Y0-CXM-385

STATISTICS

FOR BUSINESS AND ECONOMICS

William E. Becker

INDIANA UNIVERSITY

SOUTH-WESTERN College Publishing

An International Thomson Publishing Company

Reprinted, with corrections, May, 1995.

Credits are listed on page xii.

ME63AA

2 3 4 5 6 7 8 MT 1 0 9 8 7 6 5

Printed in the United States of America

ISBN: 0–538–84033–1

Acquisitions Editor: Jack Calhoun
Developmental Editor: Dennis Hanseman
Production Editor: Robin Schuster
Production House: Shepherd, Inc.
Internal Designer: Joe Devine
Cover Designer: Larry Hanes/Birdland Design
Cover Photo: ©1994, Comstock

Library of Congress Cataloging–in–Publication Data

Becker, William E.
Statistics for business and economics / William E. Becker.
p. cm.
Includes bibliographical references and index.
ISBN 0–538–84033–1
1. Statistics. 2. Commercial statistics. 3. Economics--Statistical methods. I. Title.
HA29.B3838 1995
519.5—dc20 94–18153
CIP

International Thomson Publishing
South-Western is an ITP Company. The ITP trademark is used under license.

About the Author

William E. Becker is Professor of Economics at Indiana University, Bloomington. He is editor of the *Journal of Economic Education* and also serves on the editorial board of the *Economics of Education Review*. Before joining the faculty of Indiana University in 1979 he was a tenured faculty member at the University of Minnesota. He has also served as a visiting scholar and consultant to universities around the world.

Professor Becker's research appears in the *American Economic Review, American Statistician, American Journal of Agricultural Economics, Econometric Theory, Economic Inquiry, Journal of Finance, Journal of Human Resources, Journal of Risk and Insurance, Monthly Labor Review, Review of Economics and Statistics* and numerous other journals. In addition to writing commissioned monographs on the value of human capital, and giving expert statistical and econometric testimony in Minnesota, Illinois and Indiana courtrooms, he has been a paid consultant for the United States Department of Justice, Ford Motor Company, Chrysler Corporation, Nissan, Westinghouse, as well as many smaller firms and individuals. He is the coauthor of *Business and Economics Statistics* (Addison-Wesley), and coeditor of *Academic Rewards in Higher Education* (Ballinger), *Econometric Modeling in Economic Education Research* (Kluwer-Nijhoff), *The Economics of American Higher Education* (Kluwer) and *Higher Education and National Growth* (Kluwer).

Dr. Becker earned a bachelor's degree in mathematics from the College of St. Thomas, a master's degree in economics from the University of Wisconsin, and a doctorate in economics from the University of Pittsburgh. In 1987, he received the Henry H. Villard Research Award for his work in education.

Preface

This book is designed for the first and second courses in business and economics statistics. It develops and demonstrates analytic and inferential techniques for data used in all areas of business and economics. To ensure real-life applicability, it employs numerous examples, case studies, and exercises containing information from news and business publications, government publications, and scholarly journals that publish quantitative studies in the business areas.

The text makes extensive use of firm specific data as well as aggregate economic data. The theory underlying these data applications is primarily classical inferential statistics, with attention given to alternative views. A wide range of topics is considered in both specific applications and full chapter coverage. The historical development of statistical techniques and ideas is considered throughout the text. A special feature of the presentation is the Queries, which are problems similar to those an instructor might assign as exercises or give on an exam. Through worked-out solutions to these Queries, students are able to see how to prepare their work.

Graphs, charts, tables, and other illustrations are used widely in presenting ideas and concepts. Definitions and equations are highlighted to help students identify relevant sections of the text. Footnotes, optional sections, and appendices provide the flexibility needed to accommodate the different backgrounds that students and their instructors bring to the classroom. This book is also compatible with all of the major computer statistics packages now available. A host of supplementary materials, including a data disk of over 170 files containing the data used in this book, are designed to provide students with activities that will make them aware of statistical procedures for solving problems in business and enable them to use these procedures in actual data analysis.

Applications with Data

Each chapter opens with an introduction that describes various economic and business applications of the techniques to be covered in that chapter. Short cases present the data that are used throughout the chapter to solve problems related to accounting, law, marketing, economics, finance, personnel, operations and production, and administration. New situations and data are introduced in each chapter and are used to demonstrate how statistical concepts build, with more in-depth data analysis resulting in more precise solutions.

Whenever possible, real data from clearly identified sources are employed, but in some cases artificial data had to be created. Any hypothetical data, however, were constructed to closely mimic the application described in an actual situation.

Cited Sources

Real and hypothetical data sets, together with relevant statistics, are based on actual situations described in the news and business press. These examples, case studies, and exercises were chosen from business publications such as *Business Week (BW)*, and *The*

Wall Street Journal (WSJ), government publications such as the *Monthly Labor Review*, *Economic Report of the President*, and *Federal Reserve Bulletin*, and scholarly journals that publish quantitative studies in business and economics. Many articles come from the popular press, including *USA Today*, larger city newspapers such as the *Saint Paul Pioneer Press*, and small dailies such as the Bloomington *Herald Times (HT)*.

Theory

Although this book is oriented to applications, it is rooted in the theory of classical inferential statistics. The alternative theoretical views found in Bayesian and nonparametric literature are presented for comparison purposes. Exploratory techniques in data analysis are also covered. But, at all times, the primary focus of this book is the inferential techniques of classical statistics and their application to real-life problem solving.

Theory is discussed only to the extent required in the actual situations and cases under consideration. More extensive mathematical derivations and more in-depth presentations of theory are provided for the highly motivated student in footnotes and appendices. Calculus is used in only a very few footnotes and appendices, which can be ignored by students who want only a basic understanding and working knowledge of statistics.

Topics and Content Coverage

Cases provided throughout the book describe contemporary problems that face managers within firms and government agencies. For example, Chapter 1 analyzes technological developments in small versus large firms. Examples involving quality assessment and quality improvement methods are introduced in Chapters 1 and 2, and then used throughout the book. Because of the emphasis on quality improvement in business, education, and government, a capstone chapter was needed that covered more than just control charts. Chapter 18, written by industrial engineer and Purdue University Ph.D. Julie Jacko, is devoted to the history and process of quality control.

Chapters 3 and 4 introduce probability and random variables. Chapter 4 emphasizes the use of expected values and sums of random variables in financial decisions. Chapters 5 and 6 extend the coverage of probability to specific distributions for discrete and continuous random variables.

Cases and examples that deal with selected topics in sampling, estimation, testing, modeling, forecasting, indexing, and decision theory are employed throughout the book, but there are also separate chapters devoted to these topics, as seen in Chapters 7 through 17. Extensive attention is given to estimation (Chapter 8), hypothesis testing (Chapters 9, 10, and 11), and the relationship between the two. Resampling and bootstrap techniques are demonstrated in the chapters on sampling distributions (Chapter 7) and estimation (Chapter 8). Chapter 9 reviews the controversy surrounding the usefulness of hypothesis testing in business decision making. Detailed appendices to that chapter provide a full graphical demonstration of the role of Type I and Type II errors in decision making.

Because of the central role of regression analysis in of data analysis, four chapters are devoted to this topic (Chapters 12 through 15). Special attention is given to prediction, estimation, and hypothesis testing within the regression framework. The use of data transformations and qualitative covariates is demonstrated through model building with both time-series and cross-section data.

In summary, this textbook is designed to at least mention any topic an instructor might want to cover in the first two courses in statistics. Content central to statistics is covered within the text while supporting, technical, or tangential material is presented in appendices and footnotes.

Historical Perspective

In the spirit of the works of Glenn Shafer, Stephen Stigler, and Theodore Porter, this book describes how the statistical way of thinking has developed in business and economic analysis. In Chapters 3 and 4, students get a sense of the importance of alternative interpretations of probability in the formation of Bayesian and frequentist statistical theory. Similar discussions are incorporated into the chapters on estimation, hypothesis testing, and decision theory. In applying probability to decision theory, the work of psychologists is brought to bear on the extent to which decision makers behave in accordance with expected utility theory and axioms of rational choice under uncertainty.

All major topics covered in this book feature historical perspectives. When studying regression, for example, students learn that Carl Friedrich Gauss was their age when he developed the idea of least-squares estimation. They learn how Nobel Laureate economist Milton Friedman as a young scholar used multiple regression (which required weeks of calculations without a computer) to design a metal alloy for aircraft engines during World War II. Similar historical tidbits appear throughout the book to convey a sense of history in the development of statistical ideas.

Answered Queries

In addition to the case studies that introduce each chapter, other empirical examples are used to develop and demonstrate the statistical way of thinking. These "Queries" are employed regularly to show students how exercises and exam questions should be answered.

Definitions and Formulas

Definitions of key terms are presented in the margins. Formulas and key equations are highlighted and numbered for easy identification; derivations are provided in footnotes and appendices. Although this book does not emphasize hand calculations using computational formulas, a complete understanding of statistics is not possible without at least an awareness of the mathematics on which statistics is founded. Thus, students are shown basic calculations using small data sets before they are shown the output from computer programs.

Illustrations

Students often have difficulty grasping the basic ideas of statistical analysis when those ideas are presented only through algebra and verbal explanations. This is especially true in drawing distinctions among population parameters, their estimators, and specific estimates. In this book, numerous diagrams are used to help students visualize difficult concepts. Throughout this book, diagrams and graphs show the relevant components of estimation and testing procedures. These illustrations are typically based on small data sets that students can use to easily replicate calculations. All diagrams and graphs included in this book are available from the publisher for instructors to use in the classroom.

Exercises

At key points in each chapter there are exercises for students to try. These exercises are based on material from the immediately preceding sections of the chapter. At the end of each chapter there are additional exercises that require students to draw on material from the entire chapter. Exercises are of two types: (1) problem-solving exercises that present questions using business and economic situations and data, and (2) drill exercises that emphasize the procedures and algebraic manipulations that are critical to an operational understanding of statistical principles. Chapters typically have over 50 exercises, many with multiple parts. Abridged answers to most even-numbered exercises are provided at the back of this book. Detailed answers to all exercises appear in the Instructor's Solutions Manual.

Flexibility

Students in introductory statistics classes usually bring highly diverse backgrounds in mathematics, economics, and other business courses. Similarly, instructors have greatly different experiences and expectations. To accommodate heterogeneity in student and instructor needs, this book includes footnotes, appendices, and optional sections. The body of the text is written for those who prefer verbal and graphical explanations. For the more demanding reader, mathematical derivations and proofs are provided in footnotes. Optional sections within the chapters contain material that need not be covered to proceed to the next chapter. Chapter appendices, like footnotes, provide discussions of technical material that may be of interest only to the most dedicated student. Unlike footnotes, however, coverage of material in these appendices is extensive.

Computer Usage

Surveys by textbook publishers and academics alike indicate that most instructors use computer programs in teaching statistics. Some of the more well-known programs include SAS, SPSS, SYSTAT, MINITAB, RATS, ySTAT, SHAZAM, TSP, MICROSTAT, STATISTIX, LIMDEP, and GAUSS, all of which run under DOS operating systems, with some available for Macintoshes and WINDOWS.

Some programs have very specific uses (e.g., RESAMPLING STATS for the bootstrap technique, and DEMOS for decision theory), or highly sophisticated and powerful uses (e.g., RATS for time-series analysis, LIMDEP for limited dependent variables applications and panel data. Others (e.g., MINITAB) are widely available and familiar to most instructors. New, easy, inexpensive, yet relatively powerful programs such as ET (Econometrics Toolkit) have been slower to catch on. To tie a textbook or a course to one computer package ignores the diversity in computer program offerings that are now available.

This textbook encourages the use of computer programs, but it does not require the use of any specific program or computer. To use the book effectively, instructors and students will need access to one of the better statistics computer programs now on the market. The case studies and exercises in the text are based on printouts from a variety of programs. All larger data sets used in this book are provided on the accompanying computer disk. Some exercises require students to analyze these data sets. As already noted, within the text students will see "Queries" that illustrate how to go about completing these exercises.

Supplementary Materials

Accompanying this textbook are two solutions manuals, prepared in cooperation with Choong-Geun Chung. The Instructor's Solutions Manual, which is free to adopters, contains detailed solutions to all exercises in the text. The Student Solutions Manual contains complete solutions to just the even-numbered exercises. This student manual is available for sale to interested students.

Choong-Geun Chung also prepared the more than 170 data files on the computer disk that accompanies this book. These files contain all of the larger data sets in ASCII, or in some cases spreadsheet (.WK1) readable form.

The publisher has made available transparencies for instructor use on classroom overhead projectors. These transparencies are available for all of the diagrams in this book.

Robert Toutkoushian, University of Minnesota, Minneapolis, has written a student workbook to complement the learning activities in this book. Rob designed this workbook to give students additional practice in working problems and to reinforce what is presented in the text.

James Isaac (Ike) Brannon, University of Wisconsin, Oshkosh, and Brian J. Peterson cooperated with me in preparing a Test Bank of over 1000 questions. These questions have been classroom tested at Indiana University where both Ike and Brian taught business and economics statistics.

Acknowledgments

This book was originally commissioned by George Lobell and has benefited from the editorial tutelage of Jack Calhoun, Dennis Hanseman, and Jim Sitlington. I appreciate the faith these editors have shown in the project. I also appreciate the patience Robin Schuster has shown in working with me and the production house.

I could never have undertaken or finished this project without the editorial and research assistance of my wife and strongest critic, Suzanne Becker. As also reflected in her work as assistant editor of the *Journal of Economic Education*, Sue's eye for detail and clarity in writing is unsurpassed. To the extent that this book is readable and error free, Sue deserves the credit. If there are any clunky passages or errors, the reader may safely assume that it was my stubbornness or carelessness that is to blame.

Elaine Yarde assisted Sue in preparing the manuscript. I gratefully acknowledge Elaine's skill and patience in cleaning up my word processing. She has remained a supportive friend during this project and I am indebted to her. I am also grateful for all the excellent support I have received from the secretarial staff in the Department of Economics at Indiana University.

My deep appreciation goes to the many statisticians who critiqued the manuscript in various stages of its development. I have tried to take their comments to heart, and the book is much better because of their help. These individuals include:

J. Isaac Brannon	University of Wisconsin, Oskhosh
Ronald Bremer	Texas Tech University
Daniel Christiansen	Albion College
Sangit Chatterjee	Northeastern University
Mark Eakin	University of Texas, Arlington
Eugene Enneking	Portland State University
William Greene	New York University
Bharat Kolluri	University of Hartford
Ronald Koot	Pennsylvania State University
Timothy Krehbiel	Miami University
Robert Mogull	California State University, Sacramento
Samuel Ramenofsky	Loyola University
Don Robinson	Illinois State University
Stanley Sclove	University of Illinois, Chicago
Larry Sherr	University of Kansas
Julian Simon	University of Maryland
Stanley Stephenson	Southwest Texas State University
Robert Toutkoushian	University of Minnesota
James Willis	Louisiana State University
Jimmie Woods	Hartford Graduate Center
Dean Young	Baylor University

My thanks also go to the many students at Indiana University and the University of Minnesota who studied from the preliminary manuscript and whose comments and suggestions helped make it a better book.

Finally, my thanks go to South-Western College Publishing and to the National Council on Economic Education for the support they have provided me over the years.

A Closing Note on Accuracy

Students and instructors using this text have a right to expect it to be error-free. In preparing the manuscript, I followed procedures designed to produce a highly accurate book. All of the material included here was used in my teaching of introductory courses in business and economic statistics and in econometrics. Choong-Guen Chung, Ike Brannon, and Brian Peterson all made use of material from Chapters 1 through 14 in their teaching of introductory business and economics statistics at Indiana University. In addition to this class testing, many early calculations were checked by either Claude Martine or Kevin Stroupe. Then, all calculations within the manuscript were rechecked by Choong-Geun Chung.

South-Western College Publishing and I are so confident of the accuracy of these calculations that we will pay $50 to the first person, associated with a class that adopts this book, who identifies a material calculation error in the textual body of a chapter.

In addition to checking calculations in the body of the text, Choong-Geun Chung also prepared the first draft of the answers to all the exercises in Chapters 1 through 17. I have checked and edited all those answers, along with those prepared by Julie Jacko for Chapter 18. Kaushik Mukhopadhaya then rechecked all of the exercises and answers provided in the Instructor's Solutions Manual. I am in Choong-Geun's and Kaushik's debt for their long hours of tedious work.

The collective efforts of all these individuals has resulted in a book that we believe is as accurate as humanly possible. I accept full responsibility for any errors that may remain.

William E. Becker
Bloomington, Indiana

Brief Contents

Contents

13 The Two-Variable Population Model 549

14 Multiple Regression 615

15 Time Series Analysis and Forecasting 677

Chapter 1

Collection and Presentation of Data

The age of chivalry has gone. That of sophisters, economists, and calculators has succeeded; and the glory of Europe is extinguished for ever.

Edmund Burke
1729–1797

1.1 Introduction

Statistician Stephen Fienberg tells us that 400 years ago problems involving coins, cards, dice, and the drawing of balls from urns were curiosities pursued by mathematicians trying to solve puzzles in probability. One hundred years later, concepts in probability were applied to data (measurements) from astronomy, demography, biology, and genetics—statistical analysis was born. By the 19th century, statistical analysis had spread to economics and the other social sciences. Newspaper columnist Chuck Stone lamented recently that "we have become a nation of numbers-crunching arithmomaniacs." (*Newspaper Enterprise Association,* September 5, 1992) If for no other reason than to protect yourself from the misuse of statistics, it is essential to have a knowledge of statistical application.

Advances have been made in the use of statistics within the social sciences, but in its most general form statistics continues to be a study of collecting, organizing, presenting, describing, summarizing, and interpreting data. Unlike probability, which continues to be a central branch of mathematics, statistics is an area of inquiry that is not central to mathematics. Statistics has a methodology closely tied to the discipline in which it is applied.[1] In this book, we will study statistics within the context of economics, accounting, finance, sociology, psychology, and other social science and business-related areas.

Today successful decision makers in both the public and private sectors must have a working knowledge of statistics—not as a general tool of data analysis but as a specific method for addressing issues and questions in their daily business environments. Although an arithmetic average is the same in medicine and accounting, a discussion of average accounts receivable is more pertinent and enlightening to those majoring in business and public policy areas than a discussion of average blood pressure.

As an example of the statistical way of thinking in business- and economics-related areas, consider a topic that is regularly addressed by the popular and business press: Are corporate chief executive officers of energy companies overpaid? A question from an economics perspective would be: Overpaid relative to whom and by how much? Information provided in a *Wall Street Journal* (April 22, 1992) article will help answer these questions in a statistical analysis. The *WSJ* details are for the compensation of a subset of managers taken from a survey of the country's biggest companies conducted by Towers Perrin, a compensation-consulting company. The *WSJ* used this information (sample data) to make a statement (inference) about the pay of all executives (population data).

As another example, consider the debate over sources of economic growth, technological change, and the role of patents. Harvard University economist Zvi Griliches concluded in a *Journal of Economic Literature* article that "the appearance of diminishing returns (to patents) at the cross-sectional level is due, I think, primarily to two effects: selectivity and the differential role of formal R & D and patents for small and large firms." Griliches calls attention to a problem in drawing conclusions about a population when the selected sample may provide a biased representation of the population.

This book emphasizes the collection, presentation, description, and interpretation of sample data for decision making about a population. A large share of the text is aimed at **statistical inference,** the process of making statements about a population based on sample data. Unlike books on statistics in biology, engineering, medicine, or other derivatives of the natural sciences, the populations of interest and the problems of sampling that we consider will deal with real-world situations in economics and other business areas.

Statistical Inference
The process of making statements about a population based on a sample.

This chapter introduces a way of thinking that should enable you to distinguish between a sample and a population, retrieve data from different sources, discuss alternative methods for obtaining data, display data in tabular and graphical form, and recognize problems in making statements about a population on the basis of sample information. By the end of this chapter, you will have a sense of how data are obtained, presented, and used in business and economic decision making.

1.2 Population and Sample Data

Population
The set or collection of all the observations or measurements of interest to a decision maker.

Sample
A subset of measurements taken from the population of interest to the decision maker.

Quantitative Data
Measurements that are expressed numerically, where the number has meaning on a number line (such as inches on a ruler).

Qualitative Data
Information in the form of an attribute or characteristic that is not numerical.

A **population** is the set or collection of all the observations or measurements of interest to a decision maker. The specific characteristics of a population are generally unknown. To make statements about things of interest in the population (for example, the compensation of CEOs versus other lower-level executives or the number of patents at small versus large firms), sample information can be used. A **sample** is a subset of measurements taken from the population. The process of making statements about a population on the basis of information provided in a sample is called statistical inference. Figure 1.1 illustrates the relationship between a population, a sample, and a statistical inference.

Population and sample information can be either quantitative or qualitative. **Quantitative data** are measurements that are expressed numerically, where the number has meaning on a number line, such as a ruler. For quantitative data, the arithmetic functions of addition and multiplication give sums and products that can be interpreted on the same number line as the original values. Baseball and basketball jerseys have numbers, but those numbers have no meaning on a number line; they cannot be added and subtracted in a meaningful way. Unlike quantitative data, baseball jersey numbers simply identify a player; they are like names. On the other hand, the salaries professional baseball players receive are quantitative data; all the players' salaries on a team can be added, with the sum being the total player salary budget of the team.

Qualitative data are information that is nonnumerical. A survey might ask whether the respondent has a job; whether the respondent is a male or female; and whether he or she is married. The responses to these inquiries would all be qualitative information. As we will see in later chapters, qualitative information can sometimes be coded to make it appear quantitative. For example, a male or female response can be coded 0 if male and 1 if female. If the underlying attribute (gender, in this case) has no meaning on a number line, then the assignment of numerical values is arbitrary. However, some

FIGURE 1.1 A Population, a Sample, and an Inference

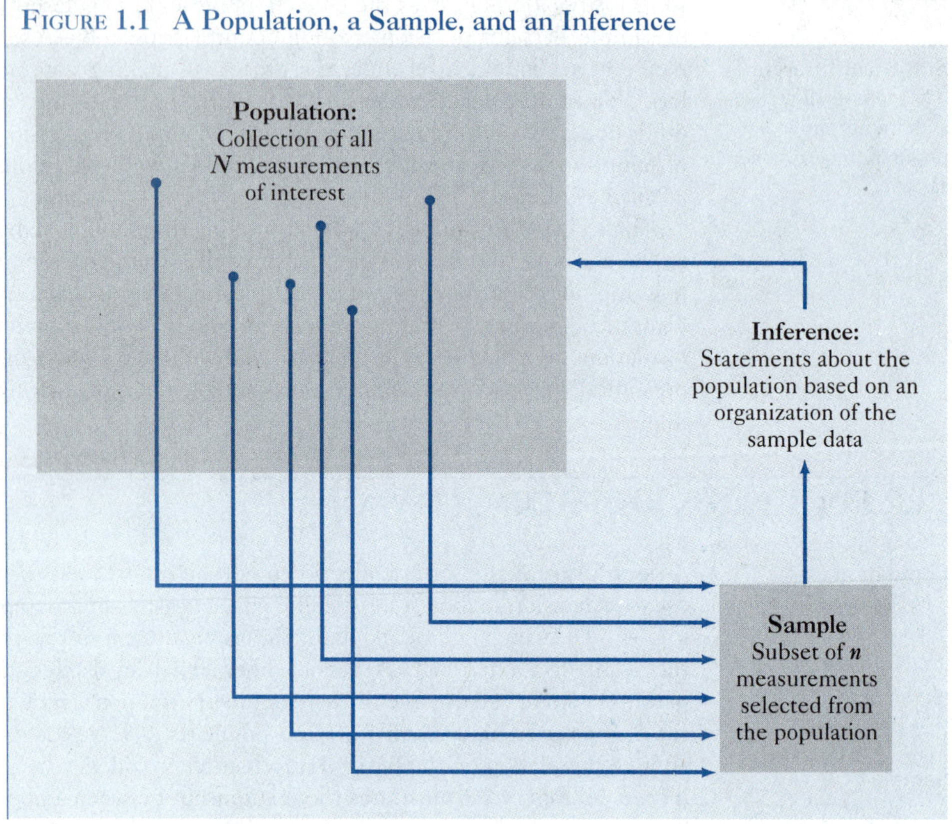

types of qualitative information can have meaning on a number line. For example, the letter grades of A, B, C, D, and F are typically coded 4, 3, 2, 1, and 0 to calculate a grade point average. Here the grade of A is clearly better than that of B, which is better than C. Thus, the numerical value given to A must exceed that given to B and C but the numerical amounts by which A exceeds B and B exceeds C are arbitrary. The measurement scale must preserve order, but the ratio by which one value exceeds the other is arbitrary. For example, a prior grading policy at Purdue University made a grade of A worth 6 points, a B worth 5, and a C worth 4.

Unlike samples, which are always finite, populations can be either finite in size or infinite. If the population has a fixed and limited size, that size is designated by the uppercase letter N. Populations that have unlimited size must be viewed as hypothetical or conceptual; after all, by definition they are not observable in a finite span of time. Shakespeare's characters, Rosencrantz and Guildenstern, as portrayed in Tom Stoppard's play, would have had to flip the coins an infinite number of times to actually observe 50 percent of the tosses resulting in heads. The idea that a coin has a 0.50

probability of coming up heads is based on the idea of an infinite number of flips that can never really be observed. Similarly, the mathematician's probabilities for card and dice tricks are based on the idea of an infinite number of attempts. Notions of probability based on infinite populations are discussed in Chapters 3, 4, 5, and 6.

In many real-world problems, a finite population can be treated as if it were infinite if the size of the sample is a small proportion of the size of the population. Sample size is designated by the lower case letter n. Ignoring the cost of acquiring the sample, bigger samples are always preferred to smaller samples. As the sample size increases, the probability that the sample represents the population increases. Sampling is not costless, however; each additional observation typically adds to the cost. As discussed in Chapters 7 and 17, when cost is considered, a smaller sample may be preferred to a larger one.

As discussed in Chapters 6, 7, 8, and 9, if relatively small samples are randomly drawn from relatively large (infinite) populations, then probability theory can be used to make inferences about the population, even though we do not know if the sample is representative of the population. This ability to make probabilistic statements about a population based on relatively little sample information is the essence of statistics.

1.3 Sampling and Statistical Analysis

Random Sample
A sample obtained by a rule involving chance as to which elements in a population will be sampled.

Line Graph
A continuous line showing the value of a series of numbers over a continuum.

If a sample was a perfect representative of the population from which it was drawn, it would be a perfect small-scale replica. Typically, however, it is impossible to know exactly how well a sample represents a population because the characteristics of the population are unknown. In some cases, however, the importance of sample selection can be assessed.

Griliches reproduced a **line chart** that originated in the research of John Bound. (See Figure 1.2.) The vertical axis is the number of patents per million dollars of research and development (R & D) expenditures. The horizontal axis is the size of the R & D program measured on a logarithmic scale that will be discussed in Chapter 6. For now, it is enough to recognize that firm size increases as we move from left to right along the horizontal axis. Smaller firms (those with smaller R & D programs) appear more efficient, receiving a larger number of patents per R & D expenditure. As firm size increases, the number of patents falls off quickly and becomes effectively constant as the size of the firm continues to increase.

Griliches argues, however, that this diagram was not constructed from sample observations that were selected by rules of probability. Rather, this sample is an "opportunity sample" based on other criteria. It consists of all manufacturing firms listed on the New York and American Stock Exchanges and also on the over-the-counter market. Almost all large firms are listed on the New York, American, or over-the-counter market. To be listed on one of these markets, a small firm has to be more successful, be more interesting to investors, or have some special appeal. A small firm listed on one of these exchanges would likely have more patents than one not listed.

FIGURE 1.2 Patents per Million R & D Dollars by R & D Size Class for Firms with both R & D and Patents

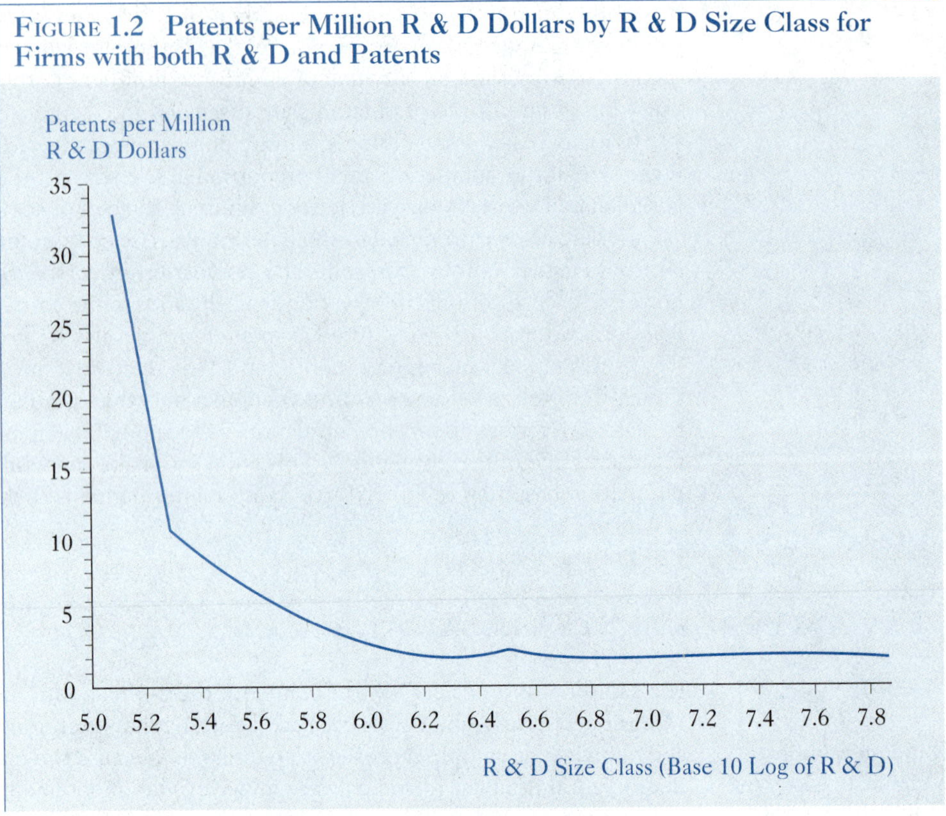

Frequency Distribution
A presentation of the number of times different values (or classes of values) in a data set occur.

Frequency
The number of times an observation (or observations in a class) occurs.

Griliches compared the number of small firms in this sample with all firms counted by the U.S. Bureau of the Census (see Table 1.1). Table 1.1 is a **frequency distribution** that shows the number of firms classified by the number of employees. There are four classes: firms with fewer than 10 employees, firms with 10 to 99 employees, firms with 100 to 999 employees, and firms with more than 1,000 employees. There were 24 firms in the sample with fewer than 10 employees, 301 with 10 to 99 employees, 952 with 100 to 999 employees, and 1,267 with more than 1,000 employees. These are the frequencies of each class. The **frequency** of the *i*th class is written f_i so the $f_1 = 24, f_2 = 301, f_3 = 952$ and $f_4 = 1{,}267$. The class frequencies must sum to the sample size, which is written $n = 2{,}544$.

The **relative frequency** of a class is the proportion (or percentage) of the total number of observations in the sample that are in that class. The relative frequency of "fewer than 10" is 0.00943, because 24 of the 2,544 firms in the sample were in this class (i.e., the relative frequency of the first class is $f_1/n = 24/2544 = 0.00943$). The other three relative frequencies (0.11832, 0.37421, and 0.49803) are shown in Table 1.1. Notice that the sum of the relative frequencies is 1.00.

Table 1.1
Selectivity of Firms

Class i	Number of Employees	Number of Firms in the Sample f_i	Relative Frequency of Firms in the Sample f_i/n	Number of Firms in the Census f_i	Relative Frequency of Firms in the Census f_i/N
1	< 10	24	0.0094340	16000	0.3883495
2	10 – 99	301	0.1183176	14300	0.3470874
3	100 – 999	952	0.3742138	9000	0.2184466
4	1,000 +	1267	0.4980346	1900	0.0461165
	Column Sum =	2544	1.0000000	41200	1.0000000

Sources: Sample is from Bound et al. (1984), and Census is from U.S. Bureau of the Census (1981), with frequencies presented in Zvi Griliches, "Patent Statistics as Economic Indicators: A Survey," *Journal of Economic Literature*, December 1990, p. 1677.

Relative Frequency
The proportion (or percentage) of the total number of observations in a data set that take a specific value (or class of values).

Relative Frequency of a class is the proportion of the total number of observations in the distribution that are in that class.

Relative frequency $= f_i/N$, for a population of size N. (1.1a)

Relative frequency $= f_i/n$, for a sample of size n, (1.1b)

where f_i is the frequency of the ith class.

Census
A survey that attempts to capture all of the items or elements in the population.

Also reported in Table 1.1 are the number of firms in the census of enterprises classified by the number of employees. Unlike a sample, a **census** is a survey that attempts to capture all of the items or elements in the population. Compiling a census is more costly than sampling. According to the U.S. General Accounting Office, for instance, the cost of counting each housing unit (in 1980 dollars) rose from $5 in 1950 to $12 in 1980, and it exceeded $17 in 1990. In many problems, a census is not possible. In a production process, for example, the durability of all items cannot be tested because such testing implies the destruction of each item.

Compared to the relative frequency of small firms in the population, small firms are underrepresented in the Bound et al. sample. About 39 percent of the census is small firms, but less than 1 percent (0.9434%) of the sample is small firms. Unfortunately, as stated by Griliches, we have no information on the firms not in the sample and, hence, cannot make an appropriate sample selectivity adjustment.

Current patents are likely to be more important to small firms than to large firms. The near-term existence of large firms is insured by their past patents, while the existence of small firms may be tied directly to their current prospects for patents. To establish a position, small firms must pursue patents more quickly and more aggressively than larger firms. Yet, small firms can be expected to report less expenditure on R & D, because research and development at small firms is an informal activity. Thus, the ratio of patents per millions of R & D dollars is driven up at small firms by the need to patent and the lack of reporting of expenditures on R & D.

Even if inventiveness rates were equal between small and large firms, the opportunity sample employed and the use of the number of patents per dollar of expenditure as the measure of inventiveness would give the impression that small firms are more efficient at inventing. Given this empirical assessment, Griliches turned to economic theory and concluded that, "This is not surprising, after all. If there were such diminishing returns, firms could split themselves into divisions or separate enterprises and escape them." (p. 1677) It is this injection of economics into the analysis that distinguishes the use of the statistics (population, sample, and relative frequencies) by an economist from that of an engineer looking at the same problem.

The purpose of the analysis is of primary importance when deciding what type of sample is needed to make an inference about a characteristic of the population. A classic example that demonstrates inappropriate sampling for the task at hand is a 60-year-old study by the now-defunct *Literary Digest.* The *Literary Digest* sent questionnaires to more than 2 million people whose names were selected from telephone directories and automobile registrations. From the responses, it predicted that Alf Landon would win the 1936 U.S. presidential election with an overwhelming victory. Instead, Franklin Roosevelt won by a substantial margin. The responses received by the *Literary Digest* were not random and could not have been representative of the population of relevant voters. They were from predominantly wealthy people who owned cars and telephones at a time when ownership of such items was beyond the reach of most people.

A more contemporary example involves the controversy over an October 14, 1992 article in the *New England Journal of Medicine* reporting on the lack of special benefits of oat bran products in lowering cholesterol. In addition to questions about the size of the sample (20 men and women), this study was criticized for using subjects not representative of the general population of the United States because the subjects were mostly dieticians.

Experient
A method of collecting sample data by observation under specified conditions that control for or remove the influence of extraneous influences.

Deciding on the type and amount of sample data needed and how these data can best be gathered is a major part of the inference-making process. The nature of the sampling will greatly affect the cost of the study. Searching existing data banks is usually the least expensive way to find useable information; the trick is knowing where to look. Although a business brainstorming session seldom passes without someone suggesting a survey as a means to collect information, conducting surveys or running **experiments** that result in meaningful data is typically an expensive way to get information. Always consider the less expensive methods for gathering data first and then consider surveys and experiments.

EXERCISES

1.1 How do a sample and a census differ?

1.2 Is it possible for a group of students in a university to be viewed as a sample for one purpose and as a population for another? Give an example.

1.3 Give a situation in which a sample would be drawn because all the elements in a population could not be tested.

1.4 Here is the breakdown of owner-occupied housing in Monroe County. (*HT,* July 2, 1993)

Less than $25,000	513
$25,000–$99,999	12,173
$100,000–$199,999	2,642
$200,000–$299,999	331
$300,000–$399,999	65
$400,000 and above	24

Source: Indiana Factbook 1992.

Construct a relative frequency distribution of owner-occupied housing in Monroe County.

1.5 In Exercise 1.4 above, is the distribution based on sample or population data? Explain.

1.6 Unlike baseball and basketball jerseys, the numbers on football jerseys are assigned on the basis of the position played. In the National Football League, quarterbacks are assigned 1 through 19, running backs 20 through 49, and offensive linemen 50 through 79. A student in Professor Becker's class observed that these jersey numbers may have meaning on a number line and, thus, represent quantitative data. To what measurement was the student referring?

1.4 Data Sources

Data can be obtained from primary sources or secondary sources. They can represent a situation at a point in time or a flow of information over time. The results that can be drawn from statistics always depend on the quality of the data.

Primary Data
Data that are published or made available from the person (or agency) responsible for collecting the data.

Primary data are observations obtained from the original generator of the data; primary source data are always preferred and are often the only data acceptable in legal proceedings. Annual reports, technical manuals, personnel files, computer files of accounting data, and other routinely filed information within a corporation may be the source of primary data on a corporation. Table 1.2, for example, illustrates the source of sales revenue of the Ford Motor Company as obtained from Ford's 1987 annual report and submitted as a deposition exhibit in a wrongful death case filed against Ford.

As corporations have moved to centralized information systems and computerized information networks, primary source data have become more easily and efficiently obtained. Production schedules, inventory levels, sales history and forecasts, and employee data are among the vast amounts of information typically found in corporate computers.

The government also collects and disseminates primary data. In addition to internal corporate data, information about a wide variety of subjects is available from numerous external sources. Consider just a few examples:

- Monthly economic indicators, such as the Consumer Price Index, housing starts, and industrial production, are available from the U.S. Bureau of Labor Statistics.
- Weekly monetary indicators, such as the money supply, banks' business loans, and free reserves, are provided by the 12 Federal Reserve Banks and the Board of Governors of the Federal Reserve System in Washington, D.C.
- A variety of business and economic indicators are published by Standard and Poor's, Moody's, Dun & Bradstreet, and numerous federal and state agencies. Almost all of this information is available on magnetic tape or on disks.
- Demographic data on the U.S. population have been collected every 10 years since 1790 by the Bureau of the Census.

Secondary Data
Data that are published or made available by someone other than the one who collected the data.

Data are also available from many **secondary data** sources. *Business Week,* for example, uses internal corporate data when it puts together its annual "Corporate Scoreboard" issue each spring. The source for the corporate data published by *Business Week* is Standard & Poor's Compustat Services, Inc. The *Business Week* publication of these data is a secondary source. It is thus not as acceptable in legal proceedings as the primary source data; its accuracy can be questioned.

Table 1.2
Ford Motor Company Annual Report Information

Total Sales Revenue (in millions)	1986	1985	1984
United States	$50,034	$43,526	$42,829
Canada	9,782	9,073	8,231
Europe	13,609	9,566	9,083
Latin America	3,286	3,107	2,764
All other (primarily Asia-Pacific)	2,046	2,028	2,590
Elimination of intercompany sales	(16,041)	(14,526)	(13,131)
Total Sales	$62,716	$52,774	$52,366

Source: December 31, 1986, Financial Report of Ford Motor Company, Coopers & Lybrand Auditors' Opinion, February 10, 1987, p. 35.

One of the best secondary sources for annual summary data on national income, expenditures, population, employment, productivity, production, wages, business activity, prices, money stock, credit, finance, profits, agriculture, and international statistics is the *Economic Report of the President,* which is issued every February. Later in the spring of each year the *Statistical Abstract of the United States* becomes available with some 1,500 tables covering everything from the amount of chicken Americans consume to the number of persons over age 40 who wear glasses. Both the *Economic Report of the President* and the *Statistical Abstract of the United States* are excellent data sources, although not always the primary source of data. Economic journalist Robert J. Samuelson, for example, stated that, "The *Statistical Abstract* has endless uses. . . . The sheer variety of statistics on so many subjects is chastening." (*Newsweek* June 1, 1987, p. 49) Both the *Economic Report of the President* and *Statistical Abstract of the United States* can be purchased from the Superintendent of Documents, U.S. Government Printing Office, Washington, D.C., 20402.

The Organization for Economic Cooperation and Development (OECD) provides international data on developing countries on computer disk and in hard copy form. These data include statistics on health, education, bank profitability, employment, productivity, and other economic indicators for Eastern and Western European, North American, and Pacific countries. This information is available from OECD, 2001 L Street, N.W., Suite 700, Washington, D.C., 20036-4910.

Besides business and government publications, trade associations, such as the American Bankers Association and the U.S. Chamber of Commerce, provide information about their areas of interest. Similarly, union and professional organizations, such as the AFL-CIO and the American Medical Association, provide information of interest to members. Numerous testing and research companies also provide data on specific products. One of the best listings of information available from associations, periodicals, libraries, companies, and other agencies—with capsule descriptions of each—is *Who Knows What: The Essential Business Resource Book* by Daniel Starer.

Data from national pollsters can also be used to investigate a diverse range of demographic, behavioral, and attitudinal topics. For example, J.D. Powers and *Consumer Reports* are famous for their survey results on consumer satisfaction with automobiles. The Roper Center and the Louis Harris Data Center are major sources of public-opinion polls. The Roper Center adds more than 500 files a year to its collection of more than 11,000 data files that cover social and political preferences and political opinions dating from 1930. Harris polls have been conducted since 1956, surveying opinions categorized into 24 topics and numerous subtopics. The Gallup Poll, which is familiar to many people, regularly contributes data files to the Roper Center. Many other public-opinion polls, such as those sponsored by the news media, contribute to the Roper Center files. The largest service for social science data is the Inter-University Consortium for Political and Social Research (ICPSR), which is a collection of more than 18,000 data files housed at the University of Michigan.

Existing data files may also provide the user with standardized items, a census, or at least samples that are representative of the entire nation and can be used as a basis for comparison. Griliches demonstrated this in his use of U.S. census data to show the sample selectivity problem in assessing small-firm efficiency in research and development. Griliches' comparisons of sample frequencies with those in the census were for one time period. When data are compared across geographical regions or among individuals at a point in time, the study is called **cross-section** analysis. On the other hand, if observations from a time period are compared with prior or subsequent time periods, then the study is called a **time-series** analysis. As demonstrated in later chapters, in a time-series analysis that covers many years it is essential to determine how important concepts are defined, how the data collection methods and the sample base may have changed over time, and how observations in one period might be connected to the next.

Cross-Section Data
Data that are collected across geographical regions or individuals, at a single point in time.

Time-Series Data
Data collected over time.

Griliches compared a national sample with a national census. National data can also be used to infer things about individual situations. For instance, in the lawsuit against Ford cited earlier, the plaintiff's expert witness stated that according to the *Statistical Abstract of the United States*, Table 1.3, the median weekly earnings of workers in the United States was $358 in 1986. This witness argued that a $2 parking ticket would inflict only a 0.55866 percent [= (2/358)(100%)] hardship on such a worker. Imposing a similar hardship on Ford would imply punitive damages of $6.738 million [= (62,716/52)(0.0055866)], assuming that the worker's weekly earnings in Table 1.3 are comparable to Ford's weekly earnings as calculated from the annual data in Table 1.2.

The author of this book, representing Ford Motor Company, questioned whether a worker's earnings should be compared to Ford's earnings in establishing punitive damages. In essence, we do not base damages and punish an entire family for an alleged misdeed of one of its members. Why, then, would we consider basing damages on the entire worldwide Ford family for an alleged misdeed at one of its transmission plants? If quantitative comparisons are made, they must be made on a like basis.

Although data exist on a wide array of topics, locating the exact types of data needed for a particular analysis and to make a particular point is always difficult. The format in which the data is available can also pose problems. Some agencies and researchers may be reluctant to share their data. In all cases, the user of existing data must check the methods by which the data were collected, recorded, and classified. Indeed, caution is needed when using any data, regardless of whether they are primary or secondary, manual or computer stored, government or privately generated. A statistical analysis can be no better than the data on which it is based. Always keep in mind the old and proven adage: garbage in, garbage out (GIGO).

TABLE 1.3
Earnings of Full-Time Wage and Salary Workers

	Median Weekly Earnings		
	1985	1986	1987
All workers	$343	$358	$373

Source: *Statistical Abstract of the United States*, 1989, p. 406.

1.5 Experiments and Randomization

Despite the existence of burgeoning data files, the questions, problems, and circumstances facing decision makers may change too quickly to allow extensive use of existing data sources. For example, final adjusted and detailed 1990 U.S. census data were scheduled to be available sometime in 1992. On August 5, 1992, the Census Bureau bowed to political pressure and put off decisions on critical adjustments until September, implying detailed data would not be available until much later. A final release of the 1990 census data in 1992 or 1993 that links demographics and dollars with specific neighborhoods is too out-of-date to be useful for business. When confronted with new situations for which there is no data available to guide in the decision-making process, government agencies and corporations are forced to collect their own data via surveys and experiments.

Surveys are typically associated with field studies, which are like opportunity samples because there is little attempt to control who gets into the sample. Market researchers use surveys to determine how products are used and how they perform, and to identify new markets. For instance, a sample survey might involve stopping people in a shopping mall to ask them what brand of soft drink they prefer. Governmental units use surveys for determining employment conditions, travel patterns, and for planning purposes.

In experiments, behavior is observed under controlled conditions. An experiment is a method of data collection in which the researcher makes every attempt to eliminate or to control for aspects of the test that are not important, or that might influence the experiment one way or another. For example, a 1990 Chrysler Corporation advertising campaign reported the results of studies comparing several models of Chrysler Corporation cars with the 1990 Honda Accord and Civic and with Toyota's Tercel and Camry. In these ads, Lee Iacocca, then Chrysler CEO, proudly stated that the Chrysler cars were judged as better than the Honda and Toyota cars by a vast majority of likely import-car buyers. The actual study used two test panels of 25 respondents each. Half of the test panel did the exterior/interior appearance, convenience, and comfort portion of the test first, followed by the riding/driving portion. The other half reversed this order. Similarly, the order of the test vehicle assigned to each respondent was rotated, as was the order of riding and driving. In this manner, the effect of order (if any) was controlled and the marketing people at Chrysler argued that they had a controlled experiment.

Experiments are designed to control for undesired effects, with subjects selected on a random basis. Simple random-sampling techniques are designed to insure that each element (and in turn, each subgroup of elements) in a population has an equal probability of being included in the sample. With simple random sampling, every sample of n elements selected from a population of N elements has a probability n/N of being selected, and each of the n elements has a probability of $1/N$ of being selected.

As discussed in detail in Chapter 7, simple random sampling may be done by a blind-draw approach (putting numbers in a hat and then drawing them out, for exam-

Random Number
A number (often a digit between 0 to 9) where there is no pattern to the numbers (each number is as likely to occur as any other number).

Simple Random Sample
A sample collected using techniques designed to insure that each element (and in turn each subgroup of elements) in a population has an equal probability of being included in the sample.

ple), a random number table (in which there is no pattern to numbered population items to be selected), or computer programs that generate **random numbers**. As you will see throughout later chapters, **simple random sampling** plays a critical part in statistical analysis. Any one sample obtained by simple random sampling may not give a perfect representation of the population. However, Chapter 7 will show that across many samples the average characteristics of these samples will represent the corresponding characteristics in the population. Chapters 3, 4, 5, and 6 provide the necessary concepts from probability theory to define when sample characteristics are close to those of the population.

Other than the pure chance factor in random sampling, an experiment will control for all other factors that can influence the results. Unfortunately, an experiment that controls for all factors other than chance is often impossible or too expensive. Econometrician Edward Leamer in "Let's Take the Con Out of Econometrics," argued that, "No one has ever designed an experiment that is free of bias, and no one ever can." (p. 33) The difference between experiments and field studies is a difference in degree, but not kind. There are always tradeoffs in considering the controls to design into a study. For instance, Chrysler ignored geographical location (possibly because it thought that preference for cars is not influenced by location); hence, it only selected participants from the greater Los Angeles area. If, however, location is important, then inferences from the Chrysler sample will yield systematic biases.

1.6 Other Types of Samples

Traditional test marketing of products calls for the introduction of a new product in a small town—such as Duluth, Erie, Indianapolis, Charleston, or Fresno—that closely matches desired aspects of U.S. census data. For example, *American Demographics* (January 1992) found Tulsa, Oklahoma to best typify the age, racial makeup, and housing value of the 555 U.S. cities with populations of at least 50,000.

Once a typical town has been identified, researchers swarm over the area collecting data on consumer and retailer reaction to the new product while they modify the package and the promotion. As stated in a *Business Week* (August 10, 1992) article, "The 12-to-18-month process can cost more than $1 million, but it's a lot better to bomb in test than in a coast-to-coast flameout. These days, though, fewer companies are bothering to see how their fledgling products play in Peoria. Instead, consumer-goods marketers such as Lever Brothers, Colgate-Palmolive, and General Mills want newer, faster, and cheaper ways to test their offerings." (See Table 1.4.)

Listening to "buyer focus groups" is a popular approach to pretesting. In this approach, a group of potential buyers are brought together and their opinions about a product are solicited. For instance, the *Wall Street Journal* (September 16, 1992) reported on the home buyers' focus groups that Marketing Directions conducted in 12 cities. From these groups, Marketing Directions learned that today's home buyers say that in houses "every inch of space has to count . . . (if a) closet is bigger than their kid's bedroom they don't like it."

TABLE 1.4
Skipping the Test

SKIPPING THE TEST

Some alternatives to the traditional 12- to 18-month market test:

PRETESTING Show a few consumers samples of new products along with ads to gauge probable response

COMPUTER MODELING Use historical data on similar products to turn small samples of data into sales projections

ROLLING THE DICE Introduce a new product region by region, fixing ads and promotions along the way to going national

FOREIGN FLING The lead country concept calls for trying out a product in an overseas market, then rolling it out globally

DATA: BW

The participants in a focus group might be shown a new product, view its ads, or just have a product concept explained. Prior to entering a focus group, recruits are asked to provide information about themselves. Once in the group, participants typically interact under the guidance of a facilitator. Finally, group members might complete a questionnaire or they might be called a few weeks later and asked about their reactions to product samples they were given. Researchers then incorporate this information, along with historical data, to form a computer-simulated market in which sales can be projected. Most data collection done through focus groups and many mail and telephone surveys do not employ the simple random-sampling techniques required for experiments.

Stratification
The act of dividing a population into subgroups that share a given characteristic or attribute.

Throughout this book, we assume that the data can at least be thought of as coming from experiments. In social research, however, simple random sampling may not be feasible or may not be efficient. In such cases, alternative sampling techniques are preferred to no controls at all. As discussed in Chapter 7, these other sampling techniques might involve the drawing of every 10th item in the population (systematic selection) or the division (**stratification**) of the population into groups with known characteristics, from which sampling then takes place. There are also a number of other techniques that may give rise to samples from which information can be packaged to make inferences about an unknown population.

EXERCISES

1.7 Find a library source where the most recent data can be obtained for each of the following.
 a. The yield on constant maturity three- and 10-year U.S. Treasury securities.
 b. Money earnings for full-time and part-time workers in the United States by years of school completed.
 c. Changes in productivity and related data on the business sector of the U.S. economy.
 d. The U.S. monthly unemployment rates by age.
 e. The U.S. rate of inflation as measured by changes in the Consumer Price Index (CPI).

f. Business failures in the United States per year.
g. The sales and profits of U.S. corporations classified by major business type.

1.8 What is the difference between a comparison of profitability of U.S. corporations among different major business types in 1994, and a comparison of profitability of U.S. corporations from 1983 to 1994?

1.9 Go to the library and locate copies (either electronically or off the shelves) of the following journals. State relevant reference information for each publication including the publisher, how often it is published, the most recent issue, and three examples of topics covered in the most recent issue.

- *Monthly Labor Review*
- *American Economic Review*
- *Survey of Current Business*
- *Federal Reserve Bulletin*
- *Business Condition Review*
- *Moody's Industrial Manual.*

1.10 Go to the library and select any yearly volume of the *American Economic Review*. Excluding the May papers and proceedings issue, how many articles in the remaining four issues can be classified as reporting on quantitative data? Of those, how many use time-series data? How many use cross-sectional data? How many use a combination of time-series and cross-sectional data? What can you conclude about the nature of data used in economic research?

1.11 A *Wall Street Journal* article (August 28, 1990) reported on the results of a survey of 167 companies by the American Society for Training and Development. According to the *Wall Street Journal*, "Some 92% of companies consider it critical to teach employees new skills, but just 23% do that to 'a great or very great extent.'"

a. What fallacy in surveying does this result highlight?
b. Why might one be suspicious of these results?

1.12 In an evaluation of pension and life insurance needs, an accountant estimated life expectancy by averaging the ages of individuals at the time of their deaths as reported on randomly selected days in the obituary column of his local newspaper. Comment on the appropriateness of this sampling and estimation procedure.

1.13 An article titled "The Science of Polling," in *Newsweek* (September 28, 1992) stated:

> Ensuring a random sample also requires something beyond statistics. During Ronald Reagan's 1984 re-election campaign, his internal 'tracking' polls showed him well ahead of Walter Mondale—except on Friday nights. 'They went into a panic every week until they figured it out,' says Frank Luntz, a political-science professor at the University of Pennsylvania . . . . The explanation: registered Republicans, on average more flush than their Democratic counterparts, were more likely to go out on the town Friday nights and not be home to answer the pollster's phone call. (p. 38)

a. This is an example of what kind of problem in sampling?
b. Instead of writing, "Ensuring a random sample . . . . " should the author have written "Ensuring a representative sample . . . ?" Explain the difference.

1.14 According to former Indiana University Kinsey Institute director June Reinisch, the 1948 Kinsey Report was based on a survey of about 5,000 men. About 10 percent of those reported having homosexual contact with another man during the previous three years. Yet, in a more recent study by the Battelle Human Affairs Research Center in Seattle, published in *Time* magazine, only about 1 percent of 3,321 men surveyed reported having sex exclusively with other men during the past 10 years. In an Associated Press release (*HT*, April 26, 1993), Reinisch said these were not comparable results because approximately

30 percent of Battelle's initial group refused to be interviewed when they were asked for their social security number and place of work. To what was she alluding?

1.15 ABC's 20/20 television broadcast on July 16, 1993 reported on a study in which individuals who had lived to be 100 years of age or more were queried in the hope of finding common characteristics. The implication was drawn that if a younger person worked at acquiring the characteristics shared by these centenarians, then the probability of reaching such an old age increased. Why was this study design inappropriate for the implication drawn?

1.7 Organizing, Condensing and Presenting Quantitative Data

Once quantitative data are collected, they must be organized and condensed so they can be reported in a way that facilitates interpretation. Griliches used a frequency distribution to organize, describe, and compare the sample and population data on firm size shown in Table 1.1. Here we address the construction of frequency distributions from the original, or raw, data collected in a sample. We will also consider the presentation of data in graphical form.

Frequency Distributions

There are four steps to organize and condense original, or raw, data into an easily understood frequency distribution:

Class Interval
A grouping of data, which occurs along a continuum, that is identified by boundary or limiting values.

Class Boundaries
The values that serve as the delineators of classes in a frequency distribution.

1. Determine the number of classes.
2. Establish the **class interval** or width for each class.
3. Set the **class boundaries** or values that form the beginning and end of each of the classes.
4. Count the number of values in the data set that fall into each class.

From just looking at raw data, the number of classes that will best represent the data may not be apparent. Terrell and Scott provide a convenient rule of thumb for determining the number of classes for a frequency distribution.

The approximate number of classes is the integer that just exceeds
$2(\text{number of values in data set})^{(0.3333)}$ (1.2)

To illustrate, consider the compensation of the chief executive officers (CEOs) of 20 energy companies surveyed by *The Wall Street Journal* (April 22, 1992). (See Table 1.5.) Comparable data were obtained on only 18 of those CEOs (see note below the table). Thus, the approximate number of classes to be employed in a frequency distribution is

$$2(18)^{(0.3333)} = 2(2.62) = 5.24$$

TABLE 1.5 Compensation of CEOs of Energy Companies

	Company	Executive	1991 Salary/ Bonus (in thousands)	% Change from 1990	Long-Term Comp. (in thousands)	Total Direct Comp. (in thousands)
1.	Amerada Hess	Leon Hess	300.0	N/A	0.0	300.0*
2.	Amoco	H. Laurance Fuller	1302.2	N/A	236.3	1538.5
3.	Ashland Oil	John R. Hall	1038.0	–6	0.0	1038.0
4.	Atlantic Richfield	Lodwrick M. Cook	1673.9	–26	2008.9	3682.8
5.	Baker Hughes	James D. Woods	916.0	–27	393.1	1309.1
6.	Chevron	Kenneth T. Derr	1385.6	2	363.8	1749.4
7.	Coastal	James R. Paul	813.8	N/A	913.4	1727.2
8.	Dresser Industries	John J. Murphy	1070.6	–3	578.7	1649.3
9.	Exxon	Lawrence G. Rawl	1812.8	17	7453.0	9265.8
10.	FINA	Ron W. Haddock	476.3	N/A	0.0	476.3
11.	Halliburton	Thomas H. Cruikshank	1200.0	9	401.5	1601.5
12.	Kerr-McGee	F.A. McPherson	542.0	–32	129.4	671.4
13.	Mobil	Allen E. Murray	2040.0	9	2624.9	4664.9
14.	Occidental	Ray R. Irani	2324.0	N/A	147.1	2471.1
15.	Pennzoil	James L. Pate	650.0	–2	0.0	650.0
16.	Sun Co.	Robert McClements Jr.	1001.1	3	0.0	1001.1
17.	Texaco	James W. Kinnear	1667.0	–14	1658.8	3325.8
18.	USX	Charles A. Corry	1313.2	–18	840.2	2153.4
19.	Unocal	Richard J. Stegemeier	1054.7	N/A	0.0	1054.7*
20.	Valero	William E. Greehey	1097.4	12	352.5	1449.9

*Salary/Bonus data not valid for comparison (e.g., partial year data).

Source: *Wall Street Journal*, April 22, 1992.

where $(18)^{(0.3333)}$ is the cube root of 18, 2.62. The number of classes to be employed in the frequency distribution is then 6, which is the integer value that just exceeds 5.24.

The second step in constructing a frequency distribution is to establish class interval widths. Often this is achieved by simply finding the highest and lowest values in the data set and dividing their difference (the range) by the number of classes.

$$\text{Approximate width of each class, after an adjustment for outliers} = \frac{\text{highest value} - \text{lowest value}}{\text{adjusted number of classes}} \tag{1.3}$$

Outlier
An extreme value.

If, however, the data set has a few extreme values, called **outliers**, then these values are best treated as special cases or as classes unto themselves. For example, among the 18 CEOs, Lawrence Rawl's compensation of $9,265,800 at Exxon is an outlier. Because this value is far above the others, it will be difficult to include in a class with lower values. It is best treated as a class of its own. The next highest compensation is that of Allen Murray's at Mobil, $4,664,900. It is nearly one-half that of Rawl's. The remaining compensations are more closely spaced down to the lowest of $476,300 for Ron Haddock at FINA. The subrange for these 17 compensations is $4,188,600.

Allocating one class to the outlier $9,265,800 leaves us with five classes to be divided into a subrange of $4,188,600. Each of these five classes is thus $837,720. Arbitrarily rounding up to the nearest $10,000 gives a class width for these five classes of $840,000. (As will be seen in a moment, this rounding up is done to insure that the classes are sufficiently wide to cover the data of interest.)

The third step is to set the class boundaries. Class boundaries are set so that the classes include all the values. The lowest boundary that begins the lowest class will typically be an arbitrary, but convenient, value slightly below the lowest value in the data set. For example, in the case of CEO compensation, the lowest compensation is $476,300. Rounding down to the nearest $10,000 gives an arbitrary lower boundary for the first class of $470,000. The upper boundary for the lowest or first class is set by adding the class width to the lower boundary. Thus, the first class includes all values from $470,000 to $1,310,000. Similarly, the upper boundary for the second class is obtained by adding the class width to the upper boundary for the first class. The second class has a lower boundary of $1,310,000 and an upper boundary of $2,150,000. Likewise, the boundaries of the third class are $2,150,000 and $2,990,000 and so on up to $4,670,000 with $9,265,800 being a special case, as shown in Table 1.6.

To include the outlier of $9,265,800 in the distribution, the prior class of $3,830,000 to $4,670,000 may be left open as $3,830,000 and higher. Ideally, open-ended classes should be avoided because of the loss of information. Sometimes they cannot be avoided conveniently, as apparently was the case for Griliches in constructing Table 1.1, where the fourth case is "1000 +."

Class boundaries must be set so that each value falls into one and only one class. The boundaries cannot overlap and must be distinct. Ideally, boundaries are set at values that do not occur in the data set, with the observed values tending to fall toward the middle of the classes. For large data sets, with values that could occur anywhere along a number line, it is critical to know the **limiting value of a class;** that is, whether the class includes the boundary value or not. We must know to which class the actual boundary value belongs. In the case of the CEO compensation, we will define the lower boundary value to belong to that class and the upper boundary value to belong to the next class. Thus, we can write the classes as:

Class Limits
The lowest and highest attainable values in a class interval.

Class 1: At least 0.470 but less than 1.310 million

Class 2: At least 1.310 but less than 2.150 million

Class 3: At least 2.150 but less than 2.990 million

Class 4: At least 2.990 but less than 3.830 million

Class 5: At least 3.830 but less than 4.670 million

Outlier: 9.2658 million

or in a shorthand notation as in Table 1.6.

Table 1.6 Frequency Distribution of Energy Company Chief Executive Officers

i	Compensation of Energy Company CEOs (in millions of $)	Frequency f_i	Relative Frequency f_i/n
1	0.470 to 1.309999	6	0.3333
2	1.310 to 2.149999	6	0.3333
3	2.150 to 2.989999	2	0.1111
4	2.990 to 3.829999	2	0.1111
5	3.830 to 4.669999	1	0.0555
	Outlier 9.265	1	0.0555
		18	1.000

If the data set consists of items that have only distinct value, as with the integer value for the number of employees in Griliches' Table 1.1, then the limiting values of the class are the lowest and highest attainable integer values in the class. The class boundaries, however, must be placed between these class limits. The class boundary that separates the class intervals given by "10–99" and "100–999" in Table 1.1 is, thus, 99.5 even though no firm could have a fractional employee.

The fourth step in constructing the frequency distribution is to count the number of observations in each class. These counts for the CEO compensation case are shown in Table 1.6. From this frequency distribution, it can be seen easily that there are six CEOs with compensation of at least $470,000 but less than $1,310,000, six CEOs with compensation of at least $1,310,000 but less than $2,150,000, and so on.

As already demonstrated for the frequency distribution of firms prepared by Griliches, Table 1.1, relative frequencies are obtained by dividing each class frequency by the total number of observations in the data set. The relative frequencies for the CEO compensation data are shown in the third column of Table 1.6. For example, the classes with the highest relative frequencies are the first and second classes (i = 1 and 2) that include values equal to or greater than $0.470 million but less than $2.150 million. Each of these classes have relative frequencies of 0.333, which is calculated as $f_1/n = 6/18$.

In summary, when constructing a frequency distribution, several general principles must be kept in mind. First, an observation must fall in one and only one class; classes cannot overlap. This first principle can never be violated. Second, classes should have widths that are the same. Sometimes data are so spread out that maintaining equal class widths is difficult; in these cases outliers should be identified and treated as special cases. Third, in deciding how many classes to have, try to avoid empty classes or classes that contain an overly large percentage of the data. Fourth, the **midpoint** (middle) of each class should be representative of the observations in that class. Ideally, observations would be distributed evenly around the class midpoint, although this ideal is seldom achievable. Finally, try to avoid open-ended intervals. As demonstrated in Table 1.1, however, this is an ideal that often is not realized.

Midpoint
The middle value of a class in a frequency distribution.

FIGURE 1.3 Histogram of Energy Company CEO Compensation

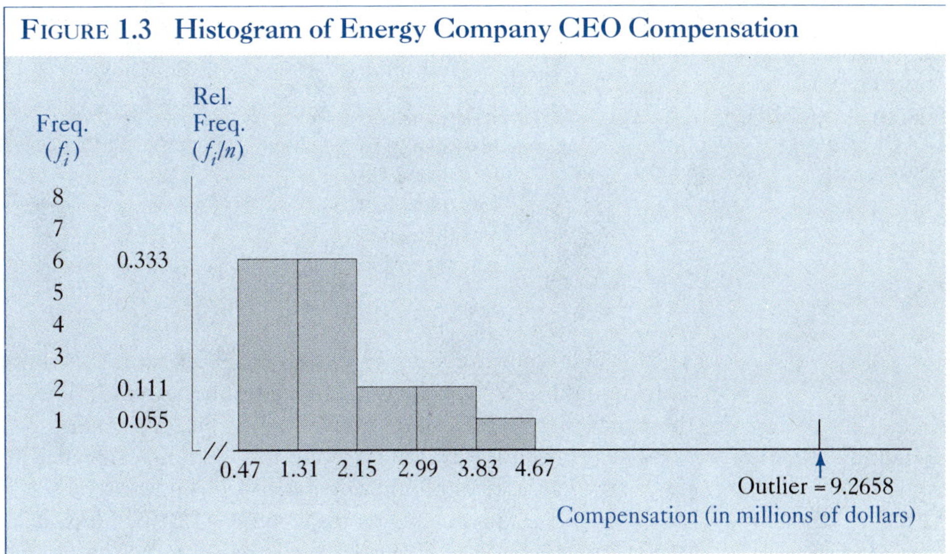

Histograms

Histogram
A graph of a frequency distribution, where class intervals are represented on a horizontal axis and the class relative or absolute frequencies are represented vertically.

The quantitative data in frequency distributions can be portrayed graphically in histograms. A **histogram** has a horizontal axis that is scaled to show all values of the measurements—in this case the compensation of energy company CEOs. It is also segmented to identify class boundaries with a vertical bar rising to the height of the frequency (or relative frequency) for each class at the identified boundary value. Figure 1.3 shows the histogram for the 18 CEOs in the sample. Notice in this figure that both the highest and lowest values in the data set are identified to eliminate any confusion about the location of these values.

Cumulative Frequency Distributions

Cumulative Frequency Distribution
A presentation of an entire set of cumulative frequencies.

Cumulative Frequency
The number of observations that have values less than or equal to a specified number.

Knowledge of the number of observations that lie below a certain value is often desired. For instance, in assessing CEO compensation, knowing the compensation level achieved by the majority of CEOs would be helpful. A **cumulative frequency distribution** shows the total number of observations that are less than a specified value. When data are in classes like those in Tables 1.1 and 1.6, the specified values are the class boundaries (10, 100, and 1,000 employees in Table 1.1 and $0.47, $1.31, $2.15, $2.99, $3.83, and $4.67 million in Table 1.6).

The **cumulative frequency** of $1.31 million for the compensation data in Table 1.6, as shown in Figure 1.3, is six, since there are six observations that are less than $1.31 million. The cumulative frequency for $2.15 million is 12: the six values less than

TABLE 1.7 Cumulative Frequency of Energy Company CEO Compensation

CEO Compensation (in thousands of dollars)	Cumulative Frequency	Cumulative Relative Frequency
less than 1310	6	0.33333
less than 2150	12	0.66667
less than 2990	14	0.77778
less than 3830	16	0.88889
less than 4670	17	0.94444
less than 9266	18	1.00000

$1.31 million plus the six values between $1.31 and $2.15 million, and similarly up to $4.67 million for which the cumulative frequency is 17. The cumulative frequency for the entire distribution, up to and including $9.2658 million, is 18.

Cumulative Relative Frequency
A ratio calculated by dividing a cumulative frequency by the total number of observations in the data set.

The **cumulative relative frequency** is a ratio calculated by dividing a cumulative frequency by the total number of observations in the data set. For example, the cumulative relative frequency for 2.15 is 0.6667 (= 12/18). Thus, we know that approximately 66.67 percent of the CEOs have compensation of less than $2.15 million. Both the cumulative frequencies and cumulative relative frequencies are shown in Table 1.7.

A graph of cumulative relative frequencies provides a good visual impression of how the data accumulate. The cumulative relative frequency function shown in Figure 1.4 was constructed from the CEO cumulative frequency distribution in Table 1.7. In Figure 1.4, the height of each bar shows the percentage of data less than the upper boundary of the class. Thus, the bar for the class "2.15" shows that 66.7 percent of these CEOs make less than $2.15 million.

Ogive
A series of lines connecting the lower boundaries of each class in a cumulative relative frequency distribution.

An **ogive** is sometimes used to smooth the graph of cumulative relative frequencies. An ogive is formed by lines connecting the lower-right corner points of the bars that make up the cumulative relative frequency graph. It is typically drawn as a smooth arching curve, as in an architectural arch of the same name. At each of these points, the ogive shows the amount of the distribution that is less than the value on the horizontal axis. Figure 1.4 is the ogive for the CEO compensation data. The ogive will be used in the next chapter to find hypothetical values that might lie within given class intervals.

QUERY:
For all eight industrial groups in the *Wall Street Journal* (April 22, 1992) data set on CEO total pay—including salary, bonus, and long-term incentives—construct a frequency distribution and a histogram to give a visual display of the data. Also, provide the cumulative frequencies associated with the frequency distribution you construct. Is there a big difference between this distribution and that of the CEOs of energy companies?

ANSWER:
Because this data set is large, a histogram cannot be constructed easily without a computer program. With a statistics package such as the ECONOMETRICS TOOLKIT (ET), this data set can be managed quickly. The data set is available in

FIGURE 1.4 Cumulative Frequency and the Ogive

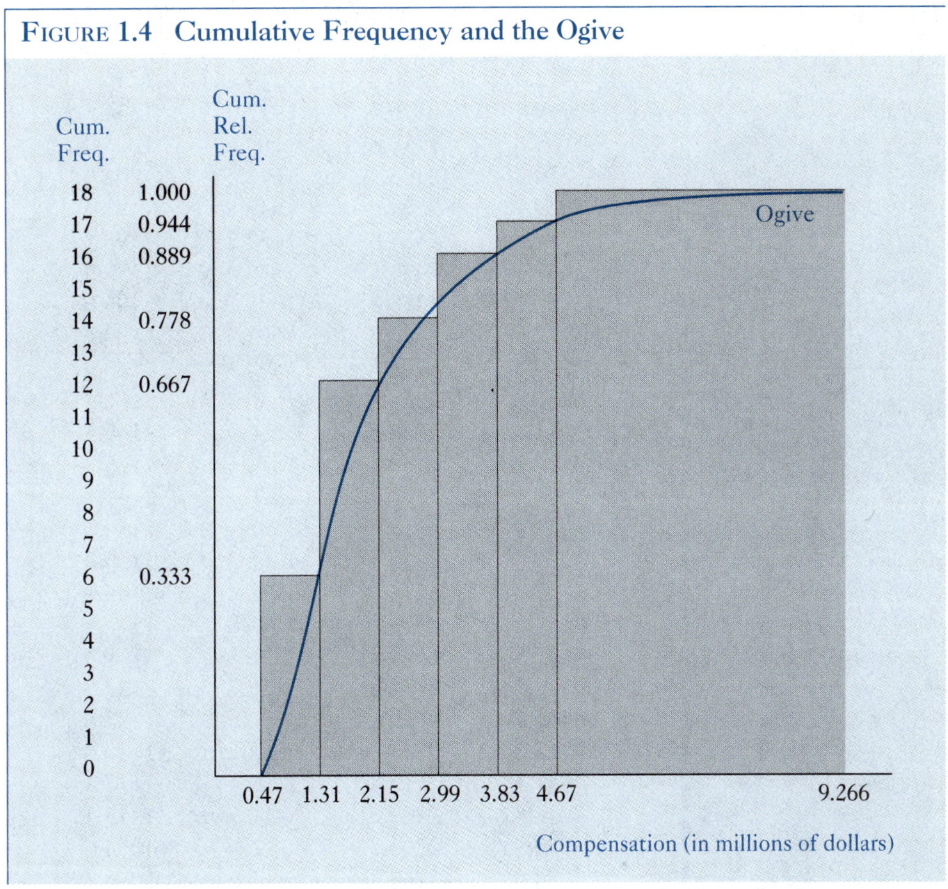

an ASCII file accompanying this book so it need not be typed into a file. It can be read into almost any computer program. For some programs, this may mean first bringing the file into a word processor to remove all the nonnumerical text and extra rows. Also, note that not all observations are comparable. There are only 317 comparable compensations, which follow in descending order.

74799.7	3221.6	1915.5	1324.5	1000.0	676.4
22563.2	3176.2	1911.9	1324.2	997.4	673.8
16913.2	3067.6	1898.6	1322.4	995.0	671.9
16778.9	3067.6	1872.8	1314.8	994.0	671.4
13812.7	3034.8	1840.4	1311.3	990.8	662.5
12178.0	3011.3	1835.2	1309.1	982.6	660.0
11436.2	2958.7	1803.6	1304.4	976.7	659.5
9725.9	2951.9	1761.5	1295.7	971.8	653.6
9265.8	2943.4	1750.8	1294.4	960.2	650.0
7620.0	2907.6	1749.4	1275.0	948.6	650.0

7211.8	2880.4	1740.0	1271.5	945.8	625.4
6654.9	2877.1	1727.2	1239.9	941.1	624.9
6411.4	2850.0	1719.4	1235.2	937.6	623.3
6299.5	2810.0	1717.0	1234.1	935.6	617.7
5955.8	2750.8	1709.6	1206.9	931.4	615.3
5829.7	2568.2	1672.3	1203.0	925.0	609.4
5630.0	2550.9	1649.3	1195.4	924.7	600.6
5497.3	2523.0	1632.2	1193.2	923.1	596.7
5441.5	2496.1	1625.0	1188.0	910.3	580.1
5299.1	2488.2	1623.7	1185.0	891.9	575.0
5173.5	2477.8	1601.5	1180.4	886.8	574.4
5100.8	2473.6	1590.4	1180.0	879.0	574.0
5030.8	2471.1	1590.2	1166.3	878.1	557.0
5011.1	2468.2	1586.9	1166.3	875.9	556.2
4887.0	2468.0	1571.6	1165.2	870.0	546.3
4664.9	2444.6	1570.0	1161.4	850.8	540.8
4450.0	2419.7	1568.4	1150.0	842.0	538.5
4320.0	2419.4	1566.2	1150.0	837.0	536.0
4302.6	2413.9	1553.2	1144.0	820.9	533.1
4243.0	2411.2	1545.8	1140.8	816.7	526.3
4207.5	2373.8	1538.5	1139.3	810.3	525.0
4156.3	2334.5	1528.7	1132.8	807.5	523.7
4151.7	2250.7	1513.5	1127.8	806.0	523.6
4112.3	2239.2	1506.7	1120.5	775.0	522.5
4047.1	2212.2	1495.1	1118.4	774.6	522.0
4046.4	2207.0	1490.7	1115.4	774.4	504.6
3999.8	2203.7	1486.9	1094.6	770.1	500.0
3950.0	2169.7	1467.2	1094.4	764.2	491.4
3949.2	2153.4	1449.9	1090.0	756.9	488.8
3905.6	2141.8	1443.8	1086.5	754.6	476.3
3880.0	2119.2	1428.0	1085.5	752.7	475.0
3836.1	2079.4	1400.7	1078.2	743.6	455.0
3747.0	2070.0	1398.6	1067.4	739.6	452.8
3682.8	2042.5	1395.1	1066.1	730.7	450.0
3681.1	2017.0	1382.9	1049.8	730.0	447.9
3616.4	2009.9	1377.5	1041.2	722.9	442.4
3600.0	2007.2	1376.7	1038.0	722.3	418.5
3576.1	2004.1	1346.9	1037.5	696.0	405.0
3423.4	2001.5	1341.1	1019.6	691.8	389.6
3353.0	2000.0	1336.7	1009.5	683.0	375.0
3344.5	1991.8	1335.0	1009.4	680.5	270.0
3325.8	1955.6	1334.0	1006.9	678.9	100.0
3308.5	1950.0	1325.3	1001.1	677.1	

Although Formula 1.2 suggests that there should be 14 classes,

$$2(317)^{(0.3333)} = 2(6.817) = 13.6$$

to make comparisons with the relative frequency distribution in Table 1.6 only five classes were employed. Using the same class boundaries as employed for the energy companies in Table 1.6, ET gives the following information on the histogram.

Histogram for CEOPAY computed using 317 observations
Observations out of range: too low = 11, too high = 25

			Frequency		Cumulative	
	Lower Limit	Upper Limit	Total	Relative	Total	Relative
0	470.000	1310.000	142	.5053	142	.5053
1	1310.000	2150.000	72	.2562	214	.7616
2	2150.000	2990.000	33	.1174	247	.8790
3	2990.000	3830.000	17	.0605	264	.9395
4	3830.000	4670.000	17	.0605	281	1.0000

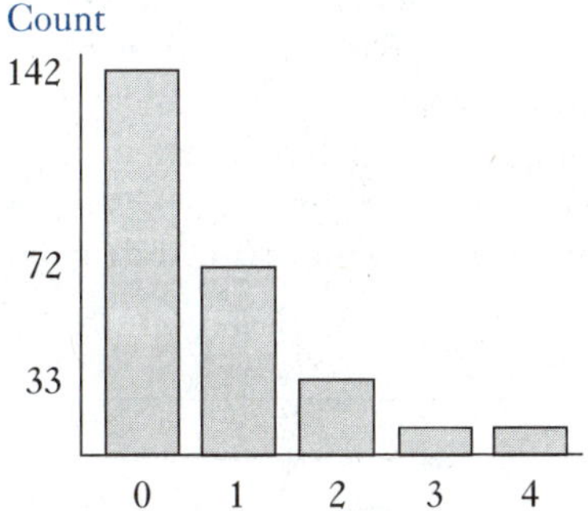

The ECONOMETRICS TOOLKIT requires that all class limits have a lower and an upper bound. Because we closed the upper class at $4.67 million, ET states that there are 25 observations higher than this value, or 7.89 percent of 317. Notice that ET does not give the histogram for the relative frequency, although our comparisons must be made in terms of the relative frequencies since the two data sets are of different sizes. This poses no problem, because ET gives the corresponding relative frequencies in tabular form immediately before printing the histogram. However, these relative frequencies are based on only the 281 observations within the $470 thousand and $4,670 thousand range.

There is some difference between the compensation of CEOs in general and those of the energy companies. For example, none of the energy company CEOs make less than $0.47 million per year while the distribution of all CEOs shows 11, or 3.47 percent, making less than $0.47 million. In addition, 71 percent [(11 + 142 + 72)/317] of the 317 CEOs make less than $2.15 million while only 67 percent of the 18 energy company CEOs make less than $2.15 million. Before concluding that energy company CEOs are among the highest paid CEOs, however, notice that at the high end of the compensation distribution, 7.89 percent of the 317 CEOs make more than $4.67 million, while only 5.56 percent of the 18 energy company CEOs do. Energy company CEOs are not among the lowest-compensated CEOs, but they are not among the highest-compensated ones either.

1.8 Stem-and-Leaf Graphics (Optional)

As an alternative to the histogram representation of a frequency distribution, a less formal means of preliminary and exploratory data analysis can be undertaken with a stem-and-leaf diagram. A stem-and-leaf diagram visually combines the frequency distribution and histogram into one chart and does not require the researcher to be concerned about the number of classes or the setting of endpoints. It is ideally suited to small data sets.

To construct a stem-and-leaf diagram with paper and pencil for the 18 energy company CEOs' compensation data, the lead digit (the stem) of each dollar figure is written in ascending order to the left of a vertical line, as in Figure 1.5, panel a. The next digit (the leaf) for each compensation is then written to the right of the vertical line on the same row as the associated stem. Computer programs can be used to construct these diagrams, as shown in Figure 1.5, panel b, where a MINITAB printout is shown. From these diagrams we can quickly see that most of the compensations are in the one to two million dollar range and that there is one high outlier.

Unlike histograms, no strict rules are required for constructing stem-and-leaf diagrams. For example, although the stem in our example was the first digit, we could have specified one-half of the first digits as the stem and the remaining digits as the leaf. This flexibility makes stem-and-leaf diagrams easy to construct with paper and pencil.

Exercises

1.16 A student is overheard saying that "all frequency distributions can be represented in histograms, which are a form of bar charts, but not all bar charts represent frequency distributions." Comment on the correctness of this quote.

1.17 Business Week (August 31, 1992) asked 401 executives: "How do you rate the President's team of economic advisers . . . . Excellent, pretty good, only fair, or poor?" They portrayed the results in the accompanying diagram.

This diagram is a/an (check those that apply):

a. Bar chart
b. Histogram
c. Line graph
d. Relative frequency distribution
e. Ogive
f. Representation of the population

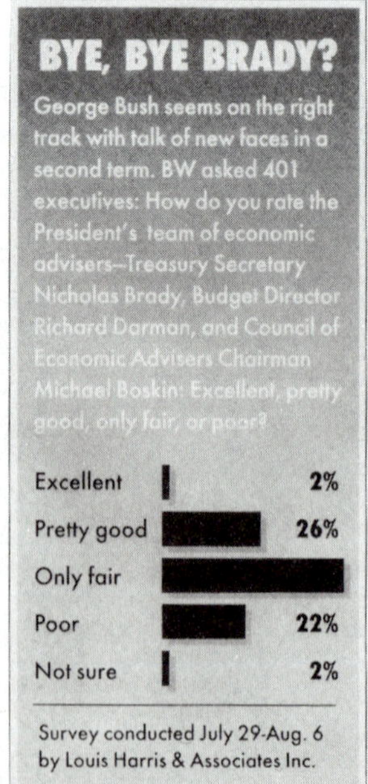

1.18 Construct a histogram for the breakdown of owner-occupied housing in Monroe County as shown in Exercise 1.4.

1.19 Construct an ogive for the breakdown of owner-occupied housing in Monroe County as shown in Exercise 1.4.

FIGURE 1.5 Stem-and-Leaf Diagrams for Energy Company CEO Compensation

Panel a: Diagram drawn with paper and pencil

The Stem	The Leaf
0	4 6 6
1	0 0 3 4 5 5 6 7 7
2	1 4
3	3 6
4	6
•	
•	
•	
9	2

Panel b: Diagram created in MINITAB, where the user's commands are lowercase and the computer's prompt and response are uppercase.

```
MTB > read 'ceo18' c6–c7        ← to read in the file ceo18, which contains two
          18 ROWS READ             rows: c6 is the company number and c7 is
                                   the CEO compensation

ROW     C6       C7            ← MINITAB's response
  1     10     476.3
  2     15     650.0
  3     12     671.4
  4     16    1001.1
  .      .     .

MTB > stem c7                   ← Command to create a stem-and-leaf diagram
                                   for c7

Stem-and-leaf of C7   N = 18    ← MINITAB's response
Leaf Unit = 100

    3    0  466
  (9)    1  0003456677          ← Stem that has the leaf middle value in the
    6    2  14                     ordered data set is identified with the count
    4    3  36                     given in parentheses
    2    4  6
    1    5
    1    6
    1    7
    1    8
    1    9  2

MTB > stop
```

1.20 Census Bureau data for 1989 show the poorest 20 percent of the population of married couples with children had a cash income under $22,700; the next 20 percent had income under $34,100; the middle 20 percent had income under $45,500; the next 20 percent had income up to $62,200; and the richest 20 percent had income over $62,200. Draw the ogive for this distribution.

1.21 Construct a pie chart from the relative frequencies you calculated from the breakdown of owner-occupied housing in Monroe County in Exercise 1.4.

1.22 Do you prefer your histogram (Exercise 1.18), ogive (Exercise 1.19), or pie chart (Exercise 1.21) representation of the breakdown of owner-occupied housing in Monroe County? Explain.

1.23 From the following pie chart taken from the *Chicago Tribune* (September 6, 1992), what percentage of employees are affiliated with Ameritech Information Systems?

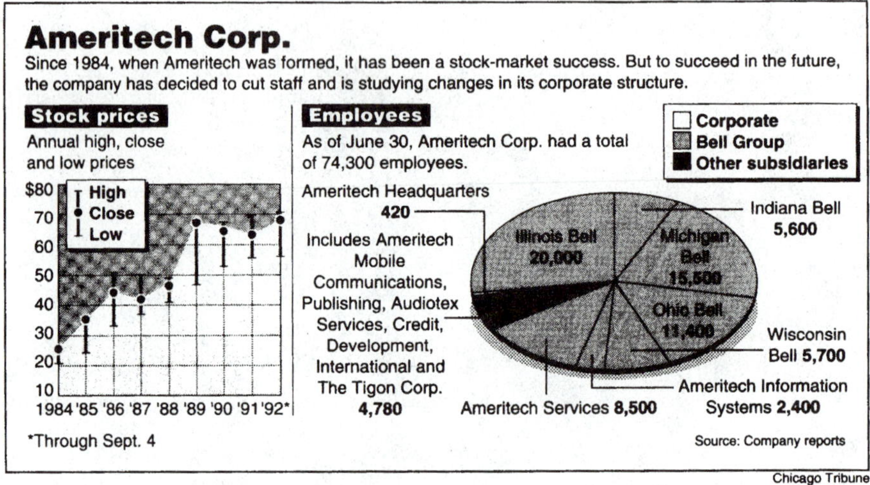

1.24 On the following page are two pie charts from the *Wall Street Journal* (September 9, 1992) that show the distribution of pension funds among mutual funds, insurance companies, banks, and other financial institutions. The article said that the total value of pension funds in 1992 was $1 trillion. Make a bar chart showing the dollar amounts of this trillion dollar pool held by the four identified institutions.

1.25 An article in the *Wall Street Journal* (November 13, 1992) reported that "Airlines Wage Fare Battle City by City, Spreading Confusion as Well as Bargains." The following table was provided to support the text:

	American	Continental	Delta	Northwest	United
Phoenix–St.Louis	$ 400	$ 248	$ 400	$ 400	$ 228
Austin–Los Angeles	248	258	360	340	248
Detroit–Kansas City	164	113	280	113	113

a. Draw three separate diagrams showing the five airline fares of each of the three cities in a bar chart.

b. Use a spreadsheet program or a graphics program to create one diagram showing a three-dimensional bar chart of fares for the three cities and the five airlines.

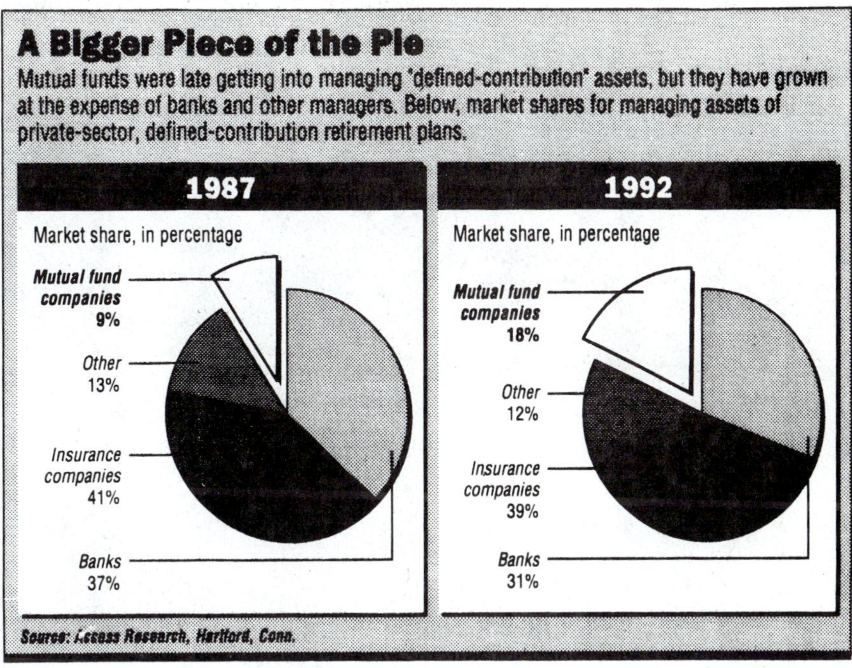

1.26 Following is information on supermarket sales and market share for the leading sports drinks from the *Wall Street Journal* (September 29, 1992). Draw a pie chart showing the market share of each product.

Sports Drink	Sales 52 Weeks to 7/12/93 (millions)	Market Share 52 Weeks to 7/12/92	Sales Growth 1/1/92–7/12/92
Gatorade	$ 288.3	85.2%	+ 7%
Gatorade Light	14.6	4.3	+ 16
10-K	12.2	3.6	—
PowerBurst	6.7	2.0	– 72
Private label	4.2	1.2	– 19
1st Ade	2.6	0.8	– 1
All Sport	1.7	0.5	+451
Joggin' in a Jug	1.6	0.5	+138
Enduro	1.2	0.4	– 45
Body Works	0.9	0.3	+ 6
Spike	0.4	0.1	+182

Source: Information Resources Inc.

1.27 Texas Instruments is cited as the source of a *PC Magazine* (May 26, 1992) bar chart designed to show that "a DSP (digital signal processor) can process a MAC (multiply and accumulate) 10 times as fast as a 486 chip." Comment on why this bar chart is poorly drawn for its stated purpose.

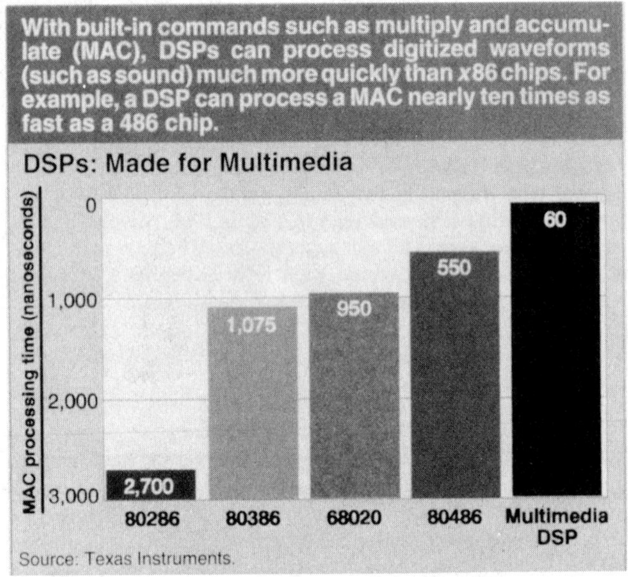

1.28 Below is a pie chart that shows the tuition cost of an executive-education program, as reported in the *Wall Street Journal* (September 10, 1993). From this information, construct the corresponding histogram and ogive. What does the ogive show that the pie chart and histogram do not?

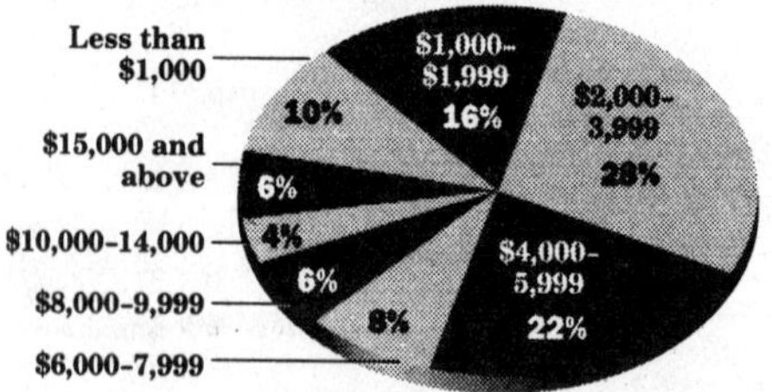

1.9 Presenting Qualitative Data

Bar Chart
A graph showing the frequencies (or relative frequencies) of some qualitative or quantitative attributes as a rectangle or bar drawn from one of the axes.

Pie Chart
A circle divided into sections to show the relative number of times a characteristic occurs.

In addition to line graphs, such as Figure 1.2, bar charts are often used to present quantitative data as in Figure 1.3. Bar charts and pie charts can be used to present qualitative data as well.

A histogram is an example of a **bar chart** in which the vertical axis (height of a bar) shows the frequency of quantitative values shown on the horizontal axis. Bar charts are not restricted to showing frequency distributions. For example, *Business Week* (August 17, 1992) used a bar chart to show a 0.2 percent drop in the U.S. Commerce Department's composite index of leading indicators for June (see Figure 1.6). This bar chart showed that "the drop was the first since December."(p. 23) Chapter 16 is devoted to the construction, presentation, and use of index numbers.

Pie charts are another graphical method of presenting qualitative data; they are especially popular for presenting financial and marketing information. For example, the *Newsday* (reprinted in *HT* June 14, 1992) press release shown in Figure 1.7 used a pie chart in the form of a bottle top to show the market share of leading soft drinks. This diagram shows that Coke Classic has 20 percent of the market as defined by sales. This 20 percent is shown in the pie chart by allocating 20 percent of the 360 degrees (72°) in the circle to Coke Classic. Similarly, Pepsi's 18.4 percent share of the market is shown by allocating 18.4 percent of the 360 degrees (66.24°) to Pepsi.

Today, computer programs are readily available to create different types of diagrams to show both quantitative and qualitative data. The use of computers in statistics is addressed throughout this book. For now it is sufficient to recognize how data are collected, organized, condensed, and reported in numerical and graphical form. The next chapter introduces summary measures that can be employed in describing data. Spreadsheet programs such as LOTUS 1-2-3, QUATTRO, and EXCEL will

Figure 1.6 A *Business Week* Bar Chart Showing that "The indicators stumbled in June"

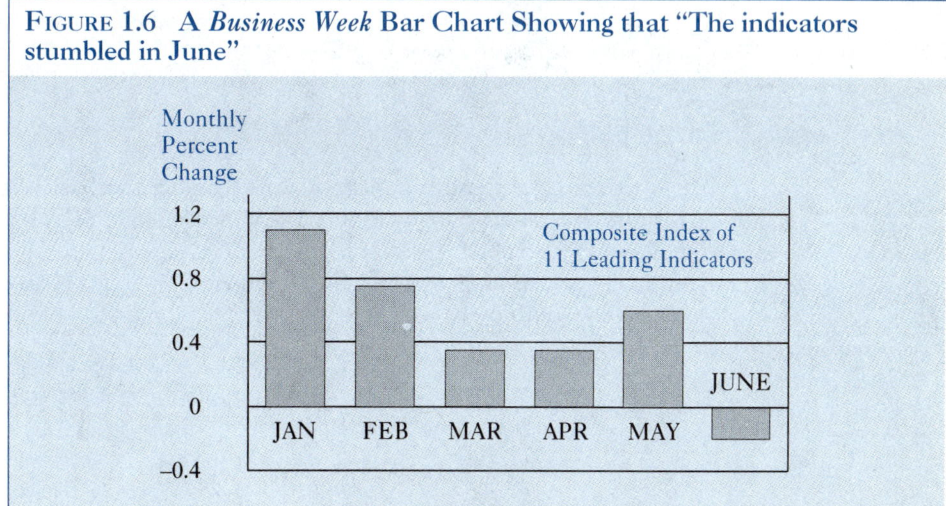

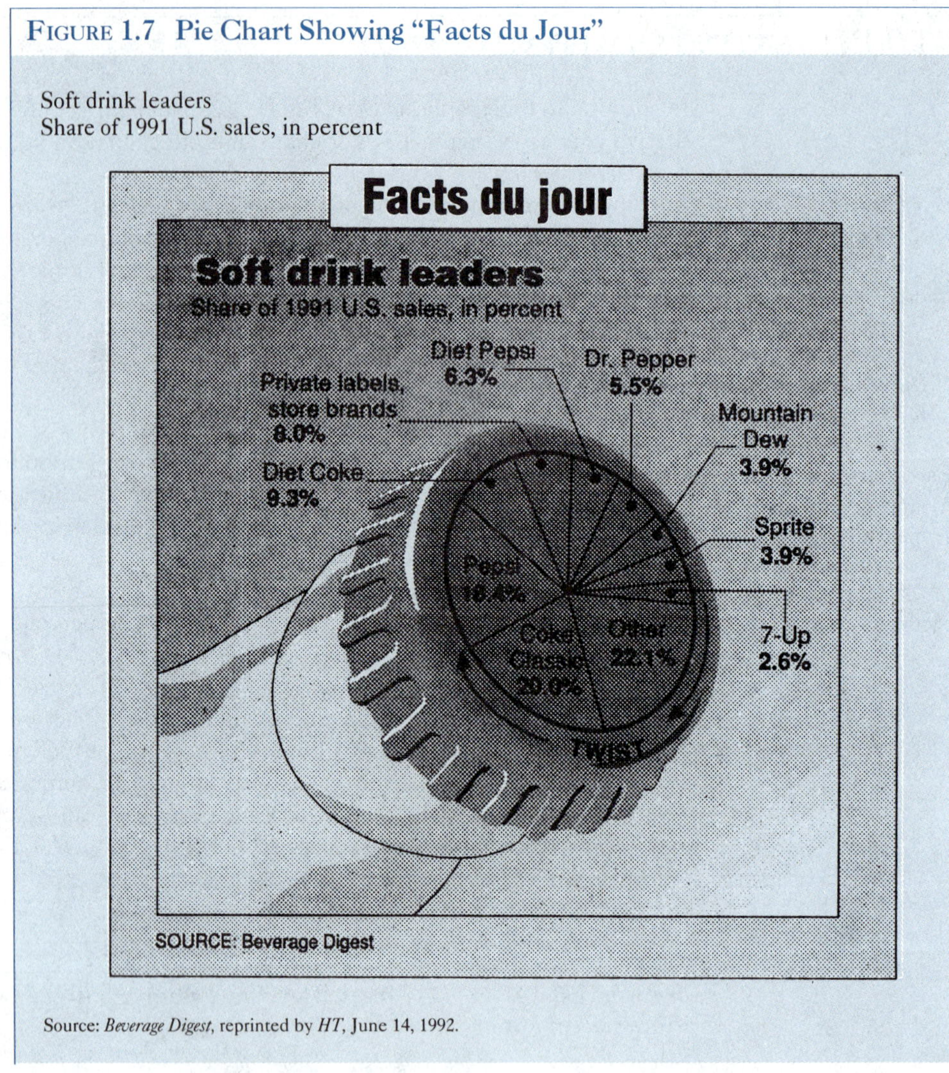

FIGURE 1.7 Pie Chart Showing "Facts du Jour"

Soft drink leaders
Share of 1991 U.S. sales, in percent

Source: *Beverage Digest*, reprinted by *HT*, June 14, 1992.

compute most of these summary measures and create basic diagrams to show results. As shown by the 12 charts in Figure 1.8 prepared by *PC Magazine* (November 10, 1992), presentation graphics packages can produce even more sophisticated data charts. Surprisingly, few programs do a fully adequate job with a histogram—some things may still be done best by hand.

FIGURE 1.8 Presentation Graphics: A Guide to Data Charting

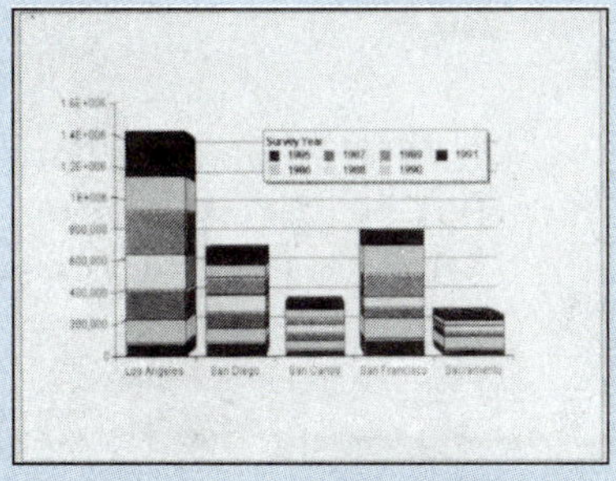

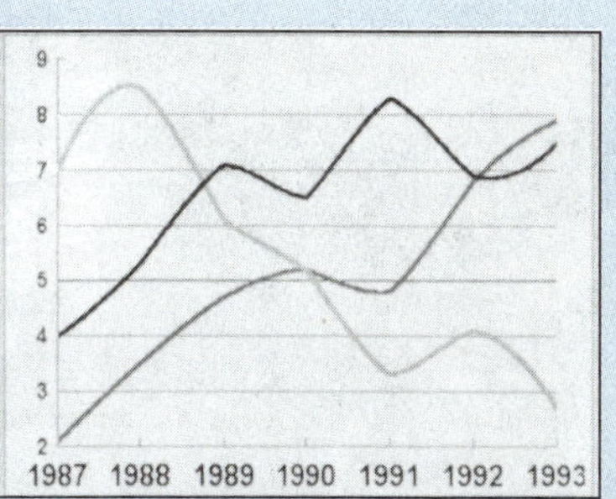

Column charts are useful for comparing discrete categories. For example, let's say you want to chart the number of participants by city in an ongoing survey. By representing each year as a different color, you can show the total and the break out by year while still keeping the display simple and attractive.

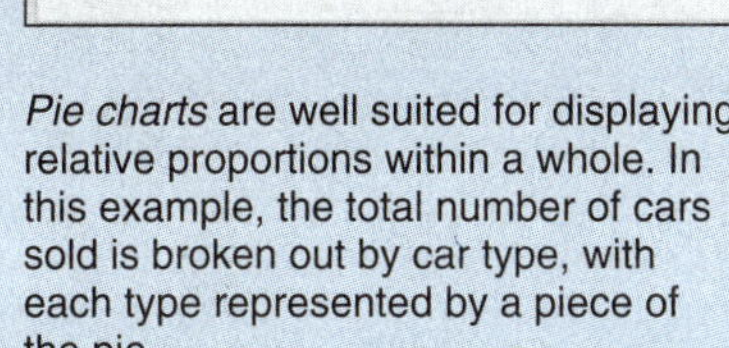

Pie charts are well suited for displaying relative proportions within a whole. In this example, the total number of cars sold is broken out by car type, with each type represented by a piece of the pie.

Everybody is familiar with the *line graph:* Today's charting tools provide options to smooth straight-line estimates into curves. Line charts are particularly useful in displaying trends (indicated by the slope of the line) over time. Anchors—lines running from the curve to the x- or y-axis—clarify exactly where a plotted point falls.

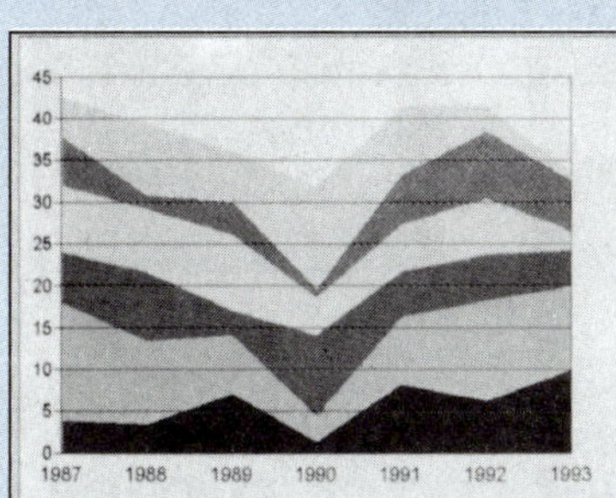

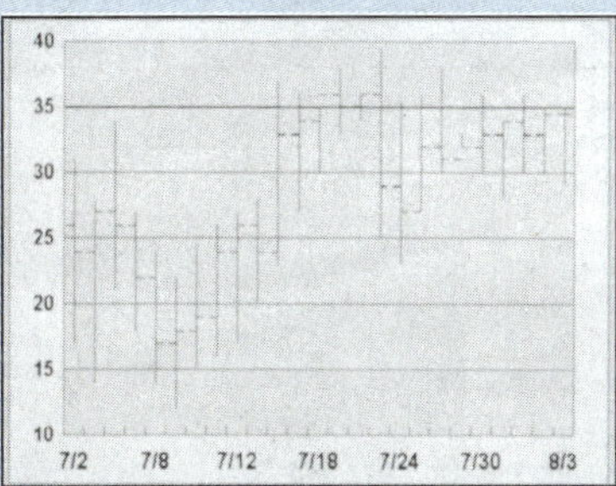

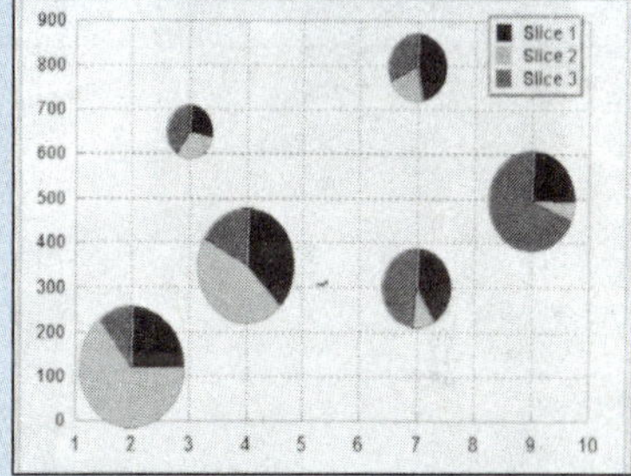

An *area plot* is useful for displaying the contribution of multiple items in making up a total over a continuous period of time. For example, you might track sales by product with each colored band representing a different product line.

High-low-close charts are commonly used to convey information about stock prices over time; in general, they can also be used to display concisely a range of observations taken over several periods of time. The vertical bars can be used to indicate the high and low, while ticks on the left and right of the bar indicate the initial and final observations, respectively. Color coding can be used to emphasize either the trend or the first or last observations.

Bubble plots are effective for displaying certain types of multidimensional data sets; bubble size is a visually intuitive way to represent relative importance. Let's say you want to look at product sales (on the *y*-axis) by region (on the *x*-axis). The diameter of the bubble could indicate the size of the potential market. By making each bubble into a pie chart, an additional variable can be introduced without making the chart visually complicated. For example, each slice could represent a different income group within the market.

(continued)

FIGURE 1.8 Presentation Graphics: A Guide to Data Charting *(continued)*

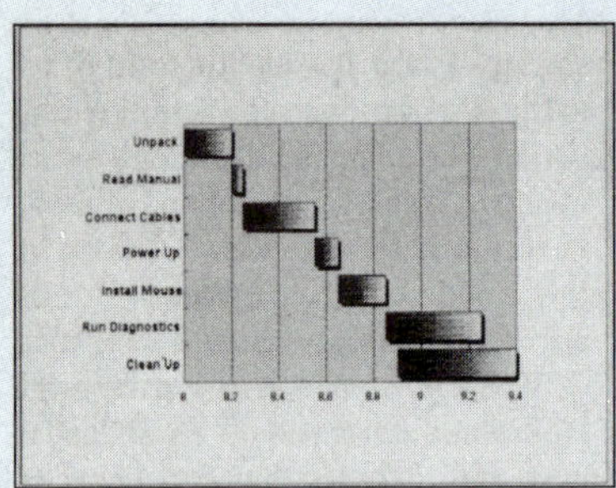

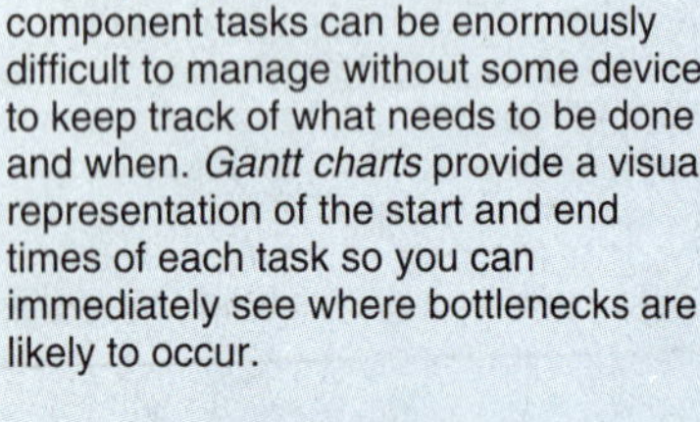

A complex project with many component tasks can be enormously difficult to manage without some device to keep track of what needs to be done and when. *Gantt charts* provide a visual representation of the start and end times of each task so you can immediately see where bottlenecks are likely to occur.

Scatter plots are an effective way to detect patterns visually in a data set. In regression analysis, a scatter plot of the errors is crucial for catching systematic patterns that would invalidate your results. The example shown includes *x,y* anchors, which make it easier to see the values associated with each point.

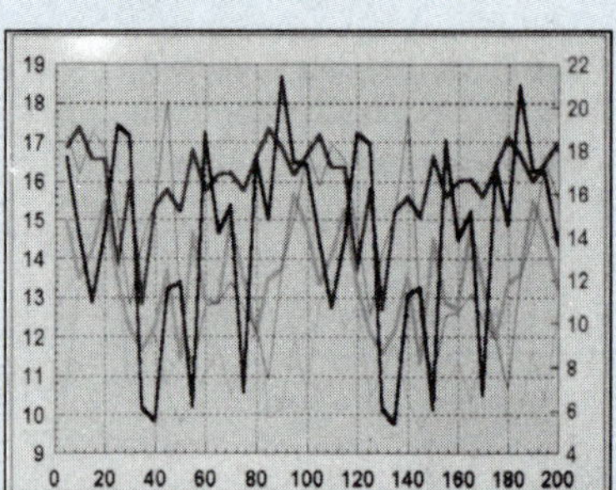

X,y plots with a double *y*-axis are useful for comparing the shapes of two curves that are scaled differently. For example, you could use an *x,y* plot to analyze seasonal sales patterns of two products with different sales volumes; the *x*-axis would be time, and each *y*-axis would be a product's sales. Similarities in seasonal patterns will be immediately apparent when the curves are plotted together on the same chart.

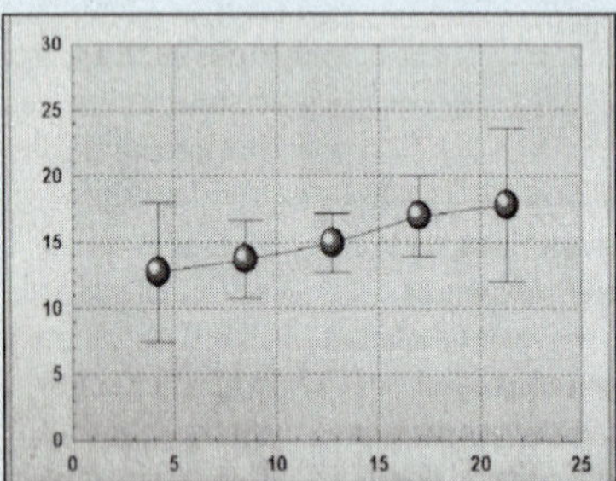

An *error plot* displays a measure of dispersion (the *standard deviation*) in addition to the mean, the give you a better sense of how well the average represents the data. Error-bar plots are commonly used to display the results of a regression analysis; an overlay of the *x,y* curve can be used to emphasize the trend.

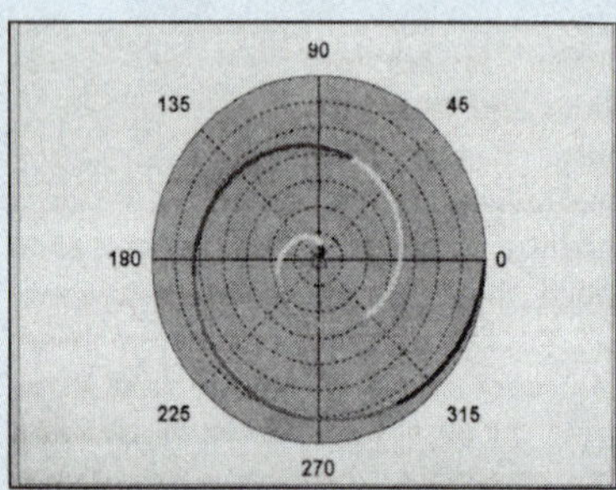

The Cartesian (*x,y*) coordinate system is useful for many things, but it is not the only way to represent a point in space. Points can also be described in terms of a circle, rather than a rectangle; each point is located according to its distance from the center and its angle from a fixed radius within the circle. *Polar plots* are a more natural way than Cartesian plots to describe certain phenomena and are commonly used in star mapping and vector analysis.

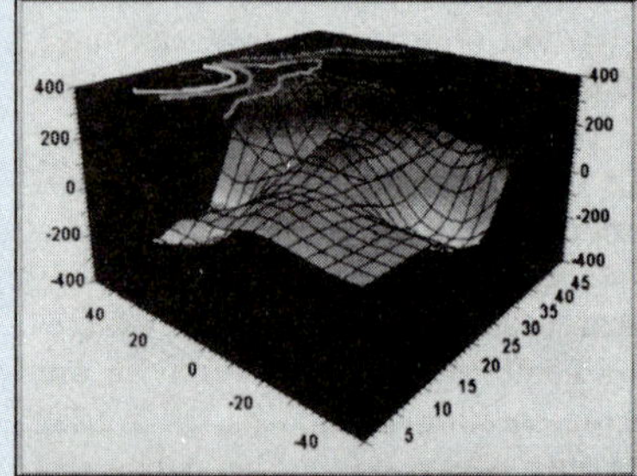

Surface plots are useful for three-variable trend analysis. They provide a quick visual summary of complex relationships such as the combined effect on sales volume of dollars spent on television advertising and in-store promotion. Overlaying a color-coded contour plot can help make the information even clearer.

Chapter Exercises

1.29 According to an article in the *Herald Times* (September 5, 1990), titled "Statistics Switch Blurs Census Data," Bloomington, Indiana, between the 1980 and 1990 census, grew by 11.5 percent by one set of figures and 15 percent by another. In 1980 the official population count was 52,044, but between 1980 and 1990 the city annexed an area that, in 1980, had a population of 1,152 people. In addition, 619 people needed to be added to the 1980 base because of an error in the 1980 count. The 1990 census figure for Bloomington was 59,998.

a. Show how the 11.5 percent and 15 percent growth rates were obtained.

b. Which percentage more accurately reflects population growth in Bloomington?

1.30 In sex and race discrimination cases, considerable courtroom time is usually devoted to defining the preselection pool (population) from which potential employees might be hired. Typically, plaintiffs (those bringing suit against an employer) argue that the appropriate preselection pool should be the entire U.S. working-age population. Defendants (employers) argue for a much more restrictive preselection pool, for example, only those in the local community who have obtained a specific education level. Why do the plaintiffs and defendants argue in this way over the appropriate preselection pool?

1.31 A *New York Times* News Service release on December 3, 1989 stated:

> The 117 executives surveyed told Runzheimer (the survey firm) they preferred the Taurus, followed by GM's Celebrity . . . . The highest rated foreign auto was a Toyota, which was selected as the most desirable by 6 percent of the managers.

a. From this quote we know that Toyota was selected by how many of the managers?

b. What else would you like to know about this sample before drawing a conclusion about the car that managers consider the most desirable?

1.32 In its second annual ranking of female-owned businesses, *Working Woman* (May 1993) reported that the number of such companies grew 20 percent in the past year to 6.5 million. Why might you question the magnitude of this increase based on data-collection considerations?

1.33 The following table is from *The Wall Street Journal* (April 19, 1990).

What's Happened to Domestic Airline Fares?

Group	Study Date	Conclusion
Economic Policy Institute	March 1990	Rate of decline in air fares has slowed since deregulation in 1978.*
Bureau of Labor Statistics	April 1990	Up 10.3% in March from a year earlier.
Transportation Department	February 1990	Down 15% in 1988 from 1984.*
American Express	April 1990	Coach fares 9% higher than a year earlier as of April 1, discount coach up 3%, advance purchase up 10%.
Air Transport Association	April 1989	Fallen on average 20% since deregulation.*
General Accounting Office	June 1989	Cost 27% more in 1988 at airports with only one or two dominant carriers than at other airports.

*Adjusted for inflation.

The accompanying article stated that, "Since free enterprise in the skies entered its second decade last year, at least 12 studies have tried to measure its impact on fares . . . the studies often clash on broad issues. Some researchers pronounce deregulation a smashing success while others say consumer savings are grossly exaggerated or nonexistent." From the above table, indicate why these studies appear to give different results.

1.34 An article on productivity in the *Wall Street Journal* (August 12, 1992) provided this line diagram showing two productivity indices, one for manufacturing and the other for the service sector. The article stated, "Manufacturing, whose share of jobs has shrunk as the service share has grown, has seen its productivity surge by a third since the late 1970s. But because of hardly any gain in the vast service sector, hourly output per employee in the overall economy has edged up less than 10% in that period . . . productivity in the service-type businesses fell at a 2.3% annual rate in the second quarter (of 1992), while manufacturing productivity rose 4.7%." Is the line diagram consistent with this quote? Why or why not?

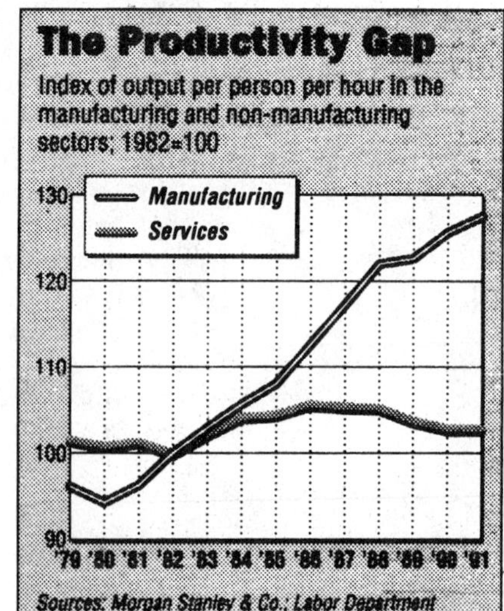

1.35 The *Wall Street Journal* (August 12, 1992) article that provided the line diagram in the previous exercise also stated that there are no productivity figures for nearly 70 percent of the people with service jobs. These include such key industries as insurance, health care, real estate, and stock and bond brokerage. Furthermore, economist Ronald Schmidt of the San Francisco Federal Reserve Bank, and Michael Darby of UCLA and the U.S. Commerce Department's chief economist, raise fundamental questions: "If productivity is so poor in service—where unionization often is low—then why are employers willing to grant consistently large pay increases? . . . capital spending in service (is) rising a brisk 8% this year on top of two years of substantial gains—hardly a sign of stagnating productivity. Over the same two years, manufacturing outlays were about flat . . . in addition foreign trade data show the U.S. with large surpluses in services for years. . . ." What does this economic reasoning suggest about measured productivity in the service sector?

1.36 According to a *Herald Times* (June 23, 1990) article, between 8 p.m. on June 17th and 5 p.m. on June 21st, 20 Indiana state troopers stopped 561 trucks and ticketed most of them to collect $42,985 in fines. Why would you be suspicious about conclusions drawn from this information regarding weight and number of violations of "the typical type of truck," on the highways?

1.37 A consultant says owners of private companies with annual sales below $100 million earned a median salary of just $51,000 in 1989. Those with sales of $20 million to $100 million have a median salary of $231,000 compared with $508,000 for heads of big public companies. The conclusion according to *Business Week* (April 16, 1990): "If you want a huge salary don't start your own business." Why doesn't this information imply that starting your own business yields small rewards?

1.38 Design a sample for the main library at your university. Assume the library wants to administer a questionnaire not only to users, but to potential users as well. Be sure to specify the population. Use as many of the techniques discussed in this chapter as possible.

Assume the budget for this study is limited. Lists of students (by class), faculty, and staff are available from the university.

1.39 The diagram on the right from the *Wall Street Journal* (August 14, 1992) shows average mental health costs per employee of U.S. companies. What kind of diagram is this? What are the approximate absolute and percentage increases in average cost per employee from 1987 to 1991?

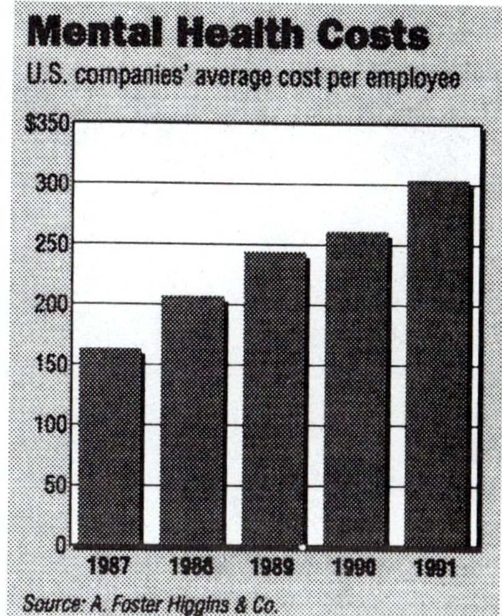

1.40 A note in the *Wall Street Journal* (October 8, 1992) reported "About 86% of U.S. units of foreign firms support the arts in the U.S. . . . says a survey by the Business Committee for the Arts. The result is based on 44 responses to the 115 survey forms sent out."

a. How high could this percentage have been had the response rate been 100 percent?

b. How low could this percentage have been had the response rate been 100 percent?

1.41 In sailboat racing, rule violations are handled by the local yacht club. Those involved in protests who are not satisfied with the results at the local level may appeal to the U.S. Sailing Association. Below are the frequency distributions for the time needed to process those appeals in 1989, 1990, and 1991, from *American Sailor* (September 1992).

Draw the histogram for each year of data. What does a comparison of these histograms show?

Time to Process an Appeal

Time in weeks	FY 89	FY 90	FY 91	Total
0 to < 5	5	2	4	11
5 to < 10	1	8	1	10
10 to < 15	9	4	1	14
15 to < 20	7	8	2	17
20 to < 25	3	3	3	9
25 to < 30	4	0	3	7
30 or more	0	1	4*	5
Total	29	26	18	73

*Only two appeals received before 10/31/91 were still open as of 4/30/92, and in both cases the Committee is waiting for information from the protest committee. Both will have taken more than 30 weeks to process and are included in the statistics.

1.42 On the following page is a pie chart from the *Wall Street Journal* (September 29, 1992) showing the amount of federal money going to small business for innovative research. Convert this pie chart into a bar chart showing the percentage from each type of government agency.

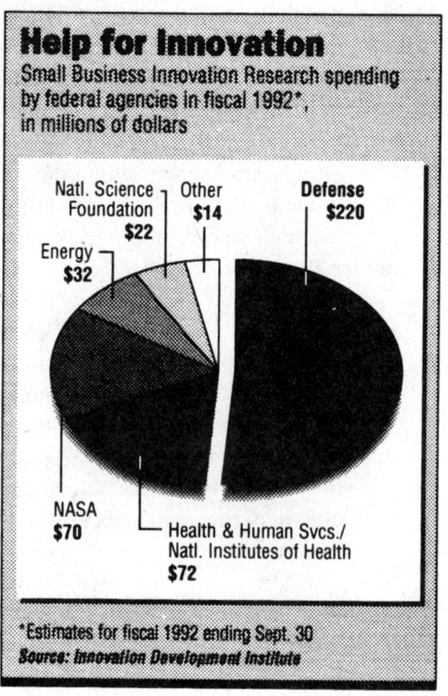

1.43 Following is a diagram and relative frequency distribution for U.S. home sales by price in the first quarter of 1990 as prepared by the *Chicago Tribune* (reported in *HT*, June 24, 1990).

a. Construct the histogram for this relative frequency data.

b. Construct the ogive for this relative frequency data.

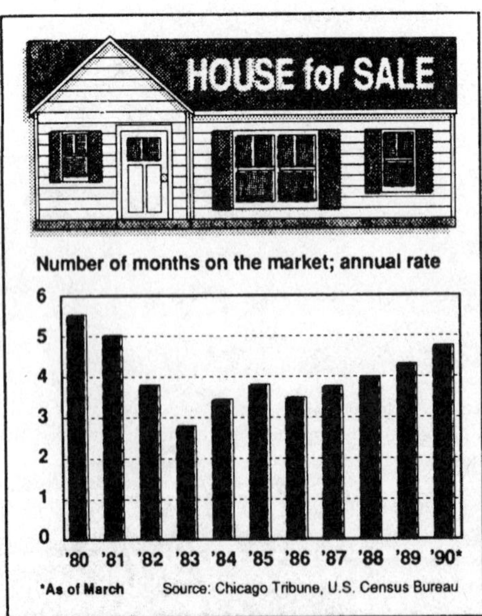

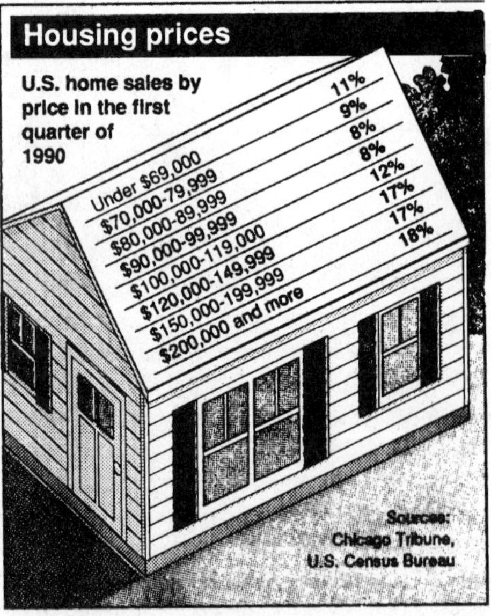

1.44 The article "Resort Ads Caught Snowing the Ski Set" (*WSJ*, December 22, 1992), reported on Killington Ski Resort's hiring of an engineering firm to measure its ski terrain and that of other New England ski areas in an attempt to show that "dishonesty and hype are rampant." The aerial survey conducted by W. Byrd LaPrade Inc. for Killington showed that while Killington "over-reported its mileage by some 9%," the other areas had "degrees of hype" much greater, as shown. Demonstrate why the "degree of hype" should be even greater than reported if it is a measure of over-reporting of the actual miles of trails.

Ski Area	Actual Miles of Trails	Miles Claimed by Resort	Degree of Hype
Killington	70	77	9%
Sugarbush	46	53	13%
Sunday River	29	36	19%
Sugarloaf	36	45	20%
Okemo	25	32	22%
Smuggler's Notch	22	32	31%

1.45 During World War II many economists, mathematicians, and statisticians were members of Columbia University's Statistics Research Group, which did high-level consulting work for the armed services. As part of this group's work, statistician Abraham Wald was asked where to place armor on planes. It seemed obvious to the aircraft engineers that armor was needed at the places most frequently hit, as found in a large sample of battle-proven airplanes. After studying the bullet holes of a sample of returning planes, Wald's conclusion was to place the armor where bullet holes were least frequently found in these planes. What was wrong with the aircraft engineers' sampling design? What did they overlook?

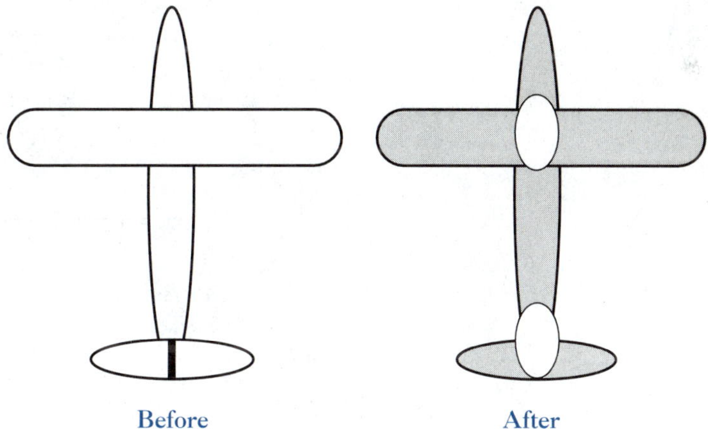

Source: Howard Wainer, "Graphical Visions from William Playfair to John Tukey," *Statistical Science*, August 1990: p. 343.

1.46 Below is a pie chart from an article in the *Wall Street Journal* (August 2, 1993) reporting on the smokeless tobacco market share. What percentage of the market did each of these four producers have? What is wrong with the *WSJ* diagram?

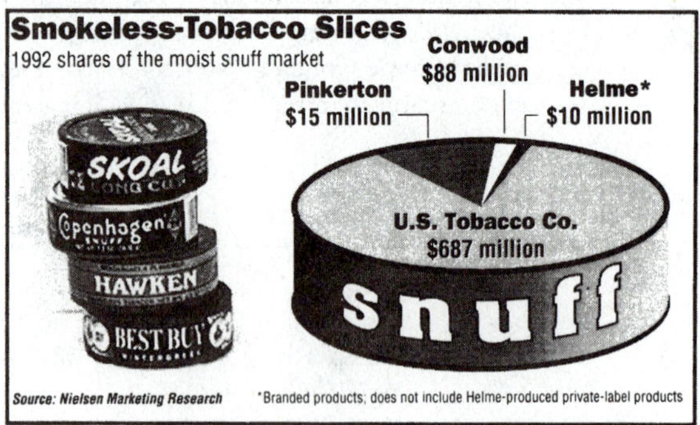

1.47 Below is a diagram from *Business Week* (August 9, 1993) giving a breakdown of voters, and the general population, by age. What does this diagram tell you?

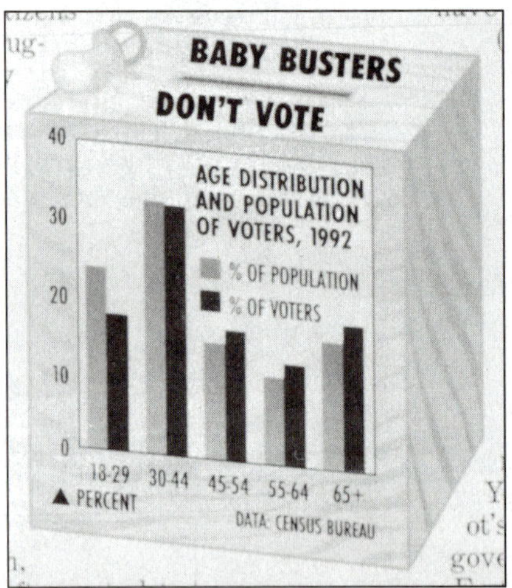

1.48 Redraw the *Business Week* diagram in the previous exercise as two separate and properly drawn histograms: one age histogram for the general population and the other for voters. Why doesn't the *Business Week* diagram properly represent these frequency distributions? To assist you in your construction, the approximate relative frequencies follow:

	Relative Frequencies	
Age	Population	Voters
18–29	24.0%	18.0%
30–44	32.5	32.0
45–54	15.0	17.5
55–64	11.5	13.5
65 +	17.0	19.0

1.49 The computer ASCII file "EX1-49.PRN" contains the 51 average salaries for teachers by state and the District of Columbia. Make up a frequency distribution, and draw the histogram for these salaries.

1.50 The following pie chart shows the length of time that executive-education programs are conducted, as reported in *The Wall Street Journal* (September 10, 1993). From this information, construct the corresponding cumulative relative frequency distribution. What does the cumulative relative frequency distribution show you that the pie chart does not?

TAKING THEIR TIME

▶ The average executive-education program in 1992 lasted 2½ weeks. A breakdown:

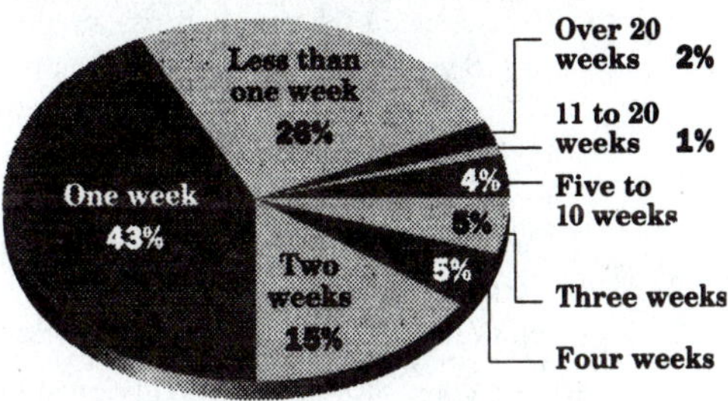

NOTE: Percentages total more than 100% due to rounding

Source: Bricker's International Directory

1.51 In the 1980s the Minnesota Association of Commerce and Industry (MACI) was concerned about the number of firms leaving the state. Many theories were put forward to explain the alleged departures, including high state taxes and cold weather. The MACI staff proposed a survey of MACI member firms soliciting information on the reasons for firms leaving the state. What was wrong with this proposed study design; what did the MACI staff overlook? How does this sampling problem relate to that faced by Wald in Exercise 1.45?

Academic References

Clogg, Clifford. "The Impact of Sociological Methodology on Statistical Methodology." *Statistical Science,* May 1992, pp. 183–195.

Fienberg, Stephen. "A Brief History of Statistics in Three and One-Half Chapters: A Review Essay." *Statistical Science,* May 1992.

Fisher, Ronald. *Statistical Methods for Research Workers.* New York: Hafner Publishing Company, 1970.

Griliches, Zvi. "Patent Statistics as Economic Indicators: A Survey." *Journal of Economic Literature,* December 1990, pp. 1661–1707.

Leamer, Edward. "Let's Take the Con Out of Econometrics." *American Economic Review,* March 1983, p. 33.

Starer, Daniel. *Who Knows What: The Essential Business Resource Book.* New York: Henry Holt, 1993.

Terrell, George R. and David W. Scott. "Oversmoothed Nonparametric Density Estimates." *Journal of the American Statistical Association,* Vol. 80, March 1985, pp. 209–214.

ENDNOTES

1. Sir Ronald Fisher, in the 1970 14th edition of possibly the most influential book in statistics, *Statistical Methods for Research Workers,* wrote:

> The science of statistics is essentially a branch of Applied Mathematics, and may be regarded as mathematics applied to observational data. (p. 1)
>
> Statistical methods are essential to social studies, and it is principally by the aid of such methods that these studies may be raised to the rank of sciences. This particular dependence of social studies upon statistical methods has led to the unfortunate misapprehension that statistics is to be regarded as a branch of economics, whereas in truth methods adequate to the treatment of economic data, insofar as these exist, have mostly been developed in biology and the other sciences. (p. 2)

Fisher's view, although traceable to the first version of his book in 1925, is still held today by many scholars in the natural sciences and departments of mathematics. Econometricians (economists who apply statistical methods in economics), psychometricians, cliometricians, and other "metricians" in the social sciences have different views of the process by which statistical methods have developed. Given that Fisher's numerous and great contributions to statistics were in applications within biology, genetics, and agriculture, his view is understandable, although it is refuted by sociologist Clifford Clogg (1992) and the numerous commenters on his article.

FORMULAS

- Relative frequency $= f_i/N$, for a population of size N.
- Relative frequency $= f_i/n$, for a sample of size n
  where f_i is the frequency of the ith class.
- The approximate number of classes is the integer that just exceeds
  2 (number of values in data set)$^{(0.3333)}$.
- The approximate width of each class is highest value number of classes.

Chapter 2

Description and Summary of Data

The purpose of computing is insight, not numbers.

Richard Hamming
b. 1915

2.1 Introduction

Frequency distributions alone cannot tell the whole story about a data set. We need more information to describe the central values, variability, and shape of the distribution, whether population or sample. The following situations depict problems encountered in business and economics that require knowledge of data position and data dispersion.

- A *Wall Street Journal* (April 22, 1992) article showed that "median CEO total pay—including salary, bonus and long-term incentives—came to $1.3 million in 1991." But what is the mean total pay for these chief executive officers? A stockholder of an energy company may ask, "What do these values imply about overcompensation or undercompensation of my company's CEO?"
- A *Wall Street Journal* (September 23, 1991) article on bank closings reported that, "A House Banking Committee study found that regulators kept critically ill banks open an average of 10 to 12 months before failure." Is this average correct? Why might it be misleading?
- According to journalist Joyce Lain Kennedy, in a syndicated column on merit pay policies that appeared in the *Herald Times* (February 26, 1989), a *Newsweek* survey showed that roughly 80 percent of American workers believe they are above average in merit. She quotes a Hays management consultant who said Americans suffer from the "Lake Woebegone myth," where all the children are above average. Is it possible for nearly everyone to be above the average? Or must 50 percent be above and 50 percent be below the average?
- A university administrator asked an assistant to calculate the faculty salaries that defined the median and third quartile of the salary distribution. The assistant reported the values for the 50th and 75th percentile. The administrator criticized the assistant, saying that the 50th and 75th percentiles are the same as the median and the third quartile only for bell-shaped distributions. He said the distribution of faculty salaries is skewed and thus the median is not the 50th percentile and the third quartile is not the 75th percentile. Who is correct, the administrator or the assistant?
- Suppose that in the immediate past, a corporate stock has yielded a net quarterly average return of $100 per $1,000 invested, with a standard deviation of $50. Moreover, the distribution of the net return has been bell shaped (or normal). If we assume the future will be like the past, what is the likelihood that an investor of $1,000 will lose money on this investment?
- An executive training firm wants to design a CPA preparatory course that is long enough to ensure that at least 84 percent of the participants will pass. How long should the course be?

By the end of this chapter, you will be able to calculate and use the three measures of central location—mean, median and mode—and other measures of location, such as percentiles and quartiles. You also will be able to compute and interpret three measures of dispersion—the range, the variance, and the standard deviation. Using the mean and standard deviation, you will be able to make statements about the proportions of a distribution that are associated with deviations from the mean.

2.2 Descriptive Summary Measures

For large data sets, a listing of the data in the original order of collection (the raw data) or even a frequency distribution representation may provide too much information to identify the "middle" of the data and their "spread." What we need are summary measures that represent a typical middle value and typical deviations from that middle value. These descriptive summary measures are of two types: measures of central tendency and measures of dispersion.

Measure of Central Tendency or Central Location
A value that identifies the middle of a distribution; it is a measure of the "typical" value.

Measures of central tendency or central location are numbers that represent the typical or middle values in a distribution. To illustrate, consider the issue of CEO compensation introduced in the first chapter. Table 2.1 reproduces the compensation of the 20 CEOs of the energy companies surveyed by *The Wall Street Journal.* We want to know the distance of each CEO's total compensation (salary and bonus) from the typical compensation, but first, the "typical" must be defined. The typical, middle, or central compensation can be defined in different ways, depending on the purpose of the inquiry and the nature of the data, as discussed in Sections 2.3 through 2.10.

Table 2.1 Compensation of CEOs of Energy Companies

	Company	Executive	1991 Salary/ Bonus (in thousands)	% Change from 1990	Long-Term Comp. (in thousands)	Total Direct Comp. (in thousands)
1.	Amerada Hess	Leon Hess	300.0	N/A	0.0	300.0*
2.	Amoco	H. Laurance Fuller	1302.2	N/A	236.3	1538.5
3.	Ashland Oil	John R. Hall	1038.0	–6	0.0	1038.0
4.	Atlantic Richfield	Lodwrick M. Cook	1673.9	–26	2008.9	3682.8
5.	Baker Hughes	James D. Woods	916.0	–27	393.1	1309.1
6.	Chevron	Kenneth T. Derr	1385.6	2	363.8	1749.4
7.	Coastal	James R. Paul	813.8	N/A	913.4	1727.2
8.	Dresser Industries	John J. Murphy	1070.6	–3	578.7	1649.3
9.	Exxon	Lawrence G. Rawl	1812.8	17	7453.0	9265.8
10.	FINA	Ron W. Haddock	476.3	N/A	0.0	476.3
11.	Halliburton	Thomas H. Cruikshank	1200.0	9	401.5	1601.5
12.	Kerr-McGee	F.A. McPherson	542.0	–32	129.4	671.4
13.	Mobil	Allen E. Murray	2040.0	9	2624.9	4664.9
14.	Occidental	Ray R. Irani	2324.0	N/A	147.1	2471.1
15.	Pennzoil	James L. Pate	650.0	–2	0.0	650.0
16.	Sun Co.	Robert McClements, Jr.	1001.1	3	0.0	1001.1
17.	Texaco	James W. Kinnear	1667.0	–14	1658.8	3325.8
18.	USX	Charles A. Corry	1313.2	–18	840.2	2153.4
19.	Unocal	Richard J. Stegemeier	1054.7	N/A	0.0	1054.7*
20.	Valero	William E. Greehey	1097.4	12	352.5	1449.9

*Salary/bonus data not valid for comparison (e.g., partial year data).

Source: *Wall Street Journal*, April 22, 1992.

Measure of Dispersion
A value that indicates how "spread out" the distribution is.

Measures of dispersion are numbers that represent the spread or variability of values in a distribution. The distance of each CEO's compensation from a measure of central tendency, for instance, reflects the dispersion of the distribution. A measure of dispersion, however, is a single summary number that reflects these deviations. Several summary measures of dispersion will be presented later in this chapter. As introduced in Section 2.11 and the sections that follow, the choice of a specific measure of dispersion will depend on the goal of the inquiry and type of data considered.

2.3 The Mean as a Measure of Central Tendency

Mean
A measure of central location defined by the arithmetic average, which is the sum of the values in a data set divided by the number of values.

The arithmetic average is also called the **mean.** For the mean to be useful, all the data used in its calculation must be comparable. For instance, Table 2.1 reports different forms of pay for 20 CEOs of the energy companies, but comparable data are available for only 18 of those CEOs. The mean total direct compensation for those 18 CEOs is $2,245,861, which is calculated from the data in Table 2.1 in thousands of dollars as

$$\frac{1538.5 + 1038.0 + \ldots + 2153.4 + 1449.9}{18} = \frac{40425.5}{18} = 2245.861$$

Because it is inconvenient to list all 18 comparable compensations that form the sum 40425.5, the notation 1538.5 + ... + 1449.9 is used to indicate that all of the 18 comparable compensations from 1,538.5 to 1,449.9 in the last column of Table 2.1 are included in the sum of 18 compensations. In a more general form, the calculation of the mean of the 18 comparable values is written

$$\text{Mean} = \frac{x_1 + x_2 + \ldots + x_{17} + x_{18}}{18}$$

where x_1 is the value of the first observation (that is, $x_1 = 1538.5$), x_2 is the second value ($x_2 = 1038.0$), and so on, until the final value ($x_{18} = 1449.9$).

Even more compact sigma notation can be used to represent the sum of x_1 to x_{18}. Capital sigma Σ is used to indicate a summation, and the symbol x_i indicates the variable over which the addition is to occur. The subscript i identifies the ith value in a string of values. For instance, in the energy companies' CEO compensation, $i = 1$, or $i = 2$, or any number to $i = 18$. The beginning and ending x_i to be added are identified by a starting number of i below the sigma and the stopping number above the sigma. Thus, the notation

$$\sum_{i=1}^{18} x_i$$

symbolizes the summation of values from x_1 to x_{18}; that is,

$$\sum_{i=1}^{18} x_i = x_1 + x_2 + \ldots + x_{17} + x_{18}$$

Using this sigma notation, the mean of the CEO energy company compensation is written

$$\text{Mean} = (1/18)\sum_{i=1}^{18} x_i = (1/18)(x_1 + x_2 + \ldots + x_{17} + x_{18})$$

As discussed in Chapter 1, observable data sets consist either of all N values in the population of interest or an n observation sample (or subset) of those N values. Hypothetical populations obviously can be of infinite size but observed data must be countable. Here we are concerned with only observable data sets that are either a sample or a population.

Population Mean (μ) A parameter of the population that is the arithmetic average; it is the measure of central location identified by the Greek letter "mu," μ.

The **population mean** is typically designated by the Greek letter μ (verbalized "mu"), or μ_x (verbalized "mu sub-ex"). If the population values are labeled y, then the mean of y would be designated μ_y. The x and y subscript identifiers are helpful when comparing means from different populations but they may be deleted when there is no confusion about the population to which the mean applies.

The **population mean** of x is the sum of the N x values divided by N. In shorthand notation this population mean is designated μ_x and is calculated as

$$\mu_x = \frac{1}{N}\sum_{i=1}^{N} x_i = (1/N)(x_1 + x_2 + \ldots + x_N) \quad (2.1a)$$

For an n observation sample of x values, the **sample mean** is written $\bar{x}$, which is read "ex bar." A sample of y values would have a mean identified by $\bar{y}$, and likewise for any other letter capped with a bar to identify a sample mean.

The **sample mean** of x is the sum of the n x values divided by n. In shorthand notation, this sample mean is designated $\bar{x}$ and is calculated as

$$\bar{x} = \frac{1}{n}\sum_{i=1}^{n} x_i = (1/n)(x_1 + x_2 + \ldots + x_n) \quad (2.1b)$$

As a general rule, Greek letters are used to identify population measures and Roman letters are used to signify sample measures. Whether a data set is a population or a sample depends on the type of data and the use to which they are to be put. For instance, if we are only interested in comparing how far any given energy company CEO's compensation is from the mean of the 20 CEOs who head energy companies, then the 20 compensation values are a population. But we do not have the 20 compensations; we have only 18 comparable values. We thus have a subset of all 20 CEOs; these 18 usable records are a sample.

Even if *The Wall Street Journal* had complete information on all 350 companies in its survey, the resulting data set would not be a population if we are interested in all corporations in the United States. Typically, smaller data sets are samples from larger sets, but even large data sets may be samples of still larger sets of data. Remember, only a complete census of the group of interest will yield the population.

QUERY 2.1:
In reviewing the salary increases of six employees at a small firm, the office manager noticed that the mean percentage increase (4.46 percent) was not equal to the percentage increase in the mean salary (4.63 percent). Does this suggest a mistake in the calculation of the means?

	1992 Salary	Percentage Increase	1993 Salary
	$60,000	4.26%	$62,557
	60,000	4.26%	62,557
	58,000	12.07%	65,000
	52,883	1.94%	53,908
	51,250	2.00%	52,275
	45,800	2.24%	46,825
Mean	$54,656	4.63%	$57,187
Mean percentage increase in salaries		4.46%	

ANSWER:
No mistake was made in the calculation of the means. The percentage increase in the mean salary (4.63 percent) is the same as the actual percentage increase in the total budget for the six employees:

$$6(\$54656) = \$327936$$
$$6(\$57187) = \$343122$$
$$(343122 - 327936)/327936 = 0.0463$$

although it was calculated on the basis of the change in the mean salary in each year

$$\$54656)$$
$$\$57187)$$
$$(57187 - 54656)/54656 = 0.0463$$

The mean percentage increase (4.46 percent) was calculated as the average of the six percentage increases in salaries.

$$\frac{4.26\% + 4.26\% + 12.07\% + 1.94\% + 2.00\% + 2.24\%}{6} = 4.46\%$$

The mean percentage increase has no relationship to the actual budget for the six employees. It can be decreased, for example, by giving larger dollar raises to the higher-paid employees and smaller dollar raises to the lower paid. It is an artifact of the way in which dollar raises are allocated.

2.4 The Weighted Mean

Data are often reported in frequency distributions. If the frequency of each value in the original data set is not known, and only the frequencies of class intervals are provided, then the mean may be approximated by "weighting" representative class values by their respective frequencies. If, however, the frequency or relative frequency of each value in the data set is known, then the exact mean may be calculated by weighting actual values by their respective relative frequencies.

Approximation Method

Once values are grouped into class intervals, their individual identities are lost. This can be seen in Table 2.2, where risky banks were grouped by the time interval before they failed. For instance, 92 banks failed in the 10- to 12-month interval after being placed on the high-risk list. Although we know the frequency of bank failures for the 10- to 12-month interval, we cannot determine the exact length of time before each of the 92 banks failed because of the grouping. *The Wall Street Journal* did not provide the original data, and we can work only with the frequency distribution provided.

For data that are grouped in a frequency distribution, a true mean cannot be calculated if the actual values are unavailable. An approximate mean can be calculated by assuming that the observations in each class fall evenly throughout the class. This assumption enables us to use a class midpoint to represent the average value in the class. For instance, if the 92 banks failed evenly between 10 and 12 months, then the class midpoint of 11 months is the average time to failure for these 92 banks.

Table 2.2 Lingering Death

Length of time banks with "extremely high" risk of failure stayed open before failing, January 1, 1985 to May 10, 1991

Months	Number of Banks
0–3	74
4–6	53
7–9	62
10–12	92
13–18	95
19–30	51
31–58	10

Source: *Wall Street Journal,* September 23, 1991.

Table 2.2 shows seven classes, so $i = 1, 2, \ldots$ or 7. In general, the maximum number of classes is designated by c (in Table 2.2, $c = 7$). Let the ith class midpoint be m_i. The 10- to 12-month class is then the fourth class, and the midpoint of this class is $m_4 = 11$. The frequency of each m_i midpoint is denoted by f_i, so the frequency of the fourth class is $f_4 = 92$. The frequency f_i is the weight given to the class midpoint value m_i in the calculation of the data set mean. An approximate **weighted mean** may be calculated by multiplying each midpoint value (m_i) by its frequency weight (f_i). The sum of these c products is then divided by the sum of frequencies (which is the total number of observations, N or n).

A general statement of the population weighted mean for grouped data is

Population weighted mean

$$\mu_x = \frac{1}{N}\sum_{i=1}^{C} f_i m_i = (1/N)(f_1 m_1 + f_2 m_2 + \ldots + f_C m_C) \tag{2.2a}$$

The sample weighted mean for grouped data is, likewise

Sample weighted mean

$$\bar{x} = \frac{1}{n}\sum_{i=1}^{C} f_i m_i = (1/n)(f_1 m_1 + f_2 m_2 + \ldots + f_C m_C) \tag{2.2b}$$

The bank failure data in Table 2.2 might be viewed as population data because they represent all 437 banks that failed between January 1, 1985, and May 10, 1991. On the other hand, this is but a slice of time, so that bank failures during this period are only a subset of all such bank failures that could have occurred to the present day. Since we will be interested in making statements about all bank failures, these data will be treated as a sample because they do not include every bank failure to the present or into the future.

The data and calculations required for the approximate weighted mean in the case of bank failures are given in tabular form in Table 2.3, where all calculations were done using the computer spreadsheet program LOTUS 1-2-3. These calculations could have been done with any other computer spreadsheet (such as EXCEL or QUATTRO) or a hand-held calculator. The relevant quantities also can be arranged in algebraic form:

$$\begin{aligned}\bar{x} = (1/437)\sum_{i=1}^{7} f_i m_i &= (1/437)[1.5(74) + 5.0(53) + 8.0(62) + 11.0(92) \\ &\quad + 15.5(95) + 24.5(51) + 44.5(10)] = 11.55835\end{aligned}$$

In some cases, we may not even know the size of the data set but we may know the relative frequencies of the classes. In such cases, alternatives to Formulas (2.2a) and (2.2b) can be employed. These alternatives are based on the relative frequency of each class. For the population, this formula is

Table 2.3 Calculation of a Weighted Mean for the Bank Failure Data

m_i Class Midpoints (months)	(f_i) Frequency of Banks Failing (number)	$f_i m_i$ Product of Frequency Times the Midpoint
1.5	74	111.0
5.0	53	265.0
8.0	62	496.0
11.0	92	1012.0
15.5	95	1472.5
24.5	51	1249.5
44.5	10	445.0
	$n = \Sigma f_i = 437$	$5051.0 = \Sigma f_i m_i$

Mean = $11.55835 = \Sigma f_i m_i / n$

Population weighted mean

$$\mu = \sum_{i=1}^{c} (f_i / N)(m_i) = [(f_1 / N)(m_1) + (f_2 / N)(m_2) + \ldots (f_c / N)(m_c)] \quad (2.3a)$$

For the sample weighted mean, the frequency formula is

Sample weighted mean

$$\bar{x} = \sum_{i=1}^{c} (f_i / n)(m_i) = [(f_1 / n)(m_1) + (f_2 / n)(m_2) + \ldots (f_c / n)(m_c)] \quad (2.3b)$$

Notice that each term in these formulas already contains a division by the total number of observations because the relative frequency is the weight; thus, no additional division by N or n is involved after summing the c terms. As demonstrated later, to use these alternative formulas, we do not need to know the number of observations in the data set.

Comments on Approximating a Mean

The data in Table 2.2 and the weighted mean of 11.56 months in Table 2.3 support the House Banking Committee study that found that "regulators kept critically ill banks open an average of 10 to 12 months before failure," as reported in *The Wall Street Journal.* Our calculation is not exact, however.

First, we had to assume the boundaries of the class intervals, since they were not given. The class intervals are not of equal width, so the boundaries cannot be equally spaced. Assuming that the class boundaries are at the half-month points between identified class values implies that the boundaries of the "0–3" month class are –0.5 to 3.5 months, and the next "4–6" month class has boundaries of 3.5 to 6.5 months, and so on. In the case of the first class, however, this looks a bit strange, since –0.5 is not possible. On the other hand, if we assume that the bank failures occurred only at the monthly integer values in the classes, then these class boundaries are irrelevant. But why would bank failures occur only in exact monthly intervals?

Either of the above sets of assumptions are consistent with the 11.56-month average as an approximation of the mean that could be calculated with the raw data. As long as the data are only reported in frequency form, as they were in *The Wall Street Journal*, the correctness of any assumption regarding the placement of the data remains open to question.

Second, even if the House Banking Committee had access to the raw data, an average calculated from bank failures would be for those banks that failed within 58 months. We do not know what happened after 58 months. We do know from *The Wall Street Journal* article that "90% of the banks that received extended credit failed," but what about the other 10 percent? They are not included in the calculation of length of time before failure because they did not fail within the 58 months. But what if some of them failed at 158 months, and some at 258 months, and . . . ? This would greatly raise the average time to failure, yet the House Banking Committee study's average of 10 to 12 months before failure does not reflect these extremes.

Many widely cited averages involving data collected over time suffer from a truncation of the data before the underlying process has exhausted all possibilities. The average duration of unemployment, the expected life span, the mean time to complete a college education, and the mean time before tire failure are all examples of averages calculated from data for which the length of time or duration in a given condition may be cut off before extreme values are realized. Econometrician Nicholas Kiefer (1988) discusses problems in calculating means and other descriptive statistics when time-series data are "censored" by beginning the study period after the spell or process has started, by making observations at time intervals too wide to observe the spells, or by ending the study prior to the completion of spells. Means calculated with such data are always susceptible to these sampling problems.

Exact Method

If the actual values in a data set are reported with their respective frequencies or relative frequencies, then an exact mean may be calculated by weighting values by their respective relative frequencies. As an example, consider an instructor who assigns grades so that the top 13 percent of the students get A's (4 points), the next 25 percent get B's (3 points), the next 50 percent get C's (2 points), and the bottom 12 percent equally split D's (1 point) and F's (0 points).

The number of students who will finish the course is unknown in the first week of classes, so the mean grade at the end of the course cannot be calculated with frequencies as in Formula (2.2a). Because the relative frequencies of grades are known,

TABLE 2.4 Weighted Mean of Grades

Letter Grade	Numerical Grade (m_i)	f_i/N	$(f_i/N)m_i$
A	4	0.13	0.52
B	3	0.25	0.75
C	2	0.50	1.00
D	1	0.06	0.06
F	0	0.06	0.00
			Mean = 2.33

however, the mean grade can be determined by the alternative Formula (2.3a), where the calculations are

$$\mu = \sum_{i=1}^{5}(f_i/N)m_i = (.13)(4) + (.25)(3) + (.50)(2) + (.06)(1) + (.06)(0) = 2.33$$

Or, if done in a spreadsheet, they would appear as in Table 2.4.

No matter whether 100, 200, or 1,000 students complete the course, the average numerical grade will be 2.33. Furthermore, as long as the A, B, C, D, and F letter grade equivalents of 4, 3, 2, 1, and 0 do not represent midpoints of intervals (brought about by + and – letter grades), 2.33 is the exact mean.

2.5 APPLICATIONS OF THE MEAN TO QUALITY CONTROL

Now that we have examined the mean and weighted mean in some detail, let's put these concepts to practical use in the field of quality control.

Every production process involves time. Data collected over time in sequential order are called time-series data. The number of award-winning Cadillac Sevilles produced per day at the General Motors Detroit–Hamtramck plant from the early days of production in August 1991 (when half the cars were pulled off the line to correct defects) to the time when full production of 480 cars a day was reached in February 1992 is a time series of interest to the GM executives directly responsible for sales. Similarly, the daily time series showing the sequential fall in defects from a list of 700 potential problems, when production started, to a list of 200 by February 1992, and to 5.1 defects per car in June 1992, is of great interest to quality control experts.

Control Chart
A graph showing a time series of measures of central location, or variability in the output of a process.

$\bar{x}$ Chart
A control chart showing the time series of average output levels in a production process.

As discussed in *Business Week* (June 29, 1992), corporations track time series on quantity and quality measures through audits targeted at specific goals. **Control charts** are often constructed to study variations in the process outcomes or results as they evolve over time. The most commonly used control chart is the **$\bar{x}$ chart,** which shows how the mean of a process changes over time. It is constructed by graphing the mean of item measurements produced by a production process at sequential time periods. Chart 2.1 shows two $\bar{x}$ charts: the first for average student course grades and the second for test scores on the 44-item final exam administered every semester at

Indiana University–Bloomington in the introductory business and economics statistics course.

Common Cause
Variability showing up in a control chart that is common to each time observation, such as the randomness inherent in sampling.

Special or Assignable Cause
Variability in a control chart that is identifiable with something known to affect observations over time in a unique manner and for which corrective steps can be taken.

An examination of both panels in Chart 2.1 finds quite a bit of random variability in these $\bar{x}$ series. As discussed in detail in Chapter 18, a **common cause** of this variability is the randomness inherent in the type of testing done (students guessing on multiple-choice questions) and the randomness inherent in sampling (each semester has a different mix of students and instructors). **Special** or **assignable causes** of variations (those for which corrective steps can be taken) can be discerned in identifiable patterns in these $\bar{x}$ series. For instance, the average course grade shot up between fall 1987 and spring 1989, when it hit a high of 2.818. In approximately the same period, records of final exam scores are missing. In fall 1988, the average score on the final went "out of control," hitting an all-time high of 31 out of 44 questions correct. An assigna-

CHART 2.1 $\bar{x}$ Charts for Student's Grades and Test Scores

Panel a: Average course grade (A = 4, B = 3, . . . to F = 0) for all sections of Introductory Business and Economics Statistics at Indiana University–Bloomington, 1986–1992

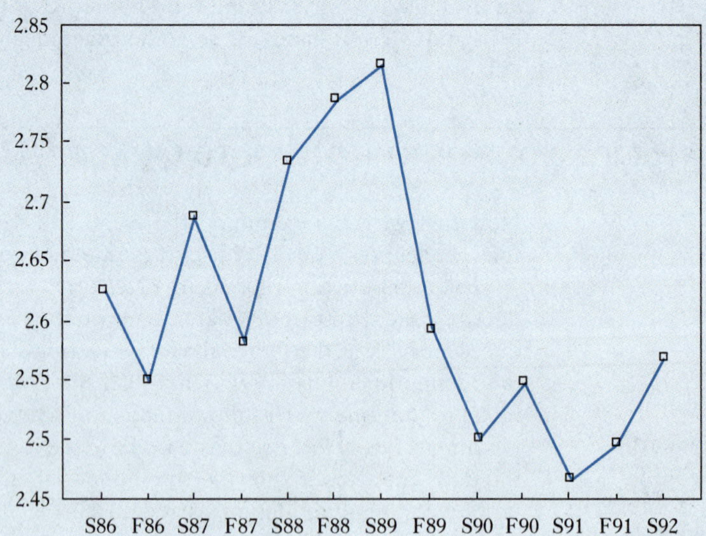

Average grade for all sections

Spring 1986	2.629	Fall 1989	2.596
Fall 1986	2.553	Spring 1990	2.510
Spring 1987	2.692	Fall 1990	2.550
Fall 1987	2.587	Spring 1991	2.470
Spring 1988	2.739	Fall 1991	2.500
Fall 1988	2.790	Spring 1992	2.572
Spring 1989	2.818		

(continued

CHART 2.1 $\bar{x}$ Charts for Student's Grades and Test Scores *(continued)*

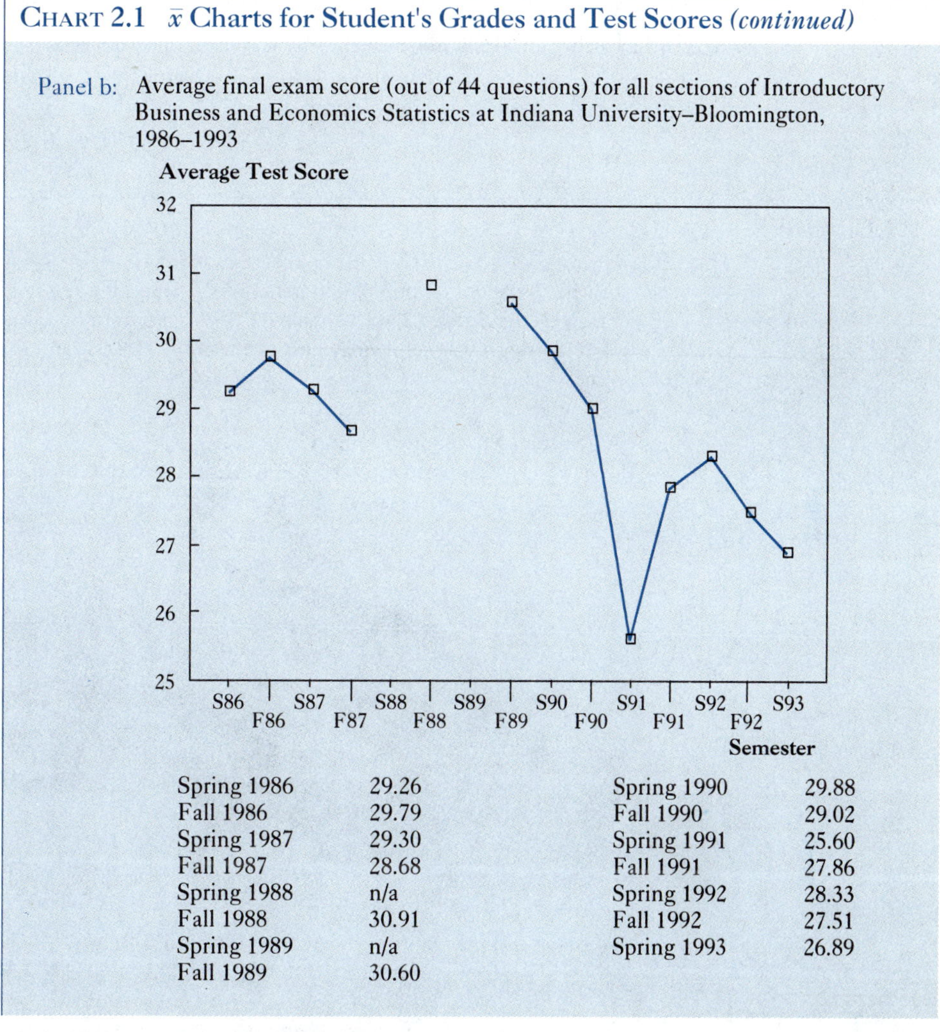

Spring 1986	29.26	Spring 1990	29.88
Fall 1986	29.79	Fall 1990	29.02
Spring 1987	29.30	Spring 1991	25.60
Fall 1987	28.68	Fall 1991	27.86
Spring 1988	n/a	Spring 1992	28.33
Fall 1988	30.91	Fall 1992	27.51
Spring 1989	n/a	Spring 1993	26.89
Fall 1989	30.60		

Upper Control Limit The upper extreme above which the process monitored in a control chart is said to be out of control and in need of change.

Lower Control Limit The lower extreme below which the process monitored in a control chart is said to be out of control and in need of change.

ble cause may be that an interim faculty member coordinated the multisection course for this two-year period. Attention to detail in constructing the final exam and in policing the grades given by the many course instructors apparently were not of high importance to this interim coordinator.

In addition to the plot of the average score, **control limits** are often added to control charts. Control limits show the extremes of $\bar{x}$ for which the process is said to go from "controlled" to "out of control." Chapter 18 addresses the manner in which an **upper control limit** (UCL) and **lower control limit** (LCL) may be set. In the case of the Cadillac Seville, for example, the UCL for average defects was given as "a target of 1.0 per car," while "an internal quality audit in early June found 5.1 defects per car," so the process was still not in control. *Business Week* (June 29, 1992) reported, "J.D. Power &

Associates Inc.'s initial quality survey found 0.46 defects per Lexus LS400 last fall—and 21 other models, none of them Cadillacs, had less than one defect per car." Thus, one defect per car is an achievable upper control limit. Chapter 18 addresses the setting of control limits in more detail.

EXERCISES

2.1 In what ways do the population parameter μ and sample statistic $\bar{x}$ differ?

2.2 An article in *Business Week* (August 17, 1992) reported on the success of the Saturn automobile. It gave the following chart of consumer satisfaction, based on J.D. Power's survey of new cars:

Lexus	Infiniti	Saturn	Acura	Mercedes-Benz	Toyota	Industry Average
179	167	160	148	145	144	129

Why is it acceptable to say that Saturn purchasers may be more satisfied than Acura purchasers, but it is not correct to say that Saturn buyers are 24 percent more satisfied than the average car buyer?

2.3 An August 4, 1992, article in *The Wall Street Journal* reported that "although there haven't been any major accidents since the Valdez disaster, about 4,700 gallons of oil or other liquids considered hazardous have been spilled along the (Alaskan pipe) line in 351 minor mishaps, according to data compiled by Alaska's Department of Environmental Conservation." What is the average spill per mishap? Why shouldn't information on the average spill be of great interest to environmentalists? What should they be interested in?

2.4 A *Wall Street Journal* (August 25, 1992) article reported that of the 17,000 members surveyed by the Texas State Teachers Federation, the average amount of their own money spent to buy classroom supplies the prior year was \$242. What was the total size of the subsidy these teachers contributed for classroom supplies?

2.5 According to the *Economic Report of the President,* in 1991 the population of the United States was 252.688 million people. Disposable income was $4,209.6 billion. What was average disposable income per person in 1991?

2.6 In its second annual ranking of women-owned businesses, *Working Woman* (May 1993) reported that there are 6.5 million such companies and that these women-owned businesses employed 12 million people, which is more than the roughly 11.7 million employed by the Fortune 500 companies.

   a. What is the average workforce of the women-owned companies?
   b. What is the average workforce of the Fortune 500 companies?

2.7 W. E. Deming, statistician and quality control expert, tells the story of a company president who routinely looked at the average weekly error rate of employees. To reduce the error rate, he then had "motivational talks" with those whose error rates were above the average. Deming responded to the company president who did this by saying, "I thereupon told him (the president) about a recent letter to the editor of the *Times* of London. The writer had been studying a report from the Ministry of Health, where it was obvious that about half the children in the United Kingdom were below the average weight. A disgrace on the nation. We must do something about the nourishment of our children. My listener laughed at this joke, but his understanding thereof was not deep enough to see that his style of supervision fell in to the same trap, causing more errors, not fewer." *Out of the Crisis* (MIT Press, 1986, p. 389). What was the trap into which this president had fallen?

2.6 The Median as a Measure of Central Tendency

Most American workers believe they are better than average, according to Kennedy's column cited at the beginning of this chapter. Is it possible for more than 50 percent be above the mean? To answer this question, we must look at another summary measure of central location: the **median.**

Median
The value that divides an ordered set of data into two parts, with the values of at least half of the data being less than or equal to the median and at least half of the data being greater than or equal to the median.

Locating the Median in the Original Data Set

The most common method of locating the median involves an ordering of the data from lowest value to highest. The median is then the middle value that splits the data into two equal halves. For example, suppose we want to find the middle value of the following seven observations:

9.0 lb, 2.5 lb, 5.5 lb, 6.0 lb, 2.0 lb, 1.5 lb, and 2.0 lb

The first step is to order the data from lowest to highest values:

1.5, 2.0, 2.0, 2.5, 5.5, 6.0, 9.0

The middle value in this ordered array is the fourth number, 2.5; thus, 2.5 lb is the median because three observations are less than 2.5 lb and three observations are greater than 2.5 lb.

The middle value can always be found by adding one to the total number of observations in the set of data and then dividing by two. That is, if there are N observations, the middle observation is $(N + 1)/2$. For $N = 7$, the middle observation is the fourth value in the ordered array since $(7 + 1)/2 = 4$.

For an *even* number of observations, there is no unique middle value. For an eight-observation set of data, for example, the middle observation would be the 4.5th, which does not exist; any value between the fourth and the fifth ordered value could work. In the ordered data set \$1, \$2, \$2, \$6, \$7, \$8, \$9, \$10, any value between \$6 and \$7 could serve as a median, even though there is no actual observation there. The value halfway between the two values that flank the median (\$6.50) is typically picked.

In both of the above examples, the medians of 2.5 lb and \$6.50 split their respective data sets into two equal pieces. The means of these data sets, however, do not equal these medians. In the first example, the mean is 4.07 lb, which is greater than the median. In the second example, the mean is \$5.625, which is less than the median. Thus, in the first example the majority of observations (four out of seven) are below the mean, and in the second the majority of observations (five out of eight) are above the mean. Clearly, the majority of observations (50 percent or more) can be above or below the mean if the median and mean are not the same value.

If the mean exceeds the median, the majority of observations will have values lower than the mean. If the median exceeds the mean, then most of the values will be greater than the mean.

The median is not well suited to describing the central location of some small data sets. For instance, the data set \$1, \$1, \$3 does not have a median that splits the data into two equal halves, although \$1 fulfills the definition of a median. At least half the values are equal to or less than \$1, and at least half the values are greater than or equal to \$1. Computer programs will force a median onto a data set even when the efficacy of imposing a median is dubious. Users of these programs are left with the problem of making sure the concept of the median is meaningful before attempting to compute one.

QUERY 2.2:

What is the median for the CEO compensation data in Table 2.1? How does it compare to the mean? Why might the median be preferred as a measure of the "typical" compensation?

ANSWER:

Using the data sort command in QUATTRO, the 18 comparable values are quickly ordered from lowest to highest.

Company #	Total Pay (in thousands of \$s)
10	476.3
15	650.0
12	671.4
16	1001.1
3	1038.0

(continued)

Company #	Total Pay (in thousands of $s)	
5	1309.1	
20	1449.9	
2	1538.5	
11	1601.5	
8	1649.3	← 1625.4 Approximate Median
7	1727.2	
6	1749.4	
18	2153.4	
14	2471.1	
17	3325.8	
4	3682.8	
13	4664.9	
9	9265.8	

The two middle ordered values are 1601.5 and 1649.3; the median could be any value in between. We can therefore consider 1625.4, which is midway between, as the median compensation for CEOs of the energy corporations. A median salary of $1,625,400 is well below the mean of $2,245,900 for these CEOs. The majority of CEOs make less than the mean compensation of $2,245,900 because the mean is being pulled up by a few very high compensation packages, such as the $9,265,800 total paid to Lawrence G. Rawl at Exxon. For this reason, the median might be preferred to the mean as a measure of the typical compensation of these CEOs.

QUERY 2.3:

As a check on how your statistics computer program responds to a command for the median, request the descriptive statistics for the extremely small data set $1, $1, and $3.

ANSWER:

To get descriptive statistics from the MINITAB computer program, for example, enter the following statements (as given in lowercase after each uppercase prompt from MINITAB).

```
MTB> set c1
DATA> 1 1 3
DATA> end
MTB> describe "c1"
```

```
         N     MEAN    MEDIAN    TRMEAN    STDEV    SEMEAN
C1       3    1.667     1.000     1.667    1.155     0.667
       MIN      MAX        Q1        Q3
C1   1.000    3.000     1.000     3.000
```

In this printout, we have not yet discussed the STDEV (standard deviation) = 1.155, and the SEMEAN (standard error of the mean) = 0.667; these will be considered later in this chapter. The TRMEAN is the "trimmed mean," a mean that MINITAB calculates after discarding the top 5 percent and bottom 5 percent of the data to mitigate the effect of extreme values. This data set is so small, however, that this feature is not operational as seen by the equality of the TRMEAN and the MEAN = 1.667. Similarly, the program did not produce meaningful quartiles (Q1 and Q3), which will be discussed later in this chapter as the values that place 25 percent and 75 percent below them. The median produced by the program, MEDIAN = 1.000, is likewise of little value because the data set is so small. Ideally, half the values would be above and half the values below the median, but here no values are smaller than $1 and one value ($3) is larger.

Locating the Median in Grouped Data

Median Class
The first class to have a cumulative relative frequency of 50 percent or more; it is the class that contains the median.

Determination of a median can be especially trying if only frequency distributions are available and the original raw data values cannot be retrieved. A **median class** in a frequency distribution is easy to find; it is the first class to have a cumulative relative frequency of 50 percent or more. Determining the actual median value within the median class, however, is impossible without the original values; it must be approximated.

In the case of unemployment duration shown in Table 2.5, the median class is five to 15 weeks. This class is the first interval for which the cumulative relative frequency is 50 percent or more (78.12 percent of the distribution is below 15 weeks). That is, the five- to 15-week class contains the middle value, which is between the 3,437th and 3,438th ordered observation [(N+1)/2 = (6874 + 1)/2 = 3437.5].

The actual median value that is buried within the median class interval in Table 2.5 is not apparent because the original values are not given. By assuming that the values are spread evenly throughout the median class, however, we can approximate a median by algebraic interpolation. An alternative procedure to approximate the median is to use the ogive graph. Both these procedures are illustrated.

Algebraic Approximation of the Median

The cumulative frequency up to the lower boundary of the median class was 3,169 persons (or 46.1 percent of the data). The median, as already determined, is the value of

TABLE 2.5 Unemployment Duration in 1990 (thousands of persons)

Duration	Frequency of Duration	Relative Frequency of Duration	Cumulative Relative Frequency of Duration	
less than 5 weeks	3169	0.461012	0.461012	
5 to 15 weeks	2201	0.320192	0.781204	← Median Class
15 to 27 weeks	809	0.117689	0.898894	
27 weeks and over	695	0.101105	1.000000	
	6874	1.000000		

Source: *Economic Report of the President*, February 1992, p. 342.

the 3,437.5th observation. Thus the median value must be 268.5 observations (that is, 3,437.5 – 3,169) into the median class. There are 2,201 observations in the median class of five to 15 weeks. If these 2,201 observations are evenly distributed within the class, then the median is 12.2 percent (268.5/2201) of the way into the median class. Because the median class is 10 weeks wide (15 – 5), the approximate median is found about 1.22 weeks (that is, 0.122 × 10) into the median class. An approximate median is thus 6.22 weeks (5 + 1.22).

From this example a formula can be derived for finding the median in grouped data:

$$\text{Median} \cong L + (o_1/o_2)(U - L) \tag{2.4}$$

where L is the low boundary of the median class (e.g., 5 weeks),
U is the upper bound of the median class (e.g., 15 weeks),
o_1 is the number of observations into the median class until the median is found (e.g., 268.5), and
o_2 is the number of observations in the median class (e.g., 2201).

To assess the magnitude of error in using Formula (2.4), compare the approximate median of 6.2 weeks with the actual median duration of unemployment of 5.4 weeks as reported by the Department of Labor, U.S. Bureau of Labor Statistics. The difference between this reported median calculated from raw data and the approximation from Formula (2.4) reflects the critical nature of the assumption that the observations within the median class (o_2) are evenly distributed. In the case of unemployment duration, apparently they are not; the observations in the median class must be more concentrated at the lower end.

Graphical Approximation of the Median

Instead of assuming an even distribution of observations within the median class, an alternative graphical method is to use the cumulative relative frequency distribution and its ogive. From a starting point at 0.5 on the vertical axis, a horizontal line is drawn to the ogive with a vertical line then drawn to the horizontal axis. The connecting value on the horizontal axis is the approximate median. This procedure is shown in Figure 2.1, where the approximate median value read from the horizontal axis is about 5.4.

There are other methods of approximating medians when the data are in frequency form. If extremely accurate measures of the median are required, however, the analyst should get the original raw data.

QUERY 2.4:
The Wall Street Journal article on bank failures summarized at the beginning of this chapter reported that "90% of the banks that received extended credit failed" within a 58-month time period; the frequency of failures is given in Table 2.2. Construct a frequency distribution showing the number of months before failure of all the troubled banks that received extended credit. What are the mean and the median of

FIGURE 2.1 Median Location on the Ogive for the Distribution of Unemployment Duration

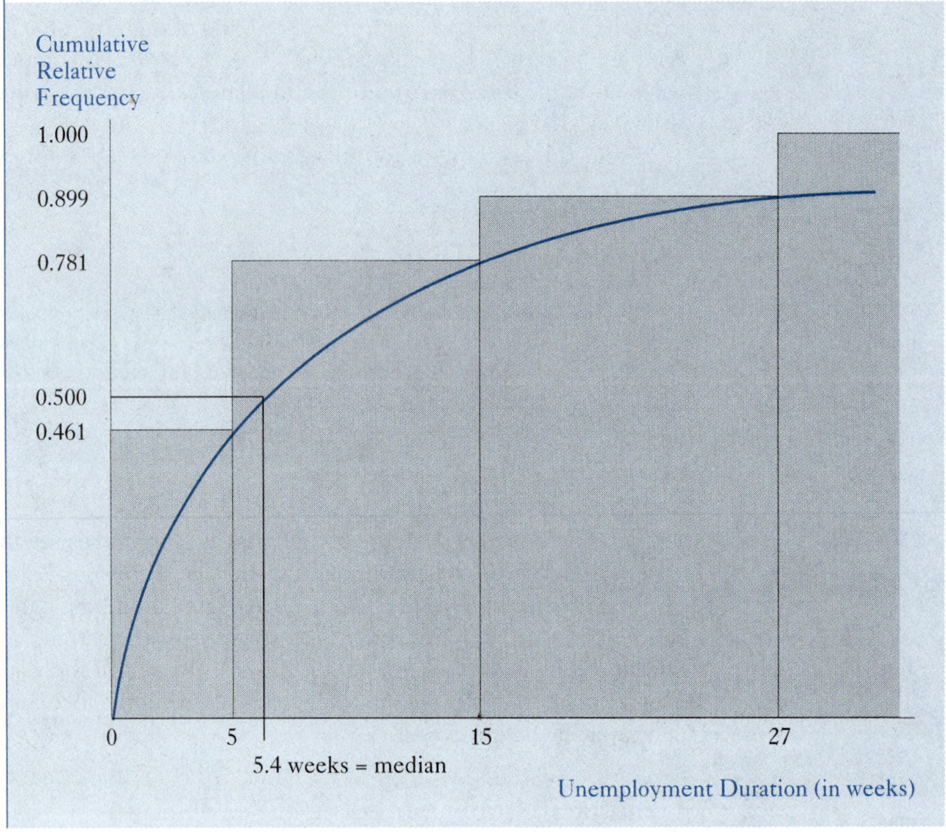

this distribution? Why is one of these measures of central tendency preferred to the other in this case?

ANSWER:

Because 437 banks represent 90 percent of all these troubled banks, there must have been some 486 such banks ($0.9x = 437$, so $x = 437/0.9 = 485.55$). Because the actual number and time until failure of the 49 banks that had not failed in 58 months is unknown, the mean cannot be approximated for all 486 banks. Thus, the mean is not relevant in this case. The complete frequency distribution, the portion of the cumulative frequency distribution needed to approximate the median, and the approximate median are

Months to Failure	Number of Banks	Cumulative Frequency	
0–3	74	74	
4–6	53	127	
7–9	62	189	
10–12	92	281	⟵ Median class: The median is the value of the 243.5 observation [(486+1)/2]. Assuming an even distribution of data within this class, the approximate median is located 59.2 percent of the way into the class. The approximate median is 11.3 months [(0.592)(12.5 – 9.5) + 9.5].
13–18	95		
19–30	51		
31–58	10		
59 or more or never failed	49		
	486		

From this frequency distribution, we know that about 50 percent of all the troubled banks failed within 11.3 months after being designated as troubled and receiving emergency extended credit.

QUERY 2.5:

Acting under the Oil Pollution Act of 1990, the National Oceanic and Atmospheric Administration invited six economists, including two Nobel Prize winners, to assist in writing damage-assessment rules for oil spills. The economists concluded that contingent-value measures of passive use of the environment by those who may never see or use an affected area (such as Prince William Sound prior to the Exxon Valdez oil spill) can produce initial estimates for a judicial process of damage assessment. As reported in *The New York Times* (September 6, 1993), the state of Alaska commissioned a $3 million survey of 1,043 randomly selected people in 50 states to see what Americans would pay to avoid a repeat of the 1989 Valdez disaster. The sample median response was $31, from which the state of Alaska inferred that all households in the United States collectively would be willing to pay $2.8 billion—the damages it ascribed to Exxon. The sample mean of $94, implied even higher damages of $8.6 billion. Although this contingent survey was never tested in the courts, because the liability suit was settled for $1 billion before the survey was introduced as evidence, what is the difference in the interpretation that can be given to the median and mean in these damage calculations?

ANSWER:

Interpreted as the answer to a popularity survey or a referendum, the $2.8 billion inferred median represents the amount that the majority of U.S. households would vote to pay to prevent another oil spill. Interpreted as a market survey or market outcome, and the mean suggests that households would collectively pay $8.6 billion to buy a spill-free environment for Prince William Sound.

2.7 The Mode as a Measure of Central Tendency

Mode
The mode is the most frequently occurring value.

The **mode** is a third measure of central tendency or central location of the data. It is the value that occurs most frequently. It provides the answer to questions such as: What fan belt size is most often requested in an auto parts store? What is the most common teaching load at the University of Wisconsin? What is the customary workweek for a real estate agent?

In the set of data 9.0 lb, 2.5 lb, 5.5 lb, 6.0 lb, 2.0 lb, 1.5 lb, and 2.0 lb, the mode is 2.0 lb because it occurs most often. For this seven-item data set, the median is 2.5 lb and the mean is 4.07 lb; thus, the mode is less than the median, which in turn is less than the mean. As will be shown later in this chapter, this is the anticipated ordering of these three measures of central tendency for large data sets, with continuous measurements along a number line where there is only one modal value and only a few high values pull up the mean. If a few low values are pulling the mean down, then the mean is expected to be less than the median, which in turn is expected to be less than the mode.

For data sets with discrete values, as in all the data sets described so far in this chapter, or for those with more than one mode, the position of the mode(s) relative to the median and mean need not be consistent with the above anticipated pattern. In the data set $1, $2, $2, $6, $7, $8, $9, $10, a median is $6.50 and the mean is $5.625. Although the mean is being pulled down by a few low values, those two low values of $2 happen also to be the mode.

Because the mode does not take account of the magnitude of the values involved, it is not a good indication of the middle of a set of data. In the above two data sets, the number 2 was the mode in each even though its relative contribution to the sum of the eight values was different in the two sets. Reporting the mode by itself can be misleading. For instance, in the spring of 1992, an administrator at the University of Wisconsin told a legislator that the most frequent teaching load on the Madison campus was two courses a semester. Such a fact could be consistent with a mean and median teaching load well above or below two courses a semester.

Unimodal Distribution
A distribution with one mode.

Bimodal Distribution
A distribution with two modes.

Trimodal Distribution
A distribution with three modes.

Distributions that have a single mode are called **unimodal.** Those with two modes are **bimodal,** and those with three modes are **trimodal.** A frequency distribution can have only one mean but there is no limit to the number of modes that it may have. Some data sets do not even have one well-defined mode. For instance, the CEO compensation data in Table 2.1 does not have a well-defined mode because each value occurs only once.

Modal Class
The class in a frequency distribution with the most occurrences in it.

Finally, if data are grouped into classes, a **modal class** is defined as that with the greatest frequency. In Table 2.2, for example, 95 banks failed in 13 to 18 weeks, which is the interval with the highest frequency and is thus the modal class. Without having the raw data, however, we cannot know if this interval contains the exact mode or if there are one, two, three, or even more modes.

Historical Perspective

Strange as it may seem today, arithmetic and graphical techniques for describing data once had a terrible reputation. They were viewed as tools used by social reformers to denigrate the doctrines of free will and individual responsibility and by scoundrels to mislead and obscure truth.

According to Stephen Stigler (1986, pp. 169–172), one of the most controversial concepts has been "the average man," or, as his creator Adolphe Quetelet labeled him, "l'homme moyen." Quetelet was born in Belgium in 1796; he became a mathematician, an astronomer, and then a sociologist. He created the notion of the "average man" in 1835 as a fictional being, never imagining that this idea would take on a life of its own and live on in newspaper headlines and corporate board rooms. It has become the means to disguise the identity of individuals. More important for the social sciences, Quetelet provided the key notion and framework for separating the fixed (that felt in common by all) from the random (individual) determinants of behavior.

William Playfair (1759–1823) introduced and developed the fundamental rules for graphical representation of quantitative data. His goal was to replace tables of numbers with visual representations of his "linear arithmetic." However, graphing involves many diverse skills. The visual needs of art, the accuracy of drafting, the rigors of mathematics, and the fuzziness of social science concepts did not mix well in the minds of established scholars of the time.

Famed nursing pioneer and hospital reformer Florence Nightingale (1820–1910) was a leading advocate of Quetelet's "social physics" and Playfair's visual representation of data. She invented the "coxcomb" diagram, the forerunner of a pie chart, in which the magnitude being represented is proportional to the area of a wedge in a circular diagram. To her, average mortality rates in the general public were the measures against which average death rates in the military should be compared. She used data and graphs in defiance of the prevailing doctrine of John Stuart Mill and Charles Dickens, who viewed statistics as dehumanizing and characteristic of "a deadly statistical clock" that was an anathema of individualism.

The proper use of numbers and graphs to represent social phenomena continued to be debated well into the twentieth century. As reflected in the popular book by Darrell Huff, *How to Lie With Statistics* (1954), much of post-World War II academic interest in descriptive statistics and graphs focused on debunking false calculations and improperly drawn graphs that were designed to fool naive readers. Descriptive statistics and graphs were viewed as devices either for showing the obvious to the ignorant or obscuring the truth from the informed. Credit for demonstrating the absurdity of these myths goes to world-renowned statistician John Tukey.

In the 1970s, Tukey demonstrated how graphs can be used as tools of analysis, and he created new types of graphs to explore large and complex data sets. The advent of inexpensive computer programs in the 1980s enabled the further development of graphical techniques in data analysis. Today, descriptive statistics and graphical analysis are key components in the use of nonparametric statistical techniques, a few of which are introduced in later chapters of this book.

In the world of manufacturing, W. Edwards Deming used histograms and quality control charts demonstrated in this chapter to convince Japanese engineers and executives that statistical methods could be used to overcome the post-World War II belief that "made in Japan" meant "cheap junk." Deming was a statistician on the faculty at New York University when he advanced the process control tools that helped Japanese manufacturers achieve a new reputation. Curiously, however, major U.S. manufacturers were not willing to follow the example of Japan until the late 1980s. For years, Japanese firms have been competing among themselves to win the annual Deming Medal. It wasn't until 1988 that American firms began competing for the Malcolm Baldrige National Quality Award, given annually to the company making the greatest advances in improved product quality and customer satisfaction.

Total Quality Management (TQM) is no longer viewed as a panacea by which companies can dash past their competitors, as reflected in numerous articles with titles like "The Cost of Quality: Faced with hard times business sours on 'Total Quality Management'" (*Newsweek*, September 7, 1992). Florida Power & Light, winner of Japan's Deming Prize for quality management slashed its program because of excessive paperwork. Less than two years after winning the 1990 U.S. Commerce Department's Baldrige award, the Houston oil-supply company Wallace Co. filed for Chapter 11 bankruptcy protection. These disappointments have not discouraged the numerous quality control gurus who continue to crisscross the United States selling their total quality management techniques to large and small firms alike. Clearly, a knowledge of statistics is required to ferret out the useful from the wasteful business practices being advanced in the name of TQM. The philosophies of TQM and many of the different statistical techniques of process control are described in detail in Chapter 18 of this book.

2.8 Percentiles, Deciles, and Quartiles

pth Percentile
The pth percentile is the value of the observation for which p percent of the observed values are equal to or less than it.

First Quartile (or lower quartile)
The 25th percentile defines the lowest quartile of a distribution; the value for which one quarter of the values are equal to or lower.

Third Quartile (or upper quartile)
The 75th percentile of a distribution.

Second Quartile
The 50th percentile is the second quartile of a distribution.

The use of proportions in the description of data is not restricted to qualitative data. We already have seen that cumulative relative frequencies describe the proportion of quantitative observations up to a numerical value. Percentiles, deciles, and quartiles identify special values by their relative positions in the set of data. The **pth percentile** is the value for which p percent of the data set is less than or equal to the value. The typical percentiles cited in research are the 25th, 50th, and the 75th percentiles. These three percentiles are referred to as the quartiles. The **first quartile** is the 25th percentile; the **third quartile** is 75th percentile. The **second quartile** is the 50th percentile, which for large data sets is also the approximate median. Although not commonly reported, **deciles** are defined by the 10th, 20th, 30th, . . . and 90th percentiles.

Like the median, the percentile, decile, and quartile values for small data sets and grouped data may not be unique and well defined. For example, Table 2.6 gives the relative frequencies and cumulative relative frequencies for the U.S. Census age groups in Figure 2.2. The 25th percentile (and first quartile) can be seen to be an age of slightly less than 18 years, because the cumulative relative frequency up to age 18 years is 0.255737. The 50th percentile (fifth decile and second quartile), as with the median, is in the age interval 25 to 45 years, but we cannot determine it exactly because the original data are camouflaged by the grouping. (From other current population reports of census data, the median age in 1990 is known to be 32.9 years.) Similarly, the 75th percentile (third quartile) is in the 45–55 age interval, but without the original data

Table 2.6
1990 U.S. Census by Age

Age Group	1990 Population	Relative Frequency	Cumulative Relative Frequency
less than 5	18,354,443	0.073798	0.073798
5 to 18	45,249,989	0.181938	0.255737
18 to 25	26,737,766	0.107505	0.363243
25 to 45	80,754,835	0.324694	0.687938
45 to 55	25,223,086	0.101415	0.789353
55 to 65	21,147,923	0.085030	0.874384
65 to 75	18,106,558	0.072801	0.947186
75 to 85	10,055,108	0.040429	0.987615
85 +	3,080,165	0.012384	1.000000
	248,709,873	1.000000	

Decile
The first decile is the value that puts one-tenth of the values at or below it; the second decile puts two-tenths of the values at or below it, etc.

we can only approximate it by making an assumption about the position of observations within this class (as evenly distributed or following an ogive, for example).

Query 2.6:
Find the first, second and third quartiles, for the CEO compensation data in Table 2.1, and assess their distances from the median.

Answer:
Using the data sort command in LOTUS 1-2-3, the 18 comparable observations on CEO compensation can be ordered quickly.

Original Company #	Total Pay (in thousands of $s)	
9	9265.8	
13	4664.9	
4	3682.8	
17	3325.8	
14	2471.1	← Third Quartile
18	2153.4	
6	1749.4	
7	1727.2	
8	1649.3	
11	1601.5	← Second Quartile
2	1538.5	
20	1449.9	
5	1309.1	
3	1038.0	← First Quartile
16	1001.1	
12	671.4	
15	650.0	
10	476.3	

The median, as determined in Query 2.2, is $1,625,400, which is also the 50th percentile and second quartile. The 25th and 75th percentiles are the values of the 4.5 and 13.5 observations, which of course do not exist. As approximations, either $1,019,550 or $1,038,000 may be specified as the first quartile and $2,312,250 or $2,471,100 may be set as the third quartile.

2.9 Skewness

Skewness
A measure of the degree of asymmetry of a distribution.

Positively (or right) Skewed Distribution
An asymmetric distribution in which the mean is greater than the median.

Negatively (or left) Skewed Distribution
An asymmetric distribution in which the mean is less than the median.

Symmetric Distribution
A distribution in which the mean and median are the same value.

Knowing all three measures of central tendency can suggest the shape of a frequency distribution. A relationship between the measures of central location and the shape of the distribution is described by **skewness.** In general, distributions that have a mean value greater than the median are **positively (or right) skewed** and those with a mean value less than the median are **negatively (or left) skewed.** If the mean is the same value as the median, the distribution is said to be **symmetric.**

If a data set is large (infinite) with no breaks in the data (continuous), and if the distribution is right-skewed, with only a single mode, then the mean will exceed the median, and both the mean and median will be greater than the mode. As stated earlier, the mean exceeds the median because a few large values, in the upper tail pull up the mean. Such a right-skewed distribution is Figure 2.2, panel a. For a left-skewed unimodal distribution, the mean will be less than the median and the mode will exceed both the mean and median. The mean is less than the median because a few low values in the lower tail pull the mean down, as depicted in Figure 2.2, panel b. Because the median is not influenced by extreme values in the tails, it is often preferred to the mean as a measure of central tendency for highly skewed data sets.

A unimodal, continuous, and symmetric distribution will have a bell shape for which the mean, median, and mode will all be equal, as shown in Figure 2.2, panel c. For symmetric distributions, whether the mean or median is used as the measure of central tendency is irrelevant since they are the same value.

2.10 Measures of Dispersion

As already discussed in assessing skewness, differences in the mean, median, and mode convey information about the distribution. The distance between certain data points and an average measure of dispersion is relevant for decision makers. In fact, a 1992 television ad for the Lexus automobile portrayed the uniformity of the distance between the fenders and trunk lid, engine hood, and the doors by having a marble roll around the car as the car was turned in a giant vise. Because the assembly process of the Lexus is near perfect, very little variation in the clearance was evident as the marble moved.

Obviously, door-fender and hood-fender clearances do not have to be as precise as on a Lexus for daily operation. At some point, however, the width of these clearances

FIGURE 2.2 Measures of Central Tendency and Skewness

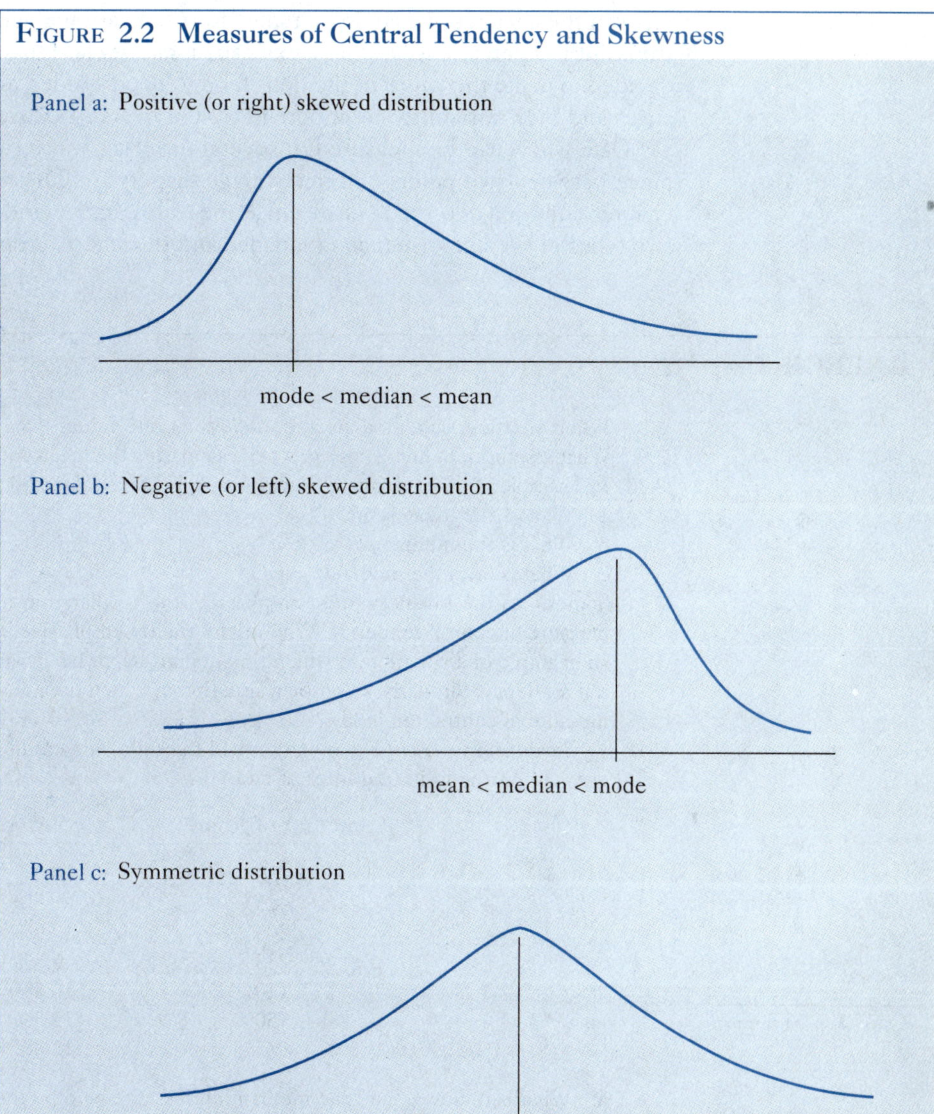

on even the cheapest car could be too wide or too narrow to permit the doors, hood and trunk lid to operate easily. The Lexus ads make it clear that Lexus management has decided to make the width of the door-fender, hood-fender, and trunk lid-fender clearances and their variability an important part of overall Lexus quality assessment.

Data width can be measured in several different ways; two such ways are the distance between two points and the average dispersion. The remaining sections of this chapter consider two types of distance measures (range and interquartile range) and two types of average dispersion (variance and the standard deviation).

EXERCISES

2.8 When are the mean, median, and mode the same value?

2.9 What assumptions are necessary when calculating the mean and median with grouped data?

2.10 In a sample of nine job qualifying exams, applicants scored 50, 35, 40, 65, 80, 90, 95, 85, 70.

a. What is the mean score?

b. What is the median score?

c. What is (are) the mode(s)?

2.11 Sometimes the business press emphasizes the median and other times the mean as the measure of central tendency. Why might one be emphasized rather than the other? Give an example of a situation in which the median might be an inappropriate measure of central tendency. Similarly, describe a situation in which the mean might be an inappropriate measure of central tendency.

2.12 A salesperson for a large corporation had the following credit card balances on her different cards at the indicated interest rate.

Amount of Loan	Interest Rate
$ 250	19.0%
425	18.5%
1,200	17.0%
235	16.5%
110	16.0%
750	15.5%
935	15.0%

a. What is the mean and the median interest rate paid by this person?

b. What is the shape of the distribution of interest rates?

2.13 The *DeVoe Report* (June 2, 1980) quoted then U.S. President Jimmy Carter as saying "half the people in this country are living below the median income—and this is intolerable." What is disputable and what is true in this quote?

2.14 In sailboat racing, rule violations are handled by the local yacht club. Anyone involved in a protest who is not satisfied with the results may appeal to the U.S. Sailing Association. Below are the frequency distributions for the time to process appeals in 1989, 1990, and 1991, from *American Sailor* (September 1992).

Approximate the median time to process appeals for each of the three years. What does a comparison of medians show? Why wouldn't a comparison of means be appropriate or even possible from these data?

	Time to Process an Appeal		
Time in weeks	FY 89	FY 90	FY 91
0 to < 5	5	2	4
5 to < 10	1	8	1
10 to < 15	9	4	1
15 to < 20	7	8	2
20 to < 25	3	3	3
25 to < 30	4	0	3
30 or more	0	1	4*
Total	29	26	18

*Only two appeals received before 10/31/91 were still open as of 4/30/92, and in b cases the Committee is waiting for information from the protest committee. Both have taken more than 30 weeks to process and are included in the statistics.

2.15 Below is information on supermarket sales and market share for the leading sports drinks, from *The Wall Street Journal* (September 29, 1992). What are the mean and median sales for these 11 products? Which is a better representation of sales of a representative product in the sports drink market? Explain.

Sports Drink	Sales 52 Weeks to 7/12/93 (millions)	Market Share 52 Weeks to 7/12/92	Sales Growth 1/1/92–7/12/92
Gatorade	$288.3	85.2%	+7%
Gatorade Light	14.6	4.3	+ 16
10-K	12.2	3.6	—
PowerBurst	6.7	2.0	–72
Private label	4.2	1.2	–19
1st Ade	2.6	0.8	–1
All Sport	1.7	0.5	+451
Joggin' in a Jug	1.6	0.5	+138
Enduro	1.2	0.4	–45
Body Works	0.9	0.3	+6
Spike	0.4	0.1	+182

Source: Information Resources Inc.

2.16 On the following page is a Computer Associates' advertisement that appeared in *PC Magazine* (June 30, 1992). Determine a median for each of the three spreadsheets. Why is the median better than the average as a measure of central tendency for this data?

2.17 *The Wall Street Journal* (October 27, 1992) reported that of the 94 patients treated for skin cancer (melanoma lesions), "the average time from their surgery and injection to the recurrence of melanoma was 30 months." Is it clear whether "the average time" pertains to the mean or the median? Why might the mean time from injection to recurrence be higher than 30 months, regardless of whether "the average time" pertains to the mean or the median? Why might the median time before recurrence be a better measure of successful treatment than the mean?

VARBUSINESS
Spreadsheet Report Card

	CA-SuperCalc	Lotus 1-2-3	Microsoft Excel
Ease of use	7.30	6.50	7.11
Memory Requirement	7.00	5.41	6.14
Ease of programming	6.48	5.86	6.26
Ability to manipulate data	7.31	6.71	7.00
Sorting capabilities	7.50	6.64	6.68
Provision for software security	6.96	5.25	5.10
Report writing capabilities	6.78	5.33	6.17
Ease of use of interface	7.45	6.19	6.77
Software integration capabilities	7.30	6.23	6.78
Ease of data retrieval	7.50	6.78	7.00
Satisfaction with product profitability	6.81	5.75	6.42
Overall quality of product	7.70	7.18	7.53
Provision for customer support	7.52	5.79	6.22
Charges for training time	6.43	5.60	5.71
Provision for technical support	7.34	5.55	5.95
Provision for marketing support	6.69	5.71	5.93
Documentation & product information	6.90	6.70	6.98
Frequency of updates & revisions	6.59	5.75	6.15
OVERALL AVERAGE	7.09	6.05	6.44

2.18 *The Wall Street Journal* (October 27, 1992) article on skin cancer quoted Dr. Donald L. Morton as saying "that of the 136 patients who were vaccinated, half have survived for 23 months or longer compared with an average survival of 7.3 months for unvaccinated melanoma patients in similar advanced states of the disease." Do you suspect that Morton is using the "average survival" wording to refer to the mean or median survival?

2.19 Exercise 1.20 required the drawing of the ogive for the Census Bureau data showing that in 1989 the poorest 20 percent of the population of married couples with children had cash income under \$22,700; the next 20 percent had income under \$34,100; the middle 20 percent had income under \$45,500; the next 20 percent had income up to \$62,200; and the richest 20 percent had income over \$62,200. Use this ogive to estimate the median income.

2.11 The Range

The simplest measure of variability is the range. The range is the difference between the largest and smallest values in a set of data. After the data are ordered from lowest to highest, the range is the difference between the two extremes. In the energy company CEO pay example, Table 2.1, the lowest comparable yearly compensation was that of Ron W. Haddock of FINA (\$476,300) and the highest compensation was received by Lawrence G. Rawl of Exxon (\$9,265,800). Thus, the range of comparable compensation packages was \$8,789,500.

In quality control work, the range is plotted in the so-called "R-chart." For instance, the top panel in Chart 2.2 shows the range of student test scores on the final

CHART 2.2 R-Chart for Student Scores

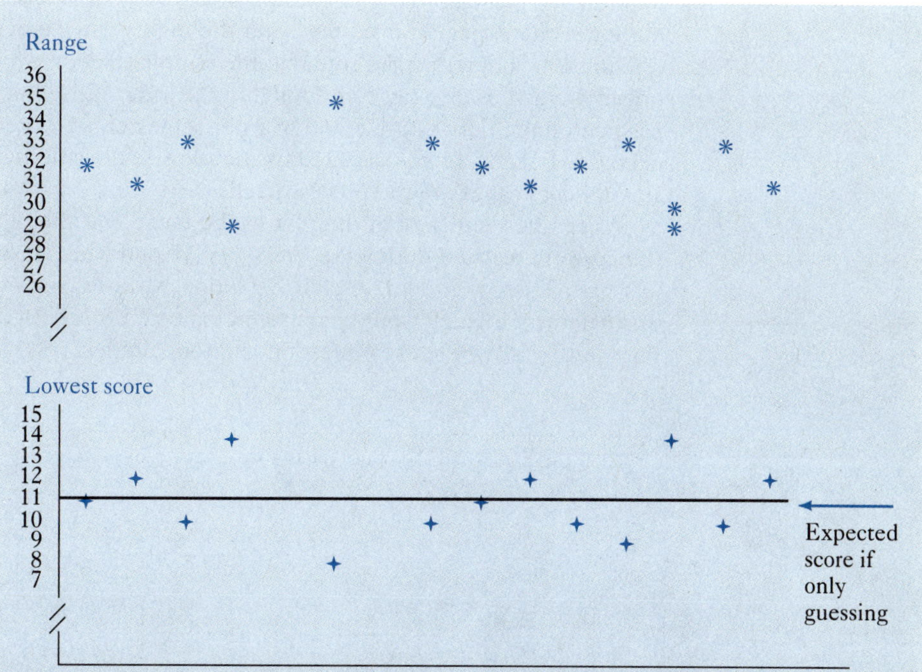

Semester	Final Exam Scores			Enrollment
	Highest	Lowest	Range	
Spring 1986	43	11	32	962
Fall 1986	43	12	31	758
Spring 1987	43	10	33	905
Fall 1987	43	14	29	761
Spring 1988	n/a	n/a	n/a	901
Fall 1988	43	8	35	640
Spring 1989	n/a	n/a	n/a	863
Fall 1989	43	10	33	839
Spring 1990	43	11	32	747
Fall 1990	43	12	31	712
Spring 1991	42	10	32	792
Fall 1991	42	9	33	489
Spring 1992	43	13	30	689
Fall 1992	43	10	33	475
Spring 1993	43	12	31	652

exams in the statistics course at Indiana University discussed earlier. Maximum and minimum values are likely to be influenced by sample sizes (the smaller the sample, the lower the expected maximum and the higher the expected minimum); therefore, care must be taken to make sure that the sample sizes on which an R-chart is based are (ideally) equal or else large and roughly the same. Enrollment in the statistics course has been high, between 475 and 692 per semester, so extreme test scores can be expected to be relatively unaffected by sample size deviations. Observed extremes probably reflect changes in the process itself.

As can be seen in both the plot of the range and the plot of the lowest score, extremes were reached during the 1987–1989 period when an interim coordinator was directing this large-enrollment course. Curiously, in several semesters, the lowest score was below that which would have been expected from just guessing (11 correct questions out of 44, where each question had four alternatives).

FIGURE 2.3 Same Range with Different Dispersions

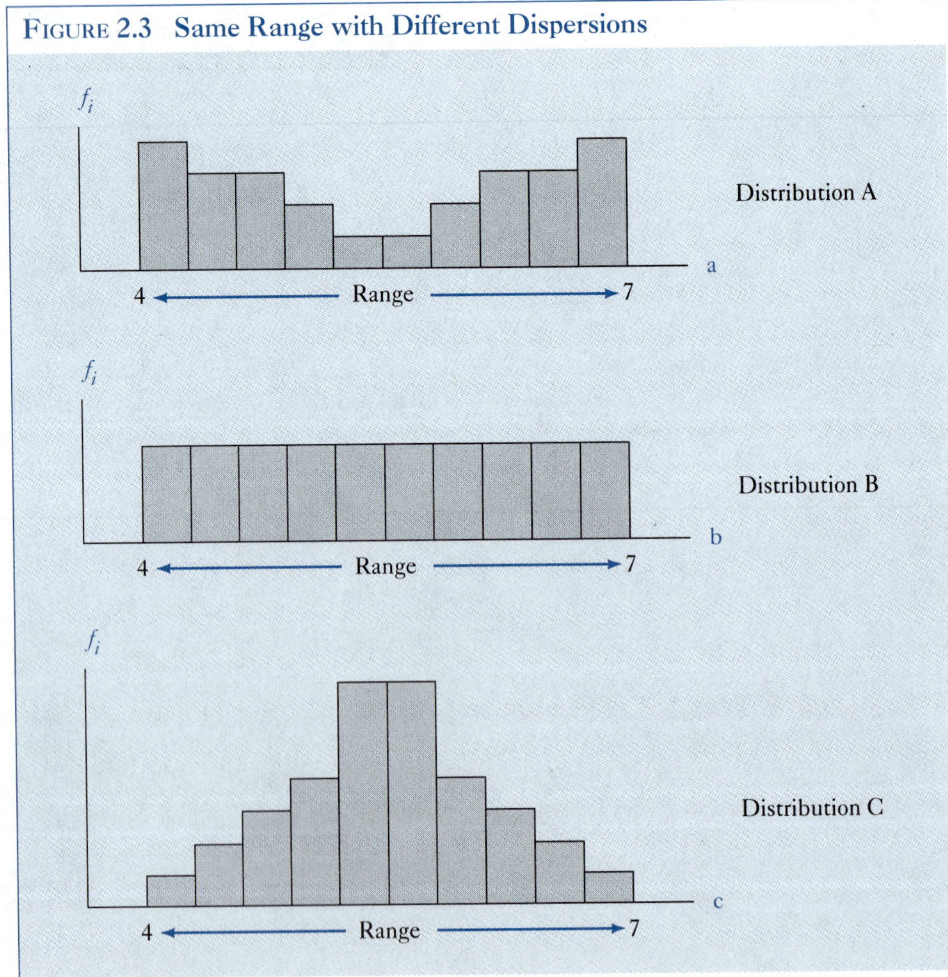

Although the range is easy to calculate, it is not used much outside of quality control work. One reason why the range is often ignored is that distributions may have the same range but yet appear different in terms of observable dispersion. All three distributions in Figure 2.3, for instance, have the same mean ($\mu = 5.5$) and the same range ($7 - 4 = 3$). Yet, distribution A appears to be more "variable" or "spread out" or "diverse" than distribution B, and distribution B appears more diverse than distribution C.

A second reason why the range is not used extensively is that it is sensitive to the sample size, as noted earlier. As the sample size increases, the range tends to grow, as well. The influence of sample sizes may be removed by dividing the range by the sample size n. As discussed in the next sections, however, the range divided by the sample size is not a common measure of average dispersion. It is still sensitive to extreme values and insensitive to other values.

2.12 The Interquartile Range and the Box Plot

Another measure of distance between data points is the interquartile range, which is the difference between the third and first quartiles. Because the third quartile is the 75th percentile and the first quartile is the 25th percentile, the interquartile range gives the middle 50 percent of the data. An advantage of using the interquartile range is that it can be calculated from frequency data that have open-ended class intervals, whereas the range can not. For instance, Figure 2.4 contains a histogram from the *Occupational Outlook Handbook* (first published by the U.S. Bureau of Labor Statistics). It shows that

Figure 2.4

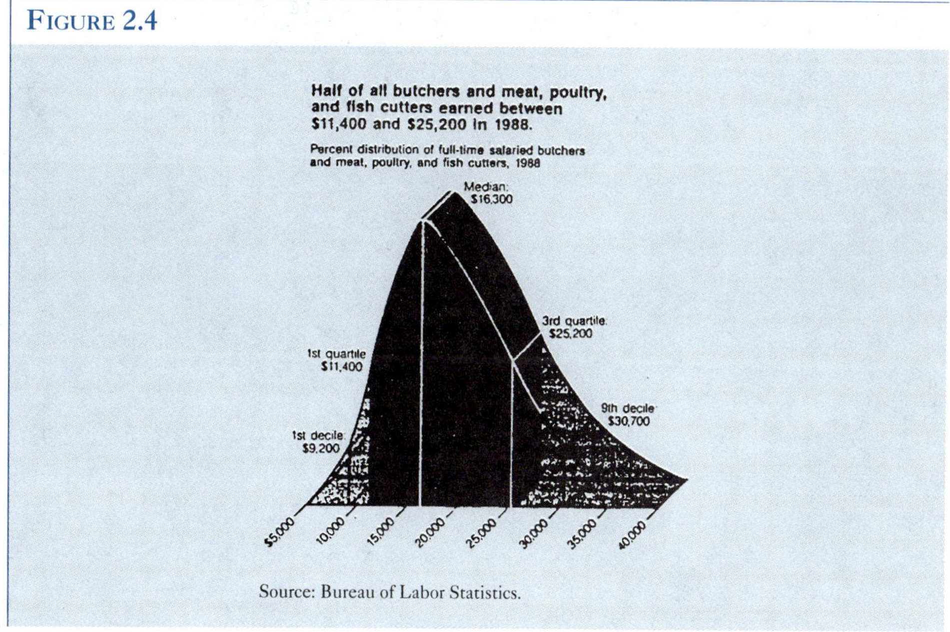

Source: Bureau of Labor Statistics.

"half of all butchers and meat, poultry, and fish cutters earned between \$11,400 and \$25,000 in 1988."

The *Handbook* also uses interquartile ranges to show how annual salaries differ across and within occupations. For instance, Figure 2.5 provides a *Handbook* box plot that shows the interquartile range of salaries for eight levels of engineers, five levels of

FIGURE 2.5 Box Plot for Salaries

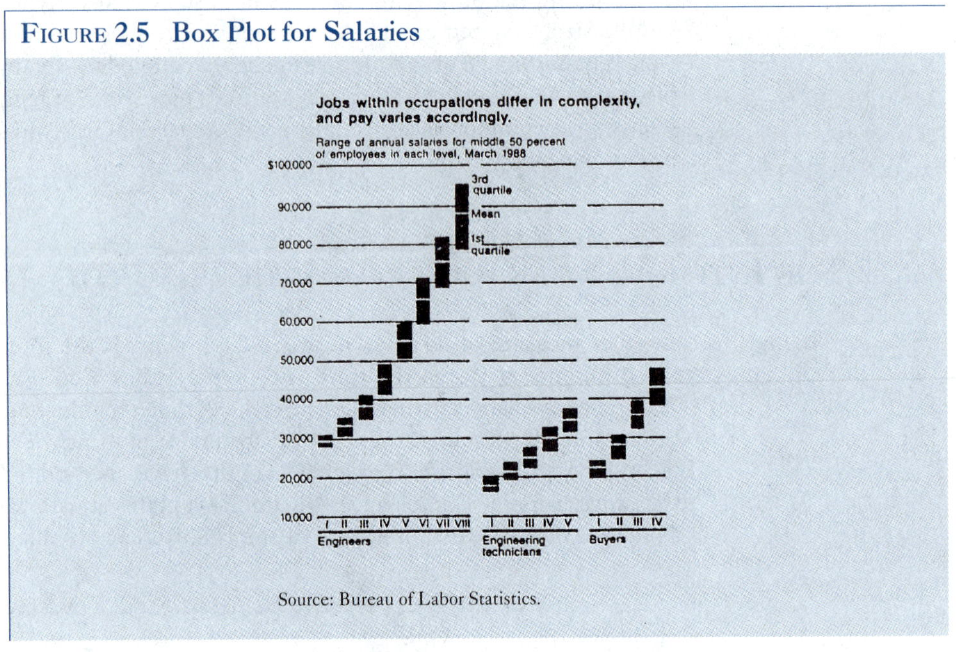

Source: Bureau of Labor Statistics.

FIGURE 2.6 Box Plot for CEO Compensation

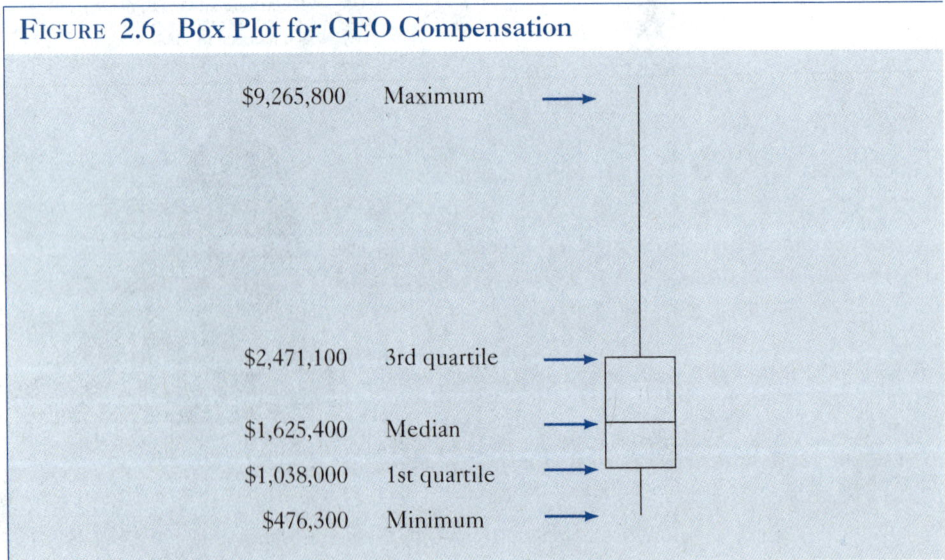

engineering technicians, and four levels of buyers. These interquartile salary levels, as indicated by each boxed area, reflect differences in experience, starting with entry-level jobs and continuing up the career ladder to the most complex and responsible supervisory positions within the occupation.

Because the mean is pulled in the direction of the extreme tail values, box plots are often centered on the median. In fact, in 1952 when Mary Eleanor Spear invented the "range bar," which John Tukey popularized as the box plot, she used the median as the center of the data. This more standard box plot is shown in Figure 2.6 for the CEO compensation data for which we derived the quartiles in Query 2.6. The right skewness of these data is apparent in the long "whisker" extending from the top of the box and the median being closer to the first quartile than to the second quartile in the box.

2.13 Variance and Standard Deviations

Variance
The average of squared differences between values and their mean.

The most common average measures of dispersion are the **variance** and the standard deviation. Although the conceptual definition of a variance is the same for populations and samples, the method of calculation is slightly different. Calculation of both population and sample variances begins by subtracting the mean from each value in the data set; those individual differences, which are called deviations, are then squared and summed. For the population, the sum of squared deviations is divided by the population size (N). For a sample, the sum of squared deviations is divided by the sample size minus one ($n - 1$). The reason for the division by $n - 1$ for sample data versus division by N for population data will be addressed briefly in this chapter but in more detail in Chapter 8.

To illustrate the calculation of a variance, consider the 18 comparable values of CEO compensation in Table 2.1. The mean of this data set was calculated to be \$2,245,861, or, in the units of measurement used in *The Wall Street Journal*, 2245.861 thousands of dollars; thus, the 18 deviations from the mean are

$x_i - \bar{x}$		
1538.5	− 2245.861 =	−707.361
1038.0	− 2245.861 =	−1207.861
3682.8	− 2245.861 =	1436.939
1309.1	− 2245.861 =	−936.761
1749.4	− 2245.861 =	−496.461
1727.2	− 2245.861 =	−518.661
1649.3	− 2245.861 =	−596.561
9265.8	− 2245.861 =	7019.939
476.3	− 2245.861 =	−1769.561
1601.5	− 2245.861 =	−644.361
671.4	− 2245.861 =	−1574.461
4664.9	− 2245.861 =	2419.039
2471.1	− 2245.861 =	225.239

(continued)

$x_i - \bar{x}$ *(continued)*
650.0 − 2245.861 = −1595.861
1001.1 − 2245.861 = −1244.761
3325.8 − 2245.861 = 1079.939
2153.4 − 2245.861 = −92.461
1449.9 − 2245.861 = −795.961
0.000

Because of the magnitude of these chief executives' compensation, the 18 deviations from the mean cannot be easily calculated with a hand-held calculator. A spreadsheet program, such as LOTUS 1-2-3 or EXCEL, is best suited to confirm the above arithmetic. Notice that as a check of the arithmetic, the sum of deviations from the mean must sum to zero, assuming no rounding error. That is,

$$\text{Sum of deviations} = \sum_{i=1}^{18}\left(x_i - \bar{x}\right) = \left(1538.5 - 2245.861\right) + \ldots$$

$$\ldots + \left(1449.9 - 2245.861\right) = 0$$

Because the sum of deviations is always zero (regardless of the shape of the distribution), this sum is not meaningful as a measure of dispersion. Squaring each of these deviations, however, removes the negative signs. Therefore, summing the *squared* deviations from the mean does provide a measure of variability that will differ from distribution to distribution—the more variability in the distribution, the larger the sum of squared deviations from the mean. As shown in Table 2.7, the sum of squared devia-

Table 2.7 Sample Variance of the Energy Companies CEO Compensation

	x_i	$(x_i - \bar{x})$	$(x_i - \bar{x})^2$
	1538.5	−707.361	500359.742
	1038.0	−1207.861	1458928.464
	3682.8	1436.939	2064793.370
	1309.1	−936.761	877521.379
	1749.4	−496.461	246473.635
	1727.2	−518.661	269009.348
	1649.3	−596.561	355885.159
	9265.8	7019.939	49279542.004
	476.3	−1769.561	3131346.526
	1601.5	−644.361	415201.242
	671.4	−1574.461	2478927.790
	4664.9	2419.039	5851749.146
	2471.1	225.239	50732.557
	650.0	−1595.861	2546772.686
	1001.1	−1244.761	1549430.224
	3325.8	1079.939	1166268.004
	2153.4	−92.461	8549.057
	1449.9	−795.961	633554.090
Column sum =	40425.5	0.000	72885044.423

Variance of x = 72885044.423/17 = 4287355.554

tions from the mean for the CEO compensation data is 72885044.423; again, this is a sum easily obtained with a computer spreadsheet program:

$$\text{Sum of squared deviations} = \sum_{i=1}^{18}\left(x_i - \bar{x}\right)^2 = \left(1538.5 - 2245.861\right)^2 + \ldots + \left(1449.9 - 2245.861\right)^2 = 72885044.423$$

Sample Variance (s^2) The average of squared deviations of values in a sample that are free to vary after the sample mean is fixed at its value in a specific sample; a statistic designated by a letter s squared, s^2.

Population Variance (σ^2) The average of squared deviations of population values from the population mean for a population; a parameter designated by the Greek letter "sigma" squared, σ^2.

Parameter A numerical value that describes a characteristic of a population (e.g., the mean (μ) is a parameter that gives central location).

Sample Mean ($\bar{x}$) The arithmetic average of a sample; a measure of central location identified by the symbol "ex-bar," $\bar{x}$.

Statistic A numerical value describing a characteristic of a sample as a function of the sample values.

Finally, the sample variance is obtained by dividing the sum of squared deviations by 17 (that is, by $n - 1$):

$$\text{Sample variance} = \frac{\text{Sum of squared deviations}}{\text{Sample size minus one}} = \frac{72885044.423}{17} = 4287355.554$$

Unlike the variance of the population that has a single and fixed mean μ, the mean of a sample $\bar{x}$ varies from sample to sample. That is, two samples from the same population need not have the same mean. As demonstrated in Chapter 8, when the sample mean is treated as a unique value within a single sample, only $n - 1$ of the values in that sample can be treated as if they were freely determined.

A **sample variance** is designated by s^2. The **population variance** is identified by the symbol σ^2, which is the square of the lowercase Greek letter sigma. They are calculated as

$$\text{Population variance}: \ \sigma^2 = \frac{1}{N}\sum_{i=1}^{N}\left(x_i - \mu\right)^2 \qquad (2.5a)$$

$$\text{Sample variance}: \ s^2 = \frac{1}{n-1}\sum_{i=1}^{n}\left(x_i - \bar{x}\right)^2 \qquad (2.5b)$$

The population variance is calculated the same way as the sample variance, except that the sum of squared deviations around the mean μ is divided by the population size N. Because the population is all of the values of x, μ does not vary; its one value is the population **parameter** that characterizes a central tendency of the population data. Similarly, the σ^2 has one and only one value that represents a characteristic of the population; it is a parameter of the population that measures dispersion. On the other hand, as $\bar{x}$ varies from sample to sample, so does s^2. The **sample mean** and sample variance are called sample **statistics** because they give the status of the sample at hand; sample statistics report different values from sample to sample even though the samples are all drawn from the same population. In the case of the CEO compensation, the one sample variance we calculated is written

$$s^2 = 4287355.554 \text{ (thousands of dollars squared)}$$

If we randomly selected another sample of CEO compensation, s^2 would be a different value; it is a sample statistic.

The variance is measured in square units. It is thus not possible to relate it directly to the original units of measurements. To return to the original units of measure, the

Standard Deviation Square root of the variance.

positive square root of the variance is typically reported. The positive square root of the variance is called the **standard deviation.**

The **standard deviation** is the positive square root of the variance.

The population standard deviation is $\sigma = +\sqrt{\sigma^2}$ (2.6a)

For the sample, the standard deviation is $s = +\sqrt{s^2}$ (2.6b)

For the energy company CEO compensation data, the sample standard deviation is 2.070593 million dollars.

$$s = \sqrt{4287355.554} = 2070.593 \text{ thousands of dollars}$$

What should be apparent from Formulas (2.5) and (2.6) is that the variance and standard deviation are larger for distributions with many values far away from their mean (such as in Figure 2.3a) than for distributions with many values close to their mean (as in Figure 2.3c). Notice also that while the variance is an "average squared deviation," there is no simple verbal definition for the standard deviation. The best we can say is that the standard deviation relates directly to the proportion of values that are grouped around the mean, as will be shown later in this chapter.

QUERY 2.7:

The weights of a sample of seven pennies were measured and reported to the nearest 0.02 gram to be 3.11, 3.03, 3.05, 3.15, 3.11, 2.99, and 3.01 grams. What is the approximate mean and standard deviation? Why aren't your calculated mean and standard deviation exact?

ANSWER:

The mean and standard deviation are easily calculated in a spreadsheet program (such as LOTUS 1-2-3) as follows:

Weight	Deviations	Deviations Squared
3.11	0.04571	0.002089
3.03	–0.03429	0.001175
3.05	–0.01429	0.000204
3.15	0.08571	0.007346
3.11	0.04571	0.002089
2.99	–0.07429	0.005518
3.01	–0.05429	0.002946
Sum = 21.45		0.021371 = Sum
Mean = 3.064286		0.003562 = Sample variance
		0.059681 = Sample standard deviation

These calculations are not exact because the original weights were reported after rounding to the nearest 0.02 grams.

QUERY 2.8:

Treating the eight industrial groups in *The Wall Street Journal* (April 22, 1992) data set on CEO total compensation as the population, what are the mean and standard deviation of the CEO compensation? Is this distribution skewed?

ANSWER:

Before attempting to obtain the desired population parameters, notice that this is a large data set and calculations cannot be made easily without a computer program. With a spreadsheet program such as LOTUS 1-2-3, however, or a statistics package such as the ECONOMETRICS TOOLKIT, this data set can be managed quickly. The data set is available in an ASCII file accompanying this book, so it need not be typed into a file. It can be read into almost any computer program; for some programs, this may mean first bringing the file into a word processor to remove all the nonnumerical text and extra rows. Keep in mind that not all observations in this data set are comparable; there are only 317 comparable compensations. ET gives the following computer printout for these values.

Variable	Mean	Std. Dev.	Skew.	Minimum	Maximum	Cases
CEOPAY	2325.6	4763.9	11.690	100.0	.7480E+05	317

From this computer printout, we see that for the 317 CEOs, the lowest compensation level is \$100,000 and the highest is \$74.8 million. (Notice that .7480E+05 is notation for $0.7480*10^5 = 0.7480*10{,}000 = 74{,}800$.) Because the formula for the mean involves the same calculations for both population and sample data, the mean compensation for this population of 317 CEOs is \$2.3256 million.

This computer printout, and most others produced by statistical packages, automatically calculates sample statistics. If population data are read in, the program does not recognize this and makes incorrect calculations of the variance and standard deviation. To convert to populations values reference must be made to the formulas employed. To get the population standard deviation, we must retrieve the sum of square deviations from the 4763.9 figure given in the printout. To facilitate this with a hand calculator, we first put the standard deviation in millions of dollars as 4.7639. The computer had done the following calculation, which we need to undo by hand.

$$\sqrt{\frac{\Sigma\left(x_i - \text{mean}\right)^2}{\left(n-1\right)}} = \sqrt{\frac{\Sigma\left(x_i - \text{mean}\right)^2}{\left(316\right)}} = 4.7639$$

By multiplying both sides of this expression by $\sqrt{316}$ and then squaring both sides we get the sum of squared deviations

$$\Sigma\,(x_i - \text{mean})^2 = 7171.54$$

Now dividing this sum of squared deviations by the population size 317 gives the variance of the population.

$$\frac{\Sigma(x_i - \text{mean})^2}{N} = \frac{7171.54}{317} = 22.62315$$

and the population standard deviation is 4.75638 million dollars. Because the data set is large, whether the standard deviation formula for a sample or population was used is of little consequence here (4.7639 versus 4.75638 million). For a small data set, however, such is not the case. We must know whether a data set is a population or a sample and how our computer program is treating it.

A box and whisker plot can be produced showing the information on the quartiles, expressed below in millions of dollars:

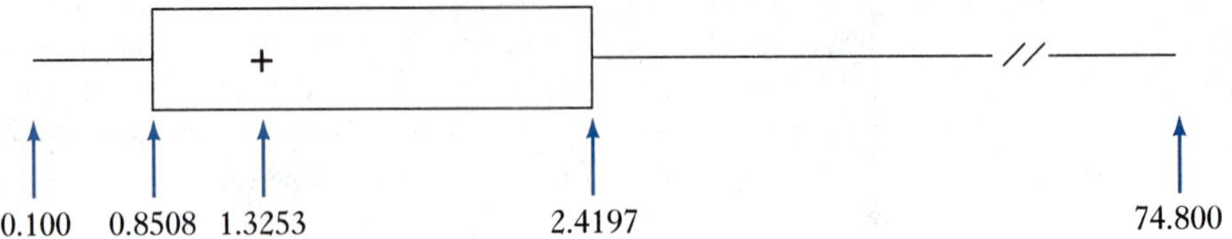

Clearly this distribution is highly right skewed, with the median compensation 1.3253 million dollars closer to the first quarlite, 0.8508, than the second, 2.4197. It is much more skewed than the sample of 18 energy company CEOs, where the box and whisker plot given in Figure 2.6 is reproduced here for scale comparison:

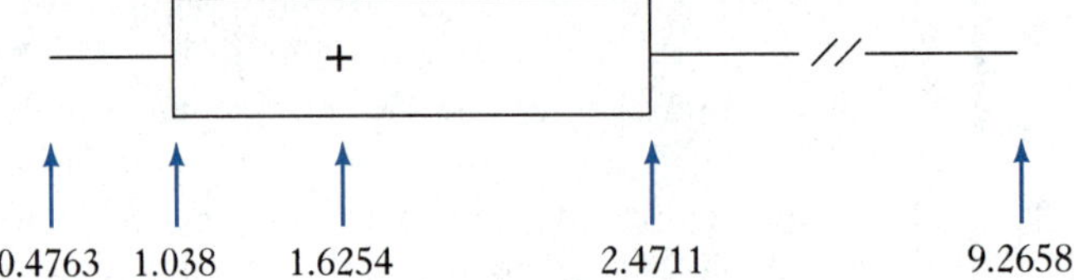

The right or positive skewness of the population of all 317 CEOs is also reflected numerically in the "coefficient of skewness" of 11.690, which was reported earlier in the computer printout in the column titled "Skew." We will not go into the details of the coefficient of skewness here. It is sufficient to know that a positive coefficient of skewness indicates a distribution that is skewed right and a negative value indicates left skewed.[1]

2.14 The Variance and Standard Deviation in Frequency Form

The procedures for calculating a variance after data have been grouped into a frequency distribution are similar to those for the ungrouped raw data except the value or class midpoint (m_i) at which multiple observations occur is used instead of x_i, and deviations of m_i from the mean are weighted by the respective frequencies (f_i). Formulas (2.7a, b and c) give formulas for determining the variances for population and sample data after the original values have been placed in a frequency distribution.

Weighted variances for grouped data:

Population variance:

$$\sigma^2 = \frac{1}{N}\sum_{i=1}^{c} f_i\left(m_i - \mu\right)^2 \qquad (2.7a)$$

where μ is the mean of the population, m_i is the value or class midpoint at which multiple observations occur, f_i is the frequency of the ith class, c is the number of classes, and N is the number of items in the population $N = \Sigma f_i$

Alternative population variance:

$$\sigma^2 = \sum_{i=1}^{c}\left(f_i / N\right)\left(m_i - \mu\right)^2 \qquad (2.7b)$$

where f_i/N is the relative frequency of ith class

Sample variance:

$$s^2 = \frac{1}{n-1}\sum_{i=1}^{c} f_i\left(m_i - \bar{x}\right)^2 \qquad (2.7c)$$

where $\bar{x}$ is the sample mean and the sample size is $n = \Sigma f_i$

Notice that the Formula (2.7b) for the population variance is simply Formula (2.7a) rewritten in terms of the relative frequencies. By focusing on the relative frequency, we do not need to know the actual population size; only the relative frequencies of occurrences are needed. As in our discussion of mean calculations, Formulas (2.7a, b, and c) give approximate variances if the data have been grouped and placed in frequency form. As in the approximation of the mean, if we do not know the original values, we simply assume that the class midpoint represents the class.

As with the calculation of a mean, if we are fortunate and have the frequencies or relative frequencies of each of the original values in the population, then Formula (2.7a) or its alternative (2.7b) can be used to get the exact variance of the population. Table 2.8 illustrates the calculation of the true population variance for the population grade distribution in Table 2.4. The population variance is

$$\sigma^2 = \sum_{i=1}^{5}\left(f_i / N\right)\left(m_i - 2.33\right)^2 = 0.9611$$

and the population standard deviation is

$$\sigma = \sqrt{\sigma^2} = \sqrt{0.96} = 0.980357$$

TABLE 2.8 Weighted Population Variance Calculation of Grades

Letter Grade	Numerical Grade (m_i)	f_i/N	$(f_i/N)m_i$	$(m_i - \mu)$	$(m_i - \mu)^2$	$(f_i/N)(m_i - \mu)^2$
A	4	0.13	0.52	1.67	2.7889	0.362557
B	3	0.25	0.75	0.67	0.4489	0.112225
C	2	0.50	1.00	−0.33	0.1089	0.054450
D	1	0.06	0.06	−1.33	1.7689	0.106134
F	0	0.06	0.00	−2.33	5.4289	0.325734
			$\mu = 2.33$			$0.9611 = \sigma^2$

TABLE 2.9 Weighted Sample Variance Calculation of Grades in One Course Offering by an Instructor Who Has Taught Many Courses

Letter Grade	Numerical Grade (m_i)	f_i	$(f_i)m_i$	$(m_i - \bar{x})$	$(m_i - \bar{x})^2$	$(f_i)(m_i - \bar{x})^2$
A	4	13	52.00	1.67	2.7889	36.2557
B	3	25	75.00	0.67	0.4489	11.2225
C	2	50	100.00	−0.33	0.1089	5.4450
D	1	6	6.00	−1.33	1.7689	10.6134
F	0	6	0.00	−2.33	5.4289	32.5734
			233.00			96.110
			$\bar{x} = 2.33$			$s^2 = 96.11/99$
						$= 0.970808$

For sample data, there is no corresponding relative frequency form of Formula (2.7b), because the division of the sum of squared deviations is by $n - 1$ and not by n. For instance, if at the end of a semester, 100 students complete a course for which 13 receive an A grade, 25 a B, 50 a C, six a D, and six an F, and if this course is viewed as a subset of all the courses this instructor has taught, then grades in this one course are a sample of all grades given in all of this instructor's courses. As such, the variance must be calculated by Formula (2.7c), as shown in Table 2.9. This sample variance is

$$s^2 = \frac{1}{99}\sum_{i=1}^{5} f_i\left(m_i - 2.33\right)^2 = 96.11/99 = 0.970808$$

and the sample standard deviation is

$$s = \sqrt{s^2} = \sqrt{0.970808} = 0.985296$$

EXERCISES

2.20 According to an article in *The Wall Street Journal* (May 3, 1988), advertisers bought \$1.59 billion of commercial time on network sports programs in 1987, a 10 percent increase from 1986. The top 10 advertisers are given below, where advertising expenditures are in millions of dollars and identified by the variable "ADVERTISE."

General Motors	111.400	AT&T	34.300
Anheuser-Busch	102.000	Tandy	30.600
Phillip Morris	78.600	IBM	29.400
U.S. Air Force	42.200	Sears	27.800
Chrysler	37.600	Ford	27.000

Interpret the following computer output by showing how the identified quantities were calculated for this data

Variable	Mean	Standard Deviation	S.E. of Mean	Cases
ADVERTISE	52.09000	32.54380	10.29125	10

How much time did the other advertisers buy on network sports programs in 1987?

2.21 Is there skewness in the data in Exercise 2.20 dealing with advertising expenditures on network sports programs?

2.22 Calculate the mean and variance for the following two samples:

Set A is \$1, \$4, –\$6, and \$2.

Set B is \$100, \$400, –\$600 and \$200.

What is the relationship between the means of these sets and the relationship between the variances? Would this relationship be different if these were population data?

2.23 A larger sample than that reported in Query 2.7 about the weight of newly minted pennies was undertaken by W. J. Youden, as reported by S. B. Vardeman, ("What About the Other Intervals?" *American Statistician*, August 1992, p. 195). The weight of 100 pennies was measured to the nearest 10^{-4}g but reported to the nearest 0.02g. What is the approximate mean and standard deviation?

Penny weight	Frequency
2.99	1
3.01	4
3.03	4
3.05	4
3.07	7
3.09	17
3.11	24
3.13	17
3.15	13
3.17	6
3.19	2
3.21	1

2.24 In a study of used car prices, a sample of dealerships in a large city showed a wide discrepancy in final sales prices. For 50 cars of identical specifications, the following prices were recorded.

Prices	Relative Frequency
At least \$9,800 but under \$10,000	.10
At least \$9,600 but under \$ 9,800	.20
At least \$9,400 but under \$ 9,600	.40
At least \$9,200 but under \$ 9,400	.30

Approximate the mean and standard deviation for this sample.

2.25 Is there skewness in the distribution of car prices in Exercise 2.24?

2.26 The three distributions in Exercise 2.14 deal with the time to process appeals in 1989, 1990, and 1991, from *American Sailor* (September 1992). In which of the three years is the distribution of time to process an appeal most skewed? Why?

2.27 An article in *The Wall Street Journal* (October 12, 1992) reported the mean compensation of CEOs in seven countries. An earlier *WSJ* (September 24, 1992) article reported information on the world's 100 largest public companies. Below are summary data on CEO compensation and number of headquartered companies in the identified countries.

	CEO Compensation	Number of Headquarter Companies in Country
United States	\$717,237	49
France	\$479,772	2
Italy	\$463,009	1
Britain	\$439,441	16
Canada	\$416,066	0
Germany	\$390,933	4
Japan	\$390,723	23
Other	not avail	5

a. What are the mean and standard deviation of CEO compensation per country, ignoring the number at headquarters?

b. What are the mean and standard deviation of number of headquartered companies per country for the seven identified countries?

c. What are the approximate mean and standard deviation of CEO compensation for the 95 companies, ignoring the country?

d. How does your answer in part b influence the difference in your answers in parts a and c?

2.28 For the data on CEO compensation and company headquarters per country, in Exercise 2.27, as the number of headquartered companies deviates from its mean what tends to happen to deviations in CEO compensation? What can you conclude from this?

2.15 Data Location and the Standard Deviation

Rule of Thumb for Bell-Shaped Distributions

For most bell-shaped distributions, the interval $\mu \pm \sigma$ contains about 68 percent of the distribution, $\mu \pm 2\sigma$ contains about 95 percent of the distribution and $\mu \pm 3\sigma$ contains nearly all of the distribution.

In later chapters, we will examine exact probability calculations associated with several different distributions.[2] For now, it is helpful to see how the standard deviation relates to the area on either side of the mean of a symmetric bell-shaped distribution. The **rule of thumb** for these bell-shaped distributions, such as those in Figure 2.7, is that

about 68 percent of the distribution is between

$$\mu - 1\sigma \text{ and } \mu + 1\sigma$$

about 95 percent falls between

$$\mu - 2\sigma \text{ and } \mu + 2\sigma$$

nearly all of the distribution is between

$$\mu - 3\sigma \text{ and } \mu + 3\sigma$$

FIGURE 2.7 Rule of Thumb for Bell-Shaped Distributions

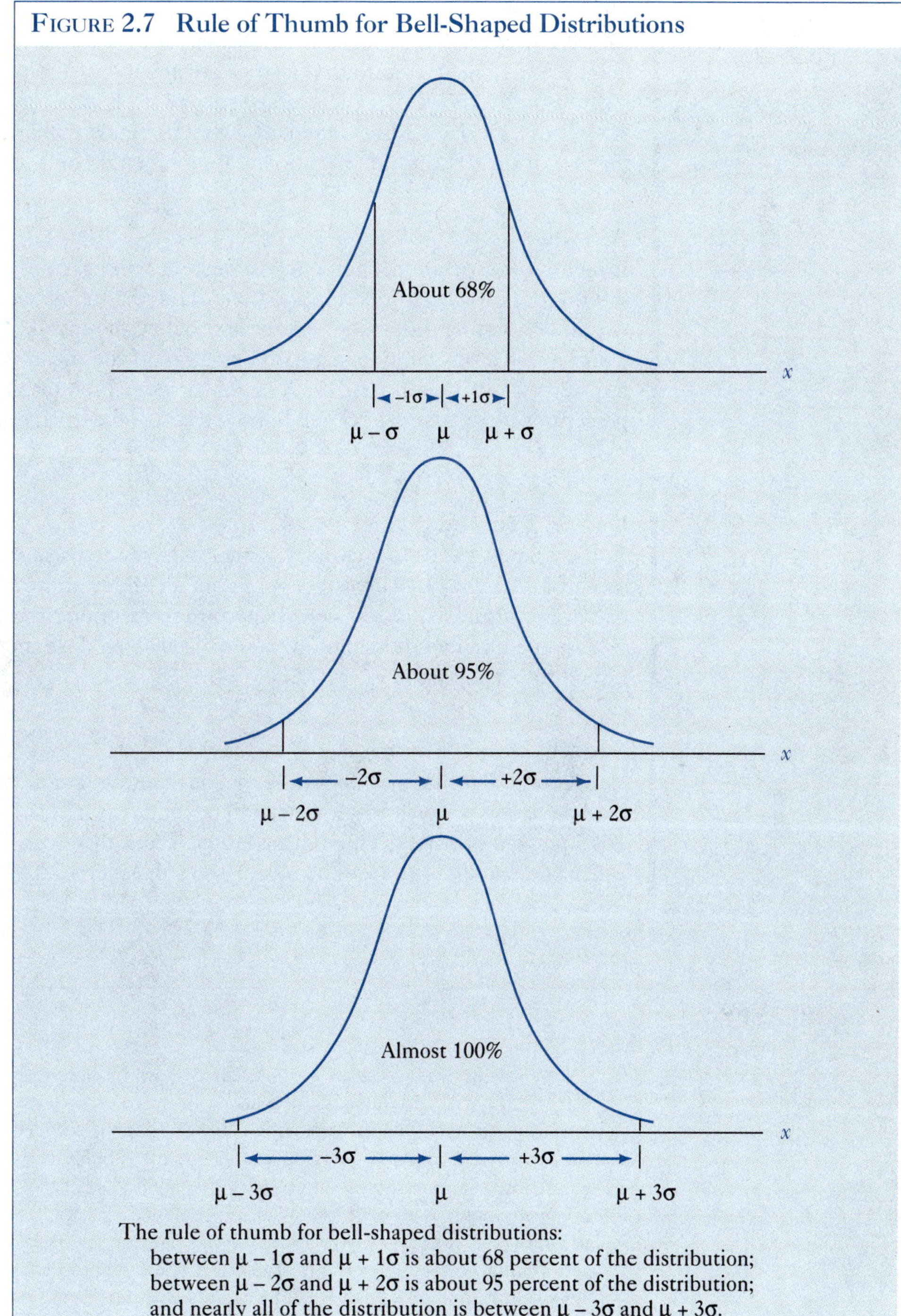

The rule of thumb for bell-shaped distributions:
between $\mu - 1\sigma$ and $\mu + 1\sigma$ is about 68 percent of the distribution;
between $\mu - 2\sigma$ and $\mu + 2\sigma$ is about 95 percent of the distribution;
and nearly all of the distribution is between $\mu - 3\sigma$ and $\mu + 3\sigma$.

QUERY 2.9:
An executive training firm wants to design a CPA review study course that is of a length sufficient to ensure that at least 84 percent of the participants will pass the CPA exam. It knows that the average review study time required to pass this exam is 40 hours, with a 9.5 hour standard deviation. How much time should be devoted to study if the distribution of time is "approximately bell shaped"?

ANSWER:
If 40 hours is specified, then only 50 percent will pass the exam. From the rule of thumb for bell-shaped distributions, we know that 68 percent of the distribution is within one standard deviation above and below the mean. Thus, 32 percent of the distribution is beyond one standard deviation of the mean, with 16 percent in each tail. Thus, if 49.5 hours [$\mu + \sigma = 40 + 9.5$] are specified for the study course, only 16 percent of the course takers will fail the exam and 84 percent will pass.

QUERY 2.10:
In the immediate past, a corporate stock has yielded a net semiannual average return of $100 per $1,000 invested, with a standard deviation of $50. If the distribution of the net return has been bell shaped, then assuming the future is like the past, what is the likelihood that an investor of $1,000 will lose money on this investment if held for a half year?

ANSWER:
Between $100 return and no return at all is two standard deviations, since each standard deviation is worth $50. In each tail beyond two standard deviations of the mean is 2.5 percent of a bell-shaped distribution. Thus, this investment will result in losses 2.5 percent of the time. (The return also will exceed $200 about 2.5 percent of the time.)

EXERCISES

2.29 A study of the loss payment per insurance claim by homeowners showed an average of $810, with a standard deviation of $90.
   a. If the distribution of loss payments is bell shaped, then what percentage of payments is above $990? What percentage of payments is below $720?
   b. *If the shape of the distribution of loss payments is unknown, then what percentage of payments is above $990? What percentage of payments is below $720?

2.30 According to a note in *The Wall Street Journal* (December 30, 1987), telephone calls that are put on hold average 55 seconds on hold. If the standard deviation is 15 seconds, what is the smallest number of calls that are handled within 25 and 85 seconds, assuming a bell-shaped distribution?

2.31 How, if at all, would your answer in Exercise 2.30 change if you now learned that the distribution of hold time in the above question was not bell shaped?

2.32 *According to a fact sheet in *The Wall Street Journal* (September 10, 1993), the minimum, maximum, and average salary of students in executive training programs in 1992 were approximately $20,000, $400,000, and $80,000. If the standard deviation of this distribution was $40,000, what proportion of students had salaries above $140,000?

2.16 Looking Ahead

Many key concepts and important ideas for almost any statistical analysis have been surveyed in this chapter. The remaining chapters in this book continue to come back to these basic notions and greatly expand on them. Of particular importance is recognition that the median may be preferred to the mean as a measure of central location if a distribution is skewed. For measures of dispersion, the standard deviation is usually preferred to the range.

Essential throughout this book is the notion that randomness and chance are inherent in sampling. Both the measures of central location and the measures of dispersion vary from sample to sample. Chapters 7 and 8 deal explicitly with the implications of chance for estimating population parameters. To appreciate the implications of the errors in estimation that may result from random chance events, an understanding of probability, expected values, and probability distributions is required.

The basics of probability and expected value calculations are discussed in Chapters 3 and 4, and specific probability distributions (including the binomial, normal, and other discrete and continuous distributions) are considered in Chapters 5 and 6. If you have already had a course in probability, you may be able to skip over Chapters 3 through 6 and go directly to the application of descriptive statistics in estimation and hypothesis testing, Chapters 7, 8, and 9. In the absence of such a course, however, Chapters 3 through 6 are necessary before you continue your study of inferential statistics.

Chapter Exercises

2.33 An article on airplane crashes in the *Saint Paul Pioneer Press* (July 9, 1992) stated, "Less than one person died for every 100 million miles flown in 1990 . . . . But the death rate climbed up to 6 persons per 100 million airline miles flown in years where there are big airline disasters . . . . Between 1990 and 1991 an average of about 4 people died for every 100 million airline miles flown." How is this a different use of the word "average" than what is used in this chapter? Could you apply the concept of a median and a mode to these data?

2.34 Below are data on average delivery times after 8 a.m. for a *Business Week* (June 6, 1988) experiment involving the sending of 12 letters via each of seven overnight delivery services.

* Read Endnotes to answer this part/problem.

Can the mean, median, and mode delivery time for all 84 letters be determined from this data? Can the standard deviation and variance be determined? If possible, do so.

Emery Worldwide	1 hour, 31 minutes
Federal Express	1 hour, 44 minutes
Airborne Express	2 hours
DHL Worldwide	3 hours, 2 minutes
Purolator Courier	3 hours, 12 minutes
U.S. Postal Service Express Mail	3 hours, 15 minutes
United Parcel Service	3 hours, 31 minutes

2.35 The following distribution is the word processing speed of a large pool of clerical workers:

Words Per Minute	Frequency
35–44	1
45–54	7
55–64	12
65–74	15
75–84	6
85–94	3
95–104	4
105–114	1

a. What is the mode?
b. What is the median?
c. What is the mean?
d. What is the variance?
e. What is the standard deviation?
f. What is the range?
g. What proportion of the clerical workers type 64 or fewer words per minute?
h. Do your calculations in parts a through f depend on whether these are population or sample data? In what type of problem might these be population data, and in what type of problem might these be sample data?

2.36 On the following page is a *Wall Street Journal* (August 10, 1992) article reporting to show that "High-Priced Stocks Are Hardest Hit." It shows the percentage of stocks, within a "P/E decile grouping of stocks," whose price fell by 30 percent or more. For example, in the tenth decile the median price/earnings ratio was 80.5, and about 35 percent of the stocks in this highest decile had price drops of 30 percent or more. From this diagram, what are the fifth, 15th, 25th, . . . 95th percentiles for the P/E ratios?

2.37 Regarding the life of new nickel-metal hydride batteries, a Scripps Howard News Service (*HT*, July 19, 1992) press release stated that, "Most passenger cars don't need a range of much more than 300 miles," with an average range of maybe 350 miles. "If the battery can be charged in 15 minutes, that's about the time it takes to get a snack and stretch your legs." What do you know about the distribution of battery life range from these statements?

2.38 A *Wall Street Journal* (May 6, 1988) article on airline safety stated that, "The average age of U.S. jetliners has risen 21% since 1979 to 12.53 years, and more than half of the

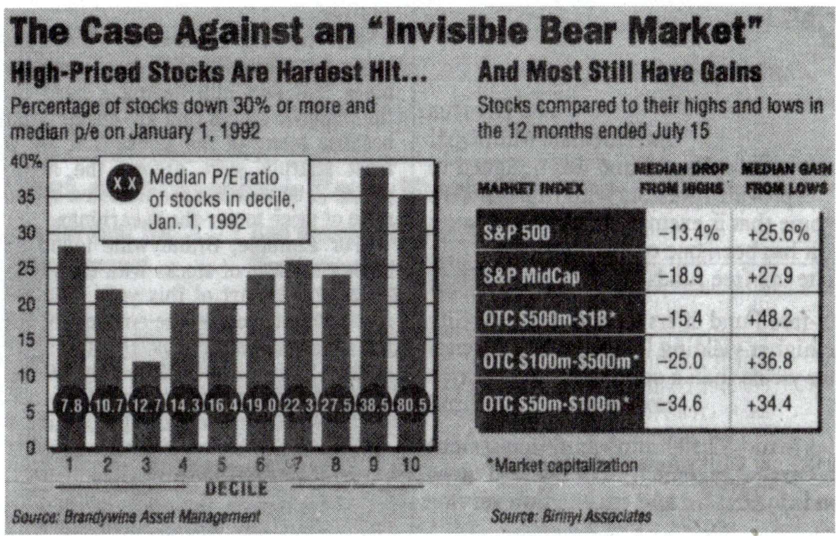

MARKET INDEX	MEDIAN DROP FROM HIGHS	MEDIAN GAIN FROM LOWS
S&P 500	−13.4%	+25.6%
S&P MidCap	−18.9	+27.9
OTC $500m-$1B*	−15.4	+48.2
OTC $100m-$500m*	−25.0	+36.8
OTC $50m-$100m*	−34.6	+34.4

2,767 jets in service at major airlines are 16 years old or older . . . . Manufacturers figure the 'economic'—or profitable—life of a jet at 20 years . . . . The carriers insist that a jet can fly safely practically forever . . . . "Assuming the distribution of age is unimodal, draw the distribution implied by this quote.

2.39 The following frequency distribution from *Business Week* (September 21, 1992) shows the share of credit card accounts being charged the indicated interest rates. The accompanying article also stated that "only 2% are paying under 12%."

a. What percentage of card holders are paying less than 16 percent but at least 12 percent?

b. Assuming that the mean rate paid by those who are paying less than 12 percent is 6 percent, and that 25 percent is the mean for those paying more than 19 percent, what is the approximate overall mean rate? What is the approximate standard deviation?

Card Rate	Share of Accounts
Over 19%	36%
18–19	8
17–18	15
16–17	12
Less than 16	29

2.40 For the information in Exercise 2.39 on the rates charged card holders, what is the approximate median rate? What did you have to assume to calculate this median?

2.41 An A&E television special, "First Flights" (May 19, 1993), featured astronaut Neil Armstrong saying that "in the early years of World War II, the average number of missions B17 bombers survived was the unlucky 13." He also said that "200 planes per week were lost during the major campaigns of World War II." Comment on the appropriateness of these averages.

2.42 A survey conducted by American Sports Data found that "people who ski fewer than 12 times a year, for example, account for half of all ski sales. People who play tennis fewer than 25 times a year buy nearly half of all rackets. And people who golf fewer than 25 times a year make up about half of the market for golf club sets."(*WSJ*, June 2, 1993) From this quote, what, if anything, do you know about the approximate mean, median, or modal number of times per year people who a) buy skies ski, b) buy rackets play tennis, or c) buy golf clubs play golf?

2.43 "Most of Indiana's top officials spend less than one-tenth of what President Clinton paid ($200) to have his hair cut and styled last month in Los Angeles," according to an Associated Press article (*HT*, June 1993). From this quote, can you provide an approximate mean, median, or modal value for the amount paid by Indiana's top officials for haircuts and stylings?

2.44 In reporting on health care costs, *Business Week* (March 15, 1993) stated, "Only 5.7% chose such coverage (low-cost but high-satisfaction health care plans), while 54.2% are in plans with above-average premiums." Is there a mistake here? How can 54.2 percent be above the average? Explain.

2.45 The information in Exercise 2.39 from *Business Week* (September 21, 1992) on the rates charged card holders, was from a survey of the top 100 institutions in the credit card industry, by the *Nilson Report*, a Santa Monica, California, newsletter firm that tracks the industry. These top 100 institutions account for 139 million card accounts.

a. In calculating the variance of rates from these data, is the population or sample size 100 million or 139 million?

b. Should the population or the sample variance formula be used?

c. Why wouldn't it make much difference if the population or sample formula were used to calculate the variance?

2.46 Below are the automobile output plans for the fourth quarter of 1991 and 1992, as reported in *The Wall Street Journal* (October 8, 1992):

	1992	1991	% Change
GM	647,000	674,000	–4.0
Ford	307,000	251,000	+22.3
Chrysler	118,837	151,000	–21.3
Honda	100,000	112,174	–10.9
Mazda	64,000	47,000	+36.2
Nissan	60,000	34,000	+76.5
Toyota	58,903	47,000	+25.3
Nummi	43,000	51,000	–15.7
Diamstar	32,500	39,712	–18.2
Sub-ISU	14,000	15,000	–6.7

What is the mean output of automobiles from these 10 companies in each year? What is the median in each year? Why aren't the mean and median equal in 1992? Why aren't they equal in 1991?

2.47 In Exercise 2.46 what is the average percentage change in output? What is the percentage change in average output? Why aren't these two measures of change the same?

2.48 In Exercise 2.46, which is a better measure of change: the percentage change in the median or the percentage change in the average? Explain.

2.49 Below is a diagram and relative frequency distribution for U.S. home sales by price in the first quarter of 1990, as prepared by the *Chicago Tribune*. (*HT*, June 24, 1990)

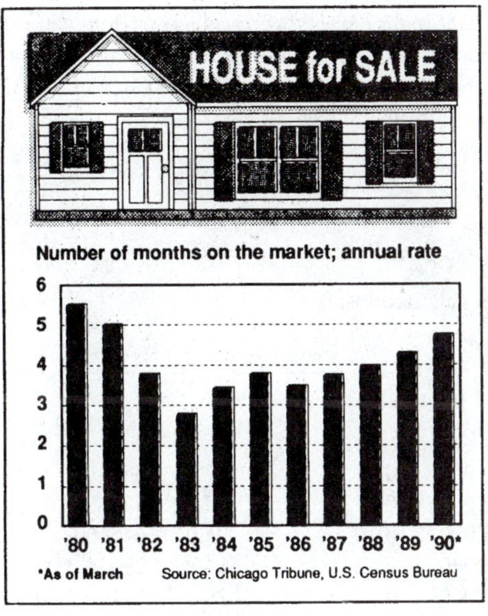

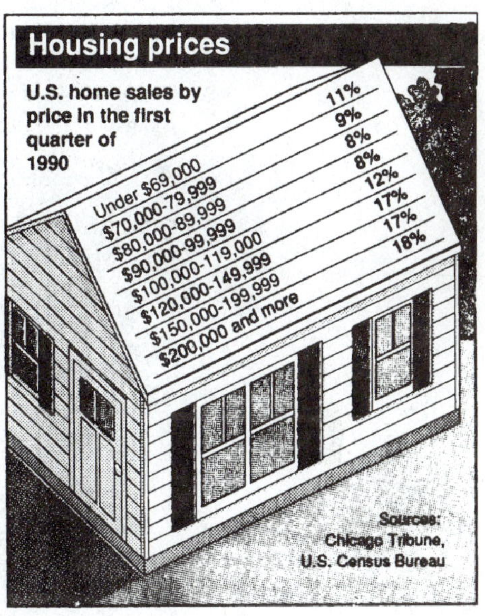

a. What is the approximate mean and median price of home sales?
b. What did you have to assume to obtain your mean and median in part a?

2.50 The 125th edition of the *World Almanac and Book of Facts* (Pharos Books, 1993) tells of British doctors recommending that beer mugs be made of tougher glass, so they do less damage in bar fights. "Mugs are the weapon of choice in 3,400 to 5,400 bar fights a year in England and Wales."(*Parade*, October 25, 1992) Which of the following can be determined from this quote?

a. The median is 4,400.5 fights.
b. The mean is 4,400 fights.
c. The range is 2,000 fights.
d. If the number of times mugs are selected in fights is approximately normal, the approximate standard deviation is 285.7 fights.

2.51 An article in *The Wall Street Journal* (November 13, 1992) reported, "Airlines Wage Fare Battle City by City, Spreading Confusion as Well as Bargains." The following table was provided to support the text:

	American	Continental	Delta	NorthWest	United
Phoenix-St. Louis	$400	$248	$400	$400	$228
Austin-Los Angeles	248	258	360	340	248
Detroit-Kansas City	164	113	280	113	113

For which city is there the largest dispersion in price? Is your answer sensitive to the measure of dispersion used? Explain.

2.52 Which airline has the median price for the three cities identified in the above exercise?

2.53 An article in *The Wall Street Journal* (November 25, 1992) reported that "women doctors in private practice earned median incomes 34% less than male colleagues in 1991, according to *Medical Economics* magazine." To check this statement and related issues, samples of 100 female and 200 male doctors were selected. Below are computer-generated descriptive statistics and worksheet calculations made from these data, where the mean earnings are reported in thousands of dollars for FEMALES and MALES. Use these printouts to answer the next eight questions.

Variable	Mean	Standard Deviation	S.E.Mean	Cases
FEMALE	117.180	89.17444	.89174	100
MALE	185.730	143.3742	.71687	200

Some worksheet calculations:

Earnings: thousands of dollars	mid pt m	Female: n = 100 Rel Frq (f/n)		(f/n)m	$f(m-\bar{x})^2$	Male: n = 200 Rel.Frq. (f/n)		(f/n)m	$f(m-\bar{x})^2$
400 plus	700.000	0.01		7.0	334546.6	0.05		35.0	521220.5
300–400	350.000	0.01		3.5	52166.6	0.06		21.0	51520.5
200–300	250.000	0.10		25.0	164865.6	0.21		52.5	73205.0
100–200	150.000	0.36		54.0	29036.2	0.43	158.13	64.5	112338.0
60–100	80.000	0.33	97.575	26.4	57108.5	0.18		14.4	791356.5
below 60	30.000	0.19		5.7	159420.6	0.07		2.1	966729.5
		1.00		121.60	797144	1.00		189.50	2516370

a. The mean earnings of the 100 female doctors is ________?

b. The median earnings of the 100 female doctors is ________?

c. If each female doctor who is above the median earnings level for female doctors in the sample had her earnings increased by $60,000, then what would happen to the median earnings of the 100 female doctors?

d. If each female doctor who is above the median earnings level for female doctors in the sample had her earnings increased by $60,000, then what would happen to the mean earnings of the 100 female doctors?

e. If each female doctor who is above the median earnings level for female doctors in the sample had her earnings increased by $60,000, then what would happen to the mode earnings of the 100 female doctors?

f. What proportion of the total earnings of these 300 female and male doctors goes to the female doctors?

g. What is the variance of the earnings of the 100 female doctors?

h. Is the distribution of female earnings skewed?

2.54 (Use a computer spreadsheet and the data file "EX2–54.PRN" on the computer disk that accompanies this book for this problem.) An Associated Press story (July 31, 1992) reported on the number of farms by state, as given below. Calculate three measures of central location and three measures of dispersion for these farm data. What difference does it make in your calculation of each of the measures of central location and each measure of dispersion if you treat these data as a population versus sample? What would determine whether they are population or sample data?

State	Number of Farms
Alabama	46,000
Alaska	540
Arizona	8,000
Arkansas	46,000
California	81,000
Colorado	25,500
Connecticut	3,900
Delaware	2,700
Florida	39,000
Georgia	46,000
Hawaii	4,500
Idaho	21,000
Illinois	81,000
Indiana	65,000
Iowa	102,000
Kansas	67,000
Kentucky	91,000
Louisiana	30,000
Maine	7,100
Maryland	15,600
Massachusetts	6,900
Michigan	54,000
Minnesota	88,000
Mississippi	38,000
Missouri	107,000
Montana	24,600
Nebraska	56,000
Nevada	2,500
New Hampshire	2,900
New Jersey	8,500
New Mexico	13,500
New York	38,000
North Carolina	60,000
North Dakota	33,000
Ohio	78,000
Oklahoma	71,000
Oregon	37,500
Pennsylvania	52,000
Rhode Island	700
South Carolina	24,500
South Dakota	35,000
Tennessee	88,000
Texas	183,000
Utah	13,200
Vermont	6,900
Virginia	44,000
Washington	38,000
West Virginia	20,000
Wisconsin	79,000
Wyoming	9,200
TOTAL	2,095,740

2.55 (Use a computer spreadsheet and the file "EX2–55.PRN" on the data disk that accompanies this book for this problem.) For the following data on different types of taxes paid by tax-payers, calculate the average tax across the states paid in each of the four categories. Add up the averages for the four categories. Calculate the average tax for the total paid within a state. Why is there a difference between the sum of the averages across the four categories and the average of the total? (Data are from *USA Today* October 10, 1991.)

State	Sales Tax	Indiv. Income	Gas Tax	Tags	Total Tax
Alabama	257	278	73	31	945
Alaska	0	0	75	37	2811
Arizona	523	290	91	63	1194
Arkansas	357	314	93	33	962
California	458	565	46	36	1459
Colorado	251	407	99	33	932
Connecticut	743	186	94	53	1603
Delaware	0	685	95	35	1696
Florida	633	0	60	37	1027

(continued)

State	Sales Tax	Indiv. Income	Gas Tax	Tags	Total Tax
Georgia	407	443	68	12	1093
Hawaii	1062	627	48	18	2107
Idaho	381	400	107	56	1131
Illinois	357	375	80	54	1128
Indiana	460	377	102	32	1101
Iowa	340	458	120	78	1193
Kansas	352	346	91	39	1077
Kentucky	295	328	98	40	1156
Louisiana	299	175	95	17	968
Maine	415	473	112	39	1271
Maryland	329	599	94	31	1349
Massachusetts	325	816	50	42	1557
Michigan	343	422	80	52	1220
Minnesota	427	658	105	73	1559
Mississippi	423	167	120	24	931
Missouri	371	350	69	39	965
Montana	0	350	140	46	1073
Nebraska	322	314	133	36	959
Nevada	667	0	93	56	1317
New Hampshire	0	37	73	46	537
New Jersey	426	382	54	44	1350
New Mexico	552	238	109	67	1329
New York	334	850	30	30	1591
North Carolina	267	511	120	34	1186
North Dakota	362	165	105	60	1060
Ohio	331	380	90	36	1054
Oklahoma	268	318	105	85	1105
Oregon	0	643	83	76	980
Pennsylvania	356	271	63	36	1113
Rhode Island	396	425	73	40	1229
South Carolina	415	396	103	22	1128
South Dakota	359	0	115	29	719
Tennessee	481	21	129	31	870
Texas	449	0	89	42	866
Utah	410	375	77	24	1026
Vermont	242	446	95	66	1183
Virginia	219	498	101	40	1067
Washington	919	0	99	36	1525
West Virginia	426	288	116	41	1243
Wisconsin	406	537	108	33	1341
Wyoming	358	0	83	93	1348

2.56 A *Wall Street Journal* article (April 18, 1988) reported the following percentage of time that managers spent in development programs and how much time they thought they should be spending.

Workdays Per Year	Percent Who Currently Spend	Percent Who Prefer To Spend
0	10	2
1–2	34	6
3–5	39	48
6–10	13	39
11+	4	5

a. What are the approximate means and medians for the two distributions? How do they differ?
b. Why isn't the above data in a proper form for a frequency distribution? What are the implications of this for the approximation of the mean and median?
c. Is the approximate mean or median the preferred measure of central location for the above data? (Explain, using your answer in part b.)

2.57 Surveys to determine product quality are extremely popular. Consider, for example, the following evaluation data for two automobiles. The results are based on new purchaser's response to the survey question.

"The overall quality of your new car is ________ ?"

9 8 7 6 5 4 3 2 1 0
High Low

Summary Data	Car A	Car B
Number of new purchasers in survey	360	300
Number responding	60	50
Mean response value	4.0	5.0
Standard deviation of responses	1.0	2.0
Median response value	5.0	4.0

a. What are the response rates for each car as a percentage of new purchasers surveyed?
b. If there had been a 100 percent response rate, how high could each car's mean score have been?
c. About how many new purchasers who responded rated car A between 7 and 1?
d. A salesperson reviewed this data and decided that car B is clearly better. This person argued, "The majority of new car buyers gave car B an average score higher than A. The higher variability in car B's rating is more than made up by the higher mean score of car B. The majority of new car buyers cannot be wrong; thus, I am confident in the superiority of car B." Point out at least four errors or shortcomings in this reasoning.

2.58 *Business Week* (August 31, 1992) reported the following costs for a "night at home," which included McDonald's Quarter-Pounders with cheese for two, a six-pack of either Budweiser or Miller Lite, and a new Monopoly board game.

City	Cost
New York City	\$27.44
Anchorage	\$26.69
Tulsa	\$18.42
Orlando	\$17.97
Cleveland	\$17.57

What are the mean and median? What are the variance and standard deviation for this sample data?

2.59 Below are data on the percentage of gross domestic product that each industrialized nation spent on education, according to the Organization for Economic Cooperation and Development publication "Education at a Glance (1992)," which are also in the ASCII computer file "EX2–59.PRN." Use a computer statistics package to determine the mean, median, and mode of the total column. Comment on the appropriateness of each of these measures as a description of the central tendency of these data.

School Spending

Country	Public	Private	Total
Denmark	6.8	0.1	6.9
Finland	6.8	.	6.8
Norway	6.6	.	6.6
Canada	6.4	0.8	7.2
Netherlands	6.3	0.3	6.6
Belgium	6.1	.	6.1
Luxembourg	6	.	6
Ireland	5.8	0.4	6.2
Sweden	5.7	.	5.7
Austria	5.6	.	5.6
Switzerland	5.1	.	5.1
France	5.1	0.7	5.7
United States	5	0.7	5.7
Australia	4.8	.	4.8
Italy	4.8	.	4.8
United Kingdom	4.7	.	4.7
Portugal	4.7	0.2	4.9
Germany	4.3	1.9	6.2
Spain	3.9	1.1	5
Japan	3.8	1.2	4.9
Average	4.8	0.9	5.7

Source: *Herald-Times* (September 24, 1992).
Missing data and zero entries are indicated as "."

2.60 Below are data on Corvette sales, from *Corvette Quarterly* (Winter 1992), which are also in the ASCII computer file "EX2–60.PRN" that accompanies this text. Use a computer spreadsheet program or computer statistics package to check if the average yearly production run for each body style has increased since the Corvette's introduction in 1953. Why would or wouldn't this be relevant information in planning the introduction of a new body style later in the 1990s?

	Type of Production			
Year	Convertible	Coupe	Total	Body Style
1953	300	0	300	1
1954	3640	0	3640	1
1955	700	0	700	1

(continued)

Year	Type of Production: Convertible	Coupe	Total	Body Style
1956	3467	0	3467	2
1957	6339	0	6339	2
1958	9168	0	9168	2
1959	9670	0	9670	2
1960	10261	0	10261	2
1961	10939	0	10939	2
1962	14531	0	14531	2
1963	10919	10594	21513	3
1964	13925	8304	22229	3
1965	15376	8186	23562	3
1966	17762	9958	27720	3
1967	14436	8504	22940	3
1968	18630	9936	28566	4
1969	16633	22129	38762	4
1970	6648	10668	17316	4
1971	7121	14680	21801	4
1972	6508	20496	27004	4
1973	4943	25521	30464	4
1974	5474	32028	37502	4
1975	4629	33836	38465	4
1976	0	46558	46558	4
1977	0	49213	49213	4
1978	0	46776	46776	4
1979	0	53807	53807	4
1980	0	40614	40614	4
1981	0	40606	40606	4
1982	0	25407	25407	4
1983	0	43	43	4
1984	0	51547	51547	5
1985	0	39729	39729	5
1986	7315	27794	35109	5
1987	10625	20007	30632	5
1988	7407	15382	22789	5
1989	9749	16663	26412	5
1990	7630	16016	23646	5
1991	5672	14967	20639	5

style code: 1: introductory model
2: side panel insert model
3: original Stingray model
4: shark shape model
5: wide stance model

2.61 The data set from *The Wall Street Journal* (September 24, 1992) on the world's 50 largest insurers is in the spreadsheet file "EX2–61.WK1," which is readable by any LOTUS compatible computer spreadsheet. Use a spreadsheet program to calculate 1) the average capital asset ratio and 2) the ratio of average capital to average assets. Why aren't these the same?

2.62 The data set from *The Wall Street Journal* (September 24, 1992) on the world's 100 largest banks is in the spreadsheet file "EX2–62.WK1," which is readable by any LOTUS compatible computer spreadsheet. Use a spreadsheet program to calculate 1) the average capital asset ratio and 2) the ratio of average capital to average assets. Why aren't these the same?

2.63 The data set from *The Wall Street Journal* (September 24, 1992) on the world's 25 largest security firms is in the spreadsheet file "EX2–63.WK1," which is readable by any LOTUS compatible computer spreadsheet. Use a spreadsheet program to show how the capital growth rates of 0.8 percent, 23.2 percent, 0.9 percent and 10.7 percent were calculated respectively for Japan, the United States, the other countries, and the top 25 average.

2.64 The data set from *The Wall Street Journal* (September 24, 1992) on the world's 100 largest public companies is in the spreadsheet file "EX2–64.WK1," which is readable by any LOTUS compatible computer spreadsheet. Use a spreadsheet program to show how the sales and profit growth rates compare for the Netherlands/United Kingdom, the United States, Japan, Germany, France, Italy, Switzerland, and the other countries.

2.65 In a television show "The Birth of Europe," A&E Journey (August 11, 1991), the commentator stated that "during World War I (1914), the average survival rate of an English flier was two weeks." Why should you suspect that this was an incorrect statement of the facts? What did the two weeks most likely pertain to?

2.66 Using the salary data on the men and the women bank executives at the midwestern bank in the data file "EX2–66.PRN," calculate the mean and median salaries for each. Are women receiving much less than men? By what measure and by how much? If the 10 highest paid women each received $50,000 raises, what would happen to your measures of discrimination? Which measure of central tendency is best for assessing the cost of removing salary differences between men and women?

2.67 Construct an $\bar{x}$ chart, based on hour of the day, for the average time each of the 196 patients waits in the emergency room for the hospital data set provided in the "EX2–67.PRN" file on the computer disk accompanying this book. Is wait time related to time of day?

2.68 Construct an R chart for the 196 patients in the emergency room data found in Exercise 2.67. Place time of arrival on the horizontal axis and the time in the emergency room on the vertical axis. Does time of arrival influence the range of wait time?

2.69 The computer ASCII file "EX1–49.PRN" was introduced in Exercise 1.49. The Scripps Howard News Service (*HT*, April 23, 1993) article that accompanies this data stated that the average teacher's salary for the 50 states and District of Columbia was $35,334. What is the average salary for teachers in the data set? Why don't these averages agree?

Academic References

Deming, W. Edwards. *Out of the Crisis.* Cambridge, Mass.: MIT Center for Advanced Engineering Study, 1986.

Kiefer, Nicholas. "Economic Duration Data and Hazard Functions." *Journal of Economic Literature,* June 1988, pp. 646–681.

Spear, Mary Eleanor. *Charting Statistics.* New York, McGraw-Hill, 1952.

Stigler, Stephen. *The History of Statistics.* Harvard University Press, 1986, pp. 169–172.

Tukey, John. *Exploratory Data Analysis.* Reading, Mass., Addison-Wesley, 1977.

Tukey, John. "Some Graphic and Semigraphic Displays," in T. A. Bancroft, ed., *Statistical Papers in Honor of George W. Snedecor.* Ames, Iowa: Iowa University Press, 1972, pp. 293–316.

Tukey, John and Martin Wilk. "Data Analysis and Statistics: Techniques and Approaches," in Edward R. Tufte, ed., *The Quantitative Analysis of Social Problems.* Reading, Mass.: Addison-Wesley, 1970, pp. 370–390.

ENDNOTES

1. Computer programs typically calculate the coefficient of skewness for sample data as

$$\frac{\dfrac{\sum\left(x_i - \text{mean}\right)^3}{n-1}}{\left[\dfrac{\sum\left(x_i - \text{mean}\right)^2}{n-1}\right]^{(3/2)}}$$

For population data, this formula would have to be rewritten for divisions by N instead of n – 1. Regardless of whether the data are for a population or a sample, the denominator of the coefficient of skewness is always positive; the numerator dictates the sign of skewness. If low values of x are close to the mean, while high values extend far beyond the mean, then the coefficient of skewness will be positive; in this case, "right skew" and "positive skew" are used as synonyms. If the distribution is skewed left (mean < median), then the coefficient of skewness will be negative, and "left skew" and "negative skew" are used as synonyms. If a distribution is symmetric (mean = median), then the coefficient of skewness will be zero.

2. If the shape of the distribution is unknown, then the percentage of values contained within a distance of $\pm k$ standard deviations of the mean will be at least $[1 - (1/k)^2]$, as determined by Russian mathematician Pafnuty Chebychev more than a century ago. Unfortunately, Chebychev's theorem does not convey much information . For instance, within two standard deviations of the mean (an interval $\mu \pm 2\sigma$) we know only there are at least 75 percent $[1 - (1/2)^2]$ of the observations. If $k = 2.5$, then at least 84 percent $[1 - (1/2.5)^2]$ is inside the interval $\mu \pm 2.5\sigma$. If $k = 3$, then at least 88.9 percent $[1 - (1/3)^2]$ of the values fall in the interval $\mu \pm 3\sigma$.

FORMULAS

- Population mean of x is the sum of the N x values divided by N. In shorthand notation, this population mean is designated μ_x and is calculated as

$$\mu = \frac{1}{N}\sum_{i=1}^{N} x_i = (1/N)(x_1 + x_2 + \ldots + x_N)$$

- Sample mean of x is the sum of the n x values divided by n. In shorthand notation, this sample mean is designated $\bar{x}$ and is calculated as

$$\bar{x} = \frac{1}{n}\sum_{i=1}^{n} x_i = (1/n)(x_1 + x_2 + \ldots + x_n)$$

- Population frequency weighted mean

$$\mu = \frac{1}{N}\sum_{i=1}^{c} f_i m_i = (1/N)(f_1 m_1 + f_1 m_2 + \ldots + f_c m_c)$$

or

$$\mu = \sum_{i=1}^{c} (f_i / N)(m_i)$$

- Sample frequency weighted mean

$$\bar{x} = \frac{1}{n}\sum_{i=1}^{c} (f_i m_i) = (1/n)(f_1 m_1 + f_2 m_2 + \ldots + f_c m_c)$$

- Population relative frequency weighted mean

$$\mu = \sum_{i=1}^{c} (f_i / N)(m_i) = [(f_1/N)(m_1) + (f_2/N)(m_2) + \ldots + (f_c/N)(m_c)]$$

- Sample relative frequency weighted mean

$$\bar{x} = \sum_{i=1}^{c} (f_i / n)(m_i) = [(f_1/n)(m_1) + (f_2/n)(m_2) + \ldots + (f_c/n)(m_c)]$$

- Median ~ L + (o_1/o_2)(U – L)
  where L is the low boundary of the median class
  U is the upper boundary of the median class
  o_1 is the number of observations into the median class until the median is found, and
  o_2 is the number of observations in the median class.
- Population variance

$$\sigma^2 = \frac{1}{N}\sum_{i=1}^{N} (x_i - \mu)^2$$

- Sample variance

$$s^2 = \frac{1}{n-1}\sum_{i=1}^{n} (x_i - \bar{x})^2$$

- Population standard deviation

$$\sigma = +\sqrt{\sigma^2}$$

- Sample standard deviation

$$s = +\sqrt{s^2}$$

- Weighted population variance:

$$\sigma^2 = \frac{1}{N} \sum_{i=1}^{c} f_i \left(m_i - \mu\right)^2$$

where μ is the mean of the population, m_i is the value or class midpoint at which multiple observations occur, f_i is the frequency of the ith class, c is the number of classes, and N is the number of items in the population $N = \Sigma f_i$

- Alternative weighted population variance

$$\sigma^2 = \sum_{i=1}^{c} \left(f_i / N\right)\left(m_i - \mu\right)^2$$

- Weighted sample variance:

$$s^2 = \frac{1}{n-1} \sum_{i=1}^{c} f_i \left(m_i - \bar{x}\right)^2$$

where $\bar{x}$ is the sample mean and the sample size is $n = \Sigma f_i$

- Coefficient of skewness

$$\frac{\dfrac{\Sigma\left(x_i - \text{mean}\right)^3}{n-1}}{\left[\dfrac{\Sigma\left(x_i - \text{mean}\right)^2}{n-1}\right]^{(3/2)}}$$

Chapter 3

Probability

Chance is a nickname for Providence.

Sebastien Roch Nicolas Chamfort
1741–1794

3.1 INTRODUCTION

Data collection, summary, and presentation, as discussed in the first two chapters, are essential for reducing the guesswork in decision making. But risk and uncertainty in decisions cannot be avoided completely. The following represent the different types of problems involving chance events that are considered in this chapter.

- A firm hires three people from a labor force that is 50 percent female. What is the probability that this firm hires no women, if its hiring process is equivalent to a random draw? If the firm is observed hiring no women, then what could we conclude about the hiring process?
- The "morning line" or starting odds appear in local newspapers wherever there is horse racing. These odds represent the best guess of experts (handicappers) about horses' relative chances of winning that day. A handicapper's "8" means that the horse has an eight to one chance of winning and a "5–2" means that the horse has a five to two chance of winning. But what do these odds imply about the probability of a horse winning?
- In state lotteries, there are typically more than one type of game, but at least one game is designed to identify winners and losers quickly while others are more involved and ongoing games. In the Indiana state lottery, for example, "based on the 75 million tickets available for each six- to eight-week game, the odds of winning an instant cash prize are:
  - $1, 1 in 12.50 tickets
  - $2, 1 in 31.25
  - $5, 1 in 64.94
  - $10, 1 in 135.14
  - $25, 1 in 384.62
  - $500, 1 in 7,692.00
  - $5,000, 1 in 100,000.00"(*HT*, October 8, 1989).

  What is the probability of having a winning ticket in this instant cash game?
- A company that tests athletes and prospective employees for drugs, announces that its test is "95 percent accurate," by which it means that 95 percent of drug users are accurately identified and 95 percent of nondrug users are accurately identified. From government statistics, we know that approximately 5 percent of the adult population in a community are drug users. A human resource manager of a large firm in this community wants to know whether it would be a good idea to start testing prospective employees.
- According to a story in *Newsweek* (June 8, 1992), promoting drugs in medical journals is big business, with American drug companies spending $350 million a year for advertising. A study by UCLA researchers showed 100 of 109 ads in medical journals had factual errors in areas where the Food and Drug Administration (FDA) has established explicit standards. The FDA checks only about 15 percent of all these drug ads. What is the probability of the FDA finding deficiencies?

Subjective Probability Probability that is based on personal belief or judgment.

Objective Probability Probability that at least in some sense is based on observable relative frequencies.

Each of these problems involves chance factors that requires the determination of probabilities of uncertain events. By the end of this chapter, you will be able to identify other economics and business situations in which probability calculations are needed. You will be able to distinguish between **subjective** and **objective probabilities** and give a historical perspective on their developments. You will know the implication of dependent, independent, and mutually exclusive events, and know how to calculate the probabilities of unions and intersections of events. Using the basic rules of probability, you will be able to make probability calculations for everyday business and economic decisions involving compound probabilities.

3.2 Relative Frequency and Probability

Before addressing the problems identified at the beginning of this chapter, let's make sure we have an idea of what probability is. One way to do this is to flip a coin 10 times, record the results, and then flip it 20 times, record the results, and, finally, flip it 40 times and record the results. You can record your results as in Table 3.1, where both the frequency and relative frequencies are calculated for each of the three sample sizes of $n = 10$, $n = 20$, and $n = 40$. You should recall from Chapter 1 that the randomness inherent in sampling should make your results different from those in Table 3.1.

Probability The long-term relative frequency with which a single outcome (or set of outcomes called an event) occurs.

Notice in Table 3.1 how the relative frequency of heads is approaching or converging to 0.50. If we could continue to flip the coin in samples of $n = 80$, $n = 160$, $n = 320$ and so on, then as we devoted more and more time to generating larger and larger samples, this relative frequency would settle down on the limiting value 0.50. This long-term limit to which relative frequency converges is called a **probability**.

3.3 Experiments, Outcomes, and Probability

The occurrence or realization of a result, a "head" or a "tail" in Table 3.1, is called an outcome. For a coin toss, these are the only two outcomes, and each is equally likely. For the roll of a die there are six outcomes: a face value of 1, 2, 3, 4, 5, or 6, and each is equally likely. In the lottery's instant cash game discussed at the beginning of this chapter there are eight outcomes: No instant cash, \$1, \$2, \$5, \$10, \$25, \$500, and \$5,000, but each of these outcomes is not equally likely.

Table 3.1 Outcome of Coin Toss

Outcome	Frequency	Relative Frequency	Frequency	Relative Frequency	Frequency	Relative Frequency
Heads:	3	0.3	12	0.6	21	0.525
Tails:	7	0.7	8	0.4	19	0.475
$n =$	10		20		40	

Experiment
Trials or testing situations that can be repeated a large (infinite) number of times where the chance factor is the only source of difference in outcomes for each administration.

The calculation of probability requires knowledge of how the outcome is determined. For instance, shaking a die in a cup for two seconds and then dumping it on the table is an experiment in which each face value (an outcome) has an equal one in six chance of occurring. As introduced in Chapter 1, an **experiment** is a situation that can be repeated a large (infinite) number of times, with the random chance factor being the only source of difference in outcomes for each running of the experiment.

In the state lottery instant cash game, the experiment entails the identification of winning contestants (tickets). This experiment may be replicated month after month with many different contestants or with the same people entering the lottery over and over. In other experiments, actual replication may not take place or it may not be perfect. For example, the conditions under which robberies take place are not identical, but yet they may be viewed as experiments that give rise to a randomly determined outcome: the dollar value of things taken.

Many uncertain events involve outcomes that are not numerical. Whether or not a lottery contestant or a slot machine player has a match on three out of six "windows" is an event that may be deemed a success or a failure with no regard to the dollar value of the outcome. Whether a statistics test is "passed" or failed; whether a job applicant is rated unacceptable, satisfactory, or excellent; or whether a customer likes or does not like the service are all examples of outcomes that are not quantitative. They are qualitative attributes that are determined with a chance factor (guessing on the exam, for example). As we will see, probability calculations can be made on any attribute that occurs by chance.

Sample Space (or Outcome Set)
The set of all possible outcomes in an experiment.

In the lottery's instant cash game the prizes are: No instant cash, $1, $2, $5, $10, $25, $500, and $5,000. These are the only eight outcomes that are possible in one administration of a game in which there are 75 million equally likely ways these outcomes could occur. This is the **sample space** or **outcome set** for the experiment that determines the "immediate redemption value of ticket."[1] The sample space includes all the outcomes and ways in which they could occur in an experiment.

One of the outcomes in the sample space must occur as an outcome of an experiment, but which one is uncertain. Before running an experiment, any one of the elements could be selected. After the experiment, however, the outcome selected is known with certainty. The amount of a winning ticket in last month's lottery was observed and is public knowledge, but the amount of a winning ticket eight weeks from now is in doubt. Identification of the sample space and the nature of the chance factor that determines the outcomes are the keys in our work to determine probability.

In the instant cash state lottery, the chance of winning is based on the 75 million tickets. This qualifier and the stated chances of winning show how the probability is related to the different ways the outcomes in the sample space can occur and their relative frequency of occurrences. In Chapter 1, relative frequencies were defined in terms of the number of times outcomes had actually been observed. The concept of probability pertains to the relative frequency of outcomes that have not yet occurred but that would occur over a large (infinite) number of future experiments that could be conducted in a sufficiently long (infinite) time period.

In the instant cash lottery game, if one in 12.5 tickets is worth $1 and there are 75 million tickets, then we know that there must be six million of the $1 tickets

(= 75000000/12.5) per running of the game. This *does not say* that a contestant who plays the game 25 times (where each play is in a new lottery) will get two tickets worth $1 each. It *does not say* that a contestant who plays the game 50 times (in a new lottery) will get four tickets worth $1 each. A chance or probability of 0.08 (=1/12.5) says that if a contestant plays the instant cash game an extremely large (infinite) number of times, then 8 percent of the time he or she will win $1, where each play is a new lottery.

For the $2 prize, the chance of winning is one in 31.25, which for 75 million tickets implies 2.4 million tickets worth $2 per running of the lottery. Again, this does not say that a contestant who plays the game 125 times will get four tickets worth $2 each. A likelihood or probability of 0.032 (=1/31.25) says that if a contestant plays the instant cash game an extremely large (infinite) number of times, then 3.2 percent of the time he or she will win $2. And similarly for the other prizes shown in Table 3.2, where the probability of a winning amount is calculated as the number of equally likely ways an outcome may occur divided by the number of equally likely ways all outcomes may occur.

$$\textbf{Probability of an outcome} = \frac{\text{Number of equally likely ways the outcome may occur}}{\text{Total number of equally likely ways all outcomes may occur}} \qquad (3.1a)$$

Exhaustive Outcomes
All of the possible outcomes; all of those outcomes that make up the sample space.

There are several things to be emphasized in Table 3.2. First, the chance of winning must be interpreted as an average win per number of occurrences (for example, a $1 winning ticket occurs an average of once in 12.5 tickets). There are no assurances that one winning dollar ticket will appear in 12.5 tickets; this only happens on average. Similarly, the decimals in the probability column must be interpreted as a relative frequency over a large (infinite) number of plays. Finally, notice that the probabilities of all eight outcomes in the sample space sum to one, as with relative frequency. The eight outcomes are **exhaustive** since there are no other possible outcomes.

TABLE 3.2
Chances of Winning the Instant Cash Lottery

Prize	Number of winning tickets in 75 million	Chance of winning	Probability of winning
$ 0	64,684,610	1 in 1.1595	0.862461
$ 1	6,000,000	1 in 12.50	0.080000
$ 2	2,400,000	1 in 31.25	0.032000
$ 5	1,154,912	1 in 64.94	0.015399
$ 10	554,980	1 in 135.14	0.007400
$ 25	194,998	1 in 384.62	0.002600
$ 500	9,750	1 in 7,692.0	0.000130
$5,000	750	1 in 100,000	0.000010
	75,000,000		1.000000

Source: *HT*, October 8, 1989.

Mutually Exclusive
Outcomes or events that cannot occur together.

The outcomes that form the sample space must also be **mutually exclusive;** they cannot occur together. That is, one outcome (ticket value) cannot be both a \$1 and a \$5,000 winner at the same time.

The sample space must be formed by outcomes that are mutually exclusive and exhaustive so that the probabilities of all the outcomes sum to one. For example, in a coin toss

$$P(\text{head}) + P(\text{tail}) = (1/2) + (1/2) = 1.00$$

For the throw of a die,

$$P(1) + P(2) + P(3) + P(4) + P(5) + P(6) = (1/6) + (1/6) + (1/6) + (1/6) + (1/6) + (1/6) = 1.00$$

And in the instant cash game,

$$P(\$0) + P(\$1) + P(\$2) + P(\$5) + P(\$10) + P(\$25) + P(\$500) + P(\$5{,}000) = 0.862461 + 0.080000 + 0.032000 + 0.015399 + 0.007400 + 0.002600 + 0.000130 + 0.000010 = 1.000000$$

3.4 Events and Probability

Event
A subset of the sample space that is formed by outcomes that share an attribute of interest.

In addition to the probability of an outcome, the probability of a subset of the sample space can be defined. The subset of outcomes of interest is called an **event**; it can consist of one or more outcomes in the sample space. In the instant cash game, we form the event "contestant wins some instant cash" with seven of the eight outcomes (\$1, \$2, \$5, \$10, \$25, \$500, and \$5,000); this event is fulfilled if any one of these seven outcomes occurs.

The probability of an event is the sum of the probabilities of the outcomes that form the event. For example, the probability that a contestant has a ticket that turns out to be an instant winner is the sum of the probabilities that the contestant receives \$1, \$2, \$5, \$10, \$25, \$500, or \$5,000:

$$P(\text{win }\$\text{s}) = P(\$1) + P(\$2) + P(\$5) + P(\$10) + P(\$25) + P(\$500) + P(\$5000) = 0.080000 + 0.032000 + 0.015399 + 0.007400 + 0.002600 + 0.000130 + 0.000010 = 0.137539$$

This adding-up process gives rise to another formula for the probability of an event: the number of equally likely ways that the outcomes making up an event may occur divided by the total number of equally likely ways all outcomes may occur.

Probability of an event

$$= \text{Sum of the probabilities of the outcomes in the event}$$

$$= \frac{\text{Number of equally likely ways outcomes in an event may occur}}{\text{Total number of equally likely ways all outcomes may occur}} \quad (3.1b)$$

Because the eight outcomes $0, $1, $2, $5, $10, $25, $500, and $5,000 are exhaustive in the instant cash game with 75 million equally likely tickets, the events "win no instant dollars" and "win some instant dollars" are also said to be exhaustive. They are also mutually exclusive because they share no outcomes and cannot occur together; that is, one ticket cannot be both a "loser" and a "winner." If events are mutually exclusive and exhaustive, then their probabilities must sum to one. In the instant cash game,

$$P(\text{win \$s}) + P(\text{win no \$s}) = 0.137539 + 0.862461 = 1$$

EXERCISES

3.1 Probability may be referred to as relative frequency in the long term. What does this mean?

3.2 What is the difference between an outcome and an event in the calculation of probability?

3.3 The National Basketball Association used to flip a coin to determine which of the two worst teams would get the first draft choice. For many years, heads dominated, and sportscasters began to question the fairness of the process. How could heads keep coming up if this was a fair procedure?

3.4 An article in *The Wall Street Journal* (August 6, 1992) on risk analysis stated that, Risk analysis begins with scientific studies . . . . The data from the studies are then run through computer models of bewildering complexity, which produce results of implausible precision . . . risk analysis of the space shuttle showed a chance of failure so miniscule that it could probably fly every day for 300 years without an accident. Then the Challenger blew up." Comment on the reporter's understanding of probability.

3.5 With the notable exception of problems involving dice, coins, card tricks, and the drawing of balls from urns, why are the conditions of classical probability seldom fulfilled?

3.6 If the probability is 0.2 that an event will occur and if in four attempts we do not observe the event, then do we know how many times it will occur in the next six attempts? Explain.

3.7 If the probability is 0.1 that an event will occur and if in nine attempts the event is not realized, then what will happen on the next attempt?

3.8 A denigrating statement about the creative genius of artists sometimes heard at business meetings and dinner parties is that "if you had 700 monkeys with 700 typewriters, in 700 years at least one of them would have typed a play by Shakespeare." This statement is false, but it is true that "there is almost a 100 percent chance that if you had an infinite number of monkeys and typewriters for an infinite number of years, then at least one of them would type a play by Shakespeare." What is the difference between these two statements? Why is one false and the other true?

3.9 From the following probabilities $P(A_1) = 0.62$, $P(A_2) = 0.09$, $P(A_3) = 0.29$, $P(A_1 \,\&\, A_2) = 0.24$, determine if A_1, A_2, and A_3 are mutually exclusive and exhaustive.

3.5 DETERMINING THE SAMPLE SPACE

Determination of the sample space and its subsets is not always straightforward. A tree diagram may be helpful in identifying the different ways in which outcomes occur. To see this, consider a firm hiring three people. Assume that the firm sequentially and randomly draws these three people from a labor force that continually renews itself so that it is always 50 percent female. That is, each time a woman (or a man) leaves the labor force, a like woman (or man) enters so the probability of drawing a woman on the first hiring draw is 0.50; the probability that the second draw is a woman is 0.50, and the probability that the third is a woman is also 0.50. Also assume that each hiring situation is identical and unrelated to the previous one or the subsequent one. (Cases in which these probabilities and hiring situations change from hire to hire are considered later in this chapter and in more detail in following chapters.)

A tree diagram shows the likelihood of the hiring outcomes. It is thus a model or description of the way in which hiring may progress. Starting on the left, the first branch in Figure 3.1 represents the first hire, where either a woman (W) or man (M) may be hired. Movement up the top branch represents a woman hired; movement down the bottom branch represents a man hired. At each of the three branching positions, the same routing is followed. Each of the eight paths, from the left starting point to the stopping point on the right, represents an outcome. Because each of these eight paths is equally likely, we know the probability of each path is 1/8 = 0.125.

The outcome WMW, for example, is the path marked in Figure 3.1. The probability of the outcome WMW can be visualized by starting at the left-most point. Although we are interested only in one firm, to visualize probability here it is advantageous to think of 40 firms starting at this left-most point. We would expect 20 of these 40 firms to get a women on the first draw (assuming a continually updated labor force that is always 50 percent women). Of those 20, only 10 would get a man on the second draw and

FIGURE 3.1 A Tree Diagram as a Model of Hiring

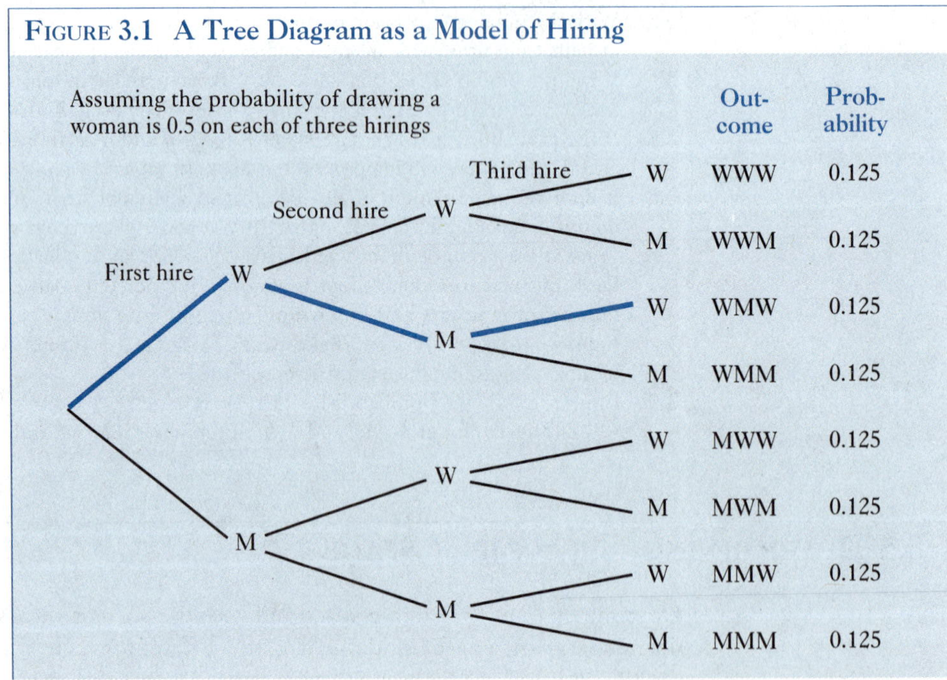

of those 10, only five would get a woman on the third. Thus, only five of the original 40 firms (or one-eighth) would have the outcome WMW.

In general, tree diagrams provide a good way to represent sequential experiments. These diagrams are used in many probability examples in this and later chapters.

QUERY 3.1:
Give two different representations of the outcome set for a coin toss involving one flip of a nickel and a flip of a dime. Show the probability of each outcome.

ANSWER:
Below are a tree and a rectangular or tabular array of the outcomes and probabilities associated with the two coin tosses.

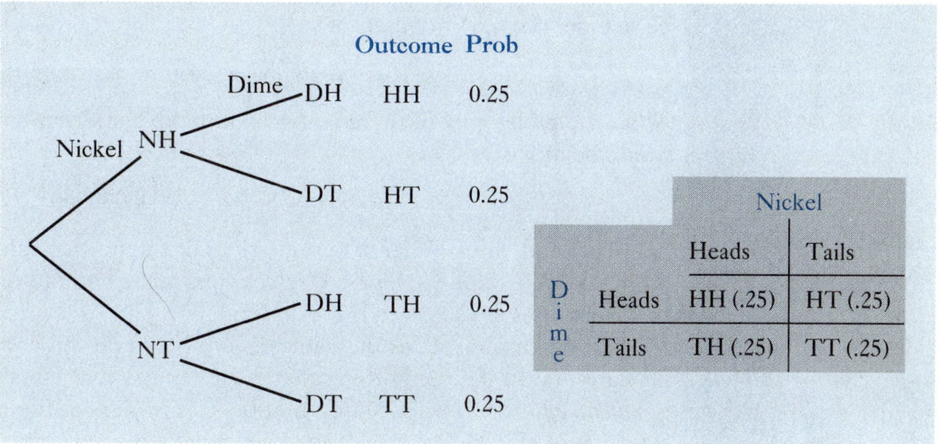

3.6 Counting Techniques

Basic rules of counting also may be helpful in determining the sample space and its subsets. In the instant cash lottery example, the total number of ways all possible outcomes occur was given as 75 million, the sample space. If a winning ticket number were selected, but no one had purchased the identified ticket number, the game would still end and a new one would begin with no change in future cash prizes or the probabilities.

In big, ongoing lotteries, the total number of ways all outcomes can occur is determined by a fixed number of draws from a known population. These big games differ from the instant cash game because the payoffs are not immediate; they accumulate into big amounts (multimillion-dollar pots) until there is a winner, and they are shared by anyone holding a purchased ticket with the correct numbers.

Most big lotto games involve the random selection of six numbers from a pool of about 40 to 54 numbers. If the six numbers are selected from 54, without replacement, each sequence of six numbers is called a **permutation.** The sequence 1,2,3,4,5,6 is a permutation; the sequence 6,5,4,3,2,1, is another. Similarly, 16,51,4,32,22,1 is a permutation, as is 51,16,32,4,1,22. For permutations, the order matters.

Permutation
Any one of the different sequences of numbers that can be formed from a set of numbers.

The first number in a permutation of six numbers can be any one of the 54 numbers that can be drawn from the population. The second number, however, is restricted to the 53 numbers that were not chosen on the first selection. Thus, for any number selected on the first draw, there are 53 that could be matched with it on the second. Thus, the number of permutations of two numbers that can be selected from 54 is 2,862 = (54)(53). If a third number is restricted to the 52 numbers not selected on the first and second draws, the number of permutations of three numbers that can be selected from 54 is 148,824 = (54)(53)(52). By the sixth selection, there are only 49 numbers left in the population. Thus, the number of permutations of six numbers that can be selected from 54 numbers is

$$18{,}595{,}558{,}800 = (54)(53)(52)(51)(50)(49)$$

Thus, the probability of there being a winner in a multimillion-dollar lotto of this type would be only

$$1/18595558800 = 0.000000000053776$$

In general, the number of permutations of x items that can be selected from n items is $n(n{-}1)(n{-}2)\ldots(n{-}x{+}1)$, where selection is made without replacement and the order in which items are selected is meaningful.

Winning in most big lotto games is not based on the order in which the six numbers are drawn. The only thing that is relevant is that the six numbers on a winning ticket match with the six chosen numbers, regardless of the order. If the ordering in a selection of objects is not relevant, then what we need is the number of combinations that can be formed. That is, while AB and BA are two permutations of A and B, they are a single combination. The number of different ways of selecting x items from n items, ignoring the order of selection, is the number of **combinations** that can be formed from n items taken x at a time. This number of combinations is

Combination
A group of elements or items selected from a set with no regard to the order in which they are selected.

$$\frac{n(n-1)(n-2)\ldots(n-x+1)}{x(x-1)(x-2)\ldots(2)(1)}$$

An equivalent for this combination count can be written

$$\frac{n(n-1)(n-2)\ldots(2)(1)}{[x(x-1)(x-2)\ldots(2)(1)][(n-x)(n-x-1)(n-x-2)\ldots(2)(1)]}$$

Factorial of n ($n!$)
$n! = n(n{-}1)(n-2)\ldots(3)(2)(1)$

A shorthand notation for the product $n(n{-}1)(n{-}2)\ldots(2)(1)$ is the **factorial $n!$.** For example, $5! = 5\cdot4\cdot3\cdot2\cdot1$, where $0! = 1$. Using the factorial notation, the number of combinations of x items selected from n items is identified by $C(n,x)$ and calculated as

Combination of x items selected from n items is

$$C(n,x) = \frac{n!}{x!(n-x)!} \tag{3.2}$$

For $n = 54$ and $x = 6$,

$$C(n,x) = \frac{n!}{x!(n-x)!} = C(54,6) = \frac{54!}{6!48!}$$

$$= \frac{54 \cdot 53 \cdot 52 \cdot 51 \cdot 50 \cdot 49}{6 \cdot 5 \cdot 4 \cdot 3 \cdot 2 \cdot 1}$$

$$= \frac{18595558800}{720} = 25,827,165$$

So there are 25,827,165 different ways of randomly drawing six numbers from 54, with no regard to the order of selection. (Inexpensive hand calculators will not be accurate in the above division. They will show 18595558800/720 approximately equal to 25827.16+, where the + may be any digit depending on the manner in which the calculator does rounding. The number on the screen (25827.16+) must be multiplied by 1,000 to approximate the correct answer. A spreadsheet program, such as LOTUS 1-2-3, can be programmed to accurately do this division.)[2]

Only one combination of six numbers on a purchased ticket will match the six selected numbers. The appropriate probability of winning this big jackpot in which the order of selected numbers does not matter is thus 1/25827165 = 0.0000000387189, which is far greater than the probability of winning when order matters.

QUERY 3.2:
In the hiring example where three people were to be hired from a labor force in which 50 percent were female, how many different ways can two women be hired?

ANSWER:
For $n = 3$ and $x = 2$,

$$C(n,x) = \frac{n!}{x!(n-x)!} = C(3,2) = \frac{3!}{2!1!}$$

$$= \frac{3 \cdot 2 \cdot 1}{2 \cdot 1 \cdot 1} = 3$$

So there are three different ways of getting two women hired in three attempts. These three outcomes are represented in Figure 3.1 as WWM, WMW, and MWW.

QUERY 3.3:

If a technician tests four computers per hour, with three of four expected to work properly, what is the probability of any two in a batch of four computers selected at random working properly? Determine your answer algebraically and show the outcome set.

ANSWER:

If the sequence in which the computers are tested is not important, the size of the sample space is

$$C(n,x) = \frac{n!}{x!(n-x)!} = C(4,2) = \frac{4!}{2!2!} = 6$$

Let G_1, G_2, and G_3 be the good computers and B_1 the bad one. The outcome set, assuming that sequencing is irrelevant, is then the six combinations

$$\{G_1G_2,\ G_1G_3,\ G_2G_3,\ G_1B_1,\ G_2B_1,\ G_3B_1\}$$

Again, if the sequence in which the computers are tested is unimportant, there are three outcomes that form the event "two good," namely:

$$\{G_1G_2,\ G_2G_3,\ G_1G_3\}$$

Algebraically, these three outcomes are calculated as

$$C(n,x) = \frac{n!}{x!(n-x)!} = C(3,2) = \frac{3!}{2!1!} = 3$$

Thus, the probability of selecting two of four computers at random and having both work properly on the first try is 3/6 = 0.5. (Note: If the effects of sequencing were to be considered, then the outcome set would be twice as large, with 12 permutations, and the event space would be twice as large, with six permutations, but the probability of two of four computers working properly would still be 0.5.)

The Best of Marilyn on Queries about Probability

The following are excerpts from Marilyn vos Savant's column "Ask Marilyn" in *Parade Magazine.*

QUERY 3.4:

After each holiday gift exchange, my five nieces and nephews write their names on slips of paper and put them into a basket, from which they then draw the name of the person for whom they'll buy a gift the following year. Last year, for the first time, each of the five drew his or her own name. What are the chances of such an occurrence taking place?

— Richard Coffey
Newington, Conn.
September 13, 1992

ANSWER:

It's unlikely, of course, but the chances aren't as low as you might think—one in 120.

—Marilyn vos Savant
September 13, 1992

QUERY 3.5:

A briefcase has a three-wheel combination lock, and each wheel is numbered from 0 to 9. How many possible combinations are there? And what are the odds against finding the correct combination?

—Bill Shannon
Memphis, Tenn.

ANSWER:

There are 1,000 combinations involved, so you could say the chances are only one in 1,000 that you'd find the right one the first try. But it actually would be fairly easy to try them all.

But if a briefcase has two of those locks, which many do, each with its own combination, the chances are one in a million, and finding the correct set may seem 1,000 times more laborious but again it really isn't. You would just open the first lock by rolling through the 1,000 combinations, then leave it open while you roll through the 1,000 combinations on the second lock.

—Marilyn vos Savant
November 1, 1992

QUERY 3.6:

This problem appeared in our local paper: "If each of 10 Little League baseball teams played the other teams twice, how many games would there be?" The National Council of Teachers of Mathematics says the answer is 90. Shouldn't it be 180?

—John L. Lowden
Wilmington, N.C.

ANSWER:

Ninety is correct. Let's reduce the number of teams to four (a total of 12 games, not 24) to see the reason more clearly:

A plays B	A plays D	B plays D
A plays B	A plays D	B plays D
A plays C	B plays C	C plays D
A plays C	B plays C	C plays D

The reason the number is only half what we'd expect is that when A plays B (etc.), B plays A at the same time—that is, during the same game. So one game suffices for both teams.

—Marilyn vos Savant
June 27, 1993

Exercises

3.10 Two portable computers are to be selected from a pool of five. Unknown to the potential users is that computer number one is defective. It is known that computers one (A_1) and two (A_2) are the same make and model (A) and computers three (B_1), four (B_2), and five (B_3) are the same make and model (B), which is different from that of computers one and two. Use a tree diagram to show the possible outcomes in the sample space.

3.11 If each of the 20 outcomes in Exercise 3.10 is equally likely, what is the probability that at least one of the two selected computers is defective and that at least one of the computers is a make and model A?

3.12 A department consists of 10 members. Three are to be selected for different jobs. The order in which assignments are made is important, because the three jobs are viewed as being of different status. In how many different ways can the jobs be assigned?

3.13 From the information in Exercise 3.12, if the assignment order is not important (all three jobs are considered of equal status, for example), then in how many different ways can the jobs be assigned?

3.14 It is sometimes argued that one of the consequences of affirmative action legislation is that hiring becomes random. If six applicants for a position are not of equal ability but, unknown to the prospective employer, could be ranked from most able (1) to least able (6), then if two of these applicants are hired at random what is the probability that the most able is hired?

3.15 There are three private contractors competing for two government contracts. The government agency could assign each contract to any one of the contractors or assign them both to the same contractor. If all outcomes are equally likely, what is the probability that both contracts go to the third contractor? (Identify the contractors and show the possible ways in which the contracts could be assigned.)

3.16 A specialty shop can produce five engines per day. If the selection of each is equally likely, what is the probability that the selection of two engines will result in both starting on the first try if only three of the five are properly timed to start?

3.17 A preselection pool consists of seven applicants for two jobs. If all seven applicants are of equal ability and the two jobs are the same, then in how many ways can the jobs be assigned?

3.18 If the two jobs in Exercise 3.17 are not the same, then in how many ways can the jobs be assigned?

3.7 Multiple Events

Multiple sets of outcomes may be of interest to decision makers. For instance, if 50 percent of the available labor force is female, the U.S. Employment and Equal Opportunity Commission (EEOC) might be suspicious of a firm that hires fewer than two women for three jobs or one that hires all three of the same sex. This event can be written as

"fewer than two women *or* all the same sex"

Compound (or Joint) Event
Events that must occur together.

The EEOC and an employer will definitely be interested in the probability of this **compound** or **joint event**. Compound or joint events are formed by two or more events in one of two ways: either as a union of two or more events or as an intersection of two or more events. Below we consider such compound events together with complements and conditional statements involving multiple events.

Unions and Intersections

Union of Two Events
All of the outcomes that make up the two events.

The probability calculation for "fewer than two women *or* all the same sex" is referred to as the probability of a **union of the two events** "fewer than two women" and "all the same sex." From Figure 3.1, it can be seen that there are four outcomes that have "fewer than two women": WMM, MWM, MMW, and MMM. This event is represented by the shaded area in Figure 3.2, panel a. There are two outcomes that have "all the same sex": WWW and MMM, as represented by the shaded area in Figure 3.2, panel b. Panel c shows the five outcomes that make up the compound event "fewer than two women or all the same sex." (Notice that adding the caveat "or both events" to the description of this compound event would not change the shaded area in panel c or the probability of this compound event.)

The probability of fewer than two women or all the same sex is five-eighths, or 0.625. This probability cannot be obtained by simply adding the probability of "fewer than two women" and the probability of "all the same sex" because these two events are not mutually exclusive; they share the outcome MMM. This double counting must be eliminated by subtracting it from the sum of the probability of these two events. That is,

$$\begin{aligned}&P(\text{fewer than two women or all the same sex})\\&= P(\text{fewer than two women}) + P(\text{all same sex})\\&- P(\text{fewer than two women and all same sex})\\&= \frac{4}{8} + \frac{2}{8} - \frac{1}{8} = \frac{5}{8} = 0.625\end{aligned}$$

Intersection of Two Events
The outcomes that fulfill the conditions for both events.

In general, if A or B are the two events of interest, then the probability of A or B is the probability of the union of A or B. The probability of A or B is calculated as the probability of A plus the probability of B minus the probability of A and B. The probability of the **intersection**, $P(A \text{ and } B)$, is subtracted from the sum of the probability of A and the probability of B because it is included in each of these probabilities and it cannot be counted twice, so one of the intersection probabilities is removed by subtracting it from the total.

Probability of ***A*** or ***B***

$$P(A \text{ or } B) = P(A) + P(B) - P(A \text{ and } B) \tag{3.3}$$

Complement of an Event
The set of outcomes that is not in the event.

Complements

If we define the event "all of the same sex" as event A, then those outcomes that are not in A are called the **complement** of A. The complement of A in the hiring case is represented by the unshaded area in Figure 3.2, panel b: WWM, WMW, WMM, MWW, MWM, and MMW. The complement of any set A is written $\bar{A}$ and is the set of all outcomes not in A.

Because the outcomes in A and the outcomes in $\bar{A}$ make up 100 percent of the sample space, $P(A) + P(\bar{A}) = 1.00$, we have the formula for the probability of the complement of A given by

Probability of the Complement of *A*

$$P(\bar{A}) = P(\text{not } A) = 1 - P(A) \tag{3.4}$$

The probability that not all hires are of the same sex is thus 0.75 (= 6/8). As another example, let event B be "fewer than two women." The complement is then "two or more women," which is the unshaded area in Figure 3.2, panel a. The probability of the complement of *B* is thus 0.50 (= 4/8).

FIGURE 3.2 Venn Diagrams for Events in the Hiring of Men and Women

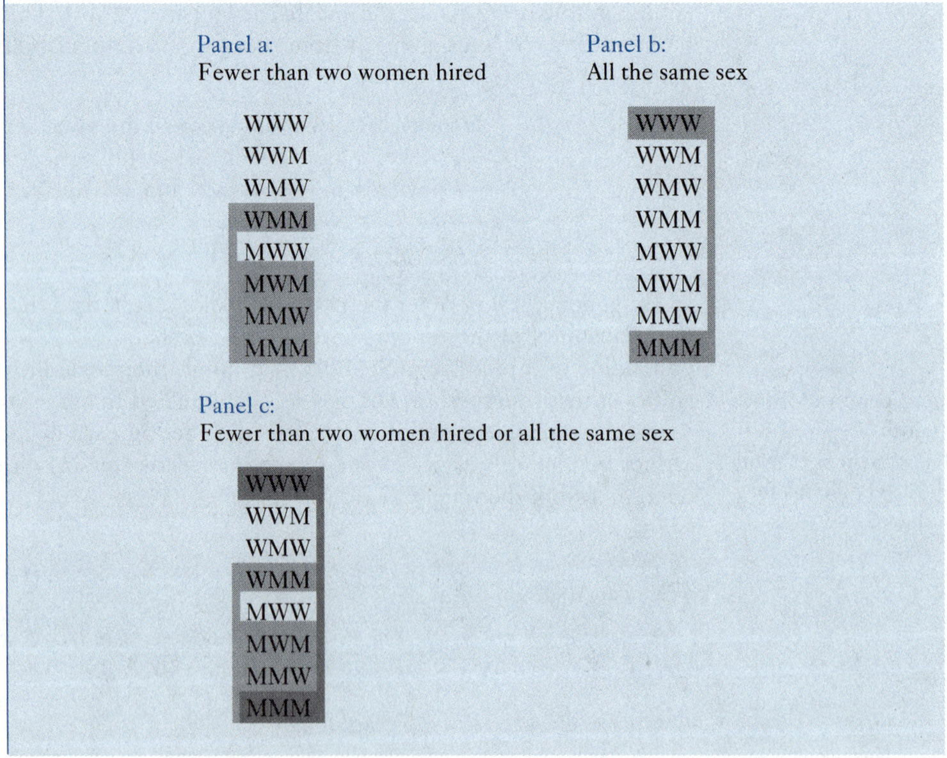

QUERY 3.7:

A firm has only six applicants (three men and three women) for three jobs. If hiring is random, what is the probability of at least one woman being hired? Why does your answer here differ from that obtained from Figure 3.1?

ANSWER:

First, the size of the sample space is

$$C(n,x) = \frac{n!}{x!(n-x)!} = C(6,3) = \frac{6!}{3!3!} = 20$$

Second, notice that the probability of at least one woman is the complement of all men hired. Thus,

$$P(\text{at least one woman}) = 1 - P(\text{all three men})$$

Only one of the outcomes involve the hiring of all three men.

$$C(n,x) = \frac{n!}{x!(n-x)!} = C(3,3) = \frac{3!}{3!0!} = 1$$

Thus, the probability of selecting at least one woman is

$$P(\text{at least one woman}) = 1 - (1/20) = 19/20 = 0.95$$

In Figure 3.1, the probability of hiring at least one woman was 0.875 (= 1.000 – 0.125), because the probability of hiring a woman on each trial was assumed fixed at 0.5. This assumption was reasonable as long as the pool from which individuals were drawn was extremely large or renewed over the sequence of hiring. In this query, that is not the case since the number of applicants is only six and there is no replacement. Thus, the probability of hiring a woman must change from trial to trial.

Conditional Probability

Conditional Probability The likelihood of one event given that another event occurs.

Decision makers are often interested in the probability of one event given another. This is a **conditional probability**. For example, the EEOC might have documentation of an employer saying that it would hire "at most one woman." It might then ask the question: What is the probability of no woman being hired given that at most one woman is to be hired?

Suppose again that Figure 3.2, panel a, is the relevant sample space for this employer. We can see that there are four outcomes that make up the event "at most one woman": WMM, MWM, MMW, and MMM. Of these, only one constitutes the event "no women hired," MMM. Thus the conditional probability of no women hired, given that at most one woman is to be hired, is 1/4 or 0.25.

In general terms, if A is one event and B another, then the probability of A given B is calculated by the formula

Conditional probability of A given B

$$P(A \mid B) = \frac{P(A \text{ and } B)}{P(B)} \tag{3.5}$$

where $P(B) > 0$

The formula for conditional probability facilitates the determination of the probability of the intersection. In many problems, the decision maker will need to know the probability of the intersection (A and B) but he or she will only know a conditional probability and the probability of the event on which the conditional probability is based. As demonstrated by the next Query 3.8, Formula (3.5) provides the key to solving for the intersection of the two events.

QUERY 3.8:

According to a story in *Newsweek* (June 8, 1992), promoting drugs in medical journals is big business, with American drug companies spending $350 million a year for advertising. A study by UCLA researchers showed 100 of 109 ads in medical journals had factual errors in areas for which the Food and Drug Administration (FDA) has established explicit standards. The FDA checks only about 15 percent of all drug ads. What is the probability of the FDA checking and finding an error for which it has standards?

ANSWER:

Let the probability of finding an error if checked be

$$P(\text{error} \mid \text{check}) = 100/109 = 0.917431$$

The probability of FDA checking is $P(\text{check}) = 0.15$. Thus, the probability of the FDA checking and finding an error is 0.1376. From a rewriting of Formula (3.5)

$$\begin{aligned} P(\text{error and check}) &= P(\text{error} \mid \text{checked})*P(\text{check}) \\ &= (0.917431)(0.15) \\ &= 0.137615 \end{aligned}$$

QUERY 3.9:

A drug-testing company claims that its screening tests are 95 percent accurate, meaning that 95 percent of drug users are accurately identified and 95 percent of those who don't use drugs are accurately identified. From government statistics, we know that approximately 5 percent of the adult population in a community are drug users. A human resource manager of a large firm in this community asks if it would be a good idea to start testing prospective employees.

ANSWER:

Unless the manager wants to waste company resources, these tests should not be purchased. If a prospective employee is identified as a drug user (tests positive), there is only a 0.5 probability that he or she is actually a drug user. This can be seen by constructing the following table:

	Drug user	Nondrug user	
test positive	0.0475	0.0475	0.0950
test negative	0.0025	0.9025	0.9050
	0.0500	0.9500	1.0000

where $P(\text{test positive} \mid \text{drug use}) = 0.95$, as claimed by the drug company, and $P(\text{drug use}) = 0.05$, as found in government records for your community, imply that $P(\text{test positive and drug use}) = (0.95)(0.05) = 0.0475$ and $P(\text{test negative and nondrug use}) = (0.95)(0.95) = 0.9025$, from which the other two intersections are obtained by subtracting from the relevant column probabilities at the bottom of the table. The probability of someone actually being a drug user given that he/she tests positively is now calculated as

$$\begin{aligned} P(\text{drug user} \mid \text{positive}) &= P(\text{drug user and positive})/P(\text{positive}) \\ &= 0.0475/0.095 = 0.5000 \end{aligned}$$

Thus, there is only a 50-50 chance (as in a coin toss) that someone in the community who tests positively for drugs is actually a drug user. A manager thus has little reason to make use of the drug tests.

3.8 INDEPENDENCE

Two events are independent if information about the occurrence of one does not affect the probability of the other occurring. Coin tosses are typically independent: Whether a head or tail occurs on the first toss has no influence on a head or tail on the second. The probability that a tennis player wins a tennis tournament, however, is likely related to the seed or ranking assigned prior to the tournament; winning is not independent of seed.

If two events A and B are independent then

$$P(A \mid B) = P(A)$$

That is, the probability of A given B is the same as the probability of A; the occurrence of B does not affect the probability of A. From the definition of conditional probability, we also know that independence must imply that

$$\frac{P(A \text{ and } B)}{P(B)} = P(A)$$

Multiplying both sides of this expression by the probability of B gives

$$P(A \text{ and } B) = P(A)P(B)$$

which is a helpful formula for calculating the probability of the intersection of two **independent events.**

Independent Events Two events that do not influence each other so that the probability of the first occurring, given that the second occurs, is the probability of the first in isolation.

QUERY 3.10:

The probability of rolling a one with either of two dice is one-sixth. What is the probability of rolling snake eyes (two ones with a pair of dice)?

ANSWER:

The outcomes of the dice rolls are independent; thus, the probability of snake eyes is 1/36 [=(1/6)(1/6)].

QUERY 3.11:

G. Howard (1987) describes a system that was alleged to be able to "beat" big, ongoing lotto games such as that in Pennsylvania. It is based on the so-called balanced game theory, which says that numbers that occur randomly should have their sums distributed approximately as a bell-shaped distribution. A lotto game that involves the selection of six numbers out of 54, for example, should yield a sum of those six numbers that is bell shaped with a mean, median, and mode of 165. [The median of the 54 numbers is 27.5; with six numbers drawn, the numbers with the highest probability must thus sum to 165 = (6)(27.5).] With the sum of the six selected numbers forming a bell-shaped distribution around a mean, median, and mode of 165, feasible ranges of values for the sum of the six selected numbers can be calculated using the rule of bell-shaped distributions. For example, a 68 percent interval could be computed with the explanation that 68 percent of all sums will fall within this range of values. If a ticket with a sum of the six numbers is chosen outside this range, then those numbers have only a 32 percent chance of being selected. This is alleged to imply that tickets with numbers that sum to 165 should be purchased. Is this theory correct?

ANSWER:

R. Paulson (1992) provides a good answer. He states that the information about sums being approximately bell shaped around a theoretical mean is correct, as discussed in Chapter 7 when the Central Limit Theorem is introduced. Paulson also states, however, that it is irrelevant in the prediction of future lottery winning numbers. As we will see in Chapter 7, the Central Limit Theorem explains the mathematics of what is expected to happen, but it doesn't tell a person which particular combination that adds up to a given sum will occur.

The balanced game system ignores the quantity of various combinations of numbers that all add up to the same total. Paulson considers a lottery game in which six numbers are selected from 40, so the most probable sum for those six numbers is 123 [=(6)(40+1)/2]. The probability of winning, however, is not altered by having your chosen numbers add up to 123. There are 56,512 different combinations that sum to 123, and there are 3,838,380 possible combinations for this particular game. Let the event A be the winning numbers sum to 123; let event B be your ticket numbers sum to 123, and event C is you win. If your numbers sum to 123, then your probability of winning is $P(A \cap C) = P(C \mid A)P(A)$, or (1/56,512)(56,512/3,838,380) = 1/3,838,380. The following conditional probabilities show that the probability of winning is not affected by the ticket numbers summing to 123:

$$P(C|B) = \frac{P(B \cap C)}{P(B)} = \frac{P(B|C)P(C)}{P(B)}$$

$$= \frac{(56{,}512/3{,}838{,}380) \times (1/3{,}838{,}380)}{56{,}512/3{,}838{,}380}$$

This is the same as your probability of winning if your numbers do not add up to 123; events B and C are independent.

The Best of Marilyn on Conditional Probability

The following are excerpts from Marilyn vos Savant's column "Ask Marilyn," in *Parade Magazine.*

QUERY 3.12:

I'll come straight to the point. In the following question and answer, you blew it!

"Suppose you're on a game show and you're given a choice of three doors. Behind one door is a car; behind the others, goats. You pick a door—say, No. 1—and the host, who knows what's behind the doors, opens another door—say, No. 3—which has a goat. He then says to you, 'Do you want to pick door No. 2?' Is it to your advantage to switch your choice?"

You answered, "Yes, you should switch. The first door has a 1/3 chance of winning, but the second door has a 2/3 chance."

Let me explain: If one door is shown to be a loser, that information changes the probability to 1/2. As a professional mathematician, I'm very concerned with the general public's lack of mathematical skills. Please help by confessing your error and, in the future, being more careful.

— Robert Sachs, Ph.D.
George Mason University
Fairfax, VA
December 2, 1990

You blew it, and you blew it big! I'll explain: After the host reveals a goat, you now have a one-in-two chance of being correct. Whether you change your answer or not, the odds are the same. There is enough mathematical illiteracy in this country, and we don't need the world's highest IQ propagating more. Shame!

—Scott Smith, Ph.D.
University of Florida
December 2, 1990

Your answer to the question is in error. But if it is any consolation, many of my academic colleagues also have been stumped by this problem.

—Barry Pasternack, Ph.D.
California Faculty Association
December 2, 1990

You are in error . . .

—Frank Rose, Ph.D.
University of Michigan
February 17, 1991

. . . in this matter, in which I do have expertise, your answer is clearly at odds with the truth.

—James Rauff, Ph.D.
Millikin University
February 17, 1991

May I suggest that you obtain and refer to a standard textbook on probability before you try to answer a question of this type again?

—Charles Reid, Ph.D.
University of Florida
February 17, 1991

Your logic is in error . . .

—W. Robert Smith, Ph.D.
Georgia State University
February 17, 1991

You are utterly incorrect about the game-show question . . . . How many irate mathematicians are needed to get you to change your mind?

—E. Ray Bobo, Ph.D.
Georgetown University
February 17, 1991

I am in shock that after being corrected by at least three mathematicians, you still do not see your mistake.

—Kent Ford
Dickinson State University
February 17, 1991

Maybe women look at math problems differently than men.

—Don Edwards
Sunriver, OR
February 17, 1991

I still think you're wrong. There is such a thing as female logic.

—Don Edwards
Sunriver, OR
July 7, 1991

You are the goat!

—Glenn Calkins
Western State College
February 17, 1991

You're wrong, but look at the positive side. If all those Ph.D.s were wrong, the country would be in very serious trouble.

—Everett Harman, Ph.D.
U.S. Army Research Institute
February 17, 1991

You are indeed correct. My colleagues at work had a ball with this problem, and I dare say that most of them—including me at first—thought you were wrong!

—Seth Kalson, Ph.D.
Massachusetts Institute of Technology
February 17, 1991

I also thought you were wrong, so I did your experiment, and you were exactly correct . . . .

—William Hunt, M.D.
West Palm Beach, FL
July 7, 1991

I put my solution of the problem on the bulletin board in the Physics Department office here, following it with a declaration that you were right. All morning I took a lot of criticism and abuse from colleagues, but by late in the afternoon most of them came around. I even won a free dinner from one overconfident professor.

—Eugene Mosca, Ph.D.
U.S. Naval Academy
Annapolis, MD
July 7, 1991

After considerable discussion and vacillation here at the Los Alamos National Laboratory, two of my colleagues independently programmed the problem, and in one million trials, switching paid off 66.7% of the time. The total running time on the computer was less than one second.

—G.P. DeVault, Ph.D.
Los Alamos National Laboratory
Los Alamos, NM
July 7, 1991

Now fess up. Did you really figure all this out, or did you get help from a mathematician?

—Lawrence Bryan
San Jose, CA
July 7, 1991

ANSWER:

Good heavens! With so much learned opposition, I'll bet this one is going to keep math classes all over the country busy on Monday.

My original answer is correct. But first, let me explain why your answer is wrong. The winning odds of 1/3 on the first choice can't go up to 1/2 just because the host opens a losing door. To illustrate this, let's say we play a shell game. You look away, and I put a pea under one of three shells. Then I ask you to put your finger on a shell. The odds that your choice contains a pea are 1/3, agreed? Then I simply lift up an empty shell from the remaining two. As I can (and will) do this regardless of what you've chosen, we've learned nothing to allow us to revise the odds on the shell under your finger.

The benefits of switching are readily proved by playing through the six games that exhaust all the possibilities. For the first three games, you choose No. 1 and switch each time; for the second three games, you choose No. 1 and "stay" each time, and the host always opens a loser. Here are the results (each row is a game):

DOOR 1	DOOR 2	DOOR 3
AUTO	GOAT	GOAT
Switch and you lose.		
GOAT	**AUTO**	GOAT
Switch and you win.		
GOAT	GOAT	**AUTO**
Switch and you win.		
AUTO	GOAT	GOAT
Stay and you win.		
GOAT	**AUTO**	GOAT
Stay and you lose.		
GOAT	GOAT	**AUTO**
Stay and you lose.		

When you switch, you win two out of three times and lose one time in three; but when you don't switch, you only win one in three times.

You can play the game with another person acting as host with three playing cards—two jokers for the goats and an ace for the auto. Doing it a few hundred times to get valid statistics can get a little tedious, so perhaps you can assign it for extra credit—or for punishment. (*That'll* get their goats!)

—Marilyn vos Savant
December 2, 1990

. . . I'm receiving thousands of letters, nearly all insisting that I'm wrong, including one from the deputy director of the Center for Defense Information and another from a research mathematical statistician from the National Institute of Health! Of the letters from the general public 92% are against my answer, and of the letters from universities, 65% are against me . . . . But math answers aren't determined by vote . . . . The original answer is still correct . . . .

—Marilyn vos Savant
February 17,1991

NOTE: *The New York Times* gave front-page coverage to this "Let's Make a Deal" controversy. Outside her column, however, Marilyn vos Savant responded only to an article in the *American Statistician* (November 1991). Her letter to the authors (J.P. Morgan, N.R. Chagantry, R.C. Dahiya, and M.J. Doviak) and their rejoinders demonstrate that issues of probability depend critically on the nature of the question asked and the sample space defined.

QUERY 3.13:
A shopkeeper says she has two new baby beagles to show you, but she doesn't know whether they're male, female or a pair. You tell her that you want only a male, and she telephones the fellow who's giving them a bath. "Is at least one a male?" she asks. "Yes," she informs you with a smile. What is the probability that the *other* one is a male?

—Stephen I. Geller
Pasadena, CA
October 13, 1991

ANSWER:
One out of three.

—Marilyn vos Savant
October 13, 1991

QUERY 3.14:
This is regarding the problem where a shopkeeper has two baby beagles, but she doesn't know if they're male, female or a pair. You come into the shop to buy a pup, but you want only a male, so she phones the fellow who is giving the beagles a bath and asks, "Is at least one a male?" After he gives her the answer, she tells you, "Yes!" What's the probability that the other one is a male?

You answered, "One out of three." There are only three possible explanations for this:

1. You are considering the possibility that the second puppy has been neutered, but because you did not make this clear, you leave open the next explanation.
2. You are (still) confused about probability. If so, I suggest you brush up on this information.
3. Your incorrect responses are intentional, done as a means to solicit mail. If so, there has to be a better way.

—James Larsen, Ph.D.
Wright State University
January 5, 1992

I disagree with your answer. The observer knows the first dog is a male; the probability that the other is a male is 50-50.

—Richard Jones, Ph.D.
University of New Haven
January 5, 1992

I believe you're wrong! Before you know that one of the baby beagles is a male, the chances are one out of three. But once the above fact is known, the chances change to 50-50.

—Edward Weiss, Ph.D.
Bethel College
January 5, 1992

Okay. But here's what is puzzling to me: If there were only one beagle pup, the probability of it being male would be one out of two. What can explain why the presence of another beagle pup should affect the probability that the pup in question is a male? I know this is cute, but how far away does the second beagle have to be in order for it not to affect the sex of the first beagle?

—Steve Marx
Worcester, MA
January 5, 1992

ANSWER:
The original answer is correct. We didn't define a "first beagle" or a "second beagle," and so either beagle can be in a doghouse on the moon, and it still would affect the outcome.

If we could shake a pair of puppies out of a cup the way we do dice, there are four ways they could land: male/female or female/male or male/male or female/female. So there are three ways in which at least one of them could be a male. And since the partner of a male in those three is either a female, a female, or another male, the chances of that partner being a male are one out of three.

The key is that we didn't specify which beagle was a male, so it can be either one. If we'd said instead, "The one nibbling your ankle is a male—what are the chances that the one sleeping is a male?" the chances would be 50-50. But we just specified that "at least one" was a male, so we don't know which it is and whether its partner is awake or asleep. And this means the chances of that partner being a male are only one out of three.

—Marilyn vos Savant
January 5, 1992

QUERY 3.15:
Three prisoners on death row are told that one of them has been chosen at random for execution the next day, but the other two are to be freed. One privately begs the warden to at least tell him the name of one other prisoner who will be freed. The warden relents: "Susie will go free." Horrified, the first prisoner says that because he is now one of only two remaining prisoners at risk, his chances of execution have risen from one-third to one-half! Should the warden have kept his mouth shut?

—Marvin M. Kilgo, III
Camden, SC
July 5, 1992

ANSWER:
It didn't matter. Even though there are only two remaining prisoners at risk, the first prisoner still has only a one-third chance of execution. Oddly enough, however, things don't look so good for the other one, whose chances have gone up to two-thirds!

—Marilyn vos Savant
July 5, 1992

EXERCISES

3.19 Identify the following as either the probability of an event, a joint event (unions or intersections), or a conditional event.

a. The probability the next president of the United States is under 50 years of age and female.
b. The probability it rains tomorrow given that it rained today.
c. The probability that the stock market rises and the prime interest rate declines next week.
d. The probability a person earns a grade of B in a statistics course if he or she earned a B in microeconomics.
e. The probability that either of two machines in a production process produces a defect.
f. The probability that a real estate agent is a member of the "ten million dollar club" and is under 35 years old.
g. The probability that an unidentified executive is earning more than $100,000 per year regardless of his or her age.

3.20 State whether each of the following is true or false and give your reason.

a. If $P(A_1 \mid A_2) = 0$, then A_1 and A_2 are independent.

b. If $P(A_1 \mid A_2) = 0$, then A_1 and A_2 are mutually exclusive.

c. If A_1 and A_2 are independent, then $P(A_1) = P(A_2)$.

3.21 Two independent machines produce plastic decals, with defect rates of 0.015 and 0.025. If 100 decals are produced, what is the probability of three or fewer defects?

3.22 A *Business Week*/Harris Poll (June 15, 1992) of 400 female executives at companies with $100 million or more in sales found that 27 percent had been harassed, but of those who had been harassed, only 25 percent reported it to their employers. What is the probability of a female executive having been harassed and reporting it?

3.23 Purdue University Cooperative Extension Service reports that 50 percent of fatal tractor accidents are from tractor rollovers and 18 percent are from run overs. "Indeed, tractor accidents are the leading cause of farm-related fatalities in Indiana, accounting for 53 percent of all farm-related deaths, according to the American Society of Agricultural Engineers."(*HT*, June 14, 1992)

a. Given a fatal farming accident, what is the probability it was from a tractor rollover?

b. Given a fatal farming accident, what is the probability it was from something other than tractor rollover or tractor run over?

3.24 A *WSJ* (July 15, 1992) article reported that institutional investors are getting more assertive in an attempt to curb excessive executive pay. Of the 56 institutions surveyed, 66 percent (37 of 56) said they favored the creation of compensation committees to review or even set executive pay. In addition, 50 percent said such committees should have the power to hire and pay compensation consultants directly. From this information, can you calculate the probability that an institutional investor would favor the creation of a compensation committee that has the power to hire and pay compensation consultants? Why or why not? Explain.

3.25 A *Wall Street Journal* article (August 11, 1992) reported that "Most American men—eight out of 10—use a hair dryer, according to industry estimates . . . . And while men purchase just 20 percent of the 20 million hair dryers each year, Frank Lindsey, vice president of marketing for Conair Corp., a leading maker of blow dryers, says that's not important." If use and purchase of hair dryers are independent, what is the probability that a man does not buy but uses a hair dryer?

3.26 A *Wall Street Journal* (July 14, 1992) article, suggested that "Only 7%" of companies have a "very complete" understanding of the Americans With Disabilities Act. A *WSJ* (August 11, 1992) article reported that "76% worry" about possible lawsuits under this new disability law. What is the probability of a company being worried and having a complete understanding of this act, assuming independence?

3.27 An arms dealer accepts or rejects shipments of bullets on the basis of testing a few bullets sampled from boxes selected from the shipment. Suppose we know that an inspector has accepted 99 percent of all good shipments and has incorrectly rejected 1 percent of the good shipments. In addition, the inspector accepts 94 percent of all shipments. It is also known that 10 percent of all shipments are of inferior quality. Find the probability that a shipment is

a. rejected.

b. good.

c. good and accepted.

d. inferior or rejected.

e. inferior if it is accepted.

f. good if it is rejected.

3.28 A financial analyst states that the probability "a real estate stock rises is 0.75, if the stock market rises." The probability that the stock market will rise is 0.2. What is the probability of both the stock market in general rising and the real estate stock rising?

3.29 Steven C. Salop, in "Evaluating Uncertain Evidence With Sir Thomas Bayes: A Note For Teachers," *Economic Perspectives* (Summer 1987), provides a discussion of the reliability of evidence in a wrongful injury suit against the Blue Cab Company for a hit-and-run accident alleged to involve one of its blue taxis. Salop states that the eyewitness was ". . . 80 percent reliable in identifying the color of taxi. That is, he was able to identify the correct color of taxis 80 percent of the time . . . ." Salop provided the following "matrix" to show how many of the city's 15 blue taxis and 85 green taxis this eyewitness could be expected to identify correctly—12 of the 15 blue and 68 of the 85 green. Salop stated: "Therefore, when the eyewitness testifies that a particular taxi is blue, the likelihood that the taxi is blue in fact is only 12/29, or 41 percent . . . the plaintiff should have sued the Green Cab company, on the strength of the Registrar's testimony (of 15 blue and 85 green cabs in the city)." In coming to this conclusion, Salop made a major error confusing conditional with unconditional probability. What is his error, and what are its implications for the analysis?

		Color Perceived by Eyewitness		
		Blue	Green	Total
Color in fact	Blue	12	3	15
	Green	17	68	85
	Total	29	71	100

3.30 A computer manufacturer uses disk drives from two suppliers. The first supplier accounts for 80 percent of the drives. It is known that 5 percent of all drives are defective and that 6 percent of the drives supplied by the first supplier are defective. If a disk drive is found to be defective, what is the probability it came from the first supplier?

3.31 In a retirement community, 60 percent of the population are coffee drinkers and 30 percent are smokers. Moreover all the smokers are coffee drinkers. What is the probability that a randomly selected person from the community would be a smoker, given that the person is a coffee drinker?

3.9 Subjective Versus Objective Probability

Odds
A statement of chance that gives the number of times an event will occur versus the number of times it will not occur over a large (infinite) number of trials.

The words **odds**, probability, chance, and likelihood are typically used as synonyms. But they may have different meanings in different circumstances, and they may have different meanings even though they may appear the same.

For instance, a TV sportscaster who says that the odds of Michigan beating Michigan State are "3 to 4," is saying that if the two teams play each other many (infinite) times, on the average three of seven times Michigan will win. On the other hand, in horse racing the "morning line odds" (or simply morning line) on a horse are written as "5–2" or "5:2." These *do not* say that the expert handicapper believes this horse will win the race five of seven times. These odds pertain to the money wager: if you bet \$2 on this horse to win and it does then you will get \$7 back—the \$2 bet plus \$5 net return. Alternatively, if this horse could run a large (infinite) number of races against the same field, and each time you bet \$2 on it to win, five of seven times you would get nothing back; two of seven times you would get \$7 back. That is, the handicapper is saying that only two out of seven times will this horse win, on the average, for an approximate probability of winning of 0.2857.

Recognize that the sportscaster's odds of a team winning and a handicapper's odds of a horse paying off need not be based on a direct reference to any relative frequency distribution. The sportscaster's odds are his or her belief about a team's chance of winning. The handicapper's odds are his or her best guess about the probabilities of a horse "being in the money." Similarly, when stock market forecasters predict what is going to happen to stock prices, they need not be basing their likely scenario of a rise or fall on actual past data.

The sportscaster's and handicapper's odds and the forecaster's likely scenario are not the same as statements about probabilities in the state-run lotteries, wheels of chance (e.g., a 38-number roulette wheel or a 360-number wheel of fortune), dice games (e.g., craps or Yahtzee), card games (e.g., blackjack or stud), or flipping coins. In those games of chance, the likelihoods of outcomes are fixed in accordance with physical, time, or rule constraints. Because a roulette wheel is numbered one through 38, the number 15 will come around once each turn of the wheel. Once in 38 spins, on the average, the ball will land on number 15. It makes sense to talk about repeated spins of the wheel. The imagination is stretched, however, in thinking about replicating basketball games in Michigan, horse races at Churchill Downs, and days at the New York Stock Exchange.

In the case of the stock market, horse racing, or other investment opportunities involving uncertain events that cannot be replicated, the relative frequentist's interpretation of probability can be approximated by finding groups of comparable stocks or horses to see what fraction of them had price appreciation or actually won races under like conditions. The obvious question is how to find groups of comparable stocks or horses.

Peter Asch and Richard Quandt (1991, p.111) demonstrated how race-time odds can be used to group horses. Race-time odds, unlike the handicapper's morning odds, are calculated as the amount bet on a horse divided by the total amount bet on all horses in the race. This relative frequency measure is still a subjective probability measure, because it is based on the opinions of the betting public, but it may be viewed as "more objective" because it reflects what racetrack patrons are betting. The "most objective measure" of the probability of winning might be to count how many horses in each group actually won as a percentage of the total number of horses in the group. Even this measure, however, is not truly objective in the sense that it represents a count of replicable results; it overlooks events that might have taken place during the races that may have influenced the results in a nonreplicable way. Similarly, in studies of the stock market, identical daily replications are not possible.

The notion that there is a truly objective probability that can be interpreted as a long-term relative frequency is an idea that is never fully realized in the real world. It is a concept that is fulfilled in degrees. It is not an absolute.

Historical Perspective

As demonstrated by questions asked of Marilyn vos Savant, the concept of probability is not easily applied to everyday problems. Part of the reason for this is that the issues are difficult to formulate and are open to different interpretations. Another is that the definition and meaning of probability are not unique and universally accepted. There is no universally accepted and fully adequate definition of probability.

Galileo (1564–1642), mathematician and astronomer, may have been the first to formally attempt and apply a definition of probability. In his high-paying academic position at the University of Pisa, Galileo had no explicit duties other than to answer queries from his Highness Cosimo II of Tuscany. Cosimo, an avid dice player, reportedly had noticed that throwing three six-sided dice resulted in a sum of nine less often than a sum of 10; yet, he reasoned that there are six outcomes (3 + 3 + 3, 5 + 3 + 1, 4 + 3 + 2, 5 + 2 + 2, 6 + 2 + 1, and 4 + 4 + 1) that sum to nine and six that yield 10 (4 + 3 + 3, 4 + 4 + 2, 5 + 4 + 1, 5 + 3 + 2, 6 + 3 + 1, and 6 + 2 + 2), suggesting that the sum of nine and 10 should have the same number of occurrences. What is amiss, he asked Galileo?

Galileo responded that the six outcomes identified by Cosimo are not each equally likely. There are 216 (= 6 · 6 · 6) ways of forming a sum from the outcomes on the faces of three dice; of these, 25 form a sum of nine and 27 sum to 10. Galileo then formulated the classic definition of probability:

> The probability of an outcome (or event) is the number of ways an outcome may occur divided by the total number of ways all outcomes may occur; thus,

$$P(9) = 25/216 = 0.11574 < P(10) = 27/216 = 0.125$$

This classical definition of probability suffers from two shortcomings. First, the concept is defined in terms of its measurement (circularity). Second, it cannot easily handle a large (infinite) number of outcomes and large (infinite) number of ways of obtaining those outcomes. Jacques Bernoulli (1654–1705) evaded the circularity problem by defining probability as a "degree of confidence" that may vary from person to person but could be objectively determined by looking at observable frequencies. This frequency approach was later placed on an axiomatic footing (in which generalities are deduced from assumptions) when John Venn defined probability as the limiting value of the percentage of favorable outcomes in an infinite sequence of independent replications (Richard von Mises, 1964).

The axiomatic definition of probability suffers from three weaknesses. First, an exact replication is impossible—if the rolling of a die is replicated perfectly, then wouldn't it always result in the same outcome? Second, because infinity is never realized, probability is never a precise or exact number; it is always an estimate. Third, and possibly most important in applications, the sample space is typically unclear. When taking a road trip, is the objective probability of your car crashing determined by your experiences on all prior trips? Your trips only to the same location? The trips of all other drivers? Or only those drivers with your kind of car; or, . . . ?

Economist John Maynard Keynes (1921) and mathematician Harold Jeffreys (1948) advanced a third definition of probability. Although Jeffreys tried to distance himself from Keynes, they both set out an approach to probability based on personal degrees of belief and linked those to the notion of utility. Both approaches view objective probability as a measure of the strength of the connection between a set of evidence and the truth of some hypothesis, when viewed by a "rational person." However, the establishment of the hypothesis believed true (e.g., person is innocent) against which the probability of another hypothesis is assessed (person is guilty) requires a prior subjective judgment so the resulting probability calculation cannot be uniquely objective.

Finally, there is the subjective or personal school of probability, as codified by Leonard Savage (1954). Subjective or Bayesian probabilities can be degrees of beliefs about one-time events (e.g., the introduction of a new series of Cadillacs or the destruction of the center structure of the Golden Gate Bridge in an earthquake) or events that can be replicated. The only requirement is that the belief system is consistent, so that multiple bets cannot be made with the agent winning under all outcomes. The question to be raised, however, is: How are these consistent subjective probabilities formulated by the agent?

As pointed out by Paul Schoemaker (1982) in his review of different types of probability, "The important point is that probability is not a simple construct . . . . Its measurement is obviously difficult in real-world settings." Casual students of statistics and serious scholars alike have learned that there may be no unique and unquestionable answer to complex problems involving probabilities.

EXERCISES

3.32 In what way do odds and probability measure the same thing?

3.33 If the odds on the next Indiana University versus Purdue University basketball game are given as 3-2 in IU's favor, what does this mean in terms of the probability that IU wins?

3.34 Based on the following information on the horses scheduled to run in the 1993 Preakness, which horses were the favorites to win (first), place (second) and show (third)?

PREAKNESS STAKES FIELD

118th running, 13/16 miles, 4:32 p.m., Saturday

PP		Jockey	Trainer	Owners	Odds
1.	Personal Hope	Stevens	Mark Henning	Debbie & Lee Lewis	9-2
2.	El Bakan	Perret	Alfredo Callejas	Robert Perez	20-1
3.	Prairie Bayou	Smith	Tom Bohannan	Loblolly Stable	3-1
4.	Hegar	Ferrer	Penny Lewis	Huntington Point Stable	20-1
5.	Too Wild	McCauley	Nick Zito	Bill Condren/Joe Cornacchia	20-1
6.	Union City	Valenzuela	D.Wayne Lukas	Overbrook Farm	15-1
7.	Woods of Windsor	Wilson	Ben Perkins Jr.	Mrs. Augustus Riggs IV	8-1
8.	Rockamundo	Prado	Oris J. Glass Jr.	Gary & Mary West	20-1
9.	Sea Hero	Bailey	Mack Miller	Rokeby Stable	7-2
10.	Wild Gale	Sellers	Mike Doyle	Little Fish Stables	8-1
11.	Koluctoo Jimmy Al	McCarron	Bruce Levine	Basil J. Plasteras	10-1
12.	Cherokee Run	Day	Frank Alexander	Jill Robinson	12-1

3.35 Below are some of the odds cited by Jack Saylor in the *Detroit Free Press* (June 24, 1993) for the 78-player field in the Senior Players Championship at the Tournament Players Club of Michigan. Which of these 14 players is most likely to win the championship according to these odds? Which is least likely to win? What are the two implied probabilities of the most likely and least likely winning the championship?

Player	Odds
Jim Albus	28–1
George Archer	30–1
Bob Charles	16–1
Jim Ferree	50–1
Al Geiberger	40–1
Larry Gilbert	18–1

(continued)

Player	Odds
Bob Murphy	6–1
Jack Nicklaus	8–1
Chi Chi Rodriguez	12–1
J.C. Snead	25–1
Dave Stockton	10–1
Lee Trevino	4–1
Tom Weiskopf	5–1
Bob Wynn	60–1

3.10 Concluding Comments

This chapter provided an introduction to the application of probability for decisions involving outcomes and events associated with experiments. You now know how to define uncertainty in the context of these outcomes and events and their associated probabilities. You are able to distinguish between subjective and objective probabilities and give a historical perspective on their developments. With the basic rules of probability, you are able to make probability calculations for everyday business and economic decisions involving unions and intersections of events, complementary events, and independent events. You know how to work with conditional probability and relate this concept to the idea of independence. The next chapter extends these concepts and procedures to similar topics involving random variables.

Chapter Exercises

3.36 What are the similarities and differences between relative frequency and probability?

3.37 Since it was introduced in England in 1985, DNA testing has been a most powerful means of identification in legal matters. In July 1992, for example, Leonard Callace was released from a New York State prison for a rape he did not commit. DNA testing of spots on the victim's clothing and the DNA of Callace did not match. When the results of DNA testing point to their clients' guilt, defense attorneys bring in expert witnesses who question the probability of a match. "In a Manhattan murder case, one method yielded a one in 500 chance that the DNA might belong to someone other than the defendant. Figured differently, the odds were one in 739 billion . . . . While the chance of a certain pattern occurring in the general population may be one in 5 million, . . . the chance of a match among Polish-Americans may be one in 200,000."(*Parade Magazine,* April 25, 1993)

a. Why weren't these differences in the probability of a match relevant for the Callace acquittal?

b. Why do these probabilities differ by so much?

3.38 A corporate board consists of 10 members. Suppose that the 10 members are to be assigned to three committees: three to each of the first two committees and four to the third committee. In how many ways can three committees be formed?

3.39 In Exercise 3.38, if three of the 10 board members are female, what is the probability that all three would be assigned to the first committee if assignments are random?

3.40 Light bulbs are typically sold in packages of three. If every package has one defective bulb, what is the probability that three bulbs randomly selected from two packages will have no defects?

3.41 Courts now confine guilty parties to house arrest, where the individual can travel back and forth from work to home and, with the exception of this travel, must always be either at home or at work. Private contractors are hired to monitor the process. If each morning an inspector must select one from 10 parties to monitor for the day, what is the probability that the same party is selected twice in a seven-day week?

3.42 There are eight candidates, six female and two male, for a committee that will consist of two people.

a. What is the probability that the first person selected is a male and the second is a female?

b. What is the probability of two males being selected?

c. What is the probability of selecting at least one male?

d. What is the probability the second person selected is a male, given that the first selected is male?

3.43 You know that 20 percent of the applicants for positions with your firm will be females and members of minority races. Twenty-five percent of the applicants are female. If the next applicant is known to be a female, what is the probability this person is of a minority race?

3.44 Two independent components of a computer system are known to have average failure rates of 0.001 and 0.002 in 10,000 hours of operation, and either failure will make the system inoperable. If 50 systems are purchased, what is the probability that two or fewer will fail in the first 10,000 hours as a result of a component failure?

3.45 Following the runway fire at Kennedy Airport in New York of a TWA airplane bound for San Francisco on July 28, 1992, a commentator said "the survivors should not be fearful of continuing their travel to San Francisco by air on July 31, 1992, because few people are ever involved in an airplane accident and since these people had already been involved in one it was even more unlikely that they would be involved in another." Comment on the correctness of this statement.

3.46 Television commercials are designed to appeal to the most likely viewing audience of the sponsored program. However, S. Ward, in a 1972 *Journal of Advertising Research* article, notes that children often have a very low understanding of commercials, even of those designed to appeal especially to children. Ward's studies show the following percentages of children who do or do not understand TV commercials, for the age groups listed:

	Ages		
	5–7	**8–10**	**11–13**
Do not understand (%)	55	65	30
Understand (%)	45	35	70

An advertising agent has shown a television commercial to a 12-year-old and another to a 9-year-old child in a laboratory experiment to test their understanding of the commercials.

a. What is the probability that the message of the commercial is understood: 1) by the 9-year old? 2) by the 12-year old?

b. What is the probability that both children demonstrate an understanding of the TV commercials?

c. What is the probability that one or the other children demonstrates an understanding of the TV commercials?

d. What did you have to assume in Parts b and c? Is this assumption reasonable? Why or why not?

3.47 A new method of drug testing for drunken driving can identify 95 percent of all drunks. Five percent of the time, however, the test will identify a driver as drunk when such is not the case. If 3 percent of the drivers on the road are drunk, what is the probability that a driver who is identified as drunk by this new test is not? What is the relevance of this probability?

3.48 The following diagram from *Popular Science* (June 1993) shows the percentage of computer users who work away from the office. The accompanying text stated that "while 75 percent of today's professional workforce does some part of its job away from the office, fewer than a third of those 27 million people are the frequent-flyer 'Globetrotters' or highway-

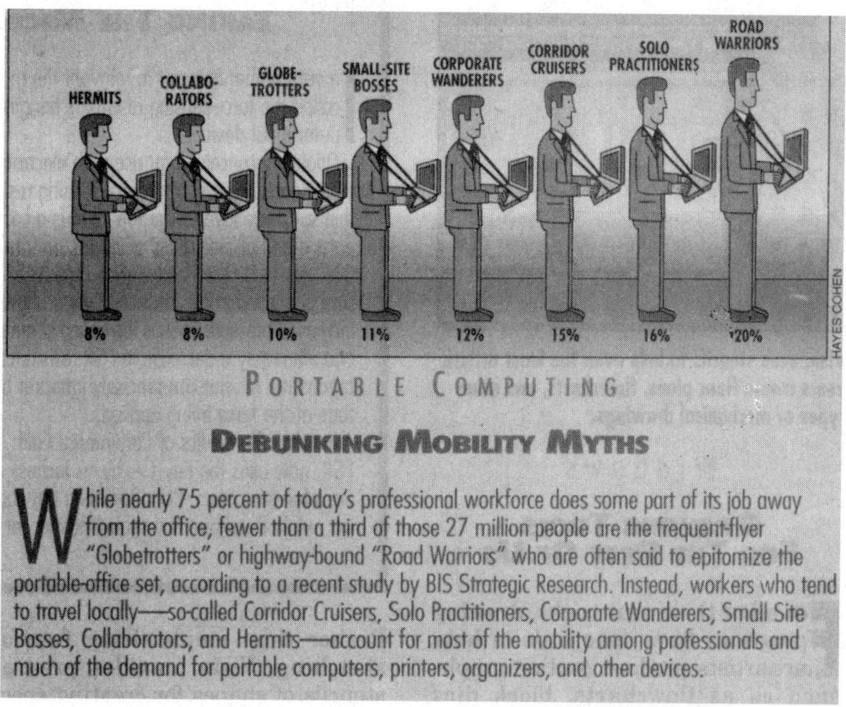

PORTABLE COMPUTING

DEBUNKING MOBILITY MYTHS

While nearly 75 percent of today's professional workforce does some part of its job away from the office, fewer than a third of those 27 million people are the frequent-flyer "Globetrotters" or highway-bound "Road Warriors" who are often said to epitomize the portable-office set, according to a recent study by BIS Strategic Research. Instead, workers who tend to travel locally—so-called Corridor Cruisers, Solo Practitioners, Corporate Wanderers, Small Site Bosses, Collaborators, and Hermits—account for most of the mobility among professionals and much of the demand for portable computers, printers, organizers, and other devices.

bound 'Road Warriors' who are often said to epitomize the portable-office set, according to a recent study by BIS Strategic Research." Redo the percentages in the above diagram to show the fraction of the total professional workforce that would fall into each of the eight categories. What is the probability of randomly selecting from the total professional workforce one person who is either a "Globetrotter" or "Road Warrior?"

3.49 At the Porter County Fair, a church-run carnival game involved the placing of money on one of six colored squares. A wheel was spun to determine a winning color. If blue is one of the colors, what is the probability of it being a winner? What are the odds of it being a winner?

3.50 There are five used cars on a lot: two are okay and three have major problems. A prospective buyer selects two of these cars to test drive. What is the probability that

a. the two are okay?
b. at least one is okay?
c. both are okay, given that one of those selected is okay?
d. both are not okay, given that one those selected is not okay?

3.51 An Indiana basketball player is expected to complete nine of 10 free throws, on the average. If such is the case, what is the probability that an Indiana player will make two free throws in the next three attempted? What did you have to assume to calculate this probability?

3.52 A financial analyst states that "there is a 0.75 probability that a real estate stock will rise, if the Dow Jones Industrial average (DOW) rises." The probability that the DOW will rise is given as 0.25. What is the probability that both the DOW and the real estate stock will rise together?

3.53 The probability that an employee smokes is 0.30 and the probability that an employee misses more than five days a year due to illness is 0.45. Further, the probability that an employee misses more than five days given that the employee smokes is 0.8. What is the probability that a randomly chosen worker smokes and misses more than five days a year due to illness?

3.54 The probability that a smoke detector will sound an alarm at the first indication of a fire is 0.9. As a safety measure, a hotel room is designed with two smoke detectors. What is the probability that one or the other will sound an alarm at the first indication of a fire? What did you have to assume about these detectors to determine this probability?

3.55 What is the probability of at least one daughter in a family with

a. two children?
b. four children?
c. six children?

3.56 In planning an advertising budget that includes the purchase of TV time during the World Series, questions that need to be answered concern the probability that the series will last four, five, six, or seven games. What are these probabilities, assuming the winning of each game is an independent event and the probability of either team winning any one game is 0.50?

Academic References

Asch, Peter, and Richard E. Quandt. *Racetrack Betting: The Professors' Guide to Strategies.* New York: Praeger, 1991.

Howard, G. "Balanced Game Theory." *Gambling Times,* January 1987, pp. 36–39.

Keynes, John Maynard. *A Treatise on Probability.* London: Macmillan, 1921.

Jeffreys, Harold. *Theory of Probability.* Oxford: Clarendon Press, Second Edition, 1948.

Von Mises, Richard. *Mathematical Theory of Probability and Statistics.* New York: Academic Press, 1964.

Morgan, J.P., N.R. Chagantry, R.C. Dahiya, and M.J. Doviak. "Let's Make a Deal: The Player's Dilemma." *American Statistician,* November 1991.

Paulson, R. "Using Lottery Games to Illustrate Statistical Concepts and Abuses." *American Statistician,* August 1992, pp. 202–204.

Savage, Leonard. *The Foundations of Statistics,* New York: John Wiley, 1954.

Schoemaker, Paul. "The Expected Utility Model: Its Variants, Purposes, Evidence and Limitations." *Journal of Economic Literature,* June, 1982.

Endnotes

1. The Indiana state lottery has six window tickets that are scratched open by contestants. Three matching windows wins the holder \$1 to \$5,000 in "instant cash." There is also a more involved and ongoing game in which 150,000 ticket purchasers out of 75 million get a chance to be on television, if they mail their tickets with the matching three windows that identify them as entrants. Of those 150,000 tickets, about 75 are planned for selection to be on the TV show. Each week, six finalist TV contestants pick unmarked doors for \$3,000 to \$30,000 prizes; the big prize winner also tries for \$1 million. "But the odds of a person winning one of the \$1 million jackpots are about 1 in 24 million." (*HT*, October 8, 1989)
2. To appreciate how a hand calculator does arithmetic with large numbers, try one with the typical eight-digit display. Enter the number 4321, which gives a calculator display of 00004321. Depressing the "x square" key gives a display that reads 18671041. Depressing the x square key again gives an error message, if the calculator does not use scientific notation. If the calculator uses scientific notation, then 3.4861 14 appears, which is scientific notation for 3.4861 times 10^{14}. Only five digits are presented accurately. when a number gets too large to be displayed, calculators that use scientific notation automatically shift to powers of ten. Statistical programs typically use floating point calculations when working with extremely large or small numbers. Floating point is similar to scientific notation except that it is based on powers of two instead of 10.

Formulas

- Probability of an outcome

$$= \frac{\text{Number of equally likely ways the outcome may occur}}{\text{Total number of equally likely ways all outcomes may occur}}$$

- Probability of an event

$$= \text{Sum of the probabilities of the outcomes in the event}$$

$$= \frac{\text{Number of equally likely ways the outcome in an event may occur}}{\text{Total number of equally likely ways all outcomes may occur}}$$

- Combination of x items selected from n items is

$$C(n,x) = \frac{n!}{x!(n-x)!}$$

- Probability of A or B

$$P(A \text{ or } B) = P(A) + P(B) - P(A \text{ and } B)$$

- Probability of the complement of A

$$P(\bar{A}) = P(\text{not } A) = 1 - P(A)$$

- Conditional probability of A given B

$$P(A|B) = \frac{P(A \text{ and } B)}{P(B)}$$

CHAPTER 4

RANDOM VARIABLES

I shall never believe that God plays dice with the world.

Albert Einstein
1879–1955

4.1 Introduction

When working with outcomes that can be denominated in dollars, or any other quantifiable unit of measurement, knowledge of the event probabilities, as presented in the previous chapter, is not sufficient. We also need to have a central or expected value and a gauge of variability around that expected value. We need a concept that can represent all the values that might occur, along with their probabilities of occurrence. The following situations describe business and economics situations in which such knowledge will prove helpful, if not financially profitable:

- An article in *Newsweek* (December 9, 1991) on "fame and filthy lucre" reported that the average take in heists in the previous year was $3,244 but that 85 percent of the robbers were caught. What is the expected return from a robbery? What magnitude of retribution would be a deterrent to would-be thieves?
- According to an article in *Business Week* (June 29, 1992), Cadillac is having trouble capitalizing on its award-winning Seville STS. "Part of the problem is that Cadillac misjudged the market. It thought 75% of Hamtramck's production would be the Seville's sister car, the two-door Eldorado (retailing at $35,500), 15% would be the basic Seville (at $37,000), and only 10% would be the STS, which retails for $38,500. Instead 58% of the output is the Seville, half of that the STS." (p. 88) What is the probability distribution of projected retail sales of these new Cadillacs? What is the value of retail sales originally expected? Why and by how much does this differ from what was realized?
- During the 1992 America's Cup yacht race in San Diego, American industrialist Bill Koch and Italian tycoon Raul Gardini had to assess how much the other was spending on the race. Each took a different strategy in trying to throw off the other. Gardini was quoted as saying "whatever it takes," while Koch was forever complaining about cost. If Gardini is believed to be spending $165 million, what is the probability that Koch is spending $100 million?
- CNN MoneyLine (July 1, 1992) featured a story on failing interest in and falling state revenues from lotteries because of too few winners and rising administrative costs. In response, a social activist proposes an alternative to the Indiana instant cash game that he claims would raise the payout and prizes without adding to the administrative cost because it would make use of the existing structure. Instead of only a fraction (one in 24 million) of nonwinners in the instant cash game being eligible to enter the bigger but more distant $1 million prize game, he proposes that each of the 75 million ticket holders have an independent chance at winning $1 million.[1] If the chance of winning a $1 million jackpot is one in 24 million and if the games are independent drawings from the same ticket pool, what is the probability that a ticket is a winner in either? What is the overall expected win for a ticket in these combined games?
- An article in *The Wall Street Journal,* "Putting Together Right Mix of Funds in Your Portfolio," stated "the idea is to own a variety of funds that tend to do well at different times, so that the overall value of your portfolio doesn't jump around too much." The question is: How does diversification reduce the risk of price fluctuation?

By the end of this chapter, you will be able to define a random variable to represent all of the chance values that are relevant to a situation. You will distinguish discrete random variables from continuous random variables. You also will be able to work with sums of random variables in economics and business situations where probability calculations are to be made. You will be able to calculate and interpret the expected value and variance of sets of discrete random variables and to relate these measures to risk assessment. You will know how to use the variance, covariance, and coefficient of variation in the assessment of the uncertainty associated with random variables.

4.2 Random Variables and Probability

Random Variable
A mapping or function that assigns one and only one numerical value to each outcome in an experiment.

When an experiment gives rise to numerically valued outcomes that occur by chance, those numerical values are said to be the realizations of a **random variable.** This and later chapters are devoted to the properties of random variables. At this stage, however, four points must be made clear:

- The notion of a random variable requires that an experiment can be conducted, with outcomes determined by chance.
- The outcomes of the experiment constitute a sample space, where the outcomes are both mutually exclusive and exhaustive.
- The outcomes are identified by unique numerical values that occur in accordance with a probability distribution.
- The numerical values are either discrete numbers or a continuous range of values along a number line.

Thus, a random variable is a rule for assigning numbers to outcomes that occur in an experiment by chance. Although experiments that give rise to nonnumerical outcomes are not associated with random variables, we can often transform qualitative outcomes into meaningful quantitative outcomes, and the experiments thus can be associated with a random variable. The rating of job candidates, for example, can be represented by the discrete numerical scale: 0 = unacceptable, 1 = satisfactory, and 2 = excellent. Pass/fail, success/failure, and yes/no outcomes and the like can be quantified by arbitrarily assigning the number one to the first option (pass, success, yes) and a zero to the other (fail, failure, no).

In the case of employment, the hiring of one type of person (say a woman) might be assigned the number 1 and the hiring of another type (a man) the number 0. Each hiring can now be viewed as a random variable resulting in either a 1 or 0. In addition, we might be interested in the total number of women hired in n hires, in which case the hiring of x women in n tries is a random variable that equals 0, 1, 2, 3, 4, . . . n, as discussed further in Chapter 5. For now, consider the small sample hiring situation depicted in Figure 3.1. Three people were to be hired by random draw, where the probability of selecting (hiring) a woman was fixed at 0.50 on each attempt. How many women might be hired in three attempts? The answer to this question can be interpreted as the outcome or realization of a random variable, denoted by the letter x.

$$x = \text{number of women hired}$$

The values of x are $x = 0, 1, 2$, and 3. The likelihood of each of these four values can be determined from the sample space in Figure 3.1, reprinted in Figure 4.1. The probability of each of the eight outcomes in the sample space is $(0.5)(0.5)(0.5) = 0.125$; thus, the probability of each of the four values of the random variable are

$$P(x = 0) = P(\text{MMM}) = 0.125$$

$$P(x = 1) = P(\text{WMM}) + P(\text{MWM}) + P(\text{MMW})$$
$$= 0.125 + 0.125 + 0.125 = 0.375$$

$$P(x = 2) = P(\text{WWM}) + P(\text{WMW}) + P(\text{MWW})$$
$$= 0.125 + 0.125 + 0.125 = 0.375$$

$$P(x = 3) = P(\text{WWW}) = 0.125$$

These probabilities make up the probability distribution of x, as shown on the right-hand side of Figure 4.1.

The sample space on the left-hand side of Figure 4.1 is written more compactly as the probability distribution on the right. The original sample space on the left is required to obtain the probabilities shown for each of the values of the random variable x on the right. Once the probability distribution $P(x)$ is obtained, however, the original sample space of male and female outcomes can be disregarded because the probability distribution contains all the information required to answer probability questions about the values of x. In Figure 4.1, the random variable x is a rule that maps each outcome in the original sample space into a numerical value x.

Discrete Random Variable
A random variable that has only a finite number of specific and distinct values.

In this chapter and the next, random variables that give rise to discrete values (as in $x = 0, 1, 2, 3$) are emphasized. Such a random variable is called a **discrete random variable,** because it can take only distinct values along a number line. In contrast to a discrete random variable is a continuous random variable, which is one that can have any value along a number line. For example, the continuous random variable x might

FIGURE 4.1 Sample Space and Probability Distribution

Outcome	Probability
MMM	0.125
WWM	0.125
WMW	0.125
WMM	0.125
MWW	0.125
MWM	0.125
MMW	0.125
WWW	0.125

Probability Distribution

x	$P(x)$
0	0.125
1	0.375
2	0.375
3	0.125

be the height of students in a dorm and its realization x might be any value between four feet and eight feet. Chapter 6 has a detailed discussion of continuous random variables.

In the case of horse racing, slot machines, lotteries, and similar experiments where the names of the winners are the basic outcomes, a discrete random variable can be defined as the dollar value of the winning prize. Remember, in these games of chance, it is the dollar value that defines the random variable and not the name of the winner. Random variables are always defined in terms of their numerical values and the probability of the occurrence of those values.

4.3 Probability Mass

In Chapter 1, relative frequency distributions were presented in tabular lists and in histograms. In a like manner, tables and diagrams can be used to present the probability associated with given values of a random variable. These tables and diagrams are called "probability distributions." When drawn as in a histogram, the vertical height at each discrete random variable value is the probability of that value.

In the case of Cadillac production at the Detroit-Hamtramck plant, according to *Business Week,* GM executives planned on 75 percent of production to be Eldorados (retailing at \$35,500), 15 percent basic Sevilles (at \$37,000), and 10 percent Seville STSs (at \$38,500). The probability distribution for the random variable "retail price of Hamtramck-produced Cadillac" is shown in Figure 4.2, where each vertical bar is one dollar wide. The vertical bar at \$35,500, for instance, has a height of 0.75 and an implied width from \$35,499.50 to \$35,500.50. The mass of this bar is its height (0.75) times the one unit horizontal distance, which is just the (long-term) relative frequency 0.75 given by the height. Because this bar actually represents an area, even though we need only know its height, it is called the **probability mass** at \$35,500. The probability mass is 0.15 at \$37,000 and 0.10 at \$38,500. Although a diagram such as that in Figure 4.2 is commonly called a "probability function," it is more appropriately referred to as a "probability mass function."

Probability mass function (p.m.f.)
A tabular list, a graph, or a formula that gives the probability for each possible value of a discrete random variable, where the sum of probability masses is one.

An alternative representation of probability mass is shown as a list in Table 4.1 for the instant cash lottery game introduced in Chapter 3. These probabilities were formed by the first and last columns of Table 3.2. The probability of a winning \$1 instant cash ticket is written

$$P(x = 1) = P(1) = 0.08,$$

of a \$2 ticket is

$$P(x = 2) = P(2) = 0.032,$$

and so on.

The probabilities in this lottery example are not easily represented in a diagram because of the wide range in x values from \$0 to \$5,000. A diagram is a helpful way to visualize probabilities only if there is not a lot of dispersion in the random variable.

FIGURE 4.2 Probability Mass Function of Retail Prices of Hamtramck-Produced Cadillacs

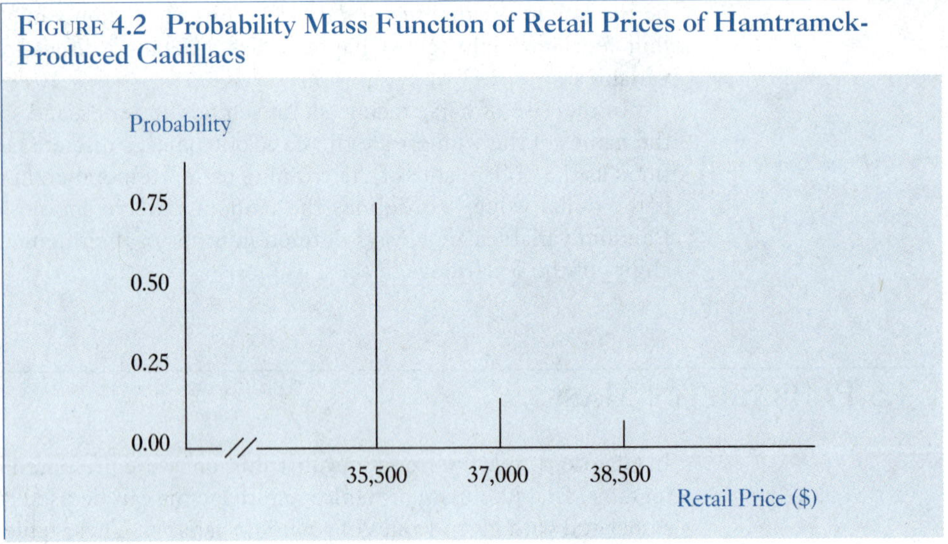

TABLE 4.1 Probability Mass Function of Instant Cash Lottery

Prize won (x)	Probability of winning $P(x)$
$ 0	0.862461
$ 1	0.080000
$ 2	0.032000
$ 5	0.015399
$ 10	0.007400
$ 25	0.002600
$ 500	0.000130
$5,000	0.000010
	1.000000

4.4 The Expected Value of a Random Variable

Calculations of probabilities associated with a random variable say nothing about the cost and benefit of making an investment, placing a bet, or making a decision in any other uncertain situation. One possible question is: How much money can we expect to make or lose as the result of an action?

In Chapter 2, the mean was calculated to describe the central location of a set of numbers. Calculations were made on the relative frequency of the observed values. A random variable represents a set of numerical values whose occurrences are determined by probability. These probabilities can be used as long-term relative frequencies in the calculation of a mean. The mean of a random variable is called its **expected value.** The expected value will provide an answer to questions about how much can be won or lost in uncertain investments and games of chance.

Expected Value of x, $E(x)$
Mean (μ) of the random variable x.

As with any population mean μ, the expected value of a random variable x is calculated by multiplying each value x by its (long-term) relative frequency, and then these products are summed, $\Sigma xP(x)$. Unlike any other weighted mean calculations in Chapter 2, however, the use of the expected value designation indicates that the relevant relative frequency is the probability of x, which is $P(x)$. Thus, we have Formula 4.1.

Expected value of x

$$E(x) = \mu = \Sigma_x xP(x) \tag{4.1}$$

where $P(x)$ is the probability of an x value,

$\sum_x xP(x)$ is the sum of x and $P(x)$ products, for all the values of x,

$E(x)$ is the expected value of x, and

μ is the mean of x.

Formula (4.1) is similar to Formula (2.2a) for a mean calculated with relative frequencies, except that here each value of x is weighted by its probability. Formula (4.1) provides a way of computing the mean of x based on the long-term relative frequencies. Notice that the summation is over all the values of x and their respective probabilities, which is indicated by the x under the sigma summation sign. As an example, consider Table 4.2; the first column again contains the x values for the instant cash lottery, and the second gives their probabilities. The third column is formed by the products $xP(x)$. The sum of the eight entries in the third column is the expected value of x,

$$E(x) = \sum_x xP(x) = 0.475$$

TABLE 4.2
Calculation of the Expected Value in the Instant Cash Lottery

Prize won (x)	Probability of winning $P(x)$	$xP(x)$
0	0.862461	0.000
1	0.080000	0.080
2	0.032000	0.064
5	0.015399	0.077
10	0.007400	0.074
25	0.002600	0.065
500	0.000130	0.065
5000	0.000010	0.050
		$0.475 = \Sigma xP(x) = E(x) = \mu$

The expression $E(x) = 0.475$ says that, prior to knowing if a ticket is a winner or loser, each ticket has an expected win of $0.475 in the instant cash lottery game.[2] That is, for a large number of ticket purchases, the mean amount received per ticket will be 47.5 cents in this game. No contestant will actually receive 47.5 cents from any one ticket, since this value isn't even possible; 47.5 cents is an average value.

A contestant who could buy all 75 million tickets in one administration of the game would get an average return of 47.5 cents per ticket. Alternatively, if he or she bought one ticket every time the game was run (every six to eight weeks), then he or she could expect to receive 47.5 cents per ticket over a long (infinite) period of time. No matter which interpretation is used, the average return to a contestant is 47.5 cents per ticket in this instant cash lottery game.[3]

QUERY 4.1:

Newsweek (December 9, 1991) reported that the average take in bank robberies was $3,244 but that 85 percent of the robbers were caught. What is the expected return from a bank robbery? What magnitude of retribution would be a deterrent to would-be thieves?

ANSWER:

Assuming that robbers who are caught lose their take, the expected value of a robbery is $486.60.

$$E(x) = \mu = \Sigma x P(x) = (.15)(3244) + (.85)(0) = 486.6$$

To serve as a deterrent, any punishment would have to be viewed by the criminal as imposing a cost on him or her that exceeds $486.60, assuming he or she makes the decision only on expected values.

QUERY 4.2:

In the case of Cadillac production at the Detroit–Hamtramck plant, according to *Business Week* (June 29, 1992), GM executives planned on 75 percent of production to be Eldorados (retailing at $35,500), 15 percent basic Sevilles ($37,000), and 10 percent Seville STSs ($38,500). As it turned out, 58 percent of the output was in the Seville, 29 percent in the STS, and 13 percent in the Eldorado. Why did the expected and actual percentages differ, and what are the consequences in terms of expected and actual return per car?

ANSWER:

Initially the executives could only be working with subjective probability, since they had no experience with this new series of Cadillacs. However even if they had a more objective notion of probability based on long-term relative frequencies, there is no reason why the initial year of production should match those long-term values. The expected value of the initial retail returns (x), based on probabilities $P(x) = 0.75$, 0.15, and 0.10, from the Cadillac production is $36,025 per car. The actual retail returns (y), with realized relative frequencies $f/n = 0.13$, 0.58, and 0.29, was $37,240 per car. Thus, on the average, Cadillac retail sales are $1,215 per car more than what was initially expected.

x	$P(x)$	$xP(x)$
35500	0.75	26625
37000	0.15	5550
38500	0.10	3850
		$\mu_x = 36025$

y	f/n	$y(f/n)$
35500	0.13	4615
37000	0.58	21460
38500	0.29	11165
		$\mu_y = 37240$

Exercises

4.1 What is the difference between the information in the sample space and outcome set and that in a random variable?

4.2 *New York Times* syndicated columnist Marilyn Geewax, in an article titled "Financial plight of aging Catholic nuns should be lesson for all women"(*Saint Paul Pioneer Press,* June 3, 1993), stated that the median age of a U.S. nun is 65 years. What is the approximate probability of selecting a nun at random who is 65 years or older?

4.3 In Exercise 4.2, if three nuns are selected at random, what is the implied probability mass function of the number of nuns 65 years or older? Use a tree diagram to show the determination of the relevant probabilities.

4.4 How does the notion of an expected value differ from that of an arithmetic average calculated with relative frequency data?

4.5 A *Business Week* (March 12, 1984) article described a violent crime insurance policy that costs $210 and pays $50,000 to a victim. If police records in your community suggest a one in a 1,000 chance of being victimized, what is your expected return in buying this policy? What did you assume in your calculations?

4.6 Would you buy the violent crime insurance described in the above question? Why or why not?

4.7 In the summer of 1992, the author of this text received an "Application for Service Protection Plus" from General Electric for insurance on a television set to cover the cost of major service work. The television set cost about $500, and the service insurance was being offered at "the special 20% discount." The cost was listed at $119.95 for one year's coverage, less $23.99 for a discounted one-year contract price of $95.96. Ignoring the use that is obtained from the TV set prior to a claim, what would have to be the probability of total destruction to break even on this policy?

4.8 Why would someone buy the General Electric service insurance policy described in Exercise 4.7?

4.9 An open-air theater promoter is considering taking out an insurance policy to cover possible losses if it rains. If the event in question is a complete failure because of rain, the promoter feels that a loss of $80,000 will be incurred. If it is moderately successful, with a little drizzle occurring, a loss of $25,000 would be incurred. Insurance actuaries have determined that the probabilities of total failure or moderate success are 0.01 and 0.05, respectively. Assuming the promoter would be willing to ignore all other possible losses, what premium should he be willing to pay the insurance company for the policy?

4.10 The local Schwinn dealer has three models of men's 27-inch bicycles: a $220 model, a $240 model, and a $260 model. The Schwinn dealer makes a profit of $35 on the first model, $45 on the second model, and $60 on the third model. For the coming year, the Schwinn dealer estimates that for customers who buy the 27-inch size, 60 percent will want the lowest-priced model, 30 percent will purchase the $240 model, and 10 percent will buy the $260 model. Thus, the probability of each profit is:

$$P(\$35\text{ profit}) = 0.60,\ P(\$45\text{ profit}) = 0.30,\text{ and } P(\$60\text{ profit}) = 0.10$$

a. Define the random variable for this situation.
b. Find the expected value of this random variable.
c. Find the variance of this random variable.
d. What are the expected profits for selling 100 of the 27-inch bikes?

4.5 The Dispersion of a Random Variable

Variance of a Random Variable The mean squared deviation of outcome values from the mean of the random variable, where each squared deviation is weighted by its probability.

Just as the expected value of a random variable is a long-term average value, measures of dispersion around this mean can be defined in probability terms. The **variance of a random variable** is a measure of the spread of the distribution. The variance of a random variable is the mean squared deviation of values from their mean. Unlike the sample variance discussed in Chapter 2, however, now probability is used for the weighing of those squared deviations. The variance of a random variable x is thus the expected squared deviation of the values of x around the expected value of x. In shorthand notation,

$$\text{Var}(x) = \sigma^2 = E[x - E(x)]^2 = E(x - \mu)^2$$

As with any finite population variance σ^2, the variance of a random variable x is calculated by multiplying each squared deviation of values of x from their mean by the associated (long-term) relative frequency $P(x)$, and then summing these products, $\Sigma(x - \mu)^2 P(x)$. Thus, we have the formula.[4]

Variance of random variable x

$$\sigma^2 = E\left[x - E(x)\right]^2 = E(x - \mu)^2 = \sum_x (x - \mu)^2 P(x) \qquad (4.2)$$

where $P(x)$ is the probability of an x value,

$\sum_x (x - \mu)^2 P(x)$ is the sum of products, of squared deviations multiplied by respective probabilities, for all the values of x,

$E(x)$ is the expected value of x, which is also the mean of x.

Formula (4.2) is similar to Formula (2.7a), with the exception that probability or long-term relative frequency is used for weighting the squared deviations. A variance calculated by Formula (4.2) starts with the mean of x, as calculated in Table 4.2 and re-

TABLE 4.3 Calculation of the Variance in the Instant Cash Lottery

x	$P(x)$	$xP(x)$	$x - \mu$	$(x - \mu)^2$	$(x - \mu)^2 P(x)$
0	0.862461	0.000	−0.475	0.2256	0.1945874
1	0.080000	0.080	0.525	0.2756	0.0220506
2	0.032000	0.064	1.525	2.3256	0.0744207
5	0.015399	0.077	4.525	20.4757	0.3153016
10	0.007400	0.074	9.525	90.7258	0.6713464
25	0.002600	0.065	24.525	601.4760	1.5638187
500	0.000130	0.065	499.525	249525.2323	32.4395778
5000	0.000010	0.050	4999.525	24995250.2926	249.9525029
	1.000000	0.475			285.2336060
		= μ			= $\sum_x (x - \mu)^2 P(x)$

peated in Table 4.3, followed by the calculation of deviations from this mean, as shown in the fourth column of Table 4.3. Next, the deviations from the mean are squared, as in column five, and multiplied by the probability of getting this squared deviation, column six. Finally, the probability weighted squared deviations are added, with the resulting variance of $\sigma^2 = 285.233606$.

Standard Deviation of a Random Variable x
The standard deviation of random variable x is the positive square root of the variance.

As with the data sets in Chapter 2, the **standard deviation of a random variable x** is the positive square root of the variance. So for the instant cash lottery, the standard deviation is $16.89.

$$\sigma = \sqrt{\sum_x (x - \mu)^2 P(x)} = \sqrt{285.233606} = 16.89$$

Standard deviation of random variable x

$$\sigma = \sqrt{E[x - E(x)]^2} = \sqrt{E(x - \mu)^2} = \sqrt{\sum_x (x - \mu)^2 P(x)} \qquad (4.3)$$

If there was no variability in x (it was not a random variable and had only one value), then the variance and the standard deviation would be zero. On the other hand, if x can take values far away from its mean with great likelihood, then the variance and standard deviation will be large because both factors of the product $(x - \mu)^2 P(x)$ will then be large.

QUERY 4.3:

In the case of Cadillac production at the Detroit-Hamtramck plant, where GM executives planned on 75 percent of production to be Eldorados (retailing at $35,500), 15 percent basic Sevilles ($37,000), and 10 percent Seville STSs ($38,500), but realized 58 percent of the output in the basic Seville, 29 percent in the Seville STS, and 13 percent in the Eldorado, what is the difference in variability in actual retail prices from what was expected initially?

ANSWER:

The expected value of the initial retail returns (x) from the Cadillac production is \$36,025 per car. The actual retail returns (y) average \$37,240 per car. As a measure of variability around these means, the initial standard deviation based on probabilities is \$980.75, and the standard deviation based on realized relative frequencies is \$1,537.41. Thus, there is less variability in actual sales than what was expected initially.

x	$P(x)$	$xP(x)$	$(x-\mu)^2$	$(x-\mu)^2P(x)$	
35500	0.75	26625	275625	206718.75	
37000	0.15	5550	950625	142593.75	
38500	0.10	3850	6125625	612562.50	
		$\mu_x = 36025$		961875.00	$980.75 = \sigma_x$

y	f/N	$y(f/N)$	$(x-\mu)^2$	$(x-\mu)^2(f/N)$	
35500	0.13	4615	3027600	393588	
37000	0.58	21460	57600	33408	
38500	0.29	11165	1587600	46404	
		$\mu_y = 37240$		887400	$942.02 = \sigma_y$

4.6 RISK ASSESSMENT

The standard deviation is often used in finance as a measure of the riskiness of an investment. For instance, if two investments each have the same expected monetary return, but the first has a standard deviation smaller than the second, then the first is less risky than the second. To see the implications of these two different standard deviations and their ramifications for risk, we consider a bell-shaped distribution of returns.

Standard Deviation Comparison

From the rule of bell-shaped distributions introduced in Chapter 2, if monetary returns are bell shaped, then approximately 68 percent of the time dollar returns will be within one standard deviation (either direction) of the expected dollar return; about 95 percent of the time those returns will be within two standard deviations (either side) of the expected return. Thus, for two investments, if the standard deviation of the second is twice the magnitude of the first, then within the same dollar bounds the second (with the larger standard deviation) has about 68 percent of its returns and the first (with the smaller standard deviation) will have about 95 percent of its returns.

As an illustration, consider two corporate stocks that both have an expected net return (dividend and price appreciation) of \$150 on a \$1,000 investment. From historical records, suppose that the standard deviation of the net return of the first stock is calculated to be \$20, and the standard deviation of the second is \$40. Assume returns are bell shaped, as in Figure 4.3. From the rule of normality, 95 percent of the time the first stock will have a net return between \$110 and \$190, because for this stock the distance from \$150 to \$110 or from \$150 to \$190 is two standard deviations below and two standard deviations above the mean. Because the distance from \$150 to \$110 or from \$150

FIGURE 4.3 Probability Distributions of Net Returns to Two Stocks With Different Standard Deviations

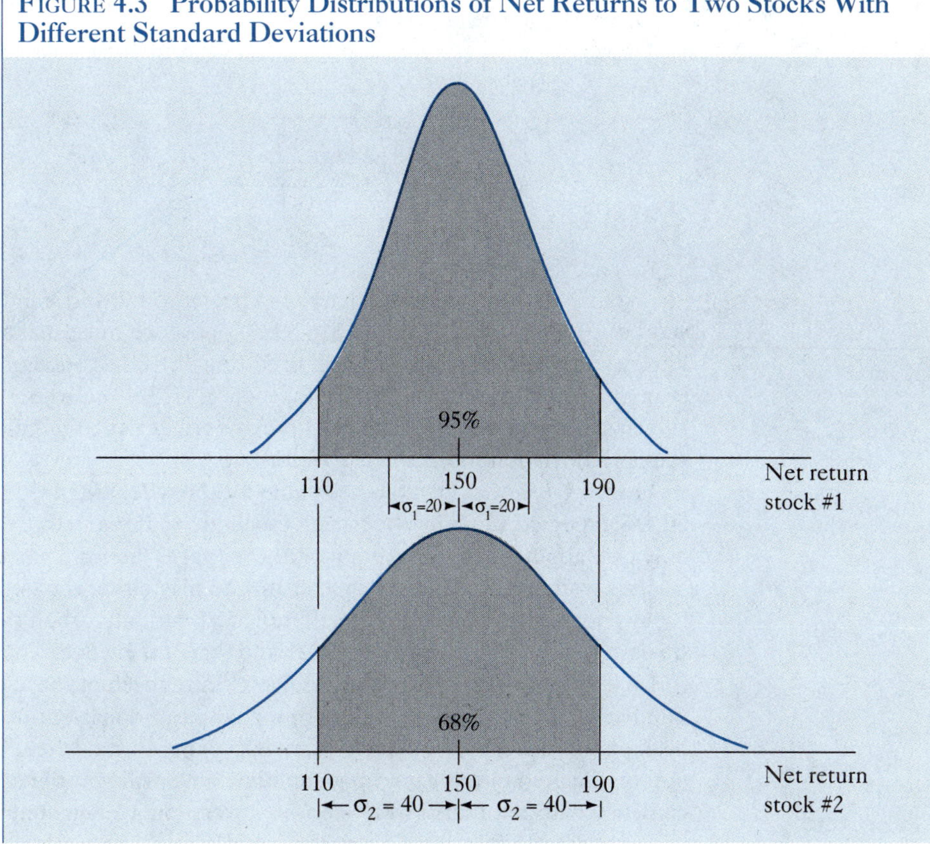

to $190 represents only one standard deviation below and one standard deviation above the mean in the second stock, only about 68 percent of the net return for this stock will be within the $110 and $190 interval. Clearly, for stocks expected to have similar net returns, the more risky stock will have the larger standard deviation.

Coefficient of Variation Comparison

Coefficient of Variation
Ratio of the standard deviation of a random variable to its mean.

To compare riskiness among distributions that have different means and standard deviations, a measure often employed is the coefficient of variation. The **coefficient of variation** is calculated by dividing the standard deviation by the expected value. Because the standard deviation typically gets larger as the magnitude of values involved in its calculation get larger, an unadjusted comparison of standard deviations between stocks with greatly different expected returns is not a meaningful way to assess risk. Dividing standard deviations by their respective means results in the cancellation of the units of measurement and removes the influence of the size of the variable's values. The coefficient of variation is a measure of dispersion free of the measurement unit of the original values.

Coefficient of variation

The coefficient of variation is a ratio of the standard deviation to the expected value or mean of a variable.

$$C = \frac{\sqrt{E\left[x - E(x)\right]^2}}{E(x)} = \frac{\sqrt{E(x-\mu)^2}}{\mu} = \frac{\sigma}{\mu} \tag{4.4}$$

From Table 4.3, we know that the expected value and standard deviation of the instant cash game are \$0.475 and \$16.89. Suppose we must make a choice between this lottery and a dollar bet on a "wheel of fortune" that has 360 numbers on which a clicker might rest after the wheel is spun; it pays \$325 if the wheel stops on your number. Which should we choose? Table 4.4 provides the calculation of the expected value and standard deviation for this wheel of fortune.

Table 4.4 shows the expected value for the wheel to be \$0.903, which exceeds the \$0.475 expected value for the instant cash lottery. If we are not concerned with risk and only care about what will happen on the average, then the wheel is preferred.

It is worth recognizing that the chance to play either the lottery or the wheel of fortune is not costless. If each were offered at \$1 per entry, then the expected net win for the lottery would be a negative \$0.525 and that of the wheel would be a negative \$0.097 (= 1 – 0.903). That is, if we paid a dollar to enter each of these games a large (infinite) number of times, then each time we would pay a dollar but on average we would get back only 47.5 cents in the lottery and 90.3 cents on the wheel. Most times we will get nothing back in either game, but sometimes we will win the amounts specified in accordance with the odds given—for the lottery, on average, one in 1 million times you will even win \$5,000, but on average you will lose 52.5 cents per entry.

If we are concerned with the risk of winning or losing, then we must consider the dispersion around the expected value. The standard deviation of the wheel (\$17.11) exceeds that of the lottery (\$16.89), suggesting that the wheel of fortune is more risky. But after deflating the standard deviations by the magnitudes of the expected value involved, the lottery continues to show itself as more risky. The coefficient of variation for the lottery is

$$C = \sigma/\mu = \$16.89/\$0.475 = 35.56$$

The coefficient of variation for the wheel is only

$$C = \sigma/\mu = \$17.11/\$0.903 = 18.95$$

TABLE 4.4 Mean, Variance, and Standard Deviation for a 360-Number Wheel of Fortune that Pays \$325 on One Number

x	$P(x)$	$xP(x)$	$x-\mu$	$(x-\mu)^2$	$(x-\mu)^2P(x)$
0	0.997222	0.000	–0.903	0.8150	0.8127438
325	0.002778	0.903	324.097	105039.0095	291.7750263
	1.000000	0.903 = μ			292.5877701 = σ^2
					σ = 17.10519717

Further consideration of risk and investors' behavior toward risk are provided in later chapters. The analysis of risk has become a critical part of government regulations. For instance, U.S. Presidential Executive Order 12291 (*Federal Register*, 1981, pp. 13193–98) states that all federal regulatory action must be accompanied by a cost/benefit analysis that includes risk assessments. Modern-day financial analysis is founded on cost/benefit risk analysis. Liability considerations in product development require risk analysis. The pricing of insurance policies is based on expected payout and risk of disaster. In short, all decision making should incorporate an assessment of expected gains and losses and the risks involved.

EXERCISES

4.11 The following investments are available at a price of $10,000 each:

Investment:	x	y
Expected net returns:	$ 1,000	$ 900
Standard deviation	$ 1,000	$ 450

a. What are the coefficients of variation for the returns of x and y?
b. If the distributions of returns of both x and y are continuous and bell shaped, what is the probability that x is less than zero and what is the probability that y is less than zero?
c. How do your answers in parts a and b relate?

4.12 Graph the probability mass function associated with the random variable x:

x	0	1	2	3	4	5	6
P(x)	.02	.09	.23	.32	.23	.09	.02

a. What is the expected value of x?
b. What is the standard deviation of x?
c. What proportion of the distribution of x is within two standard deviations of the expected value of x?
d. Is your answer in part c consistent with the "rule of thumb" for a bell-shaped distribution?

4.13 These three investments are available at a price of $12,000 each.

Investments:	*A*	*B*	*C*
Expected net returns	$ 1,000	$ 1,500	$ 2,000
Standard deviation	$ 500	$ 700	$ 1,150

Which investment has the least risk? By what measure?

4.14 If you ignore risk, which one of investments *A*, *B*, and *C* is the preferred purchase in Exercise 4.13? Why?

4.15 The following investments are a vailable at a price of $100,000 each:

Investment:	*A*	*B* (in thousands of dollars)
Expected net returns	10	15
Standard deviation	3	4

Which investment has the more risk? By what measure?

4.16 If an investor ignores risk, and will purchase only one of the investments in Exercise 4.15 but not both, which one should he or she buy? Why?

4.17 One of the following investments is available to you at a price of $10,000 each:

Investment:	x	y
Expected net returns	$ 2,000	$ 1,500
Standard deviation	$ 1,000	$ 750

If you ignore risk, which investment would you make? Which of the above investments of $10,000 in x or y is the least risky? By what measure?

4.7 Multiple Random Variables

In the instant cash game discussed in the previous chapter, the random variable was the immediate redemption value of a ticket (No instant cash, $1, $2, $5, $10, $25, $500, and $5,000). Experiments may involve more than one variable, however. For example, a variation of the Indiana state lottery, introduced at the beginning of this chapter, has two random variables associated with each ticket: one for the instant cash game (which we analyzed) and another for the millionaire game (which we will consider in the next section). An insurance company interested in the age, gender, education level, and marital status of a driver with a propensity to be involved in accidents provides another example of a multivariable probability situation in which accidents are divided into four random variables for age, gender, education, and marital status.

The remainder of this chapter focuses on questions related to situations involving multiple random variables. First, the key role of independence is shown in the calculation of the probability of joint occurrence of two random variables. Next, attention is given to expected values and variances of multiple random variables. Consideration is given to the role of the variance in risk assessment. The importance of the covariances is demonstrated in investment decisions.

4.8 Joint Probability Distributions

Analysis of multiple random variables often requires the construction of joint probability distributions that show the probabilities of combined sets of the variables. As a special case, consider the construct of a bivariate probability distribution for analysis of two discrete random variables. A table for a bivariate probability distribution contains the probabilities of all combined outcomes of the two random variables. For instance, what we know about the two random variables in the proposed millionaire and instant cash lotteries is represented in the bivariate probability distribution in Table 4.5.

The probabilities in the instant cash game, from Table 3.2 and Table 4.1, are shown down the right-hand margin, $P(x)$. The chances of winning (one in 24 million) and losing in the proposed millionaire game are given across the bottom margin of the table, where

$$P(y = 0) = 0.99999995833 \text{ and } P(y = 1000000) = 0.00000004167$$

TABLE 4.5
Incomplete Bivariate Probability Distribution for Instant Cash and Millionaire Games

		Proposed Millionaire Game (y): no million $0	win million $1,000,000		
Instant Cash Game (x)	$ 0			0.86246146270	Probability of winning in the Instant Cash Game P (x)
	$ 1			0.08000000000	
	$ 2			0.03200000000	
	$ 5			0.01539882969	
	$ 10			0.00739973361	
	$ 25			0.00259996880	
	$ 500			0.00013000520	
	$5,000			0.00001000000	
		0.99999995833	0.00000004167	1.00000000000	
		Probability of winning in this Millionaire Game P (y)			

Marginal Probability
The sum of joint probabilities over all values of one variable in a multivariate probability distribution.

In keeping with their placement in this table, $P(x)$ and $P(y)$ are called **marginal probabilities;** as in Chapter 3, they represent probabilities of the occurrence of values for one random variable with no regard to the other variable. Notice that each marginal probability is between zero and one and that the marginal probabilities for each random variable sum to one.

Missing in Table 4.5 are the joint probabilities that reflect the likelihood that the values of one variable will occur with the values of the other variable. For instance, the first empty cell in Table 4.5 is the probability that a ticket is worth nothing in the instant cash game *and* worth nothing in the millionaire game. It must be a probability between the row marginal probability 0.86246146270 and the column marginal probability 0.99999995833. The last empty cell in the lower right-hand corner is the probability that a ticket is worth $5,000 in the instant cash game *and* $1 million in the millionaire game. It must be a probability between 0.000010 and 0.00000004167. All 16 cells in the table will be exhaustive, mutually exclusive, and have probabilities that sum to unity.

Before attempting to determine these joint probabilities, we need to address how the two variables are related. The social activist's proposals for the millionaire game as given at the beginning of this chapter call for a contestant's winning or losing in the instant cash game to not affect his or her chances of winning or losing in the millionaire game. Winning tickets in the instant cash game are chosen randomly, and winners in the millionaire game are selected randomly from the 75 million tickets: outcomes in the two games are independent. As in the previous chapter, Section 3.8, independence means that if a contestant did not win in the instant cash game, the probability of winning the million dollars is not influenced; or, a contestant winning $1 instant cash does not alter the probability of winning the million; or, given that a contestant won $5,000 does not change the probability of winning the million, and similarly for the other instant cash winnings. This critical assumption of independence between the two games is summarized in the following statements of conditional probabilities.

$$P(y = 1000000) = P(y = 1000000 \mid x) = 0.00000004167$$

which says the probability of winning the million is equal to the probability of winning the million given x in the instant cash game.

$$P(y = 0) = P(y = 0 \mid x) = 0.99999995833$$

which says the probability of not winning the million is equal to the probability of not winning the million given x in the instant cash game.

where x =\$0, \$1, \$2, \$5, \$10, \$25, \$500, and \$5,000

In this enumeration $P(y = 0) = 0.99999995833$ and $P(y = 1000000) = 0.00000004167$ are the two probability masses that make up the two marginal probability distributions $P(y = 0)$ and $P(y = 1000000)$. There are eight conditional random variables of the form "$y \mid x$"—one for each of the eight values of x in the instant cash game. For each of those values of x, y is either 0 or 1000000, with probabilities determined by the conditional probabilities $P(y = 0 \mid x)$ or $P(y = 1000000 \mid x)$. But because of independence, these conditional probabilities are equal to their marginal probabilities, regardless of the value of x.

Independent Random Variables Random variables y and x are independent if the conditional probability distribution of y, given any value of x, is the same as the marginal probability distribution of y, and if the conditional probability distribution of x given any value of y is the same as the marginal probability distribution of x.

If all occurrences of the two random variables x and y are **independent**, then the two random variables are said to be independent. Two random variables are independent if and only if the marginal and conditional probability distributions are identical, as is the case above.

If two variables are independent, then the joint probability of any pair of occurrences from the two variables is the product of the two marginal probabilities. The reason for this can be seen by considering the breakdown of the 75 million tickets as given in Table 4.1. There are 64,684,610 tickets that yield nothing in the instant cash game, but one in 24 million of these will be a winner in the proposed millionaire game. That is, the number of tickets, on the average, that will result in \$0 in the instant cash game *and* \$1 million in the millionaire game are 2.695 out of 75 million tickets:

$$\text{Frequency of } x = 0 \text{ and } y = 1000000 = (64{,}684{,}610)(1/24{,}000{,}000) = 2.695192$$

The joint probability of \$0 in the instant cash game *and* \$1 million in the millionaire game is thus 0.00000003594 = 2.695192/75000000. Or in terms of probability

$$\begin{aligned} P(x = 0 \text{ and } y = 1000000) &= (64{,}684{,}610/75{,}000{,}000)(1/24{,}000{,}000) \\ &= (0.8624614627)(0.00000004167) \\ &= P(x = 0) \times P(y = 1000000) \\ &= 0.00000003594 \end{aligned}$$

Similarly, by multiplying the marginal products for the other 15 y and x pairs in Table 4.5, the associated joint probabilities can be obtained. For instance, the joint probability of \$5 in the instant cash game *and* \$0 in the proposed millionaire game is 0.01539882905:

$$
\begin{aligned}
P(x = 5 \text{ and } y = 0) &= P(x = 5) \times P(y = 0) \\
&= (0.01539882969)(0.99999995833) \\
&= 0.01539882905
\end{aligned}
$$

as shown in Table 4.6.

Notice in Table 4.6 that the joint probability of winning \$5,000 in the instant cash game *and* \$1 million in the millionaire game is given as zero. This probability is not zero; it is just smaller than the decimal places shown. It can be expressed, however, in scientific notation as 4.167×10^{-13}, which is 4.167 in 10 trillion:

$$
\begin{aligned}
P(x = 5000 \text{ and } y = 1000000) &= P(x = 5000) \times P(y = 1000000) \\
&= (0.000010)(0.00000004167) \\
&= 0.0000000000004167 = 4.167 \times 10^{-13}
\end{aligned}
$$

Marginal probabilities can be multiplied together to determine joint probabilities only if the variables are independent. If the variables are dependent, then from a rewriting of Formula (3.5), we know only that a joint probability can be expressed as the product of a conditional probability times the associated marginal probability. That is, if A and B are two random variables for which independence has not been established, then the joint probability of a specific value of A, $\boldsymbol{a}$, and a specific value of B, $\boldsymbol{b}$, is the probability of $A = \boldsymbol{a}$ given $B = \boldsymbol{b}$ times the probability of $\boldsymbol{b}$.[5]

TABLE 4.6 Bivariate Probability Distribution for Instant Cash and Proposed Millionaire Games

		Proposed Millionaire Game *(y)*: no million \$0	Proposed Millionaire Game *(y)*: win million \$1,000,000	Probability of winning in the Instant Cash Game *P (x)*
Instant Cash Game *(x)*	\$ 0	0.86246142676	0.00000003594	0.86246146270
	\$ 1	0.07999999667	0.00000000333	0.08000000000
	\$ 2	0.03199999867	0.00000000133	0.03200000000
	\$ 5	0.01539882905	0.00000000064	0.01539882969
	\$ 10	0.00739973330	0.00000000031	0.00739973361
	\$ 25	0.00259996869	0.00000000011	0.00259996880
	\$ 500	0.00013000519	0.00000000001	0.00013000520
	\$5,000	0.00001000000	0.00000000000	0.00001000000
Probability of winning in this Millionaire Game *P (y)*		0.99999995833	0.00000004167	1.00000000000

If A and B are two random variables, which may take values a and b, then the joint probability of a and b is

$$P(A{=}a \text{ and } B{=}b) = P(A{=}a \mid B{=}b) \times P(B{=}b) \tag{4.5}$$

If, however, A and B are independent, then and only then does

$$P(A{=}a \mid B{=}b) = P(A{=}a)$$

and the probability of $A{=}a$ and $B{=}b$ can be calculated as the product of the two marginal probabilities.[6]

Probability of $A = a$ and $B = b$, if the random variables A and B are independent, is written

$$P(A{=}a \text{ and } B{=}b) = P(A{=}a) \times P(B{=}b) \tag{4.6}$$

QUERY 4.4:
Below is a bivariate probability distribution for the amounts of money America's Cup competitors Bill Koch and Raul Gardini were alleged to be spending. These probabilities were obtained from a cursory review of what TV commentators, journalists, and yachting experts were saying the two were spending. They are thus subjective probabilities. What does treating these probabilities as objective imply about the independence of spending? If Gardini is believed to be spending \$165 million, what is the probability that Koch is spending \$100 million?

Probabilities of Expenditures by the Challenger and Defender in 1992 America's Cup

Bill Koch's spending on the defense	Raul Gardini's spending on the challenge				
	\$105,000,000	\$135,000,000	\$165,000,000	\$195,000,000	
\$ 50,000,000	0.00	0.01	0.01	0.00	0.02
\$ 75,000,000	0.01	0.19	0.20	0.09	0.49
\$100,000,000	0.00	0.17	0.24	0.08	0.49
	0.01	0.37	0.45	0.17	1.00

ANSWER:
If Bill's spending of \$50 million and Raul's spending of \$105 million were independent, then

$$\text{P(Bill spends 50)} \times \text{P(Raul spends 105)} = (.02)(.01) = 0.0002$$

which would have to equal

$$\text{P(Bill spends 50 and Raul spends 105 million dollars),}$$

but this is not the case; the latter probability is zero. These two variables are dependent. None of the joint probabilities are equal to the products of the respective marginal probabilities. It only takes one inequality to show dependence, while all of the joint probabilities must equal the products of their respective marginal probabilities for independence. Finally, from a rewriting of Formula (4.5), if Raul Gardini is believed to be spending $165 million (***b***), the probability that Bill Koch is spending $100 million (***a***) is 0.533.

$$P(A = \text{a} \mid B = \text{b}) = \frac{P(A = \text{a and } B = \text{b})}{P(B = \text{b})} = \frac{0.24}{0.45} = 0.5333$$

QUERY 4.5:

A stock analyst predicts that the stock price of a new startup pharmaceutical corporation, currently selling at $100, may rise by $30 if the corporation's diet drug is approved by the FDA, but may fall by $30 if it is not. On questioning by the interviewing TV host, the analyst stated that the probability of the $30 increase in the next six months is 60 percent, the likelihood of a price decrease is 30 percent, and the chance of no noteworthy price change is 10 percent. For comparison, the analyst also stated that $100 worth of a mature pharmaceutical corporation's stock would likely rise in value by $15 with a 40 percent chance, fall by $15 with a 10 percent chance, and not change with a 50 percent chance. From these probabilities, determine the bivariate probability distribution of the price change of the startup and the mature corporation, assuming independence. Is this assumption of independence reasonable?

ANSWER:

The bivariate probability distribution is

		Startup Corp.			
		$ –30	0	30	
	$ 15	0.12	0.04	0.24	0.40
Mature Corp.	0	0.15	0.05	0.30	0.50
	–15	0.03	0.01	0.06	0.10
		0.30	0.10	0.60	1.00

The assumption of independence is reasonable if the new startup is to be unaffected by the same economic forces that influence the mature corporation. This might be true if the new startup's patent is truly unique and revolutionary, but, in general, all stock prices in an industry tend to move together.

4.9 ADDITIVE PROBABILITY

In the consideration of a single random variable, we calculated the probabilities that an instant cash contestant wins $1, $2, $5, $10, $25, $500, or $5,000: $P(x > 0) = 0.137539$.

TABLE 4.7 Abridged Probability Table for Instant Cash and Proposed Millionaire Games

		Proposed Millionaire Game (y)			
		no million $0	win million $1,000,000		
Instant Cash Game (x)	No $s	0.86246142676	0.00000003594	0.86246146270	Probability of winning in Instant Cash Game $P(x)$
	Some $s	0.13753853157	0.00000000573	0.13753853730	
		0.99999995833	0.00000004167	1.00000000000	
		Probability of winning in this Millionaire Game $P(y)$			

We also know the probability that a contestant wins $1 million in the millionaire game: $P(y = 1000000) = 0.00000004167$. We can now ask and answer the question: What is the probability of winning in either the instant cash game or the proposed millionaire game?

Again, because there are two random variables involved, a bivariate probability table is needed, Table 4.7. As in Table 4.6, the joint probabilities in Table 4.7 were calculated as products of the marginal probabilities, because the two games represent independent random variables.

In essence, calculation of the probability of winning in either game is the same as the calculation of the probability of a union presented in the previous chapter. If ***a*** and ***b*** are values of two variables A and B, then the probability of ***a*** or ***b*** is the probability of ***a*** plus the probability of ***b*** minus the probability of ***a*** and ***b***. As in our discussion of the probability of the union of two events in the previous chapter (where the probability of the intersection of the two events is subtracted from the sum of the two event probabilities), here both marginal probabilities include the joint occurrence of the values, so one of the joint probabilities is removed by subtracting it from the sum of the two marginal probabilities.

Probability of $A = \boldsymbol{a}$ or $B = \boldsymbol{b}$, for the random variables A and B, is written

$$P(A=\boldsymbol{a} \text{ or } B=\boldsymbol{b}) = P(A=\boldsymbol{a}) + P(B=\boldsymbol{b}) - P(A=\boldsymbol{a} \text{ and } B=\boldsymbol{b}) \tag{4.7}$$

The probability of winning in either the instant cash or the millionaire game is

$$\begin{aligned} P(x > 0 \text{ or } y = 1000000) &= P(x > 0) + P(y = 1000000) - P(x > 0 \text{ and } y = 1000000) \\ &= 0.13753853730 + 0.00000004167 - 0.00000000573 \\ &= 0.13753857324 \end{aligned}$$

This says that in fewer than 14 out of 100 times will a contestant win some amount in either of these two independent lottery games. In a like manner, other additive probabilities associated with these two games can be determined.

QUERY 4.6:

Using the information on the joint probabilities of Bill Koch's and Raul Gardini's spending on the 1992 America's Cup yacht race, what is the probability that either one is spending at their respective maximum ($100 million for Koch or $195 million for Gardini)?

ANSWER:

Let Koch's spending be $\boldsymbol{a}$ and Gardini's $\boldsymbol{b}$. The probability of either spending at his maximum is 0.58.

$$P(\boldsymbol{a} \text{ or } \boldsymbol{b}) = P(\boldsymbol{a}) + P(\boldsymbol{b}) - P(\boldsymbol{a} \text{ and } \boldsymbol{b})$$

$$= 0.49 + 0.17 - 0.08 = 0.58$$

4.10 LINEAR FUNCTIONS OF RANDOM VARIABLES

Instead of being independent, two random variables might be related linearly. A random variable y is a linear function of another random variable x if

$$y = \alpha + \beta x$$

where α (alpha) and β (beta) are constants and y and x are realized in matched pairs. For instance, y might be annual profits of a car dealership and x might be number of cars sold. Ian Ayers (1991) reports on how car dealerships extract different premiums per car sold, depending on the size of the dealership, type of car, and the age, race, and sex of the customers and salespersons. Thus, the number of cars sold x is a random variable that influences dealership profits y.

If the number of cars sold in a fixed time period is a random variable with a known expected value, then the expected profit per car is likewise known and written as[7]

$$E(y) = \alpha + \beta E(x)$$

Continuing the car dealership example, suppose the fixed cost of the dealership is $220,000 per year. The gross profit from each car sold is the selling price less the variable cost associated with the car (invoice plus shipping, preparation, salesperson's commission, etc.); say this gross profit is $800. The number of cars sold through the dealership is a random variable with an expected value of 300. The expected value of this dealership's annual profit is $20,000: $y = -220000 + 800x$, so

$$E(y) = -220000 + 800(300) = 20000$$

In Chapter 6 on the normal distribution, a linear transformation of one random variable is introduced to create another normal random variable. In Chapters 12, 13, and 14, regression analysis is introduced to do much more with the linear relationships that exist among economic and business variables. For now, it is sufficient to recognize that models such as $y = \alpha + \beta x$ can be constructed and that the distribution of y is a function of that of x.

EXERCISES

4.18 For the following bivariate probability table,

	A_1	A_2	A_3
B_1	0.12	0.03	0.10
B_2	0.21	0.02	0.18
B_3	0.11	0.09	0.14

a. give the six marginal probabilities.
b. find $P(A_1 \text{ and } B_1)$.
c. determine $P(A_1 \text{ or } B_2)$.
d. calculate $P(A_1 | B_1)$.
e. are A_1 and B_2 independent or dependent?

4.19 Northwest Airlines runs a commuter service from Chicago to Minneapolis, with several flights per day. In the winter months, bad weather influences arrival time. Let $y = 1$ if the flight is on time and 0 otherwise. Let $x = 1$ if the flight is made during bad weather and 0 if not. The bivariate probability distribution for y and x is as follows:

		x (weather)	
		0 (good)	1 (bad)
y (arrival)	0 (late)	0.02	0.14
	1 (on time)	0.80	0.04

a. What are the marginal probabilities?
b. Are weather conditions and arrival times independent?

4.20 According to an article in the *WSJ* (February 24, 1993), only 0.9 percent of some 113,829,200 tax returns get audited. Only about 20 percent of those audited get a field audit, in which an agent comes to the taxpayer's home or place of business. The article states, however: "When the IRS goes looking for a problem, it usually finds one . . . .The average payment: $15,801 after a field audit, and $2,348 after an office audit." If a taxpayer is notified that he or she is going to be audited but does not yet know whether it will be a field or office audit, what is the additional payment that he or she can expect?

4.21 Using only the information in Exercise 4.20, what is the additional payment that any individual taxpayer can expect after completing his or her return if auditing is random? Why doesn't this expected value have much meaning for low-income students?

4.11 EXPECTED VALUE AND VARIANCE OF SUMS OF RANDOM VARIABLES

Random variables can be related in many different ways. Thus, multiple variable analysis is tricky but essential to business and economic problem solving. Later chapters are devoted to multivariate analysis in a linear equation (regression) setting; what is important here is to grasp how random variables interact in the calculation of measures of central tendency and dispersion in the sums and differences of the variables.

Consider Query 4.5, involving a stock analyst's predictions. The analyst states that the stock price of a new startup pharmaceutical corporation, currently selling at $100, may rise by \$30 if the corporation's diet drug is approved by the FDA, but may fall by \$30 if it is not. The analyst also states that the probability of the \$30 increase is 60 percent; the likelihood of a price decrease is 30 percent, and the chance of no noteworthy price change is 10 percent. On the other hand, the analyst states that \$100 worth of a mature pharmaceutical corporation's stock would likely rise in value by \$15 with a 40 percent chance, fall by \$15 with a 10 percent chance, and not change with a 50 percent chance. The query involved the joint probability distribution of price changes for the two stocks. Now we ask about the expected value of the price change in these two stocks.

We already know how to calculate the expected values of the net returns to the two stocks (\$9 for the startup and \$4.50 for the mature pharmaceutical corporation) and the standard deviations (\$27 and \$9.605).

$$\text{Startup}: \quad \mu_s = E(s) = \sum sP(s) = 9.00$$

$$\sigma_s = \sqrt{\sum (s - \mu_s)^2 P(s)} = \sqrt{729} = 27$$

$$\text{Mature}: \quad \mu_m = E(m) = \sum mP(m) = 4.50$$

$$\sigma_m = \sqrt{\sum (m - \mu_m)^2 P(m)} = \sqrt{92.25} = 9.605$$

Is there any advantage in splitting an investment by putting, say, 50 percent in the startup and 50 percent in the mature corporation? That is, is there any advantage to diversification?

If, instead of putting \$100 into the startup or \$100 into the stock of the mature corporation, \$50 was placed in each, then the expected net return for this diversified portfolio would be \$6.75.

$$E(0.5s + 0.5m) = 0.5\mu_s + 0.5\mu_m = 0.5(9.00) + 0.5(4.50) = 6.75$$

The expected value of the sum of random variables is equal to the sum of the expected values, adjusted for any proportional shares, which in this case is 50 percent in each stock.

In general, the expected value of any sum of random variables formed by a multiple β_1 of a random variable x plus a multiple β_2 of a random variable y plus a multiple β_3 of a random variable z plus . . . can be represented as

Expected value of a weighted sum of random variables

$$E(\beta_1 x + \beta_2 y + \beta_3 z + \ldots) = \beta_1 E(x) + \beta_2 E(y) + \beta_3 E(z) + \ldots \qquad (4.8)$$

Where the $x, y, z, \ldots$ are the random variables and the β's are fixed coefficients.

Notice that Formula (4.8) holds regardless of the relationship among the random variables. No assumption of independence is required to calculate the expected value of a sum of random variables. In terms of an investment strategy, forming a portfolio of different investments simply yields an expected value that is a weighted sum of the expected returns of the investments, where the weights are the proportion of funds invested.

The advantage of diversification shows up in the dispersion to be expected. If the prices of the two pharmaceutical stocks that form the above portfolio are independent, then the variance of the return to the portfolio is

$$\text{Var}(0.5m + 0.5s) = 0.25\text{Var}(m) + 0.25\text{Var}(s) =$$
$$= 0.25(92.25) + 0.25(729) = 205.3125,$$

where the 0.5 is squared because the dollar values in the random variables m and s are squared. But therein lies the advantage of diversification. The standard deviation is a measure of risk; it is the square root of the variance. Because the square root is based on a multiplicative process while the variance is calculated as an additive process, gains through diversification are possible. The standard deviation of this portfolio is $\$14.33 = \sqrt{205.3125}$, which is less than half-way between the standard deviation of the startup (\$27) and the mature stock (\$9.61), even though 50 percent of the total investment was in each.

Another way to see the advantage of diversification for risk reduction is to consider the coefficients of variation for a \$100 investment in the startup, the mature corporation, and a portfolio consisting of \$50 in each.

Startup: $\sigma_s/\mu_s = 27/9.00 = 3$

Mature: $\sigma_m/\mu_m = 9.605/4.50 = 2.134$

Diversified: $\sigma_d/\mu_d = 14.3287/6.75 = 2.123$

The diversified portfolio (with 50 percent of the total investment in each stock) has an expected return half-way between a portfolio that consists of only the startup and another that has only the mature corporation, but the coefficient of variation (2.123) for the diversified portfolio is even smaller than that of the mature stock (2.134).

In general, the variance and standard deviation of any sum of independent random variables that is formed by a multiple β_1 of a random variable x plus a multiple β_2 of a random variable y plus a multiple β_3 of a random variable z plus . . . can be represented as

Variance of a weighted sum of independent random variables

$$\text{Var}(\beta_1 x + \beta_2 y + \beta_3 z + \ldots) = \beta_1^2\sigma_x^2 + \beta_2^2\sigma_y^2 + \beta_3^2\sigma_z^2 + \ldots \quad (4.9)$$

Where the σ's are standard deviations of the x, y, z . . . variables and the β's are fixed coefficients.

Standard deviation of a weighted sum of independent random variables

$$\text{St. Dev.}\left(\beta_1 x + \beta_2 y + \beta_3 z + \ldots\right) = \sqrt{\beta_1^2\sigma_x^2 + \beta_2^2\sigma_y^2 + \beta_3^2\sigma_z^2 + \ldots} \quad (4.10)$$

4.12 Covariance and Dependent Random Variables (Optional)

As already hinted in Query 4.5, the prices of two stocks in the same industry are likely to be related. As the demand for products within an industry increases, all companies can expect to benefit to some degree. Similarly, as the demand for the industry's products falls, all companies within it can expect to suffer. As the industry expands or contracts, however, competition within the industry may favor one company to the detriment of another. The assumption of independence between the stock prices of a startup and a mature corporation within the pharmaceutical industry is questionable.

Without the assumption of independence, the probability that the price of the startup rises when the price of the mature corporation falls cannot be obtained, unless the joint probability distribution for the two random variables is known. In the real world, joint probabilities are seldom known. At best, we know marginal probabilities and conditional probabilities from which we can infer joint probabilities. For multiple variables, however, it is unlikely that we would have sufficient information to infer all the joint probabilities that go into forming the entire multivariate probability distribution. Later chapters will address the estimation of these missing joint probabilities. For now, it is important to recognize these implications for the calculation of the variance of two random variables.

If two random variables x and y tend to move in the same direction, then many of the products of x deviations $(x - \mu_x)$ and y deviations $(y - \mu_y)$ should be positive. When x is above its mean, for example, y would usually be above its mean. More specifically, consider a bivariate distribution in which x_i is the ith valuc of variable x and y_j is the jth value of variable y, and the joint probability $P(x_i, y_j)$ is relatively large whenever the product $(x_i - \mu_x)(y_j - \mu_y)$ is positive. Likewise, the joint probability $P(x_i, y_j)$ is relatively small whenever the product $(x_i - \mu_x)(x_j - \mu_y)$ is negative. For such a bivariate distribution, the probability weighted sum of (i,j) cross products should be positive:

$$\sum_i \sum_j (x_i - \mu_x)(y_j - \mu_y)P(x_i, y_j) > 0$$

Alternatively, if x and y tend to move in opposite directions, then the products of deviations in x and deviations in y times their respective probabilities should tend to be negative:

$$\sum_i \sum_j (x_i - \mu_x)(y_j - \mu_y)P(x_i, y_j) < 0$$

When there is no relationship (independence) between x and y, positive and negative deviations will tend to cancel so

$$\sum_i \sum_j (x_i - \mu_x)(y_j - \mu_y)P(x_i, y_j) = 0$$

Covariance of x and y
A measure of the joint or "co" variability of two variables x and y.

This sum of deviation cross products weighted by probabilities is called the **covariance of x and y,** abbreviated Cov (x,y). It shows how, on the average, deviations of x are related to deviations of y. In the case of the pharmaceutical companies, a positive covariance indicates that as the price of the startup rises (or falls), the price of the mature corporation tends to rise (or fall). A negative covariance says that as the price of the startup rises (or falls), the price of the mature corporation stock falls (or rises). If the two are independent, then the covariance is zero and there is no relationship between changes in the price of the two companies. Unfortunately, we seldom know a covariance with certainty; as discussed in later chapters, it typically must be estimated.

Covariance for random variables x and y

$$\sum_i \sum_j (x_i - \mu_x)(y_j - \mu_y) P(x_i, y_j) \quad (4.11)$$

When the covariance between two random variables x and y is known to be nonzero, the variance for the sum of the two variables must take account of this. If the two variables are positively related, then the positive covariance will add to the variability in the sum. If the two variables are negatively related, then the negative covariance will reduce variability.[8]

In general, the variance and standard deviation of any sum of random variables that is formed by a multiple β_1 of a random variable x plus a multiple β_2 of a random variable y plus a multiple β_3 of a random variable z plus . . . is represented as

Variance of a weighted sum of any random variables

$$\text{Var}(\beta_1 x + \beta_2 y + \beta_3 z + \ldots) = \beta_1^2\sigma_x^2 + \beta_2^2\sigma_y^2 + \beta_3^2\sigma_z^2 + \ldots \quad (4.12)$$
$$+ 2\beta_1\beta_2\text{Cov}(xy) + 2\beta_1\beta_3\text{Cov}(xz) + 2\beta_2\beta_3\text{Cov}(yz) + \ldots$$

Where the σ's are standard deviations of the x, y, z . . . variables and the β's are fixed coefficients.

Again the standard deviation of the sum of the random variables is the square root of the variance.

Standard deviation of any weighted sum of random variables

$$\text{St. Dev.}\left(\beta_1 x + \beta_2 y + \beta_3 z + \ldots\right) = \sqrt{\text{Var}\left(\beta_1 x + \beta_2 y + \beta_3 z + \ldots\right)} \quad (4.13)$$

Notice in Formula (4.12) that if the covariances are zero, as is the case for independent variables, then Formula (4.9) and Formula (4.12) are the same. To the extent that covariance is positive, with β's > 0, the variance of the sum of random variables is increased. To the extent that covariance is negative, the variance of the sum of random variables is decreased. It is for this reason that financial managers advise investors to put offsetting investments in their portfolios to reduce risk while attempting to maintain expected returns. "The idea is to own a variety of funds that tend to do well at different times, so that the overall value of your portfolio doesn't jump around too much," as stated in a *Wall Street Journal* article on "Putting Together Right Mix of Funds in Your Portfolio."

EXERCISES

4.22 For the three investments listed below, what is the expected net return for a portfolio that consists of $4,000 in each of the three investments (for a total portfolio investment of $12,000)?

Investments:	*A*	*B*	*C*
Expected net returns	$ 1,000	$ 1,500	$ 2,000
Standard deviation	$ 500	$ 700	$ 1,150

4.23 In Exercise 4.22, would your answer be affected if you learned that investments *A*, *B*, and *C* are independent? Explain.

4.24 What is the standard deviation of the return in Exercise 4.23, assuming the returns are independent? What is the significance of the assumption of independence?

4.25 The following investments are available at a price of $100,000 each:

Investments:	*A*	*B* (in thousands of dollars)
Expected net returns	11	16
Standard deviation	4	5

A third investment may be formed by placing $50,000 in investment *A* and $50,000 in investment *B*. What is the standard deviation of this third investment, assuming *A* and *B* are independent? What is the significance of the assumption of independence? If an investor ignores risk, which investment would he or she select [*A*, *B*, or the third option (.5*A* + .5*B*)]? Why?

4.26 Let x and y be two *independent discrete random variables.* The probability distributions of x and y are given below:

x	1	2	3	4
$P(x)$	.4	.1	.3	.2

y	4	3	2	1
$P(y)$	.2	.3	.1	.4

a. What is $E(x)$?
b. What is $E(y)$?
c. What is $E(x - y)$?

4.27 From the x and y distributions described in Exercise 4.26, form the bivariate probability distribution. What is the probability that $x = 2$ and $y = 3$?

4.28 Use the data in Exercise 4.26 and a computer spreadsheet package to calculate the
a. variance of x
b. variance of y
a. covariance of x and y

4.29 Let x be the annual return on a share of GM's stock and y be the annual return on a share of Chrysler's stock. $E(x)$ = $20, Var($x$) = 10 square dollars, $E(y)$ = $15, Var($y$) = 5 square dollars, Cov(x,y) = 2 square dollars.
a. What is the expected return for a portfolio that contains one unit of each type of stock?
b. What is the variance of return for the portfolio?

4.13 Concluding Comments

You are now sufficiently versed in the notion of probability and its application to business and economics problems involving random variables. The importance of knowing whether variables are independent should be clear. Using the basic rules of probability, you should be able to make probability calculations for everyday business and economic decisions involving joint probabilities. Finally, you should be able to calculate and interpret the expected value and variance of sets of random variables and relate these measures to the notion of risk. In Chapter 17, we extend the use of these concepts in a formal consideration of decision theory and risk analysis. Some students may want to explore Chapter 17 now to see how the ideas presented in Chapters 3 and 4 are used in decision theory. Others will charge into the next two chapters, which extend our discussion of random variables and probability distributions. Chapters 5 and 6 introduce well-known and widely applied discrete and continuous probability distributions.

Chapter Exercises

4.30 Exercise 3.49 describes a carnival game involving the placing of money on one of six colored squares. The spinning of a wheel determines the winning color. If you place a $10 bet on blue (for which you either lose your $10 if blue is not spun or win $10 if blue is spun), what is your expected win?

4.31 The probability distribution for the number of sales calls (x) made by a salesperson is given in the following table.

x	0	1	2	3	4 Calls per day
$P(x)$	.10	.40	.25	.15	.10

a. What is the expected value for the random variable x?

b. What is the standard deviation of x?

c. What is the approximate shape of the above distribution?

4.32 An article in the *Sunday Herald Times* (June 28, 1992) quoted John Gordon, chief financial officer of Domino's Pizza, in Dayton, Ohio, as saying that he did not understand the hassles pizza delivery drivers may get from insurance companies. He cited "a company study that found drivers who deliver fast food have fewer accidents per mile driven than the national average." Why is this an irrelevant fact to the insurance company when it is considering expected loss and the risk associated with a policy?

4.33 In a malpractice case following the death of a man during a cardiac operation, a medical doctor stated that even if the cardiac procedure had been a complete success, the deceased could not have lifted more than 25 pounds and there would have been a 30- to 40-percent likelihood that the arteries would have narrowed again, leaving little or no lifting ability after six months. The plaintiff's attorney argued that had the operation been a success, the deceased could have worked in retail sales. Professor Becker wrote for the defense that since two gallons of milk or two gallons of paint or two gallons of most anything else weighs in excess of 17.5 pounds, and since the *Occupational Handbook* stated that most retail sales workers are required to bag and package purchases and stock shelves and racks, prior to the operation the deceased could not realistically have expected to work in retail sales even if the operation had been a success. On what basis was Becker basing his opinion?

4.34 "Pepsi has said it expects to grab more than 2% of the $48 billion soft-drink market with Crystal Pepsi in its first year as a national product," according to *The Wall Street Journal* (December 15, 1992). In statistical terms, what does the word "expects" mean here?

4.35 A note in *The Wall Street Journal* (December 30, 1987) reported that, "Last year, 52 percent of all calls (to the Internal Revenue Service) were put on hold for an average of 55 seconds." What is the expected wait time for someone calling in to the IRS?

4.36 At a dinner party, an economist was heard saying that "the state lottery is a tax on stupid people." What did he mean by this?

4.37 If the economist's statement in Exercise 4.36 is true, would you expect the state lottery to be a progressive or regressive tax with respect to personal income?

4.38 When state lotteries became popular in the 1970s, proponents argued that the revenue could go to education, roads, and other infrastructure needs. Some pointed to the University of Delaware's use of state lottery money to build College Hall in 1834 as the earliest example of higher education's involvement with lotteries. As argued by Duke University economics professor Charles Clotfelter in his book on lotteries, however, the cost of administering lotteries is extremely high (e.g., in 1991, $151 million was turned over to the state of Indiana at a cost of $68 million) and revenues are erratic (e.g., between the first quarter of 1990 and the last of 1991, the Indiana state lottery has seen its profits cut in half); thus, lotteries have not delivered what was expected.

a. As more and more states introduced lotteries, why did they fail to deliver the expected profits?

b. Had expected profits been realized and placed into more effective education, why might this have led to future reductions in lottery revenues?

Academic References

Ayers, Ian. "Fair Driving: Gender and Race Discrimination in Retail Car Negotiations." *Harvard Law Review,* February, 1991.

Chernoff, H. "How to Beat the Massachusetts Number Game." *Mathematical Intelligencer,* 1981, pp. 166–172.

Endnotes

1. As described in the previous chapter, the Indiana state lottery has six window tickets that are scratched open by contestants. Three matching windows win the holder $1 to $5,000 in instant cash. Alternatively, the holder may lose or have three matching windows that enable him or her to mail in an entry to an ongoing and more involved TV contest. Only about 75 out of 75 million tickets will place holders in a pool from which six go onto a weekly TV show, where they pick from unmarked doors for $3,000 to $30,000 prizes. Each week, the big prize winner on this TV show gets to try for $1 million. (*HT,* October 8, 1989) The variant proposed here allows holders of each of the 75 million tickets to be one of those selected for a chance at the $1 million prize, regardless of the outcome in the instant cash game. There

need be no TV show for awarding the $1 million; three of the six windows match for $1 million along with a match (or no match) on the other three windows.

2. This expected value of $0.475 is close to that suggested by published estimates. "If all 75 million tickets are sold, there will be more than 10 million cash prize winners who will pocket more than $35 million." This quote implies an expected value of $0.467 per ticket (= 35000000/75000000). An alternative estimate of the mean return can be obtained from actual payouts. In 1991, 54 percent of total revenue was paid in prize money, implying an average return to a $1 ticket of 54 cents, which is slightly higher than the expected return.
3. In the case of big, ongoing lotteries, as with horse racing, a caveat must be added to the interpretation of the expected value of a winning entry. Unlike fixed payoff games of chance involving dice (for example, craps), cards (for example, blackjack), the Indiana instant cash game, or even the stock market, horse racing and some lotteries offer a parimutuel betting system, in which the payoff is really not fixed. If you place $10 on number 15 in roulette and the wheel stops with the marker resting on 15, you receive the same amount ($380) no matter how many other players bet on number 15, assuming the house has the funds. In most big, ongoing lotteries, the payoff to the winning ticket is shared equally by all holders of tickets with the winning six numbers. We will not go into the calculation of the expected value for these types of random variables. The interested reader is referred to H. Chernoff (1981, pp. 166–172), who describes the calculation of expected returns in these lotto games and uses basic probability concepts to analyze alleged strategies to beat the Massachusetts lotto.
4. As an alternative computation formula, some textbooks emphasize the use of the variance formula $\sigma^2 = \sum_x x^2 P(x) - \mu^2$, which is algebraically equivalent to Formula (4.2).

$$\begin{aligned}\sigma^2 &= E[x - E(x)]^2 = E(x - \mu)^2 = \Sigma\,(x - \mu)^2 P(x)\\ &= \Sigma\,(x^2 - 2x\mu + \mu^2)P(x) = \Sigma x^2 P(x) - 2\mu\,\Sigma x P(x) + \mu^2 \Sigma P(x)\\ &= \Sigma x^2 P(x) - 2\mu^2 + \mu^2 = \Sigma x^2 P(x) - \mu^2\end{aligned}$$

5. In our discussion of a random variable, the distinction between the variable and the occurrence of one of its values has been made. Here we are letting the uppercase *A* be the random variable and the bold-faced lowercase ***a*** the realization of a specific value. As in scholarly publications, a random variable may be represented by either a lowercase or uppercase letter. Here bold-faced lowercase letters will be used for the realization of the random variable.
6. If two occurrences are mutually exclusive, then they must be dependent. For example, if the marginal probability of each is 0.5, but the probability of their intersection is 0, then

$$P(A = \mathrm{a}/B = \mathrm{b}) = \frac{P(A = \mathrm{a} \text{ and } B = \mathrm{b})}{P(B = \mathrm{b})} = \frac{0}{0.5} = 0 \neq 0.5 = P(\mathrm{a})$$

7. This result follows from the following proof:

$$E(y) = \sum_y yP(y)$$

$$E(y) = \sum_x (a + bx)P(x)$$

$$E(y) = \sum_x \left[aP(x) + (bx)P(x)\right]$$

$$E(y) = a\sum_x P(x) + b\sum_x xP(x)$$
$$= a + bE(x)$$

since $\Sigma P(x) = 1$ and $\Sigma xP(x)$ is the mean of x.

8. Let x and y be two random variables, and let A and B be two constants; then,

$$\text{var}(Ax + By) = E\left[(Ax + By) - E(Ax + By)\right]^2$$
$$= E\left[A(x - \mu_x) + B(y - \mu_y)\right]^2$$
$$= E\left[A^2(x - \mu_x)^2 + B^2(y - \mu_y)^2 + 2AB(x - \mu_x)(y - \mu_y)\right]$$
$$= A^2\text{Var}(x) + B^2\text{Var}(y) + 2AB\text{Cov}(xy)$$

FORMULAS

- Expected value of x

  $$E(x) = \mu = \sum_x xP(x)$$

  where $P(x)$ is the probability of x,

  $\sum_x xP(x)$ is the sum of x and $P(x)$ products, for all the values of x,

  $E(x)$ is the expected value of x, and μ is the mean of x.

- Variance of random variable x

  $$\sigma^2 = E\left[x - E(x)\right]^2 = E(x - \mu)^2 = \sum_x (x - \mu)^2 P(x)$$

  where $P(x)$ is the probability of x,

  $\sum_x (x - \mu)^2 P(x)$ is the sum of products of squared deviations multiplied by respective probabilities, for all the values of x, $E(x)$ is the expected value of x, and μ is the mean of x.

- Standard deviation of random variable x

  $$\sigma = \sqrt{E\left[x - E(x)\right]^2} = \sqrt{E(x - \mu)^2} = \sqrt{\sum_x (x - \mu)^2 P(x)}$$

  The standard deviation of random variable x is the positive square root of the variance.

- Coefficient of variation
The coefficient of variation is a ratio of the standard deviation to the expected value or mean of a variable.

$$C = \frac{\sqrt{E\left[x - E(x)\right]^2}}{E(x)} = \frac{\sqrt{E(x-\mu)^2}}{\mu} = \frac{\sigma}{\mu}$$

- Probability of $A = a$ and $B = b$, if A and B are dependent

$$P(A = a \text{ and } B = b) = P(A = a \mid B = b) \times P(B = b)$$

- Probability of $A = a$ and $B = b$, if A and B are independent

$$P(A = a \text{ and } B = b) = P(A = a) \times P(B = b)$$

- Probability of $A = a$ or $B = b$

$$P(A = a \text{ or } B = b) = P(A = a) + P(B = b) - P(A = a \text{ and } B = b)$$

- Expected value of a weighted sum of random variables

$$E(\beta_1 x + \beta_2 y + \beta_3 z + \ldots) = \beta_1 E(x) + \beta_2 E(y) + \beta_3 E(z) + \ldots$$

where the $x, y, z, \ldots$ are the random variables and the β's are fixed coefficients.

- Variance of a weighted sum of independent random variables

$$\text{Var}(\beta_1 x + \beta_2 y + \beta_3 z + \ldots) = \beta_1^2\sigma_x^2 + \beta_2^2\sigma_y^2 + \beta_3^2\sigma_z^2 + \ldots$$

where the σ's are standard deviations of the $x, y, z \ldots$ variables and the β's are fixed coefficients.

- Standard deviation of a weighted sum of independent random variables

$$\text{St. Dev.}(\beta_1 x + \beta_2 y + \beta_3 z + \ldots) = \sqrt{\beta_1^2\sigma_x^2 + \beta_2^2\sigma_y^2 + \beta_3^2\sigma_z^2 + \ldots}$$

- Covariance for random variables x and y

$$\sum_i \sum_j (x_i - \mu_x)(y_j - \mu_y)P(x_i, y_j)$$

- Variance of a weighted sum of any random variables

$$\text{Var}(\beta_1 x + \beta_2 y + \beta_3 z + \ldots) = \beta_1^2\sigma_x^2 + \beta_2^2\sigma_y^2 + \beta_3^2\sigma_z^2 + \ldots$$
$$+ 2\beta_1\beta_2\text{Cov}(xy) + 2\beta_1\beta_3\text{Cov}(xz) + 2\beta_2\beta_3\text{Cov}(yz) + \ldots$$

where the σ's are standard deviations of the *x*, *y*, *z*. . . variables and the β's are fixed coefficients.

- Standard deviation of any weighted sum of random variables

$$\text{St. Dev.}\left(\beta_1 x + \beta_2 y + \beta_3 z + \ldots\right) = \sqrt{\text{Var}\left(\beta_1 x + \beta_2 y + \beta_3 z + \ldots\right)}$$

CHAPTER 5

DISTRIBUTIONS OF DISCRETE RANDOM VARIABLES

If thou must choose
Between the chances,
choose the odd;
Read the New Yorker;
trust in God;
And take short views.

Wystan Hugh Auden
1907–1973

5.1 Introduction

The following seven problems share in common a probability calculation of a discrete random variable involving a count.

- The defense for boxer Mike Tyson in his 1992 trial for rape in Indianapolis claimed that the pool of 50 persons from which a jury of 12 members was selected was not representative of the community at large. With 22 percent of the community black, what is the probability of randomly selecting a pool of 50 with seven or fewer blacks?
- A *Wall Street Journal* (April 20, 1992) article reported on a study by University of Delaware Professor Valerie P. Hans of jurors' attitudes toward business in wrongful death and injury cases. "Of the 18 cases, 14 were decided in favor of plaintiffs, higher than the national norm of about 50%." The question is: How likely was a sample with 14 out of 18 decisions for the plaintiff, if the expected outcome was nine out of 18 by the national norm?
- A *Wall Street Journal* (August 29, 1991) article reported on the problems that General Motors is having at its plant in Orion Township, Michigan. According to the article, "On some days, the cars have more than three times the company goal of two defects per car." For 20 key checkpoints, what is the probability that a randomly selected car from the Orion Township plant will have six or more defects, if the plant is producing cars that average only two defects per car?
- In "Why Investors Shouldn't Listen to Economists," (*WSJ*, January 22, 1993), statistics compiled by Robert Beckwitt, a portfolio manager at Fidelity Investment, showed "only five of 34 economists who participated in 10 or more surveys managed to guess correctly which way long-term bond yields were going more than half of the time." The conclusion drawn was that economists cannot forecast as well as flipping a coin. But what is the probability of five of 34 individuals predicting correctly six or more yield movements in 10 tries? What are the implications of this probability for the accuracy of economists' forecasts?
- A claim of sex discrimination in hiring involves an employer who has hired 10 workers, only one of whom was a woman. In the preselection pool of 20 equally qualified applicants, eight were women. What is the probability of selecting one woman in 10 random draws from this pool?
- In the lawsuit of *Ultramares Corp. v. Touche*, a creditor claimed that the accountants had been negligent in certifying financial statements of a failed company. The accountants had checked invoices in a sample of 100 (drawn from 1,000 invoices), but none of 17 fraudulent invoices were in the sample. The question is: Should the accountants be held liable?
- At a large midwestern university, each semester thousands of students take the final exams in the principles of microeconomics and macroeconomics. Over the last 10 years, an average of 1.2 students per semester were caught cheating. The question: How likely is it to catch someone cheating in the administration of the exam?

The number of blacks, number of favorable decisions, number of defects, number of accurate forecasts, number of women, number of fraudulent invoices, and number of cheaters are the outcomes of processes that can result only in a whole number count; in these examples, outcomes involving fractional measurements are not possible. A rule or probability model for assigning or generating only distinct values is called a discrete random variable. (Continuous random variables involving height, weight, and other measures, in which any value along a number line is a possible outcome, are discussed in the next chapter.)

Drawings from three different probability distributions are represented by the discrete random variables in the above seven examples. The first four (blacks, jurors' decisions, defects, and forecasts) make use of the binomial distribution. The two random variables for number of women hired and number of fraudulent invoices entail the hypergeometric probability distribution. The random process determining the number of cheaters is a Poisson probability distribution.

This chapter provides the skills needed to use the binomial, hypergeometric, and Poisson distributions in probability calculations. By the end of this chapter, you will be able to state when the binomial, hypergeometric, or Poisson distributions are appropriate representation of the process underlying a discrete random variable. You will be able to make probability calculations with these distributions using either a formula, a computer program, or a table in the back of the book.

5.2 The Binomial Distribution

Each of the 20 checkpoints on a car produced at the Orion Township plant represents a quality characteristic, factor, or attribute that either is conforming to a standard set by GM or is not. At each of these checkpoints, a random variable can be defined that is equal to one if the check conforms to the standard or that is equal to zero if it does not. The sum of these individual random variables then forms a random variable that is the number of checkpoints that conform to the standard. Similarly, in the jury selection case, each selection of a juror can be viewed as an experiment in which the random variable is equal to one if a black is selected and to zero if not. The sum of these individual random variables is the number of blacks in the preselection pool. The composite random variable x in either of these two cases assumes only the values 0, 1, 2, 3, . . . , n, where n is the sample size ($n = 20$, for the car and $n = 50$, in the jury preselection pool).

Binomial Probability Distribution
A discrete probability distribution for the sum of a series of n independent and identical trials, where each trial results in one of the two possible outcomes, (0 or 1) that occur with fixed probabilities.

The **binomial probability distribution** is formed by a probability mass function that may be used when the random variable x can only take on values $x = 0$, $x = 1$, $x = 2$, $x = 3$, and so on. The unique feature of the binomial process, however, is that these nonnegative whole number values are obtained as the sum of n identical and independent trials in which each trial yields either a one or zero as an outcome. That is, each trial itself is a random experiment yielding either a one or zero as its outcome. In the car example, each checkpoint can be viewed as a trial, with the outcome either defect or no defect. If we are interested in the total number of defects (x), then on each of the 20 trials (checkpoints), a one is recorded ($x_i = 1$) if the checkpoint does not conform to the standard and a zero is recorded if it does ($x_i = 0$). The total number of defects is then the sum of the outcomes on the 20 trials.

$$x = x_1 + x_2 + \ldots + x_{20} = \sum_{i=1}^{20} x_i$$

Similarly, each person selected for the preselection pool can be thought of as the outcome of a trial in which the outcome is either black or nonblack, with the total number of blacks determined by the outcome ($x_i = 1$ if black or $x_i = 0$ if nonblack) on each of the 50 trials.

$$x = x_1 + x_2 + \ldots + x_{50} = \sum_{i=1}^{50} x_i$$

For the binomial distribution, one trial cannot influence or be associated with the outcome of another; the outcome of each of the n trials must be independent of one another; the probability of each outcome cannot change from trial to trial. Each trial must be an exact replication of the others. For example, on the first trial in the hiring of 10 workers from an applicant pool of eight women and 12 men applicants, the probability that a woman is hired is 0.40 = 8/20. If a woman is hired, then the probability that a woman is hired on the second trial falls to 0.368 = 7/19. The probability of hiring a woman does not remain fixed from trial to trial, and the binomial distribution cannot be used to calculate probability because the trials are not independent. (As shown later in this chapter, in this hiring example where the probabilities change from trial to trial, the hypergeometric distribution may be used.)

In laboratory experiments or drawing balls from an urn, the probability of "success" or "failure" on each trial can be held fixed by replacement. That is, we can draw a ball from an urn, record the color, place the ball back in the urn, and on the next draw the probability of getting that color ball is the same as on the first draw, regardless of the number of balls in the urn.

In real-world problems in business and economics, seldom will the probability of an event remain fixed from trial to trial. In the case of jury selection, for example, when the first of 50 people is selected, there is one fewer person remaining in the pool, thus the probability of selecting a black on the second trial must be different than on the first, but the change is trivial. In the Indianapolis area, where the Tyson trial took place, there are more than 400,000 registered voters. Selecting one, two, or 49 blacks from this base is not going to greatly influence the probability that the next person selected is black. Without replacement, the probability of selecting a black cannot remain exactly fixed from trial to trial, but any change in probability is minuscule because the sample size (n = 50) is extremely small relative to the population size (N = 428,318). Unlike the hiring example in which 10 applicants are to be selected from 20, selecting only 50 potential jurors from a population that exceeds 400,000 allows us to use the binomial in probability calculations.

In the car example, we do not know for sure from the quote if the probability of a defect is fixed at 0.10 for each checkpoint, which represents the average of two defects over 20 checkpoints. The likelihood of a defect at each checkpoint could be different. Each checkpoint could constitute a totally different trial. If such is the case, then the binomial could not be used. In addition, the consequence (dissatisfied customer, expense of repair, lost sale, etc.) could differ depending on the defect—in which case, the likelihood of a defect might be weighted by its cost. Here we will assume that the likelihood of a defect at any one of 20 checkpoints is fixed at 0.10 and that each checkpoint involves an identical visual inspection. We will also assume that the consequence of the defect is identical across defects.

In summary, the binomial distribution is applicable only if

1. the random variable of interest takes on the values of 0, 1, 2, 3, . . . n, where this value is the event formed by the sum of n random variables;
2. there are n trials that give rise to the n random variables;
3. only one of two outcomes (0 or 1) is possible for each trial;
4. the probability of either outcome on each trial remains fixed (or at least approximately fixed) from trial to trial. That is, the trials are independent and identically distributed.

5.3 The Binomial Probability Mass Function

The probability of one of the two outcomes on each of the n trials is a parameter of the binomial distribution. There is no universally agreed on symbol for this binomial parameter, but in this book it will be designated by the letter π. In the car example, if the goal of only two defects at 20 checkpoints is achieved, then the probability of finding a defect at any one of the 20 checkpoints might be $\pi = 0.10$. In the case of jury selection, 22 percent of the Indianapolis population is black, so the probability of a black on each of the 50 selections is $\pi = 0.22$, assuming draws are random selections from the population.

From the discussion of independence in Chapter 4, the probability that all 20 checkpoints on a car are not conforming to the standard is the product of the probability of a defect at each checkpoint, if the checkpoints are independent. That is, the probability of 20 defects in 20 trials, with the probability of a defect on any trial fixed at $\pi = 0.10$, is $(0.10)^{20}$.

$$
\begin{aligned}
P(20 \text{ defects}) &= P(\text{defect at 1st checkpoint}) \text{ times} \\
&\quad\; P(\text{defect at 2nd checkpoint}) \text{ times} \\
&\quad\; P(\text{defect at 3rd checkpoint}) \text{ times} \\
&\quad\; \ldots \\
&\quad\; P(\text{defect at 19th checkpoint}) \text{ times} \\
&\quad\; P(\text{defect at 20th checkpoint}) \\
&= (\pi)(\pi) \ldots (\pi)(\pi) \\
&= (0.10)(0.10) \ldots (0.10)(0.10) = (0.1)^{20} \\
&= 0.00000000000000000001 = 1.0(10)^{-20}
\end{aligned}
$$

The probability of 20 perfect checkpoints is likewise determined to be the probability of zero defects, which is 0.121577.

$$\begin{aligned} P(0 \text{ defects}) &= P(1\text{st no defect}) \times P(2\text{nd no defect}) \times \ldots \\ &\quad \ldots \times P(19\text{th no defect}) \times P(20\text{th no defect}) \\ &= (1-\pi)(1-\pi)\ldots(1-\pi)(1-\pi) \\ &= (0.90)(0.90)\ldots(0.90)(0.90) = 0.1215766546 \end{aligned}$$

These are extreme possibilities. Joint probabilities of the different combinations of defects can be calculated, but attention must be given to the different ways in which the defects can occur over the 20 checkpoints. One defect, for example, can occur separately at each checkpoint, so the probability of one defect over the 20 checkpoints is

$$\begin{aligned} P(1 \text{ defect}) &= 20 \times P(\text{one defect}) \times P(\text{no defects at 19 points}) \\ &= 20(\pi)(1-\pi)^{19} \\ &= 20(0.10)(0.90)^{19} = 0.270170 \end{aligned}$$

Two defects can occur over the 20 checkpoints in 190 different ways. As presented in Chapter 3, this is calculated by the combination

$$20!/2!18! = (20)(19)/(2)(1) = 190$$

Factorial
The product of all positive integers from 1 up to and including a given integer; as in $5! = 5 \times 4 \times 3 \times 2 \times 1 = 120$, where $1! = 0! = 1$.

where again 20!,18!, and 2! are **factorials** standing for the products

$$20! = 20 \times 19 \times 18 \times \ldots \times 3 \times 2 \times 1,\ 18! = 18 \times 17 \times \ldots 3 \times 2 \times 1, \text{ and } 2! = 2 \times 1$$

Thus, the probability of two defects is

$$\begin{aligned} P(2 \text{ defects}) &= 190 \times P(\text{two defects}) \times P(\text{no defects at 18 points}) \\ &= 190(\pi)^2(1-\pi)^{18} \\ &= 190(0.10)^2(0.90)^{18} = 0.285180 \end{aligned}$$

Three defects can occur over the 20 checkpoints in 1,140 different ways [20!/3!17! = (20)(19)(18)/(3)(2)(1) = 1140]; thus, the probability of three defects is

$$\begin{aligned} P(3 \text{ defects}) &= 1140 \times P(3 \text{ defects}) \times P(\text{no defects at 17 points}) \\ &= 1140(\pi)^3(1-\pi)^{17} \\ &= 1140(0.10)^3(0.90)^{17} = 0.190120 \end{aligned}$$

In general, the binomial probability of x "ones" and $(n-x)$ "zeros," in n trials, where π is the probability of a "one" on any of n trials, is given by the following formula.

Binomial Probability Mass Function

$$P(x) = \frac{n!}{x!(n-x)!}\pi^x(1-\pi)^{n-x} \tag{5.1}$$

The GM executive stated "On some days, the cars have more than three times the company goal of two defects per car." Although it is not clear from this quote how many cars had six or more defects in a day, if the production process is "in control" and producing a mean of only two defects per car, the probability of randomly drawing a car with exactly six defects is calculated with Formula 5.1 to be 0.008867.

$$\begin{aligned} P(6 \text{ defects}) &= (20!/6!14!)(.1)^6(.9)^{14} \\ &= 38760(0.1)^6(0.9)^{14} = 0.008867 \end{aligned}$$

The probability of exactly seven defects is

$$\begin{aligned} P(7 \text{ defects}) &= (20!/7!13!)(.1)^7(.9)^{13} \\ &= 77520(0.1)^7(0.9)^{13} = 0.001970 \end{aligned}$$

These are rare events and might be taken as evidence that something is wrong with the production process. That is, if the production process was operating with a mean of only 2 defects at 20 checkpoints it is unlikely (0.008867 probability) that we would randomly select a car with six defects. It is more unlikely that we would select one with seven defects (0.00197) and even more unlikely that we would get one with eight defects (0.000356 probability, which the reader should verify).

As discussed in Chapter 3, if events are mutually exclusive, then the probability of a union of the events can be obtained as the sum of the probabilities. For one car, if six defects are found then no other count is possible. If seven defects are found then no other count is possible, and so on. Thus, the probability of six or more defects is the sum of the probability of six defects plus the probability of seven defects . . . plus the probability of 20 defects, which is

$$\begin{aligned} P(x \geq 6) &= P(6) + P(7) + \ldots + P(20) \\ &= 0.008867 + 0.001970 + 0.000356 + \ldots \\ &= 0.0112 \end{aligned}$$

Only about 112 in 10,000 cars (or approximately one in a hundred cars) should have six or more defects if the production process is operating with an average of only 2 defects at 20 checkpoints per car. Thus if we observe 6, 7, 8 or more defects we have evidence that the cars are not being produced at the company goal of two defects per car.

Determined with a Computer

Calculations with Formula (5.1) are easy for small sample sizes but, as should be apparent from the car defect case, even for a sample of $n = 20$ the arithmetic exceeds what can be done easily with an inexpensive hand calculator. Computer programs such as STATABLE, MICROSTAT, and the ECONOMETRICS TOOLKIT make calculations with the binomial mass function easy where the user need only specify n and π. For example, below is the code required to have MINITAB calculate binomial probabilities for $n = 20$ and $\pi = 0.1$.

```
MTB > pdf; [hit enter key]
SUBC > binomial n = 20 p = 0.1. [hit enter key]
```

Minitab responds with the following probabilities

BINOMIAL WITH N = 20 P = 0.1000

K	P(x = K)
0	0.1216
1	0.2702
2	0.2852
3	0.1901
4	0.0898
5	0.0319
6	0.0089
7	0.0020
8	0.0004
9	0.0001
10	0.0000

Spreadsheet programs such as LOTUS 1-2-3, QUATTRO, and EXCEL also can be programmed easily to make calculations with Formula (5.1). Alternatively, a binomial table, such as that in the back of this book, can be used.

Determined with a Binomial Table

Tables of the binomial distribution for $n = 1, 2, 3, 4, 5, 6, 7, 8, 9, 10, 20$, and 50 are given in Appendix Table A.1, part of which is reproduced below. The numbers in Table 5.1 represent the middle of the binomial values for $n = 20$.

By putting a decimal point in front of each four-digit number in the body of Table 5.1, the probabilities of $x = 1, 2, 3, \ldots 20$ are obtained. The two-digit numbers across the top and bottom are the values of π, again with the decimal point omitted.

The values along the left margin designate the x-values when π is less than 0.50. In the car defect case, where $\pi = 0.10$, we are interested in $x = 6$, for six defects. We first

Table 5.1 Some Binomial Values for $n = 20$ from Appendix Table A.1

$n = 20$

x π	01	02	03	04	05	06	07	08	09	10	
0	8179	6676	5438	4420	3585	2901	2342	1887	1516	1216	20
1	1652	2725	3364	3683	3774	3703	3526	3282	3000	2702	19
2	0159	0528	0988	1458	1887	2246	2521	2711	2828	2852	18
3	0010	0065	0183	0364	0596	0860	1139	1414	1672	1901	17
4	0000	0006	0024	0065	0133	0233	0364	0523	0703	0898	16
5	0000	0000	0002	0009	0022	0048	0088	0145	0222	0319	15
6	0000	0000	0000	0001	0003	0008	0017	0032	0055	0089	14
7	0000	0000	0000	0000	0000	0001	0002	0005	0011	0020	13
8	0000	0000	0000	0000	0000	0000	0000	0001	0002	0004	12
9	0000	0000	0000	0000	0000	0000	0000	0000	0000	0001	11
	99	98	97	96	95	94	93	92	91	90	π x

look down the left-hand column to the row identified as 6. Next we look across the top of the table to the column headed 10, which corresponds to $\pi = 0.10$. The intersection of row 6 and column 10 gives the probability of $x = 6$, after putting a decimal point in front of this intersection value of 0089,

$$P(6 \text{ defects} \mid n = 20, \pi = 0.10) = 0.0089$$

Other than rounding, this is the same probability we calculated above with Formula (5.1). Similarly, other values of interest in the car defect case can be obtained by looking down the column headed 10. For instance, the probability of $x = 7$, with $n = 20$ and $\pi = 0.10$, is

$$P(7 \text{ defects} \mid n = 20, \pi = 0.10) = 0.0020$$

Notice that no probabilities are shown for $x > 9$ and $01 < \pi < 10$ because these probabilities are approximately zero.

For π greater than 0.50, we find the π of interest (with the decimal point omitted) in the bottom row of the binomial table, then we find the x value in the far right column. For example, 10 π values are shown across the bottom of Table 5.1 (.99, .98, .97, . . . , .90). If $\pi = 0.91$, then the probability of $x = 12$ is found in the cell formed by the intersection of column with the footer 91 and the right-margin row labelled 12.

$$P(x = 12 \mid n = 20, \pi = 0.91) = 0.0002$$

Again, values with obvious zero probabilities may not be shown in the binomial table to save on space.

Query 5.1:

A multiple-choice test has four unrelated questions. Each question has five possible choices, but only one is correct. Thus, a person who guesses randomly has a probability of 0.2 of guessing correctly. Draw a tree diagram showing the different ways in which a test taker could get 0, 1, 2, 3, and 4 correct answers. Sketch the probability mass function for this test. What is the probability a person who guesses will get two or more correct?

Answer:

Letting Y stand for a correct answer and N a wrong answer, where the probability of Y is 0.2 and of N is 0.8 for each of the four questions, the probability tree diagram is as shown on the following page.

This probability tree diagram shows the branches that must be followed to show the calculations captured in the binomial mass function for $n = 4$ and $\pi = 0.2$. For example, the tree diagram shows the six different branch systems that yield two correct and two wrong answers (which corresponds to $4!/2!2! = 6$). The binomial mass function shows the probability of two correct answers as

$$P(x = 2 \mid n = 4, \pi = 0.2) = 6(.2)^2(.8)^2 = 6(0.0256) = 0.1536 = P(2)$$

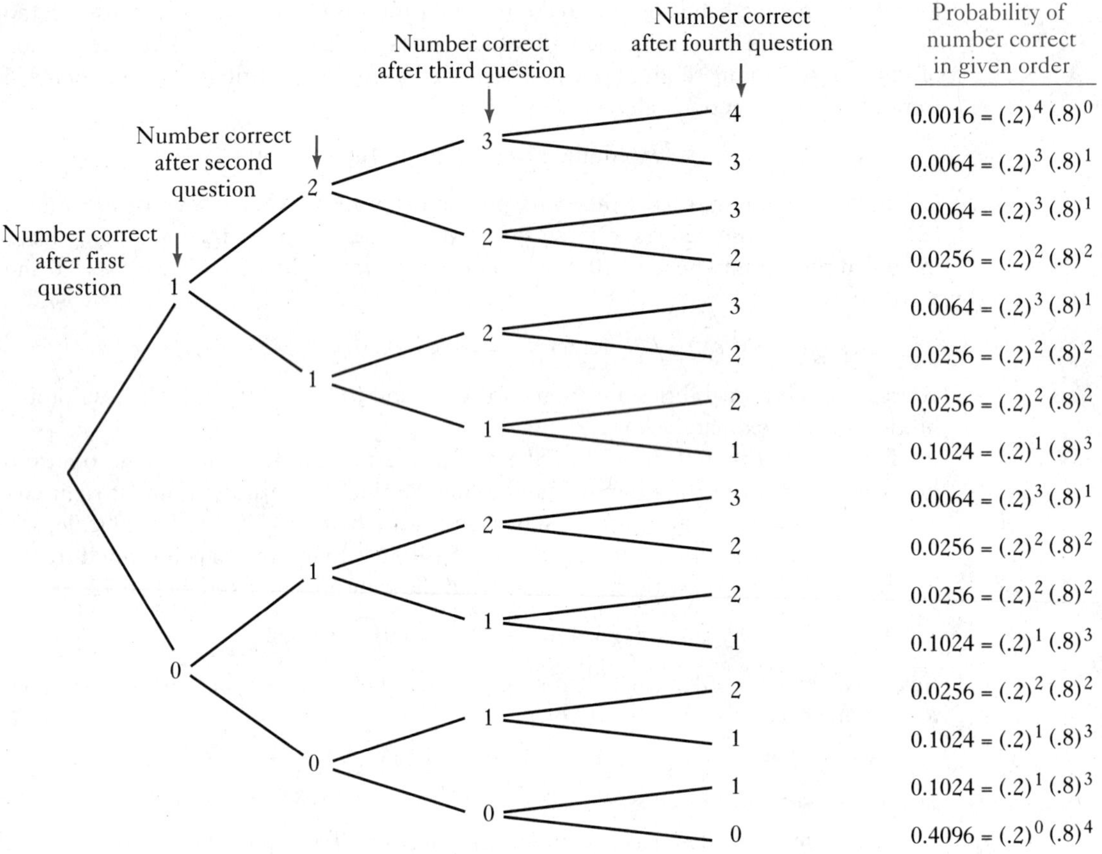

The probability of three correct answers is

$$P(x = 3 \mid n = 4, \pi = 0.2) = 4(.2)^3(.8) = 4(0.0064) = 0.0256 = P(3)$$

The probability of four correct answers is

$$P(x = 4 \mid n = 4, \pi = 0.2) = (.2)^4 = 0.0016 = P(4)$$

Thus, the probability of two or more correct answers by guessing is

$$\begin{aligned} P(x \geq 2 \mid n = 4, \pi = 0.2) &= P(2) + P(3) + P(4) \\ &= 0.1536 + 0.0256 + 0.0016 \\ &= 0.1808 \end{aligned}$$

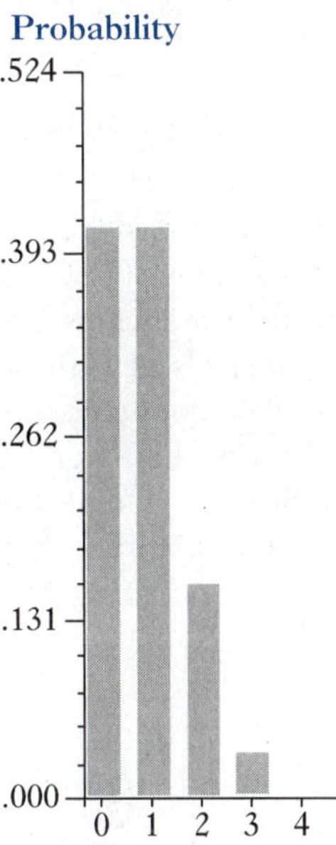

QUERY 5.2:

Following the development of the nationally normed *Test of Economic Literacy* (TEL), authors William Walstad and John Soper reported that the typical student in their selected sample of 2,483 students averaged about 20 out of 46 questions correct, in a midschool year administration of the TEL. Later, in a session at the 1988 annual meeting of the American Economic Association, Walstad stated that guessing would, on average, yield 11.5 questions correct, because each question had four choices. In *Vital Speeches of the Day* (March 25, 1989), Walstad stated:

> So typical high school students score only 15 percentage points above a chance score on this test. Clearly 40 percent correct represents a failing grade under even the most liberal grading standards. This level of economic knowledge among most high school students is shocking!

Are these results really shocking? How likely is a score of 20 questions correct if a student is just guessing, assuming questions are independent? Assuming questions are independent, calculate the probability of 20 correct if students are just guessing, and state whether you think they are just guessing or actually know some economics. State why you would question the assumption of independence and its implications.

ANSWER:

The probability of 20 correct out of 46 independent multiple-choice questions, with four choices, is

$$P(x = 20 \mid n = 46, \pi = 0.25) = \frac{46!}{20!26!}(0.25)^{20}(0.75)^{26} = 0.003$$

which is most easily determined with a statistics computer program such as STATABLE or the ECONOMETRICS TOOLKIT or by writing a calculation routine in a spreadsheet such as EXCEL or LOTUS 1-2-3. This says that only three in 1,000 times, over a large (infinite) number of tests, will a student get exactly 20 correct answers by random guessing. Binomial probabilities approach zero for guesses of more than 20 correct answers. That is, there is little chance of a student guessing 20 questions correctly. The fact that students averaged around 20 questions correct suggests that something other than chance is at work here.

Before concluding that students know economics, however, we should call attention to the manner in which the TEL was constructed. Each question was put on

Binomial Probability Mass Function for 46 Test Questions and $\pi = 0.25$

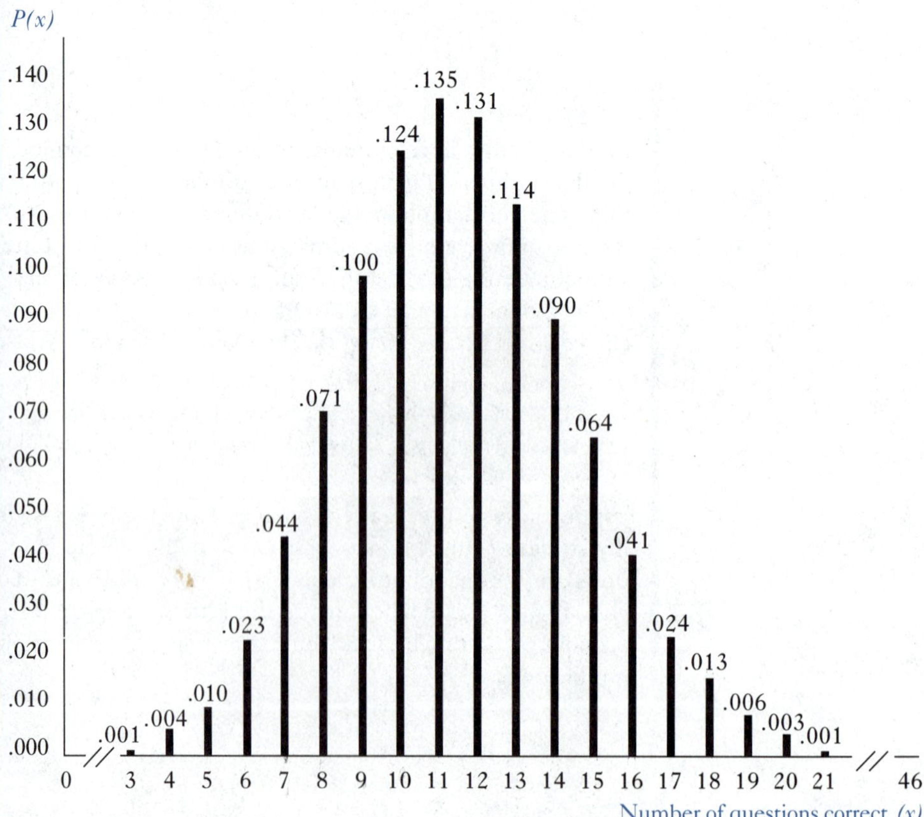

the test because it tended to identify students who know economics and those who do not. Reliability was built into the test by using questions that differentiate the same students in the same way—students with higher scores on the test should get any question correct with higher probability than those with lower scores. As concluded by Becker, Greene, and Rosen,"This rule for inclusion makes questions dependent, and the probability of successful guessing on each question is not fixed at 0.25."(1990, p. 14) In addition, it is likely that some questions made use of information in others, making them related and thus dependent.

QUERY 5.3:

The defense for boxer Mike Tyson in his 1992 trial for rape claimed that the pool of 50 persons from which a jury of 12 members was to be selected was not representative of the community at large. With 22 percent of the community black, what is the probability of randomly selecting a pool of 50 with seven or fewer blacks?

ANSWER:

From Appendix Table A.1, with $n = 50$, $\pi = .22$, the probability of at most seven is

$$P(x \leq 7 \mid n = 50, \pi = .22) = .0000 + .0001 + .0004 + .0018 + .0059 + .0152 + .0322 + .0571 = .1127$$

which is not an unusual occurrence. That is, observing seven or fewer blacks in a 50-person group selected at random from a population with 22 percent blacks would occur by chance approximately 11 out of 100 times over many (infinite) draws of samples of 50.

EXERCISES

5.1 Use Appendix Table A.1 to find the following binomial probabilities
   a. $P(X \leq 7 \mid n = 20, \pi = 0.3)$
   b. $P(Y = 3 \mid n = 10, \pi = 0.7)$
   c. $P(2 < W \leq 9 \mid n = 30, \pi = 0.2)$

5.2 Eric von Hippel, an MIT professor, and Cornelius Herstatt, of the Swiss Federal Institute of Technology, in their monograph "An Implementation of the Lead User Market Research Method In a 'Low-Tech' Product Area: Pipe Hangers" (Alfred P. Sloan School of Management, MIT, 1992), describe their work with Hilti AG, a leading European manufacturer of fastening, hanging, and shelving related products. To test whether designers were too far ahead of what customers wanted, the authors tested the designers' ideas on a sample of 12 regular customers: 10 preferred the new product and said they would pay up to 20 percent more for it. If less than the majority of customers (49 percent) would prefer the new product and be willing to pay up to 20 percent more for it, what is the probability of getting a sample of 12 in which 10 or more say they prefer it and are willing to pay up to 20 percent more for it? What are the implications of this probability?

5.3 A headline in the *Wall Street Journal* (June 12, 1992) reported: "Sears Is Accused of Billing Fraud at Auto Centers." The article reports on a year-long study by the California Department of Consumer Affairs in which undercover agents established that 89.5 percent of the owners of cars with worn brakes but no other mechanical defects were told that additional and more expensive repairs were needed. "In a second phase, the department informed Sears of the investigation in December, 1991, and then conducted 10 more tests in January, of which seven resulted in unnecessary repairs." What is the probability of seven or fewer unnecessary repairs if, in the population of all Sears repairs, 89.5 percent of the cars were still getting unnecessary repairs?

5.4 *Business Week* (June 29, 1992) reported in an article about Cadillac production at its Detroit-Hamtramck plant that, "Today paint jobs are right on 82% of the cars after the first pass, a 90% improvement and that has all but eliminated costly end-of-the-line paint repairs." (p. 89) If this is a true statement, what is the probability of randomly selecting 10 Cadillacs coming off the line and finding fewer than four defective paint jobs?

5.5 In an attempt to curb excessive executive pay, a *Wall Street Journal* (July 15, 1992) article reported that institutional investors are getting more assertive. Of the 56 institutions surveyed, 66 percent (37 of 56) said they favored the creation of compensation committees to review or even set executive pay. If this sample was representative of the population, what is the probability of drawing another random sample of 56 institutions for which 50 percent or less of the institutions would favor the creation of compensation committees?

5.6 A *Wall Street Journal* article (August 11, 1992) reported that "Most American men—eight out of 10—use a hair dryer, according to industry estimates." If you ask five of your male friends whether they use a hair dryer, what is the probability that four will say they do? What did you assume to reach your answer?

5.7 A fire detection device utilizes three temperature-sensitive cells acting independently of each other in such a manner that any one or more may activate an alarm. Each has a probability of 0.8 of activating the alarm when the temperature reaches 100 degrees. What is the probability that the alarm will function if the temperature reaches 100 degrees?

5.8 If the nine-member United States Supreme Court is hearing a case for which a verdict is upheld or overturned, what is the probability that five or more jurists will uphold a guilty verdict if there is a 95 percent chance that any one of them will do so? What do you have to assume to make this probability calculation? Why might this assumption be unreasonable?

5.9 Below are the 10 biggest down days for the Dow Jones Industrial Average, as reported in the *Wall Street Journal* (October 6, 1992). For 21 business-day months, assuming that potential big down days are distributed as a binomial random variable, what is the expected number of 10 big down days occurring in October? What is the standard deviation? How many standard deviations are there between what is expected and what actually occurred?

The 10 Biggest Down Days

Percentage changes in the Dow Jones Industrial Average since 1926

Date	Decline
October 19, 1987	22.6%
October 28, 1929	12.8
October 29, 1929	11.7
November 5, 1929	9.9
August 12, 1932	8.4
October 26, 1987	8.0
July 21, 1933	7.8
October 18, 1937	7.8
October 5, 1932	7.2
September 24, 1931	7.1

Source: U.S. General Accounting Office.

5.10 Use the information in Exercise 5.9 to calculate the probability of getting six big down days in October for the sample of 10. What conclusion could you draw from this?

5.11 Dollar Rent A Car System has a catchy phrase for its ads: "Right on the airport. Right on the money." A *Wall Street Journal* (June 4, 1993) feature reported, however, "Dollar isn't in 15 of the top 100 airport markets. Instead, it operates off airport or not at all in those other locations." If the *WSJ* is correct, what is the probability of finding a random sample of 20 top airport markets for which at least 17 have on-site Dollar operations?

5.4 Location and Dispersion

Random variables have means (that measure central location), variances (that measure dispersion), and probability distributions that may be symmetric or skewed. For the binomial π and n determine central location, dispersion, and skewness.

Chapter 4 showed the mean or expected value for any random variable x to be calculated by the formula

$$\mu = E(x) = \Sigma\, xP(x)$$

This formula can now be used to determine the average value for a specific random variable that conforms to the binomial distribution. For example, the expected number of questions guessed correctly on a four-question, multiple-choice test, where each independent question has five choices, is

$$\mu = 0(0.4096) + 1(0.4096) + 2(0.1536) + 3(0.0256) + 4(0.0016) = 0.80$$

Guessing on this test will yield an average of 0.8 questions correct, over a large (infinite) number of tests where each test has four independent questions with five alternatives. Notice that this mean value could have been obtained by simply multiplying the sample size n by the probability of success, π;

$$\mu = (4)(0.2) = 0.8$$

Thus, the mean or expected value of a binomial random variable can be obtained by the shortcut Formula (5.2).[1]

Mean of a Binomial Random Variable

$$\mu = E(x) = n\pi \tag{5.2}$$

The variance, or expected squared deviation, was given in Chapter 4 by the formula

$$\sigma^2 = E(x - \mu)^2 = \Sigma(x - \mu)^2P(x)$$

For the four-question multiple-choice test, where the expected number correct is 0.80, the variance is

$$\sigma^2 = (0 - 0.8)^2(0.4096) + (1 - 0.8)^2(0.4096) + (2 - 0.8)^2(0.1536) + (3 - 0.8)^2(0.0256) + (4 - 0.8)^2(0.0016) = 0.64$$

But this variance value could have been calculated by multiplying the sample size n by the probability of success π times the probability of failure $(1 - \pi)$;

$$\sigma^2 = (4)(0.2)(0.8) = 0.64$$

The variance of a binomial random variable can be obtained by the shortcut Formula (5.3).[2]

Variance of a Binomial Random Variable

$$\sigma^2 = E(x - \mu)^2 = n\pi(1 - \pi) \tag{5.3}$$

Notice that if $\pi = 0.5$, the variance of a binomial random variable is at its maximum. With $\pi = 0.5$, the binomial distribution is also symmetric. If $\pi < 0.5$, then the binomial distribution is right skewed. If $\pi > 0.5$, then the binomial distribution is left skewed. As n gets larger, however, the skewness in the binomial distribution vanishes; even with $\pi \neq 0.5$ it will appear symmetric.

Figure 5.1 shows four different binomial distributions for $n = 4$, 8, 16, and 32, and $\pi = 0.2$. In panel a, where $n = 4$, the right skewness is apparent. In panel d, where $n = 32$, the distribution looks more symmetrical than skewed. Additional increases in n would blur the skewness further, to the point that the binomial mass function would appear to be a bell-shaped curve. In fact, as described in the next chapter, it was this observation that gave rise to the normal distribution.

QUERY 5.4:

When an experimental munitions production process is first set up, it is thought to have promise for future refinement if 70 percent of the bullets in a magazine fire properly. Samples of $n = 50$ are to be tested. What is the expected number of bullets that should fire properly if the process has promise? What is the likelihood of getting a sample of bullets one or two standard deviations from either side of this mean? To what use can you put this information in deciding when the production process does or does not have promise?

ANSWER:

The expected number of bullets that should fire properly if the production process has promise is 35, assuming the firing of each of the 50 bullets is an independent and identical trial where the probability of success on each trial is 0.7. That is,

$$\mu = \pi n = (0.7)(50) = 35$$

This, of course, is the average number of properly firing bullets over a large (infinite) number of samples of size 50. For any given sample of 50, the observed number of properly firing bullets may be somewhat higher or lower, even if the production process has promise. Below is the binomial mass function that shows the probabilities of getting a sample of 50 bullets with 29 through 42 good bullets.

FIGURE 5.1 Vanishing Skewness as the Sample Size is Increased

Panel a: $\pi = 0.2$, and $n = 4$

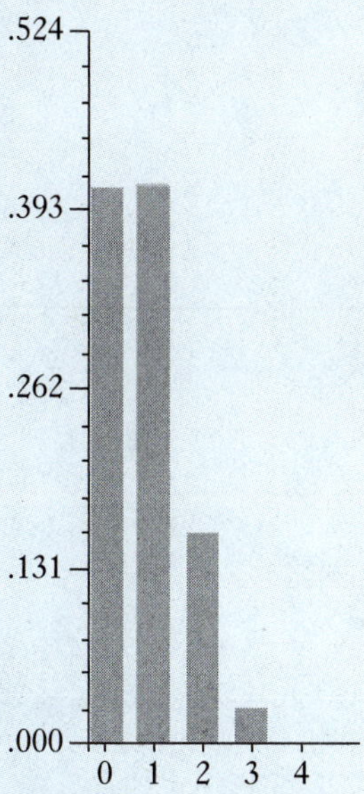

Panel b: $\pi = 0.2$, and $n = 8$

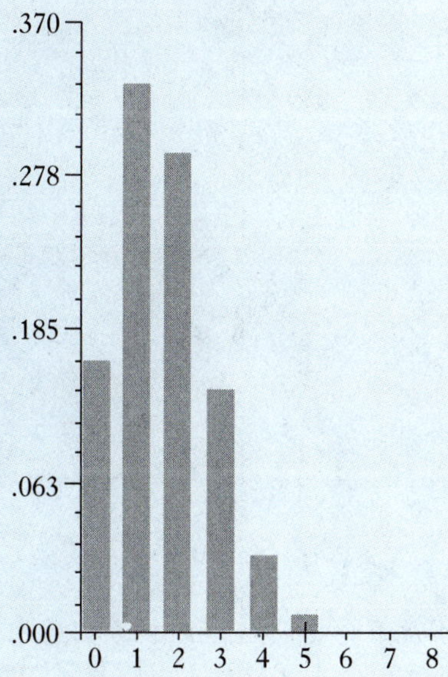

(continued)

FIGURE 5.1 Vanishing Skewness as the Sample Size is Increased (*continued*)

Panel c: $\pi = 0.2$, and $n = 16$

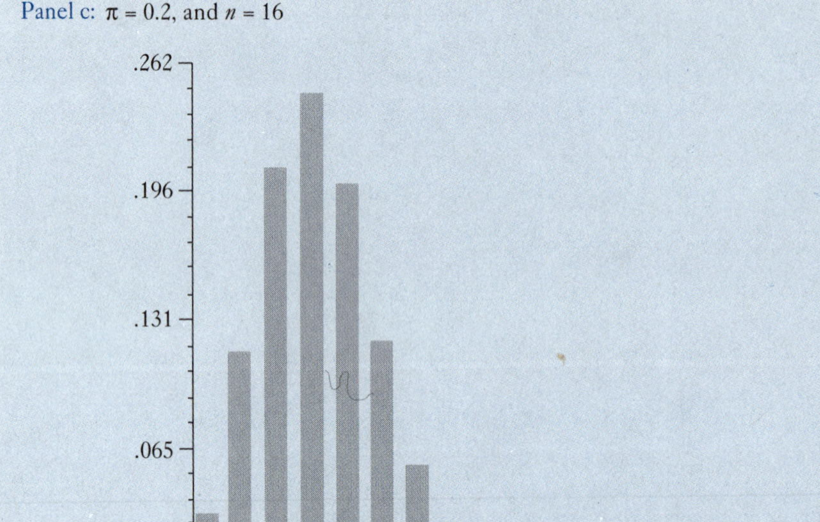

Panel d: $\pi = 0.2$, and $n = 32$

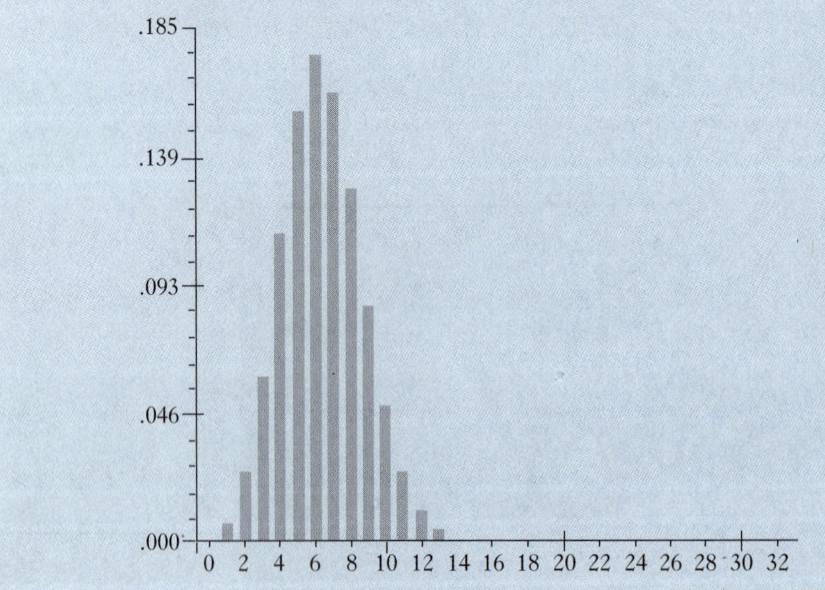

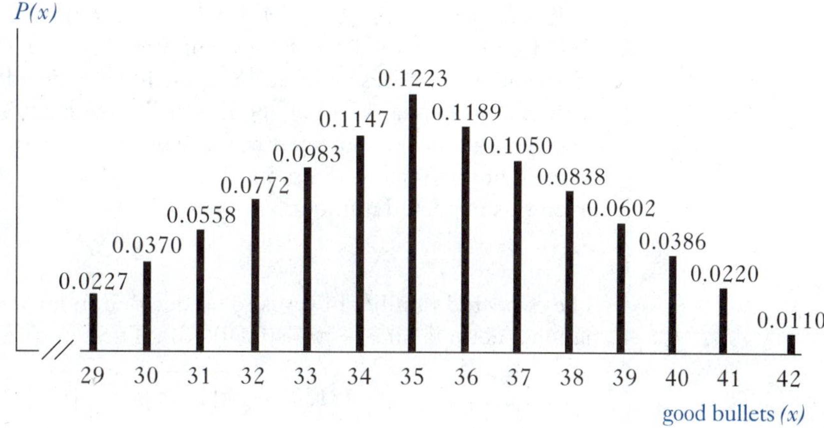

The variance and standard deviation for this binomial are

$$\sigma^2 = n(\pi)(1-\pi) = 50(.70)(.30) = 10.5 \text{ bullets squared}$$

$$\sigma = \sqrt{10.5} = 3.24 \text{ bullets}$$

If the distribution of bullets was perfectly bell shaped, then by the rule of thumb for bell-shaped distributions presented in Chapters 3 and 4, approximately 68 percent of the distribution should be within one standard deviation of the mean, which is

$$\mu \pm \sigma = 35 \pm 3.24, \text{ or } 31.76 \text{ to } 38.24$$

From the above probability mass function, the sum of the probabilities for x values between 31.76 and 38.24 is 0.7202.

$$P(x = 32) + P(x = 33) + \ldots + P(x = 37) + P(x = 38)$$

$$= 0.0772 + 0.0983 + 0.1147 + 0.1223 + 0.1189 + 0.1050 + 0.0838$$

$$= 0.7202$$

The difference between the exact binomial probability of 0.7202 and the bell-curve approximation of 0.68 reflects the fact that this binomial is not perfectly symmetric and it is not continuous.

By the rule of thumb for bell-shaped distributions, approximately 95 percent of the observations should be within two standard deviations of the mean. That is, 95 percent of the observations should be between 28.52 and 41.48 [$\mu \pm 2\sigma = 35 \pm 2(3.24)$]. The exact sum of the probabilities of values in this interval is 0.9565, which again is close to that suggested by the rule of thumb.

Since only about 5 percent of the time will a sample of 50 bullets yield less than 28.5 or more than 41.5 firing properly, when these extremes are realized the production process should be reviewed to see what is wrong (when $x < 28.5$) or what has changed to cause the improvement (when $x > 41.5$).

QUERY 5.5:

A *Wall Street Journal* (April 20, 1992) article reported on a study by University of Delaware Professor Valerie P. Hans of jurors' attitudes toward business in wrongful death and injury cases. "Of the 18 cases, 14 were decided in favor of plaintiffs, higher than the national norm of about 50%." How many standard deviations are there between 14 and the expected number of cases favoring the plaintiffs based on the national norm? What is the probability of 14 or more cases decided for the plaintiffs, given this expected number?

ANSWER:

The expected number of cases to be decided in favor of the plaintiffs based on the national norm is nine [$= \pi n = (0.50)(18)$]. The standard deviation is

$$2.12132\left[=\sqrt{\pi(1-\pi)n}=\sqrt{(0.5)(0.5)(18)}\right]$$

Thus, the observed number of 14 cases is 2.357 standard deviations higher than the expected number of nine cases. The probability of observing 14 or more out of 18, when the expected number is nine, is calculated by the STATABLE computer program as follows:

$$P(x = 14 \mid \pi = 0.5, n = 18) = 0.01167$$
$$P(x = 15 \mid \pi = 0.5, n = 18) = 0.00311$$
$$P(x = 16 \mid \pi = 0.5, n = 18) = 0.00058$$
$$P(x = 17 \mid \pi = 0.5, n = 18) = \underline{0.00007}$$
$$P(x \geq 14 \mid \pi = 0.5, n = 18) = 0.01543$$

With $\pi = 0.50$ the binomial distribution is symmetric and is close to being a smooth bell-shaped distribution with just 18 cases. By the rule of thumb for bell-shaped distributions, the probability of being two standard deviations above the expected number is approximately 0.025, which in this case is

$$P(x > \mu + 2\sigma) = P(x > 9 + 4.24264) = P(x > 13.24264) \cong 0.025$$

which is not far from

$$P(x \geq 14 \mid \pi = 0.5, n = 18) = 0.01543$$

QUERY 5.6:

The introduction to this chapter included a report that "only five of 34 economists who participated in 10 or more surveys managed to guess correctly which way long-term bond yields were going more than half of the time." It asked: Are these results as likely if done by 34 people each flipping a coin 10 times to predict yields? It also asked about the implications of this probability for the accuracy of economists' forecasts.

ANSWER:

The probability of five of 34 individuals predicting six or more yield movements in 10 tries by flipping a coin is obtained by first using the binomial distribution to find the probability of a coin tossed 10 times giving six or more heads (corresponding to six or more correct yield predictions, $x \geq 6$). Using a statistics package such as ECONOMETRICS TOOLKIT this probability is 0.377:

$$P(x \geq 6 \mid n = 10, \pi = 0.5) = 0.377$$

Next, assuming the 34 individuals operate independently, the probability of five individuals, $y = 5$, predicting six or more yield movements correctly by each flipping a coin 10 times is 0.0023:

$$P(y = 5 \mid n = 34, \pi = 0.377) = 0.0023$$

That is, only about 23 out of 10,000 times would we expect to see five individuals predict six or more yield movements correctly if they were predicting by 50-50 guessing, as in flipping a coin. Observing five of 34 economists predicting correctly more than half of the long-term bond yields in 10 tries is an extremely rare occurrence if based on 50-50 coin-toss-type guessing. Contrary to the implications of the *WSJ* article, economists appear to predict long-term bond yields better than what would be expected if they were just guessing.

EXERCISES

5.12 Find the mean and standard deviation of the number of successes in the binomial distribution as specified by

a. $\pi = 0.2, n = 10$

b. $\pi = 0.5, n = 17$

c. $\pi = 0.75, n = 30$

5.13 In Exercise 5.11, what is the expected number and standard deviation of the number of airports in a sample of 20 of the 100 top airports for which Dollar has on-site operations?

5.14 In May 1992, NordicTrack was advertising that its exerciser was still being used on a regular basis (at least three hours per week) by seven out of 10 buyers five years after purchase. If this claim is true:

a. What is the expected number still using the exerciser five years later in a random sample of three purchasers? What does this number mean?

b. What is the probability of drawing a random sample of three with no regular users five years later?

5.15 For the data and conditions in Exercise 5.14, if you select a sample of three buyers and find no regular users, what could you conclude (and why) about the correctness of NordicTrack's ad?

5.16 If only 50 percent of the buyers of the NordicTrack exerciser are regular users five years later, and you select a sample of 10 buyers, what is the probability of selecting a sample that seven or more are regular users five years later? What can you conclude from this and your answer in Exercise 5.15?

5.17 According to *Newsweek* (June 8, 1992), "American drug companies are justly famous for their largesse . . . they spend $550 million a year on advertising in medical journals . . . (but) of the 109 ads reviewed (by UCLA researchers), 100 contained 'deficiencies in areas for which the FDA has established explicit standards for quality.'" If the expected number of deficiencies was 55 ads, what is the probability of finding 100 or more deficiencies? What did you have to assume to calculate this probability with the binomial distribution?

5.18 During the Persian Gulf operation in 1991, 53 of the 291 Americans who lost their lives were black. About a quarter of those who took part in Desert Storm were black.
   a. Did fewer or more blacks die than would be expected by chance during Desert Storm?
   b. What was the probability of getting 53 or fewer blacks losing their lives if proximity to risk was random?

5.5 The Hypergeometric Distribution

As with the binomial distribution, the hypergeometric distribution is used when

1. the random variable of interest takes on the values of 0, 1, 2, 3, . . . *n*, where this value is the event formed by the sum of *n* random variables;
2. there are *n* trials that give rise to the *n* random variables; and
3. each trial has only one of two outcomes (0 or 1).

But unlike the binomial, the hypergeometric distribution does not require that the probability of either outcome on each of the *n* trials remain fixed from trial to trial. That is, the trials are dependent.

Consider the case of claimed sex discrimination in hiring. The employer hired 10 workers from a preselection pool of eight women and 12 men, but only one woman was hired. The question is: What is the probability of selecting one woman in 10 random draws from this pool?

Here again there are two outcomes (a person is hired or is not hired) on each of 10 trials. Now the probability of each outcome changes from trial to trial. With random selection, the probability of hiring a woman on the first trial is 8/20 = 0.40. If a woman is hired, then the probability that the person hired on the second trial is a woman is 7/19 = 0.368421, because, of the 19 applicants left, seven are women. The probability of hiring a woman changes from the first trial to the second trial, which means the binomial is inappropriate. The appropriate probability mass function is the hypergeometric.

Hypergeometric Probability Distribution
A discrete probability distribution for the sum of a series of *n* trials, where each trial results only in one of two outcomes (0 or 1), and the outcome of each trial influences the outcome of the next trial.

Applications of the **hypergeometric probability distribution** require that the population (all applicants) is divided into two categories: those that share one attribute (women) and those that share another (men). For a population of size N, let N_1 = the number sharing the one attribute and let N_2 = number sharing the other attribute. Because there were eight women and 12 men,

N_1 = number in group one (= 8 women),
N_2 = number in group two (= 12 men), and
$N = N_1 + N_2$ (= 20 applicants).

For a sample of size n, let

x_1 = number realized from group one (= 1 woman hired),
x_2 = number realized from group two (= 9 men hired), and
$n = x_1 + x_2$ (= 10 hired).

We are interested in calculating the probability of x_1 and x_2 in a sample of size n. In our case, we are interested in the probability that $x_1 = 1$ (one woman is hired) and $x_2 = 9$ (nine men are hired). That is, $P(x_1 = 1 \text{ and } x_2 = 9)$, which is given by Formula (5.4).

Hypergeometric Probability Mass Function

$$P(x_1 \text{ and } x_2) = \frac{\dfrac{N_1!}{x_1!(N_1 - x_1)!} \dfrac{N_2!}{x_2!(N_2 - x_2)!}}{\dfrac{N!}{n!(N-n)!}} \tag{5.4}$$

Hand calculations with Formula (5.4) are feasible for small populations, but for large N a computer statistics package is needed or calculation routines can be written for a spreadsheet. The following calculations can be verified by either method, or by hand calculation if need be. Substituting values into Formula (5.4) gives

$$P(x_1 = 1, x_2 = 9) = \frac{\dfrac{8!}{1!(8-1)!} \dfrac{12!}{9!(12-9)!}}{\dfrac{20!}{10!(10!)}}$$

$$= \frac{8 \times 12 \times 11 \times 10 \times 9 \times 8 \times 7 \times 6 \times 5 \times 4 \times 10}{20 \times 19 \times 18 \times 17 \times 16 \times 15 \times 14 \times 13 \times 12 \times 11} = .00953$$

If being hired has nothing to do with a person's sex, then the probability that exactly one woman and nine men are hired by the equivalent of a random draw from a preselection pool of eight women and 12 men applicants is 0.00953. This means that fewer than one in 100 times will we see a random draw of 10 individuals from this pool of 20 applicants result in only one woman being selected—an extremely unlikely event.

To see how unlikely the hiring of one woman is relative to other possibilities when 10 people are randomly drawn from the preselection pool of 20, consider the hypergeometric probability mass function in Figure 5.2. The most likely outcome is four women, which occurs with a probability of 0.350083. Some 83 percent of the time we should see three, four, or five women hired.

The hiring of only one woman does not appear consistent with what we could expect from a random draw from the preselection pool. In particular, substituting the probabilities in Figure 5.2 into the formula for the mean in Chapter 4, gives the expected number of women to be hired as four.

FIGURE 5.2 Hypergeometric Probability Mass Function for Women Hired

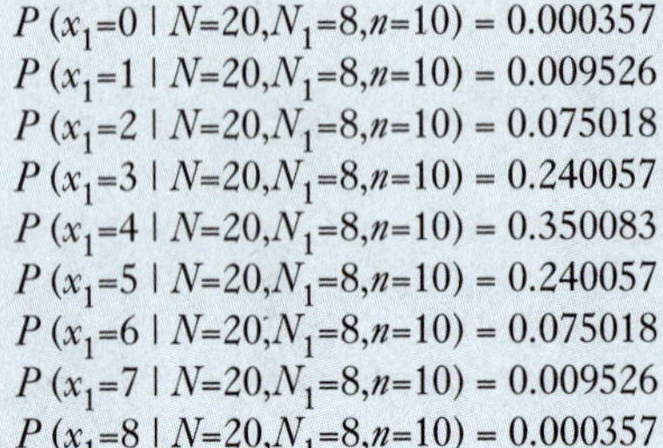

$P(x_1=0 \mid N=20, N_1=8, n=10) = 0.000357$
$P(x_1=1 \mid N=20, N_1=8, n=10) = 0.009526$
$P(x_1=2 \mid N=20, N_1=8, n=10) = 0.075018$
$P(x_1=3 \mid N=20, N_1=8, n=10) = 0.240057$
$P(x_1=4 \mid N=20, N_1=8, n=10) = 0.350083$
$P(x_1=5 \mid N=20, N_1=8, n=10) = 0.240057$
$P(x_1=6 \mid N=20, N_1=8, n=10) = 0.075018$
$P(x_1=7 \mid N=20, N_1=8, n=10) = 0.009526$
$P(x_1=8 \mid N=20, N_1=8, n=10) = 0.000357$

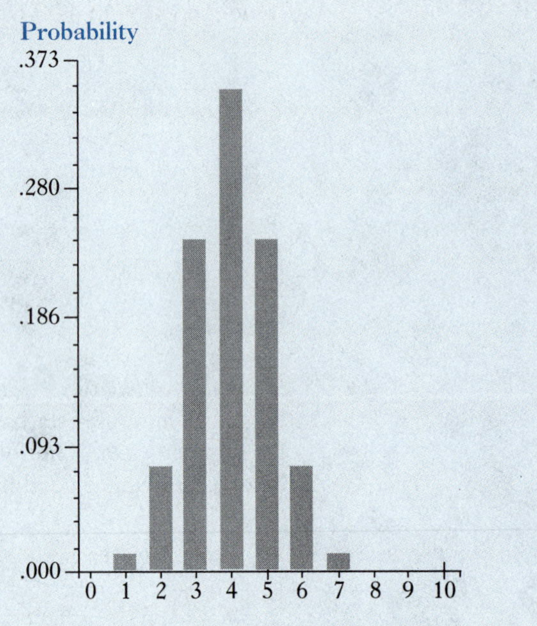

$$\mu = E(x_1) = \sum_{x_1=0}^{8} x_1 P(x_1 \mid N = 20,\ N_1 = 8,\ n = 10)$$

$$= (0)(.000357) + (1)(.009526) + (2)(.075018) + (3)(.240057)$$

$$+ (4)(.350083) + (5)(.240057) + (6)(.075018) + (7)(.009526)$$

$$+ (8)(.000357) = 4.00$$

This mean value, however, can be calculated more easily by using shortcut Formula (5.5), which is the expected value of any random variable distributed as a hypergeometric.

Mean of Hypergeometric Random Variable

$$\mu_1 = E(x_1) = \frac{nN_1}{N} \tag{5.5}$$

In our case, the expected number of women to be hired is

$$\mu_1 = nN_1/N = (10)(8)/20 = 4$$

Notice that, although the probability of success from trial to trial changes in the hypergeometric distribution but not in the binomial distribution, the formula for the expected value of a hypergeometric (Formula 5.5) is the same as for the binomial (Formula 5.2), since $N_1/N = \pi$. The variance of a hypergeometric (Formula 5.6) can likewise be compared to that of a binomial (Formula 5.3), however, with one difference.

Variance of a Hypergeometric Random Variable

$$\sigma^2 = \frac{N-n}{N-1}\left[\frac{nN_1}{N}\right]\left[1-\frac{N_1}{N}\right] \tag{5.6}$$

With the exception of the ratio $(N - n)/(N - 1)$, the variance of the hypergeometric in Formula (5.6) is the same as that of the binomial in Formula (5.3), again since $N_1/N = \pi$. In general, the variance of the hypergeometric is less than a corresponding binomial, because $(N - n)/(N - 1) < 1$. If the population size N is large relative to the sample size n, however, then $(N - n)/(N - 1) \cong 1$, and the variance of the hypergeometric and the binomial are approximately the same. It is for this reason that the binomial distribution can be used to approximate hypergeometric probabilities when the population is large and the sample is small. This is true even though there is no replacement to maintain a fixed probability of success from trial to trial. When N is large relative to n, the probability of success changes little from trial to trial, making binomial calculations possible although there is no replacement.

QUERY 5.7:

Michael Finkelstein and Bruce Levin (1990) report on a lawsuit involving *Ultramares Corp. v. Touche.* The accountants (Touche) had checked invoices in a sample of 100 (drawn from 1,000 invoices), but none of 17 fraudulent invoices were in the sample. When the company failed, a creditor sued, claiming the accountants had been negligent in certifying the financial statements. Were the accountants negligent?

ANSWER:

The probability of having no fraudulent invoices in a random sample of 100, drawn from a population of 1,000 with 17 fraudulent invoices, is

$$P(x_1 = 0,\ x_2 = 100) = \frac{\dfrac{17!}{0!(17-0)!}\ \dfrac{983!}{100!(983-100)!}}{\dfrac{1000!}{100!(900!)}}$$

$$= \frac{983 \times 982 \times \ldots \times 884}{1000 \times 999 \times \ldots \times 901} = 0.164$$

Thus, 16.4 percent of the time a sample of 100 invoices with no fraudulent invoices may show up, even though 17 of the 1,000 invoices in the population are fraudulent. That is, the random error in sampling could account for not getting any fraudulent invoices—the accountants need not have done anything wrong.

To the above argument, however, the creditors could argue that the accountants should have drawn a larger sample that would have reduced the random error and reduced the probability of getting a sample with no fraudulent invoices. But even for a sample of 200, there is still a 0.022 chance that the sample would not contain a fraudulent invoice.

$$P(x_1 = 0, x_2 = 200) = \frac{\dfrac{17!}{0!(17-0)!}\;\dfrac{983!}{200!(983-200)!}}{\dfrac{1000!}{200!(800!)}}$$

$$= \frac{983 \times 982 \times \ldots \times 784}{1000 \times 999 \times \ldots \times 801} = 0.022$$

Creditors and investors alike must recognize that when accountants use statistical sampling (as described by Donald Roberts, 1987, for example), there is always a chance of sampling error, as discussed in more detail in later chapters.

Exercises

5.19 There are 12 portable computers available for loan, three of which are defective. If three computers are randomly selected for a traveling sales team, what is the probability that all three are the defective ones?

5.20 A bank has 15 assistant vice-presidents who have applied for three openings at the vice-president levels. Ten of the assistant VPs are women and five are men. What is the probability that all three openings are filled by men if selection is random?

5.21 What is the expected number of women selected to fill the open positions in Exercise 5.20? What does this imply about the filling of these three positions?

5.22 In Exercises 5.20 and 5.21, why can't you assume that there is replacement so that the binomial distribution could be used?

5.23 A dishonest bookkeeper has altered five entries in an accounts receivable file that has a total of 20 entries. An auditor is about to randomly select four of the 20 entries to check if they are still outstanding or if they have been paid. What is the probability that at least one of the four accounts selected will be one of those altered by the dishonest bookkeeper?

5.6 The Poisson Distribution

Like the binomial and the hypergeometric, we use the **Poisson distribution** when we are interested in the probability of a count (0, 1, 2, 3, . . .). Unlike the binomial and

Poisson Probability Distribution
A discrete probability distribution for the occurrence of an event $x = 0, 1, 2, \ldots,$ when the probability of occurrence is small but the number of opportunities for occurrence is large.

hypergeometric distributions, for the Poisson there is no specific upper limit to the count, although a finite count is expected. For instance, a production manager might want to determine the probability that an assembly line is stopped x times for repairs during an eight-hour shift. At least theoretically, the line could be stopped an unlimited or infinite number of times, but in all likelihood it will only be stopped a few times. Conceptually, there could be any number of blisters on the bottom of a new fiberglass boat, but in all likelihood there will be few, if any. These are examples of situations in which the Poisson distribution may apply because rare events occur randomly in a fixed time period or over a finite space in which many occurrences are possible.

Although the Poisson distribution is defined for an unlimited number of occurrences, it is often used in situations where the maximum number of occurrences is large but known to be finite. In these situations, the important point is that an extremely low count is expected when a large number is possible. That is, the Poisson distribution may be applicable when the number of possible occurrences in a period of time or space is really not without bound but is known to be large, while the number of expected realizations is small.

For instance, at a large midwestern university, the number of students who could cheat during final exams in micro- and macroeconomics is large (about 2,000 per semester) but the actual number caught is small (mean of 1.9), and thus the Poisson distribution may be considered for probability calculation. As another example, the U.S. Centers for Disease Control found that 13 people were killed in North Carolina by falling or being thrown from a horse in the ten-year period between 1979 and 1989; thus, the mean number is 1.3 per year. Although there is clearly a finite number of horse riders it is extremely large relative to the number of occurrences observed, and the Poisson distribution may be considered for probability calculation. In fact, the first application of the Poisson distribution, by Ladislaus von Bortkiewicz, involved the calculation of deaths from horse kicks in the Prussian army in 1898.

In addition to the Poisson random variable being based on an unlimited number of possible occurrences, the probability of an occurrence must be the same for all (time or space) intervals of equal length or size. The occurrence of the event in any interval must also be independent of the occurrence in any other interval. A Poisson random variable is defined by these conditions, and probabilities for a Poisson random variable are calculated by Formula (5.7).

Poisson Probability Mass Function

$$P(x) = \frac{e^{-\lambda}\lambda^{x}}{x!}, \text{ for } x = 0,\ 1,\ 2,\ 3 \ldots \text{ and } \lambda > 0$$

$$P(x) = 0, \text{ otherwise} \tag{5.7}$$

where $e = 2.71828$ and

λ is the mean (and also the variance) of x

Poisson probabilities are easily determined by a statistics computer program or by writing a routine in a spreadsheet or by using a table such as that in Appendix Table A.2. To determine a Poisson probability by any of these methods, the mean or expected number of occurrences λ (lambda) must first be known. This mean is typically estimated from historical data, as in the exam cheating case where $\lambda = 1.9$ or in the horseback death case where $\lambda = 1.3$. A unique feature of the Poisson distribution is that these means are also the variances of their respective distributions.

In specifying λ, it is important to make sure that the intervals are standardized. For example, in the case of cheaters on final exams, all the classrooms in which the test are administered should be of roughly equal size. It might be that cheating in large classrooms is more likely than cheating in small classrooms. In his classic study of deaths from horse kicks, Bortkiewicz followed 14 cavalry corps for 20 years and discarded four corps that were considerably larger than the others. Presumably there were more opportunities for deaths from horse kicks in these larger corps than in the typical corps.

Appendix Table A.2 provides the probability mass function for selected values of λ between 0.1 and 10. For example, the panel containing the probability mass function for $\lambda = 1.3$ is reproduced in Table 5.2. From this we learn, for example, that the probability of no one falling off a horse and dying is 0.2725 in the 10-year period followed by the U.S. Centers for Disease Control. The probability of one person falling off a horse and dying is 0.3543 in a 10-year period, and so on, to an approximate zero probability of nine people falling off a horse and dying in a 10-year period. It is similar for cheaters, where the probability of catching no cheaters in a given semester is 0.1496.

Notice in Table 5.2 that the Poisson distribution is right skewed. This skewness is more pronounced for smaller values of λ. If $\lambda < 1.0$, as in panel a of Figure 5.3 where $\lambda = 0.6$, the Poisson distribution is highly right skewed with the probability of x decreasing throughout the range of x. If n is large and π is small, so that $\lambda = n\pi$ is only moderately small, then the Poisson and the binomial have approximately the same right-skewed shape. As λ increases, as with the binomial, the Poisson approaches a bell-shaped curve, becoming almost symmetric by the time $\lambda = 10$. For larger λ, probabilities from the Poisson distribution may be approximated with probabilities determined from the normal distribution, which is the focus of the next chapter.

TABLE 5.2
Poisson Probabilities

	λ									
x	1.1	1.2	1.3	1.4	1.5	1.6	1.7	1.8	1.9	2.0
0	.3329	.3012	.2725	.2466	.2231	.2019	.1827	.1653	.1496	.1353
1	.3662	.3614	.3543	.3452	.3347	.3230	.3106	.2975	.2842	.2707
2	.2014	.2169	.2303	.2417	.2510	.2584	.2640	.2678	.2700	.2707
3	.0738	.0867	.0998	.1128	.1255	.1378	.1496	.1607	.1710	.1804
4	.0203	.0260	.0324	.0395	.0471	.0551	.0636	.0723	.0812	.0902
5	.0045	.0062	.0084	.0111	.0141	.0176	.0216	.0260	.0309	.0361
6	.0008	.0012	.0018	.0026	.0035	.0047	.0061	.0078	.0098	.0120
7	.0001	.0002	.0003	.0005	.0008	.0011	.0015	.0020	.0027	.0034
8	.0000	.0000	.0001	.0001	.0001	.0002	.0003	.0005	.0006	.0009
9	.0000	.0000	.0000	.0000	.0000	.0000	.0001	.0001	.0001	.0002

FIGURE 5.3 Poisson Probability Mass Functions for Different Means

Panel a: Mean λ = 0.6

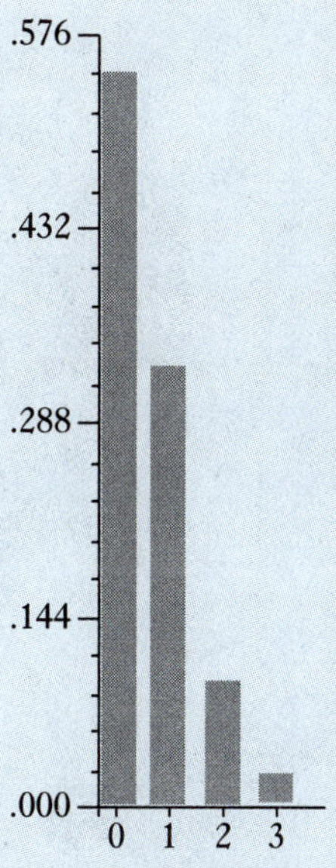

Panel b: Mean λ = 1.3

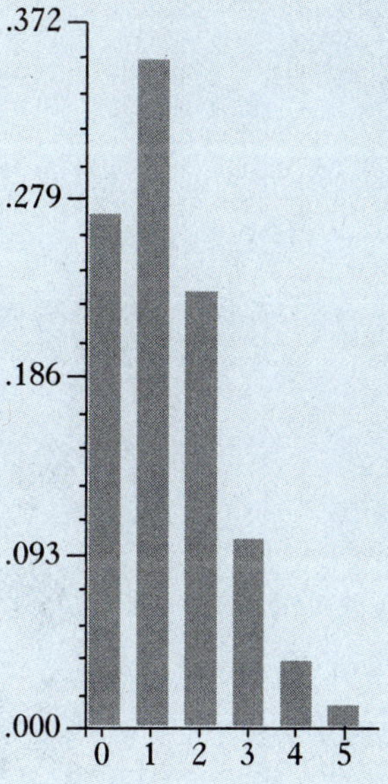

(continued)

FIGURE 5.3 Poisson Probability Mass Functions for Different Means (*continued*)

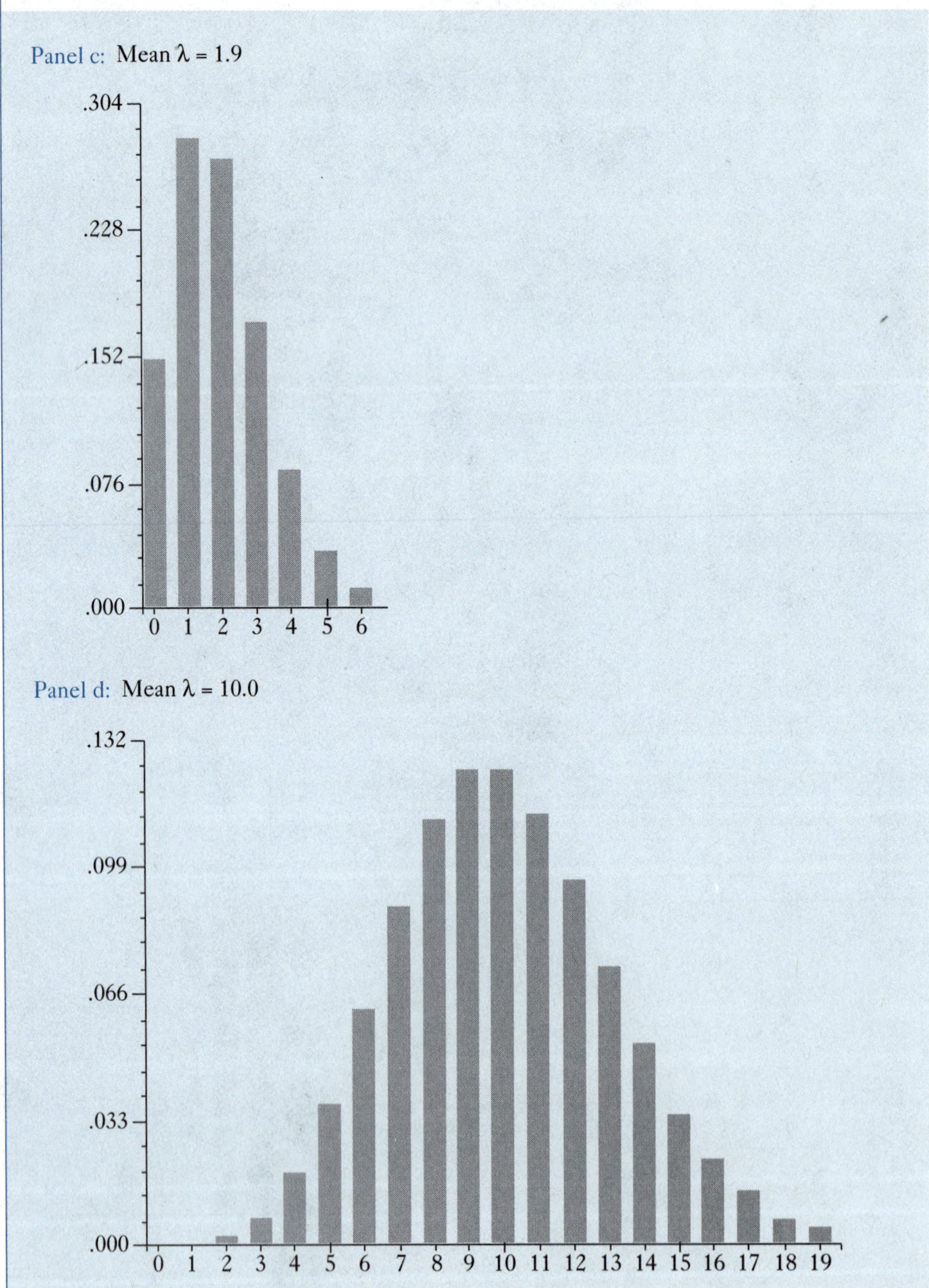

QUERY 5.8:
The average breakdown rate of a change machine in a dorm food vending room is 0.1 per weekday. What is the probability there will be a breakdown on any given week day?

ANSWER:
The Poisson distribution may be applicable here because the machine can break down an unspecified large number of times on each day but the probability of a breakdown is small. The probability of a breakdown is given as being the same from day to day, with occurrences independent across days. Either Appendix Table A.2 or a computer program can be used to determine the probabilities of 1, 2, 3, or more breakdowns. This probability is easiest to compute as 1 – P(no breakdowns), which is

$$P(x = 1, 2, 3 \ldots \mid \lambda = 0.1) = 1 - P(x = 0 \mid \lambda = 0.1) = 1 - 0.9048 = 0.0952$$

Thus, approximately 10 percent of the time, there will be a breakdown of the machine on any given week day.

EXERCISES

5.24 Use Appendix Table A.2 to find the following Poisson probabilities.
   a. $P(x = 2 \mid \lambda = 3)$
   b. $P(y \leq 5 \mid \lambda = 2.7)$
   c. $P(1 < w < 4 \mid \lambda = 3)$

5.25 The number of blisters in the gelcoat of fiberglass boats produced at Melges Boat Works is known to be Poisson distributed, with a mean rate of 0.01 blisters per square foot. In the layup (production) of the size of a hull that measures 24 feet by 5 feet, what is the probability that the gelcoat will have no blisters?

5.26 In an article titled "Murder, By the Numbers: Will Statistics Prove a Nurse Guilty of Homicide?" *Newsweek* (May 20, 1988) reported that nurse Jane Boulding's patients suffered 57 cardiac arrests, while the average for her co-workers was less than five during a 15-month period. By the Poisson distribution, what is the probability of 57 or more cardiac arrests if the expected number of cardiac arrests is five? Is the Poisson distribution appropriate in this case? Would you conclude nurse Boulding may have been a contributing factor in her patients' deaths?

5.27 The article in *Business Week* (June 29, 1992) on the production of Cadillac Sevilles reported that "an internal quality audit in early June found 5.1 defects per car vs. a target of 1.0." If the target is being maintained on average (the process is in control), then what is the probability that a randomly selected car will have five or more defects, assuming defects follow the Poisson distribution?

5.28 In steel mills in less-developed countries, an average of three serious injuries per week is not unusual. These injuries tend to occur independently of one another and at a relatively constant rate. At such mills, what is the probability of no serious injuries in a week?

5.29 During peak hours, the average number of telephone calls transferred by a receptionist at a major auto dealership is three per minute. What is the probability that the receptionist will have to transfer no calls in a minute, assuming calls follow a Poisson distribution?

5.30 During the breakfast, lunch, and dinner periods, restaurant customers arrive at a register at an average of 10 per hour. Arrivals have a Poisson distribution. In a given hour, what is the probability

a. of 10 customers arriving at the register?
b. of five customers arriving at the register?
c. of no customers arriving at the register?

Historical Perspective

The Bernoulli family produced no fewer than five scholars who contributed to the study of probability and statistics, and several other academics whose contributions to other areas of science are noteworthy. Jacob Bernoulli (1654–1705) advanced the study of combinatorial probabilities, as found in the binomial and hypergeometric distributions of today. His primary objects of analysis were the fabled urn (from which white or black pebbles were drawn), coins (flipped for a desired side), and dice (rolled for a specific face value).

Bernoulli was a testy and discontented person who spent more time competing with and denigrating his academic brother John than collaborating with him. After his death, John wanted nothing to do with Jacob's yet-unpublished papers. It wasn't until eight years later, when nephew Nicholas edited them in *Ars Conjectandi,* that the papers became a seminal work on the mathematics of probability and introduced the first law of large numbers and the "Bernoulli numbers." In this work, the vocabulary of "fertile cases" or "sterile cases" was used for successful and unsuccessful trials that could actually be enumerated in the context of urns, coins, and dice. Bernoulli argued that simple extensions to the study of disease, weather, or games of skill "would be a sign of insanity," because the underlying processes in real-world applications were hidden, making the enumeration of cases impossible. For such situations, he recognized the common-sense idea that the more observations there were on the unknown proportion of "cases," the closer the calculated proportion would be to the true proportion, but he could not formalize this notion mathematically.

Mathematician Siméon Denis Poisson (1781–1840) derived his distribution in its cumulative form as a limit to Bernoulli's binomial distribution when the number of trials is large but the chance of success is small. Poisson, a fun-loving and socially minded Frenchman, was an innovator in applying algebra and probability to analyses of guilty verdicts and juror errors. When it came to deriving the distribution that would later bear his name, however, he was not concerned with real-world applications; as were the Bernoullis before him, he was intrigued with its mathematical properties.

Antonine-Augustin Cournot (1801–1877), whose early work was largely ignored by contemporary mathematicians but which later provided many concepts in economics, popularized the Poisson distribution. In 1898, Polish mathematician/statistician/economist Ladislaus von Bortkiewicz (1868–1931) published "The Law of Small Numbers," in which he introduced the now-famous example of soldiers' deaths from horses' kicks following the Poisson distribution.

5.7 Concluding Comments

Within almost all areas of business practices, legal proceedings, and the science of economics generally, the binomial, hypergeometric, and Poisson distributions have become key tools of statistical analysis. As concluded by Donald Roberts, 275 years after Jacob Bernoulli's death, applied statisticians were able to show that while the simple urn, coin, and dice models of the early years of probability were viewed as insufficient to support the needs of business and economic applications, they did provide the focus for debate on the mathematics of probability theory. Nearly three centuries later, desktop computers put the power of these distributions at the fingertips of analysts for everyday use in problems never envisioned by the mathematicians who invented them.

In the next chapter, we will see how further extensions of the binomial probability distributions gave rise to the most famous of all distributions: the normal distribution. Curiously, the normal distribution, as with the Poisson distribution, was first proposed as a way to approximate the cumulative binomial probabilities. It was not until well into the eighteenth century that applications of the normal distribution came into their own.

Chapter Exercises

5.31 A *Wall Street Journal* (August 14, 1987) article reported that, "IBM samples 50 circuit boards out of every 270-board lot; if one is defective, every board in the lot is rejected." (p. 12) Why is this a demanding standard?

5.32 In the above problem, if there is only one defective board in the lot, what is the probability of getting it in a sample of 50?

5.33 As reported in the *American Economic Review* (May 1992), a sample of 600 economists was surveyed by mail to assess their reactions to certain economic policies. One proposition was "Wage-Price Controls are a useful policy option in the control of inflation." (p. 204) Out of the 464 responding economists, 73.9 percent said they disagreed. What is the probability of getting this response if three quarters of economists did not favor wage-price controls? Do your calculations with the binomial and the hypergeometric. Which distribution is the most appropriate?

5.34 An Associated Press release (*HT,* June 22, 1992) reported on research by psychologist Marsha Kaitz of the Hebrew University in Jerusalem in which "seventy-two blindfolded people in the study tried to distinguish their romantic partner from two decoys of similar age, weight and height . . . the blindfolded people were correct 58 percent of the time in the forehead test."

a. What is the expected number of correct identifications if only guessing is involved?
b. What is the probability of correct identification of 42 or more persons if participants were just guessing?
c. Do you think they were just guessing?

5.35 In the research reported in Exercise 5.34 about the blindfolded identification of partners, "women identified their man's hand 69 percent of the time." What is the probability that 72 women will correctly identify 50 or more men if the women were just guessing? Do you think the women were just guessing?

5.36 In *The Art of Conjecture* (Basic Books, 1967), Bertrand de Jonvenel wrote:

> If there is a choice between only two decisions, and if one consults eleven experts who each have 95 chances in 100 of arriving at the right verdict, a simple calculation shows the odds are 1,000 to 1 that a simple majority will be in favor of the right choice. (p. 204)

Is there a "simple calculation" that gives the odds 1,000 to 1 or is it 10,000 to 1?

5.37 There was some discussion on a computer bulletin board about the correctness of the simple calculation in the quote in Exercise 5.36 regarding an assumption that had to be made to derive the associated probability.

a. What was the assumption?

b. Under what conditions would it or would it not be true?

5.38 An individual involved in the computer bulletin board discussion of the probability presented in Exercise 5.37 stated that the relevant probability is not only six out of 11 but should include 7, 8, 9, 10, and 11 correct, as well. Is this person correct? How does this change the probability?

5.39 An article in *The Wall Street Journal* (June 18, 1987) stated, "Drugstore pharmacists catch an average of more than four prescription errors a week." If the "catch rate" was exactly 4.0 per seven-day week, what is the Poisson probability that an error is found on the first day of the week? Why might a critic question the use of the Poisson distribution in this case?

5.40 An article in *The Wall Street Journal* (May 10, 1991) reported on Nucor Steel's efficiency in increasing output. The article reported, however, "Its worker death rate since 1980 is the highest in the industry—in fact, more than double the industry average. Since 1980 11 Nucor workers have died . . . ." If the industry average death rate in this 11-year period was 0.4 deaths per year, what is the likelihood of observing 11 or more deaths in 11 years?

5.41 A manufacturer of automobile spark plugs knows that 10 percent of its plugs are defective. A random sample is drawn. Use this printout from ECONOMETRICS TOOLKIT to answer the following questions.

Distributions	x Prob[X = x]	x Prob[X = x]	x Prob[X = x]	x Prob[X = x]
1 Binomial	0 .042391	5 .102305	10 .000365	15 .000000
2 Poisson	1 .141304	6 .047363	11 .000000	16 .000000
3 Geometric	2 .227656	7 .018043	12 .000000	17 .000000
4 Hypergeom.	3 .236088	8 .005764	13 .000000	18 .000000
5 Neg. Binom	4 .177066	9 .001565	14 .000000	19–30 not 0

Success probability π	.100	$0 < \pi < 1$
Sample size, n	30	$1 \le n \le 99$

Mean= 3.000, Std. Deviation= 1.643

Enter P,C,R, or I for the type of computation

Prob P[X = x]	what is x?3	P=	.236	1 – P=	.764
Cumul P[X ≤ x]	what is x?3	P=	.647	1 – P=	.353
Range P[a ≤ X ≤ b]	what is a?3	b?30		P=	.589
Invrs: x for P[X ≤ x]= P:		what is P?		X=	

a. What is the probability that at least three spark plugs are defective?
b. What is the probability that three spark plugs are effective?
c. Which of the following statements is true?
   1. The average number of defective spark plugs over an infinite number of large random samples ($n > 30$) is 3.00.
   2. The average number of defective spark plugs per sample, over an infinite number of random samples, is 3.00.
   3. In each sample of size 30, the average number of defective spark plugs is 3.00.
   4. In this sample, the number of defective spark plugs is 3.00.
   5. In this sample, the average number of defects is $p = 3/30$.

5.42 Assume a quality control expert randomly selects three items from the end of an assembly line, for which it is claimed that there is a 0.8 probability, π, that an item is totally error free. If the number of items that are error free is a binomial random variable x, then
a. the expected number of error-free items is calculated by the shorthand formula $E(x) = n\pi$, which yields a numerical value of ________ .
b. by summing over the two values and their probabilities for each of the three trials, show how the resulting six terms form the above expected value.
c. explain what your numerical expected value means.
d. the variance is calculated by the shorthand formula $E(x - \mu)^2 = n\pi(1 - \pi)$, which yields a numerical value of _________ .
e. by summing over the numerical values and their probabilities for each of the three trials, show how the six resulting terms form the above value of the variance.

5.43 In testing the items in Exercise 5.42, the items are destroyed and cannot be replaced. Thus, for this process to be treated as a binomial, what do you have to assume?

5.44 In testing the three items in Exercise 5.42, if you find one totally error-free item, are you willing to conclude that the assembly line is running in a way that the probability of any given item being error free is 0.8? Explain and justify your answer in terms of probabilities.

5.45 (computer software required) In a survey of 105 female judges, 54 percent said sexual bias was widespread in California state courts, according to the *Wall Street Journal* (August 20, 1992). How likely was this sample result if, in fact, the majority of female judges (50 percent) do not think that sexual bias was widespread?

5.46 On CBS's TV program "Shame On You" (June 2, 1993), the announcers described their experiment in which five VCR repair centers were randomly selected and brought new VCR units that had a blown fuse inserted. Three of the five repair centers said major repairs were needed, with the cost of repair exceeding $90; two of the centers correctly diagnosed the $1 blown fuse. What is the probability of getting these results if the majority (50 percent) of all VCR repair centers are honest and good at diagnostics?

5.47 Cars enter a parking lot for Indiana University basketball games at an average rate of four per minute. If this process conforms to a Poisson distribution, what is the probability of more than five cars entering in a specific minute?

5.48 In 1974, the national maximum speed limit of 55 miles per hour was enacted. D. B. Kamerud (*Transportation Research*, 1983) reported that the fatality rate for interstate highways in 1975 was approximately 16 per 100 million vehicle miles, but had the 55 mph limit not been enacted it would have been 25 deaths per 100 million vehicle miles.
a. What are the probabilities of at most 10 deaths per 100 million vehicle miles with and without the 55 mph limit in 1975?
b. What are the probabilities of at most 20 deaths per 100 million vehicle miles with and without the 55 mph limit in 1975?

c. What are the probabilities of at most 30 deaths per 100 million vehicle miles with and without the 55 mph limit in 1975?

d. What did you have to assume to do the probability calculations in parts a, b, and c? Are these reasonable assumptions to make? Why or why not?

5.49 If the courts place the average value of a life at $2 million, what is the expected savings in 1975 from the enactment of 55 miles per hour speed limit in 1974, using the data and required assumptions of Exercise 5.48?

5.50 Before concluding that the enactment of the 55 miles per hour speed limit was economically justified, what other costs and benefits must be considered?

5.51 As a student in graduate school, Professor Becker had an old V8 Buick convertible that overheated, and the engine seized. An experienced mechanic informed Becker that it sounded as if one piston on the right bank of four cylinders was cracked, but he could not tell which one. Becker gambled and replaced only three of the four pistons. What is the probability that the bad piston was not replaced?

5.52 "Child resistent lighters" are so designated if a child cannot ignite the lighter in 15 minutes of trying. If in a classroom of 30 students, 15 could figure out how to ignite a lighter, what is the probability of selecting five students at random and finding none of them able to perform the task?

5.53 *The Wall Street Journal* (December 17, 1992) reported that "12 out of the 31 analysts that track IBM had a 'buy' on the company as of September 1, when the stock had already fallen to $87.625. Only three analysts had an outright 'sell' . . . ." By December 21, 1992, IBM's price had fallen to an 11-year low of $51.875, while Intel's stock price rose to a record $83.50 and Microsoft's value likewise rose (*WSJ*, Decemb :r 21, 1992). If over the last major turning point in the price of IBM stock, only three of . 1 analysts were correct in their "sell" orders, would you conclude that they are using some hing other than chance in their orders?

ACADEMIC REFERENCES

Becker, William, William Greene, and Sherwin Rosen. "Research on Economic Education." *American Economic Review*, May 1990, pp. 14–22.

Finkelstein, Michael, and Bruce Levin. *Statistics for Lawyers*, New York: Springer-Verlag, 1990.

Roberts, Donald. *Statistical Auditing*. American Institute of Certified Public Accountants, 1987.

ENDNOTES

1. $\mu = \Sigma\, xP(x)$, which for the binomial yields

$$\mu = \underbrace{[(1)(\pi) + (0)(1-\pi)]}_{\text{1st trial}} + \underbrace{[(1)(\pi) + (0)(1-\pi)]}_{\text{2nd trial}} + \ldots + \underbrace{[(1)(\pi) + (0)(1-\pi)]}_{n\text{th trial}}$$

$$\mu = \pi \qquad + \pi \qquad + \ldots + \pi$$

$$\mu = n\pi$$

2. For the binomial random variable $x = x_1 + x_2 + \ldots + x_3$, the variance of x is

$$\sigma^2 = \text{Var}(x_1 + x_2 + \ldots + x_3) = \text{Var}(x_1) + \text{Var}(x_2) + \ldots + \text{Var}(x_3)$$

because the trials are independent $\text{Cov}(x_i x_j) = 0$

Thus, for the binomial random variable x,

$$\sigma^2 = \underbrace{[(1-\pi)^2\pi + (0-\pi)^2(1-\pi)]}_{\text{1st trial}} + \underbrace{[(1-\pi)^2\pi + (0-\pi)^2(1-\pi)]}_{\text{2nd trial}} + \ldots$$

$$= [\pi - 2\pi^2 + \pi^3 + \pi^2 - \pi^3] + [\pi - 2\pi^2 + \pi^3 + \pi^2 - \pi^3] + \ldots$$

$$= \pi(1-\pi) + \pi(1-\pi) + \ldots$$

$$= n\pi(1-\pi)$$

FORMULAS

- Binomial Probability Mass Function

$$P(x) = \frac{n!}{x!(n-x)!}\pi^x(1-\pi)^{n-x}$$

where

$$x = x_1 + x_2 + \ldots + x_n = \sum_{i=1}^{n} x_i$$

$$x_i = 0 \text{ or } 1$$

$$P(x_i = 1) = \pi \text{ and } P(x_i = 0) = 1 - \pi$$

- Mean of a Binomial Random Variable

$$\mu = E(x) = n\pi$$

- Variance of a Binomial Random Variable

$$\sigma^2 = E(x-\mu)^2 = n\pi(1-\pi)$$

- Hypergeometric Probability Mass Function

$$P(x_1 \text{ and } x_2) = \frac{\dfrac{N_1!}{x_1!(N_1 - x_1)!}\;\dfrac{N_2!}{x_2!(N_2 - x_2)!}}{\dfrac{N!}{n!(N-n)!}}$$

where N_1 = number in group one
N_2 = number in group two
$N = N_1 + N_2$
x_1 = number realized from group one
x_2 = number realized from group two
$n = x_1 + x_2$
! indicates a factorial, as in
$N! = (N-1)(N-2)(N-3)\ldots(2)(1)$

- Mean of a Hypergeometric Random Variable

$$\mu_1 = E(x_1) = \frac{nN_1}{N}$$

- Variance of a Hypergeometric Random Variable

$$\sigma^2 = \frac{N-n}{N-1}\left[\frac{nN_1}{N}\right]\left[1-\frac{N_1}{N}\right]$$

- Poisson Probability Mass Function

$$P(x) = \frac{e^{-\lambda}\lambda^x}{x!}, \text{ for } x = 0, 1, 2, 3\ldots \text{ and } \lambda > 0$$

$$P(x) = 0, \text{ otherwise}$$

where $e = 2.71828$ and

λ is the mean (and also the variance) of x

CHAPTER 6

DISTRIBUTIONS OF CONTINUOUS RANDOM VARIABLES

A single death is a tragedy;
a million deaths is a statistic.

Joseph Stalin
1879–1953

6.1 Introduction

The height of men and women, length of ears of corn, and many other variables involving measures of time and weight have been shown to be approximately normally distributed. The bell-shaped curve commonly identified as the normal distribution in the popular press has even played a key role in Michael Crichton's suspense novel *Jurassic Park:*

> Arnold was pushing buttons. Another screen came up.

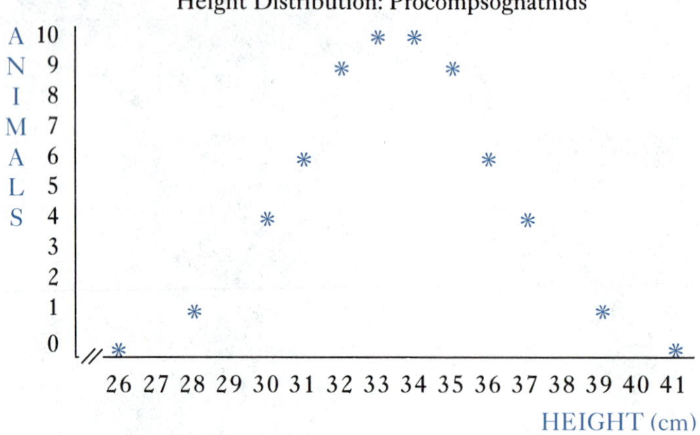

> . . . You see here we have a normal . . . . It shows that most of the animals (Procompsognathids) cluster around an average central value, and a few are either larger or smaller than the average, at the tails of the curve. (p. 132)

Although this bell-shaped curve is possibly the most well-known shape within and outside of academe, its origin and properties are often misrepresented. For instance, the historical perspective in this chapter presents Abraham De Moivre as the 1733 originator of the normal distribution, which he advanced as a way to approximate the binomial distribution. In addition, as you will learn in this chapter, because the normal distribution is continuous (with an infinite number of possible outcome values), the probability of any one value occurring is zero. Thus, the vertical axis in Arnold's diagram *does not* represent the frequency (or relative frequency) of values appearing on the horizontal axis. These are typical errors made in the popular use of the normal distribution.

This chapter presents the normal distribution as a continuous distribution for use in problems of the following type.

- An Associated Press (*HT*, December 21, 1991) news release reported that a "new British regulation will require bigger beer glasses that can accommodate a full 20-ounce British pint." According to the Institute of Trading Standards Administration, 80 percent of the "pints" were short and beer drinkers were losing $420 million a year. The questions are: How short was the average fill? How large will glasses need to be to ensure that most all the fills are at least 20 ounces?

- *The Wall Street Journal* (February 26, 1992) reported that the average tenure of marketing directors employed by law firms in 1987 was about six months. By 1990 it had increased to 2.1 years, but almost one-third of the marketing directors had been on the job for a year or less. The question: Could tenure be normally distributed?
- By U.S. law, compensating arrangements must be made for the handicapped. A problem facing personnel directors is how much additional time must be given to the mentally disabled to enable them to complete a corporate qualifying test. If the length of time required to complete such an examination by those with this disability is normally distributed with a mean of 65 minutes and a standard deviation of 15 minutes, then when can the exam be terminated so that 99 percent of those with the disability have sufficient time to complete the exam?

By the end of this chapter, you will be able to answer these and similar questions involving continuous random variables. You will be able to specify the conditions required to use the normal distribution in probability calculations. You will be able to make probability calculations using the normal distribution. You will know the characteristics of the normal distribution that distinguish it from other distributions (such as the uniform and the lognormal). You will know when to use the normal versus these other probability distributions and the general procedure of determining probabilities for any continuous distribution.

6.2 Continuous Random Variables and Probability Distributions

Continuous Random Variable
A random variable that may take on any value along an uninterrupted interval of a number line.

Within the limits of the glass, the volume of beer in the glass is a **continuous random variable**: the amount of beer can be any value between zero and the maximum volume of the glass. Time to complete a test is likewise continuous, as is the height and weight of animals and numerous other variables that can be expressed as decimals (e.g., 19.135 ounces, 1.25 hours, or 5.6 feet).

Economic and financial variables such as interest payments and wages may also be continuous, even though transactions are typically reported only to a cent. Folklore has a dishonest bank computer programmer in Minneapolis capitalizing on the one-cent reporting limit in calculating interest payments. He programmed the bank's computer to place the truncated fractions of cents in his own bank account when interest calculations were made for thousands of bank customers. A senior citizen doing cumulative calculation on a hand-held calculator caught the discrepancy in her account. Her repeated inquiries finally got the bank management's attention.

For a discrete random variable, the probability that the variable equals one of its limited number of values is given by the mass of the associated bar in a histogram. We have already seen in the case of the binomial distribution that if the population proportion of successes is $\pi = 0.50$, then the probability that x equals 0, 1, 2, 3, or 4 is the mass (= height times the one unit width) of the identified bar in Figure 6.1, panel a. For $n = 8$, the probability of $x = 0, 1, 2, \ldots 7$ or 8 is the mass of the associated bar in panel b, and so on for $n = 16$ in panel c, and $n = 32$ in panel d. Because each bar associated with

$x = 1, 2, 3, \ldots$ in a binomial distribution is of equal unit width, with $\Sigma_x P(x) = 1$, the probability of any discrete x is simply its bar's height.

Notice as n increases from $n = 4$ to $n = 32$ in Figure 6.1, the binomial probability mass function looks more and more like a smooth bell-shaped curve. As the number of bars increase, they appear to get narrower, because the total sum of their masses must continue to be unity, $\Sigma_x P(x) = 1$. In the limit, for the truly smooth bell-shaped curve, the mass associated with each value on the horizontal axis would be infinitesimal (zero). That is, the probability of realizing a specific x value would be zero.

FIGURE 6.1 Binomial Probability Mass Functions: as n Increases the Binomal Approaches the Normal Probability Density Function

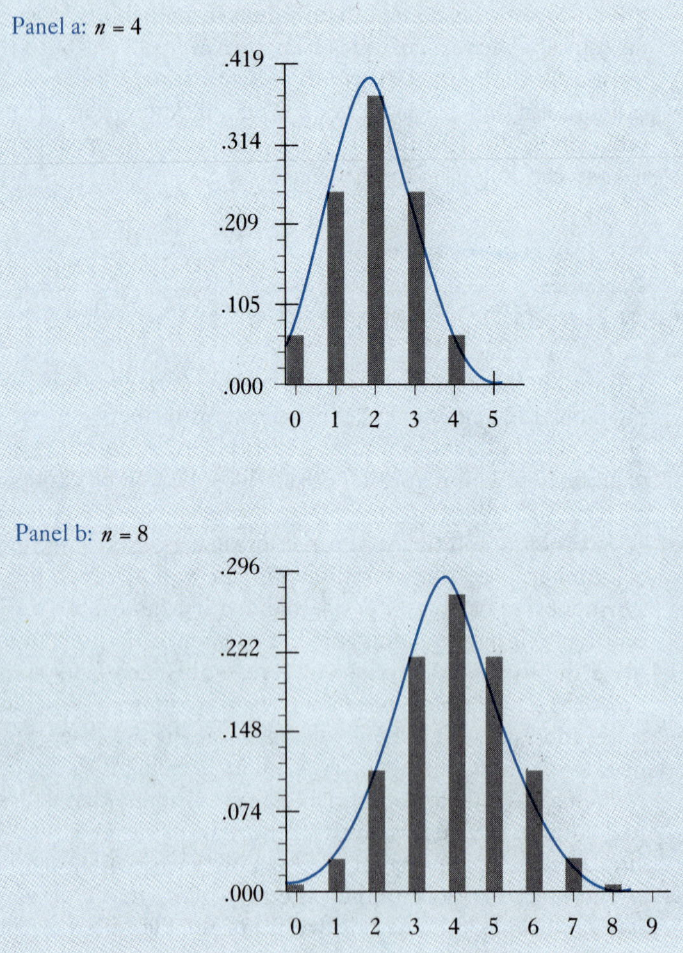

Panel a: $n = 4$

.419
.314
.209
.105
.000
0 1 2 3 4 5

Panel b: $n = 8$

.296
.222
.148
.074
.000
0 1 2 3 4 5 6 7 8 9

(continued)

FIGURE 6.1 **Binomial Probability Mass Functions: as n Increases the Binomial Approaches the Normal Probability Density Function** ***(continued)***

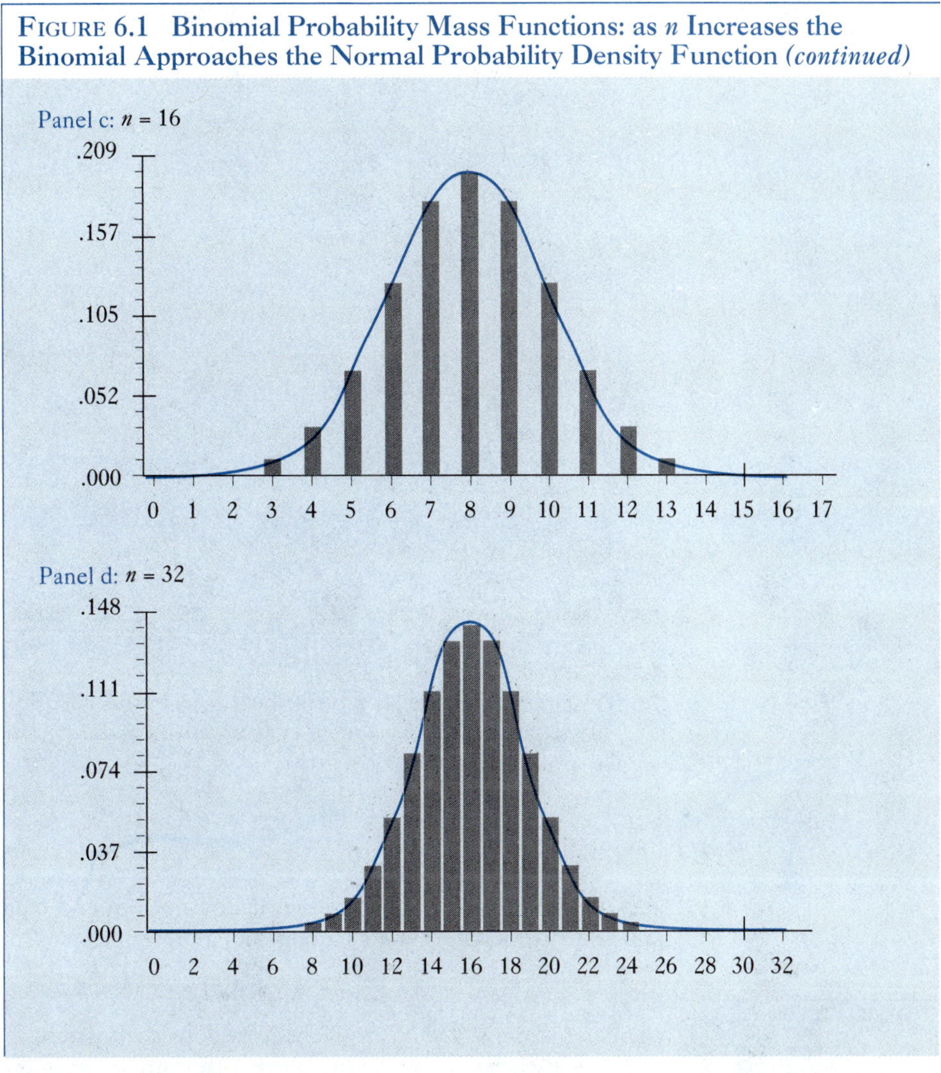

Probability Density Function (pdf)
A function that, for any interval of values, gives the probability that the associated random variable, x, takes a value in the interval.

A random variable x is continuous if it can take any value along a number line but the probability of any one of these infinite values is zero. Because the probability mass at any one value is zero, only a range of x values have a meaningful non-zero probability. The curve showing the area or density of this range of x values is called the **probability density function (pdf)** of x. The probability that x takes a value in a range between a and b is $P(a < x < b)$, which is the area under the probability density function between a and b. For a normal probability density function, Figure 6.2 shows such an area.

The normal probability density function is not the only continuous distribution used in statistics. All continuous distributions have probability density functions—examples include the rectangular (panel a) and the right-skewed (panel b) distributions in Figure 6.3. In each of these cases, the probability that a continuous random variable

FIGURE 6.2 Normal Probability Density Function

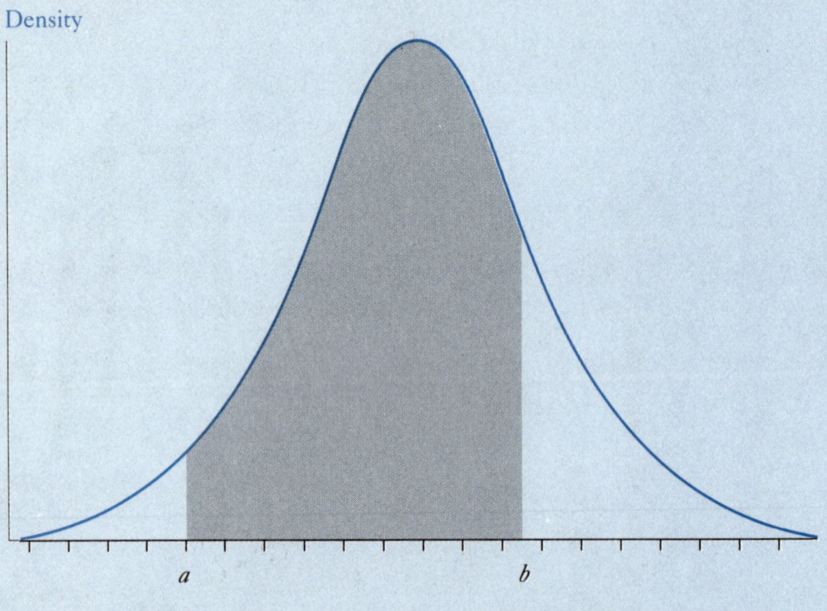

Note: The shaded area under the bell-shaped normal density function is the probability that the continuous random variable x takes a value between a and b.

takes a value between $\boldsymbol{a}$ and $\boldsymbol{b}$ is the shaded area under the pdf. Because there is a zero probability that a continuous random variable x equals the single value $\boldsymbol{a}$ (or $\boldsymbol{b}$),

$$P(\boldsymbol{a} < x < \boldsymbol{b}) = P(\boldsymbol{a} \le x \le \boldsymbol{b})$$

regardless of the shape of the probability density function.

If $\boldsymbol{a}$ is the smallest value for which a continuous random variable is defined and $\boldsymbol{b}$ is any larger value, then the curve or function giving the area under the density function between $\boldsymbol{a}$ and $\boldsymbol{b}$ is called the cumulative distribution function or simply the distribution function. For instance, the density function for the uniform distribution in Figure 6.3, panel a, is

$$\text{Density of } u = f(u) = 1/h, \text{ for } 0 < u < h$$

and the distribution function is

$$P(u < \boldsymbol{b}) = \boldsymbol{b}/h,$$

which is shown in Figure 6.4, panel a, as the continuous straight line.[1]

FIGURE 6.3 Some Continuous Nonnormal Probability Density Functions

Panel a: Density Function for a Uniform or Rectangular Distribution, where the continuous random variable u can take any value from 0 to h (called the support of u) and the hatched area is the probability of u taking a value between a and b. The total area under this density function is one.

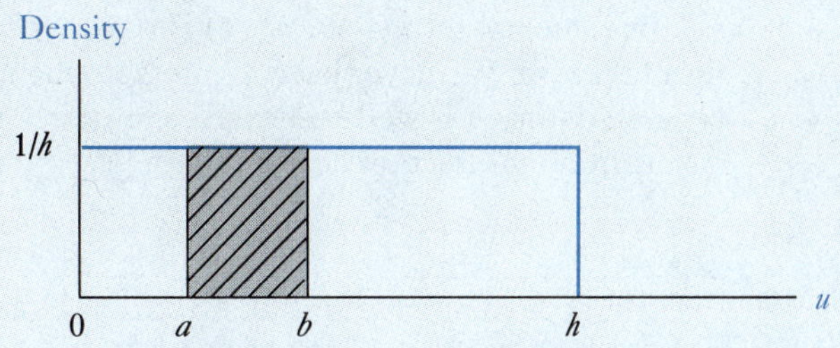

Panel b: Density function for a Right-Skewed Distribution, where the continuous random variable v can take any value and the hatched area is the probability of v taking a value between a and b. The total area under this density function is one.

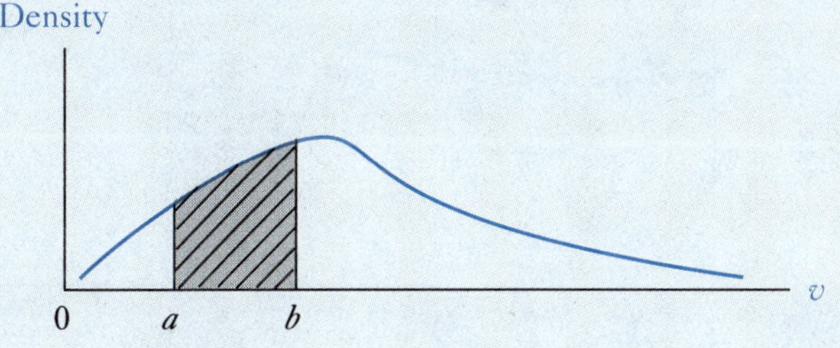

The density function for a normal random variable x is

$$f(x) = \frac{1}{\sigma\sqrt{2\pi}} e^{-(x-\mu)^2/2\sigma^2}$$

where $-\infty < x < +\infty$

μ is the mean of x

σ is the standard deviation of x

π is 3.14159

e is 2.71828

The probability of normal random variable x yielding a value between a and b is

$$P(a < x < b) = \int_a^b \frac{1}{\sigma\sqrt{2\pi}} e^{-(x-\mu)^2/2\sigma^2}\,dx$$

Probability Distribution Function
A function giving the probability that a random variable takes on values less than a specific upper limit.

which is the area under the density function between a and b. Notice that this probability is not defined at a point for the normal distribution, as $\boldsymbol{a}$ goes to $\boldsymbol{b}$ in the above integral, $P(\boldsymbol{a} < x < \boldsymbol{b})$ goes to 0.

The integral for $P(-\infty < x < b)$ is typically referred to as the **probability distribution function.** We never have to integrate the normal probability distribution function because its shape is well known. As shown in Figure 6.4, panel b, the normal probability distribution function is a serpentine-type line that rises ever faster from negative

FIGURE 6.4 Density and Distribution Functions

Panel a: The distribution function for a uniform distribution is a straight line whose uniform rise shows an accumulation of probability as u increases from 0 to its maximum value of h in the density function.

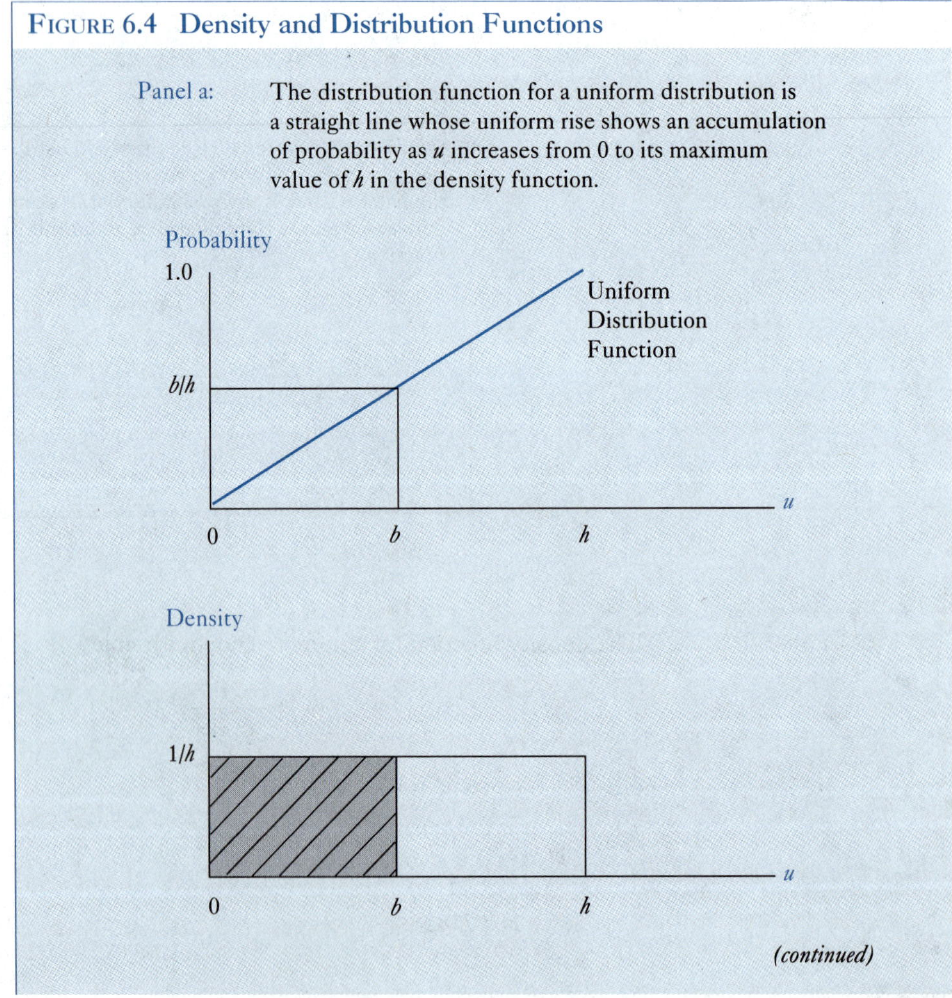

(continued)

infinity to the mean, where 50 percent of the distribution has accumulated. It then rises ever slower to a 100 percent cumulative probability as x approaches a positive infinity. Probability is read from the vertical axis. For example, approximately 16 percent of the distribution is in the lower tail of the normal density function below $\mu - \sigma$. This is the probability that is shown on the vertical axis of the normal probability distribution function for the area under the density function between $-\infty$ and $\mu - \sigma$.

$$P(-\infty < x < \mu - \sigma) = 0.16$$

Other probabilities can be obtained from the vertical axis of the normal distribution function in a similar fashion. As we will see, tables and computer programs make reading of this vertical axis easy and more precise than what can be accomplished by visual inspection.

FIGURE 6.4 Density and Distribution Functions *(continued)*

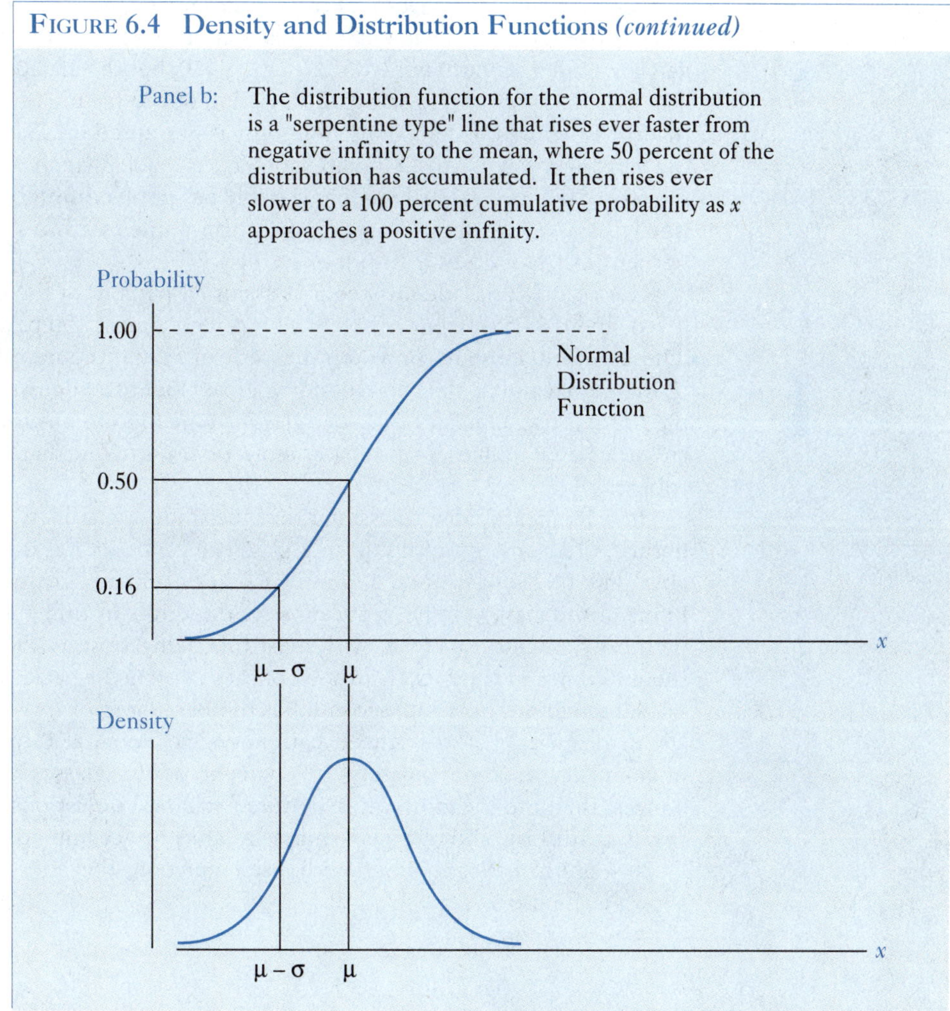

6.3 Normal Distributions

The distribution of any normal random variable requires the specification of two parameters: the mean μ and the standard deviation σ (or variance σ^2). The mean or expected value gives the central location of the variable, and the standard deviation is a measure of its variability. Because normal random variables can have different means and standard deviations, there is a family of normal distributions where each member is identified by its mean and standard deviation.

The mean of the normal distribution simply centers the distribution on the horizontal axis; the higher the mean, the farther to the right the center of the distribution. Differences in the size of a standard deviation, however, do not affect the probability of being within one standard deviation of the mean. Figure 6.5 shows three normal density functions with different means and different standard deviations. The area within one standard deviation of the mean remains the same, regardless of the magnitude of the standard deviation and position of the mean. That is, distribution *s* is less variable than distribution *w* ($\sigma_s = 2.1 < \sigma_w = 4.9$), but the area under the respective density functions within one standard deviation of the mean is fixed. It is for this reason that the rules of bell-shaped distributions, as presented in Chapter 2, hold regardless of the variability in a normal random variable. As a bell-shaped distribution, regardless of the size of μ or σ, a normal random variable has **approximately 68 percent of the values within $\mu \pm \sigma$, about 95 percent of the values within $\mu \pm 2\sigma$, and almost 100 percent of the values within $\mu \pm 3\sigma$.**

Notice that this rule implies that about 84 percent of the distribution is less than $\mu + \sigma$, about 97.5 percent is below $\mu + 2\sigma$, and almost 100 percent is beneath $\mu + 3\sigma$. These percentages are shown on the vertical axis of the normal distribution function in Figure 6.6 and as the corresponding areas under the density function. As an example of their use, consider the beer glass size needed to ensure that nearly all the fills are at least 20 ounces. We must assume first, however, that "fills" are normally distributed.

It seems reasonable to assume that the underlying distribution of glass fills (x) is normal. The amount of liquid in a glass can be treated as a continuous random variable. The fill is not restricted to integer values, and any decimal is possible within the limits of the glass volume. Furthermore, the errors in filling a glass should vary symmetrically around the mean, with most fills being close to the center value. (Look at the difference in fluid levels in soda bottles in an eight-pack, for example.)

Although the press release cited at the beginning of this chapter did not give the standard deviation, experienced bartenders can set σ at 0.15 ounces. From the rule of normality, if approximately 100 percent of the glasses are to have 20 or more ounces, then the mean fill must be three standard deviations higher, at 20.45 ounces [= 20 + 3(0.15)], and the glass must be able to accommodate at least 20.9 ounces [= 20 + 6(0.15)]. For this normal density function, the relevant values and areas are shown in Figure 6.7.

FIGURE 6.5 Different Variability in Normal Densities

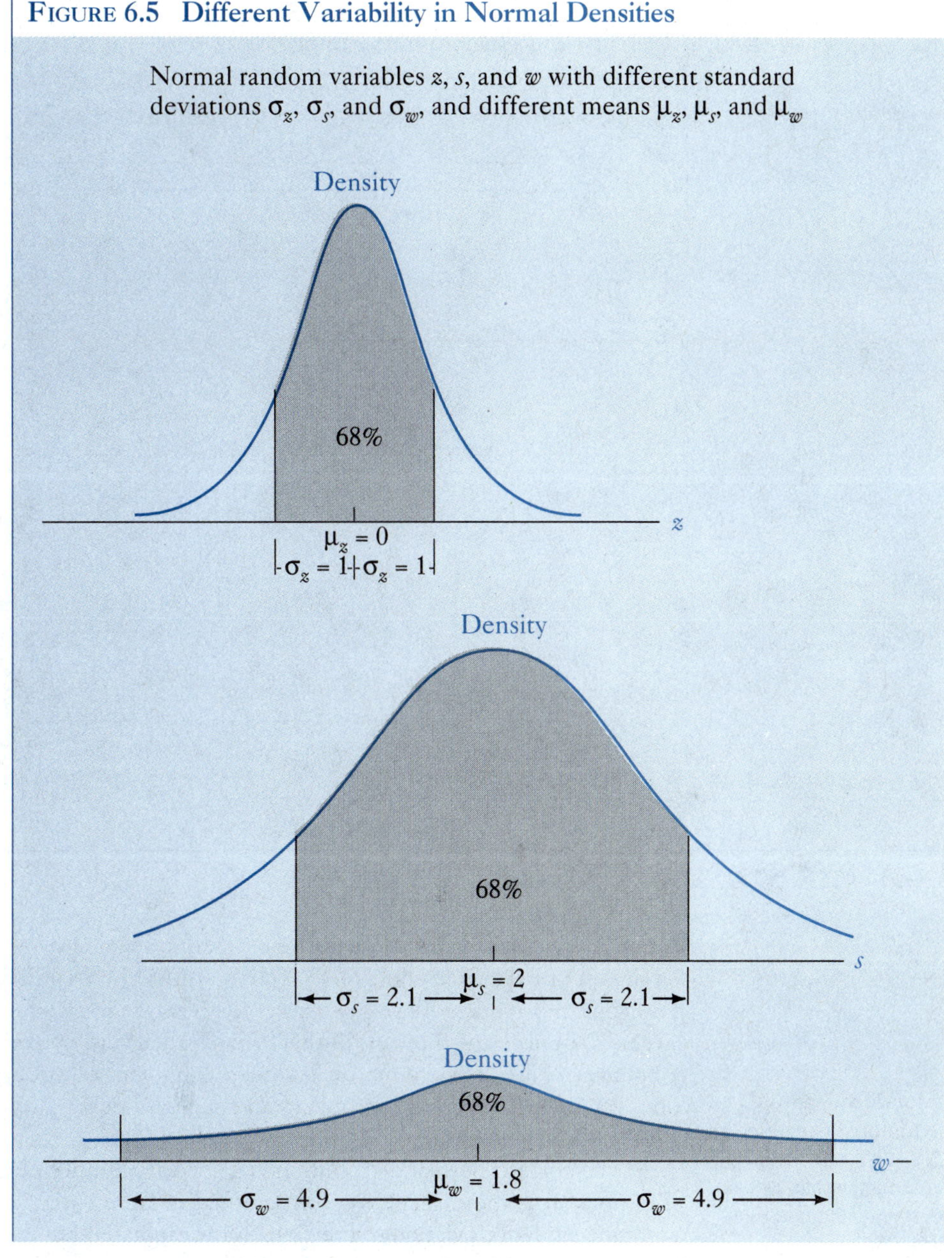

FIGURE 6.6 The Rule for Bell-Shaped Distributions as Related to the Normal Density and Distribution Function

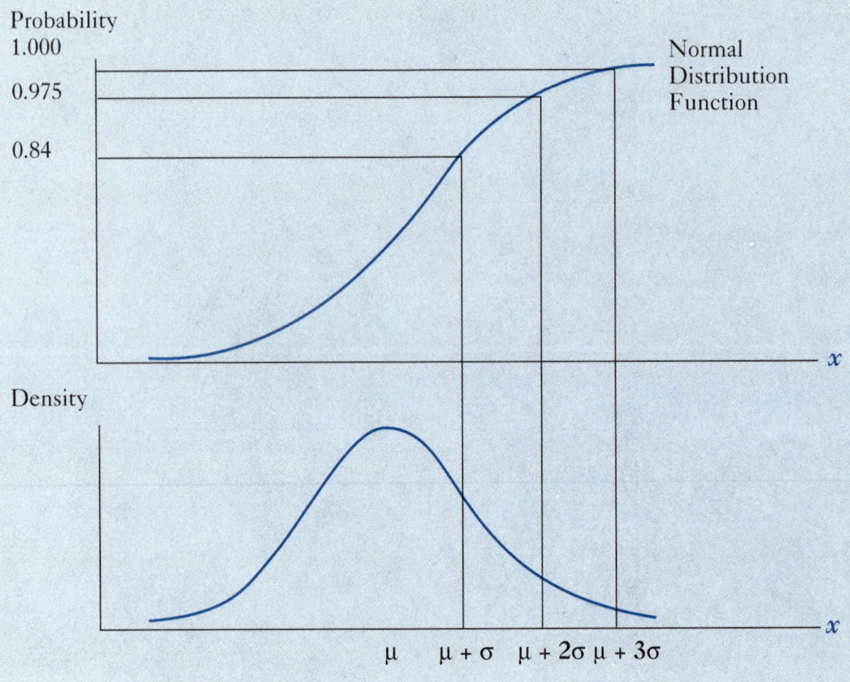

6.4 THE STANDARD NORMAL DISTRIBUTION

Standard Normal Random Variable A normal random variable, z, that has a mean of 0 and standard deviation of 1.

According to Great Britain's Institute of Trading Standards Administration, 80 percent of glasses of beer were short of a 20-ounce fill in the past. To answer the question of how short the average fill was, we need to know the number of standard deviations between 20 ounces and a mean fill that is consistent with 80 percent of the fills less than 20 ounces. This can be done by "standardizing" or "transforming" fills to form a random variable that has a mean of zero and standard deviation of one—a **standard normal random variable.**

The standard normal distribution is a special member of the family of normal distributions. It is the distribution for a normal random variable that has a mean of zero and the standard deviation of one, which is designated by the letter z. That is, z is the standard normal random variable that is distributed normally with mean zero and standard deviation of one. In mathematical notation, this is written

$$z \sim N(\mu = 0, \sigma = 1), \text{ or simply } z \sim N(0, 1)$$

As shown in Figure 6.5, changing the mean of a normal random variable only changes the position of the distribution without changing the probabilities assigned to

FIGURE 6.7 The Normal Density for Desired Ounces of Beer in a Glass

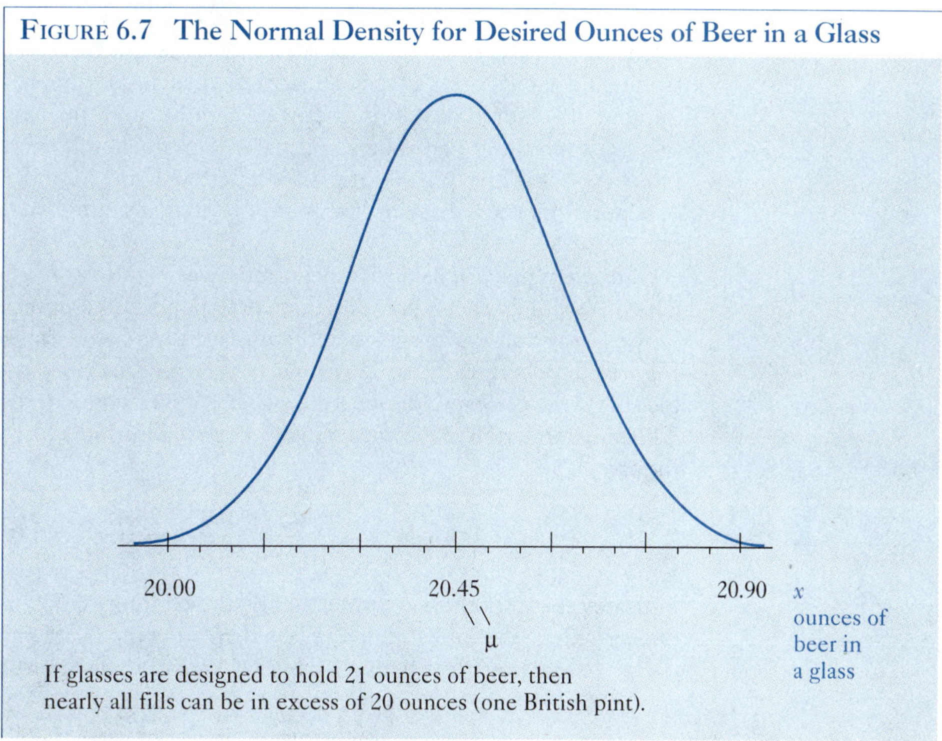

being a certain number of standard deviations from the mean, panel a. In addition, changing the size of a standard deviation does not affect the probability of being within a given number of standard deviations of the mean, panel b. These observations are captured in an important theorem: **any linear transformation of a normal random variable is also a normal random variable.** Any normal random variable x can be transformed into a standard normal random variable z by Formula (6.1).

Standard normal random variable: $z = \dfrac{x - \mu}{\sigma}$ (6.1)

where x is a normal random variable with mean μ and standard deviation of σ so z has a mean of zero and standard deviation of one. That is, if $x \sim N(\mu, \sigma)$, then $z \sim N(0, 1)$.

Although it may not be obvious in Formula (6.1), z is a linear transformation of the random variable x because it can be written as a straight line function of x:

$$z = (1/\sigma)x - (1/\sigma)\mu$$

If x is distributed normally, with $E(x) = \mu$ and standard deviation σ, then z is distributed normally with $E(z) = 0$ and standard deviation of one.[2]

In essence, the standard normal random variable z gives the number of standard deviations a particular value x is above (or below) the mean of x. The standard normal distribution is a redrawing of the x distribution such that the vertical axis for z shows the number of standard deviations between x and μ, and the area under the density function between these two values is equal to the probability of getting a value of z in the interval $z - 0$. Thus, probabilities associated with any normal random variable x can be obtained from the z distribution by simply "transforming" or "standardizing" x to form a z.

For example, Figure 6.7 can be redrawn in terms of "standardized" values of x, which are simply the values of z that correspond to the number of standard deviations between the value of x considered and its mean. These transformations are shown in Figure 6.8, where the x scale (which is measured in ounces, in panel a) has been replaced by the corresponding z scale (which counts standard deviations in x, in panel b). The transformation of x values into z is done via Formula (6.1). For example, the transformation

$$z = \frac{x - \mu}{\sigma} = \frac{20.00 - 20.45}{0.15} = -3.00$$

indicates that 20 ounces is three standard deviations below a mean glass fill of 20.45 ounces. Thus, the probability of x less than 20, *if* $\mu = 20.45$ and $\sigma = 0.15$, is the same as the probability of z less than –3, which is known to be approximately 0. Similarly,

$$z = \frac{x - \mu}{\sigma} = \frac{20.15 - 20.45}{0.15} = -2.00$$

shows that 20.15 ounces is two standard deviations below the mean of 20.45 ounces and the probability of x less than 20.15 is the same as the probability of z less than –2, which is known to be approximately 0.025. The reader is left to show that 20.30 is one standard deviation below an assumed mean of 20.45 ounces. The probability of x being less than 20.30 is the same as the probability of z being less than –1, which is approximately 0.16.

But we do not know the old mean fill of beer glasses. We want to find the old mean fill for which 80 percent of the glasses had less than 20 ounces. In essence, this is done by sliding the normal distributions in Figure 6.5 to the left until 80 percent of the glasses have less than 20 ounces of beer. Figure 6.9 shows this point as a "standardized" z value of 0.84, which is discussed in more detail in the next section. Here it is only necessary to recognize that the probability of z less than 0.84 is 0.80 and this probability is all of the area under the standard normal density function to the left of a z of 0.84; this is also the area under the density function to the left of x at 20 ounces. We needed z to obtain the probabilities associated with x:

$$P(z < 0.84) = P(x < 20) = 0.80$$

FIGURE 6.8 Normal Density for Beer Glasses and the Corresponding Standard Normal z

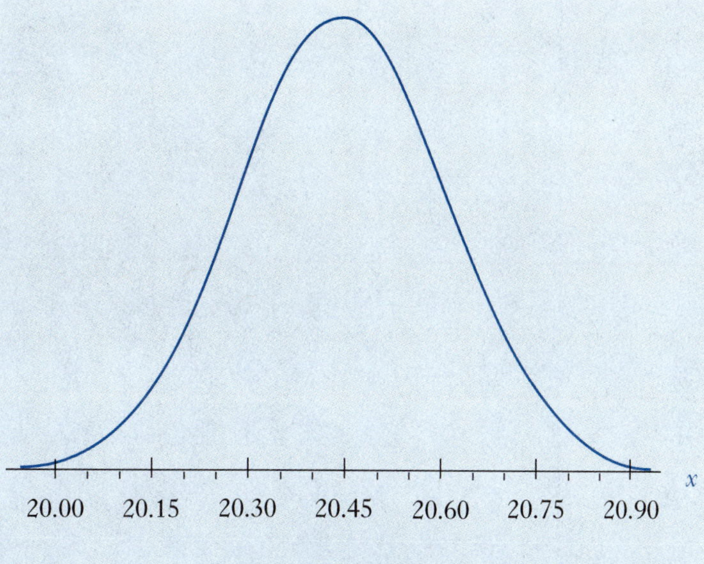

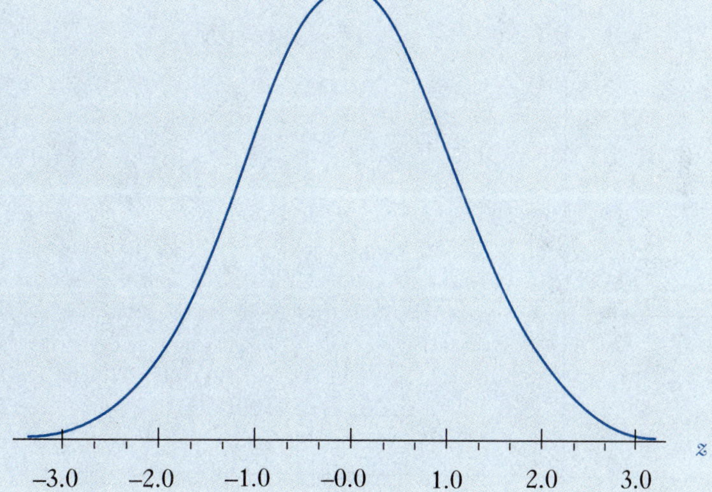

Notice that $z = 3.0$ indicates that 20.90 ounces is 3 standard deviations above the mean of 20.45. Similarly $z = -3.0$ states that 20 ounces is 3 standard deviations below the mean of 20.45. A z of 2.0 shows that 20.75 is 2 standard deviations above the mean of 20.45 and a z of –2.0 shows that 20.15 is 2 standard deviations below the mean of 20.45, and so on.

FIGURE 6.9 Normal Density for Fills of the Old Beer Glasses

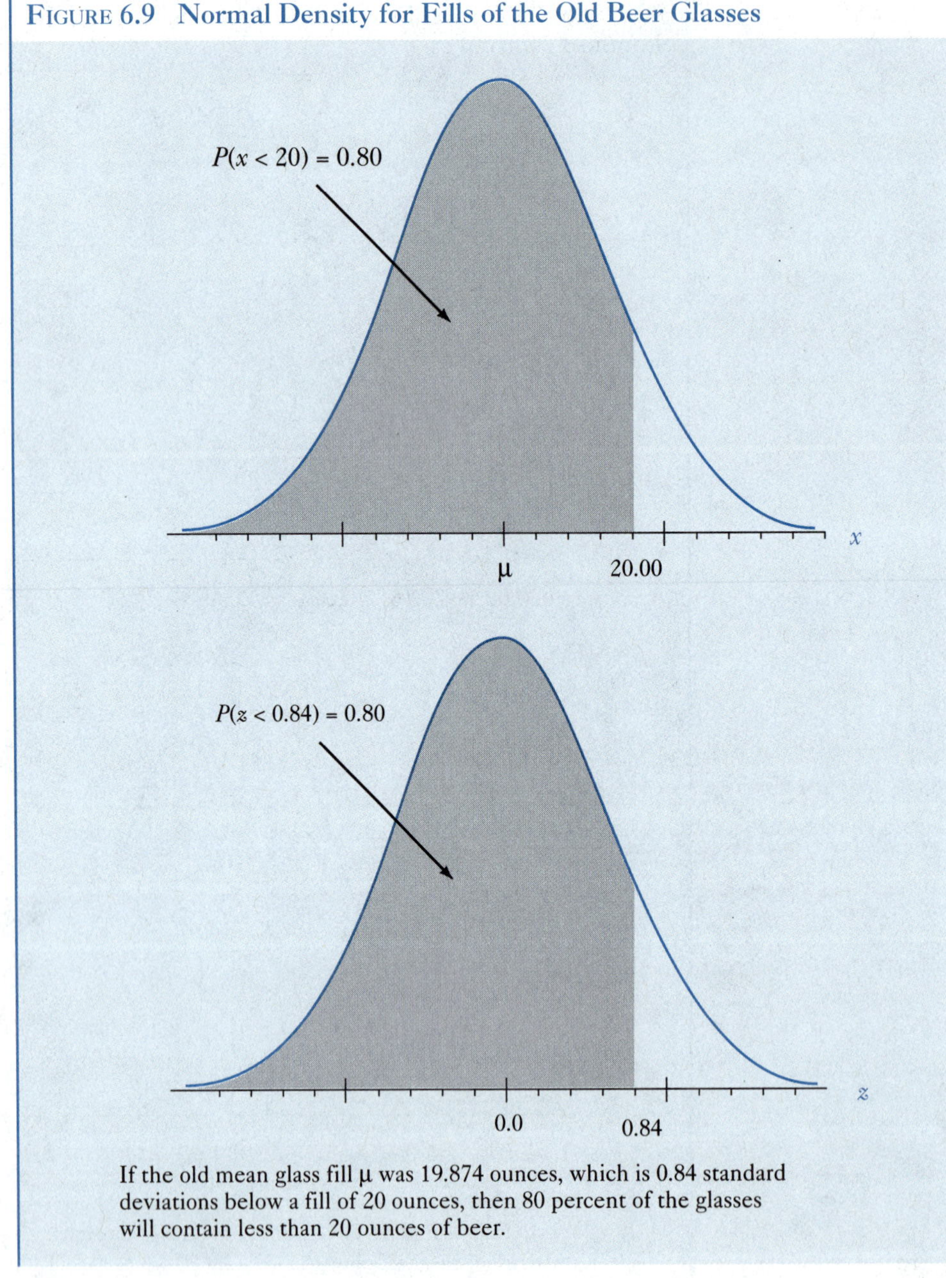

If the old mean glass fill μ was 19.874 ounces, which is 0.84 standard deviations below a fill of 20 ounces, then 80 percent of the glasses will contain less than 20 ounces of beer.

The old mean fill now can be found, because 20 ounces must be 0.84 standard deviations above the mean glass fill; otherwise 80 percent of the fills could not be less than 20 ounces. Thus, the old mean glass fill was 19.87 ounces, which is found by Formula (6.1):

$$z = \frac{x - \mu}{\sigma}$$

$$0.84 = \frac{20.00 - \mu}{0.15}$$

$$19.874 = \mu$$

6.5 Calculations with the Standard Normal Table

Appendix Table A.3 is a cumulative probability table for the standard normal distribution. The body of this table provides the probabilities of z less than the indicated positive z value $\boldsymbol{b}$, that is, $P(-\infty < z < b)$. The use of this table can be illustrated by first finding a probability we already know, such as $P(-\infty < z < b) = P(z < 1.00) \cong 0.84$.

For bell-shaped density functions, about 68 percent of the distribution is within one standard deviation on either side of the mean. Thus, about 16 percent is above one standard deviation of the mean and about 84 percent is between $-\infty$ and $\mu + \sigma$, as shown in the distribution function in Figure 6.6. This is a probability that can be obtained exactly from the standard normal tabulations in Appendix Table A.3, a part of which is reproduced in Table 6.1. Look down the left-hand margin of this standard normal table until 1.0 is reached, and then move one column to the right (to the column headed .00): the number in the corresponding cell is .8413. This is the cumulative probability $P(z < 1.00) = 0.8413$, which we already knew to be approximately 84 percent from the rule of bell-shaped densities.

From Appendix Table A.3, and as reproduced in Table 6.1, the area corresponding to other z values can be obtained. For example, the probability of $z < 0.84$ can be seen to be 0.7995, which is approximately the 0.80 area shown in Figure 6.9. Notice, however, that the distribution for a normal random variable is still not exhausted completely, even at three standard deviations above the mean; for example, $P(z < 3.49) = 0.9998 \neq 1.00000$. The normal distribution has a range that extends from a negative infinity to a positive infinity. Nevertheless, most of its mass is within three standard deviations of the mean, as suggested by the rule for bell-shaped distributions.

Cumulative probabilities for negative z values are not provided in Appendix Table A.3. Because the normal distribution is symmetric, however, they can be determined indirectly from the information provided. To illustrate, consider $P(z < -1.08)$. The reader should be able to identify this probability in Appendix Table A.3; it is graphed in panel a, Figure 6.10. Panel b shows that the area below -1.08 is the same as the area above $+1.08$; i.e., $P(z < -1.08) = P(z > 1.08)$. The area above 1.08 is one minus the area below 1.08, $P(z > 1.08) = 1 - P(z < 1.08)$. From Appendix Table A.3, the cumulative

TABLE 6.1 Reading the Standard Normal Table

To find the cumulative probability associated with any z value, accurate to the tenths position as in $z = 1.0$, go down the left margin until the corresponding tenths are found, then read probability in the column titled .00. Thus $P(z < 1.0)$ is 0.8413. To find the probability for z values accurate at the hundredths position, say $P(z < 0.84)$, go down the first column to the row .8 and then go across this row to the column headed .04 to find $P(z < 0.84) = 0.7995$. Notice that 50 percent of the distribution is below $z = 0$ and the distribution is still not exhausted completely, even at $z = 3.49$.

z	.00	.01	.02	.03	.04	.05	.06	.07	.08	.09
.0	.5000	.5040	.5080	.5120	.5160	.5199	.5239	.5279	.5319	.5359
.1	\\									
.	$P(z < 0.00)$									
.										
.					$P(z < 0.84)$					
.7					//					
.8	.7881	.7910	.7039	.7967	.7995	.8023	.8051	.8078	.8106	.8133
.9										
1.0	.8413	.8438	.8461	.8485	.8508	.8531	.8554	.8577	.8599	.8621
1.1	\\									
.	$P(z < 1.00)$									
.										
.										$P(z < 3.49)$
3.3										\\
3.4	.9997	.9997	.9997	.9997	.9997	.9997	.9997	.9997	.9997	.9998

value $P(z < 1.08)$ is 0.8599, panel c, which upon subtraction from one gives the area under the density function above 1.08.

$$
\begin{aligned}
P(z > 1.08) &= 1 - P(z < 1.08) \\
&= 1 - 0.8599 \\
&= 0.1401
\end{aligned}
$$

Because the normal distribution is symmetric, 0.1401 is also the area under the density function below –1.08.

$$P(z < -1.08) = P(z > 1.08) = 0.1401$$

QUERY 6.1:

If the time required to complete an examination by those with a certain learning disability is believed to be distributed normally, with a mean of 65 minutes and a standard deviation of 15 minutes, then when can the exam be terminated so that 99 percent of those with the disability can finish?

ANSWER:

Because the average and standard deviation are known, what needs to be established is the amount of time, above the mean time, such that 99 percent of the distribution is lower. This is a distance that is measured in standard deviations, as given

FIGURE 6.10 Cumulative Probabilities for Negative z Values

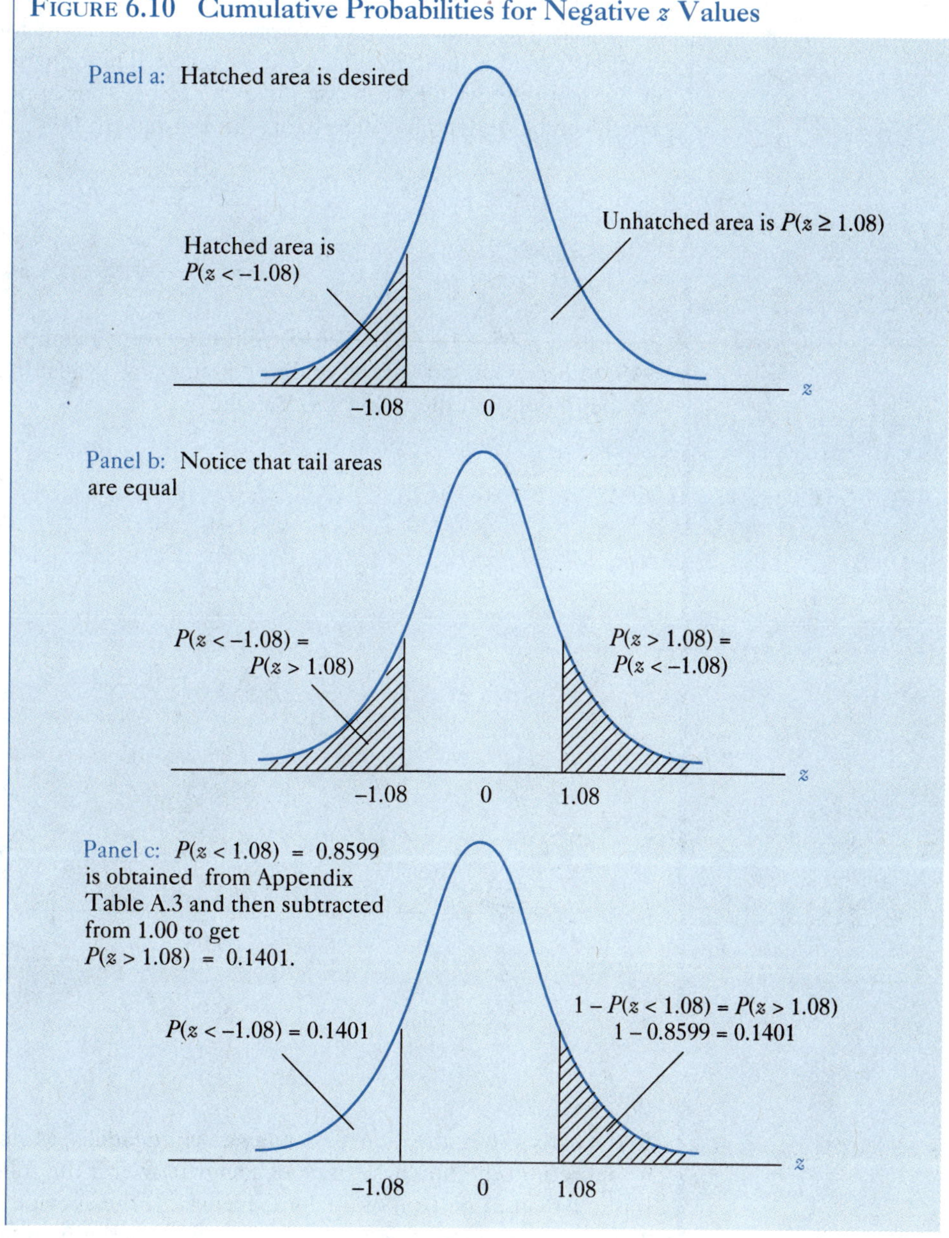

by the z value corresponding to the 0.99 probability found in the body of Appendix Table A.3. The closest cumulative probability that can be found is .9901, in the row labeled 2.3 and column headed by .03, $z = 2.33$. The time needed so that 99 percent of the learning disabled can complete the test is thus approximately 100 minutes, which can be derived by solving for x, in Formula (6.1).

$$z = \frac{x - \mu}{\sigma}$$

$$2.33 = \frac{x - 65}{15}$$

Thus, $x = 65 + 15(2.33) = 99.95$ minutes. Using a computer program such as the SAS, SPSS, or any one of the other programs now available, the relevant quantities can be found as in the following diagram.

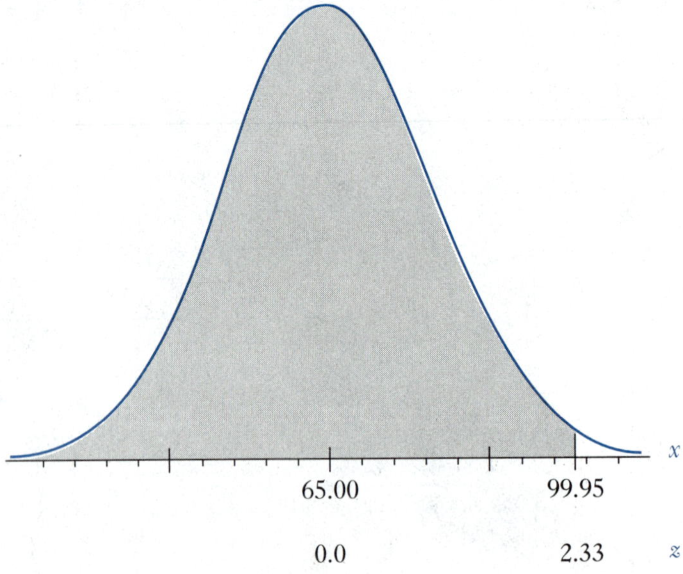

QUERY 6.2:

In the early 1980s, the Toro Company of Minneapolis, Minnesota, advertised that it would refund the purchase price of a snowblower if the following winter's snowfall was less than 21 percent of the local average. If the average snowfall is 45.25 inches, with a standard deviation of 12.2 inches, what is the likelihood that Toro will have to make refunds?

ANSWER:

Within limits, snowfall is a continuous random variable that can be expected to vary symmetrically around its mean, with values closer to the mean occurring most often. Thus, it seems reasonable to assume that snowfall (x) is approximately normally distributed with a mean of 45.25 inches and standard deviation of 12.2 inches. Nine and a half inches is 21 percent of the mean snowfall of 45.25 inches and, with a standard deviation of 12.2 inches, the number of standard deviations between 45.25 inches and 9.5 inches is z:

$$z = \frac{x - \mu}{\sigma} = \frac{9.50 - 45.25}{12.2} = -2.93$$

Now $P(x < 9.50) = P(z < -2.93)$, where the hatched area below is the area of interest

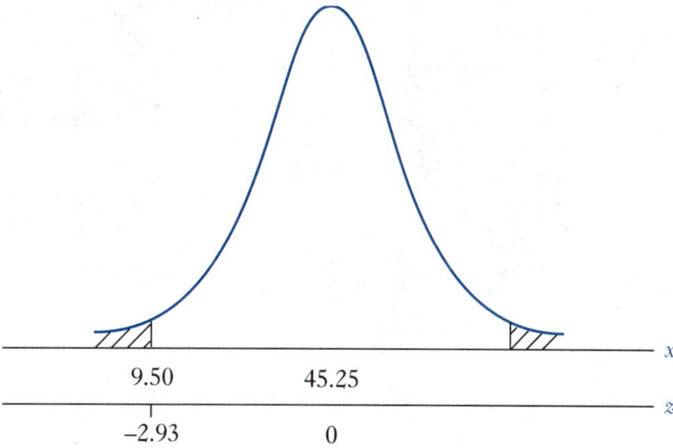

Using Appendix Table A.3, $P(z > 2.93) = 1 - P(z < 2.93)$ $= 1 - 0.9983 = 0.0017$, which is also $P(z < -2.93)$

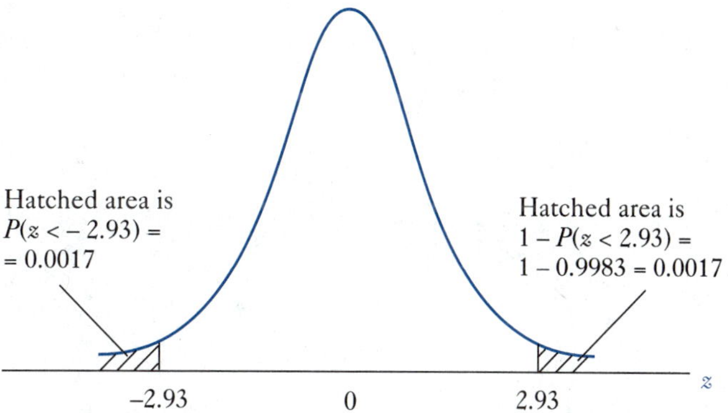

Thus, there is a 0.17 percent chance that the snowfall next winter will be less than 9.5 inches and that Toro will have to make refunds.

QUERY 6.3:

When workers are paid by the piece, the average time required to complete the task is usually well publicized by the employer. Suppose that a time study shows that a task takes 2.4 minutes to complete, on the average, with a standard deviation of 0.6 minute. If time of task is normally distributed, how often is the task completed in one to three minutes?

ANSWER:

At the lower limit, where $x = 1.0$,

$$z = \frac{x - \mu}{\sigma} = \frac{1.0 - 2.4}{0.6} = -2.33$$

At the upper limit, where $x = 3.0$,

$$z = \frac{x - \mu}{\sigma} = \frac{3.0 - 2.4}{0.6} = 1.00$$

Thus, the desired area is

$$P(1.0 < x < 3.0) = P(-2.33 < z < 1.00)$$

From Appendix Table A.3, we get

$$P(z < 1.00) = 0.8413$$

which graphically is the undotted area below

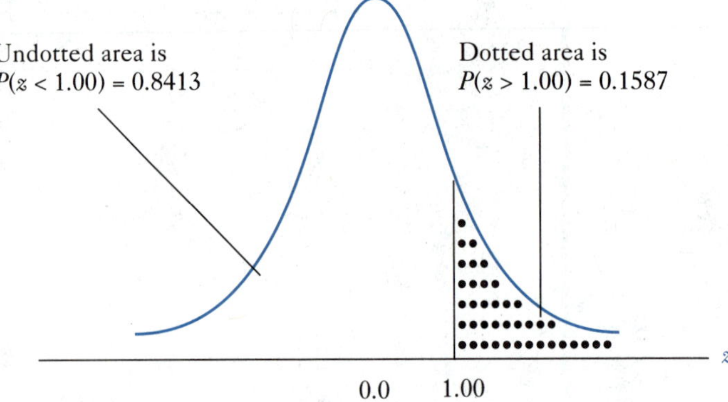

Using the symmetry property of the normal distribution and Appendix Table A.3, we can determine

$P(z < -2.33) = 1 - P(z < 2.33) = 1 - 0.9901 = 0.0099$, which graphically is

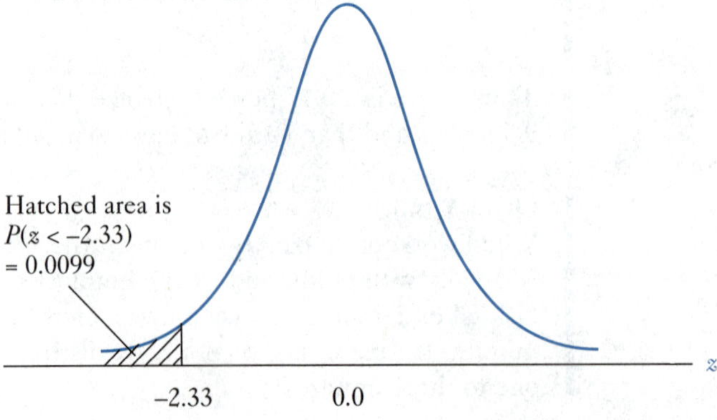

Subtracting the hatched area from the undotted area above gives

$$P(-2.33 < z < 1.00) = P(z < 1.00) - P(z < -2.33)$$
$$= 0.8413 - 0.0099 = 0.8314$$

Thus, 83.14 percent of the time, the task is completed between one and three minutes. Below is the graphical representation of this percentage and time frame. This graph and corresponding information was produced with the ECONOMETRICS TOOLKIT computer program by simply following the menus and entering the mean $\mu = 2.4$, standard deviation $\sigma = 0.6$, and range of values $(1.0 < x < 3.0)$ for the desired probability calculation; no transformation to the standard normal was required.

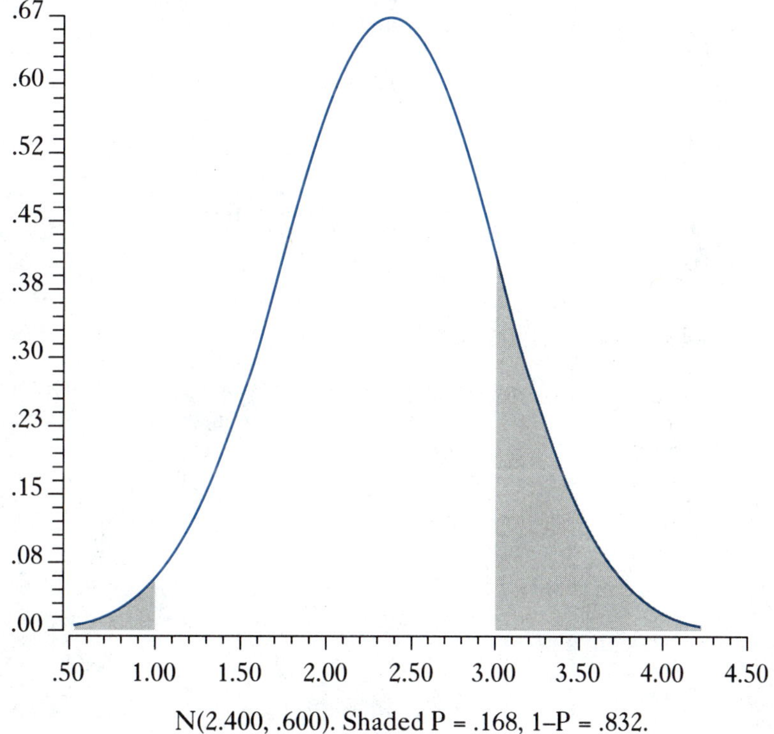

N(2.400, .600). Shaded P = .168, 1–P = .832.

QUERY 6.4:

The Wall Street Journal (February 26, 1992) reported that in 1990 the job average tenure of marketing directors at law firms was 2.1 years but that one-third had been on the job for a year or less. The question: Could marketing director tenure be approximately normally distributed?

ANSWER:

What we know in this problem is reflected in the following diagram, where x is job tenure, which we tentatively assume is normally distributed.

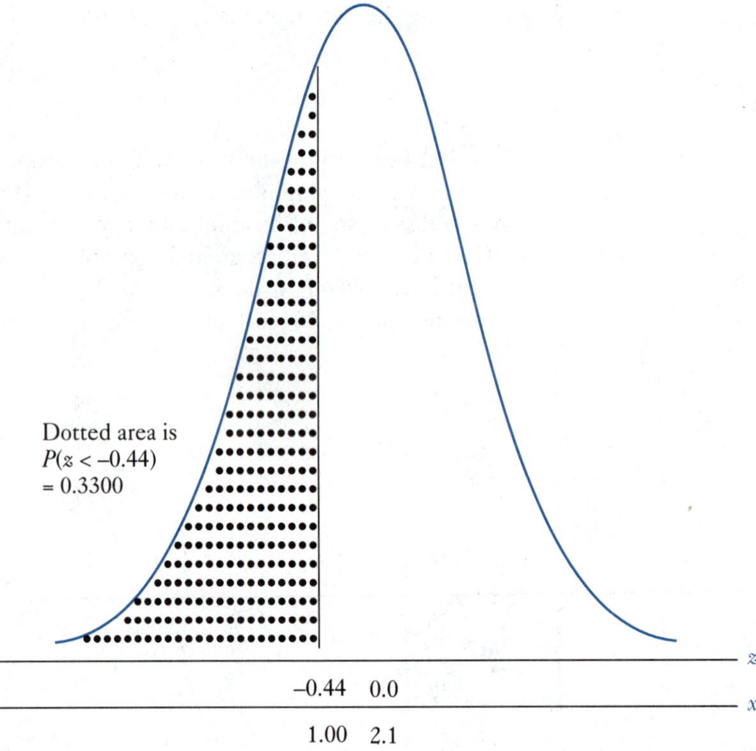

What we do not know is the standard deviation of x, but this can be solved for in the Formula (6.1), assuming x is normally distributed. Substituting for x, μ, and z and solving for σ gives

$$z = \frac{x - \mu}{\sigma}$$

$$-0.44 = \frac{1.0 - 2.1}{\sigma}$$

$$\sigma = 2.5$$

If the distribution of tenure is normally distributed, then the standard deviation is 2.5 years. But, with a mean of 2.1 years, $\mu - \sigma$ would be a negative 0.4 years, which is impossible. Thus, it would not be appropriate to assume that job tenure of marketing directors is approximately normally distributed.

6.6 The Lognormal Distribution (Optional)

The normal distribution cannot be used directly to address many problems in business and economics that involve random variables with highly skewed distributions. Many

HISTORICAL PERSPECTIVE

Many students think that scholarship and the development of scientific ideas has always been harmonious and civilized. The historical development of the normal distribution, however, is filled with controversy and turbulence. Germans, for example, credit the great scientist Carl Friedrich Gauss (1777–1855) for deriving the normal distribution; the French cite the equally renowned Pierre Simon Laplace (1749–1827). But Abraham De Moivre (1667–1754), a French expatriate living in London, invented the normal distribution, and Englishman Thomas Simpson (1710–1761) popularized it.

In 1733, De Moivre proposed what is now called the normal density function as an equation that, when integrated, would approximate Bernoulli's binomial distribution for large sample sizes. He calculated the probability of the value of a variable falling within one, two, and three standard deviations of its mean. The precision of his equation and method of calculation, and his emphasis on measuring distance from the mean in terms of standardized units, met with indifference from his contemporaries, possibly because of the absence of applications in De Moivre's work and his highbrow style. Empirical scientists were not interested in mathematical nuances; they wanted techniques or methods that would lead to generalizations from observable data, and De Moivre failed to provide any.

In 1740, Thomas Simpson published two books that presented ideas and concepts similar to those of De Moivre but in a manner that was readable by a wider audience. Simpson's early interest in astrology led him to the study of astronomy and then mathematics, which he learned on his own. At the age of 19, he married a 50-year-old widow with two children. He supported his family by weaving and teaching mathematics. It may have been his need for money that led him to write books that appealed to a readership beyond the highbrows for whom De Moivre wrote.

Simpson cited De Moivre appropriately, but De Moivre was envious of Simpson's success. De Moivre may have felt that the sale of his earlier books would suffer, because in 1743 he came out with a second edition to compete with Simpson's. In the preface, De Moivre blasted Simpson for writing on the same subject and mutilating his work. More to the point, however, is De Moivre's condemnation of Simpson for selling his book at a very moderate price. Needless to say, a war of words between Simpson and De Moivre followed.

Stephen Stigler (1986, pp. 88–94) asserts that, unlike De Moivre, Simpson opened the door to practical applications of the normal distribution for assessing uncertainty by making it the distribution of errors in measurement, as introduced in the next chapter. Lagrange, Laplace, Bayes, and then Gauss slipped through this door and advanced the normal distribution as the most important error distribution in statistics (as discussed in the next chapter and applied throughout the remainder of this book).

of these problems can be handled, however, with the logarithmic-normal distribution. The so-called lognormal distribution is well suited to quantities that are formed by a multiplicative process instead of an additive process.[3]

To appreciate the distinction between an additive versus a multiplicative process, consider a young entry-level worker whose earnings start at $100 per week. If in each subsequent year of employment, earnings rise to $110, $120, $130, $140 . . . , then earnings are growing by the constant additive amount of $10 per year of experience.

FIGURE 6.11 Earnings Growing at 10 Percent Per Year

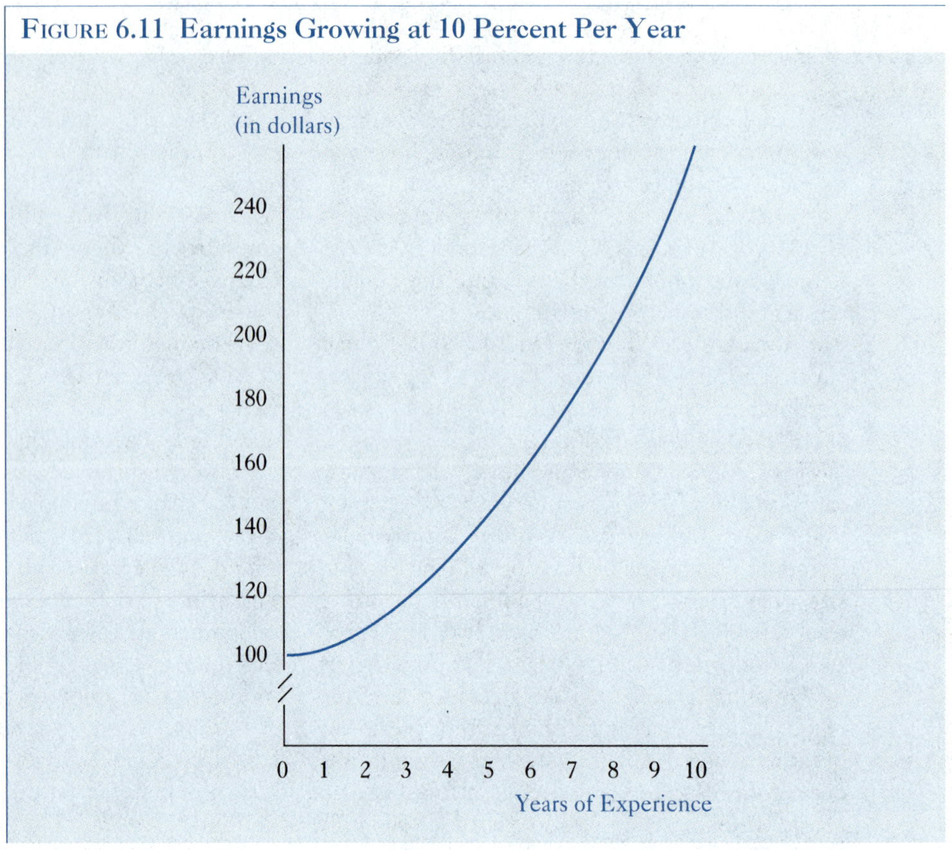

Alternatively, if raises are at 10 percent per year, then the earnings series is \$110, \$121, \$133.10, \$146.41 . . . . This latter compensation policy has earnings determined by the multiplicative process known as compounding. Compounding gives rise to a series that grows exponentially over time, as shown in Figure 6.11.[4]

Labor economists have found that earnings tend to grow exponentially, with a few older workers having very high earnings because of compounding (think of the multimillion-dollar yearly compensation packages enjoyed by many of the 50- to 65-year-old chief executive officers of the Fortune 500 companies). The distribution of earnings is right skewed; a few workers have extremely high earnings that pull up the mean. As a result, the majority of workers make less than the mean earnings. The lognormal distribution reflects such a right-skewed distribution.

In their classic book, J. Aitchison and J.A.C. Brown (1957) state, "in economic data (right) skewed frequency curves are the rule rather than the exception . . . each generation must learn anew the lessons taught by Galton, Kapteyn, Gibrat, and many other investigators of the properties of the lognormal distribution."(p. xvii) As discussed later in the chapters on regression analysis, random variables involving right-skewed distributions may be more common in economics and finance than those involving the symmetric, bell-shaped density of the normal distribution.

Many random variables that have right-skewed distributions can be transformed into variables that are symmetric and bell shaped. Variables for which a logarithmic transformation yields a normal random variable are said to be distributed lognormally. For instance, a natural logarithmic transformation of the sample data in the second column of Table 6.2 results in the third column of values headed "ln$x$." [These log transformations are easily performed in a spreadsheet program; for example, in LOTUS 1-2-3, the command for the log transformation is "@ln( )".]

Unlike the original variable x, Figure 6.12, panel a, ln$x$ is symmetric and bell shaped, panel b. With the exception of the discreteness of the data, ln$x$ appears to be approximately normal.

Figure 6.12 also demonstrates that if data are initially reported in the natural logarithmic scale, they may be converted back to the original measurements by an antilog or an exponential. For instance, if $\ln x = 1.50$, then x is the $e^{1.50}$, which is 4.5 as in the original data of Table 6.2. [Antilog transformations are easily performed in a spreadsheet program; for example, in LOTUS 1-2-3, the command for the antilog exponential transformation is "@exp( )".]

Finally, the lognormal distribution has some handy properties. One property that is extremely helpful in working with large data sets is that the median of x is the exponential of the mean of ln$x$. For the earnings data in Figure 6.12, the median of x is 20, which is the antilog of the mean of ln$x$, $20 = exp(3)$. This property enables the location of the median without first ranking the values. There are other attributes of the lognormal distribution that are considered in more detail in Crow and Shimizu (1988). In later chapters, when regression and correlation are presented, logarithmic and other data transformations will be shown to be essential features of modern-day data analysis.

TABLE 6.2 Earnings and Natural Logarithm of Earnings

Frequency	Earnings	Log of Earnings
f	x	$\ln x$
1	4.50	1.504077
2	7.39	2.000127
4	12.20	2.501435
7	20.00	2.995732
4	33.11	3.499835
2	54.60	4.000033
1	90.00	4.499809

FIGURE 6.12 Earnings Data and Log of Earnings

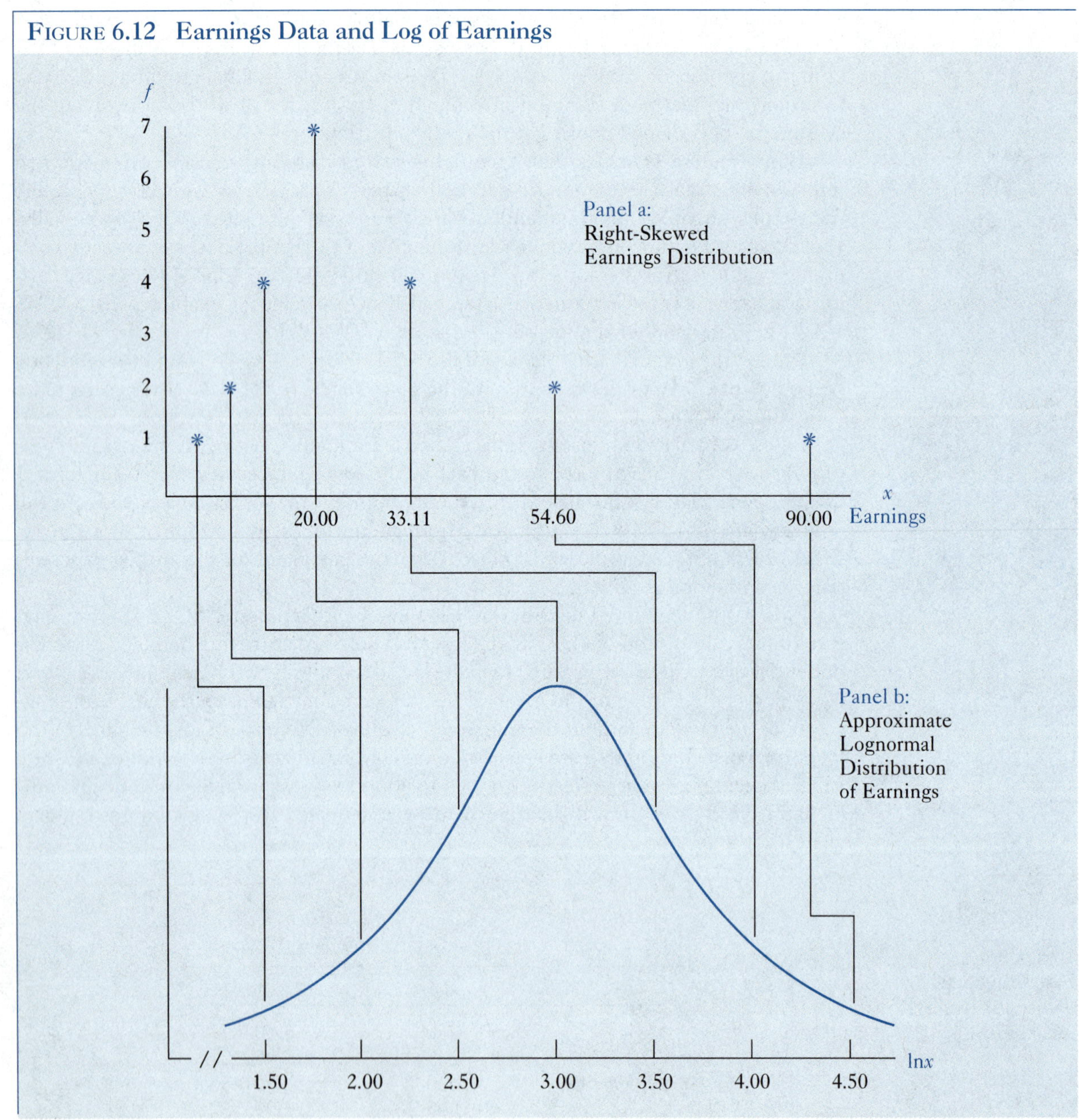

6.7 Concluding Comments

The bell-shaped density of the normal distribution is the most widely used curve in statistics. Many variables have been shown or assumed to behave as a normal random variable. As demonstrated in the previous section, even when a variable is highly skewed, a data transformation often can be found to create a related variable that is normally distributed. This chapter presents the properties of the normal distribution for use in a wide range of problems.

The normal distribution can be used as a way of describing the behavior of data. Although helpful in many types of problems, it does not address the central role of the normal distribution in statistical inference. It does not address how the normal distribution is used to make inferences about an unknown population based on sample observations. That is the focus of the next chapter.

Chapter Exercises

6.1 Classify the following variables as continuous or discrete:
   a. the gender of a chief executive officer
   b. the time it takes to be serviced at Sears
   c. the number of new housing starts in Tucson in June
   d. the age at which a person purchases his or her first car

6.2 State the differences and similarities between a standard normal distribution and any other normal distribution.

6.3 The random variable z is normally distributed with zero mean and variance of one. Find the following probabilities:
   a. $P(z < 2.01) =$ _____
   b. $P(z > 2.95) =$ _____
   c. $P(z > -2.05) =$ _____
   d. $P(z < -1.45) =$ _____
   e. $P(-2.03 < z < 1.06) =$ _____
   f. $P(1.04 < z < 1.99) =$ _____

6.4 The random variable z is normally distributed with zero mean and variance of one. Find the following values of b:
   a. $P(z > b) = 0.9946$
   b. $P(z < b) = 0.8186$
   c. $P(z > b) = 0.0003$
   d. $P(-b < z < b) = 0.673$
   e. $P(z < b) = 0.0228$
   f. $P(0 < z < b) = 0.1664$

6.5 The random variable x is normally distributed with mean 5.00 and standard deviation 2.00. Determine the following.
   a. $P(x < 2.00) =$ _____
   b. $P(2.00 < x < 6.4) =$ _____
   c. $P(-1.00 < x < 0) =$ _____
   d. $P(x < 6.40) =$ _____
   e. $P(x > -2.00) =$ _____
   f. $P(x > 6.23) =$ _____

6.6 The random variable Y is normally distributed with mean of 25.00 and standard deviation of 9.0. Find the following values of b:
   a. $P(y > b) = 0.0007$
   b. $P(y > b) = 0.0550$
   c. $P(y < b) = 0.9948$
   d. $P(b > y > 25) = 0.032$
   e. $P(b < y < 25) = 0.5429$
   f. $P(y < b) = 0.2123$

6.7 As director of a local welfare agency, you are responsible for the allocation of food stamps. You are to provide food stamps only to those households in the lowest 12 percent of the income distribution in your area. Although most income distributions are skewed right, your area's income distribution is approximately normal, with mean income of $21,340 and standard deviation of 4,980. Below what income level should food stamps be provided?

6.8 From time studies, it is determined that a task should take an average of 2.05 minutes to complete, with a standard deviation of 0.4 minutes. What is the probability that a randomly selected employee will take between 1.45 minutes and 2.95 minutes? (Sketch and calculate the area, assuming completion time is normally distributed.)

6.9 A recruiting firm states that it is only interested in students from the top 5 percent of a graduating class. The average grade of the graduating class is well known to be 2.65, with a standard deviation of 0.45. Nevertheless, a student with a grade point of 3.61 applies for an interview and is turned down because the cutoff is set at 3.6265 based on the following calculation:

$$2.17 = z = (b - \mu)/\sigma = (b - 2.65)/0.45 = > b = 3.6265$$

Is this calculation justified? That is, is it appropriate to assume the normal distribution? Explain.

6.10 According to a *Wall Street Journal* (April 1, 1983) article, Americans have voluntarily contributed to reducing the $1.2 trillion public debt with checks totaling $135,575.15. "'Nearly half of the contributions are for less than $10,' says IRS spokeswoman Ellen Murphy, 'but a very generous soul' forwarded $25,000." Is it reasonable to assume that the distribution of contributions to reduce the debt is normal? Why or why not?

6.11 (Optional section required) In Exercise 6.10, is it reasonable to assume that the distribution of contributions to reduce the debt is lognormal? Why or why not?

6.12 "Sears Is Accused of Billing Fraud at Auto Centers" was a headline in *The Wall Street Journal* (June 12, 1992). The article reported on a year-long study by the California Department of Consumer Affairs in which "its agents were overcharged nearly 90% of the time, by an average of $223." If the other 10% average $25 in under charges, and if the distribution of the amount of over, under, and correct charges is normally distributed, what is the standard deviation of those charges?

6.13 Using your standard deviation estimated in Exercise 6.12, what proportion of overcharges were above $585?

6.14 In *The Wall Street Journal* article about Sears billing fraud, "The worst example was in the San Francisco suburb of Concord, where an agent was overcharged $585 to have the front brake pads, front and rear springs, and control arm bushings replaced." Is this consistent with the assumption of normality in Exercise 6.12 and the probability calculated in Exercise 6.13? Explain.

6.15 A soft-drink machine can be regulated so that it discharges an average of μ ounces per cup. If the ounces of fill are normally distributed with standard deviation equal to 0.3 ounces, what is the setting of μ so that 8-ounce cups will overflow only 1 percent of the time?

6.16 The achievement scores for a college entrance examination are normally distributed with a mean of 75 and a standard deviation of 10. What fraction of the scores would you expect to lie between 70 and 90?

6.17 The random variable y is normally distributed with a mean of 90 and a standard deviation of 18. How are the areas between 72 and 90 and between 90 and 108 related?

6.18 A radio manufacturer knows the length of time until a set needs servicing is normally distributed with a mean of 36 months and standard deviation of eight months. The probability that a randomly selected radio will require servicing within the first six months after purchase is approximately ________.

6.19 An article in *The Wall Street Journal* (September 2, 1992) reported on the ability and willingness of insurance companies to pay customer claims following disasters. The article stated: "Customers receive an average of 87 cents for every dollar they claim in losses." If the standard deviation is 13 cents, why is it unreasonable to assume that the amount received per dollar claimed is normally distributed?

6.20 A *Wall Street Journal* (November 11, 1992) article on the clothing sizes that retailers carry stated, "In good times retailers use a sort of bell curve of sizes, with the greatest number of items in the midrange. For an estimated 35% to 40% of women, that means sizes 6, 8, and 10." Assuming sizes are normally distributed, with a mean size of 8, what values for the standard deviation would give 35 percent to 40 percent of the distribution in the size 6 to 10 interval?

6.21 The article on clothing sizes in Exercise 6.20 continued: "For men, it's waist sizes 30–36 inches" which constitute the midrange of 35 percent to 40 percent. If men's waist sizes are normally distributed with a mean of 33, what must the magnitude of the standard deviation be so that 35 percent to 40 percent of the distribution is in the 30 to 36 interval?

6.22 A note in *The Wall Street Journal* (December 9, 1992) stated "the average American worked 81 days this year just to pay federal taxes." Assume the distribution of days worked to pay taxes is normally distributed, with a mean of 81 days and standard deviation of 10 days. What is the probability of selecting a taxpayer who worked more than 100 days to pay taxes?

6.23 According to *The Chronicle of Higher Education* (December 2, 1992), the average time in office of university presidents was seven years; 14 percent of the presidents left office within the first year of appointment. If presidential job tenure is normally distributed, what is the standard deviation of this distribution?

6.24 Using the standard deviation you calculated in Exercise 6.23 and the other information provided, what is the probability that a president stays in office more than 10 years?

6.25 A section of the *DEMOS Tutorial: An Introduction to DEMOS*, Version 3.0b2 (June 1992), which is a computer program for use in decision making that will be introduced in Chapter 17 of this book states:

> Probability Density. The height of the curve at a given value along the horizontal axis denotes the relative probability that the variable has that value. (p. 15)

What is wrong with this definition of probability density?

6.26 In response to reports of needles being found in Pepsi-Cola soda cans in June 1993, Pepsi reported that tampering with cans in plants was unlikely. Pepsi officials stated that the lines run at a rate that fills 11,000 to 12,000 cans per minute, with each fill and seal taking less than eight seconds. If, on average, the line fills and seals 11,500 cans per minute, with a standard deviation of 142 cans per minute, what is the probability of randomly checking the speed of the line and finding an average line speed of 11,250 to 11,750 cans per minute? What do you have to be able to assume to calculate this probability? Is this assumption reasonable?

6.27 Members of a large church congregation are known to have yearly salaries between \$20,000 and \$330,000, with an average of \$81,240. Why would it be inappropriate to use the normal distribution to calculate the proportion who have salaries within one standard deviation of the mean?

6.28 (Optional section required) In Exercise 6.27, why would it be appropriate to use the lognormal distribution to make probability calculations? In your explanation, make use of the minimum, maximum, and average salaries of the church congregation members.

6.29 A *Wall Street Journal* (December 17, 1992) article on razors reported that per shave, "the average U.S. woman shaves about 412 square inches of skin, while the average man shaves 48 square inches, according to Gillette. Yet women shave less often than most men and buy far fewer blades per year—10 compared with an average of 30 for men." If the distribution of blades purchased by men is lognormal, and if the average number of blades purchased by women (10) almost equals the median number of blades purchased by men, what is the mean of this lognormal distribution?

Academic References

Aitchison, J. and J.A.C. Brown. *The Lognormal Distribution, With Special Reference to Its Uses in Economics.* Cambridge: Cambridge University Press, 1957.

Crow, E.L. and K. Shimizu. (eds.) *Lognormal Distributions: Theory and Applications.* New York: Marcel Dekker, 1988.

Stigler, Stephen. *The History of Statistics: The Measurement of Uncertainty before 1900.* Cambridge: Harvard University Press, 1986.

Endnotes

1. The probability density function for the uniform or rectangular distribution shown in Figure 6.3, panel a, is

$$f(x) = 1/h, \text{ for } 0 < x < h$$

The probability distribution function gives the area between a and b as

$$F(a < x < b) = \int_a^b (1/h)dx = (1/h)(b) - (1/h)(a)$$

A probability density function for a right-skewed distribution as shown in Figure 6.3, panel b, is

$$f(x) = 12x(1 - x)^2, \text{ for } 0 < x < 1$$

The probability distribution function gives the area between a and b as

$$F(x) = \int_a^b 12x(1-x)^2\,dx$$

$$= 3b^4 - 8b^3 + 6b^2 - (3a^4 - 8a^3 + 6a^2)$$

2. From Chapter 4, we have the following:

$$E(z) = (1/\sigma)E(x) - (1/\sigma)\mu = 0, \text{ if the mean of } x \text{ is } \mu.$$

$$\text{Also, } E[z - E(z)]^2 = E(z - 0)^2 = E[(1/\sigma)x - (1/\sigma)\mu]^2$$

$$= E[(1/\sigma^2)x^2 - 2(1/\sigma^2)\mu x + (1/\sigma^2)\mu^2]$$

$$= (1/\sigma^2)E(x - \mu)^2 = 1, \text{ if the variance of } x \text{ is } \sigma^2$$

3. For the lognormal distribution, the probability of x between 0 and b is

$$P(0 < x < b) = \int_0^b \frac{1}{x\sigma\sqrt{2\pi}} e^{-(\ln x - \mu)^2 / 2\sigma^2} dx$$

where $0 < x < +\infty$
$\ln x$ is the natural (or base e) logarithm of x
μ is the mean of $\ln x$
σ is the standard deviation of $\ln x$
π is 3.14159
e is 2.71828

4. Algebraically, if the starting earnings of the ith person in year 0 are Y_{i0}, then annual raises at a rate of b imply earnings in year t of

$$Y_{it} = (1 + b)^t Y_{i0}$$

With continuous compounding at rate b, earnings grow exponentially by

$$Y_{it} = e^{bt} Y_{i0}$$

where $e = 2.718$, Euler's constant.

FORMULA

- Standard normal random variable: $z = \frac{x - \mu}{\sigma}$

  where x is a normal random variable with mean μ and the standard deviation of σ so z has a mean of zero and standard deviation of one. That is, if $x \sim N(\mu, \sigma)$, then $z \sim N(0, 1)$.

CHAPTER 7

SAMPLING AND SAMPLING DISTRIBUTIONS

Statistical figures referring to economic events are historical data. They tell us what happened in a nonrepeatable historical case.

Ludwig Edler von Mises
1881–1973

7.1 Introduction

The previous chapter was devoted to situations in which only one item was drawn from a population. Rewriting the three introductory problems makes the single draw explicit.

- If there is an 80 percent probability that *one* randomly selected glass will have less than 20 ounces of beer, and fills are normally distributed, what is the mean fill?
- If the average job tenure of a marketing director at a law firm is 2.1 years and the probability of randomly selecting *one* director who has been in the job for a year or less is one-third, could tenure be normally distributed?
- If the average length of time required to complete an examination is normally distributed, with a mean of 65 minutes and a standard deviation of 15 minutes, then how much time must be given so there is a 99 percent probability that *one* randomly selected test taker will have completed the exam?

This chapter is concerned with the drawing of more than one item from a population. It is devoted to probability calculations based on the average of items in a sample. In contrast to the previous chapter, problems of the following type are considered.

- The Associated Press (December 21, 1991) reported on the new regulation in Great Britain requiring beer glasses to accommodate a full 20-ounce British pint. To check for compliance, an agent visits a pub at nine random times ($n = 9$), ordering a draft on each visit with the following results: 19.77, 20.10, 19.95, 19.89, 20.09, 20.08, 19.89, 19.67, and 20.01 ounces. What is the probability of a mean as low as that calculated from this sample if the pub is serving glasses of beer averaging 20 ounces?
- If the average job tenure of a marketing director at a law firm is 2.1 years, how many directors ($n = ?$) would have to be sampled for the sample mean tenure to be approximately normally distributed? Why would we want the sample mean to be normally distributed?
- If the length of time required to complete an examination is normally distributed, with a mean of 65 minutes and a standard deviation of 15 minutes, then what is the sample mean time for which 99 percent of the samples of 60 test takers ($n = 60$) would have a lower average completion time?

This chapter first describes how sampling is done and then addresses the calculation of probabilities based on multiple-item samples. By the end of this chapter, you will be able to describe simple random sampling, systematic sampling, stratified sampling, and cluster sampling. You will be able to use a random number table and a computer program to generate simple random samples. You will be able to describe and work with the sampling distribution of the sample mean and sample proportion and to use the z distribution in probability calculations.

7.2 Probability Samples

The idea of drawing samples of one or more items to make statements about the population has been used throughout this book. For example, in Chapter 5, we asked how likely was a sample with 14 out of 18 decisions for the plaintiff, if the expected count was nine out of 18 by the national norm? In calculating this probability, we acknowledged that each sample of 18 decisions could give a different count for the plaintiff. Similarly, not new to this chapter are the notions that any two randomly selected numbers need not be the same and that no two collections of randomly selected sets of numbers need have the same distribution. The fact that no single sample can be expected to be a perfect representation of its population is likewise not new.

What is introduced in this chapter is the manner in which probability is assigned to the chance of getting any given sample. The probability or likelihood of getting a sample result is introduced as the measure of "how close" the sample result is to the result expected in the population. Before attempting to calculate this probability, consideration must be given to the different ways in which sampling might give rise to mathematically defensible probabilities.

7.3 Simple Random Samples

Simple Random Sampling
A procedure in which items are drawn or selected from a population in such a way that each item (and every subset of items) in the population is equally likely to be chosen.

Simple random sampling techniques are designed to ensure that each element (and, in turn, each subgroup of elements) in a population has an equal probability of being included in the sample. With simple random sampling, if a population consists of N elements, then each and every sample of n elements (where n may equal 1, 2, 3, . . . $N - 1$) should have a probability $1/N^n$ of being selected.

From Chapters 3, 4, and 5, we know that the size of the sample n relative to the size of the population N and whether sampling is done with or without replacement are critical pieces of information if every possible subset of n items is to be assumed to have an equal probability of being selected. For a finite population of size N, the probability of selecting the first sample of size n, without replacement, is $n!(N - n)!/N!$; the probability of the second sample is $n!(N - 2n)!/(N - n)!$. Only with replacement of the items after each draw would the probability of getting n items from sample to sample remain fixed at $1/N^n$.

In the laboratory, replacement might seem possible; in actual business and economics practices, replacement is seldom possible. If, however, the population is extremely large relative to the sample size, then even though the probability of each sample in repeated sampling does not remain fixed, the changes in probability could be trivial. For instance, if $n = 30$ and $N = 30{,}000$, then for the first sample the probability of drawing 30 items is

$$n!(N - n)!/N! = 0.1307 \times 10^{-101}$$

On the second sampling,

$$n!(N - 2n)!/(N - n)! = 0.1347 \times 10^{-101}$$

using the computer program MAPLE V for calculations.

As long as N is large relative to n, whether there is or is not replacement will not be critical in the drawing of simple random samples. In practice, randomization based on the notion of fixed sampling probabilities is more a matter of degree than an absolute. As a crude rule of thumb, if the population size is at least 10 times greater than the sample size, then sampling with or without replacement should be of little consequence.

The most frequently used methods of simple random sampling are blind draws, random number tables, and random number generators within statistics programs.

Blind Draw

In cases where the elements or items in a population may be numbered easily, a blind draw may be used as a way of selecting a simple random sample. A blind draw may entail selecting cards from a shuffled deck, choosing names out of a hat, picking Bingo numbers from a revolving drum, or letting numbered ping-pong balls float or fall into a bin, as is common in many state lotteries.

In the blind-draw approach, the analyst might place the names of each of the 400 customers who entered a store contest on separate but identical cards, and then thoroughly shuffle the cards in a revolving drum. Then 10 customer cards might be blindly drawn, so that there is no reference to the information on the cards. The cards would not be replaced, since the same customer might be selected more than once, but again, since the sample is small relative to the population, this is of little concern. We do not know, however, whether this procedure will result in a sample of customers that is truly representative of the customers who shop in the store or only the customers who enter the contest.

Putting name cards, numbered ping-pong balls, or the like in a revolving drum may give the appearance of good mixing or randomizing, but the results of selection need not produce a sample that is a perfect representation of the desired population. In 1970 during the Vietnam War, the order in which birthdays were selected from a drum was used to determine the order of being taken in the military draft into the U.S. armed service. A total of 366 capsules were placed in a drum, each representing a birthday. The Selective Service director and delegates selected capsules "at random," after the drum was rotated several times.

Although the above method looked impressive on television, as pointed out by S. Fienberg (1971), the results did not produce a good representation of birthdays. Randomization suggests that among the first 183 birthdays selected, about 30 days should have been from November and December. Of the first 183 days selected, however,

November and December birthdays were overrepresented with 46 capsules. Thus, Fienberg questioned the randomness of the process.

If the capsules were put in the drum in monthly order and then not thoroughly mixed, the days from the sequential months would have tended to come up together. This would be an example of **nonsampling error** because it was a mistake that could have been avoided by spinning the drum more times to more fully mix the capsules. On the other hand, it is possible that the drum was turned enough to adequately mix the capsules but that the noncontrollable random error inherent in the sampling process produced these results. Because a nonsampling error is identifiable, however, the pure sampling error of a randomization process is in doubt.

Nonsampling Error
A discrepancy between a sample value and a population parameter that might result from such things as sampling from the wrong population, improperly worded questions, or adding a column of numbers incorrectly.

The job of the analyst is to avoid nonsampling errors. Intrinsic in randomization, however, is sampling error that makes it impossible to know for sure whether a particular sample result is representative of an unknown population. Reviewing the method of sampling, and reviewing the results themselves to see if they are consistent with what would be expected from a random process, may shed light on the adequacy of the process used for randomizing. This exploration, however, will never tell you whether the sample is or is not representative of an unknown population.

Random Number Table

Because of the possibility of nonsampling error in the blind draw approach (caused by inadequate mixing, clerical errors, or even fraud), the better method for simple random sampling is to use a table of random numbers. Appendix Table C.1 (partly reproduced in Table 7.1) consists of rows and columns of five-digit random numbers. Each of the digits in the five-digit number, as well as the five-digit number itself, are random numbers because they appear with no identifiable order or pattern. Random number tables were first produced by the Rand Corporation in 1955. As discussed in the next section, random numbers can be generated by most statistical software programs for microcomputers .

To use the random number table, each item or element in the population is first given a number. To illustrate, the 1992 special bonus issue of *Business Week* lists 1,000 of "America's Most Valuable Companies," ranked and identified by the numbers one through 1,000. These 1,000 companies and their market values are reproduced in Appendix Table B.1.

To minimize the occurrence of superfluous numbers, the number of digits in the *BW* list of 1,000 companies must be renumbered 0 to 999. To select a simple random sample of 10 companies, an arbitrary starting point is now selected in the random number table by dropping a pencil on the page and taking the closest three-digit number hit by the tip (or by dropping a coin and taking the number under the coin or any like procedure).

Assume the pencil tip hits in the seventh column of grouped numbers, on the 12th five-digit number from the top. This number is 93551. Because a three-digit number (0 to 999) is desired, the first three digits of 93551 are all that is needed as a starting point. The number 935 indicates that the first company selected for the sample is actually numbered 936 on the *BW* list.

TABLE 7.1 Random Numbers and Sample of *Business Week's* Most Valuable Companies

Random Number 0–999	*Business Week* Rank 1–1,000	Company Name	Market Value $Mil.
935	936	Weingarten Realty	522
588	589	Kelly Services	1069
345	346	Blockbuster Entertainment	2215
075	76	Scecorp	8963
859	860	Cypress Semiconductor	610
896	897	Teradyne	572
771	772	Banponce	722
008	9	Bristol-Myers Squibb	41431
632	633	Hartford Steam Boiler Inspection	959
601	602	Hannaford Brothers	1029
			Sum = 58092.0
			Average = 5809.2

Any systematic plan for selecting the next nine three-digit numbers may be used as long as the plan was determined without first looking at the table. The easiest planned sequence is to read down the nine five-digit numbers. The next three-digit number is 588, indicating that the second company in the sample is 589 in the *BW* list. The remaining eight numbers are 345, 075, 859, 896, 771, 008, 632, and 601, which identify the *BW* companies numbered 346, 76, 860, 897, 772, 9, 633, and 602, as shown in Table 7.1.

If a selected sample number occurs more than once in the random number table sequence, the repeat numbers are skipped, unless sampling is with replacement. If sampling is with replacement, then each number is used in the order selected, even if it appears more than once. In the selection of the 10 companies from *BW,* sampling is without replacement because 10 separate companies are desired. In most real-world situations, it may make no sense to include the same company twice, or to repeat an interview with the same person, or test the same part more than once. Again, if the sample size is small relative to the population size, the question of whether sampling is with or without replacement should be of minor importance.

If a population size is between 100 and 999, say $N = 400$, then three-digit numbers are needed, but it is possible to get three-digit numbers from the random number table that exceed the population size. In such cases, these values are simply skipped. For instance, if $N = 400$ and the random number table calls for observation 567 to be included, then this overly large number is ignored and the next number in the random number systematic sequence is selected.

Computerized Random Number Generator

Random numbers are generated by programs that are keyed to a computer's internal clock. Typical microcomputer programs require the user to enter the sample size (10) and population size (1,000); the random generator then gives a screen display of the random numbers. For example, Table 7.2, panel a, shows the commands that are

Table 7.2 Random Number Generation in RESAMPLING STATS

Panel a: RESAMPLING STATS commands that generate 10 random numbers, from 1 to 1,000, and place them in a file A, which is then printed on screen.

```
GENERATE 10 1,1000 A [enter]
PRINT A [enter]
F5
```

Panel b: RESAMPLING STATS screen display of the 10 numbers generated in panel a.

```
Start execution

A = 460  90  445  735  30  578
    705   4  239  102

Successful compilation (0.2 seconds)
```

Panel c: The market values of the 10 *Business Week* companies identified in panel b.

Random Number and *Business Week* Rank 1–1,000	Company Name	Market Value $Mil.
460	Premark International	1523
90	Apple Computer	7583
445	Parker Hannifin	1592
735	Equitable Resources	772
30	Southwestern Bell	17449
578	Penn Central	1092
705	Commerce Clearing House	819
4	Wal-Mart Stores	60954
239	Ingersoll-Rand	3183
102	Allied-Signal	6769
	Sum =	101736
	Average =	10173.6

needed in the computer program RESAMPLING STATS to generate 10 random numbers out of 1,000; the resulting 10 observation numbers then appear as the screen printout shown in Table 7.2, panel b. The market values of the corresponding companies from *Business Week* are shown in panel c.

7.4 Systematic Sampling

Systematic random sampling involves the selection of every *k*th element from a list of elements, starting with any randomly selected element before the *k*th element. Systematic sampling is a popular way of generating samples in accounting. It is typically less expensive to select every *k*th element than to search for the *n* randomly

Systematic Random Sampling
Selection of every kth element on a list of all elements in the population, starting with any randomly selected element before the kth element.

determined items. In systematic sampling, however, only the first of n items can be considered as randomly determined.

Systematic sampling is not the same as simple random sampling. With systematic sampling every element and, in turn, each subgroup of elements in the population does not have the same opportunity of being included in the sample after the first element is selected—only the starting point is randomly determined. For example, if every 100th item is selected from the *Business Week* list of 1,000 most valuable companies, then consecutively numbered companies will never be included after the first company is randomly selected. If a phone book is the source of a list, systematic sampling could be used to minimize the likelihood that relatives would be included in a sample—after the first name is selected, systematic sampling leads to a name down the alphabetized list.

Systematic sampling is convenient for populations formed by lists, stacks, or series. For example, a sample of 10 companies from the *Business Week* list is quickly obtained by selecting every 100th company (after randomly starting within the first 100 names). Table 7.3 shows this systematic sample for a random starting point of the 35th company, followed by companies 135 to 935.

Quality control analysts regularly use systematic sampling for periodic monitoring of production processes. Accountants also employ systematic sampling to check for fraudulent shipping statements, billing statements, and other financial records that follow an orderly posting. When the quantity of interest is in a continuous series or is in a rank order (as with the *Business Week* data on market value), a systematic sample gives a good estimate of the true population average.

There are situations, however, in which systematic sampling should be avoided. If the data have cyclical components, then systematic sampling may be an inappropriate method of sampling. For instance, an article in *The Wall Street Journal* (May 18, 1992) said that people are most likely to move in August. If every 12th month was selected, then the result would not give a good representation of average revenue of a moving

Table 7.3 Systematic Sample of *Business Week's* Most Valuable Companies

Business Week Rank Order Number	Company Name	Market Value $Mil.
35	Boeing	15828
135	Chemical Banking	5679
235	Dow Jones	3223
335	Florida Progress	2336
435	Marriott	1630
535	National Service Industries	1257
635	Amsouth Bancorporation	958
735	Equitable Resources	772
835	Hon Industries	638
935	Delta Woodside Industries	514
		Sum = 32835.0
		Average = 3283.5

company for the years under study. Either revenue would be overestimated, if August was selected, or underestimated, if August was not selected as the starting month. As another example, consider the Chapter 5 Query 5.7 that presented the case of *Ultramares Corp. v. Touche.* In that case, accountants used systematic sampling, but the sample failed to turn up the pattern in the fraudulent invoices.

Many variations on systematic sampling have been designed to overcome the periodicity in data. For instance, instead of selecting only one element at a fixed interval, a group of elements may be selected at a fixed interval. This is known as block sequence sampling. Block sequence selection requires taking a group of observations that are next to each other in a series. This is systematic sampling because the group, or "blocks," are at fixed intervals in the series. As with systematic sampling, block sequence sampling gives a probability sample as long as the starting point is random. This process is not equivalent to simple random sampling, because, once the starting point is selected, the remaining elements do not have an equal chance of getting into the sample.

7.5 Stratified Sampling

When important characteristics of the population are known prior to sampling, the sampling process can be made more efficient if the population is divided into subgroups of those with common characteristics. For example, in a study of starting salaries at a college, graduates might be grouped, or stratified, by their majors. In a study of higher education costs, schools might be placed in one of two subgroups: public or private. Heterogeneous populations can always be divided into subgroups that are more homogeneous if those in the population can be identified by the characteristic. These more homogeneous subgroups are referred to as **strata.**

Strata
Subgroups of a population that share a given characteristic or are homogeneous with respect to some attribute.

The Gallup and Harris polls and other national surveys use stratification techniques to get a better representation of the population than what would be possible by simple random sampling for a given same size. For a given amount of time and money invested in sampling, more information is gleaned from using stratified samples than simple random samples. Stratified sampling is efficient. That is, the added information about characteristics of the population reduces the sample error found in simple random sampling because we have more information. This, of course, assumes that easily identifiable and correct information about population strata are available.

Proportional Stratified Sampling
A form of sampling in which items are randomly drawn from known subgroups of the population (called *strata*), where items in each subgroup have at least one shared characteristic.

If an analyst can identify strata, and knows the proportion of the population contained in these strata, then small, but highly representative, samples can be chosen by randomly drawing from each strata in accordance with the population proportions. Such sampling is sometimes called **proportional stratified random sampling.** For example, suppose the list of graduates mentioned above is classified by four majors in which 1,500 are liberal arts majors, 1,000 are business majors, 850 are education majors, and 150 are natural science majors. A proportional stratified random sample of 100 graduates would reflect these same proportions, with 43 liberal arts majors, 29 business majors, 24 education majors, and four natural science majors.

7.6 Cluster Sampling

There are many combinations of sampling and groupings that can give rise to random samples. For instance, in a marketing survey, if face-to-face communication is required for sampling and the members of a population are at distant locations, then it would be very costly to visit randomly selected persons. Instead, the sampling design might be restricted to a few randomly selected cities. Within these cities, a surveyor visits specific neighborhoods, which might be randomly selected, where they then might interview an adult from every fourth house on every sixth street. Such sampling is called **cluster sampling,** because groups or clusters of elements are first selected and then elements within a cluster are chosen.

Cluster Sampling
A form of sampling in which the population is divided into groups or clusters. These clusters are randomly selected, and elements within the selected clusters are then chosen randomly.

In cluster sampling, the idea is to form groups of diverse subjects; whereas, in stratified sampling, a group's members are supposed to be similar. In cluster sampling there is little difference between groups, while in stratified sampling differences between the groups are desired.

In addition to distance between possible subjects being an impetus for cluster sampling, it is also useful when all the members of a population cannot be individually identified. For instance, we might want to survey students at certain types of colleges on their attitudes about free enterprise. There is no single list identifying and classifying the more than 12 million students currently in U.S. higher education institutions. As discussed in Becker and Lewis (1992, pp. 3–4), however, the Carnegie Foundation for the Advancement of Teaching has classified the 3,400 accredited institutions in American Higher Education under 20 major categorical types. Colleges within a major categorical type are supposedly alike, but students within each college are different. Thus, each college within a major category is a cluster of students for which a student list could be obtained. Colleges to be sampled within a major category could be selected randomly, and then students within the selected colleges could be selected randomly to form a cluster sample for surveying attitudes.

The disadvantage of cluster sampling is that the sampling error is larger than in simple or stratified random sampling. Each step in the cluster sampling process increases sampling error. As we will learn in the next sections, however, sampling error can always be decreased by increasing the sample size.

Exercises

7.1 What alternative methods of sampling (simple random, systematic, block sequence, etc.) may be used in the following situations:

a. to check the daily attendance record of students?
b. to audit weekly average accounts receivable?
c. to monitor error rates in computer chips?

7.2 How does stratification cut the cost of national surveys and increase the precision of predictions?

7.3 A graduate student is heard to say that "a good random sample is one that is a perfect representation of the population, and a bad one is one that does not." Comment on the appropriateness of this student's understanding of sampling.

7.4 In March 1991, J. D. Power & Associates mailed a six-page questionnaire to 73,000 owners of 1990 model cars. Recipients received $1 for responding. According to *Business Week* (June 10, 1991), about 23,000 of the questionnaire recipients responded. Why wouldn't these 23,000 respondents be a simple random sample of 1990 car owners?

7.5 For the hospital data set provided in the file "EX7-5.PRN," use a computer random number generator or table of random numbers to select a simple random sample of size $n = 10$, with replacement, from the 235 patients. Calculate the average wait in the emergency room for your sample.

7.6 From the data in Exercise 2.54, select a systematic sample of 10 states and calculate the mean number of farms. Now use a random number generator to select 10 states without replacement and calculate the sample mean number of farms. Which sample yielded a mean closer to the mean of all the states in Exercise 2.54?

7.7 Sampling Distribution of the Sum of Random Variables

Classical statistics is based on simple random sampling. We have seen in Figures 5.1 and 6.1 that when the size of the simple random sample is large, the distribution of the binomial random variable x, which is formed as the number of "successes" in n trials, can be approximated by the normal distribution. That is,

$$\text{if } x = x_1 + x_2 + \ldots + x_n$$

$$\text{where } x_i = 1 \text{ or } 0, \text{ with } P(x_i = 1) = \pi$$

$$\text{and } p(x_i = 0) = 1 - \pi, \text{ then}$$

$$x \to N\left[\mu = n\pi,\ \sigma = \sqrt{n\pi(1-\pi)}\right] \text{ as } n \to \infty$$

Although not emphasized in Chapters 5 and 6, this result can be explained by constructing the distribution of x from repeated sampling, where each sample is of size n. For instance, Figure 7.1 shows the code written to generate 13 samples in the computer program RESAMPLING STATS. Each of the 13 samples shown represents 32 independent and identical trials. The outcome of each trial is either a one or a zero, $x_i = 1$ or 0, where the probability of a one is 0.20 and the probability of a zero is 0.80, $P(x_i = 1) = 0.2$ and $P(x_i = 0) = 0.8$. The sum of the 32 outcomes, for each of the 13 samples, yields one of the values for x, where $x = \sum_{i=1}^{32} x_i$. As seen in the histogram for x, the distribution of x has a pattern that is bell shaped.

Increasing the number of samples from 13 to 130 and then to 1,000, as shown in Figures 7.1, 7.2, and 7.3, simply makes the bell shape more defined as an approximately symmetric and smooth curve that approximates the normal distribution. It is

FIGURE 7.1 Printout from RESAMPLING STATS for 13 Samples, Each of Size $n = 32$, and the Probability of Success on Each Trial is $\pi = 0.2$

```
PROGRAM:  1: REPEAT 13              (produce 13 samples)
          2: GENERATE 32 1,5 x      (sample is x, where there is a 1 in 5 chance of success on
                                    on each of 32 trials)
          3: PRINT x                (print x)
          4: COUNT x = 1 J          (sum l's in x and place count in J)
          5: SCORE J Z              (keep track of successes in each sample and place in Z)
          6: END                    (end the loop)
          7: HISTOGRAM Z            (graph histogram of number of successes per sample of
                                    size 4)
```

13 SIMPLE RANDOM SAMPLES, EACH OF SIZE n = 32, WITH CHANCE OF SUCCESS π = 0.2

```
x: 0 1 0 0 0 0 0 0 1 0 0 0 0 1 0 0 0 0 1 0 1 0 0 1 1 0 0 0 1 0 0 0    8 = x
x: 0 0 0 0 0 1 0 0 0 0 1 1 0 0 1 0 0 1 0 0 1 0 0 0 0 0 0 0 0 0 0 1    7 = x
x: 1 0 0 0 0 0 1 0 0 0 0 0 0 0 0 0 1 1 0 0 0 0 0 0 0 0 0 0 1 0 0 0    5 = x
x: 0 0 1 0 0 0 0 1 0 0 0 1 0 0 0 1 0 0 0 0 0 0 0 1 0 0 0 0 0 0 0 0    5 = x
x: 0 1 0 0 0 1 1 0 0 1 0 0 0 1 0 0 0 0 1 0 0 0 0 0 0 0 0 0 0 0 0 0    6 = x
x: 1 0 0 1 0 0 0 0 0 0 0 0 1 0 0 1 0 1 0 0 0 0 0 1 0 0 0 0 0 0 0 0    6 = x
x: 0 0 1 0 1 0 0 1 0 0 0 0 0 0 1 0 0 0 0 0 0 0 0 1 1 0 0 0 0 0 0 0    6 = x
x: 0 0 0 1 0 0 0 1 1 1 0 0 0 0 0 1 1 0 0 0 0 1 0 0 0 1 0 0 0 0 0 0    8 = x
x: 0 0 0 0 0 0 0 0 0 0 0 1 1 0 1 0 1 1 0 1 0 1 1 0 0 0 1 0 0 0 1 0   10 = x
x: 0 0 0 0 0 0 0 0 1 0 0 0 0 1 0 0 1 0 0 0 0 0 0 0 0 1 0 0 0 0 0 0    4 = x
x: 0 0 0 0 1 1 0 0 1 0 0 0 0 0 0 0 0 0 0 0 1 0 1 0 0 1 0 0 0 0 0 1    7 = x
x: 1 0 0 0 0 0 0 0 0 0 0 0 0 0 0 0 0 0 0 1 0 1 0 0 0 0 0 0 0 0 0 0    3 = x
x: 0 1 1 1 1 0 0 0 1 0 0 0 0 0 0 0 1 0 0 0 1 0 0 1 0 0 1 0 0 0 0 0    9 = x
```

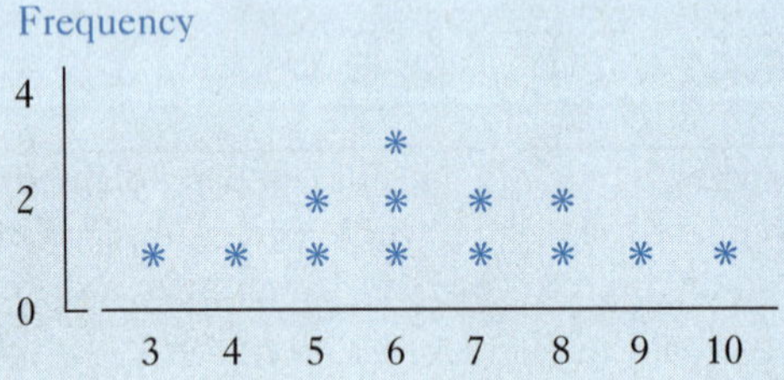

```
Frequency
4 |
  |               *
2 |           *   *   *   *
  |   *   *   *   *   *   *   *   *
0 L_________________________________
      3   4   5   6   7   8   9   10
```

Bin Center	Freq	Pct	Cum Pct
3	1	7.7	7.7
4	1	7.7	15.4
5	2	15.4	30.8
6	3	23.1	53.8
7	2	15.4	69.2
8	2	15.4	84.6
9	1	7.7	92.3
10	1	7.7	100.0

FIGURE 7.2 Printout from RESAMPLING STATS for 130 Samples, Each of Size $n = 32$, and the Probability of Success on Each Trial is $\pi = 0.2$

Bin Center	Freq	Pct	Cum Pct
1	1	0.8	0.8
2	3	2.3	3.1
3	10	7.7	10.8
4	14	10.8	21.5
5	17	13.1	34.6
6	32	24.6	59.2
7	22	16.9	76.2
8	13	10.0	86.2
9	11	8.5	94.6
10	5	3.8	98.5
12	2	1.5	100.0

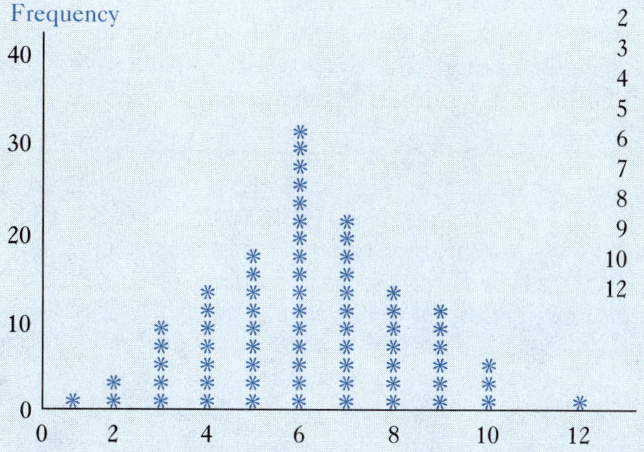

Frequency
40
30
20
10
0
0 2 4 6 8 10 12

FIGURE 7.3 Printout from RESAMPLING STATS for 1000 Samples, Each of Size $n = 32$, and the Probability of Success on Each Trial is $\pi = 0.2$

Bin Center	Freq	Pct	Cum Pct
0	1	0.1	0.1
1	5	0.5	0.6
2	27	2.7	3.3
3	62	6.2	9.5
4	108	10.8	20.3
5	163	16.3	36.6
6	180	18.0	54.6
7	160	16.0	70.6
8	127	12.7	83.3
9	83	8.3	91.6
10	48	4.8	96.4
11	22	2.2	98.6
12	10	1.0	99.6
13	4	0.4	100.0

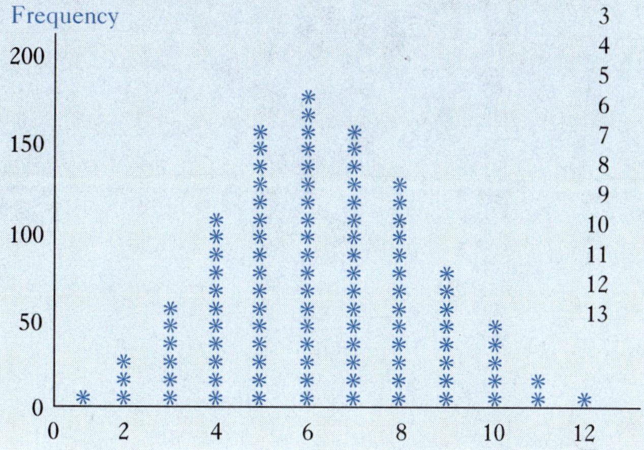

Frequency
200
150
100
50
0
0 2 4 6 8 10 12

important to recognize that it is the relatively large sample size of $n = 32$ that is producing the bell-shaped distribution for x and not the number of samples (13, 130, and 1,000). Increasing the number of simple random samples simply smooths the pattern.

To fully appreciate that the large n is producing the bell-shaped distribution of x, consider a relatively small sample of $n = 4$. Figures 7.4, 7.5, and 7.6 show the distributions generated by the RESAMPLING STATS program that correspond to Figures 7.1, 7.2, and 7.3, but with a smaller n of 4. Notice that as the number of samples, each of size $n = 4$, increases from 13 to 130 to 1,000, the distribution of x does not become bell shaped. A sample size of $n = 4$ is too small to produce the symmetric bell-shaped distribution of x, regardless of the number of samples.

FIGURE 7.4 Printout from RESAMPLING STATS for 13 Samples, Each of Size $n = 4$, and the Probability of Success on Each Trial is $\pi = 0.2$

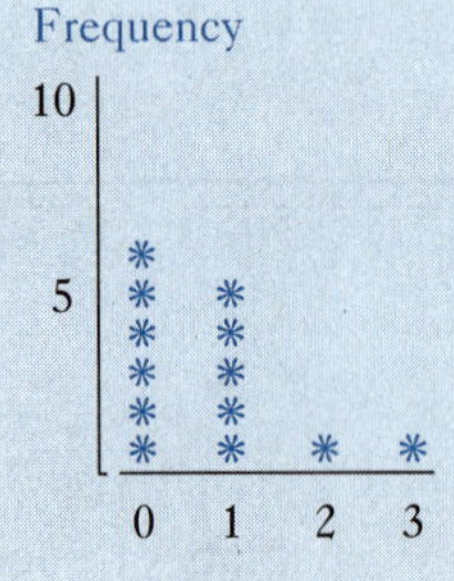

```
Frequency
10 |
   |
   |
   | *
 5 | *  *
   | *  *
   | *  *
   | *  *
   | *  *  *  *
   +------------
     0  1  2  3
```

Bin Center	Freq	Pct	Cum Pct
0	6	46.2	46.2
1	5	38.5	84.6
2	1	7.7	92.3
3	1	7.7	100.0

FIGURE 7.5 Printout from RESAMPLING STATS for 130 Samples, Each of Size $n = 4$, and the Probability of Success on Each Trial is $\pi = 0.2$

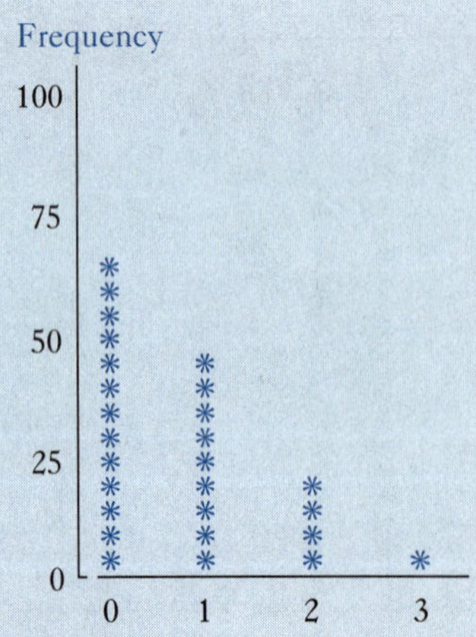

```
Frequency
100 |
    |
    |
 75 |
    |
    | *
    | *
 50 | *
    | *    *
    | *    *
    | *    *
    | *    *
 25 | *    *
    | *    *    *
    | *    *    *
    | *    *    *
    | *    *    *    *
  0 +-------------------
      0    1    2    3
```

Bin Center	Freq	Pct	Cum Pct
0	64	49.2	49.2
1	43	33.1	82.3
2	20	15.4	97.7
3	3	2.3	100.0

FIGURE 7.6 Printout from RESAMPLING STATS for 1000 Samples, Each of Size $n = 4$, and the Probability of Success on Each Trial is $\pi = 0.2$

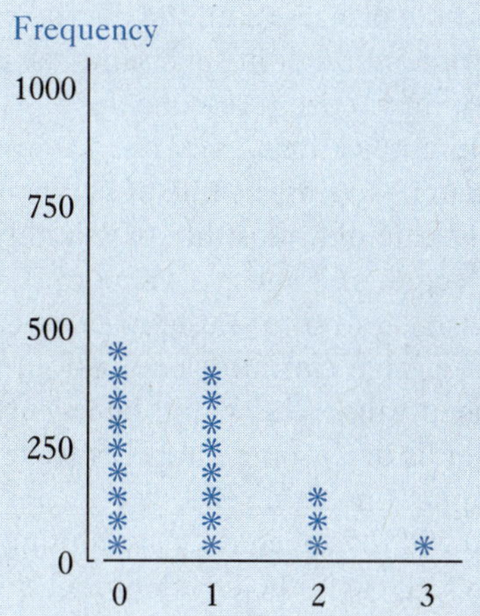

```
Frequency
1000 |
     |
 750 |
     |
 500 |
     |  *
     |  *     *
     |  *     *
 250 |  *     *
     |  *     *
     |  *     *     *
     |  *     *     *
   0 |  *     *     *     *
     ---------------------------
        0     1     2     3
```

Bin Center	Freq	Pct	Cum Pct
0	428	42.8	42.8
1	392	39.2	82.0
2	152	15.2	97.2
3	28	2.8	100.0

The distributions of x, as portrayed by the patterns in the histograms in Figures 7.1 through 7.6, are called sampling distributions of x. When the shape of the sampling distribution is known, it is used as the probability density function for probability calculations associated with specific sums. As in the case of the binomial random variable, if the sample size is large, then the sampling distribution of x can be approximated by the normal distribution. As demonstrated in the next sections of this chapter, any sampling distribution representing the sum of random variables can be approximated by the normal distribution if the sample size is sufficiently large. The normality of the sampling distribution forms the basis of estimation and hypothesis testing in all the remaining chapters of this book. It is thus a most important concept.

7.8 SAMPLING DISTRIBUTION OF THE SAMPLE MEAN

Sampling Distribution of the Sample Mean, $\bar{x}$
The probability distribution of the means of all possible samples of fixed size randomly drawn from a given parent population.

In the case of sampling from populations in which the mean is of interest, the random error intrinsic to sampling causes the mean calculated from any one sample to differ from the population mean. It also causes the mean of one sample to differ from that of another. Unlike the population mean (μ), which has a specific fixed value, the mean of a sample ($\bar{x}$) is a random variable that takes on different values from sample to sample. The random variable $\bar{x}$ has a probability distribution that is called the **sampling distribution of the sample mean,** or, in its short-hand form, the distribution of $\bar{x}$.

The conceptualization of the distribution of $\bar{x}$ is the most important notion in statistics. It provides the logical foundation for estimation and hypothesis testing, as discussed in the following chapters. If you can learn to visualize the theoretical construction of the sampling distribution of the sample mean, then the statistical estimation and testing procedures introduced in the next three chapters will make sense. If you cannot grasp the difference between obtaining a specific value for $\bar{x}$, as calculated from a single simple random sample of size n, and the distribution of all the values of $\bar{x}$ that could be calculated from a large number of such samples, each of size n, then estimation and hypothesis testing may never seem reasonable.

In most real-world applications, only one sample of size n is drawn from which a sample mean is calculated—a single value of $\bar{x}$. If another sample could be drawn, however, then a different value of $\bar{x}$ most likely would be obtained. If yet another sample of size n could be selected, still another value of this sample mean might be realized. Because there are a large (infinite) number of possible samples, each of size n, that could be drawn from a large (infinite) population, theoretically there are an infinite number of sample mean values that could be calculated. To begin to visualize the resulting distribution, let us draw some samples of size $n = 20$ from the *Business Week* data on market value in Appendix Table B.1.

Samples of $n = 20$ will be selected by a sampling plan tailored to the random number table of Appendix C.1. As in the development of Table 7.1, a random starting point is selected, with the next 10 three-digit numbers taken but then a random skip is enacted, with the next 10 three-digit numbers taken to make a total of 20 random numbers. For example, building on the 10 random numbers in Table 7.1 (as reproduced in the first three lines of Table 7.4), a random skip to a new part in the random number table gives the next 10 three-digit numbers, shown in the second three lines of Table 7.4, where the associated information on firm rank and market value is taken from the *Business Week* data in Appendix B.1. The next 12 samples of size $n = 20$ were constructed in a like manner. Notice that, although the individual companies in a sample of 20 were selected without replacement, once a sample of 20 is drawn and the mean market value calculated, all 20 companies are returned to the population, so that the probability of drawing any sample of $n = 20$ remained fixed from sample to sample. This process of repeated sampling is called resampling.

The 13 sample means in Table 7.4 are summarized in a frequency distribution and represented in a histogram in Figure 7.7. Although not shown here, the population of 1,000 companies had a mean market value of 3375 million dollars and a median of 1390 million dollars. This population is highly right skewed. On the other hand, the overall mean of sample means is 3045.426 and the median of the sample means is 2958.65; this distribution of sample means is almost symmetric.

Notice also that the mean of the sample means ($\Sigma\bar{x}/13 = 3045.426$) is close to the population mean ($\mu = 3375$), with six of the 13 sample means within \$1,000 of the population mean. In addition, approximately 68 percent of the 13 observations are within plus or minus one standard deviation of the overall mean of the sample means; that is, nine means are within 3045.426 ± 1447.673, which is consistent with 8.84 (or 68 percent of 13) to be expected from the rule of normality. In summary, the mean of sample means appears to be close to the mean of the population from which samples are drawn, and the distribution of the means around the mean of sample means appears to be almost symmetric, bell shaped, and approximately normal.

TABLE 7.4 Thirteen Samples of Size $n = 20$ of *Business Week*'s Most Valuable Companies (Market Value in Millions of Dollars)

First simple random sample of size $n = 20$											Mean ($\bar{x}$)
BW Rank	936	589	346	76	860	897	772	9	633	602	
Sample x	522	1069	2215	8963	610	572	722	41431	959	1029	
BW Rank	870	273	400	492	930	964	4	690	988	883	
Sample x	601	2804	1830	1425	529	498	60954	850	476	587	6432.30
Second simple random sample of size $n = 20$											
BW Rank	68	423	379	986	55	660	933	737	624	845	
Sample x	9808	1681	1965	476	10964	908	525	771	977	627	
BW Rank	462	546	26	532	933	583	811	473	964	375	
Sample x	1516	1197	20229	1261	525	1081	677	1492	498	1995	2958.65
Third simple random sample of size $n = 20$											
BW Rank	286	553	844	676	543	304	721	814	297	379	
Sample x	2680	1157	628	876	1210	2519	786	671	2573	1965	
BW Rank	968	454	790	775	594	938	300	279	187	421	
Sample x	494	1542	702	717	1046	519	2564	2763	4204	1685	1565.05
Fourth simple random sample of size $n = 20$											
BW Rank	26	428	100	549	958	303	467	88	996	294	
Sample x	20229	1664	7091	1182	505	2520	1504	8007	469	2620	
BW Rank	958	303	567	88	996	294	50	24	789	513	
Sample x	505	2520	1123	8007	469	2620	12423	21039	704	1340	4827.05
Fifth simple random sample of size $n = 20$											
BW Rank	516	536	288	950	629	627	476	152	298	969	
Sample x	1318	1248	2666	512	968	974	1482	5160	2570	493	
BW Rank	417	216	358	496	870	579	166	707	827	531	
Sample x	1715	3463	2104	1412	601	1089	4801	817	654	1266	1765.65
Sixth simple random sample of size $n = 20$											
BW Rank	531	681	849	64	898	946	345	610	330	131	
Sample x	1266	863	619	9877	570	514	2218	1012	2359	5739	
BW Rank	605	95	284	729	259	566	999	727	66	705	
Sample x	1023	7237	2701	779	2943	1126	466	783	9843	819	2637.85
Seventh simple random sample of size $n = 20$											
BW Rank	578	964	334	235	544	507	172	808	316	968	
Sample x	1092	498	2337	3223	1208	1361	4737	681	2458	494	
BW Rank	536	413	74	777	433	908	882	723	518	631	
Sample x	1248	1742	9309	715	1636	562	587	785	1314	962	1847.45

(continued)

TABLE 7.4 Thirteen Samples of Size n = 20 of *Business Week's* Most Valuable Companies (Market Value in Millions of Dollars) *(continued)*

											Mean ($\bar{x}$)
Eighth simple random sample of size n = 20											
BW Rank	720	270	746	561	93	337	142	564	64	486	
Sample x	789	2821	752	1132	7307	2321	5535	1130	9877	1439	
BW Rank	61	820	418	416	921	507	725	781	46	419	
Sample x	10088	665	1708	1723	543	1361	784	710	13835	1706	3311.30
Ninth simple random sample of size n = 20											
BW Rank	301	259	101	380	922	462	546	26	532	933	
Sample x	2547	2943	7009	1963	542	1516	1197	20229	1261	525	
BW Rank	729	259	566	990	727	66	705	20	591	302	
Sample x	779	2943	1126	474	783	9843	819	21962	1063	2540	4103.20
Tenth simple random sample of size n = 20											
BW Rank	862	676	510	544	213	214	531	422	596	39	
Sample x	608	876	1351	1208	3547	3530	1266	1682	1043	15619	
BW Rank	80	150	551	249	118	461	290	359	791	591	
Sample x	8715	5253	1169	3008	6043	1516	2653	2097	699	1063	3147.30
Eleventh simple random sample of size n = 20											
BW Rank	968	917	604	405	591	562	374	290	889	642	
Sample x	494	554	1024	1789	1063	1132	2001	2653	580	944	
BW Rank	369	713	302	592	596	520	619	919	588	258	
Sample x	2027	805	2540	1058	1043	1303	984	549	1073	2950	1328.30
Twelfth simple random sample of size n = 20											
BW Rank	326	679	883	669	777	664	96	451	164	675	
Sample x	2405	866	587	894	715	900	7225	1563	4809	878	
BW Rank	608	842	696	222	859	257	226	134	200	849	
Sample x	1018	629	832	3366	611	2958	3345	5701	3881	619	2190.10
Thirteenth simple random sample of size n = 20											
BW Rank	781	699	850	43	153	227	678	206	417	216	
Sample x	710	829	618	14308	5157	3345	868	3709	1715	3463	
BW Rank	254	207	86	54	369	713	302	592	596	520	
Sample x	2969	3688	8155	11217	2027	805	2540	1058	1043	1303	3476.35

Median of the 13 sample means = 2958.65 Mean of the 13 sample means = 3045.426

FIGURE 7.7 Frequency Distribution and Histogram for Means of 13 Samples, Each of Size n = 20, of *Business Week's* Most Valuable Companies (Market Value in Millions of Dollars)

Summary measures of the 13 sample means
Median = 2958.650 Mean = 3045.426 Standard Deviation = 1447.673

Frequency distribution of the 13 sample means

Sample means of market values		Frequency		Cumulative	
Lower Limit	Upper Limit	Total	Relative	Total	Relative
8.000	1600.000	2	.153846	2	.1538
1600.000	2400.000	3	.230769	5	.3846
2400.000	3200.000	3	.230769	8	.6154
3200.000	4000.000	2	.153846	10	.7692
4000.000	4800.000	1	.076923	11	.8461
4800.000	5600.000	1	.076923	12	.9231
5600.000	6433.000	1	.076923	13	1.0000

Histogram of the frequency distribution of the 13 sample means

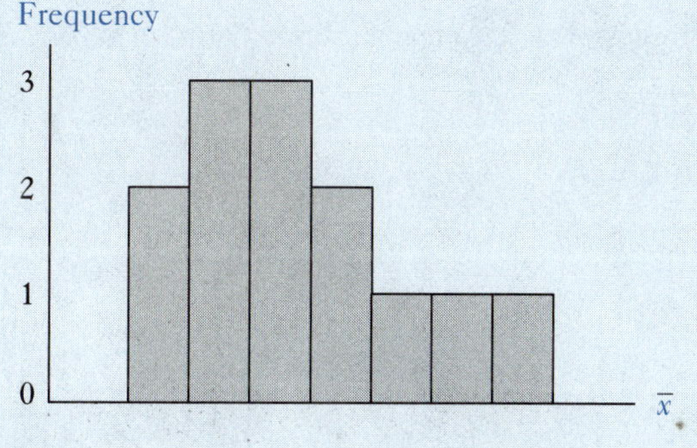

If we increased the sample size from n = 20 to n = 40, the distribution of $\bar{x}$ would become even more bell shaped and symmetric than that shown in Figure 7.7. This can be seen in the change in shape from panel a to panel b in Figure 7.8. The critical recognition here is that increasing the sample size n makes the distribution of the sample mean more bell shaped and symmetric.

If we now increase the number of samples from 13 to a larger number, where each continues to be of size n = 40, the distribution of $\bar{x}$ would become smoother and appear more continuous, but it would continue to appear bell shaped and symmetric. The smoothing of the distribution of $\bar{x}$ is shown by the movement from panels b to d in Figure 7.8. Remember, it was the increase in the sample size from n = 20 to n = 40 that made the distribution of $\bar{x}$ more symmetric and bell shaped. Increasing the number of

FIGURE 7.8 Sampling Distribution of the Sample Mean

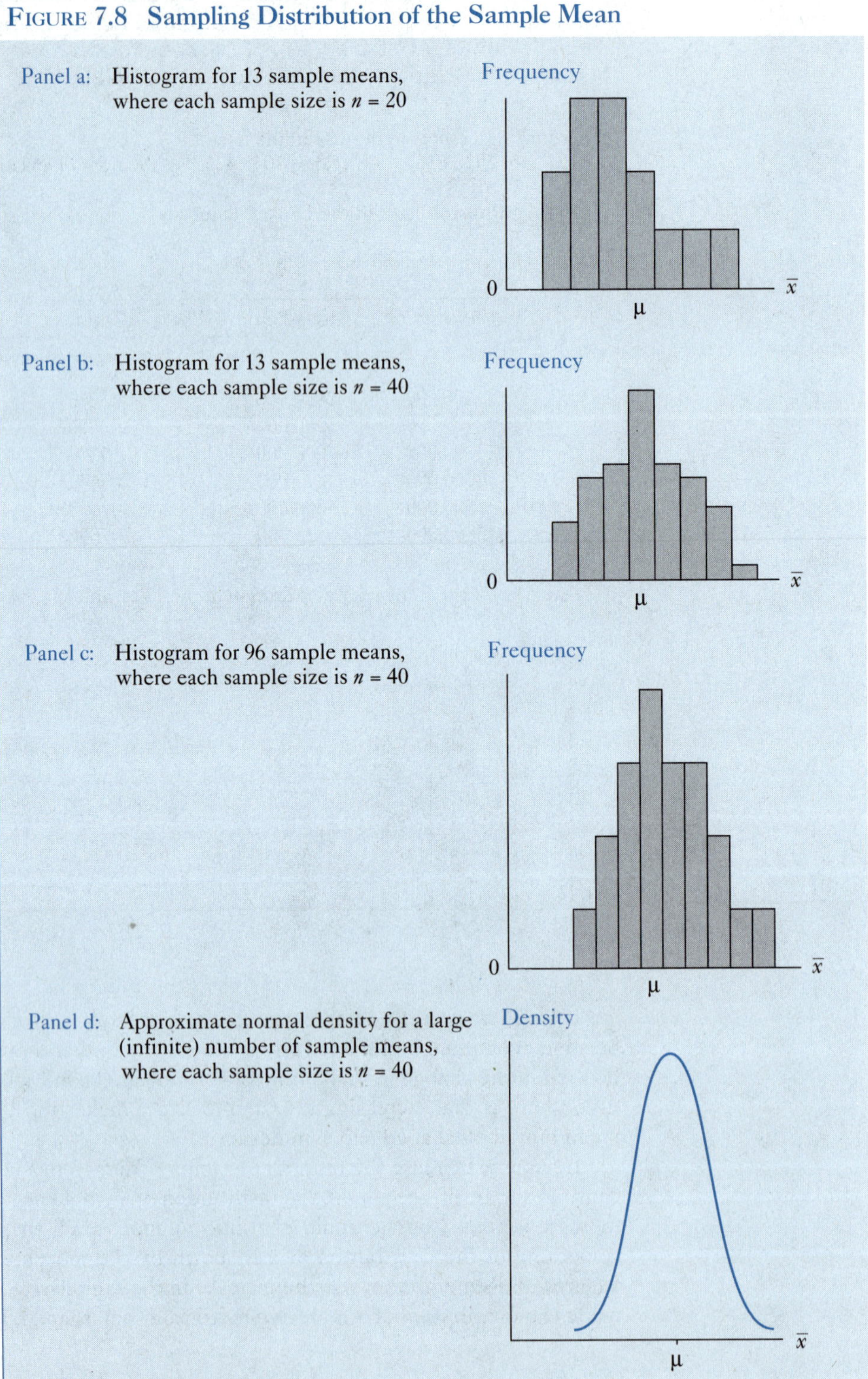

samples only smoothed the distribution of the sample mean. In what follows, we will always assume that, at least theoretically, an infinite number of sample means can be calculated from an infinite number of samples where each sample is of size n.

Possibly the most important theorem in statistics is the Central Limit Theorem. In the case of the sample mean, it says that as the sample size n increases (as from $n = 20$ to $n = 40$) the distribution of the infinite $\bar{x}$ values that theoretically could be calculated with resampling will take on the symmetric and bell shape of the normal distribution. More precisely,

Central Limit Theorem (for sample mean $\bar{x}$)

If $x_1, x_2, x_3, \ldots x_n$ is a simple random sample of n elements from a large (infinite) population, with mean μ and standard deviation σ, then the distribution of $\bar{x}$ takes on the bell-shaped distribution of a normal random variable as n increases and the distribution of the ratio

$$\frac{\bar{x} - \mu}{\sigma/\sqrt{n}}$$

approaches the standard normal distribution as n goes to infinity.

The precise statement of the Central Limit Theorem is based on an infinite number of simple random samples, each of size n, so that the expected value of the sample mean, $E(\bar{x})$ is equal to the population mean μ. This *does not* say that any one sample mean is the population mean. It says that the overall average of an extremely large (infinite) number of $\bar{x}$'s is equal to the population mean. That is, if we could continue to resample as in Table 7.4 and Figure 7.7, and calculate the overall mean of all the $\bar{x}$'s from these samples, then that overall mean would eventually equal the population mean.

Notice in Table 7.4 and Figure 7.7, that with only 13 samples, each of size $n = 20$, we are already close to the population mean as reported in *Business Week*. The sampling distribution of $\bar{x}$ is based on an infinite number of samples, each of which is of size n. As a rule of thumb, n should be at least 30 to invoke the Central Limit Theorem to establish the shape of the distribution of $\bar{x}$. That is, each of an infinite number of samples must be at least of size $n = 30$ to assert that the sampling distribution of their means is approximately normal.

Finally, the Central Limit Theorem requires that the population variable x has a fixed standard deviation σ so that the sampling distribution of the sample mean $\bar{x}$ can have a standard deviation of $\sigma_{\bar{x}} = \sigma/\sqrt{n}$.[1] Other names statisticians give to this standard deviation of the sample mean are the **standard error of the mean** or **standard deviation of $\bar{x}$,** which is calculated by dividing the standard deviation of the population σ by the square root of the sample size $\sqrt{n}$; i.e., $\sigma_{\bar{x}} = \sigma/\sqrt{n}$. The ratio $\sigma/\sqrt{n}$ is a measure of the sampling error. It is a measure of dispersion of the sample means around their expected value $E(\bar{x})$. As the sample size n increases, the distribution of $\bar{x}$ collapses on the point at which $E(\bar{x}) = \mu$, because $\sigma/\sqrt{n}$ decreases as n increases. This is portrayed in Figure 7.9 as a movement from a sample size of $n = 30$, in panel a, to $n = 120$, in panel d.

FIGURE 7.9 The Effect of Increasing the Sample Size on the Distribution of the Sample Mean

Panel a: Approximate normal density for a large (infinite) number of sample means, where each sample size is $n = 30$

Panel b: Approximate normal density for a large (infinite) number of sample means, where each sample size is $n = 60$

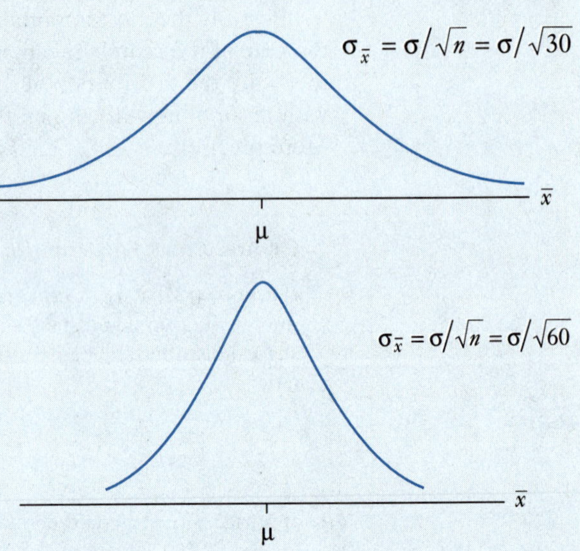

Panel c: Approximate normal density for a large (infinite) number of sample means, where each sample size is $n = 90$

Panel d: Approximate normal density for a large (infinite) number of sample means, where each sample size is $n = 120$

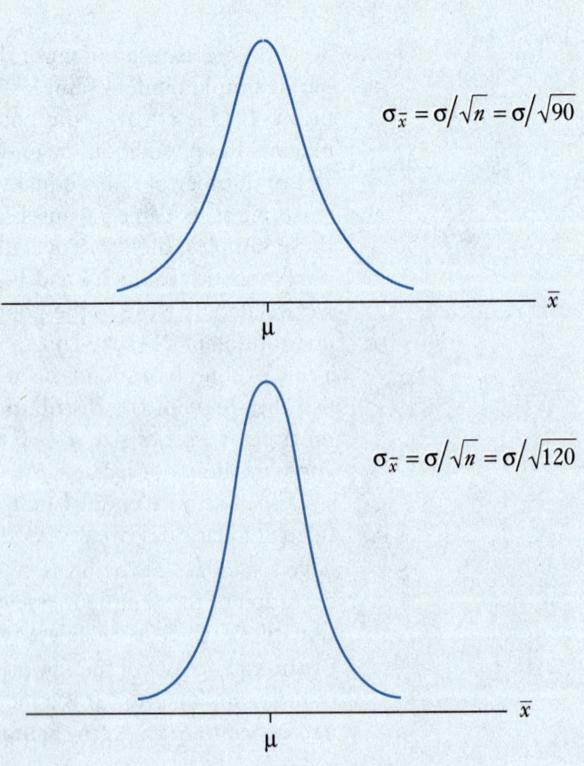

HISTORICAL PERSPECTIVE

In statistics, as in all walks of life, credit is not always given to the right person. That is, even in the history of statistical thought, errors have not been eliminated.

Abraham De Moivre (1667–1754) invented the normal distribution, and Thomas Simpson (1710–1761) opened the door for its use in practice problems. The great German scientist Carl Friedrich Gauss (1777–1855), however, gets credit for its development in the context of the Central Limit Theorem and its role as the most important error distribution in statistics. But Gauss was not the originator of the Central Limit Theorem. (As discussed in later chapters of this book, Gauss's contribution was the application of the normal distribution to least-squares.)

Pierre Simon Laplace (1749–1827) formulated the Central Limit Theorem in 1810. Laplace extended the work of De Moivre and Simpson to a major generalization: Most any random variable formed by a sum (such as a mean) will be approximately normally distributed, if the number of terms in the sum is large. From the early formulation of the Central Limit Theorem, there was confusion about the sum to which it applies and the use to which it might be put.

To this day, it is common to hear students say: "As the sample size gets large, *its* distribution becomes normal." But to what does the pronoun "its" refer? The distribution of a sample obviously will approach that of the population from which it is drawn as the sample size increases. If the population is normal, then the sample will approach the normal distribution as the sample size increases. If the population is not normal, then the sample will not approach the normal distribution as the sample size increases. The Central Limit Theorem, however, does not apply to the distribution of the sample; it pertains to the distribution of all the sample means that could be calculated from an infinite number of samples where each sample is of size n.

In the nineteenth century, "curve fitting" and the finding of "normality in nature" and the characteristic of the "average man" were popular, but this had little to do with the implications of the Central Limit Theorem. As discussed in Chapter 2, for example, Adolphe Quetelet (1797–1874) made his early reputation by applying a form of cluster sampling in a failed attempt to make probabilistic statements about the population. He never made the connection between the mean of one simple random sample and the distribution of means from all such samples.

Quetelet had not been educated in statistics, but he finally learned the errors in his ways and then discarded all notions of sampling and probability analysis in favor of a complete census. Although he was an extremely productive scholar, he never fully grasped the difference between (1) simple random sampling and the use of the Central Limit Theorem and (2) his nonrandom sampling techniques and his inability to make mathematically defensible probabilistic statements from such samples. Gauss and Laplace made the Central Limit Theorem operational as a mathematical concept for assessing uncertainty through probability and making an inference about a parameter of a large population from a single simple random sample.

EXERCISES

7.7 If the population standard deviation is \$25 and the sample size is 25, then the standard error of the mean is __________.

7.8 If the sample size increases by a multiple of four, then the standard error of the mean _______ by a multiple of _______.

7.9 The average yearly salary of retail store managers in a particular city is known to be \$22,000, with a standard deviation of \$120. A random sample of 100 store managers reveals a sample average salary of \$23,000 and sample standard deviation of \$122. From this information, what is the distribution of the sample mean and what is its mean and standard deviation?

7.10 Writing in the *British Medical Journal* (March 15, 1986), Martin Gardner and Douglas Altman state:

> The distinction between two widely quoted statistics—the standard deviation and the standard error—is, however, often misunderstood. The standard deviation is a measure of the variability between individuals in the level of the factor being investigated, such as blood alcohol concentrated in a sample of car drivers, and is thus a descriptive index. By contrast, the standard error is a measure of the uncertainty in a sample statistic. For example, the standard error of the mean indicates the uncertainty of the mean blood alcohol concentration among the sample of drivers as an estimate of the mean value among the population of all car drivers (p. 748).

Comment on the correctness of this quote.

7.11 A court case, *Cuomo et al. v. Baldrige: 2629,* was brought by New York State and others to require the U.S. Census Bureau to adjust the 1980 Census numbers for alleged undercounting. In 1987, Judge Sprizzo ruled against New York. At one point in the trial, J Sprizzo said, "I take it your standard error should be a fixed statistical number which you then subtract from your results and you get what is left, basically which is supposed to measure the accuracy of what you are measuring?" The witness (David Freedman), responded, "I hate to argue with you, but it isn't quite like that." Explain the problem with the judge's understanding of the standard error and its relationship to accuracy.

7.12 An article in *Newsweek* on "The Science of Polling" (September 28, 1992) implied that the standard error of estimation varies inversely with the sample size, but increasing the sample size does not result in proportional reductions in the standard error. Is this correct? Explain.

7.13 Sherlock Holmes, in A.C. Doyle's *The Sign of the Four* (1890), said: "You can, for example, never foretell what any one man or woman will do, but you can say with precision what an average number will be up to." How does this quote relate to the notion of precision in statistics and the sampling distribution of the mean?

7.9 Sampling Distribution of the Sample Mean for Large Samples

By the Central Limit Theorem, the distribution of $\bar{x}$ is approximately normal and the ratio

$$z = \frac{\bar{x} - \mu}{\sigma/\sqrt{n}}$$

is an approximate standard normal random variable, where

1. the population variable x can have most any distribution provided it has a fixed mean, $E(x) = \mu$, and variance σ^2
2. the sample size n is large ($n \geq 30$), and
3. the distribution of $\bar{x}$ has $E(\bar{x}) = \mu$ and $Var(\bar{x}) = \sigma^2/n$

The development of the Central Limit Theorem from Table 7.4 and Figure 7.7 was unusual because the population mean was known and we were able to resample with replacement. Typically the value of the population mean is something under study, and only one simple random sample of size n is available. As long as n is sufficiently large, however, the Central Limit Theorem tells us that the sampling distribution of the mean will be approximately normally distributed. Only one sample is needed to make probability calculations.

The Central Limit Theorem enables us to calculate probabilities associated with the sample mean. For instance, an article (*HT* May 26, 1992) on the use of hot rod Ford Mustangs and Chevrolet Camaros by Indiana State Police quoted an officer as saying, "It is not uncommon for the average daily speed of people stopped to be 83 miles an hour." If a sample of 36 vehicles stopped on a given day shows a mean speed of 83 miles per hour, what is the probability of observing this or a higher sample mean speed if the overall mean speed of vehicles stopped is 65 miles per hour, assuming a population standard deviation of 12 mph?

Because the sample size is large ($n = 36$), and assuming it was based on simple random sampling techniques, the $\bar{x}$ distribution is approximately normal and the z ratio is close to the standard normal. Thus, from the information provided we have

$$z = \frac{\bar{x} - \mu}{\sigma/\sqrt{n}} = \frac{83 - 65}{12/\sqrt{36}} = 9.0$$

The mean of the $\bar{x}$ distribution is the same as the mean of the population. If the population mean is 65 miles per hour, then the mean of all possible sample means must also equal 65; that is, $E(\bar{x}) = \mu = 65$. The standard deviation of the sampling distribution of all the possible $\bar{x}$ values is $\sigma_{\bar{x}} = \sigma/\sqrt{n} = 12/\sqrt{36} = 2.0$. Thus, $z = 9.0$ indicates that the sample mean of 83 miles per hour is nine standard deviations above the expected sample mean of 65 miles per hour, as shown in Figure 7.10.

The highest z value given in Appendix Table A.3 is for 3.49, for which only 0.0002 of the values are higher. To be practical, we treat the probability of observing values nine standard deviations above the mean as zero because the probability is minute. If vehicles stopped by the state police were averaging 65 mph, it is next to impossible that a simple random sample of 36 vehicles would be observed averaging 83 miles per hour. Observing this 83 mile per hour average is thus extremely strong evidence that vehicles stopped by the state police are not averaging 65 miles per hour, Figure 7.10. (More tests of this type are done with the sampling distribution of $\bar{x}$ in Chapter 9.)

FIGURE 7.10 Probabilities of Vehicle Speed

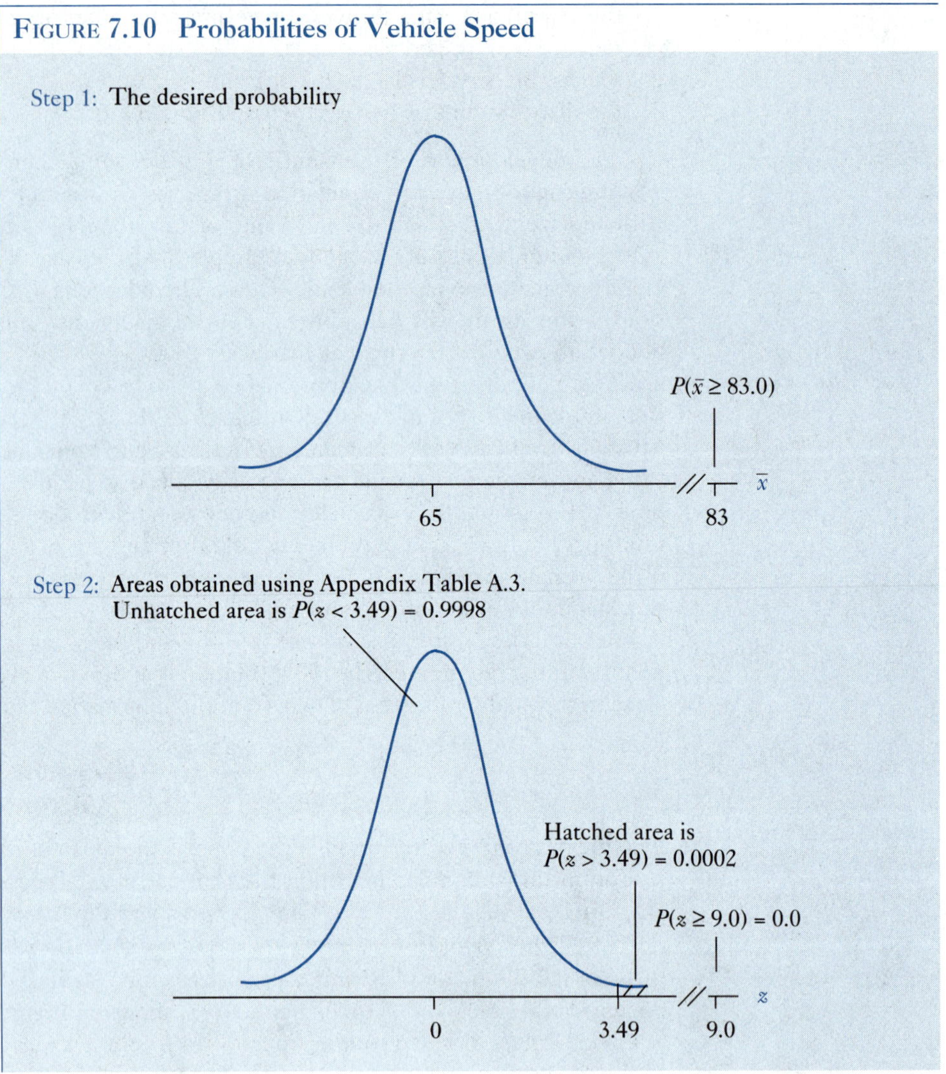

QUERY 7.1:

If the length of time required to complete an examination is normally distributed, with a mean of 65 minutes and a standard deviation of 15 minutes, then what is the sample mean time for which 99 percent of the samples of 60 test takers would have a smaller mean completion time?

ANSWER:

The population mean is 65 minutes, so the mean of all possible sample means must also equal 65 minutes; that is, $E(\bar{x}) = \mu = 65$. The standard deviation of the sampling distribution of all the possible $\bar{x}$ values is $\sigma_{\bar{x}} = \sigma/\sqrt{n} = 15/\sqrt{60} = 1.9365$. We also know that the $\bar{x}$ value for which we are looking puts 99 percent of the distribution below it. As shown in the diagram below, the z value that puts 99 percent of the distribution below it is 2.33. Thus, we know that $\bar{x}$ value we want is 2.33 standard deviation above the mean, or

$$z = \frac{\bar{x} - \mu}{\sigma/\sqrt{n}} = \frac{\bar{x} - 65}{1.9365} = 2.33$$

from which we can solve for $\bar{x}$ as

$$\bar{x} - 65 = (1.9365)(2.33) = 4.512$$

$$\bar{x} = 65 + 4.512 = 69.512$$

The 99 percent of the samples of 60 test takers will yield a mean of 69.512 minutes or less to complete the test.

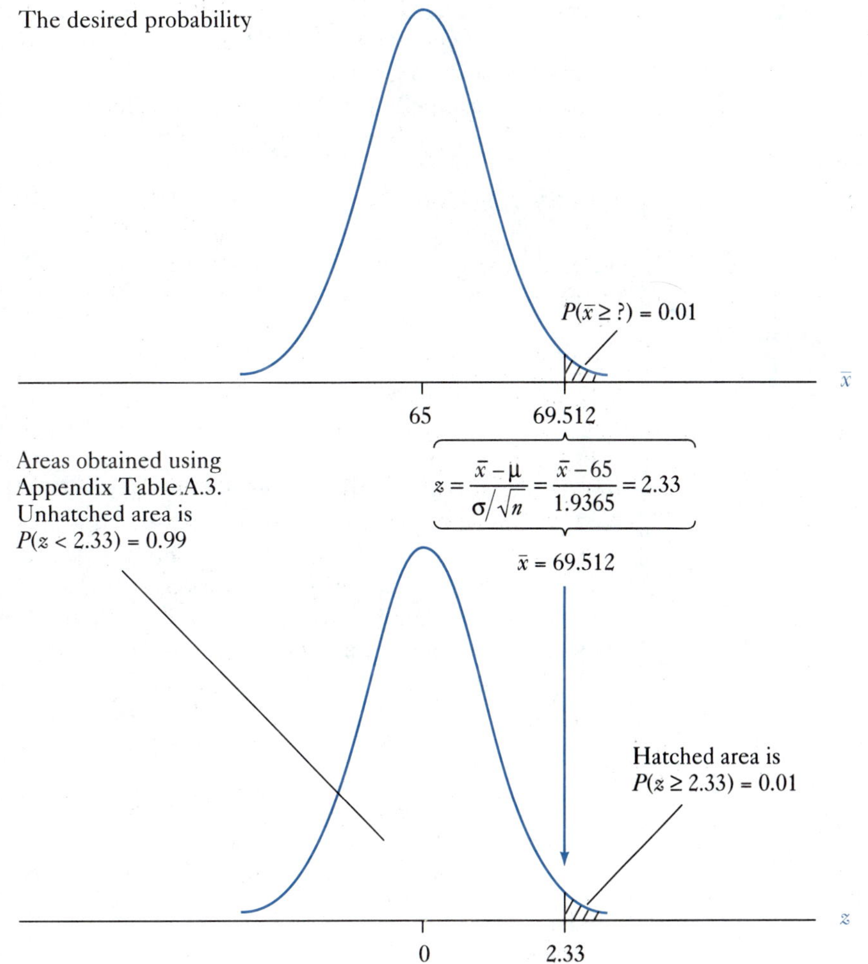

7.10 Sampling Distribution of the Sample Mean for Small Samples

The question to be asked here is what is the shape of the sampling distribution of the sample mean if the sample size n is relatively small ($n < 30$)? To answer this question, notice that the z formula used to standardize x in the previous chapter is a special case of the more general z formula for standardizing $\bar{x}$. This can be seen by setting $n = 1$

$$z = \frac{\bar{x} - \mu}{\sigma_{\bar{x}}} = \frac{\bar{x} - \mu}{\sigma/\sqrt{n}} = \frac{\bar{x} - \mu}{\sigma/\sqrt{1}} = \frac{x - \mu}{\sigma}$$

For $n = 1$, the mean of the sample is the value of the single element selected, so $\bar{x} = x$.

To make probability calculations with z in Chapter 6, where $n = 1$, we had to first establish that the population x was normally distributed. The same is true here when $n = 1$ because $x = \bar{x}$. Clearly, for $n = 1$, if x is normally distributed, then $\bar{x}$ is normally distributed. If x is normally distributed, then $\bar{x}$ must also be normally distributed for $n = 2, 3, 4$, and so on, since increasing the sample size does not make the sampling distribution of the sample mean "less normal." As long as x is normally distributed, $\bar{x}$ will be normally distributed, regardless of the sample size.[2]

Query 7.2:
If the average job tenure of a marketing director at a law firm is 2.1 years, how many directors would have to be sampled for the sample average tenure to be approximately normally distributed?

Answer:
Query 6.4 in the previous chapter addressed a question about the normality of job tenure of marketing directors and concluded that job tenure could not be normally distributed and that it was right skewed. Thus, the only way that samples drawn from this population distribution could give rise to a sampling distribution of sample means that is approximately normally distributed would be if the samples were large enough to involve the Central Limit Theorem. As a rule of thumb, the Central Limit Theorem should become operational if the sample size n is at least 30. This general rule of thumb, however, holds only if the population is not extremely skewed. If the population of job tenure is highly skewed right, then a sample larger than 30 may be needed for the sampling distribution of the sample mean to be approximately normal.

Query 7.3:
A press release (December 21, 1991) reported on a new regulation in Great Britain that requires bigger beer glasses that can accommodate a full 20-ounce British pint and a creamy head; brewers and pub landlords are to be fined for selling less. To test

for compliance, an agent visits a pub at nine random times, ordering a draft on each visit with these results:

19.77, 20.10, 19.95, 19.89, 20.09, 20.08, 19.89, 19.67, and 20.01 ounces.

What is the probability of a mean as small as that calculated from this sample if the pub is serving glasses of beer with a mean of 20 ounces?

ANSWER:

Although this query did not provide information on the dispersion in the population, from Chapter 6 we know that experienced bartenders would produce the population fill that is normal, with a standard deviation σ of 0.15 ounces. The sample average is calculated as

$$\bar{x} = \sum_{i=1}^{9} x_i / 9 = \begin{pmatrix} 19.77 + 20.10 + 19.95 + 19.89 + 20.09 \\ +20.08 + 19.89 + 19.67 + 20.01 \end{pmatrix} / 9 = 19.9389$$

Thus the z ratio is

$$z = \frac{\bar{x} - \mu}{\sigma_{\bar{x}}} = \frac{\bar{x} - \mu}{\sigma/\sqrt{n}} = \frac{19.9389 - 20}{0.15/\sqrt{9}} = -1.22$$

The sample average fill of 19.9389 ounces is 1.22 standard deviations less than the expected average fill of 20 ounces, as portrayed in the diagram below. The probability of observing a sample of $n = 9$ with a mean of 19.9389 or less if the actual population average fill is 20 ounces is determined using Appendix Table A.3 as follows

$$\begin{aligned} P\left(\bar{x} < 19.9389 \,\middle|\, \mu = 20, \sigma = 0.15\right) &= P\left(z < -1.22\right) \\ &= P\left(z > 1.22\right) = 1 - P\left(z < 1.22\right) \\ &= 1 - 0.8888 = 0.1112 \end{aligned}$$

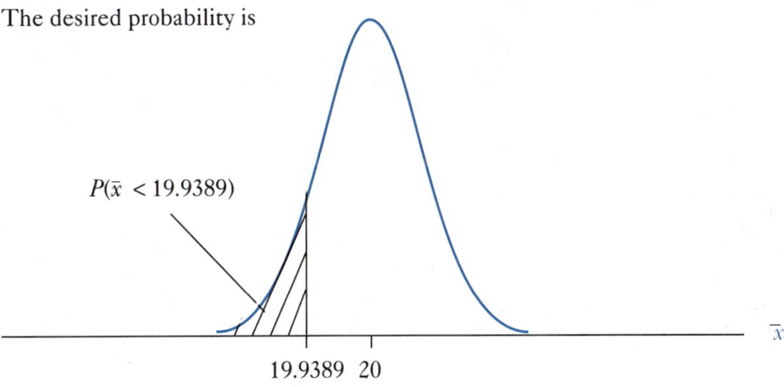

which in the standard normal is

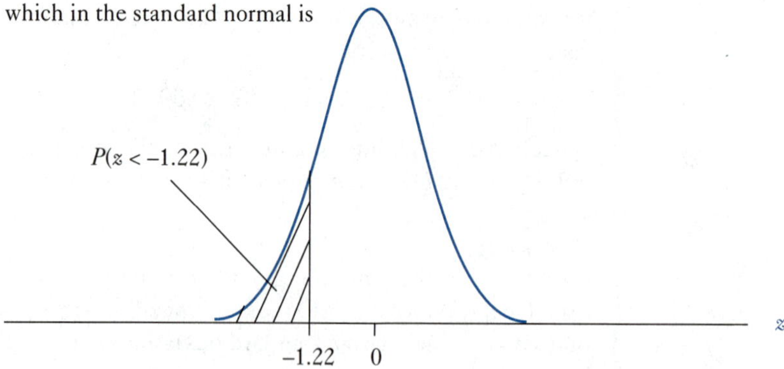

and obtained from Appendix Table A.3 as the area above $z = 1.22$

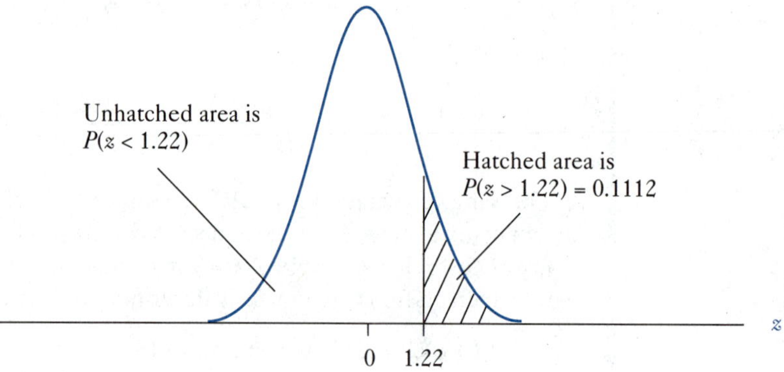

Something that occurs with a probability of 11.12 percent is not that unusual. So even though this sample gave an average fill of less than 20 ounces, it is possible that in the population the average fill is 20 ounces. The difference could result from the random error inherent in sampling.

Exercises

7.14 What is required to assert that the sampling distribution of the sample mean is approximately normally distributed?

7.15 Are the following statements true or false? Why?

a. The population mean is equal to a sample mean.
b. The population mean is equal to the sample mean if and only if the sample size exceeds 30.
c. The population mean is equal to the mean of all sample means, if sampling is random.
d. The variance of the mean is the same as the variance of x.
e. The variance of the distribution of the mean is s^2.
f. The variance of the distribution of the mean is σ_x^2/n.

7.16 A bakery sells an average of 24 loaves of bread per day. Sales (x) are normally distributed with a standard deviation of four.

a. If a random sample of size $n = 1$ (day) is selected, what is the probability this x value will exceed 28?

b. If a random sample of size $n = 4$ (days) is selected, what is the probability that $\bar{x}$ will exceed 28?

c. Why does the answer in part a differ from that in part b?

7.17 The state police believe that the average speed of trucks exceeds the 55 mph speed limit. A random sample of 100 trucks shows a mean speed of 61.8 mph. What is the probability of observing a sample average speed of 61.8 mph or more if the population average speed is 55 mph, assuming a population standard deviation is six mph?

7.18 Kathryn Walker and William Gauger, in "Time and Its Dollar Value in Household Work" (*Family Economic Review*, Fall 1973), report that "time spent on household work by husbands averages about 1.5 hours per day . . . ." A small sample of four families showed an average of 1.25 hours per day spent by the husband in household work. The question is: What is the probability of observing a sample average of 1.25 hours or less if the true population average is 1.5 hours, with a standard deviation of 0.4 hours? In your answer to this question, what did you assume about the population of time in household work and the distribution of the mean time? What did you have to assume about the sample of four families?

7.19 A given type of TV set can be expected to be service free for 36 months, with a standard deviation of six months. The distribution of required service is also known to be normally distributed. If four televisions of this type are purchased, the probability that the average service-free period will exceed two years for this random sample of four sets is approximately ________.

7.20 It is known that students graduating from a certain large university have a mean grade point average of 2.80, with a standard deviation of 0.80, and that all complete the same number of credits for graduation. Furthermore, it is known that these grade-point averages are not normally distributed. If a sample of 36 students is selected, what is the approximate probability that the mean grade point average is at least 2.60?

7.11 Sampling Distribution of the Sample Proportion

The sample proportion $p = x/n$ is a type of sample mean; it is the average number of "successes" in a sample of size n, where $x = \Sigma x_i$ and $P(x_i = 1) = \pi$ and $P(x_i = 0) = 1 - \pi$. Like $\bar{x}$, the sample proportion p varies from sample to sample. The shape of the sampling distribution of p can be determined by the Central Limit Theorem.

Central Limit Theorem (for sample proportion p)
If $x_1, x_2, x_3, \ldots x_n$ is a simple random sample of n elements from a binomial population, with a proportion of success π, then the distribution of $p = \Sigma x_i/n$ takes on the bell-shaped distribution of a normal random variable as n increases and the distribution of the ratio

$$\frac{p - \pi}{\sqrt{\pi(1 - \pi)/n}}$$

approaches the standard normal distribution as n goes to infinity.

Because p is just a sample mean, its expected value in simple random sample is π, the population proportion.[3] The standard error of p (standard deviation of p) likewise can be shown to be $\sqrt{\pi(1-\pi)/n}$.

Although the sample proportion $p = x/n$ can be treated as a mean in applying the Central Limit Theorem, there is no accurate way of getting normality on p for small samples, since p's numerator x is a binomial random variable. As first shown in Figure 5.1, and recast here as a sampling distribution in Figures 7.4, 7.5, and 7.6, when the sample size n is small, the binomial random variable x does not approach the normal; it only approaches the normal for large sample sizes. When n is small, probability calculations are best done with x and the binomial distribution itself.

QUERY 7.4:

During the Indianapolis 500 time trials, TV Channel 13 in Indianapolis telecast a show called "Speed Quest." (May 23, 1993) The show was devoted to the new role of microcomputers in auto racing. The announcer stated that IBM supplied computers to all but a few of the competitors because it believed that 60 percent of racing fans are prospective computer purchasers. For a simple random sample of 30 racing fans, what is the probability that 15 or fewer would be prospective computer purchasers if IBM is correct in its belief that the true proportion is 60 percent?

ANSWER:

This sample is sufficiently large to invoke the Central Limit Theorem and assume that the sampling distribution of the sample proportion (p) of fans who are prospective computer purchasers is approximately normal. The mean of the sampling distribution of p is $E(p) = \pi = 0.60$, if IBM is correct, and the standard error of p is

$$\sigma_p = \sqrt{\pi(1-\pi)/n} = \sqrt{(.6)(.4)/30} = 0.0894427$$

The probability of p less than or equal to 0.50 ($p = x/n = 15/30$) is now calculated as

$$P(p < .5 \mid \pi = .6,\ n = 30) = P\left(z < \frac{0.5 - 0.6}{0.0894427}\right) = P\left(z < -1.12\right) = 0.1314$$

This probability calculation is shown graphically as the hatched area in the sampling distribution of p.

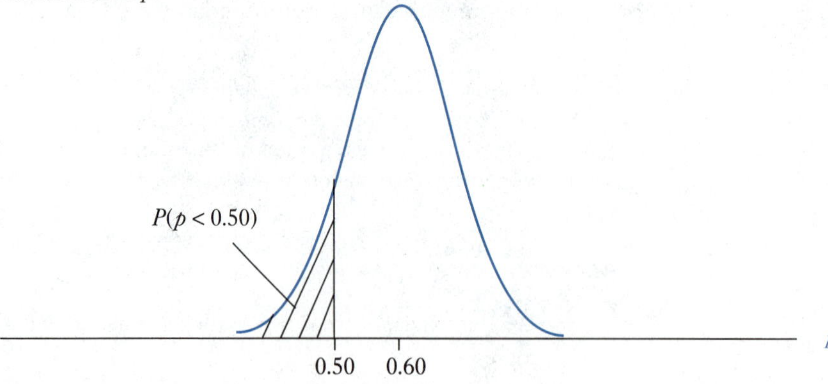

In the standard normal distribution this area is

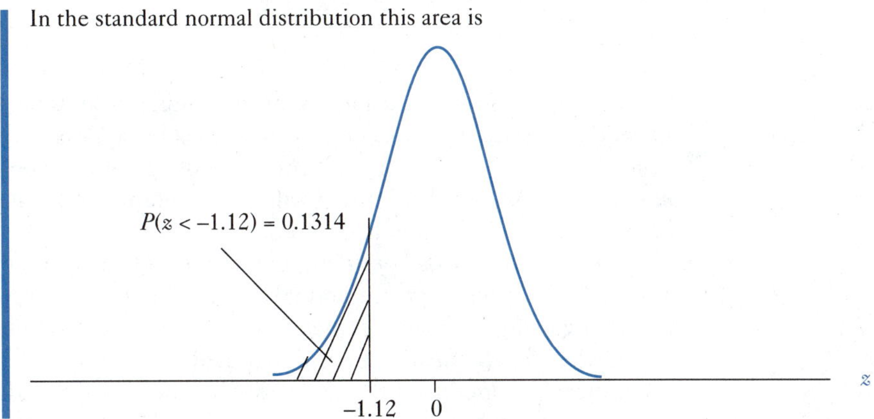

EXERCISES

7.21 What is required to assert that the sampling distribution of the sample proportion is approximately normally distributed?

7.22 Are the following statements true or false? Why?
   a. The population proportion is equal to a sample proportion.
   b. The population proportion is equal to the sample proportion if and only if the sample size exceeds 30.
   c. The population proportion is equal to the mean of all sample proportions, if sampling is random.
   d. The variance of the sample proportion is the same as the variance x.
   e. The variance of the distribution of the proportion is s^2/n.
   f. The variance of the distribution of the proportion is $\pi(1 - \pi)/n$.

7.23 According to a survey of 111 CPA firms, conducted by Professor David Doran of Pennsylvania State University, 18 percent wrongly thought they were shielded from disclosure to the IRS, as clients' talks with lawyers are privileged. (*Wall Street Journal,* May 26, 1993) If most (50 percent) CPA firms think they have such privileges, when they do not, what is the probability of getting 18 percent or fewer in a simple random sample of 111 CPA firms saying they do? Draw the sampling distribution of the sample proportion and label all points of interest.

7.24 In her nationally syndicated column, Ann Landers stated, "Your story (on poor nursing care) is a chilling one . . . (however) I would say at least 90 percent of the nurses are compassionate and caring. Too bad you ran into a few who were not." (*HT,* May 28, 1993) If the true proportion of compassionate and caring nurses is 90 percent, for 40 randomly selected nurses what is the probability that 34 or fewer will be compassionate and caring?
   a. using the binomial distribution?
   b. using the standard normal approximation?

7.25 A note in *The Wall Street Journal* (December 9, 1992) stated, "One year ago, 37 of 42 economists surveyed by *The Wall Street Journal* predicted a tax cut in 1992—and 36 of 42 predicted President Bush would be re-elected." Of course, neither of these events took place. Is this evidence that the distribution of the sample proportion was not normally distributed? Discuss.

7.12 Concluding Comments

The idea of a sampling distribution for the sample mean is possibly the most important concept in classical statistics. The sample mean is approximately normally distributed if the sample size is large (regardless of the shape of the population from which samples are drawn). This distribution will also be normal (regardless of the sample size) if the population is normally distributed.

The normality of the sampling distribution of the sample mean must first be established before the standard normal distribution can be used to calculate the probabilities involving the sample mean. In addition, these probability calculations will require that the population mean, standard deviation, and sample size are known. When these parameters of the population are not known, estimation techniques must be employed. Estimation is the subject of the next chapter in this book.

Chapter Exercises

7.26 In Chapter 9, reference is made to the attempts of Bloomington (Indiana) Hospital to assess employee satisfaction. In 1991, Bloomington Hospital hired a consulting firm to survey 35 of 1,700 employees. The 35 employees were hand picked by division managers to determine whether or not employees were satisfied (and would or would not want to unionize). Why can't these 35 employees be considered a simple random sample of the 1,700 employees?

7.27 Why won't increasing the sample size necessarily reduce nonsampling error while it will reduce sampling error?

7.28 A random sample of size n people has been drawn from a population distributed around a mean weight of 320 pounds with a population standard deviation of 60 pounds.

a. If $n = 144$, the sample average will be distributed with a mean and a standard deviation of approximately what?

b. With what approximate probability will the average weight for the 144 people lie beyond 335 pounds?

c. If the sample size had been 36 rather than 144, what would have been the standard deviation for the sampling distribution of the average weight?

7.29 In the early days of microcomputers, hard disk drives had a mean time between failures (MTBF) of 5,000 to 20,000 hours. Today, drives should last 20,000 to 40,000 hours, making disk drive failure rare in the first several years of service. If the average time between failures for a new type of drive is 40,000 hours, with standard deviation of 5,000 hours, what is the probability that a sample of 10 drives will have a mean time between failures of less than 35,000 hours? less than 30,000 hours? less than 25,000 hours? less than 20,000 hours? Why don't these probabilities fall off in the same proportion as the hours?

7.30 A production process turns out washers with an inside average diameter of 0.25 inches and a standard deviation of 0.00001 inches. Every half hour, a lot of 16 washers is checked. The latest sample had a mean of 0.25007 inches. Approximately how likely is it to observe this sample mean or one larger if the population average diameter is in control producing washers that average 0.25? In order to calculate this probability, what needs to be assumed?

7.31 Sheet steel is produced to automobile company specifications that call for an average thickness of 0.20 inches, with a standard deviation of 0.021 inches. A random sample of 25 different locations on a roll of steel showed a mean width of 0.191 inches. Past steel shipments have had normally distributed thicknesses. If the entire roll actually averages 0.20 inches, how likely is a sample average of 0.191, or less?

7.32 The amount of beer in a bottle is normally distributed with a mean of 11.7 ounces, and a standard deviation of 0.40 ounces. An independent research firm takes a sample of 16 bottles and finds the value of the sample mean, which it fails to report. If the probability of getting a sample mean $\bar{x}$ less than the unreported sample mean is 0.025, what is the value of the unreported sample mean?

7.33 Chuck Stone, in his Newspaper Enterprise Association syndicated column (*HT*, September 5, 1992) wrote: "Increases of one or two points are monumentally insignificant because the SAT's standard error of measurement is 18 points." If this standard error is correct, what is the probability of the average SAT score rising by one or more points from its previous mean?

7.34 An article in *The Wall Street Journal* (August 31, 1989) reported on the release of U.S. government statistics that later have to be revised substantially. The article mentioned that one widely reported series of statistics that measures the monthly change in sales of new homes showed a huge 5.5 percent increase, but a footnote stated that two standard errors were equal to approximately nine percentage points. What does this imply about the likelihood that sales decreased?

7.35 A *Wall Street Journal* (November 11, 1992) article on the clothing sizes that retailers carry stated, "In good times retailers use a sort of bell curve of sizes, with the greatest number of items in the midrange. For an estimated 35% to 40% of women, that means sizes 6, 8 and 10." If the average population size is 8, what are the limits on the implied standard error of the mean for a sample of size 4, if in the population 35 to 40 percent of the distribution is in the size 6 to 10 interval? Why is it reasonable to assume that the sampling distribution of the mean is normally distributed?

7.36 *The Wall Street Journal* (November 11, 1992) article on the clothing sizes referenced in Exercise 7.35 also stated that "For men, it's waist sizes 30–36 inches," which constitutes the midrange of 35 percent to 40 percent. If the population average men's waist size is 33, what are the limits on the implied standard error of the mean for a sample of size 4, if 35 to 40 percent of the population distribution is in the 30 to 36 interval?

7.37 A *Wall Street Journal* (August 14, 1987) article reported that "IBM samples 50 circuit boards out of every 270-board lot; if one is defective, every board in the lot is rejected." Why is this a demanding standard?

7.38 A Lanier TV advertisement on CNN (April 20, 1993) stated, "Lanier copiers are up and running 98 percent of the time." On thirty randomly selected days, the operating status of a Lanier copier is checked; it is found to be "up and running" 29 times. Why wouldn't it be appropriate to calculate the probability of getting a sample of 30, containing 96.67 percent or less "successes," by assuming that $p = x/n$ is normally distributed, with mean of 98 percent? How should this probability be calculated?

7.39 Chapter 5 investigated the correctness of the article "Why Investors Shouldn't Listen to Economists." *WSJ* (January 22, 1993) The article reported that "only five of 34 economists who participated in 10 or more surveys managed to guess correctly which way long-term bond yields were going more than half of the time." We found that the probability of five individuals each predicting six or more yield movements correctly, where each flipped a coin 10 times, was 0.0023. Now use the normal approximation, where appropriate, to find the probability of 14.7059 percent or fewer of the 34 individuals predicting six or more yield movements correctly by each flipping a coin 10 times.

7.40 Exercise 1.44 reported on "Resort Ads Caught Snowing the Ski Set." (*Wall Street Journal*, December 22, 1992) Below are the data, along with the differences between claimed and actual miles of these New England area ski areas. Calculate the mean difference between claimed and actual miles. What is the probability of at least this mean difference if the true population mean difference is zero, assuming the population is normally distributed with standard deviation of two miles?

Ski Area	Actual Miles of Trails	Miles Claimed by Resort	Claimed Less Actual
Killington	70	77	7
Sugarbush	46	53	7
Sunday River	29	36	7
Sugarloaf	36	45	9
Okemo	25	32	7
Smuggler's Notch	22	32	10

7.41 Using the data and test in Exercise 7.40, if an area owner questioned if this was a representative sample and a legitimate test, how would you respond?

7.42 In June 1993, there were reports of needles being found in Pepsi-Cola soda cans. Pepsi reported that tampering with cans in plants was unlikely, because the lines run at a rate that fills 11,000 to 12,000 cans per minute, with each fill and seal taking less than eight seconds. If, on average, the line fills and seals 11,500 cans per minute, with a standard deviation of 142 cans per minute, what is the probability of randomly checking the speed of the line

a. once and finding an average line speed of only 11,250 or fewer cans per minute? What do you have to be able to assume to calculate this probability? Is this assumption reasonable?

b. on 36 different occasions and finding an average line speed of only 11,250 or fewer cans per minute? What do you have to be able to assume to calculate this probability?

7.43 (Appendix 7 required) Ann Landers was headlined "Jury is in: Engineers are 'different' breed." (*HT*, September 13, 1993) To support this conclusion, she stated, "I've been swamped with letters from wives, daughters, husbands, mothers, and sisters of engineers." She also reprinted nine of these letters as examples of how different engineers are. Comment on the sampling technique implied by Ann's receipt of letters and whether or not only sampling error is involved. Comment on the validity and reliability of interpreting words in the letters.

7.44 (Appendix 7 required) According to a Scripps Howard News Service (*HT*, September 18, 1992), Charles McDowell, a supervisory special agent in the Air Force Office of Special Investigation in Washington, D.C., distributes a 57-item questionnaire that is meant to establish whether a rape complainant is telling the truth. Dean Kilpatrick, a clinical psychologist and director of Crime Victims Research and Treatment Center at the Medical University of South Carolina, is quoted in the article saying, "From a scientific perspective, there is absolutely no documentation of the validity of those test items." What was Kilpatrick saying about the questions?

7.45 (Appendix 7.0 required) An article in *Newsweek* (May 6, 1985) quotes William Meyers, author of *The Image-Makers: Power and Persuasion on Madison Avenue,* as saying that image is as important as taste in the soft-drink field. The article also quotes Faith Popcorn of Brainreserve as saying that Pepsi has the image in the youth market. If this is true, what problems will Coca-Cola have in drawing conclusions about future market success of its new syrups from blind taste tests in which labels are not displayed? Is this a problem of validity, sample size, or randomness in the experiment design? Explain.

Academic References

Becker, W., and D. Lewis. (eds.) *The Economics of American Higher Education.* Boston: Kluwer Academic Publishers, 1992.

Fienberg, S. "Randomization and Social Affairs: The 1970 Draft Lottery." *Science,* January 22, 1971.

Endnotes

1. For the variance of $\bar{x}$, we need to determine $E[\bar{x} - E(\bar{x})]^2$
   For simple random samples we know that $E(\bar{x}) = \mu$. Thus, $E[\bar{x} - E(\bar{x})]^2 = E(\bar{x} - \mu)^2$, which can be expanded as

$$E(\bar{x}-\mu)^2 = E\left[n^{-1}\sum_{i=1}^{n}(x_i-\mu)\right]^2 = n^{-2}E\left[\sum_{i=1}^{n}(x_i-\mu)\right]^2$$

$$= n^{-2}E\left[\sum_{i=1}^{n}(x_i-\mu)^2 + \sum_{\substack{i=1\\ i\neq j}}^{n}\sum_{j=1}^{n}(x_i-\mu)(x_j-\mu)\right]$$

$$= n^{-2}\left[\sum_{i=1}^{n}E(x_i-\mu)^2 + \sum_{i=1}^{n}\sum_{j=1}^{n}E(x_i-\mu)(x_j-\mu)\right]$$

$$= n^{-2}(n\sigma^2)$$

$$= \sigma^2/n$$

The variance of x

As long as x's are independent, the expected value of a crossproduct is zero

   We now know that $E(\bar{x} - \mu)^2$ is σ^2/n. That is, the variance of the sampling distribution of the sample mean is the population variance divided by n.
2. The normal density function for the sample mean with $n = 2$, for example, is

$$f(\bar{x}) = \frac{e^{-[(x_1+x_2)/2-\mu]^2/\sigma^2}}{\sigma\sqrt{2\pi}}$$

where $x \sim N(\mu, \sigma)$, and x_1 and x_2 are random variables formed by independent and identical draws from x. The expected value of the sum of the two random variables,

$$(x_1/2) + (x_2/2), \text{ is } E[(x_1/2) + (x_2/2)] = (\mu/2) + (\mu/2) = \mu$$

For the variance of $\bar{x}$, we need to determine the variance of the sum of the two random variables $(x_1/2)$ and $(x_2/2)$. From Chapter 4, Formula (4.12),

$$\text{Var}[(x_1/2) + (x_2/2)] = (1/2)^2\text{Var}(x_1) + (1/2)^2\text{Var}(x_2) + 2(1/2)(1/2)\text{Cov}(x_1x_2)$$

If values of x_1 and x_2 are drawn independently from the same distribution (simple random sampling), then $\text{Cov}(x_1x_2) = 0$

$$\text{and Var}[(x_1/2) + (x_2/2)] = (1/4)\text{Var}(x_1) + (1/4)\text{Var}(x_2)$$
$$= (1/4)\sigma^2 + (1/4)\sigma^2 = \sigma^2/2$$

Thus, $\bar{x} \sim N(\mu, \sigma/\sqrt{2})$, for $n = 2$, and similarly for $n = 3, 4 \ldots$ to the most general large sample case, as shown in the first endnote of this chapter.

3. The expected value of a binomial random variable x is

$$E(x) = n\pi$$

which upon dividing through by n yields

$$E(x/n) = n\pi/n = \pi$$

Thus, $E(p) = \pi$

The variance of a binomial random variable x is

$$E[x - E(x)]^2 = n\pi(1 - \pi)$$

which upon dividing through by n^2 yields

$$E[(x/n) - E(x/n)]^2 = n\pi(1 - \pi)/n^2 = \pi(1 - \pi)/n$$

Thus, $E(p - \pi)^2 = \pi(1 - \pi)/n$

FORMULAS

- Formula for transforming the normal random variable $\bar{x}$ to a standard normal random variable z

$$z = \frac{\bar{x} - \mu}{\sigma/\sqrt{n}} = \frac{\bar{x} - \mu}{\sigma_{\bar{x}}}$$

where the population random variable x has a mean of μ and a standard deviation of σ, and the sample mean $\bar{x}$ has a mean of μ and a standard deviation of $\sigma_{\bar{x}} = \sigma/\sqrt{n}$.

- Formula for transforming the normal random variable p to a standard normal random variable z

$$z = \frac{p - \pi}{\sqrt{\pi(1 - \pi)/n}}$$

where $x_1, x_2, x_3, \ldots x_n$ is a simple random sample of n elements from a binomial population, with a proportion of success π, and $p = \sum x_i/n$

Appendix 7 Avoiding Errors in Sampling

Sampling Error
The discrepancy between a sample value and the population parameter resulting from the drawing of only a subset from the population.

The essence of statistics is making inferences about a population using a sample. A statistical analysis must be concerned with error, because even if a perfect experiment could be conducted, some samples will not perfectly represent the population from which they are drawn. The chance element inherent in simple random sampling is called experimental error or **sampling error** or random error.

Although sampling error cannot be avoided, increasing the sample size will reduce its magnitude, as shown in this chapter. There are errors, however, that are not affected by sample size. These systematic and persistent errors are known as nonsampling errors or bias. The Chapter 1 discussion of Griliches's argument that diminishing returns may not exist in inventiveness, even though the sample data of Figure 1.2 suggest otherwise, is an analysis of nonsampling error that was caused by the exclusion of less profitable small firms from the sampling process. Other nonsampling errors may result from invalid and unreliable measurement instruments, measurement errors, and errors due to omitted or missing observations.

Validity

Validity
The degree to which a survey instrument measures what it purports to measure.

In surveys, errors caused by lack of **validity** are a form of nonsampling error that may result because a respondent does not understand the question. For example, the U.S. Bureau of Labor Statistics has found that poorly defined questions aimed at establishing a person's work status might be misunderstood. Retired persons might classify themselves as "unemployed" in a survey.

Another error of validity may occur if the respondent does not know the information with certainty. In completing a dental questionnaire, for example, a respondent might not remember the exact date of the last check-up and thus record the wrong date. David Mingay, a cognitive psychologist at the National Center for Health Statistics warns in *Business Week* (February 17, 1987) that, "Most of us just don't have a good internal calendar for dating events."

Problems of validity make it difficult or impossible to interpret results. For example, from personal correspondence with *Business Week,* the author learned that the respondent percentages in Appendix Table 7.1 were to a questionnaire given to participants at a conference of executives of smaller enterprises. They were asked for yearly compensation to the approximate values indicated; the respondents only had the option of choosing among the seven alternatives provided.

But how did an executive who makes $950,000.01 respond when confronted with compensation options "$900,000" or "$1,000,000 or more"? If he/she thinks "$1,000,000 or more," then would he or she have expected the class bound in a frequency distribution to have been "$950,000 or more" and not "$1,000,000 or more." Alternatively, he/she may think that $900,000 is closer to his/her $950,000.01 compensation than is the "or more" portion of the "$1,000,000 or more" option and thus select the $900,000 option. Similarly, at the lower end, an executive making $149,999.99 may say that the "$200,000" option is closer to his/her compensation than the midpoint ($50,000) or the lower limit (zero) in the "$100,000 or less" option.

Appendix Table 7.1 What the Boss Makes at the Up-and-Comers

A new survey of 235 small and midsize companies, most with annual sales of $2 million to $50 million, sheds some light on the boss's paycheck.

Approximate Annual Compensation (salary and bonus)	Respondents (percent)
$ 100,000 or less	40 %
200,000	39
300,000	9
400,000	6
500,000	3
900,000	1
1,000,000 or more	1
No Answer	1

Source: *Business Week*, June 27, 1992, p. 76.

The possible confusion regarding the interpretation of responses at the extremes makes this survey invalid. Given that 79 percent of the responses were in the two lowest options, the issue of validity is not trivial.

Reliability

Closely tied to the need for valid data collection instruments is the need for reliable instruments. Reliable surveys are ones that give stable results.

Reliability
The degree to which a survey instrument yields stable results.

There are generally two types of **reliability** that researchers consider. Reliability over time implies that the respondents would respond the same if asked the same questions at another time. For example, a response that is preferred on one day is still preferred on another. Internal reliability pertains to the degree of similarity between the manner in which each question on a survey characterizes responders and the way in which the entire survey does. For example, one multiple-choice question on a 20-question exam is said to be reliable if those who received high scores on the exam also tended to get that one question correct. Internal reliability also implies that if on one question a respondent says A is preferred to B which in turn is preferred to C, then on another like question the respondent does not reverse this order.

In Chapter 2, Query 2.5 introduced the national survey that the state of Alaska commissioned to value what Americans who may never visit Prince William Sound would pay to ensure that another oil spill like the Exxon Valdez will not occur. *The New York Times* (September 6, 1993) reported that the six economists the National Oceanic and Atmospheric Administration invited to assist in writing damage-assessment rules for oil spills concluded that such contingent-value measures of passive use of the environment by those who may never see or use an affected area can produce valid and reliable initial estimates for a judicial process of damage assessment. The article also reported, however, that many economists question this conclusion.

On the issue of validity, Harvard econometrician and Exxon consultant Zvi Griliches was quoted as saying, "Economists just don't want to admit there is stuff that cannot be measured." On reliability, several economists were cited for pointing out the difference between what people say they would do and what they have actually done—economists traditionally have relied on market transactions to reveal value. Peter Diamond and Jerry Hausman of the Massachusetts Institute of Technology pointed to studies showing the public's inclination to be overly generous with hypothetical outlays. Walter Mead of the University of California at Santa Barbara stated that contingent studies give too wide of range; he cited studies that put the nonuse value of saving old-growth forest in the Pacific Northwest at $119 billion to $359 billion, while sparing the whooping crane from extinction was valued at $51 billion to $715 billion. William Desvousges of the Research Triangle Institute in North Carolina identified the irrationality of respondents who say they are willing to pay the same amount to save 2,000 migratory birds from oil-coated ponds as they are willing to pay to save 20,000 or 200,000 birds.

Although these comments raise questions about the validity and reliability of contingent surveys, Nobel Prize-winning economist Robert Solow, one of the advisers to the National Oceanic and Atmospheric Administration, still concluded, "You can learn quite a lot from a well-done contingent-value study." Ohio State University economist Alan Randall added, "If passive uses are ignored, it would encourage the riskiest enterprises in the most pristine environments."

Measurement

Measurement errors are another form of nonsampling error. They are often caused by mistakes in reading instruments or in the instruments themselves, or in recording data. For example, a researcher may make an inaccurate observation, or there may be a keypunching or typographical error in recording the data, or someone may merely add a column of numbers incorrectly or transpose digits. Serious measurement error can destroy the value of an entire data set.

Errors involving validity, reliability, and measurement can be avoided by taking time and care in constructing instruments, reporting, and recording data. Typically the least expensive methods of collecting data are also the most likely to have nonsampling errors. Appendix Table 7.2 provides a brief review of the costs and possible nonsampling errors that might occur as a result of using 1) direct measurement, 2) personal interview, or 3) self-enumeration via questionnaire.

Omitted and Missing Observations

The last category of Appendix Table 7.2, self-enumeration, indicates that the survey method of asking individuals to respond to questionnaires does not lead to a high response rate. People do not like to take the time to fill out questionnaires. A *Business Week* article (February 17, 1987) stated, for example, "The public is so barraged by surveyors that people refuse to be guinea pigs (for the surveyors) about 40% of the time." "Missing observations" can lead to incorrect inferences about the population if those who do take the time to complete the questionnaire are not representative of the population under investigation.

The Wall Street Journal (August 4, 1987) reported on how the U.S. Census Bureau took elaborate steps to figure out how many people it missed and where those people live. "By cross-referencing all sorts of records, and double-checking national census with small-scale post-census surveys, bureau officials think they can estimate, with a high degree of accuracy, the extent of the national undercount." But because these techniques are based on theory instead of actual observations, there is a never-ending debate as to whether the actual count is more accurate than an estimate that adjusts for these unenumerated Americans.

Researchers can now attempt to account for "sample selection bias" that is caused by missing observations; however, all sample selection adjustment methods require some assumptions about the distribution that is then used to generate the missing values. The implied validity and reliability of such assumptions can always be questioned.

Appendix Table 7.2 Sampling Procedures

Direct observation: Sampling may be by direct observation.

In a time study, a researcher periodically checks on the activities of a manager (reading reports, giving dictation, talking on the telephone, etc.) and records this activity on a time sheet.
In a personnel training study, a researcher records the interaction between trainer and trainee on a hidden video recorder.
In a marketing study of automobile purchasing behavior, a researcher observes and records the exchange between the salesperson and potential customers.

Advantages

1. Data can be collected at various times or even continuously.
2. There are few limitations on what is observed and recorded.
3. Distorted or incomplete recall by subjects is not a problem.

Disadvantages

1. Participants may be aware of the observer and alter their behavior as a result.
2. Observers, because of personal biases or a lack of training, may not record precisely what they observe.
3. The method is expensive because of the cost of training and paying the observers.

Personal interview: In a personal interview, predetermined questions are asked. These questions are written on a form, and the responses are recorded by the interviewer. The actual questioning may be either face to face or over the telephone. Telephone interviews, for example, are often used in political polls.

Advantages

1. People are more likely to respond when they are contacted directly. Thus, personal interviews tend to yield high response rates.
2. Direct contact enables the interviewer to clarify any questions or misunderstandings that the interviewee might have about the questions asked. Thus, personal interviews are more likely to result in reliable and valid responses from those contacted.
3. Supplemental information, such as a respondent's facial reaction to given questions, can be recorded in personal interviews.

(continued)

APPENDIX TABLE 7.2 Sampling Procedures *(continued)*

Disadvantages

1. The interviewer may not follow directions. In any study, it is essential that the sample be collected using a method that yields a good representation of the population. It is not uncommon, however, for interviewers to disregard the procedures specified. For example, assume an interviewer is told to survey the father in each household. In some cases the father may not be available, so the interviewer decides (incorrectly) to interview the mother.
2. Interviewers may influence respondents by the manner in which they ask a question or the reaction they give to a response.
3. Personal interviews are a relatively expensive method of data collection because it is usually necessary to pay for the time of the interviewers.

Self-enumeration via questionnaires: Possibly the most often used method of data collection is a questionnaire the participants fill out themselves. Such questionnaires may be distributed by mail, handed out in a shopping mall, stuffed in newspapers, or distributed by some other methods.

Advantages

1. This method is relatively inexpensive, since no observer or interviewer is required.
2. The questionnaire can be put aside until it is convenient for the respondent to answer it.

Disadvantages

1. There is no control over who completes the questionnaire. Thus, the data collected may not be representative of the population.
2. Most people do not like to fill out questionnaires. Thus, a low response rate can be expected from self-enumeration. To increase the response rate, follow-up letters, telephone calls, and even personal interviews are often necessary. Such follow-up work adds to the cost of the study.
3. The time period between the date the questionnaires are issued and the date they are returned may be relatively long. Thus, timely data may be difficult to collect by self-enumeration.

Source: William E. Becker and Donald L. Harnett, *Business and Economics Statistics with Computer Applications*, Addison-Wesley Publishing Company, 1987, pp. 12–13.

Chapter 8

Estimation

What good are analytical skills if the information you're analyzing is skimpy or misleading?

Tom Peters
September 14, 1994

8.1 INTRODUCTION

We have already learned that random variables have probability distributions from which data are assumed to be generated. As will be seen, both the observed sample data and the assumed probability distribution enable us to 1) estimate population parameters, 2) test specific parameter values, and 3) make predictions. Estimation is discussed in this chapter; later chapters address testing and prediction. Here we consider problems of the following types.

- An Associated Press release (*HT*, February 5, 1992) stated that "A majority of adults surveyed in Japan complained of fatigue and emotional stress . . . 53 percent said they felt somewhat tired . . . and stressed. . . ." According to the news release, the survey was based on a random sample of 3,000 people over the age of 20, but "the poll did not give a margin of error." Of what value is this estimate?
- In the survey of Japanese workers, it was also found that "the average Japanese employee worked 2,044 hours, including 185 hours of overtime, in fiscal 1990, which ended March 31, 1991, the equivalent of 51 40-hour weeks a year." Again the article stated that no margin of error was given for inferences about the total Japanese population.
- A recruiting firm's survey of 64 "highly mobile" executives shows that the average raise received by these executives was $25,320, with a standard deviation of $5,429. From this, the recruiting firm concluded that a 95 percent interval estimate for the mean raise received by all such executives ranges from $23,990 to $26,650.
- An article in *PCWEEK* (February 11, 1991) reports on a study of the number of PCs shipped that failed in the first year after installation. In a sample of 47 Compaq computers, 17 percent failed in the first year. The question: With 95 percent confidence, what proportion of all Compaq computers would you expect to fail within the first year?
- An article in *The Wall Street Journal* (June 20, 1991) stated: "According to Wall Street lore, a headquarters move tends to be an ominous sign. If a company is 'spending a lot of money on a trophy headquarters, it's a negative indicator' for the stock, says Charles Biderman, editor of the newsletter *Market Trim Tabs*." The *WSJ* did a "minisurvey" of the stock prices of 10 companies that moved between 1986 and the end of 1988: five stocks beat the market and five underperformed. To do a better job of estimating the effect of moving, what size sample is needed?

Statistical inference in examples such as these requires the construction of mathematical models that represent the populations from which the observed data are thought to originate. Chapters 5 and 6 dealt with some common population distributions from which data are often generated. Estimation of the parameters of these probability distributions as incorporated in statistical models requires the calculation of sample statistics. A **sample statistic** is a formula or rule by which a value is obtained as a descriptive measure of the sample. If the sample statistic is then used as a predictor of a population parameter, it is called an estimator of the parameter.

Sample Statistic
A function of the observations in a sample that does not depend upon any unknown parameter.

For a specific sample, a statistic has one value. Across all possible samples, however, the statistic may take an infinite number of values; the likelihood of any given value is determined by a probability distribution. That is, a statistic is a random variable. From the Central Limit Theorem (Chapter 7), we know that statistic $\bar{x}$ (which is determined by the rule $\bar{x} = \Sigma x_i/n$, where x_i is the ith observed value in a sample of size n) is approximately normally distributed for large sample sizes. Similarly, the statistic p (which is the sample proportion $p = x/n$, where x is the number of successes in a sample of size n) is approximately normally distributed for large sample sizes.

Point Estimate
The single value, calculated from sample data, used to infer a parameter (characteristic) of the population from which the sample data are assumed to be generated.

In the case involving the survey of Japanese workers, the sample proportion (p = 0.53) is used to estimate the population proportion (π) of the Japanese workers feeling tired or stressed. In the second case involving Japanese workers, the sample mean ($\bar{x}$= 2,044) is the statistic used to estimate the population mean (μ) hours worked per year. The specific numerical values p = 0.53 and $\bar{x}$ = 2,044 are called the **point estimates** of the respective parameters π and μ. A point estimate is the single value calculated from a specific sample statistic (the estimator) that is used to predict the population parameter of interest.

Interval Estimate
A range or set of values, calculated from sample data, that is used to infer a parameter (characteristic) of the population from which the sample data are drawn.

The chance factor inherent in sampling makes it unlikely that a point estimate for any given sample will exactly equal the parameter value being estimated. It is unlikely, for example, that 53 percent of Japanese workers feel tired or stressed. To capture this uncertainty, a range of values may be calculated as the estimate of a population parameter. A specific range of values is called an **interval estimate.** For example, the recruiting firm's prediction that highly mobile executives are getting raises between \$23,990 and \$26,650 is an interval estimate of the true population average raise. When interval estimates are associated with specific confidence levels, they are called **confidence intervals.**

Confidence Interval
An interval that, with stated probability, includes a parameter value.

By the end of this chapter, you will be able to calculate and use the point and interval estimates for the population parameters μ and π. You also will be able to approximate an interval estimate for the population median, if you complete the optional sections at the end of this chapter. You will be able to use the idea of an interval estimator to determine the sample size needed for statistical inferences involving either a population mean or a population proportion.

8.2 Point Estimation

A point estimate is a single numerical value that is calculated from a sample and used to make an inference about a population parameter. In the example of pay raises for mobile executives, the sample average of \$25,320 is a point estimate of the population mean raise. Similarly, 53 percent is a point estimate of the population proportion of Japanese workers who feel tired and stressed. In both examples, however, had different samples been drawn, different values of $\bar{x}$ and p would likely have resulted and, thus, different estimates of μ and π would have been obtained.

The actual value that is calculated from sample data as an estimate of a population parameter is the outcome of the sampling process. This value is the realization of a

Point Estimator
A formula applied to sample data and used to infer the value of a population parameter.

statistic (or random variable) that represents all the estimated values that could be calculated from a large number of samples, each of size n. This statistic is called the **point estimator,** and the observed value is called the point estimate. The following five point estimators are considered in this chapter:

1. the sample mean ($\bar{x}$), which is a point estimator of the population mean (μ)
2. the sample proportion (p), which is a point estimator of the population proportion (π)
3. the sample variance (s^2), which is a point estimator of the population variance (σ^2)
4. the sample standard deviation (s), which is a point estimator of the population standard deviation (σ)
5. the sample median, which is a point estimator of the population median.

Remember that a point estimator is a statistic and that every statistic is a random variable having its own probability distribution. The point estimate is the realization of the point estimator.

8.3 Properties of Estimators

Although we would like an estimator that always yields the exact value of the population parameter of interest, the chance factors in sampling make this impossible. At a minimum, it seems reasonable that an estimator should have an expected value equal to the population parameter estimated and have minimal dispersion around that mean value. Estimators with these two properties are said to be "unbiased" and "efficient."

Unbiasedness
The property of any estimator whose expected value equals the population parameter being estimated.

Unbiasedness says that the mean of the estimator is the value of the population parameter being estimated. Unbiasedness does not say that any one estimate will equal the population parameter. The sample mean ($\bar{x}$) is an unbiased estimator of the population mean μ, if simple random samples are drawn, where each sample is of size n. Another way to say this is that the expected value of $\bar{x}$ equals μ or simply $E(\bar{x}) = \mu$.[1] Recognize that unbiasedness says nothing about the outcome of one sample of size n. For an infinitely large population, an infinite number of samples of size n can be drawn and an infinite number of $\bar{x}$ values can be calculated. It is the mean of all these $\bar{x}$ values that equals the population mean.

In the case of the mobile executive pay raises, $\bar{x}$ is the estimator of the population mean (μ) raise for all such executives; the point estimate is $\bar{x}$ = \$25,320 for our one sample. If many (infinite) samples of size $n = 64$ could be selected, then the average of all these sample means would equal the population mean. The point estimate \$25,320, however, need not equal the population mean. It is but one value that $\bar{x}$ may realize; only the expected value or mean of $\bar{x}$ is equal to μ.

The sample mean is an unbiased estimator of the population mean. The sample median is an unbiased estimator of the population median. The sample proportion is an unbiased estimator of the population proportion. But not all sample statistics are unbiased estimators of their population counterparts. For example, the population variance is

$$\sigma^2 = (1/N)\Sigma(x - \mu)^2$$

which suggests the formula $(1/n)\Sigma(x - \bar{x})^2$ as its estimator. The formula $(1/n)\Sigma(x - \bar{x})^2$, however, is a biased estimator of σ^2. (Recall that once a value for a sample mean is determined, only $n - 1$ of the observations can be treated as free to vary.) For this reason, the sample variance is calculated by the formula[2]

$$s^2 = [1/(n - 1)]\Sigma(x - \bar{x})^2$$

The sample variance s^2 is an **unbiased** estimator of the population variance σ^2; the average of all possible values of s^2 equals σ^2. The sample variance for the mobile manager pay raises is $s^2 = (\$5{,}429)^2$, but notice that since the variance is in squared units of measurement, the sample standard deviation ($s = \$5{,}429$) is reported. Again \$5,429 is not necessarily equal to the population's standard deviation; it is only an estimate. Different samples would give different estimates. Only the expected value (mean) of s^2 is equal to the population variance σ^2.

If the value of the population mean is known, then bias could be measured as the difference between the expected value of the estimator and the population mean. But if the population mean is known, we would not need to estimate it. Typically, we have no way of measuring biasedness; it is a theoretical concept.

Every estimator has some dispersion around its expected value. The "best" estimators will not only be unbiased but will also have minimal variability. To appreciate this idea, consider the analogy of two guns used during the American Revolution. Englishmen equipped with smooth bored muskets were able to shoot 100 yards, but with a three-foot side error (six-foot range) for five out of 10 shot accuracy. Colonials with Pennsylvania flintlock 58-inch grooved barrel rifles could sharp shoot (with little side error for nine out of 10 shot accuracy) at well over 100 yards. A soldier equipped with either of these guns can be viewed as an estimator or finder of a target. Both sets of armed soldiers were accurate (unbiased), in the sense that on average they could hit a target at 100 yards. The colonials armed with Pennsylvania rifles were more "precise" or **efficient**, in the sense that their average shots had less dispersion. In statistics, one estimator is said to be more efficient (or more precise) than another if its variance is smaller than the other.

Efficiency
The property of an estimator whose variance is smaller than that of any other estimator using the same sample size.

As introduced and discussed in the previous chapter, the sample mean $\bar{x}$, has a variance of σ^2/n, where σ^2 is the variance of the population from which samples of size n are drawn.[3] The standard deviation of the estimator is obtained by taking the square root of the variance, $\sigma/\sqrt{n}$. The standard deviation of the estimator is called the **standard error.** The standard deviation of the estimator $\bar{x}$ should not be confused with the standard deviation of the sample. The standard deviation of the sample is

Standard Error of an Estimator
The standard deviation of the estimator.

$$s = \sqrt{\left[1/(n-1)\right]\Sigma\left(x - \bar{x}\right)^2}$$

which reflects variability in a sample around the sample's mean $\bar{x}$. The standard deviation or standard error of $\bar{x}$ is

$$\sigma_{\bar{x}} = \sigma/\sqrt{n}$$

FIGURE 8.1 The Population and Sampling Distribution of $\bar{x}$

Panel a: Population of x

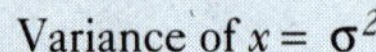

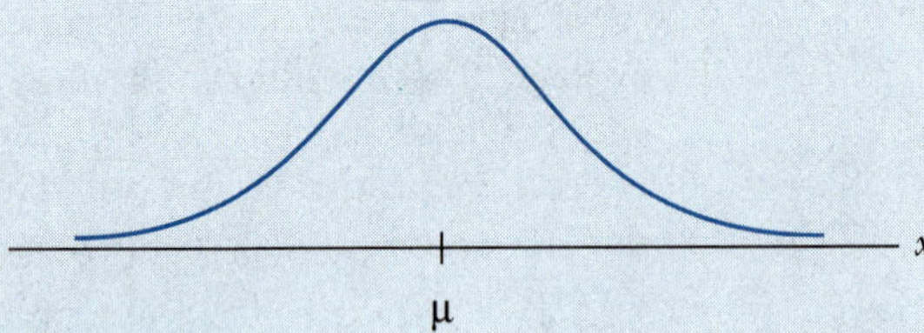

Panel b: Sampling distribution of $\bar{x}$, for $n = 1$

Variance of $\bar{x} = \sigma^2/1$
$= \sigma^2$

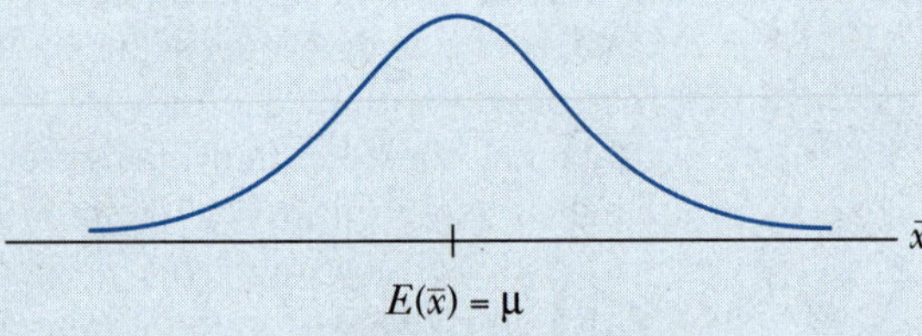

Panel c: Sampling distribution of $\bar{x}$, for $1 < n < \infty$, say $n = 25$

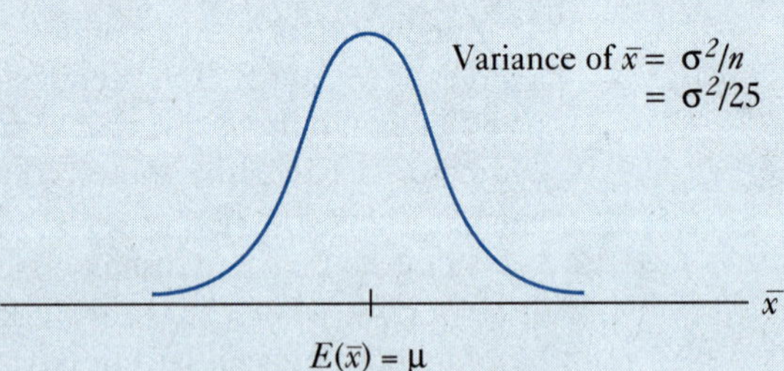

Panel d: Sampling distribution of $\bar{x}$ as n goes to infinity

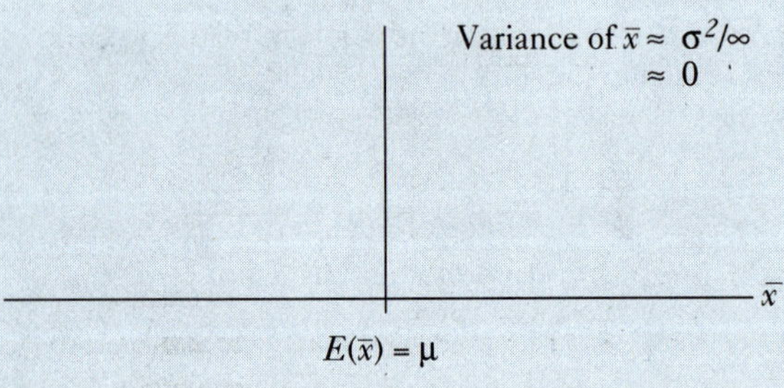

The standard error of $\bar{x}$ measures variability in the $\bar{x}$'s around their expected value $E(\bar{x})$. Although s is an estimator of σ, and thus can be used in the estimation of the standard error of $\bar{x}$ (as $s/\sqrt{n}$), it should not be confused with σ or $\sigma/\sqrt{n}$.

Along with its mean and standard error, every estimator has a probability distribution that determines its values. This distribution, which is called the sampling distribution of the estimator, is determined by the model. For instance, if the population of x is normal, with mean μ and variance σ^2, then the sampling distribution of $\bar{x}$ is also normal but its variance is σ^2/n, where again n is the sample size. Under these assumptions, the model for $\bar{x}$ is

$$\bar{x} \sim N\left(\mu, \sigma/\sqrt{n}\right)$$

which says that $\bar{x}$ is distributed normally with mean μ and standard deviation (standard error) of $\sigma/\sqrt{n}$. Alternatively, and as discussed in the previous chapter, if each random sample is large ($n > 30$), then from the Central Limit Theorem we know that the sampling distribution of the estimator is approximately normally distributed, regardless of the population distribution.

Critical in our understanding of the sampling distribution is the recognition that the variance of the sampling distribution is not the same as the variance of population from which samples are drawn. In fact, $\bar{x}$ will have a variance equal to the population variance only in the extreme case of $n = 1$. If $n = 1$, then $\bar{x}$ is the one value in the sample. At the other extreme, if each sample is extremely large (infinite), then the variance of $\bar{x}$ approaches zero; that is, the sampling distribution of the sample mean is a spike at its mean. These extremes are shown in Figure 8.1, where for simplicity the population is pictured as being normal, which implies that the sampling distributions of $\bar{x}$ are likewise normal, regardless of the sample size. Notice in Figure 8.1 that as n increases (from panel a to panel d), the deviations from the expected value of $\bar{x}$ (that is, μ) decrease to the point of no noticeable dispersion when n is extremely large.

Consistency
The property of an estimator by which the probability that its value is near the parameter's value approaches unity as the sample size increases.

The property of an estimator that reflects the influence of sample size on precision is **consistency**. An estimator is consistent if the probability that its value is near the population parameter's value increasingly approaches unity as the sample size increases. As seen in Figure 8.1, because the sampling distribution of $\bar{x}$ collapses on μ as n increases (movement from panel b to d), $\bar{x}$ is a consistent estimator of μ.

8.4 Confidence Intervals

A point estimate is a single number; it provides no idea of the error associated with the estimator. For this reason, estimates are often reported with a "margin of error." Unfortunately, it is not always clear what this expression means. For example, what does a reporter on the six o'clock news mean when he or she talks about the outcome of a poll and its margin of error? The implication is that there is a point estimate to which one may add or subtract some measure for chance events. Such interval estimates that provide information on variability in sampling may be preferred to point estimates.

The interpretation of the error in sampling, however, requires knowledge of the probability distribution underlying the sampling distribution of the estimator, which the newscaster seldom provides.

Because it is only one number, a point estimate is sharp or more precise than an interval estimate, but it is also more often wrong. An interval estimate, on the other hand, provides a range of values. Its center value is typically the point estimate of the population parameter being estimated, and its upper and lower end values are determined by the margin of error. The margin of error is based on probability considerations.

For a random variable L less than a certain number θ, we write $Prob(L < \theta)$; similarly, we know that $Prob(\theta < U)$ is the probability that the random variable U is greater than the number θ. Thus, the

$$Prob(L < \theta < U) \tag{8.1}$$

is the probability that a population parameter θ is between a lower limit L and an upper limit U, where L and U are still the random variables and the parameter θ is still the certain number. The range L to U is an interval estimate of the parameter θ.

The interval $L < \theta < U$ is called a confidence interval, where the random values L and U are defined by the probability of the interval containing the population parameter. The population parameter θ to be estimated might be either the population mean μ or population proportion π.

8.5 Confidence Interval for the Population Mean

The confidence interval for the population mean is written $L < \mu < U$. If, prior to sampling, there is a 95 percent chance that this L–U interval will contain the true μ, then after sampling the resulting interval is called a 95 percent confidence interval for μ. This means that if 100 samples are randomly selected, and for each sample a value of L and a value of U are determined, then the population mean will be between the L and U values for 95 of the 100 intervals, on the average.

L and U are random variables that yield different values from sample to sample, but the population mean μ is a certain number. Thus, the probability statement

$$P(L < \mu < U) = 0.95 \tag{8.2}$$

does not say that there is a 95 percent probability that μ is between specific values of L and U. The interpretation is rather that for a large (infinite) number of samples, each sample would yield different values of L and U, but 95 percent of these L–U intervals will contain μ and 5 percent of the L–U intervals will not contain μ. This interpretation says nothing about any specific interval.

Consider the case in which $\bar{x}$ is normally distributed, with mean μ and variance σ^2/n. From the previous chapter, we know that there is a 0.95 probability that an $\bar{x}$ value is obtained between $\mu - 1.96\sigma/\sqrt{n}$ and $\mu + 1.96\sigma/\sqrt{n}$. That is,

$$Prob\left(\mu - 1.96\sigma/\sqrt{n} < \bar{x} < \mu + 1.96\sigma/\sqrt{n}\right) = 0.95 \tag{8.3}$$

Notice the difference between Equations (8.3) and (8.2). To express (8.3) as written in (8.2), both μ and $\bar{x}$ must be subtracted from each side of the inequalities in Equation (8.3), yielding

$$Prob\left(-\bar{x} - 1.96\sigma/\sqrt{n} < -\mu < -\bar{x} + 1.96\sigma/\sqrt{n}\right) = 0.95 \tag{8.4}$$

which upon multiplying through by a negative one gives

$$Prob\left(\bar{x} + 1.96\sigma/\sqrt{n} > \mu > \bar{x} - 1.96\sigma/\sqrt{n}\right) = 0.95$$

or

$$Prob\left(\bar{x} - 1.96\sigma/\sqrt{n} < \mu < \bar{x} + 1.96\sigma/\sqrt{n}\right) = 0.95 \tag{8.5}$$

Equation (8.5) can be seen to be identical to (8.2), where

$$L = \bar{x} - 1.96\sigma/\sqrt{n} \text{ and } U = \bar{x} + 1.96\sigma/\sqrt{n}$$

As in the interpretation of Equation (8.2), (8.5) says that before sampling, when $\bar{x}$ has not yet been calculated, there is a 95 percent chance that the interval between $L = \bar{x} - 1.96\sigma/\sqrt{n}$ and $U = \bar{x} + 1.96\sigma/\sqrt{n}$ will include the population mean. Here $\bar{x}$ and thus L and U are still random variables whose values have not yet been determined. Now if samples are drawn one after another, each of size n, then 95 out of 100 L–U intervals constructed will contain μ, on the average.

A confidence interval is based on a probability statement about $\bar{x}$ and not μ. As shown in Figure 8.2, the value of $\bar{x}$ will differ from sample to sample, but it will always be the middle of the confidence interval. The 20 theoretical intervals in the lower part of Figure 8.2 show only one interval that does not cover μ, which is what should be expected over a large (infinite) number of samples, on the average. Unfortunately, once an actual value of $\bar{x}$ is calculated from a sample, we do not know whether the resulting interval does or does not include μ. After all, μ is the parameter being estimated; if it were known, there would be no need for its estimation.[4]

The assignment of 95 percent probability is not relevant after a sample is actually drawn and after an $\bar{x}$ value is substituted into the formula $\bar{x} \pm 1.96\sigma/\sqrt{n}$. After the 64 "highly mobile" executives were sampled and a 95 percent confidence interval for the mean raise was calculated to be \$23,990 to \$26,650, it is not correct to say that there is a probability that μ is between \$23,990 and \$26,650. Either the interval \$23,990 to \$26,650 includes the population mean raise or it does not. Because there is no random variable in the interval

$$\$23{,}990 < \mu < \$26{,}650$$

there is no probability. We can say only that under repeated sampling, 95 percent of similarly constructed intervals would contain the true population mean, and the interval $\$23{,}990 < \mu < \$26{,}650$ may (or may not) be one of those correct intervals.

FIGURE 8.2 Interpreting a 95 Percent Confidence Interval

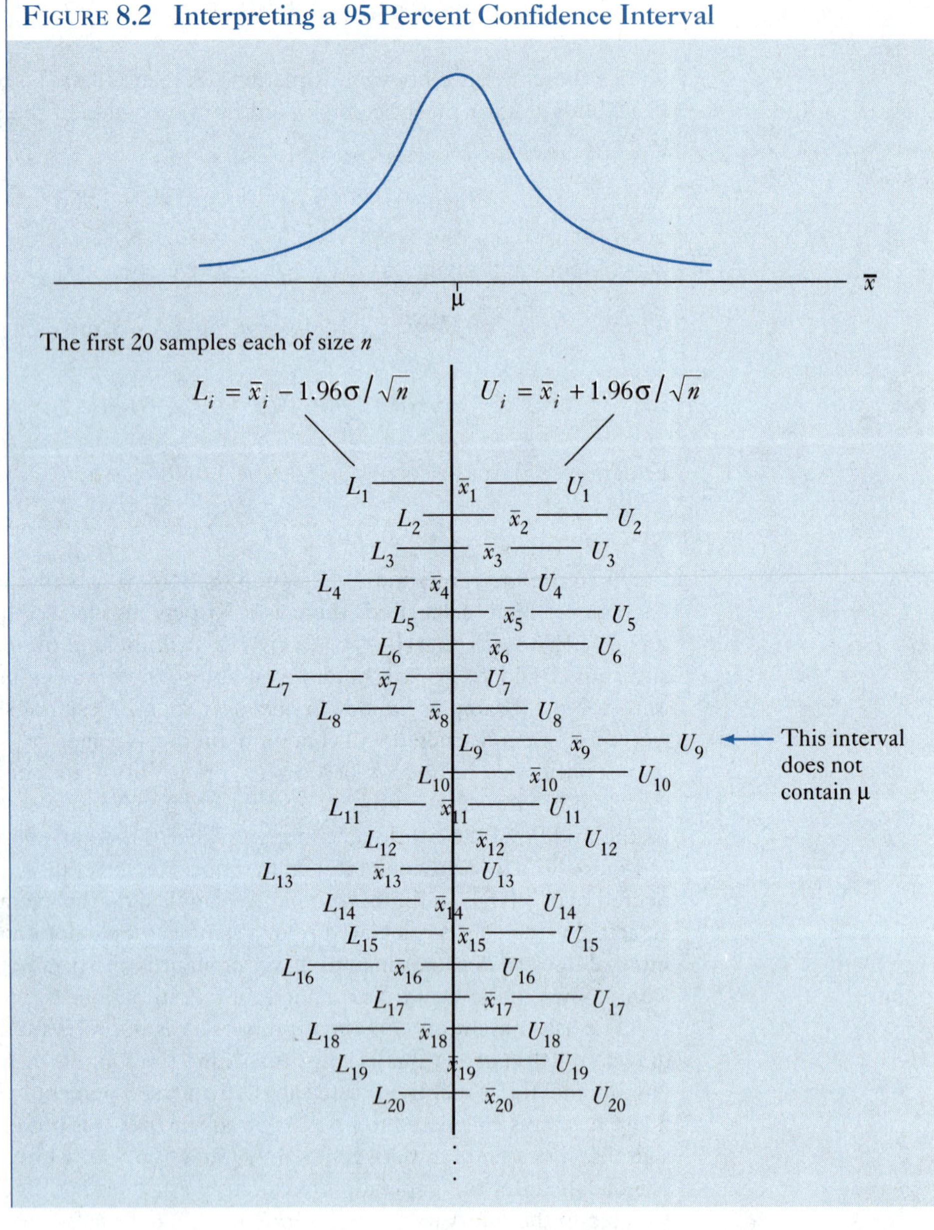

8.6 Level of Confidence

Typically 90, 95, or 99 percent confidence intervals are constructed. For these three levels of confidence, the probability that the *L–U* interval does not contain μ is thus 0.10, 0.05, or 0.01. The probability of the *L–U* interval not containing μ is designated by the Greek letter alpha (α), where for the typical three confidence intervals, α is either the decimal 0.10, 0.05, or 0.01, and the level of confidence is then given by $(1 - \alpha)100\%$. If $\alpha = 0.01$, for example, then the confidence interval is 99 percent: $(1 - \alpha)100\% = (1 - 0.01)100\% = 99\%$. In the general case, a $(1 - \alpha)100\%$ confidence interval for μ is

$(1-\alpha)100\%$ confidence interval for μ

$$\bar{x} - z_{\alpha/2}\left(\sigma/\sqrt{n}\right) < \mu < \bar{x} + z_{\alpha/2}\left(\sigma/\sqrt{n}\right) \qquad (8.6)$$

where $z_{\alpha/2}$ is the value of the standard normal variable z that puts $\alpha/2$ percent in each tail of the distribution.

As an application of Equation (8.6), consider the mobile executive pay raise example of a 95 percent confidence interval for the mean raise where $\alpha = 0.05$ and $z_{.025} = 1.96$. For a 90 percent confidence interval, $\alpha = .10$ and $z_{.05} = 1.645$. For $\alpha = .01$, $z_{.005} = 2.575$.

Alpha typically is set by considering the tradeoff between the desired level of confidence and the required accuracy. If alpha is low ($\alpha = 0.01$), then the width of the interval will be wide and 99 out of 100 of the *L–U* intervals will, on the average, contain the population parameter being estimated. Such an interval, however, may not be highly informative because of its width. In the mobile executive pay raise example, a 99 percent confidence interval for μ is

$$\bar{x} - z_{\alpha/2}\left(\sigma/\sqrt{n}\right) < \mu < \bar{x} + z_{\alpha/2}\left(\sigma/\sqrt{n}\right)$$

$$25320 - \left(2.575\right)\left(5428.57/\sqrt{64}\right) < \mu < 25320 + \left(2.575\right)\left(5428.57/\sqrt{64}\right)$$

$$\$23{,}572.68 < \mu < \$27{,}067.32$$

The width of this 99 percent confidence interval is $3,494.64, versus the $2,660 width for the 95 percent confidence interval. That may be too wide for some purposes. Increasing the confidence level further, by decreasing alpha, would simply widen the interval still further.

At a fixed level of confidence, and for a given population variance, the width of the confidence interval can be narrowed by increasing the sample size. As the sample size is increased, the standard deviation of the sampling distribution of the estimator is reduced and the margin of error $\pm z_{\alpha/2}\left(\sigma/\sqrt{n}\right)$ is likewise decreased. It is for this reason that larger samples are preferred to smaller samples in the construction of confidence intervals.

EXERCISES

8.1 Are the following statements true or false, and why?
   a. The sample mean is an unbiased estimator of the population mean because they are equal for large samples.
   b. The expected value of the sample standard deviation is equal to the population standard deviation; therefore, it is an unbiased estimator of the population standard deviation.
   c. The sample variance is equal to the population variance for large samples; therefore, it is an unbiased estimator of the population variance.

8.2 At what stage in the construction of a confidence interval are the words "confidence" and "probability" interchangeable? At what stage are they not?

8.3 After surveying 100 new Cadillac owners, a marketing analyst at General Motors has established a 99 percent confidence interval for the mean income of all new Cadillac owners: $\$65,345 < \mu < \$123,145$.
   a. What was the alpha for this interval?
   b. What value of $z_{\alpha/2}$ was used in this interval?
   c. What was the point estimate of the mean income?
   d. What was the value of the standard deviation used in this interval?

8.4 Suppose two demographers independently take random samples of size $n = 64$ to estimate the mean age in a geographical region and then construct a 95 percent confidence interval based on their separate samples.
   a. Would the two demographers necessarily have the same point estimate for μ?
   b. Would you expect the margin of error for the two demographers to be the same? Explain.

8.5 If the sample size is increased fourfold, what happens to the width of the confidence level, assuming σ is known?

8.6 A simple random sample of size 16 is drawn from a normal population of prices, with $\sigma =$ \$12. A confidence interval for the mean price is constructed and found to have a width from lower to upper end of \$18.60. What was the level of confidence?

8.7 A *Wall Street Journal* survey just prior to the launching of the Desert Storm action in 1991 reported that the majority of Americans were hawkish on Iraq and that "chances are 19 to 20 that if all registered voters in the U.S. had been surveyed . . . the findings would differ from these poll results by no more than 3.2 percentage points in either direction." Comment on this interpretation of the margin of error.

8.8 In a Gallup telephone poll of 767 adults, as reported in *Newsweek* (March 2, 1992), 73 percent said George Bush was not doing enough for the economy, with a margin of error of plus or minus 4.13 percent. What value of z would give this margin of error?

8.9 For the Gallup/*Newsweek* survey reported in Exercise 8.8, what is the implied level of confidence and the confidence interval for the population proportion that would say Bush was not doing enough on the economy?

8.10 The Gallup/*Newsweek* survey reported in Exercise 8.8 involved many questions. Would the margin of error be the same for each question, as suggested by the reporting of a single 4.13 percent margin? Explain.

8.7 WHEN σ IS UNKNOWN

The formula for the upper or lower limits of a confidence interval, $\mu = \bar{x} \pm z_{\alpha/2}(\sigma/\sqrt{n})$, requires that the population standard deviation σ be known. In most situations, this is not the case. In the mobile executive pay raise case, for example, σ is unknown; only the sample standard deviation s = \$5,428.57 is actually known. We used s^2 as our estimator of σ^2. . . and simply substituted s for σ. But this substitution introduces more variability into the confidence interval than what is assumed in z. In particular, the standard normal transformation z of the sampling distribution of the mean involves only one statistic $\bar{x}$:

$$z = \frac{(\bar{x} - \mu)}{\sigma/\sqrt{n}}$$

Substituting s for σ results in a ratio involving two statistics $\bar{x}$ and s:

$$\frac{(\bar{x} - \mu)}{s/\sqrt{n}}$$

At the turn of the century, William Gosset developed the distribution for this ratio. He called his ratio the t statistic and used it to calculate confidence intervals for μ.

To standardize $\bar{x}$, when σ is unknown:

$$t = \frac{\bar{x} - \mu}{s/\sqrt{n}} \quad (8.7)$$

For small samples, the population x must be normal for $\bar{x}$ to be normal.

The confidence limits for μ, when σ is unknown:

$$\mu = \bar{x} \pm t_{\alpha/2}\, s/\sqrt{n}$$

where $t_{\alpha/2}$ is the value of t that puts α/2 in each tail, and

$$s = \sqrt{\frac{\sum(x - \bar{x})^2}{n - 1}} \quad \text{and } \bar{x} = \frac{1}{n}\sum x_i$$

The t distribution has $n - 1$ degrees of freedom.

t Distribution
A probability distribution formed by the ratio of a normal random variable to the square root of an independently distributed Chi-square random variable divided by its degrees of freedom.

The ***t* distribution** in Equation (8.7) depends on randomness in both $\bar{x}$ and s.[5] The sample standard deviation s depends on the number of observations in a sample of size n that are free to vary around the sample mean $\bar{x}$. If the sample size is $n = 1$, for example, then there is no variability around the mean—the value of this single observation is also the mean of the sample. If the sample size is $n = 2$, then one of the values is free to vary. Once either of the two values is selected, the other value is determined by the mean and the first value. To illustrate, let a be the first value and b be the second. For any $\bar{x}[= (a + b)/2]$, b depends on the value of a and, hence, is not free to vary. For example, if $\bar{x} = 12$ and a is 15, then b must be 9. Similarly, when $n = 3$, any two values are free to vary, but once two are selected, then the third is fixed. In general, for any $\bar{x}$ value and sample size of n, once $n - 1$ values are determined, the remaining value is no longer free to vary. Thus, "the degrees of freedom" are said to be $n - 1$.

Degrees of Freedom for the Mean (df) The number of observations that are "free to vary" around the mean of a sample.

The **degrees of freedom** (abbreviated df) represent the number of observations in a sample that are "free to vary" around the mean of the sample. For the sample mean, the degrees of freedom for use in the t distribution are df $= n - 1$.

With the notable exception of the degrees of freedom adjustment, the t distribution is similar to the standard normal distribution. Both z and t are continuous, unimodal, bell-shaped distributions, with means of zero. They both require that the sample mean $\bar{x}$ be normally distributed. The t distribution differs from the z in that the sample standard deviation s is used as an estimator for the population standard

FIGURE 8.3 Examples of the Family of t Distributions

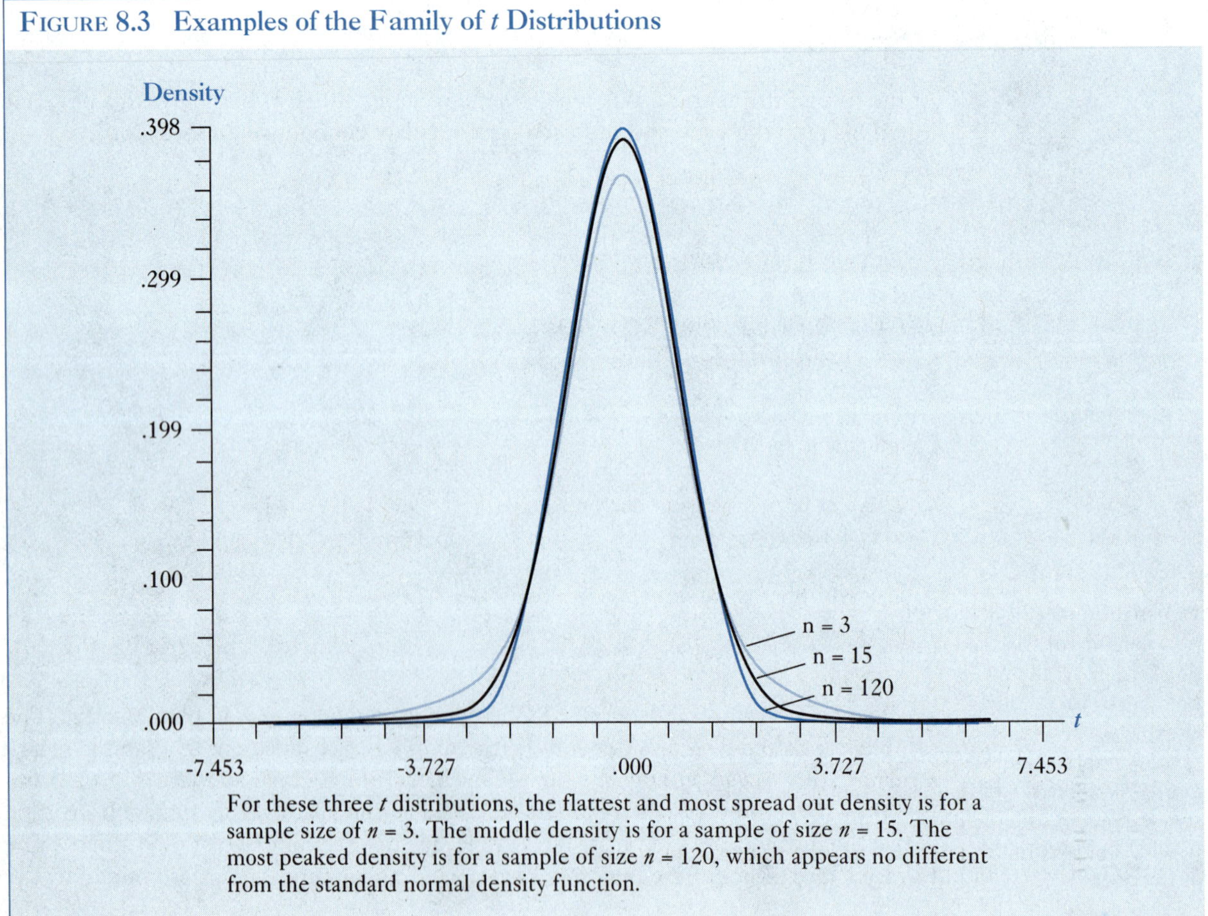

For these three t distributions, the flattest and most spread out density is for a sample size of $n = 3$. The middle density is for a sample of size $n = 15$. The most peaked density is for a sample of size $n = 120$, which appears no different from the standard normal density function.

deviation σ. Unlike σ, values of s vary from sample to sample, which tends to make the t more spread out than the z, with heavier or larger areas in the tails of the t.

Although the expected value of s^2 is σ^2, the probability that any individual s value is close to σ depends on the sample size n—the larger the sample size, the more likely it is that an individual s value will be close to σ. For large samples, little difference between s and σ should be expected. Thus, for relatively large samples, it should be of minor importance whether the t or z ratio is used to "standardize" $\bar{x}$, or whether t or z is used in a confidence interval.

Because the variability in t depends on the degrees of freedom, the t distribution can be thought of as a family of distributions—a separate member for each degree of freedom. Three of these member distributions are shown in Figure 8.3. With 120 degrees of freedom, the t distribution is indistinguishable from the z. Even with only 15 degrees of freedom, the t distribution does not appear greatly different from the z. At three degrees of freedom, a difference is easily discernible. Use of the t distribution for small samples is critically important.

Historical Perspective

Few people think of the beer industry as employing path-breaking scientists. Yet, William S. Gosset (1877–1937) invented the t distribution while working for Guinness brewery in Ireland.

Gosset recognized that in most applications involving the estimation of the population mean, the value of the population variance is unknown. In 1908, he published "The Probable Error of a Mean" under the pseudonym "Student." Gosset's introduction of the t distribution eventually revolutionized statistical work with small samples. At first, however, the wide applicability of Student's t was not recognized. It was not until 1925, when Sir Ronald A. Fisher called attention to them in his textbook, that applications became popular in scientific inquiry.

Gosset was a student of both chemistry and mathematics. He learned statistics from Karl Pearson (1857–1936), one of a few distinguished English statisticians at the turn of the century. Unlike most scientists of his time, however, Gosset did not pursue a life in academe.

Gosset made his major scholarly contribution while working for private industry. His employer, Guinness, employed many university educated scientists but did not encourage scholastic publishing. So, Gosset published under his pseudonym.

In the United States, Chemical Rubber Co., a private firm in Cleveland, Ohio, assumed a key role in the preparation, maintenance, and dissemination of probability tables for statistical research. Not all contributions in statistics have come from academic institutions and think tanks.

Although Gosset discovered the t distribution and Fisher perfected and popularized it, it was Edward Paulson (1942) who provided a convenient algorithm for approximating probabilities based on Student's t. Paulson showed that the area more extreme than the absolute value of t, with $n - 1$ degrees of freedom, could be approximated with the area in the upper tail of the standard normal distribution. Provided that the degrees of freedom are greater than three ($n - 1 > 3$), the approximate relationship between $|t|$ and the value of z is

$$z^* = \frac{\left[1 - \dfrac{2}{9(n-1)}\right] t^{2/3} - \dfrac{7}{9}}{\sqrt{\dfrac{2}{9(n-1)} t^{4/3} + \dfrac{2}{9}}}$$

The accuracy of this algorithm can be assessed in the following table. The exact absolute value of

t that put 0.10 and 0.025 in *each of the tails* of the t distribution was used to calculate the associated z^* values that put approximately 20 and 5 percent in the upper tail of the standard normal distribution. These approximate z^* values compare favorably with the exact z values of 0.843 and 1.646 that put 0.20 and 0.05 in the upper tail of the standard normal distribution.

$n-1$	t	z^*	Exact z
5	1.476	0.845962	0.843
5	2.571	1.650117	1.646
10	1.372	0.848719	0.843
10	2.228	1.661954	1.646
20	1.325	0.850333	0.843
20	2.086	1.667223	1.646

Today, inexpensive statistical computer programs can calculate probabilities associated with the entire family of t distributions, regardless of the degrees of freedom considered. W.I. Kennedy and J.E. Gentile (1980) have reviewed some of the alternative algorithms that are used by programmers to achieve extremely high accuracy.

8.8 Confidence Interval for the Mean, σ Unknown (and n small)

Whether the standard normal distribution or the t distribution is used in the construction of a confidence interval only affects the calculation of the margin of error. Let's consider that calculation now for the t distribution.

From Equations (8.6) and (8.7), the $(1 - \alpha)100\%$ confidence interval for μ when σ is unknown is

$$Prob\left(\bar{x} - t_{\alpha/2} s / \sqrt{n} < \mu < \bar{x} + t_{\alpha/2} s / \sqrt{n}\right) = 1 - \alpha \tag{8.8}$$

where $t_{\alpha/2}$ is the value of t that puts $\alpha/2$ percent of the t distribution in the upper tail and $-t_{\alpha/2}$ puts $\alpha/2$ percent of the distribution in the lower tail, with $n - 1$ degrees of freedom. These values of t can be obtained in several different ways, including microcomputer programs such as ECONOMETRICS TOOLKIT, STATABLE, and MICROSTAT. Appendix Table A.4 also presents some key cumulative probabilities for a limited number of members of the t distribution, which are identified by the degrees of freedom in the left-hand column. Only 34 different t distributions are presented for df = 1 through 30 and for df = 40, 60, 120, and infinity.

To appreciate the influence of the degrees of freedom, consider the value of t that puts 2.5 percent in the upper tail. If the degrees of freedom are infinite, then this t value is 1.96, which is the last value in the column headed by the cumulative probability 0.975. The t for df = ∞ is the same as the z value that puts 2.5 percent above it and 97.5 percent below. If the degrees of freedom fall to 120, in the second to the last row, the t value that puts 2.5 percent in the upper tail rises to 1.98. With df = 60, the critical t that puts 2.5 percent above it is 2.00; with df = 40, this t rises to 2.021. Notice that even with only 30 degrees of freedom, the t that puts 2.5 percent in the upper tail is 2.042, which is still close to the z of 1.96.

For a 95 percent confidence interval, the critical t values for df = ∞, 120, 60, 40, and 30 are ±1.96, ±1.98, ±2.00, ±2.021, and ±2.042. Clearly the width of a confidence interval is not going to be greatly affected if the z value of ±1.96 is used instead of the more appropriate t value. For extremely small samples, however, use of the appropriate t is essential. With only one degree of freedom, for example, the critical t that puts 2.5 percent in the upper tail is 12.706, which is not close to the 1.96 value for z.

Critical t values for different confidence levels are determined in Appendix Table A.4 by looking across the row specified by the degrees of freedom. Suppose, for example, that we have a sample of n = 23 observations and n – 1 = 22 degrees of freedom; row 22 from Appendix Table A.4 is reproduced in the top part of Figure 8.4. The

FIGURE 8.4 Cumulative Probabilities for the t Distribution, df = 22

From Appendix Table A.4, for 22 degrees of freedom,

df	0.75	0.90	0.95	0.975	0.99	0.995	0.9995
•							
•							
•							
22	0.686	1.321	1.717	2.074	2.508	2.819	3.792
•							
•							
•							

the cumulative probabilities given across the top of the table are associated with the t values within the table in the following way

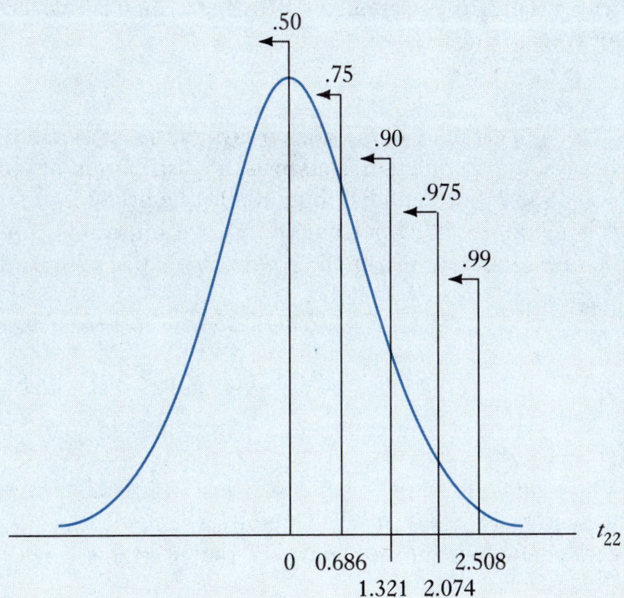

relationship between the cumulative probability, given as the column headers, and the t value in the body of the table is shown in the lower part of Figure 8.4.

The first cumulative probability in Figure 8.4 is 0.75, and its corresponding t value is 0.686. As indicated in the lower part of Figure 8.4, 75 percent of the distribution is below $t = 0.686$, with 22 degrees of freedom. A 50 percent confidence interval for μ, with 22 df, is now given by

$$Prob\left(\bar{x} - 0.686s/\sqrt{23} < \mu < \bar{x} + 0.686s/\sqrt{23}\right) = 0.50 \tag{8.9}$$

The second cumulative probability in Figure 8.4 is 0.90, and the corresponding t value is 1.321. This indicates that with 22 degrees of freedom, 90 percent of the distribution will be below t = 1.321; thus, –1.321 and +1.321 are the t values for an 80 percent confidence interval:

$$Prob\left(\bar{x} - 1.321s/\sqrt{23} < \mu < \bar{x} + 1.321s/\sqrt{23}\right) = 0.80 \tag{8.10}$$

And since 95 percent of the distribution is less than 1.717, a 90 percent confidence interval is

$$Prob\left(\bar{x} - 1.717s/\sqrt{23} < \mu < \bar{x} + 1.717s/\sqrt{23}\right) = 0.90 \tag{8.11}$$

and so on.

QUERY 8.1:

A mill produces sheet metal that it says will average 0.20 inch in thickness, with deviations determined by the normal distribution. A truckload is received for which a sample of 23 different locational measurements yields a mean width of 0.192 inch, with a standard deviation of 0.0187 inch. Should you accept the shipment as having an average thickness of 0.20 inch?

ANSWER:

The point estimate of the thickness of the sheet metal in the shipment is 0.192 inch. Because the population is normal, the distribution of sample average thickness $\bar{x}$ is also normal, but the t distribution must be used for probability calculations because σ is unknown. A 95 percent confidence interval is constructed to reflect the randomness inherent in sampling, where $t = 2.074$ with 22 degrees of freedom.

$$.192 - 2.074\left(.0187/\sqrt{23}\right) < \mu < .192 + 2.074\left(.0187/\sqrt{23}\right)$$

$$.192 - .0081 < \mu < .192 + .0081$$

$$.1839 < \mu < .2001$$

The claimed 0.20 inch thickness is just inside this interval, suggesting that the true μ could be that high, as estimated by a 95 percent confidence interval. For a 90 percent confidence interval, however, the interval is

$$.192 - 1.717\left(.0187/\sqrt{23}\right) < \mu < .192 + 1.717\left(.0187/\sqrt{23}\right)$$

$$.192 - .0067 < \mu < .192 + .0067$$

$$.1853 < \mu < .1987$$

which does not include the claimed 0.20 thickness. If it is critical that the steel average thickness be 0.20, then this shipment should be rejected. On the other hand, if the exact thickness is not that critical, it can be accepted.

QUERY 8.2:

It is common for business travelers not to submit receipts for the cost of a taxi ride from airports to city centers; they "estimate the fare to have been . . . ." The accounting department of a large firm wants to know when to question these "estimates."

ANSWER:

An interval estimate of the mean cost of taxi service for a given city can be estimated from a random sample of taxi costs from the airport to the city center. Below is such a sample reported by eight travelers:

Traveler 1	\$28.00	Traveler 5	\$22.00
Traveler 2	\$27.00	Traveler 6	\$15.00
Traveler 3	\$24.50	Traveler 7	\$14.00
Traveler 4	\$24.00	Traveler 8	\$ 7.00

By typing the following four command lines into the MINITAB computer program, a 99 percent confidence interval for the mean taxi fair will be calculated.

```
MTB  > set into c4
DATA > 28 27.0 24.5 24 22 15 14 7
DATA > end
MTB  > tinterval 99 percent c4
```

MINITAB provides the following printout:

	N	MEAN	STDEV	SE MEAN	99.0 PERCENT C.I.
C4	8	20.19	7.40	2.62	(11.03, 29.34)

MINITAB did the calculations for this 99 percent confidence interval with the t distribution. This computer program did not know if the sample mean was or was not normally distributed; it simply followed the commands. Here we justify the normality of $\bar{x}$ based on the assumed normality of the population.

Because all values are rounded to the nearest cent, it is worthwhile to see the calculations MINITAB did:

$$20.188 - 3.499(2.615) < \mu < 20.188 + 3.499(2.615)$$

where the estimated standard error of the mean is

$$s/\sqrt{n} = 7.397/\sqrt{8} = 2.615$$

Thus, our interval estimate of the average cost of a taxi ride between the airport and the city center is \$11.03 to \$29.34. The accounting department might be wise to question any "estimated" charges outside this interval—a reported cost higher than \$29.34 might indicate exaggerated costs, and a reported cost less than \$11.03 might indicate that a taxi was never taken.

EXERCISES

8.11 In what ways do the z distribution and the t distribution differ? In what ways are they the same?

8.12 Between the t and the z, which has more area beyond the standardized value of 1.00? What determines the magnitude of this difference in area beyond the value 1.00?

8.13 Find the probabilities for Student's t distribution and value of b.

a. $P(t < 1.721 \mid \text{df} = 21) =$ ____
b. $P(t > 3.707 \mid \text{df} = 6) =$ ____
c. $P(t > -2.306 \mid \text{df} = 8) =$ ____
d. $P(t > b \mid \text{df} = 13) = 0.995, b =$ ____
e. $P(t < b \mid \text{df} = 3) = 0.9995, b =$ ____
f. $P(t > b \mid \text{df} = 23) = 0.005, b =$ ____

8.14 What is the consequence of using the t distribution versus the z distribution in the construction of a confidence interval?

8.15 What assumptions are involved in the use of the t distribution to form a confidence interval for μ?

8.16 When is the z distribution more appropriate than the t distribution for constructing a confidence interval? When is the distinction irrelevant?

8.17 An audit of a retailer's inventory was conducted by randomly selecting the purchase invoices from $n = 100$ unsold units in stock. The average purchase price per unit was found to be \$17.50, with a standard deviation of \$6.75.

a. Find a 95 percent confidence interval for the mean purchase price of all units of inventory held by the retailer.
b. Audit results are interpreted in terms of the stated precision and the materiality of the derived estimate. The precision of the estimate is the level of confidence in the confidence interval, while materiality is synonymous with accuracy—the absolute difference between population and sample mean. How does changing σ, n, and α change the level of confidence and width of the interval?

8.18 In the 1992 "Red Book" that most banks and credit unions use to assess car loans, the average retail price of 1989 Corvette convertibles was \$21,750. A sample of attempted purchases of these cars, as advertised in the local "Wheels & Deals" periodical, resulted in the following sampling of "firm and final prices":

\$19,500, \$25,550, \$21,995, \$22,500, \$23,500, and \$21,000

What is the probability of getting the mean for this sample of size 6, or one with a higher mean, if the average retail purchase price is \$21,750, with a population standard deviation approximately equal to that of this sample? What assumptions did you make in your analysis?

8.19 In the early 1980s, a Baltimore directory-assistance operator was suspended because she averaged three seconds more than the 30-second standard. If this suspension was based on the random sample of the length of 12 different calls given below, would a 99 percent confidence interval suggest that this operator had exceeded the 30-second standard?

Call	Time (seconds)	Call	Time (seconds)
1	29	7	37
2	28	8	28
3	26	9	35
4	39	10	30
5	44	11	32
6	40	12	26

8.20 Below is a computer output of descriptive statistics on profits (in cents) per dollar of sales. (The numbers under the minimum and maximum headings are the lowest and highest profits in the sample.)

NAME	n	MEAN	STD. DEV.	MINIMUM	MAXIMUM
profit	23	6.28	1.34	2.96	11.21

a. What is the point estimate of average profit per dollar of sales?
b. What is the estimated standard error of average profit per dollar of sales?
c. Construct a 95 percent confidence interval for average profits per unit of sales.

8.9 Estimating the Population Proportion π

Recall from the discussion of the binomial distribution in Chapter 5 that if π is the proportion of computers that do not fail in the first year, then $n\pi$ is the expected number of failure-free machines in a sample of n computers.

$$E(x) = n\pi \tag{8.12}$$

When π is not known, it is estimated by $p = x/n$, where p is the proportion of successes x in a sample of size n. From Equation (8.12), we can see that

$$E(x/n) = \pi \tag{8.13}$$

Thus, as shown in Chapter 7, the expected value of p is π.

Equation (8.13) states that the sample proportion of successes is an unbiased estimator of the population proportion of successes. Within any one sample, p need not equal π; only over a large number of samples will the average p equal π. The values of p will vary around its expected value π in accordance with a variance given by[6]

$$\sigma_p^2 = E(p - \pi)^2 = \pi(1 - \pi)/n \tag{8.14}$$

Thus the distribution of p has a mean of π and standard deviation of $\sqrt{\pi(1-\pi)/n}$. From the Central Limit Theorem, we know that the distribution of p will be approximately normal if n is sufficiently large. Thus, for large sample sizes, we have the model

$$p \sim N\left(\pi, \sqrt{\pi(1-\pi)/n}\right) \tag{8.15}$$

which enables us to define the $(1 - \alpha)100\%$ confidence interval for π as

$$p - z_{\alpha/2}\sqrt{\pi(1-\pi)/n} < \pi < p + z_{\alpha/2}\sqrt{\pi(1-\pi)/n} \tag{8.16}$$

where $z_{\alpha/2}$ is the value of the standard normal variable z that puts $\alpha/2$ percent in the tail of the distribution. Since π must be between 0 and 1, if a confidence interval for π includes values outside this range, then the sample size was clearly not sufficiently large to invoke the Central Limit Theorem.

Equation (8.15) cannot be used directly to calculate a confidence interval for π, because it requires knowledge of π. The estimator p is substituted for π, which is legitimate as long as the sample size n is large. Regardless of how large the sample is, however, this substitution results in only an approximate $(1 - \alpha)100\%$ confidence interval of π. Thus, the approximate computational formula for the confidence interval is

$(1 - \alpha)100\%$ confidence interval for π:

$$p - z_{\alpha/2}\sqrt{p(1-p)/n} < \pi < p + z_{\alpha/2}\sqrt{p(1-p)/n} \tag{8.17}$$

where $z_{\alpha/2}$ is the value of the standard normal variable z that puts $\alpha/2$ percent in each tail of the distribution.

QUERY 8.3:

In a sample of 47 Compaq computers, 83 percent did not fail in the first year. The question: With 95 percent confidence, what proportion of all Compaq computers would you expect not to fail within the first year?

ANSWER:

The point estimate of the population proportion of Compaq computers that did not fail in the first year is 83 percent. A 95 percent confidence interval estimate of this proportion is

$$.83 - 1.96\sqrt{.83(.17)/47} < \pi < .83 + 1.96\sqrt{.83(.17)/47}$$

$$.7226 < \pi < .9374$$

which means that our interval estimate of the percentage of Compaq computers that will not fail in the first year is 72.26 percent to 93.74 percent.

QUERY 8.4:

"Statisticians Occupy Front Lines in Battle Over Passive Smoking" (*Wall Street Journal,* July 28, 1993), reports that Kenneth Brown, a statistical consultant on whom the U.S. Environmental Protection Agency has relied, "produced the controversial 90% confidence interval, or 90% probability that the lung-cancer-risk range is between 4% and 35% higher for passive smokers than those who aren't so exposed." Gio Gori, a medical doctor and frequent expert witness for the tobacco industry, was cited as stating "that the standard in such studies is to calculate the range within which it is 95% certain that the true answer lies . . . (but) it would be so wide it might even hint that passive smoking actually reduced the risk of lung cancer." Comment on the correctness of these comments and on the use of a 95 versus a 90 percent confidence interval.

ANSWER:
First, the interpretations of the 90 percent confidence intervals as having a "90% probability that the lung-cancer-risk range is between 4% and 35%" and the 95 percent interval as "the range within which it is 95% certain that the true answer lies" are both wrong. Probability for a confidence interval is only defined before the sample is drawn; after the calculations are made, there is no longer a random variable on which probability is defined. Second, there is nothing sacred about a 90 percent or a 95 percent confidence interval; selection of either depends on the intended use. Finally, going from a 90 to a 95 percent confidence interval will lead to a wider confidence interval. If a resulting lower limit yields a percentage less than zero and such is viewed as meaningless, then this is evidence of the incorrectness of the assumptions and model upon which both the 90 and 95 percent confidence intervals were constructed.

EXERCISES

8.21 An article in the *Herald Telephone* (May 2, 1985) provided survey results showing that 75 percent of the 811 adult Indiana residents polled said they supported a state-controlled lottery. The margin of error for this survey was reported to be plus or minus 4 percent at the 95 percent confidence level. The article quoted Jon Masland as saying, "This means we are 95 percent confident that we would have gotten results within 4 percent either way of these results (0.75 ± 0.04) if we had interviewed every adult Indiana resident who has a telephone." Does this statement provide a correct interpretation of a confidence interval? Explain.

8.22 Indianapolis TV Channel 8 reported on February 24, 1992, that it had selected a random sample of 100 school bus drivers and found 32 had "blemished" driving records within the past five years. Construct a 95 percent confidence interval for the population proportion of Indiana school bus drivers with blemished driving records.

8.23 According to an article in *The Wall Street Journal* (February 24, 1992), 75 percent of the 134 current first-year M.B.A. students at the University of Pennsylvania's Wharton School said they expected to be running businesses 10 years hence. Construct a 90 percent confidence interval for the population proportion of students who expect to be running businesses ten years from now. To whom does your confidence interval apply?

8.24 A note in *The Wall Street Journal* (February 24, 1992) stated that 86 percent of the 120 women executives polled by Exec-U-Net, a western Connecticut networking group, said they once had a boss who was a "significant helper." Construct a 95 percent confidence interval for the population proportion of bosses who are significant helpers to women who become executives.

8.25 In car-buying situations, a common notion is that dealers steer customers to certain types of salespersons. This phenomenon was studied by Ian Ayers in "Fair Driving: Gender and Race Discrimination in Retail Car Negotiations," *Harvard Law Review* (February 1991).

He found, in a sample of 119 encounters, that black salesmen were paired with black male buyers 11.1 percent of the time. What is the 90 percent confidence interval for the proportion of times this pairing can be expected in the population?

8.26 Effective in 1991, interstate truck drivers were required to maintain national driver licenses. Unaddressed, however, was the state of the equipment they were using. A random sampling of 114 trucks showed 69 with faulty brakes. Construct a 95 percent confidence interval for the population proportion with faulty brakes.

8.10 Selecting the Sample Size

To this point in the analysis, we have assumed that a sample has already been drawn and n is known. At the beginning of any analysis, however, the sample size is unknown and must be determined before any statistical analysis can be undertaken. To determine the "best" sample size prior to actually drawing the sample, three factors must be considered:

1. The purpose of the sample. If it is for estimation purposes, then the reliability or confidence level must be set; that is, α must be specified. (The next chapter presents hypothesis testing as an alternative purpose for drawing a sample. Hypothesis testing provides a different set of considerations for sample selection, as discussed in Appendix 9D.)
2. The "margin of error" to be tolerated in estimation. In confidence interval estimation, this margin is called M.
3. The cost of sampling. Typically the more expensive the sampling, the smaller the sample and the larger the margin of error. (In Chapters 1 and 7, alternative methods and problems of sampling were considered.)

Here we will consider two estimation situations: sample selection for estimation of either a population mean μ or a population proportion π. A researcher might want to estimate many other parameters, but the associated sample selection procedures are just variations on the two procedures demonstrated below. (As an alternative to procedures based on the confidence interval, the careful reader will want to consider the appendices to the next chapter, where sample selection criteria are based on tradeoffs in hypothesis testing.)

Selecting a Sample Size to Estimate μ

The margin of error, M, in estimating μ when the population variance is known, is $\pm z_{\alpha/2}\left(\sigma/\sqrt{n}\right)$; it is expressed as:

$$M = \pm z_{\alpha/2}\left(\sigma/\sqrt{n}\right)$$

After specifying M and α, this equation can be solved for n, which yields

Sample size for estimating μ, σ known

$$n = \left[\frac{z_{\alpha/2}\sigma}{M} \right]^2 \tag{8.18}$$

In most situations, the population standard deviation σ is unknown. After all, if we do not know μ and need to select a sample to estimate it, how could we know σ?

In some situations, a guess about the value of σ is possible. For instance, if it is reasonable to assume that the population is normally distributed, then there should be at least four to six standard deviations between the highest and lowest observable values. A reasonable guess for the standard deviation in Equation (8.18) thus might be the highest observed value minus the lowest divided by 4, 5, or 6. In the absence of any better prior information, we will arbitrarily use a divisor of 5.

Rule of thumb for guessing σ value prior to sampling

$$\sigma^* = \frac{\text{highest minus lowest observed value}}{5} \tag{8.19}$$

This rule of thumb estimate of σ can be substituted into the formula for n so that

$$n = \left[\frac{z_{\alpha/2}\sigma^*}{M} \right]^2$$

From this formula for n, it should be apparent that decreasing the desired margin of error will increase the needed sample size. Similarly, increasing the level of confidence (by lowering α and increasing z) will increase the required sample size. Finally, the more variability in the population, the larger the sample size that will be required.

QUERY 8.5:

An accountant wants to estimate the average time between the issuing of the first billing statement and the receipt of payment, for those bills paid before a second billing statement is sent the following month. How large a sample should be drawn from the records of a large retailer? Assume the accountant wants to be 99 percent confident that her estimate of the mean is within plus or minus two days of the true mean time.

ANSWER:

The accountant can quickly check to see that some bills are paid immediately, within a day of billing, and a like number are paid just before the next billing statements are sent (31 days). Thus, a rule of thumb guess at the population standard deviation, as given in Equation (8.19), is

$$\sigma^* = \frac{\text{highest} - \text{lowest}}{5} = \frac{31 - 1}{5} = 6 \text{ days}$$

The desired sample size, using Equation (8.18), is thus calculated to be

$$n = \left[\frac{z_{\alpha/2}\sigma}{M}\right]^2 = \left[\frac{2.575(6)}{2}\right]^2 = 59.7$$

The best guess of the approximate sample size is now found by rounding up to the next highest integer. As a guess at the appropriate sample size, the accountant should sample 60 accounts receivable.

Selecting a Sample Size to Estimate π

In Equation (8.17), the margin of error in a confidence interval for the population proportion π was

$$M = \pm z_{\alpha/2}\sqrt{p(1-p)/n}$$

This equation cannot be solved for n, because the sample proportion of successes $p = x/n$ has not yet been determined—remember no sample has been drawn, so x is unknown. If we can make a reasonable guess about the value of p, we can substitute this guess (denoted p^*) for p. Thus for estimating the population proportion, we can approximate the sample size using Equation (8.20).

Sample size for estimating π:

$$n = \left[\frac{z_{\alpha/2}}{M}\right]^2 p^*(1 - p^*) \qquad (8.20)$$

where p^* denotes a guess for p and M is the margin of error

From past experience, we might have an intuitive idea or hunch about a possible value for p; this value can serve as p^* in Equation (8.20). If we have no idea about p, then set $p^* = 1/2$ in Equation (8.20). A value of $p^* = 1/2$ results in the highest value of n at any given confidence level and is thus a good starting point. The required sample size will also be increased if we lower the margin of error (M) or increase the level of confidence and reliability (which lowers α and increases z).

QUERY 8.6:

As stated in the introduction to this chapter, Wall Street lore suggests that a headquarters move tends to be an ominous sign. A *WSJ* survey of 10 companies that moved between 1986 and the end of 1988 found that the price of stocks for five companies beat the market and the other five underperformed. What size sample would we have to draw to estimate the proportion of companies whose stock fell after the move to within plus or minus 0.06 of the true proportion, with 90 percent reliability?

ANSWER:

From the *WSJ*'s minisurvey, our best guess of p prior to sampling is 0.5. The margin of error is ±0.06 and α is 10 percent. Thus

$$n = \left[\frac{z_{\alpha/2}}{M}\right]^2 p^*\left(1 - p^*\right) = \left[\frac{1.645}{0.060}\right]^2 (.5)(.5) = 187.9$$

We must therefore consider the stock prices of a random sample of 188 firms that moved headquarters if we are to be 90 percent confident that our estimate is within 0.06 of the true proportion whose stock price fell after the move.

Almost every statistics textbook discusses formulas similar to Equations (8.18) and (8.20) for determining sample size. Yet, Lawrence Kupper and Kerry Hafner (1989) have questioned these formulas in small-sample situations. Kupper and Hafner conclude that these confidence-interval-based formulas for determining sample size "behave so poorly in all instances that their future use should be strongly discouraged." (*Am Stat,* May 1989 p. 101) Unfortunately, without more information about the population, these formulas may be the only available indication of minimum sample requirements. As demonstrated in Appendix 9D in the next chapter, and as supported by the work of Kupper and Hafner, better sample selection rules can be determined if we have information on possible alternative values of the population parameters.

EXERCISES

8.27 A computer chip manufacturer produces large quantities of ROM chips to control engine emission. The production process has been running with a 30 percent defect rate. To check if a new method of production is generating fewer defects, the manufacturer plans to draw a random sample of chips. The company wants to be 95 percent confident that the estimated defect rate is within 0.05 of the true rate. How large a sample should be drawn?

8.28 A student in statistics determines that an adequate sample size to estimate a population mean dollar earnings is 100.

a. If she had set $\alpha = 0.05$, but now decides that $\alpha = 0.01$ is required, how is the sample size affected?

b. If she learns that the population standard deviation is twice as large as what she assumed originally, what happens to the desired sample size? (Ignore the changes in part a.)

c. If she originally was working with a margin of error of \$125 but now decides that a margin of error of \$200 is sufficient, how is the desired sample size influenced? (Again, ignore the changes in parts a and b.)

8.29 According to a *Wall Street Journal* (April 1, 1983) article, Americans have voluntarily contributed to reducing the \$1.2 trillion public debt with checks totaling \$135,575.15. "'Nearly half of the contributions are for less than \$10,' says IRS spokeswoman Ellen Murphy, but 'a very generous soul' forwarded \$25,000."

a. How large a sample size would be required to re-estimate the proportion of those who gave less than $10, with 90 percent confidence that you are within 3 percent of the true proportion giving less than $10? (Specify the assumptions you must make.)
b. How large a sample size would be required to re-estimate the average contribution of those who gave, with 90 percent confidence that you are within $3? (Specify the assumptions you must make.)
c. Why do your answers in parts a and b differ?

8.30 The owner of a large auto repair service claims that 80 to 90 percent of customers believe his mechanics do an excellent job. You wish to estimate the proportion of customers who actually hold this belief to within 0.05, with 90 percent confidence. How large a sample is needed?

8.31 The average daily withdrawals from new savings accounts are to be estimated. You know that the maximum withdrawal from these accounts is $2,400. You want to be 90 percent confident that your estimate is accurate to within $55. How many accounts should you sample?

8.11 Estimating the Population Median by Iteration (Optional)

Although we can rely on the Central Limit Theorem to provide normality to the sampling distribution of the mean, our estimator of the population mean, in some situations we may not want to use the mean as our measure of central tendency. For instance, if we know the population is highly skewed, then an estimate of the population median may be a better measure of central location. Cases in which a few extreme observations may skew the distribution include:

1. Individual incomes, where a few high incomes pull the mean up.
2. Time to deliver a letter, where a few letters take a very long time because of loss or misdirection.
3. Hours worked at a fast food restaurant, where a few employees work only a couple of hours a week and, thus, lower the mean.

The sample median is an unbiased estimator of the population median. A confidence interval based on the sample median can be constructed as an interval estimator of the population median. The median is the 50th percentile. It is associated with $\pi = 0.50$. Thus, for large sample sizes, confidence limits for the population median can be based on the normality of the sample proportion, where the expected value of p is 0.50. Unlike the confidence interval for the population proportion, however, an initial confidence level is set to estimate the interval, but then the confidence level is itself estimated based on the estimated interval. The confidence level is thus itself estimated as the second step in an iterative process. This procedure is possibly more easily understood with an example.

We want to estimate the central location of the population lease price of imported cars. A sample of 36 cars and their lease prices was obtained, Table 8.1. These prices suggest a skewed population price distribution because a few imports are extremely expensive, pulling up the mean. Because of this skewness, we choose to use the median as our measure of central tendency.

The sample median in this case is a value between the observations holding the 18th and 19th rank order. After ordering the observations from smallest to largest, as

Table 8.1
Import Automobile Lease Prices

Imports	Amount
Acura Integra	$146.00
Acura Legend	$248.00
Audi 80	$273.00
BMW 325	$279.00
BMW 528	$333.00
BMW 735	$679.00
Honda Accord Coupe	$155.00
Honda Accord	$149.00
Honda Civic	$120.00
Honda CRX	$131.00
Honda Prelude	$170.00
Isuzu Trooper	$179.00
Jaguar XJ-S	$658.00
Jaguar XJ6	$579.00
Mazda MX6 or 626	$152.00
Mazda 929	$258.00
Mazda RX-7	$169.00
Mercedes 190E	$361.00
Mercedes 300E	$539.00
Mercedes 420SEL	$586.00
Mercedes 560	$749.00
Nissan 300ZX	$288.00
Porsche 911	$519.00
Porsche 928	$888.00
Porsche 944	$399.00
Saab 900	$205.00
Saab 9000	$318.00
Saab Convertible	$388.00
Sterling 825S	$277.00
Toyota Camry	$177.00
Toyota Celica	$138.00
Toyota Cressida	$239.00
Toyota Supra	$289.00
Volvo 240DL	$210.00
Volvo 740	$276.00
Volvo 760	$358.00

Approximate Sample Median = 276.5

120 131 138 146 149 152 155 169 170 177 179 205 210 239 248 258 273 276
277 279 288 289 318 333 358 361 388 399 519 539 579 586 658 679 749 888

shown at the bottom of Table 8.1, we can see that the sample median is a value between \$276 and \$277. Such a value is \$276.50, which we use as a point estimate of the population median.

To get the lower and upper confidence limits, we must set a target confidence level. As with the mean, the target confidence level is typically set at 90, 95, or 99 percent. Here we will use a 95 percent target level, so $z = 1.96$. The rank order of the lower confidence limit is now given by

$$L = \left(n - z_{\alpha/2}\sqrt{n}\right)/2 = \left(36 - 1.96\sqrt{36}\right)/2 = 12.12$$

After rounding up to the nearest whole number, this is the 13th rank-ordered value, or \$210. The rank order of the upper limit is calculated to be

$$U = n - L + 1 = 36 - 13 + 1 = 24$$

which is \$333. The interval estimate of the population median is therefore

$$\$210 < \text{population median} < \$333$$

Because of the rounding and approximations involved in the entire process, the actual confidence level of this interval is not necessarily 95 percent. To approximate the true confidence level, we must recompute the probability by calculating

$$z = \frac{2(L) - n - 1}{\sqrt{n}} = \frac{2(13) - 36 - 1}{\sqrt{36}} = -1.833$$

The $Prob(z < -1.833) = 0.0336$. Thus, the confidence interval

$$\$210 < \text{population median} < \$333$$

is best described as a 93.28 percent confidence interval for the population median lease price for imported cars, where the 93.28 percent figure was obtained by $1 - 2(.0336)$.

8.12 Estimating the Population Median by the Bootstrap (Optional)

An alternative to the interactive estimation procedure of Table 8.1 is to construct the confidence interval for the median via resampling, as introduced in the previous chapter and made possible with programs such as RESAMPLING STATS. After selecting a simple random sample of size n, this program will automatically construct a hypothetical population that has exactly the same relative frequencies as values in your sample. Conceptually, this is done by copying each value in the sample, say a million times, and then mixing n million values thoroughly. Next, the program goes back and repeatedly draws simple random samples (each of size n) from this hypothetical population. Finally, the sampling distribution of the median is constructed, from which the $(1 - \alpha)$ percent confidence interval is determined by the $\alpha/2$ percentile and the $[1 - (\alpha/2)]$ percentile.

TABLE 8.2 The Bootstrap Procedure for the Confidence Interval of a Median

RESAMPLING STATS commands to bootstrap an 80 percent confidence interval for the median market value (based on Table 7.2 sample):

COPY (1523 7583 1592 772 17449 1092 819 60954 3183 6769)A	To enter the 10 market values in A.
MEDIAN A MEDA	Find the median and call it meda.
REPEAT 1000	Draw 1,000 samples.
SAMPLE 10 A AA	Keep results of 1,000 samples, each of size $n = 10$, in AA.
MEDIAN AA MED$	Find median of each sample.
SCORE MED$ Z	Put results of medians in Z.
END	Stop loop.
HISTOGRAM Z	Make histogram of medians.
PERCENTILE Z (10 90) K	Put 10th and 90th percentile of histogram of medians in K.
PRINT MEDA K	Print the initial sample median and the 10th and 90th percentile of the histogram of medians.

RESAMPLING STATS results:

Bin Center	Freq	Pct	Cum Pct
1000	66	6.6	6.6
1500	334	33.4	40.0
2000	13	1.3	41.3
2500	155	15.5	56.8
3000	94	9.4	66.2
4000	63	6.3	72.5
4500	15	1.5	74.0
5000	82	8.2	82.2
5500	24	2.4	84.6
7000	104	10.4	95.0
7500	24	2.4	97.4
9500	3	0.3	97.7
10500	3	0.3	98.0
12000	2	0.2	98.2
12500	12	1.2	99.4
17500	6	0.6	100.0
	1000		

MEDA = 2387.5 ← Initial sample median

K = 1307.5 6769 ← 10th and 90th percentiles of histogram of the medians

The (1 – α) confidence interval estimator obtained from resampling is called the bootstrap estimator—after the saying about pulling one's self up by one's bootstraps. In the statistical bootstrap, it is the distribution of a statistic that is being pulled or inferred from a sample.

The bootstrap is a fairly new nonparametric method, introduced and developed by Julian Simon and Bradley Efron in the late 1960s and early 1970s. The basic idea is that the relative frequency distribution of a simple random sample is itself an estimator of the relative frequency distribution of the population. As such, a simple random sample can be "blown up" to an (infinitely) large population from which resampling takes place. A sampling distribution can then be constructed; it is no longer just a theoretical construct.

Table 8.2 provides the RESAMPLING STATS commands and results from the bootstrap estimation of an 80 percent confidence interval of the median market value, starting from the 10 sample companies in Table 7.2. The resulting end values of the 80 percent confidence for the median market value are shown at the bottom of Table 8.2. First, notice that the median market value of the sample in Table 7.2 is 2387.5 million dollars, which is not the center of the 1307.5- to 6769.0-million-dollar bootstrap confidence interval in Table 8.2; the bootstrap is based on the estimated sampling distribution of the median and not the median in one sample. Second, the bootstrap 80 percent confidence interval for the median includes the true population median, which from *Business Week* is known to be 1390 million dollars.

Computer programs such as RESAMPLING STATS make the construction of confidence intervals for medians easy, although the idea of resampling is not universally accepted by statisticians. It is thus presented here as an option.

8.13 Concluding Comments

This chapter introduced the notion of a model for the estimation of the population mean, proportion, and median. In essence, it was the first step in inferring population parameters with sample values. The next two chapters deal with testing specific parameter values based on these sample measures. Later chapters will return to the idea of prediction when regression analysis is considered. Unlike armchair predictions based on hunches, all our work with estimation and hypothesis testing relies on probability theory and sample data.

Chapter Exercises

8.32 Is the t or z distribution more appropriate for calculating probabilities associated with the sample mean under the following conditions?

a. A small sample from a skewed population with known standard deviation

b. A small sample from a normal population with known standard deviation

c. A small sample from a skewed population with an unknown standard deviation

d. A small sample from a normal population with an unknown standard deviation

e. A large sample from a skewed population with an unknown standard deviation
f. A large sample from a skewed population with known standard deviation

8.33 Individual A constructed a 90 percent confidence interval for the proportion of graduates who do not get jobs within a year of graduation and reported the interval to be $0.15 < \pi < 0.19$. Using the same data, individual B constructed a 95 percent confidence interval and reported the interval to be $0.16 < \pi < 0.18$. Could both individual A and individual B be right? Explain.

8.34 A study in *Parade* (June 10, 1984) magazine reported that the median pay in the United States was $18,700. Why might this be a better point estimate of central location than the mean pay level?

8.35 A *Wall Street Journal* (February 4, 1992) article reported on a survey of 466 corporate-benefit managers by consultant Towers Perrin. The survey showed "77% support tax credits for people who buy their own health insurance." What is the margin of error associated with a 95 percent confidence interval of the true proportion who favor such tax credits? How wide would the 95 percent confidence interval be?

8.36 An article in *The Chronicle of Higher Education* (December 1984) reported on a survey of faculty members showing that their average yearly salary was $29,700. The article stated "the 'confidence interval' for that figure is $410, meaning that there is a 95 percent probability that the estimated average varies from the true average of the sample population by no more than plus or minus $410." Does this statement provide a correct interpretation of a confidence interval? Explain.

8.37 A Scripps Howard News Service (*HT*, February 9, 1992) release reported on a test carried out by the Insurance Institute for Highway Safety. The test showed that at five miles per hour, crashes of 24 cars resulted in repair bills ranging from $769 to $3,864. To be 95 percent confident that your estimated mean is within $200 of the true average cost of repair, how large a sample need be drawn?

8.38 An Associated Press (*HT*, February 11, 1992) release stated, "A leading test critic accused the makers of the Scholastic Aptitude Test in January of 'covering up' evidence that coaching can improve scores by an average of 100 points or more." The article went on to say that coaching is big business since "at least 100,000 students pay $500 or more each year to commercial coaching firms." (Each section of the two-part test is scored on a scale of 200 to 800.) To be 95 percent confident that your estimate of the mean improvement is within 15 points of the true mean, how large a sample is needed to estimate the improvement in test scores?

8.39 A note in *The Wall Street Journal* (February 18, 1992) describes a National Car Rental system that allows customers to rent and return cars with a machine similar to an automatic teller. But the system does not provide travelers with information on mileage. To estimate the mileage on National cars, you walk through one of its large lots and record the following mileage on a random sample of nine cars.

9,876	11,346	12,321
8,564	12,590	3,210
7,487	8,886	9,389

a. What is your point estimate of the average mileage on National cars?
b. What is a 95 percent confidence interval estimate of the average mileage on National cars?
c. What does your answer in part b tell you?

8.40 In economic recessions, small businesses tend to suffer because accounts receivable rise as the time between billing and payment increases. For a small plumbing and heating supply firm, the number of days since billing for 25 sample accounts receivable is:

42, 59, 30, 35, 40, 39, 101, 55, 92, 55, 81, 37, 73, 87, 12, 55, 63, 23, 72, 66, 71, 41, 20, 31, and 82.

a. What is the point estimate of the average days to payment?
b. What is the 90 percent interval for the average days to payment?
c. How are your estimators in parts a and b "biased" because of bills that are never paid?

8.41 Below is a MINITAB computer printout of descriptive statistics for a sample of the starting salaries of 31 graduating seniors. What is the margin of error? What is the value of the *t* statistic?

	N	MEAN	STDEV	SE MEAN	95.0 PERCENT C.I.
C4	31	24342.0	3246.0	583.0	(23151.5, 25532.5)

8.42 An article in *The Wall Street Journal* (February 24, 1992) stated that "for Biotech (companies), pure genius isn't enough." It said that companies had to be adequately capitalized. Below are stock market capitalizations (in millions of dollars) for a sample of biotech companies. Use a computer program to calculate a 95 percent confidence interval for the average capitalization of all biotech firms.

Amgen	$8,309.7
Genentech	3,080.2
Chiron	1,509.6
Centocor	1,296.8
Synergen	1,205.4
Genzyme	935.9
Biogen	879.8
Gensia	860.1
Immunex	791.8
Alliance	644.0
US Bioscience	570.8
Zoma	489.5
Medimmune	448.8
Affymax	409.6
Immune Response	395.2
Genetics Institute	379.6
Molecular Biosystems	373.8
Cytogen	352.3
Immunomedics	337.1
Liposome Co.	334.9

8.43 An article in *The Wall Street Journal* (February 3, 1992), titled "Depressed Job Market Prompts Paradox: Upsurge of Interest in M.B.A. Programs," reported on the number of applications to 10 programs: Dartmouth, Harvard, Stanford, U. of Pennsylvania, MIT, Columbia, Northwestern, U. of Virginia, U. of Chicago, and UCLA, for the past 11 years.

Year	Applications for fall of identified school year, in thousands
1981	32.578
1982	30.498
1983	28.916
1984	25.701
1985	28.365
1986	27.686

1987	31.569
1988	33.522
1989	33.357
1990	35.944
1991	36.531

a. What is the average number of applications for these 11 years at these 10 schools?
b. Why, in this case, is the average a biased estimator of the average number of applications to MBA programs?

8.44 The average price of Canadian fir, the lumber industry pricing benchmark, is to be estimated to within \$10 with 95 percent confidence. From *The Wall Street Journal* (February 27, 1992), you learn that this fir has "fetched from \$164 to \$206 per 1,000 board feet." How many contracts should you sample to estimate this mean with the desired accuracy?

8.45 Writing on "Confidence Intervals Rather Than P Values: Estimation Rather Than Hypothesis Test," *British Medical Journal* (March 15, 1986), Martin Gardner and Douglas Altman provided the following interpretation of a 95 percent confidence interval for the difference in means:

> Put simply, this means that there is a 95% chance that the indicated range includes the 'population' difference in mean blood pressure levels—that is, the value which would be obtained by including the total population of diabetics and non-diabetics at which the study is aimed. (p. 747)

Comment on the correctness of this interpretation.

8.46 An article in *The Wall Street Journal* (April 14, 1992) reported, "a study of 515 companies acquired in recent years found that 65% of their CEOs had departed, mostly in the first year after the transaction." Form a 95 percent confidence interval for the mean proportion of CEOs who depart after a takeover.

8.47 An article in *Business Week* (May 11, 1992) on mail-order computers, quoted Michael S. Dell, CEO of Dell Computer, saying "Our people handle an average of 40 to 60 calls a day." How large a sample is needed to estimate the mean number of calls Dell's people are handling to within plus or minus two calls, with 99 percent confidence?

8.48 Eric von Hippel, an MIT professor, and Cornelius Herstatt, of the Swiss Federal Institute of Technology, in their monograph "An Implementation of the Lead User Market Research Method In a 'Low-Tech' Product Area: Pipe Hangers" (Alfred P. Sloan School of Management, MIT, 1992), describe their work with Hilti AG, a leading European manufacturer of fastening, hanging, and shelving related products. To test whether designers were too far ahead of what customers wanted, the authors tested the designers' ideas on a sample of 12 regular customers: 10 preferred the new product and said they would pay up to 20 percent more for it.
a. Using Formula (8.17), construct a 95 percent confidence interval for the population proportion π of customers that would prefer the new product and be willing to pay up to 20 percent more for it.
b. What is wrong with this interval estimate in part a, and why did you get these strange results?

8.49 A *Business Week*/Harris Poll (June 15, 1992) of 400 female executives at companies with \$100 million or more in sales found that 27 percent had been harassed. Give a 95 percent confidence interval for the proportion of female executives who have been harassed.

8.50 According to the U.S. Bureau of Labor Statistics, *Occupational Outlook Handbook* (April 1990), "When listing property for sale, agents and brokers make comparisons with similar property being sold to determine its fair market value . . . parties to the transaction usually seek the advice of real estate appraisers, objective experts who do not have a vested interest in the property. . . . An appraisal is an unbiased estimate of the quality, value, and best use of specific property."(p. 234) Compare and contrast the meaning of the word "unbiased" as used in this quote versus its use in statistical estimation.

8.51 A headline in *The Wall Street Journal* (June 12, 1992) read "Sears Is Accused of Billing Fraud At Auto Centers." The article reports on a year-long study by the California Department of Consumer Affairs in which "Undercover agents took 38 cars with worn brakes but no other mechanical defects to 27 Sears automotive centers throughout California. . . . In 34 instances . . . agents were told of other defects. . . ." Construct a 95 percent confidence interval for the true proportion of cars for which nonexistent defects could be expected.

8.52 According to a *WSJ* (July 14, 1992) article, "Only 7% of 434 companies surveyed by the International Foundation of Employee Benefit Plans have a 'very complete' understanding of the new Americans With Disabilities Act." What is a point estimate of the population proportion of companies that have a very complete understanding of this act and a point estimate of the standard deviation of the distribution of the sample proportion of companies that have a very complete understanding of this act?

8.53 A *WSJ* (August 11, 1992) article reported that "of 104 personnel managers surveyed by Bureau of National Affairs, a newsletter firm, 76% worry about possible lawsuits under the new disability law." What is a point estimate of the population proportion (and its standard deviation) of personnel managers who are worried about lawsuits under this law?

8.54 An IRS agent questions the daily receipts reported by Healthy House Foods. Healthy claims receipts averaging \$517.40 per day. A random sample of receipts for five days yielded totals of \$558, \$501, \$478, \$490, and \$579.

a. What is the probability of getting this sample average, or larger value, if Healthy's claim is true?

b. What assumption is needed to do the probability calculation in part a?

c. Should the IRS continue to pursue Healthy House Foods based on this sample information? (Give reasons.)

8.55 A machine that is set to produce two-inch bolts is known to turn them out with deviations in length that appear to follow the normal distribution. The process is monitored with a sample of 10 bolts selected at random every 10 minutes. The last lot showed the following lengths:

Sample	Length	Sample	Length
1	1.99992	6	2.00021
2	2.00014	7	2.00007
3	2.00001	8	2.00014
4	2.00010	9	1.99996
5	1.99995	10	2.00042

a. What are the mean and standard deviation of this sample?

b. What is the probability of getting a mean as large or larger than the one calculated in part a, if the process is in control?

c. Based on your answer to part a, does it appear the machine is working properly?

8.56 A simple random sample of size 25 is drawn from a normal population of weights, with $\sigma = 15$ pounds. A confidence interval for the mean weight is constructed and found to have a width from end to end of 11.76 pounds. What was the level of confidence?

8.57 A simple random sample of size 9 is drawn from a population of hourly wages, with an unknown variance. A 98 percent confidence interval for the mean is constructed and found to have a margin of error of ±\$9.67/hr.

a. What was the standard error of the mean?

b. What was the sample's standard deviation?

8.58 In a survey of 105 female judges, 54 percent said sexual bias was widespread in California state courts, according to *The Wall Street Journal* (August 20, 1992). Construct a 90 percent confidence interval for the true proportion of female judges who think sexual bias is widespread in California state courts. Does this interval include the majority of judges?

8.59 *Newsweek* (September 28, 1992), in an article on polling, said:

> To be fair, polls that at first blush look contradictory may not be. A *Miami Herald* poll last week had George Bush ahead of Bill Clinton by 7 points (48–41) in Florida; in a *New York Times*/CBS News poll, Bush trailed (42–48). But since the polls had margins of error of about 4 points, they overlapped (add and subtract 4 from each number to get the possible range). The margin of error, calculated according to a textbook statistical formula, varies inversely with the sample size. In general, about 500 responses gives a possible error of 5 percent either way, 2,500 responses decreases it to 1 percent. (p. 38)

Comment on the correctness of this quote.

8.60 Exercise 7.38 describes a Lanier TV advertisement on CNN (April 20, 1993) that stated "Lanier copiers are up and running 98 percent of the time," and the results of a status check of a Lanier copier on thirty randomly selected days in which the copier was "up and running" 29 times. Why wouldn't it be appropriate to calculate a 99 percent confidence interval (based on the z distribution) for the true population proportion of up time in this problem?

8.61 A *Wall Street Journal*/NBC News poll (January 29, 1993) of the views of 1,009 adults toward President Clinton included a statement that "Chances are 19 of 20 that if all adults with telephones in the U.S. had been surveyed, the findings would differ from these poll results by no more than 3.1 percentage points in either direction." Is this wording a correct representation of the implications of the margin of error? Explain.

8.62 The statement accompanying a *Wall Street Journal* (February 19, 1993) report on a Roper Organization survey of 500 black business owners said: "According to Roper, the odds are 19 to 20 that if the survey of entrepreneurs had been expanded to include the entire SBA list of black-owned businesses, with annual revenue of at least $100,000, the findings would differ from these poll results by no more than five percentage points in either direction." Comment on the correctness of this interpretation of the margin of error.

8.63 (Requires Endnotes) In the March 1991 issue of the *Journal of the American Statistical Association,* there was an article on "Bayesian and Frequentist Predictive Inference for the Patterns of Care Studies." A year earlier, in the March 1990 issue, there was an article on "Approaches to Bayesian Confidence Intervals." What distinction between classical (frequentist) and Bayesian estimation is called to your attention by these article titles?

Academic References

Fisher, Ronald A. *Statistical Methods for Research Workers,* first edition. New York: Hafner Publishing Company, Inc., 1925.

Kennedy, W.I., and J.E. Gentile. *Statistical Computing.* New York: Marcel Dekker, 1980.

Kupper, Lawrence, and Kerry Hafner. "How Appropriate are Population Sample Size Formulas?" *American Statistician.* May 1989, pp. 101–105.

Gosset, William (under pseudonym Student). "The Probable Error of a Mean." *Biometrica* VI, 1908, pp. 1–25.

Paulson, Edward. "An Approximate Normalization of the Analysis of Variance Distribution." *Annals of Mathematical Statistics* 13, 1942, pp. 233–235.

Zehna, Peter. "On Proving That $\bar{x}$ and s^2 Are Independent." *American Statistician,* May 1991, pp. 121–122.

Endnotes

1. To demonstrate that the expected value of the sample mean $\bar{x}$ is μ we have

$$E(\bar{x}) = En^{-1}\sum_{i=1}^{n} x_i = n^{-1}\sum_{i=1}^{n} Ex_i = n^{-1}n\mu = \mu$$

2. A natural estimator of σ^2 to consider is

$$s_*^2 = n^{-1}\sum_{i=1}^{n}(x_i - \bar{x})^2$$

But this is a biased estimator of σ^2. The bias can be demonstrated by deriving the expected value s_*^2.

$$s_*^2 = n^{-1}\sum_{i=1}^{n}(x_i - \mu)^2 - (\bar{x} - \mu)^2$$

Now to determine the expected values of the right-hand terms, we have

$$E(s_*^2) = n^{-1}\sum_{i=1}^{n} E(x_i - \mu)^2 - E(\bar{x} - \mu)^2$$

$$= n^{-1}n\sigma^2 - E(\bar{x} - \mu)^2 = \sigma^2 - E(\bar{x} - \mu)^2 \neq \sigma$$

To get an unbiased estimator, we need to determine $E(\bar{x} - \mu)^2$, which is the variance of the sampling distribution of $\bar{x}$.

$$E(\bar{x} - \mu)^2 = E\left[n^{-1}\sum_{i=1}^{n} x_i - \mu\right]^2 = n^{-2}E\left[\sum_{i=1}^{n}(x_i - \mu)\right]^2$$

$$= n^{-2}E\left[\sum_{i=1}^{n}(x_i - \mu)^2 + \sum_{i=1_{i \neq j}}^{n}\sum_{j=1}^{n}(x_i - \mu)(x_j - \mu)\right]$$

$$= n^{-2}\left[\sum_{i=1}^{n} E(x_i - \mu)^2 + \sum_{i=1}^{n}\sum_{j=1}^{n} E(x_i - \mu)(x_j - \mu)\right]$$

$$= n^{-2}(n\sigma^2)$$

$$= \sigma^2/n$$

The variance of x

As long as x's are independent, the expected value of a crossproduct is zero

We now know that $E(\bar{x} - \mu)^2$ is σ^2/n. Thus,

$$E(s_*^2) = n^{-1}n\sigma^2 - E(\bar{x} - \mu)^2 = \sigma^2 - E(\bar{x} - \mu)^2 = \sigma^2 - \sigma^2/n$$

Therefore,

$$E\left[ns_*^2/(n-1)\right] = \sigma^2$$

Since s_*^2 is $n^{-1}\sum_{i=1}^{n}\left(x-\bar{x}\right)^2$, an unbiased estimator of σ^2 is

$$s^2 = \left(n-1\right)^{-1}\sum_{i=1}^{n}\left(x_i-\bar{x}\right)^2$$

3. For this derivation, see the latter part of endnote 2.
4. In contrast with the classical view in which the population parameter is fixed and its estimator is a random variable, Bayesians treat the population parameter as if it is the random variable and the estimator is fixed. Bayesians argue that since the calculated sample statistic is known with certainty, a distribution should be assumed for what is unknown—the population parameter. Thus, Bayesians can calculate the probability that the population parameter is within a specific numerical interval. The Bayesian framework for estimation and testing is considered in more detail in the next chapter, in optional section 9.13.
5. "Student's t" is defined by the ratio of a normal random variable to the square root of an independently distributed chi-square random variable divided by its degrees of freedom. For this statistic, $= \bar{x} = \Sigma x/n$ and $s^2 = \Sigma(x-\bar{x})^2/(n-1)$ must be independent and $(n-1)s^2/\sigma^2$ must have a chi-square distribution with $n-1$ degrees of freedom. This relationship is demonstrated by Peter Zehna (1991).
6. For a binomial random variable x, the variance was given by

$$E(x-\pi n)^2 = \pi(1-\pi)n$$

which can be rewritten as

$$n^2E[(x/n)-\pi]^2 = \pi(1-\pi)n$$

Dividing both sides by n^2 yields the variance of p:

$$E[(x/n)-\pi]^2 = \pi(1-\pi)/n$$

FORMULAS

- $(1-\alpha)100\%$ confidence interval for μ, where σ is known:

$$\bar{x} - z_{\alpha/2}\left(\sigma/\sqrt{n}\right) < \mu < \bar{x} + z_{\alpha/2}\left(\sigma/\sqrt{n}\right)$$

where $z_{\alpha/2}$ is the value of the standard normal variable z that puts $\alpha/2$ percent in each tail of the distribution.

- $(1-\alpha)100\%$ confidence interval for μ, where σ is unknown:

$$\bar{x} - t_{\alpha/2}\left(s/\sqrt{n}\right) < \mu < \bar{x} + t_{\alpha/2}\left(s/\sqrt{n}\right)$$

where $t_{\alpha/2}$ is the value of t that puts $\alpha/2$ in each tail, and

$$s = \sqrt{\frac{\Sigma\left(x-\bar{x}\right)^2}{n-1}} \text{ and } \bar{x} = \frac{1}{n}\Sigma x_i$$

and the t distribution has $n-1$ degrees of freedom. For small samples, the population x must be normally distributed so $\bar{x}$ can be normally distributed.

- The t statistic

$$t = \frac{\bar{x} - \mu}{s/\sqrt{n}}$$

where

$$s = \sqrt{\frac{\sum\left(x - \bar{x}\right)^2}{n - 1}}$$

and

$$\bar{x} = \frac{1}{n}\sum x_i$$

For small samples, the population x must be normally distributed so $\bar{x}$ can be normally distributed.

- $(1 - \alpha)100\%$ approximate confidence interval for π:

$$p - z_{\alpha/2}\sqrt{p(1-p)/n} < \pi < p + z_{\alpha/2}\sqrt{p(1-p)/n}$$

where $z_{\alpha/2}$ is the value of the standard normal variable z that puts $\alpha/2$ percent in each tail of the distribution.

- Sample size for estimating μ, σ known

$$n = \left[\frac{z_{\alpha/2}\sigma}{M}\right]^2$$

where M is the margin of error.

- Sample size for estimating μ, σ unknown

$$n = \left[\frac{z_{\alpha/2}\sigma^*}{M}\right]^2$$

where M is the margin of error, and σ^* is an educated guess for σ, which might be given by

$$\sigma^* = \frac{\text{highest minus lowest observed value}}{5}$$

- Sample size for estimating π:

$$n = \left[\frac{z_{\alpha/2}}{M}\right]^2 p^*\left(1 - p^*\right)$$

where M is the margin of error, and p^* denotes an educated guess for p.

Chapter 9

Single Sample Hypothesis Testing

Who can refute a sneer?

William Paley
1743–1805

9.1 Introduction

In addition to parameter estimation, we want to test for specific parameter values in a model. We have already seen how a confidence interval can suggest a range of values that a parameter might assume, but what about the correctness of any one value? That is the question for this and the next two chapters.

A claim or an assumption about the value of a population parameter is called a **hypothesis.** Making a decision about the correctness of one hypothesis versus another is called **hypothesis testing.** For example,

Hypothesis
A claim or assumption about the value of a population parameter.

Hypothesis Testing
The process of judging which of two contradictory statements is correct.

- The task of filing a 1040 tax form is supposed to take a person, on average, a mere nine hours and five minutes, according to the Internal Revenue Service. Several professional tax preparers cited anecdotal evidence and personal experiences, in a *Wall Street Journal* article, that implies an average time much longer. A hypothesis test might shed light on this apparent contradiction.
- An article in *Popular Mechanics* reported that a major rubber company had found that the mean (alternator, fan, power steering, or air conditioning) accessory-drive belt lasts at least four years and 11 months, while mechanics typically use a rule of thumb based on an expected belt life failure before that point. A test could determine the reasonableness of the rubber company's claim versus that of the mechanics.
- An Associated Press release (January 3, 1991) stated that for the 14 states participating in Lotto America as of January 1991, Wisconsin residents had won the most jackpots (14 out of 51), but they also had bought the most tickets (31 percent of tickets purchased). Lottery spokesperson Bret Vorhees was quoted as emphasizing that "a person's residence has nothing to do with the chance of winning." A hypothesis test could be used to see if the proportion of Wisconsin winnings is significantly different from the 31 percent expected based on ticket sales.
- Andy Rooney on "60 Minutes" (May 2, 1993) asked: "How do manufacturers decide on how much to put in a package?" In the case of paper toweling, a marketing expert says 60 feet, "because that's the average amount used by a family of four between pay checks." A hypothesis test could assess whether or not the mean amount of paper toweling used by such families is 60 feet per two-week pay period.
- A *Wall Street Journal* (February 12, 1991) reported that the median housing price in Houston, Texas, declined from $77,430 in 1982 to $70,656 in 1990. A real estate agent can use a hypothesis test to see if the median is currently less than $70,656.

A decision about the correctness of one hypothesis versus another alternative hypothesis could be based on a hunch, an anecdote, the flip of a coin, or even the positions of the stars. Unlike decisions that are based on isolated incidents, the statistical hypothesis testing procedures demonstrated in this chapter require the use of sample information and decision rules founded in probability models that fit the situation.

This chapter considers business and economic decision-making situations in which, at least conceptually, repeat simple random sampling is possible, as in an experiment. The rules for choosing among alternatives will take into account the uncertainty inherent in such sampling. Decision rules that identify, control, and minimize the chance of errors in reaching a conclusion will be emphasized.

By the end of this chapter, you should be able to identify hypothesis testing situations, specify the hypotheses to be tested, use a probability based decision rule for rejecting or accepting a hypothesis based on observed sample data, and reach a conclusion. You will be able to test statements about means, proportions, and medians. If you complete the optional section, you will be able to articulate the difference between hypothesis testing in a classical versus a Bayesian framework. The appendices will enable the more advanced students to address the importance of sample size and power through the analysis of operating characteristic curves.

9.2 Hypotheses: An Illustration

Null Hypothesis
A statement about a population parameter that is assumed true until sufficient contrary sample evidence is presented.

Alternative Hypothesis
A statement about a population parameter that is being tested; it is true if the null hypothesis is false.

To assess the correctness of the rubber company's claim that the mean life of accessory-drive belts is at least four years and 11 months requires the specification of an alternative statement, which is supplied by the mechanics. Statistical hypothesis testing requires a pair of hypotheses: one which is initially assumed true, called the **null hypothesis,** and the other representing a challenging view, called the **alternative hypothesis.** If the rubber company is wrong, then the mechanics' experience would suggest that the average belt life is less than 59 months (4 years and 11 months).

The null and alternative hypotheses typically are written as mutually exclusive and exhaustive statements about the value of a parameter of a population model. As in Chapter 4, mutually exclusive means that both the null and alternative hypotheses cannot be accepted as true; exhaustive means that there are no other options. For the drive belts, any one of four outcomes is thus possible.

1. The mean time to failure is actually at least 59 months, and we conclude that this is so by not rejecting the null hypothesis. That is, we draw the correct inference.
2. The mean time to failure is actually at least 59 months, but we conclude that this is not the case by rejecting the null hypothesis. That is, we draw the wrong inference.
3. The mean time to failure is actually less than 59 months, and we reject the null hypothesis. That is, we draw the correct inference.
4. The mean time to failure is actually less than 59 months, but we do not reject the null hypothesis. That is, we draw the wrong inference.

Similarly, there are four possible outcomes in any other hypothesis testing situation.

Because the null and alternative hypotheses form a mutually exclusive and exhaustive set of options, a choice between them is either correct (options 1 or 3 above) or one of two errors is made (options 2 or 4 above). The null hypothesis may be rejected when it is true, as in case 2, or the null hypothesis may not be rejected when it is false, as in case 4.

Types and Cost of Errors

The chance factors in sampling can lead to the wrong conclusion in hypothesis testing. When this occurs, an error of inference is said to occur. As depicted in Table 9.1, the

Table 9.1
The Four Outcomes in Hypothesis Testing

Conclusion reached based on sample data		Population Situation: Null hypothesis is actually true	Population Situation: Null hypothesis is actually false
	Null hypothesis is rejected	Type I error	correct conclusion
	Null hypothesis is not rejected	correct conclusion	Type II error

Type I error is
rejecting the null hypothesis when it is true.

Type II error is
not rejecting the null hypothesis when it is false.

Type I Error
Rejecting the null hypothesis when it is true.

Type II Error
Not rejecting the null hypothesis when it is false.

Maintained Assumptions
A model and its related assumptions that are taken to be true during a hypothesis test.

Specification Error
A situation in which the maintained assumptions in a hypothesis test are, in fact, wrong.

first type of error of inference (rejecting the null when it is true) is called a **Type I error.** The second type of error (not rejecting the null when it is false) is called a **Type II error.** Later in this chapter, we will assign probabilities to these errors; for now, we need only keep in mind these two types of errors of inference that could occur whenever a decision is made.

Because hypothesis testing is based on sample information, an inference about the population requires a model of the population and the sampling distribution of the statistic used to estimate the population parameter of interest. The model and its related assumptions are called **maintained assumptions** because they are assumed true during the test. If these maintained hypotheses are wrong and lead to incorrect inferences about the population, then a **specification error** is said to have occurred.

We never want to make an error. The only way the possibility of an error could be eliminated, however, would be if we knew the population and its parameters with certainty, but then there would be no need for the test. Typically we have to assume a model of the population and estimate the parameters of the model. As long as sample information is used to make an inference about the population, errors are possible. It is the researcher's obligation to set up the test to minimize the likelihood of errors, given the budget available and the cost of the errors.

The costs or the seriousness of the two different types of errors of inference are not equal. If we conclude, for example, that the average life of a drive belt is at least four years and 11 months (and do not replace the belt earlier), when in fact the average life is much less, then an error could be costly. Relative to the cost of replacing a belt, a belt that breaks 50 miles from the nearest service station right before an important business meeting could be extremely costly. As discussed later in this chapter, ideally the probability of incurring such costs is minimized by the criterion or decision rule for choosing between the null and alternative hypotheses.

Stating the Hypotheses

The null hypothesis is typically denoted by the symbol H_O and the alternative hypothesis by H_A. In the case of the drive belt illustration, the null hypothesis is

$$H_O: \mu \geq 59$$

where μ is the population mean belt life (time before breakage) measured in months. This null hypothesis says that the population mean is greater than or equal to 59 months (four years and 11 months). Often along with the null hypothesis, there are other maintained assumptions that are not tested. For instance, if a spokesperson for the rubber company was quoted saying that "the average belt goes for four years and 11 months (or more), with a standard deviation of 3.333 months and almost all of the distribution ranging from just over four years to just under six," then the null hypothesis H_O: $\mu \geq 59$ says nothing about the observable range being from about 49 months to 69 months or the standard deviation being 3.333 months. This information about the dispersion of belt life is given or maintained during the test of the mean. These maintained assumptions are considered to be true during the test of the mean. Of course, if they are false, then the test of the mean may not be valid.

The alternative hypothesis, as suggested by the mechanics' statements, is that consumers should expect a belt life of less than 59 months:

$$H_A: \mu < 59$$

This alternative hypothesis says that the population mean is less than 59 months. It allows for the population mean to differ from the value in the null hypothesis in only one direction and is thus called a one-tailed alternative. The null and alternative hypotheses are mutually exclusive and exhaustive because the null hypothesis is consistent with any value that *is* at least 59 months and the alternative includes any value that *is not* at least 59 months.

The null hypothesis is assumed correct until sufficient sample information is obtained to refute it. Thus, the null hypothesis is the statement that is tested for possible rejection in favor of the alternative. By implication, the alternative hypothesis is accepted when the null is rejected. Because the truth of either hypothesis is never known with certainty, some statisticians prefer to say only that "H_O can (or cannot) be rejected" instead of saying that "H_A is (or is not) accepted."

In the belt life example, $\mu \geq 59$ is not rejected until sample information suggests that it can be rejected in favor of $\mu < 59$. If that sample information is not forthcoming, technically we should not say that $\mu \geq 59$ is accepted. It is more correct to say that $\mu \geq 59$ cannot be rejected. This subtle distinction between not rejecting the null hypothesis and accepting it is often overlooked in the statement of the conclusion.

The subtlety in wording in the statistical hypothesis testing procedure is similar to the legal process in the United States, where the defendant is assumed to be innocent until the evidence indicates guilt beyond a reasonable doubt. If the jury determines that the weight of the evidence is not sufficient to conclude guilt, the defendant is not proven innocent; he or she is said to be "not guilty" and "not convicted" or simply "freed."[1]

9.3 Decision Rule: An Illustration

A statistical hypothesis test requires sample information. In the case of the test of drive belts, time until failure might be obtained from a local taxicab company, a limousine service, or any other company that maintains fleets of motor vehicles. The National Association of Fleet Administrators also may be contacted for data, assuming there is no difference in belt life between vehicles used in fleets and those in individual use.

The sample we are going to use here consists of 36 drive belts that averaged 57.8 months before failure. This sample evidence ($\bar{x} = 57.8$) is used to decide whether H_O: $\mu \geq 59$ should be rejected in favor of H_A: $\mu < 59$. That is, we will try to establish if the sample result $\bar{x} = 57.8$ is sufficiently below $\mu = 59$ to be reasonably sure that the null hypothesis should be rejected.

p-Value

The difference between the observed $\bar{x}$ value of 57.8 and the smallest value of $\mu = 59$ in the null hypothesis H_O: $\mu \geq 59$ is associated with a probability; namely, the probability that a sample of size $n = 36$ results in a $\bar{x}$ value between 57.8 and 59, Figure 9.1, panel a. The bigger the probability of a $\bar{x}$ value between 57.8 and 59, the smaller the probability of one at or below 57.8. These probability calculations are the same type of probability determinations we did in Chapters 7 and 8. Now, however, the probability of obtaining a sample $\bar{x}$ value at or below 57.8, if $\mu = 59$, is called a **"*p*-value."** That is, the *p*-value is $P(\bar{x} \leq 57.8 \mid \mu = 59)$. As can be seen in Figure 9.1, panel a, the *p*-value is the area under the density function and below $\bar{x} = 57.8$. It is all in one tail because this is a one-tailed test.

_p_-Value
The probability that a test statistic assumes a value more extreme than the observed value, assuming the null hypothesis is true.

The *p*-value requires the calculation of probabilities for values of the estimator in the tail (or tails) of the sampling distribution.[2] In the drive belt example, the sampling distribution is centered at $\mu = 59$, which is called the "limiting value" in the null hypothesis because it represents the lower limit of the values $\mu \geq 59$. If the observed $\bar{x}$ value is below the limiting value in this null hypothesis, then the probability of values less than the observed $\bar{x}$ value (e.g., $\bar{x} < 57.8$) is the *p*-value. The *p*-value is the probability of values "beyond" or "more extreme than" the observed $\bar{x}$ value. If the observed $\bar{x}$ value is above the limiting value in H_O, then values "beyond" or "more extreme than" the observed $\bar{x}$ value are those values that are greater than the observed $\bar{x}$ value, and it is the probability of these values that is reflected in a *p*-value. (In Section 9.6, a *p*-value based on areas in both tails of the sampling distribution is introduced when "two-tailed tests" are considered. For now, we are dealing with *p*-values defined only for a one-tailed test.)

To calculate a *p*-value, the probability density function for the estimator must be known. If the population distribution of belt life is normal, with mean μ and variance

σ^2, then the sampling distribution of $\bar{x}$ will be normal and our model for $\bar{x}$ is

$$\bar{x} \sim N\left(\mu, \sigma/\sqrt{n}\right)$$

Alternatively, if the sample size is large, then from the Central Limit Theorem, $\bar{x}$ is known to be approximately normally distributed for simple random sampling.

Because H_0 is assumed true until contrary information is obtained, for the limiting value ($\mu = 59$) for which the null hypothesis is true, the expected value of $\bar{x}$ is 59, as shown in Figure 9.1. Assuming the population standard deviation is $\sigma = 3.333$, the standard deviation of the $\bar{x}$ distribution (also called the standard error of the mean, $\sigma_{\bar{x}}$) is

$$\sigma_{\bar{x}} = \sigma/\sqrt{n} = 3.333/\sqrt{36} = 0.5555$$

To calculate $P(\bar{x} \le 57.8 \mid \mu = 59)$, $\bar{x}$ can be standardized as in Chapter 7. Now, however, the calculated value of z is denoted z_c.

$$z_c = \frac{\bar{x} - \mu}{\sigma_{\bar{x}}} = \frac{\bar{x} - \mu}{\sigma/\sqrt{n}} = \frac{57.8 - 59.0}{3.333/\sqrt{36}} = -2.16$$

From Appendix Table A.3 or a computer program,

$$P(z \le -2.16) = 1 - 0.9846 = 0.0154$$

Thus, the p-value is 0.0154. There is a 1.54 percent chance of observing a mean of 57.8 or less, in a sample of 36 belts, when the expected mean is 59.0 and $\bar{x}$ is normally distributed with mean of 59 and standard error of 0.5555. That is, if the population model is true, as assumed, then in only 154 out of every 10,000 randomly selected samples, on the average, will the sample mean be less or equal to 57.8.

The p-value is a conditional probability that can be used to show how far the sample result is from that assumed in the null hypothesis. An unlikely sample result (an $\bar{x}$ value far away from the μ value in H_0) is signified by a small p-value. The observation of sample data that is highly unlikely casts doubt on the correctness of the null hypothesis. On the other hand, a large p-value is the result of the sample mean being "close to" the limiting value of μ in H_0 and thus does not cast doubt on the correctness of the null hypothesis.

Decision Rule

Determining what constitutes a small or large p-value requires reference to the probability of a Type I error, which will be discussed further in the next section. Clearly, though, extremely small p-values indicate a sample result far away from that expected from the null hypothesis. An extremely small p-value is strong evidence against the null hypothesis. A large p-value indicates that the sample results are close to that expected when the null hypothesis is true.

FIGURE 9.1 Calculating the p-value in the Drive Belt Life Illustration

Panel a: Hatched area is the p-value for testing $\mu = 59$ versus $\mu < 59$.

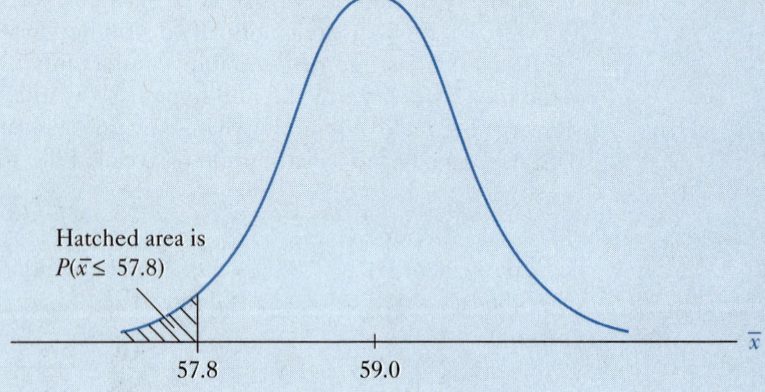

Panel b: Standardizing $\bar{x}$ gives the desired hatched area in the z distribution.

$$z_c = \frac{\bar{x} - \mu}{\sigma/\sqrt{n}} = \frac{57.8 - 59.0}{3.333/\sqrt{36}} = -2.16$$

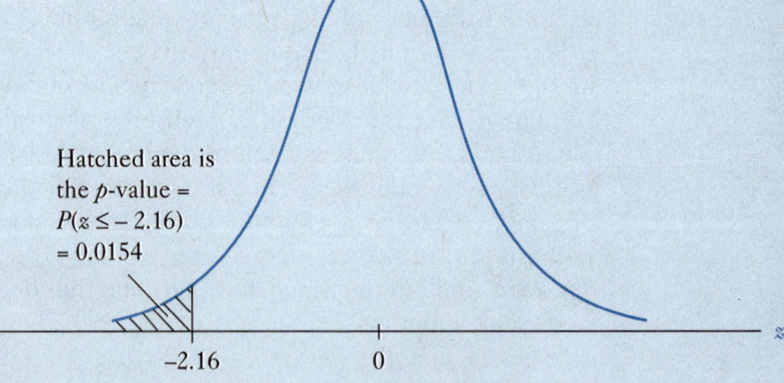

Recall from Chapter 7 that $P(z \le -2.16) = P(z \ge 2.16)$.
Using Appendix Table A.3 or a computer program,
$P(z \ge 2.16) = 1 - 0.9846 = 0.0154$.

Decision Rule for selecting between H_0 and H_A:

Reject the null hypothesis (H_0) if the p-value is small.
Do not reject the null hypothesis if the p-value is large.

As in a legal proceeding, where the defendant is assumed innocent unless sufficient evidence is presented for a guilty verdict, H_0 is assumed true unless there is sufficient sample evidence (as reflected in a small p-value) to reject H_0. In a court of law, sometimes a defendant that seems guilty is not convicted because of lack of evidence. If the defendant is guilty but incorrectly acquitted, then in statistical language a Type II error is made. A Type II error occurs in statistical testing when the p-value is too big to reject H_0 even though H_0 is false, although unknowingly so by the researcher. On the other hand, an innocent person may be found guilty—in hypothesis testing, there is always the possibility of incorrectly rejecting H_0 when it is true, and the Type I error is made. As with the law, in a statistical decision there is always a tradeoff between the cost and probability of making a Type II versus a Type I error.

The p-value of 0.0154 in the drive belt illustration would be a small probability in the eyes of most any jurist. Something that occurs with such a low probability (0.0154) is so unlikely that when it does occur, there is good reason to suspect that the null hypothesis is wrong. Replacing alternator, fan, power steering, and air conditioning drive belts more frequently than every four years and 11 months seems advisable since the average life of these belts may be less than 59 months. In reaching this conclusion, however, we must acknowledge that there is a chance we might be wrong. The population standard deviation might not be 3.333; the sampling process might be biased toward an understatement of the mean because it was drawn from fleet vehicles[3] or, because of the chance factors inherent in a random sample, the sampling error could have brought about an unlikely sample. The cost of being wrong, however, seems trivial compared to an error on the other side.

HISTORICAL PERSPECTIVE

Ronald A. Fisher (1890–1962), Jerzy Neyman (1894–1981), and Egon S. Pearson (1895–1980) are credited with codifying and advancing the objective science of statistical testing. This development was not accomplished in a spirit of cooperation or appreciation for alternative views; even to this day heated debate about the merits of classical hypothesis testing continues.

Sir Ronald Fisher was a determined, feisty, and combative scholar. He overcame his handicap of poor eyesight, developing the almost microscopic handwriting that was a mark of distinction in pre-word-processing scholarship. He was schooled in mathematics and physics, and made major contributions to genetics and eugenics. Yet, he denigrated theoretical mathematicians and stressed the need to avoid contact with those in the natural sciences who did not know statistics. Possibly because of his interest in applying statistics to the biological sciences, he was unable to appreciate developments of a statistical methodology within economics and the other social sciences.

Being human, Fisher was known for making some major errors in practical empirical matters. In the service of the tobacco industry, for example, Fisher erred greatly in assessing the evidence against cigarette smoking. Based in part on faulty calculations, Fisher ascribed the harmful effects of smoking exclusively to genetic disposition.

Fisher's approach to statistical testing emphasized observed phenomena and not explicit theoretical constructs; his arguments were developed through examples. To Fisher, the observation of highly unlikely sample evidence was sufficient to reject a null hypothesis, but observed likely sample evidence had no informational value. That is, sample evidence could lead only to the rejection of the null hypothesis. Researchers had to be concerned only about the probability of a Type I error, because the speculative nature of a Type II error made it irrelevant.

In contrast to Fisher's emphasis on the Type I error, statisticians Jerzy Neyman and Egon Pearson viewed testing as a contest in which either the null or alternative hypothesis was rejected at the expense of the other: If the null was rejected, then by implication the alternative was not rejected and vice versa. The Neyman-Pearson theory of hypothesis testing emphasized both the probabilities and costs of Type I and Type II errors. They sought tests that had the "power" to differentiate between the null and alternative hypotheses. To them, unlikely sample information supported the alternative hypothesis and likely sample information supported the null hypothesis. Fisher tried to dismiss Neyman and Pearson as mere mathematicians without experience in the natural sciences and without real scientific problem-solving skills.

The distinction between the approaches of Fisher and Neyman–Pearson are lost in most modern-day studies. The logic and efficacy of hypothesis testing in general, however, continues to be debated, as reflected in the exchange of Spielman (1974), Carlson (1976), and Spielman (1978) on "The Logic of Tests of Significance." Of particular interest is the philosophical importance of Spielman's assertion that the rejection of the null hypothesis is only justified, when a rare event has been observed, if it is also reasonable to believe that the significant outcomes are not rare when the null hypothesis is false. According to Spielman, for classical hypothesis testing to be logically consistent, it also must be true that prior to sampling and testing "it is not incredible that the (null) hypothesis is false." (For instance, if one was convinced that the null hypothesis was true—from past experience or divine intervention—how could one rare occurrence justify its rejection?) This prior subjective judgment, however, is viewed as contrary to the objective nature of the classical methods popularized by Fisher, his followers, and his critics.

EXERCISES

9.1 What is the meaning of a p-value in a hypothesis test in which the alternative hypothesis calls for a lower-tailed null hypothesis rejection region?

9.2 Why does a small p-value imply that the sample does not represent the population assumed in the null hypothesis?

9.3 Prior to the introduction of its "guaranteed 80,000 mile tire"(*HT* July 19, 1992), the engineering department at Bridgestone tire company had to demonstrate that it did indeed have a better tire that would last more than 80,000 miles, on the average.

a. Specify H_0 and H_A, and state the consequences of a wrong conclusion.

b. If you work for *Consumer Reports,* how would you prefer to specify H_0 and H_A? If you were a salesperson for the tire company, how might you prefer to specify the null and alternative hypotheses? Could the way the hypotheses are specified make a difference in the conclusion reached on the basis of sample information? Explain.

c. Let H_0: $\mu \leq 80{,}000$ and H_A: $\mu > 80{,}000$, and assume a sample of size $n = 36$ yields a mean of 76,380. Calculate a p-value, assuming the population standard deviation is 5,234.

d. What conclusion can you draw in part c at $\alpha = 0.01$?

9.4 If the statement on the side of a light bulb package is as follows:

Watts	Light Output (Lumens)	Expected Life (Hours)	Stand. Dev.
60	855	1000	50

How could the correctness of the expected life claim be tested?

9.5 A *Wall Street Journal* (July 1, 1991) on P&G development of a low-calorie fat for candy bars quoted Joan Steuer, president of Chocolate Marketing, Inc., a Los Angeles consulting firm, saying, "It's one of the last indulgences people have. The average U.S. chocolate consumption is 11 pounds per person a year." State the null and alternative hypotheses implied by this quote, and describe the nature of the Type I and Type II errors for a test of these hypotheses.

9.4 General Procedures for Hypothesis Testing

Statistical testing can involve many different types of hypotheses and different test statistics. The procedure for conducting statistical tests of hypotheses, however, follows roughly the same steps, regardless of what is tested. Although not explicitly identified in the drive belt example, the steps are as follows.

1. Specify the model and the null and alternative hypotheses.
2. Formulate a decision rule to reject or not reject the null hypothesis based on tolerable error levels.
3. Select a test statistic.
4. Collect the sample data.
5. Calculate the value(s) of the test statistic and the associated *p*-value.
6. Reach a conclusion (reject or do not reject H_0).

The next sections will provide a discussion of these steps in more detail than was provided in the introductory drive belt illustration. The drive belt illustration, however, provides an ideal example of this testing procedure in which the reader does not have to get bogged down in technical matters. What follows is a more general, yet more technical, discussion of the hypothesis testing procedure.

9.5 The Null and Alternative Hypotheses (Step 1)

The determination of the null and alternative hypotheses and the model in which the test will be performed is not always easy, for there may be several different ways of formulating the problem. In tests of the population mean μ based on the sample mean $\bar{x}$, the model typically will be

$$\bar{x} \sim \text{Normal}\left(\mu, \sigma_{\bar{x}}\right), \text{ for } \sigma_{\bar{x}} = \sigma/\sqrt{n}$$

where σ is the population standard deviation.

Justification for assuming that $\bar{x}$ is approximately normally distributed is based either on the Central Limit Theorem or information that the population is normally distributed. Other models and hypotheses also can be specified. As will be discussed later in this chapter, for example, a test of the population proportion μ based on the sample proportion $p = x/n$ can be based on the model[4]

$$p \sim \text{Approximately Normal}\ (\pi, \sigma_p),\ \text{for}\ \sigma_p = \sqrt{\pi(1-\pi)/n}$$

where now only the Central Limit Theorem can be used to justify this model since x is a binomial random variable. That is, the sample must be large ($n > 30$).

Even when such models are not explicitly identified with the null and alternative hypotheses, they are implicit, since a test of a population parameter cannot be done without an assumption about the distribution of the estimator. Models contain all of the assumptions that are maintained throughout the test. The appropriateness of these maintained assumptions will receive more attention in later chapters; our primary concern here is to acknowledge that there are implicit assumptions about the model that are required to test the null and alternative hypotheses.

Rules of thumb for specifying the null and alternative hypotheses are often based on the Fisherian idea that a new theory is required to dispute an older one. In accordance with this notion of testing, the null hypothesis will reflect the old theory and the new theory will be captured in the alternative. More generally, the alternative hypothesis contains the statement being tested.

For example, when Barbara A. Metz, a researcher at Battelle Memorial Institute in Columbus, Ohio, claimed to "have an antidote that could make tires last longer" (*Business Week*, September 28, 1992), the average life of the old tires goes in the null hypothesis and the alternative is that the average life of new tires exceeds the old. Unfortunately, not all testing situations are as clear-cut.

In claiming that they were underpaid, the female employees who brought suit against a large midwestern bank for sex discrimination in the 1980s would have found it advantageous to have the null hypothesis be that the average salary of women is less than that of men, since the null is assumed true until contrary evidence is provided. The bank, on the other hand, would want the null to be the reverse (average salary of women is greater than or equal to that of men).[5] As will be shown, and as discussed in the 1991 deposition of the author as an expert witness in this case, a statistical test of discrimination is possible if the latter is used but is not possible if the former is used.[6]

As another example, consider the debate over the use of Retin-A as a wrinkle-prevention ointment. (*Wall Street Journal*, June 12, 1991) Johnson & Johnson (the producer of Retin-A) was alleged to be promoting Retin-A for removing wrinkles (the null hypothesis). The U.S. Food and Drug Administration was not convinced and asked the U.S. Justice Department for a decision on whether to proceed with a civil lawsuit against the giant pharmaceutical company. FDA Commissioner David Kessler was quoted by the *WSJ* as saying, "We do not have data to assess the long-term safety of Retin-A for chronic use in wrinkles." The FDA wanted the null hypothesis to be that Retin-A does not work and the alternative hypothesis to be that it does. Although a pharmaceutical company might like the FDA to assume that a new treatment works until proved otherwise, the FDA insists that the null hypothesis be that it does not work with the alternative accepted only when sufficient sample data supports such a conclusion.

As an alternative to the rules of putting that which is new and challenging in the alternative, a Neyman-Pearson approach suggests that the null hypothesis should contain the statement that implies the highest cost if it is rejected wrongly. This is a good rule of thumb, and it may help shed light on the appropriate statement of the null and the alternative hypotheses in some situations. In the case of Retin-A, however, whose costs should be considered? The person with wrinkles? Johnson & Johnson's? The FDA's? Or society's? In the case of the women's suit against the bank for sex discrimination in salary determination, whose cost should be used to determine the null hypothesis? The women's? The bank's? Or society's? Specifying the null and alternative hypotheses on the basis of the cost of making an error clearly depends on whose costs are considered.

There are exceptions to the rules of thumb commonly used to determine the null and alternative hypotheses. The null hypothesis, however, can never be that the population parameter is not equal to a specific value. In tests of the population mean or population proportion, the null must contain an equal sign, a less than or equal to sign, or a greater than or equal to sign. To perform a statistical hypothesis test of the type considered here, the sampling distribution must be centered at a specific value; thus, the null hypothesis must always include equality with that value.

Formulating the Null Hypothesis

To test a population mean or a population proportion, the null hypothesis must contain an equal sign, a less than or equal to sign, or a greater than or equal to sign.

The null hypothesis can never be that the population parameter is not equal to a specific value, is only less than a specific value, or is only greater than a specific value.

One-Tailed Tests

The tax preparers cited in *The Wall Street Journal* article at the beginning of this chapter claimed the average time required to complete a tax form is higher than that claimed by the IRS. They argued that the Internal Revenue Service had understated the time required to find and check documents required to complete a tax form and that the average time for completion was greater than the IRS's nine hours and five minutes. Thus the null and alternative hypotheses are

$$H_O\text{: } \mu \leq 9.0833 \text{ and } H_A\text{: } \mu > 9.0833$$

Unlike the drive belt illustration, the alternative hypothesis in this tax illustration allows for the population average to be greater than that in the null hypothesis. The alternative hypothesis for belt life illustration was "one tailed," but it stated that the population average would be less than that given in the null hypothesis.

$$H_O\text{: } \mu \geq 59 \text{ and } H_A\text{: } \mu < 59$$

But like the belt life illustration, the test of the time to complete a tax form is a unidirectional test. Thus, they are both said to be "one-tailed" tests.

One-Tailed Hypothesis Test
A test in which the alternative hypothesis allows for departure from the value given by H_O in only one direction.

A **one-tailed hypothesis test** is appropriate whenever direction is indicated by the use of words such as "greater than," "less than," "inferior to," "superior to," "at least," or "at most." The alternative hypothesis in a one-tailed test will always contain either a < or > sign. The implied null hypothesis will then contain either a ≥ sign or a ≤. For example, if H_A: μ > 9.0833 is the alternative hypothesis, then the implied null hypothesis is the reverse statement H_O: μ ≤ 9.0833.

Although the null hypothesis H_O: μ ≤ 9.0833 includes all the values at or below 9.0833, only μ = 9.0833 need be used in a one-tailed test against the alternative μ > 9.0833. If μ = 9.0833 is rejected in favor of the alternative μ > 9.0833, then any value of μ < 9.0833 likewise would be rejected, regardless of its actual magnitude. Thus, the null and alternative in a one-tailed test are often written as H_O: μ = 9.0833 versus H_A: μ > 9.0833, where the equal sign in the null is understood to imply equal to or less than. It is the alternative hypothesis that establishes the test as one-tail.

Two-Tailed Tests

Two-Tailed Hypothesis Test
A test in which the alternative hypothesis allows for departure from the values given by H_O in either direction.

In contrast to one-tailed tests, in a **two-tailed hypothesis test** the alternative hypothesis allows for values on either side of a value for which equality is claimed in the null hypothesis. For example, in deciding on the amount of paper toweling to place on a roll, the marketing expert said 60 feet. This expert could be wrong if μ is greater than 60 or less than 60, μ ≠ 60. In the absence of any other information, however, we would have to assume he is correct, μ = 60. Putting the assumed equality in the null hypothesis now gives

$$H_O: \mu = 60 \text{ and } H_A: \mu \neq 60$$

In addition to means, we will be testing statements about the population proportion, π. As an example, consider the Lotto America situation in which the official claimed that a person's residence has nothing to do with the chances of winning. If this is true, then 31 percent of the jackpots can be expected to go to Wisconsin residents since that is the proportion of tickets they purchased. Any deviation in either direction from this percentage might be used to cast doubt on its validity. Thus, if π is the population proportion expected by chance, the null and alternative hypotheses are

$$H_O: \pi = 0.31 \text{ and } H_A: \pi \neq 0.31$$

If no direction is suggested, and departure from the value in the null hypothesis can go in either direction, then a two-tailed test is required. Often, in questions words such as "equal to," "similar to," and "no different from" are indications that a two-tailed test may be needed. For a two-tailed test, the null hypothesis will contain an equal sign and the alternative hypothesis will have a not-equal-to sign.

9.6 The Probabilities of Type I and II Errors (Step 2)

As already presented in Table 9.1, there are four outcomes associated with a choice between the null and alternative hypotheses; two lead to mistakes. A Type I error occurs if the null hypothesis (H_O) is rejected when it is true. A Type I error may also be stated as accepting the alternative hypothesis (H_A) when it is false. A Type II error occurs if the null hypothesis (H_O) is not rejected when it is false. A Type II error may also be stated as rejecting the alternative (H_A) when it is true. Unfortunately, since the true value of a population parameter is typically unknown, we never know for sure whether a conclusion based on sample information is correct or incorrect.

Before sampling, a decision rule should be established for determining when H_O is to be rejected and when it is not rejected. This decision rule is based on the probability of a Type I error versus the probability of a Type II error. The Greek letter **alpha,** (α) is used to represent the probability of making a Type I error and the letter **beta,** (β) represents the probability of a Type II error.

Alpha, α
The probability of making a Type I error.

Beta, β
The probability of making a Type II error.

Ideally both the probabilities of a Type I error and a Type II error would be made as small as possible in a hypothesis test. Values of α can be specified easily, but, as shown in two appendices to this chapter, determining β is an involved process. Because a decision rule for rejecting (or not rejecting) the null hypothesis can be formulated with only knowledge of the probability of a Type I error, many analysts rely on the fact that there is an inverse relationship between α and β and never attempt to actually determine β. This **α–β tradeoff** and related issues are also discussed in the appendices to this chapter.

α–β Tradeoff
Decreasing the probability of a Type I error (α) increases the probability of a Type II error (β), holding the sample size constant.

The researcher sets α in accordance with the cost of making a Type I error. If rejecting the null hypothesis when it is true would result in an extremely costly error, then α is set relatively small, at $\alpha = 0.01$, for example. (The α–β tradeoff implies, however, that β is then relatively large.) If rejecting the null hypothesis when it is true would result in little cost, then α is set relatively large, at $\alpha = 0.10$, for example. (The α–β tradeoff implies, however, that β is then relatively small.) The α–β tradeoff prevents the analyst from setting $\alpha = 0$: if α is decreased to zero, then β is increased to its maximum. Recall from our earlier discussions, however, that assignment of cost to Type I and Type II errors is not an easy task.

In the drive belt illustration, suppose that the probability of a Type I error had been preset at $\alpha = 0.05$. This **level of statistical significance** means that 5 percent of the time we are willing to be wrong when the null hypothesis H_O: $\mu \geq 59$ is rejected and the alternative H_A: $\mu < 59$ is accepted. This probability of a Type I error is shown in Figure 9.2. The shaded area that is 5% of the area under the density function is associated with the $\bar{x}$ values for which H_O will be rejected. The H_O rejection region is in the lower tail because the alternative hypothesis is true only if the population mean is less than 59.

Level of Statistical Significance
The (lowest) level of α for which the null hypothesis can be rejected.

In Figure 9.2, the value of $\bar{x}$ for which 5 percent of the $\bar{x}$ values are lower is $\bar{x} = 58.09$, which is called the critical value. The critical $\bar{x}$ value is identified by the letter *b*

FIGURE 9.2 Probability of a Type I Error and the Critical $\bar{x}$ Value in the Drive Belt Illustration

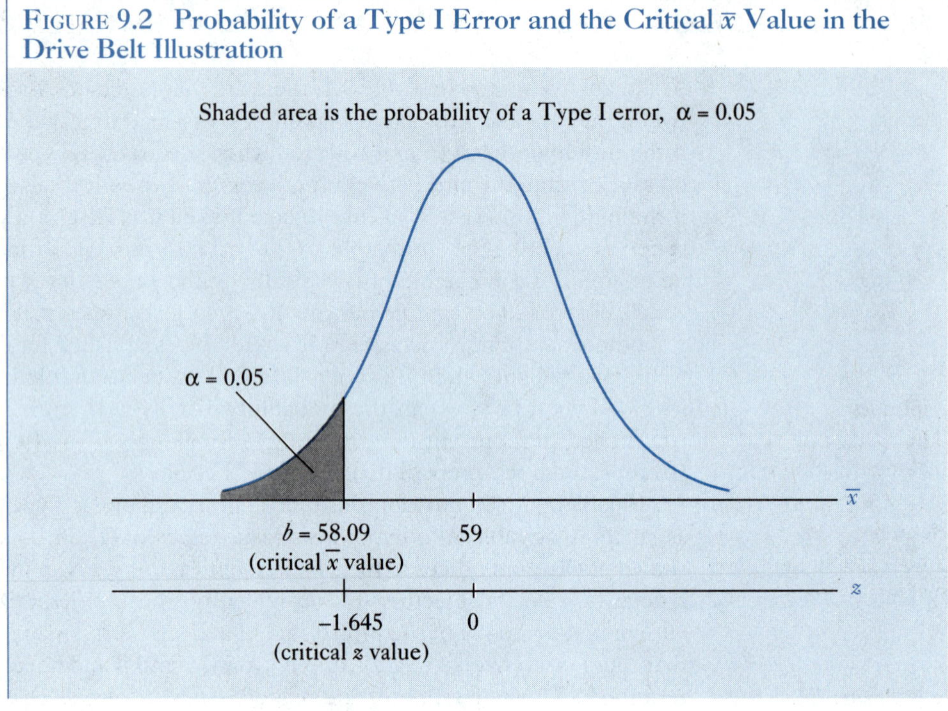

in Figure 9.2. The critical $\bar{x}$ value, b = 58.09, is found by substituting b for $\bar{x}$ in the z formula in Chapter 7, which gives Formula (9.1).

Critical $\bar{x}$ value is b, which is standardized as z_b

$$z_b = \frac{b - \mu_0}{\sigma/\sqrt{n}} \tag{9.1}$$

Corresponding to the critical $\bar{x}$ value, b, is its corresponding standardized value; this critical value of z is z_b. From Appendix Table A.3 or a computer program, the critical value of z that puts 5 percent of the distribution in the lower tail of the z distribution is –1.645.

$$P(z \le -1.645) = 0.05$$

z_b = –1.645 is called the critical z value; its relationship to the critical $\bar{x}$ value b is shown in Figure 9.2.

Substituting $z = -1.645$, $\mu_0 = 59$, $\sigma = 3.333$, and $n = 36$ into Formula (9.1) yields:

$$-1.645 = \frac{b - 59}{3.333/\sqrt{36}}$$

$$\text{and } b = 58.09$$

This critical value of b is shown in Figure 9.2, along with the corresponding critical z value of -1.645. As shown, 58.09 months is 1.645 standard deviations ($1.645\sigma/\sqrt{n}$) below 59 months in the $\bar{x}$ distribution.

Calculated Test Statistic
The value of the statistic that is calculated from observed sample data and then used to test a hypothesis about the parent population.

Figure 9.1 and Figure 9.2 provide all the ingredients for a statistical decision between rejecting or not rejecting the null hypothesis. We already had decided that the observed sample mean of 57.8 months, the **calculated test statistic** value, was significantly below the null hypothesis value of 59 because 57.8 was 2.16 standard deviations below 59, Figure 9.1, which was reflected in a small p-value of 0.0154. Now in Figure 9.2, the critical $\bar{x}$ value of 58.09 and critical z of -1.645 and α-risk of 0.05 can be used to define what constitutes "significantly below" and "a small p-value." Since the observed sample values ($\bar{x} = 57.8$ and $z = -2.16$) are beyond or more extreme than their critical values (58.09 and -1.645), the null hypothesis should be rejected because the observed sample values are "significantly far away." Another way to say this is that since the p-value of 0.0154 is smaller than the α-level of 0.05, the null hypothesis should be rejected. Our earlier decision rule for choosing between the null and alternative hypotheses now can be restated with more rigor.

Decision rule for selecting between H_0 and H_A:

Reject the null hypothesis (H_0) if the p-value is smaller than α.

Do not reject the null hypothesis if the p-value is larger than α.

One-Tailed Hypothesis Tests

For one-tailed tests, the decision rule based on an α-risk and a p-value comparison has already been illustrated in the drive belt example. The general principle is summarized in Figure 9.3 and Figure 9.4. In both graphs, H_0 is not rejected if the p-value $\geq \alpha$ and H_0 is rejected if the p-value $< \alpha$. These are the same conclusions that would be reached by comparing the observed $\bar{x}$ value (and its calculated z) with the critical $\bar{x}$ value (and critical z).

Two-Tailed Hypothesis Tests

For a two-tailed hypothesis test, the decision rule continues to be the same: reject H_0 if the p-value is less than α. Now, however, account must be taken of the fact that the

FIGURE 9.3 Decision Rules for Lower One-Tailed Hypothesis Tests

Panel a: In a lower-tailed alternative hypothesis test where the population mean μ is being tested as

$H_O: \mu \geq \mu_0$
$H_A: \mu < \mu_0$

the decision is to "not reject H_O" if the p-value $\geq \alpha$ as shown below

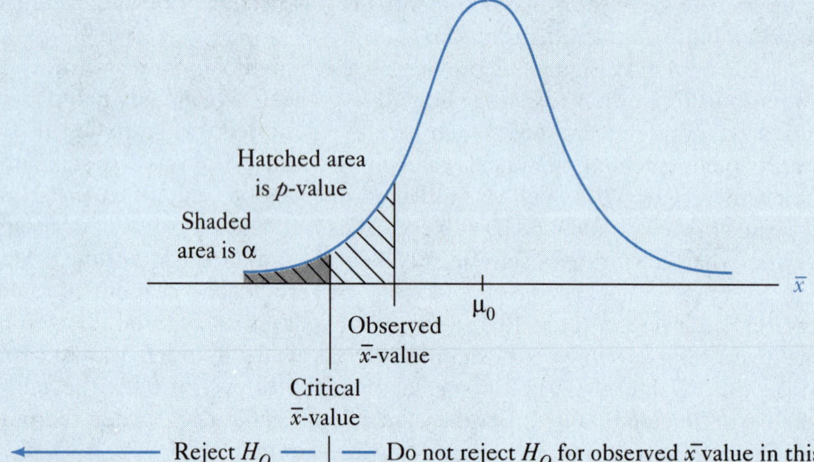

Reject H_O — | — Do not reject H_O for observed $\bar{x}$ value in this region

Panel b: In a lower-tailed alternative hypothesis test where the population mean μ is being tested as

$H_O: \mu \geq \mu_0$
$H_A: \mu < \mu_0$

the decision is to "reject H_O" if the p-value $\leq \alpha$ as shown below

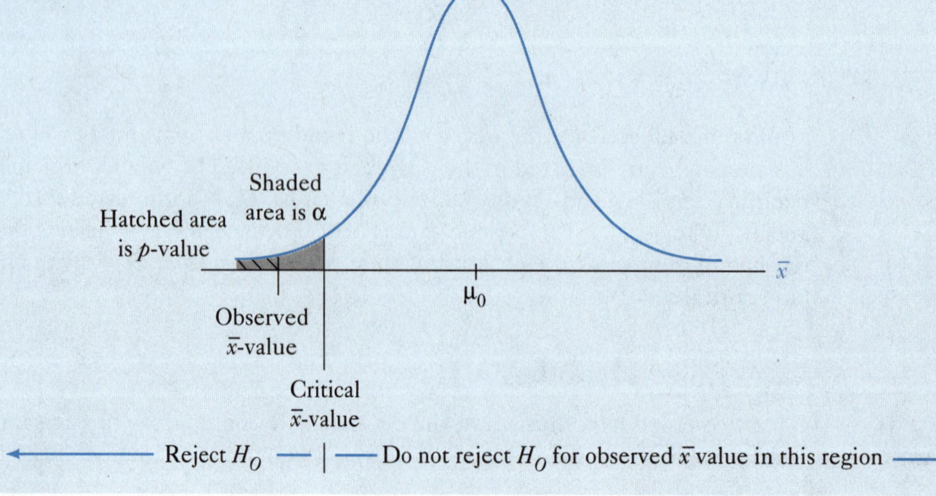

Reject H_O — | — Do not reject H_O for observed $\bar{x}$ value in this region

FIGURE 9.4 Decision Rules for Upper One-Tailed Hypothesis Tests

Panel a: In an upper-tailed alternative hypothesis test where the population mean μ is being tested as

H_O: $\mu \leq \mu_0$
H_A: $\mu > \mu_0$

the decision is to "not reject H_O" if the p-value $\geq \alpha$ as shown below

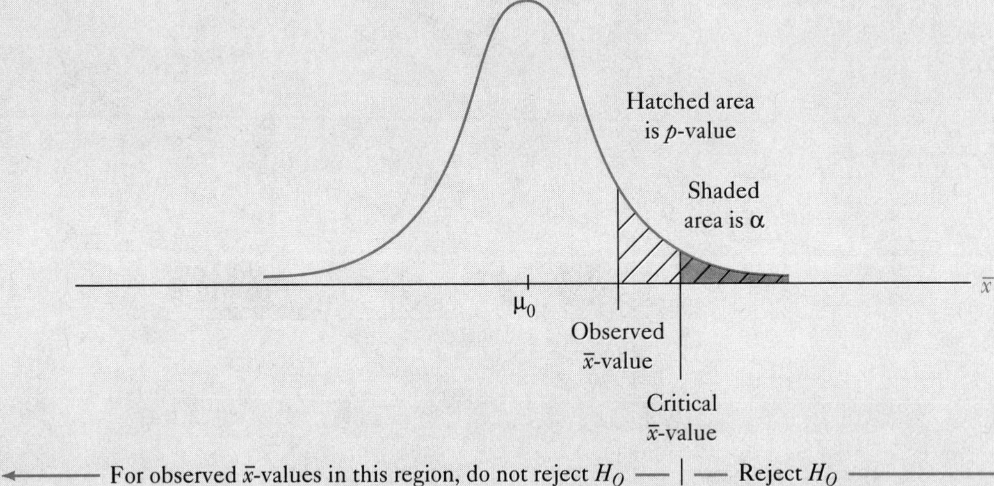

Panel b: In an upper-tailed alternative hypothesis test where the population mean μ is being tested as

H_O: $\mu \leq \mu_0$
H_A: $\mu > \mu_0$

the decision is to "reject H_O" if the p-value $\leq \alpha$ as shown below

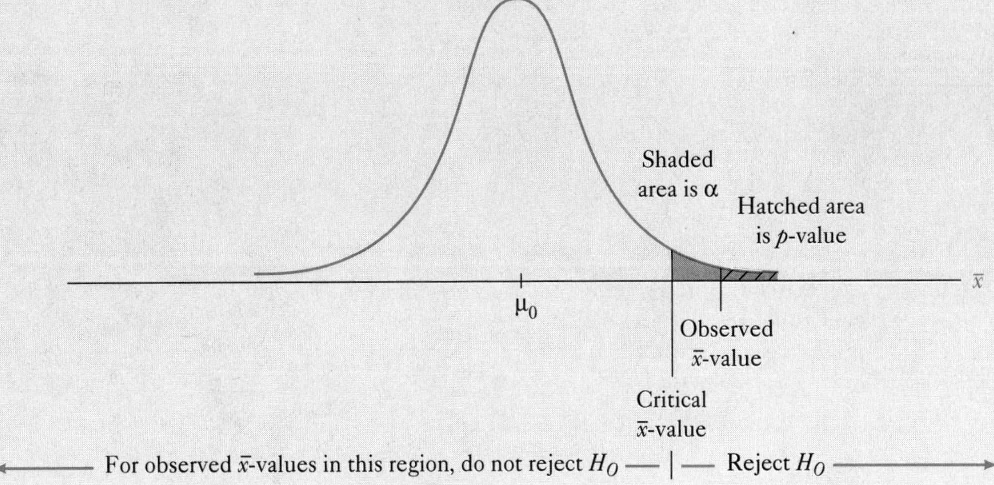

Figure 9.5 Decision Rules for Two-Tailed Hypothesis Tests Where $H_O: \mu = \mu_0$ and $H_A: \mu \neq \mu_0$

Panel a: When H_O cannot be rejected.

Because the alternative hypothesis allows for μ to differ from μ_0 in either direction, this graph shows shaded areas in both tails as α so that one tail contains only half of α. The hatched area beyond the observed $\bar{x}$-value is now only half of the p-value; thus, the hatched area must be doubled to obtain the p-value. Since half the p-value is greater than $\alpha/2$, the p-value is greater than α, and H_O is not rejected.

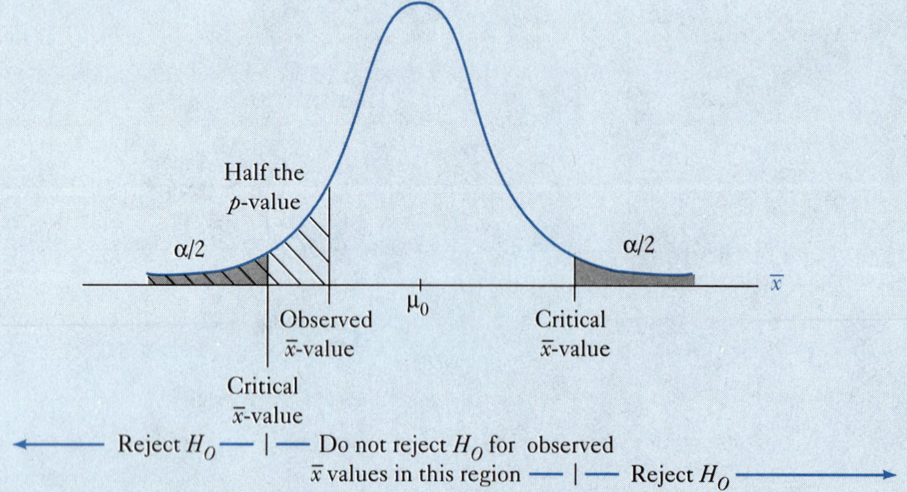

Panel b: When H_O can be rejected.

Because the alternative hypothesis allows for μ to differ from μ_0 in either direction, this graph shows shaded areas in both tails as α so that one tail contains only half of α. The hatched area beyond the observed $\bar{x}$-value is now only half of the p-value; thus, the hatched area must be doubled to obtain the p-value. Since half the p-value is less than $\alpha/2$, the p-value is less than α, and H_O is rejected.

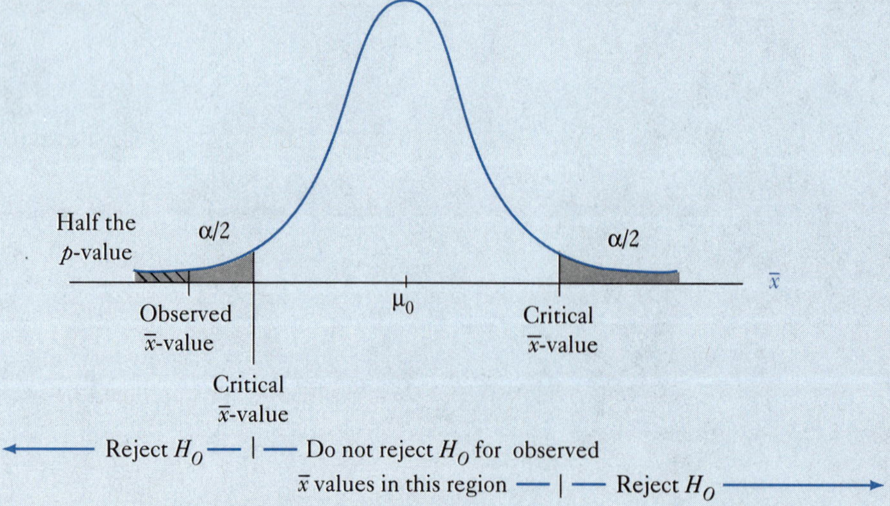

alternative hypothesis permits values on either side of the single value specified in the null hypothesis. The probability of a Type I error is not put in one tail of the distribution, but rather it is split equally between the two tails. That is, $\alpha/2$ is put in each tail beyond the critical test statistic value, as shown in Figure 9.5.

Students often have difficulty distinguishing between the determination of the critical test statistic value for a given α-level in a two-tailed test and the calculation of the p-value based on the observed $\bar{x}$ value. Unlike dividing α in two to determine the critical test statistic value, as shown in Figure 9.5, the calculation of the p-value in a two-tailed test requires a doubling of the area under the density function for $\bar{x}$ values more extreme than the observed $\bar{x}$. Thus, for a two-tailed test, only half of α is the probability of $\bar{x}$ values more extreme than the critical test statistic value on just one side of the mean. For a two-tailed test, only half of the p-value is the probability of $\bar{x}$ values more extreme than the observed $\bar{x}$ value.

EXERCISES

9.6 State the null and alternative hypotheses for each of the following, using the notation given in this chapter.

a. The average number of applicants for a training program in sales at IBM for the past five years has been 87.4. Has this average changed?

b. An econometric analysis was obtained to determine if the rate of inflation in the area had changed from last year, when the rate was 5 percent.

c. The personnel director at GM believes that sick leaves have decreased since a stricter policing program was implemented. In the past, employees averaged three sickdays per month.

d. A machine stamps parts at a defect rate of 10 percent, when it is operating correctly. When is the defective rate much too high (out of control)?

e. A sailboat racer believes that a new line stretches no more than the old line he had been using. The old line stretched 1.8 percent of its length at 15 percent of its working load.

f. A management study was to determine if the cost of piece work had changed since last year, when the cost was $378 per piece.

g. An engineer thinks a new composite has an average breaking strength that is greater than the old composite. The old breaking strength averaged 1,000 pounds per square inch.

h. The claim is made that at most 7 percent of the population will be unemployed in the next recession.

i. An accounting study is to assess whether or not there has been a decrease in the percentage of overdue accounts receivable. At the last audit, 6 percent of these accounts were overdue.

j. A marketing study is to dete,mine if the whitening power of a new bleach is at least equal to that of the old. The old bleach registered a 7 on a 0-to-10 scale, where 10 was the whitest.

9.7 For each of the above, state the nature of the Type I error and Type II error.

9.8 Find the critical z or t value for each of the following.

a. For a two-tailed test, where H_O: $\mu = \mu_0$, $\alpha = 0.02$, $n = 36$, and the population variance is known.

b. For a one-tailed test, where H_A: $\mu > \mu_0$, $\alpha = 0.05$, $n = 16$, and the population variance is unknown but the population is normal.

c. For a two-tailed test, where H_O: $\pi = \pi_0$, $\alpha = 0.05$, and $n = 40$, and π_0 or its estimate is not close to 0 or 1.

9.9 What are the different rules of thumb for deciding whether a hypothesis test is one- or two-tailed? What are the rules for determining which statement is to go into the null versus the alternative hypothesis?

9.10 Why doesn't the probability of a Type II error equal one minus the probability of a Type I error?

9.11 A *Wall Street Journal* article (January 18, 1985) on computer product design systems quoted Henry Eichfield, an official at Computervision Corp., as saying, "Thirty percent down time isn't atypical for the world of data processing." Design an experiment to test this claim based on sample information. What are the costs you need to consider?

9.12 An expert witness in a labor dispute concluded that there was no significant difference between the wage paid at the firm (\$13.45/hour) and the national average (\$14.78/hour). The sample standard deviation was \$3.99 for a sample of 36. What was the maximum Type I error level at which this expert could have been testing?

9.7 Selecting a Test Statistic (Step 3)

The values of a test statistic behave according to a known probability distribution. In this chapter, we use two familiar test statistics introduced in Chapters 7 and 8—the z and the t statistics. In the case of the sample mean, both of these test statistics are based on the model

$$\bar{x} \sim \text{Normal}\left(\mu, \sigma_{\bar{x}}\right), \text{ for } \sigma_{\bar{x}} = \sigma/\sqrt{n}$$

In this model, and as discussed in Chapter 8, if the population standard deviation σ is known, then the z test statistic can be used to calculate probabilities for the $\bar{x}$ distribution, assuming the expected value of $\bar{x}$ is μ_0, as given in the null hypothesis, and either the parent population is normal or the sample size is large ($n > 30$) so that $\bar{x}$ is approximately normal. This formula for this z statistic is[7]

z test statistic for $\mu = \mu_0$

$$z = \frac{\bar{x} - \mu_0}{\sigma/\sqrt{n}} \tag{9.2}$$

Formula (9.2) gives the standardized value of $\bar{x}$ around its expected value of $\bar{x}$ as given in H_O. This formula was used in the drive belt illustration, where the z was calculated as

$$z_c = \frac{\bar{x} - \mu}{\sigma/\sqrt{n}} = \frac{57.8 - 59.0}{3.333/\sqrt{36}} = -2.16$$

Usually the population standard deviation is unknown. When the population standard deviation is unknown, as discussed in Chapter 8, the Student's t test statistic is used to standardize $\bar{x}$ around its expected value μ_0, as given in the null hypothesis.

t test statistic for $\mu = \mu_0$

$$t = \frac{\bar{x} - \mu_0}{s/\sqrt{n}}$$

$$\text{where } s = \sqrt{\Sigma(x_i - \bar{x})^2/(n-1)} \qquad (9.3)$$

Formula (9.3) requires that $\bar{x}$ is normally distributed, which for small samples is assured if the population x is normal. Unlike Formula (9.2), in (9.3) the population standard deviation σ is not used, since it is unknown. σ is estimated by the sample standard deviation. The denominator of Formula (9.2), $\sigma/\sqrt{n}$, is the standard deviation of the distribution (also called the standard error of the mean). The denominator of the Formula (9.3), $s/\sqrt{n}$, is called the **estimated standard error of the mean;** it is the estimator for the standard deviation of all the $\bar{x}$'s that could be obtained under repeated sampling.

Estimated Standard Error of the Mean
The estimated value of the standard deviation of x; computed as s divided by the square root of the sample size, n.

To illustrate the use of the t test statistic, consider the claim of the marketing expert who said the mean amount of paper toweling used by a family of four is 60 feet per pay period. The null and alternative hypotheses were given as

$$H_O: \mu = 60 \text{ and } H_A: \mu \neq 60$$

For this test, suppose a simple random sample of 31 families was selected; each family was paid to have its paper towel use monitored. The sample mean amount of toweling used was 64.1 feet, and the sample standard deviation was 13.452 feet. The population standard deviation is unknown, so the appropriate test statistic is t, and its calculated value is

$$t_c = \frac{\bar{x} - \mu_0}{s/\sqrt{n}} = \frac{64.1 - 60}{13.452/\sqrt{31}} = 1.697$$

Even if the population was not normally distributed in this case, approximate normality of $\bar{x}$ can be justified in this case by the Central Limit Theorem, since n is relatively large. The calculated t statistic can be used to determine the p-value, as done in the section following our discussion of the sample.

9.8 The Sample (Step 4)

Throughout this book we assume that simple random sampling is possible. Recall that this does not imply that samples are "representative" of the population; after all, if a sample was representative, there would be no need for hypothesis testing, since the sample mean would be the population mean. Classical hypothesis testing requires simple random sampling, as discussed in Chapter 7.

Typically, data collected by simple random sampling are not available. For example, in March 1991, J.D. Power & Associates mailed a six-page questionnaire to 73,000 owners of 1990 model cars. Recipients received $1 for responding, and, according to *Business Week* (June 10, 1991), about 23,000 did. Whether these 23,000 represent a simple random sample of 1990 car owners is debatable. Are the less wealthy more likely to respond? Are those who are pleased (or upset) with their cars more likely to respond? Are those who bought a certain make or model more likely to respond? Were owners more likely to respond to different types of problems?

As another example of problems caused by nonrandom sampling, consider the case of Bloomington (Indiana) Hospital and its attempts to assess employee satisfaction. In 1991, Bloomington Hospital hired the consulting firm Alpha Company (of Minneapolis) for $12,000 to survey 35 of 1,700 employees. The 35 employees were picked by division managers, and the Alpha Group provided probability estimates as to whether or not employees were satisfied (and would or would not want to unionize). Because the 35 employees were hand picked and not selected by simple random sampling, the use of classical statistics to make inferences is dubious.

If we have reason to believe that simple random sampling was not employed, then the validity of the testing procedure must be called into question. Discussions of biases that are introduced through sample selection were presented in Chapters 1 and 7 and the Chapter 7 appendix. Sample selection problems and the absence of a plausible probability model, however, do not necessarily imply that probability-based testing is worthless. University of Chicago Professor of Statistics Paul Meier (1986) argues that

> such testing serves a useful purpose as a benchmark: if the observed association would not be counted statistically significant had it arisen from a randomized study, it could not be counted as persuasive, when even that foundation is lacking. If the observed association is highly statistically significant, however, the extent of its persuasiveness depends on many uncertain judgments about background factors, and its persuasive value is not at all reflected in the significance level itself. (p. 271)

What one can learn from nonrandom samples clearly depends on the situation and the conclusion suggested by the data.

9.9 Determining *p*-Values (Step 5)

We have already calculated the *p*-value for one-tailed tests involving the z test statistic. Those *p*-values involved one of the following probability calculations: $P(z \geq z_c)$ or $P(z \leq z_c)$. Determining *p*-values using the t test statistic for a one-tailed test is the same as for the z except that the t distribution is used. As shown in Figure 9.5, calculating the p-value in a two-tailed test is not as straightforward because the alternative hypothesis allows for a difference from the value in the null hypothesis in either direction.

In a two-tailed test, the absolute difference between the observed $\bar{x}$ value and the value in the null hypothesis is the measure of interest; the direction (– or +) of the difference is irrelevant. Because the absolute value of the test statistic is of interest, the area under the density function for values of $\bar{x}$ beyond or more extreme than the observed $\bar{x}$ value in the one tail is only half of the p-value (as the area beyond a critical test statistic is only half of the α-level). Doubling the area under the density function for $\bar{x}$ values beyond the observed $\bar{x}$ value gives the p-value in a two-tailed test. Thus we have the following rules for p-values.

Rules for determining *p*-values

If H_A is $\mu > \mu_0$, then

$$p\text{-value} = P(z \geq z_c)$$

If H_A is $\mu < \mu_0$, then

$$p\text{-value} = P(z \leq z_c)$$

If H_A is $\mu \neq \mu_0$, then

$$p\text{-value} = P(z \geq |z_c|)$$
$$= 2P(z \geq z_c), \text{ if } z_c > 0$$

or

$$= 2P(z \leq z_c), \text{ if } z_c < 0$$

Determining the *p*-value for a One-Tailed *t* Test

A p-value for a one-tailed t test follows the same steps as used in a z test, except probabilities are calculated from the t distribution in a computer package or as contained in Appendix Table A.4. As an example, consider a test to determine whether or not a computer disk drive has an average access time of no more than nine milliseconds versus an alternative that the average access time is greater than nine milliseconds; that is,

$$H_0: \mu \leq 9 \text{ and } H_A: \mu > 9$$

Assume the model is

$$\bar{x} \sim \text{Normal}\left(\mu, \sigma_{\bar{x}}\right), \text{ for } \sigma_{\bar{x}} = \sigma/\sqrt{n}$$

but that the population standard deviation σ is unknown. A sample of 16 disk drives yields a sample mean of $\bar{x} = 11.21$ and a sample standard deviation of $s = 3.40$. Assuming that population distribution of computer disk access time is normally distributed (to justify the assumption that $\bar{x}$ is approximately normal), the t test statistic is appropriate and its calculated value is 2.60.

$$t_c = \frac{\bar{x} - \mu_0}{s/\sqrt{n}} = \frac{11.21 - 9.00}{3.40/\sqrt{16}} = 2.60$$

FIGURE 9.6 The p-value for an Upper-Tailed t Test

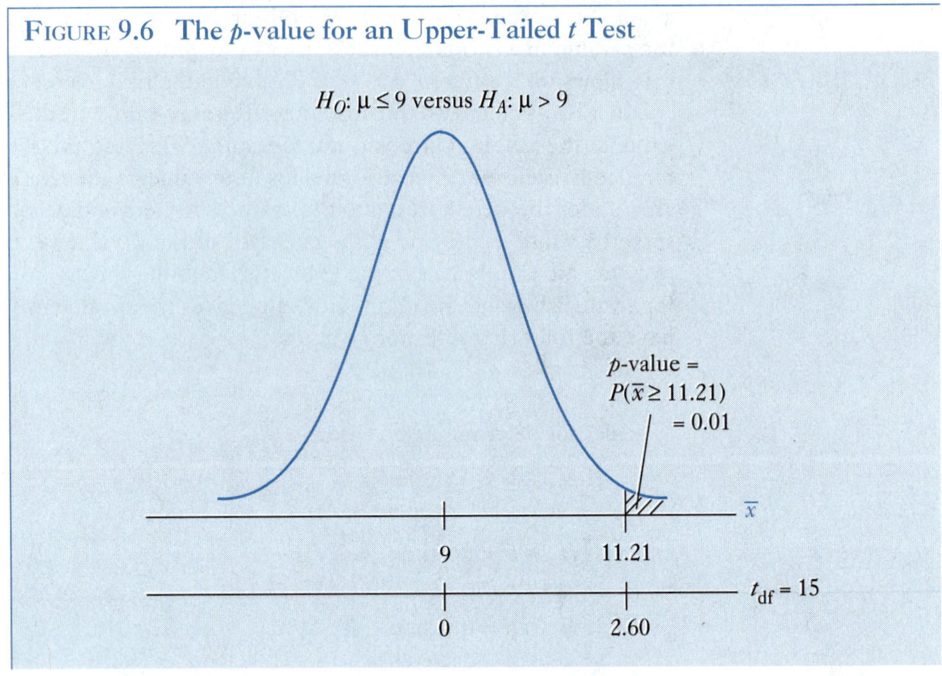

Because this is an upper-tail test, the p-value is $P(t \geq 2.60)$. This probability is determined by looking in Appendix Table A.4 across the row for 15 degrees of freedom. (Recall from Chapter 8 that the degrees of freedom are df = $n - 1$.) From Appendix Table A.4, the closest t value to 2.60, with 15 degrees of freedom, is 2.602—that is, $P(t \leq 2.602 \mid \text{df} = 15) = 0.99$. Thus, the approximate p-value is $P(t \geq 2.6) = 1 - 0.99 = 0.01$. This area is represented in Figure 9.6. (As discussed in earlier chapters, the exact probability for every value of t cannot always be obtained from Table A.4. When this happens, a computer program must be used to get exact probabilities.)

QUERY 9.1:

A *Wall Street Journal* (April 1, 1991) article reported that because of price wars, the average retail price for a gallon of self-serve regular unleaded gasoline was only $1.04, with some places charging less than 90 cents per gallon, even though market analysts say prices should be higher. As additional information on gasoline prices, the article reported prices in ten cities:

Atlanta	$ 0.94
Los Angeles	0.94
Dallas	0.98
Miami	1.03
San Francisco	1.05
Philadelphia	1.08
Las Vegas	1.12

Long Island	1.15
Boston	1.15
Chicago	1.19

But this sample has an average of \$1.063, suggesting that it is not representative of the population or that the national average might be higher than \$1.04 or that sampling error caused the difference. Do these data refute or support the idea that the national average is greater than \$1.04?

ANSWER:

The null and alternative hypotheses are

$$H_O: \mu \le 1.04 \text{ and } H_A: \mu > 1.04$$

The t test statistic is calculated using LOTUS 1-2-3 as

x	$(x-\bar{x})$	$(x-\bar{x})^2$
0.94	–0.123	0.015129
0.94	–0.123	0.015129
0.98	–0.083	0.006889
1.03	–0.033	0.001089
1.05	–0.013	0.000169
1.08	0.017	0.000289
1.12	0.057	0.003249
1.15	0.087	0.007569
1.15	0.087	0.007569
1.19	0.127	0.016129

$$\sum x = 10.63 \qquad 0.07321 = \sum(x-\bar{x})^2$$

$$\bar{x} = 1.063 \qquad s^2 = 0.00813 = \sum(x-\bar{x})^2/(n-1)$$

$$s = 0.09019$$

$$s/\sqrt{n} = 0.02852$$

$$t = 0.8064 = (\bar{x}-\mu)/s/\sqrt{n}$$

We are using t here because σ is unknown and the sample is small. Furthermore, assuming that the population is normally distributed does not seem unreasonable. At any of the typical α-levels, the null hypothesis cannot be rejected because the calculated $t = 0.8064$ is so small. (For example, at $\alpha = .10$ the critical t is 1.383, with nine degrees of freedom.) Similarly, using a computer program such as MINITAB, the p-value can be determined to be 0.22:

$$P(\bar{x} > 1.063 \mid \mu = 1.04) = P(t > 0.8064 \mid \text{df} = 9) = 0.22$$

This p-value is relatively large, suggesting that \$1.063 is "close to" \$1.04, given the variability in sampling. Thus, these ten sample observations on the price of gasoline are not inconsistent with the claim that the national average is \$1.04, and they do not support the claim that the national average is greater than \$1.04. The p-value and the related statistics are shown in the diagram below.

(Note that without a computer program, exact p-values cannot be obtained. From the t table in Appendix A.4, for example, there is no information for $t = 0.8064$, with nine degrees of freedom. In this table, however, 25 percent of the distribution can be seen to be beyond $t = 0.703$, df = 9. In the absence of accurate p-values, we are restricted to comparing calculated and critical test statistics values to complete a statistical hypothesis test.)

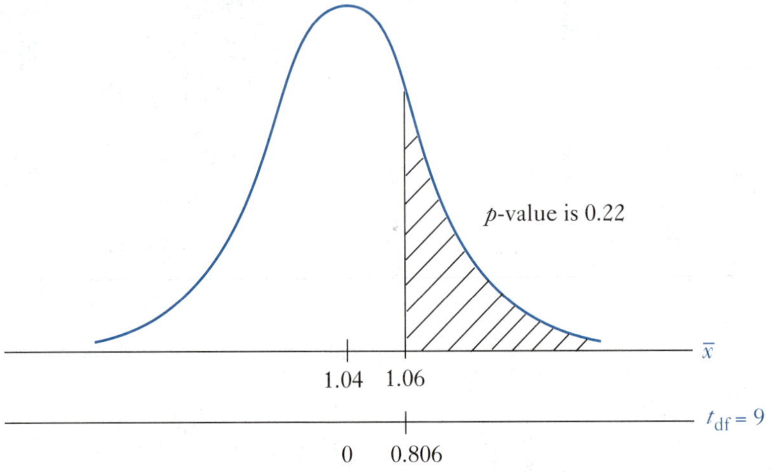

Determining a p-value for a Two-Tailed Test

Determining the p-value for a two-tailed alternative hypothesis requires attention because the sample value can fall on either side of the value in the null hypothesis. The absolute value of the difference between the value in the null hypothesis and the observed $\bar{x}$ value is of interest. Thus, the area beyond the observed $\bar{x}$ value (as given by the corresponding area beyond the calculated test statistic value) is only half of the p-value. For comparisons with α, the area beyond the observed $\bar{x}$ value (when it is in either tail) must be doubled to obtain the absolute value of being beyond this value. That is, the p-value in a two-tailed test of μ is either

$$P(z \geq |z_c|) \text{ or } P(t \geq |t_c|)$$

In the paper towel example, the null and alternative hypotheses were $\mu = 60$ and $\mu \neq 60$. For a sample of size 31, the mean was 64.1, the estimated standard error was 2.416, and the calculated t test was 1.697. The area beyond 1.697 can be obtained from the 30th row in Appendix Table A.4, because the degrees of freedom are 30. The area beyond 1.697 is 0.05—that is, $P(t \geq 1.697) = 1 - 0.95 = 0.05$. But this is only the p-value for an upper-tailed test. A two-tailed test accommodates $\bar{x}$ values on both sides of

FIGURE 9.7 The p-value in the Two-Tailed Test of Paper Towel Use

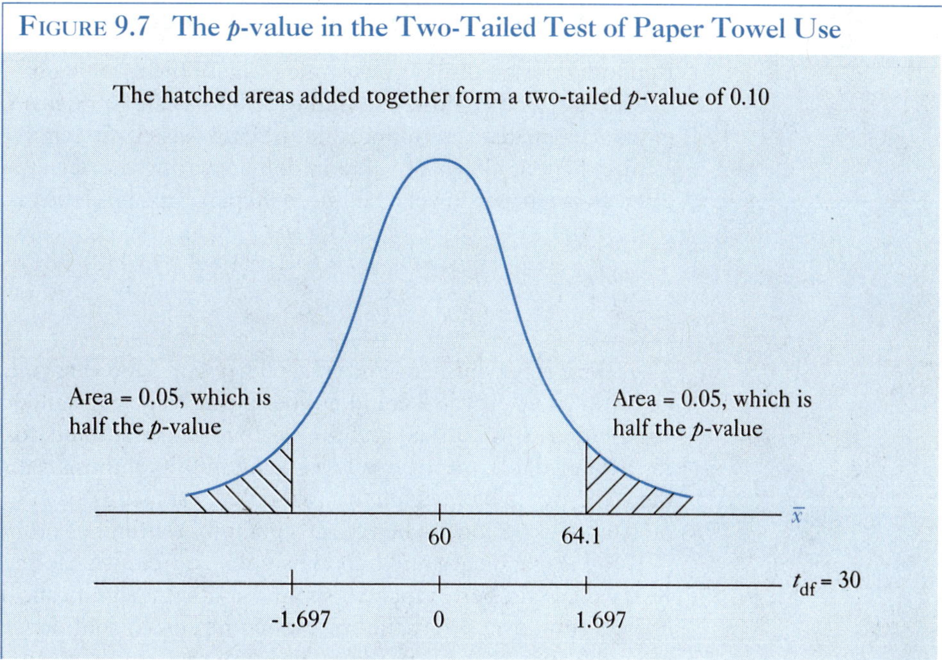

the value in the null hypothesis. Thus, the p-value for this two-tailed test is twice as large as the one-tailed p-value. In Figure 9.7, the p-value is 0.10.

9.10 REACHING A CONCLUSION (STEP 6)

As we have seen, the observation that a sample mean (or sample proportion) differs from the expected value contained in the null hypothesis does not necessarily imply that this expected value is wrong. The chance factors inherent in simple random sampling could account for the difference. What matters is the probability of observing sample values at or beyond this distance by chance (the p-value). If this probability is small, then the null hypothesis is rejected because the sample mean is "not close" to that expected. When the null hypothesis is rejected, the sample information is said to be "statistically significantly different" from that expected. When it is not rejected, the sample mean is said to be "not significantly different" from that expected from the null hypothesis.

Whether sample data is judged to be statistically significant or not can be sensitive to the α-risk at which the test is conducted. In the paper towel illustration, for example, the p-value was 0.10, which implies that H_0 is rejected if $\alpha = 0.15$, but is not rejected if $\alpha = 0.05$; it is "statistically significant" at exactly the 0.10 level.

Using Confidence Intervals in Conclusions

Reporting the *p*-value is worthwhile because the *p*-value gives the lowest probability of a Type I error for which the null hypothesis can be rejected. Reporting a confidence interval estimate, as introduced in the previous chapter and discussed in Appendix 9B of this chapter, also is of value in demonstrating the sensitivity of results to sampling error. In the paper towel example, a 90 percent confidence interval for μ is

$$\bar{x} \pm t(s/\sqrt{n}) = 64.1 \pm (1.697)(13.452/\sqrt{31}) = 64.1 \pm 4.1$$

$$60 \leq \mu \leq 68.2$$

The 90 percent confidence interval $60 \leq \mu \leq 68.2$ indicates that any null hypothesis value between 60 and 68.2 could not be rejected in a two-tailed test, for $\alpha = 0.10$ and this sample data. Both $\mu = 60$ and $\mu = 68.2$ are the end points for statistical significance at the 10 percent Type I error level—any null hypothesis value less than 60 or greater than 68.2 would be rejected in a two-tailed test using this information. (Remember that probability is defined to be zero at a point for continuous probability density functions.)

Notice that once a null hypothesis of no difference has been rejected in a two-tailed test, we can do better than to simply say that a significant difference exists at the α Type I error level. From the corresponding $(1 - \alpha)$ confidence interval, we can say that the population mean is in the direction of the sample mean. In the case of paper towels, for example, the 90 percent confidence interval suggests that the population mean is greater than 60 feet. Researchers often overlook this connection between a two-tailed test and the implications of the corresponding confidence interval.

Maintaining Uncertainty in Conclusions

Statistician John W. Tukey (1991, pp. 100–101) warns that the "most dangerous feature of 'accepting the null hypothesis' is the giving up of explicit uncertainty." Reporting an interval that is centered on the null hypothesis and has end points of other null hypotheses that also could not be rejected at the stated Type I level provides a method for acknowledging the uncertainty attached to the "acceptability" of the null hypothesis. Such an interval is demonstrated in Query 9.2 and its answer and more fully discussed in Appendices 9A and 9B.

Rejecting the null hypothesis and concluding that something is statistically significant does not necessarily imply that it is of practical importance. Concluding, for example, that the average access time for the computer disk drives is significantly greater than nine milliseconds (because the *p*-value = 0.01 was small, Figure 9.6) may have little practical consequence if the programs to be run do not require repeated disk access or if an inexpensive disk caching system can be used to speed up access time. As another example, consider training programs where one group of trainees is instructed by one method and a second group is instructed by another method. If the average scores on a multiple-choice test show a statistically significant difference between the learning of the two groups, but this difference is only two multiple-choice questions out of

50, what is the practical importance of this finding? Statistical significance is not synonymous with practical significance.

Finally, rejecting the null hypothesis does not "prove" that it is false. Failing to reject the null hypothesis does not "prove" that it is true. As long as conclusions are based on sample data, there is a chance that an error could be made, regardless of the conclusion reached. The testing procedures outlined here can detect situations in which it is unlikely that the sample data could have occurred if the null hypothesis was true. As long as there is a chance that the sample data could have occurred if the null hypothesis was true, there is no assurance that the results of the test are correct—Type I and Type II errors are always possible. Two appendices to this chapter explore the trade-off between Type I and Type II errors in much more detail.

QUERY 9.2:

The president of a small firm asks the personnel manager to justify her projection that new college graduates will have average starting salaries of $600 per week. To do this, she contacted a simple random sample of her colleagues at other firms and obtained the following information on starting salaries they intend to offer.

$ 650	580	622	685	614	543	586	610
615	548	600	590	612	547	663	605

The question: "Are these salaries similar to her projected average offer of $600 per week?"

ANSWER:

If "similar to" is defined by a sample mean being significantly close to $600, at the 1 percent Type I error level, then we have the following hypothesis testing situation.

H_O: $\mu = 600$; that is, the industry average salary is $600, which is the personnel officer's claim.

H_A: $\mu \neq 600$; that is, the industry average salary is something other than $600.

The following calculations are made with LOTUS 1-2-3, where we assume that the population of starting salaries, and thus $\bar{x}$, is normally distributed.

x	$x - \bar{x}$	$(x - \bar{x})^2$
650	45.625	2081.640
580	−24.375	594.1406
622	17.625	310.6406
685	80.625	6500.390
614	9.625	92.64062
543	−61.375	3766.890
586	−18.375	337.6406
610	5.625	31.64062

x	$x - \bar{x}$	$(x - \bar{x})^2$
615	10.625	112.8906
548	–56.375	3178.140
600	–4.375	19.14062
590	–14.375	206.6406
612	7.625	58.14062
547	–57.375	3291.890
663	58.625	3436.890
605	0.625	0.390625

$\sum x = 9670$ $\qquad$ $24019.74 = \sum(x - \bar{x})^2$

$\bar{x} = 604.375$ $\qquad$ $40.01045 = \sqrt{\sum(x - \bar{x})^2 / (n-1)} = s$

$s/\sqrt{n} = 10.004$ $\qquad$ $0.43732 = (\bar{x} - \mu)/s/\sqrt{n} = t$

The critical t ($\alpha = .01$ with 15 degrees of freedom) is ±2.947, since this is a two-tailed test. Because the calculated t is not beyond the critical t, the null hypothesis is not rejected. Using the computer program STATABLE, the p-value is determined to be 66.82 percent (in excess of 50 percent if you only have access to Appendix Table A.4), which is greater than the probability of a 1 percent Type I error, as can be seen in the following diagram.

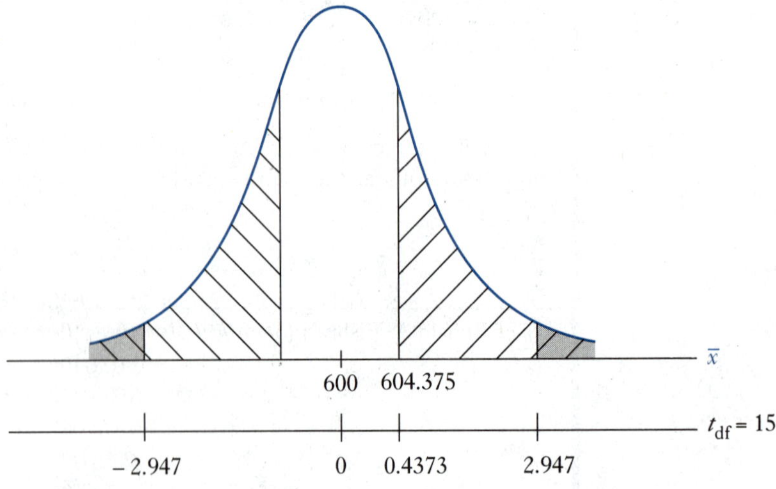

An alternative to LOTUS 1-2-3 for calculations and STATABLE for probability determination is to use a statistical program such as MINITAB for both steps. After bringing up the MINITAB program, data to be stored in the MINITAB files are entered with a "set" command following the MTB> prompt. The routine below shows the 16 observations stored in the first column of the data set, as designated by the name "c1" following the set command.

```
MTB>   set c1
DATA>  650  580  622  685  614  543  586  610
DATA>  615  548  600  590  612  547  663  605
DATA>  end
```

The t test for H_0: $\mu = 600$ is requested with the command "ttest," as follows:

```
MTB>   ttest 600 c1
```

MINITAB responds with the information:

TEST OF MU = 600.000 VS MU N.E. 600.000

	N	MEAN	STDEV	SE MEAN	T	P-VALUE
C1	16	604.375	40.016	10.004	0.44	0.67

A p-value of 0.67 is extremely large, which the personnel manager might want to interpret as strong evidence in favor of her notion that starting salaries will average \$600. After reading the appendices to this chapter, however, the president could point out that her sample does not enable her to reject a null hypothesis as high as μ = \$633.85 or as low as μ = \$574.89, in a 0.01 two-tailed test. This interval is wide and, as a practical matter, makes budgeting for new hires uncertain and extremely difficult.

Exercises

9.13 Test the following null and alternative hypothesis, at the 5 percent Type I error level.

H_0: $\mu_x \geq 29$

H_A: $\mu_x < 29$

The population standard deviation is 16.

a. for a random sample of 64 observations, with a mean of 26

b. for a random sample of 64 observations, with a mean of 23

9.14 How does the interpretation of a confidence interval (as presented in Chapter 8) compare to a two-tailed hypothesis test?

9.15 D.A.S. Fraser, writing in the *Journal of the American Statistical Association* (June 1991), tells of Ronald A. Fisher's (*Statistical Methods for Research Workers*, 1925) use of data that had been previously analyzed by Student (William S. Gosset, 1908) to test for an improvement in health associated with a new drug therapy. The ten observations used as a measure of improvement were

1.2 2.4 1.3 1.3 0.0 1.0 1.8 0.8 4.6 1.4

Fraser assumed the population standard deviation was known to be 1.5.

a. Why is it critical to know if the population variance is known or unknown?

b. Do you think Fisher knew that the population variance was 1.5? Why or why not?

c. Does the assumption of a known population variance affect the conclusion to be drawn here?

d. Go to the library to find Fisher's original presentation of this example. (Note: in 1954 Fisher's book was in its twelfth revised edition, but this example continued to appear on page 121.)

9.16 A note in *The Wall Street Journal* (February 18, 1992) describes a National Car Rental system that allows customers to rent and return cars with a machine similar to an automatic teller used by banks. But the system does not provide travelers with information on mileage, which customers want to know to ensure that their rental car has low mileage. The note quotes a company representative as saying that this is not a problem because "... on average its cars have fewer than 10,000 miles ...."

a. State the null and alternative hypotheses to test the company representative's claim.

b. Based on the following random sample of the mileage on nine rental cars, can the company's claim be accepted as true? At what level of significance?

9,876	11,346	12,321
8,564	12,590	3,210
7,487	8,886	9,389

9.17 *The Wall Street Journal* (October 27, 1992) reported that of the 94 patients treated for skin cancer (melanoma lesions), "the average time from their surgery and injection to the recurrence of melanoma was 30 months." In the past, the population average for other groups of melanoma patients receiving alternative types of treatment had been 17.5 months.

a. State the null and alternative hypotheses to test whether the new treatment results in a greater average time before the recurrence of the melanoma. (Define all notation.)

b. Draw a diagram showing the null hypothesis rejection region for a test at a 5 percent Type I error level. (Make sure to label all points of interest.)

c. If the p-value for this test was 0.0002, what was the population standard deviation?

d. Given your answer in part c, how do you know that the population of time before melanoma recurrence cannot be normally distributed? What does the population look like?

e. Why might you question the use of average time before melanoma recurs as a measure of successful treatment?

9.18 A new production process is being introduced. Prior to its introduction, the piece rate per worker was 80 units per hour. For a sample of 47 workers, the average piece rate was 96 units per hour, with a standard deviation of 21. Is the new process better than the old? (Test at the 5 percent Type I error level.)

9.19 Associated Press (December 21, 1991) reported on a new regulation in Great Britain that requires bigger beer glasses that can accommodate a full 20-ounce British pint and a creamy head; brewers and pub landlords are to be fined for selling less. To test for compliance, an agent visits a pub at nine random times, ordering a draft on each visit with the following results:

19.77, 20.10, 19.95, 19.89, 20.09, 20.08, 19.89, 19.67, and 20.01 ounces.

Would you conclude that on average this pub is serving glasses of beer with less than 20 ounces? (Test at the 5 percent level.)

9.11 Testing a Population Proportion

As already stated, hypothesis testing is not restricted to tests of means. The proportion of items possessing a given attribute or characteristic can be tested in exactly the same way as a mean; only a different model and statistics are required. The population proportion π might represent the proportion of females in a workforce, the percentage of executives flying commercial airlines during a year, the anticipated ratio of defective items to total monthly production, Coca-Cola's market share next year, or the fraction of lottery jackpots that Wisconsin residents can expect to win.

Consider the Lotto America situation in which 31 percent of the tickets were purchased by Wisconsin residents and 27.45 percent (14 of 51) of the jackpots were won by Wisconsin residents. We wanted to test whether a person's residence had anything to do with the chances of winning. The null and alternative hypotheses are

$$H_O: \pi = 0.31 \text{ and } H_A: \pi \neq 0.31$$

where π is the population proportion of Wisconsin winners expected by chance and the model of its sample estimator is

$$p \sim \text{Normal}\left(\pi, \sigma_p\right), \text{ for } p = \left(x/n\right) \text{ and } \sigma_p = \sqrt{\pi\left(1-\pi\right)/n}$$

Although x is the number of winners from Wisconsin, which is a discrete count distributed as a binomial, the sample proportion p can be approximately normal because the sample size n is relatively large.[8] For this model, the z test statistic is

z Test Statistic for a Population Proportion π

$$z = \frac{p - \pi_0}{\sqrt{\pi_0\left(1-\pi_0\right)/n}} \tag{9.4}$$

where π_0 is the limiting proportion in the null hypothesis, and there are x successes in a large sample of size n, and the estimator of π is $p = x/n$

A calculated value for this test statistic is determined by inserting the proportion of successes in the sample ($p = x/n = 14/51$) and setting π equal to its expected value ($\pi_0 = 0.31$) from the null hypothesis.

$$z = \frac{\left(x/n\right) - \pi_0}{\sqrt{\pi_0\left(1-\pi_0\right)/n}} = \frac{0.2745 - 0.31}{\sqrt{0.31\left(0.69\right)/51}} = \frac{-0.0355}{0.06476} = -0.548$$

FIGURE 9.8 Test of Wisconsin Residency and Winning the Lotto America Jackpots

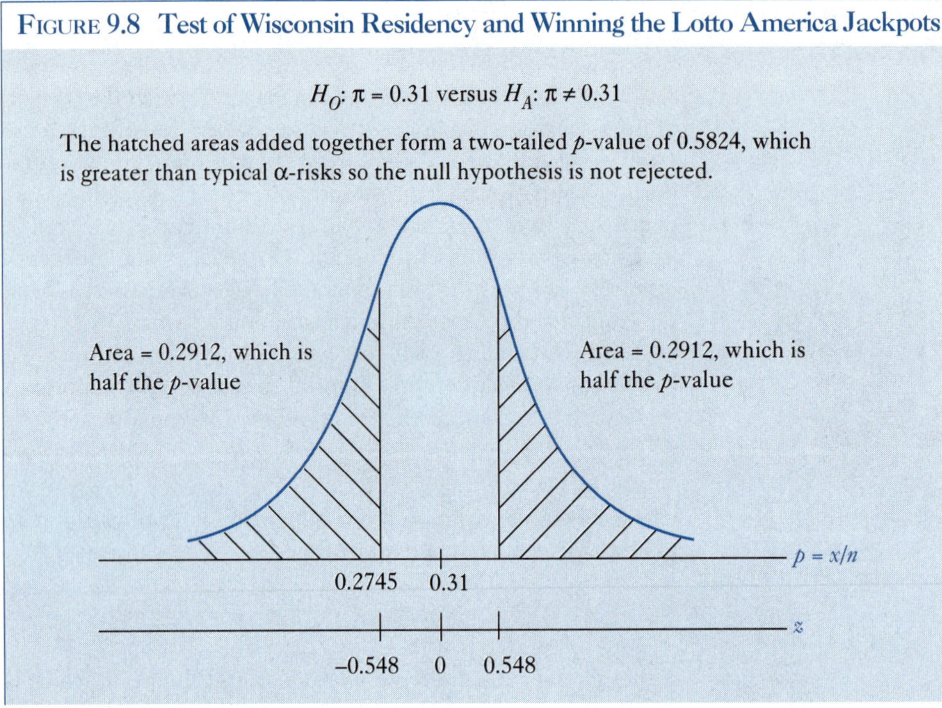

The use of the z test statistic is justified here by the Central Limit Theorem, because n is large, the distribution of p is approximately normal. As before, a p-value is the area in the tails of the distribution. Because it is a two-tailed test, the p-value in this case is $2P(z \leq -0.55) = 2(0.2912) = 0.5824$, which is very large. Thus, for typical α-levels, the null hypothesis cannot be rejected and the lottery spokesman may be correct—Wisconsin residency is not related to winning. This testing process is summarized in Figure 9.8. Remember, however, that in our failure to reject the null hypothesis, we have not proven it. There are many other values that one could assume for π that could not be rejected with this sample data.

If the sample size is small, then the z-statistic cannot be used to test a proportion. As demonstrated in the next section for a test of the median (the value of x that puts 50 percent of the values below it), small sample tests of proportions must be done with the binomial distribution. The definition of the p-value for the binomial distribution is the same as when the normal approximation is employed. Because the binomial distribution is discrete, however, care must be taken to include the probability of the observed value of x plus the probability of any more extreme values in the calculation of the one-tailed p-value.

QUERY 9.3:

According to Fancorp Inc. (a consulting firm), "more than 42 percent of franchises are purchased by husband-and-wife teams." (*Business Week*, May 6, 1991, p. 70.) If this is true, a house maintenance service aimed at these couples may be profitable. In your area, are more than 42 percent of the franchises owned by husband-wife teams?

ANSWER:
The null hypothesis is $\pi \leq 0.42$, and the alternative is $\pi > 0.42$. A simple random sample of 36 franchise owners shows that 19 are owned by husband-wife teams. The calculated z statistic for this relatively large sample is

$$z = (0.5278 - 0.42) \Big/ \sqrt{(0.42)(0.58)/36} = 1.31$$

The p-value is 0.0951, which at the 5 percent Type I error level implies that the null hypothesis should not be rejected but at the 10 percent Type I error level it should. That is, the conclusion is sensitive to the α-level. The cost of making a Type I versus a Type II error is practically important here.

QUERY 9.4:
In "Statisticians Occupy Front Lines in Battle Over Passive Smoking" (*Wall Street Journal*, July 28, 1993), statistical consultant Kenneth Brown was cited for the calculation "that there were only two chances in 100—a probability of 0.02—that the 19% figure was a matter of happenstance. That more than meets the standard of 0.05 (five chances out of 100) at which most scientific studies are considered statistically significant." Brown also produced a 90 percent confidence interval for the increased risk of cancer as being approximately 4 percent to 35 percent higher for passive smokers than those who aren't so exposed. Are Brown's hypothesis test and confidence interval consistent?

ANSWER:
Brown appears to be reporting a one-tailed hypothesis test with a p-value of 0.02 and α-level of 0.05. He also calculated a two-tailed confidence interval with $\alpha = 0.10$, which is not comparable to his one-tailed hypothesis test. Had he done a two-tailed hypothesis test, then estimation with a corresponding two-tailed hypothesis test would be appropriate. (Recall from Query 8.4 in the previous chapter, however, that it is doubtful whether the normality assumption underlying this test is true if passive smoke cannot reduce the risk of lung cancer.)

EXERCISES

9.20 A *Business Week* (June 29, 1992) article had a headline that read: "It's Not All That Easy Having a Hit. Just Ask Caddy." The article was about the success General Motors is having with the Cadillac Seville made at its Detroit–Hamtramck plant. According to the article, "Still, an internal quality audit in early June found 5.1 defects per car vs. a target of 1.0." What would you need to know to assess whether this was a statistically significant difference?

9.21 A *Wall Street Journal* (July 26, 1984) article reported that a survey of 1,200 hourly workers found "only 45% of hourly workers think their firm is a good place to work." At the 5 percent significance level, can you conclude that less than 50 percent of all hourly workers think their firm is a good place to work?

9.22 A *Wall Street Journal* (October 27, 1992) article stated that the "past is no guarantee of (Money) manager's future." It cited a study by Frank Russell Co., a pension-consulting firm, that "tested whether an investor could predict future performance from the track records of 106 hand-picked managers monitored by the firm. It found that the winners' chances of repeating their success were 'statistically indistinguishable (in either direction) from a proportion of 50% (the coin flipping chance of success).'"

a. State the null and alternative hypotheses for this test. (Make sure to define all notation.)
b. Draw a diagram showing the null hypothesis rejection region, assuming the test was done at a 5 percent Type I error level. (Make sure to label all points of interest.)
c. If the null hypothesis is assumed true, what is the value of the standard error of the estimator?
d. In the limiting case, how high (or low) could the estimate have been?

9.23 As stated in Exercise 7.24, Ann Landers, in her nationally syndicated column, wrote: "Your story (on poor nursing care) is a chilling one . . . (however) I would say at least 90 percent of the nurses are compassionate and caring. Too bad you ran into a few who were not . . . ."(*HT*, May 28, 1993).

a. For a test of Ann's claims, what are the null and alternative hypotheses?
b. If a random sample of 40 nurses yields 34 compassionate and caring, what conclusion would you draw at the 5 percent Type I error level?

9.24 A soft drink bottler is considering introducing one of three new plastic bottles. The bottler believes that if customers could select among the three types, more than one-third would pick bottle A. As a check, the bottler surveys 100 consumers and finds that 36 percent prefer bottle A, 29 percent prefer bottle B, and 35 percent prefer bottle C. Do more than one-third of the customers prefer bottle A over bottles B and C? (Test at the 0.1 Type I error level.)

9.25 As a mandate for change, a union leader wants to get more than 60 percent of the membership vote. She surveys 64 of the voting members and finds that 40 intend to vote for her and her platform of change. At the 1 percent significance level, should she expect to get her victory with the mandate for change?

9.12 Test of the Median (Optional)

If a sample is small ($n < 30$) and there are "outliers" that pull the mean toward these extreme values, then the median may be a preferred measure of central location. A highly skewed sample suggests a nonnormal population, which makes the t and z ratios inappropriate statistics for a small sample test of the population mean. A test of the population median, however, uses the number of observations greater than (or less than) a hypothesized median as the test statistic. The exact distribution of this statistic is the binomial distribution, for the initial assumption that the null hypothesis is true.

The population median is the value of the 50th percentile—half the values are greater than the median, and half are less. The expected (or average) sample median is the population median. Thus, if it were not for sampling error, half the values in a sample should be above the population median stated in the null hypothesis, if the null hypothesis is true. The chance factors inherent in simple random sampling, however, may cause the proportion of values above the expected median to be slightly less than or slightly greater than 50 percent in any given sample. Only on average will 50 percent of the sample values be above the purported population median, if it is true. But if many more than 50 percent of the sample values are above (or below) the median value in the null hypothesis, the null hypothesis must be rejected. Again what constitutes "many more" is defined by a p-value—the probability of observing no more than the number below (or above) the hypothesized median, as determined from the binomial distribution with $\pi = 0.5$.

To illustrate the test of the median, consider the real estate example from *The Wall Street Journal* (February 12, 1991) in which the agent wants to know if the median selling price of homes in the Houston, Texas, area is less than its past value of $70,656. For this test, the null and alternative hypotheses are

H_O: Median ≥ $70,656 (at least 50 percent of the observations are above $70,656)

H_A: Median < $70,656 (less than 50 percent of the observations are above $70,656)

She randomly selects 20 houses that recently sold in the area and finds that 14 of them sold for less than $70,656 and six sold for more. The p-value for this test is determined from the binomial table Appendix A.1 or a computer program to be

$$p\text{-value} = P[x \leq 6 \mid \pi = 0.5 \text{ (that is, median is 70,656)}, n = 20] = 0.0577$$

Notice that here x represents the number at or above the median, as called for in the null hypothesis.

At a 5 percent Type I error level, the null hypothesis cannot be rejected, but at the 6 percent level it can. Clearly this test is sensitive to the α-level selected. We can say that the median is statistically significantly less than $70,656 down to the 0.0577 Type I error level.

As another example, consider a test of grade points for students who have completed 60 credit hours at a large state university. The school newspaper claims that the overall average GPA is 2.3 but that the majority of students have GPAs above 2.5. The school administration says this is not true. You have what you believe to be a simple random sample of 10 students, of which seven students have GPAs above 2.5. The sample data are

3.55	3.16	3.08	2.61	2.56
2.54	2.51	2.10	1.89	1.00

The sample mean is 2.50, and the sample median is 2.55. As with the population, the sample is left skewed because its mean 2.50 is less than its median 2.55. Because the sample size is small and because it, as well as the population of GPAs, is skewed to the left, a test of the population mean with either the z or t distribution is not appropriate.

The statements about the population median can be tested with the binomial distribution. The null and alternative hypotheses are

H_O: Median ≤ 2.5 (at least 50 percent of the observations are below 2.5)

H_A: Median > 2.5 (less than 50 percent of the observations are below 2.5)

The p-value for this test is determined from the binomial table Appendix A.1 or a computer program to be

$$p\text{-value} = P[x \leq 3 \mid \pi = 0.5 \text{ (that is, median is 2.5)}, n = 10] = 0.1719$$

Here x is the number at or below the median, as called for by the null hypothesis. At even the 10 percent Type I error level, however, the null hypothesis cannot be rejected. The median is not significantly greater than 2.5 until the 0.1719 Type I error level, which is considered too high a level for statistical significance in most situations. The null hypothesis should not be rejected.

In concluding our discussion of the test of the median, recognize that this test is identical to any test of the proportion $\pi = 0.5$, where x is distributed as a binomial and the ratio x/n is approximately normal if n is large. The alternative hypothesis may indicate either a one- or two-tailed test. Although not as prevalent as the one-tailed test, the two-tailed test of the median can be performed as in any other two-tailed tests of a proportion or mean. For a two-tailed test of the median, with a small sample, the procedure just demonstrated for the one-tailed test is repeated, except that the mass (probabilities) in the tail of the binomial distribution is doubled to determine the p-value. Again, if the p-value is small (less than α), then the null hypothesis (median equals value) is rejected and the alternative hypothesis (median does not equal the value) is accepted. If the p-value is large (greater than α), then the null hypothesis cannot be rejected.

9.13 Bayesian Hypothesis Testing (Optional)

In contrast with the classical view in which the population parameter is fixed and its estimator is a random variable, Bayesians treat the population parameter as if it is the random variable and the estimator is fixed. Bayesians argue that because the calculated sample statistic is known with certainty, a distribution should be assumed for what is unknown—the population parameter. Bayesians calculate the probability that the population parameter is equal to hypothesized values; the higher this probability, the more likely the hypothesized values.

In the drive belt test, for example, the null hypothesis was $\mu \geq 59$. If forced to comply with the classical starting point, a Bayesian will assume that the initial opinion or belief about μ is that the probability of $\mu \geq 59$ is unity—that is, H_O is assumed true as in the classical world. This starting position is called a "noninformative prior."

With only knowledge that $\bar{x} = 57.8$, the expected value of the random variable μ is set at 57.8, in the Bayesian world. As seen in Figure 9.9, treating $\bar{x}$ as fixed and μ as random "reverses" or "inverts" the probability distribution and its interpretation from that

FIGURE 9.9 A Bayesian Framework for Testing Drive Belt Life

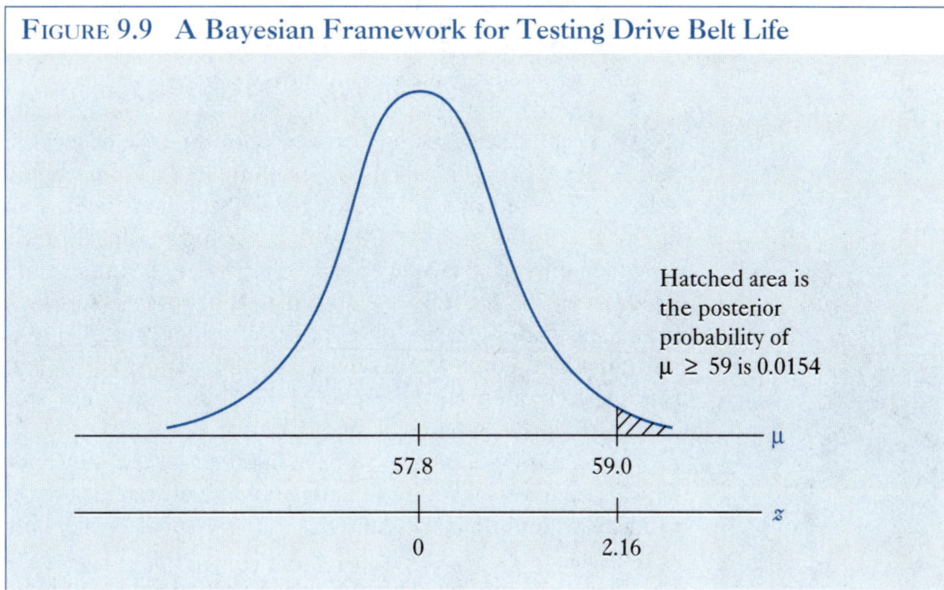

found in classical hypothesis testing. The Bayesians ask: Given the sample mean 57.8, what is the probability that $\mu \geq 59$? This probability is called the posterior probability; it is shown in Figure 9.9 to be 0.0154, which is the same size as the *p*-value in the classical hypothesis test of Figure 9.1. Here, however, the Bayesian says that the posterior probability of $\mu \geq 59$ is too small to believe that the mean belt life is at least 59 months. (Based on Figure 9.1, the classicalist says the probability of a sample mean less than 57.8 is too small to believe $\mu \geq 59$.)

If the researcher has no prior belief about μ (a noninformative prior), then the (posterior) probability of the null hypothesis is always the same size as the *p*-value of classical statistics. As alluded end of the historical perspective earlier in this chapter, the difference between Bayesian and classical answers do not arise until differences in the prior subjective judgment about the null hypothesis are entertained. When the null hypothesis is that of a single point equality (e.g., $\mu = 59$) and informative priors are entertained, classical and Bayesian interpretations are no longer the same, as discussed by James Berger and M. Delampady (1987).

9.14 CONCLUDING COMMENTS

University of Chicago Business School Professor Harry Roberts (1990) argues that "in business the major emphasis is in solving immediate problems, while in science the major emphasis is on broad empirical generalizations." (p. 372) In business, he asserts, "the most serious misunderstanding is interpreting tests of sharp null hypotheses as business decision procedures. It is very difficult to find a realistic business decision that is clarified by formulation as a statistical test . . . significance levels are irrelevant to the manager who must make the business decision." (p. 382)

There are many illustrations one can cite to document the misuse of concepts from statistical hypothesis testing, especially in the case of legal proceedings where "the 5 percent level of significance" has been treated as something magical and unquestionable, as discussed by Paul Meier (1986). Numerous books and articles have been written on the use of statistical hypothesis testing in legal proceedings aimed at settling disagreements over business decisions. For an example, see Michael Finkelstein and Bruce Levin (1990). The use of statistical hypothesis testing in legal matters involving personnel matters (salary determination, hiring, firing, and promotion decisions), acquisitions and corporate takeover attempts, antitrust, safety, production and quality control, promotion and marketing, and numerous other areas cannot be ignored by managers who may be held liable for their actions in court. In this day and age of litigation, no top-level manager can be ignorant of the concepts underlying classical statistical hypothesis testing. An understanding of statistical significance is relevant to the manager who must make decisions with personal, social, political, and economic consequences.

Even in day-to-day decision making, a knowledge of hypothesis testing is needed. The idea of Type I and II errors and the cost of making one or the other certainly should influence what is assumed true until contrary information is obtained. In quality control, samples of output are periodically taken to assess whether production is in control, as discussed in Appendix 9C of this chapter. In modern-day accounting, sampling is used to determine whether or not fraud has occurred. It is very easy to find realistic business decisions that are clarified by the ideas of a statistical test. It may be true that a manager may never have to base his or her decision on the outcome of a single statistical test. However, no good manager is going to ignore the accumulation of evidence against one hypothesis, in favor of the other, when there are profits to be made.

Although many business managers may never have to formulate null and alternative hypotheses in the formal and rigorous manner discussed here, the six-step procedure for decision making has applications well beyond one or two courses in statistics. As suggested in the discussion of Bayesian decision theory, this basic six-step procedure can be modified to incorporate a decision maker's prior beliefs, prejudices, and assessments of costs. But without an understanding of hypothesis testing process, none of these extensions will make much sense.

Chapter Exercises

9.26 A *Business Week* (July 8, 1991) note reported that steelmakers test each batch of high-quality alloys by drawing a sample of molten metal. This sample is then heated in a vacuum and the vapors tested. If the composition of the vapors does not deviate too greatly from the norm, the batch is accepted as being satisfactory (the process is in control).

Describe the null and alternative hypotheses in this test and the nature of the Type I and Type II errors.

9.27 The U.S. Supreme Court in *International Brotherhood of Teamsters v. United States* (1977) stated, "It is ordinarily to be expected that nondiscriminatory hiring practices will in time result in a work force more or less representative of the racial and ethnic composition of the population in the community from which employees are hired." In one area, 25 percent of the accountants available for new accounting jobs are black. Thus, a company that has not discriminated against black applicants should hire a proportion of blacks that is not significantly different from 0.25. A company in this area hired 80 new accountants over the past several years, 16 of whom were black. Does this indicate discriminatory hiring at the 5 percent significance level?

9.28 A *Wall Street Journal* (August 29, 1991) article reported on the problems that General Motors is having at its plant in Orion Township, Michigan. According to the article, "On some days, the cars have more than three times the company goal of two defects per car."

a. To test whether the average is more than six defects per car, 16 cars are selected at random with the following results:

3, 6, 8, 9, 5, 7, 10, 4, 5, 12, 8, 4, 6, 9, 1, and 7 defects.

What distribution can be used for this test?

b. What conclusion can you draw from this data? (Test at the 5 percent Type I error level.)

9.29 In a *Wall Street Journal* (August 17, 1987) article on selling machine tools to U.S. firms, Toyoda representatives claimed that "Toyoda machines would operate 95% of the time." In 47 randomly selected eight-hour shifts, it was found that the Toyoda machines were down because of breakage an average of a half hour per shift, with a standard deviation of ten minutes. Can you reject the Toyoda claim?

9.30 A *Wall Street Journal* (June 20, 1991) article reported that there was no clear pattern to whether corporate relocation influenced stock prices. In some cases, the corporate moves were associated with spending a lot of money on a trophy headquarters, which tended to lower the company's stock price, but in other cases the move was to a less expensive place. A sample of 30 companies that recently moved showed the stock price of 17 falling after the move. Does this contradict the *WSJ* hypothesis?

9.31 An article by John Cunniff (Associated Press, January 21, 1992) reported on a survey of American workers, which was conducted by Towers Perrin, a human resource consultant. It was conducted for the National Association of Manufacturers and included its 4,000 member companies. Cunniff states:

> While the response rate was only 10 percent, the findings were significant. The survey revealed, for example, that the average manufacturer rejected five out of every six candidates for a job . . . more than half the companies reported major employee skill deficiencies . . . . American companies are crippled in applying technological advances and in otherwise improving productivity.

Comment on the "significance" of these findings.

9.32 As a result of the deregulation of U.S. airlines, regulators claimed that many airlines would go bankrupt, even if the industry expanded. They predicted that following deregulation, the airline industry would be represented by average revenue-generating passenger miles of 42 billion. The 1990 revenue passenger miles, shown on the following page from the *Philadelphia Inquirer*, are from the Air Transportation Association, in billions of miles. Has the regulators' claim been realized?

Airline	Miles
American	76.9
United	76.0
Delta	59.0
Northwest	51.5
Continental	41.2
USAir	35.6
Trans World	34.2
America West	11.1
Southwest	10.0
Alaska	4.5

9.33 The defense for boxer Mike Tyson in his 1992 trial on a rape charge claimed that the pool of 50 persons from which a jury of 12 members was to be selected was not representative of the community at large. Twenty-two percent of the community was black, while only 14 percent of the pool was black. At the 5 percent Type I error level, is it reasonable to assume that the jury was selected randomly from the community at large?

9.34 Procter & Gamble wants to know if Folgers' market share has increased since it cut the price of Folgers coffee "an average of 10% to 15% since 1991" (*Business Week*, July 19, 1993). P&G knows that its market share last year was 0.19 (19 percent of the market) in test areas. In 10 randomly selected market test sites, P&G now has the following shares of the market:

.20 .19 .17 .21 .15 .20 .25 .24 .22 .18

a. Specify a model that can be used for testing here.
b. Test whether P&G market share can be expected to exceed 19 percent this year.

9.35 Mothers Against Drunk Driving claims that the majority of crimes committed in the United States involve alcohol. This view is challenged by an Orange County beer distributor who noted that of the 621 crimes recorded in Orange County last year, only 306 involved drunken driving, public intoxication, or charges against minors for illegal use or possession of alcohol. Comment on the distributor's challenge, and provide statistical evidence to support your position.

9.36 A consumer protection group interprets the claim on the side of the General Electric light bulb package to imply that more than 50 percent of GE's 60-watt Soft-White light bulbs should last longer than 1,000 hours. A sample of 36 bulbs is tested, and 23 are found to last longer than 1,000 hours. Is this sufficient evidence to conclude that 50 percent of all GE 60-watt bulbs will last more than 1,000 hours? (Test at the 5 percent Type I error level.)

9.37 In the week prior to Super Bowl XXVI in Minneapolis in 1992, the "Super Bowl Theory of Stock Prices" was circulating. It stated that if the Washington Redskins (of the NFC conference) won, then a bull market (prices up) would characterize 1992. If the Buffalo Bills (of the AFC) won, then a bear market (prices down) would prevail in 1992. Empirical support for this theory was that for 22 of the previous 25 Super Bowls, conference affiliation of the winner predicted the stock market correctly. Pure randomness suggests that the relationship should have been 50 percent. Thus, a finance analyst calculated the test statistic

$$z = \left(0.88 - 0.50\right) \Big/ \sqrt{\left(0.5\right)\left(0.5\right)/25} = 3.80$$

and concluded that the Super Bowl Theory of Stock Prices is correct. Comment on this test and conclusion.

9.38 The Washington Redskins (NFC) won the 1992 Super Bowl and the Dallas Cowboys (NFC) won the 1993 Super Bowl. The stock market finished in both 1992 and 1993 with higher prices than at the start of each of these years. How would this information affect the financial analyst's calculations of the z statistic and your comments on the conclusion to be drawn from this test in Exercise 9.37?

9.39 In a study of the way individuals are treated by car dealers, "testers" were sent into dealerships as would-be buyers. Author Ian Ayers, in "Fair Driving: Gender and Race Discrimination in Retail Car Negotiations" (*Harvard Law Review*, February 1991), stated, "When the seller did reveal his cost, the represented cost was substantially higher than independent estimates of seller cost for the same model, as seen (below) . . . they systematically overstate their costs. The greatest misrepresentation was made to white female testers."

Seller Misrepresentation of Cost Data

	Average Misrepresentation
White Male Testers	$ 849
White Female Testers	1046
Black Male Testers	752
Black Females	–

a. Comment on the statistical significance of the white male and female difference in misrepresentation.

b. Comment on the statistical significance of the white and black male difference in misrepresentation.

9.40 An article titled "The Bloodbath In Market Research" in *Business Week* (February 11, 1991) stated that "the validity of surveys is being jeopardized by the growing refusal of Americans to participate. Annoyed by the intrusiveness, 36 percent of consumers declined to answer a phone query in 1990."

a. How does refusal to participate jeopardize validity of surveys?

b. How could you test the correctness of the claim that 36 percent of Americans called would not respond to questions?

9.41 A *Wall Street Journal* (January 4, 1991) article reported that "The average American consults with a doctor 5.4 times a year. Most contacts are at the doctor's office, an average of 3.2 visits a person each year." To test whether the average is 3.2 visits at the doctor's office, a local group of physicians pooled data on the average number of visits for those who actually came to their offices and concluded that their sample average was 6.5 visits a person each year, well above the claimed average of 3.2 visits. Comment on their sample design and the validity of their conclusion.

9.42 A *Business Week* (May 28, 1990) article claimed that "In the past two years, Grano (head of retail sales at Paine-Webber) has hired 150 brokers, each of whom had gross commissions of at least $500,000." To test the accuracy of this statement, a competitor interviewed a sample of 25 of these brokers as potential candidates for a fictitious position. The average reported by these brokers was $450,000, which was close to *Business Week's* average, with a standard deviation of $50,000.

a. What conclusion would you draw regarding the correctness of the *Business Week* statement about gross commissions?

b. What might you question about the sampling technique employed by the competitor? What are the implications of this for your test in part a?

9.43 Writing for the *Washington Post* (*HT*, February 9, 1992) Frank Swoboda reported on the sixth annual health-care benefits survey by Foster Higgins & Co. Foster claimed that "the average cost to employers for health care rose 12.1 percent per employee in 1991—to \$3,605—as more and more companies shifted costs of medical coverage to their employees." This cost seemed high, so a personnel manager sampled 10 of his counterparts at other companies and found the following costs:

\$4,145	\$2,950	\$3,320	\$2,978	\$3,576
\$3,423	\$3,398	\$3,975	\$3,552	\$3,502

At the 5 percent Type I error level, does this sample data support a population average cost of \$3,605 or is the population average less?

9.44 Foster, Higgins & Co. also claimed, following its survey of health-care benefits, that health-care costs average 12 percent of total payroll cost. For the ten similar companies contacted in Exercise 9.43, the following health-care costs as a percentage of payroll were recorded:

14.1%	10.2%	12.1%	10.3%	11.8%
11.6%	12.1%	13.7%	11.5%	11.9%

At the 5 percent Type I error level, does this sample data support a population average of 12 percent?

9.45 A headline in *The Wall Street Journal* (February 25, 1992) stated, "The Temporary Help Business is Likely to Rebound This Year." Yet, the article cited a survey of 600 companies by TempForce, Inc., a temporary personnel firm, as showing "37% plan staff increases for 1992, and 48% don't see any change, while 15% expect cuts." Comment on the correctness of this headline based on the reported data.

9.46 Although law firms are now free to advertise, it is generally believed that the majority of firms do not employ marketing specialists. A *Wall Street Journal* (February 26, 1992) article reported that a survey of 400 firms "found that 48% had marketing directors." Is there a significant difference here between the observed 48 percent and the hypothesized 50 percent?

9.47 There is a claim of sex discrimination in hiring. In the relevant period, the employer has hired 64 workers, only eight of whom were women. In the preselection pool of equally qualified applicants, 20 percent were women. An expert witness for the employer states that the probability of observing the eight or fewer women in a random sample of 64 when drawn from this pool was 6.68 percent, implying that the employer could have been randomly sampling from this pool with no consideration to the sex of the applicants.

a. State the null and alternative hypotheses the expert was testing.
b. Conduct the statistical test performed by the expert. Is the conclusion of no sex discrimination supported? At what level of significance?
c. (Requires appendices) At the same level of significance used by the expert witness, what is the lowest proportion of women in a preselection pool that likewise could not be rejected? What are the practical implications of this?

9.48 A *Wall Street Journal* (April 20, 1992) article reported on a study by University of Delaware Professor Valerie P. Hans of jurors' attitudes toward business in wrongful death and injury cases. "Of the 18 cases, 14 were decided in favor of plaintiffs, higher than the national norm of about 50%." Is there a significant difference between Hans' sample and the national norm?

9.49 In the mail-order computer business, a 30 percent failure rate for systems is considered average. An article in *Business Week* (May 11, 1992) reported on an order of 50 Gateways placed by Thomas Walsh, executive director for Houston law firm Mayor, Day, Caldwell

& Keaton: "3 wouldn't work out of the box, and 14 monitors and 10 more computers soon failed." Do these 27 failures from this sample of 50 support the result that Gateway's average defect rate is 30 percent?

9.50 A report by the U.S. Centers for Disease Control (*HT*, May 15, 1992) examined the blood-alcohol content of 13 people killed in North Carolina by falling or being thrown from a horse between 1979 and 1989. Five, or 39 percent, had been drinking, prompting a government warning against drinking and riding. Dr. Thomas Cole stated, "It makes as little sense to drink and ride as it does to drink and drive." Comment on the correctness of the conclusion that drinking alcohol prior to riding increases the risk of falling or being thrown from the horse and dying.

9.51 A *Newsweek* (September 7, 1992) article on "Total Quality Management" stated that "most U.S. companies don't think much of TQM's impact on their ability to dash past competition." (p. 48) In a survey of 500 companies by Boston-based Arthur D. Little, 36 percent said the process of TQM was having "a significant impact" on their ability to quash competition. Are these survey results sufficient to support *Newsweek's* claim? At what "significance level"?

9.52 (Appendices helpful but not required for basic answer) Chuck Stone, in his Newspaper Enterprise Association syndicated column (*HT*, June 6, 1992), wrote:

> Of the 739,763 inmates in federal and state prisons, a minimum of 5 percent—36,000—are innocent, according to estimates of several nationally prominent legal experts whom I have interviewed. Now, 95 percent accuracy is eminently respectable. But when policemen, prosecutors, judges, and juries systematically make errors 5 percent of the time—and authorities refuse to budge when new evidence indicates a miscarriage of justice—the cumulative effect makes a mockery of fair trials.

Comment on the statement that the legal system has a 95 percent accuracy rate.

9.53 (Appendices required) A *Wall Street Journal* (January 22, 1992) article on the health care field cited the Pennsylvania Health Care Cost Containment Council suggesting that Harrisburg Hospital's death rate is excessive. The article states that, in addition to questioning the correctness of the numbers, federal Health Care Financing Administration analysts say "there are 90 chances in 100 that Harrisburg Hospital's 1989 death rate is truly worse than those of other hospitals rather than being a statistical fluke; the hospital insists only a higher standard—95 chances in 100—is valid."

a. Draw sampling distributions of the death rates to show the issue being debated regarding the "90 chances in 100" versus the "95 chances in 100."
b. Comment on the correctness of the hospital's rebuttal.

9.54 (Appendices required) A. Tversky and D. Kahnemann ("Belief in the law of small numbers." In *Judgment Under Uncertainty: Heuristics and Biases*, D. Kahnemann, P. Slovic, and A. Tversky, eds., pp. 23–31, Cambridge University Press, 1982) provide a demonstration that even many scientists do not understand the importance of power and sample size in defining successful replications. They distributed a questionnaire at a meeting of psychologists, with the following inquiry:

> An investigator has reported a result that you consider implausible. He ran 15 subjects, and reported a significant value, $t = 2.46$. Another investigator has attempted to duplicate his procedure, and he obtained a nonsignificant value of t with the same number of subjects. The direction was the same in both sets of data. You are reviewing the literature. What is the highest value of t in the second set of data that you would describe as a failure to replicate? (p. 28)

Tversky and Kahnemann reported the following results:

> The majority of our respondents regarded $t = 1.70$ as a failure to replicate. If the data of two such studies ($t = 2.46$ and $t = 1.70$) are pooled, the value of t for the combined data is about 3.00 (assuming equal variances). Thus, we are faced with a paradoxical state of affairs, in which the same data that would increase our confidence in the finding when viewed as part of the original study, shake our confidence when viewed as an independent study. (p. 28)

a. Demonstrate how Tversky and Kahnemann determined that the "t for the combined data is about 3.00."
b. Why does this experiment demonstrate a lack of understanding of power and sample size?

9.55 (Appendices required) Jessica Utts ("Replication and Meta-Analysis," *Statistical Science*, 1991, 4, pp. 363–378) describes a question she posed at a History of Philosophy of Science Seminar at the University of California at Davis:

> Two scientists, Professors A and B, each have a theory they would like to demonstrate. Each plans to run a fixed number of Bernoulli trials and then test H_0: $p = 0.25$ versus H_A: $p > 0.25$. Professor A has access to large numbers of students each semester to use as subjects. In his first experiment, he runs 100 subjects, and there are 33 successes ($p = 0.04$, one-tailed). Knowing the importance of replication, Professor A runs an additional 100 subjects as a second experiment. He finds 36 successes ($p = 0.009$, one-tailed).
>
> Professor B only teaches small classes. Each quarter, she runs an experiment on her students to test her theory. She carries out ten studies this way, with the results in Table 1.

Table 1 Attempted Replications by Professor B

n	Number of Successes	One-Tailed p-Value
10	4	0.22
15	6	0.15
17	6	0.23
25	8	0.17
30	10	0.20
40	13	0.18
18	7	0.14
10	5	0.08
15	5	0.31
20	7	0.21

> "I asked the audience by a show of hands to indicate whether or not they felt the scientists had successfully demonstrated their theories. Professor A's theory received overwhelming support, with approximately 20 votes, while B's theory received only one vote." (p. 367)

a. What are each professor's aggregate results for the 200 experiments conducted by each?
b. What is the one-tail p-value for the combined trials of each professor?
c. Jessica Utts found the aggregate results in parts a and b to be in direct contrast to those reached by the 20 audience participants who voted that Professor A's theory was supported. Why?

9.56 (Appendices required) Jessica Utts, in the reference in Exercise 9.55, describes another scenario she presented to the same seminar:

> "In December of 1987, it was decided to prematurely terminate a study on the effects of aspirin in reducing heart attacks because the data were so convincing (J.A. Greenhouse and S.W. Greenhouse, 1988, "An Aspirin a Day . . . ?" *Chance,* 1, 24–31; and R. Rosenthal, 1990, "How Are We Doing in Soft Psychology?" *American Psychologist,* 45, pp. 775–777). The physician-subjects had been randomly assigned to take aspirin or a placebo. There were 104 heart attacks among the 11,037 subjects in the aspirin group, and 189 heart attacks among the 11,034 subjects in the placebo group . . . $p < .00001$."
>
> "After showing the results of that study, I presented the audience with two hypothetical experiments conducted to try to replicate the original result, with outcomes in Table 2.

Table 2 Hypothetical Replications of the Aspirin/Heart Attack Study

	Replication #1 Heart attack		Replication #2 Heart attack	
	Yes	No	Yes	No
Aspirin	11	1156	20	2314
Placebo	19	1090	48	2170
Chi-square	2.596,	$p = 0.11$	13.206,	$p = 0.0003$

> "I asked the audience to indicate which one they thought was a more successful replication. The audience chose the second one, as would most journal editors, because of the 'significant *p*-value.'"(p. 367)

a. What is the proportion of heart attacks in the two replications? How do these proportions compare to the original study? On this comparison, which replication is closer to the original?
b. Construct a 95 percent confidence interval for the relative risk for the original study. Which replication has a relative risk proportion contained in this confidence interval?
c. What can you conclude about significance levels from this scenario?

9.57 (Appendices required) Which is more dramatic and implies more assurance, rejecting the null hypothesis with a small or a large sample? (Hint: See Richard Royall, "The Effect of Sample Size," *American Statistician,* November 1986.)

Academic References

Berger, James, and M. Delampady. "Testing Precise Hypotheses (with Discussion)." *Statistical Science,* 2, 1987, pp. 317–352.

Carlson, Roger. "The Logic of Tests of Significance." *Philosophy of Science,* 43, 1976, pp. 116–128.

Edelman, David. "A Note on Uniformly Most Powerful Two-Sided Tests." *American Statistician,* August 1990: pp. 219–220.

Finkelstein, Michael, and Bruce Levin. *Statistics for Lawyers.* New York: Springer Verlag, 1990.

Fisher, Franklin. "Statisticians, Econometricians, and Adversary Proceedings." *Journal of the American Statistical Association* 81, June 1986: pp. 277–286.

Kiefer, Nicholas. "Economic Duration Data and Hazard Functions." *Journal of Economic Literature* 26, June 1988, pp. 646–679.

Kupper, Lawrence, and Kerry Hafner. "How Appropriate Are Population Sample Size Formulas." *American Statistician,* May 1989, pp. 101–105.

Meier, Paul. "Damned Liars and Expert Witnesses." *Journal of the American Statistical Association* 81, June 1986, pp. 269–276.

Roberts, Harry V. "Business and Economic Statistics." *Statistical Science,* November 1990, pp. 372–390.

Shilling, Edward G. *Acceptance Sampling In Quality Control.* New York: Marcel Dekker, 1982.

Spielman, Stephen. "The Logic of Tests of Significance." *Philosophy of Science* 41, 1974, pp. 211–226.

Spielman, Stephen. "The Logic of Tests of Significance," *Philosophy of Science* 45, 1978, pp. 120–135.

Tukey, John W. "The Philosophy of Multiple Comparisons." *Statistical Science,* February 1991, pp. 100–101.

Wald, Abraham. *Sequential Analysis.* New York: John Wiley & Sons, 1947.

ENDNOTES

1. Kirk Bloodsworth, after two trials and almost nine years in prison for raping and murdering a 9-year-old girl in Maryland, "walked out of prison a free man thanks to DNA testing." (Associated Press, *HT:* June 29, 1993) County State's Attorney Sandra O'Connor was quoted saying: "I am not prepared to say that he is innocent . . . but we do not have enough evidence to convict him beyond a reasonable doubt." Notice the newspaper story simply said Bloodsworth was a "free man," and attorney O'Connor would not say that he was innocent beyond a reasonable doubt. Innocence is assumed, not proved. Innocence is rejected by the evidence, and errors are sometimes made.
2. Recall from Chapter 7 that $P(\bar{x} \leq 57.8) = P(\bar{x} < 57.8)$ because probability at a single value is defined to be zero for a continuous probability density function. The area in the tail of the sampling distribution $P(57.8 < \bar{x} < 59 \mid \mu = 59)$ is of interest because $P(57.8 = \bar{x})$ is zero, regardless of the population mean assumed. The occurrence of any specific value, even when drawn from a discrete distribution, can be viewed as rare or unusual, regardless of its distance from the mean. Calculating the area beyond or more extreme than the observed value eliminates this problem.
3. Any assessment of the life of an item (be it a light bulb, tire, or human life) may involve a truncation of the period of observation before all items have ceased working. If these long-life items are ignored (deleted from the sample), then the mean will be understated. If the time for these long-life items is recorded for only the span of the period of observation, the mean again will be understated but not as badly as when the observations are deleted. Guessing at the remaining life of these long-life items may result in still less understatement, but it might also lead to an overstatement—we can never know what goes on outside the observation period. These and other "censoring problems" in "duration data" are discussed by Nicholas Kiefer (1988).

4. Instead of testing for a value of π, based on a value of p, the expected value of x, given by πn, could be tested with the number of successes in the sample, pn. The model would then be

$$x \sim \text{Binomial}\left(\pi n, \sigma_x\right), \text{ for } \sigma_x = \sqrt{\pi\left(1-\pi\right)n}$$

As long as n is large, and π is not too close to 0 or 1, the normal and the binomial models will give similar results. If n is small, however, the binomial model must be used.

5. On May 1, 1989, the U.S. Supreme Court made it easier for plaintiffs to prevail in a major category of lawsuits involving sex, race, and age discrimination in personnel matters. In *Hopkins v. Price Waterhouse,* the Court rejected the accounting firm's argument that Ann Hopkins should be required to prove that it was sex discrimination rather than legitimate judgment that led to her lack of a promotion to partner. When Hopkins brought the charges, the firm was assumed innocent (the null hypothesis), but the Court said the derogatory comments made by firm partners were enough to establish the existence of illegal discrimination (the alternative hypothesis). The Court said that once this prima facia case was made, it was then the firm's obligation to show it was innocent (the null hypothesis becomes guilt and the alternative innocence). But the Court ruled that the firm had to show only by "a preponderance of the evidence" that the firm's reasons for denying promotion were legitimate. This standard of proof is the least onerous. It is equivalent to a large probability of a Type I error level in which the firm's reasons were "more likely than not" legitimate ones. A lower court had required "clear and convincing" evidence, which is equivalent to a small probability of a Type I error level.

6. For a discussion of how courts have viewed statistical hypothesis testing, see Franklin Fisher (1986) and Paul Meier (1986).

7. We have been and will continue to assume that samples are drawn from large (infinite) populations. If the population is not large relative to the sample, then the standard error of $\bar{x}$ is

$$\frac{\sigma}{\sqrt{n}}\left(1-\frac{n}{N}\right)$$

where the factor $(1 - \frac{n}{N})$ is called the finite population correction factor, and n/N is the fraction of the population included in the sample. If n is small relative to N, then the finite population correction factor is approximately 1. If the n is large relative to N, however, then the z test statistic is

$$z = \frac{\bar{x} - \mu_0}{\dfrac{\sigma}{\sqrt{n}}\left(1-\dfrac{n}{N}\right)}$$

As a rule of thumb, as long as $n < 0.10N$, the finite population correction factor is ignored.

8. For p to be approximately normal, the expected value of p $[E(p) = \pi]$ as well as its realization $(p = x/n)$ cannot be overly close to the extreme values of 0 or 1, and n must be sufficiently large to ensure that the interval $p \pm 3.5\sigma_p$ does not include 0 or 1. If these conditions are not met, then the binomial distribution must be used to calculate probabilities for x based on the sample of size n.

FORMULAS

- To find the critical $\bar{x}$ value, b

$$z = \frac{\bar{x} - \mu_0}{\sigma/\sqrt{n}} = \frac{b - \mu_0}{\sigma/\sqrt{n}}$$

- The z test statistic for $E(\bar{x}) = \mu_0$

$$z = \frac{\bar{x} - \mu_0}{\sigma/\sqrt{n}}$$

- The t test statistic for $E(\bar{x}) = \mu_0$

$$t = \frac{\bar{x} - \mu_0}{s/\sqrt{n}}$$

$$\text{where } s = \sqrt{\Sigma\left(x_i - \bar{x}\right)^2/\left(n-1\right)}$$

- The z test statistic for a population proportion π_0

$$z = \frac{\left(x/n\right) - \pi_0}{\sqrt{\pi_0\left(1 - \pi_0\right)/n}}$$

where $p = x/n$

APPENDIX 9A THE TYPE I AND TYPE II ERROR TRADEOFF AND THE EFFECT OF SAMPLE SIZE

A tradeoff exists between the probability of a Type I error and a Type II error. As the probability of a Type I error α is decreased, the probability of a Type II error β increases. This inverse relationship can be illustrated in the drive belt test, where $\bar{x}$ = 57.8 was used to decide whether H_O: $\mu \geq 59$ should be rejected in favor of H_A: $\mu < 59$, at $\alpha = 0.05$.

According to the null hypothesis, and the claim of the rubber company, the mean life of drive belts will be centered at 59 months, Figure 9.A1, panel a. If H_O is true, 5 percent of the $\bar{x}$ values from samples of size $n = 36$ will lie below the critical value b = 58.09. If H_O is not true, then the true average belt life might be a number such as 57.7, Figure 9.A1, panel b. If the true mean belt life is 57.7 months, then the probability that H_O is incorrectly accepted (Type II error) is given by the dotted area in Figure 9.A1, panel b. Thus the probability of a Type II error is 0.2420, if $\mu = 57.7$. This probability

FIGURE 9.A1 The Relationship Between α and β

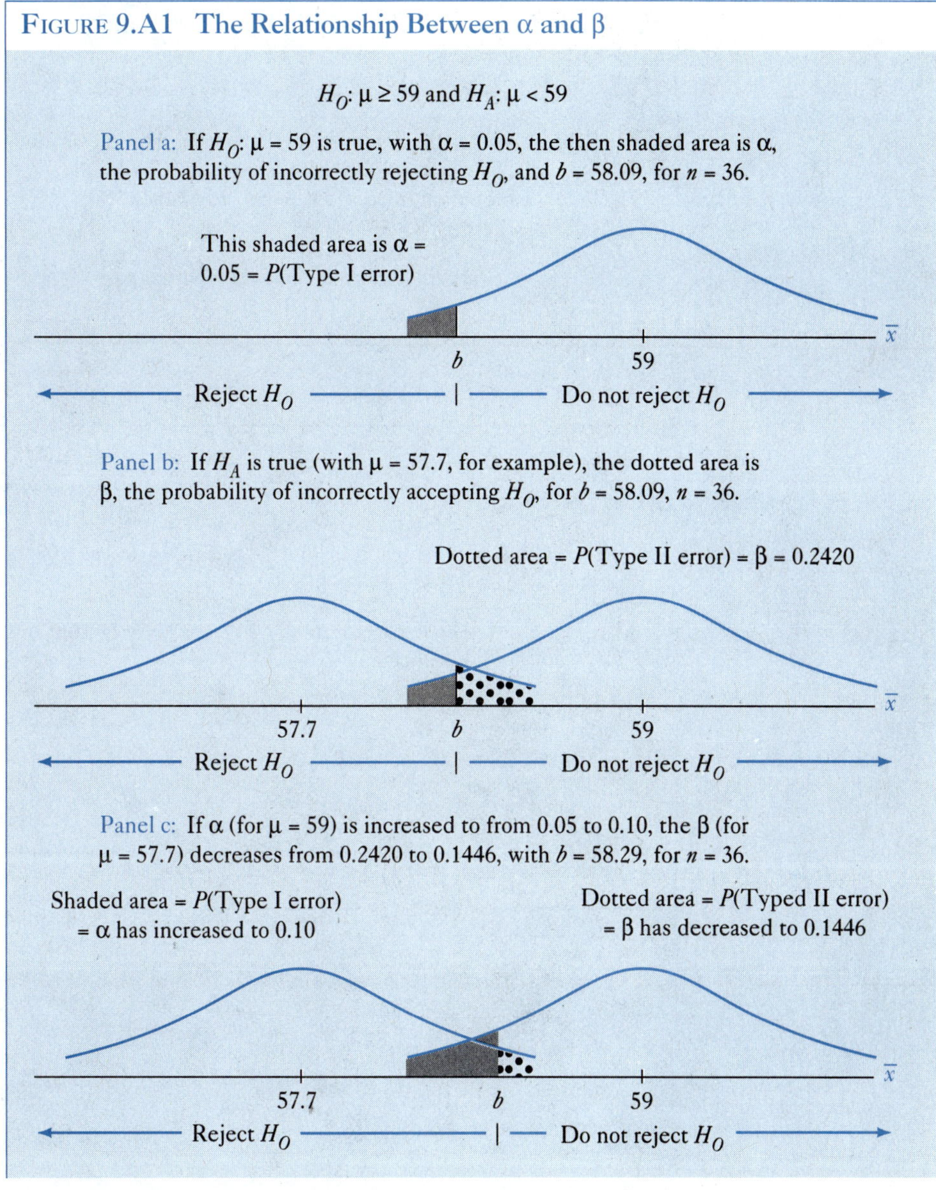

is calculated as $P(\bar{x} \geq 58.09 \mid \mu = 57.7) = P(z \geq 0.70) = 0.2420$, where

$$z = \frac{58.09 - 57.7}{3.333/\sqrt{36}} = 0.70$$

FIGURE 9.A2 The Relationship Between α and β

H_O: μ ≥ 59 and H_A: μ < 59

Panel a: If H_O: μ = 59 is true, with α = 0.05, then the shaded area is the probability of incorrectly rejecting H_O (a Type I error), with n =36. If H_A is true (with μ = 57.7, for example), then the dotted area is the probability of incorrectly accepting H_O (a Type II error), n = 36.

Shaded area = α Dotted area = β

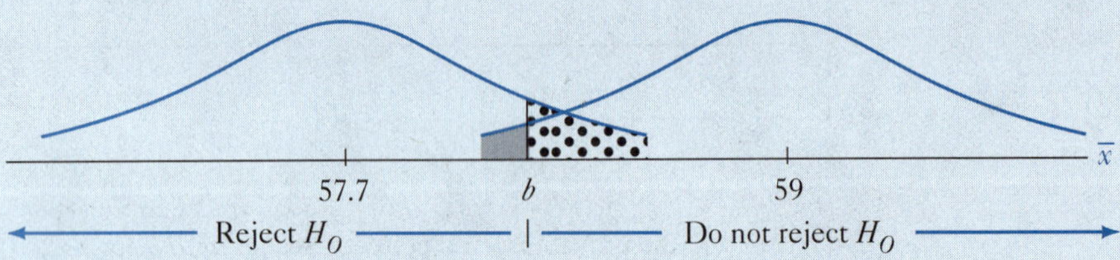

Panel b: If n increases from 36 to 144, with α fixed at 0.05, the standard deviation of $\bar{x}$ decreases from 0.556 to 0.278.

Shaded area = α is unchanged

Dotted area = β has decreased

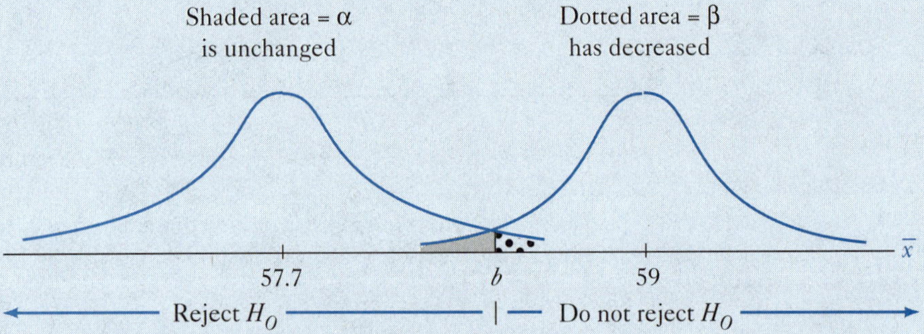

Panel c: If α and β can vary but b is fixed as in Panel a, both α and β decrease as n increases from 36 to 144.

Shaded area = α has decreased

Dotted area = β has decreased

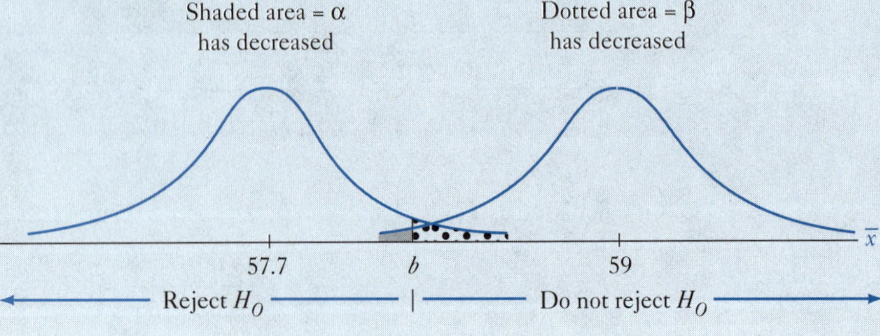

If the probability of making a Type I error α is increased, from 0.05 to 0.10, the critical $\bar{x}$ value (b) will rise from 58.09 to 58.29, as in Figure 9.A1, panel c. As b is moved to the right, the set of $\bar{x}$ values leading to the rejection of H_0 becomes larger, with the shaded area increasing from 5 to 10 percent. This implies that the dotted area representing β is smaller, as seen by comparing panels b and c. In particular, with $\mu = 57.7$, the probability of a Type II error falls from 0.2420 to 0.1446, since $P(\bar{x} \geq 58.29 \mid \mu = 57.7) = P(z \geq 1.06) = 0.1446$, where

$$z = \frac{58.29 - 57.7}{3.333/\sqrt{36}} = 1.06$$

In general, for a given sample size, raising (lowering) the probability of a Type I error will lower (raise) the probability of a Type II error.

Either or both the probabilities of a Type I error and a Type II error can be reduced by increasing the sample size. Figure 9.A2, panel a, shows the initial probabilities of Type I and Type II errors when μ equals 59 versus 57.7. If the sample size is increased to $n = 144$, however, the standard deviation of the $\bar{x}$ distribution is cut in half, going from 0.556 $(= 3.333/\sqrt{36})$ to 0.278 $(= 3.333/\sqrt{144})$. The effect of this decrease in the standard deviation is shown in Figure 9.A2, panel b. The increase in the sample size moves the critical $\bar{x}$ value of b closer to the null hypothesis value of μ, if α remains at 0.05. The probability of incorrectly accepting H_0 falls as b moves closer to 59 so the β area is smaller in panel b than in panel a. If, on the other hand, the critical value of b remains fixed, as in panel c, then α decreases along with β. Increasing the sample size can decrease both the probability of a Type I error and a Type II error.

Appendix 9B The Operating-Characteristic Curve and the Power of a Test

The decision rule for rejecting or accepting the null hypothesis is based on a predetermined and fixed level of α. But how large is β, the probability of a Type II error? In Figure 9.A1 of Appendix 9A, the calculation of β for an alternative value of μ was demonstrated. The null hypothesis was $\mu = 59$, and the alternative value for which β was calculated was $\mu = 57.7$. At $\mu = 57.7$, $\beta = 0.2420$. As μ falls further below 59, β will decrease. For example, at $\mu = 57$, with H_0: $\mu = 59$ and $\alpha = 0.05$, $\beta = 0.0250$ [because $\alpha = 0.05 = P(z \leq -1.645) = P(\bar{x} \leq 58.09 \mid \mu = 59)$ and $P(\bar{x} \geq 58.09 \mid \mu = 57.0) = P(z \geq 1.96) = 0.0250$].

Operating-Characteristic Curve The graph showing the probability of Type II errors as the population parameter takes values other than that specified in the null hypothesis.

Figure 9.B1 shows the declining values of β as μ falls below the null hypothesized value of $\mu = 59$. Panel a has H_0: $\mu = 59$ and assumes $\mu = 59$. In panel b, H_0: $\mu = 59$ and $\alpha = 0.05$ but μ is set equal to 57.7, so $\beta = 0.2420$. Panel c has H_0: $\mu = 59$ and $\alpha = 0.05$ but with $\mu = 57$, $\beta = 0.0250$. Plotting each value of β against the corresponding μ, as a continuous function, gives Figure 9.B2, which is called the **operating-characteristic curve** for this test.

FIGURE 9.B1 Correspondence Between μ and β

H_O: μ ≥ 59 and H_A: μ < 59

Panel a: If H_O: μ = 59 is true, with α = 0.05, then the shaded area is α.

Shaded area is α = 0.05

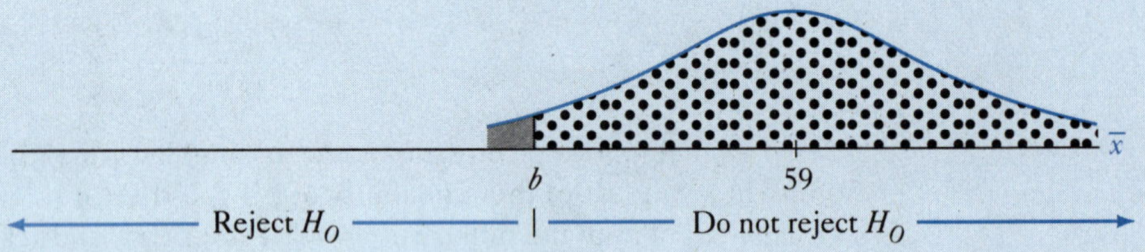

Panel b: If H_A is true (with μ = 57.7, for example), the dotted area is β, the probability of incorrectly accepting H_O, for b = 58.09, n = 36.

Shaded area is α = 0.05 Dotted area = β = 0.2420

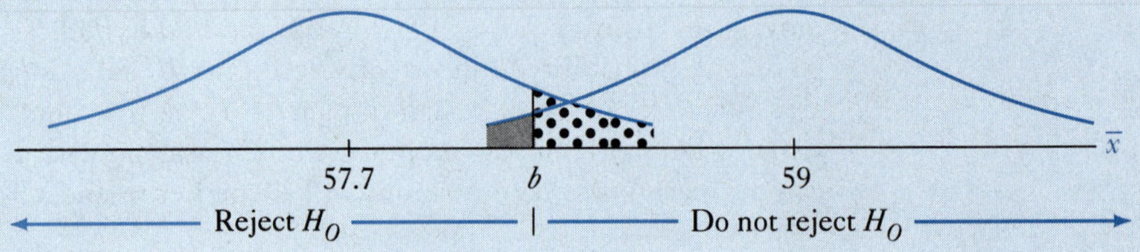

Panel c: If H_A is true (with μ = 57.0, for example), the dotted area is β, the probability of incorrectly accepting H_O, for b = 58.09, n = 36.

Shaded area is α = 0.05 Dotted area = β = 0.0250

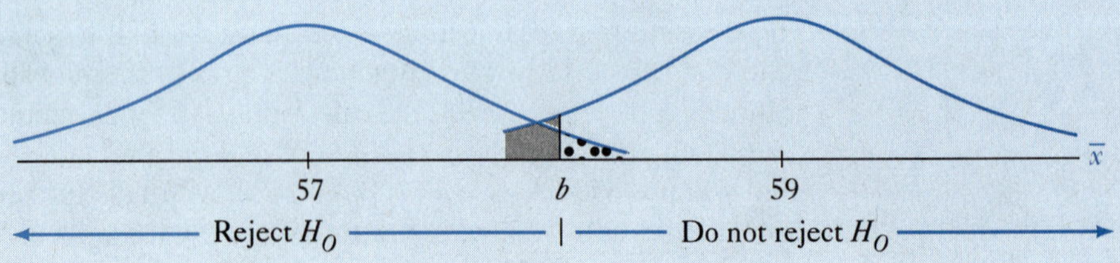

FIGURE 9.B2 Operating-Characteristic Curve of Test that Mean Drive Belt Life is 59 Months, for $n = 36$

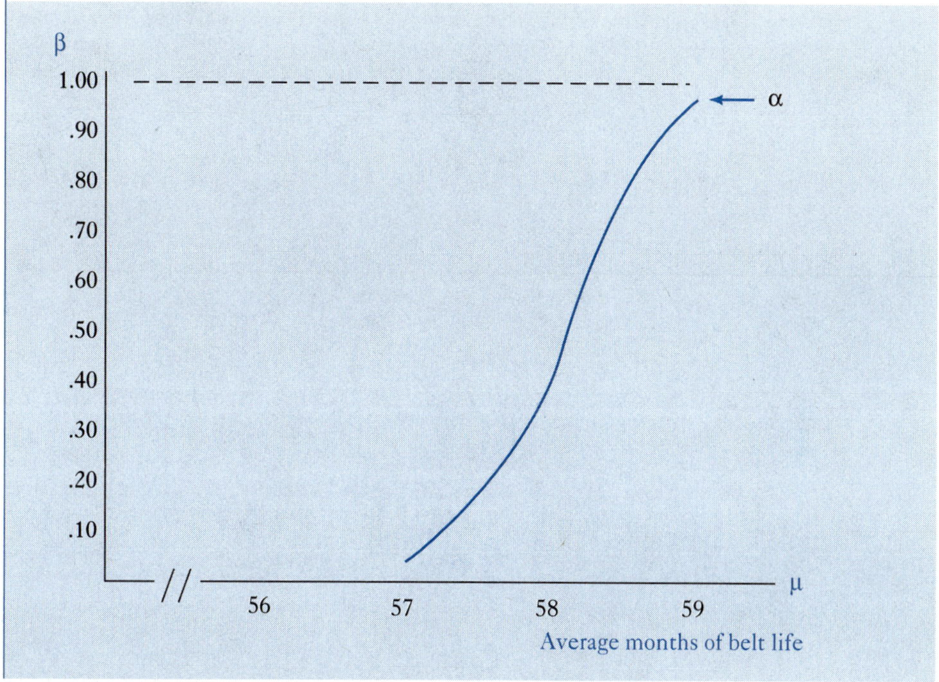

As shown in Figure 9.B3, increasing the sample size shifts the entire operating-characteristic curve to the right, for a fixed critical $\bar{x}$ value b. The reason for this already has been demonstrated in Figure 9.A2, panel c, where for a fixed critical $\bar{x}$ value b, both α and β were shown to decrease for an increase in sample size from 36 to 144. It is worth pointing out, however, that if μ truly equals 59, then an increase in the sample size will reduce α but increase β, for a fixed critical $\bar{x}$ value b. This can be seen in Figure 9.B3 by the higher β value at $\mu = 59$ for the operating-characteristic curve for $n = 144$. If α is held constant at 0.05 and the critical $\bar{x}$ value b is left free to adjust, however, then both the operating-characteristic curves for $n = 36$ and $n = 144$ would show $\beta = 0.95$ at $\mu = 59$.

Power of a Test
The probability of rejecting the null hypothesis in favor of a specific-valued alternative when the alternative is true.

Econometricians, psychometricians, biometricians, cliometricians, and statisticians in general often talk about the **power of a test.** The power curve for a test shows the value of $(1 - \beta)$, whereas the operating-characteristic curve shows the value of β at each value of μ. From the operating-characteristic curve in Figure 9.B3, we can see how the power of the test increases when the sample size increases. Sample sizes can become so large that the slightest deviation from the value in the null hypothesis results in the rejection of the null hypothesis. On the other hand, a population could have so much variability, that a test of its mean with a given sample size would have little power to differentiate among alternative values of its mean.

FIGURE 9.B3 Operating-Characteristic Curve of Test that Mean Drive Belt Life is 59 Months, for n = 36 Versus n = 144

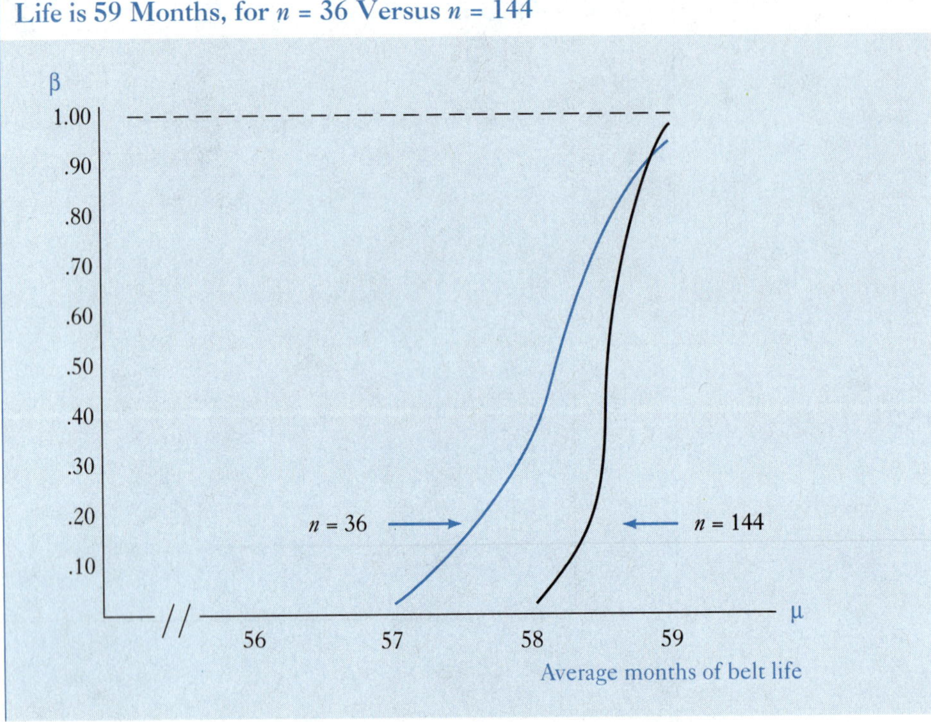

The shape of the operating characteristic curve depends on the null and alternative hypotheses being tested. If the test is

$$H_O: \mu \leq \mu_0 \text{ versus } H_A: \mu > \mu_0$$

then the operating characteristic curve will be like that in Figure 9.B4. If the test is

$$H_O: \mu = \mu_0 \text{ versus } H_A: \mu \neq \mu_0$$

then the operating-characteristic curve will be like that in Figure 9.B5. Tests based on populations with more variability or those using smaller sample sizes will have flatter operating-characteristic curves.

Ideally, either operating-characteristic curves or power curves would be constructed for all tests. This is seldom done, possibly because of the computational cost. At a minimum, however, the implications of the test's power should be checked when the null hypothesis is not rejected. For example, in the paper towel illustration, the null and alternative hypotheses were

$$H_O: \mu = 60 \text{ and } H_A: \mu \neq 60$$

and the test statistic and its p-value were

$$t_c = \frac{\bar{x} - \mu_0}{s/\sqrt{n}} = \frac{64.1 - 60}{13.452/\sqrt{31}} = 1.697$$

FIGURE 9.B4 Operating-Characteristic Curve for a Test of an Alternative $\mu > \mu_0$ Versus the Null $\mu = \mu_0$

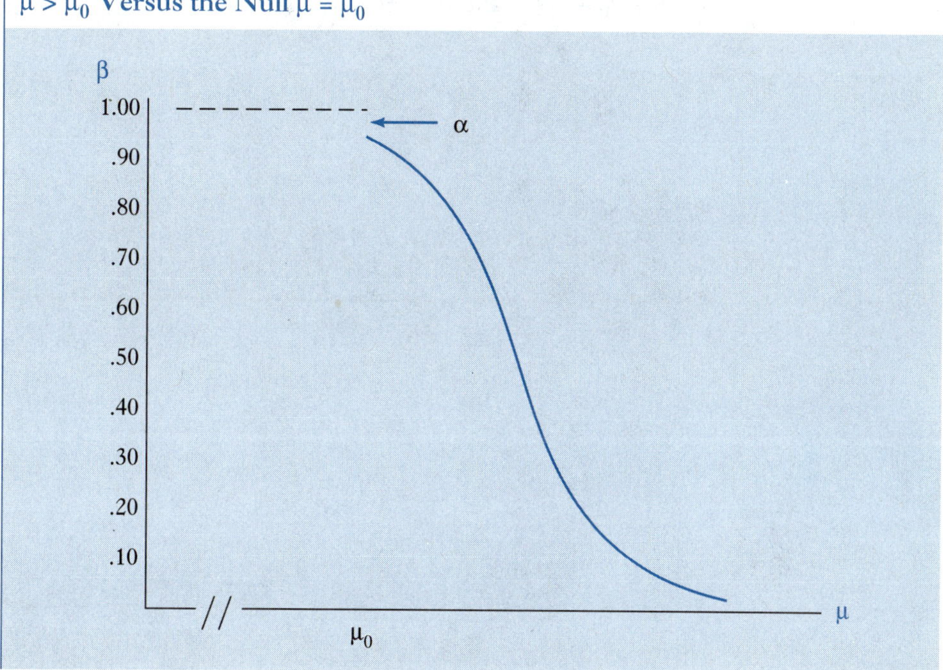

$$p\text{-value} = 2P(t \geq 1.697) = 2(0.05) = 0.10$$

Thus, the null hypothesis could not be rejected for any α level at or below 10 percent. But at the 10 percent level of significance, μ values as high as 68.2 likewise cannot be rejected because

$$t_c = \frac{\bar{x} - \mu_0}{s/\sqrt{n}} = \frac{64.1 - 68.2}{13.452/\sqrt{31}} = -1.697$$

$$p\text{-value} = 2P(t \leq -1.697) = 2(0.05) = 0.10$$

Not being able to reject either $\mu = 60$ or $\mu = 68.2$ may seem paradoxical, but it simply reflects the lack of power of the test. The more powerful the test, the closer non-rejectable values will be to each other. When the null hypothesis is not rejected, remember there is always another parameter value that can likewise not be rejected. Not rejecting the null hypothesis is not equivalent to accepting the parameter value it contains.

Statisticians generally agree that the usual one-tailed test for the mean of a normally distributed random variable (with known variance) is uniformly most powerful, in the sense that for a preset Type I error it minimizes the Type II error for any alternative value of the mean. Two-tailed tests are generally regarded as having no power, although David Edelman (1990) shows that in a restricted sense, even two-tailed tests can have power.

FIGURE 9.B5 Operating-Characteristic Curve for a Test of an Alternative $\mu \neq \mu_0$ Versus the Null $\mu = \mu_0$

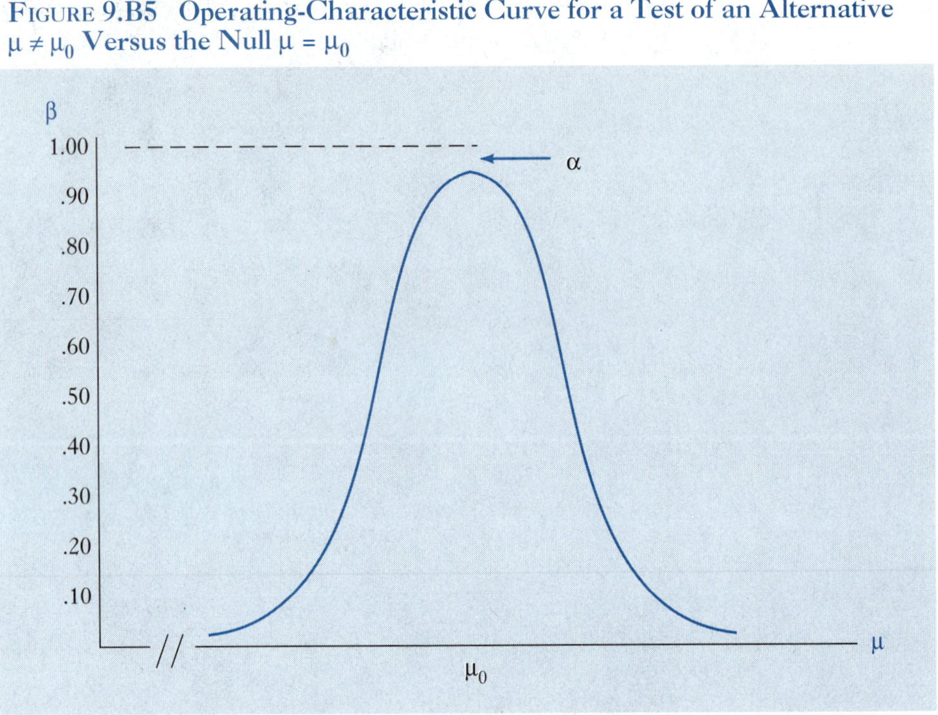

APPENDIX 9C THE OPERATING-CHARACTERISTIC CURVE AND ACCEPTANCE SAMPLING

Acceptance Sampling
Sampling a lot, batch or shipment to determine if it has too many defects.

Operating-characteristic (OC) curves have become important tools in statistical quality control. **Acceptance sampling**, as it is known today, originated during World War II as a problem presented to mathematician/statistician Abraham Wald by Nobel Laureate economist Milton Friedman and statistician W. Allen Wallis. At issue was the number of rounds of ammunition firing needed to determine whether one method of ordnance handling was superior to another. Wald's classic book *Sequential Analysis* set the foundation for the statistical quality control procedures popularized by W. E. Deming and others in the 1980s.

The role of the OC curve in quality control can be appreciated by letting x be the number of defects in a sample of size n so that $p = x/n$ is the estimator of the proportion of defects π in the entire lot, shipment, or batch under analysis. If b is the critical number of defects in the sample for which the lot is rejected, then if the sample proportion of defects p is observed above b/n, when only π_0 defects were expected, the shipment is rejected. As a hypothesis test this can be stated

$$H_O: \pi \leq \pi_0 \text{ (The lot does not have too many defects.)}$$

$$H_A: \pi > \pi_0 \text{ (The lot has too many defects.)}$$

Decision rule: If $p = x/n > b/n$, then H_O is rejected and the entire lot, batch, or shipment is unacceptable.

Now for alternative defect rates ($\pi > \pi_0$), the operating-characteristic curves can be constructed. The resulting OC curves would have the general shape shown in Figure 9.C1, with the population proportion of defects $0 < \pi < 1$ measured on the horizontal axis.

Typically in acceptance sampling, values of π, p, and b/n close to zero for small samples are of interest. For instance: What are the probabilities of obtaining no more than one defect in a sample of six items, if the lot actually contains 2 percent or 4 percent defects? In this case, the null hypothesis can be thought of as $\pi = 0$ versus the alternative $\pi > 0$. The sample proportion p cannot be assumed to be normally distributed, however, since the critical value $b/n = 1/6$ is too close to zero. The binomial distribution

FIGURE 9.C1 Operating-Characteristic Curve for a Test in Which a Lot is Accepted, if $\pi = 0$ is Not Rejected, and the Lot is Rejected if $\pi > 0$ is Accepted

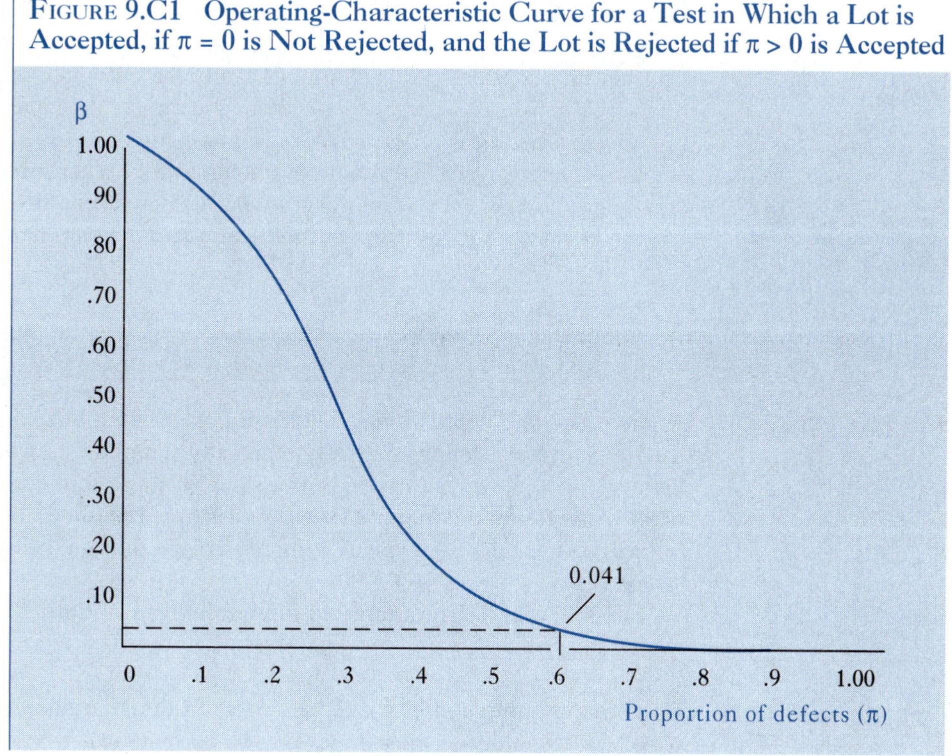

must be employed. By laborious hand calculations, or from an extended binomial table, or ideally from a computer program, the following probabilities of Type II errors can be obtained.

$$P(x = 0 \text{ or } 1 \mid n = 6, \pi = .02) = (6!/0!6!)(.02)^0(.98)^6 + (6!/1!5!)(.02)^1(.98)^5 = 0.994$$
$$P(x = 0 \text{ or } 1 \mid n = 6, \pi = .04) = (6!/0!6!)(.04)^0(.96)^6 + (6!/1!5!)(.04)^1(.96)^5 = 0.978$$

and similarly for ever increasing values of π to

$$P(x = 0 \text{ or } 1 \mid n = 6, \pi = .90) = 0.000$$

The calculated β values are plotted to form an operating-characteristic curve as in Figure 9.C1. This OC curve states that 99.4 percent of the time a lot will be accepted (because the null hypothesis $\pi = 0$ is not rejected) when, in fact, 2 percent of the lot is defective ($\pi = 0.02$ is true). Not until the majority of items in the lot are defective can this test be relied on to lead to a rejection of the shipment. When 60 percent of the lot is in fact defective, there is less than a 5 percent chance (0.041, as shown in Figure 9.C1) that it will be erroneously accepted. To obtain a more powerful test, the sample size must be increased or the critical value of b must be reduced from one to zero.

The idea behind quality control based on acceptance sampling is simple: If the manufacturing process is producing too many defects "it is out of control" and needs to be fixed. To make this judgment for continuous processes that are associated with mass production or for products that can be checked only through their own destruction, sampling is the only means of testing. Similarly, for those purchasing these goods, the decision to accept or reject a shipment can be based only on statistical tests. Conceptually, the power of the test $(1 - \beta)$ can be maximized by increasing the sample size or by reducing b subject to the cost of sampling and the cost of erroneously rejecting a shipment. There typically is no unique solution to this constrained maximization problem, however. Edward G. Shilling (1982) discusses acceptance plan solutions currently used in business and government production and procurement.

Appendix 9D Type I and Type II Errors and Sample Size Selection

As discussed in Chapter 8, formulas based on the idea of a confidence interval can be used to suggest a sample size before actually sampling. Lawrence Kupper and Kerry Hafner (1989), however, question the value of these formulas for small-sample situations. They found that sample size formulas based on the tradeoff between the probabilities of Type I and Type II errors "perform amazingly well even for very small sample sizes."

Suppose that we are interested in performing a test of

$$H_O\colon \mu \le \mu_0 \text{ and } H_A\colon \mu > \mu_0$$

for a random sample of size n to be selected from a normal distribution with variance σ^2. The question is: What sample size should be drawn?

We know that the critical $\bar{x}^*$ is related to the critical z value, for a test in which the probability of a Type I error is α, by the formula

$$z_\alpha = \left(\bar{x}^* - \mu_0\right)/\sigma/\sqrt{n}$$

where z_α is the value of z that puts α in the upper tail. Alternatively, the relationship between this value of $\bar{x}^*$ and the value of z_β, if μ is truly equal to μ_1 (where $\mu_1 > \mu_0$), is

$$z_\beta = \left(\bar{x}^* - \mu_1\right)/\sigma/\sqrt{n}$$

where β is the probability of a Type II error. Rearranging both equations to obtain $\bar{x}^*$ as a function of the other quantities and substituting for $\bar{x}^*$ from one equation into the other gives

$$\mu_0 = z_\alpha \sigma/\sqrt{n} = \mu_1 + z_\beta \sigma/\sqrt{n}$$

from which we can solve for n as

$$n = \left[\left(z_\alpha - z_\beta\right)\sigma/\left(\mu_1 - \mu_0\right)\right]^2$$

Although not considered by Kupper and Hafner, a sample size formula based on a one-tailed test of

$$H_0\colon \pi \le \pi_0 \text{ and } H_A\colon \pi > \pi_0$$

can be derived in a similar manner. In particular, the sample size n is given by

$$n = \left\{\left[z_\alpha \sqrt{\pi_0\left(1 - \pi_0\right)} - z_\beta \sqrt{\pi_1\left(1 - \pi_1\right)}\right]/\left(\pi_1 - \pi_0\right)\right\}^2$$

Whenever possible these formulas for n are preferred to those based on the confidence interval formulations. Their use, however, requires some knowledge of alternative values for μ and π. Finally, there are computer programs such as NCSS-PASS that can aid in power analysis and sample size calculations. It is not necessary to do these calculations by hand.

Chapter 10

Two-Sample Hypothesis Testing

Moreover, it sometimes appears that statistics are less likely to be relied upon if they challenge one's own interest.

Mary W. Gray
(*Statistical Science*, 1993)

10.1 Introduction

So far our estimation and testing has involved a single random variable. This chapter extends those ideas to tests involving two variables. Later chapters will continue this extension to estimation and predictions involving multiple variables.

Tests of the difference between two means, the difference between two proportions, the difference between two medians, as well as the independence of two random variables are presented in this chapter in the following situations.

- Women managers brought suit against a large midwestern bank for salary discrimination. At a deposition, an expert witness presented data in which the average annual salary of 219 women managers was \$34,593, while that of 392 male managers was \$43,331. The question: Are women managers receiving significantly less than men?
- Participants are tested at the beginning of a corporate training program; they are retested at the end of the program. The question: Was the value added by this program statistically significant?
- *PCWEEK* (February 1991) reported results from an International Data Corporation study of computer sites. It found that 92 percent of the 106 responding managers of sites using Hewlett Packard equipment were "satisfied" or "very satisfied" with HP, while only 84 percent of the 115 responding Compaq site managers were so satisfied with Compaq. Can a hypothesis test determine whether Compaq computer site managers in general are less satisfied than HP site managers?
- The *American Statistician* (August 1990) reported on a study showing that, "The difference between mean exam scores for students in the computer use group and the control group was not statistically significant." What is the practical importance of this finding?

The hypothesis testing procedure described in Chapter 9 can be used in all these personal and business situations. Because many different statements of hypotheses are possible, however, care must be taken to correctly specify the null and alternative hypotheses to test for specific numerical values of the parameters (Step 1). This chapter introduces new test statistics that must be employed when testing for differences between two values (Step 3). The t and z statistics are used to test differences in means, proportions, and medians. In testing for differences between two measures of central tendency, potential differences in measures of dispersion must also be considered, because sampling may now be from two different populations (Step 4). The F distribution is introduced for testing differences between variances.

The specification of Type I error probabilities (Step 2), the determination of a p-value (Step 5), and the process of reaching a conclusion (Step 6) typically vary little from one problem to another. These steps were discussed in detail in Chapter 9, and little need be added in this chapter. By the end of this chapter, you should fully recognize why only Steps 1, 3, and 4 in the hypothesis testing procedures need to be changed to fit the context of the situation, while Steps 2, 5, and 6 generally remain the same from situation to situation.

10.2 Difference Between Means

Possibly the most common tests in statistics are those involving differences between means. For example, we might want to compare revenues from sales in two different cities, or the average number of units sold in two different months, or two factory shifts might be compared on the basis of defective items produced. These situations may involve hypotheses tests of the difference between two population means based on samples that share no common elements. The samples may be completely independent of each other and may be of unequal size. The objective is to determine if there is a significant difference between the two populations based on the two sets of independent randomly selected samples. The test statistic may be a z or a t test, with the familiar form

$$\text{Test statistic} = \frac{\text{Estimator} - \text{Expected value of estimator}}{\text{Standard deviation of estimator}}$$

where the expected value of estimator is again given by the null hypothesis, and the estimator's standard deviation is called the standard error. Now, however, the estimator and its standard error will be based on the difference between two random variables.

Let the mean of the first of two possibly different populations be μ_1, and that of the second population be μ_2. If the theory is that there is no difference between these two means, then the null and alternative hypotheses are

H_O: $\mu_2 - \mu_1 = 0$

H_A: $\mu_2 - \mu_1 \neq 0$, if μ_1 is suspected to be different than μ_2.

If the theory suggests that one population mean is less than the other, the null and alternative hypothesis for a one-sided, two-sample test may be specified as

H_O: $\mu_2 - \mu_1 \geq 0$

H_A: $\mu_2 - \mu_1 < 0$, if μ_1 is suspected to be greater than μ_2;

or H_O: $\mu_2 - \mu_1 \leq 0$

H_A: $\mu_2 - \mu_1 > 0$, if μ_1 is suspected to be less than μ_2.

Conceptually, in a test of means using two samples, any numerical difference between the population means may be tested, but usually the value in the null hypothesis is zero.

The point estimator of the population difference in means $(\mu_2 - \mu_1)$ is the difference in sample means $(\bar{x}_2 - \bar{x}_1)$. This is an unbiased estimator, since on average it will equal the population difference in means. To use either the t or z statistics, the distribution of the estimator $(\bar{x}_2 - \bar{x}_1)$ must be normally distributed, which it will be if each sample mean is itself normally distributed or if the sample sizes are large. Whether the t or z statistic is appropriate will once again depend on knowledge of the variances and sample sizes.

When Variances Are Known (or Samples Are Large)

If the population variances are known, then the standard error of the estimator will be related positively to these two individual population variances, σ_1^2 and σ_2^2, and negatively to the two sample sizes, where[1]

$$\sigma_{(\bar{x}_2 - \bar{x}_1)} = \sqrt{\frac{\sigma_1^2}{n_1} + \frac{\sigma_2^2}{n_2}}$$

Because s_2^2 and s_1^2 are unbiased estimators of σ_2^2 and σ_1^2 and because $s/\sqrt{n}$ approaches $\sigma/\sqrt{n}$ quickly as n is increased, the sample variances provide good approximations for the population variances for large sample sizes. Thus, if the sample sizes are large, tests of differences in means can be conducted using the z statistic, even if the population variances are unknown.

z test statistic for two independent samples

$$z = \frac{(\bar{x}_2 - \bar{x}_1) - D_O}{\sqrt{\frac{\sigma_1^2}{n_1} + \frac{\sigma_2^2}{n_2}}} \tag{10.1}$$

where D_O is the difference specified by the right-hand side of H_O. D_O is typically zero.

FOR USE WHEN:

1. The two parent population variances are known, the random samples are independent of each other, and
   a) each is at least of size $n = 30$, or
   b) each is drawn from a normal parent population.

OR 2. Each independent sample is sufficiently large ($n > 30$) that the sample variances are good approximations of their respective population variances.

Although population variances are seldom known, Formula (10.1) is extremely useful for the initial check of mean differences.[2] For instance, Table 10.1 provides the sample data presented by the econometrician in the suit brought by the women managers at the midwestern bank for discrimination in the salary determination process. Data are provided for the beginning and end of the liability period, which was predetermined by the court. In the beginning of the liability period, for example, the sample of 219 women averaged \$34,593, with a standard deviation of \$6,229; the highest salary paid a woman was \$66,800, and the lowest was \$21,240. A sample of 392 men in the same time period averaged \$43,331, with a standard deviation of \$15,842; the highest salary paid a man was \$195,900, and the lowest was \$25,500. The question: Do women receive significantly less than men?

For this question, the null and alternative hypotheses are

H_O: $\mu_m - \mu_w \le 0$

H_A: $\mu_m - \mu_w > 0$

where the subscripts designate men (m) or women (w). Although the population variances are unknown, the sample sizes are large, so the z in Formula (10.1) can be used. At the beginning and end of the liability period, the calculated values of the test statistics are:

Beginning of liability period

$$z = \frac{(\bar{x}_2 - \bar{x}_1) - 0}{\sqrt{\frac{\sigma_1^2}{n_1} + \frac{\sigma_2^2}{n_2}}} \approx \frac{(43331 - 34593) - 0}{\sqrt{\frac{(15842)^2}{392} + \frac{(6229)^2}{219}}} = 9.665$$

End of liability period

$$z = \frac{(\bar{x}_2 - \bar{x}_1) - 0}{\sqrt{\frac{\sigma_1^2}{n_1} + \frac{\sigma_2^2}{n_2}}} \approx \frac{(45809 - 36840) - 0}{\sqrt{\frac{(17601)^2}{376} + \frac{(6998)^2}{199}}} = 8.671$$

Each of the two calculated values of the z test statistic ($z = 9.665$ and 8.671) is well above the critical values associated with the typical Type I error levels (at $\alpha = 0.05$ in a one-tailed test, the critical z is 1.645). Thus, the null hypothesis is rejected, and the conclusion is that women are paid significantly less than men. To jump from this "statistical significance" to the practical conclusion that women are discriminated against on the basis of their sex is not justified, however.

Recall that classical hypothesis testing requires simple random samples. These samples of salaries are not simple random samples; there are many factors involved here besides just sampling error. Notice, for instance, that the standard deviations for male salaries are over two and a half times larger than those of females, and that at the high end of the salary ranges, males are receiving over three times what the highest paid females are receiving. The source of this additional variability in male salaries must be assessed before any firm conclusions are advanced. Consideration of the sources and consequences of this additional variability will have to wait until later chapters, when multivariate analysis is introduced.

Table 10.1 Salaries at Midwestern Bank

LIABILITY PERIOD	SIZE	AVERAGE	ST. DEV.	LOW	HIGH
Beginning of liability period					
MEN	392	43331	15842	25500	195900
WOMEN	219	34593	6229	21240	66800
MEN – WOMEN	173	8738		4260	129100
End of liability period					
MEN	376	45809	17601	26700	211000
WOMEN	199	36840	6998	24000	71400
MEN – WOMEN	177	8969		2700	139600

QUERY 10.1:

During a period in the late 1980s, automobile ads claimed that "Leasing a New Car Is Now Smarter and More Advantageous Than Buying," as stated, for example, in a Bay Leasing ad in the *Chicago Tribune* (March 8, 1988). In particular, a sample of domestic vehicles was presented as substantially less expensive to lease than imports. Based on the Bay Leasing data given in its ad (as reproduced in the table below), are domestic vehicles less expensive to lease than imports?

Domestics	Amount	Imports	Amount
Buick Electra	$237.00	Acura Integra	$147.00
Buick Reatta	$357.00	Acura Legend	$247.00
Buick Regal	$177.00	Audi 80	$267.00
Cadillac Brougham	$327.00	BMW 325	$277.00
Cadillac Deville	$317.00	BMW 528	$327.00
Cadillac Eldorado	$337.00	BMW 735	$677.00
Cadillac Seville	$357.00	Honda Accord Coupe	$157.00
Chevy Astrovan	$147.00	Honda Accord	$147.00
Chevy Blazer	$157.00	Honda Civic	$117.00
Chevy Beretta	$147.00	Honda CRX	$127.00
Chevy Celebrity	$157.00	Honda Prelude	$167.00
Chevy Corsica	$147.00	Isuzu Trooper	$177.00
Chevy Corvette	$387.00	Jaguar XJ-S	$657.00
Chevy Corvette Conv	$447.00	Jaguar XJ6	$577.00
Chrysler Lebaron Coupe	$177.00	Mazda MX6 or 626	$147.00
Chrysler Fifth Avenue	$257.00	Mazda 929	$257.00
Dodge Caravan	$167.00	Mazda RX-7	$167.00
Eagle Premier	$217.00	Mercedes 190E	$357.00
Ford Aerostar	$167.00	Mercedes 300E	$537.00
Ford F-250	$187.00	Mercedes 420SEL	$587.00
Ford Mustang	$127.00	Mercedes 560	$747.00
Ford Thunderbird	$167.00	Nissan 300ZX	$287.00
Ford Taurus	$157.00	Nissan Maxima	$227.00
Jeep	$177.00	Porsche 911	$517.00
Lincoln Town Car	$337.00	Porsche 928	$887.00
Lincoln Continental	$347.00	Porsche 944	$397.00
Mercury Sable	$157.00	Saab 900	$207.00
Olds Ciera	$157.00	Saab 9000	$317.00
Olds Delta 88	$217.00	Saab Convt	$387.00
Olds Regency 98	$247.00	Sterling 825S	$277.00
Pontiac Bonneville	$207.00	Toyota Camry	$177.00
Pontiac Grand Am	$147.00	Toyota Celica	$137.00
Pontiac Grand Prix	$187.00	Toyota Cressida	$237.00
Pontiac Trans Am	$207.00	Toyota Supra	$287.00
		Volvo 240DL	$207.00
		Volvo 740	$277.00
		Volvo 760	$357.00

ANSWER:
The average monthly cost of leasing is $226.71 for domestic vehicles and $325.65 for imports. Within the sample, domestic cars are $98.94 less expensive than imports to lease, on average. If these data are viewed as simple random samples of all such leasing expenses, then from a hand calculation or a computer printout the following can be obtained:

SAMPLE	IMPORTS	DOMESTIC
Mean	325.65	226.71
Variance	37695.35	7530.21
St. Dev.	194.15	86.78

Because the samples are relatively large, the z statistic can be used for testing; its calculated value for a test of $\mu_2 - \mu_1 \le 0$ versus $\mu_2 - \mu_1 > 0$ is

$$z = \frac{(\bar{x}_2 - \bar{x}_1) - 0}{\sqrt{\frac{\sigma_1^2}{n_1} + \frac{\sigma_2^2}{n_2}}} \approx \frac{(325.65 - 226.71) - 0}{\sqrt{\frac{37695.35}{37} + \frac{7530.21}{34}}} = \frac{98.94}{35.22} = 2.81$$

The p-value is 0.0025, which is small. It implies that the null hypothesis $(\mu_2 - \mu_1 \le 0)$ can be rejected, and the conclusion is that domestic cars are significantly less expensive than imports to lease.

When Variances Are Unknown but Assumed Equal

The two-sample z test in the last section typically is used only when the samples are large, because population variances are seldom known. Here we consider a two-sample test that may be appropriate when the samples are small and the population variances are *unknown*. This two-sample t test requires the following assumptions:

1. The population variances are unknown but reasonably assumed to be equal.
2. The samples are independent of each other and are drawn from normal parent populations.

These assumptions are maintained through the testing procedure.[3] In essence, they imply that the two populations are identical except for their means, the difference of which is being tested.

Unlike a z statistic, the t statistic is based on an estimate of the standard error of the estimator $(\bar{x}_2 - \bar{x}_1)$. As discussed in Chapters 8 and 9, when the standard error of one mean is estimated, the degrees of freedom are $n - 1$. For two samples of identical size, the degrees of freedom would be $2(n - 1)$. If the samples are of different sizes n_1 and n_2, then the degrees of freedom are $n_1 + n_2 - 2$.

The test statistic for this two-sample test is:[4]

t test statistic for difference between means, with equal population variances and independent samples

$$t = \frac{(\bar{x}_2 - \bar{x}_1) - 0}{s_{\bar{x}_2 - \bar{x}_1}} \tag{10.2}$$

where

$$s_{\bar{x}_2 - \bar{x}_1} = \sqrt{\left(\frac{s_1^2(n_1 - 1) + s_2^2(n_2 - 1)}{n_1 + n_2 - 2}\right)\left(\frac{n_1 + n_2}{n_1(n_2)}\right)}$$

with $n_1 + n_2 - 2$ degrees of freedom and 0 is the value on the right-hand side of H_O.

FOR USE WHEN:

1. The two parent population variances are unknown but equal, and
2. Both independent samples are small but drawn from populations that are normally distributed.

As an illustration of this two-sample test, consider a quality control problem in which there are two shifts of workers at a large restaurant. Patrons are provided with questionnaires that ask them to assess service on a scale from 0 (bad) to 100 (great). Suppose that the following ratings are obtained:

Day shift scores:
91.0 55.0 62.0 64.0 41.0 63.0 35.0 57.0 94.0 68.0 48.0 72.0

Night shift scores:
81.0 86.0 73.0 77.0 90.0 91.0 75.0 48.0 97.0 74.0

The question is whether there is a statistically significant difference in ratings between the two shifts. That is, the null and alternative hypotheses are

$$H_O: \mu_2 - \mu_1 = 0$$

$$H_A: \mu_2 - \mu_1 \neq 0$$

where μ_2 is the average night shift service rating, and μ_1 is the average day shift service rating.

Using a spreadsheet computer program (such as EXCEL, LOTUS 1-2-3, or QUATTRO), the calculations in Table 10.2 can be made easily. Substituting into Formula (10.2), we obtain the calculated test statistic value of $t = 2.44$, where we assume the day shift and night shift populations of ratings are normally distributed.

Table 10.2 Calculated Values for a Pooled, Two-Sample t Test with Equal but Unknown Variances

DAY SHIFT (x_1)

Scores

91.0	55.0	62.0	64.0	41.0	63.0	35.0	57.0	94.0	68.0	48.0	72.0

$\bar{x}_1 = 62.50$

Squared Deviations

812.25	56.25	0.25	2.25	462.25	0.25	756.25	30.25	992.25	30.25	210.25	90.25

$s^2{}_1 = 313.00$

$s_1 = 17.69$

NIGHT SHIFT (x_2)

Scores

81.0	86.0	73.0	77.0	90.0	91.0	75.0	48.0	97.0	74.0

$\bar{x}_2 = 79.20$

Squared Deviations

3.24	46.24	38.44	4.84	116.64	139.24	17.64	973.44	316.84	27.04

$s_2^2 = 187.07$

$s_2 = 13.68$

Estimated Standard Error = 6.86

$t = 2.44$

$$t = \frac{(\bar{x}_2 - \bar{x}_1) - 0}{\sqrt{\left(\frac{s_1^2(n_1 - 1) + s_2^2(n_2 - 1)}{n_1 + n_2 - 2}\right)\left(\frac{n_1 + n_2}{n_1(n_2)}\right)}} = \frac{(79.2 - 62.5) - 0}{\sqrt{\left(\frac{(313)(11) + (187.07)(9)}{12 + 10 - 2}\right)\left(\frac{12 + 10}{12(10)}\right)}}$$

$$t = \frac{16.7}{6.86} = 2.44$$

The p-value for this calculated statistic is obtained as in the past with a computer program, where again the area beyond 2.44 is doubled because this is a two-tailed test.

$$p\text{-value} = 2P(t > 2.44, \text{ with df} = 20) = 0.024$$

This p-value indicates that the null hypothesis can be rejected for a Type I error probability as low as 0.024; that is, the level of significance is 0.024. If the probability of a Type I error is preset at the 5 percent level (critical t of 2.086), then the null hypothesis is rejected, but at the 2 percent Type I error level (critical t of 2.528), the null hypothesis is not rejected. The sensitivity of the conclusion to these critical values is depicted in Figure 10.1.

As an alternative to calculating from scratch on a spreadsheet program, most statistical packages do a two-sample t test on command once the data have been entered.

FIGURE 10.1 Decision Rules for a Two-Tailed, Pooled Two-Sample t Test

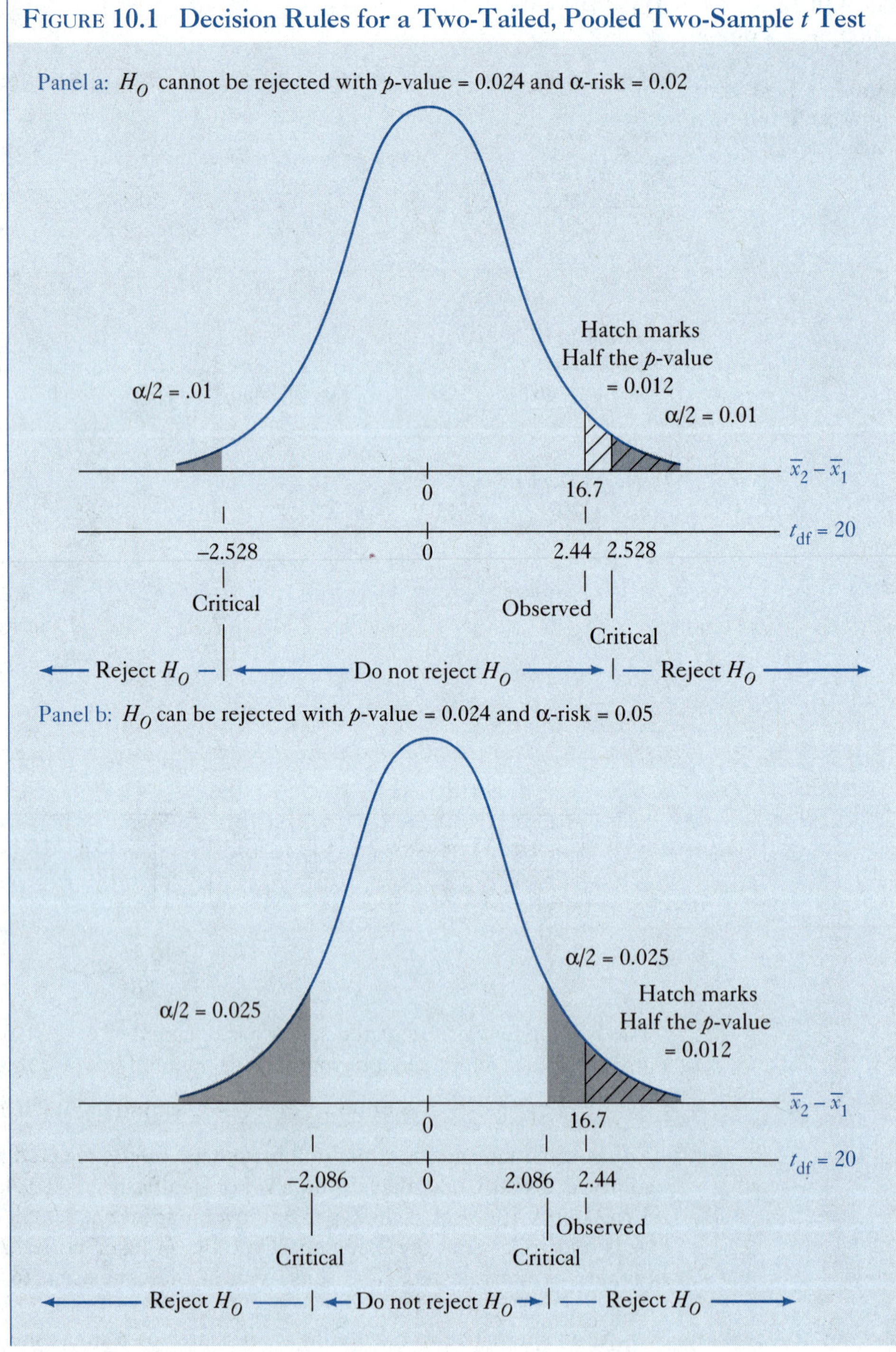

TABLE 10.3 MINITAB Output for Restaurant Service Ratings

MINITAB commands for data are columns named "DAY" and "NIGHT" in a file named "REST":

```
MTB > RETRIEVE "REST"
MTB > TWOSAMPLE 'DAY' 'NIGHT'
SUBC > ALTERNATIVE = 0
```

MINITAB output:

TWOSAMPLE T FOR NIGHT VS DAY

	N	MEAN	STDEV	SE MEAN
DAY	12	62.50	17.69	5.11
NIGHT	10	79.20	13.68	4.33

TEST MU DAY = MU NIGHT (VS NE): T = 2.44 P = 0.024 DF = 20

For instance, using MINITAB one can easily obtain the output in Table 10.3 after entering the data and the appropriate commands. These prepackaged computer routines are easy to use, typically highly accurate, and thus worth using. Make sure, however, that you know what the program is doing and that you have checked its accuracy. To check the accuracy of a program, take a small data set (such as one found here) and work through a test procedure with a hand calculator or spreadsheet program and then have the package do the same routine. In addition, if you have access to more than one program, you can check one against the other.

EXERCISES

10.1 An article in *The Wall Street Journal* (March 27, 1992) reported that tests show U.S. students have only a basic understanding of science. "Moreover, students at parochial and other private schools scored better than public school students at all grades, though test officials said that for twelfth-graders the gap wasn't statistically significant."

a. What are the null and alternative hypotheses to which the "statistical significance" refers?

b. What is the meaning of "statistical significance" in this context?

10.2 A study by the EPA compared the highway fuel economy of domestic and imported passenger cars. In the 3,000-pound weight class, 18 domestic cars showed an average of 34.3 mpg (miles per gallon) and 14 imported cars yielded an average of 34.8 mpg. The population gas mileage is known to be normally distributed in both cases and the variances are known to be $\sigma_1^2 = 2.1$ (domestic) and $\sigma_2^2 = 1.9$ (imported). At the the 5 percent Type I error level, are the means the same?

10.3 An article in *The Wall Street Journal* (January 29, 1991) reported on Domino's plans to introduce a new pizza that will have 45 percent more cheese than the company's old pizza and a thicker crust. "The average pizza will cost about 7 percent more as a result of the extra ingredients. That will lift the price of the average Domino's pizza, currently $7.84 including discounts, by about 50 cents." On the next page are old and new pizza prices from a sample obtained at 10 outlets. Was Domino's projection that the average price would rise by 50 cents reasonable?

Outlet	1	2	3	4	5	6	7	8	9	10
Old($)	8.00	9.00	7.50	6.75	8.50	7.20	6.50	7.77	8.00	9.20
New($)	8.25	10.00	8.00	7.50	8.25	7.90	7.00	8.88	8.50	9.75

10.4 A software house is experimenting with a new font size that is expected to increase the reading speed of older people. After testing thousands of people with the original font, it was established that a certain passage took older people an average of $\mu_1 = 0.16$ minutes and younger readers an average of $\mu_2 = 0.12$ minutes, with standard deviations of $\sigma_1 = 0.015$ minutes for older folks, and $\sigma_2 = 0.011$ minutes for the younger. For a sample of 16 older people, the average time to read the passage printed in the new font is now 0.12 minutes. For the sample of 12 younger readers, it is 0.11 minutes.

a. Is there now a difference in the reading times, assuming no change in measures of dispersion? (Test at the 5 percent Type I error level.)

b. Are the assumptions (the model) you had to assume to do your test reasonable?

10.5 An article in *The Wall Street Journal* (November 25, 1992) suggested women doctors in private practice have incomes that average 34 percent less than male colleagues. Below are descriptive statistics for analysis of this issue, where the mean earnings are reported in thousands of dollars for females and males.

Variable	Mean	Standard Deviation	S.E.Mean	Cases
FEMALE	117.180	89.17444	.89174	100
MALE	185.730	143.3742	.71687	200

a. What is the approximate value of the standard error of the differences in means? Explain why you used the formula you did.

b. At the 0.01 Type I error level, do women doctors, on the average, earn less than male doctors? State the assumptions you had to make to conduct this test.

10.6 A recent study suggests that licensed engineers serving as tenured university faculty members have higher incomes (university salary plus earnings from consulting) than their industry counterparts. Use the following information to test this claim at the 5 percent level. Specify the assumptions you needed to make to do this test.

Academic Engineers	Industry Engineers
$\bar{x}_1 = \$95,000$	$\bar{x}_2 = \$84,000$
$s_1 = \$9,000$	$s_2 = \$7,000$
$n_1 = 12$	$n_2 = 13$

10.3 Test of Equality of Variances

In the preceding two-sample t test, the sample variances (s_1^2 and s_2^2) were pooled to form an estimate of the assumed common population variance σ^2. But do the two populations from which the samples were drawn have the same variances? Does σ_1^2 equal σ_2^2, with both equal to the common variance σ^2 as assumed?

A new distribution must be introduced to test for equality of variances. It is called the F distribution. As with the t distribution, the F is a "family" of distributions. Each **F distribution** is formed as a ratio of one sample variance, s_1^2, to that of another, s_2^2, where the population variance of one random variable, σ_1^2, equals that of the other, σ_2^2.

The two underlying random variables must be normal and independently distributed, and they must have been standardized to have means of zero and variances of one.[5]

F test statistic for a test of a difference in variances, given a null hypothesis of no difference

$$F = \frac{s_1^2}{s_2^2} \tag{10.3}$$

s_1^2 and s_2^2 are sample variances, where $s_1^2 > s_2^2$, and there are $n_1 - 1$ degrees of freedom in the numerator, and $n_2 - 1$ degrees of freedom in the denominator.

The numerator and denominator involve independent simple random samples that may be of different sizes and variabilities. Critical values of *F* depend on the degrees of freedom in the numerator and denominator, where the larger sample variance is in the numerator (i.e., $s_1^2 > s_2^2$). This book provides tables with these *F* values for different degrees of freedom and four upper-tail probabilities, Appendix Tables A.5. As with the family of values shown in a *t* table, in the body of an *F* table are the *F* values, but now across the top edge are the degrees of freedom in the numerator. Down the left-hand margin are the degrees of freedom used in the denominator of the *F* ratio, where four different levels of significance are grouped together for each degree of freedom. For example, to find the value of *F* that puts 0.05 of the distribution above it, with $n_1 - 1 = 11$ degrees of freedom in the numerator and $n_2 - 1 = 9$ in the denominator, look across the top of the table until you get to the column marked "11" and look down the left margin until you get to the row marked "9." The group of four *F* values that is in the column headed "11" and row marked "9," are 1.58, 2.40, 3.10, and 5.18. The relationships between these *F* values and the *F* distribution probabilities are depicted in Figure 10.2.

One-Tailed Test

The *F* values in Figure 10.2 are the critical *F* values that put 25 percent, 10 percent, 5 percent, and 1 percent of the distribution in the upper tail, with 11 and 9 degrees of freedom. They define the rejection regions for a one-tailed test of

$$H_O: \sigma_1 - \sigma_2 \leq 0$$

$$H_A: \sigma_1 - \sigma_2 > 0$$

where again the decision rule is to reject the null hypothesis if the calculated *F* ratio is greater than the critical value at the specified (1, 5, 10, or 25 percent) Type I error level.

The calculated value of the *F* test statistic is obtained by simply substituting in the values for s_1^2 and s_2^2. In the restaurant service example, from Table 10.2 we have

$$F = \frac{s_1^2}{s_2^2} = \frac{313.00}{187.07} = 1.67$$

FIGURE 10.2 Critical Values of F Statistic for a Test of H_O: $\sigma_1 - \sigma_2 \leq 0$ versus H_A: $\sigma_1 - \sigma_2 > 0$ with 11 and 9 Degrees of Freedom

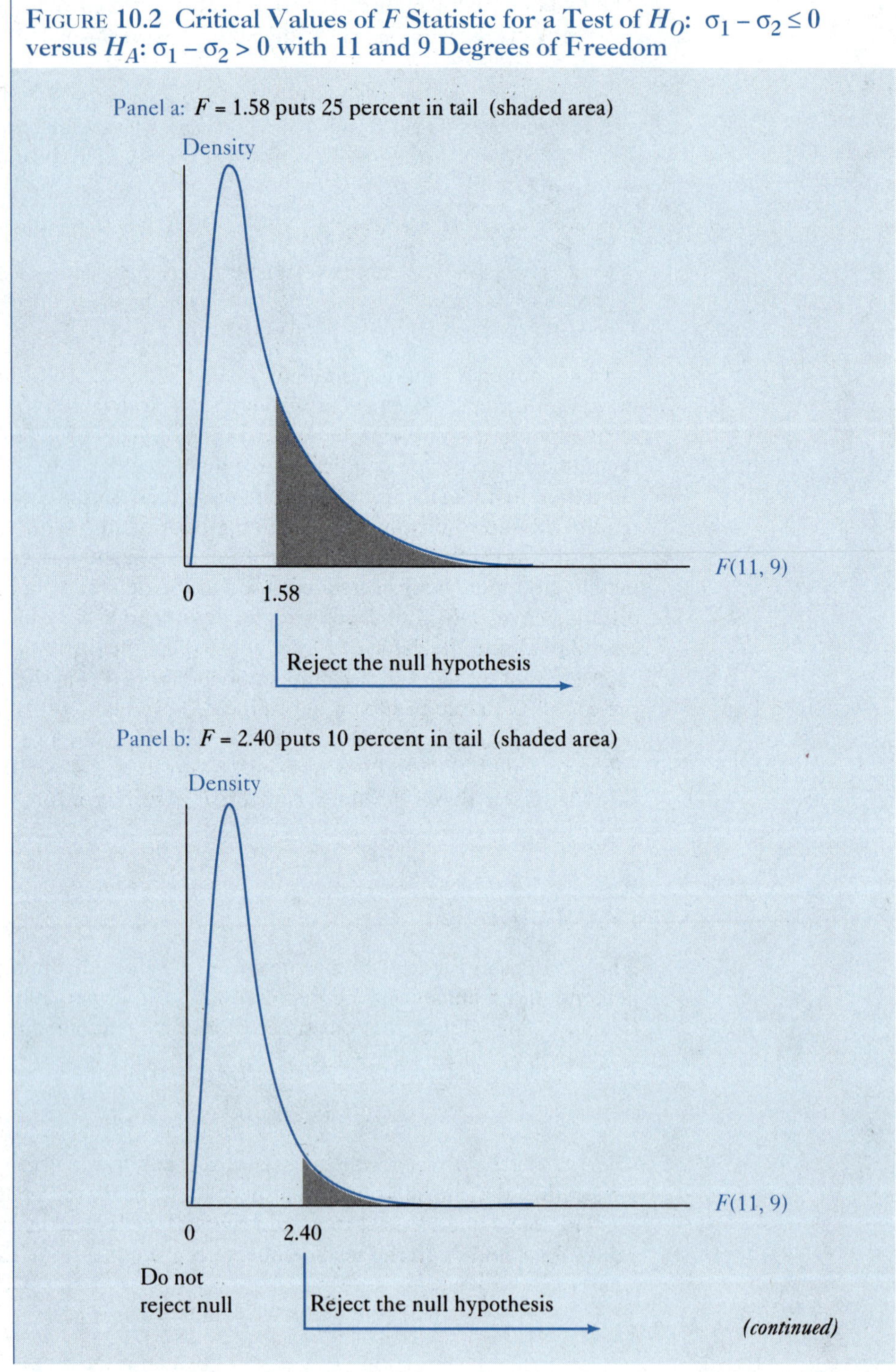

(continued)

FIGURE 10.2 Critical Values of F Statistic for a Test of H_O: $\sigma_1 - \sigma_2 \leq 0$ versus H_A: $\sigma_1 - \sigma_2 > 0$ with 11 and 9 Degrees of Freedom (*continued*)

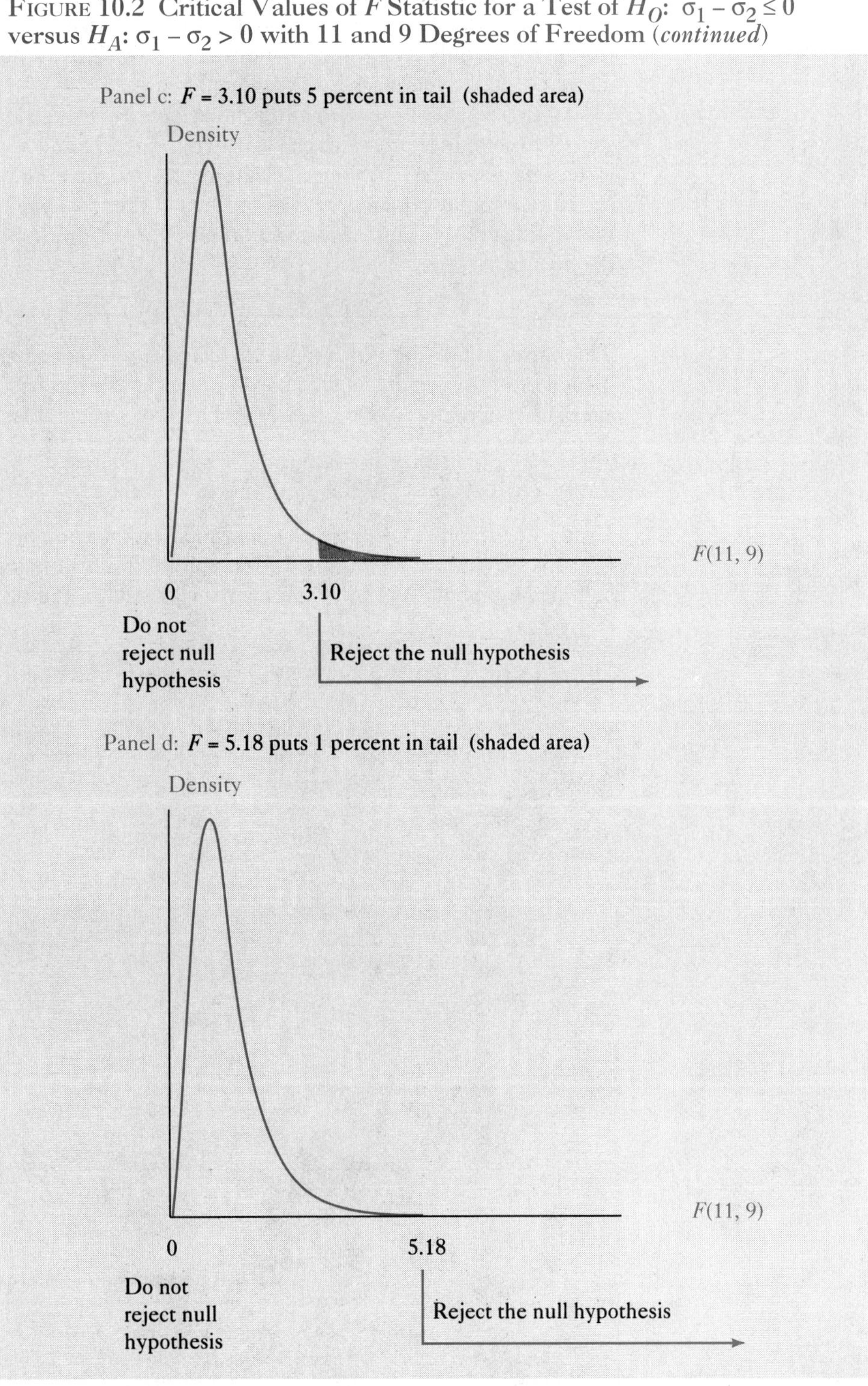

In this example the degrees of freedom are 11 for the numerator and 9 for the denominator. The calculated F of 1.67 is not large. For instance, it is in the null hypothesis rejection region only for $\alpha = 0.25$, panel a in Figure 10.2, but not for $\alpha = 0.10$, 0.05, or 0.01. Thus, the null hypothesis is not rejected.

As an alternative to comparing the calculated F with its critical value, those with statistical computer programs can calculate p-values. The p-value for the one-tailed test is the area associated with F values more extreme than the calculated F value. Because the larger sample variance is always placed in the numerator of the F statistic, the area that defines the p-value is always in the upper tail. In the restaurant service example, the p-value is

$$P(F > 1.67, \text{ with } df_1 = 11, df_2 = 9) = 0.225$$

This means that the null hypothesis cannot be rejected unless the probability of a Type I error was set above 0.225. At the typical α levels of 1, 5, or even 10 percent, the alternative hypothesis of σ_1 greater than σ_2 is unacceptable.

Two-Tailed Test

If, prior to sampling, we have no idea which variance might be larger, then a two-tailed F test is needed. In a two-tailed F test of a difference in variances, as in Figure 10.3, half of the probability of a Type I error is placed in the upper tail. The critical F for

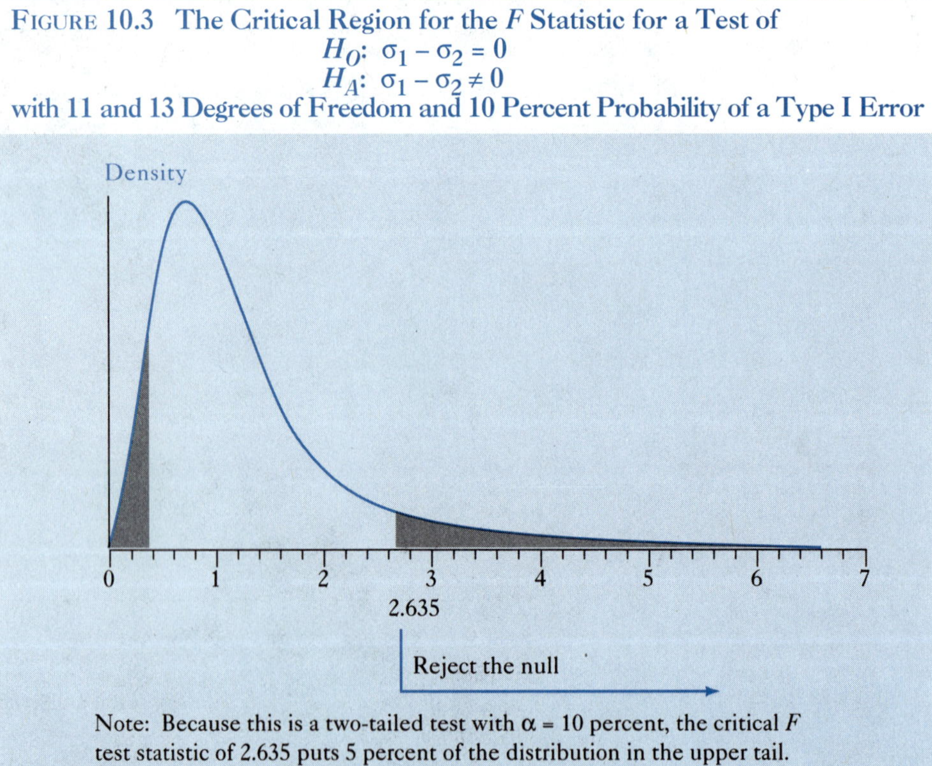

FIGURE 10.3 The Critical Region for the F Statistic for a Test of H_O: $\sigma_1 - \sigma_2 = 0$, H_A: $\sigma_1 - \sigma_2 \neq 0$ with 11 and 13 Degrees of Freedom and 10 Percent Probability of a Type I Error

Note: Because this is a two-tailed test with $\alpha = 10$ percent, the critical F test statistic of 2.635 puts 5 percent of the distribution in the upper tail.

rejecting the null hypothesis H_O: $\sigma_1 - \sigma_2 = 0$ is the value of F that puts $\alpha/2$ in the upper tail. Correspondingly, only half of the p-value is the area above the calculated F.

Once the sample data have been collected, the larger sample variance s_1^2 is again put in the denominator of the calculated F statistic in Formula (10.3). Of course, a philosophical issue can be raised as to why sample information is used in setting up the F statistic but not in respecifying the test as a one-tailed test. Instead of pursuing this issue, a demonstration of the two-tailed test might be of more value at this stage of our discussion. The two-tailed test procedure is shown as part of Query 10.2.

QUERY 10.2:

As part of the process of monitoring cash register receipts at two Subway restaurants, samples of 12 weeks at one location and 14 weeks at the other were obtained. The following dollar shortages were recorded:

Store One: $1.34 0.00 5.00 1.17 0.00 0.52 7.23 1.01 0.00 0.10 0.00 2.48

Store Two: $2.30 6.75 3.79 1.52 0.45 0.00 2.56 3.67 4.35 0.33 5.31 0.00 0.13 1.12

The question: Are there differences between the two restaurants?

ANSWER:

Assuming we are interested in checking for differences in the means and variances of the two underlying populations, and assuming those populations are normally distributed, the "proc ttest" routine in the SAS computer program may be employed to get the following statistics.

TYPE	N	Mean	Std Dev	Std Error	Minimum	Maximum
one	12	1.57083333	2.30174028	0.66445519	0	7.2300
two	14	2.30571429	2.18072576	0.58282347	0	6.7500

Variance	T	DF	Prob < |T|
Unequal	–0.8315	22.9	0.4143
Equal	–0.8351	24.0	0.4119

For H_O: Variances are equal, F'=1.114 Df=(11,13) Prob>F'=0.8422

The last line of this SAS printout contains the information for testing

$$H_O: \sigma_1 - \sigma_2 = 0$$
$$H_A: \sigma_1 - \sigma_2 \neq 0$$

where the calculated F is

$$F = \frac{s_1^2}{s_2^2} = \frac{(2.30174028)^2}{(2.18072576)^2} = 1.114$$

and half of the p-value for this F test of variances is

$$P(F > 1.114, \text{ with df}_1 = 11, \text{df}_2 = 13) = 0.4211$$

As shown in the SAS printout, the full p-value is 0.8422, which is well above any of the Type I error levels at which we typically test. Thus, the null hypothesis of equal variances cannot be rejected.

With an assumption of equal variances, the appropriate t statistic is shown in the SAS printout in the row labeled "Equal," $t = 0.8351$. It was calculated via Formula (10.2). For the hypotheses

$$H_O: \mu_1 - \mu_2 = 0$$

$$H_A: \mu_1 - \mu_2 \neq 0$$

half the p-value for this t test of means is

$$P(t > 0.8351, \text{ with df} = 24) = 0.20595$$

As shown in the SAS printout, the full p-value for this two-tailed t test is 0.4119, which is well above any of the typical Type I error levels. Thus, the null hypothesis of equal means cannot be rejected.

EXERCISES

10.7 Determine the following:
   a. $P(F > 5.22, \text{ with df}_1 = 11, \text{df}_2 = 3) = \underline{?}$
   b. $P(F > \underline{?}, \text{ with df}_1 = 8, \text{df}_2 = 19) = 0.05$
   c. $P(F > 3.37, \text{ with df}_1 = 10, \text{df}_2 = 20) = \underline{?}$
   d. $P(F > \underline{?}, \text{ with df}_1 = 12, \text{df}_2 = 3) = 0.01$

10.8 In Exercise 10.6, was it reasonable to assume equality of variances, as was required to do the test with Formula (10.2)?

10.9 In horse racing, the subjective probability of a horse winning is the total amount bet on the horse divided by the total amount bet on all the horses in the race. The objective probability is the fraction of times that the horses actually win. Peter Asch and Richard Quandt, *Racetrack Betting: The Professors' Guide to Strategies.* New York: Praeger, 1991, p. 112) provide the following data on these probabilities for races in Atlantic City, N.J.

Groups	Objective Probability	Subjective Probability
Favorites	0.361	0.325
Second-lowest odds	0.218	0.205
Third-lowest odds	0.170	0.145
Fourth-lowest odds	0.115	0.104
Fifth-lowest odds	0.071	0.072
Sixth-lowest odds	0.050	0.048
Seventh-lowest odds	0.030	0.034
Eighth-lowest odds	0.017	0.025
Ninth-lowest odds	0.006	0.018

If bettors are able to accurately predict, then there should be no difference between objective and subjective probabilities other than chance. At the 0.05 Type I error level, is there a difference between the means of the subjective and objective probabilities? (Assume normality as needed, but test for equality of variances.)

10.10 Daniel Seligman (*Forbes*, March 8, 1982) argued that federal deficit projections were extremely tenuous. He provided official deficit projections (by the administrations in office) between 1970 and 1982. These data (in billions of dollars) are:

Fiscal year	Projected Deficit	Actual Deficit
1970	–3.4	2.8
1971	–1.3	23.0
1972	11.6	23.4
1973	25.5	14.9
1974	12.7	4.7
1975	9.7	45.2
1976	51.9	66.4
1977	59.1	57.9
1978	47.0	48.8
1979	60.6	27.7
1980	29.9	59.6
1981	15.8	57.9
1982	27.5	98.6

Assuming the underlying populations are normal, is there a significant difference between the projected and the actual deficits, as Seligman claims? Test for differences between variances and differences between means at the 0.01 Type I error level.

10.11 In the early days of microcomputers, hard disk drives had a mean time between failures (MTBF) of 5,000 to 20,000 hours. As evidenced by a *Wall Street Journal* (December 22, 1983) ad placed by Leading Edge Products, Inc., in which the claim was made that its drives had a MTBF of 20,000 while IBM's was only 8,000 hours, producers really tried to tout their drive life. Test the Leading Edge claim using the following data, at the 5 percent significance level.

	MTBF($\bar{x}$)	St.Dev.(s)	(n)
LE	20,000	3,200	6
IBM	8,000	1,900	4

10.4 Small Samples of Different Size and Different Variances

Barry Moser and Gary Stevens (1992) argue that the practice of preliminary variance testing is unwarranted. They show that if the sample sizes are close to each other, the homogeneity of the population variances ($\sigma_2 = \sigma_1$) is irrelevant. If both the sample sizes and the sample variances are greatly different, however, a pooled t test, as in Formula (10.2), is not appropriate.

For situations involving small samples of disparate sizes and sample variances, Moser and Stevens propose the use of the method advanced by F. E. Satterthwaite and others. The calculated test statistic for the null hypothesis $\mu_1 - \mu_2 = 0$ is as in the z formula (10.1), but is now appropriately labeled t^* to reflect the fact that the population variances are unknown. In addition, a special formula is used to determine the degrees of freedom.

test statistic for difference between means from small independent samples

$$t^* = \frac{(\bar{x}_2 - \bar{x}_1) - D_O}{\sqrt{\frac{s_1^2}{n_1} + \frac{s_2^2}{n_2}}} \tag{10.4}$$

degrees of freedom are

$$\text{df} = \frac{(F/n_1 + 1/n_2)^2}{F^2/n_1^2(n_1 - 1) + 1/n_2^2(n_2 - 1)}$$

where $F = s_1^2/s_2^2$, for $s_1^2 > s_2^2$.

Calculation of the degrees of freedom in Formula (10.4) is computationally burdensome. Fortunately, recent versions of the more complete statistical packages automatically make these calculations when two-sample tests are requested. For example, in Query 10.2 the SAS program printout shows the degrees of freedom for the t^* test, allowing for unequal variances, of 22.9.

$$\text{df} = \frac{(F/n_1 + 1/n_2)^2}{F^2/n_1^2(n_1 - 1) + 1/n_2^2(n_2 - 1)} = \frac{[1.114/12 + 1/14]^2}{(1.114)^2/(12)^2(11) + 1/(14)^2(13)} = 22.9$$

Rounding to the nearest whole number of 23 degrees of freedom yields the area beyond the $t^* = 0.8315$ of 0.20715. That is, half the p-value for a two-tailed test of means, assuming a difference in variances, is

$$P(t > 0.8315, \text{ with df} = 23) = 0.20715$$

As shown in the SAS printout in Query 10.2, the full p-value for this two-tailed t test is 0.4143, which is well above any of the typical Type I error levels. Thus, the null hypothesis of equal means cannot be rejected, even when we assume differences in the variances of the two populations.

10.5 Two-Sample Tests without the Assumption of Normality

All of the above small sample t tests require that the samples come from populations that are distributed normally. There are ways to test whether this assumption of normality is reasonable, but these tests have little power when samples are small.[6] Many

HISTORICAL PERSPECTIVE

The F distribution, as named by George Snedecor in honor of R.A. Fisher who discovered it, was an impressive achievement in formula computation. Fisher derived the probabilities of what is now called the F distribution in the early 1920s, long before computers were invented. It was 20 years before Edward Paulson (1942) provided a convenient algorithm for approximating probabilities for the F distribution that can be done easily with even a hand calculator. He showed that an F, with $n_1 - 1$ degrees of freedom in the numerator, and $n_2 - 1$ degrees of freedom in the denominator, could be approximated with a standard normal random variable. Provided that the degrees of freedom in the denominator exceed three ($n_2 - 1 > 3$), the relationship between F and z is

$$z^* = \frac{\left[1 - \frac{2}{9(n_2 - 1)}\right] F^{1/3} - \left[1 - \frac{2}{9(n_1 - 1)}\right]}{\sqrt{\frac{2}{9(n_2 - 1)} F^{2/3} + \frac{2}{9(n_1 - 1)}}}$$

The accuracy of this algorithm can be assessed in the following table. The exact F that puts 0.05 in the tail of the F distribution was used to calculate the approximate z^* value. This approximate z^* compares favorably with the exact z of 1.645 that puts 0.05 in the tail of the standard normal distribution.

$n_1 - 1$	$n_2 - 1$	F	z^*
4	8	3.84	1.6436
20	9	2.94	1.6427
5	20	2.71	1.6466
30	20	2.04	1.6452
15	40	1.92	1.6397

As a demonstration of this approximation method, consider the calculated F of 1.67 from the restaurant service example. The approximate p-value for this test is

$$P(F > 1.67, \text{ with } df_1 = 11, df_2 = 9)$$
$$= P(z^* > 0.76) = 0.224$$

Today, inexpensive statistical computer programs can calculate probabilities associated with the entire family of F distributions, regardless of the degrees of freedom considered. W.I. Kennedy and J.E. Gentile (1980) review some of the alternative algorithms that are used by programmers to approximate tail areas in the F distribution.

Nonparametric Test
A statistical test that does not require the normal probability distribution or another well-defined distribution for derivation of the sampling distribution. Also known as a distribution free test.

statisticians therefore prefer to use so called "distribution free" or **nonparametric tests** of differences in population means whenever samples are small. One such test, which makes use of the t distribution, requires information on only the rank order of observations. This "Wilcoxon-Mann-Whitney rank sum test" in its exact form does assume that the variances of both populations are the same, but it is not as sensitive to violations of this assumption as is the t test based on the two sample means. More important, it does not require that the populations be normal.

The Wilcoxon-Mann-Whitney rank sum test of population differences, as with its t test counterpart discussed earlier, starts with the assumption that the two random samples were drawn independently. But now, the sample observations from both samples must be ranked from lowest to highest in a single array. The combined ranking of the observations in the restaurant service example is shown in Table 10.4. Notice that the rank of each observation is based on all 22 observations in both the night and day shifts. The lowest value is 35, so its rank is 1; the highest value is 97, so its rank is 22. Identical observation values, such as 48, are each given the average of their ranks.

TABLE 10.4 Wilcoxon-Mann-Whitney Rank Sum Test

SHIFT	OBSERVATION	RANK r_1	$(r_1 - \bar{r}_1)^2$
day	35	1.0	57.5069
day	41	2.0	43.3403
day	48	3.5	25.8403
day	55	5.0	12.8403
day	57	6.0	6.6736
day	62	7.0	2.5069
day	63	8.0	0.3403
day	64	9.0	0.1736
day	68	10.0	2.0069
day	72	11.0	5.8403
day	91	19.5	119.1736
day	94	21.0	154.1736
		Mean = 8.583	39.1288 = Variance of ranks

SHIFT	OBSERVATION	RANK r_2	$(r_2 - \bar{r}_2)^2$
night	48	3.5	132.2500
night	73	12.0	9.0000
night	74	13.0	4.0000
night	75	14.0	1.0000
night	77	15.0	0.0000
night	81	16.0	1.0000
night	86	17.0	4.0000
night	90	18.0	9.0000
night	91	19.5	20.2500
night	97	22.0	49.0000
		Mean = 15.0	25.5000 = Variance
			32.9958 = Pooled variance
			2.4595 = Estimated standard error
			2.609 = t

The two-sample t test is calculated on the ranks in each sample (r_2 and r_1), where these ranks are from an ordering of the observations in the combined samples. All the required quantities shown in Table 10.4 were obtained from a spreadsheet program. Substituting into Formula (10.2), we obtain the calculated test statistic value of $t = 2.609$.

$$t = \frac{(\bar{r}_2 - \bar{r}_1) - 0}{\sqrt{\left(\frac{s_1^2(n_1 - 1) + s_2^2(n_2 - 1)}{n_1 + n_2 - 2}\right)\left(\frac{n_1 + n_2}{n_1(n_2)}\right)}} = \frac{(15.000 - 8.583) - 0}{\sqrt{\left(\frac{(39.1288)(11) + (25.5)(9)}{12 + 10 - 2}\right)\left(\frac{22}{120}\right)}}$$

$$t = \frac{6.417}{2.4595} = 2.609$$

The p-value for this calculated static is again obtained with a computer program. Because the alternative hypothesis is still H_A: $\mu_2 - \mu_1 \neq 0$, the area beyond 2.609 is doubled for a two-tailed test. The p-value is

$$2P(t > 2.609 | \text{df} = 20) = 0.0168$$

This p-value indicates that the null hypothesis H_O: $\mu_2 - \mu_1 = 0$ can be rejected for a Type I error probability as low as 0.0168. If the probability of a Type I error is preset at the 5 percent level (critical t of 2.086), then the null hypothesis is rejected. Even at the 2 percent Type I error level (critical t of 2.528), the null hypothesis is rejected. At the 1 percent level (critical $t = 2.845$), the null hypothesis could not be rejected, but this is possibly an overly-low Type I error level.

An alternative form of the Wilcoxon-Mann-Whitney rank sum test is the U test for difference in population location. This nonparametric test is not as powerful, in terms of its ability to reject the null hypothesis, as the t test of differences. But, it does not require strict equality of the population variances. Once again, it requires that the random samples be combined into a single ordered array, using each individual sample member's rank from this combined array to get the following:

$$T_A = n_A n_B + \frac{n_A(n_A+1)}{2} - \Sigma r_A \text{ and } T_B = n_A n_B - T_A$$

where the symbol Σr_A denotes the sum of the ranks in sample A, and n_A and n_B are the sizes of samples A and B. The statistic for the Wilcoxon-Mann-Whitney test is U, where U is defined as

$$U = \min[T_A, T_B]$$

It can be shown that U has an approximately normal distribution with a mean of $n_A n_B/2$ and variance $n_A n_B(n_A + n_B + 1)/12$. For samples of at least size 8, the z test statistic can be formed

$$z = (U - n_A n_B)/2 \Big/ \sqrt{n_A n_B (n_A + n_B + 1)/12}$$

Wilcoxon-Mann-Whitney U test

$$T_A = n_A n_B + \frac{n_A(n_A+1)}{2} - \Sigma r_A$$

$$T_B = n_A n_B - T_A \tag{10.5}$$

where Σr_A denotes the sum of the ranks in the A sample, and n_A and n_B are the sample sizes

$$U = \min[T_A, T_B]$$

$$z = (U - n_A n_B)/2 \Big/ \sqrt{n_A n_B (n_A + n_B + 1)/12}$$

As in all z statistics, there is only one random variable; in this test it is U. If the value realized for U in a pair of samples is large relative to U's variance, the calculated z will be large, and the null hypothesis of no difference between the A and B populations is rejected. In the restaurant service example, Formula (10.5) is used to obtain T(day) = 95 and T(night) = 25, with U being the minimum of the these two, U is 25. Thus, the calculated z is –2.3, and again the null hypothesis of no difference between the shifts is rejected, at a significance level of 0.02.

Nonparametric tests of rank cannot pinpoint the population differences in normal populations as sharply as a t test based on sample observations. But the Wilcoxon-Mann-Whitney test of rank, when used in situations in which the t test on observation values is valid, is only slightly less powerful than the t test, as shown by the early work of A.M. Mood and many other later researchers. Because the Wilcoxon-Mann-Whitney tests do not require stringent assumptions about the parameters of a distribution, the validity of rejecting the null hypothesis is not suspect when these tests are used. For these reasons, many statisticians prefer the nonparametric tests in all situations.

QUERY 10.3:

Business Week (January 12, 1987) reported on recent changes in the dollar volume of mergers and acquisitions for selected industries. The article implied that these activities would continue to increase or at least stay the same, since, according to Deborah Allen Oliver (president of Claremont Economics Institute), stocks were not "fully valued" at that time. Curiously, *BW*'s supporting table of the ten industries with the most merger and acquisition activity in 1985 and 1986 seems to suggest the reverse. The question is: If the test was done on the basis of the top ten industries in 1985 versus the top ten in 1986, would the conclusion of no decrease continue to be supported?

Industries with the Most Merger and Acquisition Activity

1985	Dollar Volume Billions	1986	Dollar Volume Billions
1 Mining, oil, and gas	15.3	1 Communication	13.4
2 Food	14.6	2 Electrical machinery	12.0
3 Transportation	12.9	3 Food	11.9
4 Chemicals	12.3	4 Retailing	9.2
5 Banking	7.1	5 Nonelectrical machinery	8.6
6 Communication	7.1	6 Utilities	5.7
7 Retailing	6.3	7 Chemicals	5.3
8 Utilities	6.2	8 Banking	5.1
9 Electrical machinery	5.9	9 Transportation	3.8
10 Photo, dental, and optical instruments	5.4	10 Mining, oil, and gas	3.7

ANSWER:

Let μ_1 and μ_2 be the population average dollar volume of mergers and acquisition activity in 1985 and 1986. Similarly, let $\bar{x}_2$ and $\bar{x}_1$ be the sample average dollar volumes of mergers and acquisition activity in 1985 and 1986. The null and alternative hypotheses are then

$$H_O: \mu_2 - \mu_1 \geq 0 \text{ and } H_A: \mu_2 - \mu_1 < 0$$

Both a parametric and a nonparametric two-sample t test can be employed to test these hypotheses. The quantities needed to assess statistical significance can be calculated with a spreadsheet program. The quantities for each test are

	x_1	$(x_1 - \bar{x}_1)^2$	r_1	$(r_1 - \bar{r}_1)^2$	x_2	$(x_2 - \bar{x}_2)^2$	r_2	$(r_2 - \bar{r}_2)^2$
1	15.3	35.8801	20	60.84	13.4	30.5809	18	84.64
2	14.6	27.9841	19	46.24	12	17.0569	15	38.44
3	12.9	12.8881	17	23.04	11.9	16.2409	14	27.04
4	12.3	8.9401	16	14.44	9.2	1.7689	13	17.84
5	7.1	4.8841	10.5	2.89	8.6	0.5329	12	10.24
6	7.1	4.8841	10.5	2.89	5.7	4.7089	6	7.84
7	6.3	9.0601	9	10.24	5.3	6.6049	4	23.04
8	6.2	9.6721	8	17.64	5.1	7.6729	3	33.64
9	5.9	11.6281	7	27.04	3.8	16.5649	2	46.24
10	5.4	15.2881	5	51.84	3.7	17.3889	1	60.84
	9.31	15.67877	12.2	28.566	7.87	13.23566	8.8	38.844

	$\bar{x}_2 - \bar{x}_1$	$\bar{r}_2 - \bar{r}_1$
difference	–1.44	–3.4
pooled variance	14.457	33.705
estimated standard error	1.700	2.596
t ratio	–0.847	–1.310

For the t test of means, both populations must be normal and have the same variances. The calculated F statistic is

$$F = \frac{s_1^2}{s_2^2} = \frac{15.679}{13.236} = 1.18$$

There are nine degrees of freedom in the numerator and denominator, so at the 10 percent Type I error level the critical F is 2.44 for a one-tailed test. Thus, assuming independence of samples, the calculated F of 1.18 is not large, and the null hypothesis that the population variances are the same cannot be rejected.

For the two tests of central location, the critical t statistic is –1.33, at the 10 percent Type I error level, with 18 degrees of freedom. Neither calculated t (–0.847 nor –1.310) is in the null hypothesis rejection region, and the conclusion is that merger and acquisition volume has not decreased.

EXERCISES

10.12 In addition to the assumption of equality of variances, what other assumptions did you have to make to conduct your test of engineers' salaries, in Exercise 10.6? Why might these additional assumptions be unreasonable? What alternative test might you perform? Can you do the proposed test with the data provided? Why or why not?

10.13 If the relevant populations in Query 10.2 about the two Subway restaurants are not normal, how would you test for differences in cash register shortages? Do this test at the 5 percent Type I error level.

10.14 The test of the difference between projected and actual deficits, Exercise 10.10, is based on a model in which the samples are assumed to come from populations that are normal. Repeat this test assuming that this assumption is false. Does your conclusion differ?

10.15 Below is the SAS printout for the test of difference of means in day and night shift service rates for the data in Table 10.2. Is the appropriate assumption for the *t* test of means one of variance equality or inequality? With the quantities provided, show how the relevant statistic for a test of variance equality was calculated by SAS. Conduct the relevant test of variance equality.

TYPE	N	Mean	Std Dev	Std Error	Minimum	Maximum
one	12	62.5000000	17.69180601	5.10718448	35.0000	94.0000
two	10	79.2000000	13.67723169	4.32512042	48.0000	97.0000

Variance	T	DF	Prob > |T|
Unequal	2.4953	19.9	0.0215
Equal	2.4361	20.0	0.0243

For H0: Variances are equal, F' = 1.67 Df = (11,9) Prob > F' = 0.4487

10.6 Matched Pairs

Consider a large corporate training program in which a random sample of trainees is tested at the beginning of the program (pretest) and another random sample is tested at the end (posttest). Using the testing procedure for independent samples, the average pretest score is then subtracted from the average posttest score to form an average "change score," which is used as a measure of the "value added" by the program. As we have seen, statistics can be calculated from this sample information to test the null hypothesis that the difference in average scores is zero versus the alternative that it is not. Symbolically, these null and alternative hypotheses are written

$$H_O: \mu_2 - \mu_1 = 0 \qquad H_A: \mu_2 - \mu_1 \neq 0$$

where μ_2 is the population average posttest score, and μ_1 is the population average pretest score.

Because the pretesting and posttesting do not involve the same individuals, there is variability in these test scores that is caused by the different individuals involved; this variability may have nothing to do with the training program. For example, by chance, less able trainees could have been in the pretest sample and more able in the posttest sample; this would give rise to a large difference in sample averages, even if the program added nothing.

There is another study design, however, that can remove some of the variability that is caused by individual differences in the pretesting and posttesting. Because the intent of the study is to assess the effect of the training program on the performance of individuals, the same individuals who are tested at the start of the program should be tested at the end. The unwanted variability assignable to the difference in individuals tested is thus removed. Any difference in the pair of scores for each individual now may reasonably be interpreted as the outcome of the common cause (the training program) each experienced. These paired observations are called **matched pairs**. The average of the difference between matched pairs, at least for large sample sizes, can be assumed to be normal so that a paired t test is possible. For small sample sizes, the assumption of normality is debatable since the Central Limit Theorem does not apply.

Matched Pairs
A sample consisting of paired observations where the pairing is based on a common attribute.

Table 10.5 gives pretest and posttest scores for eight individuals in a training program discussed by R.W. Mee and T.C. Chua (1991). Each individual's difference or change score (d) is calculated by subtracting the pretest from the posttest. Then the average difference score ($\bar{d}$) and its standard error $s_d/\sqrt{n}$) can be calculated by hand or, better yet, by any one of a number of spreadsheet programs (e.g., LOTUS 1-2-3, EXCEL, or QUATTRO). Once this average difference score is calculated, a t statistic is defined, where 0 is the expected value of $\bar{d}$, as assumed in the null hypothesis $\mu_2 - \mu_1 = 0$. Once again, to use the t distribution, for small n, we must be able to assume that the distribution of the estimator ($\bar{d}$) is approximately normally distributed.

Matched pairs t statistic

$$t = \frac{\bar{d} - \mu_0}{s_d / \sqrt{n}} \tag{10.6}$$

where $s_d = \sqrt{\Sigma\left(d_i - \bar{d}\right)^2 / (n-1)}$

The value of t in our matched pairs test is calculated using the values in Table 10.5.

$$t = \frac{\bar{d} - \mu_0}{s_d / \sqrt{n}} = \frac{3 - 0}{4.24264 / \sqrt{8}} = 2.0$$

For a two-tailed test, the p-value is obtained from a computer program as p-value = $2P(\bar{d} > 3 \mid \mu_0 = 0) = 2P(t > 2$, with df = 7) = 0.0856. For a two-tailed test, with a 0.05 probability of a Type I error, the critical values are obtained from Appendix Table A.4,

$$P(t > 2.365, \text{ with df} = 7) = 0.025$$

If there is 5 percent chance of a Type I error, Figure 10.4, panel a, shows that the null hypothesis cannot be rejected. If the probability of a Type I error is 10 percent, however, then the null hypothesis can be rejected, since, as shown in panel b, Figure 10.4, $P(t > 1.895$, with df = 7) = 0.05. Clearly, the conclusion is sensitive to the α-risk.[7] We can conclude that the difference in test scores is significantly different from zero down to the 0.086 Type I error level, for a two-tailed test.

TABLE 10.5 Calculation of Difference (or Change) Scores and Associated t Statistic for Matched Pairs

pretest	posttest	d	$d-\bar{d}$	$(d-\bar{d})^2$
45	49	4	1	1
52	50	−2	−5	25
63	70	7	4	16
68	71	3	0	0
57	53	−4	−7	49
55	61	6	3	9
60	62	2	−1	1
59	67	8	5	25

$\Sigma d = 24$ $\qquad$ $126 = \Sigma(d-\bar{d})^2$

$\bar{d} = \Sigma d/n = 3$ $\qquad$ $s_d = 4.24264 = \sqrt{\Sigma(d-\bar{d})^2/(n-1)}$

$t = (\bar{d}-\mu_0)/s_d/\sqrt{n} = 2$ $\qquad$ Standard error $= 1.5 = s_d/\sqrt{n}$

Similar t tests of matched pairs can be defined for one-tailed hypothesis tests using the same procedures we have used throughout this and the previous chapter. In a one-tailed test, the p-value is only the area beyond the calculated value of the estimator. Thus, for example, if the above t-test of the training program effect had been a one-tailed test, the p-value would have been

$$P(\bar{d} > 3 \mid \mu_0 = 0) = P(t > 2, \text{ with df} = 7) = 0.0428$$

and an alternative hypothesis H_A: $\mu_2 - \mu_1 > 0$ would be supported because the null hypothesis is rejected at the 5 percent Type I error level.

Scatterplot (or Scatter Diagram) A two-dimensional graph showing the paired (x, y) observations with x measured along the horizontal axis and y measured along the vertical.

In addition to testing hypotheses with matched pairs, a **scatterplot** or **scatter diagram** can be used to show the sample relationship between the paired observations. A scatterplot is a two-dimensional graph showing the paired observations with their coordinates on the vertical (Y) and horizontal (X) axes. Figure 10.5 shows the scatterplot for the posttest (Y) and pretest (X) scores. If the training program was totally ineffective (the null hypothesis $\mu_2 - \mu_1 = 0$ is not rejected), then the eight points would have been scattered randomly around the "45-degree" line. To the extent that the program confused people and actually decreased their understanding, the points should be below the 45-degree line, where posttest scores are less than pretest scores. As shown, six of the eight points are above the 45-degree line, suggesting that the training program had a positive effect. The question to be raised later in the chapters on regression analysis is whether this effect resulted from the education in the training program or some other aspect of it.

EXERCISES

10.16 In the *American Statistician* article "Regression Toward the Mean and the Paired Sample *t* Test" cited in this chapter, the authors (R.W. Mee and T.C. Chua) identify a sample selection problem inherent in the pretesting and posttesting of students who have the

FIGURE 10.4 Decision Rules for a Two-Tailed Matched Pairs t Test

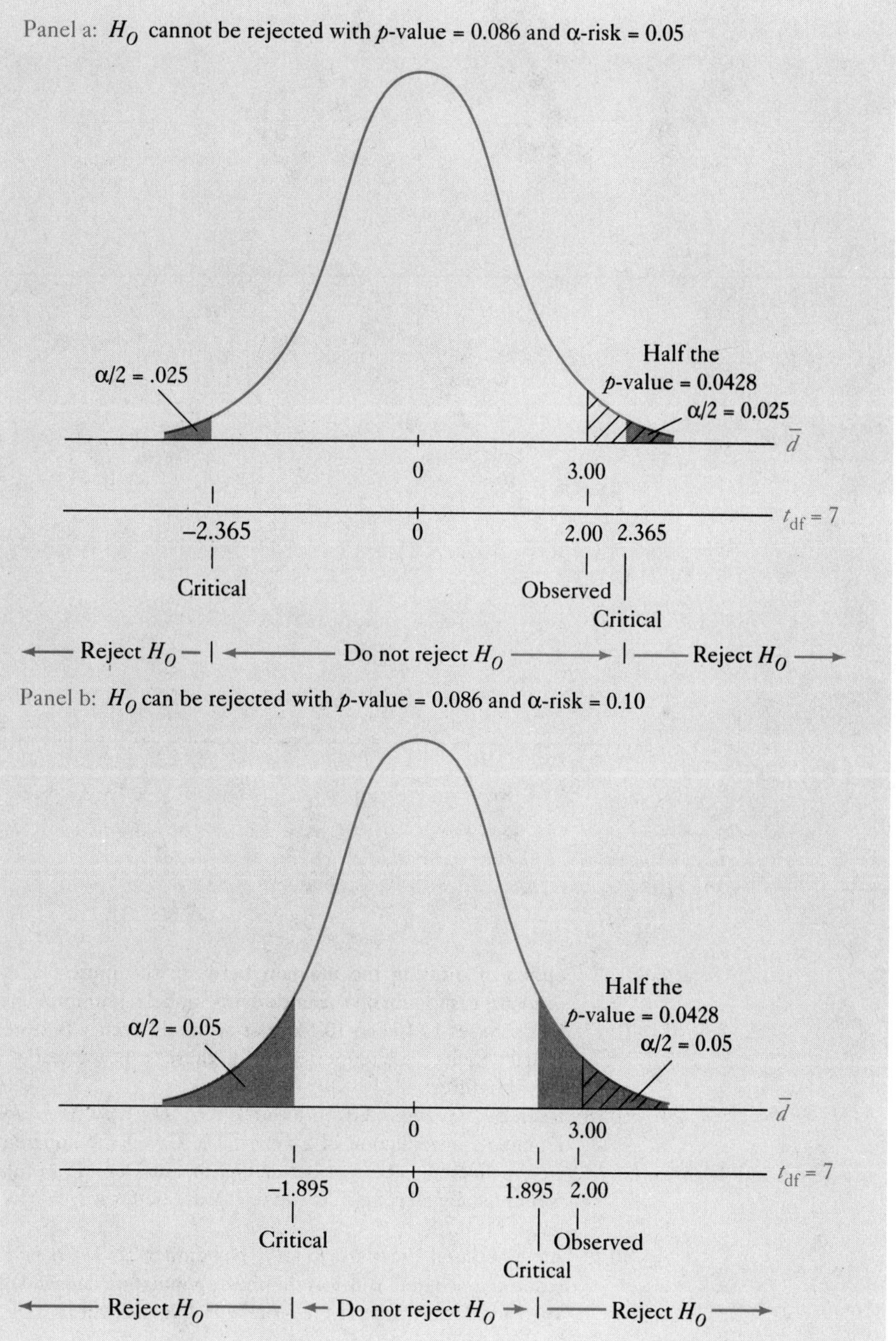

Panel a: H_0 cannot be rejected with p-value = 0.086 and α-risk = 0.05

α/2 = .025

Half the
p-value = 0.0428
α/2 = 0.025

$\bar{d}$: 0, 3.00

$t_{df} = 7$: −2.365, 0, 2.00, 2.365

Critical (−2.365); Observed (2.00); Critical (2.365)

← Reject H_0 | ← Do not reject H_0 → | Reject H_0 →

Panel b: H_0 can be rejected with p-value = 0.086 and α-risk = 0.10

α/2 = 0.05

Half the
p-value = 0.0428
α/2 = 0.05

$\bar{d}$: 0, 3.00

$t_{df} = 7$: −1.895, 0, 1.895, 2.00

Critical (−1.895); Critical (1.895); Observed (2.00)

← Reject H_0 | ← Do not reject H_0 → | Reject H_0 →

FIGURE 10.5 Scatterplot of Pretest and Posttest Pairs

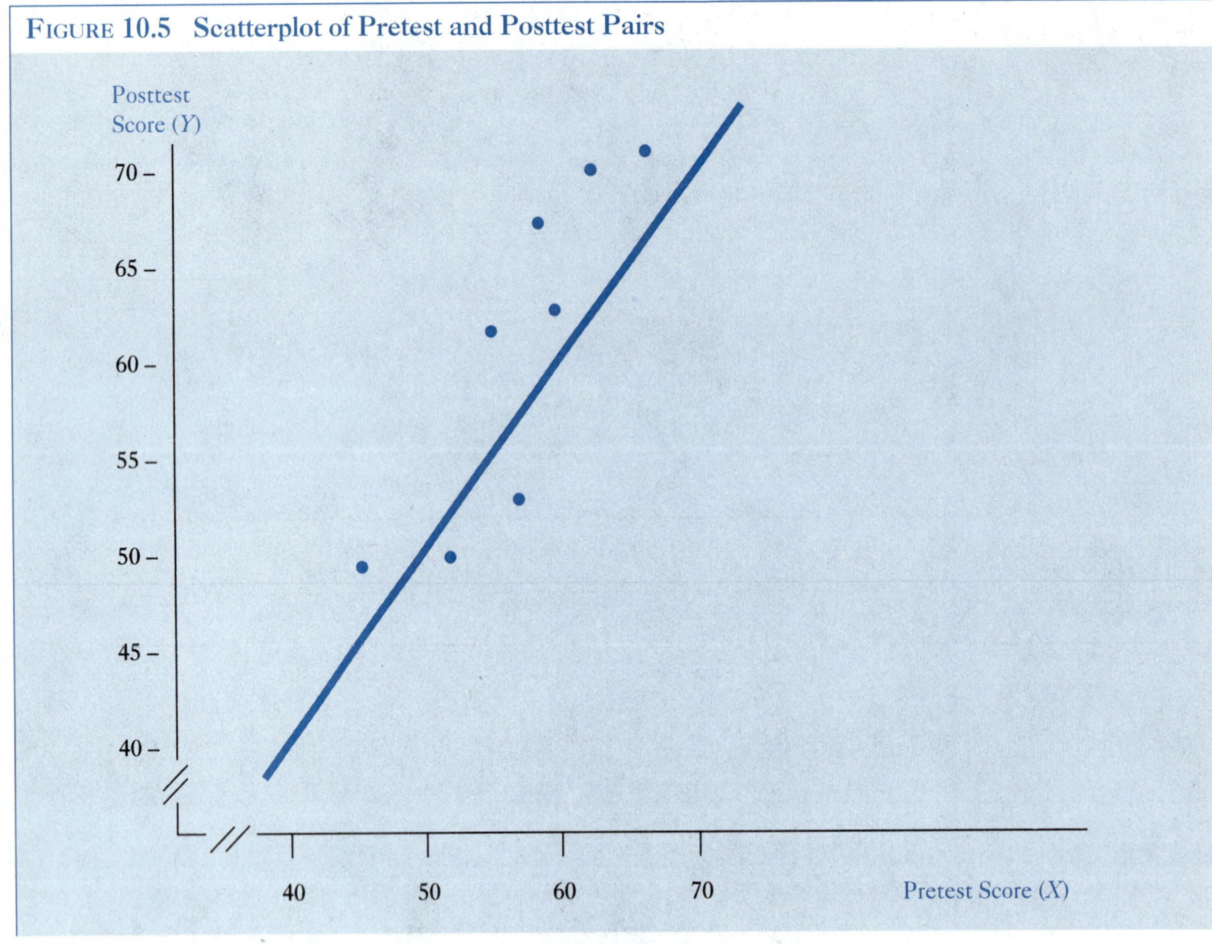

option of quitting the program between the pretest and posttest. Why might these dropouts cast doubt on the underlying model (assumptions) required for the paired *t* test?

10.17 In Exercises 10.10 and 10.14, tests of the difference between the mean of projected and actual deficits were requested. Redo these tests as a test of matched pairs. Do your conclusions differ?

10.18 Exercise 10.3 reported on an article in *The Wall Street Journal* (January 29, 1991) on Domino's introduction of a new pizza. Given the apparent manner in which the data were collected, is there a better way to test Domino's projection that the average price would rise by 50 cents? Do the test your "better way." Why is this way better in this situation?

10.19 An Associated Press article (*HT*, November 22, 1989) said that "nearly half the top 25 cities lost people." But was the mean population change significantly different between 1988 and 1986? (Test at the 5 percent Type I level.)

	1988	1986
New York	7353000	7263000
Los Angeles	3353000	3259000
Chicago	2978000	3010000
Houston	1698000	1729000
Philadelphia	1647000	1643000
San Diego	1070000	1015000
Detroit	1036000	1086000
Dallas	987000	1004000
San Antonio	941000	914000
Phoenix	924000	894000
Baltimore	751000	753000
San Jose	738000	712000
San Francisco	732000	749000
Indianapolis	727000	720000
Memphis	645000	653000
Jacksonville	635000	610000
Washington	617000	626000
Milwaukee	599000	605000
Boston	578000	574000
Columbus	570000	566000
New Orleans	532000	554000
Cleveland	521000	536000
El Paso	511000	492000
Seattle	502000	486000
Denver	492000	505000

10.20 After Iraq's invasion of Kuwait in 1991, there was concern about the effect of the Persian Gulf conflict on world oil prices. Runzheimer International, a management consulting firm based in Rochester, Wisconsin, compiled figures in U.S. dollars per gallon for the least expensive gasoline available as of December 1990 and 1991 in 18 cities around the world. (*Parade Magazine*, March 29, 1992)

Location	12/90	12/91
Atlanta	$1.26	$ .98
Auckland	2.42	2.02
Bogota	.53	.59
Caracas	.25	.29
Dublin	4.17	3.88
Frankfurt	2.91	3.29
Hong Kong	3.42	3.34
Jerusalem	2.74	2.72
Johannesburg	2.14	1.92
Kingston	1.60	1.32
London	3.11	3.22
Mexico City	.91	1.35
Milan	4.86	4.69
New Delhi	2.53	2.14

(*continued*)

Location	12/90	12/91 (*continued*)
Paris	4.02	3.66
Tokyo	4.45	4.27
Toronto	2.13	1.74
Washington, DC	1.47	1.16

a. Was there a significant change in the average price of gasoline in 1991?
b. Are the conditions for classical hypothesis testing fulfilled in this problem?

10.21 According to Ralph Rush of Technomic Inc., a Chicago-based food consulting firm, the price increases that will accompany Domino's introduction of a new pizza is "a calculated gamble" that could alienate customers. Design an experiment to test whether customers are alienated. Be explicit about what data you would collect and how you would obtain the data.

10.7 Testing Matched Pairs without the Assumption of Normality

The t test for matched pairs requires that the small samples come from normal populations. Although there are ways to test whether this maintained assumption of normality is reasonable, these tests have little power when samples are small, which is precisely when they are needed. The assumption of normality can be abandoned and a "distribution free" or nonparametric test can be employed. One such test is the Wilcoxon signed-ranks test, which requires only that the population be continuous and symmetric.[8]

The Wilcoxon test starts from the same point as the paired t test but requires that the absolute value of the differences $|d|$ be ranked (in ascending or descending order), with this ranking then "signed" in accordance with the original sign of the difference. That is, if the original difference was negative, then a negative sign is attached to the rank of the absolute value of this difference, Table 10.6. Notice in Table 10.6 that when differences are equal, average rank is assigned to each of the differences. (If a difference d is zero, it is removed from the analysis and the sample size is reduced by each zero difference removed.)

The test statistic has a distribution that is approximately the t distribution, but now the t ratio is calculated as

$$t = \frac{\bar{r} - 0}{\sqrt{\dfrac{\Sigma\left(r - \bar{r}\right)^2 / (n-1)}{n}}} = \frac{\bar{r} - 0}{s/\sqrt{n}} = \frac{3 - 0}{4.326/\sqrt{8}} = 1.961$$

The calculated value of this t is only a little lower than that obtained in the earlier t test of observed values, so the results represented in Figure 10.4 are not materially altered. The null hypothesis H_O: $\mu_2 - \mu_1 = 0$ is rejected in favor of the alternative H_A: $\mu_2 - \mu_1 \neq 0$, at the 10 percent Type I error level but not at the 5 percent level.

TABLE 10.6 Calculation of Difference (or Change) Scores for Wilcoxon Signed-Ranks Test

pretest	posttest	d	$\mid d \mid$	Rank	Signed Rank (r)	$(r - \bar{r})^2$
52	50	–2	2	1.5	–1.5	20.25
60	62	2	2	1.5	1.5	2.25
68	71	3	3	3	3	0
57	53	–4	4	4.5	–4.5	56.25
45	49	4	4	4.5	4.5	2.25
55	61	6	6	6	6	9.00
63	70	7	7	7	7	16.00
59	67	8	8	8	8	25.00
					24	131.00

Average Signed Rank ($\bar{r}$) = 3

$s^2 = \Sigma (r - \bar{r})^2 / (n - 1) = 18.714$

$s = 4.326$

$s/\sqrt{n} = 1.529$

$t = 1.961$

QUERY 10.4:

Use the *Business Week* (January 12, 1987) data in Query 10.3 to do a t test on the industry-matched pairs of the dollar volume of mergers and acquisition in 1985 and 1986. On an industry-by-industry basis, was there a significant decrease in activity between 1985 and 1986?

ANSWER:

With the exception of the photo, dental, and optical instruments industry in 1985 and nonelectrical machinery in 1986, these data can be treated as matched pairs. A matched pairs t test can assess significance with the remaining nine industries, assuming the individual populations (and thus the difference score d) are normally distributed. The importance of the normality assumption in any conclusion reached can be assessed by also performing the Wilcoxon signed-ranks test. For both tests, the null and alternative hypotheses are

- H_O: Population mean difference in volume between 1985 and 1986 is at least zero
- H_A: Population mean difference in volume between 1985 and 1986 is less than zero

The quantities needed to assess statistical significance by the two tests can be obtained easily with a spreadsheet program. The resulting quantities are

Matched Pairs	1985	1986	d	$(d - \bar{d})^2$	$\mid d \mid$	Rank	Signed Rank (r)	$(r - \bar{r})^2$
Utilities	6.2	5.7	–0.5	2.119	0.5	1	–1	0.444
Transportation	12.9	3.8	–9.1	51.043	9.1	8	–8	40.111
Retailing	6.3	9.2	2.9	23.576	2.9	4	4	32.111
Mining, oil, and gas	15.3	3.7	–11.6	93.015	11.6	9	–9	53.778

(*continued*)

Matched Pairs *(continued)*	1985	1986	d	$(d-\bar{d})^2$	$\lvert d \rvert$	Rank	Signed Rank (r)	$(r-\bar{r})^2$
Food	14.6	11.9	−2.7	0.554	2.7	3	−3	1.778
Electrical Machinery	5.9	12.0	6.1	64.892	6.1	5	5	44.444
Communications	7.1	13.4	6.3	68.154	6.3	6	6	58.778
Chemicals	12.3	5.3	−7.0	25.446	7.0	7	−7	28.444
Banking	7.1	5.1	−2.0	0.002	2.0	2	−2	0.111
Column sum divided by degrees of freedom =	9.74	7.78	−1.95	41.100			−1.666	32.500

Paired t test

$$t = -1.95\Big/\sqrt{41.1/9} = -0.915$$

Wilcoxon signed-ranks test

$$t = -1.666\Big/\sqrt{32.5/9} = -0.877$$

By either test statistic, the volume in 1986 is not significantly less than in 1985. For example, at the 10 percent Type I error level, the critical t is −1.397, with 8 degrees of freedom. Neither of the calculated t's of −0.915 and −0.877 is close to this critical value, so the null hypothesis is not rejected.[9]

10.8 Testing the Difference between Population Proportions

As we saw in the previous chapter, a test of population proportion can be undertaken with a normal distribution approximation if the sample size is large. This is also true for the difference between population proportions, such as the difference between the proportions of men and women who qualify for promotions or the proportion of defects in one assembly plant versus another. For the test of differences between proportions, the appropriate statistic and its value depend on the hypothesized difference between the two population proportions $\pi_2 - \pi_1$ and the difference in sample proportions $p_2 - p_1$.

Typically, the difference in population proportions is hypothesized to be zero in one of three testing situations:

$$H_O\colon \pi_2 - \pi_1 = 0 \qquad H_O\colon \pi_2 - \pi_1 \geq 0 \qquad H_O\colon \pi_2 - \pi_1 \leq 0$$

$$H_A\colon \pi_2 - \pi_1 \neq 0 \qquad H_A\colon \pi_2 - \pi_1 < 0 \qquad H_A\colon \pi_2 - \pi_1 > 0$$

where π_2 is the proportion possessing the attribute of interest in one population and π_1 is the proportion in the other population. In all three cases, the null hypothesis involves the assumption that $\pi_2 - \pi_1 = 0$, which implies that the populations are identical, so we can define $\pi = \pi_2 = \pi_1$. Identical populations also imply that the two independent samples of sizes n_1 and n_2 can be pooled into one sample of size $n_1 + n_2$. If one of these experiments yielded a sample of x_1 realizations of the attribute of interest (with a sample proportion of $p_1 = x_1/n_1$), while the other yielded x_2, such "successful" trials (with a sample proportion of $p_2 = x_2/n_2$), then the pooled proportion of successes is

$$p = \frac{x_1 + x_2}{n_1 + n_2}$$

The point estimator of π is p and the expected value of p is π. The point estimator of the difference in population proportions $\pi_2 - \pi_1$ is the difference in sample proportions $p_2 - p_1$. If $\pi_2 - \pi_1 = 0$ (the null hypothesis), the expected value of $p_2 - p_1$ is 0 and the variance of the $p_2 - p_1$ distribution is approximately[10]

$$\sigma^2(p_2 - p_1) \approx \frac{p(1-p)}{n_1 n_2}(n_1 + n_2)$$

Thus, the approximate z test statistic for large samples is

z test statistic for $\pi_2 - \pi_1$

$$z \approx \frac{(p_2 - p_1) - 0}{\sqrt{\dfrac{p(1-p)}{n_1 n_2}(n_1 + n_2)}} \tag{10.7}$$

where: H_O: $\pi_2 - \pi_1 = 0$
$p_2 = x_2/n_2$, $p_1 = x_1/n_1$ and $p = (x_1+x_2)/(n_1+n_2)$

This is illustrated by an International Data Corporation study commissioned by Hewlett Packard (and obtained from HP by this author). The study shows that at 106 sites using Hewlett Packard computers and at 115 sites using Compaq computers, 92 percent of the HP site respondents and 84 percent of the Compaq site respondents were "satisfied" or "very satisfied" with the vendor. A hypothesis test to determine if Compaq site respondents are less satisfied than HP site respondents can be performed with the z statistic, assuming two simple random samples from binomial distributions.

Letting π_2 be the proportion satisfied at the HP sites and π_1 the proportion satisfied at the Compaq sites, the null and alternative hypotheses are

H_O: $\pi_2 - \pi_1 \leq 0$

H_A: $\pi_2 - \pi_1 > 0$, HP sites more satisfied than Compaq's

For $\pi_2 - \pi_1 = 0$, the pooled estimate of π is

$$p = \frac{x_1 + x_2}{n_1 + n_2} = \frac{0.84(115) + 0.92(106)}{115 + 106} = \frac{97 + 98}{221} = 0.88$$

The variance of the distribution of $p_2 - p_1$ is approximately

$$\sigma^2(p_2 - p_1) \approx \frac{p(1-p)}{n_1 n_2}(n_1 + n_2) = \frac{0.88(0.12)}{115(106)}(221) = 0.0019$$

Thus, the calculated value of the test statistic is

$$z \approx \frac{(p_2 - p_1) - 0}{\sqrt{\frac{p(1-p)}{n_1 n_2}(n_1 + n_2)}} = \frac{0.92 - 0.84 - 0}{\sqrt{0.0019}} = 1.818$$

The p-value for this one-tailed test is

$$p\text{-value} = P(p_2 - p_1 > .08 \mid \pi_2 - \pi_1 = 0) = P(z > 1.818) = .035$$

This p-value indicates that the null hypothesis $\pi_2 - \pi_1 = 0$ can be rejected and the alternative hypothesis $\pi_2 - \pi_1 > 0$ accepted for a Type I error probability as low as 0.035. This 3.5 percent level of statistical significance is depicted in Figure 10.6. (Note that if this had been a two-tailed test, then the p-value would have been 0.07.)

If the samples of the two computer sites truly represented the outcomes of independent experiments, then a p-value of 0.035 would be strong evidence in favor of the conclusion that those at Hewlett Packard sites were more satisfied than those at Compaq sites. International Data Corporation stated clearly on page one of its report, however, that those sites it surveyed were not chosen by simple random sampling

FIGURE 10.6 p-value for Hewlett-Packard versus Compaq Computer Satisfaction

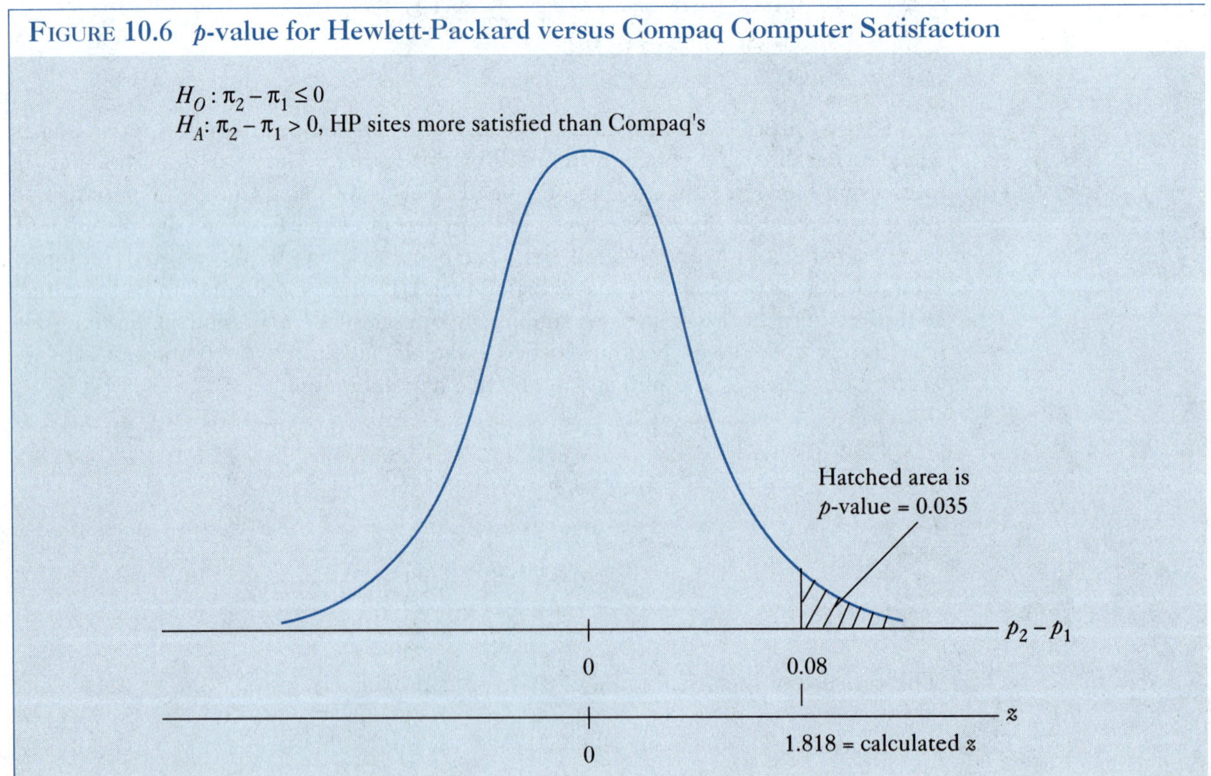

methods. Instead, they were selected on the basis of specific criteria. Furthermore, there is no reason to believe that those who responded represented a random sample of those surveyed. Because IDC had some discretion on who got into the samples, surely IDC would not bias the sample against Hewlett Packard—recall that HP commissioned the study. An article in *The Wall Street Journal* (July 31, 1991) on conflicting results in survey rankings of computers also identified another area of concern. The money that Hewlett Packard spends publicizing the IDC results advances IDC because IDC is identified as the surveying expert in all of HP's ads. In designing a study and identifying winners, an astute pollster is not going to ignore the subjects' ability to promote the pollster. These sampling and motivational considerations weaken the practical importance of the 0.035 p-value.

Exercises

10.22 In 1989, a J.D. Power's quality survey named the Buick LeSabre the most trouble-free domestic model, with only 89 defects per 100 cars. But according to a *Wall Street Journal* (September 5, 1991) article, the Ford Crown Victoria had just 91 defects per 100 cars, "a difference Power says was within the margin of error." Buick heavily advertised its victory. If you were considering the purchase of one of these cars, how seriously should you take these results?

10.23 A *Wall Street Journal* (January 22, 1992) article stated that, in 1989, the Pennsylvania Health Care Cost Containment Council counted 12 deaths at Harrisburg Hospital in 368 bypass operations and six deaths at Lancaster General Hospital in 447 bypass operations. At the 0.01 Type I error level, is this difference due to only randomness? Besides randomness, what other factors might be influencing this difference?

10.24 A *Wall Street Journal* (February 4, 1992) article reported the results from two surveys of employers' intentions to discontinue placing retirement plan assets in Guaranteed Investment Contracts. For a sample of 40 companies, consultant Foster Higgins found that 50 percent will not reinvest or will only partly reinvest in GIC, while 38 percent of the 183 employers in a similar poll by the International Foundation of Employee Benefit Plans are reducing or ending GIC holdings. Is there a significant difference in the population proportions represented by these two surveys? (Test at the 5 percent level.)

10.25 A rental car system representative claims that compared to last year, compact cars now make up a smaller percentage of the total fleet housed at airports. On this day last year, your lot had 40 cars, including 11 compacts. Today you have 48 cars, 12 of which are compacts.

a. State the null and alternative hypotheses involved in the rental car system's claim.
b. Treating your daily observations as random samples, do you believe that nationally the proportion of compacts has fallen? At what level of significance?
c. Comment on the validity of your test.

10.26 According to an article in *The Wall Street Journal* (February 26, 1992), a survey by Sally Schmidt, a marketing consultant in Burnsville, Minnesota, in 1991 of some 400 law firms "found that 48% had marketing directors, down from 66% in 1990." Was this a statistically significant drop in the proportion of law firms with marketing directors? How sensitive is your test to the probability of a Type I error?

10.27 As stated in the Exercise 10.11, in the early days of microcomputers, hard disk drives had a mean time between failures (MTBF) of 5,000 to 20,000 hours. Because there are 8,760 hours in a year, disk failures in the first year of operation were common. Today, drives should last from 20,000 to 40,000 hours, making disk drive failure rare in the first several years of service. The question to be asked, however, is whether there is a difference between failure rates in the first five years of service for two different types of experimental drives. Below is information on the failure rates of two types of disk drives. At the 5 percent significance level, is there a difference between the failure rates?

	Proportion failing (p)	Sample size (n)
Type A	30%	30
Type B	42%	50

10.28 The newspaper article "Hoosier Lottery targets minorities in marketing plan"(*HT*, September 1, 1992) describes a poll taken in November and December of 1991 that "showed 42.8 percent of blacks surveyed had never played the lottery, compared to 38.6 percent of whites. While those numbers are statistically equal, lottery officials believe many minorities don't trust the lottery." Write the null and alternative hypotheses tested here. What is the apparent meaning of the statement "those numbers are statistically equal," and why is this poor wording?

10.29 Management consultant Tom Peters (in his Tribune Media Service column of November 11, 1992, *HT*) presented an Australian study of 48 cancer patients. After seeing 24 of these patients, a letter was sent by the physician summarizing the visit; the other 24 received no communication. Of the 24 who got the letter, 13 stated that they were "completely satisfied with the physician's care; 4 of the 24 who did not get a letter said they were completely satisfied." At what level of significance could the hypothesis of no difference between the two groups be rejected?

10.9 The Median Test for Two Samples (Optional)

In Chapter 9, a single sample test of the population median was introduced. This test can be extended to a two-sample test of the differences between population medians.

The test of differences between medians makes use of the z statistic approximation introduced in the last section for testing the differences between proportions. As in the rank-sum test, the observations from two simple random samples must be combined in a rank ordering, as if they came from the same population (the null hypothesis). Then the difference in the proportions of observations below this overall median in each sample is calculated and standardized as a normal random variable.

Suppose the data in Table 10.7 are two simple random samples of employee seniority (in years) at two different plants. We want to test whether or not the median seniorities are the same.

$$H_O\text{: Median}_2 - \text{Median}_1 = 0 \text{ versus } H_A\text{: Median}_2 - \text{Median}_1 \neq 0$$

The 20 observations from plant one and the 19 observations from plant two already have been ranked in a 39-observation array, using a spreadsheet program and the sort command. The median was found to be 22 years, with 12 values below it at plant one

Table 10.7 Employee Seniority

	PLANT	SENIORITY	RANK
	1	1	1.5
	1	1	1.5
	1	3	4
	1	5	6
	1	8	8
	1	11	10
	1	12	11
	1	15	13
	1	16	14.5
	1	16	14.5
	1	17	16
	1	20	19
Median →	1	22	20
	1	25	22.5
	1	28	25
	1	30	27
	1	30	27
	1	31	29
	1	37	31
	1	40	34.5
	2	2	3
	2	4	5
	2	7	7
	2	10	9
	2	13	12
	2	18	17
	2	19	18
	2	23	21
	2	25	22.5
	2	27	24
	2	30	27
	2	33	30
	2	38	32
	2	39	33
	2	40	34.5
	2	42	36
	2	45	37
	2	46	38
	2	47	39

	Below Median	
Plant	#	Proportion
1	12	$0.6000 = p_1$
2	7	$0.3684 = p_2$
		$-0.2316 = p_2 - p_1$

and seven below it at plant two. Thus the relevant individual sample proportions below the overall median are $p_1 = 12/20 = 0.6$ and $p_2 = 7/19 = 0.3684$.

According to the null hypothesis, the two plants are identical; if so, then 50 percent of the observations in each sample are expected to be below the median of 22 years. The null hypothesis thus implies that $\pi_1 = \pi_2 = \pi = 0.50$, and the expected value of the pooled proportion p is 0.50; that is, $E(p) = \pi = 0.50$. Thus, the test of differences between proportions in Formula (10.7) now can be written

$$z \approx \frac{(p_2 - p_1) - 0}{\sqrt{\frac{\pi(1-\pi)}{n_1 n_2}(n_1 + n_2)}} = \frac{0.3684 - 0.6 - 0}{\sqrt{\frac{0.5(0.5)}{(20)(19)}(20+19)}}$$

$$= \frac{-0.2316}{\sqrt{0.02566}} = -1.45$$

The p-value for this two-tailed test is

$$p\text{-value} = 2P(p_2 - p_1 < -.2316 \mid \pi = .5) = 2P(z < -1.45) = 0.147$$

This p-value indicates that the null hypothesis of no difference between the median seniority at each plant ($\text{median}_2 - \text{median}_1 = 0$) cannot be rejected at typical Type I error probability levels. (As in the case of a one-sample test of the median described in Chapter 9, a one-tailed test of the difference in medians would have yielded a p-value of 0.0735 in this example if the alternative hypothesis was $\text{median}_2 - \text{median}_1 > 0$.)

Unlike the rank-sum test, in this test of medians the only relevant information is whether the observations are above, at, or below the median of the combined array. As a result, this test is not as powerful as the rank-sum test. This lack of power could be important in some practical applications. For instance, suppose initially there was no difference between the salaries of employees but then a new management team gave substantial salary raises to all the younger, white males who were above the median salary. Because the median did not change, a test of differences in median salaries across the races, sexes, or ages would not detect this change. A test of means or rank would capture differences brought about by this new salary policy.

10.10 Concluding Comments

Two sample tests of means and proportions are popular in business, education, medicine, law, and most other applied areas of investigation. Typically, they are part of the initial investigation of data that accompanies presentations of the descriptive statistics. Violations in the required assumptions of normality, independence of simple random samples, or other modeling considerations require more complex methods of analyses.

In principle, one should test for violations of the required population normality; in practice, existing tests for violations are weak. One is always wise to consider results from both a parametric test of differences and one of its nonparametric counterparts. Consistent results from both types of tests indicate that the assumption of normality is not crucial. Lack of consistency across tests raises questions of power: Which test is more powerful? Although this question is easy to ask, there may be no unique answer. One test may be more powerful than another at different levels of α and different values consistent with the alternative hypothesis. As shown by Stephen Spielman (1978) in his discussion of tests of differences in the effects of vaccines and types of operations on mortality rates, it is extremely unlikely that any one test is uniformly more powerful than another across all the values contained in (diffuse) alternative hypotheses.

Even when results are consistent for different forms of tests, however, this consistency says little about other aspects of the research model employed. Two-variable models of differences, from which two independent samples are obtained, cannot capture the influence of systematic, nonrandom factors that might be influencing both parametric and nonparametric test results. Alternatives to these simple bivariate models are thus considered in the following chapters.

Chapter Exercises

10.30 The test of the difference between subjective and objective probabilities in horse racing, Exercise 10.9, starts with a model in which the samples are assumed to come from normal populations. Repeat this test without making this assumption.

10.31 An article in *Business Week* (August 9, 1993) reported on how the layout and size of computer keyboards is believed to affect typing speed and accuracy. To test this belief, one secretarial pool, with seven members, is equipped with one type of keyboard while another, with six members, is equipped with the other. After the secretaries had become accustomed to working on the boards, they were tested for speed and accuracy with the following results.

	Speed (words/minutes)							Accuracy (errors/100 words)						
Pool One:	100	90	67	101	88	91	75	2	1	2	4	0	1	2
Pool Two:	68	102	84	70	81	93		2	2	1	1	1	3	

Assuming that the speeds, errors, and the secretaries themselves are drawn from normal populations, with equal (respective) variances, is there a difference between the keyboards (in terms of speed and accuracy)? At what levels of significance?

10.32 Exercise 10.31 involves the assessment of typing speed and accuracy. Is it reasonable to assume that the respective population variances are equal? (Test for equality of variances, at the 0.01 Type I error level.)

10.33 Repeat the typing speed and accuracy test in Exercise 10.31, assuming the populations are not normal. What are your conclusions regarding the two keyboards now?

10.34 What alternative experiment could you propose to test one keyboard against the other in Exercise 10.31? Why would your alternative be preferred to that described in Exercise 10.31?

10.35 In *Cox v. Conrail* (1987), the plaintiff alleged that Conrail discriminated against blacks in hiring. At issue was the fairness and job-relatedness of the knowledge test for engineer trainees. The court opinion was based on the following data gathered from 46 training classes.

	Number Who				
	Started	Terminated	Tested	Passed	Failed
Blacks	85	15	70	42	28
Whites	533	36	497	432	65
Total	618	51	567	472	93

Because the plaintiff did not provide an expert witness to testify on the implications of these numbers, Judge Flannery stated that the court was not able to evaluate the relevance and importance of these data or the appropriateness of any statistical analysis applied to them. He wondered if a sample size of 70 blacks was sufficient to draw conclusions about any disparate impact of the test. What expert guidance could you have provided? (Hint: See Joseph L. Gastwirth, "Statistical Reasoning in the Legal Setting," *American Statistician*, February 1992, pp. 55–69.)

10.36 In the case of *Cox v. Conrail* (1987), described in Exercise 10.35, test to see if blacks and whites have the same termination rates. Test at the 5 percent Type I error level. What are the practical implications of your test for this case?

10.37 In the case of *Cox v. Conrail* (1987), described in Exercise 10.35, test to see if blacks and whites have the same pass rates (based on those who took the test). Test at the 5 percent Type I error level. Does this test have a disparate impact on blacks?

10.38 What is the importance of testing for differences in termination rates before testing for differences in test scores between blacks and whites in the case described in Exercises 10.36 and 10.37?

10.39 In the early 1990s, a few Japanese leaders made derogatory comments about the work ethic of Americans. A *Wall Steet Journal* (February 6, 1992) article, however, had the headline "If the U.S. Work Ethic is Fading, 'Laziness' May Not Be the Reason." In the article, University of Michigan economist Thomas Juster was cited as saying that U.S. and Japanese men each spend about 56 hours a week working. "What sets them apart," according to Juster, "is that Japanese men spend 52 hours per week working for pay and only 3.5 hours—a mere half hour a day—working around the house . . . . In the U.S. men work 44 hours a week for pay and 14 hours a week at home." How could you test Juster's claim of no difference between hours worked in the United States versus Japan?

10.40 (Appendix 9D would be helpful) A survey by the National Association of Fleet Administrators found that 57 percent of 1992 model cars purchased by fleet buyers specified the installation of air bags, but only 60 percent required antilock brakes. *The Wall Street Journal* (February 10, 1992), reporting on the results, did not give sample sizes. At the 1 percent Type I error level, what is the smallest sample size that would lead to the rejection of the null hypothesis of no difference between the population proportion ordering air bags and antilock brakes?

10.41 A *Wall Street Journal* (August 29, 1991) article reported on the problems that General Motors was having at its plant in Orion Township, Michigan. According to the article "On some days, the cars have more than three times the company goal of two defects per car." A *Business Week* (June 29, 1992) article had headlines saying that "It's Not All That Easy Having a Hit. Just Ask Caddy." The article was about the success General Motors was having with the Cadillac Seville made at its Detroit-Hamtramck plant. According to this *BW* article, however, "an internal quality audit in early June found 5.1 defects per car vs. a target of 1.0." But an average of slightly more than six defects versus slightly more than five defects doesn't seem to be much of a difference. How could you go about testing if there is a significant difference between the defect rate at the Orion Township Oldsmobile plant and the Detroit-Hamtramck Cadillac plant? What would you need to know to test whether the Cadillac target of one (or fewer) defect per car is resulting in fewer defects than Oldsmobile's goal of two (or fewer) defects?

10.42 A short note in *Business Week* (June 29, 1992) reported on surveys of some 500 CEOs and some 500 board members to gauge the strength of their agreement with a series of statements on corporate management. The closest agreement was on the proposition "Stock options should be extended to all employees"—17 percent of the CEOs agreed and 20 percent of the board members agreed. But do these results show agreement? (Test at the 5 percent Type I error level.)

10.43 A note in *The Wall Street Journal* (June 23, 1992) reported that "since Smith Kline Beecham's pharmaceutical division gave laptop computers and portable printers to all 1,800 of its sales staff, each sales person now makes, on average, one-and-a-half more calls per week." At the 5 percent Type I error level, would the conclusion that there was a significant increase in number of calls be greatly influenced by the size of the standard deviation in a matched pairs t test?

10.44 Chuck Stone, in his Newspaper Enterprise Association column (*HT,* September 5, 1992) wrote: "Increases of one or two points are monumentally insignificant because the SAT's standard error of measurement is 18 points. That means you can take the SAT 10 times, and your score might change 10 times, but your score would still fall within that range of 18 points." Comment on the correctness of the ideas expressed in this quote.

10.45 Use the data from *PC Magazine* (June 30, 1992) on spreadsheets in Exercise 2.16. Perform both a parametric and nonparametric test for differences in the measure of central tendencies of the CA SUPER CALC and the Microsoft EXCEL programs. Do the tests give the same results? Which test is better in this situation and why?

10.46 The data set from *The Wall Street Journal* (September 24, 1992) on the world's 100 largest public companies is in the spreadsheet file "EX10-46.WK1," which is readable by LOTUS and any compatible spreadsheet. Do a Wilcoxon signed-ranks test of the differences in rank in market value between the years 1992 and 1991.

10.47 The data set from *The Wall Street Journal* (September 24, 1992) on the world's 100 largest public companies is in the spreadsheet file "EX10-47.WK1," which is readable by LOTUS compatible spreadsheets. Calculate the SALES for 1990 from the 1991 data on SALES and the percentage change from 1990. Do a Wilcoxon-Mann-Whitney rank sum test of the differences between these two years, using only companies for which there is complete data.

10.48 The data set from *The Wall Street Journal* (September 24, 1992) on the world's 100 largest public companies is in the spreadsheet file "EX10-48.WK1," which is readable by any spreadsheet that is LOTUS compatible. Calculate the PROFIT for 1990 from the 1991 data on PROFIT and the percentage change from 1990. Do a Wilcoxon-Mann-Whitney rank sum test of the differences between these two years, using only companies for which there is complete data.

10.49 "Crash tests perfomed by safety regulators showed that six of 11 GM pickups tested performed flawlessly or as well as competitors," according to Harry Peace, GM's general counsel as quoted in *The Wall Street Journal* (April 30, 1993). State the null and alternative hypotheses for a test of no difference between GM and competitors' trucks in crash tests.

10.50 In Exercise 10.49 involving GM trucks, if nine of 11 competitors' trucks performed flawlessly, could the null hypothesis of no difference be rejected at the 5 percent level of significance? What is the practical significance of this result?

10.51 Courts typically put more weight on published than on unpublished research that finds statistical differences at the 5 percent significance level. A *Newsweek* (March 22, 1993) article cited Kenneth Rothman of Boston University as saying that significance testing is "a clumsy substitute for thought." He argues that a finding of no significant difference could actually mask a very large difference. Use the appendices to Chapter 9 to show how Rothman could be correct.

10.52 "Workers plan to retire at 62, on average, an EBRI-Gallup Organization survey of 1,000 people shows. Last year, respondents in a similar poll . . . planned to retire at an average age of 61."(*Wall Street Journal,* July 13, 1993) If both the sample size and population standard deviations were the same in each year, how small would this standard deviation have to be to conclude that there was a significant increase in the average retirement age, at the 0.05 Type I error level?

10.53 Is the standard deviation calculated in Exercise 10.52 a reasonable magnitude? (Explain why or why not.) Is it thus okay to conclude that a sample average increase from 61 to 62 years is significant? (Explain.)

10.54 Exercise 2.16 presented a Computer Associates' advertisement that appeared in *PC Magazine* (June 30, 1992). Why is the arithmetic mean an inappropriate statistic to use for a comparison of the three spreadsheet programs?

Academic References

Bera, A., and C. Jarque. "Model Specification Tests: A Simultaneous Approach." *Economics Letters* 6, 1980, pp. 255–259.

Iman, R.L. "Graphs for Use with the Lilliefors Test for Normal and Exponential Distributions." *American Statistician* 36(2), 1982, pp. 108–112.

Kennedy, W.J., and J.E. Gentile. *Statistical Computing.* New York: Marcel Dekker, 1980.

Kiefer, N., and M. Salmon. "Testing Normality in Econometric Models," *Economics Letters* 11, 1983, pp. 123–128.

Lilliefors, H.W. "On the Kolmogorow-Smirnov Test for Normality with Mean and Variance Unknown," *Journal of the American Statistical Association* 62, 1967, pp. 399–402.

Mee, R.W., and T.C. Chua. "Regression Toward the Mean and the Paired Sample *t* Test." *American Statistician,* February 1991, pp. 39–42.

Moser, Barry, and Gary Stevens. "Homogeneity of Variance in the Two-Sample Means Test." *American Statistician,* February 1992, pp. 19–21.

Paulson, Edward. "An Approximate Normalization of the Analysis of Variance Distribution." *Annals of Mathematical Statistics* 13, 1942, pp. 233–235.

Spielman, Stephen. "Statistical Dogma and the Logic of Significance" *Philosophy of Science* 45, 1978, pp. 120–135.

Endnotes

1. The variance of the sample means $(\bar{x}_2 - \bar{x}_1)$ is the variance of the difference between two random variables. Thus we know from Chapter 4 that

$$\begin{aligned}\mathrm{Var}(\bar{x}_2 - \bar{x}_1) &= \mathrm{Var}(\bar{x}_2) + \mathrm{Var}(\bar{x}_1) - 2\mathrm{Cov}(\bar{x}_2\bar{x}_1) \\ &= (\sigma_2 / \sqrt{n_2}) + (\sigma_1 / \sqrt{n_1}) - 2\mathrm{Cov}(\bar{x}_2\bar{x}_1) \\ &= (\sigma_2 / \sqrt{n_2}) + (\sigma_1 / \sqrt{n_1})\end{aligned}$$

   if the samples are independent, which implies zero covariance.

2. The point estimator of $(\mu_2 - \mu_1)$ is $(\bar{x}_2 - \bar{x}_1)$. The interval estimator is a $(1 - \alpha)$ 100% confidence interval defined by

$$\left(\bar{x}_2 - \bar{x}_1\right) \pm z_{\alpha/2}\sqrt{\frac{\sigma_1^2}{n_1} + \frac{\sigma_2^2}{n_2}}$$

3. The first assumption implies that $\sigma_1 = \sigma_2 = \sigma$ so that the standard error of the difference in sample means can be written

$$\sqrt{\frac{\sigma_1^2}{n_1} + \frac{\sigma_2^2}{n_2}} = \sigma\sqrt{\frac{1}{n_1} + \frac{1}{n_2}}$$

To estimate σ^2 the two samples are "pooled" and treated as if they came from the same population, which is assumed under the null hypothesis. Given the maintained assumptions, it can be shown that this pooled estimator of σ^2 can be used to obtain an unbiased estimator of the standard error of the difference in sample means; that is,

$$E\left[\left(\frac{s_1^2(n_1 - 1) + s_2^2(n_2 - 1)}{n_1 + n_2 - 2}\right)\left(\frac{n_1 + n_2}{n_1(n_2)}\right)\right] = \sigma^2\left[\frac{1}{n_1} + \frac{1}{n_2}\right]$$

4. The point estimator of $\mu_2 - \mu_1$ is $(\bar{x}_2 - \bar{x}_1)$. The interval estimator is given by the $(1 - \alpha)$ 100% confidence interval defined by

$$\left(\bar{x}_2 - \bar{x}_1\right) \pm t_{\alpha/2}\sqrt{\left(\frac{s_1^2(n_1 - 1) + s_2^2(n_2 - 1)}{n_1 + n_2 - 2}\right)\left(\frac{n_1 + n_2}{n_1(n_2)}\right)}$$

where the degrees of freedom are $n_1 + n_2 - 2$.

5. The F ratio is formed by two independent normal random variables x_1 and x_2 in the following way

$$F = \frac{\Sigma\left(x_{1i} - \bar{x}_1\right)^2 / (n_1 - 1)\sigma_1^2}{\Sigma\left(x_{2i} - \bar{x}_2\right)^2 / (n_2 - 1)\sigma_2^2} = \frac{s_1^2/\sigma_1^2}{s_2^2/\sigma_2^2} = \frac{s_1^2}{s_2^2}$$

If $\sigma_1^2 = \sigma_2^2$

6. For tests of normality, see, for example, H. W. Lilliefors (1967), R. L. Iman (1982), N. Kiefer and M. Salmon (1983), and A. Bera and C. Jarque (1980).

7. A $(1 - \alpha)$ 100 percent confidence interval for μ is

$$\bar{d} \pm t_{\alpha/2} s_d / \sqrt{n}$$

Thus, a 95 percent confidence interval is calculated as

$$3 \pm (2.365) 4.24264 / \sqrt{8}$$

$$-0.55 < \mu < 6.55$$

which includes the null hypothesized value of $\mu = 0$.

On the other hand, the 90 percent confidence interval is

$$3 \pm (1.895)4.24264/\sqrt{8}$$

$$0.1575 < \mu < 5.8425$$

which does not include the null hypothesized value of $\mu = 0$.

8. Rankings are not normally distributed; they are uniformly distributed with the median rank and mean rank being the same. This symmetry is sufficient for the sampling distribution of average rankings to be approximately normal.
9. In concluding that mergers and acquisition activity volume has not fallen between 1985 and 1986, an interesting philosophical question can be raised regarding the manner in which the data from 1985 and 1986 can be considered a simple random sample. These years cannot be repeated for another sampling, so what is the meaning of the sampling distribution of means?
10. The exact variance of $(p_2 - p_1)$ is

$$\frac{\pi_1(1-\pi_1)}{n_1} + \frac{\pi_2(1-\pi_2)}{n_2} \underset{\uparrow}{=} \pi(1-\pi)\left[\frac{1}{n_1} + \frac{1}{n_2}\right] \underset{\uparrow}{=} E(p)[1-E(p)]\left[\frac{1}{n_1} + \frac{1}{n_2}\right]$$

$$\text{If } \pi = \pi_1 = \pi_2 \qquad E(p) = \pi$$

If the difference between population proportions $(\pi_1 - \pi_2)$ is assumed equal to something other than zero, then this convenient reduction in the variance of $(\pi_2 - \pi_1)$ is not possible.

FORMULAS

- z test statistic for two independent samples

$$z = \frac{(\bar{x}_2 - \bar{x}_1) - D_O}{\sqrt{\dfrac{\sigma_1^2}{n_1} + \dfrac{\sigma_2^2}{n_2}}}$$

where D_O is the difference specified by the right-hand side of H_O. D_O is typically zero.
FOR USE WHEN:

1. The two parent population variances are known, the random samples are independent of each other, and
   a) each is at least of size $n = 30$, or
   b) each is drawn from a normal parent population.

OR 2. Each independent sample is sufficiently large ($n > 30$) that the sample variances are good approximations of their respective population variances.

- t test statistic for difference between means, with equal population variances and independent samples

$$t = \frac{(\bar{x}_2 - \bar{x}_1) - 0}{s_{\bar{x}_2 - \bar{x}_1}}$$

where

$$s_{\bar{x}_2-\bar{x}_1} = \sqrt{\left(\frac{s_1^2(n_1-1)+s_2^2(n_2-1)}{n_1+n_2-2}\right)\left(\frac{n_1+n_2}{n_1(n_2)}\right)}$$

with $n_1 + n_2 - 2$ degrees of freedom and 0 as the value on the right-hand side of H_O.
FOR USE WHEN:

1. The two parent population variances are unknown but equal, and
2. Both independent samples are small but drawn from populations that are normally distributed.

- *F* test statistic for a test of a difference in variances, given a null hypothesis of no difference

$$F = \frac{s_1^2}{s_2^2}$$

s_1^2 and s_2^2 are sample variances, where $s_1^2 > s_2^2$, and there are $n_1 - 1$ degrees of freedom in the numerator, and $n_2 - 1$ degrees of freedom in the denominator.

- test statistic for difference between means from small independent samples

$$t^* = \frac{(\bar{x}_2 - \bar{x}_1) - D_O}{\sqrt{\frac{s_1^2}{n_1} + \frac{s_2^2}{n_2}}}$$

Degrees of freedom are

$$\text{df} = \frac{(F/n_1 + 1/n_2)^2}{F^2/n_1^2(n_1-1) + 1/n_2^2(n_2-1)}$$

where $F = s_1^2/s_2^2$, for $s_1^2 > s_2^2$.

- Wilcoxon-Mann-Whitney *U* test

$$T_A = n_A n_B + \frac{n_A(n_A+1)}{2} - \Sigma r_A$$

$$T_B = n_A n_B - T_A$$

where Σr_A denotes the sum of the ranks in the *A* sample, and n_A and n_B are the sample sizes

$$U = \min[T_A, T_B]$$

$$z = (U - n_A n_B/2)\Big/\sqrt{n_A n_B(n_A + n_B + 1)/12}$$

- matched pairs t statistic

$$t = \frac{\bar{d} - \mu_0}{s_d / \sqrt{n}}$$

where $s_d = \sqrt{\Sigma\left(d_i - \bar{d}\right)^2 / (n-1)}$

- z test statistic for $\pi_2 - \pi_1$

$$z \approx \frac{\left(p_2 - p_1\right) - 0}{\sqrt{\frac{p(1-p)}{n_1 n_2}\left(n_1 + n_2\right)}}$$

where H_O: $\pi_2 - \pi_1 = 0$

$p_2 = x_2/n_2$, $p_1 = x_1/n_1$ and $p = (x_1+x_2)/(n_1+n_2)$

Chapter 11

Analysis of Variance and Contingency Tables

That's not an experiment
you have there,
that's an experience.

Sir Ronald A. Fisher
1890–1962

11.1 Introduction

The previous two chapters were devoted to inferences about a single population and differences between populations. This chapter extends the analysis to differences among several populations. Tests of the differences among population means, population proportions, and population medians are presented in this chapter in the following situations.

- The gas mileage of three different makes of compact cars was established by testing them in different environments. The average mileage of each car was then computed based on the performance achieved in the tests. The question: Can we conclude that the cars really differ in their mileage?
- The discrimination suit against the large midwestern bank discussed in the previous chapter claimed discrimination in the hiring and retention of female employees. The question here is whether the bank attempted to change the composition of its staff of managers after the filing of the sex discrimination suit.
- An instructor assigned 28 students in a course to four project study groups on the basis of an alphabetical ordering of their names. After the first exam, she wants to know if student performance in the four groups was sufficiently close to believe that this assignment method was similar to a simple random sampling.

The six-step hypothesis testing procedure described in the previous chapters continues to apply here. After specifying the null and alternative hypotheses (Step 1), this chapter introduces new test statistics that are employed in testing for differences among the parameter values of different populations (Step 3). An analysis of variance (abbreviated ANOVA) is introduced, for use with the F statistic (Step 4), to test for differences among means. The chi-square (χ^2) test statistic is introduced to test for the independence of attribute proportions among populations. The chi-square distribution is also used to test for differences among population medians. The use of a p-value, as determined by a computer program for both the F and χ^2 distributions, is demonstrated (Step 5). The location of the critical F and χ^2 values, for a specified Type I error probability level, are also shown (Step 2), and this critical value is used in reaching a conclusion (Step 6). Thus, by the end of this chapter, the reader will be able to conduct a classic analysis of variance via an F test, a chi-square test of independence, and the Kruskal-Wallis nonparametric ANOVA.

11.2 Variability Between and Within Samples

Because of possible differences in driver ability, different road conditions, or the general driving environment, testing agencies such as *Consumer Reports* typically have cars test driven more than once to assess gas mileage. The typical Fisherian experimental design involves assigning "subjects" randomly to "treatments." For example, 15 drivers might be randomly assigned to each of the three compact cars.

Historical Perspective

Along with the development of test statistics and their distributions, articulating the role played by "random arrangements" in experimental design was another major contribution to statistics made by Sir Ronald A. Fisher.

For many years, Fisher was associated with an agricultural experimental station in England. A typical question was whether different types of fertilizers gave different yields. The fertilizer type was a categorical treatment variable or factor that had a binary outcome of yes, if the given fertilizer was used, and no, if not. Yield was a continuous variable whose measurement had meaning on a number line. Because yield was made a function of the type of treatments, it was labeled the dependent variable.

Fisher designed experiments in which plots of land were randomly assigned to different fertilizer treatments. By measuring the mean yield on each of several different randomly assigned plots of land, Fisher eliminated, or "averaged out," the effects of nontreatment influences (such as weather and soil content) so only the effect of the fertilizer was reflected in differences among the mean yields.

Unfortunately, in many applications in business and economics, strict adherence to the principles of Fisherian experimental design is impossible. We usually have opportunistic data that do not come from well-defined experiments employing simple random sampling. The difference between the handling of experimental data and opportunistic data is a difference in degree, not kind. The computations for an analysis of means, for example, are the same regardless of the manner in which the data are collected. Our ability to extract causal implications from the analysis, however, is not the same. There may be unknown factors that are biasing results when the experiment employs nonrandom sampling. But, as stated by econometrician Edward Leamer, "No one has ever designed an experiment that is free of bias, and no one ever can."(1983, p. 33) We typically have to work with the data at hand. Cochran (1983) provides a detailed discussion of the implications of working with data from experiments and opportunistic data obtained from nonrandomized studies.

Finally, Fisher was interested in analyzing differences in mean yields among fertilizers. We will be interested in different mean gas mileage among cars. By historical convention, we refer to this analysis as a comparison "between means," even though more than two means are involved. The method of analysis is called "analysis of variance," even though we are interested in differences in the means. The name "analysis of variance" comes from the use of ratios of variances to establish differences in means. It is typically referred to by the acronym ANOVA.

Table 11.1 shows the 15 gas mileages achieved by the drivers of the three cars. Because of random assignment, only the chance factors inherent in sampling should cause differences in the sample averages unless there really is a difference in the gas mileage of the three cars. That is, the sample average mileages of the three cars can be expected to differ (which they do: $\bar{y}_1 = 31$, $\bar{y}_2 = 29$, and $\bar{y}_3 = 36$), but the question is whether these sample differences are sufficiently large to believe that the population mean mileage differs among the three cars.

Notice that we are representing the gas mileage random variable by the letter y. This is done to make it clear that gas mileage may depend on the type of car. Dependent variables are typically plotted on the vertical axis, which is usually designated by the letter y, although any letter could be used.

TABLE 11.1 Miles Per Gallon for Three Cars Achieved in Five Testings

Car 1:	35	31	30	30	29	Mean $\bar{y}_1 = 31$ mpg
Car 2:	28	32	29	31	25	Mean $\bar{y}_2 = 29$ mpg
Car 3:	36	35	37	39	33	Mean $\bar{y}_3 = 36$ mpg

The null hypothesis is that there is no difference between the population means. The alternative hypothesis is at least one of the means differs from the others. Symbolically, these null and alternative hypotheses are written

$$H_O\colon \mu_3 = \mu_2 = \mu_1$$

$$H_A\colon \text{at least one of the } \mu_i \text{ differs}$$

where μ_i is the population average miles per gallon

for the ith car, $i = 1, 2, 3$

Grand Mean The mean of the total sample with no regard to treatment factor groupings.

This test requires an analysis of the variability "between" sample means relative to the variability "within" the samples. Variability between i samples is reflected in the way that each sample mean $\bar{y}_i$ deviates from the **grand mean** ($\bar{\bar{y}} = 32$), which is the mean of y ignoring the treatment groupings ($\bar{y} = 32$). Variability within samples is reflected in the way that each observation within a sample deviates from the specific sample mean $\bar{y}_i$. If there is a lot of variability among the sample means and little variability within the samples, then the null hypothesis is rejected in favor of the alternative.

To appreciate the idea behind the decision to reject or not reject the null hypothesis on the basis of a comparison of within and between variability, consider the scatterplot in Figure 11.1. In panel a, for each of the three treatments (cars or farmland plots, for example) all five of the observations (mileage or plant yield, for example) are close to their respective means, $\bar{y}_i$, $i = 1, 2, 3$. In panel b, the observations within the samples are more diverse, although the sample means are the same as in panel a; that is, the between sample variability of means is the same in both panels. Thus, it seems more likely that the mean differences in panel b are associated with chance events than the same mean differences in panel a.

Figure 11.2 provides a scatterplot of the data in Table 11.1. The F distribution is employed to assess whether the between variability is sufficiently larger than the within variability to reject the null hypothesis of no difference between population mean mpg for the three cars. A ratio of the between to within variability forms the test statistic. This statistic follows the F distribution if the underlying populations are approximately normal, with similar variances.

Notice in Figure 11.2 that the cars are identified on the horizontal or x axis. Each car is viewed as a treatment; it is an attribute that has no meaning along a number line. Thus, the positioning of the car along the x axis is irrelevant. The gas mileage data on car 1, for example, could be plotted vertically to the left or right of cars 2 and 3 with no loss in meaning.

FIGURE 11.1 Scatterplot of Sample Observations and Their Means

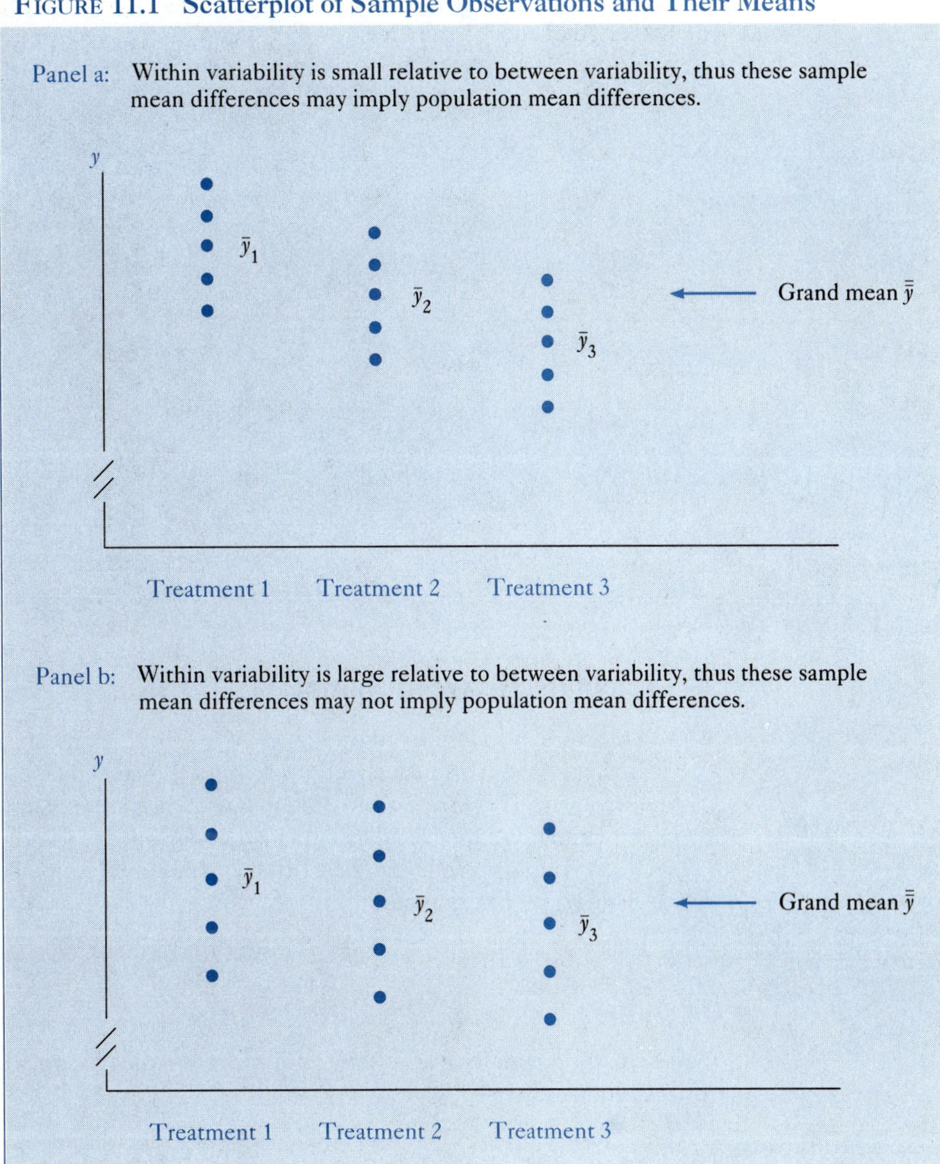

FIGURE 11.2 Scatterplot of Sample Observations and the Mean Gas Mileage of the Three Cars

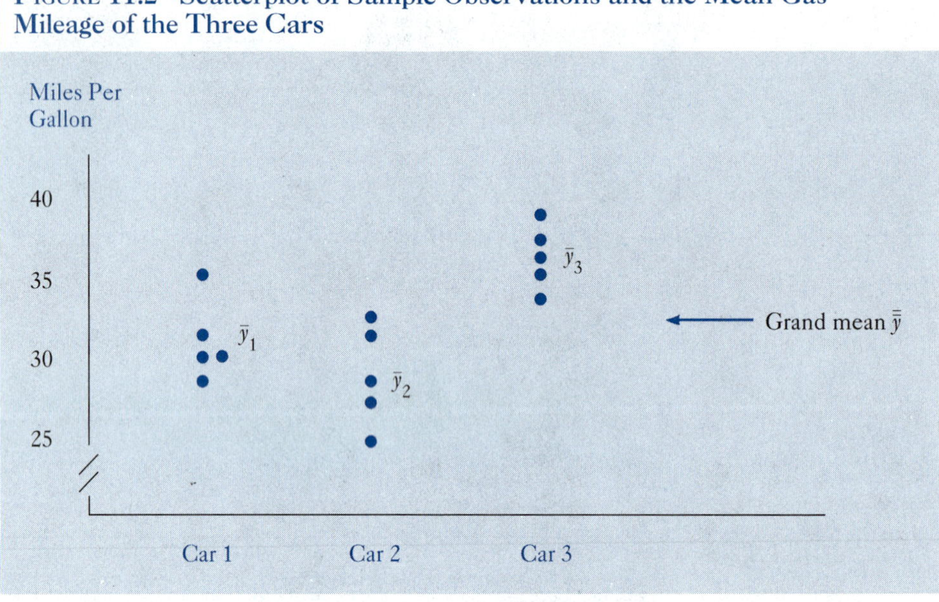

Variability Between Samples

As a measure of variability, the variance of the three car means in Table 11.1 can be calculated easily with a spreadsheet program, as shown in Table 11.2. The variance of the sample means ($\bar{y}_i$ = 31, 29, and 36) around their grand mean ($\bar{\bar{y}}$ = 32) is called the between sample variance ($s^2_{\bar{y}}$ = 13). As long as each of the samples are of equal size, the between sample variance can be written as a modified form of the unweighted variance Formula (2.5b) in Chapter 2:

$$s^2_{\bar{y}} = \frac{\Sigma\left(\bar{y}_i - \bar{\bar{y}}\right)^2}{k-1} \tag{11.1}$$

where, for k sample means, the grand mean is a modification of the unweighted mean in Formula (2.1b).

$$\bar{\bar{y}} = \frac{\Sigma\,\bar{y}_i}{k}$$

The most efficient and typical design of an experiment is to have equal sample sizes for each treatment. Unequal sample sizes can be accommodated by calculating weighted means and weighted variances, although, for brevity, the formulas are not presented here. The MINITAB printout, for an example involving unequal sample sizes, however, is provided in Query 11.1. Gudmund Iversen and Helmut Norpoth (1987) provide an excellent and easy to follow guide to these "unbalanced designs," although other good guides can be found in most any library.

Table 11.2 Calculation of the Between Sample Variance

						$\bar{y}_i$	$\bar{y}_i - \bar{\bar{y}}$	$(\bar{y}_i - \bar{\bar{y}})^2$
Car 1	35	31	30	30	29	31	−1	1
Car 2	28	32	29	31	25	29	−3	9
Car 3	36	35	37	39	33	36	4	16
						96	0	26
						$\bar{\bar{y}} = 96/3 = 32$		$s_{\bar{y}}^2 = 26/2 = 13$

Variability Within Samples

The variability between samples needs to be compared to a measure of the variability within the samples. As a measure of the variability within each sample, we calculate the sum of squared deviations within the sample (Table 11.3). The sum of squared deviations was calculated for the first car as

$$\Sigma(y_{1j} - \bar{y}_1)^2 = (35-31)^2 + (31-31)^2 + (30-31)^2 + (30-31)^2 + (29-31)^2 = 22 \quad (11.2)$$

where y_{1j} is the jth observation for the first car.

Adding the sum of squared deviations for the three samples, and then dividing this sum by the total degrees of freedom for the three samples ($n - 1 = 4$, for each), gives the pooled within sample variance ($s_w^2 = 6$). The pooled within sample variance is just an extension of the variance formula used in the two-sample case in the previous chapter. In general, for k samples, where each is of size n, the pooled within sample variance is

$$s_w^2 = \frac{\Sigma\left(y_{1j} - \bar{y}_1\right)^2 + \Sigma\left(y_{2j} - \bar{y}_2\right)^2 + \ldots + \Sigma\left(y_{kj} - \bar{y}_k\right)^2}{k(n-1)} \quad (11.3)$$

Notice that if the null hypothesis is correct, the five observations on each of the three cars represent three random samples of gas mileage from one population. Formula (11.3) provides an estimator of the variance of that population (σ^2).

The standard error of the mean (that is, the variance of the sampling distribution of $\bar{y}$) is estimated by $s_{\bar{y}}^2$ in Formula (11.1). The relationship between the standard error of the mean and the population variance is known (from Chapters 8 and 9) to be

$$\sigma_{\bar{y}}^2 = \sigma^2 / n \quad (11.4)$$

Thus, if the null hypothesis is true,

$$s_{\bar{y}}^2 \approx s_w^2 / n \quad (11.5)$$

and the F ratio

$$F = \frac{s_{\bar{y}}^2}{s_w^2 / n} \quad (11.6)$$

Table 11.3 Calculation of the Within Sample Variance

Car 1:

y_{1j}	35	31	30	30	29	
$y_{1j}-\bar{y}_1$	4	0	–1	–1	–2	
$(y_{1j}-\bar{y}_1)^2$	16	0	1	1	4	$22 = \Sigma_j (y_{1j}-\bar{y}_1)^2$

Car 2:

y_{2j}	28	32	29	31	25	
$y_{2j}-\bar{y}_2$	–1	3	0	2	–4	
$(y_{2j}-\bar{y}_2)^2$	1	9	0	4	16	$30 = \Sigma_j (y_{2j}-\bar{y}_2)^2$

Car 3:

y_{3j}	36	35	37	39	33	
$y_{3j}-\bar{y}_3$	0	–1	1	3	–3	
$(y_{3j}-\bar{y}_3)^2$	0	1	1	9	9	$20 = \Sigma_j (y_{3j}-\bar{y}_3)^2$
						72
						$s_w^2 = 72/12 = 6$

will be approximately one. That is, if the null hypothesis is correct, with the exception of random chance, the between variance should equal the within variance. If the null hypothesis is not correct, however, then the numerator of the F ratio in (11.6) will exceed the denominator as the difference in population means gives a wider dispersion in the sample means. That is, the between variance will exceed the within variance if means differ.

In our example, the calculated F statistic in (11.6) is

$$F = \frac{s_{\bar{y}}^2}{s_w^2/n} = \frac{ns_{\bar{y}}^2}{s_w^2} = \frac{5(13)}{6} = \frac{65}{6} = 10.833$$

which seems well above an approximate value of one that is expected if the null hypothesis is correct. As with all statistical tests, however, there is a need here to have a critical value against which this calculated value can be compared.

11.3 Comparing Critical and Calculated Values

As in all statistical tests, to assess the correctness of the null hypothesis we can define the critical value of F, for a set probability of a Type I error. If the computed value of the F statistic exceeds the critical value, then the null hypothesis should be rejected.

FIGURE 11.3 Critical Values of *F*-Test Statistic for 2 and 12 Degrees of Freedom

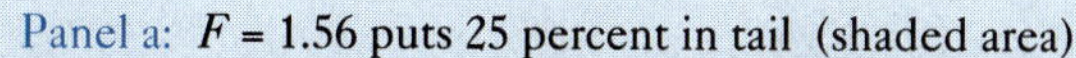

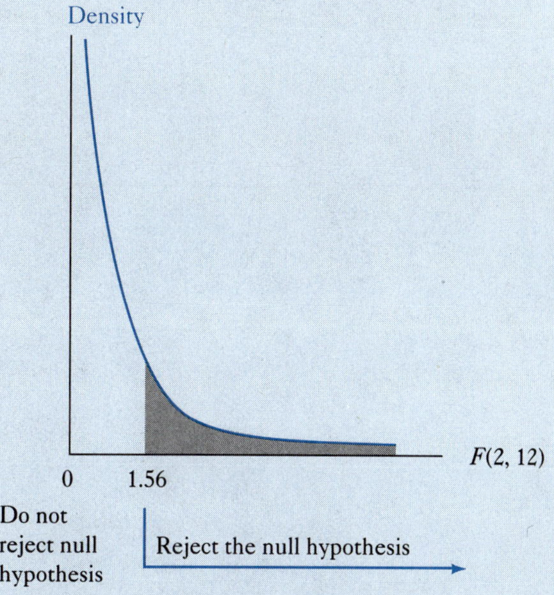

Panel b: F = 2.81 puts 10 percent in tail (shaded area)

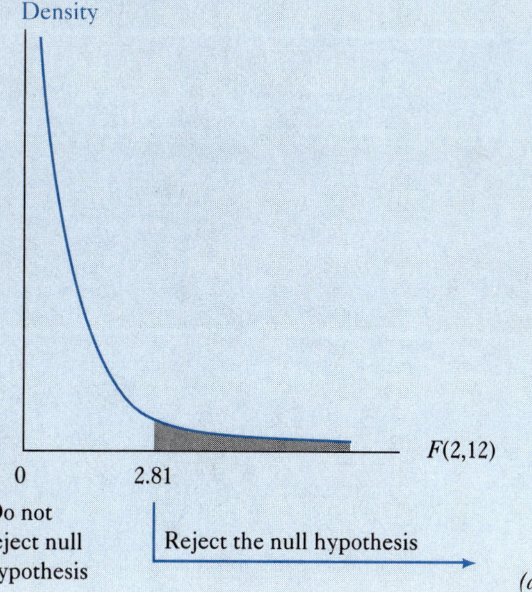

(continued)

FIGURE 11.3 Critical Values of F-Test Statistic for 2 and 12 Degrees of Freedom (*continued*)

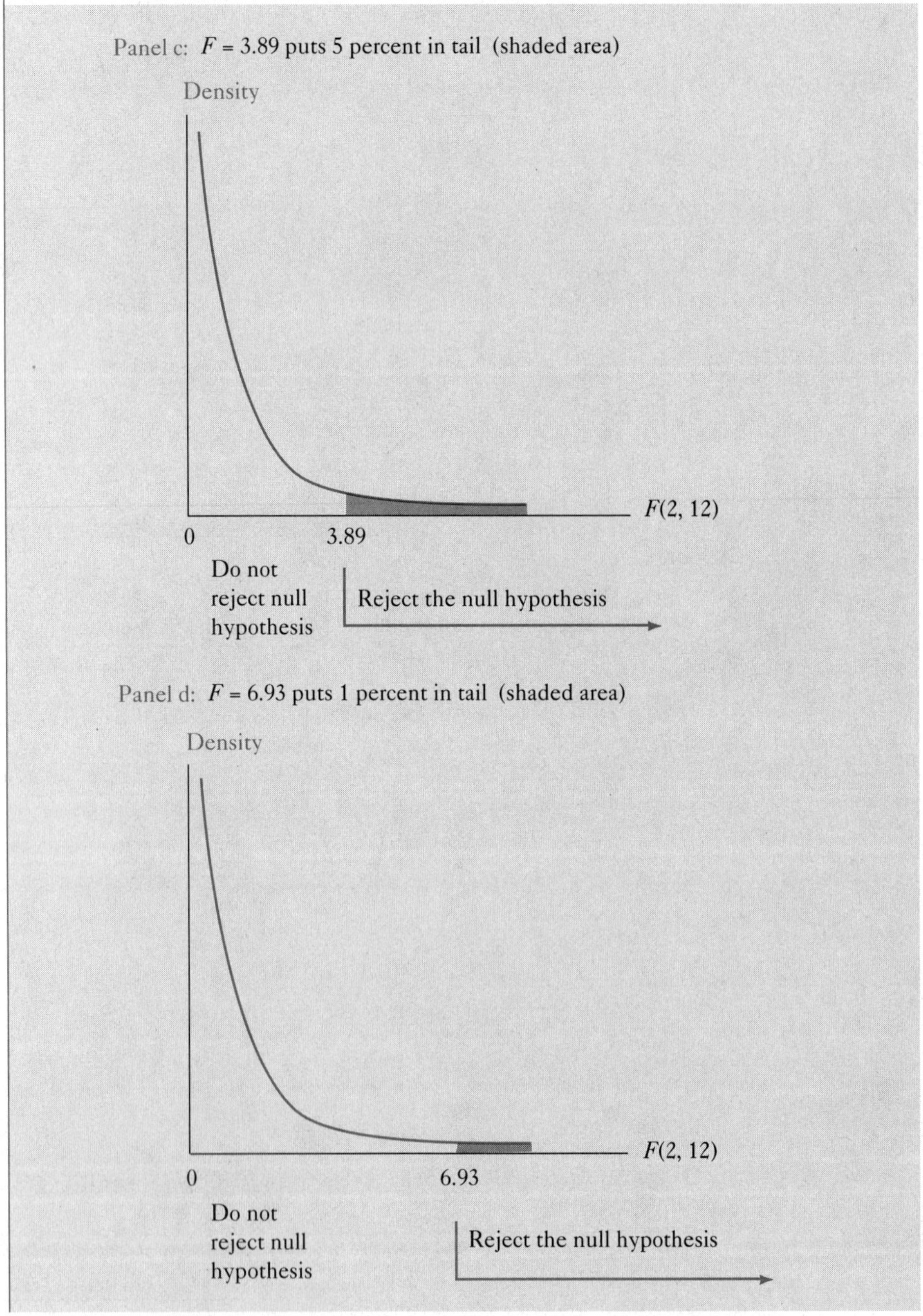

Recall that the probability of a Type I error is the area beyond the critical value in the test statistic in a one-tailed test, assuming the null hypothesis is true. This area and the associated critical value of F can be obtained from Appendix Table A.5 or by a computer program such as STATABLE.

As introduced in Chapter 10, the F distribution depends on the degrees of freedom in the numerator [here being $df_1 = (k - 1)$] and the degrees of freedom in the denominator [now being $df_2 = k(n - 1)$]. Appendix Table A.5 provides values for a family of F distributions. In the body of the table are the F values; across the top edge of the table are the degrees of freedom in the numerator, and down the left-hand margin are the degrees of freedom in the denominator of the F ratio. In the body of the F table, four different F values for four different levels of significance are grouped together for each set of degrees of freedom.

To find the critical value of F that puts 0.05 of the distribution above it, with $df_1 = k - 1 = 3 - 1 = 2$ in the numerator and $df_2 = k(n - 1) = 3(5 - 1) = 12$ in the denominator, as called for in our test of car gas mileage, look across the top of the table until you get to the column marked "2" and look down the left margin until you get to the row marked "12." The group of four F values that are in the second column and twelfth row are 1.56, 2.81, 3.89, and 6.93. As shown in Figure 11.3, these are the critical F values that put 25 percent, 10 percent, 5 percent, and 1 percent of the distribution in the tail, with 2 and 12 degrees of freedom.

For our car mileage example, the calculated F value of 10.833 is well beyond any of the critical values in Figure 11.3. The null hypothesis of no difference in the average mileage of the three cars can be rejected at even the 0.01 Type I error level.

A p-value can be determined with a computer program such as STATABLE. The p-value can be interpreted as the smallest Type I error level for which the null hypothesis can be rejected. The p-value is the area beyond the calculated F value, which is

$$P(F > 10.833, \text{ with } df_1 = 2, df_2 = 12) = 0.002$$

This p-value implies that the null hypothesis of no difference among the population means would be rejected as long as the probability of a Type I error is not set below 0.002. We can be relatively confident in drawing the conclusion that the three cars do achieve different gas mileage, assuming that the only source of randomness in our samples is sampling error.

11.4 The Analysis of Variance (ANOVA) Table

ANOVA tables simply provide a convenient way of presenting the sum of square calculations and associated adjustments needed to obtain the F statistic. In Table 11.4, for example, the first row shows the calculations for the numerator of the F ratio ($ns_{\bar{y}}^2$), and the second row shows the denominator (s_w^2).

The bottom row in an ANOVA table is the sum of squared deviations for all the observations around the grand mean; it is the sum of squared deviations of all the values,

Table 11.4 ANOVA Table for General Spreadsheet Calculation

Source of Variation	Sum of Squares	Degrees of Freedom	Mean Square	*F* Ratio
Factor One: Differences between sample means; that is, differences explained by the cars	$\text{FSS} = n\sum_{i=1}^{k}(\bar{y}_i - \bar{\bar{y}})^2 = 130$	$\text{df}_1 = (k - 1) = 2$	$\text{MSF} = \text{FSS}/\text{df}_1 = ns_{\bar{y}}^2 = 65$	$F = \text{MSF}/\text{MSE} = ns_{\bar{y}}^2 / s_w^2 = 10.83$
Residual Error: Differences within samples; that is, differences not explained	$\text{ESS} = \sum_{i=1}^{k}\sum_{j=1}^{n}(y_{ij} - \bar{y}_i)^2 = 72$	$\text{df}_2 = k(n - 1) = 12$	$\text{MSE} = \text{ESS}/\text{df}_2 = s_w^2 = 6$	
Total: Overall difference; variability to be explained	$\text{TSS} = \sum_{i=1}^{3}\sum_{j=1}^{5}(y_{ij} - \bar{\bar{y}})^2 = 202$	$\text{df}_t = kn - 1 = 14$		

Total Sum of Squares (TOTALSS or TSS)
The sum of squared deviations of a set of observations about their grand mean.

regardless of the sample to which they belong. For the 15 values in Table 11.1, the **total sum of squares (TOTALSS or TSS)** was calculated as

$$\sum_{i=1}^{3}\sum_{j=1}^{5}\left(y_{ij} - \bar{\bar{y}}\right)^2 = (35-32)^2 + \ldots + (29-32)^2 + (28-32)^2 + \ldots + (25-32)^2 + (36-32)^2 + \ldots + (33-32)^2 = 202 \tag{11.7}$$

The total sum of squares is designated TotalSS or just TSS. As should be apparent from the calculations shown in (11.7), TSS is the numerator of the variance of y as given in Formula (2.5b) in Chapter 2. The formula for TSS in an ANOVA table simply introduces new notation.

TSS is sometimes called the variability in y to be "explained" by the treatments or nonchance factors. It is a measure of the squared distance between sample observations of y and the mean of y, as shown for one such y point in Figure 11.4.

Analysis of Variance (ANOVA)
A procedure for testing the statistical significance of the differences among the means of two or more groups defined by treatments and factors.

As portrayed in Figure 11.4, **analysis of variance (ANOVA)** is based on the idea that TSS can be partitioned into two components, or sources, of variability: that which is being "explained" by $\bar{y}_i$ and that which is not. Using these two sources of variability, each observation on the dependent variable y then can be decomposed into three terms. Any observation y_{ij} thus can be written:

$$\begin{aligned} y_{ij} = {} & \text{grand mean} \\ & + \text{deviation of the } i\text{th treatment mean from grand mean} \\ & + \text{deviation of observation from treatment mean} \end{aligned} \tag{11.8}$$

FIGURE 11.4 Decomposition of the Sums of Squares

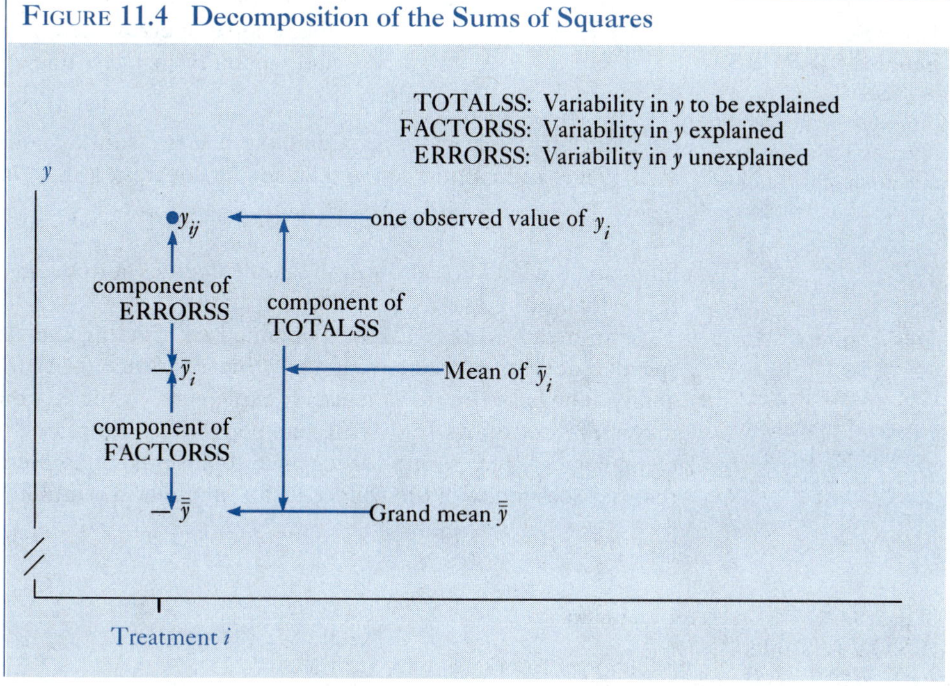

Or in terms of the notation in Figure 11.4

$$y_{ij} = \bar{\bar{y}} + \left(\bar{y}_i - \bar{\bar{y}}\right) + \left(y_{ij} - \bar{y}_i\right) \tag{11.9}$$

which, upon rearranging terms, gives

$$y_{ij} - \bar{\bar{y}} = \left(\bar{y}_i - \bar{\bar{y}}\right) + \left(y_{ij} - \bar{y}_i\right) \tag{11.10}$$

Squaring both sides of (11.9) and summing over i and j yields the decomposition of the total sum of squares into sources of variability associated with the treatment or non-chance factor and that associated with unexplained chance factor. That is,

$$\Sigma_i \Sigma_j \left(y_{ij} - \bar{\bar{y}}\right)^2 = \Sigma_i \Sigma_j \left(\bar{y}_i - \bar{\bar{y}}\right)^2 + \Sigma_i \Sigma_j \left(y_{ij} - \bar{y}_i\right)^2 \tag{11.11}$$

or

Factor Sum of Squares (FACTORSS or FSS) The sum of squared deviations of the individual factor group means from the grand mean.

$$\Sigma_i \Sigma_j \left(y_{ij} - \bar{\bar{y}}\right)^2 = n\Sigma_i \left(\bar{y}_i - \bar{\bar{y}}\right)^2 + \Sigma_i \Sigma_j \left(y_{ij} - \bar{y}_i\right)^2 \tag{11.12}$$

where: $\Sigma_i \Sigma_j \left(y_{ij} - \bar{\bar{y}}\right)^2$ is the total sum of squares SSTOTAL or TSS—the variability in y to be explained.

$n\Sigma_i \left(\bar{y}_i - \bar{\bar{y}}\right)^2$ is called the **factor sum of squares FACTORSS or FSS**—the variability in y explained by the treatment factor.

Error Sum of Squares (ERRORSS or ESS)
In ANOVA, the sum of squared deviations of the group observations from their group mean.

$\Sigma_i \Sigma_j \left(y_{ij} - \bar{y}_i\right)^2$ is called the residual or **error sum of squares (ERRORSS or ESS)**—the variability in y left unexplained by the treatment factor.[1]

Notice in Table 11.4 that adding the first two sums of squares (FSS = 130 and ESS = 72) yields a sum equal to the total sum of squares (TSS = 202).[2]

$$TSS = FSS + ESS \tag{11.13}$$

Similarly, the between and within sample degrees of freedom ($df_1 = 2$ and $df_2 = 12$) sum to the total ($df_t = 14$).

Mean Square (MS)
A variance for the identified source of variability; obtained by dividing the sum of squares by the respective degrees of freedom.

The **mean square (MS)** is obtained by dividing the sum of squares by the respective degrees of freedom. Each MS is a variance for the identified source of variability. The between car variance is "explained" by the fact that the rows in Table 11.1 may represent draws from different populations (different types of car gas mileage technologies). The within car variance represents random error or residual variability that is associated with the chance factor inherent in sampling—it is "unexplained by"

TABLE 11.5 ANOVA Table with MINITAB Computer Program Input and Output

User input:

```
MTB>      READ INTO  C1  C2  C3
DATA>      35  28  36
DATA>      31  32  35
DATA>      30  29  37
DATA>      30  31  39
DATA>      29  25  33
DATA>      END
           5 ROWS READ
MTB>       AOVONEWAY  IN  C1,  C2,  C3
```

MINITAB output:

```
ANALYSIS OF VARIANCE
SOURCE   DF       SS        MS       F        p
FACTOR    2   130.00     65.00   10.83    0.002
ERROR    12    72.00      6.00
TOTAL    14   202.00
                                      INDIVIDUAL 95 PCT CI'S FOR MEAN
                                      BASED ON POOLED STDEV
LEVEL     N     MEAN     STDEV       ----+---------+---------+---------+--
C1        5   31.000     2.345            (------*------)
C2        5   29.000     2.739       (------*------)
C3        5   36.000     2.236                     (------*------)
                                     ----+---------+---------+---------+--
POOLED STDEV  2.449                    28.0      31.5      35.0      38.5
```

Unexplained Variance
The pooled within sample variance that represents the random error or residual variability.

or has nothing to do with different technology of the three cars. Thus, the F ratio can be expressed in a more intuitive form than that in Formula (11.6):

$$F = \frac{ns_{\bar{y}}^2}{s_w^2} = \frac{\text{Explained variability}}{\text{Unexplained variability}}\left(\text{with an adjustment for } n\right) \quad (11.14)$$

Computer programs will automatically do all of the calculations we have done so far and report the results in an ANOVA table. When an ANOVA table is obtained from a computer program, as shown in the MINITAB printout in Table 11.5, the formulas are not shown. It is assumed that the user knows how the computer program does the calculations and that the user knows how to interpret the results provided. The MINITAB output automatically provides 95 percent confidence intervals for the means of each sample. Unlike the confidence intervals discussed in Chapters 8 and 9, each of these confidence intervals uses the same "pooled standard deviation" in the determination of the standard error of the identified means. The margins of error for each of the three-treatment means is set equal to $\pm t_{\alpha/2}\, s/\sqrt{n} = \pm(2.776)2.449/\sqrt{5} = \pm 3.04$.

Query 11.1:

In reviewing the service ratings given to the two shifts of workers in a restaurant illustration of Chapter 10, Tables 10.2 and 10.3, a manager claimed that the wrong test was used. The manager claimed that the test of mean differences in rates required a one-way analysis of variance and an F test. Was the manager correct?

Answer:

Whether this test is done with the t distribution or the F distribution is irrelevant because there are only two treatment factors: day shift and night shift; thus, the manager was wrong. To see this, let us repeat the analysis in Table 10.3, only this time we will use the "ONEWAY" routine instead of the "TWOSAMPLE" routine in MINITAB. For the "ONEWAY" routine, the data must be entered in two columns, with the second column identifying the shift.

In reviewing the MINITAB printout on the following page, first notice that the means and standard deviations reported at the bottom of the printout are exactly the same as those reported in Table 10.2 and Table 10.3. The 95 percent confidence intervals are calculated on the basis of this information. As already stated, each of these confidence intervals uses the same "pooled standard deviation" in the determination of the standard error of the identified means. Unlike the case of equal sample sizes, however, here the margins of error are

$$\pm t_{\alpha/2}\, s/\sqrt{n} = \pm\left(2.201\right)16.01/\sqrt{12} = 10.17 \text{ and}$$

$$\pm t_{\alpha/2}\, s/\sqrt{n} = \pm\left(2.262\right)16.01/\sqrt{10} = 11.45$$

for the respective sample means of 62.5 and 79.2.

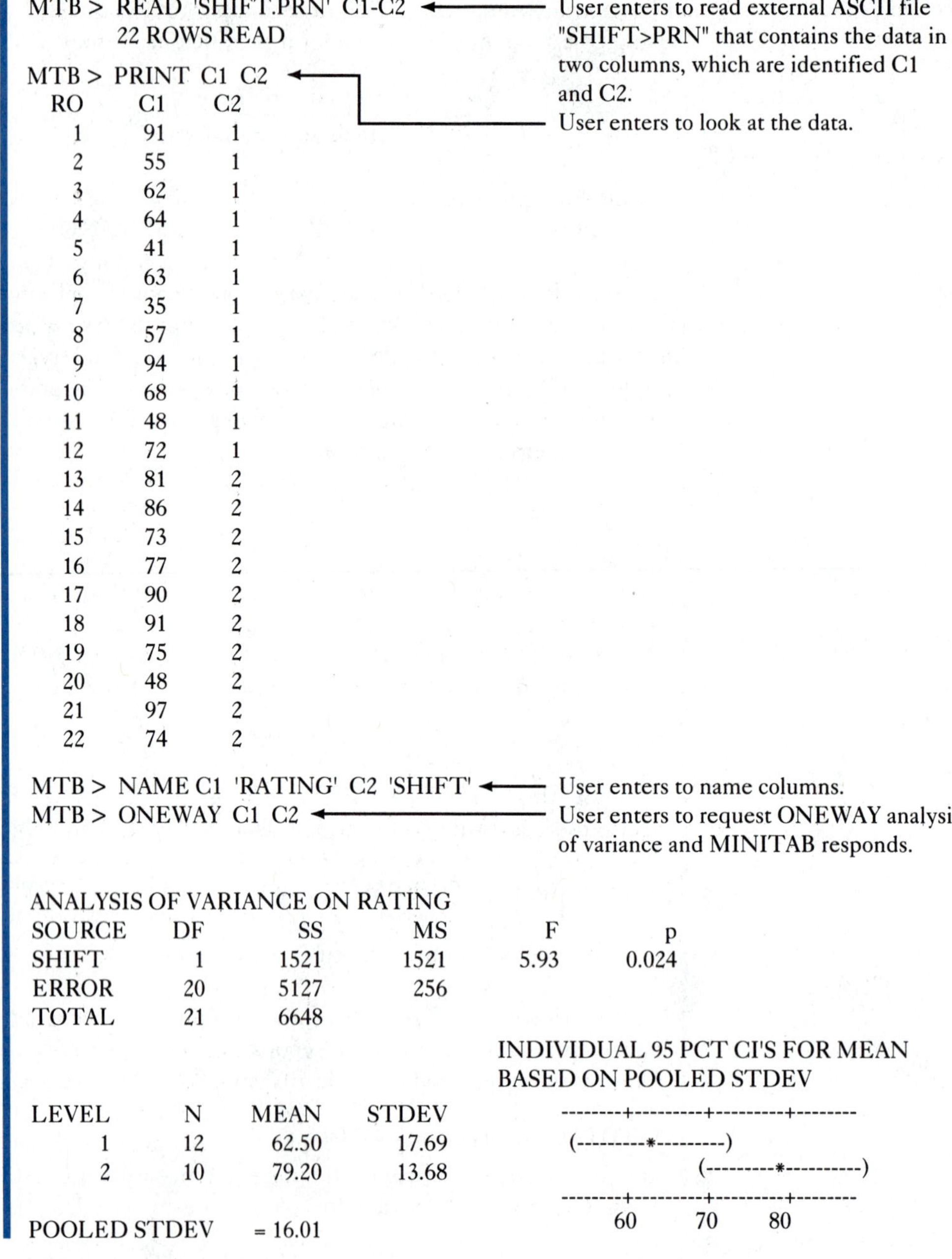

```
MTB > READ 'SHIFT.PRN' C1-C2
      22 ROWS READ
MTB > PRINT C1 C2
 RO    C1    C2
  1    91     1
  2    55     1
  3    62     1
  4    64     1
  5    41     1
  6    63     1
  7    35     1
  8    57     1
  9    94     1
 10    68     1
 11    48     1
 12    72     1
 13    81     2
 14    86     2
 15    73     2
 16    77     2
 17    90     2
 18    91     2
 19    75     2
 20    48     2
 21    97     2
 22    74     2

MTB > NAME C1 'RATING' C2 'SHIFT'
MTB > ONEWAY C1 C2

ANALYSIS OF VARIANCE ON RATING
SOURCE    DF      SS      MS       F       p
SHIFT      1    1521    1521    5.93   0.024
ERROR     20    5127     256
TOTAL     21    6648
                                  INDIVIDUAL 95 PCT CI'S FOR MEAN
                                  BASED ON POOLED STDEV
LEVEL      N    MEAN   STDEV     --------+---------+---------+--------
    1     12   62.50   17.69     (---------*---------)
    2     10   79.20   13.68                (---------*----------)
                                 --------+---------+---------+--------
POOLED STDEV = 16.01                    60        70        80
```

User enters to read external ASCII file "SHIFT>PRN" that contains the data in two columns, which are identified C1 and C2.

User enters to look at the data.

User enters to name columns.

User enters to request ONEWAY analysis of variance and MINITAB responds.

Next, notice in the analysis of variance part of the printout that the calculated F statistic is 5.93, with a p-value of 0.024. This is exactly the same p-value reported in Table 10.3. The reason for these identical p-values is that the F distribution with one degree of freedom in the numerator is exactly the same as the t distribution. In fact, with one degree of freedom in the numerator, the F distribution is just the square of the t distribution. From Table 10.3, the calculated t is 2.44, which is the square root of the 5.93 calculated F.

Whether the t distribution or the F distribution is used for this test, or any other test involving only two treatment factors, is irrelevant. As long as the numerator of the F ratio has only one degree of freedom, $F = t^2$.

EXERCISES

11.1 Fill in the missing values in the following ANOVA table

Source of variability	Sum of Squares	Degrees of Freedom	Mean Square	F	p-Value
FACTOR	_____	2	23	___	____
ERROR	260	___	___		
TOTAL	_____	14			

11.2 In the above ANOVA table, is the factor significant at the 5 percent Type I error level?

11.3 Fill in the missing values in the following ANOVA table

Source of variability	Sum of Squares	Degrees of Freedom	Mean Square	F	SIGNIFICANCE
BETWEEN	172	4	____	___	____
WITHIN	_____	15	____		
TOTAL	662	___			

11.4 In the above ANOVA table, is the between variability making a significant contribution to the total sum of squares? (Test at the 5 percent Type I error level.)

11.5 For the following ANOVA table, what is the value of F and the p-value? Does the factor identified as the MODEL make a significant contribution?

SOURCE	DF	SUM OF SQ	MEAN SQ
MODEL	10	457.50	45.75
ERROR	1	4.50	4.50
TOTAL	11	462.00	

11.6 Below are change scores for students who were randomly assigned to three different instruction programs. Are these programs equally effective, on the average? (Test at the 5 percent Type I error level.) What do you have to assume to do this test?

Program 1	Program 2	Program 3
10	12	8
8	7	9
11	10	10
7	14	11

(continued)

Program1	Program 2	Program 3 *(continued)*
9	8	9
13	11	7
10	9	13
12	13	10

11.7 Four different assembly processes were under consideration. Thirty-two workers were randomly assigned to the four processes, eight per process. The number of correctly assembled units in an eight-hour work shift were recorded:

Process 1	Process 2	Process 3	Process 4
31	29	28	32
36	32	36	33
36	35	29	33
34	32	31	31
30	25	30	27
29	30	26	28
24	31	32	30
35	33	27	29

At the 0.01 Type I error level, is there a significant difference between the four processes?

11.5 Two-Way ANOVA

So far we have assumed that the five tests of each of the three cars were conducted independently of each other. We have conducted what is called "a one-way analysis of variance" in which one factor (the design efficiency of the cars) is suspected of causing differences in the gas mileage of three different makes of cars. We have been assuming that the three samples were generated by simple random sampling with 15 independently assigned drivers.

An alternative experimental design would be to assign only five drivers to each of the three cars. An obvious advantage is less expense. As with the matched pairs test discussed in Chapter 10, there is also a reduction in sampling error because the driver differences have been removed from the comparison. On the negative side, we could be building in bias in that the order in which the cars are driven might affect the results if the drivers acquire skills while driving. We will assume, however, that our five drivers are highly skilled from the start and do not increase driving skill during the tests. Furthermore, the order in which cars are driven is assumed to be randomly determined.

For computational convenience and to make comparisons with earlier results, let us assume that the five drivers of the three cars achieved the same results as in Table 11.1. Now, however, the fact that there are only five drivers will reduce the variability in gas mileage that could not be explained by only knowing the make of the car. Reduction in the unexplained residual error will increase the calculated value of the F ratio and provide stronger evidence against the null hypothesis of equality of population means.

There is clearly variability in the drivers' average miles per gallon achieved across the three cars. In Table 11.6, driver five averaged only 29 miles per gallon while driver

Table 11.6 Miles Per Gallon for Three Cars Achieved in Five Testings

	Driver					
	One	Two	Three	Four	Five	Car Mean
Car 1:	35	31	30	30	29	31
Car 2:	28	32	29	31	25	29
Car 3:	36	35	37	39	33	36
Driver Mean	33	32.67	32	33.33	29	32 mpg

Table 11.7 Sample Miles Per Gallon Classified by Car *i* and Driver *j*

	Dr 1	Dr 2	Dr 3	Dr 4	Dr 5	$\bar{y}_i$	$\bar{y}_i - \bar{\bar{y}}$	$(\bar{y}_i - \bar{\bar{y}})^2$
Car 1:	35	31	30	30	29	31	–1	1
Car 2:	28	32	29	31	25	29	–3	9
Car 3:	36	35	37	39	33	36	4	16
						32		26 = $\Sigma(\bar{y}_i - \bar{\bar{y}})^2$

$\bar{y}_j$	33.00	32.67	32.00	33.33	29.00	32.00
$\bar{y}_j - \bar{\bar{y}}$	1.00	0.67	0.00	1.33	–3.00	
$(\bar{y}_j - \bar{\bar{y}})^2$	1.00	0.44	0.00	1.78	9.00	12.22 = $\Sigma(\bar{y}_j - \bar{\bar{y}})^2$

four averaged 33.33 miles per gallon. This source of variability is captured in the ANOVA table by introducing it as a second factor. That is, we are now interested in the average mileage of each car ($\bar{y}_i$) and average mileage of each driver ($\bar{y}_j$), as well as variability around these means, as calculated in Table 11.7.

In the one-way ANOVA of Table 11.4 there were two sources of variability in the total variability in gas mileage: 1) the one factor (car) that was used to explain variability in gas mileage (SSF = 130); and 2) the error or residual variability that was left unexplained (SSE = 72). In two-way ANOVA, as shown in Table 11.8, the total variability in gas mileage is the same as in the one-way ANOVA in Table 11.4 (SST = 202). Now, however, there are three sources of variability.

First, the variability associated with the car factor continues to be the same as in Table 11.4, but in the two-way ANOVA in Table 11.8 it is identified as FSS1 = 130. The second, and new computation, has to do with the variability that can be explained by the differences in drivers (FSS2 = 36.67). FSS2 is the sum of squared deviations of the five drivers' means ($\bar{y}_j$) around the grand mean ($\bar{\bar{y}}$), as calculated in Table 11.7, multiplied by the number of such means k, where k is three for each of the cars. As shown in Table 11.8,

TABLE 11.8 Two-Way ANOVA Table for General Spreadsheet Calculation

Source of Variation	Sum of Squares	Degrees of Freedom	Mean Square	F Ratio
Factor One: Differences between sample means explained by the cars	$FSS1 = n\sum_{i=1}^{k}(\bar{y}_i - \bar{\bar{y}})^2$ $= 5(26) = 130$	$(k-1)$ $= (3-1)$ $= 2$	$MSF1 = \frac{FSS1}{(k-1)}$ $= 65$	$F1 = \frac{MSF1}{MSE}$ $= 14.7$
Factor Two: Differences between sample means explained by the drivers	$FSS2 = k\sum_{j=1}^{n}(\bar{y}_j - \bar{\bar{y}})^2$ $= 3(12.22) = 36.67$	$(n-1)$ $= (5-1)$ $= 4$	$MSF2 = \frac{FSS2}{(n-1)}$ $= 9.17$	$F2 = \frac{MSF2}{MSE}$ $= 2.08$
Residual Error: Differences within samples; that is, differences not explained	$ESS = TSS - (FSS1 + FSS2)$ $= 202 - (130 + 36.67) = 35.33$	$(k-1)(n-1)$ $= 8$	$MSE = \frac{SSE}{(k-1)(n-1)}$ $= 4.42$	
Total: Overall difference; variability to be explained	$TSS = \sum_{i=1}^{3}\sum_{j=1}^{5}(\bar{y}_{ij} - \bar{\bar{y}})^2 = 202$	$kn-1$ $= 14$		

$$FSS2 = k\sum_{j=1}^{n}\left(\bar{y}_j - \bar{\bar{y}}\right)^2 = 3\left(12.22\right) = 36.67 \tag{11.15}$$

The third source of variability is the error or residual sum of squares left unexplained by the car and driver factors. It is calculated by simply extending the equality in Equation (11.13) to take account of the two explanatory factors. That is, from Equation (11.13) we have

$$TSS = FSS + ESS$$

Now, however, FSS consists of two factors: FSS1 and FSS2. Thus, the sum of squares equality is written:

$$TSS = FSS1 + FSS2 + ESS \tag{11.16}$$

With TSS = 202, FSS1 = 130, FSS2 = 36.67, we can solve for ESS.

$$202 = 130 + 36.67 + ESS$$

$$202 - 130 - 36.67 = ESS$$

$$35.33 = ESS$$

Similarly, the degrees of freedom can be determined. The total degrees of freedom remains unchanged at 14. The first car factor continues to use 2 degrees of freedom.

The new driver factor has 4 degrees of freedom (5 – 1 = 4). Thus, the residual error must have 8 degrees of freedom [14 – (2 + 4) = 8]. As in the one-way ANOVA, the mean squares (MSF1, MSF2, and MSE) are obtained by dividing the sum of squares by the respective degrees of freedom. Two F ratios have been calculated in Table 11.8. Both of these F ratios were calculated using the general formula that is appropriate for each factor in ANOVA, regardless of the number of factors employed, namely:

F Statistic for Multiple Factor Anova

$$F = \frac{\text{Variability explained (adjusted for a factor's degrees of freedom)}}{\text{Variability left unexplained (adusted for degrees of freedom)}} \tag{11.17}$$

As with one-way ANOVA, modern-day statistical computer programs will automatically do multifactor ANOVA. Users need only input the data and issue the appropriate commands. All of these major computer programs will provide screen displays and printouts containing the information in Table 11.8. We still need to know what to do with that information once obtained.

11.6 Testing in Multiple Factor ANOVA

Once the total sum of squares has been broken down into its component parts, the significance of the factors used to explain total variability can be tested. The significance of each factor in an ANOVA can be tested via the F statistic, Equation (11.17).

To test whether there is a difference in the mean miles per gallon achieved by the three cars, after accounting for driver differences, the calculated F is obtained from Table 11.8.

$$\text{F} = \text{MSF1/MSE} = 65/4.42 = 14.7 \tag{11.18}$$

If the null hypothesis of no difference among the cars is true, then this ratio has an F distribution, with 2 and 8 degrees of freedom. From Appendix Table A.5, the critical F that puts 1 percent in the tail is 8.65. Our calculated F of 14.7 lies well beyond even this critical F for a small Type I error of 0.01, so it seems reasonable to reject the null hypothesis of no difference among the cars. (From a computer program the p-value can be determined to be 0.002, for a calculated F = 14.7, with 2 and 8 degrees of freedom.)

We can now also test to see if there is a difference between the gas mileage that is achieved by the five different drivers. The null hypothesis is that there is no difference between the population mean miles per gallon achieved by the drivers and the alternative is that the mean of at least one driver differs from the others. The calculated F is approximately

$$F = \text{MSF2/MSE} = 9.17/4.42 = 2.08 \tag{11.19}$$

If the null hypothesis of no difference among the drivers is true, then this ratio has an F distribution, with 4 and 8 degrees of freedom. From Appendix Table A.5, the critical F that puts 10 percent in the tail is 2.81. Our calculated F of 2.08 is well below even this critical F for a large Type I error of 0.10; the null hypothesis of no difference among the drivers should not be rejected. (From a computer program, the p-value can be determined to be 0.175, for a calculated $F = 2.08$, with 4 and 8 degrees of freedom.)

Both of the above F tests require that the assumptions of the model underlying the test are true. The population model implicit in the decomposition of the total sum of squares, as derived from Equation (11.11), is

$$y_{ij} = \mu_y + \lambda_1 + \lambda_2 + \ldots + \varepsilon_{ij}$$

where μ_y is the population mean of y,
λ's are the effects on an observation of being in treatment group 1, 2 . . . as indicated by the subscript,

and ε_{ij} is the random effect on the jth observation in the ith treatment group, for each of the groups.

The assumptions maintained throughout the test are:

1. With the possible exception of the identified λ factors, only random sampling error ε is involved in the formation of the total sum of squares.
2. Each ε_{ij} is normally distributed, which implies that each y_{ij} is normally distributed.
3. Each ε_{ij} has the same and independent distribution, with fixed mean and constant variance, implying the same for each y_{ij}.
4. There is no relationship between assignment to the treatments and the random effects.

Many books have been written on the design of experiments and the specification of the underlying population models. These experimental designs and model specifications have acquired many different names across the disciplines in which they are employed. We have only provided an introduction to that work here. Later chapters will deal in more detail with the extension of the analysis of variance framework in correlation analysis and regression analysis.

11.7 Multiple Factor ANOVA with a Computer

The computational chore for ANOVA increases with the size of the data set and the number of factors considered. The general statistical computer programs such as MINITAB, SYSTAT, and SPSS will all do multifactor ANOVA. These programs require that the data be presented in a matrix form, where elements of each factor are identified by a numerical value.

In the automobile gas mileage case, the dimensions of the matrix are defined by the driver (1, 2, 3, 4, and 5) and the car (1, 2, and 3), as shown in Table 11.9. The data in the matrix form of Table 11.9 is entered into a computer program for a two-way ANOVA. (Unlike one-way ANOVA, two-way ANOVA cannot be conducted directly on data entered as arrayed in Table 11.7.) For the MINITAB program, this data entry is shown in the top part of Table 11.10.

TABLE 11.9 Block Data Entry Design for Two-Way ANOVA

mpg	Driver	Car
35	1	1
28	1	2
36	1	3
31	2	1
32	2	2
35	2	3
30	3	1
29	3	2
37	3	3
30	4	1
31	4	2
39	4	3
29	5	1
25	5	2
33	5	3

By a comparison of the calculations in Table 11.8 and the ANALYSIS OF VARIANCE part of Table 11.10, it can be seen that the computer program did the same arithmetic that we did earlier. Notice, however, that the MINITAB printout also provides a 95 percent confidence interval for each of the factor means. These confidence intervals were calculated as discussed in Chapters 8, 9, and 10. The assumption here is that within the treatment factor group, only random sampling error is accounting for differences in observed values.

EXERCISES

11.8 Fill in the missing values in the following ANOVA table

Source of variability	Sum of Squares	Degrees of Freedom	Mean Square	F	p-Value
FACTOR1	____	2	___	1.00	___
FACTOR2	____	3	___	___	___
ERROR	____	10	26		
TOTAL	400	__			

11.9 In the above ANOVA table, are each of the factors significant at the 5 percent Type I error level?

11.10 Fill in the missing values in the following ANOVA table

Source of variability	Sum of Squares	Degrees of Freedom	Mean Square	F	SIGNIFICANCE
FACTOR1	61	2	___	___	___
FACTOR2	____	3	___	___	___
ERROR	____	__	10		
TOTAL	825	15			

TABLE 11.10 Two-Way ANOVA Computer Output

```
MTB>   READ C1 C2 C3
DATA>      35     1      1
DATA>      28     1      2
DATA>      36     1      3
DATA>      31     2      1
DATA>      32     2      2
DATA>      35     2      3
DATA>      30     3      1
DATA>      29     3      2
DATA>      37     3      3
DATA>      30     4      1
DATA>      31     4      2
DATA>      39     4      3
DATA>      29     5      1
DATA>      25     5      2
DATA>      33     5      3
DATA>    END

MTB > NAME C1 'MPG' C2 'DRIVER' C3 'CAR'
  ROW    MPG   DRIVER   CAR

MTB > TWOWAY C1 C2 C3;
SUBC> ADDITIVE;
SUBC> MEANS C2 C3.

ANALYSIS OF VARIANCE    MPG
SOURCE             DF        SS        MS
DRIVER              4     36.67      9.17
CAR                 2    130.00     65.00
ERROR               8     35.33      4.42
TOTAL              14    202.00

                    Individual 95% CI
DRIVER       Mean   ------+---------+----------+---------+----
   1         33.0                (----------*----------)
   2         32.7              (-----------*-----------)
   3         32.0            (----------*----------)
   4         33.3                  (----------*-----------)
   5         29.0    (----------*----------)
                    ------+---------+---------+---------+-----
                       27.5      30.0      32.5      35.0

                      Individual 95% CI
CAR          Mean   -+---------+--------+---------+---------+
   1        31.00          (------*-------)
   2        29.00    (-------*------)
   3        36.00                          (------*------)
                    -+---------+--------+---------+---------+
                  27.00     30.00    33.00     36.00     39.00
```

11.11 In the above ANOVA table, are the two between variability measures (factor1 and factor2) making a significant contribution to the total sum of squares? (Test each at the 5 percent Type I error level.)

11.12 For the following ANOVA table, what are the values of F and the respective p-values for factors C3 and C4? Do these factors make significant contributions?

SOURCE	DF	SS	MS
C3	3	382.00	127.33
C4	2	58.50	29.25
ERROR	6	21.50	3.58
TOTAL	11	462.00	

11.13 For the following ANOVA table, what are the values of F and the respective p-values for factors X1 and X2? Do these factors make significant contributions?

SOURCE	DF	SS	MS
X1	3	470.67	156.89
X2	2	43.17	21.58
ERROR	6	36.83	6.14
TOTAL	11	550.67	

11.14 For the following data, is hourly wage explained by the region of the country in which the person lives and the job classification? (Test at the 1 percent Type I error level.)

Hourly Wage	Region of Country	Job Classification
$5	1	1
4	1	2
4	1	3
7	1	4
11	2	1
10	2	2
2	2	3
8	2	4
15	3	1
14	3	2
8	3	3
10	3	4
22	4	1
20	4	2
21	4	3
18	4	4

11.15 After conducting the one-way test in Exercise 11.7, it was learned the last four observations were all from the afternoon shift and the first four were from the morning shift. How does this information alter your desired testing procedure? Conduct a test to see if the process and the shift are important contributors to explaining the number of correctly assembled units processed.

11.8 The Chi-Square Test for Independence

Tests of differences in the proportion of observations possessing a given attribute versus the proportion possessing the attribute in a second sample can be generalized to multiple attributes that might be defined for two different variables. The chi-square test of independence is designed to test whether two variables are related on the basis of the different attributes they have. For example, in the case of the midwestern bank mentioned in the introduction to this chapter, we might expect that as the liability period progresses, the bank might attempt to change the gender mix in its workforce. Thus, one variable is the gender of managers and the other variable is the time in the liability period. The question is whether the gender mix of managers is independent of the beginning, middle and end of the liability period?[3] The hypotheses for this test are

H_O: gender and time are independent

H_A: gender and time are dependent

The Chi-Square Statistic

Development of the statistic for the chi-square test requires the introduction of a new conceptual framework. First, the test for these hypotheses is said to determine the "goodness-of-fit" between an observed joint occurrence of these two variables and a set of expected joint occurrences. The observed frequencies of these joint occurrences are presented in a **contingency table.** Table 11.11 is the contingency table for the midwestern bank data on gender and time period.

Contingency Table A table showing the frequency of joint occurrences of two variables.

Let the symbol O_{ij} represent the observed frequency in the ith row and the jth column of the contingency table. Table 11.11 then gives $O_{11} = 392$, indicating that at the beginning of the liability period there were 392 men employed; $O_{21} = 219$ indicates that 219 women were employed at the beginning of the liability period. In the next column, $O_{12} = 394$ indicates that 394 men were at the bank in the middle of the period, and so on until the last cell in the second row and third column, $O_{23} = 199$.

The expected frequency of the ith row and jth column (E_{ij}) is calculated by assuming that the two variables are independent; that is, H_O is true.[4] As discussed in

Table 11.11 Women and Men Employed at the Beginning, Middle and End of the Liability Period

Number	Begin	Middle	End	
Men	392	394	376	1162
Women	219	223	199	641
	611	617	575	1803

Relative Frequency	Begin	Middle	End	
Men	0.217415	0.218524	0.208541	0.644481
Women	0.121464	0.123682	0.110371	0.355518
	0.338879	0.342207	0.318912	1.00000

Chapters 3 and 4, when two events are independent, the product of their marginal probabilities equals the probability of their intersection. If H_O is true, the probability of randomly drawing a man *and* selecting the beginning of the liability period is the product of the probability of a man *times* the probability of being at the beginning of the liability period. These two marginal probabilities are estimated by the observed first row relative frequency (1162/1803) and the first column relative frequency (611/1803). Thus, the estimated joint probability of drawing a man *and* the beginning of the liability period is 0.217868 = (1162/1803)(611/1803). The expected frequency of men in the first period is then this estimated joint probability times the total number of executives. That is,

$$E_{11} = \frac{1162}{1803} \times \frac{611}{1803} \times 1803 = 0.218401 \times 1803 = 393.7781 \tag{11.20}$$

In formula form, the expected frequency of any one of the cells in the ith row and jth column can be written as

$$E_{ij} = \hat{p}_i \hat{p}_j n \tag{11.21}$$

where $\hat{p}_i$ is the estimated marginal probability of the ith row (as calculated from the relative frequency data in the ith row of the contingency table);

$\hat{p}_j$ is the estimated marginal probability of the jth column (as calculated from the relative frequency data in the jth column of the contingency table); and n is the number of observations in the contingency table.

Expected frequencies (the E_{ij}'s) are easily calculated with a spreadsheet program such as LOTUS 1-2-3. Table 11.12 gives the joint probabilities for each of the six cells (i.e., $\hat{p}_i\hat{p}_j$), and Table 11.13 gives the expected frequencies. Movement from Table 11.12 to Table 11.13 shows the multiplication of each cell in Table 11.12 by n = 1803. As a rule of thumb, each E_{ij} must be at least five for the chi-square distribution to be appropriate. Next, the chi-square goodness-of-fit test requires a numerical comparison of the observed frequencies (as in Table 11.11) with the expected frequencies (as in Table 11.13). If the "fit is good" (meaning the frequencies in the two tables are fairly close), then the null hypothesis of independence is not rejected. If the fit is not good, then the null hypothesis is rejected in favor of the alternative hypothesis of dependence. The

TABLE 11.12 Estimated Joint Probabilities of Women and Men Employed at the Beginning, Middle, and End of the Liability Period, Assuming Independence

	Beginning	Middle	End	$\hat{p}_i$
Men	0.218401	0.220546	0.205533	0.644481
Women	0.120478	0.121661	0.113379	0.355518
$\hat{p}_j$	0.338879	0.342207	0.318912	1.000000

Table 11.13 Expected Number of Women and Men Employed at the Beginning, Middle and End of the Liability Period, Assuming Independence

	Beginning	Middle	End	
Men	393.7781	397.6450	370.5768	1162
Women	217.2218	219.3549	204.4231	641
	611	617	575	1803

numerical measure for making this decision is the chi-square (the Greek letter χ, squared) test statistic, which is determined by the following formula:

Chi-Square Statistic

$$\chi^2 = \sum_{i=1}^{r} \sum_{j=1}^{c} \frac{\left(O_{ij} - E_{ij}\right)^2}{E_{ij}} \tag{11.22}$$

where O_{ij} is the observed frequency in ijth cell; E_{ij} is the expected frequency in ijth cell ($E_{ij} \geq 5$); and the degrees of freedom are df = $(r - 1)(c - 1)$, for a contingency table with r rows and c columns.

Using the information in Tables 11.11 and 11.13, the chi-square test statistic is calculated using Formula (11.22) as follows:

$$\chi^2_{\text{calc}} = \sum_{i=1}^{2} \sum_{j=1}^{3} \frac{\left(O_{ij} - E_{ij}\right)^2}{E_{ij}} = 0.3398 \tag{11.23}$$

$$= \frac{(392 - 393.7781)^2}{393.7781} + \frac{(394 - 397.6450)^2}{397.6450} + \frac{(376 - 370.5768)^2}{370.5768}$$

$$+ \frac{(219 - 217.2218)^2}{217.2218} + \frac{(223 - 219.3549)^2}{219.3549} + \frac{(199 - 204.4231)^2}{204.4231}$$

We will learn that this calculated chi-square value of $\chi^2_{\text{calc}} = 0.3398$ is relatively small, indicating that there is not much difference between the observed and expected values from the contingency table. As with other hypothesis tests, small calculated test statistic values indicate that the null hypothesis should not be rejected. That is, the calculated $\chi^2_{\text{calc}} = 0.3398$ is small, so we conclude that there is a "good fit" between the observed and expected relative frequencies—gender and the time within the liability period appear to be independent.

The *p*-value

What constitutes a small chi-square value can be determined by getting the *p*-value that is associated with the calculated chi-square test statistic. This requires knowledge of the degrees of freedom and the shape of the chi-square distribution. The degrees of freedom for the chi-square are

$$\text{df} = (r - 1)(c - 1) \tag{11.24}$$

where r is the number of rows in the contingency table, and

c is the number of columns in the contingency table.

In the case under consideration, there are two rows and three columns so

$$\text{df} = (2 - 1)(3 - 1) \tag{11.25}$$

The *p*-value in a chi-square test is the probability

$$P(\chi^2 > \chi^2_{\text{calc}}) \tag{11.26}$$

which in the bank employment case is the probability of a chi-square value larger than $\chi^2_{\text{calc}} = 0.3398$, with 2 degrees of freedom. As with other test statistics, if the calculated test statistic is large, then its *p*-value will be small. If the *p*-value is less than a preset probability of a Type I error (α), then H_0 is rejected. On the other hand, if the calculated test statistic is small, then its *p*-value will be large. If the *p*-value is greater than α, then H_0 is not rejected.

The *p*-values are easily determined using a computer program. A chi-square computer routine gives the probability of obtaining a chi-square value of 0.3398 or larger [i.e., $P(\chi^2 \geq 0.3398)$, with 2 degrees of freedom] equal to 0.844. Because this *p*-value is extremely large, for any typical probability of a Type I error (say $\alpha = 0.10$), the null hypothesis of independence is not rejected, and the conclusion is that gender and the time of observation during the liability period may be independent.

11.9 Critical Values and the Chi-Square Distribution

An alternative to finding the *p*-values with a computer program is to determine the chi-square critical value for $(r - 1)(c - 1)$ degrees of freedom. Critical values for the chi-square distribution are displayed in Appendix Table A.6. Like other tests, if the calculated χ^2_{calc} value is less than the critical value in Appendix Table A.6, for $(r - 1)(c - 1)$ degrees of freedom, then H_0 is not rejected. If χ^2_{calc} is greater than the critical value, then H_0 is rejected.

The chi-square distribution is a continuous distribution. Its shape depends on the degrees of freedom, $\text{df} = (r - 1)(c - 1)$. It is thus a family of distributions, one for each degree of freedom. A few of these distributions are shown in Figure 11.5, for df = 2, df = 16, and df = 30. For df = 2, the chi-square density function appears to be decreasing

FIGURE 11.5 Chi-Square Distribution

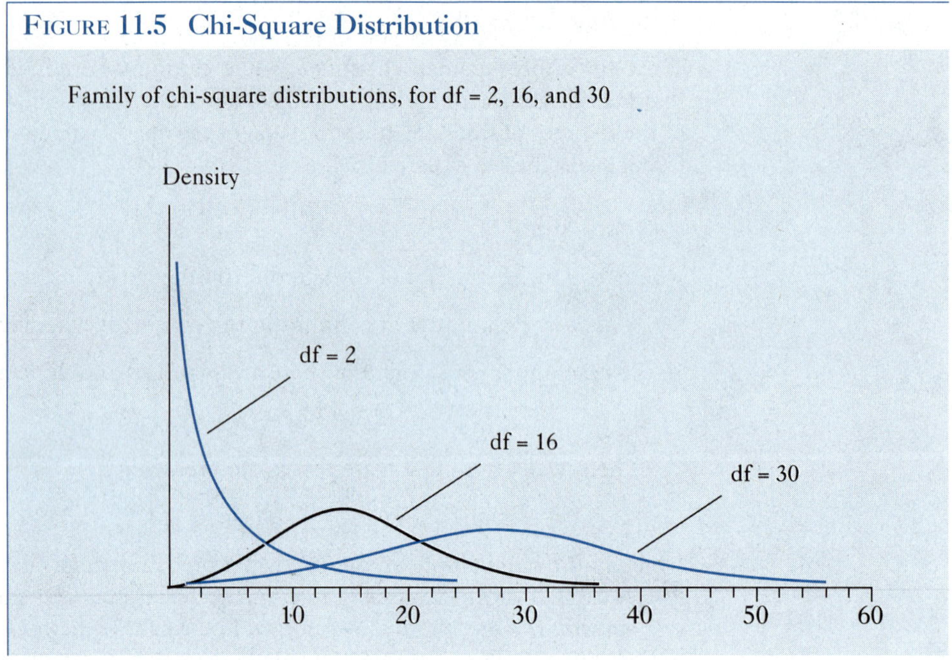

(which is also the case for df = 1, although not shown here). If df increases, as seen in Figure 11.5 for df = 16, the chi-square distribution is at first noticeably right skewed. It becomes more symmetrical, however, as the degrees of freedom increase further. For df = 30, the chi-square distribution is very close to the normal distribution. As degrees of freedom increase and approach infinity, the chi-square distribution becomes indistinguishable from the normal distribution.

If the probability of a Type I error is $\alpha = 0.10$, the critical chi-square value, with 2 degrees of freedom (as shown in Figure 11.6 with values obtained from Appendix Table A.6) is 4.61. In the midwestern bank case, the calculated chi-square value (χ^2_{calc} = 0.335415) is less than this critical value (4.61), as shown in Figure 11.6. Thus, H_0 is not rejected; the conclusion is that the gender and time within the liability period are independent.

Most of the large computer programs such as SAS and SPSS have routines for doing chi-square tests of independence. They may be listed as "goodness-of-fit tests," "test of homogeneity," or "cross-tab tests." Regardless of the name given, the procedures described here for the chi-square test of independence are appropriate for an analysis of the relationship between the rows and columns of a contingency table.

FIGURE 11.6 Chi-Square Test for Bank Employment

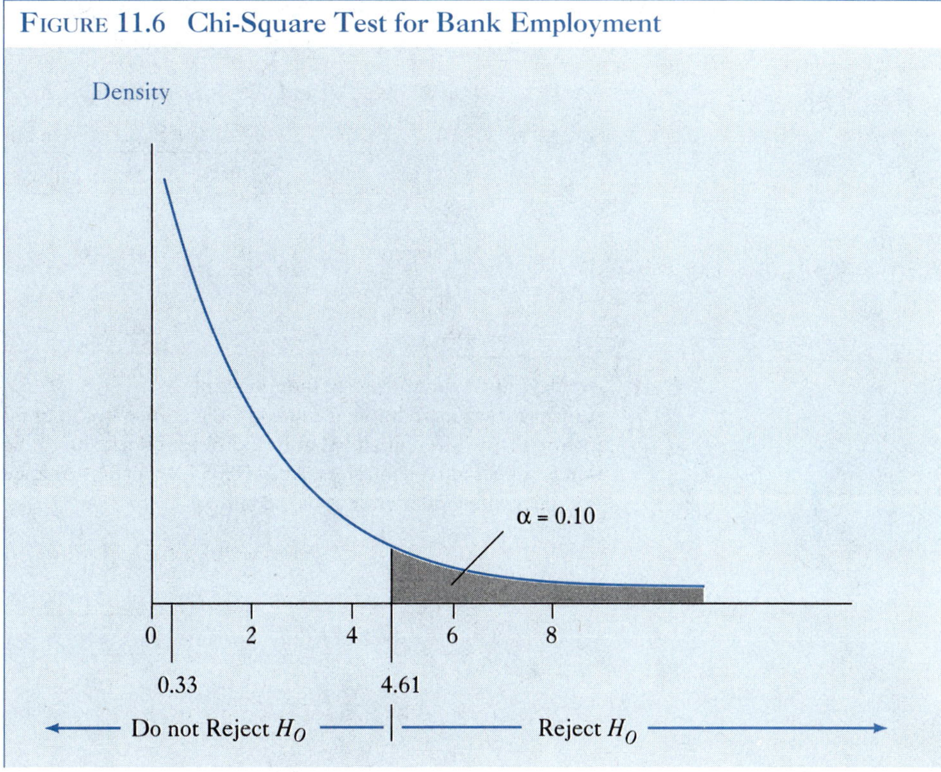

EXERCISES

11.16 If two events X and Y are not related and the probability of event X is 0.30 and the probability of Y is 0.25, then for a sample of 200 what is the expected frequency of X and Y?

11.17 What is the null hypothesis rejection region in the chi-square test of independence for the following contingency table sizes and Type I error levels?

a. Contingency table is $r = 3$ by $c = 4$ and $\alpha = 0.05$.

b. Contingency table is 3 by 3 and $\alpha = 0.10$.

c. Contingency table has 5 rows and 7 columns and the probability of a Type I error is 0.01.

11.18 Absenteeism among assembly line workers is of concern. In particular, there may be a relationship between the day of the week on which absences occur and the type of job performed by the worker. Workers are classified as skilled, semiskilled, and laborer. A random sample of days absent was selected from records kept over the past several years; workers were never included more than once in the following results:

Days Absent Among Assembly Line Workers

	Type of Worker		
Days	Skilled	Semiskilled	Laborer
Mondays	18	23	21
Tuesdays	10	16	17
Wednesdays	12	11	9
Thursdays	17	16	16
Fridays	22	26	20

Are days and type of worker independent?

11.19 A manager wants to know if there is a difference in the number of trucks unloaded during the three different shifts, at four different dock sites. The following sample information is collected on dock sites. At the 0.05 level of significance, are trucks unloaded at the four dock sites independent of the shifts?

Number of Trucks Unloaded

	Dock Site			
Time of Shift	1	2	3	4
7:30– 3:30	8	12	7	10
3:30–11:30	11	9	8	9
11:30– 7:30	10	12	10	9

11.20 *The Business Failure Record*, compiled by Dun and Bradstreet, reports business failures by industry type and period of observation. For the four periods below, is the type of business related to the period of observation in which failures occurred? (Use a spreadsheet program.)

	Number of Failures in Period			
Industry	1st	2nd	3rd	4th
Mining and Manufacturing	2,455	2,074	2,612	2,035
Wholesale Trade	1,316	1,016	1,473	984
Retail Trade	8,495	4,429	7,386	4,650
Construction	760	912	2,607	1,687
Commercial Service	593	731	1,367	1,392

11.21 An article in *The Wall Street Journal* (December 23, 1993) reported that although President Clinton had only 27 confirmed appointments to the U.S. District and Appeals Courts, only 72.3 percent (20 of 27) were white compared to Bush's 89.2 percent (165 of 185) and Reagan's 93.5 percent (344 of 368). Is the race of the judicial appointments independent of the president in power? (Test at the 0.05 Type I error level.)

11.10 Nonparametric ANOVA (Optional)

In addition to the test of independence, the chi-square distribution can also be used in a nonparametric ANOVA. Recall that the classical F test ANOVA required that the populations be approximately normally distributed with similar variances. There are situations in which the sample information makes such assumptions highly questionable.

Table 11.14 shows, for example, the exam scores for 28 students who had been placed in four study groups on the basis of an alphabetic ordering of their names.

Table 11.14 Student Exam Scores in Four Study Groups

Student Number	Study Group	Exam Score
1	1	96
2	1	75
3	1	80
4	1	10
5	1	57
6	1	61
7	1	56
8	2	45
9	2	73
10	2	71
11	2	65
12	2	66
13	2	100
14	2	61
15	3	55
16	3	67
17	3	63
18	3	50
19	3	90
20	3	68
21	3	83
22	4	15
23	4	50
24	4	94
25	4	86
26	4	68
27	4	64
28	4	58

Notice the unusually low score of 10 for student number 4 in study group one, the extremely high ceiling exam score of 100 in study group two, and an atypical score of 15 in study group four. The occurrence of the uncommonly low "outlier" scores and the ceiling score suggest that the underlying populations may not be normally distributed. To test for differences among the four groups, a classical ANOVA may be questionable. As an alternative, the Kruskal-Wallis nonparametric analysis of variance method may be employed.

The Kruskal-Wallis ANOVA tests differences among population medians. The null and alternative hypotheses are

H_O: The four population medians scores are equal

H_A: At least one of the population medians differs

The Kruskal-Wallis test statistic is based on the rank ordering of values and is thus not sensitive to the magnitude of outliers, although it does require that the underlying populations be continuous and similar in shape.

Provided that no more than one-fourth of the values are equal to each other, the Kruskal-Wallis test statistic is calculated as

$$H = \frac{12}{T(T+1)} \sum_{t=1}^{k} \frac{R_t^2}{n_t} - 3(T+1) \tag{11.27}$$

where: T is the total sample size, $T = kn$, for each of the k groups being of equal size $n_t = n$,
R_t is the sum of ranks within the tth group,
n_t is the sample size of the tth group
$t = 1, 2, \ldots k$

The Kruskal-Wallis H is approximately chi-square distributed, with $k - 1$ degrees of freedom, provided that there are five or more observations in each group ($n_t \geq 5$). With $k - 1 = 4 - 1 = 3$ degrees of freedom, the critical chi-square value that puts 10 percent in the upper tail is 6.25.

The calculation of the Kruskal-Wallis test statistic starts with a rank ordering of the total sample, with no regard to the group they are in. The ranking of the 28 students shown in Table 11.15 was done easily with the LOTUS 1-2-3 spreadsheet. Notice that there were scores tied at 50, 61, and 68. Each of these three tied scores received its average rank.

Once the scores have been ranked, Formula (11.27) can be employed to calculate the Kruskal-Wallis test statistic. Again, the calculations are made easily on a spreadsheet such as LOTUS 1-2-3, as shown in Table 11.16.

From Table 11.16, $H = 0.2650$, which is far below the critical chi-square value of 6.25, for a 10 percent Type I error level. Thus, we cannot reject the null hypothesis of no differences among the groups. Although these groups were formed by an alphabetic ordering of names, the exam test scores are such that they could have come from a simple random assignment.

TABLE 11.15 Rank Ordering of Student Exam Scores

Student Number	Study Group	Exam Score	Rank
1	1	96	27
2	1	75	21
3	1	80	22
4	1	10	1
5	1	57	8
6	1	61	10.5
7	1	56	7
8	2	45	3
9	2	73	20
10	2	71	19
11	2	65	14
12	2	66	15
13	2	100	28
14	2	61	10.5
15	3	55	6
16	3	67	16
17	3	63	12
18	3	50	4.5
19	3	90	25
20	3	68	17.5
21	3	83	23
22	4	15	2
23	4	50	4.5
24	4	94	26
25	4	86	24
26	4	68	17.5
27	4	64	13
28	4	58	9

QUERY 11.2:

Distributions of salaries are often positively skewed with the mean exceeding the median. In salary discrimination suits, why isn't the median used as the measure for comparing salaries among groups for which disparity is alleged? For instance, why wouldn't the median salaries of blacks, whites, and Hispanics be compared in a Kruskal-Wallis test? If a classical ANOVA requires that the populations are normally distributed, but the samples suggest otherwise, how can an ANOVA be justified?

ANSWER:

The median could be used for comparing salaries among groups, but a comparison of medians might not capture pay increases that were targeted at just those in the top end of a subgroup distribution. Furthermore, the cost of removing discrimination based on medians cannot be transformed directly into the total wage bill, but comparisons of means can be (average wage times workforce equals wage bill). As

TABLE 11.16 Calculation of the Kruskal-Wallis Test Statistic for Student Exam Scores

Student Number	Study Group	Exam Score	Rank	R_t	R_t^2	$R_t^2/7$
1	1	96	27			
2	1	75	21			
3	1	80	22			
4	1	10	1			
5	1	57	8			
6	1	61	10.5			
7	1	56	7	96.5	9312.25	1330.321
8	2	45	3			
9	2	73	20			
10	2	71	19			
11	2	65	14			
12	2	66	15			
13	2	100	28			
14	2	61	10.5	109.5	11990.25	1712.892
15	3	55	6			
16	3	67	16			
17	3	63	12			
18	3	50	4.5			
19	3	90	25			
20	3	68	17.5			
21	3	83	23	104	10816	1545.142
22	4	15	2			
23	4	50	4.5			
24	4	94	26			
25	4	86	24			
26	4	68	17.5			
27	4	64	13			
28	4	58	9	96	9216	1316.571
						$\sum R_t^2/7 = 5904.928$

Kruskal-Wallis Test Statistic:

$$H = \frac{12}{T(T+1)} \sum_{t=1}^{k} \frac{R_t^2}{n_t} - 3(T+1) = 0.0147783(5904.928) - 3(29) = 0.2650$$

discussed in Chapter 6, if the salary distribution is highly right skewed, it can be "normalized" by a logarithmic transformation. A classical ANOVA via an F test can then be employed on log-normal data. If the data is only slightly right skewed, this transformation may not be needed because the validity of the classical ANOVA is not greatly influenced by small departures from normality.

11.11 LOOKING AHEAD

This chapter is the third in a three-chapter series on hypothesis testing that provides a structure for testing differences among means, proportions, and medians. Classical analysis of variance with the F statistic, and the chi-square test of independence, continue to be two of the most widely used methods in statistics. As we will see in the remaining chapters of this book, however, from a mathematical modeling standpoint, there are more intuitive approaches to analyzing the relationship between a dependent variable and more than two explanatory factors.

The next three chapters introduce you to multivariate analysis within a regression framework. In the third of these, Chapter 14, we return to the analysis of variance ideas, but in regression applications. If you decide to continue with the study of statistics you will find that the concepts employed in ANOVA keep reappearing.

CHAPTER EXERCISES

11.22 For four outlets in 16 randomly selected days, the value of pilfered merchandise is estimated to be as follows (in thousands of dollars).

Outlet Number	1	2	3	4
Mean	1.243	0.892	1.312	1.157
Standard Deviation	0.351	0.263	0.365	0.319

a. What is the grand mean?

b. What are the between and within variances?

11.23 In Exercise 11.22 at the 0.05 Type I error level, is there a difference in the mean level of pilfering among the four stores? Why would a regional manager of these four stores be interested in the results of this test?

11.24 In the study of quality control, there is a famous lesson known as the red bead experiment. The experiment involves each of several employees sampling 50 beads from a box of 4,000 beads, of which 18 percent are red (representing defects) and 82 percent are white (representing good quality). The lesson is designed to show that some employees may look as if they are big producers of defects (drawing many red beads) even though the process is totally random. Based on the following red bead outcomes, are the five individuals equal producers? (Each of the five persons drew 50 beads on each of three days.)

Day	Ann	David	Frank	Molly	Sally
1	8	13	12	15	11
2	10	12	9	6	14
3	6	9	13	7	5

11.25 For the above data from a red bead experiment, is either the day or the person making a significant contribution to explaining total variability?

11.26 On five consecutive days the word processing speeds achieved at six randomly selected times were recorded with the following results, in words per minute:

Monday	Tuesday	Wednesday	Thursday	Friday
80	85	92	101	83
77	79	86	85	84
65	70	67	68	68
86	85	87	89	94
67	78	76	75	77
90	92	94	94	92

Is there a difference in word processing speed among the five days? (Test at the 5 percent level of significance.)

11.27 If you now discover that the speeds in Exercise 11.26 were from six different employees, as represented by each row, how would your analysis in 11.26 change? How would your conclusion be influenced?

11.28 For the situation described in Exercises 11.26 and 11.27, do the six word processors make a significant contribution to explaining variability?

11.29 The following data on dog anesthesia reports milliseconds between heartbeats; it is from P.H. Westfall and S.S. Young, *Resampling Based Multiple Testing,* John Wiley & Sons, Inc., 1993, p. 135. Use a classical ANOVA to determine if the four anesthesia methods (columns) produce the same mean time between heartbeats.

	Halothane Absent		Halothane Present	
Dog	CO_2 High	CO_2 Low	CO_2 High	CO_2 Low
01	426	609	556	600
02	253	236	392	395
03	359	433	349	357
04	432	431	522	600
05	405	426	513	513
06	324	438	507	539
07	310	312	410	456
08	326	326	350	504
09	375	447	547	548
10	286	286	403	422
11	349	382	473	497
12	429	410	488	547
13	348	377	447	514
14	412	473	472	446
15	347	326	455	468
16	434	458	637	524
17	364	367	432	469
18	420	395	508	531
19	397	556	645	625

11.30 Why might a critic question the use of the F statistic to test for mean differences in time between heartbeats in Exercise 11.29?

11.31 Use a computer spreadsheet program to conduct a Kruskal-Wallis nonparametric analysis of variance to test for mean time differences between heartbeats in Exercise 11.29.

11.32 Lloyd's Register of Shipping keeps track of minor, but costly, damage to the forward section of cargo vessels (from contact with large waves, debris at sea, docks, and the like) for purposes of setting insurance premiums and standards for hull construction. The data below identify four types of ships and the number of damage incidents. At the 0.05 Type I error level, is there a difference between damage incidents among the four types of ships?

Ship Type	Number of Damage Incidents
1	39
1	30
1	58
1	54
1	14
1	45
1	2
1	18
1	10
2	1
2	3
2	5
2	9
2	18
2	0
2	11
2	10
2	6
3	0
3	1
3	0
3	1
3	2
3	12
3	1
3	2
3	3
4	1
4	1
4	2
4	1
4	5
4	3
4	2
4	1
4	0

11.33 The data set in the computer ASCII file "EX11-33.PRN" has information on the day (numbered 40 to 46) a patient entered the emergency room of the hospital (first column), the time of entering (1 to 1440 minutes past midnight, as recorded in the second column), the time of leaving the emergency room, which could extend into the next day (third column), the admitting doctor (1, 2, 3, or missing data point ".", as seen in the second to last column), and other variables.

a. Create a variable to show the minutes that each person was in the emergency room. (Make sure to correctly calculate the time for those whose stay extended from one day to the next.)

b. Is the day on which the person enters the emergency room significantly related to the time spent in the emergency room?

11.34 Using the data described in Exercise 11.33, do a two-way ANOVA to determine if the day and the doctor are each significant in explaining time in the emergency room.

11.35 In Fisher's testing of fertilizer, a common test design involved the use of a "Latin square," which was so named because the fertilizer treatments were symbolized by the Latin letters A, B, C, . . . instead of the Greek letters. Below is a Latin square for farmland that has been divided into three plots (Plot 1, Plot 2, and Plot 3). For the three seasonal plantings (Season 1, Season 2, and Season 3), three different fertilizers (A, B, and C) are to be applied in the order shown. What is special about this design and its use in a test of mean yield?

	Plot 1	Plot 2	Plot 3
Season 1	A	C	B
Season 2	B	A	C
Season 3	C	B	A

In the test of differences in mean yields among the three fertilizers, what is the role of the Latin square design?

11.36 In contrast to the Latin square, Exercise 11.35, is the randomized-block design, in which subjects are matched on a variable that the researcher can control. Say we have 60 students to divide into three classes for which instruction will be given by different methods. From a pretest, we know each student's knowledge going into the course. We divide them into blocks: the three highest-scoring students are block one, the next highest three are block two, and so on to the 20th block. In the randomized-block design we now randomly assign the students in each block to one of the three classes. In our subsequent test of differences among the mean posttest scores of the three classes, what is the role of the blocks?

11.37 A study of bank teller efficiency yielded the following:

		Teller Type
Observation Period	**Average Minutes To Complete Transaction**	**2 If only human 1 If combined human and auto 0 If fully automated**
1	2.0	2
2	1.0	0
3	.9	1
4	1.1	2
5	2.5	2
6	2.3	0
7	2.1	2

(continued)

Observation Period	Average Minutes To Complete Transaction	Teller Type *(continued)* 2 If only human 1 If combined human and auto 0 If fully automated
8	2.2	0
9	1.2	2
10	2.0	2
11	.9	1
12	.9	2
13	1.7	2
14	1.1	0
15	2.0	2
16	1.9	0
17	1.1	1
18	.9	2
19	1.5	0
20	1.0	1

a. Is average time to complete a transaction a function of the type of teller? (Set up test and discuss.)

b. What other factors, in addition to type of teller, might influence transaction time? How does recognition of this missing factor influence your answer in part a?

11.38 The computer ASCII data file "EX11-38.PRN" contains the records of 154 cars tested by the National Highway Traffic Safety Administration for damages resulting from 35 mph frontal crashes into a fixed barrier. Within the file, the first line identifies the variable:

"Car" is the first column; it identifies the car.

"Drs" is the second column; it is the number of doors.

"yr" is the third column; it is the last two digits of the model year of the car.

"Cls" is the fourth column; it is size class, where 1 is light, 2 is compact, 3 is medium, and 4 is heavy.

"lbs" is the fifth column; it is the weight of the car.

"Pro" is the sixth column; it is the protection type, where 1 is manual seat belts, 2 is motorized belts, 3 is passive belts, 4 is driver air bag, and 5 is driver and passenger air bags.

"HD" is the seventh column; it is a driver head injury index, where a score below 1000 indicates serious injury is unlikely.

"HP" is the eighth column; it is a passenger head injury index, where a score below 1000 is not critical.

"CD" is the ninth column; it is a driver chest injury index, where a score below 60 indicates serious injury is unlikely.

"CP" is the 10th column; it is a passenger chest injury index, where a score below 60 is not critical.

"DFL" is the 11th column; it is femur load to the driver's left leg, where loads below 2,250 pounds indicates that serious injury is unlikely.

"DFR" is the 12th column; it is the femur load to the driver's right leg, where loads below 2,250 pounds are not critical.

"PFL" is the 13th column; it is femur load to the passenger's left leg, where loads below 2,250 pounds indicates that serious injury is unlikely.

"PFR" is the 14th column; it is the femur load to the passenger's right leg, where loads below 2,250 pounds are not critical.

For crash protection types 1, 2, 3, and 4, randomly pick 15 cars from each type.

a. For your samples, what is the average driver head injury index score for each of the four crash protection types?

b. Based on your sample information, is average driver head injury index score the same across the four types?

11.39 For your sample of 15 cars from each of the four crash protection types identified in Exercise 11.38, what is the average driver chest injury index score for each of the four protection types? Is average driver chest injury index score the same across the four types?

11.40 For your sample of 15 cars from each of the four crash protection types identified in Exercise 11.38, calculate the average femur load for the left and right leg of each car driver, FDA. What is the average FDA for each of the four protection types? Is the driver femur average load the same across the four types?

11.41 The computer ASCII data file "EX11-33.PRN", presented in Exercise 11.33, contains information from a hospital emergency room. Is time in the emergency room related to the treating emergency room doctor, as identified by numbers 1, 2, and 3, in the second to last column of information?

Academic References

Becker, William, and Arlington Williams. "Assessing Personnel Practices in Higher Education: A Case Study in the Hiring of Females." *Economics of Education Review*, Vol. 5, No. 3, 1986, pp. 265–272.

Cochran, W.G. *Planning and Analysis of Observational Studies*. New York: Wiley, 1983.

Conway, Delores. "Can Statistics Tell Us What We Do Not Want to Hear?" *Statistical Science*, May 1993, pp. 158–171.

Iversen, Gudmund, and Helmut Norpoth. *Analysis of Variance*. Newbury Park: Sage University Papers, Second Edition, 1987.

Leamer, Edward. "Let's Take the Con Out of Econometrics." *American Economic Review*, March 1983, pp. 31–43.

Endnotes

1. From Figure 11.4,

$$y_{ij} - \bar{\bar{y}} = \left(\bar{y}_i - \bar{\bar{y}}\right) + \left(y_{ij} - \bar{y}_i\right)$$

Squaring both sides yields

$$\left(y_{ij}-\bar{\bar{y}}\right)^2=\left(\bar{y}_i-\bar{\bar{y}}\right)^2+\left(y_{ij}-\bar{y}_i\right)^2+2\left(\bar{y}_i-\bar{\bar{y}}\right)\left(y_{ij}-\bar{y}_i\right)$$

Before summing over all *i* and *j*, note that

$$\Sigma_i\,\Sigma_j\left(\bar{y}_i-\bar{\bar{y}}\right)\left(y_{ij}-\bar{y}_i\right)=\left(\bar{y}_1-\bar{\bar{y}}\right)\Sigma_j\left(y_{1j}-\bar{y}_1\right)+\left(\bar{y}_2-\bar{\bar{y}}\right)\Sigma_j\left(y_{2j}-\bar{y}_2\right)+\ \ldots$$
$$=\left(\bar{y}_1-\bar{\bar{y}}\right)(0)+\left(\bar{y}_2-\bar{\bar{y}}\right)(0)+\ \ldots=0$$

where $\Sigma_j(y_{1j}-\bar{y}_1)$ and $\Sigma_j(y_{2j}-\bar{y}_2)$, and so on, are each the sum of deviations from the mean within the identified treatment group. Each of these sums is zero.
Thus summing over all *i* and *j* yields

$$\Sigma_i\,\Sigma_j\left(y_{ij}-\bar{\bar{y}}\right)^2=\Sigma_i\,\Sigma_j\left(\bar{y}_i-\bar{\bar{y}}\right)^2+\Sigma_i\,\Sigma_j\left(y_{ij}-\bar{y}_i\right)^2$$

or TotalSS = FactorSS + ErrorSS.

2. The use of the TSS, FSS, and ESS notation is unfortunate because it is not standardized across books or computer programs. You should be aware that some use TSS or SST to represent the treatment sum of squares while others use ESS or SSE to represent the "explained sum of squares," or BSS or SSB to represent the "between sum of squares." The treatment sum of squares, explained sum of squares, and between sum of squares are all the same quantity, which we label FactorSS or just FSS when its meaning should be clear. In addition, the error sum of squares is sometimes called the residual sum of squares (RSS or SSR), or within sum of squares (WSS or SSW), or unexplained sum of squares (USS or SSU). Here we label the error sum of squares ErrorSS or just ESS when its meaning should be clear. You should take care to check the definition of this sum of squares notation in using computer programs and other printed material.
3. Employers in discrimination cases often argue that the minority in question is not interested in the jobs being considered. For example, in *EDOC v. Sears* [839 F.2d 302 (7th cir. 1988)], Sears argued that women were not interested in certain sales positions, but shortly after the charges were filed the fraction of females hired into those positions increased greatly. Professor Delores Conway (1993) comments that this "A Funny Thing Happened on the Way to the Courtroom" phenomenon is a major reason plaintiffs have not prevailed in several cases.
4. A maintained hypothesis that is often overlooked is that the outcomes or elements of one event of a variable are not shared by other events of the variable. In the case of a workforce, this implies that at the beginning of the liability period there is a drawing of *n* individuals and in the middle of the period there is another drawing. That is, the first *n* individuals do not automatically continue to the middle of the period. If the time between the beginning and middle of the period is short, this assumption is untenable since typically only a small fraction of employees terminate and only a small number are hired. An intertemporal model of these flows and the way in which the total workforce stock changes over time is provided by William Becker and Arlington Williams (1986).

FORMULAS

- Between sample variance

$$s_{\bar{y}}^2 = \frac{\Sigma\left(\bar{y}_i - \bar{\bar{y}}\right)^2}{k-1}$$

where, for k sample means ($\bar{y}_i$, $i = 1, 2, \ldots k$)
and the grand mean ($\bar{\bar{y}}$) is

$$\bar{\bar{y}} = \frac{\sum_{i=1}^{k} \bar{y}_i}{k}$$

- Pooled within sample variance:

$$s_w^2 = \frac{\sum_{i=1}^{k}\sum_{j=1}^{n}\left(y_{ij} - \bar{y}_i\right)^2}{k\left(n-1\right)}$$

where there are k samples, each of size n, for each treatment factor subgroup

- Sum of squares equality

$$\text{TSS} = \text{FSS1} + \text{FSS2} + \ldots + \text{ESS}$$

where TSS is the total sum of squares

FSS1, FSS2, . . . are sums of squares associated with the identified factors 1, 2 . . .

ESS is the error sum of squares

- F statistic for multiple factor ANOVA

$$F = \frac{\text{Variability explained (adjusted for a factor's degree of freedom)}}{\text{Variability left unexplained (adjusted for degrees of freedom)}}$$

where the numerator and denominator degrees of freedom depend on the quantities considered.

Assumptions:

1. With the possible exception of the identified factors, only random sampling error is involved in the formation of the total sum of squares. That is, the observations within samples are statistically independent.
2. The populations and thus the sampling distributions of the means are approximately normally distributed.
3. The populations are identical with the possible exception of their central location. That is, the populations have identical variances even if their means differ.

- Chi-square statistic for the test of independence

$$\chi^2 = \sum_{i=1}^{r} \sum_{j=1}^{c} \frac{\left(O_{ij} - E_{ij}\right)^2}{E_{ij}}$$

where O_{ij} is the observed frequency in *ij*th cell;
E_{ij} is the expected frequency in *ij*th cell ($E_{ij} \geq 5$);
and the degrees of freedom are df = $(r-1)(c-1)$,
for a contingency table with r rows and c columns.

Assumption:
The null hypothesis is always that the two variables of interest are independent and normally distributed.

- The Kruskal-Wallis test statistic

$$H = \frac{12}{T(T+1)} \sum_{t=1}^{k} \frac{R_t^2}{n_t} - 3(T+1)$$

where T is the total sample size, $T = kn$, for each of the k groups being of equal size $n_t = n$,
R_t is the sum of ranks within the *t*th group,
n_t is the sample size of the *t*th group $t = 1, 2, \ldots k$

Assumption:
H is approximately chi-square distributed, with $k - 1$ degrees of freedom.

CHAPTER 12

CORRELATION AND REGRESSION ANALYSIS WITHIN A SAMPLE

Too many use statistics as a drunk man uses a lamp post—for support and not illumination.

Keely v. Westinghouse Elec. Comp. (E.D. Mo. 1975)

12.1 Introduction

Knowing the algebraic relationship between two variables enables predictions for one of the variables based on values of the other. For example:

- How do highway deaths change when the speed limit is changed? A letter to the editor of *The New York Times* (November 29, 1986) by an official of the Sports Car Club of America suggests that there is no relationship.
- A *Wall Street Journal* (June 16, 1987) article suggests that credit card balances and interest rate charges are related inversely, but do observed data support this claim?
- A *Business Week* (June 22, 1987) article argued that as the percentage of women among those holding Master in Business Administration (MBA) degrees has increased, the percentage of women in management and administration jobs has also increased. Is this relationship supported by the data? Can future changes be predicted from this relationship?
- According to a *Wall Street Journal* (July 8, 1987) article, financial analysts, bankers, attorneys, and academics see consulting services provided by CPA firms as an impairment to their objectivity in auditing. How can the relationship between consulting and objectivity be described?
- A *Business Week* (June 18, 1990) article asked what would be the effect on cigarette consumption if the cigarette tax were increased by 10 percent? The regression analysis of Nobel Laureate economist Gary Becker and fellow economists Kevin Murphy and Michael Grossman suggests that current consumption will fall by 4 percent in the short run and 7.5 percent in the long run.

This chapter is devoted to statistical techniques for describing the association between paired sample observations. It presents covariance, correlation, and regression for this purpose. The principle of least squares is developed and demonstrated. By the end of this chapter, you will be able to calculate and interpret all these descriptive measures of the sample relationship between two variables. The next chapter introduces the ideas of population modeling and parameter estimation and testing based on the descriptive statistics presented in this chapter.

12.2 Speed and Death: An Illustration

Questions and issues involving the relationship between variables appear regularly in the popular press, business and trade publications, and scholarly journals. For example, in a letter to the editor of *The New York Times* (November 29, 1986, p. 26), David Rivkin provided data on two variables: highway deaths per 100 million vehicle miles and highway speed limits for 10 countries, Table 12.1. He states:

> From this we can see that five countries with the same speed limit have very different positions on the safety list. Britain . . . with a speed limit of 70 is demonstrably safer than Japan, at 55. Clearly, speed has little to do with safety.

Table 12.1
Death Rates and Speed Limits

	Deaths Per 100 Million Vehicle Miles	Speed Limit (mph)
1. Norway	3.0	55
2. United States	3.3	55
3. Finland	3.4	55
4. Britain	3.5	70
5. Denmark	4.1	55
6. Canada	4.3	60
7. Japan	4.7	55
8. Australia	4.9	60
9. Netherlands	5.1	60
10. Italy	6.1	75

Although Rivkin argues that there is no sample relationship between speed and death, his argument is susceptible to statistical analysis. Better measures are needed of the sample relationship between Rivkin's notion of highway safety and the speed limit. Those measures must describe both central tendencies and dispersion.

12.3 Conditional Means and Deviations

Rivkin does not define exactly what he means by "a safety list." He does state that the five countries with 55 mph speed limits have very different positions on this list. In other words, there is considerable variability in the rate of deaths at 55 mph. The conditional mean death rate at a speed limit of 55 mph is 3.7 deaths per 100 million vehicle miles.

$$\text{Mean death rate, at 55 mph speed limit} = \frac{3.0 + 3.3 + 3.4 + 4.1 + 4.7}{5} = 3.7$$

Deviations from this mean or average are calculated by subtracting the conditional average from each of the five death rates. The deviations in death rates at a 55 mph speed limit are $3.0 - 3.7 = -0.7$ for Norway, and, similarly, –0.4 for the United States, –0.3 for Finland, +0.4 for Denmark, and +1.0 for Japan.

We now have a quantitative safety list for the sample of five countries with 55 mph speed limits. As suggested by Rivkin, countries on this safety list have very different positions. Three are below this speed-specific average death rate of 3.7, and two are above. Death rates vary from 0.7 deaths below the average to one death above the average, per 100 million vehicle miles.

A similar listing for deaths at 60 mph can be obtained, where the conditional mean death rate is 4.767. For the three sample observations at a speed limit of 60 mph, deviations from this conditional mean death rate are –0.467 for Canada, +0.133 for Australia, and +0.333 for the Netherlands. Now only one country is below the 60 mph conditional mean death rate of 4.767, and the death rate only varies from 0.467 deaths below this mean to 0.333 deaths above the mean, per 100 million vehicle miles.

At each of the speed limits of 70 mph and 75 mph, there is no variability in deaths within the sample, because only one country is listed at each speed. The importance of these single observations at 70 mph and 75 mph versus the multiple observations at the slower speed limits can be seen by constructing a scatterplot and connecting the conditional means to form a **regression** line. A regression line shows the central tendency (in our example the mean) of y at each value of x in the sample.

Regression of y on x The relationship between a measure of central tendency of one variable, y, across values of another variable, x. Because y is said to depend on x, y is called the dependent variable and x is the independent variable.

Sample regression lines can be obtained in many different ways. One way is to connect the conditional means of the dependent variable at each value of the independent variable in a scatterplot or scatter diagram. As introduced in Chapters 10 and 11, a scatterplot measures one variable along the vertical axis and the other variable along the horizontal axis. It is conventional to place the variable to be explained, called the dependent variable (y), on the vertical axis. The variable used to do the explaining, called the independent variable (x), is placed on the horizontal axis. In Figure 12.1, death rate is the dependent variable; it is measured along the vertical axis. The speed limit is the independent variable; it is measured along the horizontal axis.

Using the data in Table 12.1, point I in Figure 12.1 gives the x and y values for Italy. It is plotted at the intersection of a vertical line drawn from 75 mph and a horizontal line drawn from 6.1 deaths per 100 vehicle miles. Point B is for Britain and is the intersection of a vertical line from 70 and a horizontal line from 3.5. The remaining eight countries are identified and plotted in a similar manner. (Note that this is the same type of two-dimensional scatterplot drawn for the pretest, x, and posttest, y, in Chapter 10.)

Each point in Figure 12.1 is marked with a letter that identifies the country. Point I, for example, gives the x and y values for Italy: 75 mph and 6.1 deaths per 100 vehicle miles. Figure 12.1 also shows a line that connects the conditional mean death rates for the different speed limits for which there are data. This is a regression line. It shows how the central tendency (conditional mean) of y changes between x values. It does not show a uniformly positive or negative slope.

The single observations at 70 and 75 miles per hour are more influential in determining the slope of the regression line than any one of the observations at 55 or 60 miles per hour because there are multiple observations at 55 and 60 mph. For instance, if the single death rate at 70 mph were higher, then the slope of the line might be positive throughout its range. If one of the observations at 55 mph were higher (or lower), however, the slope of the line need not change because this observation enters the conditional mean with only a one-fifth weight.

Comparing one country's death rate at 55 mph with another at 70, as Rivkin does, overlooks the variability in death rates at 55 mph. Comparing the conditional mean death rate at 55 mph with the mean deaths at 60 mph or the death rates at 70 mph or 75 mph overlooks how variability in deaths changes across speed limits. What is needed is a single equation or regression line that can be used to assess how deaths deviate from their overall sample mean as the speed limit deviates from its overall sample mean.

FIGURE 12.1 **Scatterplot of Death Rates and Speed Limits**

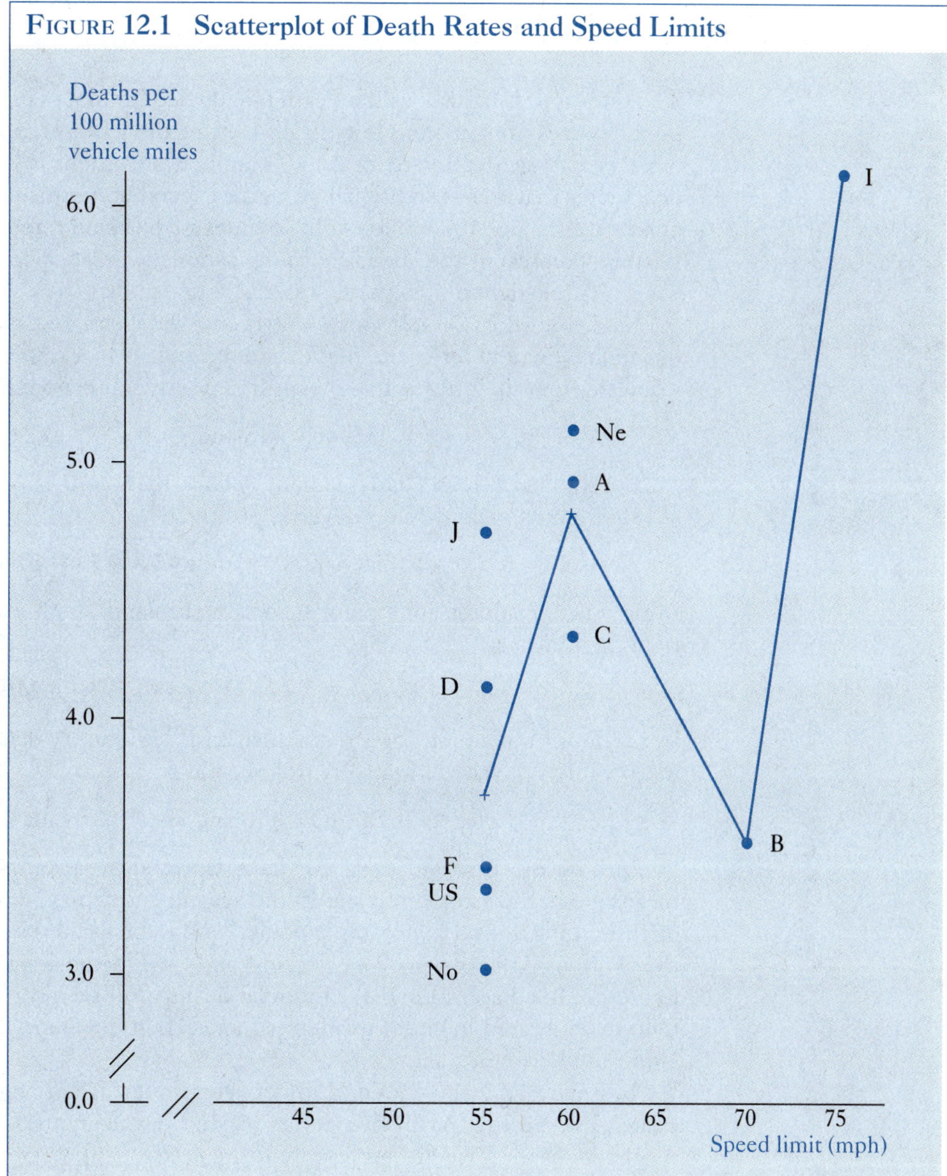

12.4 Deviations from the Means

The question is: How does the death rate deviate from its sample average as the speed limit deviates from its sample average? Letting the letter y represent the death rate, the mean or average death rate for the 10 sample observations is $\bar{y} = 4.24$. Unlike the mean death rates that were conditional on a given speed limit, this sample average death rate of 4.24 deaths per 100 million vehicle miles involves no reference to the speed limits. Letting x represent the speed limit, the sample average speed limit for all 10 cases is $\bar{x} = 60$; its calculation involves no reference to deaths.

In Figure 12.2, $\bar{y} = 4.24$ and $\bar{x} = 60$ divide the death and speed scatterplot into four quadrants. For a point in the first quadrant, such as that representing Italy, both x and y deviate from their respective means in a positive direction; that is,

$$x - \bar{x} = 75 - 60 = +15 \text{ and } y - \bar{y} = 6.1 - 4.24 = +1.86$$

For a point in the second quadrant, such as that representing Japan, x deviates negatively from its mean while y deviates positively:

$$x - \bar{x} = 55 - 60 = -5 \text{ and } y - \bar{y} = 4.7 - 4.24 = +0.46$$

In the third quadrant, for a point such as F (Finland), both x and y deviate negatively from their means:

$$x - \bar{x} = 55 - 60 = -5 \text{ and } y - \bar{y} = 3.4 - 4.24 = -0.84$$

In the fourth quadrant, for a point such as B (Britain), x deviates positively from its mean while y deviates negatively from its mean:

$$x - \bar{x} = 70 - 60 = +10 \text{ and } y - \bar{y} = 3.5 - 4.24 = -0.74$$

In both the first and third quadrant, deviations in x and y from their respective means are in the same direction. In the second and fourth, deviations in x and y are in opposite directions. If the sample points tend to be located in the quadrants I and III, the relationship between deviations in x and y will be positive. If the points tend to be located in quadrants II and IV, the relationship will be negative. If the points do not tend to be located in either quadrants I and III, or quadrants II and IV, there is no relationship between x and y.

In Figure 12.2, the relationship between x and y is not negative. If anything, it appears to be positive, with five points located in quadrants I and III and only two in quadrants II and IV. We can represent the relationship between x and y by drawing a single straight line through the point for the mean of x and the mean of y and into the quadrants with the most points. This single straight line regression would be the one that appears to "fit the data best." For the data in Figure 12.2, this line would pass through $\bar{y} = 4.24$ and $\bar{x} = 60$ and into quadrants I and III. It would have a positive slope, as drawn in Figure 12.3.

The positive relationship between x and y in Figure 12.3 is not strong, since three sample points show y deviating from its mean when x did not deviate from its mean, and two points are in quadrants II and IV. Curiously, Rivkin only emphasized these

FIGURE 12.2 Deviations in Death Rates and Speed Limits

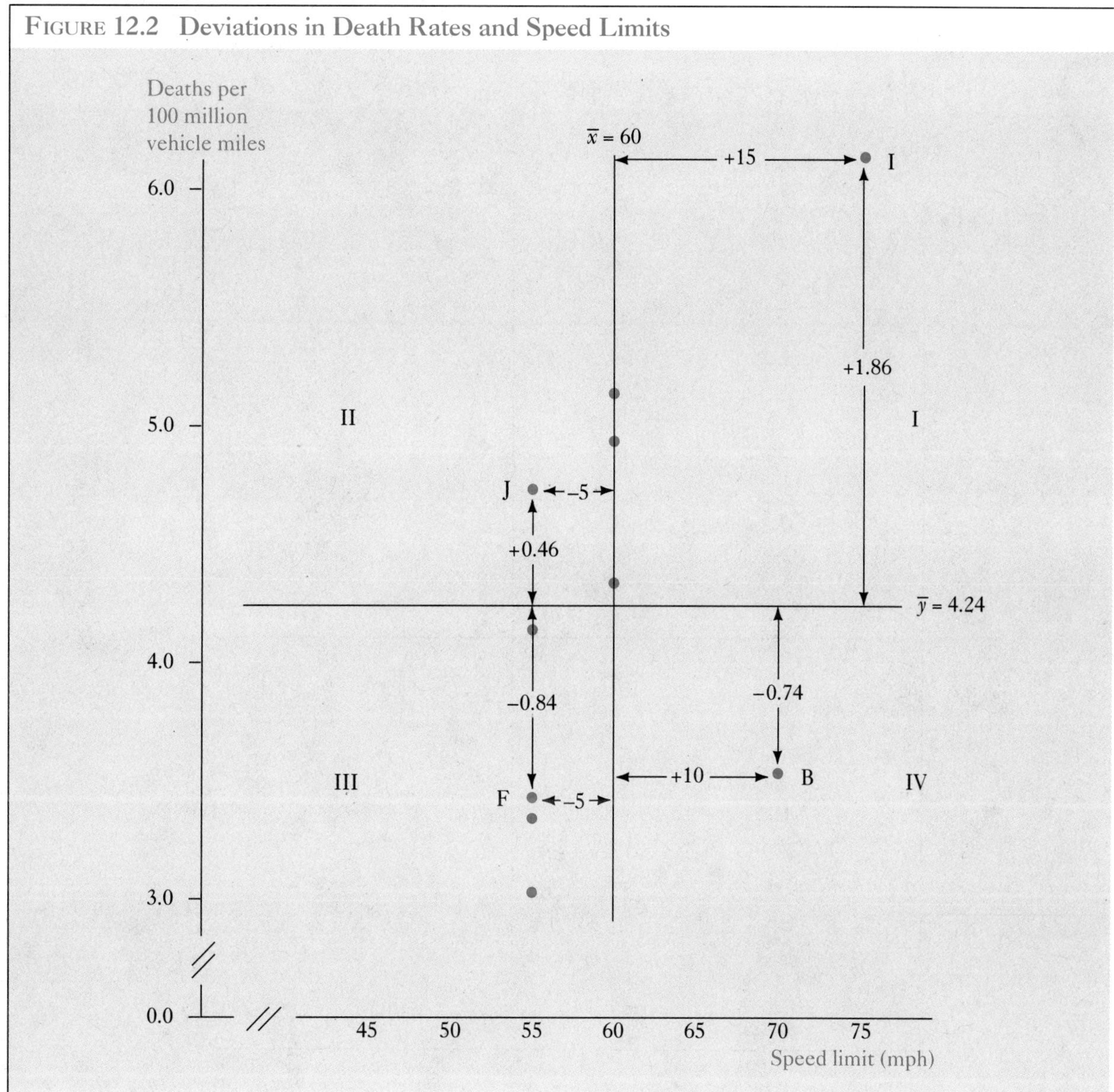

FIGURE 12.3 Scatterplot of Death Rates and Speed Limits With Continuous, Straight Regression Line

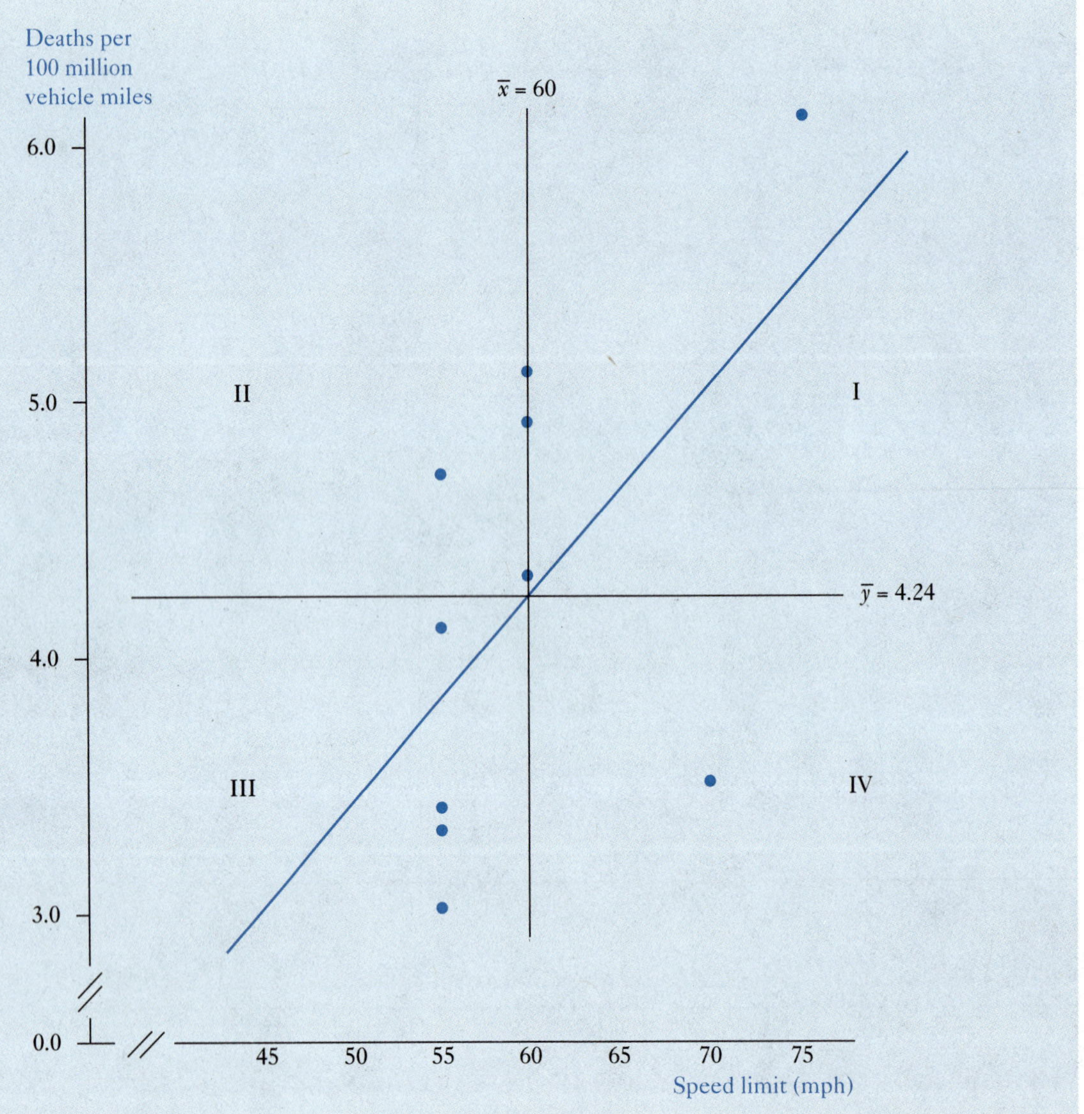

two points (Japan and Britain) in quadrants II and IV; he never mentioned that five death and speed limit pairs show a positive relationship in accordance with our positively sloped, single, and continuous straight regression line.

QUERY 12.1:

A *Wall Street Journal* (June 16, 1987) article provided the following data on outstanding credit card balances and interest rates charged at 10 selected banks.

Bank	Outstanding balances (in billions)	Most common interest rate
1	\$9.10	19.8%
2	5.30	19.8
3	4.50	17.5
4	3.30	19.8
5	2.50	17.8
6	2.00	17.9
7	1.94	20.0
8	1.27	19.8
9	1.20	19.8
10	0.99	17.7

The article suggested that banks with smaller outstanding credit card balances charge lower interest rates. Is this true? (Calculate the deviations from the mean interest rate and mean outstanding balance, and draw a scatterplot to determine how deviations from the mean credit balance are related to deviations from the mean interest rate. Draw in the single straight line regression line.)

ANSWER:

Let x be credit card balance and y be interest rate. The mean credit card balance, mean interest rate, deviations from the means, and the scatterplot are

$$\bar{x} = 3.21 \text{ and } \bar{y} = 18.99$$

Bank	$x - \bar{x}$	$y - \bar{y}$
1	5.89	0.81
2	2.09	0.81
3	1.29	–1.49
4	0.09	0.81
5	–0.71	–1.19
6	–1.21	–1.09
7	–1.27	1.01
8	–1.94	0.81
9	–2.01	0.81
10	–2.22	–1.29

The scatterplot, with the regression line drawn in, follows.

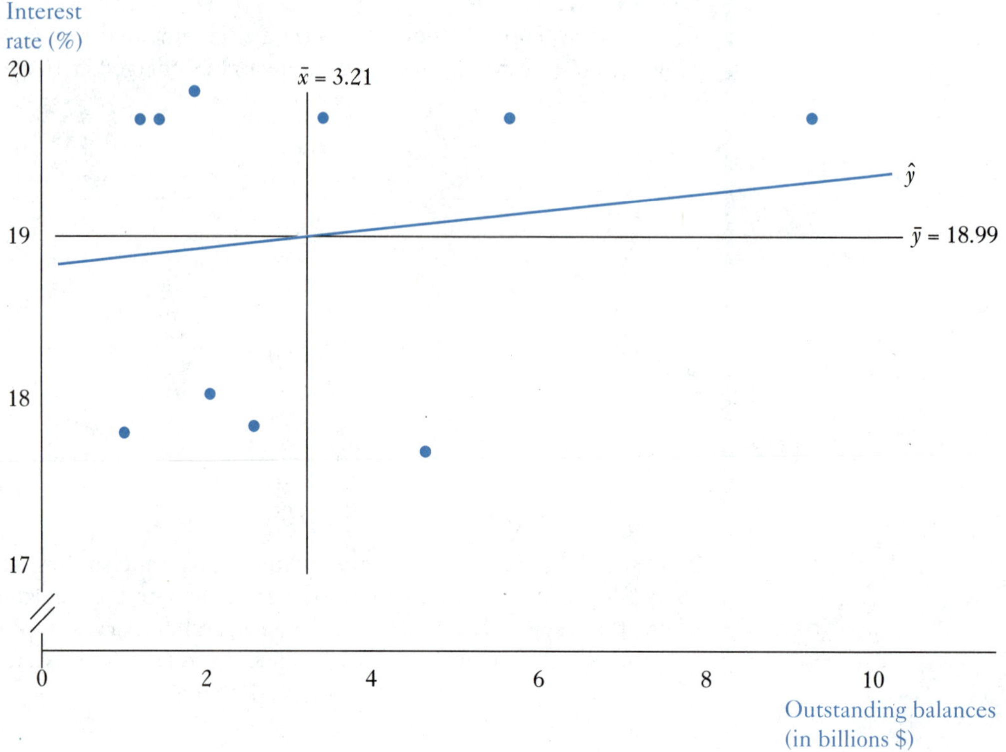

There appears to be a slight positive relationship between interest rates charged on credit cards and outstanding balances, since six of the 10 points lie in quadrants I and III. For these six points, deviations from the interest rate mean are in the same direction as deviations from the mean of outstanding balances. At best, this relationship is weak, however, because the other four points lie in quadrants II and IV, where deviations are in the opposite directions.

12.5 Measures of Covariance

Covariance is a measure of the strength of the linear relationship between two variables. It provides a quantitative answer to the question, "As the values observed for one variable rise (or fall), what tends to happen to the values observed on another variable?"

Reconsider Figure 12.2, where $\bar{y} = 4.24$ and $\bar{x} = 60$ divide the death and speed scatterplot into four quadrants. For a point in the first quadrant, both x and y deviate from their respective means in a positive direction, so the products of $(x_i - \bar{x})$ deviations and $(y_i - \bar{y})$ deviations will be positive. For a point in the second quadrant, x deviates negatively from its mean, while y deviates positively, and the product of deviations is negative. In the third quadrant, both x and y deviate negatively from their means, so the product of deviations is positive. In the fourth quadrant, x deviates positively from its mean, while y deviates negatively from its mean, and the product of deviations is negative.

If x and y tend to move in the same direction, then most products of deviations in x and deviations in y will be positive, and the sum of all such cross products should be positive; that is,

$$\sum_{i=1}^{n}(x_i - \bar{x})(y_i - \bar{y}) > 0$$

Alternatively, if x and y tend to move in opposite directions, then the products of deviations in x and deviations in y will tend to be negative and

$$\sum_{i=1}^{n}(x_i - \bar{x})(y_i - \bar{y}) < 0$$

When there is no relationship between x and y, with the observations appearing evenly in all four quadrants, positive and negative deviations from means will tend to cancel, and

$$\sum_{i=1}^{n}(x_i - \bar{x})(y_i - \bar{y}) \approx 0$$

The quantity $\Sigma(x_i - \bar{x})(y_i - \bar{y})$ is a measure of variability between x and y. In the case of death rates and speed limits, Table 12.2 shows that $\Sigma(x_i - \bar{x})(y_i - \bar{y}) = 34$, which is consistent with our earlier assessment that death rates and speed limits were positively related in Rivkin's sample.

As with the calculation of the sum of squared deviations in x, $\Sigma(x_i - \bar{x})^2$ or sum of squared deviations in y, $\Sigma(y_i - \bar{y})^2$, the magnitude of the sum of x and y deviation

TABLE 12.2 Sum of Death Rate and Speed Limit Deviation Cross Products

i	y_i	x_i	$y_i - \bar{y}$	$x_i - \bar{x}$	$(y_i - \bar{y})(x_i - \bar{x})$
1	3.0	55	–1.24	–5	6.2
2	3.3	55	–0.94	–5	4.7
3	3.4	55	–0.84	–5	4.2
4	3.5	70	–0.74	10	–7.4
5	4.1	55	–0.14	–5	0.7
6	4.3	60	0.06	0	0.0
7	4.7	55	0.46	–5	–2.3
8	4.9	60	0.66	0	0.0
9	5.1	60	0.86	0	0.0
10	6.1	75	1.86	15	27.9
Sum =	42.4	600	0.00	00	34.0

products, $\Sigma(x_i - \bar{x})(y_i - \bar{y})$, depends on the sample size. Like the variance, a measure of variability between x and y based on the sum of deviation cross products must involve an adjustment for the degrees of freedom $n - 1$. The **covariance of x and y** is defined as the average variability between x and y; it is represented by the symbol s_{xy} and calculated by the formula

Covariance of x and y
A measure of average cross variability between x and y.

$$s_{xy} = \frac{\sum_{i=1}^{n}(x_i - \bar{x})(y_i - \bar{y})}{n-1} \tag{12.1}$$

For death rates and speed limits, the covariance is $s_{xy} = 34/9 = 3.78$.

Together with the individual variances of x and y, the concept of the covariance between x and y forms the basis for correlation and regression analysis. As already discussed, a regression line is used to visualize changes in the central tendency of y as x changes. It can be defined by the line that connects the conditional means of y at each value of x in a scatterplot. The idea of covariance (and correlation), however, is based on deviations from $\bar{x}$ and $\bar{y}$. The calculation of covariance implicitly assumes that one continuous straight line is drawn through the intersection of $\bar{x}$ and $\bar{y}$ and into the two opposing quadrants containing the bulk of data.

12.6 Correlation

The sign of the covariance gives the direction of the relationship between two variables; its size gives some idea of the strength of that relationship. The actual magnitude of the covariance, however, depends on the units in which x and y are measured. (Suppose death rates had been given in one million vehicle miles instead of 100 million vehicle miles; y values would have been 100 times larger and hence s_{xy} would have been 100 times larger.) This problem is removed by dividing each deviation from the variable's mean by the standard deviation of the variable; that is,

$$\frac{(x_i - \bar{x})}{s_x} \text{ and } \frac{(y_i - \bar{y})}{s_y} \tag{12.2}$$

where

$$s_y = \sqrt{\frac{\sum_{i=1}^{n}(y_i - \bar{y})^2}{n-1}} \qquad s_x = \sqrt{\frac{\sum_{i=1}^{n}(x_i - \bar{x})^2}{n-1}}$$

The covariance based on these standardized variables is now written

$$\left[\frac{1}{n-1}\right]\sum_{i=1}^{n}\left[\frac{x_i - \bar{x}}{s_x}\right]\left[\frac{y_i - \bar{y}}{s_y}\right] = \frac{\sum_{i=1}^{n}(y - \bar{y})(x - \bar{x})}{(n-1)s_x s_y}$$

This standardized covariance is called the sample correlation and is typically denoted by the symbol r. Both s_x and s_y have the square root of $n - 1$ in their denominators, which will cancel with the $n - 1$ multiplicative factor in the denominator of r. Thus, the **coefficient of correlation** can now be written

Coefficient of Correlation
A standardized measure of the strength of the linear relationship between two variables.

$$r = \frac{\sum_{i=1}^{n}(x_i - \bar{x})(y_i - \bar{y})}{\sqrt{\sum_{i=1}^{n}(x_i - \bar{x})^2 \sum_{i=1}^{n}(y_i - \bar{y})^2}} \tag{12.3}$$

Table 12.2 provides the value for the numerator of r, for the speed limit and death data. Table 12.3 has the appropriate sum of squared deviations for the denominator. Substituting these quantities into 12.3 gives

$$r = \frac{34}{\sqrt{(450)(8.544)}} = \frac{34}{\sqrt{3844.8}} = \frac{34}{62} = .0548$$

An appreciation of the sign and magnitude of the correlation coefficient can be gained from Figure 12.4. In panel a, where there is a perfect positive relationship, the sum of products of x and y deviations from their respective means is always positive. In addition, $\Sigma(x - \bar{x})(y - \bar{y})$ is at its maximum and, hence, r is at its maximum. In panel b, the products of x and y deviations are always negative and r is at its minimum. As will be shown in the next section, the maximum value of r is +1 and its minimum is –1.

An r value of 0 indicates no straight line relationship between x and y, because the sum of positive products of x and y deviations are offset by the sum of negative products, panel c. Notice in panel d, however, that an r of 0 need not imply a lack of a relationship between x and y; an $r = 0$ only indicates no straight line association.

Correlation of x and y
A standardized measure of the average cross variability between x and y.

For r values between 0 and +1, and for r values between –1 and 0, some **correlation between x and y** is indicated, panels e and f. The closer an r value is to +1, the stronger the positive relationship between x and y; the closer to –1, the stronger the negative relationship. An r value of +0.548 indicates some positive correlation between x and y.

TABLE 12.3 Sum of Squared Deviations in Death Rates and Speed Limits

i	y_i	x_i	$y_i - \bar{y}$	$x_i - \bar{x}$	$(y_i - \bar{y})^2$	$(x_i - \bar{x})^2$
1	3.0	55	–1.24	–5	1.5376	25
2	3.3	55	–0.94	–5	0.8836	25
3	3.4	55	–0.84	–5	0.7056	25
4	3.5	70	–0.74	10	0.5476	100
5	4.1	55	–0.14	–5	0.0196	25
6	4.3	60	0.06	0	0.0036	0
7	4.7	55	0.46	–5	0.2116	25
8	4.9	60	0.66	0	0.4356	0
9	5.1	60	0.86	0	0.7396	0
10	6.1	75	1.86	15	3.4596	225
Sum =	42.4	600	0.00	00	8.5440	450

FIGURE 12.4 Scatterplots of Different Correlation Values

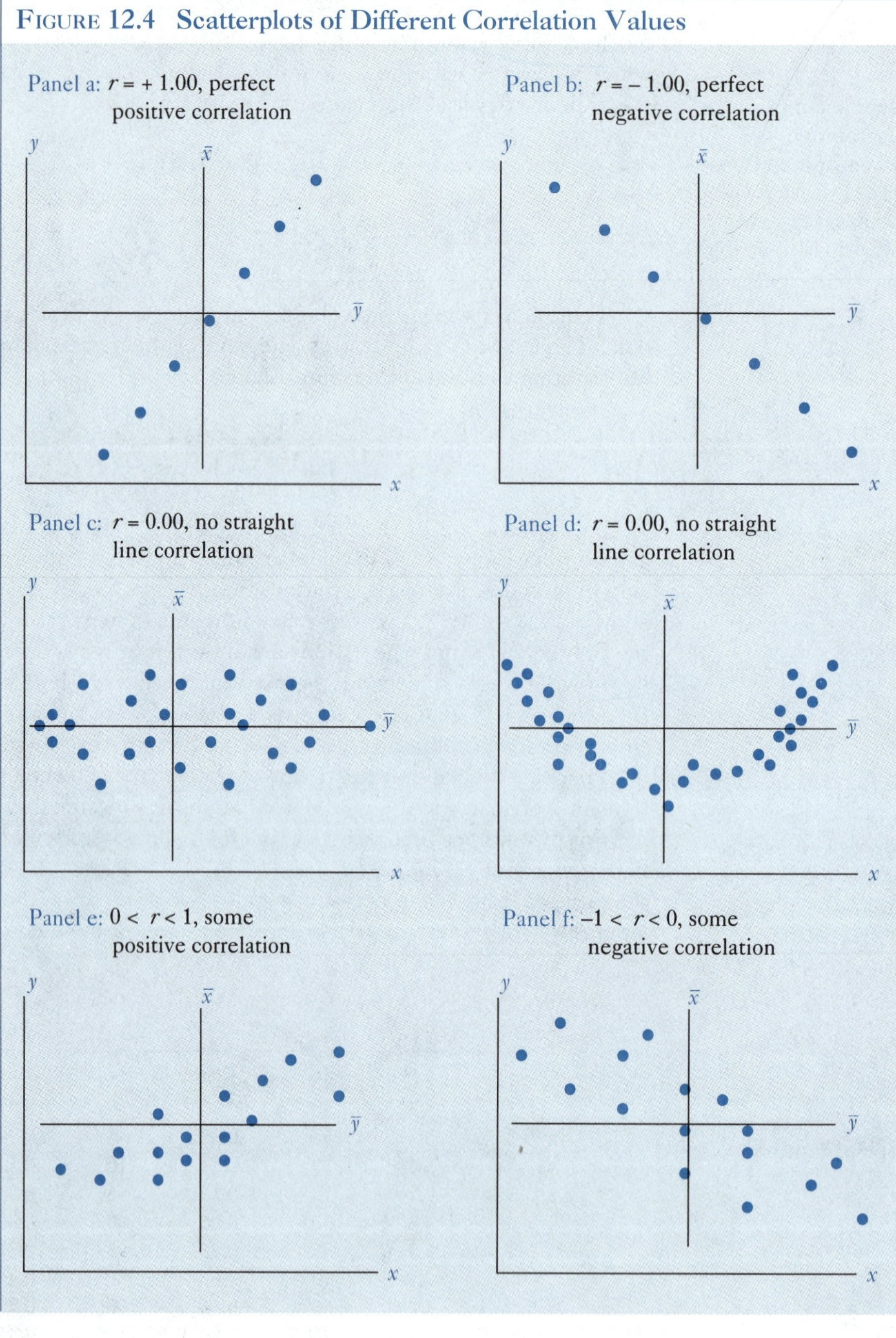

Although this correlation between deaths and speed is not perfect, it is not negative, as suggested by Rivkin in his *NYT* letter to the editor.

In closing this discussion of correlation, note that correlation does not imply causality. Although we call y (death rate) the dependent variable and x (speed limit) the independent variable, and conclude that in the sample of 10 observations there is a positive relationship between y and x, we did not prove that x causes y. First, we are working only with sample data, and we have not provided any means to assess what could happen in other samples. Second, we have not ruled out the possibility that y causes changes in x. It is possible that as highway death rates fall (or rise) countries increase (or decrease) their speed limits. Interestingly, later in his letter to the editor, Rivkin asserted that in Britain the Chiefs of Police Association petitioned the government to raise the speed limit from 70 mph to 80 mph, because highway death rates were relatively low. If other countries behaved in a similar manner, then it could be argued that lower highway death rates would lead to higher speed limits. Issues of correlation and causality will be addressed in later chapters.

QUERY 12.2:

For the *WSJ* data on outstanding credit card balances and interest rates charged by the 10 banks in Query 12.1, what is the sample variance for each variable, the covariance between the variables, and the correlation coefficient? (In your answer, relate this correlation to the scatterplot in Query 12.1.)

ANSWER:

We will need the following sums.

Balances x	Interest y	$x-\bar{x}$	$y-\bar{y}$	$(x-\bar{x})^2$	$(y-\bar{y})^2$	$(x-\bar{x})(y-\bar{y})$
9.10	19.8	5.89	0.81	34.6921	0.6561	4.7709
5.30	19.8	2.09	0.81	4.3681	0.6561	1.6929
4.50	17.5	1.29	–1.49	1.6641	2.2201	–1.9221
3.30	19.8	0.09	0.81	0.0081	0.6561	0.0729
2.50	17.8	–0.71	–1.19	0.5041	1.4161	0.8449
2.00	17.9	–1.21	–1.09	1.4641	1.1881	1.3189
1.94	20.0	–1.27	1.01	1.6129	1.0201	–1.2827
1.27	19.8	–1.94	0.81	3.7636	0.6561	–1.5714
1.20	19.8	–2.01	0.81	4.0401	0.6561	–1.6281
0.99	17.7	–2.22	–1.29	4.9284	1.6641	2.8638
32.10	189.9	0	0	57.0456	10.7890	5.1600

The variances are

$$s_y^2 = \frac{\sum_{i=1}^{n}\left(y_i - \bar{y}\right)^2}{n-1} = \frac{10.7890}{9} = 1.1988$$

$$s_x^2 = \frac{\sum_{i=1}^{n}\left(x_i - \bar{x}\right)^2}{n-1} = \frac{57.0456}{9} = 6.3384$$

The covariance is

$$s_{xy} = \frac{\sum_{i=1}^{n}(x_i - \bar{x})(y_i - \bar{y})}{n-1} = \frac{5.16}{9} = 0.5733$$

The coefficient of correlation is

$$r = \frac{\sum_{i=1}^{n}(x_i - \bar{x})(y_i - \bar{y})}{\sqrt{\sum_{i=1}^{n}(x_i - \bar{x})^2 \sum_{i=1}^{n}(y_i - \bar{y})^2}} = \frac{5.16}{\sqrt{(57.0456)(10.789)}} = 0.21$$

This correlation coefficient of +0.21 indicates a weak positive relationship between x and y. This is consistent with the observation that only six of the 10 points lie in quadrants I and III in the scatterplot of Query 12.1. For these six points, deviations from the interest rate mean are in the same direction as deviations from the mean of outstanding balances. The sum of these six positive products is slightly larger than the sum of the four negative products associated with the four points in quadrants II and IV.

EXERCISES

12.1 Answer true or false; if false, state why.

a. The sum of deviations around the mean is always zero.
b. The calculation of the sample variance involves a division by $n - 1$, instead of n, because there are only $n - 1$ deviations.
c. The variance is a measure of the average squared deviation of values around their mean.
d. If the x and y covariance is negative, then the standard deviation of either x or y must be negative.
e. If the correlation coefficient is negative, then the covariance is negative.
f. The bigger the coefficient of correlation, the stronger the relationship between two variables. Thus, $r = +0.75$ shows a stronger relationship than $r = -0.85$.
g. A positive correlation coefficient indicates that as values of one variable rise, values of the other tend to rise, and as values of the one variable fall, values of the other tend to fall.
h. A variable is "standardized" by dividing deviations from its mean by the standard deviation.
i. A negative covariance indicates that for negative deviations from the mean of one variable there tend to be negative deviations from the mean of the other variable.
j. If the covariance is zero, then we know that the coefficient of correlation is zero and that there is no relationship between the variables.
k. If the variance of one variable is zero, then the correlation coefficient is infinite because its denominator is zero.
l. If the correlation coefficient is zero, then all the y values must be equal to $\bar{y}$.

12.2 What is the value of the sum of squared errors if $r = +1$, if $r = -1$, and if $r = 0$?

12.3 Does the magnitude of r^2 or r give any information about the number or percentage of sample points above, below, or on the least squares regression line?

12.4 In an experimental executive training program, 60 trainees were chosen at random and given a test before the program began (pre) and were then given the test at the middle of the program (mid). The scatterplot of their scores is as follows:

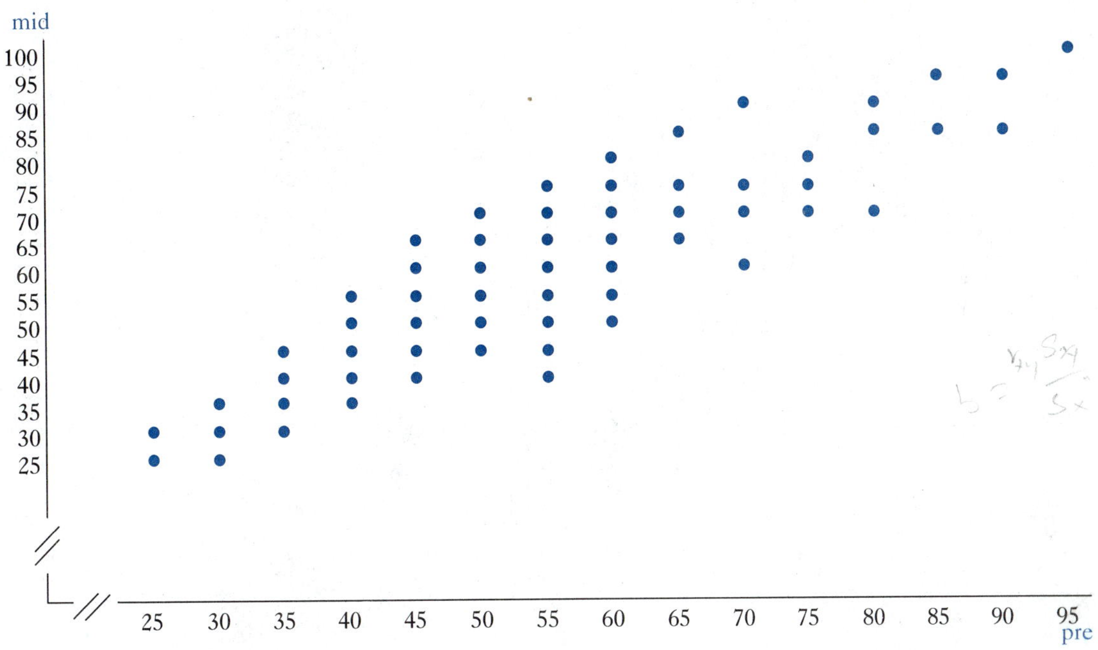

a. On the pretest and midtest frequency distributions on the next page, those who scored a 95, 90, and 85 on the pretest are matched with their midtest score, as obtained from the scatterplot. Complete this matching for the remaining 55 trainees.

b. Calculate the mean of the pretest score.

c. Calculate the change in score (change score equals midtest score minus pretest score) for each of the 60 trainees.

d. Calculate the average change score for those who scored below the average on the pretest.

e. Calculate the average change score for those who scored above the average on the pretest.

f. If you now learn that those who scored below the average on the pretest received special attention and assistance, are you willing to conclude that this special effort resulted in the larger gain scores for those below the average? If, in the second part of the course, all trainees received special attention and assistance, would you expect the change in scores from the middle of the program to the end of the program to be approximately equal for those below and above the average of the midtest? Explain why or why not.

Pretest Score	Frequency	Percent	Cumulative Frequency	Cumulative Percent
25	2	3.33	2	3.33
30	3	5.00	5	8.33
35	4	6.67	9	15.00
40	5	8.33	14	23.33
45	6	10.00	20	33.33
50	6	10.00	26	43.33
55	8	13.33	34	56.67
60	7	11.67	41	68.33
65	4	6.67	45	75.00
70	4	6.67	49	81.67
75	3	5.00	52	86.67
80	3	5.00	55	91.67
85	2	3.33	57	95.00
90	2	3.33	59	98.33
95	1	1.67	60	100.00

Midtest Score	Frequency	Percent	Cumulative Frequency	Cumulative Percent
25	2	3.33	2	3.33
30	3	5.00	5	8.33
35	3	5.00	8	13.33
40	4	6.67	12	20.00
45	5	8.33	17	28.33
50	5	8.33	22	36.67
55	5	8.33	27	45.00
60	5	8.33	32	53.33
65	5	8.33	37	61.67
70	7	11.67	44	73.33
75	5	8.33	49	81.67
80	2	3.33	51	85.00
85	4	6.67	55	91.67
90	2	3.33	57	95.00
95	2	3.33	59	98.33
100	1	1.67	60	100.00

12.5 For the following data set, construct a scatterplot, show the lines representing the x and y means, and calculate the 10 deviations from these means. Without doing any other calculations, do x and y appear to be related? Why or why not?

y:	2	−3	7	8	−2	−5	−7	4	−6	2
x:	0	4	−5	−7	8	9	9	−3	5	−1

12.6 Draw in the lines representing the average pretest and midtest scores in the scatterplot given in Exercise 12.4. How do deviations from the means of the pretest and midtests tend to be related?

12.7 The following summary statistics were calculated from the data in Exercise 12.4, where y is the midtest score and x is the pretest score:

$$\Sigma(x_i - \bar{x})^2 = 17490 \quad \Sigma(\bar{y}_i^2 - y) = 21471.25 \quad r = 0.8881$$

a. What is the standard deviation of x?
b. What is the standard deviation of y?
c. What is the value of the covariance of x and y?

12.8 Most statistical computer programs run best on microcomputers that have fast disk drives (low disk seeking and accessing time to retrieve and store data) and fast floating point calculation time (low time requirements to complete calculations). A microcomputer sales representative states that since these two items are highly correlated, and since floating point calculations are hard to time, one need only look at the disk seeking and access time to decide on appropriate machines for running statistical programs. *PC Magazine* (September 29, 1987) provided information on 12 different 80386 based microcomputers. Does this information support the sales rep's claim?

	Floating Point Calculation	Disk Seek
PC Discount Noble 383	17.4	110.4
CCC 386c	15.5	45.2
IBM PS/2 Model 80	15.6	33.6
ALR 386/2	17.1	32.7
Acer 1100	14.9	32.3
CCI ST 386	17.4	32.7
Tandy 4000	17.1	30.9
Compaq Deskpro 386	15.5	29.2
Kaypro 386	17.3	28.8
Laser Digital Pacer 386	17.3	26.4
Corvus 386	17.3	26.1
PC's Limited 386-16	13.4	19.0

12.9 For the following data set, calculate the coefficient of correlation and draw the scatterplot.

i	y	x
1	2	–2
2	5	0
3	4	–1
4	7	2
5	5	1

12.10 For the following data set, calculate the coefficient of correlation and draw the scatterplot.

i	y	x
1	–1	5
2	2	0
3	0	3
4	3	–2
5	6	–6

12.11 Without doing any calculations, determine the covariance and coefficient of correlation for the following two scatterplots.

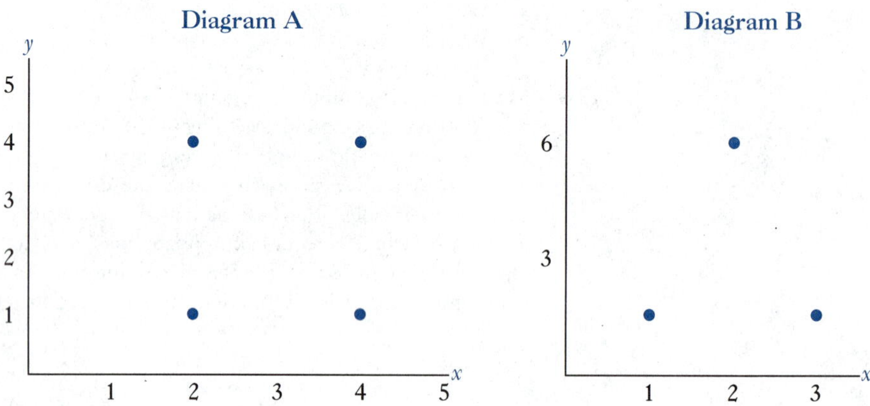

12.12 Based on the pretest and midtest scores in 12.4, suppose that only the trainees with pretest scores exceeding the average were selected for executive positions (unknown to the trainees until after the program was completed). Calculate the coefficient of correlation between the pretest and midtest scores for those selected for executive positions and those not selected. Also calculate the coefficient of correlation for the total group of 60 trainees. Are the two correlation coefficients for the subsamples greater or less than the correlation for the entire sample? If so, why; if not, why not?

12.13 Below is a bivariate frequency distribution that shows the number of times each x and y pair of observations occurred. For example, the first entry indicates that $x = 11$ and $y = 18$ occurred five times.

a. Graph the scatterplot represented by these data.
b. Calculate the variance of both x and y.
c. Calculate the covariance of x and y.
d. Calculate the coefficient of correlation.

	y									
x	18	19	20	21	22	23	24	25	26	Total
11	5	2	1	1	1	1	0	0	0	11
13	1	3	2	1	0	2	1	0	0	10
15	1	1	2	3	0	1	1	0	0	9
17	0	2	3	1	1	0	1	0	0	8
19	0	0	3	2	1	1	0	0	0	7
21	0	0	0	0	2	1	0	1	1	5

12.14 On the following page are the results of a city council race in a small town in Indiana. Is there a relationship between the number of votes obtained by candidates in the primary and the general election?

Candidate	Number of Votes in Primary	General
A	1,346	1,509
B	979	1,125
C	1,378	1,278
D	1,125	1,978
E	874	378
F	1,002	1,200

12.15 From the time of Irving Fisher, economists have been doing empirical studies of the relationship between the nominal rate of interest (i) and the expected rate of change in the general price (p_e), where the difference between these two rates is called the real rate of interest ($r_e = i - p_e$). If investors form price expectations "rationally," then they should be correct, on average, and according to a discussion by Santoni and Stone, "The Fed and the Real Rate of Interest" (Federal Reserve Bank of St. Louis, December 1982), the observed nominal rate of interest at time t should be positively related to observed rate of price inflation at time $t + 1$. While there may be a positive relationship between nominal rate at time t and inflation rate at time $t + 1$, comparing a nominal rate at time t with an inflation rate at time t may be misleading since they need not be related. Use the following Santoni and Stone data to calculate the two coefficients of correlation for the

a. relationship between the nominal rate of interest at time t, and rate of inflation at time $t + 1$.

b. nominal rate of interest at time t and the rate of inflation at time t.

(Note, for comparison purposes, the maximum number of observations is 22, after associating each nominal interest rate at time t with the inflation rate at time $t + 1$.)

Year	90-day T-Bill Rate	Inflation Rate	Year	90-day T-Bill Rate	Inflation Rate
1960	3.2	1.6	1972	4.4	4.2
1961	2.0	0.9	1973	8.7	5.8
1962	2.7	1.8	1974	10.5	8.8
1963	3.2	1.5	1975	5.8	9.3
1964	3.5	1.5	1976	5.0	5.2
1965	4.1	2.2	1977	5.5	5.8
1966	5.1	3.2	1978	7.9	7.4
1967	4.2	3.0	1979	11.2	8.6
1968	5.6	4.4	1980	13.4	9.3
1969	8.2	5.1	1981	16.4	9.4
1970	7.2	5.4	1982	13.3	4.8
1971	4.7	5.0			

12.7 The Regression Line

Knowing the coefficient of correlation does not enable us to predict values of the dependent variable from values of the independent variable. The single straight line that appears to describe the relationship between two variables in a scatterplot, however, could be used for such predictions. The discussion of Figures 12.2 and 12.3, for example, suggests that the line that best describes the relationship between the speed limits and average death rates at each speed goes through the $\bar{y}$ and $\bar{x}$ intersection and into quadrants I and III, illustrating a positive relationship between death rates and the speed limit.

The equation of the straight line drawn in the speed limit and death scatterplot in Figure 12.3 and extended down to the y axis in Figure 12.5 is of the form

$$\hat{y} = a + bx \tag{12.4}$$

where $\hat{y}$ is the predicted dependent variable (death rate) at a specific value of x,
x is the independent variable (speed limit),
a is the y-axis intercept, and
b is the slope.

Once the values of the slope b and intercept a are determined, this line can be used to describe the relationship between deaths and speed and to predict death rates at various speed limits. These values could be determined by visual inspection using a ruler or preferably by an algebraic method.

Predicted Dependent Variable
The value of the variable being explained in a regression.

Interpreting the slope b and y-axis intercept a requires recognition of the subtle difference between the actual death rates observed and those predicted. The symbol y represents the actual value of the dependent variable. The **predicted dependent variable** values are represented by the symbol $\hat{y}$; these values are calculated from Equation (12.4) for the line that "best fits" the scatterplot. Calculation of a $\hat{y}$ is conditional on the x value specified. In Figure 12.5, for example, beginning on the horizontal axis at a speed limit of 70 mph, a vertical line may be drawn to the positively sloped $\hat{y}$ line. Next, a horizontal line may be drawn to the y axis to obtain predicted deaths, which with a highly accurate ruler and a magnifying glass could be read as 4.996 per 100 million vehicle miles at 70 mph.

Slope
For a regression equation, the change in the predicted dependent variable resulting from a one unit change in the independent variable.

The **slope** b is measured as the change in predicted death rate divided by the change in speed limit. For example, at a speed limit of 70 mph, predicted deaths are 4.996 per 100 million vehicle miles. At 75 mph, predicted deaths are approximately 5.374 per 100 million vehicle miles. Thus, the slope is

$$b = \frac{5.374 - 4.996}{75 - 70} = 0.0756$$

Similar calculations between any other points on the line would yield a slope of 0.0756, because the line is straight. This slope indicates that for a one mile per hour increase in the speed limit, predicted deaths rise by 0.0756 deaths per 100 million vehicle miles. For a five mile per hour increase, predicted deaths rise by 0.378 (= 0.0756 times 5) per 100 million vehicle miles. Contrary to Rivkin's claim about the data in Table 12.1, the speed limit appears to have something to do with safety; as the speed limit increases, predicted deaths rise.

The y-axis intercept (which is also called the "constant term") is the point where the line crosses the vertical axis; that is,

$$a = -0.293$$

This point corresponds to a speed limit of zero miles per hour and a predicted death rate of negative 0.293 deaths per 100 million vehicle miles. A negative death rate is im-

FIGURE 12.5 Death Rates and Speed Limit Relationship

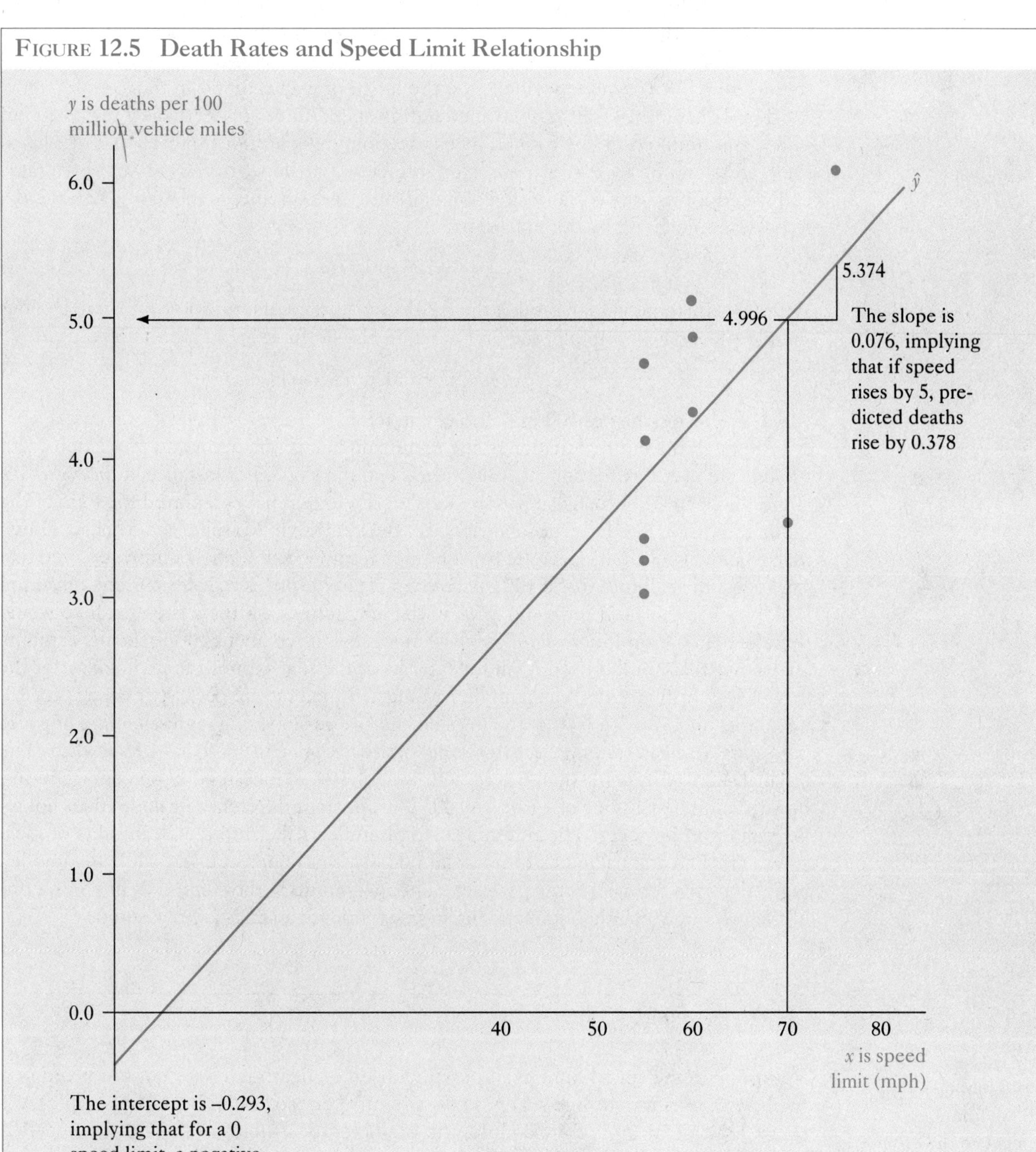

possible; here the y intercept is an artifact of fitting a straight line to a sample with only relatively high values. As will be discussed later, the y intercept may have little practical meaning if it is a value well outside the range of values found in the sample.

The equation for the straight line relating speed limits to deaths is a regression of y on x. But, unlike the line formed by connecting the conditional means of y, at each x, the regression line $\hat{y} = a + bx$ has a fixed intercept and slope. In the case of death rates and speed limits, our visual inspection and ruler measurements in Figure 12.5 yield a regression line given by the equation

$$\hat{y} = -0.293 + 0.076x \tag{12.5}$$

With this equation, we can predict the deaths for any speed limit within the range of our sample values. For example, the predicted death rate for a speed limit of 55 mph is

$$\hat{y} = -0.293 + 0.076(55) = 3.862$$

which is 3.86 deaths per 100 million vehicle miles.

Although the entire equation of the regression line is important and predictions with the line are useful, attention typically focuses on the slope. For instance, a *Business Week* (June 18, 1990) article on cigarette smokers' desire to quit never reported the form of the entire regression line fit by economists Gary Becker, Kevin Murphy, and Michael Grossman to annual per capita cigarette consumption from 1955 to 1985. It simply reported that "the economists found that a 10% increase in cigarette prices reduces current consumption by 4% in the short term and 7.5% in the long term . . . a 16¢ a pack tax hike would therefore cut consumption significantly. Americans puffed about 100 packs per capita in 1989—but that number would fall to 96 packs in the first year of the tax hike . . . at the end of four or five years, consumption would fall to about 92 packs per capita."(p. 20) From this quote we know that the implied short-term slope of a regression line of packs per capita on cigarette price, at a 160-cents-per-pack price in 1989, is $-4/16 = -0.25$. The original short-term regression, where both variables were measured in percentage terms, however, implied a slope of $-4/10 = -0.40$. The apparent difference in slopes does not reflect an error, however. The difference simply reflects the differences in units of measurement. The original regression was defined in terms of percentages, while the implied regression is in terms of cents per pack. The magnitude of the slope is defined in terms of the units in which the independent and dependent variables are measured.

12.8 Errors in Predictions

Prediction Error
The difference between the observed value of the dependent variable and the value predicted by the regression line.

Returning to the speed limit example, the observed death rates are not predicted perfectly with our straight line regression at the different speed limits in the sample. At a speed limit of 55 mph, for example, the predicted death rate is 3.862 deaths per 100 million vehicle miles while the observed deaths are 4.7, 4.1, 3.4, 3.3, and 3.0 deaths per 100 million vehicle miles. The **prediction errors** for these five observations are the differences between the observed death rates and the predicted death rate. They are calculated by subtracting the predicted death rate from the observed death rates.

for Japan, the error is 4.7 – 3.862 = 0.838

for Denmark, the error is 4.1 – 3.862 = 0.238

for Finland, the error is 3.4 – 3.862 = –0.462

for the United States, the error is 3.3 – 3.862 = –0.562

for Norway, the error is 3.0 – 3.862 = –0.862

At a speed of 60 mph, the predicted death rate is 4.24 deaths per 100 million vehicle miles. That is,

$$\hat{y} = -0.293 + 0.076(60) = 4.24$$

Again, there is variability in observed death rates around this predicted value. The three errors made in predicting death rates at 60 mph are

for the Netherlands, the error is 5.1 – 4.24 = 0.860

for Australia, the error is 4.9 – 4.24 = 0.660

for Canada, the error is 4.3 – 4.24 = 0.060

Similarly, the remaining two errors in predicting death rates at 70 and 75 mph can be calculated:

for Britain, the error is 3.5 – 4.996 = –1.496

for Italy, the error is 6.1 – 5.37 = 0.727

Although there is only one reported death rate at both 70 mph and 75 mph, our straight line regression suggests variability even at these speed limits as reflected in the difference between the actual and predicted death rates.

Residual
A prediction error measured as the vertical difference between the observed value of the dependent variable and the value predicted with the regression equation.

The errors made in predicting the sample values of the dependent variable reflect variability in observed values across all the values of the independent variable. These errors reflect the amount of y that is "left over" or "left unexplained" by the regression line; they are thus also called **residuals**. Each error or residual value is represented graphically as the vertical distance between the observed sample point and the y value predicted by the regression line at the specified x value. The 10 residuals for the straight line death rate regression are depicted in Figure 12.6.

12.9 The Method of Least Squares

The graphical method of fitting a straight line to a scatterplot is not very precise because it relies on the researcher's ability to eyeball the data and accurately read a rule. For a large data set, it would be extremely difficult, if not impossible, to use. A more reliable method of fitting a line is to use formulas that minimize the sum of the squares of the errors (or residuals). To derive these formulas, it is first necessary to introduce some notation.

FIGURE 12.6 Residuals for Straight Line Death Rate Regression as Drawn in Figure 12.5

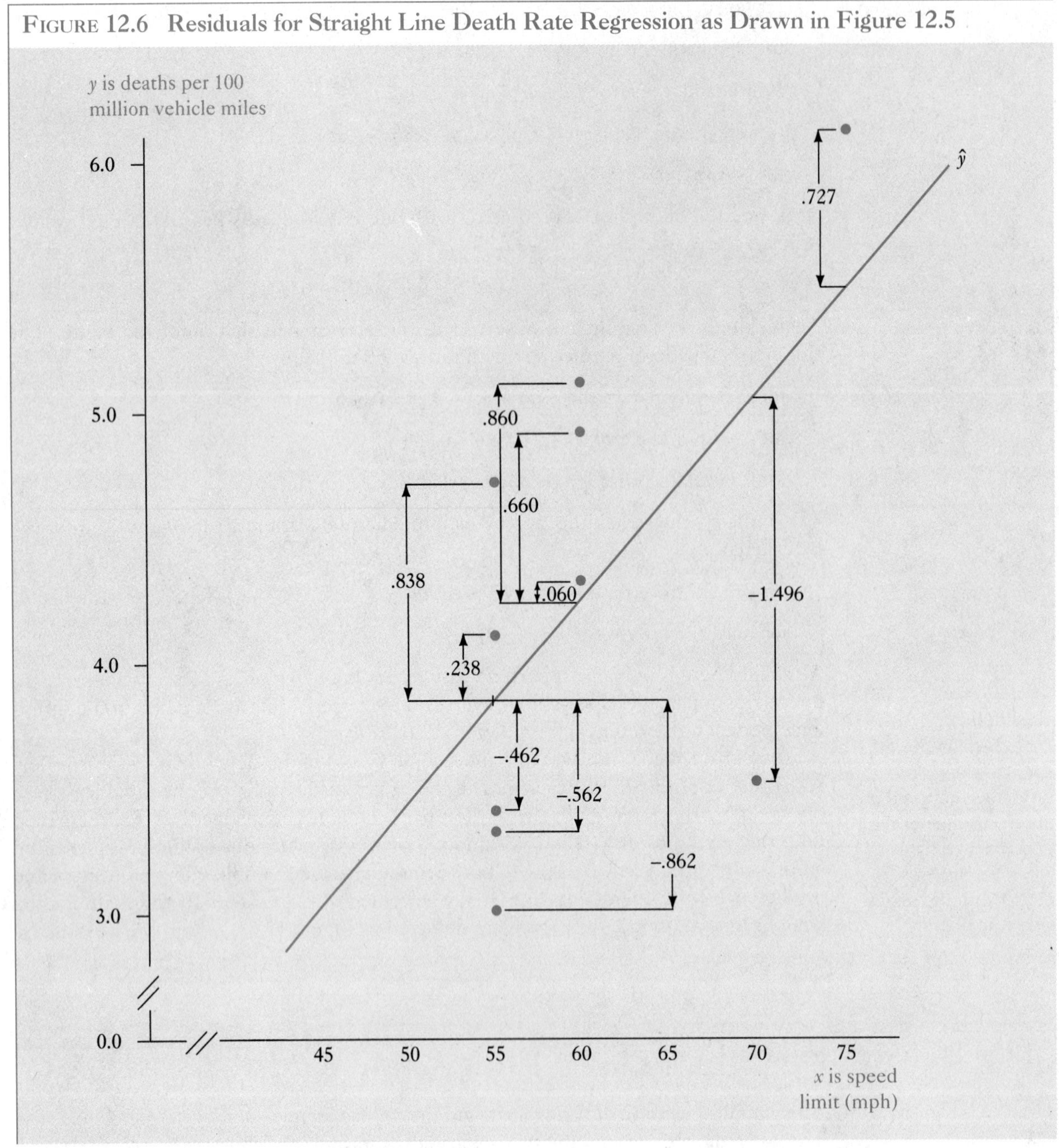

Let the residuals or prediction errors associated with the regression line be denoted by the letter *e*. The subscript *i* is used to identify a residual associated with a given point; e_i is the error for the *i*th observation on *y*.

$$e_i = y_i - \hat{y} \tag{12.6}$$

For example, in Table 12.1, Norway is the first country listed in the sample; the residual for Norway was previously calculated as

$$e_1 = y_1 - \hat{y}_1 = -0.862$$

Italy is the tenth country listed in the sample; its error was previously calculated as

$$e_{10} = y_{10} - \hat{y}_{10} = 0.727$$

and similarly for the other eight countries. The 10 residuals calculated in this way will always sum to approximately zero. Substituting the values from the speed limit case gives

$$\sum_{i=1}^{10} e_i = -0.862 - 0.562 - 0.462 - 1.496 + 0.238$$

$$+ 0.060 + 0.838 + 0.660 + 0.860 + 0.727 \approx 0$$

If we could read values from a graph with great precision, then an equation that is accurate to many decimal places could be obtained. With such accuracy, the sum of negative residuals would exactly cancel the sum of positive residuals, and the sum of residuals would be exactly zero. All properly drawn regression lines have the property that *the residuals sum to approximately zero.*

$$\sum_{i=1}^{n} e_i = e_1 + e_2 + e_3 + \ldots + e_{n-1} + e_n = 0 \tag{12.7}$$

$$= y_1 - \hat{y}_1 + y_2 - \hat{y}_2 + \ldots + y_{n-1} - \hat{y}_{n-1} + y_n - \hat{y}_n = 0$$

Because there are other lines about which deviations from the actual observed value of *y* sum to zero, this is not a unique property of the regression line. **The unique property of the regression line is that it minimizes the sum of the squares of the errors (or residuals) with respect to the slope and intercept.** To see this, the residuals must be expressed as a function of the intercept and slope, which can be done by writing $\hat{y}$ in its general slope-intercept form; that is,

$$e_i = y_i - \hat{y}_i = y_i - (a + bx_i) = y_i - a - bx_i \tag{12.8}$$

where the *i* subscript has been added to *x*, *y*, and $\hat{y}$ to identify the *i*th paired observations and calculated value. The sum of the squares of the residuals is now written

$$\sum_{i=1}^{n} e_i^2 = \sum_{i=1}^{n} \left(y_i - \hat{y}_i\right)^2 = \sum_{i=1}^{n} \left(y_i - a - bx_i\right)^2 \tag{12.9}$$

The necessary mathematical condition for minimizing the sum of the squared errors is that the partial derivatives of Σe_i^2, with respect to *b* and *a*, are both zero. The equations for *b* and *a* derived from this calculus are:[1]

$$b = \frac{\sum_{i=1}^{n}(x_i - \bar{x})(y_i - \bar{y})}{\sum_{i=1}^{n}(x_i - \bar{x})^2} \tag{12.10}$$

$$a = \bar{y} - b\bar{x} \tag{12.11}$$

As with the numerator of the coefficient of correlation, the numerator of b reflects how y deviates from its mean as x deviates from its mean. It captures the quadrant position of points in Figure 12.2. For example, for the first x and y pair of values in Table 12.1 (Norway),

$$x_1 - \bar{x} = -5 \text{ and } y_1 - \bar{y} = -1.24$$

Since both x and y deviate in the same direction from their respective means, the relationship between x and y for this pair of values is positive (i.e., –5 times –1.24 = + 6.2). For the fourth x and y pair (Britain, also shown in quadrant IV, Figure 12.2),

$$x_4 - \bar{x} = +10, \text{ and } y_4 - \bar{y} = -0.74$$

indicating a negative relationship between x and y (i.e., +10 times –0.74 = –7.4). In Figure 12.2, all x and y pairs in quadrants I and III are positive, while all x and y pairs in quadrants II and IV are negative.

The sum of the products of x and y deviation determines the sign and the magnitude of the numerator of b. As shown in column six of Table 12.4, the sum of the 10 $(x_i - \bar{x})$ and $(y_i - \bar{y})$ products is positive.

$$\sum_{i=1}^{n}(x_i - \bar{x})(y_i - \bar{y}) = 6.2 + \ldots + 27.9 = 34$$

The denominator of b reflects the variability in x around its mean. It is always positive because each $x - \bar{x}$ deviation is squared and is simply the sum of squared deviations used in the calculation of the variance of x introduced in Chapter 2. As shown in column seven of Table 12.4, the sum of the 10 $(x_i - \bar{x})^2$ values is 450.

$$\sum_{i=1}^{n}(x_i - \bar{x})^2 = (-5)^2 + \ldots + (15)^2 = 450$$

The slope of the least squares death rate regression line is

$$b = \frac{\sum_{i=1}^{n}(x_i - \bar{x})(y_i - \bar{y})}{\sum_{i=1}^{n}(x_i - \bar{x})^2} = \frac{34}{450} = 0.075556$$

The y intercept is

$$a = \bar{y} - b\bar{x} = 4.24 - (0.075556)(60) = -0.29333$$

The least squares regression line is $\hat{y}_i = -0.2933 + 0.0756x_i$, which is a more exact version of the equation we found by visual inspection of the scatterplot and ruler measurements. It is the straight line that best fits the sample data, after rounding coefficient estimates to the nearest ten thousands.

TABLE 12.4 Calculation of the Regression Slope and Intercept

i	x_i	y_i	$x_i - \bar{x}$	$y_i - \bar{y}$	$(x_i - \bar{x})(y_i - \bar{y})$	$(x_i - \bar{x})^2$
1	55	3.0	–5	–1.24	6.2	25
2	55	3.3	–5	–0.94	4.7	25
3	55	3.4	–5	–0.84	4.2	25
4	70	3.5	10	–0.74	–7.4	100
5	55	4.1	–5	–0.14	0.7	25
6	60	4.3	0	0.06	0.0	0
7	55	4.7	–5	0.46	–2.3	25
8	60	4.9	0	0.66	0.0	0
9	60	5.1	0	0.86	0.0	0
10	75	6.1	15	1.86	27.9	225
	600	42.4			34.0	450

Means: $\bar{y} = 42.4/10 = 4.24$ and $\bar{x} = 600/10 = 60$

Slope: $$b = \frac{\sum_{i=1}^{n}(x_i - \bar{x})(y_i - \bar{y})}{\sum_{i=1}^{n}(x_i - \bar{x})^2} = \frac{34}{450} = 0.07556$$

Intercept: $a = \bar{y} - b\bar{x} = 4.24 - (0.07556)(60) = -0.2933$

The least squares regression equation is

$$\hat{y}_i = -0.2933 + 0.0756x_i$$

QUERY 12.3:

In Query 12.1, a *WSJ* article was cited as saying that banks with smaller outstanding credit card balances charge lower interest rates. What is the least squares regression that is implied by this statement, using the data from the 10 banks given in Query 12.1? In your answer, interpret the slope and graph your regression line along with the respective dependent and independent variable means in a scatterplot.

ANSWER:

The following sums are needed to obtain the regression line.

Balances x	Interest y	$x - \bar{x}$	$y - \bar{y}$	$(x - \bar{x})^2$	$(x - \bar{x})(y - \bar{y})$
9.10	19.8	5.89	0.81	34.6921	4.7709
5.30	19.8	2.09	0.81	4.3681	1.6929
4.50	17.5	1.29	–1.49	1.6641	–1.9221
3.30	19.8	0.09	0.81	0.0081	0.0729
2.50	17.8	–0.71	–1.19	0.5041	0.8449
2.00	17.9	–1.21	–1.09	1.4641	1.3189
1.94	20.0	–1.27	1.01	1.6129	–1.2827
1.27	19.8	–1.94	0.81	3.7636	–1.5714
1.20	19.8	–2.01	0.81	4.0401	–1.6281
0.99	17.7	–2.22	–1.29	4.9284	2.8638
32.10	189.9	0	0	57.0456	5.1600

The slope is $$b = \frac{\sum_{i=1}^{10}(x_i - \bar{x})(y_i - \bar{y})}{\sum_{i=1}^{10}(x_i - \bar{x})^2} = \frac{5.1600}{57.0456} = 0.090$$

The intercept is $a = \bar{y} - b\bar{x} = 18.99 - 0.090(3.21) = 18.700$
The least squares regression equation is

$$\hat{y}_i = 18.700 + 0.090x_i$$

Since outstanding balances are measured in billions of dollars and interest rates are measured as a percent, the slope of the regression line indicates that for a $1 billion increase in outstanding balances, predicted interest rates rise by 0.090 percentage points.

Query 12.4:

A *Business Week* (June 22, 1987) article argued that as the percentage of women among those holding Master in Business Administration (MBA) degrees has increased, the percentage of women in management and administration jobs has also increased. To support this thesis, *BW* presented the following diagram.

Are the data in this *BW* diagram consistent with the idea that both the percentage of women among graduating MBAs and the percentage of women in management and administrative jobs move together? Find the least squares regression that describes the relationship between the percentage of women among graduating MBAs and the percentage of women in management and administrative jobs for these data. If the percentage of women among graduating MBAs rises to 35 percent, what is the predicted percentage of women in management and administrative jobs?

ANSWER:

Let y be the proportion of women in management and administrative jobs and x be the proportion of women among graduating MBAs. The following sums are needed to obtain the regression line.

Proportion of Women					
MBAs x	Managers y	$x-\bar{x}$	$y-\bar{y}$	$(x-\bar{x})^2$	$(x-\bar{x})(y-\bar{y})$
0.04	0.20	–0.123	–0.070	0.015129	0.00861
0.12	0.24	–0.043	–0.030	0.001849	0.00129
0.33	0.37	0.167	0.100	0.027889	0.01670
0.49	0.81			0.044867	0.02660

The slope b and intercept a are

$$b = \frac{\sum_{i=1}^{3}(x_i - \bar{x})(y_i - \bar{y})}{\sum_{i=1}^{3}(x_i - \bar{x})^2} = \frac{0.02660}{0.044867} = 0.592868$$

$$a = \bar{y} - b\bar{x} = 0.27 - 0.592868(0.1633) = 0.173165$$

The least squares regression equation is

$$\hat{y}_i = 0.1732 + 0.5929x_i$$

If the percentage of women among graduating MBAs rises to 35 percent, the predicted percentage of women in management and administrative jobs rises to approximately 38 percent, because

$$\hat{y}_i = 0.1732 + 0.5929(0.35) = 0.3807$$

QUERY 12.5:

In finance, the "historical beta" is the slope of a regression line between a single stock's return and the return from all stocks. This beta (slope) is a measure of the riskiness of the single stock relative to the rest of the stock market. The higher the beta, the more volatile and risky the stock in question. Below is a scatter diagram of returns for American Telephone and Telegraph stock and for the S&P 500 (the proxy for the stock market) assembled by Burton Malkiel (1990, p. 250) before the divestiture of AT&T. What is the beta for this data and what does it mean?

ANSWER:
The slope of this regression line can be seen to be about 0.5. If the return to all S&P 500 stocks rises (or falls) by 1 percent, the predicted return to AT&T stock will rise (or fall) by half a percent; thus, the AT&T return is less prone to change than the overall market return. If we assume that the future is like the past, then this relationship could be expected to continue. But the breakup of AT&T in 1983 and the deregulation of the communication industry have made predictions about AT&T outside of this sample period meaningless. (More will be said about predicting in later chapters.)

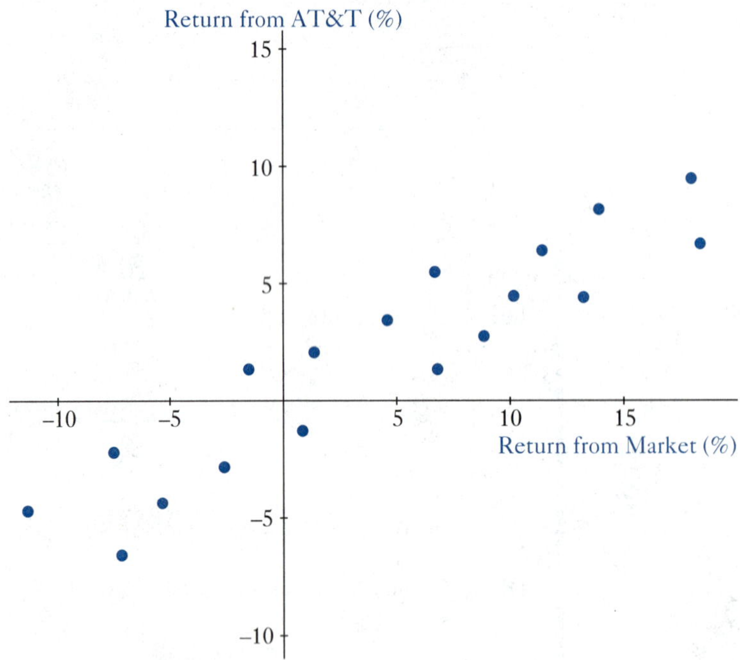

12.10 GOODNESS-OF-FIT AND CORRELATION

Total Sum of Squares (TSS)
The sum of squared differences between observed values of the dependent variable and their mean. It is the variability to be explained

Earlier in this chapter, the coefficient of correlation was introduced as a descriptive measure of the degree of association or covariance between two variables. Here the correlation coefficient is reexamined in the context of least squares regression and the analysis of variance breakdown we considered in Chapter 11.

In an analysis of variance, the measure of total or overall variability in the dependent variable is the sum of squared deviations in observed y_i values around their mean $\bar{y}$ value. This quantity is called the **total sum of squares,** abbreviated TotalSS or just TSS.

$$\text{TSS} = \sum_{i=1}^{n}\left(y_i - \bar{y}\right)^2 \qquad (12.12)$$

This quantity is the numerator of the variance of y. In the case of death rates, it was calculated to be 8.544, in Table 12.3. It represents the total sum of squared errors we would make if we used $\bar{y}$ to predict each y_i value in the sample.

Our regression line should do a better job than $\bar{y}$ in predicting the values of y_i. If it does, there will be less variability in the y_i values around the regression line $\bar{y}_i$ than there is in the y_i values around the mean line $\bar{y}$. How much less depends on how well the regression line fits the sample data.

To measure the fit of the regression line, the total sum of squares is separated into two parts. The first part represents the variability in y_i that is "unexplained" by the regression; that is, the variability in y reflected in the error when $\hat{y}_i$ is used to predict y_i. This unexplained variability is called the **error (or residual) sum of squares, ErrorSS**. ErrorSS (or just ESS) is the expression given earlier in Equation (12.9); it is the quantity that is minimized by the method of least squares.

Error (or residual) Sum of Squares (ESS)
In regression, the sum of squared prediction errors. It reflects the unexplained variability in the dependent variable.

$$\text{ErrorSS} = \sum_{i=1}^{n}\left(y_i - \hat{y}_i\right)^2 \tag{12.13}$$

The second part of the total sum of squares represents the variability in y_i that is "explained" by the regression line. It is what we called the factor sum of squares in Chapter 11. It is the difference between TSS and ESS and is now called the **regression sum of squares, RegSS (or just RSS).**

Regression Sum of Squares (RSS)
The sum of the squared differences between the predicted values of the dependent variable and their mean; also, the difference between TSS and ESS.

$$\text{RegSS} = \text{TSS} - \text{ErrorSS} = \sum_{i=1}^{n}\left(\hat{y}_i - \bar{y}\right)^2 \tag{12.14}$$

The method of least squares minimizes ErrorSS and thus maximizes RegSS. It cannot force ErrorSS to zero, since the regression line typically does not predict all y_i values perfectly.[2]

We have already seen that ErrorSS is calculated by first predicting y for each of the x values in the sample. Errors in prediction are then calculated by subtracting these predicted y values from the observed y values. Squaring the errors and summing gives ErrorSS. In column five of Table 12.5, the variability in death rates unexplained by the speed limit is calculated to be ErrorSS = 5.9751.

Table 12.5 Calculation of the Error or Residual Sum of Squares

i	y_i	$\hat{y}_i$	e_i $y_i - \hat{y}_i$	ErrorSS $(y_i - \hat{y}_i)^2$
1	3.0	3.862	–0.862	.7433
2	3.3	3.862	–0.562	.3160
3	3.4	3.862	–0.462	.2136
4	3.5	4.996	–1.496	2.2365
5	4.1	3.862	0.238	.0565
6	4.3	4.240	0.060	.0036
7	4.7	3.862	0.838	.7019
8	4.9	4.240	0.660	.4356
9	5.1	4.240	0.860	.7396
10	6.1	5.373	0.727	.5281
				5.9751

The regression sum of squares is calculated after TSS and ESS have been obtained. In particular, the regression sum of squares is

$$\text{RegSS} = \text{TSS} - \text{ErrorSS} = 8.544 - 5.975 = 2.569$$

which is the explained variance.

How well the regression line fits the data is measured by the portion of the variability in the dependent variable that is "associated with" or "explained by" the independent variable. For example, in the speed limit case, the proportion of total variability in deaths explained by the regression is

$$\frac{\text{RegSS}}{\text{TSS}} = \frac{2.569}{8.544} = 0.301$$

This indicates that 30.1 percent of the variability in deaths is explained by the speed limit in Rivkin's sample and 69.9 percent is unexplained. Denoted by r^2, the coefficient of determination measures the goodness-of-fit of the regression line.

$$r^2 = \frac{\text{RegSS}}{\text{TSS}} \tag{12.15}$$

Coefficient of Determination r²
A number between 0 and 1 that indicates the proportion of variability in the dependent variable explained by the independent variable. It is the square of the coefficient of correlation.

Ranging from 0 to 1, the **coefficient of determination, r^2**, indicates the proportion of variability in the dependent variable that is explained by the independent variable. If $r^2 = 1.00$, then 100 percent of the variability in y is explained by the regression equation and ErrorSS must be zero; all the y values are predicted perfectly. If $r^2 = 0$, then none of the variability in y is explained by x and ErrorSS = TSS; the $\hat{y}$ regression line and $\bar{y}$ line are one and the same. Since less than 50 percent of the varability in death rates is explained by speed limits, we can conclude that even the best fitting regression equation $\hat{y}_i = -0.2933 + 0.0756x_i$ does not fit the sample data very well and that there is not a good straight line relationship between death rates and speed limits.

Research reports often present the coefficient of correlation along with the regression equation. Symbolized by r, the coefficient of correlation is calculated by taking the square root of the coefficient of determination.

$$r = \sqrt{\frac{\text{RegSS}}{\text{TSS}}} \tag{12.16}$$

As discussed when the correlation coefficient was introduced in Equation (12.3), r is a measure of standardized covariance between y and x; it also measures the correlation between y and $\hat{y}$. The sign of r is positive when the regression line's slope is positive, and the sign of r is negative when the slope is negative. When ErrorSS = 0, the correlation coefficient has a maximum value of +1.00, and minimum of –1.00. It is zero if ErrorSS = TSS.[3] Our earlier discussion of Equation (12.3) continues to hold for Equation (12.16), only now the correlation coefficient is linked to the least squares regression equation.

QUERY 12.6:
From the information in Query 12.4 concerning MBA women and jobs, and Equation (12.16), what is the correlation coefficient? Comment on the appropriateness of this regression and predictions from it.

Answer:

For the correlation coefficient, we need the following sums.

y	$\hat{y}$	$y - \hat{y}$	$y - \bar{y}$	$(y - \hat{y})^2$	$(y - \bar{y})^2$
.200	.19688	.00312	−.07	9.74 (10^{-6})	.0049
.240	.24431	−.00431	−.03	18.56 (10^{-6})	.0009
.370	.36881	.00119	.10	1.42 (10^{-6})	.0100
.810	.81000	.00000	.00	29.72 (10^{-6})	.0158

The residual sum of squares is ESS = 2.972 $(10)^{-5}$, which is scientific notation for 2.972 times 1/100000, so ESS = 0.00002972. The total sum of squares is TSS = 0.0158. Thus, the regression sum of squares is RSS = TSS – ESS = 0.01577028, the coefficient of determination is r^2 = RSS/TSS = 0.9981, and the correlation coefficient is $r = 0.9991$, which is extremely high, especially for only three paired observations. Although this correlation coefficient is high, predictions from this regression equation are suspect because there are so few observations.

Linear Extrapolation Predicting values of y based on the slope of the line connecting the two known points and the x axis distance from the closest of these points to the x value for which the prediction is to be made.

Three paired observations on y and x are the smallest number of observations for which a regression can be estimated. Some computer programs give an error message for $n = 3$, but as long as there is at least one degree of freedom ($n - 2 \geq 1$), least squares calculations are feasible. If there were only two paired observations on y and x, however, then the best fitting line would be the one drawn between the two points. A prediction from this line would be done by linear extrapolation. **Linear extrapolation** only requires knowledge of the slope of the line connecting the two known points and the x axis distance from the closest of these to point to the x value for which the prediction is to be made. For instance, if one known point is at $x = 2$ and $y = 2$ and the other known point is at $x = 4$ and $y = 3$, then the linear extrapolation for $x = 5$ is $\hat{y} = 3.5$.

Historical Perspective

The widespread application of regression and correlation analysis attests to their usefulness. The origins of these techniques are traceable to French mathematician Adrien Marie Legendre (1752–1833) and German mathematician Carl Friedrich Gauss (1777–1855), who each claimed the method of least squares as his own invention. The introduction of these techniques to the social sciences, however, is associated with Sir Francis Galton (1822–1911).

Galton was trained as a physician, but he quickly became bored with the practice of medicine and turned to the study of anthropology for intellectual excitement. In *Natural Inheritance,* (London: Macmillan and Company, 1889), Galton advanced the idea of regression. His "law of universal regression" held that, "Each peculiarity of man is shared by his kinsman, but on the average in a less degree." Below is one of his famous scatter diagrams of average sibling height and mid-parent height. (Mid-parent height is the average of the father's and the mother's heights, where the latter is scaled by 1.08.) For ease in calculation and for intuitive appeal, Galton used many simple approximations: for the average, he employed the median; for dispersion, he used the median deviation from the median rather than the mean squared deviation from the mean; his regression coefficient estimates came directly from graphs, as evidenced by the following diagram, in which he calculated the slope of the regression line to be 2/3.

The single straight line that goes through the paired points in the diagram rises less steeply than the line for which the average height of children equals mid-parent height. This is the regression effect: taller (shorter) parents beget shorter (taller) children, on average. Galton called this phenomenon "regression" or "reversion" to "mediocrity," reflecting his noble upbringing and training as a physician. Galton recognized that knowing how much regression toward the mean there is for a pair of variables enabled prediction—a little regression gave good predictions, and a lot of regression gave poor predictions.

Although initially shunned by mathematicians, Galton's 100-year-old use of the word "regression," as in y regressed on x, has been accepted as an alternative to the word "function," as used in mathematics to express y as a function of x. Galton's spelling and use of the word "co-relation," however, has not survived the test of time. Generations of econometricians, biometricians, and statisticians (such as Francis Ysidro Edgeworth, 1845–1926, and Karl Pearson, 1857–1936, and Pearson's student George Yule, 1871–1951) have advanced regression and correlation analyses as alternative, but related, ways of describing the association between variables. In fact, the "r" used to symbolize Pearson's correlation coefficient was originally used to stand for regression.

What stands out in tracing the development of the least squares method, regression, and correlation is that the early work was driven more by observation and geometric demonstrations than by theory and algebraic presentations. Galton and his associates collected thousands of empirical examples of the regression effect in nature. Today, as discussed in the next chapter, many researchers, in their use of regression analysis, emphasize theory, and its algebraic representation in a model, as the engine of inquiry.

Galton's 1889 Scatter Diagram and Regression of Average (median) Height of Children on Mid (average)-Parent Height

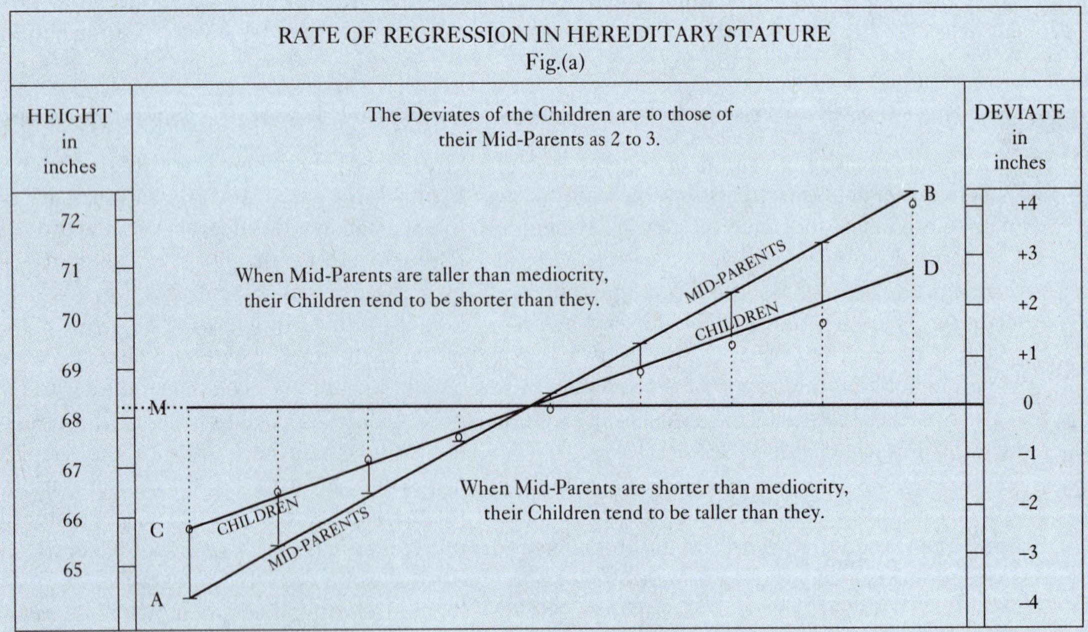

Source: Stephen Stigler (1986, p. 295).

EXERCISES

12.16 Answer true or false; if false, state why.
   a. If the means of x and y are zero, then the y intercept is zero.
   b. The slope of the sample regression line indicates how the actual value of y changes as x changes.
   c. At the mean of x, the predicted value of y is the same value as the sample mean of y.
   d. The residuals from a least squares regression are all zero.
   e. The method of least squares minimizes the residuals.
   f. If the covariance is zero between x and y, then the slope of the least squares regression line is zero.
   g. If the total sum of squares is equal to the residual sum of squares, then the slope of the regression line is zero.
   h. The coefficient of determination indicates the proportion of the total sum of squares "explained" by the regression line.
   i. If the residuals sum to zero, then the y intercept is zero.

12.17 For the data in Exercise 12.5,
   a. calculate the total sum of squares.
   b. calculate the slope and intercept of the regression line.
   c. calculate the 10 residuals.
   d. calculate the residual sum of squares.
   e. why does the total sum of squares exceed the residual sum of squares? Is this a general characteristic of least squares or specific to this data set?

12.18 For the following data set,

y:	7	−5	4	−2	1	−5
x:	−4	6	−2	−1	−3	4

   a. calculate the slope and intercept of the least squares regression line, and write the equation in slope intercept form.
   b. what is unique about this data set and your intercept value?
   c. what does the slope value indicate?

12.19 For the data sets in Exercises 12.9 and 12.10, calculate the respective least squares slopes. How do the slopes relate to the coefficients of correlation that you calculated in Exercises 12.9 and 12.10?

12.20 Use the following information to calculate the slope and intercept of the regression line.

$$\Sigma(x_i - \bar{x})(y_i - \bar{y}) = 655 \qquad \Sigma(x_i - \bar{x})^2 = 1305 \qquad \Sigma(y_i - \bar{y})^2 = 510$$

$$\Sigma x_i = 475 \qquad \Sigma y_i = 405 \qquad n = 8$$

   a. What is the equation of the regression line in slope intercept form?
   b. If $x = 40$, what is the predicted value of y?

12.21 Without doing any calculations, try to determine the slope and intercept for the two diagrams in Exercise 12.11. What problems did you encounter? How does the slope relate to the covariance and coefficient of correlation for these two diagrams?

12.22 Use the following information to determine the slope and intercept of the regression line.

$$\Sigma(y_i - \bar{y})(x_i - \bar{x}) = 1710 \qquad \Sigma(x_i - \bar{x})^2 = 3640$$

$$\Sigma x_i = 1350 \qquad \Sigma y_i = 720 \qquad n = 20$$

12.23 Using the information on variances and covariance calculated in Exercise 12.13, determine the slope of the least squares regression line for the x and y pairs.

12.24 In Exercise 12.14, for each additional vote a candidate receives in the primary, how many additional votes would you predict that he or she will receive in the general election?

12.25 According to the discussion in Exercise 12.15, we should be able to use the interest rate to predict the inflation rate in the next year. Using the data on T-bill rates and price inflation in Exercise 12.15, and a computer package, estimate the least squares regression equation between these two rates. What is your predicted inflation rate in the next year, if the T-bill rate in this year is 7 percent?

12.26 Franklin Fisher (*JASA*, June 1986, p. 278) reports on a regression he estimated in which TV station advertising revenue was the dependent variable and audience size was the independent variable. "The regression passed very close to the origin and had a slope of about $27 per year for each additional household viewing in the average half-hour of prime evening time." He reports that his estimate was attacked as incorrect until an economist at the Federal Communications Commission "took all television revenues in the United States and divided them by the number of television households multiplied by 60 percent, the fraction of prime time that the average household was found in surveys to watch television . . . (and) came up with approximately $27 per household." Demonstrate why it is not surprising that these two estimates are the same.

12.11 Rank Correlation Measures (Optional)

Questionnaires aimed at assessing opinions and attitudes often do not have scales that preserve relative magnitudes. For instance, if a survey asks whether you "generally agree," "agree with provisos," or "generally disagree" with the proposition "A large federal budget deficit has an adverse effect on the economy," then the arbitrary assigning of numerical values to these three alternative responses can only reflect the order of agreement. If "generally agree" is assigned a 3, and "agree with provisos" is assigned a 2, then "generally disagree" could be assigned any value less than 2. The number 1 seems the most natural value to be assigned to "generally disagree," but this is an arbitrary value. Any value below 2 can be assigned to "generally disagree," because the level of agreement is not a measure with a fixed starting point and a ratio preserving scale. That is, it is not a cardinal measure. A movement from "generally agree (3)" to "agree with provisos (2)" is not a 33.33 percent reduction in agreement; the movement from the numerical magnitude 3 to 2 simply reflects a decrease in agreement.

The question arises, then, as to how the responses given to a multiple-item questionnaire can be compared between groups of responders. For example, in a survey consisting of many questions about the economy, how can the correlation of responses between journalists and economists be assessed?

By design, nonparametric methods of correlation measure how strong the association is between two sets of ranked data. The most common method of rank correlation is Spearman's rho. It is the nonparametric alternative to the simple coefficient of correlation presented in the previous section.

Let d_i represent the difference between the numerical rank order of the *i*th item in two different sets of data, each consisting of *n* paired items. Spearman's rho is then determined by the following formula:

$$\text{Spearman's rho} = 1 - \frac{6\sum d_i^2}{n^3 - n} \tag{12.17}$$

If the order in two sets of data are identical, then Spearman's rho = +1. If they are perfectly related in an inverse manner, then rho = −1. If there is no agreement or "concordance," then rho = 0. So as with the simple correlation coefficient, $-1 \leq$ Spearman's rho $\leq +1$.

The d_i^2 in Formula (12.17) is the square of the difference in the rank orders of the two data sets—for example, d_1^2 can be the square of the difference between the rank order of the first question on a 28-item survey for a questionnaire given to economists and to journalists; d_2^2 is then the square of the difference in ranks of the second question, and so on to the difference in ranks on the 28th question, d_{28}^2, where $n = 28$.

R. Alston, J Kearl, and M. Vaughan (May 1992) reported the result of their national survey of the opinion of economists on 40 questions of the type cited above. W. Becker, M. Watts, and W. Walstad (1994) conducted a similar national survey of print and broadcast journalists. These two surveys shared 28 questions. Table 12.6 shows the rankings of each of these questions based on mean response to each question. Notice that the Alston, Kearl, and Vaughan numerical ordering of "generally agree," "agree with provisos," and "generally disagree" was 1, 2, and 3, while the Becker, Watts, and Walstad assignment of values to "generally agree," "agree with provisos," and "generally disagree" was the reverse, 3, 2, and 1. But this assignment of values is irrelevant, since only the rank ordering of agree to disagree is used in Spearman's rho.

The question on which most economists disagreed in the Alston, Kearl, and Vaughan survey was number 13, corresponding to the highest numerical average score of 2.65. The question on which most journalists disagreed in the Becker, Watts, and Walstad survey was the same question 13, which on their scale corresponded to the lowest numerical average score of 1.34. After ordering the questions from least agreed on to most agreed on, the magnitudes of these means are irrelevant; only the difference in rank is relevant in Spearman's rho. Column seven (d_i) in Table 12.6 shows the difference in ranks between the two groups, and column eight (d_i^2) gives the square of these differences.

Spearman's rho is calculated using Formula (12.17) as

$$\text{Spearman's rho} = 1 - \frac{6\sum d_i^2}{n^3 - n} = 1 - \frac{6(1620)}{(28)^3 - 28} = 0.556$$

Because Spearman's rho is in the middle of its positive range, there is some positive relationship between the ranking of agreement on different issues between economists and journalists, but it is clearly not strong, as would be the case if rho was close to unity.

Table 12.6 Calculating Spearman's Rho for Surveys of Economists and Journalists

Response Ranking of Economists on the 28 Questions			Response Ranking of Journalists on the 28 Questions				
Question#	Rank	Mean	#	Rank	Mean	dif	dif^2
1	24	1.57	1	27	2.65	–3	9
2	28	1.30	2	20	2.2	8	64
3	13	2.11	3	15	1.94	–2	4
4	27	1.49	4	25	2.49	2	4
5	9	2.22	5	11	1.87	–2	4
8	19	1.82	8	26	2.51	–7	49
9	16	1.98	9	21	2.22	–5	25
10	21	1.79	10	17	2.07	4	16
11	8	2.34	11	18	2.11	–10	100
12	7	2.36	12	22	2.26	–15	225
13	1	2.65	13	1	1.34	0	0
14	23	1.68	14	28	2.7	–5	25
15	2	2.63	15	4	1.58	–2	4
16	25	1.55	16	13	1.91	12	144
17	12	2.16	17	12	1.89	0	0
18	6	2.49	18	23	2.29	–17	289
19	4	2.53	19	3	1.47	1	1
20	15	2.10	20	24	2.48	–9	81
21	11	2.17	21	2	1.43	9	81
22	10	2.21	22	9	1.78	1	1
23	22	1.72	23	16	2.03	6	36
24	5	2.51	24	5	1.62	0	0
25	20	1.81	25	19	2.12	1	1
26	26	1.50	26	14	1.92	12	144
27	3	2.58	27	10	1.79	–7	49
28	14	2.11	28	6	1.66	8	64
29	18	1.85	29	8	1.71	10	100
30	17	1.91	30	7	1.67	10	100
							Sum =1620

12.12 Linear Regression and Correlation Via Computers

Computers are typically used to do the computations required to find the slope and intercept of a least squares regression. Most programs will also calculate the residuals and provide myriad other computations. For problems involving large data sets, computers provide the only practical way to do the tedious calculations.

Software programs that perform least squares regression analysis can be found at any computer center. Programs such as SAS, SPSS, SYSTAT, MINITAB, RATS, ySTAT, SHAZAM, TSP, MICROSTAT, STATISTIX, LIMDEP, and GAUSS, all of which run under DOS operating systems, are available. Many of these systems are also available for Macintoshes and Windows. All of the LOTUS 1-2-3 compatible spreadsheet programs now have least squares regression routines.

Part of a typical regression computer printout is shown in Table 12.7, where a least squares regression was performed on death rates and speed limits for the data contained in Table 12.1. There is no standard format for the order in which information is printed by the different computer programs. Most programs, however, use labels and notation similar to that found in Table 12.7, although more sophisticated programs will provide much more information than that found in Table 12.7. (For comparison, look at the SYSTAT printout that follows in Query 12.7, concerning consulting services offered by CPA firms.)

The first section of the MICROSTAT printout indicates that there were 10 observations on each of the two variables. Summary measures for these two variables are then provided. The variables are identified by name, and their means and standard deviations are given. With most programs, it is also quite easy to get a listing of the actual x and y values.

Table 12.7 MICROSTAT Computer Printout for Regression of Deaths on Speed Limits

NUMBER OF CASES: 10 NUMBER OF VARIABLES: 2

INDEX	NAME	MEAN	STD. DEV.
Indep. Var. 1	Speed	60.0000	7.0711
Dep. Var.	Deaths	4.2400	.9743

DEPENDENT VARIABLE: Deaths

INDEP. VAR.	REGRES. COEF.	STD. ERROR	T (DF = 8)
Speed	.0756	.0407	1.8575
Constant	–.2933		

STD. ERROR OF EST. = .8642
r SQUARED = .3007
r = .5483

ANALYSIS OF VARIANCE TABLE

SOURCE	SUM OF SQUARES	D.F.	MEAN SQUARE	F RATIO
REGRESSION	2.5689	1	2.5689	3.439
RESIDUAL	5.9751	8	.7469	
TOTAL	8.5440	9		

	OBSERVED	CALCULATED	RESIDUAL
1	3.000	3.862	–.8622
2	3.300	3.862	–.5622
3	3.400	3.862	–.4622
4	3.500	4.996	–1.4956
5	4.100	3.862	.2378
6	4.300	4.240	.0600
7	4.700	3.862	.8378
8	4.900	4.240	.6600
9	5.100	4.240	.8600
10	6.100	5.373	.7267

Following the descriptive data, the next section contains estimates for the regression equation. In this section, the calculated coefficient values for the slope and y intercept are provided ($b = 0.0756$ and $a = -0.2933$). Notice that the y intercept is called the "constant." As will be discussed in the next chapter, this section of the printout also contains information on the variability of the distribution of the least squares estimator, which is reported under the column titled "STD ERROR," for standard error. In the next chapter, the value reported under the column titled "T (DF = 8)" will also be discussed. As will be learned, this column gives the calculated value of the t statistic. This t statistic of 1.8575 will be used to test whether the sample slope value of 0.0756 is sufficiently different from zero to conclude that there is a population relationship between deaths and speed limits. That conclusion would be contrary to Rivkin's assertion that "speed has little to do with safety."

Moving down the printout, we find the r^2 value under the title "r SQUARED" = 0.3007 and its square root $r = 0.5483$. As will be discussed in the next chapter, the standard error of estimate, "STD. ERROR OF EST." = 0.8642, is the square root of ESS = 5.9751 divided by its degrees of freedom, D.F. = 8.

The section titled "ANALYSIS OF VARIANCE TABLE" contains the regression, residual and total sum of squares: RSS = 2.5689, ESS = 5.9751, and TSS = 8.5440. The degrees of freedom associated with each of these sums of squares is also reported, in the column headed "D.F." Other information on "MEAN SQUARE" and the "F RATIO" will be considered in a later chapter.

The last section of the printout contains the observed death rate values, their calculated or predicted values, and the residuals. Notice that the computer maintains more precision than we could in our hand calculations.

Query 12.7:

According to a survey cited in a *WSJ* article (July 8, 1987), some 50 percent of a sample of 1,000 businessmen, financial analysts, bankers, attorneys, and academics viewed consulting services done by CPA firms as an impairment to their objectivity in auditing. As accounting firms pursue revenue from consulting, they turn to their audit clients as potential customers for consulting services and ". . . the credibility gap is widening over whether the accountants can keep consulting and auditing separate." To support this thesis the *WSJ* article provided the following data.

	Consulting as % of Total U.S. Practice	% Share of Consulting Revenue from Audit Clients
Arthur Andersen & Co.	31%	36%
Peat Marwick	19	35
Ernst & Whitney	20	20
Coopers & Lybrand	19	39
Price Waterhouse	20	21
Touche Ross & Co.	20	22
Arthur Young & Co.	15	25
Laventhol & Horwath	30	12
Deloitte, Haskins & Sells	12	33

Use a computer program to establish the sample relationship between consulting revenue (x, as a percent of total practice) and the share of consulting revenue from audit clients (y). What is the equation of the regression line? What is the total variability in the share of consulting revenue from audit clients? How much variability in the share of consulting revenue from audit clients is explained by the regression? Are the accountants keeping "consulting and auditing" separate? That is, are consulting and auditing strongly related in this sample?

ANSWER:

Using the SYSTAT program, the following printout was obtained.

SYSTAT Computer Printout for Regression of Share on Consulting

DEP VAR: SHARE N: 9 MULT R: .241 SQ. MULT. R: .058
ADJ. SQ. MULT. R: .000 STANDARD ERROR OF ESTIMATE: 9.452

VARIABLE	COEFFICIENT	STD ERROR	STD COEF	TOLERANCE	T	P(2TAIL)
constant	34.314	11.568	0.000	1.0000	2.966	0.021
consult	–0.354	0.539	–0.241	1.0000	–0.657	0.532

ANALYSIS OF VARIANCE

SOURCE	SUM-OF-SQUARES	DF	MEAN-SQUARE	F-RATIO	P
REGRESSION	38.575	1	38.575	0.432	0.532
RESIDUAL	625.425	7	89.346		

Let y be the percentage share of consulting revenue from audit customers. Let x be consulting as a percentage of total U.S. practice. Then, $\hat{y} = 34.314 - 0.354x$. The total sum of squares is

$$TSS = RSS + ESS = 38.575 + 625.425 = 664.0$$

This regression explains only 5.8 percent (SQ. MULT. R = 0.058) of the variability in share of consulting revenue from audit clients. In this sample, the accounting firms are keeping "consulting and auditing" separate; that is, consulting and auditing are not strongly related. If anything, they are related in a negative direction, which is contrary to the quote from the *WSJ*.

EXERCISES

12.27 The following 60 records are the observations on pretest and midtest scores for the scatterplot in Exercise 12.4.

a. Use a computer program to calculate the slope, intercept, and coefficient of determination of the least squares regression line of the midtest score as a linear function of the pretest score.

b. Plot the least squares regression line on the scatterplot.

c. How do the slope, intercept, and coefficient of determination relate to the fit of your regression line?

Record	pre	mid	Record	pre	mid
1	25	25	31	55	70
2	25	30	32	55	75
3	30	30	33	55	65
4	30	25	34	55	45
5	30	35	35	55	40
6	35	30	36	60	50
7	35	35	37	60	55
8	35	40	38	60	70
9	35	45	39	60	65
10	40	35	40	60	80
11	40	40	41	60	75
12	40	50	42	65	75
13	40	55	43	65	65
14	40	45	44	65	85
15	45	45	45	65	70
16	45	50	46	70	70
17	45	40	47	70	60
18	45	65	48	70	90
19	45	55	49	70	75
20	45	60	50	75	70
21	50	65	51	75	75
22	50	50	52	75	80
23	50	45	53	80	70
24	50	55	54	80	85
25	50	70	55	80	90
26	50	60	56	85	95
27	55	60	57	85	85
28	55	55	58	90	85
29	55	50	59	90	95
30	60	60	60	95	100

12.28 From the following computer printout,

a. determine the slope and intercept of the fitted regression line.

b. what is the value of the total sum of squares and the residual sum of squares?

NUMBER OF CASES: 10 NUMBER OF VARIABLES: 2

NAME	MEAN	STD.DEV.
x	9.7000	5.3135
y	12.9000	6.5056

r SQUARED = .6968

12.29 Given the following computer printout,

a. determine the coefficient of determination.

b. determine the slope and intercept of the regression line.

NAME	MEAN	STD.DEV.
x	6.3333	2.1602
y	7.8333	2.3166

OBSERVED y	CALCULATED y	RESIDUAL
8.000	5.700	2.3000
10.000	8.900	1.1000
6.000	7.300	–1.3000
5.000	6.500	–1.5000
11.000	10.500	.5000
7.000	8.100	–1.1000

12.30 The manager of a utilities-included rental complex (of 200 units) is curious as to how good a predictor October heating bills are of heating bills for that winter. Having kept records for the last 15 years, she regressed winter heating bills (in thousands of dollars) on respective October heating bills (in thousands of dollars) with the following results.

Ordinary Least Squares

Dependent Variable	Y	Number of Observations	15
Mean of Dep. Variable	104.8034	Std. Dev. of Dep. Var.	34.733122
Std. Error of Regr.	29.8509	Sum of Squared Residuals	11584.0
R-squared	.31413	Adjusted R-squared	.26137
F(1, 13)	5.9540	Prob. Value for F	.02976

Variable	Coefficient	Std. Error	t-ratio	Prob|t| > x	Mean of X	Std.Dev.of X
Constant	15.9374	37.23	.428	.67557		
X	17.5466	7.191	2.440	.02976	5.06456	1.10944

a. The best point estimate of the increase in the winter heating bill if the October heating bill increases by $1,000 is ___________.

b. If next October's heating bill is $5,000, the predicted heating bill for that winter is.

c. The sample correlation coefficient is ___________.

d. In the above regression, the minimum sum of the squared vertical deviations of the points in the scatter diagram from the sample regression line is ___________.

CHAPTER EXERCISES

12.31 Exercise 10.9 provides data on the objective and subjective probability of a horse winning a race, as calculated by Peter Asch and Richard Quandt, *Racetrack Betting: The Professors' Guide to Strategies* (New York: Praeger 1991). Construct a scatter diagram for the data, where the objective probability is the dependent variable and the subjective probability is the independent variable. Calculate the least squares estimates of the intercept and slope of the "best" straight line to describe these paired observations. Predict the objective probability of a horse winning if the subjective probability is 0.22.

12.32 Calculate the *R*-square for the objective probability regression in Exercise 12.31. What does this value tell you about the quality of your predictions?

12.33 For the regression of objective and subjective probability in Exercise 12.31, predict the objective probability of a horse that has $4,000 bet on it, where the total wagered on all the horses in the race is $20,000 (which is called the pool). Why is this subjective probability of 0.2 less than the objective probability you predicted?

12.34 Peter Asch and Richard Quandt, *Racetrack Betting: The Professors' Guide to Strategies* (New York: Praeger 1991, p. 95), provide the following data on the speed of the horse for jockeys of differing years of experience.

Jockey's Experience	Horse Speed (mph)
5	37.0
2	34.8
8	39.0
6	38.0
4	37.2
1	34.5
7	39.5

a. Draw the scatter diagram for horse speed as a function of jockey experience.
b. Calculate the "best" straight line regression equation for predicting horse speed. Write the equation and give the R-square.
c. For a jockey with three years of experience, what is the predicted horse speed?

12.35 Exercise 10.19 provided data on populations for the biggest 25 cities in 1986 and 1988. Regress the 1988 Census populations for the 25 cities on their 1986 populations. Is there a negative relationship between the 1986 and 1988 populations for this sample of 25 cities?

12.36 A *Wall Street Journal* (June 28, 1990) article reported on the "Giant Pension Funds' Explosive Growth." Below are the data used to support this headline. Use regression analysis to determine the average rate of growth between 1985 and 1989 for these funds.

Top 20 Pension Funds in millions of dollars

	1985	1989
TIAA/CREF	40000	81000
Calif. Public Employees	29166	54000
General Motors	26300	40900
N.Y. State/Local	25997	44238
N.Y. City Employees	25063	45422
AT&T	24887	42700
General Electric	15812	28912
N.Y. State Teachers	15303	28130
California Teachers	15100	30335
IBM	13939	25775
N.J. Investment Div.	12494	24526
Texas Teachers	11188	23500
Ford Motor	11000	22800
Wisconsin Investment	10800	20033
Ohio Public Employees	10760	19305
Mich. Public Employees	10516	17948
Ohio Teachers	10178	18059
DuPont	9665	18681
Nynex	9447	16062
Florida State Board	8911	18700

12.37 An article in *The Wall Street Journal* (February 27, 1992) suggested that there is a relationship between "sensitivity" and the "market value" of lumber stocks. What is this correlation?

Company	Market Value Canadian $ in Millions	Sensitivity*
Siocan Forest Products	$ 103	61.6
Doman Industries	115	53.9
Intl. Forest Products	300	30.3
West Fraser Timber	400	24.8
Canfor	698	20.5
Crestbrook Forest Inds.	173	17.9
Weldwood of Canada	495	16.8
Noranda Forest	1,071	16.5
Donohue	434	13.2
Domtar	603	5.8
Canadian Pacific For. Prod.	1,115	3.9
MacMillan Bloedel	2,119	3.8
Abitibi Price	1,056	0.5

*Thousands of board feet of lumber per C$10,000 in stock-market value

12.38 An article in *Business Week* (March 30, 1992) titled "Better Schools, Not Just More School," asked the question, "Does a longer school year boost academic performance?" The following data on the days spent in school per year and average math scores of 13-year-olds on an achievement exam administered to students around the world were provided to help answer this question.

Country	Days Spent in School	Scores
China	251	80
South Korea	222	73
Former U.S.S.R.	207	71
France	210	70
Israel	215	63
Canada	188	62
Britain	192	61
U.S.	180	55
Jordan	191	40

a. What is the linear regression relationship between days of school and scores?
b. If days of school are increased from 200 to 220, by how much and in which direction will predicted scores change?
c. What is the correlation between days and scores?

12.39 The *Journal of Economic Literature* (March 1984) reported the most frequently cited journals in economics. The adjusted ranking of the top 18 journals follows:

1 American Economic Review
2 Journal of Political Economy
3 Econometrica
4 Journal of Monetary Economics
5 Journal of Economic Theory
6 Review of Economic Studies
7 International Economic Review
8 Bell Journal of Economics
9 Journal of Finance
10 Journal of Econometrics
11 Scandinavian Journal of Economics
12 Brookings Papers
13 Journal of Public Economics
14 Journal of Financial Economics
15 Review of Economics and Statistics
16 Journal of Amer. Statistical Assoc.
17 Quarterly Journal of Economics
18 Journal of Human Resources

An informal survey of economists by the author aimed at assessing the overall "quality" of journals suggests the following ranking:

1 Journal of Political Economy
2 Econometrica
3 American Economic Review
4 Journal of Monetary Economics
5 Journal of Finance
6 Review of Economic Studies
7 Journal of Financial Economics
8 Review of Economics and Stat.
9 Quarterly Journal of Economics
10 Journal of Economic Theory
11 Bell Journal of Economics
12 Journal of Econometrics
13 Journal of Amer. Statistical Assoc.
14 Brookings Papers
15 International Economic Review
16 Scandinavian Journal of Economics
17 Journal of Human Resources
18 Journal of Public Economics

a. What is the value of Spearman's rho for the journal citation ranking and the quality ranking?
b. Why should Spearman's rho be used to measure the relationship between the journal citation ranking and the journal quality ranking and not the simple coefficient of correlation?

12.40 Dun and Bradstreet compiles *The Business Failure Record,* which shows the following number of business failures by industry type and selected years.

Industry	Number of Failures in 1985	1987	1990
Agriculture, forestry, fishing	2,699	3,766	1,727
Mining	796	627	381
Construction	7,005	6,735	8,072
Manufacturing	4,869	4,273	4,709
Transportation, public utilities	2,536	2,236	2,610
Wholesale trade	4,836	4,336	4,376
Retail trade	13,494	12,240	12,826
Finance, insurance, real estate	2,676	2,550	3,881
Services	16,649	23,802	17,673

a. Calculate Spearman's rho for failures across industries in 1985 and 1987.
b. Calculate Spearman's rho for failures across industries in 1987 and 1990.
c. Compare your results in parts a and b.
d. Why wouldn't it be meaningful to calculate a simple coefficient of correlation for comparison purposes in part c?

12.41 A *Wall Street Journal* (August 7, 1992) article reported that the "cost of reaching households in prime time grew much more than the overall Olympic audience." It provided this line and bar chart shown here to support the claim. What is the approximate correlation coefficient for cost and audience reached? Is this a high correlation? Explain.

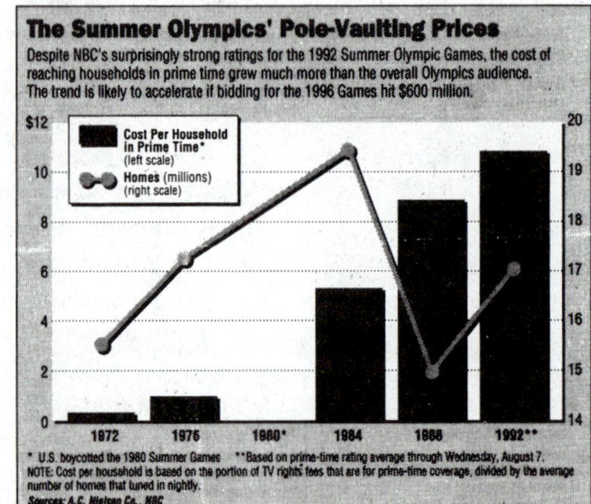

12.42 An article in *The Wall Street Journal* (October 12, 1992) reported the average compensation of CEOs in seven countries. On September 24, 1992, the *WSJ* reported information, including the home country of origin, of the world's 100 largest public companies. Below are summary data on CEO compensation and number of headquartered companies in the identified countries.

	CEO Compensation	Number of headquartered companies in country
United States	$717,237	49
France	$479,772	2
Italy	$463,009	1
Britain	$439,441	16
Canada	$416,066	0
Germany	$390,933	4
Japan	$390,723	23
Other	not avail	5

What is the correlation coefficient for CEO compensation and number of headquartered companies in each country?

12.43 For the data in Exercise 12.42, what is the least squares regression of CEO compensation on number of headquartered companies in each country? What does this regression tell you?

12.44 For each of the following scatterplots, draw in the approximate least squares regression line. (No calculations are required.)

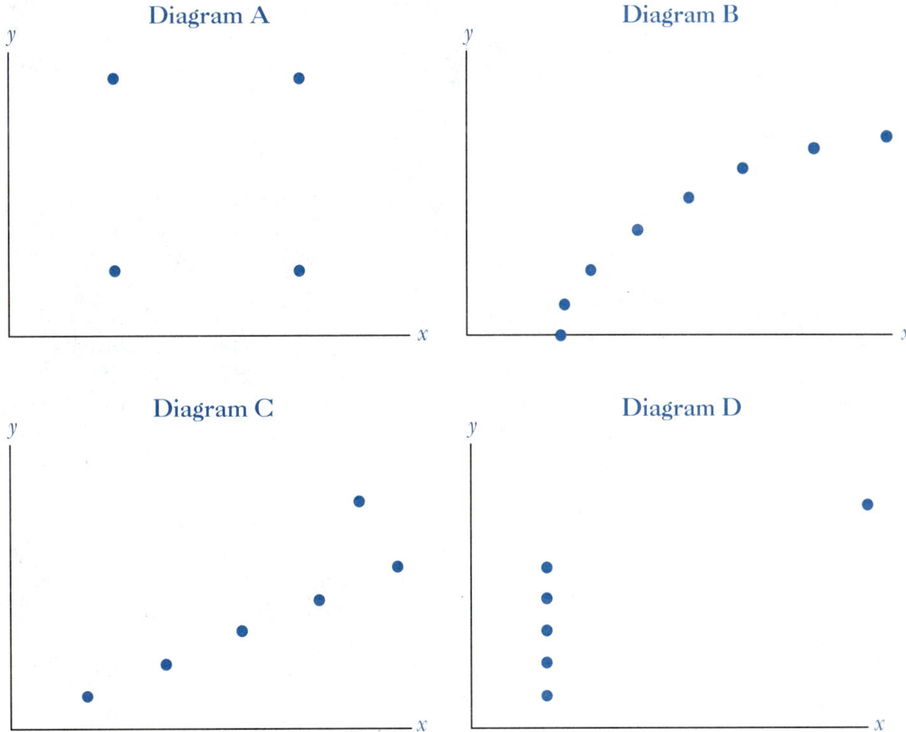

12.45 "KPMG Peat Marwich reported that cash compensation for CEOs last year rose 11.6% at 20 high-performing companies, but was up just 5% for 20 poor performers."(*WSJ*, May 11, 1993) What is the sign of the regression line slope implied by this quote?

12.46 For the data below on car weight and mileage, find the "best" fitting straight line relationship. For a car of 2,910 pounds, what is the predicted gas mileage from your equation? Why is your prediction from the least squares regression better than just taking an average of the gas mileage of the two cars that weigh 2,910 pounds?

Car	Weight	mpg	Car	Weight	mpg
1	2429	41.7	7	2700	39.1
2	2600	39.9	8	2800	39.2
3	2626	41.3	9	2731	39.9
4	2589	39.6	10	2910	37.4
5	2568	39.8	11	2875	38.9
6	2910	36.5	12	2493	42.1

12.47 Research Institute of America, New York, figures that the IRS can determine adjusted gross income for U.S. taxes on the basis of other tax deductions taken. Below is the RIA's sample data for adjusted-gross income and tax deductions calculated from preliminary 1991 data, as published in *The Wall Street Journal* (June 16, 1993).

Adjusted gross-income	Deducted taxes
$ 27500	$ 2115
35000	2540
45000	3141
62500	4161
87500	6012
150000	9616
350000	21374
750000	48289

a. What is the sample relationship betweeen income and tax deductions?

b. If someone reports adjusted-gross income of $100,000 and deducts $20,000 for other taxes paid, would you suspect the person is underreporting income or overreporting tax deductions? Explain?

12.48 According to Carnegie-Mellon University Professor M. Granger Morgan, "A simple linear relation, for instance, accurately describes the average cancer risk incurred by smokers: 10 cigarettes a day generally increase the chance of contracting lung cancer by a factor of 25; 20 cigarettes a day increase it by a factor of 50."(*Scientific American*, July 1993) Write the regression equation that Morgan has described.

12.49 The data in the computer ASCII file "EX12-49.PRN" is from *Car and Driver* (April 1993). The first column identifies the car tested and issue of *Car and Driver* in which the test appeared. The second column is the estimated price of the car tested. The third and fourth columns are 0 to 60 mph and 1/4 mile time in seconds. The fifth column is top speed. The sixth column is 70 to 0 braking distance in feet. EPA estimated fuel economy is in the seventh column. Finally, in column eight, road holding ability (measured as g force) is provided as obtained from a 300-foot skidpad. Which of the six variables (which you might name TO60, quarter, top, braking, fuel, hold) has the highest r^2 with car price?

12.50 Song Won Sohn, chief economist at Norwest Corp. in Minneapolis, asked: "Is there a reason to believe stocks and bonds move in different directions? I don't think so . . . The bond market is an excellent leading indicator of the stock market."(*Wall Street Journal*, September 13, 1993) According to the article, Sohn has studied the relative performance of the U.S. Treasury's 30-year bond and the Dow Jones Industrial Average since 1989; he predicted that the bond yield would fall from just less than 6 percent to 5.5 percent and the DOW would rise 50 to 100 points. From monthly data reported in the *Economic Report of the President*, estimate the relationship between the Dow and bond yield. Is there a strong negative correlation? Is Sohn's predicted relationship correct?

12.51 Suppose the salaries (y) of salespersons and revenue generated through sales (x) have a correlation coefficient of $r = 0.92$. What can you conclude from this value?

12.52 An article in *The New York Times* (April 11, 1994) reported on how Federal District Court Judge Clarence Newconner turned to Princeton University economics professor Orley Ashenfelter to help him decide whether to order a new election in Pennsylvania's Second State Senatorial District in Philadelphia or declare the losing candidate Bruce Marks, a Republican, the winner in the November 1993 election. In this special election to fill a Senate vacancy, Marks received 19,691 votes on voting machines and his Democrat opponent, William Stinson, received 19,127. But in absentee ballots, Stinson received 1,396 to Marks' 371. The Republicans charged that many of the absentee ballots were falsified by the Democrat-controlled County Board of Election. Professor Ashenfelter provided the author of this book with the following data used in his analysis:

		Absentee Votes			Machine Votes		
Year	District	Democrat	Republican	Difference	Democrat	Republican	Difference
82	2	551	205	346	47767	21340	26427
82	4	594	312	282	44437	28533	15904
82	8	338	115	223	55662	13214	42448
84	1	1357	764	593	58327	38883	19444
84	3	716	144	572	78270	6473	71797
84	5	1207	1436	–229	54812	55829	–1017
84	7	929	258	671	77136	13730	63406
86	2	609	316	293	39034	23363	15671
86	4	666	306	360	52817	16541	36276
86	8	477	171	306	48315	11605	36710
88	1	1101	700	401	56362	34514	21848
88	3	448	70	378	69801	3939	65862
88	5	781	1610	–829	43527	56721	–13194
88	7	644	250	394	68702	12602	56100
90	2	660	509	151	27543	26843	700
90	4	482	831	–349	39193	27664	11529
90	8	308	148	160	34598	8551	26047
92	1	1923	594	1329	65943	21518	44425
92	3	695	327	368	58480	12968	45512
92	5	841	1275	–434	41267	46967	–5700
92	7	814	423	391	65516	14310	51206
93	2	1396	371	1025	19127	19691	–564

a. Fit a regression of the difference between Democrat and Republican absentee ballots (y) to the difference between Democrat and Republican machine ballots (x) in the 21 election points between 1982 and 1992, excluding the twenty-second contested election in 1993. Write the equation for the regression. Interpret the slope intercept.

b. Show your regression line in a scatterplot. Make sure to identify the slope and intercept of the regression and to label the axis. Identify where the point would be for the contested election in 1993. Why didn't Ashenfelter include this point in the estimation of his regression?

c. What is the correlation coefficient for your regression in part a? What does it tell you?

d. Using your regression in part a, what is the predicted Democrat and Republican difference in absentee ballots in the contested election in 1993? By how many votes does this differ from the actual results?

(Note: Questions for analysis of the voting issue in Exercise 12.52 are continued in Exercises 13.50, 14.63 and 15.45.)

12.53 If $x_i^* = x_i - \bar{x}$ and $y_i^* = y_i - \bar{y}$ (i.e., x_i^* and y_i^* measure x and y in deviation form), then:

a. Prove that $b = \dfrac{\sum x_i^* y_i^*}{\sum x_i^{*2}} = \dfrac{\sum x_i y_i - \sum x_i y_i / n}{\sum x_i^2 - \left(\sum x_i\right)^2 / n}$

b. Use the data in Exercise 12.5 to calculate b by both formulas in part a. Check to make sure your results agree with the value calculated in Exercise 12.5.

c. What is the y intercept of the regression line, when x and y are measured in deviation form?

12.54 Prove that the following equalities are true.

a. Total sum of squares, $\Sigma(y_i - \bar{y})^2 = \Sigma y_i^2 - n\bar{y}^2$

b. Regression sum of squares, RSS $= \Sigma(\hat{y}_i - \bar{y})^2 = b^2\,\Sigma(x_i - \bar{x})^2$

c. Error sum of squares, ESS $=$ TSS $- b\Sigma(x_i - \bar{x})(y_i - \bar{y})$

d. Explained sum of squares, RSS $= [\Sigma(x_i - \bar{x})(y_i - \bar{y})]^2 / \Sigma(x_i - \bar{x})^2$

e. Coefficient of correlation, $r = bs_x / s_y$

Academic References

Alston, R., J. Kearl, and M. Vaughan. "Is There a Consensus Among Economists in the 1990s?" *American Economic Review*, May 1992 pp. 203–209.

Becker, W., M. Watts, and W. Walstad. "Comparison of Views of Economists, Educators, and Journalists on Economic Issues." *An International Perspective on Economics Education*, edited by W. Walstad. (Netherlands: Kluwer Academic Press, 1994).

Malkiel, Burton. *A Random Walk Down Wall Street.* (New York: Norton, 5th edition, 1990).

Stigler, Stephen M. *The History of Statistics: The Measurement of Uncertainty before 1900.* (Cambridge: Belknap Press of Harvard University Press, 1986).

Endnotes

1. Ignoring the subscript i, the sum of the squared errors is

$$\Sigma e^2 = \Sigma(y - a - bx)^2$$

Differentiating with respect to a and b gives

$$\frac{\partial\left(\Sigma e^2\right)}{\partial a} = -2\Sigma\left(y - a - bx\right) = 2\Sigma y - 2na - 2b\Sigma x$$

$$\frac{\partial\left(\Sigma e^2\right)}{\partial b} = -2\Sigma x\left(y - a - bx\right) = 2\Sigma xy - 2a\Sigma x - 2b\Sigma x^2$$

These two derivatives must equal zero for the sum of squared errors to be at a minimum. The equations that result from this necessary condition are called normal equations, so named by Carl Gauss in 1809. Because there will be one equation for each unknown, they can be solved by "ordinary methods," as shown by Adrien Legendre in 1805. That is,

$$\Sigma y = na + b\Sigma x$$

$$\Sigma xy = a\Sigma x + b\Sigma x^2$$

Solving for a in the first equation gives

$$a = (\Sigma y/n) - b(\Sigma x/n),$$

which, upon rearranging, shows that the regression line must pass through the mean of x and mean of y. Substituting for a in the second normal equation gives

$$\Sigma xy = [(\Sigma y/n) - b(\Sigma x/n)]\Sigma x + b\Sigma x^2$$

$$\Sigma xy = (\Sigma y\Sigma x/n) - b[(\Sigma x)^2/n] + b\Sigma x^2$$

$$\Sigma xy - \Sigma x\Sigma y/n = b[\Sigma x^2 - (\Sigma x)^2/n]$$

$$\Sigma xy - 2(\Sigma x\Sigma y/n) + \Sigma x\Sigma y/n = b\{\Sigma x^2 - 2[(\Sigma x)^2/n] + (\Sigma x)^2/n\}$$

$$\Sigma xy - \bar{y}\Sigma x - \bar{x}\Sigma y + \Sigma\bar{x}\bar{y} = b(\Sigma x^2 - \bar{x}\Sigma x - \bar{x}\Sigma x + \Sigma\bar{x}^2)$$

$$\Sigma(xy - \bar{y}x - \bar{x}y + \bar{x}\bar{y}) = b\Sigma(x^2 - \bar{x}x - \bar{x}x + \Sigma\bar{x}^2)$$

$$\Sigma(x - \bar{x})(y - \bar{x}) = b\Sigma(x - \bar{x})^2$$

$$b = \frac{\Sigma\left(x - \bar{x}\right)\left(y - \bar{y}\right)}{\Sigma\left(x - \bar{x}\right)^2}$$

2. The use of the RSS and ESS notation is unfortunate because it is not standardized across books or computer programs. You should be aware that some use RSS to represent the residual sum of squares and ESS the explained sum of squares. We follow the convention of using RSS for the regression sum of squares and ESS for the error sum of squares. Readers should take care to check the definition of this notation in using computer programs and text material.
3. To prove that the coefficient of correlation calculated by

$$(12.3) \qquad r = \frac{\sum_{i=1}^{n}\left(x_i - \bar{x}\right)\left(y_i - \bar{y}\right)}{\sqrt{\sum_{i=1}^{n}\left(x_i - \bar{x}\right)^2 \sum_{i=1}^{n}\left(y_i - \bar{y}\right)^2}}$$

is equivalent to the square root of the coefficient of determination calculated by the formula

$$r = \sqrt{\frac{\text{RegSS}}{\text{TSS}}} \tag{12.16}$$

it is necessary to express the regression sum of squares in terms of the least squares estimators. Ignoring the i subscript,

$$\text{RSS} = \text{TSS} - \text{ESS} = \Sigma(y-\bar{y})^2 - \Sigma(y-\hat{y})^2 = \Sigma(\hat{y}-\bar{y})^2 = b^2\Sigma(x-\bar{x})^2$$

because $\Sigma\hat{y} = \Sigma y$ ↑ ↑

because $\hat{y} = a + bx = \bar{y} - b\bar{x} + bx$

so $\hat{y} = \bar{y} + b(x-\bar{x})$

Since $b^2 = [\Sigma(x-\bar{x})(y-\bar{y})/\Sigma(x-\bar{x})^2]^2$, RSS can be written

$$\text{RSS} = \frac{\left[\Sigma(x-\bar{x})(y-\bar{y})\right]^2}{\Sigma(x-\bar{x})^2}$$

Substituting for RSS in (12.16) gives

$$r^2 = \frac{\left[\Sigma(x-\bar{x})(y-\bar{y})\right]^2}{\text{TSS}\,\Sigma(x-\bar{x})^2} = \frac{\left[\Sigma(x-\bar{x})(y-\bar{y})\right]^2}{\Sigma(y-\bar{y})^2\,\Sigma(x-\bar{x})^2}$$

which is the square of (12.3).

FORMULAS

- Covariance of x and y

$$s_{xy} = \frac{\sum_{i=1}^{n}(y_i-\bar{y})(x_i-\bar{x})}{n-1}$$

- Standardized deviation of x

$$\frac{x_i-\bar{x}}{s_x}$$

- Coefficient of correlation

$$r = \frac{\sum_{i=1}^{n}(x_i-\bar{x})(y_i-\bar{y})}{\sqrt{\sum_{i=1}^{n}(x_i-\bar{x})^2\sum_{i=1}^{n}(y_i-\bar{y})^2}}$$

- General equation of a (regression) line
  $\hat{y} = a + bx$
  where $\hat{y}$ is the predicted dependent variable,
  x is the independent variable,
  a is the y axis intercept, and
  b is the slope.
- Specific example of a regression line

$$\hat{y} = -0.293 + 0.076x$$

  where a one-unit increase in x is associated with a 0.076-unit increase in $\hat{y}$ and for $x = 0$, $\hat{y} = -0.293$.

- Residual

$$e_i = y_i - \hat{y}_i$$

- Sum of residuals

$$\sum_{i=1}^{n} e_i = e_1 + e_2 + e_3 + \ldots + e_{n-1} + e_n = 0$$

$$= y_1 - \hat{y}_1 + y_2 - \hat{y}_2 + \ldots + y_{n-1} - \hat{y}_{n-1} + y_n - \hat{y}_n = 0$$

- Residuals as a function of a and b

$$e_i = y_i - \hat{y}_i = y_i - (a + bx_i) = y_i - a - bx_i$$

- Sum of squared residuals as a function of a and b

$$\sum_{i=1}^{n} e_i^2 = \sum_{i=1}^{n} \left(y_i - \hat{y}_i\right)^2 = \sum_{i=1}^{n} \left(y_i - a - bx_i\right)^2$$

- Slope of the regression equation

$$b = \frac{\sum_{i=1}^{n} \left(x_i - \bar{x}\right)\left(y_i - \bar{y}\right)}{\sum_{i=1}^{n} \left(x_i - \bar{x}\right)^2}$$

- y intercept of the regression equation

$$a = \bar{y} - b\bar{x}$$

- Total sum of squares

$$\text{TSS} = \sum_{i=1}^{n} \left(y_i - \bar{y}\right)^2$$

- Residual or error sum of squares

$$\text{ErrorSS} = \sum_{i=1}^{n} e_i^2 = \sum_{i=1}^{n} \left(y_i - \hat{y}_i\right)^2$$

- Regression sum of squares

$$\text{RegSS} = \text{TSS} - \text{ErrorSS}$$

- Coefficient of determination

$$r^2 = \text{RegSS/TSS}$$

- Coefficient of correlation

$$r = \sqrt{\text{RegSS/TSS}}$$

- Rank Correlation

$$\text{Spearman's rho} = 1 - \frac{6\sum d_i^2}{n^3 - n}$$

where d_i represents the difference between the numerical rank order of the ith item in two different sets of data, each consisting of n paired items.

CHAPTER 13

THE TWO-VARIABLE POPULATION MODEL

If you can't generalize
from data
there's nothing else you can do
with it either.

Robert M. Pirsig
(Lila 1991, p.55)

13.1 Introduction

Chapter 12 described the sample relationships between observations on two variables: speed limits and death rates, credit card balances and interest charges, percentage of women MBAs and percentage of women managers, and so on. Here, least squares regression analysis is extended to inferences about the more general population relationship that might exist between two variables. This chapter shows how to:

- Construct a mathematical model of the population relationship between two variables.
- Use the sample least squares regression line to estimate the parameters of the population model.
- Employ the least squares estimates to test for a population relationship between the variables.
- Use the least squares regression line to estimate the expected values of the dependent variable at a certain value of the independent variable.
- Use the least squares regression line to predict a value of the dependent variable for a given value of the independent variable.
- Construct confidence intervals for the expected value and the predicted value of the dependent variable based on the values of the independent variable.

Whether a researcher should first look at the data and then construct a theoretical model of the population or specify the model and then confront it with the data is hotly debated in statistics. For classical regression analysis, the latter approach has been the dominant view, and it will receive more attention in this and the following chapters. This chapter focuses on the population relationship between just two variables. The parameters of bivariate population models will be estimated and tested using the least squares regression. The next chapter will extend this model to multiple explanatory variables.

To begin our extension of least squares regression line fitting to population inferences, consider the so-called Phillips curve relationship between inflation and unemployment.

13.2 The Phillips Curve

A negative relationship between the rate of inflation (in wages and prices) and the rate of unemployment in the United Kingdom was discovered by Alban Phillips in the 1950s. Phillips, an unassuming and relatively obscure New Zealand economist, never claimed any fixed and general relationship. He never made a statistical inference from his sample data to a population relationship. Yet, policymakers in the United States accepted the existence of such a relationship and used a "Phillips curve" to assess the relationship between inflation and unemployment in the 1960s. For example, S. Fischer and R. Dornbusch, in *Introduction to Macroeconomics* (1983, p. 292), write:

FIGURE 13.1 The Theoretical Phillips Curve

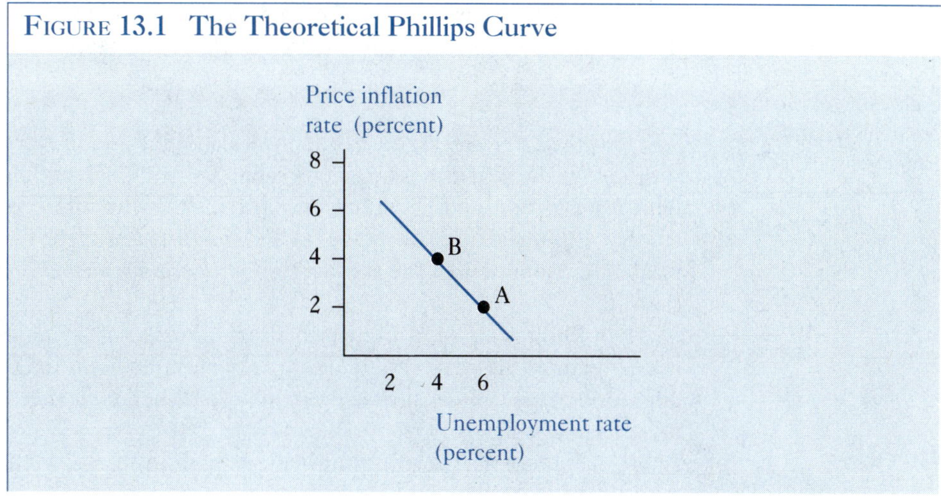

> The Phillips curve represents a trade-off between inflation and unemployment in the following sense. If aggregate demand is expanded through tax cuts, increased money, or higher government spending, output can be increased and the unemployment rate can be reduced. But with lower unemployment and less slack in the economy, there is more pressure on wages, costs, and prices. Therefore, prices will be rising at a more rapid rate the lower the rate of unemployment. . . . In the 1960s policymakers thought that a Phillips curve in the general shape of [Figure 13.1] in fact represented their policy trade-off. And they used it.

As evidenced by Alan Blinder's article in *Business Week* (November 27, 1989), in which he said the Phillips curve returned in the 1980s, and Paul Craig Roberts' *Business Week* (September 27, 1993) article, in which he claimed it had vanished, interest in the Phillips curve tradeoff continues. After three decades, the question persists: Is the observed price and unemployment relationship of the 1960s sufficient to support the hypothesis that a two percentage point increase in the unemployment rate will be associated with a two percentage point decrease in the inflation rate?

13.3 Deterministic Versus Stochastic Relationships

The theoretical Phillips curve in Figure 13.1 represents a **deterministic relationship,** because each value of the independent variable (unemployment rate) is paired with only one value of the dependent variable (inflation rate). At point A, the unemployment rate is 6 percent and the inflation rate is 2 percent. From point A to point B, the unemployment rate drops to 4 percent and the inflation rate rises to 4 percent. Thus, the slope of this theoretical relationship between points A and B is –1.00.

Deterministic Relationship
An association in which each value of the independent variable is paired with only one definite value of the dependent variable.

Letting y^* represent the inflation rate associated with the unemployment rate x, a deterministic linear relationship between these two rates can be written as

$$y^* = \alpha + \beta x \tag{13.1}$$

The Phillips curve in Figure 13.1 shows a straight line drawn through points A and B; the slope is the parameter (or fixed) value $\beta = -1.00$. Extending the straight line drawn between points A and B to the y axis implies that the intercept is the parameter value $\alpha = 8.00$. (Outside the AB range, in Figure 13.1, a straight line may not be an appropriate representation of the theory. For the moment, we will ignore this nonlinearity in the Phillips curve.)

The Phillips curve in Figure 13.1 may or may not be a good representation of the actual relationship between inflation and unemployment in the 1960s. To establish its appropriateness, we must first estimate it and then compare the estimated slope with its expected value.

If the theoretical relationship were exactly true, we would only need to get two sample points, use a ruler to join them, and then calculate the slope. Additional sample points would convey no information since they would also lie on this line. We saw in the previous chapters, however, that all the points in a scatterplot typically do not lie on a straight line. There is variability in values of the dependent variable that is not associated with the independent variable. Some of this variability is due to variables that affect inflation (e.g., the money stock, government spending, taxes, technological bottlenecks) but are not included in the study, because they are unknown or unmeasured. Some is due to purely chance occurrences beyond the control of anyone. Let all these other unknown, unmeasured, and chance factors be represented by a single random variable ε. The variable ε is called the **disturbance term.**

Disturbance Term, ε
The difference between y and its expected value at each x value in a population model.

The specification of the theoretical relationship between the observed rate of inflation and the rate of unemployment is now expanded to

$$y = \alpha + \beta x + \varepsilon \tag{13.2}$$

where y is the observed inflation rate and ε is a random variable that causes y to deviate from y^*. (Note that $\varepsilon = y - y^*$.)

The relationship between the inflation rate y and the unemployment rate x in Equation (13.2) is no longer deterministic. It is said to be **stochastic,** because for each value of x there are random disturbances that also affect y. Thus, for any given value of x, the variable y may take any one of many different values.

Stochastic Relationship
An association in which more than one value of y may be assigned to a value of x by a random process.

The distribution of ε typically is unknown, so we must infer its characteristics. In particular, ε may take negative or positive values since the net effect of any omitted and unmeasured variables may cause y to be above or below that predicted by the theory. But if the theory is correct, then there is no reason to expect a bias. Such will be the case if the mean or expected value of ε is zero at each value of x, which we will assume. This assumption about the disturbance term is written

$$E(\varepsilon \mid x) = 0 \tag{13.3}$$

where $E(\varepsilon \mid x)$ designates the expected value of the random disturbance term ε. $E(\varepsilon \mid x) = 0$ says that if we selected a large (infinite) number of ε values, the average value would be zero.

It seems reasonable that negative and positive values of ε close to zero will occur more frequently than values far from zero and that negative and positive values are equally likely to occur. That is, most values of y will be close to the deterministic component $\alpha + \beta x$ with values of ε adding little to y. In addition, in the absence of any other information, it seems reasonable to assume that the occurrence of any given value of ε is not affected by the occurrence of any other value of ε, or the value of x considered. Adding the assumption that ε has a constant variance then implies that the distribution of ε at each x value is the same. Finally, if ε is normally distributed, then the distribution of ε, and in turn y, around the deterministic line $y^* = \alpha + \beta x$ would be as shown in Figure 13.2. Notice in Figure 13.2 that only two of the ε density functions are shown. At each value of x there is an identical ε density.

Figure 13.2 shows y^* to be the conditional mean or expected value of y, at each value of x. This expected value of y, conditional on an x value, is written $E(y \mid x)$; thus,

$$E(y \mid x) = y^* = \alpha + \beta x \tag{13.4}$$

Equation (13.4) shows that $E(y \mid x)$ and y^* are the same quantity.[1] In Figure 13.2 (where $\alpha = 8.00$ and $\beta = -1.00$), $y^* = E(y \mid x) = 8 - x$ says that the expected rate of inflation is

FIGURE 13.2 Population Model of the Phillips Curve Relationship $y = \alpha + \beta x + \varepsilon$

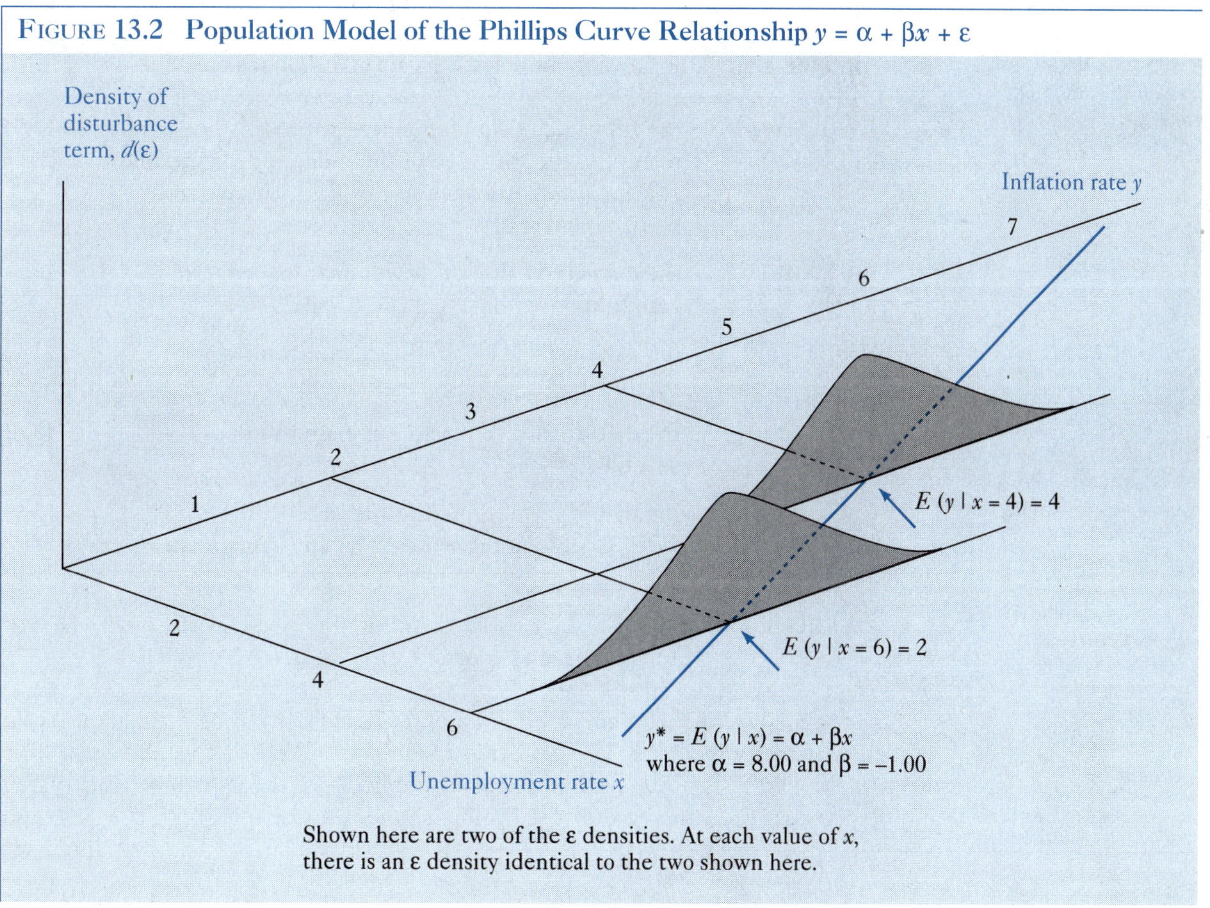

Shown here are two of the ε densities. At each value of x, there is an ε density identical to the two shown here.

4 percent when the unemployment rate is 4 percent, and the expected inflation rate is 2 percent when the unemployment rate is 6 percent. Figure 13.2 represents a redrawing of the relevant section of the Phillips curve in Figure 13.1. Here, however, the deterministic line of the Phillips curve relationship is shown on the x-y plane that is now positioned horizontally. The new idea shown in Figure 13.2 is that for each level of unemployment x, the inflation rate y can take on values other than $E(y|x)$. These random values are determined by the density function $d(\varepsilon)$. The two bell-shaped curves that rise vertically from the x–y plane show density functions for ε that are normal.

When values of ε are assumed to be generated by the normal probability distribution, the probability that y is a value within some interval of its expected value at each x is known. For instance, as drawn in Figure 13.2, at an unemployment rate of 4 percent, there is almost a 100 percent probability that the observed inflation rate will be between 2.5 percent and 5.5 percent. At an unemployment rate of 6 percent, there is almost a 100 percent likelihood that the observed inflation rate will be between 0.5 percent and 3.5 percent. Observed values of y outside these extremes should be rare if this graphical representation of the Phillips curve model is correct.

The specification of a population model of the Phillips curve is now complete. In general, the theory, the equational form used to represent it, and the assumptions about the disturbance term make up the **population model.** In the two variable case, where now the paired y and x values and the ε values are made explicit by adding the i subscript, the general population model and its six assumptions are:

Population Model
An equation used to represent a theoretical relationship among variables.

$y_i = \alpha + \beta x_i + \varepsilon_i$, the ith value of the dependent variable is linearly related to the ith value of the independent variable and the ith disturbance term value, where α and β are parameters; (13.5)

$E(\varepsilon_i) = 0$, the expected value of the ith disturbance term is zero [implying that $E(y|x) = \alpha + \beta x$]; (13.6)

$\text{Var}(\varepsilon_i) = \sigma_\varepsilon^2$, the variance of the ith disturbance term is a parameter σ_ε^2, which is the same for all i; (13.7)

$\text{Cov}(\varepsilon_i, \varepsilon_j) = 0$, the disturbance terms are unrelated so their covariances are zero; (13.8)

$\text{Cov}(\varepsilon_i, x_i) = 0$, the disturbance terms and the independent variable x are unrelated so their covariances are zero; and (13.9)

$\varepsilon_i \sim N(0, \sigma_\varepsilon^2)$, the disturbances are distributed normally with zero means and constant variance σ_ε^2. (13.10)

In what follows, the parameters of a model (α, β and σ_ε) will be estimated. Based on these parameter estimates, the expected value of y, at a specific value of x, will be estimated. Next, tests of the parameter values implied by a theory will be constructed and demonstrated. In the case of the Phillips curve, for example, both α = 8.00 and

$\beta = -1.00$ will be tested against their separate alternative hypotheses that $\alpha \neq 8.00$ and $\beta \neq -1.00$. Consideration also will be given to the ways in which the assumptions of a model may be violated.

13.4 Estimation of Coefficients

Because of the chance factors inherent in sampling, the sampling error associated with different samples give rise to different paired observations on the two variables. In turn, different intercept and slope coefficients for the regression lines that best fit each scatterplot are obtained. If the samples are randomly generated, and if the population model believed to be generating the samples is correct, then across a large (infinite) number of samples, the mean of the intercepts and the mean of the slopes will equal the respective population parameters α and β. In terms of expected values, this is written as

$E(a) = \alpha$, mean of the least squares intercepts equals the population parameter α; and (13.11)

$E(b) = \beta$, mean of the least squares slopes equals the population parameter β, where again for brevity (ignoring the i subscript): (13.12)

$y = \alpha + \beta x + \varepsilon$, $\hat{y} = a + bx$, and
y is observed and $\hat{y}$ is predicted, and a
and b values are obtained by least squares.[2]

Because $E(a) = \alpha$ and $E(b) = \beta$, a and b are said to be unbiased estimators of the respective parameters α and β. It is essential to recognize that unbiasedness *does not* imply that, for any one sample, the value of the intercept a and the slope b equal the population intercept α and slope β. Equations (13.11) and (13.12) say that if a large (infinite) number of samples could be drawn, each with n matched x-y pairs, and if for each sample a least squares regression is fit, then the mean of all the intercept values would equal α. Similarly, the mean of all the slope values would equal β, if the population model is correct.

Table 13.1 contains yearly data on the unemployment and inflation rates for the United States in the 1960s. A random sample interpretation of these data is required to show that the least squares coefficient b is an unbiased estimator of β, and a is an unbiased estimator of α.[3] As discussed in Chapter 12, drawing a line through these data, as portrayed in Figure 13.3, and hand calculations with the least squares formulas (12.10) and (12.11), yield

$$b = \frac{\sum_{i=1}^{n}(x_i - \bar{x})(y_i - \bar{y})}{\sum_{i=1}^{n}(x_i - \bar{x})^2} = \frac{-12.226}{10.776} = -1.1346$$

$$a = \bar{y} - b\bar{x} = 2.52 - (-1.1346)(4.78) = 7.9432$$

TABLE 13.1
Rates of Unemployment and Inflation in the United States, 1960-1969

Year	Inflation Rate (%)	Unemployment Rate (%)
1960	1.6000	5.5000
1961	0.9000	6.7000
1962	1.8000	5.5000
1963	1.5000	5.7000
1964	1.5000	5.2000
1965	2.2000	4.5000
1966	3.2000	3.8000
1967	3.0000	3.8000
1968	4.4000	3.6000
1969	5.1000	3.5000

Source: *Economic Report of the President;* inflation rate is GNP implicit price deflator (1972 base = 100, percentage change from previous period) and unemployment rate is for civilian work force.

FIGURE 13.3 Scatterplot of Inflation and Unemployment Rates, 1960–1969

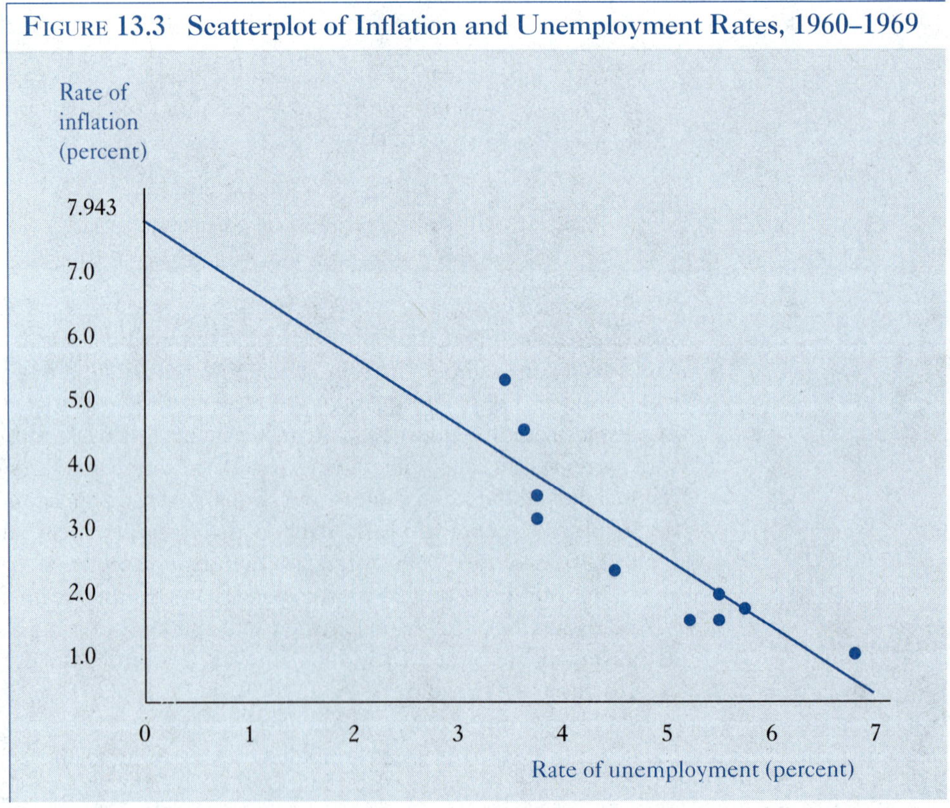

Table 13.2 Computer Output for the Least Squares Regression of the Unemployment Rate on the Inflation Rate in the 1960s

Ordinary Least Squares Estimates

Dependent Variable	INFL
Number of Observations	10.
Mean of Dependent Variable	2.52000
Std. Dev. of Dep. Variable	1.37663
Std. Error of Regression	.63096
Sum of Squared Residuals	3.1849
R-Squared	.81327

Variable	Coefficient	Std.Error	T-ratio	(Sig.Lvl)	Mean of X	Std. Dev. of X
CONSTANT	7.94319	.9402	8.449	(.00003)		
UNEM	–1.13456	.1922	–5.903	(.00036)	4.7800	1.0942

These values of b and a are also shown in the computer printout in Table 13.2 in the column headed "Coefficient."

The values $b = -1.1346$ and $a = 7.9432$ are called point estimates of the respective parameters β and α. Because of the randomness in sampling, the point estimate –1.1346 does not necessarily equal the true value of β, but it should be close. Similarly, the point estimate 7.9432 should be close to the true value of α.

We should anticipate that if the assumptions of the model are correct, then most of the x-y pairs in the sample should fall close to the deterministic part of the population model ($y^* = \alpha + \beta x$, in Figure 13.2). The best fitting least squares regression line should, in turn, give values of a and b that are close to the population α and β parameters. So until we have information to the contrary, we proceed as if the model and its assumptions are correct. That is, in the absence of any other facts, we anticipate that our calculated values of a and b will be close to the parameter values we are estimating.

"Closeness" can be defined in many different ways. For instance, the sample regression line

$$\hat{y} = 7.9432 - 1.1346x \tag{13.13}$$

can be drawn on the graph of the population model, Figure 13.4. If the sample regression line lies within the main bodies of the bell-shaped curves, as is the case in Figure 13.4, then the estimated line is consistent with the theory and the sample regression supports the Phillips curve theory that $y^* = 8.00 - x$.

The graphical demonstration of closeness in Figure 13.4 requires that the dispersion of the density function $d(\varepsilon)$ be known. But the variance of ε, σ_ε^2, is not even mentioned in Fischer and Dornbusch's discussion of the Phillips curve theory. It is a parameter to be estimated. It is estimated by s_e^2,

$$s_e^2 = \frac{\text{ErrorSS}}{n-2} \tag{13.14}$$

where ErrorSS was defined in the previous chapter by $\sum_{i=1}^{n}\left(y_i - \hat{y}_i\right)^2$

FIGURE 13.4 Population Model and Sample Regression of the Phillips Curve Relationship

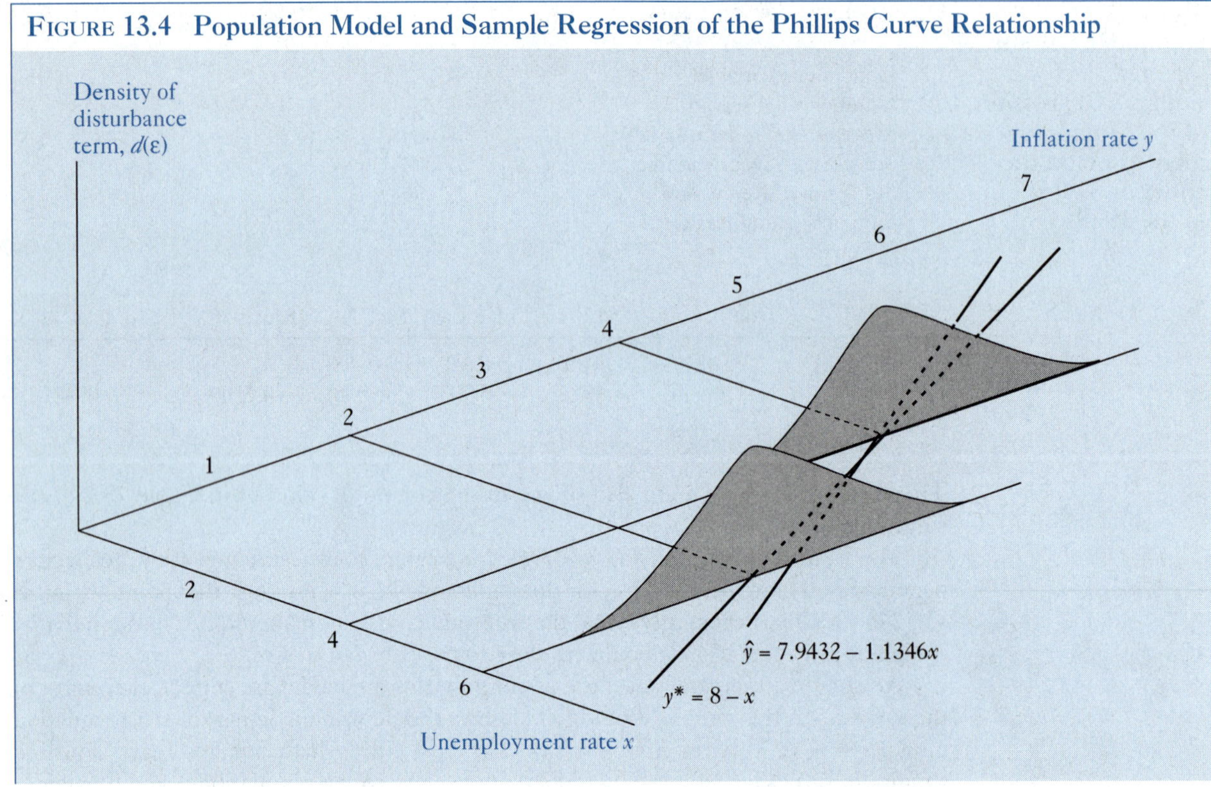

You should remember from Chapter 12 that the prediction error $e = y - \hat{y}$ and ErrorSS are calculated from values in one sample. In estimating the variance of ε using the ErrorSS, the degrees of freedom are $n - 2$, because two of the paired observations on y and x must be treated as predetermined when we use our sample specific a and b values in the calculation of $\hat{y}$. That is, two degrees of freedom are lost because the sample data is used to estimate two regressor coefficients.

If the assumed population model is correct, it is the division by $n - 2$ that makes s_e^2 an unbiased estimator of σ_ε^2. The expected value of s_e^2 is σ_ε^2, the variance of the disturbance term.[4] Remember that this does not say that for any given sample, s_e^2 equals σ_ε^2; it says that the mean of all the s_e^2 values that could be calculated will equal σ_ε^2.

The value of s_e^2 = ErrorSS/$(n - 2)$ can be calculated by hand. But as demonstrated in Chapter 12, it is typically obtained with a computer program. In Table 13.2, it is found on the line titled "Std. Error of Regression." The value of s_e^2 can also be calculated with the "sum of squared residuals,"

Standard Error of Regression
An estimator of the standard deviation of the population disturbance term in a regression.

$$s_e^2 = \frac{\text{ErrorSS}}{n - 2} = \frac{3.1849}{8} = 0.39811 = \left(0.63096\right)^2 \tag{13.15}$$

The square root of 0.39811 is called the standard error of the regression, 0.63096. The **standard error of regression,** s_e, is the estimator of the standard deviation of the population disturbance term σ_ε.

13.5 The Confidence Interval for the Expected Value of y

If the expected value of ε is zero, the expected value of y at any given x value, $E(y\,|\,x)$, is given by the deterministic equation $y^* = \alpha + \beta x$. As stated earlier, $E(y\,|\,x)$ and y^* are two different ways of saying the same thing.

The least squares regression line $\hat{y} = a + bx$ gives the point estimate of the expected value of y, at an x value. For any given sample there is no reason to believe that $\hat{y}$ will be the same as y^*, at each value of x. If the population model is correct, the expected value of $\hat{y}$ over a large (infinite) number of samples will equal y^*, for a given value of x. That is, $E(\hat{y}|x) = E(y\,|\,x) = y^*$, but randomness in sampling causes variability in $\hat{y}$, so $\hat{y}$ need not equal $E(\hat{y}\,|\,x)$.

To reflect variability in estimation, a confidence interval estimate of $E(y\,|\,x)$ can be defined. For instance, a 99 percent confidence interval for the expected value of y, at the specific x value designated by x^*, is

$$\hat{y} \pm t_{0.005} s_e \sqrt{\frac{1}{n} + \frac{\left(x^* - \bar{x}\right)^2}{\sum_{i=1}^{n}\left(x_i - \bar{x}\right)^2}} \tag{13.16}$$

where $+t_{0.005}$ is the value of t for which 0.5 percent of the area is in the upper tail and $-t_{0.005}$ is the t value for which 0.5 percent of the area is in the lower tail.[5] The z distribution could be used in Equation (13.16) if σ_ε was known and $\hat{y}$ was the only source of variability. Because σ_ε is unknown, the t distribution is used to reflect the added variability associated with s_e.

As discussed in previous chapters, the t distribution represents a family of distributions; there is a different distribution for each degree of freedom. The degrees of freedom (df) in regression analysis are determined by ErrorSS, which for two variable model, has $n - 2$ degrees of freedom. In the case of the Phillips curve regression df = $n - 2 = 8$. From Appendix Table A.4, the probability of a t value greater than or equal to 3.355, with eight degrees of freedom, is 0.005. That is,

$$P(t \geq 3.355\,|\,\text{df} = 8) = 0.005$$

At an unemployment rate of $x^* = 6$ percent, a 99 percent confidence interval for the expected inflation rate is

$$\hat{y} \pm t_{0.005} s_e \sqrt{\frac{1}{n} + \frac{\left(x^* - \bar{x}\right)^2}{\sum_{i=1}^{n}\left(x_i - \bar{x}\right)^2}} = 1.136 \pm 3.355(.631)\sqrt{\frac{1}{10} + \frac{(6 - 4.78)^2}{10.776}}$$

$$= 1.136 \pm 1.033$$

The 99 percent confidence interval for the expected rate of inflation, at an unemployment rate of 6 percent, is $0.103 \leq E(y\,|\,x = 6) \leq 2.169$, which happens to include the 2 percent inflation rate expected from the Phillips curve theory, at an unemployment rate of 6 percent.

Whenever we use Formula (13.16), there is a 0.99 probability that the resulting confidence interval will contain the expected value of y. This *does not say* that there is a 99 percent chance that the expected inflation rate is between 0.103 percent and 2.169 percent, when the unemployment rate is 6 percent. Once numerical values are substituted into Formula (13.16) there are no random variables in the expression, and no probability statement can be made. The resulting confidence interval either does or does not contain the expected value of y. The expected rate of inflation is 0.103 to 2.169 percent, for an unemployment rate of 6 percent, but the probability of 0.99 applies to the theoretical interval (13.16) and not the values calculated from this single sample.

Margin of Error
A measure of the precision of an estimator.

The quantity ±1.033 is called the **margin of error.** It reflects the imprecision in our estimate of the expected value of y. In the popular press, the words "margin of error" have come to be associated with only 95 percent confidence intervals; but all confidence intervals have margins of error. As the level of confidence decreases (e.g., from a 99 to a 95 percent confidence level) the margin of error decreases, because the t value gets smaller.

Important to notice in the margin of error for $E(y|x)$ is that it is constant at a fixed level of confidence. As x values closer to the mean value of x are considered, the margin of error gets smaller. For our 99 percent confidence interval, at $x^* = \bar{x} = 4.78$, the margin of error is only ±0.6695,

$$t_{0.005}s_e\sqrt{\frac{1}{n}+\frac{\left(x^*-\bar{x}\right)^2}{\sum_{i=1}^{n}\left(x_i-\bar{x}\right)^2}} = 3.355(.631)\sqrt{\frac{1}{10}+\frac{\left(4.78-4.78\right)^2}{10.776}} = 0.67$$

The 99 percent confidence interval for the expected value of y at a 4.78 percent unemployment level is the expected inflation rate (2.52) minus and plus the margin of error (±0.67):

$$1.85 \le E(y|x = 4.78) \le 3.19$$

which is narrower than the 99 percent confidence interval at an unemployment rate of 6 percent.

By the time x is down to an unemployment rate of 0 percent, however, the margin of error for $E(y|x)$ is a whopping ±3.15 percent. The effect on the confidence interval as the margin of error increases for values of x away from its mean can be seen in Figure 13.5. Because the margin of error is relatively large for extreme values of x, we cannot place much faith in predictions made outside the sample domain of x.

FIGURE 13.5 Confidence Interval for Rate of Inflation, 1960–1969

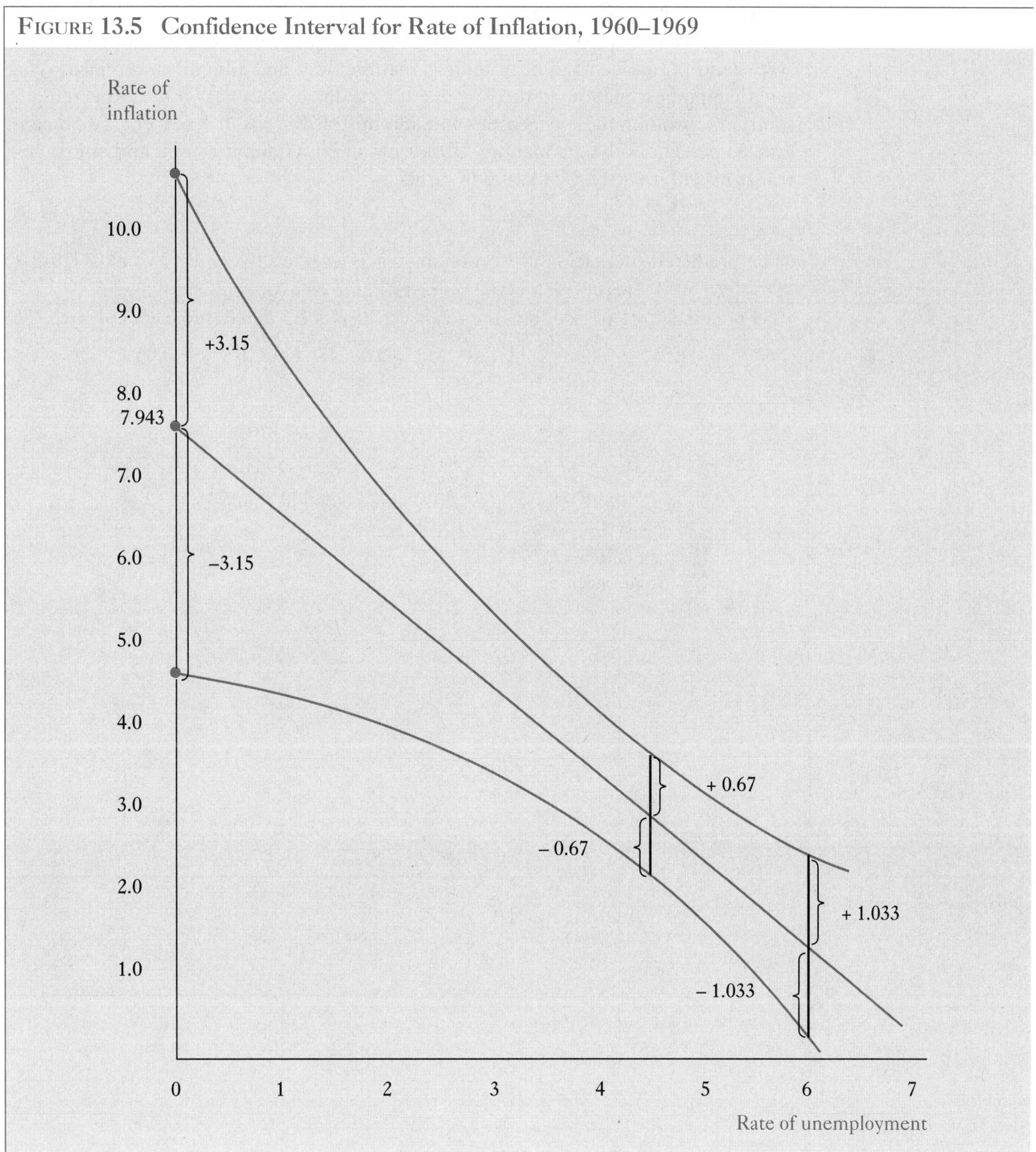

Query 13.1

Query 12.4 contained information about the percentage of women earning MBA degrees and the percentage of women in management and administrative jobs. What are the point estimates and the 95 percent confidence intervals for the expected proportion of women in management and administrative jobs, if 4 percent, 12 percent, and 33 percent of the graduating MBAs are women? Graph the point and interval estimators and identify the margin of errors.

Answer:

The predicted proportion of women managers is given by $\hat{y} = 0.1732 + 0.5929x$. At $x^* = 0.33$, $\hat{y} = 0.3688$, which is the point estimate of $E(y \mid x = 0.33)$.

For one degree of freedom, the t value that puts 2.5 percent in each tail is 12.706, and $s_e = \sqrt{\text{ESS}/n-2} = \sqrt{2.972(10^{-5})/1} = 0.00545$. Thus, $t_{0.025}s_e = 0.0692$. The mean

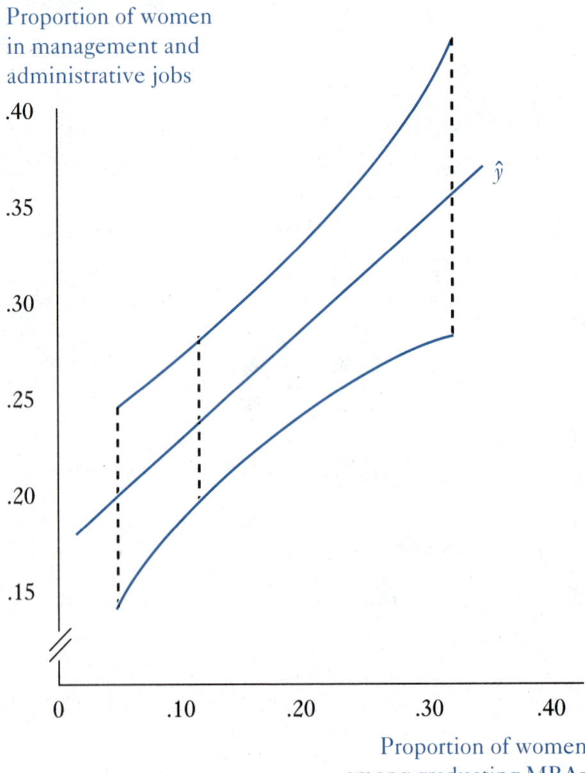

proportion of women among MBAs is $\bar{x} = 0.163$. Thus, the 95 percent confidence interval for the expected proportion of women managers, at $x = 0.33$, is

$$\hat{y} \pm t_{.025} s_e \sqrt{\frac{1}{n} + \frac{\left(x^* - \bar{x}\right)^2}{\sum_{i=1}^{n}\left(x_i - \bar{x}\right)^2}} = 0.3688 \pm 0.0692\sqrt{\frac{1}{3} + \frac{\left(0.167\right)^2}{.0449}} = 0.3688 \pm 0.0676$$

$$\text{Implying that } 0.3012 \le E(y \mid x = 0.33) \le 0.4364$$

Similarly, the point estimate and the 95 percent confidence intervals for the expected proportion of women managers, for 12 percent and 4 percent women among graduating MBAs, are

$$\text{at } x = 0.12,\ \hat{y} = 0.2443,\ 0.2020 < E(y \mid x = 0.12) < 0.2866$$

$$\text{at } x = 0.04,\ \hat{y} = 0.1969,\ 0.1402 < E(y \mid x = 0.04) \le 0.2536$$

The confidence intervals associated with these margins of error are shown in the diagram on the previous page. They are extremely wide because there are so few observations. Although a two-variable least squares regression can be calculated with only three observations, little confidence can be placed in the resulting estimates.

13.6 The Prediction Interval for an Individual Value of Y

In addition to estimating the expected value of y at given x values, $\hat{y}$ can be interpreted as providing a prediction of an individual value of y. If so, then additional variability must be recognized. The confidence interval for $E(y \mid x)$ only involves variability of $\hat{y}$ around $E(y \mid x)$, as reflected in the difference $\hat{y} - E(y \mid x)$. Any individual value of y that might occur with an x value, however, would involve the variability reflected in the difference $\hat{y} - E(y \mid x)$ plus the difference between $E(y \mid x)$ and the individual value of y that might be observed. This added variability can be seen in a comparison of Formula (13.16) and the $(1 - \alpha)$ prediction interval in Formula (13.17):

$$\hat{y} \pm t_{\alpha/2} s_e \sqrt{\frac{1}{n} + \frac{\left(x^* - \bar{x}\right)^2}{\sum_{i=1}^{n}\left(x_i - \bar{x}\right)^2} + 1} \qquad (13.17)$$

The only difference between the confidence interval for $E(y \mid x)$ and the prediction interval is that the latter is wider than the former, at any value of x^*. As with Formula (13.16), the prediction interval in Formula (13.17) reflects more imprecision in predictions the farther x^* is from $\bar{x}$. For example, at $x^* = \bar{x} = 4.78$, the 99 percent prediction interval for the individual value of y is

$$2.52 \pm 3.355(0.631)(1.049) = 2.52 \pm 2.22$$

$$\text{or} \quad 0.230 \le \text{individual value of } y \text{ at } x = 4.78 \le 4.740$$

At $x^* = \bar{x} = 6$ percent, the 99 percent prediction interval for the individual value of y is

$$1.136 \pm 2.35$$

$$\text{or} \quad -1.22 \leq \text{individual value of } y \text{ at } x = 6.0 \leq 3.49$$

EXERCISES

13.1 Using a computer program and the 60 matched pre- and midtest scores in Exercise 12.4:

a. How significant is the relationship between pretest and midtest scores? What does this significance level tell you?

b. For a pretest score of 55, what is an 80 percent confidence interval for the mean midtest score?

c. For a pretest score of 95, what is a 90 percent confidence interval for the mean midtest score?

d. Why does your answer in part b appear to make more sense than your answer in part c?

e. Why might one question the use of least squares in this situation?

13.2 In regression analysis, why is it good for prediction and estimation to have a lot of variability in the independent variable? (Hint: consider the formula for the confidence interval for $E(y \mid x)$.)

13.3 Is it possible to have a sample of paired x and y values yield a negatively sloped regression line while the deterministic part of the x and y relationship is positive in the population? Use diagrams in answering this question.

13.4 Using your information on variances and covariance calculated with the data set in Exercise 12.13, determine the point estimate of the expected value of y at $x = 5$. Form a 95 percent confidence interval for the expected value of y at $x = 5$. Form a 95 percent prediction interval for an individual value of y at $x = 5$. Why do these two interval widths differ?

13.5 A study of piece workers related the number of years of experience (to the nearest year) and the number of pieces of finished clothing returned because of imperfections. A sample of six workers is given below, where x = number of years worked, and y = number of returned clothing items.

y	13	7	5	3	3	6
x	1	2	3	3	5	4

a. For a worker with three years of experience, what is a 90 percent confidence interval for the mean returned items?

b. For a worker with three years of experience, what is a 90 percent prediction interval for returned items?

c. What is the difference between the intervals in parts a and b?

13.6 How is the confidence interval for the expected value of y at an x influenced by the size of the correlation coefficient?

13.7 The Sampling Distribution of b

Sampling Distribution of *b*
The probability density function that assigns values to the least squares estimator, b.

The "closeness" of a specific value of b to the hypothesized value β is assessed by introducing the sampling distribution of b. The **sampling distribution of *b*,** or, simply, the distribution of b, represents the probabilities of alternative values of b that could be obtained from repeated sampling. If the expected value of ε is zero, and if x and ε are not related, then the expected value of b is β. In addition, if the population disturbance term ε is distributed normally, then the distribution of b will also be normal.

If b is distributed normally, as shown in Figure 13.6, there is approximately 68 percent of the distribution within one standard deviation of the mean of b, $\beta \pm \sigma_b$. That is, the probability of obtaining a b value within the range $\beta \pm \sigma_b$ is approximately 0.68. Within two standard deviations, $\beta \pm 2\sigma_b$, there is approximately 95 percent of the distribution. Almost the entire distribution lies within three standard deviations of the mean.

Typically the magnitude of the standard deviation of the distribution of b, σ_b, is unknown. It is estimated by the standard error of b, s_b, which is calculated by[6]

$$s_b = \sqrt{\frac{s_e^2}{\sum_{i=1}^{n}(x_i - \bar{x})^2}} \tag{13.18}$$

Figure 13.6 The Sampling Distribution of *b*

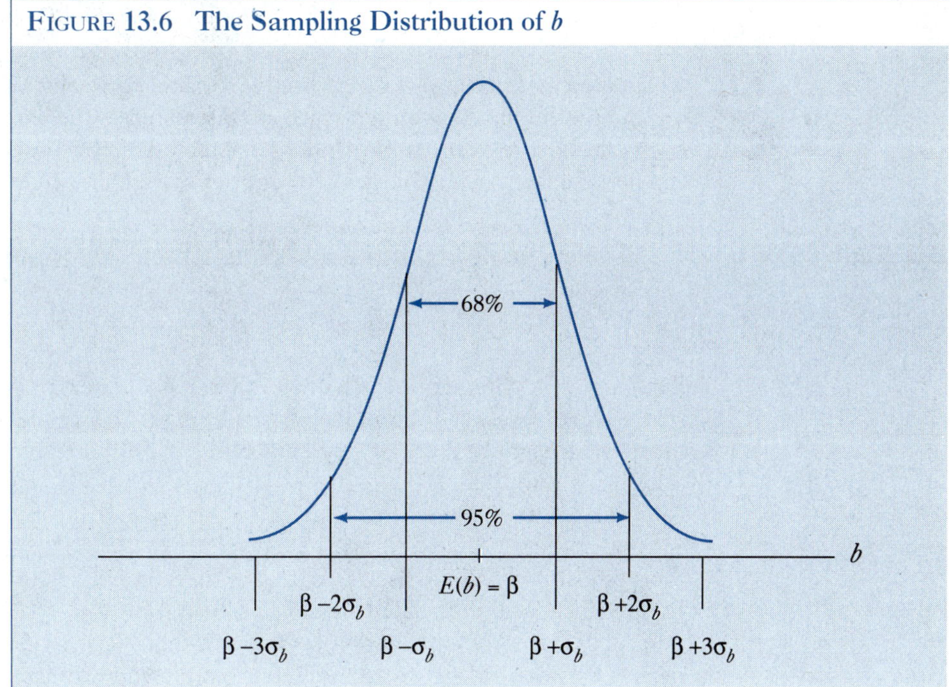

The estimated standard error of a coefficient estimator is typically calculated by computer programs. For example, in Table 13.2 the standard errors of the coefficient estimates are given in the column headed "Std. Error." For the unemployment coefficient, b, the standard error is $s_b = 0.1922$, which is calculated by

$$s_b = \sqrt{\frac{\text{ErrorSS}/(n-2)}{\sum_{i=1}^{n}(x_i - \bar{x})^2}} = \sqrt{\frac{3.1849/8}{10.776}} = 0.1922$$

When s_b is used as an estimator of σ_b, to assess how far a coefficient estimate (e.g., $b = -1.1346$) is from its hypothesized expected value ($\beta = -1.00$), it is necessary to form a "t ratio" and use the t distribution to calculate probabilities.[7] The t ratio is

$$t = \frac{b - E(b)}{s_b} = \frac{b - \beta}{s_b} \tag{13.19}$$

Computer programs calculate t ratios for an expected value of b equal to zero. For instance, in Table 13.2 under the column headed "T-ratio", the reported t ratio for the unemployment coefficient is

$$t = \frac{b - E(b)}{s_b} = \frac{-1.13456 - 0}{0.1922} = -5.903$$

↑ Assuming $E(b) = \beta = 0$

which indicates that $b = -1.13456$ is approximately 5.9 standard deviations below 0.

For the Phillips curve theory, we are interested in the closeness of $b = -1.13456$ to $\beta = -1.00$. The relevant t ratio cannot be obtained directly from a computer printout. It can be calculated using the standard error of b, which is reported on the printout.

$$t = \frac{b - E(b)}{s_b} = \frac{-1.13456 - (-1.00)}{0.1922} = -0.700$$

↑ Assuming $E(b) = \beta = -1.00$

Thus, $b = -1.13456$ is approximately 0.7 of a standard deviation below $\beta = -1.00$. To assess closeness we need the probability of being 0.7 of a standard deviation below the mean, with eight degrees of freedom. Using a computer program, such as MICROSTAT, we learn that

$$P(t \le -0.7 \mid \text{df} = 8) = 0.25$$

This probability and its associated area in the sampling distribution of b are shown in Figure 13.7. Since the shaded area for b values below $b = -1.13456$ is relatively large (25 percent of the entire distribution), we can say that $b = -1.13456$ is relatively close to the parameter value of $\beta = -1.0$. That is, our sample and estimated slope are not inconsistent with the Phillips curve theory.

FIGURE 13.7 A Sampling Distribution of b and Calculated t Value

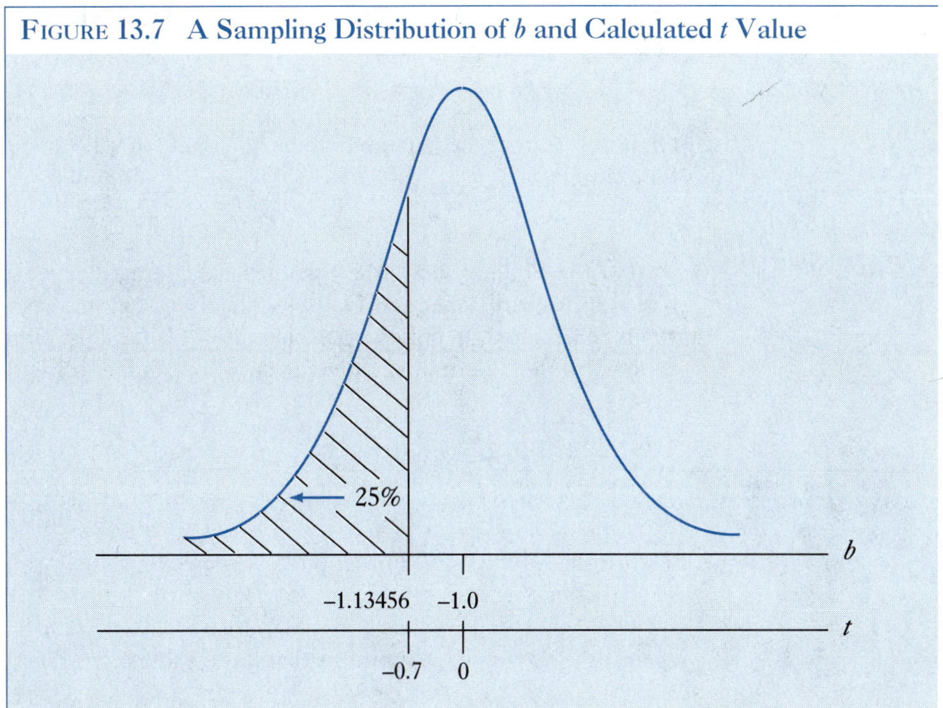

Notice that the denominator of the standard error of b is the square root of the sum of squared deviations in x. The more variability in the independent variable, the smaller the variability in the distribution of b. That is, greater variability in the independent variable makes the slope estimator more precise.

13.8 Testing Hypotheses About Individual Coefficients

Often a theory or someone's belief suggests the null and alternative hypotheses. The null hypothesis, designated as H_O, is the theory that is assumed true until contrary information is obtained. As discussed in Chapter 9 on hypothesis testing, the null hypothesis provides the value at which the sampling distribution is centered. In the case of the Phillips curve, we assumed that $E(b) = \beta = -1.0$; that is,

$$H_O: \beta = -1.0$$

The alternative hypothesis, designated H_A, is the challenging theory. It may involve a statement that allows the parameter to be any value other than that stated in the null hypothesis; for example,

$$H_A: \beta \neq -1.0$$

Or, it may restrict the parameter to be greater than the value stated in the null hypothesis; for example,

$$H_A: \beta > -1.0$$

Or, it may restrict the parameter to be less than the value stated in the null hypothesis; for example,

$$H_A: \beta < -1.0$$

For the Phillips curve, the first of these alternatives seems most appropriate (H_A: $\beta \neq -1.0$), since Fischer and Dornbusch's discussion implies that policymakers in the 1960s entertained no option other than $\beta = -1.0$. That is, either $\beta < -1.0$ or $\beta > -1.0$ is contrary to the deterministic Phillips curve presented in Figure 13.1.

13.9 Two-Tailed Test

To test the null hypothesis that β is some specific value β_O,

$$H_O: \beta = \beta_O$$

against the alternative hypothesis that β is a value other than β_O,

$$H_A: \beta \neq \beta_O$$

the ratio in Equation (13.19) is assumed to follow the t distribution, with $n - 2$ degrees of freedom. That is,

$$t = \frac{b - \beta_O}{s_b} \sim t(n-2) \tag{13.20}$$

where b is assumed to be distributed normally with mean β_O and variance σ_b^2.

As in any hypotheses testing, the probability of incorrectly rejecting the null hypothesis (Type I error) defines the calculated values of t that result in the rejection of H_O. Figure 13.8 shows the H_O: $\beta = -1.00$ and H_A: $\beta \neq -1.00$ critical rejection regions for a probability of a Type I error equal to 5 percent. According to the decision rule portrayed in Figure 13.8, if the absolute value of the calculated t ratio in Equation (13.19) exceeds 2.306, H_O: $\beta = -1.00$ is rejected in favor of H_A: $\beta \neq -1.00$. If the absolute value of the calculated t ratio does not exceed 2.306, H_O: $\beta = -1.00$ should not be rejected. In essence, this decision rule says that when the estimate of b is more than 2.306 estimated standard deviations away from the null hypothesized value of $\beta = -1.00$, we will no longer believe that β is equal to -1.00. In using this decision rule, however, we must acknowledge that 5 percent of the time we could get an estimate of b that is more than $2.306s_b$ away from $\beta = -1.00$, when in fact $\beta = -1.00$.

FIGURE 13.8 A Sampling Distribution of b and Critical t Values

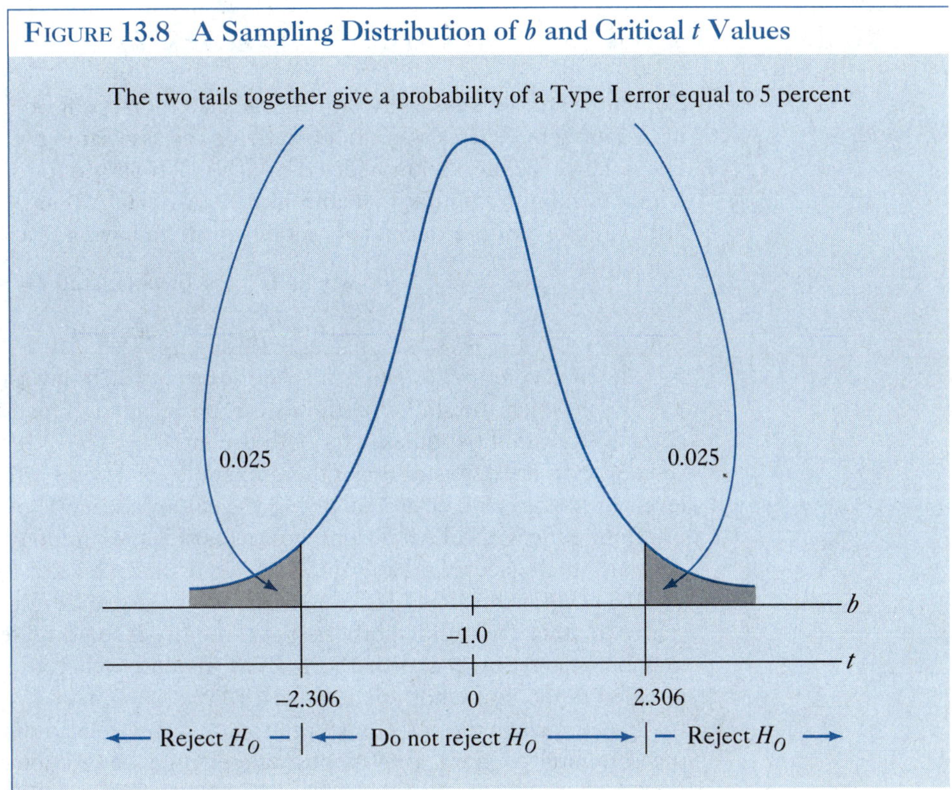

From Equation (13.20) and Figure 13.7, we know that the calculated t ratio for $b = -1.13456$ is $t = -0.7$.

$$t = \frac{b - E(b)}{s_b} = \frac{-1.13456 - (-1.00)}{0.1922} = -0.700$$

Since $b = -1.13456$ is only 0.7 estimated standard deviations below $\beta = -1.00$, while the critical t values in Figure 13.8 are minus and plus 2.306, H_0: $\beta = -1.00$ is not rejected.

We have not proven that the theoretical Phillips curve has a slope of −1.00. The sample evidence is simply not strong enough to reject this conclusion. We could be making a Type II error in not rejecting $\beta = -1.00$. In addition, any one of the other assumptions in our model (e.g., that the inflation rate is a linear function of the unemployment rate or that the disturbance term is independent of itself and the unemployment rate) could be wrong, implying that our test is biased. In the absence of any other information, however, the best we can say is that $\beta = -1.00$ cannot be rejected.

13.10 Two-Tailed Test and Computer Output

Computer printouts typically provide calculated t ratios for the test of a parameter value against zero. They also routinely provide the associated p-value for a two-tailed test. For instance, Table 13.3 provides the MINITAB output for the data in Table 13.1. As with the computer output in Table 13.2, the MINITAB output in the column headed "t-ratio" provides the calculated t ratios for the following tests

$$H_O: \alpha = 0 \text{ versus } H_A: \alpha \neq 0; \text{ calculated } t = 8.45$$

$$H_O: \beta = 0 \text{ versus } H_A: \beta \neq 0; \text{ calculated } t = -5.90$$

The next column "p" gives the significance level, or lowest probabilities of a Type I error for which the null hypotheses can be rejected. This MINITAB printout suggests that the null hypotheses for both the intercept (H_O: $\alpha = 0$) and slope (H_O: $\beta = 0$) should be rejected, regardless of the probability of Type I error, since the significance levels for both are given as "0.000," in the column headed "p" for p-value.

Unlike the MINITAB output, however, the computer output in Table 13.2 shows more decimal places. In Table 13.2, the significance level is 0.00003 for the intercept and 0.00036 for the slope. If the probability of a Type I error is set at 0.00002, for instance, neither the null hypothesis $\alpha = 0$ nor $\beta = 0$ could be rejected. This fine a distinction is not possible with the MINITAB output because significance levels are rounded to the thousandth decimal place.

Looking down the MINITAB printout in Table 13.3, at the bottom we can see that this computer program, as with most others, produces 95 percent confidence intervals for the expected value of y at the identified x values, and 95 percent prediction intervals for specific values of y at those x values. The reader can verify the correctness of the intervals shown using Formulas (13.16) and (13.17). (Notice in Table 13.3, with eight degrees of freedom, the critical t value that puts 0.025 in each tail for the 95 percent intervals is 2.306 instead of the 3.355 value used in the 99 percent intervals we calculated earlier. Discussions yet to come will address the remaining quantities in this and similar regression printouts.)

Query 13.2:
Use the SYSTAT computer printout in Query 12.7 concerning consulting services provided by CPA firms to test whether or not accountants are keeping consulting and auditing separate.

Answer:
The null and alternative hypotheses are

$$H_O: \beta = 0, \text{ consulting and auditing are not related}$$

$$H_A: \beta \neq 0, \text{ consulting and auditing are related}$$

TABLE 13.3
MINITAB Regression Output

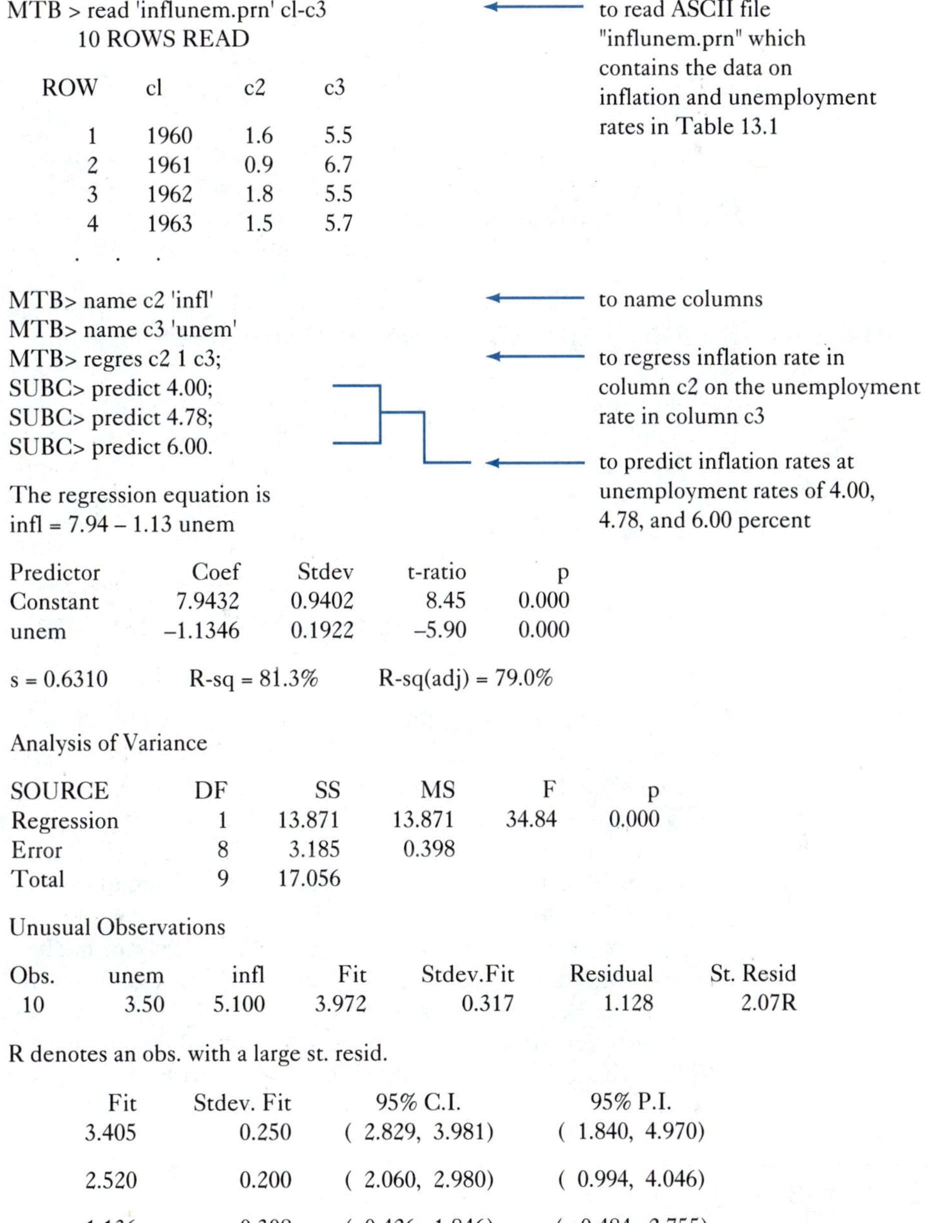

```
MTB > read 'influnem.prn' cl-c3
      10 ROWS READ

   ROW     cl      c2     c3

     1    1960    1.6    5.5
     2    1961    0.9    6.7
     3    1962    1.8    5.5
     4    1963    1.5    5.7
 .   .   .

MTB> name c2 'infl'
MTB> name c3 'unem'
MTB> regres c2 1 c3;
SUBC> predict 4.00;
SUBC> predict 4.78;
SUBC> predict 6.00.

The regression equation is
infl = 7.94 – 1.13 unem

Predictor        Coef      Stdev     t-ratio        p
Constant       7.9432     0.9402        8.45    0.000
unem          –1.1346     0.1922       –5.90    0.000

s = 0.6310        R-sq = 81.3%        R-sq(adj) = 79.0%

Analysis of Variance

SOURCE         DF        SS         MS         F         p
Regression      1    13.871     13.871     34.84     0.000
Error           8     3.185      0.398
Total           9    17.056

Unusual Observations

Obs.    unem     infl      Fit     Stdev.Fit     Residual     St. Resid
 10     3.50    5.100    3.972         0.317        1.128         2.07R

R denotes an obs. with a large st. resid.

     Fit    Stdev. Fit        95% C.I.            95% P.I.
   3.405         0.250    ( 2.829,  3.981)    ( 1.840,  4.970)

   2.520         0.200    ( 2.060,  2.980)    ( 0.994,  4.046)

   1.136         0.308    ( 0.426,  1.846)    (–0.484,  2.755)
```

The t ratio is

$$t = \frac{b - \beta_0}{s_b} = \frac{-0.354 - 0}{0.539} = -0.657$$

The $P(t < -0.657 \text{ or } t > 0.657 \mid \text{df} = 7) = 0.532$, which is far greater than typical Type I error probability levels, and H_0 is not rejected. That is, consulting and auditing do not appear to be related.

13.11 Two-Tailed Test and Confidence Intervals

As introduced in Chapter 9, reporting the significance level in hypothesis testing is helpful because it gives a sense of the strength of the evidence against the null hypothesis. Reporting a confidence interval estimate for coefficient estimates also is of value in demonstrating the sensitivity of results to sampling error. In the Phillips curve example, a 95 percent confidence interval for β is

$$b - t_{0.025}s_b < \beta < b + t_{0.025}s_b \tag{13.21}$$

$$-1.1346 - 2.306(0.1922) \le \beta \le -1.1346 + 2.306(0.1922)$$

$$-1.58 \le \beta \le -0.69$$

This 95 percent confidence interval $-1.58 \le \beta \le -0.69$ indicates that any null hypothesis value between −1.58 and −0.69 could not be rejected in a two-tailed test, for 0.05 Type I error level and this sample data. Both $\beta = -1.58$ and $\beta = -0.69$ are the end points for statistical significance at the 5 percent Type I error level—any null hypothesis value between −1.58 and −0.69 could not be rejected in a two-tailed test using this information. (Remember that probability is defined to be zero at a point for continuous probability density functions.)

Notice that once a null hypothesis of no difference has been rejected in a two-tailed test, we can do better than to simply say that a significant difference exists at the given Type I error level. From the corresponding confidence interval, we can say that β is in the direction of the sample mean. In the Phillips curve example, the 95 percent confidence interval suggests that β is negative, with −1.00 being a possible value.

Exercises

13.7 Use the data on subjective and objective probabilities given in Exercises 10.9 and cited in Exercise 12.31 to answer the following questions.

a. Test the hypothesis that there is a one-for-one tradeoff between subjective and objective probabilities.

b. What are the practical implications of your test in part a?

13.8 Use the following computer printout in parts a through d.

INDEX	NAME	MEAN	STD.DEV
INDEP.VAR.:	x	13.5000	8.3832
DEP.VAR.:	y	21.0000	12.8668

VAR.	REGRESSION COEFFICIENT	STD. ERROR	T(DF= 8)	PROB.
x	.7747	.4685	1.654	.13678
CONSTANT	10.5415			

STD. ERROR OF EST. = 11.7813 r SQUARED = .2548 r = .5047

ANALYSIS OF VARIANCE TABLE

SOURCE	SUM OF SQUARES	D.F.
REGRESSION	379.6047	1
RESIDUAL	1110.3953	8
TOTAL	1490.0000	9

a. What is the magnitude of the standard error of estimation, and how was it calculated?
b. How was $t = 1.654$ calculated?
c. Explain the relationship between $t = 1.654$ and prob = .13678.
d. Test H_0: $\alpha = 0$ versus H_A: $\alpha > 0$, at the 0.01 significance level.

13.9 Use the following computer printout in parts a through d.

Dependent Variable is	y
Number of observations is	6
Mean of Dependent Variable ...	2.00000
Std. Dev. of Dep.............	.89443
Std. Error of Regression	.29277
Sum of Squared Residuals	.34286
R-Squared	.91429

Var.	Coefficient	Std.Error	T-ratio	(Sig.Lvl)	Mean of X	Std.Dev.of X
ONE	3.14286	.2119	14.832	(.00012)	1.0000	.00000
X	–.457143	.6999E-01	–6.532	(.00284)	2.5000	1.8708

a. What is the magnitude of the coefficient of determination, and how was it calculated?
b. How was the standard error of the coefficient of X calculated?
c. Explain the relationship between $t = 14.832$ and the significance level = .00012.
d. Test H_0: $\beta = 0$ versus H_A: $\beta < 0$, at $\alpha = 0.01$.

13.10 Articles in the business press periodically discuss the relationship between the percentage drop in industrial output and the peak in jobless rates. On the following page are data on this relationship for the past 12 slumps in business activity.

Period	% Drop Industrial Output	Peak Jobless Rate
1929 – 1933	53.4	24.9
1937 – 1938	32.4	20.0
1945 – 1945	38.3	4.3
1948 – 1949	9.9	7.9
1953 – 1954	10.0	6.1
1957 – 1958	14.3	7.5
1960 – 1961	7.2	7.1
1969 – 1970	8.1	6.1
1973 – 1975	14.7	9.0
1980 – 1980	8.7	7.8
1981 – 1982	12.3	10.7
1991 – 1992*	10.0	7.0

a. Calculate the least squares regression line for the jobless rate (y) as explained by the percentage drop in industrial output (x).

b. Is there a significant relationship between the jobless rate and the drop in industrial output?

13.11 Below are data on corporate sales and the amount spent on advertising (both measured in $1,000) by different firms. Can sales be explained by advertising in a significant way?

Sales	Advertising Expense
899	61
1354	71
957	62
910	61
890	59
1055	64
1111	77
1234	79

13.12 For the data on populations in the 25 cities reported in Exercise 10.19 and cited in Exercise 12.35, is there a significant drop in populations between the two years? How does your answer here differ from the test of differences in means in Exercise 10.19? How does it differ from your answer in Exercise 12.35?

13.13 An article in *The Wall Street Journal* (February 24, 1992) stated that "for Biotech (companies), pure genius isn't enough." The article quoted Dr. Wilkerson as saying that companies must show that "their technologies work and that they can successfully turn them into effective products." On the following page are data on the number of drugs companies have in clinical trials and the number of proprietary drugs for humans on the market.

a. What is the relationship between the number of drugs in trial and the number actually in the market?

b. Is this relationship statistically significant? At what level?

	Number of Drugs in	
	Clinical Trials	Market for Human Use
Amgen	8	2
Genentech	10	3
Chiron	14	3
Centocor	5	0
Synergen	3	0
Genzyme	1	1
Biogen	3	2
Gensia	1	0
Immunex	4	2
Alliance	2	0
US Bioscience	4	1
Zoma	6	0
Medimmune	1	1
Affymax	0	0
Immune Response	1	0
Genetics Institute	3	1
Molecular Biosystems	0	0
Cytogen	1	0
Immunomedics	2	0
Liposome Co.	3	0

13.14 Explain what happens to the magnitude of s_b and $t = (b - \beta)/s_b$ as n is increased. What are the implications of this for hypothesis testing?

13.12 One-Tailed Test

Often a theory will not restrict the value of β to a specific number, or an alternative theory might be consistent with values in only one direction. For instance, in the speed limit case introduced in Chapter 12, we might wish to test whether there is a positive relationship between death rate and speed limit. For this test, the null and alternative hypotheses are

H_O: $\beta = 0$, speed limit is not related to death rate

H_A: $\beta > 0$, speed limit and death rate are positively related

Setting the probability of a Type I error at 0.05, the critical value of t, and the H_O rejection and acceptance regions are shown in Figure 13.9. Because the alternative hypothesis only allows for β values greater than that specified in the null, the entire 0.05 probability of a Type I error is put in the upper tail. (If an alternative hypothesis allows only for β values less than that specified in the null, the entire probability of a Type I error is in the lower tail.)

FIGURE 13.9 A Sampling Distribution of b and Critical t Value

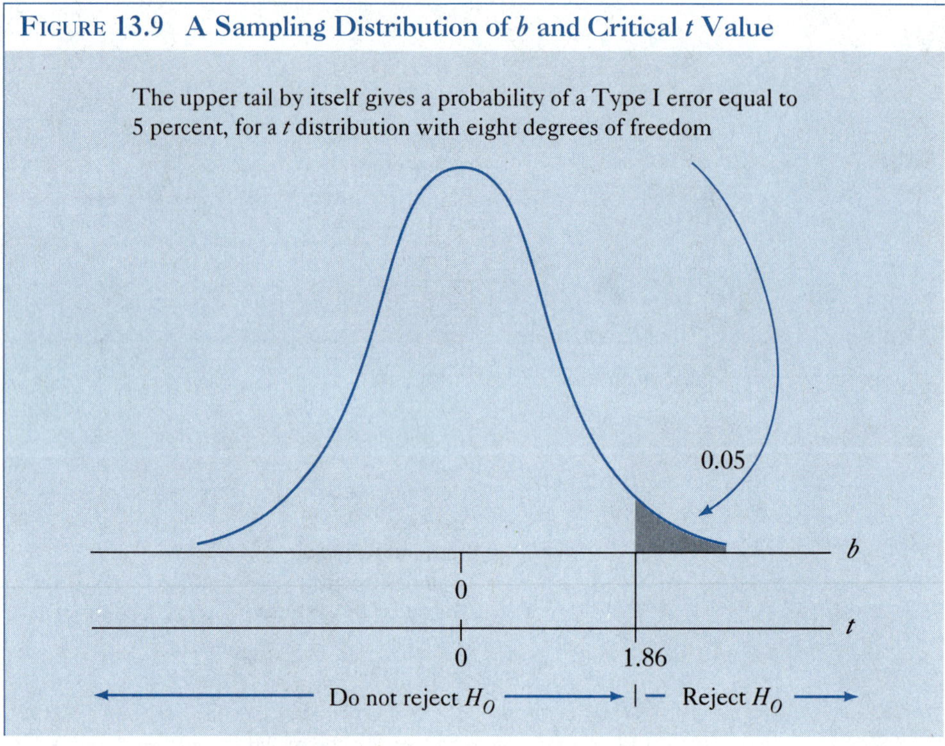

From the SAS computer printout in Table 13.4, or the MICROSTAT computer output in Table 12.7, the calculated t ratio is

$$t = \frac{b - \beta_0}{s_b} = \frac{0.075556 - 0}{0.04074} = 1.852$$

$t = 1.852$ is just short of the H_0 rejection region. The significance level or p value is 0.0504. This significance level is half the value reported in Table 13.4, because 0.1008 is the significance level for a two-tailed test where the alternative is $\beta \neq 0$. Most all computer programs report significance levels or p-values only for two-tailed tests. For our one-tailed test, we cannot reject H_0 unless the probability of a Type I error is raised to 0.0504 or higher.

As discussed in the previous chapters on hypothesis testing, raising the probability of a Type I error may allow the rejection of the null hypothesis, but at a higher likelihood of being wrong. The probability of a Type I error could be set at zero, but this would make it impossible to reject the null hypothesis when it is false and the probability of a Type II error would be at its maximum. As the probability of a Type I error is reduced, the probability of a Type II error increases. Holding the sample size fixed, the probability of a Type I error cannot be changed without inversely affecting the probability of a Type II error.

TABLE 13.4 SAS Regression Output

ANALYSIS OF VARIANCE

SOURCE	DF	SUM OF SQUARES	MEAN SQUARE	F-VALUE	PROB>F
MODEL	1	2.5689	2.5689	3.4395	0.1008
ERROR	8	5.9751	0.7469		
C TOTAL	9	8.5440			

ROOT MSE	0.8642	R-SQUARE	0.30067
DEP MEAN	4.2400	ADJ R-SQ	0.21325
C. V	0.2298		

PARAMETER ESTIMATES

VARIABLE	DF	PARAMETER ESTIMATE	STANDARD ERROR	T FOR HO: PARAMETER-O	PROB > |T|
INTERCEPT	1	–0.29333	2.460	–0.119	0.9080
SPEED	1	0.0755556	0.04074	1.852	0.1008

QUERY 13.3:

Use the information in Queries 12.4 and 12.6, concerning women with MBAs and management and administrative jobs, to test for a positive relationship between the proportion of women among MBAs and the proportion of women managers. Is there a difference in your conclusion for a probability of a Type I error of 1 percent versus 5 percent?

ANSWER:

The null and alternative hypotheses are

H_O: $\beta = 0$, the two proportions are not related

H_A: $\beta > 0$, the two proportions are positively related

With one degree of freedom, and a probability of a Type I error of 1 percent, the critical t is 31.821,

$$P(t > 31.821 \mid df = 1) = 0.01$$

With one degree of freedom, and a probability of a Type I error of 5 percent, the critical t is 6.314,

$$P(t > 6.314 \mid df = 1) = 0.05$$

The calculated t ratio is

$$t = \frac{b - \beta_O}{s_b} = \frac{0.5929 - 0}{.00545/\sqrt{.0449}} = 23$$

At the 1 percent Type I error level, the calculated t is less than the critical t, so H_O cannot be rejected. At the 5 percent Type I error level, the calculated t is greater than the critical t, so H_O can be rejected. Our conclusion is sensitive to the probability of a Type I error. Because the $P(t > 23 \mid df = 1) = 0.0276$, we can say that there is a statistically significant positive relationship between the proportion of women among MBAs and the proportion of women in management down to the 0.0276 Type I error level.

13.13 Tests of Correlation (Optional)

In bivariate analysis involving a dependent variable y and one independent variable x, the t-test of the slope of the regression line is sufficient to determine if there is a population relationship between the two variables. It can be shown that the null hypothesis

$$H_0\text{: Population correlation coefficient} = 0$$

is rejected or not rejected on the basis of rejecting or not rejecting the null hypothesis

$$H_0\text{: Population slope} = 0$$

Thus, there is no need to introduce correlation analysis in hypothesis testing within the regression framework.

There are bivariate situations, however, where the use of least squares regression analysis is questionable, but yet the correlation between variables is important. An example is the case of comparisons of survey results where the question responses are not cardinal measures. As discussed in the optional section of Chapter 12, in such a situation, Spearman's rho becomes the focus of analysis.

To test for a population relationship using Spearman's rho, the appropriate null and alternative hypotheses for rank correlation are:

H_0: the ranks in the two populations are not correlated

H_A: the ranks in the two populations are correlated

For small sample sizes ($n < 10$), tables are available for determining if Spearman's rho is significantly different from zero. If $n > 10$, the following test statistic can be used to determine if rho is significantly different from zero.

$$t = \text{rho}\sqrt{\frac{n-2}{1-\text{rho}^2}}\text{, with df} = n-2 \tag{13.22}$$

For example, the R. Alston, J. Kearl, and M. Vaughan (1992) survey of economists, and the W. Becker, M. Watts, and W. Walstad survey of print and broadcast journalists, as discussed in Chapter 12, yielded the following Spearman's rho:

$$\textbf{Spearman's rho} = 1 - \frac{6\sum d_i^2}{n^3 - n} = 1 - \frac{6(1620)}{(28)^3 - 28} = 0.556$$

as calculated from Formula (12.17), where all quantities were from Table 12.6. The t statistic is now calculated as

$$t = \text{rho}\sqrt{\frac{n \pm 2}{1 \pm \text{rho}^2}} = (0.556)\sqrt{\frac{28 \pm 2}{1 \pm (0.556)^2}} = 3.41$$

With 26 degrees of freedom and for the typical 0.05 Type I error level in a two-tailed test, the critical t is ± 2.056. Thus, the null hypothesis of no correlation between the ranking of economists and journalists on questions of economic policy can be rejected.

13.14 Violations of the Assumptions

If Assumptions (13.5) through (13.10) are true, the least squares regression line has desired properties (such as unbiased estimators with minimum variances and normally distributed sampling distributions). Assumptions (13.5) through (13.10) deal primarily with the population error term ε_i, which we never observe. We only see the residuals from the least squares regression, e_i. From the plot of e_i, however, we try to infer characteristics of the distribution of ε_i.

The plot of e_i is used to spot violations of assumptions (13.5) through (13.10). A violation of any one of these assumptions does not necessarily imply that least squares estimates are worthless. Desired properties, however, are called into question when assumptions are violated and alternative estimators must be considered. We will now deal with the identification and consequences of violations.

Normality

The construction of statistical tests and confidence intervals requires the assumption that each of the n population error terms, the ε_i, be independently and identically distributed as a normal random variable. According to Assumption (13.10),

$$\varepsilon_i \sim N(O, \sigma_\varepsilon^2)$$

If the residuals from a least squares regression are generated from a normal distribution, then a plot of the sample regression residuals $e = y - \hat{y}$ should appear random, and the histogram constructed from the frequencies of these residuals should appear normal.

Figure 13.10 shows a residual error term plot that was formed by having hypothetical yearly observations generated by resampling (as discussed in Chapter 7) from a normal density function, with a mean of zero and variance of one. The approximate normality in these sample residuals can be seen in their relative frequencies, as presented in the Figure 13.11 histogram. The larger the sample size, the more helpful the residual histogram will be in identifying randomness and normality in the residuals. It is not unusual, however, to see residuals that depart greatly from a random residual plot. Likewise, histograms of the frequencies of residuals seldom appear as normal as those in Figure 13.11.[8]

FIGURE 13.10 Plot of Residuals

−2.33 . . . . 0 . . . . 2.33

Year	Residual
1953	.996
1954	1.57
1955	.964
1956	−.643
1957	.423
1958	−.147
1959	−.339
1960	2.33
1961	.278
1962	−.655E−01
1963	−1.19
1964	.827
1965	.478
1966	−.466
1967	−.305E−01
1968	−.697
1969	1.48
1970	−2.02
1971	−.903
1972	1.48
1973	−1.70
1974	−.402
1975	.673
1976	.447
1977	−1.90
1978	−.312
1979	−.541
1980	1.22
1981	−.274
1982	−.226
1983	−1.04
1984	.386
1985	−.798
1986	−1.98
1987	1.05
1988	−.549
1989	−.132
1990	−.619
1991	1.47
1992	−.241
1993	−.934E-01
1994	1.26

FIGURE 13.11 Histogram of the Residuals in Figure 13.10

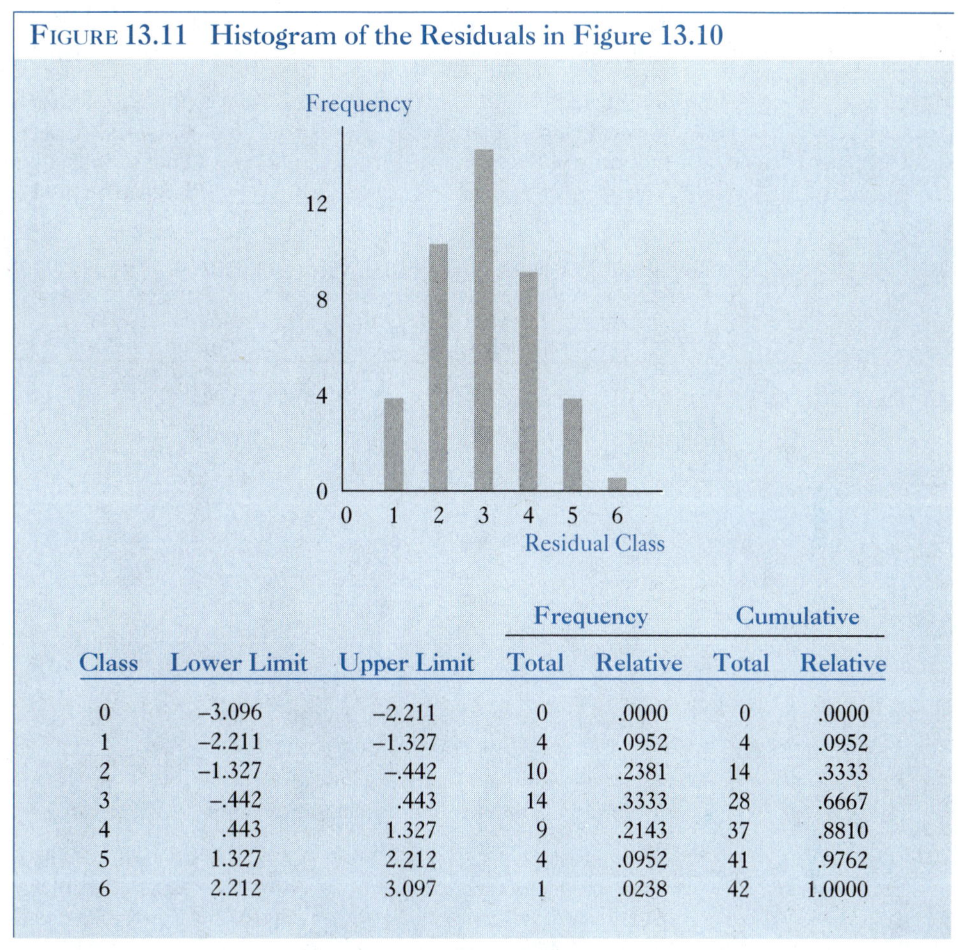

Class	Lower Limit	Upper Limit	Frequency Total	Frequency Relative	Cumulative Total	Cumulative Relative
0	−3.096	−2.211	0	.0000	0	.0000
1	−2.211	−1.327	4	.0952	4	.0952
2	−1.327	−.442	10	.2381	14	.3333
3	−.442	.443	14	.3333	28	.6667
4	.443	1.327	9	.2143	37	.8810
5	1.327	2.212	4	.0952	41	.9762
6	2.212	3.097	1	.0238	42	1.0000

Most statistical packages provide tests of normality. For instance, the ECONOMETRICS TOOLKIT has a chi-square test that is based on the skewness of the residuals. (Notice in Figure 13.11 that the residual histogram is very close to symmetric.) Unfortunately, tests of normality do not have much power; visual inspection of the plot of residuals may be just as good if not better than parametric tests.

Lack of Linearity in a Hyperbolic Scatterplot (Optional)

Assumption (13.5) states that the dependent variable y is linearly related to the independent variable x and disturbance term ε, where α and β are parameters, $y = \alpha + \beta x + \varepsilon$. The scatterplot in Figure 13.3 suggests that the relationship between the rate of inflation y and unemployment x in the 1960s is not best described by a straight line.

TABLE 13.5
Phillips Curve and Its Transformation

Panel a: Untransformed Least Squares Results, where the inflation rate is the dependent variable and the unemployment rate (UNEM) is the independent variable

R - squared	.81327	Number of Observations	10
Mean of Dep. Variable	2.5200	Std. Dev. of Dep. Var.	1.376630
Std. Error of Regr.	.6310	Sum of Squared Residuals	3.18489
Total Sum of Squares	17.056	Regression Sum of Squares	13.871

Variable	Coefficient	Std. Error	T-ratio	Prob|t|>x	Mean of X	Std. Dev. of X
Constant	7.94319	.9402	8.449	.00003		
UNEM	–1.13456	.1992	–5.903	.00036	4.78000	1.09423

PLOT OF RESIDUALS

```
UNEM    -1.13                 0                  1.13    Residual
 3.50     :                   :                 * :      1.13
 3.60     :                   :     *             :      .541
 3.80     :         *         :                   :      -.432
 3.80     :     *             :                   :      -.632
 4.50     :     *             :                   :      -.638
 5.20     :       *           :                   :      -.543
 5.50     :             *     :                   :      -.103
 5.50     :                   :  *                :      .969E-01
 5.70     :                   : *                 :      .238E-01
 6.70     :                   :      *            :      .558
```

Panel b: Transformed Least Squares Results - the inverse of the inflation rate is the dependent variable and the independent variable is the unemployment rate

R - squared	.92435	Number of Observations	10
Mean of Dep. Variable	.5149	Std. Dev. of Dep. Var.	.274319
Std. Error of Regr.	.0800	Sum of Squared Residuals	.512353E-01
Total Sum of Squares	.67726	Regression Sum of Squares	.62603 69

Variable	Coefficient	Std. Error	T-ratio	Prob|t|>x	Mean of X	Std. Dev. of X
Constant	–.637240	.1192	–5.334	.00069		
UNEM	.241028	.2438E-01	9.887	.00001	4.78000	1.09423

PLOT OF RESIDUALS

```
UNEM    -.133                 0                  .133    Residual
 3.50     :                 * :                   :      -.103E-01
 3.60     :                  *:                   :      -.319E-02
 3.80     :                   :    *              :      .338E-01
 3.80     :                   :        *          :      .547E-01
 4.50     :                   :*                  :      .746E-02
 5.20     :                   :      *            :      .506E-02
 5.50     :      *            :                   :      -.634E-01
 5.50     :*                  :                   :      -.133
 5.70     :     *             :                   :      -.700E-01
 6.70     :                   :                 * :      .133
```

Notice how the data seem to lie in a curved pattern. This curved pattern is a hyperbolic shape that may have a deterministic component,

$$\text{inflation rate} = a' + b'(1/\text{unemployment})$$

Ignoring this curvature in the data could seriously bias our representation of the relationship by a straight line regression.

The observed values of y do not occur randomly around the line $\hat{y} = a + bx$. The hyperbolic curvature in the inflation–unemployment data in Figure 13.3 can also be seen in the residuals in Table 13.5, panel a, as a "<" or (sideways) "U"-shaped pattern where residuals are first positive, then negative, and then positive again. If residuals occurred randomly around the regression line, there would be no apparent pattern. Randomly occurring residuals in time series data should shift back and forth with no more than a few positive residuals occurring and then no more than a few negative residuals occurring with no apparent pattern. In Chapter 15, statistical tests for a lack of pattern in the time series residuals will be introduced. For now, it is sufficient to recognize that there should be no apparent pattern in the residuals, as shown in Figure 13.10.

If there is a nonrandom pattern in the residuals, one way to remove the pattern is to "undo" the mathematical function that might be responsible. In the case of the hyperbolic Phillips curve data in Figure 13.3, and "<" shaped residuals in Table 13.5, panel a, a data transformation worth trying is to invert the inflation rate and then use this inversion as the dependent variable. That is, the regression

$$(1/\text{inflation rate}) = a' + b'(\text{unemployment rate})$$

should not have a "<" pattern in the residuals. Panel b of Table 13.5 provides these transformed results. Notice that the residual plot no longer has the telltale "<" shape that is a characteristic of the hyperbolic scatterplot. The slope ($b' = 0.241$) is now positive, but the relationship between the inflation rate and unemployment rate is still negative because the dependent variable is the inverse of the inflation rate. That is, if the unemployment rate goes from 4 percent to 5 percent, the dependent variable (1/inflation) is predicted to rise by 0.241, but this means the inflation rate must fall.

Researchers are always tempted to compare the coefficients of determination to assess alternative specification. Unfortunately, the r^2 values of 0.81327 and 0.92435 in Table 13.5 do not pertain to the same things. $r^2 = 0.81327$ indicates that 81.327 percent of the variability in the inflation rate is "explained by" the unemployment rate, while $r^2 = 0.92435$ states that 92.435 percent of the variability in the inverse of the inflation rate is "explained by" the unemployment rate. These coefficients of determination apply to different dependent variables and thus cannot be compared.

Lack of Linearity in Scatterplots with Multiplicative Growth (Optional)

Problems of nonlinearity often arise because economic variables tend to grow multiplicatively over time, at a constant rate but not at a constant amount. For instance, if an econometrician starts earning \$100 per hour and, with each subsequent year of experience, earnings rise to \$110, \$120, \$130, \$140 . . . , then earnings are growing by the constant amount of \$10 per year of experience. Alternatively, the raise could be at 10 percent per year of experience, giving rise to the series \$110, \$121, \$133.10, \$146.41. . . . This latter compensation policy has earnings determined by compounding, as with interest on a bond. It gives rise to a series that grows exponentially over time, as shown in Figure 13.12.

Algebraically, if the starting earnings of the person in year 0 are y_0, then annual raises at a rate of b imply earnings in year t of

$$y_t = (1 + b)^t y_0$$

With continuous compounding at rate b, earnings grow exponentially by

$$y_t = e^{bt} y_0$$

where $e = 2.718$, Euler's constant. A least squares regression of y_t on years of experience (x) would miss this compounding growth. The straight line fit to the scatterplot in Figure 13.12 underestimates future values. However, earnings could be "scaled" or transformed to create a new variable that is linear in time. Such a transformation can be accomplished by using the natural logarithms. Fortunately, computer programs will automatically do these transformations so we do not need to know much about logarithms, although some knowledge of the lognormal distribution introduced in Chapter 6 is helpful.

FIGURE 13.12 Scatterplot of Earnings, at 10 Percent Per Year Growth Rate

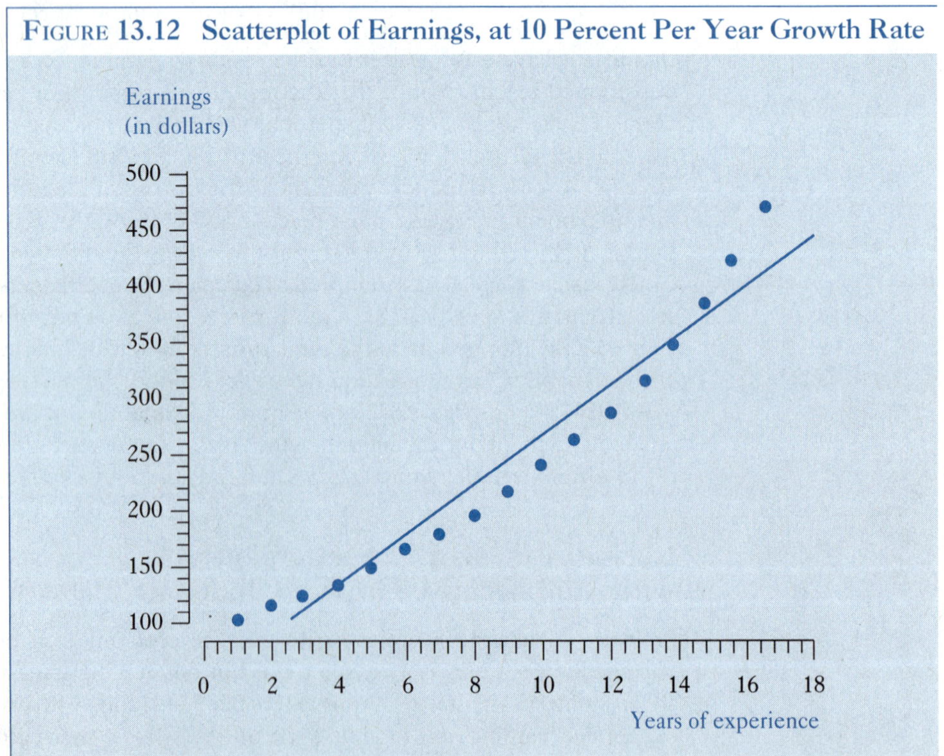

In Chapter 6, we saw that natural log transformations have a greater effect on larger values than on smaller ones; thus, this transformation will tend to straighten the pattern of values that grow exponentially. It can provide a method of pulling outliers into line with the rest of the data. In Table 13.6, for example, the earnings data have been straightened by the natural logarithm transformation. Figure 13.13 shows the straight line relationship between $\ln y$ and x (years of experience). This so-called semilogarithmic earnings model,

$$\ln(\text{earnings}) = \alpha + \beta(\text{experience}) + \varepsilon,$$

was introduced by Jacob Mincer (1974). According to Robert Willis (1986): "the Mincer earnings function has been one of the great success stories of modern labor economics. It has been used in hundreds of studies using data from virtually every historical period and country for which suitable data exist." (p. 526) Similarly, this semilogarithmic transformation has been used extensively in finance, accounting, marketing, and almost every other area of business. It is the key transformation for scaling variables that tend to change at proportional rates rather than by constant amounts.

The reason for a log transforming of y is to get an unbiased and efficient estimate of the slope coefficient β. It is not necessarily for better prediction of y. In fact, an unbiased estimate of the mean y at each x cannot be obtained directly from a regression of $\ln y$ on x, because logarithms are nonlinear transformations. Furthermore, the r^2 from a regression of $\ln y$ on x cannot be compared to a regression of y on x, because the former involves an explanation of $\ln y$ while the latter is an explanation of y. These and other peculiarities of the log transformation of the dependent variable were addressed by econometrician Arthur Goldberger (1968) more than 25 years ago, although many applied researchers continue to ignore them.[9]

TABLE 13.6 Logarithm of Earnings

Years of Experience x	Earnings y	Logarithm of Earnings $\ln(y)$
1	100.00	4.6052
2	110.00	4.7005
3	121.00	4.7958
4	133.10	4.8911
5	146.41	4.9864
6	161.05	5.0817
7	177.16	5.1771
8	194.87	5.2723
9	214.36	5.3677
10	235.79	5.4629
11	259.37	5.5583
12	285.31	5.6536
13	313.84	5.7489
14	345.23	5.8442
15	379.75	5.9395
16	417.73	6.0348
17	459.50	6.1301

FIGURE 13.13 Scatterplot of Logarithm of Earnings

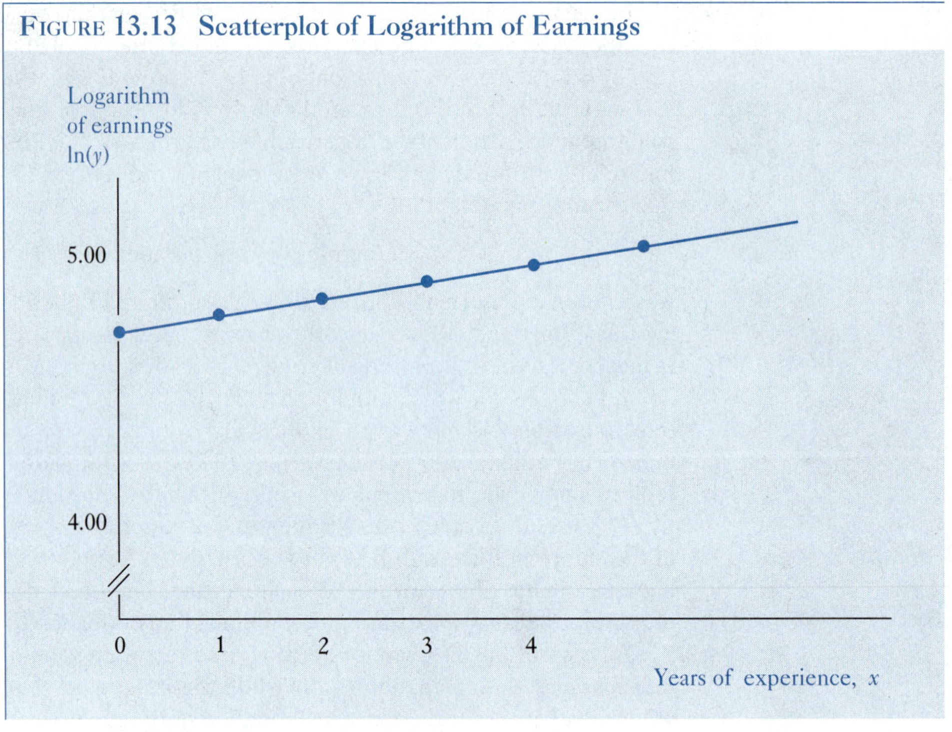

Some Other Data Transformations (Optional)

There are many other transformations that can be used to straighten the data. For instance, if the scatterplot looks like that in Figure 13.14, x can be transformed by logarithm to form the regression model

$$y = \alpha + \beta \ln(x) + \varepsilon$$

Of course, one could also fit a full logarithm specification of the form

$$\ln(y) = \alpha + \beta \ln(x) + \varepsilon$$

This so called "double log" specification is commonly associated with the Cobb-Douglas production function found in economics. In its simplest form, where capital is fixed, the Cobb-Douglas production function presents output as a function of the labor input and an error:

$$\text{Output} = \alpha \text{Labor}^{\beta} \varepsilon$$

The double log transformation yields

$$\ln(\text{Output}) = \ln(\alpha) + \beta \ln(\text{Labor}) + \ln \varepsilon$$

FIGURE 13.14 Nonlinear Scatterplot Which Could Be "Straightened" by a Logarithm Transformation of x

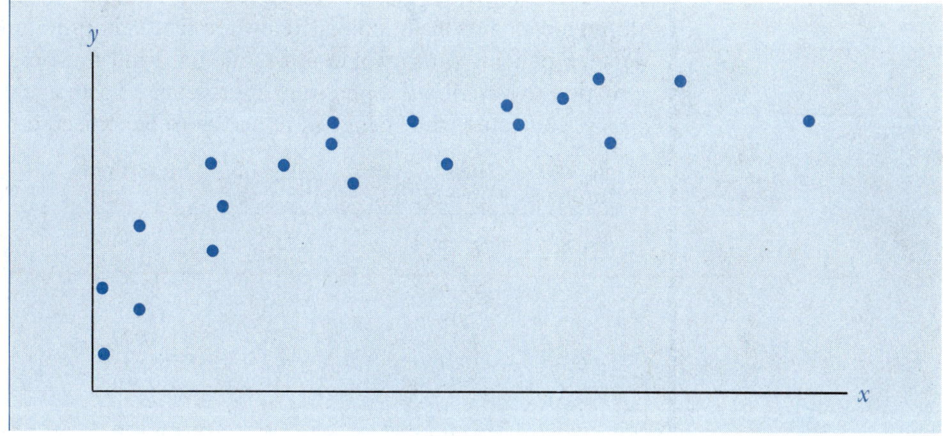

FIGURE 13.15 Nonlinear Scatterplot Which Would Require the Fitting of a Polynomial

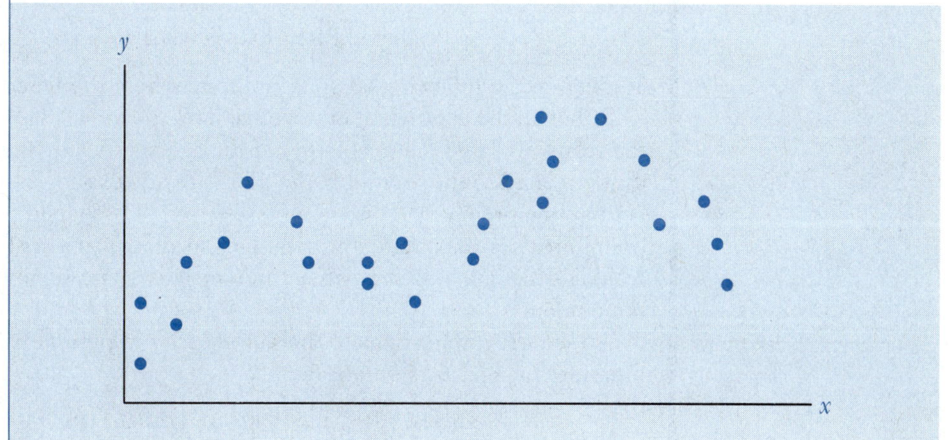

Even if the data look like that in Figure 13.15, a variable transformation can produce a linear relationship between the transformed variables. For instance, a polynomial of the form

$$y = \alpha + \beta_1 x + \beta_2 x^2 + \beta_3 x^3 + \beta_4 x^4 + \varepsilon$$

might be an appropriate specification, when curvature is highly pronounced. The fitting of polynomials is considered in more detail when multiple regression is discussed in the next chapter.

QUERY 13.4:

As evidenced by an American Insurance ad in *The Wall Street Journal* (June 14, 1988), insurance companies often use their rates on term life insurance to promote business through comparison shopping. Below are first-year premiums for three term life insurance policies with a value of \$1 million. Find the best fitting single regression equation to describe the premium at each age. From age 62.5 to 63.5, what is your best prediction of the change in premium to be expected for a \$1 million policy?

Age at Purchase	First Year Premium	Age at Purchase	First Year Premium	Age at Purchase	First Year Premium
35	590	35	470	35	460
40	720	40	690	40	685
45	920	45	1000	45	1030
50	1255	50	1455	50	1460
55	1865	55	1920	55	2130
60	3130	60	3285	60	3400
65	5320	65	4900	65	5245
70	8015	70	7500	70	7865
75	11015	75	10570	75	11790

ANSWER:

Letting PREM stand for the premium (y) and AGE for the age of the purchaser (x), the least squares regression of PREM on AGE yields

$$\text{Predicted PREM} = -9908.61 + 246.611\text{AGE}$$

as given in the following computer output, where Predicted PREM is $\hat{y}$.

Although the coefficient of correlation (R-square = 0.83346) is relatively high, and the relatively large t ratio (11.186) of the age coefficient suggests that age is significantly related to the premium, the plot of residuals shows that age is not linearly related to the premium. This "<" or (sideways) "U"-shaped residual plot suggests the premium rises exponentially with age, so that fitting a straight line gives rise to residuals that are at first positive, then negative, and finally positive again and thus not random.

To see the implications of the nonlinearity in premiums, consider the predicted premiums at ages 62.5 and 63.5:

$$\text{Predicted PREM} = -9908.61 + 246.611(62.5) = 5504.58$$

$$\text{Predicted PREM} = -9908.61 + 246.611(63.5) = 5751.19$$

These predicted premiums are far in excess of the actual premiums at ages 60 and 65.

A logarithm transformation of PREM is used to create the variable ln(PREM).

$$\text{Predicted } \ln(\text{PREM}) = 3.34960 + 0.0792908\text{AGE}$$

as given in the following computer printout.

Ordinary Least Squares		Dependent Variable	PREM
R-squared	.83346	Number of Observations	27
Mean of Dep. Variable	3655.0000	Std. Dev. of Dep. Var.	3553.770208
Std. Error of Regr.	1478.9817	Sum of Squared Residuals	.546847E+08
Total Sum of Squares	.32836E+09	Regression Sum of Squares	.27368E+09

Variable	Coefficient	Std. Error	T-ratio	Prob|t|>x	Mean of X	Std.Dev.of X
Constant	−9908.61	1246.	−7.955	.00000		
AGE	246.611	22.05	11.186	.00000	55.00000	13.15587

PLOT OF RESIDUALS

```
AGE     -.320E+04            0             .320E+04    Residual
35.0    :                   :           *       :      .187E+04
35.0    :                   :          *        :      .175E+04
35.0    :                   :          *        :      .174E+04
40.0    :                   :    *              :      764.
40.0    :                   :    *              :      734.
40.0    :                   :    *              :      729.
45.0    :                 * :                   :      -269.
45.0    :                  *:                   :      -189.
45.0    :                  *:                   :      -159.
50.0    :            *      :                   :      -.117E+04
50.0    :             *     :                   :      -967.
50.0    :             *     :                   :      -962.
55.0    :        *          :                   :      -.179E+04
55.0    :        *          :                   :      -.174E+04
55.0    :         *         :                   :      -.153E+04
60.0    :        *          :                   :      -.176E+04
60.0    :         *         :                   :      -.160E+04
60.0    :          *        :                   :      -.149E+04
65.0    :              *    :                   :      -801.
65.0    :           *       :                   :      -.122E+04
65.0    :              *    :                   :      -876.
70.0    :                   :   *               :      661.
70.0    :                   :*                  :      146.
70.0    :                   :  *                :      511.
75.0    :                   :             *     :      .243E+04
75.0    :                   :          *        :      .198E+04
75.0    :                   :                  *:      .320E+04
```

The residual plot no longer has the "<" pattern. One might argue, however, that the residuals now have a slight "S" pattern. The residuals still may be somewhat related to age, but not to the degree apparent before they were transformed. (Although tempting, we cannot compare *R*-square values here because the *R*-square value of 0.99284 pertains to the logarithm of premiums and not the premiums themselves.) Because this transformed data appear to be straighter than the original, they may be more appropriately described by a straight line drawn through them.

At ages 62.5 and 63.5, the predicted logs of the premiums are

$$\text{Predicted } \ln(\text{PREM}) = 3.34960 + 0.0792908(62.5) = 8.305275$$

$$\text{Predicted } \ln(\text{PREM}) = 3.34960 + 0.0792908(63.5) = 8.384565$$

The value 8.305275 is the predicted logarithm of the premium at age 62.5. To convert natural logarithms back to dollars, an antilogarithm transformation is required to undo the logarithm. Using a calculator with the exponent function key (usually designated by e^x) or a computer program, gives the predicted dollar premium $4,045.15, at an age of purchase of 62.5 years. That is,

$$4045.15 = \text{antilog}(8.305275) = e^{8.305275}$$

where e is 2.718, the base of the natural logarithm. Similarly, the predicted dollar premium at age 63.5 is $4,378.96.

$$4378.96 = \text{antilog}(8.384565) = e^{8.384565}$$

These premium predictions are more in keeping with the magnitudes of the actual premiums at ages 60 and 65.

The percentage change in premiums from age 62.5 to 63.5 can now be seen to be 8.25 percent:

$$\frac{(4378.96 - 4045.15)}{4045.15} = 0.082519$$

This percentage change could have been obtained without predicting the premiums since for this log specification, $e^b - 1$ is the percentage change associated with each age. That is, regardless of age

$$\frac{\text{Change in } \hat{y}}{\hat{y}} = e^b - 1 = e^{0.07929} - 1 = 0.082519$$

Typically this percentage change implied by the slope coefficient is emphasized in the estimation of a log transformed dependent variable. This emphasis is based on the nonlinearity of log transformations. The antilogs of predicted values of y involve nonlinear transformations, so the predicted values of y cannot be shown to be unbiased estimators of the expected values of y at each value of x. On the other hand, the slope coefficient b obtained from a least squares regression of $\ln y$ on x continues to provide an unbiased estimator of β in the population model $y = e^{(\alpha + \beta x + \varepsilon)}$.

Ordinary Least Squares		Dependent Variable	Ln(PREM)
R-squared	.99284	Number of Observations	27
Mean of Dep. Variable	7.7106	Std. Dev. of Dep. Var.	1.046896
Std. Error of Regr.	.0904	Sum of Squared Residuals	.204139
Total Sum of Squares	28.496	Regression Sum of Squares	28.292

Variable	Coefficient	Std. Error	T-ratio	Prob|t|>x	Mean of X	Std.Dev.of X
Constant	3.34960	.7610E-01	44.015	.00000		
AGE	.792908E-01	.1347E-02	58.862	.00000	55.00000	13.15587

PLOT OF RESIDUALS

```
AGE        -.255                 0                 .255     Residual
35.0        :                   :                  *:    .255
35.0        :                   : *                 :    .280E-01
35.0        :                   :*                  :    .645E-02
40.0        :                   :   *               :    .580E-01
40.0        :                   :*                  :    .155E-01
40.0        :                   :*                  :    .818E-02
45.0        :            *      :                   :    -.933E-01
45.0        :                  *:                   :    -.993E-02
45.0        :                   :*                  :    .196E-01
50.0        :     *             :                   :    -.179
50.0        :                *  :                   :    -.314E-01
50.0        :                 * :                   :    -.280E-01
55.0        :     *             :                   :    -.180
55.0        :       *           :                   :    -.151
55.0        :               *   :                   :    -.467E-01
60.0        :              *    :                   :    -.583E-01
60.0        :                  *:                   :    -.993E-02
60.0        :                   : *                 :    .245E-01
65.0        :                   :     *             :    .757E-01
65.0        :                  *:                   :    -.651E-02
65.0        :                   :    *              :    .615E-01
70.0        :                   :      *            :    .891E-01
70.0        :                   : *                 :    .227E-01
70.0        :                   :    *              :    .702E-01
75.0        :                   :*                  :    .106E-01
75.0        :                *  :                   :    -.306E-01
75.0        :                   :     *             :    .786E-01
```

QUERY 13.5:
A *Business Week* (March 30, 1992) article entitled "Better Schools, Not Just More School," quoted Chris Pipho, Education Commission of the States, saying, "it's difficult to extrapolate that if we added 10% to the calendar, it would increase achievement by 10%." The following data on the days spent in school per year and average

math scores of 13-year-olds on an achievement exam administered to students around the world were provided for the analysis of the relationship between the days of school and test scores. Is there a significant one for one relationship between percentage changes in school days and test scores?

Country	Days Spent in School	Scores
China	251	80
South Korea	222	73
Switzerland	207	71
Former U.S.S.R.	210	70
France	174	64
Israel	215	63
Canada	188	62
Britain	192	61
United States	180	55
Jordan	191	40

ANSWER:

As implied by Chris Pipho's quote, the relationship between test scores and days of school is not linear; Pipho's quote is about a percentage change in days of school and a percentage change in test scores. This relationship can be modeled by a regression of the logarithm of test scores on the logarithm of days of school, because the change in the logarithm of a variable is a percentage change.[10] Thus the regression to be estimated is

$$\ln(\text{SCORES}) = \beta_1 + \beta_2 \ln(\text{DAYS}) + \varepsilon$$

where ln( ) is the natural logarithm of the identified variable. The required data for estimation is now given by

	ln(DAYS)	ln(SCORES)
China	5.5255	4.3820
South Korea	5.4027	4.2905
Switzerland	5.3327	4.2627
Former U.S.S.R.	5.3471	4.2485
France	5.1591	4.1589
Israel	5.3706	4.1431
Canada	5.2364	4.1271
Britain	5.2575	4.1109
United States	5.1930	4.0073
Jordan	5.2523	3.6889

The computer results of a regression of lnSCORES on lnDAYS are

Dependent Variable: Ln(SCORES) Mean = 4.1420 Std. Dev. = .191627
Sum of Squares: Residual = .210016 Regression = .12047 Total = .33049
Std. Error of Regr. = .1620 R-squared = .36453 N = 10

Variable	Coefficient	Std. Error	T-ratio	Prob|t|>x	Mean of X	Std.Dev.of X
Constant	−1.45061	2.611	−.556	.59372		
Ln(DAYS)	1.05368	.4919	2.142	.06455	5.30768	.10980

To test if a percentage change in the school day calendar will produce an equal percentage increase in achievement, we need to test if $\beta_2 = 1.00$ versus $\beta_2 \neq 1.00$. From the above regression, we see that $b_2 = 1.05368$, with an estimated standard deviation of 0.4919. Thus, the relevant t ratio is $t = (1.05368 - 1.00)/0.4919 = 0.11$, which is far less than the absolute value of the critical t test statistic at any of the typical Type I errors levels. The null hypothesis that $\beta_2 = 1.00$ cannot be rejected. We conclude that Pipho's assertion that a percentage increase in days of school will not produce an equal percentage increase in test scores is wrong. This test suggests that the data are consistent with a 10 percent increase in days of school being associated with a 10 percent increase in test scores, but, we have not yet addressed the appropriateness of the "double log" functional form.

To assess whether logarithms of both DAYS and SCORES produces a straight line, we look either at the residuals from the regression or the scatterplot of the data. The scatter diagram for the logged data is shown below. Notice that Jordan is an "outlier" that makes the scatterplot appear nonlinear in the logarithms.

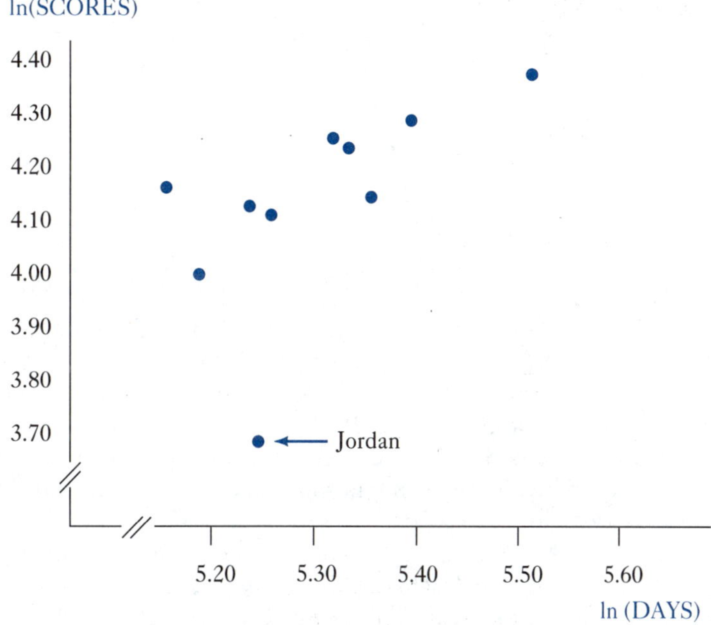

The scatterplot of the original raw data appears "more linear" than the double logged data, as can be seen in the scatter diagram on the following page:

Pipho may be wrong on two counts. First, if there is a nonlinear relationship between scores and days of school, as in percentage changes, then the hypothesis that a 10 percent increase in days of school raises scores by 10 percent cannot be rejected. Second, it appears that the relationship between days of school and test scores is better described by a linear model of the type SCORES = $\beta_1 + \beta_2$DAYS + ε. Of course, with so few sample observations, any statement about the correct model specification is highly speculative.

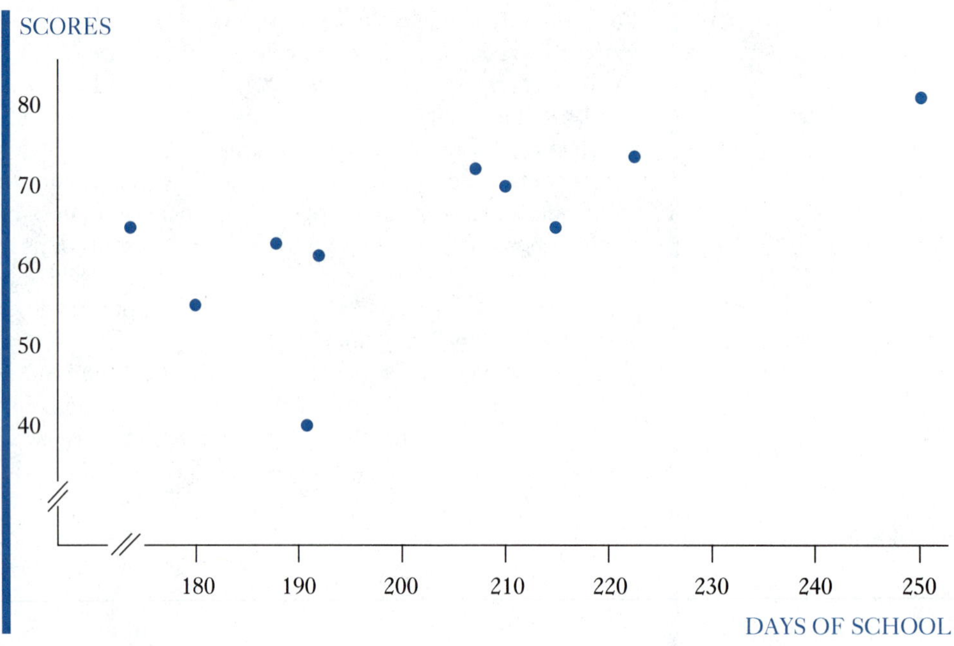

Nonconstant Error Term Variability (Optional)

Assumption (13.7) states that the variance of the disturbance term is a constant. When $\text{Var}(\varepsilon) = \sigma_\varepsilon^2$, as in Figure 13.2, ε is said to be homoscedastic. Figure 13.16 shows a situation where the variance of the disturbance term increases with *x*. (The bell-shaped distribution is wider at higher *x* values.) When Var(ε) is not a constant, ε is said to be **heteroscedastic.** Heteroscedasticity can result in least squares standard errors that are overly large and, in some situations, heteroscedasticity may bias the least squares estimators, as well.

Heteroscedasticity A situation in which the variance of ε is not fixed.

When data are collected over time, as in the inflation and unemployment case, it is not unusual for the disturbance term in one period to be related to the disturbance term in the next. This is called autocorrelation. Autocorrelation, or more generally **serial correlation,** will bias the least squares estimation of the population variance; it may thus affect least squares coefficient estimators and their standard errors.

Serial Correlation A situation in which the disturbance term in one time period is related to the disturbance term in another.

The problem of heteroscedasticity and serial correlation may result in variance estimates of the least squares estimator *b* that are overly large compared to variance estimates that could be obtained by other estimation techniques. Just as variable transformations may be used to straighten the data, transformations may be employed to make the error term homoscedastic or to remove the effects of correlation over time. These techniques are discussed in the following chapters on regression analysis.

FIGURE 13.16 Population Model with a Heteroscedastic Disturbance Term

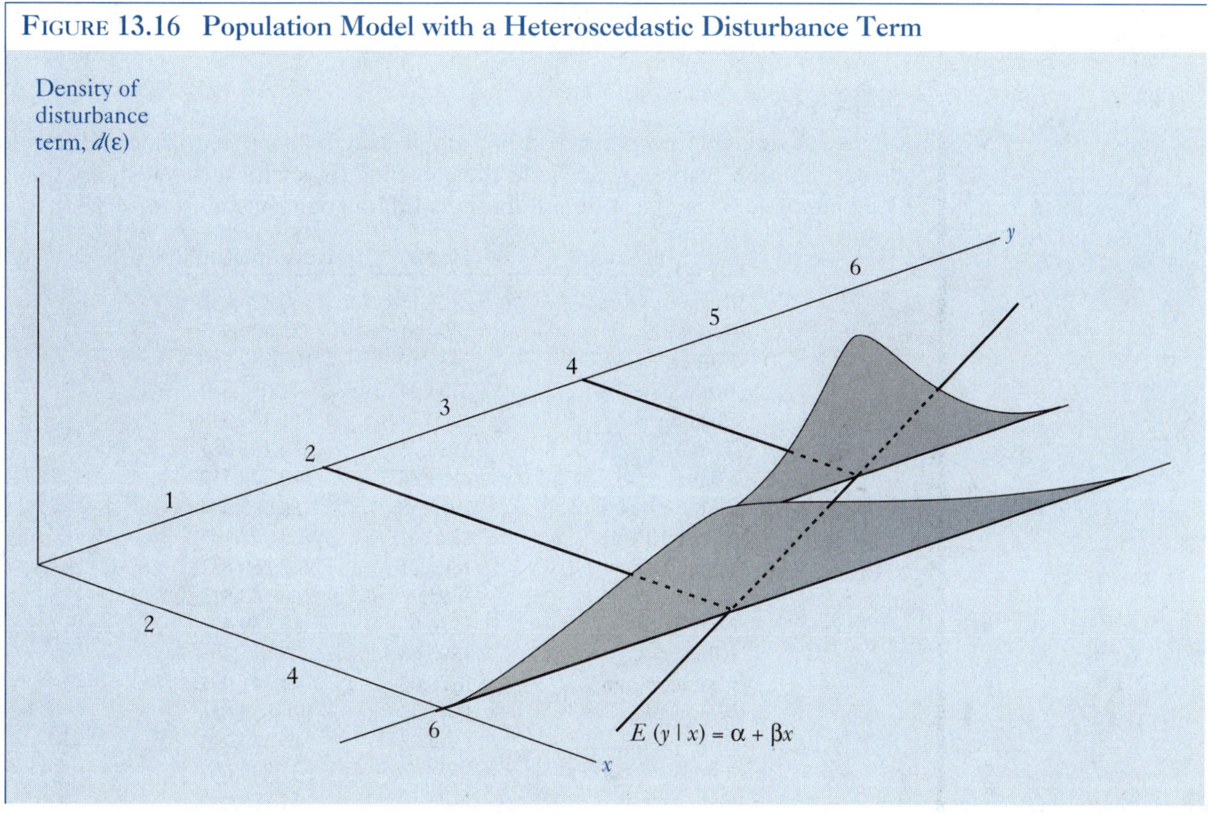

Regressor Error Term Correlation (Optional)

Assumption (13.6) states that the expected value of the disturbance term is zero and Assumption (13.9) states that it is unrelated to the regressor (or independent variable). These two assumptions play a crucial role in showing that a and b are unbiased estimators of α and β. If $E(\varepsilon) \neq 0$ or $\text{cov}(\varepsilon, x) \neq 0$, then there is no reason to believe that $E(b) = \beta$ or that $E(a) = \alpha$.

Simultaneous Relationship
A situation in which two or more variables are determined jointly, i.e., y affects x and x affects y.

A violation of Assumptions (13.6) and (13.9) occurs when there is a **simultaneous relationship** between inflation and unemployment (unemployment affects inflation and inflation affects unemployment), or because variables that are related to inflation and unemployment were left out of the equation specification. Variables that are not explicitly part of the regression specification are part of the disturbance term. Thus, the disturbance term in the inflation regression will be related to the unemployment rate to the extent that omitted variables are correlated with unemployment. As we will see in the next chapter, problems caused by omitted variables can be corrected by introducing these variables into the analysis as additional explanatory variables. Problems caused by a simultaneous relationship between y and x will require multiple step estimation procedures that are beyond the scope of this book.

QUERY 13.6:

A *Parade Magazine* (December 23, 1990) article stated, "Some magazines sell fewer ads but raise the cost per page," and thus revenue rises with fewer ads published. Based on the following data from Publishers Information Bureau, is this true?

Magazine	Pages	Revenue (in $)
Business Week	2845.15	160,326,035
People	2494.61	235,476,537
Forbes	2485.10	108,139,559
Bride's	2354.94	41,814,684
Fortune	2320.36	111,347,345
TV Guide	2186.78	236,131,981
Vogue	2159.14	78,218,413
The Economist	1953.38	16,995,384
Modern Bride	1892.42	30,851,147
Sports Illustrated	1806.71	227,289,831
Time	1793.96	246,748,515
New York	1774.58	28,759,371
Elle	1590.63	47,027,305
Cosmopolitan	1578.02	94,341,991
Newsweek	1542.73	170,638,646
Entrepreneur	1504.95	14,450,795
The New Yorker	1472.83	37,849,662
GQ	1403.82	34,220,045
Glamour	1375.27	69,992,476
Rolling Stone	1351.60	50,458,141
U.S. News & WR	1324.73	105,721,620
Architectural Digest	1306.54	34,839,841
American Way	1273.38	17,425,033
Good Housekeeping	1224.98	116,537,952
Family Circle	1218.11	99,365,045

ANSWER:

There are several ways that the statements in the article could be interpreted. Temporarily restricting the analysis to a bivariate relationship between dollars of revenue (y is REVENUE) and pages devoted to ads (x is PAGES), the article implies that there is a negative relationship between these two variables. As seen in the first of the following printouts, however, the slope of the regression line is positive ($b = 62426$) and not negative, as implied by the article. The size of the residuals, however, appears to be positively related to PAGES (and in turn REVENUE), suggesting both error term and regressor correlation and heteroscedastic errors. Error term and regressor correlation bias coefficient estimation and heteroscedasticity implies that standard errors are overly large.

A log transformation of REVENUE seems to correct the problems with the residuals, as seen in the plot of residuals in the second printout, where lnREV is the dependent variable and the log of REVENUE. Again, there is a positive relationship between revenue and pages devoted to advertisements. Because the *p*-value for a one-tailed test is 0.04082, even at the 5 percent Type I error level the alternative hypothesis of a positive relationship can be accepted.

Ordinary Least Squares		Dependent Variable is	REVENUE, n = 25
Mean of Dep. Variable	96598694.0400	Std. Dev. of Dep. Var.	75508668.810113
Sum of Squared Residuals	.116500E+18	Total Variation	.13684E+18
Regression Variation	.20337E+17	R-Squared	.14862

Variable	Coefficient	Std. Error	T-ratio	Prob|t|>x	Mean of X	Std.Dev.of X
Constant	−.138571E+08	.5693E+08	−.243	.80986		
PAGES	62426.0	.3115E+05	2.004	.05701	1769.38879	466.30786

PLOT OF RESIDUALS

```
                -.149E+09                 0                   .149E+09
PAGES                                                                        Residual
                ...................................................
.122E+04        :                        :     *                  :         .372E+08
.122E+04        :                        :        *               :         .539E+08
.127E+04        :                *       :                        :         -.482E+08
.131E+04        :                  *     :                        :         -.329E+08
.132E+04        :                        :     *                  :         .369E+08
.135E+04        :                     *  :                        :         -.201E+08
.138E+04        :                       *:                        :         -.200E+07
.140E+04        :                 *      :                        :         -.396E+08
.147E+04        :                 *      :                        :         -.402E+08
.150E+04        :             *          :                        :         -.656E+08
.154E+04        :                        :              *         :         .882E+08
.158E+04        :                        :*                       :         .969E+07
.159E+04        :                 *      :                        :         -.384E+08
.177E+04        :             *          :                        :         -.682E+08
.179E+04        :                        :                       *:         .149E+09
.181E+04        :                        :                    *   :         .128E+09
.189E+04        :            *           :                        :         -.734E+08
.195E+04        :         *              :                        :         -.911E+08
.216E+04        :                *       :                        :         -.427E+08
.219E+04        :                        :                  *     :         .113E+09
.232E+04        :                     *  :                        :         -.196E+08
.235E+04        :         *              :                        :         -.913E+08
.249E+04        :                  *     :                        :         -.331E+08
.249E+04        :                        :               *        :         .936E+08
.285E+04        :                       *:                        :         -.343E+07
.....................................................................................
```

```
Ordinary Least Squares                     Dependent Variable is          LnREV, n = 25
Mean of Dep. Variable          18.0562     Std. Dev. of Dep. Var.             .872245
Sum of Squared Residuals       15.9586     Total Variation                   18.259
Regression Variation            2.3009     R-Squared                           .12601
.......................................................................................
Variable    Coefficient    Std. Error   T-ratio   Prob|t|>x   Mean of X    Std.Dev.of X
.......................................................................................
Constant    16.8814        .6663        25.335    .00000
PAGES       .664000E-03    .3646E-03     1.821    .08164      1769.38879      466.30786
```

PLOT OF RESIDUALS

```
PAGES      -1.53                0                 1.53      Residual
.122E+04   :                   :        *          :        .724
.122E+04   :                   :           *       :        .879
.127E+04   :     *             :                   :       -1.05
.131E+04   :              *    :                   :       -.383
.132E+04   :                   :        *          :        .715
.135E+04   :                  *:                   :       -.422E-01
.138E+04   :                   :  *                :        .269
.140E+04   :            *      :                   :       -.465
.147E+04   :              *    :                   :       -.410
.150E+04   :  *                :                   :       -1.39
.154E+04   :                   :            *      :        1.05
.158E+04   :                   :     *             :        .433
.159E+04   :               *   :                   :       -.271
.177E+04   :        *          :                   :       -.885
.179E+04   :                   :               *   :        1.25
.181E+04   :                   :             *     :        1.16
.189E+04   :        *          :                   :       -.893
.195E+04   : *                 :                   :       -1.53
.216E+04   :                *  :                   :       -.140
.219E+04   :                   :           *       :        .947
.232E+04   :                   : *                 :        .106
.235E+04   :        *          :                   :       -.896
.249E+04   :                  *:                   :       -.325E-01
.249E+04   :                   :        *          :        .739
.285E+04   :                   : *                 :        .122
```

EXERCISES

13.15 The Scripps Howard News Service (*HT*, June 27, 1993) circulated an article on getting your car prepared for summer travel. The author states, "For every 10 percent increase in weight, you lose 4 percent in fuel efficiency."

a. If the dependent variable (fuel use) was measured in miles per gallon and the independent variable (vehicle weight) was measured in pounds, write an equational form that would give a slope coefficient consistent with the percentage changes cited.

b. At the extremes of vehicle weight, why would you question the percentage changes cited?

13.16 (Optional section required) An economist states that a \$100 investment in a new type of portfolio can be expected to yield the following returns over a 10-year period. (Note: at the time of the investment, year 0, the investment is worth \$100; at the end of the first year, it is worth \$110; at the end of year 2, it is worth \$120, and so on.)

Year:	0	1	2	3	4	5	6	7	8	9	10
Return:	100	110	120	130	150	160	180	200	215	240	265

To calculate the rate of return, a financial analyst obtains the following regression results, where y = return and x = year.

```
VAR.        REGRESSION COEFFICIENT   STD. ERROR   T(DF= 9)   PROB.
year               16.3636              .7755       21.100    .00000
CONSTANT           88.1818

STD. ERROR OF EST. = 8.1340      r SQUARED = .9802        r = .9900

                                              STANDARDIZED RESIDUALS

OBSERVED  CALCULATED  RESIDUAL  -2.0                 0                  2.0
100.000      88.182    11.8182    |                   |             *     |
110.000     104.545     5.4545    |                   |     *             |
120.000     120.909     -.9091    |                  *|                   |
130.000     137.273    -7.2727    |         *         |                   |
150.000     153.636    -3.6364    |               *   |                   |
160.000     170.000   -10.0000    |      *            |                   |
180.000     186.364    -6.3636    |            *      |                   |
200.000     202.727    -2.7273    |                 * |                   |
215.000     219.091    -4.0909    |               *   |                   |
240.000     235.455     4.5455    |                   |  *                |
265.000     251.818    13.1818    |                   |               *   |
```

The financial analyst concludes that "this regression fits well, because the correlation coefficient is very high, and that for each additional year the investment will return a statistically significant \$16.36." Comment on the validity of this statement.

13.17 (Optional section required) The *1980 Census* suggests that electrical engineers could expect to earn the following amounts at the indicated ages.

Age:	30	40	50	60	70
Earnings:	\$22,600	\$28,284	\$31,193	\$31,505	\$33,357

a. Fit a least squares regression to the dependent variable earnings and independent variable age; report the regression equation and all relevant statistics.

b. Construct a scatterplot of the data, and plot your regression line; show your five residuals.

c. Comment on the fit of your regression equation and the scatterplot's implications for the correctness of the least squares assumptions.

13.18 (Optional section required) An article by Padma Desai on "Reforming the Soviet Grain Economy" in the *American Economic Review* (May 1992) reported that "yields and outputs are estimated from the equation $y_t = AB^t u_t$, where y_t denotes either yield or output, t denotes time, A and B are the parameters, and u_t is the error term . . . the 1981–1989 period is characterized by the highest annual growth rate of yield, at 4.5 percent." What do the letters A and B represent in this population model?

13.19 (Optional section required) Below is a relatively famous data set constructed by F.J. Anscombe, *American Statistician*, February 1973.

	x	y_1	y_2	y_3	x_4	y_4
1	10.0	8.04	9.14	7.46	8.0	6.58
2	8.0	6.95	8.14	6.77	8.0	5.76
3	13.0	7.58	8.74	12.74	8.0	7.71
4	9.0	8.81	8.77	7.11	8.0	8.84
5	11.0	8.33	9.26	7.81	8.0	8.47
6	14.0	9.96	8.10	8.84	8.0	7.04
7	6.0	7.24	6.13	6.08	8.0	5.25
8	4.0	4.26	3.10	5.39	19.0	12.50
9	12.0	10.84	9.13	8.15	8.0	5.56
10	7.0	4.82	7.26	6.42	8.0	7.91
11	5.0	5.68	4.74	5.73	8.0	6.89

a. Obtain the descriptive statistics for each of these six variables.

b. Estimate the intercept and slope for four simple regressions of y_1 on x, y_2 on x, y_3 on x, and y_4 on x_4.

c. What are the unique features of this data set?

13.20 (Optional section required) Use a computer program and the Anscombe data set in Exercise 13.19 to get printouts of the scatterplots and fitted regression lines for the four regressions: y_1 on x, y_2 on x, y_3 on x, and y_4 on x_4. Which of the four scatterplots are consistent with the assumptions required to make an inference about the parameters in the population model $y = \beta_1 + \beta_2 x + \varepsilon$? For each scatterplot, explain which, if any, of the assumptions may be violated.

13.21 (Optional section required) Below are age-earnings profiles shown in Anil Deolalikar, "Gender Differences in the Return to Schooling and in School Enrollment Rates in Indonesia," *Journal of Human Resources*, Fall 1993, pp. 920–921. For each of the four diagrams state which, if any, of the least squares assumptions appear to be violated. For the

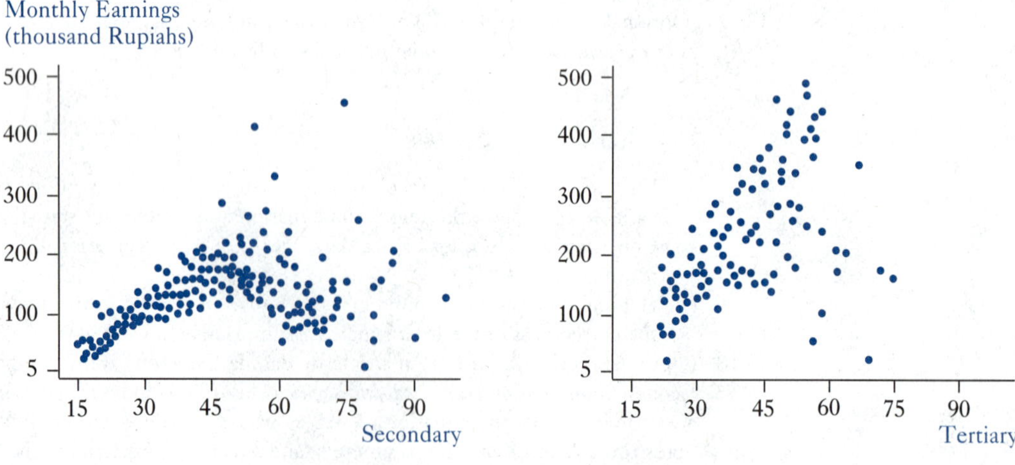

Age-Earnings profile for men, by schooling level

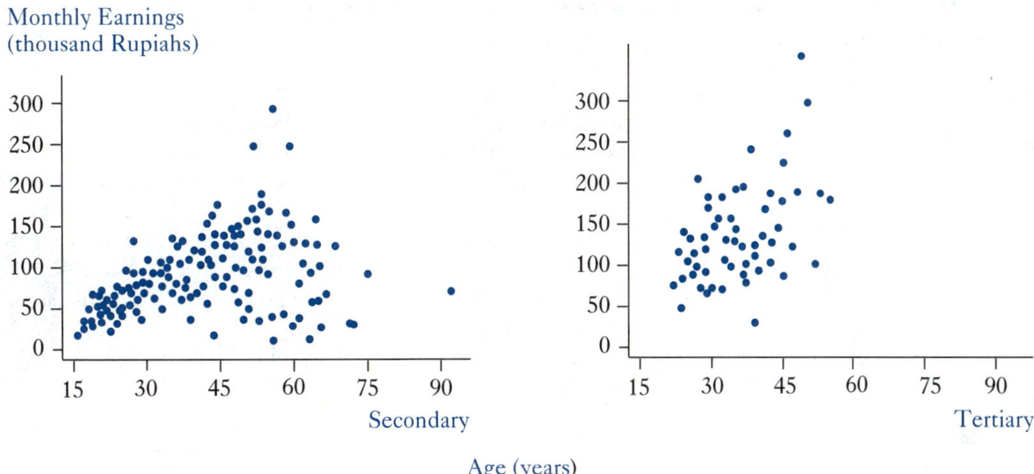

Age (years)
Age-Earnings profile for women, by schooling level

scatterplots that appear to violate the least squares assumption, write a transformed equation form that might be tried to correct the problem. State why your proposed transformed equation form might be appropriate.

13.15 Concluding Comments

Chapter 12 introduced least squares and correlation to represent and measure the strength of the relationship between paired observations in a sample. In this chapter, the techniques of classical regression analysis were extended to estimate and test population parameters. This process of statistical inference requires the building of a model to represent the population. The correctness of the assumptions that make up the model is critical to the internal validity of the inference process. As shown, analysis of the residuals of the regression provides a means for assessing the appropriateness of those assumptions.

The value of any statistical technique rests on its use in making correct inferences about the population. For predictive success, classical regression analysis involves more than just a determination of the statistical significance of an individual coefficient; consideration must be given to the overall fit of the regression, as well. The sample data and sample regression must conform to the assumptions of the population model. The coefficient of determination, while important in describing the strength of a sample relationship between two variables, has a small role in this regard. The plot of residuals, on the other hand, has a large role to play in assessing the appropriateness of the assumed population model.

The next chapters on regression analysis extend the simple regression model presented in this chapter to those involving multiple independent variables. An emphasis is placed on the building of models that are consistent with the underlying theory and sample data.

Chapter Exercises

13.22 Answer true or false to the following; if false state why.

a. The least squares estimator b is the most efficient estimator of β if the standard error of b is smaller than all alternative estimators of β.

b. If the $E(\varepsilon) = 0$ and x and ε are not related, then the least squares estimator b is an unbiased estimator of β.

c. As the standard error of b gets larger, the margin of error gets larger.

d. If a 95 percent confidence interval for the expected value of y, at $x = 20$, is $10 < E(y \mid x = 20) < 30$, then there is a 0.95 probability that the mean of y is between 10 and 30, for $x = 20$.

e. If a 95 percent prediction interval for an individual value of y is 20 ± 9, as given in the SAS computer printout

X	PREDICTED VALUE	LOWER 95% CL INDIVIDUAL	UPPER 95% CL INDIVIDUAL
4.0000	20.0000	11.0000	29.0000

then a 95 percent confidence interval for the expected value of y, at $x = 4$, could be $10 < E(y \mid x = 20) < 30$.

f. If a researcher states that there is a significant statistical relationship between x and y, then we know that the null hypothesis of no relationship is rejected irrespective of the α level specified.

g. The estimator b is a random variable for which $b = \beta$.

h. A 90 percent confidence interval for the expected value of y, at the mean of x, indicates that nine out of 10 times, on average, $\bar{y}$ will be covered by the confidence interval.

i. Hypotheses tests of the least squares slope are based on the t distribution, which requires that the sampling distribution of b be distributed normally.

j. The sampling distribution of b will be distributed normally if the population disturbance term ε is distributed normally.

k. The sole purpose of regression analysis is prediction of the dependent variable.

l. Although researchers use the t distribution to test H_O: $\beta = \beta_0$ versus H_A: $\beta \neq \beta_0$, where β_O is a value other than 0, this is not strictly correct because the standard deviation of the b is calculated under the assumption that its mean is zero.

13.23 Exercise 12.38 provides data on the days spent in school per year and average math scores on exams administered to students around the world.

a. At the 0.05 Type I error level, is there a relationship between days in school (x) and math scores (y)?

b. Construct a 95 percent confidence interval for the slope.

c. How does the confidence interval in part b relate to the hypothesis test in part a?

13.24 For the data in Exercise 12.5, use your calculations from Exercise 12.17 to test if there is a relationship between the independent and dependent variables. State the null and alternative hypothesis, and test at the 0.05 Type I error level.

13.25 Use the computer printout and your answers in Exercise 12.30 to determine a 95 percent confidence interval for the expected value of y at the sample mean value of x. Construct a 95 percent confidence interval for the expected value of y at $\bar{x} + 2s_x$; why is the second interval wider than the first?

13.26 Using the *PC Magazine* (September 29, 1987) information in Exercise 12.8 on calculation speed and disk access time, estimate the linear relationship suggested by the salesman who is cited in that exercise. What are the independent and dependent variables? As the 0.01 Type I error level, test whether there is a positive relationship between these two variables. How does your answer here differ from that in Exercise 12.8?

13.27 For the data set in Exercise 12.9, test if there is negative relationship between x and y, at the 0.01 Type I error level.

13.28 Use the information in Exercise 12.10 to calculate the standard error of the regression and the standard error of b. How could you use these two quantities?

13.29 Using the data set in Exercises 12.10, test whether each x and y relationship is significant at the 0.05 Type I error level.

13.30 Using the data in Exercise 12.14, for a candidate who receives 1,200 votes in the primary, construct a 95 percent confidence interval estimate of the expected number of votes this candidate will receive in the general election. State what this confidence interval says in a manner sufficiently clear that even the candidate can understand its interpretation.

13.31 Using the data on T-bill rates and price inflation and your computer output from Exercise 12.15, construct a 90 percent confidence interval for the expected inflation rate in the next year, if the T-bill rate this year is 7 percent?

13.32 Use the data on horse speed and jockey experience in Exercise 12.34 to answer the following questions.

a. Test the hypothesis that there is a significant relationship between horse speed and jockey experience, at the 10 percent Type I error level.
b. For a jockey with three years of experience, what is a 90 percent confidence interval for the mean horse speed?
c. For a jockey with three years of experience, what is a 90 percent prediction interval for an individual horse speed?

13.33 Use Daniel Seligman's (*Forbes*, March 8, 1982) projected versus actual deficit data in Exercise 10.10 to estimate a least squares regression of actual deficits (y) as a function of projected deficits (x). Are the projections a good prediction of the actual deficits that are realized?

13.34 In Exercise 12.39, *Journal of Economic Literature* data were provided on the most frequently cited journals in economics. Another ranking by economists of these journals in terms of their "quality" was also provided. Are the two rankings correlated? (Test at the 0.05 Type I error level.)

13.35 In Exercise 12.40, Dun and Bradstreet data was provided on the number of business failures by industry type and selected years. Based on 1985 and 1987 data, is the ranking of the business failures by industry the same across years?

13.36 Following are data on average annual percent changes in employment (employ) and industrial production (product), over the 1982–1988 period, for the United States and other G-7 nations. To assess the relationship between "employ" and "product," least squares regression results are also provided from the MICROSTAT computer package.

	employ	product
Canada	2.40%	5.90%
France	–.10	1.40
West Germany	.50	2.30
Italy	.50	2.40
Japan	1.10	4.70
United Kingdom	1.40	3.10
United States	2.40	4.90

Source: *Economic Report of the President,* February 1990, p. 45.

NAME	MEAN	STD.DEV.
product	3.5286	1.6520
employ	1.1714	.9656

VAR.	REGRESSION COEFFICIENT	STD. ERROR	T(DF= 5)	PROB.
product	.5317	.1086	4.895	.00449
CONSTANT	–.7046			

STD. ERROR OF EST. = .4395 r SQUARED = .8274 r = .9096

ANALYSIS OF VARIANCE TABLE

SOURCE	SUM OF SQUARES	D.F.	MEAN SQUARE	F RATIO	PROB.
REGRESSION	4.6286	1	4.6286	23.964	4.493E-03
RESIDUAL	.9657	5	.1931		

a. What does the slope of the regression line indicate?

b. What can be concluded about the population relationship from this slope coefficient?

13.37 For the regression in Exercise 13.36, if the percentage change in production is 3.10 percent, the predicted change in employment is approximately ___________.

13.38 From this computer printout for Exercise 13.36 we know that the total sum of squared deviations (TSS) is____________________ and the Error sum of squares (ErrorSS) is _______________. The coefficient of determination is ___________, which indicates what?

13.39 A *Wall Street Journal* (September 25, 1992) article, "Female Enrollment Falls in Many Top M.B.A. Programs," showed the percentage of women at these schools falling to an average of 25.3 percent. Use the information in Queries 12.4 and 12.6 about the percentage of women earning MBA degrees and the percentage of women in management and administrative jobs to obtain the point estimate and the 90 percent confidence intervals for the expected proportion of women in management and administrative jobs, if 25.3 percent of the graduating MBAs are women.

13.40 An article in *The Wall Street Journal* (October 12, 1992) reported the average compensation of CEOs in seven countries. Exercise 12.42 reported information on CEO compensation and number of headquartered companies in seven countries for the world's 100 largest public companies. Is there a statistically significant relationship between CEO compensation (y) and the number of headquartered companies in each country (x)?

13.41 Exercise 12.46 provides data on car weight and mileage. For a car of 2,910 pounds, what is the 95 percent prediction interval for gas mileage? Why is the reporting of this prediction interval superior to the point estimate calculated in Exercise 12.46?

13.42 Use a computer program and the data on the computer disk accompanying this book, in the ASCII file "EX13-42.PRN," to estimate the relationship between time in the emergency room (Y) and the age of the 196 patients (X). Is there a significant relationship between these two variables? What are the practical implications of your findings?

13.43 Below are the data that accompanied a *Business Week* (June 7, 1993) article on economic growth. At the 5 percent Type I error level, is there a significant relationship between the inflation rate (y) and the unemployment rate (x)?

Annual Average for 1982–1992

	Inflation Rate %	Unemployment Rate %	Misery Index %	Index of Political Stability*
Spain	7.6	18.7	26.3	4
Greece	18.0	7.6	25.6	4
Portugal	14.9	6.6	21.5	4
Italy	7.4	10.1	17.5	4
Britain	5.5	9.6	15.0	2
Belgium	3.5	10.9	14.4	3
Australia	6.4	7.8	14.2	2
Canada	4.3	9.8	14.1	2
France	4.4	9.6	14.0	3
New Zealand	7.9	6.1	14.0	2
Denmark	4.2	9.4	13.6	3
Finland	5.3	5.8	11.0	2
United States	3.8	7.1	10.9	2
Netherlands	1.9	8.7	10.6	3
Norway	5.7	3.7	9.4	2
Sweden	6.7	2.4	9.1	2
Germany	2.2	6.0	8.2	2
Austria	3.0	3.5	6.5	2
Japan	1.8	2.5	4.3	1
Switzerland	3.1	0.9	4.0	1

*Index Value: 1(most stable) to 4(least stable) based on presence of coalition government, dictatorship, representation of extreme political parties, and political conflict.
Data: DRI/McGraw-Hill, OECD, Business Week, Alberto Alesina, Harvard University.

13.44 The *Business Week* article cited in the Exercise 13.43 stated that as political stability increased, the well-being of the country increased. Based on these *BW* data, is there a significant negative relationship between the misery index (y) and political stability (x)? At what Type I error level?

13.45 Based on the data in Exercise 12.47, if someone reports adjusted-gross income of $100,000 and deducts $20,000 for other taxes paid, would you suspect him or her of underreporting income or overreporting tax deductions? Explain, using a 95 percent prediction interval.

13.46 Use the income and tax data in Exercise 12.47 to determine if there is a statistically significant relationship between adjusted-gross income and deducted taxes. Test at the 0.01 Type I error level.

13.47 The data in the computer ASCII file "EX13-47.PRN," is from *Car and Driver* (April 1993). As described in Exercise 12.49, the first column identifies the car tested and the issue of *Car and Driver* in which the test appeared. The second column is the estimated price of the car tested. The third and fourth columns are 0 to 60 mph and 1/4 mile time in seconds. The fifth column is top speed. The sixth column is 70 to 0 braking distance in feet. EPA estimated fuel economy is in the seventh column. Finally, in column eight, road-holding ability (measured as g force) is provided as obtained from a 300-foot skid-pad. Which of the variables (which you might name TO60, quarter, top, braking, fuel, hold) has the highest significance level in explaining the car price?

13.48 Exercise 12.50 cited a *Wall Street Journal* article in which Song Won Sohn stated that there was a negative relationship between bond yield and the Dow. Use your monthly data and regression estimate to test whether this relationship exists at the 0.01 Type I error level.

13.49 An article in *The Wall Street Journal* (October 29, 1993), titled "Older Drivers Pose Growing Risk on Roads," provided the following data on the relationship between number of drivers, by approximate age, involved in fatal accidents for every 100 million miles driven. A representative of the American Association of Retired Persons (AARP) fit the following regression to these data using LOTUS 1-2-3 and concluded that there was not a significant relationship between age and involvement in fatal accidents. Comment on the appropriateness of this analysis and conclusion.

AGE	MIDPT	FATAL
15 TO	17.5	9.0
20 TO	22.5	3.7
25 TO	27.5	2.9
30 TO	32.5	2.1
35 TO	37.5	1.9
40 TO	42.5	1.7
45 TO	47.5	1.7
50 TO	52.5	1.8
55 TO	57.5	1.8
60 TO	62.5	2.2
65 TO	67.5	2.5
70 TO	72.5	3.3
75 TO	77.5	4.5
80 TO	82.5	7.9
85 +	87.5	14.0

Regression Output:

Constant	1.220416
Std Err of Y Est	3.457199
R Squared	0.116930
No. of Observations	15
Degrees of Freedom	13
X Coefficient(s)	0.054214
Std Err of Coef.	0.041321

Where X is MIDPT
Y is FATAL

13.50 Exercise 12.52 provided data for the analysis of the 1993 contested election in Pennsylvania's Second State Senatorial District in Philadelphia. Republican Marks received more votes in machine voting than his Democrat opponent, Stinson, but in absentee ballots Stinson received many more votes than Marks. Use the data and the analysis in Exercise 12.52 in the following questions.

a. At the 0.05 Type I error level, is there a significant relationship between the difference in absentee votes and machine votes?
b. Construct a 95 percent prediction interval for the absentee votes in the contested election of 1993.
c. Based on your result in parts a and b do you believe that there was fraud in this 1993 election?

(Note: Questions for analysis of the voting issue are continued in Exercises 14.63 and 15.45.)

13.51 Mark Kaiser, Paul Speckman, and John Jones, "Statistical Models for Limited Nutrient Relations in Inland Waters," *Journal of the American Statistical Association* (June 1994), p. 411, provided the following four scatterplots and ordinary least squares regression lines to describe the relationship between chlorophyll (y) and phosphorus (x) in two different samples. Each of the top panels are based on logarithmic transformations of the original data shown in the respective lower panels. What does the logarithmic transformation of each of the variables accomplish? Does the log transformation of both y and x appear to work better for either the left-hand or right-hand panels? Explain.

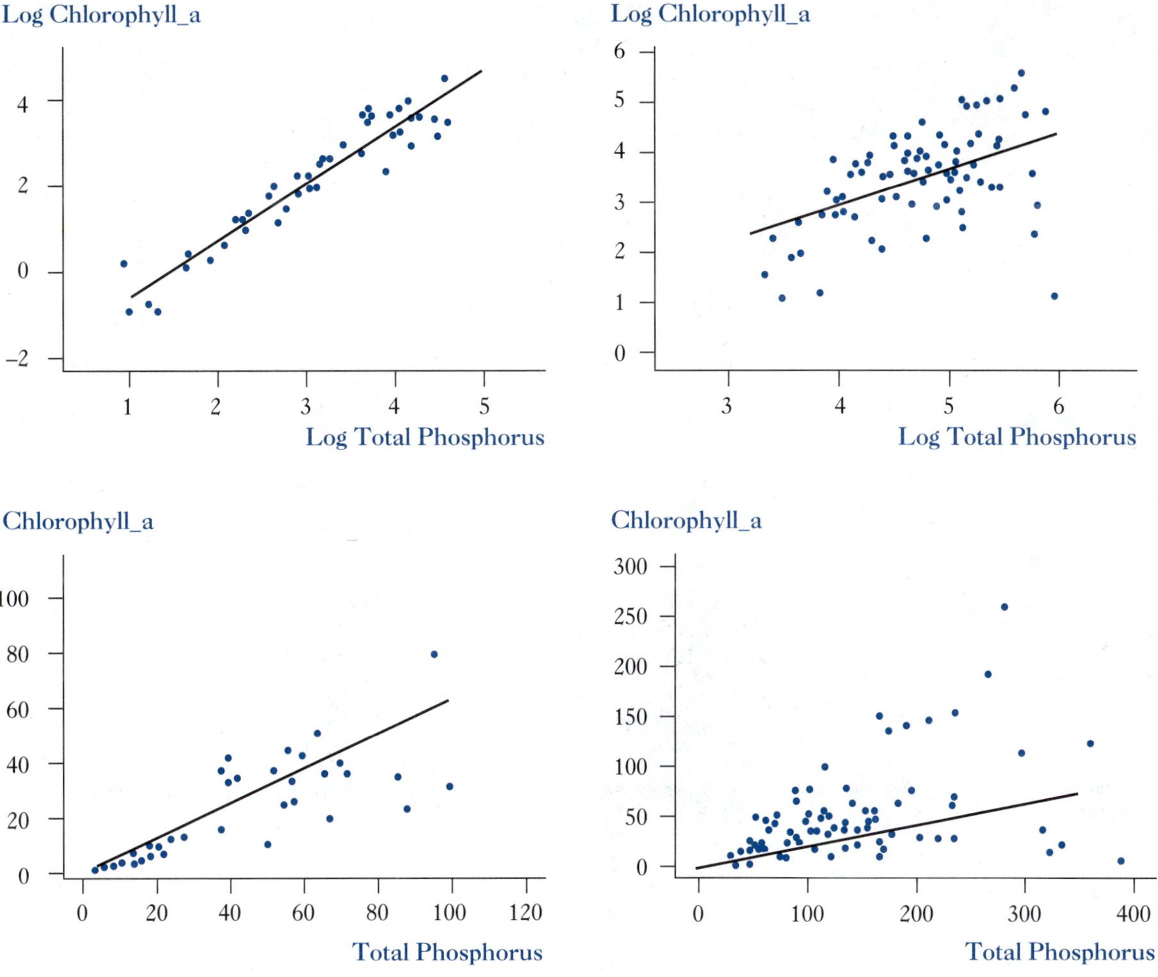

13.52 Using the endnotes, show that the standard error of b is equal to $s_b = \sqrt{\dfrac{(1 - r^2)s_y^2/(n-2)}{s_x^2}}$

13.53 Using the endnotes, show that $b = \Sigma w_i y_i$, where $w_i = \dfrac{(x_i - \bar{x})}{\Sigma(x_i - \bar{x})^2}$ and the sample mean of y is $\bar{y} \neq 0$.

Academic References

Alston, R., J. Kearl, and M. Vaughan. "Is There a Consensus Among Economists in the 1990s?" *American Economic Review,* May 1992, pp. 202–09.

Becker, William, and Robert Toutkoushian. "The Measurement and Cost of Removing Unexplained Gender Differences in Faculty Salaries." (Forthcoming).

Duan, Naihua. "Smearing Estimate: A Nonparametric Retransformation Method." *Journal of the American Statistical Association,* September 1983, pp. 606–610.

Fischer, Stanley and Rudiger Dornbusch. *Introduction to Microeconomics.* New York: McGraw-Hill, 1983.

Goldberger, Arthur S. "The Interpretation and Estimation of Cobb-Douglas Functions." *Econometrica* 35, 1968, pp. 464–472.

Hesse, R. "Too Quick and Too Dirty: Least Squares for Exponential Curves. *Decision Line,* September/October 1983, pp. 8–9.

Hesse, R. "Son of Log Transformation or Return of the Living Dead." *Decision Line,* December/January 1987, pp. 8–9.

Kennedy, Peter A. "Estimation with Correctly Interpreted Dummy Variables in Semilogarithmic Equations." *American Economic Review* 71, April 1981, p. 801.

Kennedy, Peter A. "Logarithmic Dependent Variables and Prediction Bias." *Oxford Bulletin of Economics and Statistics,* 1983, pp. 389–392.

Mincer, Jacob. *Schooling, Experience, and Earnings.* New York and London: Columbia University Press, 1974.

Willis, Robert. "Wage Determinants: A Survey and Reinterpretation of Human Capital Earnings Functions." In *Handbook of Labor Economics, Volume 1,* ed. O. Ashenfelter and R. Layard. New York: North Holland, 1986.

Endnotes

1. $E(y) = E(\alpha + \beta x + \varepsilon) = E(\alpha) + E(\beta x) + E(\varepsilon) = \alpha + \beta E(x)$

   ↑ Since α and β are parameters and the mean value of ε is zero.

   Thus, the mean value of y is equal to the mean value of x times β plus α. At any x value other than its mean, the expected value of y is written $E(y \mid x) = \alpha + \beta x$.

2. From Chapter 12, Formula (12.10) we have

$$b = \frac{\Sigma(x-\bar{x})(y-\bar{y})}{\Sigma(x-\bar{x})^2} = \frac{\Sigma(x-\bar{x})[\alpha+\beta x+\varepsilon-(\alpha+\beta\bar{x}+\bar{\varepsilon})]}{\Sigma(x-\bar{x})^2}$$

$$= \frac{\Sigma\left[\beta(x-\bar{x})^2+(x-\bar{x})(\varepsilon-\bar{\varepsilon})\right]}{\Sigma(x-\bar{x})^2}$$

$$= \beta + \frac{\Sigma(x-\bar{x})(\varepsilon-\bar{\varepsilon})}{\Sigma(x-\bar{x})^2} = \beta + \frac{\Sigma(x-\bar{x})\varepsilon}{\Sigma(x-\bar{x})^2}$$

Thus, $E(b) = \beta + E\left[\frac{\sum x\varepsilon - \bar{x}\sum\varepsilon}{\Sigma(x-\bar{x})^2}\right] = \beta$, if $\text{Cov}(x,\varepsilon) = 0$ and $E(\varepsilon) = 0$

From Formula (12.11), we have

$a = \bar{y} - b\bar{x} = \alpha + \beta\bar{x} + \bar{\varepsilon} - b\bar{x} = \alpha - (b-\beta)\bar{x} + \bar{\varepsilon}$

Thus, $E(a) = \alpha - E(b-\beta)\bar{x} + E(\bar{\varepsilon}) = \alpha$, if $E(b) = \beta$ and $E(\bar{\varepsilon}) = E(\varepsilon) = 0$

where $\bar{\varepsilon}$ is a sample mean calculated from n draws from the distribution of the population disturbance ε; it is not zero. However, its expected value can be zero if the expected value of ε is zero. Do not confuse $\bar{\varepsilon}$ with $\bar{e}$; $\bar{e}$ is the mean of the residuals of a least squares regression. It is always zero.

3. Interpreting these data as a simple random sample implies that pairs of inflation and unemployment rates are treated as if they represent equally likely draws from all combinations that could have occurred if either the 1960s are viewed as a random draw from all time periods or the 1960s could be relived an infinite number of times.

4. The sample regression error is

$$e = y - \hat{y} = \alpha + \beta x + \varepsilon - a - bx = \varepsilon - (a-\alpha) - (b-\beta)x$$

From Endnote 2, we know $a - \alpha = \bar{\varepsilon} - (b-\beta)\bar{x}$. Substituting gives

$$e = (\varepsilon - \bar{\varepsilon}) - (b-\beta)(x-\bar{x}), \text{ and squaring gives}$$

$$e^2 = (\varepsilon-\bar{\varepsilon})^2 + (b-\beta)^2(x-\bar{x})^2 - 2(b-\beta)(\varepsilon-\bar{\varepsilon})(x-\bar{x})$$

Summing over the variables that would have i subscripts gives

$$\Sigma e^2 = \Sigma(\varepsilon-\bar{\varepsilon})^2 + (b-\beta)^2\Sigma(x-\bar{x})^2 - 2(b-\beta)\Sigma(\varepsilon-\bar{\varepsilon})(x-\bar{x})$$

Note: $E(\Sigma\varepsilon^2) = n\sigma_\varepsilon^2$ and $E(\bar{\varepsilon}^2) = \text{Var}(\bar{\varepsilon}) = \sigma_\varepsilon^2/n$

From Endnote 6, which follows, $E(b-\beta)^2 = \text{Var}(b) = \sigma_\varepsilon^2/\Sigma(x-\bar{x})^2$

Using Endnote 2 and $\bar{\varepsilon}\Sigma(x-\bar{x}) = 0$, $(b-\beta)\Sigma(\varepsilon-\bar{\varepsilon})(x-\bar{x}) =$

$$\left[\frac{\Sigma(x-\bar{x})(\varepsilon-\bar{\varepsilon})}{\Sigma(x-\bar{x})^2}\right]\Sigma(\varepsilon-\bar{\varepsilon})(x-\bar{x}) = \frac{\left[\Sigma\varepsilon(x-\bar{x})\right]^2}{\Sigma(x-\bar{x})^2}$$

$$\Sigma e^2 = \Sigma\varepsilon^2 - n\bar{\varepsilon}^2 + (b-\beta)^2\,\Sigma(x-\bar{x})^2 - 2[\Sigma\varepsilon(x-\bar{x})]^2/\Sigma(x-\bar{x})^2$$

Thus, $E(\Sigma e^2) = n\sigma_\varepsilon^{\,2} - \sigma_\varepsilon^{\,2} + \sigma_\varepsilon^{\,2} - 2\sigma_\varepsilon^{\,2} = (n-2)\sigma_\varepsilon^{\,2}$

And $E[\Sigma e^2/(n-2)] = \sigma_\varepsilon^{\,2}$

5. Let $\hat{y}^*$ be the predicted value of y at $x = x^*$ and let $E(y\,|\,x^*)$ be the expected value of y at $x = x^*$. Then,

$$\hat{y}^* - E(y\,|\,x^*) = (a-\alpha) + (b-\beta)x^* \text{ and}$$

$\text{Var}(\hat{y}^*) = E[\hat{y}^* - E(y\,|\,x^*)] = E(a-\alpha)^2 + E(b-\beta)^2\,x^{*2} + 2E(a-\alpha)(b-\beta)x^*$

Note: $\text{Cov}(a,b) = E(a-\alpha)(b-\beta) = E[-(b-\beta)\bar{x} + \bar{\varepsilon})](b-\beta) = -\bar{x}E(b-\beta)^2 + E[\bar{\varepsilon}(b-\beta)] = -x\sigma_b^{\,2}$, using Endnote 8.

$$\text{Var}(\hat{y}^*) = E[\hat{y}^* - E(y\,|\,x^*)] = \sigma_\varepsilon^2\left[\frac{\bar{x}^2}{\Sigma(x-\bar{x})^2} + \frac{1}{n} + \frac{x^{*2}}{\Sigma(x-\bar{x})^2} - \frac{2\bar{x} - x^*}{\Sigma(x-\bar{x})^2}\right]$$

$$= \sigma_\varepsilon^2\left[\frac{1}{n} + \frac{(x^* - x)^2}{\Sigma(x-\bar{x})^2}\right]$$

6. Using Endnotes 2 and 4,

$$\text{Var}(b) = E(b-\beta)^2 = E\left[\frac{\Sigma(x-\bar{x})(\varepsilon-\bar{\varepsilon})}{\Sigma(x-\bar{x})^2}\right]^2 = E\frac{\left[\Sigma\varepsilon(x-\bar{x})\right]^2}{\left[\Sigma(x-\bar{x})^2\right]^2}$$

$$\text{Thus, } \sigma_b^2 = \frac{\sigma_\varepsilon^2}{\Sigma(x-\bar{x})^2}$$

$\text{Var}(a) = E(a-\alpha)^2 = E[-(b-\beta)\bar{x} + \bar{\varepsilon}]^2$
$\quad = \bar{x}^2E(b-\beta)^2 + E(\bar{\varepsilon})^2 - 2\bar{x}E[(b-\beta)\bar{\varepsilon}]$

$$= \bar{x}^2\frac{\sigma_\varepsilon^2}{\Sigma(x-\bar{x})^2} + \frac{\sigma_\varepsilon^2}{n}$$

$$\text{Thus, } \sigma_a^2 = \sigma_\varepsilon^2\left[\frac{\bar{x}^2}{\Sigma(x-\bar{x})^2} + \frac{1}{n}\right]$$

7. Endnote 2 shows b to be a linear function of ε with a mean of β. Endnote 6 shows the standard deviation of the sampling distribution of b to be

$$\frac{\sigma_{\varepsilon}}{\sqrt{\sum\left(x-\bar{x}\right)^{2}}}$$

Thus, if ε is distributed normally, then

$$\frac{b-\beta}{\dfrac{\sigma_{\varepsilon}}{\sqrt{\sum\left(x-\bar{x}\right)^{2}}}}$$

is distributed normally, with mean zero and unit variance. If σ_{ε}^{2} is replaced by its estimator s_{e}^{2}, then

$$\frac{b-\beta}{\dfrac{s_{e}}{\sqrt{\sum\left(x-\bar{x}\right)^{2}}}}$$

is not distributed normally. It is distributed as t because

$$\frac{\sum e^{2}}{\sigma_{\varepsilon}^{2}} \text{ is distributed as a chi-square with df} = n-2.$$

The t distribution is formed by the ratio of a standard normal variable to the square root of a chi-square variable divided by its degrees of freedom, where the variable in the numerator is distributed independent of the variable in the denominator. Thus,

$$t=\frac{\left(b-\beta\right)\sqrt{\sum\left(x-\bar{x}\right)^{2}}}{\sigma_{\varepsilon}}\div\frac{\sqrt{\sum e^{2}/\sigma_{e}^{2}}}{\sqrt{\left(n-2\right)}}$$

Let the notation $\sim t(n-2)$ stand for "distributed as a t, with $n-2$ degrees of freedom." Then,

$$t=\frac{b-\beta}{\dfrac{s_{e}}{\sqrt{\sum\left(x-\bar{x}\right)^{2}}}}\sim t\left(n-2\right)$$

A 95 percent confidence interval for β is

$$b\pm t_{0.025}\,s_{e}\quad\sqrt{\sum\left(x-\bar{x}\right)^{2}}$$

And for hypotheses testing, the conditional t distribution is formed based on the value in the null hypothesis. That is,

$$H_O: \beta = \beta_0$$

$$H_A: \beta \neq \beta_0$$

$$\frac{b - \beta_0}{\dfrac{s_e}{\sqrt{\Sigma\left(x - \bar{x}\right)^2}}} \sim t\left(n - 2\right)$$

More attention will be given to this t ratio in the next chapter.

8. Assumption (13.10) on the normality of ε_i also could be violated if y could only assume discrete values or was bounded. For example, if y_i was the ith person's "yes (y = 1)" or "no (y = 0)" response on a question, then y is not continuous and it could not be distributed normally. Similarly, if y represented A (4), B (3), C (2), or D (1) responses to a questionnaire or if it is a count of something, then it is not continuous. Estimation techniques for use when y and the disturbance term are discrete are beyond the scope of this book.
9. The lessons taught by Goldberger have had to be retaught in most every area of business decision making, as evidenced by Kennedy (1981, 1983), Hesse (1983, 1987), and, most recently, Becker and Toutkoushian (forthcoming).
10. ∂[ln(SCORES)]/∂SCORES = ∂SCORES/SCORES
    = Percentage change in the scores.

    ∂[ln(DAYS])/∂DAYS = ∂DAYS/DAYS
    = Percentage change in the days of school.

    Thus, for the model

    $$\ln(\text{SCORES}) = \beta_1 + \beta_2 \ln(\text{DAYS}) + \varepsilon$$

    the slope is the percentage change in scores associated with a percentage change in days:

    $$\partial\text{SCORES}/\text{SCORES} = \beta_2(\partial\text{DAYS}/\text{DAYS})$$

    Percentage change in scores = β_2(Percentage change in days)

Formulas

- Deterministic linear equation

$$y^* = \alpha + \beta x$$

where y^* is the dependent variable, and x is the independent variable

- Stochastic linear equation

$$y = \alpha + \beta x + \varepsilon$$

where y is the dependent variable,
x is the independent variable, and
ε is a random disturbance term

- Expected value of y, given x

$$E(y|x) = y^* = \alpha + \beta x$$

- Expected value of ε is zero

$$E(\varepsilon) = 0$$

- Expected value of sample intercept a

$$E(a) = \alpha$$

- Expected value of sample slope b

$$E(b) = \beta$$

- Variance of the residuals

$$s_e^2 = \frac{\sum_{i=1}^{n}\left(y_i - \hat{y}_i\right)^2}{n-2} = \frac{\sum_{i=1}^{n} e_i^2}{n-2}$$

- Variance of residuals

$$s_e^2 = \frac{\text{ErrorSS}}{n-2}$$

- $(1 - \alpha)$ percent confidence interval for $E\ (y|x)$

$$\hat{y} \pm t_{\alpha/2} s_e \sqrt{\frac{1}{n} + \frac{\left(x^* - \bar{x}\right)^2}{\sum_{i=1}^{n}\left(x - \bar{x}_i\right)^2}}$$

- $(1 - \alpha)$ percent prediction interval for an individual value of y

$$\hat{y} \pm t_{\alpha/2} s_e \sqrt{\frac{1}{n} + \frac{\left(x^* - \bar{x}\right)^2}{\sum_{i=1}^{n}\left(x - \bar{x}_i\right)^2} + 1}$$

- Standard error of the sampling distribution of b

$$s_b = \sqrt{\frac{s_e^2}{\sum_{i=1}^{n}\left(x_i - \bar{x}\right)^2}}$$

- t ratio for the slope of a regression line

$$t = \frac{b - E(b)}{s_b}$$

- *t* ratio for Spearman's rho

$$t = \text{rho}\sqrt{\frac{n-2}{1-\text{rho}^2}}$$

with degrees of freedom of $n - 2$ and

$$\text{Spearman's rho} = 1 - \frac{6\sum d_i^2}{n^3 - n}$$

where d_i represents the difference between the numerical rank order of the ith item in two different sets of data, each consisting of n paired items.

Chapter 14

Multiple Regression

Figures speak
and when they do
courts listen.

Brooks v. Beto 366 F.2d
(5th Cir. 1966)

14.1 Introduction

Applications of the two variable linear model are limited, if for no reason other than real world phenomena typically involve more than one explanatory factor. The two variable model must be extended to multiple regressors to handle problems of the following types.

- In part, a product's selling price is believed to be a function of research and development expenses, production costs, and advertising expense. The question: How important is each of these explanatory variables in determining selling price?
- The relationship between inflation and unemployment observed in the 1960s may have been affected by events in the early 1970s (e.g., Nixon's price controls, the Arab oil embargo, spread of unemployment benefits) and again in the early 1980s when interest rates skyrocketed. How might the long-term Phillips curve best be estimated over the past 35 years?
- Real estate appraisers claim that the selling price of a house depends on its location, square footage, age, construction material, and condition. How well do these factors jointly explain selling price?

Multiple Regression A method of regression analysis that uses more than one explanatory variable to predict values of a single dependent variable.

All of the statistical properties and assumptions developed in the previous two chapters may be transferred directly to applications of **multiple regression;** there are only a few new concepts to learn. Unfortunately, the computational formulas of the simple two variable model cannot be extended easily. Luckily, in this day of inexpensive computers and widely available regression programs, there is no need to burden students with these cumbersome hand-calculator routines.

Only the practical use of multiple regression analysis is emphasized here. Practical applications require that we have an awareness of how data are displayed for a multiple regression analysis, but a computer can do the calculations without our knowing the formulas employed. Thus, in this chapter we need emphasize only the

- specification of the population model and the sample regression used in a least squares estimation of the parameters of the multiple independent variables;
- testing of possible population relationship between the dependent variable and the independent variables;
- estimation of the expected values of the dependent variable at certain values of the independent variables;
- prediction of a certain value of the dependent variable for given values of the independent variables; and
- construction of confidence intervals for the expected value and the predicted value of the dependent variable based on the values of the independent variables.

14.2 The Case of Equal Pay

The United States Equal Pay Acts of 1964 and 1971 are based on the principle that a person's race or sex may not be used to establish merit, and that race or sex cannot be used to justify lack of performance and a lower salary. Remedies to sex and race discrimination have attempted to put the victim in the situation she or he would be in had there been no disparate treatment. Multiple regression has come to play an important part in assessing the fairness of the salary determination processes in both the private and public sectors of the economy. These regressions are used to detect the presence of discrimination and, if present, to estimate the magnitude of the remedy.

Although corporate executives in the United States talk about basing salaries on performance, in actual practice an employee's seniority and entry-level position may prove more important. Paul Regan, a senior vice president of Corning Inc., for example, in a *Wall Street Journal* (September 10, 1991) article on the lack of merit pay in private industry, said, "If you tell me the date someone joined a company and the starting pay, I'll tell you within $500 what (he or she) makes." This suggests that Regan had the following functional relationship in mind for the salary determination process.

$$\text{predicted salary} = \text{function(starting year, starting salary)}$$

Regan did not specify a functional form. It appears that he had in mind a multiple regression of salary on starting year and starting salary, with no explanatory variables needed for individual performance. He was confident that the error in his prediction would not exceed $500 of a person's current salary.

The use of regression analysis to study salary determination in the public sector has a slightly longer history than that in private industry. Mary Gray and Elizabeth Scott (*Academe*, May 1980, p. 180), for example, suggested a multiple regression equation to explain the salaries of college faculty. When updated for the current salaries of the men in an academic department, it corresponds to:

$$\hat{y} = 43000 + 1160x_2 + 250x_3$$

where $\hat{y}$ is the predicted salary (in dollars) of a man, x_2 is the number of years since he received the Ph.D. degree, and x_3 is the number of published books he has authored. As was the case with simple regression, the sample form fit by Gray and Scott indicates that at least implicitly they must have had the following population salary model in mind:

$$y = \beta_1 + \beta_2 x_2 + \beta_3 x_3 + \varepsilon$$

where y is a person's actual salary and ε is the population disturbance or error term that reflects the random chance factors in determining a person's salary. We do not know the population parameters β_1, β_2, and β_3. We only have the point estimates 43000, 1160, and 250.

According to Gray and Scott's sample regression, if a man in this department has a five-year-old Ph.D. and one publication, then he is predicted to receive $49,050.

$$\hat{y} = 43000 + 1160x_2 + 250x_3 = 43000 + 1160(5) + 250(1) = 49050 \qquad (14.1)$$

If a woman with like x_2 and x_3 values earned only $47,725, then, according to Gray and Scott's analysis, she is underpaid by $1,325. The question, however, is whether such a demonstration proves that a person's sex is being used in the determination of salaries. That is, is there a difference between the population models used to determine male and female salaries that is attributable solely to gender?

14.3 Interpretation of Coefficients and the Prediction of Y

The first term on the right-hand side of Equation (14.1) is the y intercept. This y intercept of 43000 is represented in Figure 14.1, panel a, as the point at which the regression line crosses the y axis. It indicates that a man who has just earned a Ph.D. degree ($x_2 = 0$) and has no publications ($x_3 = 0$) has a predicted salary of $43,000.

The line in Figure 14.1, panel a, shows the relationship between the predicted salary of males and different vintage Ph.D.'s, given that the males have never published. A male who received his Ph.D. one year ago ($x_2 = 1$), but who has no publications ($x_3 = 0$), is predicted to have a salary of $44,160. A male who received his Ph.D. two years ago ($x_2 = 2$), but has no publications ($x_3 = 0$), is predicted to have a salary of $45,320, and so on. For each of these one-year increases in x_2, the predicted salary rises by $1,160. Thus, the slope coefficient 1160 indicates that a one-unit increase in x_2 is associated with a $1,160 increase in the predicted salary, holding x_3 constant.

An increase in publications (x_3) from 0 to 1 raises the predicted salary from $44,160 to $44,410 for a male who received his Ph.D. degree one year ago ($x_2 = 1$). For someone who received his Ph.D. two years ago ($x_2 = 2$), the predicted salary rises from $45,320 to $45,570 when he publishes a book (x_3 rises from 0 to 1). The implication for salaries of a change in x_3 from 0 to 1 for all values of x_2 is represented in Figure 14.1, panel b, by a move from the lower line to the higher line. At any value of x_2, increasing x_3 from 0 to 1 increases the predicted salary by $250. Similarly, in panel c, an increase in x_3 from 1 to 2 shows predicted salary rising by $250, at each value of x_2. Thus, the coefficient 250 indicates that a one-unit increase in x_3 is associated with a $250 increase in the predicted salary, calculated at any specific value of x_2.

In general, any multiple linear regression that is fit to sample data can be represented as

$$\hat{y} = b_1x_1 + b_2x_2 + b_3x_3 + b_4x_4 + \ldots b_kx_k \tag{14.2}$$

where $\hat{y}$ is the predicted dependent variable,

x_j is the jth independent variable, $j = 1, 2, 3, \ldots k$, and

b_j is the jth coefficient, $j = 1, 2, 3, \ldots k$

Typically, but not always, x_1 is set equal to one so that b_1 is the y intercept. That is, if $x_2 = x_3 \ldots = x_k = 0$, and $x_1 = 1$, $\hat{y}$ is b_1, the regression Equation (14.2) can be written more succinctly as

$$\hat{y} = b_1 + b_2x_2 + b_3x_3 + b_4x_4 + \ldots b_kx_k \tag{14.3}$$

The remaining b's are the slope coefficients: b_2 gives the change in $\hat{y}$ for a one-unit change in x_2, holding all the other x's constant; $\hat{y}$ changes by b_3, for a one-unit change in x_3, holding all the other x's constant, and so on for each of the remaining slope coefficients.

FIGURE 14.1 Interpreting the Regression Line $\hat{y} = 43000 + 1160x_2 + 250x_3$

Panel a: The relationship between $\hat{y}$ and x_2, holding x_3 fixed at 0.

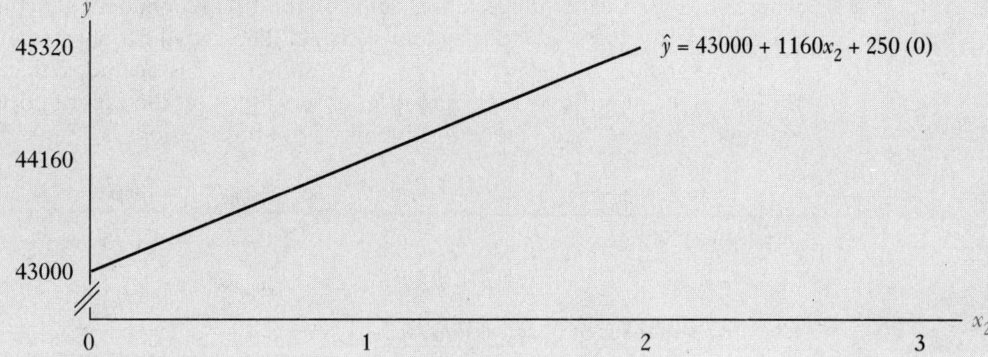

Panel b: The relationship between $\hat{y}$ and x_2, with a change in x_3 from 0 to 1.

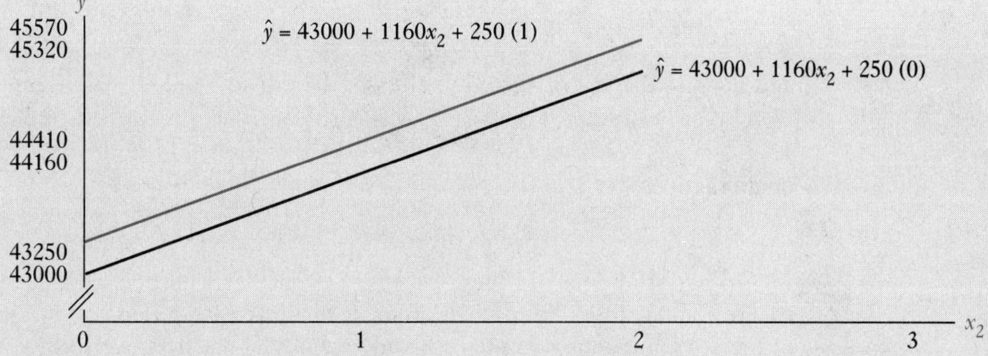

Panel c: The relationship between $\hat{y}$ and x_2, with a change in x_3 from 1 to 2.

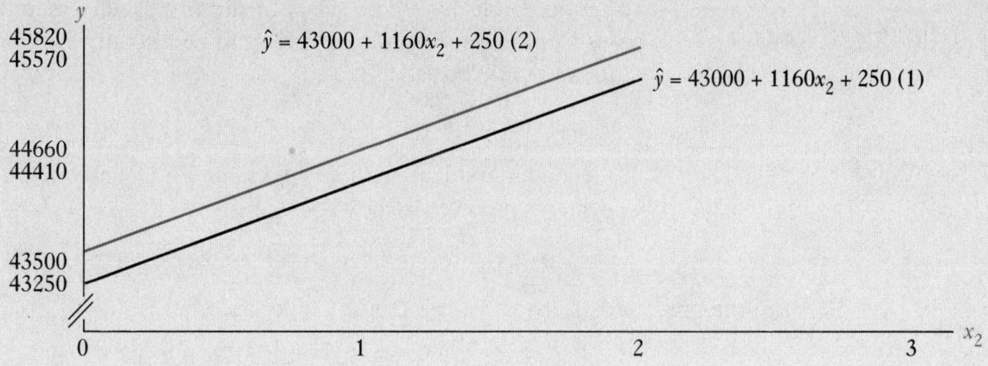

To identify specific observation values on each variable, an i subscript is added to each of the variables in Equation (14.2).

$$\hat{y}_i = b_1x_{i1} + b_2x_{i2} + b_3x_{i3} + b_4x_{i4} + \ldots b_kx_{ik} \tag{14.4}$$

where now x_{ik} indicates the ith observation on the kth independent variable and $\hat{y}_i$ is the ith predicted value of the dependent variable. The actual ith observation on the dependent variable is designated by $\hat{y}_i$. The subscript i is an index that runs from 1 through n. It identifies a set or group of observations on the identified variables. For example, the predicted y value for the first set of observations is

$$\hat{y}_1 = b_1(1) + b_2x_{12} + b_3x_{13} + b_4x_{14} + \ldots b_kx_{1k} \tag{14.5}$$

or alternatively

$$\hat{y}_1 = b_1 + b_2x_{12} + b_3x_{13} + b_4x_{14} + \ldots b_kx_{1k}$$

For the second set of observations, it is

$$\hat{y}_2 = b_1\,(1) + b_2x_{22} + b_3x_{23} + b_4x_{24} + \ldots b_kx_{2k} \tag{14.6}$$

or alternatively

$$\hat{y}_2 = b_1 + b_2x_{22} + b_3x_{23} + b_4x_{24} + \ldots b_kx_{2k}$$

and so on to the last predicted y value for the nth set of observations on the x's

$$\hat{y}_n = b_1(1) + b_2x_{n2} + b_3x_{n3} + b_4x_{n4} + \ldots b_kx_{nk} \tag{14.7}$$

or alternatively

$$\hat{y}_n = b_1 + b_2x_{n2} + b_3x_{n3} + b_4x_{n4} + \ldots b_kx_{nk}$$

To better understand these subscripts, consider a portion of updated data from a trial over alleged sex discrimination in a college at a large midwestern university. One of the academic departments involved in this trial had 10 male faculty members who were not administrators. These individuals, their current salary equivalents, the number of years since they received their Ph.D.'s, and the number of books they have published are identified in Table 14.1. The salary of the first faculty member is \$66,000; 19 years have passed since he received his Ph.D., and he has published one book. This first faculty member's predicted salary is

$$\hat{y}_1 = b_1x_{11} + b_2x_{12} + b_3x_{13} = b_1(1) + b_2(19) + b_3(1) \tag{14.8}$$

The salary of the second individual, with a 17-year-old Ph.D. and one published book, is \$63,500. This person's predicted salary is

$$\hat{y}_2 = b_1x_{21} + b_2x_{22} + b_3x_{23} = b_1(1) + b_2(17) + b_3(1) \tag{14.9}$$

and the predicted salary of the last (10th) faculty member is

$$\hat{y}_{10} = b_1x_{101} + b_2x_{102} + b_3x_{103} = b_1(1) + b_2\,(5) + b_3(2) \tag{14.10}$$

Without knowing the b values, however, no predictions can be made, which is one of the objectives of multiple regression.

Table 14.1 Male Faculty Salaries and Related Data

Individual i	Salary y_i	Constant x_{i1}	Years since Ph.D. x_{i2}	Number of Books x_{i3}
$i = 1$	$y_1 = 66000$	$x_{11} = 1$	$x_{12} = 19$	$x_{13} = 1$
$i = 2$	$y_2 = 63500$	$x_{21} = 1$	$x_{22} = 17$	$x_{23} = 1$
$i = 3$	$y_3 = 56700$	$x_{31} = 1$	$x_{32} = 11$	$x_{33} = 0$
$i = 4$	$y_4 = 56300$	$x_{41} = 1$	$x_{42} = 16$	$x_{43} = 0$
$i = 5$	$y_5 = 54900$	$x_{51} = 1$	$x_{52} = 12$	$x_{53} = 2$
$i = 6$	$y_6 = 52300$	$x_{61} = 1$	$x_{62} = 11$	$x_{63} = 0$
$i = 7$	$y_7 = 50650$	$x_{71} = 1$	$x_{72} = 7$	$x_{73} = 0$
$i = 8$	$y_8 = 50650$	$x_{81} = 1$	$x_{82} = 6$	$x_{83} = 1$
$i = 9$	$y_9 = 50500$	$x_{91} = 1$	$x_{92} = 10$	$x_{93} = 0$
$i = 10$	$y_{10} = 48000$	$x_{101} = 1$	$x_{102} = 5$	$x_{103} = 2$

Query 14.1:

Olson, Frieze and Good, in "The Effects of Job Type and Industry on the Income of Male and Female MBAs" (*Journal of Human Resources*, Fall 1987, p. 536), report the following regression of 1983 salary (y, measured in \$1,000) on the identified variables.

$$\hat{y} = 2.1114x_1 - 5.2544x_2 - 6.1665x_3 + 1.2517x_4 - 2.9075x_5$$

where x_1 is years of full-time work since receiving the MBA,
x_2 is years of part-time work since receiving the MBA degree,
x_3 is years of unemployment since receiving the MBA degree,
x_4 is years of prior full-time work, and
x_5 is one if female and zero if male.

What is the salary effect of unemployment after receipt of the MBA? What is the difference between the predicted salaries of men and women, with identical characteristics? What might be causing this difference?

Answer:

For both men and women, one additional year of unemployment (x_3) after receiving the MBA degree lowers the predicted salary by \$6,166.50. However, the predicted salary of a man is \$2,907.50 higher than for a like woman. This difference may be caused by sex discrimination in the markets for MBA's. Alternatively, it may be the result of differences in occupation selection (e.g., men tending to choose the higher-paying field of accounting), differences in industry selection (e.g., men tending to go into higher paying manufacturing industries), or the tendency of men to have line positions while females occupy staff positions. In addition to these possible systematic differences, the predicted salaries of men and women in this sample may reflect sampling errors (chance occurrences) that resulted in an unusual sample.

Query 14.2:

William Evans and W. Kip Viscusi in "Income Effects and the Value of Health" (*Journal of Human Resources*, Summer 1993), estimated and discussed the following regression (Note: E – 2 = 10^{-2} and E – 3 = 10^{-3}):

$$\hat{y} = 246.53 + 2.3E{-}2x_1 - 2.7E - 3x_2$$

where: y is the amount of money an individual will desire as compensation to accept a risk of temporary injury (monetary loss equivalence)

x_1 is yearly income (in thousands of dollars)

x_2 is $x_1^2/10{,}000$

Draw the graph of this equation in the y-x plane. Why did the authors question the estimates of monetary loss for those with incomes above $45,000?

ANSWER:

Calculation of the monetary loss equivalence requires the creation of the x_2 variable as $(x_1)^2/10{,}000$. At $5,000 intervals, it is then calculated in a spreadsheet program as

$$\hat{y} = 246.53 + 0.023x_1 - 0.0027x_2$$

x_1	x_2	$\hat{y}$	x_1	x_2	$\hat{y}$
5000	2500	354.78	35000	122500	720.78
10000	10000	449.53	40000	160000	734.53
15000	22500	530.78	45000	202500	734.78
20000	40000	598.53	50000	250000	721.53
25000	62500	652.78	55000	302500	694.78
30000	90000	693.53	60000	360000	654.53

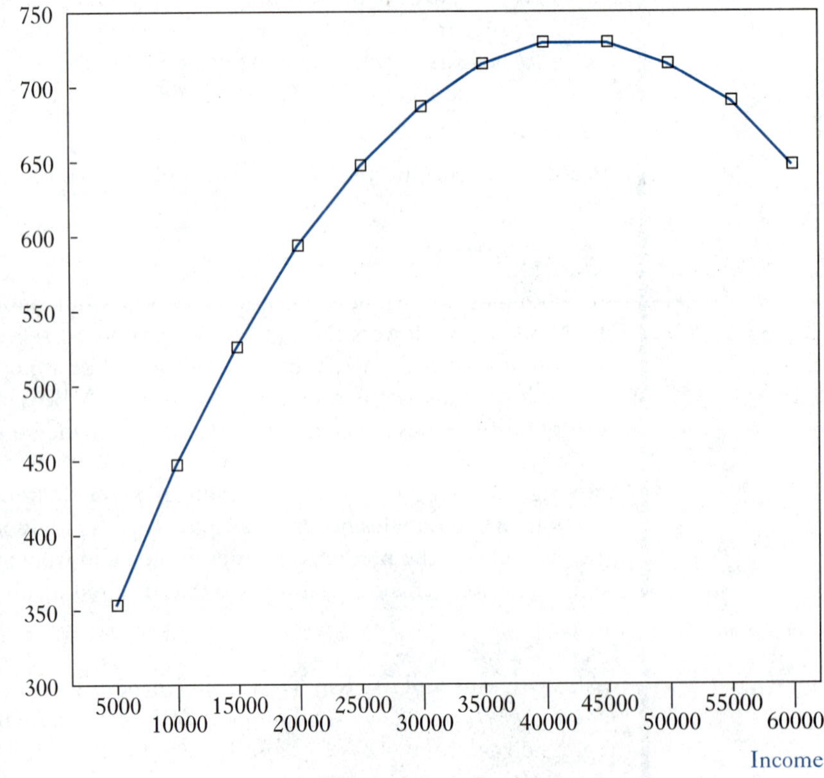

Monetary loss equivalence y is quadratic in income x; it rises with income at a decreasing rate, and then decreases, as most easily seen in the diagram on the previous page. It seems implausible that as incomes rise above $45,000, individuals would desire less compensation for additional risk of injury as their incomes grow. For this reason, predictions for individuals with incomes above $45,000 are not appropriate.

14.4 Least Squares in Multiple Regression

In multiple regression, the prediction errors or residuals are formed exactly the same as in simple regression: the observed values of y minus the corresponding predicted values $\hat{y}$. The n errors or residuals associated with the differences between an actual value of y and its corresponding predicted value is written

$$e_i = y_i - \hat{y}_i, \text{ for } i = 1, 2, \ldots n \tag{14.11}$$

Substituting for $\hat{y}$ calculated from (14.2) gives

$$e_i = y_i - b_1x_{i1} - b_2x_{i2} \ldots - b_kx_{ik}, \text{ for } i = 1, 2 \ldots n \tag{14.12}$$

The idea behind the method of least squares here is the same as in Chapters 12 and 13. The only difference is that in multiple regression the multiple b's must be selected to minimize the sum of the squares of the errors, Σe_i^2, where again

$$\sum_{i=1}^{n} e_i^2 = e_1^2 + e_2^2 + \ldots + e_n^2 \tag{14.13}$$

The ErrorSS = Σe_i^2 continues to define the sum of the squares of the errors in multiple regression, but now the predicted value of the dependent variable is based on multiple independent variables.

Explanatory Variable
Variable that is used to account for the variability in a dependent variable.

The method of least squares in multiple regression requires the minimization of the ErrorSS, jointly selecting the values for b_1 through b_k. In the one independent variable case, where only one **explanatory variable** x and a constant are used to explain y, the regression line that minimizes the error sum of squares can be seen in a two-dimensional diagram, as in Figures 12.5 and 12.6. In the two independent variable case, where $x_1 = 1$ and x_2 and x_3 are the explanatory variables used to explain y, a three-dimensional scatterplot requires the fitting of a plane or surface. Figure 14.2 shows a scatterplot and the best fitting plane that has the equation

$$\hat{y}_i = b_1 + b_2x_{i2} + b_3x_{i3} \tag{14.14}$$

The unique feature of the equation for the plane in Figure 14.2 is that there is no other plane that could be drawn in this three-dimensional scatterplot that would make the sum of squared vertical distances between the (y, x_2, x_3) points and the plane any smaller.

As more x's are added as explanatory variables, the process that leads to the optimal values of the b's that minimize the error sum of squares cannot be visualized, since three dimensions is the most dimensions that can be drawn. The k coefficients in a multiple regression equation with $k - 1$ independent variables and one constant can be

FIGURE 14.2 Three-Dimensional Scatterplot and Regression Plane

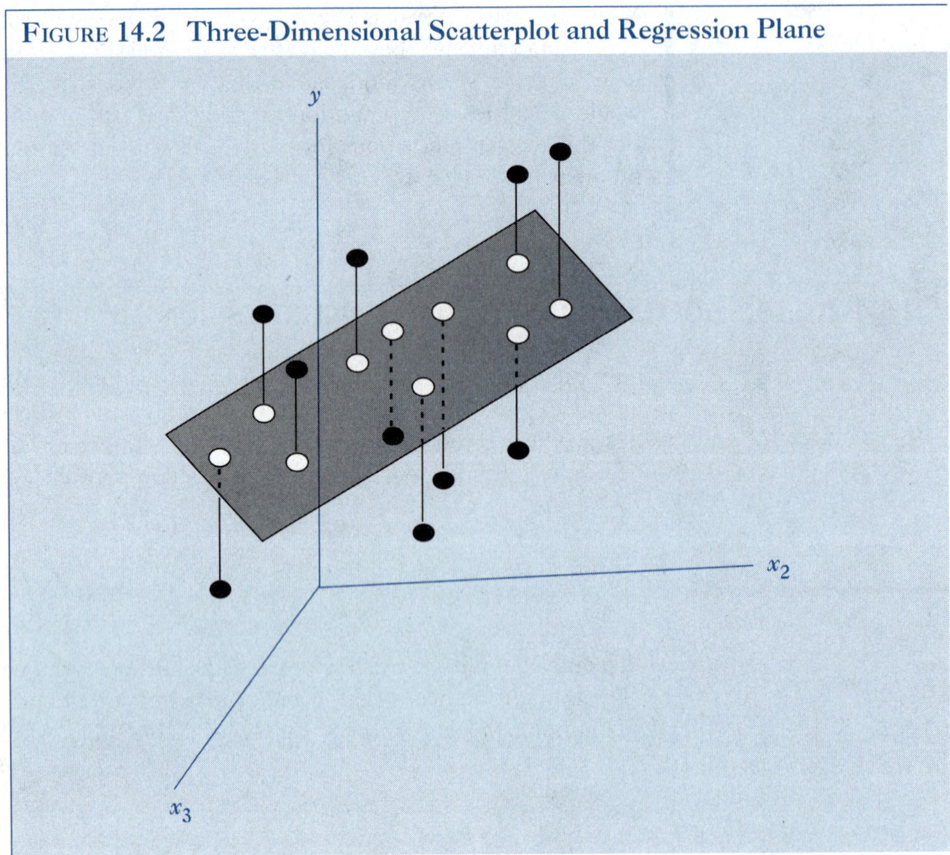

found by solving a system of k simultaneous linear equations in which $b_1, b_2, b_3, \ldots b_k$ are the unknowns. Determination of the resulting formulas and hand calculations of the b values are seldom done today.[1] The most common way to find the b coefficient is to use a computer program that has the least squares regression routine built in, as demonstrated in Query 14.3.

QUERY 14.3:

"A Scorecard on the B-Schools" (*Business Week*, January 11, 1988, pp. 164–165) provided three ratings on the top MBA programs in the United States. Michigan, the only public school among the five, had the lowest tuition but highest student/faculty ratio. Use a least squares regression of tuition (y) on the student/faculty ratio (x_2) and 0–1, public-private school variable (x_3). Predict tuition for the five schools and show the residuals sum to zero.

	y: Annual Tuition	x_2: Student/ Faculty Ratio	$x_3 = 1$, if private
Chicago	$13,500	9/1	1
Harvard	13,300	9/1	1
Michigan	11,312	18/1	0
Northwestern	13,200	9/1	1
Pennsylvania	13,354	8/1	1

ANSWER:

Below is the MINITAB computer program and results used to estimate the equation that minimizes the error sum of squares.

```
MTB > read c1 c2 c3                    ← To read in the data into
DATA> 13500  9  1                        columns 1, 2, and 3.
DATA> 13300  9  1
DATA> 11312 18  0
DATA> 13200  9  1
DATA> 13354  8  1
DATA> end
      5 ROWS READ
MTB > name c1 'y' c2 'x2' c3 'x3'      ← To name columns 1, 2, and 3.
MTB > regres c1 2 c2 c3                ← To estimate a regression where
SUBC> predict 9 1;                       the data in column 1 is the dependent
SUBC> predict 8 1;                       variable and there are two independent
SUBC> predict 18 0.                      variables in columns 2 and 3.
                                       ← To predict y for the pairs of x2 and x3
The regression equation is

y = 11684 – 21 x2 + 1835 x3

Predictor     Coef     Stdev    t-ratio      p
Constant     11684      3179       3.68   0.067
x2           –20.7     176.4      –0.12   0.917
x3            1835      1640       1.12   0.380

s = 152.8     R-sq = 98.6%     R-sq(adj) = 97.2%

Analysis of Variance

SOURCE       DF        SS         MS        F       p
Regression    2   3285682    1642841    70.41   0.014
Error         2     46667      23333
Total         4   3332349

SOURCE       DF    SEQ SS
x2            1   3256476
x3            1     29206

     Fit   Stdev.FIT        95% C.I.               95% P.I.
 13333.3        88.2   (12953.9,13712.8)    (12574.4,14092.2)
 13354.0       152.8   (12696.8,14011.2)    (12424.5,14283.5)
 11312.0       152.8   (10654.8,11969.2)    (10382.5,12241.5)
```

From the preceding MINITAB computer results, the regression equation is written

$$\hat{y} = 11684 - 20.7x_2 + 1835x_3$$

The predicted values (obtainable from the bottom of the printout) and calculated residuals are

	x_2	x_3	y	$\hat{y}$	e
Chicago	9	1	13500	13333.3	166.7
Harvard	9	1	13300	13333.3	–33.3
Michigan	18	0	11312	11312.0	00.0
Northwestern	9	1	13200	13333.3	–133.3
Pennsylvania	8	1	13354	13354.0	00.0
					000.1

As expected, with the exception of rounding error, the residuals sum to zero. Later queries will address the other quantities provided in this printout.

14.5 Assumptions, Estimation, and Hypotheses Testing

In the earlier example presented by Mary Gray and Elizabeth Scott, if a man in the department has five years of experience and one publication, then his predicted salary is $49,050, by regression (14.1). If a woman with similar attributes has an actual salary of only $47,725, then, relative to the salary predicted by the male regression, she is underpaid by $1,325. The question to be raised, however, is whether this underpayment results from chance or sex discrimination. That is, from this sample, can we conclude that the expected salary of women is different from that of comparable men?

As another example of the need for estimation and hypothesis testing, consider the data in Query 14.3 where we might now ask: Are the student/faculty ratio and type of school (private or public) significant explanatory variables in predicting MBA tuition?

Attempts to answer questions of statistical significance start with the specification of a population model. In the Gray and Scott example, a population salary model to be considered is

$$y_i = \beta_1 + \beta_2 x_{i2} + \beta_3 x_{i3} + \beta_4 x_{i4} + \varepsilon_i \tag{14.15}$$

where y_i is ith person's salary, $i = 1, 2, 3, \ldots n$,

x_{i2} is the years since the ith person completed his or her Ph.D.,

x_{i3} is the number of books published by the ith person,

x_{i4} is a 0-1 variable for the ith person's gender,

β_1, β_2, β_3, and β_4 are parameters to be estimated by the least squares estimators b_1, b_2, b_3, and b_4, and

ε_i is the population disturbance term, which reflects the random chance factors in determining the ith person's salary.

Now the variable x_{i4} is included to capture the effect of gender in the salary determination process. A person's sex is an attribute that can be represented by one of two arbitrary values. Typically, a one is assigned to represent one sex and a zero the other. An explanatory variable that represents a qualitative attribute with a 0 or 1 value is often called a **dummy variable.** Let $x_{i4} = 1$ if the ith person is female, and 0 if male. This definition of x_{i4} implies that β_4 is negative if the expected salary of females is less than that of comparable males (same vintage Ph.D. and same number of books published); it is positive if the expected salary of females is greater than that of comparable males. β_4 is zero if gender is irrelevant in the salary determination process. The hypotheses are thus

Dummy Variable
An explanatory variable that represents a qualitative attribute with a 0 or 1 value.

$$H_O\text{: } \beta_4 = 0 \text{ and } H_A\text{: } \beta_4 \neq 0$$

To test hypotheses and to construct confidence intervals, assumptions about the population model must be made.

First, as in Chapter 13, we assume the expected value of y, at any given set of values for the x's, is the deterministic part of the model; that is,

$$E(y_i|x_{i1},x_{i2}\ldots x_{ik}) = \beta_1 x_{i1} + \beta_2 x_{i2} \ldots + \beta_k x_{ik}, \text{ for } i = 1, 2 \ldots n \tag{14.16}$$

This requires that we assume that the expected value of each of the n population error terms is zero, that is,

$$E(\varepsilon_i|x_{i1},x_{i2}\ldots x_{ik}) = 0 \tag{14.17}$$

Second, now that there are more than one x, we must assume that each of these x's is independent of or linearly unrelated to the other x's. That is, the covariance between x's is zero:

$$\text{Cov}(x_{k-s},x_{k-t}) = 0,\ s \neq t \tag{14.18}$$

A violation of this second assumption is referred to as a problem of **multicollinearity,** which is discussed at the end of this chapter. The assumption of no multicollinearity is the only totally new assumption that is introduced in a movement from simple to multiple regression.

Multicollinearity
Linear dependence of explanatory variables in a multiple regression analysis.

Third, the variance of each of the n disturbance terms is σ_ε^2, which was given as Assumption (13.7) in Chapter 13,

$$\text{Var}(\varepsilon_i) = \sigma_\varepsilon^2 \tag{14.19}$$

In Chapter 13, it was also stated in Assumption (13.8) that the disturbance terms must be independent so their covariances will be zero:

$$\text{Cov}(\varepsilon_i,\varepsilon_{i-h}) = 0, \text{ for h} \neq 0 \tag{14.20}$$

Fourth, the population disturbance terms and each of the x variables must be independent so their covariances are zero. In the previous chapter, this was given by Assumption (13.9) and is now written

$$\text{Cov}(\varepsilon_i,x_{ij}) = 0 \tag{14.21}$$

Finally, as in Assumption (13.10), it is assumed that each of the n disturbance terms are distributed normally with zero mean and fixed variance σ_ε^2; i.e.,

$$\varepsilon_i \sim N(0, \sigma_\varepsilon^2) \tag{14.22}$$

If Assumptions (14.17) and (14.20) are correct, then the expected value of the least squares estimator b_4 is β_4 and $E(\hat{y}_i|x_{i1},x_{i2}\ldots x_{ik}) = E(y_i|x_{i1},x_{i2}\ldots x_{ik})$. As in the case of simple regression, the expected value of y, at the ith value of the x's, is estimated by $\hat{y}$, but now $\hat{y}$ is the multiple regression

$$\hat{y}_i = b_1x_{i1} + b_2x_{i2}\ldots + b_kx_{ik} \tag{14.23}$$

Assumptions (14.17) and (14.19) are required to define the standard error of b_4 and the other b's, as well. Assumption (14.22) implies that the least squares estimator b_4 and the other b's are each normally distributed, because they are linear combinations of ε. It is the normality of b_4 that will enable the use of the t statistic to test the hypotheses

$$H_O\colon \beta_4 = 0 \text{ and } H_A\colon \beta_4 \neq 0$$

Similarly, it is the normality of ε, and in turn the b's, that enables us to construct confidence intervals for $E(\hat{y}_i|x_{i1},x_{i2}\ldots x_{ik})$ and the β's.

Often, researchers start their analysis by simply assuming the above assumptions are true. As introduced at the end of Chapter 13, however, the correctness of these assumptions should be checked before relying on least squares estimators for predictions and hypothesis testing. The importance of these specification checks is not lessened by multiple regression, and the residual analysis techniques introduced in the optional sections of Chapter 13 continue to apply in the case of multiple regression.

Exercises

14.1 The following two data sets are based on the same observations, but the * superscript indicates that the variable is measured as a deviation from its mean. A regression fit to the second data set has the unique property that its intercept or constant term is zero. That is, only x_2^* and x_3^* are included on the right-hand side of the regression.

y	x_1	x_2	x_3	y^*	x_1^*	x_2^*	x_3^*
3	1	–2	4	2	0	–1	1
0	1	0	3	–1	0	1	0
1	1	–1	2	0	0	0	–1
2	1	–2	4	1	0	–1	1
–1	1	0	2	–2	0	1	–1

a. Use a computer program to estimate the regression

$$\hat{y}^* = b_2^*x_2^* + b_3^*x_3^*$$

b. Use a computer program to fit

$$\hat{y} = b_1 + b_2x_2 + b_3x_3$$

c. Is there a difference between b_2^* and b_2, and b_3^* and b_3? What is the difference between the y and y^* regressions?

d. Calculate the error sum of squares for both the y and y^* regressions; is there a difference? Why or why not?

14.2 Peter Asch and Richard Quandt, in *Racetrack Betting: The Professors' Guide to Strategies* (New York: Praeger 1991, p. 98), proposed the following regression to predict horse speed:

$$\text{predicted speed} = a + b(\text{jockey's experience}) + c(\text{jockey's sex}) + d(\text{weight carried on horse}) + e(\text{horse's condition})$$

a. If speed is measured in miles per hour and jockey's experience is measured in years, what is the meaning of b?

b. If the jockey's sex is equal to 1 if male and 0 if female, what is the meaning of c?

14.3 The following is the complete data set on a "A Scorecard on the B-Schools," from *Business Week* (January 11, 1988, pp. 164–165). Use the ranking on scale A and annual tuition to obtain an equation to predict starting salary. (Keep your computer printout for later use.)

School	Student/ Faculty Ratio	Annual Tuition	Appli- cations Accepted	1987 Grads	Average Starting Salary	Ranking on Scale A	B	C
Carnegie Mellon	4.6/1	$12,648	31%	134	$45,050	8	13	–
Chicago	9/1	13,500	30	692	46,000	5	5	3
Columbia	8/1	13,000	20	486	47,350	9	4	–
Dartmouth	8/1	13,000	19	165	49,100	11	11	–
Duke	12.5/1	12,500	27	241	42,000	18	10	–
Harvard	9/1	13,300	13	782	50,225	2	3	1
MIT	9/1	12,700	21	215	50,625	4	17	–
Michigan	18/1	11,312	39	400	42,750	7	6	7
Northwestern	9/1	13,200	17	480	48,550	6	1	4
North Carolina	12/1	4,533	26	156	39,625	17	18	–
Pennsylvania	8/1	13,354	18	772	48,350	3	2	5
Stanford	6.9/1	12,960	10	314	50,725	1	9	2
Texas	6.6/1	3,400	47	435	37,000	12	19	–
UCLA	9/1	5,902	21	450	44,150	13	12	–
Virginia	7.9/1	8,560	24	252	46,000	14	8	6

14.4 For the following data set, fit a first (x), second (x, x^2), and third (x, x^2, x^3) degree polynomial to the y data and comment on which best represents the sample relationship between x and y.

y	x
27.000	3.6660
28.000	3.6660
19.900	5.5830
28.000	1.5830
25.000	4.4170
25.000	4.2500
22.500	4.8330
23.500	4.8330
23.000	2.8330
19.000	2.5000

14.5 Below is an extension of Table 13.1. As in Chapter 13, fit a straight line regression to these data, but now include a dummy variable that equals 0, if the observations are from the 1960s, and 1, if from the 1970s. Write the equation for your fitted regression. Using a two-dimensional diagram, interpret the coefficient for the 0-1 dummy variable.

Year	Inflation Rate (%)	Unemployment Rate (%)	Year	Inflation Rate (%)	Unemployment Rate (%)
1960	1.6	5.5	1970	5.4	4.9
1961	0.9	6.7	1971	5.0	5.9
1962	1.8	5.5	1972	4.2	5.6
1963	1.5	5.7	1973	5.8	4.9
1964	1.5	5.2	1974	8.8	5.6
1965	2.2	4.5	1975	9.3	8.5
1966	3.2	3.8	1976	5.2	7.7
1967	3.0	3.8	1977	5.8	7.1
1968	4.4	3.6	1978	7.4	6.1
1969	5.1	3.5	1979	8.6	5.8

14.6 Labor economists assert that there is a relationship between earnings and age; as individuals age, their earnings tend to rise. Use the following data on real earnings to estimate this profile for males who are in the sales occupation under the conditions given below.

x = Age	y = Real Earnings
30	$ 8,495
45	10,517
60	9,949
30	7,756
40	10,068
50	10,467
60	10,021
70	8,957

a. Assume the "age/earnings profile" is linear.
   1. Write the equation of the regression for these data.
   2. Graph the regression equation on a scatterplot.
   3. Is the linear regression line a good representation of the "age/earnings profile"? Why or why not?

b. Assume the "age/earnings profile" is nonlinear.
   1. Write the equation of the regression fit to the above data for a third degree polynomial. (You have to create age-sq and age-cube.)
   2. Graph the regression equation on a scatterplot.
   3. Is this nonlinear regression line a good representation of the "age/earnings profile"? Why or why not?

14.7 The following regression is based on consumption (in $1,000), income (in $1,000), years of education, and age.

NAME	MEAN	STD.DEV.
income	25.3967	8.3428
ed	12.0833	3.3155
age	27.2500	9.8823
consump	22.6017	8.9233

DEPENDENT VARIABLE: consumption STD. ERROR OF EST. = 4.0027

VAR.	REGRESSION COEFFICIENT	STD. ERROR	T(DF= 8)	PROB.
income	.8463	.1712	4.943	.00113
ed	1.1606	.3792	3.061	.01556
age	–.0180	.1468	–.122	.90560
CONSTANT	–12.4259			

ANALYSIS OF VARIANCE TABLE

SOURCE	SUM OF SQUARES	D.F.
REGRESSION	747.7144	3
RESIDUAL	128.1726	8
TOTAL	875.8870	11

a. Write the regression equation in slope and intercept form.
b. What does the 0.8463 income coefficient indicate? Based on this estimate, could a $1 increase in income be associated with a $1 increase in expected consumption in the population? (Explain.)

14.8 The following data are from the Multiple Listing Service for a small midwestern college town. The prices are given in hundreds of dollars for 13 homes listed between $200,000 and $300,000. Rooms, Bedrooms, and Baths are simple room counts. Main Floor, Upper Floor, and Basement are square feet measures, with Fin Base being finished square feet in the basement. Age is the age of house in years, with zero representing a new house.

Price	2000	2050	2050	2090	2175	2175	2200	2235	2239	2295	2295	2299	2999
Rooms	12	7	6	10	7	10	10	9	8	9	10	9	8
Bedrooms	5	3	3	5	3	3	4	4	3	4	4	3	3
Baths	3.5	2.5	2.5	3	1.5	3.5	2.5	3.5	2.5	2	3.5	2.5	2
Main Fl	1478	1309	2100	1720	2050	2726	1371	1483	2850	1250	1457	2492	1898
Upper Fl	840	1064	0	1080	0	0	1079	1610	0	1000	2039	706	400
Basement	1188	1064	0	854	2050	328	1371	1443	0	1000	0	0	0
Fin Base	739	0	0	0	0	328	0	0	0	500	0	0	0
Age	5	50	3	0	27	14	0	3	1.5	71	0	1	0

a. Estimate a multiple regression of price on rooms, total square footage, and age.
b. For a new house of 2,500 square feet and eight rooms, what is the predicted price of the house?

14.9 In *Bazemore v. Friday* (1986), the U.S. Supreme Court reversed the Court of Appeals decision that rejected the petitioner's regression analyses because they did not include "all measurable variables thought to have an effect on salary levels." The Supreme Court stated that, "While the omission of variables from a regression analysis may render the analysis less probative than it otherwise might be, it can hardly be said, absent some other infirmity, that an analysis which accounts for the major factors 'must be considered unacceptable as evidence of discrimination.'"

a. Under what conditions would omitted variables bias the estimation of the included regressors? Does the U.S. Supreme Court's decision appear to address these conditions?

b. How does the omission of possible explanatory variables render the analysis less probative?

14.10 Economist Leonard Silk, in a *New York Times* (June 12, 1991) article, reported on an undergraduate's research paper aimed at explaining and predicting the relationship between a ball player's performance and pay. The paper was done by Joshua Engel in the University of Pennsylvania econometrics class of Professor Lawrence Klein, a Nobel laureate in economics. According to Silk,

> Mr. Engel found that when all other variables were held constant, every additional home run was worth $9,000 in next year's salary, and every extra run created was worth $6,000, for hitters. For pitchers, one more victory was worth $38,000, one save, $16,000, one more inning pitched, $3,000 and every one-tenth of a point off the e.r.a. is worth $12,000 on next year's pay.

a. Write the regression equation for batters; what magnitude didn't Silk supply?

b. Write the regression equation for pitchers; what magnitude didn't Silk supply?

14.11 A firm is interested in the relationship between net income Y (in $10,000 units), the number of salespeople (SALES), and the number of products produced (PRODUCT). It calculated the following multiple regression equation using existing data:

Ordinary Least Squares			
Dependent Variable	Y	Number of Observations	5
Mean of Dep. Variable	16.8000	Std. Dev. of Dep. Var.	12.214745
Std. Error of Regr.	7.2827	Sum of Squared Residuals	106.075
R-squared	.82226	Adjusted R-squared	.64452
F(2, 2)	4.6262	Prob. Value of F	.17774

Variable	Coefficient	Std. Error	T-ratio	Prob |t|>x	Mean of X	Std. Dev. of X
Constant	−17.793	11.87	−1.500	.27247		
SALES	.9121	.7630	1.196	.35441	35.00000	11.18034
PRODUCT	.1589	1.408	.113	.92047	16.80000	6.05805

a. If the company were to employ 40 salespeople and to carry 20 products, what would you predict the net income to be?

b. What is the meaning of $b_2 = 0.9121$?

14.12 Below is a rough duplication of a plot that appeared in the 1964 Surgeon General's report, *Smoking and Health.* What does this graph show regarding death rates and smoking?

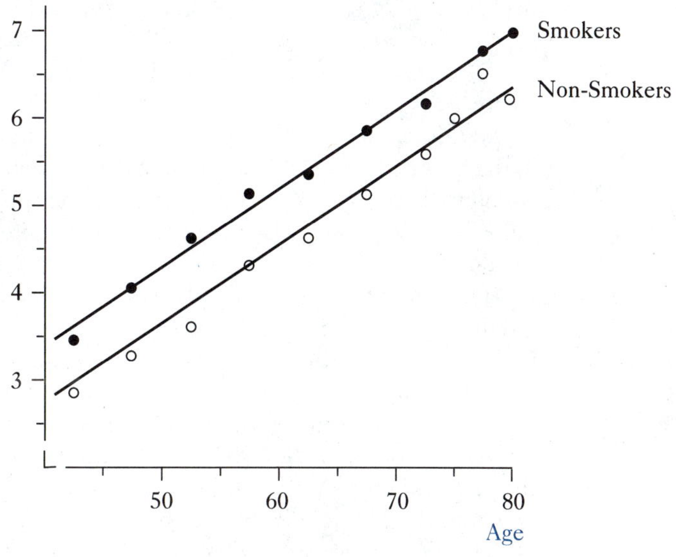

Source: Howard Wainer, "Graphical Visions from William Playfair to John Tukey," *Statistical Science,* August 1990, p. 342.

14.6 Estimation of a Conditional Expected Value of *Y*

As already stated, if the expected value of ε is zero, the expected value of y, conditional on a given set of values for the x's, is the deterministic part of the model:

$$E(y_i|x_{i1}, x_{i2} \ldots x_{ik}) = \beta_1 x_{i1} + \beta_2 x_{i2} \ldots + \beta_k x_{ik} \tag{14.24}$$

The least squares regression line

$$\hat{y}_i = b_1 x_{i1} + b_2 x_{i2} \ldots + b_k x_{ik} \tag{14.25}$$

provides the point estimate of the $E(y_i|x_{i1}, x_{i2} \ldots x_{ik})$. For any given sample, there is no reason to believe that $\hat{y}$ will be the same as the expected value of y, at the identified x values. If the population model is correct, the mean of $\hat{y}_i$, at the identified values of $x_{i1}, x_{i2} \ldots x_{ik}$, over a large (infinite) number of samples will equal $E(y_i|x_{i1}, x_{i2} \ldots x_{ik})$. That is,

$$E(y_i|x_{i1}, x_{i2} \ldots x_{ik}) = E(\hat{y}_i|x_{i1}, x_{i2} \ldots x_{ik}) \tag{14.26}$$

but randomness in sampling causes variability in $\hat{y}$, so $\hat{y}_i$, at the identified values of $x_{i1}, x_{i2} \ldots x_{ik}$, need not equal $E(y_i|x_{i1}, x_{i2} \ldots x_{ik})$.

As with simple regression involving only one explanatory variable, to reflect variability in estimation with multiple explanatory variables, a confidence interval estimate of $E(y_*|x_{*1},x_{*2} \ldots x_{*k})$ can be defined. A $(1 - \alpha)$ percent confidence interval for this expected value of y_*, at the specified x_* values, is

$$\hat{y}_* \pm t_{\alpha/2,\text{df} = n - k} s_{\hat{y}_*} \quad (14.27)$$

where $+t_{\alpha/2,\text{df} = n - k}$ is the value of t for which $\alpha/2$ percent of the area is in the upper tail of the t distribution and $-t_{\alpha/2,\text{df} = n - k}$ is the t value for which $\alpha/2$ percent of the area is in the lower tail. This t distribution has $n - k$ degrees of freedom, because k parameters are now estimated in the regression equation $(\beta_1, \beta_2, \ldots \beta_k)$. The rationale here is the same as in simple regression where two parameters were estimated and the degrees of freedom were $n - k = n - 2$.

The quantity $s_{\hat{y}*}$ is the standard error for the $\hat{y}$ surface portrayed in Figure 14.2. As in simple regression, the standard error for the $\hat{y}$ surface gets larger the farther the value of the x's are from their respective means. Unlike simple regression, the standard errors for the $\hat{y}$ surface cannot be calculated easily by hand.[2]

The standard errors for the $\hat{y}$ surface are provided in computer printouts. For example, at the bottom of the MINITAB output in the answer to Query 14.3, we see that the standard errors for the sample regression tuition plane, where the student/faculty ratio and the public/private dummy are the explanatory variables, is given in the column headed "Stdev.Fit." That is,

x_{*2}	x_{*3}	y_*	$\hat{y}_*$	Stdev.Fit
9	1	13500	13333.3	$88.2 = s_{\hat{y}_{13333.3}}$
9	1	13300	13333.3	$88.2 = s_{\hat{y}_{13333.3}}$
9	1	13200	13333.3	$88.2 = s_{\hat{y}_{13333.3}}$
8	1	13354	13354.0	$152.8 = s_{\hat{y}_{13354.0}}$
18	0	11312	11312.0	$152.8 = s_{\hat{y}_{11312.0}}$

The 95 percent confidence interval for $E(Y)$ can now be calculated using the critical $t_{0.025}$ value found in Appendix Table A.4. The probability of a t value greater than or equal to 4.303, with $n - k = 5 - 3 = 2$ degrees of freedom, is 0.025. That is,

$$P(t \geq 4.303 \mid \text{df} = 2) = 0.025$$

The 95 percent confidence interval for expected tuition at a private school with a nine to one student/faculty ratio is thus

$$\hat{y}_* \pm t_{\alpha/2,\text{df} = n - k} s_{y*} = 13333.3 \pm 4.303(88.2)$$

$$\text{or} \quad 12953.9 \leq E(y_* \mid x_{*2} = 9, x_{*3} = 1) \leq 13712.8$$

and similarly for the other 95 percent confidence interval reported at the bottom of the MINITAB printout.

Whenever we use Formula (14.27), there is a 0.95 probability that the resulting confidence interval will contain the expected value of y. This *does not say* that there is a 95 percent chance that the expected tuition is between \$12,953.9 and \$13,712.8 when the student/faculty ratio is nine to one at a private school. Once numerical values are substituted into Formula (14.27), there are no random variables in the expression, and no probability statement can be made. The resulting confidence interval either does or does not contain the expected value of y. The expected tuition is \$12,953.9 to \$13,712.8 for a private school with a nine to one student/faculty ratio, but the probability of 0.95 applies to the theoretical interval (14.27) and not the values calculated from this single sample.

The quantity ±379.52 [= ±4.303(88.2)] is called the margin of error. As in simple regression, it reflects the imprecision in our estimate of the expected value of y. As the level of confidence increases (e.g., from a 95 to a 99 percent confidence level), the margin of error increases, because the t value gets larger.

14.7 Prediction of an Individual Value of *Y*

In addition to estimating the expected value of y at given x values, $\hat{y}$ can be interpreted as providing a prediction of an individual value of y. If so, then additional variability must be recognized, as discussed in the case of simple regression in Section 13.6. This additional variability can be seen in the prediction intervals provided at the bottom of the MINITAB printout in the answer to Query 14.3.

The 95 percent prediction interval for y_* based on the prediction $\hat{y}_* = 13333.3$ is \$12,574.4 to \$14,092.2, which is wider than the corresponding 95 percent confidence interval for $E(y_* | x_{*2} = 9, x_{*3} = 1)$.[3] Prediction intervals will always be wider than the corresponding confidence intervals because the confidence interval for $E(y | x_1, x_2 \ldots x_k)$ only involves variability of $\hat{y}$ around $E(y | x_1, x_2 \ldots x_k)$, as reflected in the difference $\hat{y} - E(y | x_1, x_2 \ldots x_k)$. Any individual value of y that might occur with x values, however, would involve the variability reflected in the difference $\hat{y} - E(y | x_1, x_2 \ldots x_k)$ plus the difference between $E(y | x_1, x_2 \ldots x_k)$ and the individual value of y that might be observed.

Exercises

14.13 What influences the width of the confidence interval for the expected value of y, conditional on x values? How do these factors influence the width of this confidence interval?

14.14 Why do the prediction interval and the confidence interval for the expected value of y, conditional on x values, get wider the farther the x values are from their respective means?

14.15 For the B-school data in Exercise 14.3, for a hypothetical school ranked eighth on the A scale, with annual tuition of \$13,000, what is the 95 percent prediction interval for an individual student's starting salary? What is the 95 percent confidence interval for the expected stating salary? Which one should be of more interest to a prospective student considering an MBA program?

14.16 From your regression on the data in Exercise 14.8, for a new house of 2,500 square feet and eight rooms, what is a 95 percent prediction interval for the price of this house? What is a 95 percent confidence interval for the expected price of all such houses? What is the difference between these two interval estimates?

14.17 For the firm in Exercise 14.11, for 40 salespeople and 20 products, would a 95 percent confidence interval for net income be at its narrowest width? What is the meaning of this confidence interval?

14.8 Hypotheses Testing

Returning to the case of possible sex discrimination in the salary determination of academic men and women, the question is whether the underpayment of women was the result of chance or sex discrimination. The data in Table 14.1 are for only the men in the academic department. This same information, along with that for the five women in the department, is in Table 14.2. Notice in panel a of Table 14.2 that x_{i4} is 1 if the ith person is female and 0 if male. Panel b shows the actual ASCII file that can be read into most computer programs, as shown in Table 14.3, which uses the SAS statistics package.

Once the data are entered into a computer program, users of statistical programs such as SAS, SPSS, MINITAB, RATS, SHAZAM, and the like need only give commands to regress y (salaries) on a constant (the column of ones created by the program), x_2 (a column containing years since receipt of the Ph.D.), x_3 (a column for number of books written), x_4 (a column of 0's and 1's for gender), or any other regressors to be included in the explanation of the dependent variable. Although the location of items differs slightly, all of these computer programs provide printouts with information similar to that in Table 14.3.

The columns titled "Variable" and "Parameter Estimate" in the lower part of this SAS printout show the estimated regression to be

$$\hat{y}_i = 43036.5 + 1025.35x_{i2} + 320.74x_{i3} + 56.36x_{i4}$$

In this sample, the coefficient on the gender variable is positive, indicating that a woman is predicted to receive \$56.36 more than a man with the same publications and Ph.D. vintage. Figure 14.3 illustrates the relationship between the predicted salary and time since earning the Ph.D. for men and women who have never published a book. Upon receipt of the Ph.D. degree, men are predicted to receive a salary of \$43,036.50; comparable women are predicted to receive \$43,092.86. Because salaries are assumed to be linearly related to time since receipt of the Ph.D., this predicted \$56.36 difference continues regardless of the Ph.D. vintage considered.

Given the variability in the sampling distribution of b_4 (as reflected in its standard error of 1489.659 and t of 0.038), there is a high probability of obtaining a coefficient estimate that is \$56.36 above or below the null hypothesis value of zero. This probability of 0.9705 is given in Table 14.3 under the column titled "(Prob>|T|)" and is shown in Figure 14.4 as the area beyond ± \$56.36. It is also the area beyond the calculated t values of ± 0.038 and it is the significance level or the p-value for $b_4 = 56.36$ in a two-tailed test. Because this p-value is extremely large, and well above the usual 0.05 Type I error level,

$$H_O\colon \beta_4 = 0$$

Table 14.2
Male and Female Faculty Salaries and Related Data

Panel a: Data in matrix notation

Individual i	Salary y_i	Constant x_{i1}	Years since Ph.D. x_{i2}	Number of Books x_{i3}	Sex x_{i4}
$i = 1$	$y_1 = 66000$	$x_{11} = 1$	$x_{12} = 19$	$x_{13} = 1$	$x_{14} = 0$
$i = 2$	$y_2 = 63500$	$x_{21} = 1$	$x_{22} = 17$	$x_{23} = 1$	$x_{24} = 0$
$i = 3$	$y_3 = 56700$	$x_{31} = 1$	$x_{32} = 11$	$x_{33} = 0$	$x_{34} = 0$
$i = 4$	$y_4 = 56300$	$x_{41} = 1$	$x_{42} = 16$	$x_{43} = 0$	$x_{44} = 0$
$i = 5$	$y_5 = 54900$	$x_{51} = 1$	$x_{52} = 12$	$x_{53} = 2$	$x_{54} = 0$
$i = 6$	$y_6 = 52300$	$x_{61} = 1$	$x_{62} = 11$	$x_{63} = 0$	$x_{64} = 0$
$i = 7$	$y_7 = 50650$	$x_{71} = 1$	$x_{72} = 7$	$x_{73} = 0$	$x_{74} = 0$
$i = 8$	$y_8 = 50650$	$x_{81} = 1$	$x_{82} = 6$	$x_{83} = 1$	$x_{84} = 0$
$i = 9$	$y_9 = 50500$	$x_{91} = 1$	$x_{92} = 10$	$x_{93} = 0$	$x_{94} = 0$
$i = 10$	$y_{10} = 48000$	$x_{101} = 1$	$x_{102} = 5$	$x_{103} = 2$	$x_{104} = 0$
$i = 11$	$y_{11} = 46000$	$x_{111} = 1$	$x_{112} = 2$	$x_{113} = 0$	$x_{114} = 1$
$i = 12$	$y_{12} = 45000$	$x_{121} = 1$	$x_{122} = 2$	$x_{123} = 0$	$x_{124} = 1$
$i = 13$	$y_{13} = 62200$	$x_{131} = 1$	$x_{132} = 17$	$x_{133} = 1$	$x_{134} = 1$
$i = 14$	$y_{14} = 61000$	$x_{141} = 1$	$x_{142} = 15$	$x_{143} = 0$	$x_{144} = 1$
$i = 15$	$y_{15} = 58300$	$x_{151} = 1$	$x_{152} = 19$	$x_{153} = 1$	$x_{154} = 1$

Panel b: Data in machine readable form

```
66000  19  1  0
63500  17  1  0
56700  11  0  0
56300  16  0  0
54900  12  2  0
52300  11  0  0
50650   7  0  0
50650   6  1  0
50500  10  0  0
48000   5  2  0
46000   2  0  0
45000   2  0  1
62200  17  1  1
61000  15  0  1
58300  19  1  1
```

cannot be rejected. Even at the 0.10 Type I error level, this null hypothesis cannot be rejected because the p-value = 0.9705 is just too large. Similarly, for standard Type I error levels, books published is not significant, because its p-value is extremely large.

The p-value or significance level gives the smallest probability of a Type I error for which the null hypothesis can be rejected and the alternative hypothesis can be accepted. At the typical 0.05 probability of a Type I error level,

$$H_O\text{: } \beta_2 = 0$$

Table 14.3 Least Squares Salary Regression Using SAS

```
DATA: TABLE142;                    ← to read in file of data as presented
INFILE 'TABLE142.DAT';               in Panel b of Table 14.2
INPUT salary yrs books sex;
PROC REG;                          ← Request regression procedure
MODEL salary=yrs books sex;        ← Regression model specification
RUN;
```

The SAS System

Dependent Variable: SALARY

Analysis of Variance

Source	DF	Sum of Squares	Mean Square	F Value	Prob>F
Model	3	513200598.39	171066866.13	24.077	0.0000
Error	11	78154401.61	7104945.6009		
C Total	14	591355000			

Root MSE	2665.51038	R-square	0.8678
Dep. Mean	54800.00000	Adj R-sq	0.8318
C.V.	4.86407		

Parameter Estimates

Variable	DF	Parameter Estimate	Standard Error	T for HO: Parameter=0	Prob>|T|
INTERCEPT	1	43036.5	1686.0941934	25.524	0.0000
YRS	1	1025.350585	123.45778801	8.305	0.0000
BOOKS	1	320.743018	1001.5175425	0.320	0.7548
SEX	1	56.363139	1489.6591912	0.038	0.9705

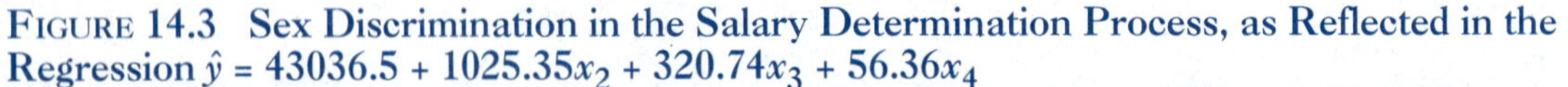

Figure 14.3 Sex Discrimination in the Salary Determination Process, as Reflected in the Regression $\hat{y} = 43036.5 + 1025.35x_2 + 320.74x_3 + 56.36x_4$

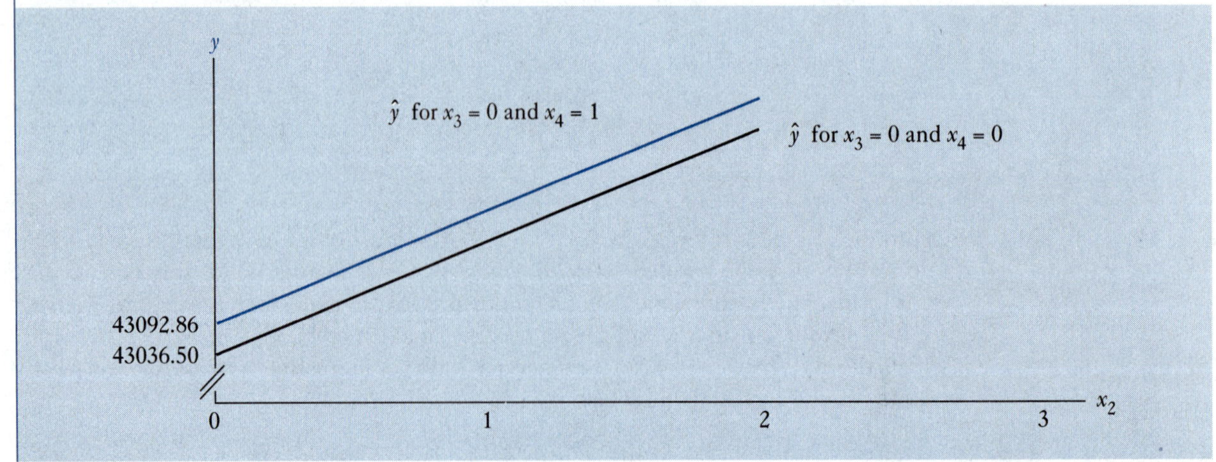

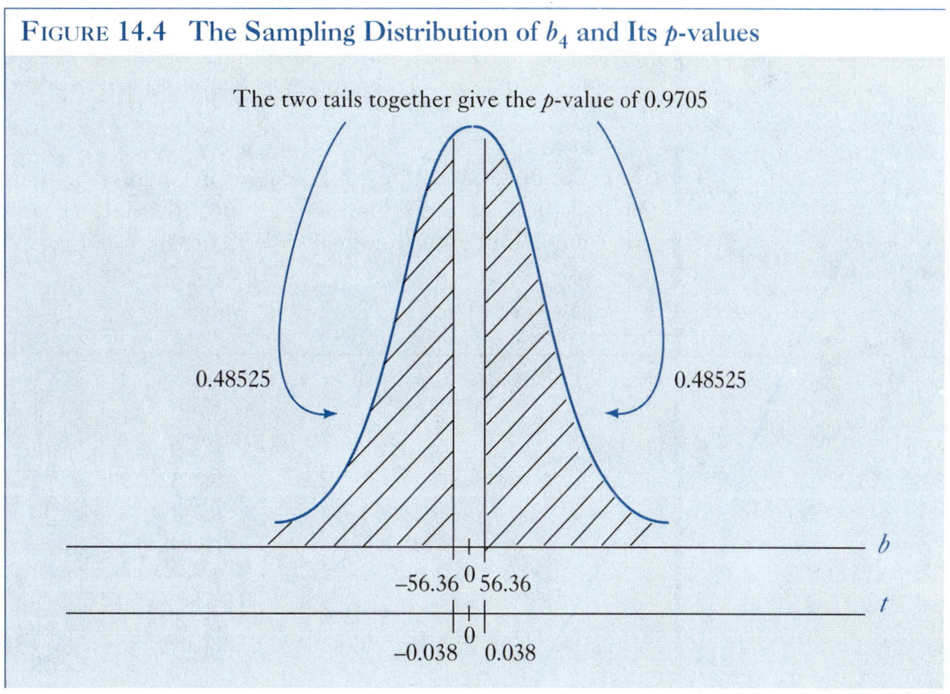

FIGURE 14.4 The Sampling Distribution of b_4 and Its p-values

can be rejected because the p-value = 0.0000 is smaller than this 0.05 level. This p-value is so small that time since the receipt of the Ph.D. degree is a significant explanatory variable at any of the typical Type I error levels.

QUERY 14.4:

A yacht broker attempted to establish the relationship between the age of a class of racing boats and their selling prices. From sales records, she obtained information on

Variable	Coefficient	Std. Error	T-ratio	(Sig.Lvl)
CONSTANT	32.4348	1.272	25.502	(.00000)
AGE	–2.91528	.3584	–8.134	(.00004)

```
                                PLOT OF RESIDUALS
           -1.40                        0                        1.40
            ..............................:...........................
PRICE       :                           :                         :  Residual
19.0 17.86  :                           :                    *    :     1.14
19.5 18.59  :                           :                *        :      .91
19.5 20.77  :  *                        :                         :    -1.27
20.0 20.77  :         *                 :                         :     -.77
22.5 22.96  :    *                      :                         :     -.46
22.5 23.69  :          *                :                         :    -1.19
23.0 23.69  :              *            :                         :     -.69
25.0 24.42  :                           :          *              :      .58
25.5 25.15  :                           :      *                  :      .35
28.0 26.60  :                           :                        *:     1.40
```

10 boats sold through the brokerage. She ran the regression shown here and concluded that age was a significant explanatory variable, with each year predicted to lower the selling price by $2,915 (price was measured in $1,000). Comment on her analysis.

ANSWER:
From our discussion of residual analysis in Chapter 13, these residuals do not appear random (boats in the middle of the price distribution tend to have negative residuals) and do not appear normally distributed, as seen from their histogram:

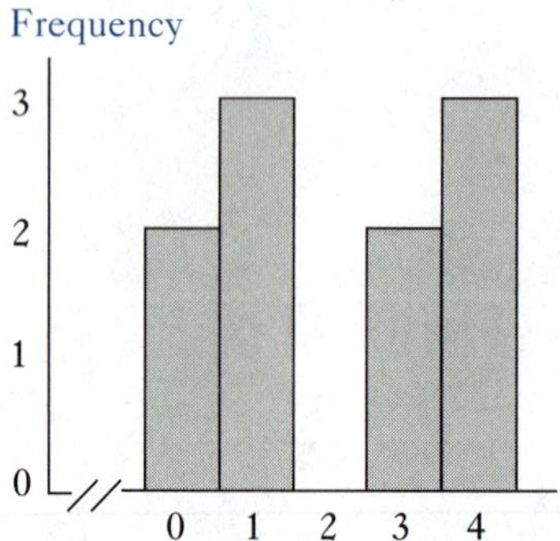

Residual Mean = 0.000, std. dev. = 0.992

Class	Lower Limit	Upper Limit	Frequency Total	Frequency Relative	Cumulative Total	Cumulative Relative
0	−1.500	−.900	2	.2000	2	.2000
1	−.900	−.300	3	.3000	5	.5000
2	−.300	.300	0	.0000	5	.5000
3	.300	.900	2	.2000	7	.7000
4	.900	1.500	3	.3000	10	1.0000

Because these residuals suggest violations of the assumption required for hypothesis testing, her test of significance is questionable. As discussed in the optional sections at the end of Chapter 13, the "<" pattern in residuals suggests a nonlinear relationship that may be captured by a polynomial of the form PREDICTED PRICE = CONSTANT + b_1AGE + b_2AGE^2. The data set is now

PRICE	AGE	AGESQ
28.000	2.0000	4.0000
25.500	2.5000	6.2500
25.000	2.7500	7.5625
23.000	3.0000	9.0000
22.500	3.0000	9.0000
22.500	3.2500	10.563
20.000	4.0000	16.000
19.500	4.0000	16.000
19.500	4.7500	22.563
19.000	5.0000	25.000

The broker had data on only price and age. We created AGE^2 by squaring the age variable. Within the statistics package, we had to identify AGE^2 by the name AGESQ because the statistics program could not make superscripts. The AGESQ variable is then included as if it were just another regressor, and the new regression results are

Variable	Coefficient	Std. Error	T-ratio	(Sig.Lvl)
CONSTANT	44.9683	2.355	19.099	(.00000)
AGE	–10.4656	1.383	–7.566	(.00013)
AGESQ	1.05818	.1925	5.498	(.00091)

PLOT OF RESIDUALS

```
                  -0.810                                   0                                      0.810
                  .. .. .. .. .. .. .. .. .. .. .. .. .. .. .. .:.. .. .. .. .. .. .. .. .. .. .. .. .. .. .. ..
PRICE             :                                         :                                     :    Residual
19.0  19.095      :                                  *      :                                     :   -.947E-01
19.5  19.132      :                                         :                        *            :        .368
19.5  20.037      :                 *                       :                                     :       -.537
20.0  20.037      :                                    *    :                                     :   -.367E-01
22.5  22.132      :                                         :                             *       :        .368
22.5  23.095      :               *                         :                                     :       -.595
23.0  23.095      :                                  *      :                                     :   -.951E-01
25.0  24.190      :                                         :                                    *:        .810
25.5  25.418      :                                         :  *                                  :    .821E-01
28.0  28.270      :                        *                :                                     :       -.270
```

Now the residuals appear to be random, but the effect of age is no longer the same across all ages. For example, boats that are one, two, and three years old have the following respective predicted selling prices (in \$1,000):

$$\text{Predicted price} = 44.9683 - 10.4656(1) + 1.0582(1) = 35.561$$
$$\text{Predicted price} = 44.9683 - 10.4656(2) + 1.0582(4) = 28.270$$
$$\text{Predicted price} = 44.9683 - 10.4656(3) + 1.0582(9) = 23.095$$

From one to two years of age, the change in predicted selling price is –\$7,291. From two to three years of age, the change in predicted selling price is –\$5,175. Students who know calculus realize that the effect of age on the predicted selling price is

$$\text{(CHANGE IN PRICE)/(CHANGE IN AGE)} = -10.4656 + 2(1.0582)(\text{AGE})$$

Thus, the effect of age is negative for boats that are less than 4.95 years old, after which the effect of age turns positive. However, there are no boats in the sample that approach the 4.95-year turning point. Finally, note that both the AGE and AGESQ effects are highly significant, suggesting that they both belong in a population model of selling price.

14.9 Confidence Interval for β_j

An alternative to a hypothesis test of a specific value of β_j is the consideration of the values in the confidence interval for β_j. The 100(1–α) percent confidence interval for β_j is again given by the estimator minus or plus the margin of error:

$$b_j \pm t_{\alpha/s} s_{bj}, \text{ for df} = n - k$$

or

$$b_j - t_{\alpha/2} s_{b_j} \leq \beta_j \leq b_j + t_{\alpha/2} s_{b_j} \tag{14.28}$$

where $t_{\alpha/2}$ again is the value of t that puts $\alpha/2$ in the tail of a t distribution, with $n - k$ degrees of freedom.

In the case of the gender coefficient, the data in Table 14.3 yield a 95 percent confidence interval for β_4 of

$$56.36 \pm (2.201)\ 1489.659$$

or

$$-3222.38 \leq \beta_4 \leq 3335.10$$

From this 95 percent confidence interval, we know that any hypothesis value of β_4 between –\$3,222 and \$3,335 could not be rejected in a two-tailed test, at the 0.05 Type I error level. This is an extremely wide interval that includes the no discrimination value of zero but also includes a sizable discrimination and reverse discrimination dollar value.

Historical Perspective

Although Sir Frances Galton receives credit for introducing the ideas of regression and correlation analysis to the social sciences, their use in economics is more closely tied to George Yule (1871–1951). Yule developed the idea of "net or partial regression" to represent the influence of one variable on another, holding other variables constant. He invented the multiple correlation coefficient R for the correlation of y with many x's. Yule's regressions looked much like those of today. In 1899, for instance, he published a study in which changes in the percentage of persons in poverty in England between 1871 and 1881 were explained by the change in the percentage of disabled relief recipients to total relief recipients (called the "out-relief ratio"), the percentage change in the proportion of old people, and the percentage change in the population.

Predicted change percent in pauperism
= –27.07 percent
+ 0.299(change percentage in out-relief ratio)
+ 0.271(change percentage in proportion of old)
+ 0.064(change percentage in population).

Stephen Stigler (1986, pp. 356–357) states that, although Yule's regression analysis of poverty was well known at the time, it did not have an immediate effect on social policy or statistical practices. In part, this lack of recognition was the result of the harsh criticism it received from the leading English economist, Arthur Pigou. In 1908, Pigou wrote that statistical reasoning could not be rightly used to establish the relationship between poverty and out-relief because even in a multiple regression (which Pigou called "triple correlation"), the most important influences, superior program management and restrictive practices, cannot be measured quantitatively.

Pigou identified the most enduring criticism of regression analysis; namely, the possibility that an unmeasured, but relevant, variable has been omitted from the regression and that it is this variable that is giving the appearance of a causal relationship between the dependent variable and the included regressors. As described by Michael Finkelstein and Bruce Levin (1990, pp. 363–364 and pp. 409–415), for example, defense attorneys continue to argue that the plaintiff's experts omit relevant market and productivity variables when they use regression analysis to demonstrate that women are paid less than men. Modern academic journals are packed with articles that argue for one specification of a regression equation versus another for everything from the demand for places in higher education to the demand for money.

Both Yule and Pigou recognized the difference between marginal and partial association, and the condensing of variables in contingency tables. Today, some statisticians try to give credit for this identification to E.H. Simpson (1951). The so-called "Simpson's paradox" says that marginal and partial association can differ even in direction so that what is true for parts of a sample need not be true for the entire sample. This is represented in the following figure, where the two separate regressions of y on the high values of x and the low values of x have positive slopes, but a single regression fit to all the data shows a negative slope. As in Pigou's criticism of Yule 90 years ago, this "paradox" may be caused by an omitted, but relevant, explanatory variable for y that is also related to x. It might be better named the "Yule-Pigou paradox" or "Yule-Pigou effect."

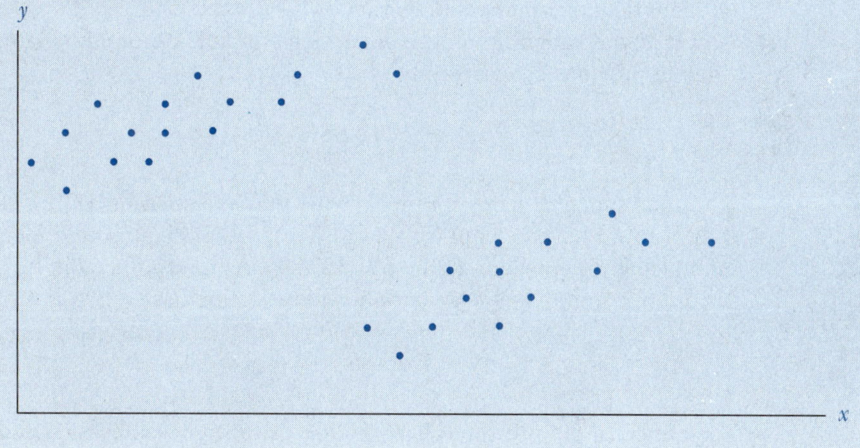

Exercises

14.18 In Exercise 14.5, was there a significant change in the Phillips curve relationship between the 1960s and 1980s? Why might you question the appropriateness of your test in this situation?

14.19 Redo Exercises 14.5 and 14.18, using a third degree polynomial to capture the curvature in the data. That is, for inflation and unemployment rates, estimate the following model.

$$\text{inflation} = \text{constant} + \beta_1 \text{ unemployment} + \beta_2(\text{unemployment})^2 + \beta_3 \text{ Dummy} + \varepsilon$$

Is $\beta_3 = 0$ or is $\beta_3 \neq 0$, at the 5 percent Type I error level?
How does this differ from your previous results?

14.20 For the consumption regression in Exercise 14.7, which, if any, of the individual regressors are significant at the 1 percent Type I error level?

14.21 For the B-school data in Exercise 14.3, what are the 95 percent confidence intervals for the A scale rank coefficient and the annual tuition coefficient in the starting salary regression? From these confidence intervals, can you tell if either of these coefficients would be significantly different from zero at the 0.05 Type I error level?

14.22 In Exercise 14.11, at what level is there a statistically significant positive relationship between net income and the number of salespeople?

14.23 For the consumption regression in Exercise 14.7, what is the largest and smallest hypothesized coefficient for education that could not be rejected, at the 0.01 Type I error level in a two-tailed test?

14.10 Multiple Coefficient of Determination

The relative amount of variability in the dependent variable explained by a multiple regression is defined the same as in simple regression; it is the ratio of the regression sum of squares (RegSS) to the total sum of squares (TSS). Now, however, the regression sum of squares, as well as the residual or error sum of squares (ErrorSS), is based on more than one independent variable.

Observed variability in the dependent variable y around its sample mean $\bar{y}$ was defined in previous chapters to be

$$\text{TSS} = \sum_{i=1}^{n}\left(y_i - \bar{y}\right)^2 \tag{14.29}$$

This total sum of squares continues to be that in Equation (12.12). Using the salary data from Table 14.2, the hand (or spreadsheet) calculation routine for TSS is demonstrated in Table 14.4, where TSS = 591,355,000. As reintroduced in this chapter, the residual or error sum of squares continues to be defined as

$$\text{ErrorSS} = \sum_{i=1}^{n} e_i^2 = \sum_{i=1}^{n}\left(y_i - \hat{y}_i\right)^2 \tag{14.30}$$

As defined in the previous chapters, the ErrorSS measures the amount of variability in sample observations on y around the regression plane, but in multiple regression $\hat{y}$ is calculated for each of the n sets of sample values on $x_2, x_3, \ldots x_k$. A hand (or spreadsheet) routine for calculating ErrorSS would start with the determination of the n values for the e vector followed by their individual squaring and summing. For the data set in Table 14.2 and the regression coefficients in Table 14.3, this process is shown in Table 14.5, where ErrorSS is calculated to be 78,154,401.

As discussed in the previous chapters, the total sum of squares is equal to the error sum of squares plus the regression sum of squares. Thus, the regression sum of squares can be obtained by subtracting ErrorSS from TSS, that is,

$$\text{RegSS} = \text{TSS} - \text{ErrorSS} \tag{14.31}$$

Table 14.4 Calculation of the Total Sum of Squares

y	$y-\bar{y}$	$(y-\bar{y})^2$
66000	11200	125440000
63500	8700	75690000
56700	1900	3610000
56300	1500	2250000
54900	100	10000
52300	–2500	6250000
50650	–4150	17222500
50650	–4150	17222500
50500	–4300	18490000
48000	–6800	46240000
46000	–8800	77440000
45000	–9800	96040000
62200	7400	54760000
61000	6200	38440000
58300	3500	12250000
822000	0	591355000 = TSS
54800 = $\bar{y}$		

From Tables 14.4 and 14.5, the regression sum of squares is

$$\text{RegSS} = 591{,}355{,}000 - 78{,}154{,}401 = 513{,}200{,}599$$

Hand or even spreadsheet calculations of TSS, ErrorSS, and RegSS are tedious. In addition, rounding errors tend to accumulate. For example, note in Table 14.5 that the sum of residuals is not exactly zero ($\Sigma e = -0.152$), as would be the case if there were no rounding error. A difference from zero of –0.152 is trivial, given the size of numbers with which we are working. A computer regression program written for statistical calculations in double precision, however, will not make an error even this small in the calculation of residuals and their squares.

Regression programs automatically provide sums of squares in an analysis of variance table. For the regression results in Table 14.3, an analysis of variance table is reproduced in Table 14.6. The measures of variability reported in Table 14.3 in the analysis of variance section of the SAS printout are large. As seen in Table 14.6, some computer programs report large numbers in scientific notation, where E+08 means multiply by 10^8. While all calculations are made in double precision, the number of digits displayed in the computer printout will be limited by the space available and the format specified by the programmer. Multiplying the values given for the sum of squares in Table 14.6 by 100,000,000 gives values extremely close to those obtained earlier in Tables 14.3, 14.4 and 14.5, namelv

$$\text{TSS} = 591{,}355{,}000 \quad \text{RegSS} = 513{,}201{,}000 \quad \text{ErrorSS} = 78{,}154{,}000$$

TABLE 14.5 Calculation of the Error Sum of Squares

y_i	$\hat{y}_i = 43036.5 + 1025.35x_{i2} + 320.743x_{i3} + 56.3631x_{i4}$	$e_i = y_i - \hat{y}_i$
66000	43036.5 + 1025.35(19) + 320.743 (1) + 56.3631 (0) = 62838.893	3161.107
63500	43036.5 + 1025.35(17) + 320.743 (1) + 56.3631 (0) = 60788.193	2711.807
56700	43036.5 + 1025.35(11) + 320.743 (0) + 56.3631 (0) = 54315.350	2384.650
56300	43036.5 + 1025.35(16) + 320.743 (0) + 56.3631 (0) = 59442.100	–3142.100
54900	43036.5 + 1025.35(12) + 320.743 (2) + 56.3631 (0) = 55982.186	–1082.186
52300	43036.5 + 1025.35(11) + 320.743 (0) + 56.3631 (0) = 54315.350	–2015.350
50650	43036.5 + 1025.35(7) + 320.743 (0) + 56.3631 (0) = 50213.950	436.050
50650	43036.5 + 1025.35(6) + 320.743 (1) + 56.3631 (0) = 49509.343	1140.657
50500	43036.5 + 1025.35(10) + 320.743 (0) + 56.3631 (0) = 53290.000	–2790.000
48000	43036.5 + 1025.35(5) + 320.743 (2) + 56.3631 (0) = 48804.736	–804.736
46000	43036.5 + 1025.35(2) + 320.743 (0) + 56.3631 (1) = 45143.563	856.437
45000	43036.5 + 1025.35(2) + 320.743 (0) + 56.3631 (1) = 45143.563	–143.563
62200	43036.5 + 1025.35(17) + 320.743 (1) + 56.3631 (1) = 60844.556	1355.444
61000	43036.5 + 1025.35(15) + 320.743 (0) + 56.3631 (1) = 58473.113	2526.887
58300	43036.5 + 1025.35(19) + 320.743 (1) + 56.3631 (1) = 62895.256	–4595.256
		–0000.152

e_i^2
9992598
7353897
5686556
9872792
1171126
4061636
190140
1301098
7784100
647600
733484
20610
1837228
6385157
21116379
78154401

$\Sigma e_i^2 = 78154401$
ErrorSS = 78,154,401

TABLE 14.6 Analysis of Variance Information for Regression Results in Table 14.3

	Sum of Squares	d.f.
Total	5.91355E+08	14
Regression	5.13201E+08	3
Residuals	.78154E+08	11

Std. Error of Regression . .	2665.51038
R-Squared	.86784
Adjusted R-Square	.83179

A measure of the goodness-of-fit of the multiple regression equation is called the multiple coefficient of determination and is designated by R^2. As with the simple coefficient of determination, r^2 in Formula (12.15), the **multiple coefficient of determination (R^2)** is

Multiple Coefficient of Determination (R^2) The proportion of variability in y explained by a set of regressors.

$$R^2 = \frac{\text{RegSS}}{\text{TSS}} = 1 - \frac{\text{ErrorSS}}{\text{TSS}} \tag{14.32}$$

The R^2 is a number between zero and one that indicates the proportion of variability in the dependent variable explained by the independent variables. If $R^2 = 1.00$, then 100 percent of the variability in the sample y's is explained by the regression equation, and ErrorSS must be zero; all the y values are predicted perfectly. If $R^2 = 0$, then none of the variability in y is explained by x_2 through x_k, and ErrorSS = TSS. From Table 14.6, we see that 86.78 percent of the variability in salaries (y) is explained jointly by Ph.D. vintage (x_2), books published (x_3), and gender (x_4).

14.11 Adjusted Coefficient of Determination

The coefficient of determination is never lowered by adding regressors, because more regressors cannot increase the sum of squares of the errors (ErrorSS). Thus, R^2 is of limited value for comparing the explanatory power of regressions with different numbers of regressors. One way around this deficiency is to adjust R^2 for the difference in degrees of freedom associated with the errors when $\hat{y}$ and $\bar{y}$ are used to predict y.

As stated in previous chapters, the variability of sample y values around their mean $\bar{y}$ is given by the total sum of squares, TSS. The associated degrees of freedom are $n - 1$. Thus, the sample variance of y is TSS divided by $n - 1$.

$$s_y^2 = \frac{\text{TSS}}{n-1} = \frac{\Sigma\left(y_i - \bar{y}\right)^2}{n-1} \tag{14.33}$$

In calculating this sample variance, one degree of freedom is given up because the value of $\bar{y}$ is fixed in its calculation. The square root of the sample variance s_y is an estimator of the population standard deviation σ_y.

In calculating the error sum of squares, ErrorSS, k degrees of freedom are given up, because the values of b_1 through b_k are fixed in the calculation. The variance of e is calculated by dividing ErrorSS by $n - k$.

$$s_e^2 = \frac{\text{ErrorSS}}{n-k} = \frac{\Sigma\left(y_i - \hat{y}_i\right)^2}{n-k} = \frac{\Sigma\left(y_i - b_1 - b_2 x_{i2} \ldots - b_k x_{ik}\right)^2}{n-k} \tag{14.34}$$

The square root of the variance of e is called the standard error of the regression. The standard error of the regression is an estimator for the standard deviation of the population disturbance term, σ_ε. It is typically reported in a regression computer printout and does not have to be calculated by hand. In Table 14.6, the standard error of the regression is $s_e = 2665.51038$ and is calculated as

$$s_e = \sqrt{\frac{\text{ErrorSS}}{n-k}} = \sqrt{\frac{.78154\text{E}+08}{15-4}} = 2665.51038$$

The adjusted R-square, $\bar{R}^2$, is defined as

$$\bar{R}^2 = 1 - \frac{s_e^2}{s_y^2} = 1 - \frac{(n-1)\text{ESS}}{(n-k)\text{TSS}} \tag{14.35}$$

The rationale behind the adjusted R-square, $\bar{R}^2$, is that k parameters have been used to explain y by $\hat{y}$ while only one parameter is estimated in explaining y by $\bar{y}$. If the additional parameter estimates do not add more to explanatory power than what is lost in degrees of freedom, then relative explanatory power should decrease. The adjusted R-square can decrease because the ratio of $n - 1$ to $n - k$ will increase even if ErrorSS is unaffected by the addition of regressors.

The adjusted R-square is typically reported in regression computer printouts. For example, in Tables 14.3 and 14.6, the adjusted R-square is $\bar{R}^2 = .83179$, which is calculated as

$$\bar{R}^2 = 1 - \frac{(n-1)\text{ ErrorSS}}{(n-k)\text{ TSS}} = 1 - \frac{(15-1)0.78154\text{E}+08}{(15-4)5.91355\text{E}+08} = 0.83179$$

14.12 Testing the Population Model

Both R^2 and $\bar{R}^2$ are descriptive measures of how well a linear regression explains sample variability in the dependent variable. They do not measure the extent to which the deterministic part of the population model is related to the dependent variable. Yet one of the first questions a researcher wants to answer is whether all of the explanatory variables in the population model jointly influence the dependent variable.

If the assumed population model is

$$y_i = \beta_1 + \beta_2 x_{i2} + \beta_3 x_{i3} + \ldots + \beta_k x_{ik} + \varepsilon_i$$

but in fact there is no linear relationship between y and all the x's, then β_2 through β_k must be zero. On the other hand, if there is a linear relationship between y and the x's, then at least one of these β's must be nonzero. Thus, the null and alternative hypotheses are

$$H_O\text{: } \beta_2 = \beta_3 = \ldots = \beta_k = 0$$

$$H_A\text{: at least one of these } \beta\text{'s is nonzero}$$

The null hypothesis implies that β_1 is the mean of y irrespective of the x_2 through x_k considered. Random sampling from a population in which $\beta_2 = \beta_3 = \ldots = \beta_k = 0$ is expected to yield an error sum of squares, adjusted for its degrees of freedom, that is relatively large, and a regression sum of squares, adjusted for its degrees of freedom, that is relatively small. Thus, the ratio of RegSS/(k – 1) to ErrorSS/(n – k) should be close to zero, if H_0 is true. [Note: Just as RegSS is determined by subtracting ErrorSS from TSS, so too are its degrees of freedom determined by subtracting $n - k$ from $n - 1$. The degrees of freedom for RegSS are $k - 1 = (n - 1) - (n - k)$.]

The question is how high does the calculated ratio of RSS/(k – 1) to ErrorSS/(n – k) have to be before H_0 is rejected? To answer this question, the probability distribution of

$$\frac{\text{RegSS}/(k-1)}{\text{ErrorSS}/(n-k)}$$

must be introduced.

The ErrorSS/σ_ε^2 is distributed as a chi-square, with $n - k$ degrees of freedom. RegSS/σ_ε^2 is distributed as a chi-square, because the b's and ε are distributed normally, but RegSS has only $k - 1$ degrees of freedom. The ratio of two independent chi-square random variables, adjusted for their degrees of freedom, forms a new random variable that is distributed as an F.

$$\frac{\text{RegSS}/(k-1)}{\text{ErrorSS}/(n-k)} \sim \text{F}\frac{(\text{df}1 = k-1)}{(\text{df}2 = n-k)} \tag{14.36}$$

As introduced in Chapter 10, the F distribution represents a family of distributions. The shape of the F is controlled by two sets of degrees of freedom (df1 and df2). When there is only one degree of freedom in the numerator, df1 = $k - 1 = 2 - 1 = 1$, the F is just the square of the t distribution. Beyond this special case, the shape of the F changes with the degrees of freedom.

The probability of a particular range of F values depends on the degrees of freedom in the numerator (df1 = k – 1) and the degrees of freedom in the denominator (df2 = n – k). Appendix Table A.5 shows the F value for different sets of degrees of freedom. For example, if there are two degrees of freedom in the numerator and three degrees of freedom in the denominator, then at the intersection of column 2 and row 3, the F value 9.55 indicates that only 5 percent of the F values are higher for this distribution. That is,

$$\text{Prob}\left(F\frac{\text{df}1 = 2}{\text{df}2 = 3} > 9.55\right) = 0.05$$

For the regression in Table 14.3, and the analysis of variance results in Table 14.6, df1 = $k - 1 = 3$ and df2 = $n - k = 11$. In Appendix Table A.5, the intersection of column 3 and row 11 shows an F value of 3.59 as the critical value that puts 5 percent of the F distribution higher. Thus, under the null hypothesis in this testing situation, H_0: $\beta_2 = \beta_3 = \ldots = \beta_k = 0$, the probability of getting an F value higher than 3.59 is 5 percent.

$$\text{Prob}\left(F\frac{\text{df}1 = 3}{\text{df}2 = 11} > 3.59\right) = 0.05$$

A test of whether H_O is correct requires the specification of the probability of a Type I error. In our testing situation, if this probability is set at 5 percent, then 3.59 is the critical F value. A calculated F value greater than 3.59 would result in the rejection of H_O, because there is a less than 5 percent chance of getting such values if H_O is true, with three and 11 degrees of freedom. On the other hand, a calculated F value less than 3.59 results in a decision not to reject H_O.

Remember that the decision to reject H_O (or to not reject H_O) says only that this sample of data is consistent with the hypothesis that is not rejected. Sample data can never prove that the null or the alternative hypothesis is true because there is always the chance of either a Type I (rejecting the null hypothesis when it is true) or a Type II error (failing to reject the null hypothesis when it is false). Reducing the probability of a Type I error increases the probability of a Type II error, and vice versa.

For the sample data on which the Table 14.3 regression and Table 14.6 analysis of variance are based, the calculated F ratio is

$$F = \frac{\text{RegSS}/(k-1)}{\text{ErrorSS}/(n-k)} = \frac{5.13201\text{E}+08/(4-1)}{0.78154\text{E}+08/(15-4)} = 24$$

For a probability of a Type I error equal to 5 percent, this calculated F value of 24 is well beyond the critical F value of 3.59, with three and 11 degrees of freedom. This implies that there is far less than a 5 percent chance of getting a calculated $F = 24$, if $\beta_2 = \beta_3 = \ldots = \beta_k = 0$. Thus, H_O is rejected in favor of H_A. That is, we conclude that at least one of these β's is nonzero; there is a relationship between y and the x's in the population.

Most computer programs print out the sums of squares RegSS and ErrorSS, and the mean sums of squares RegSS/(k–1) and ErrorSS/(n–k), which form the basis of the overall F test. In addition, most programs calculate the F and give the probability of obtaining this or a more extreme F value, if the null hypothesis is true. This information is usually contained in an analysis of variance table similar to that in the analysis of variance section of Table 14.3 or as reproduced in Table 14.7, where scientific notation is used. The significance level of 0.00004 is the smallest probability of a Type I error for which H_O: $\beta_2 = \beta_3 = \beta_4 = 0$ could be rejected and H_A: at least one β is nonzero is supported.

Worth noting is that the analysis of variance table for multiple regression contains the same information as an analysis of variance table in Chapter 11. In particular, because it reflects differences between the predicted effect ($\hat{y}$) and the sample mean ($\bar{y}$), the regression sum of squares is identical to the quantity called the treatment sum of squares or the between-group sum of squares in Chapter 11. The error sum of squares is like the within-group sum of squares in Chapter 11. Some programs use these names interchangeably, although the use of "between-group variability" versus the "within-group variability" is typically restricted to an analysis of variance done outside the regression framework, as in Chapter 11.

Table 14.7 Analysis of Variance Table for Regression Results in Table 14.3

	Sum of Squared	d.f.	Mean Square	F(Sig.Lvl.)
Total.	5.91355E+08	14		
Regression . . .	5.13201E+08	3	1.71067E+08	24.08 (.00004)
Residuals	.78154E+08	11	0.07105E+08	

EXERCISES

14.24 Based on your regression in Exercises 14.3 and 14.15, are both of the explanatory variables (rank on scale A and tuition) jointly significant at the 0.05 Type I error level? What is the difference between this test and the tests you conducted in Exercise 14.21?

14.25 For the consumption regression in Exercise 14.7, calculate the F ratio and, at the 1 percent Type I error level, test if income, education, and age are jointly significant. How does this test differ from the test you conducted in Exercise 14.20?

14.26 According to Johnson, Johnson, and Buse in *Econometrics Basic and Applied* (Macmillan, 1987, p. 121), if the computed F exceeds the critical F, for a 5 percent Type I error level, then "the probability is at least .95 that the sample came from a population where at least one of the β's is different from zero." Comment on the correctness of this statement.

14.27 Exercise 14.8 called for a multiple regression of housing prices. At what level was this overall regression "statistically significant"? What does this mean?

14.28 Why is it true that if $R = 1.00$ then 100 percent of the points are correctly predicted, but if $R = 0.9$ we do not know that 90 percent of the points are correctly predicted? What does an $R = 0.9$ tell us?

14.13 PROBLEMS IN ESTIMATION

Many applied statisticians have a favorite computer horror story involving a regression equation whose parameters could not be estimated. When the computer program spits back error messages and refuses to provide parameter estimates, or it gives some estimates that are absurd, it is likely one of two problems: insufficient variability within an explanatory variable or a linear relationship among the explanatory variables.

Insufficient Variability Within an Explanatory Variable

Throughout our discussion of regression, we have emphasized that the number of observations must exceed the number of explanatory variables plus the constant; that is, $n > k$. In addition, we need to check for adequate variability in the x data. To appreciate the importance of variability in an explanatory variable, consider the estimation of the parameters in the model $y = \beta_1 + \beta_2 x_2 + \varepsilon$, using the data in Figure 14.5. The sample size is $n = 4$ and the number of parameters to be estimated is $k = 2$ but there is no variability in the explanatory variable. The parameters of the model are not estimable if all the y values are concentrated at one value of x. Ideally, we should have a lot of variability within our explanatory variables. A lot of variability in the x's reduces the standard errors in estimation, which is a desirable property.

As another example of the need for adequate variability in the explanatory variables, consider a production process in which y is plant output and x_2 is the labor hour input. To capture diminishing returns to the labor input, the model is $y = \beta_1 + \beta_2 x_2 + \beta_3 x_2^2 + \varepsilon$, where diminishing returns implies that $\beta_3 < 0$. If the only data set available is that shown in Figure 14.6, then β_3 cannot be estimated and tested. As shown, $b_3 > 0$,

FIGURE 14.5 Insufficient Variability in an Explanatory Variable

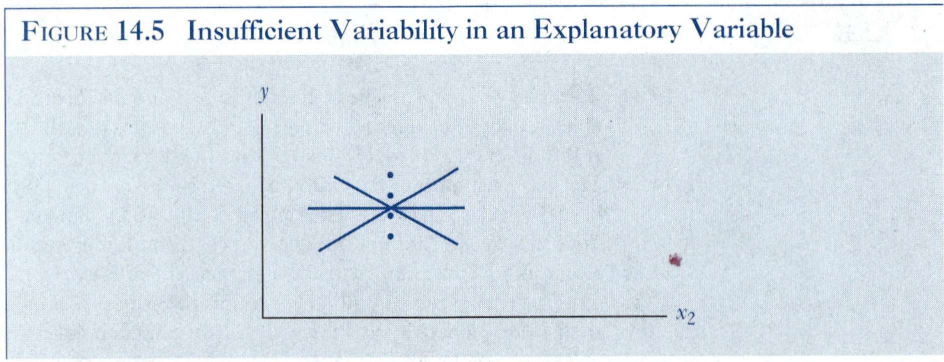

FIGURE 14.6 Insufficient Data for Fitting a Second-Order Polynomial

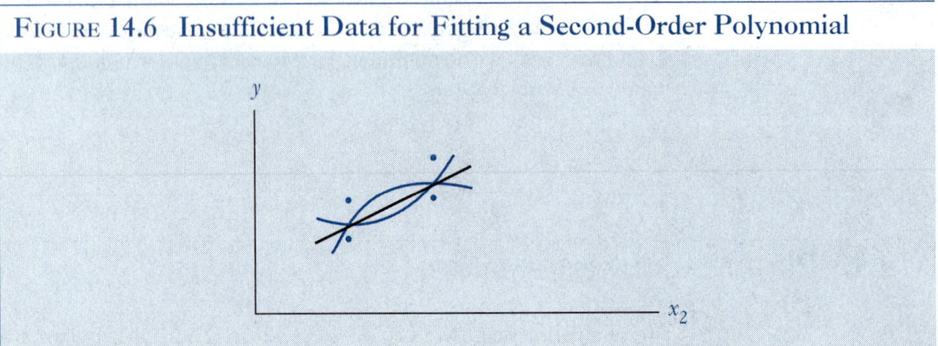

$b_3 = 0$, or $b_3 < 0$ cannot be discerned from the data even though $n = 4 > k = 3$. At least three distinct x values are needed to estimate a second-order polynomial.

Linear Relationship Among Explanatory Variables: Multicollinearity

In specifying a multiple regression, we have to be concerned that two or more of the explanatory variables might be measuring the same thing or be so highly correlated that they are almost measuring the same thing. For example, in a time-series study of the demand for new houses in a large city over a 10-year period, including an explanatory variable for the weekly 15-year conventional mortgage rate at each bank in the city would result in redundancy since competition among banks tends to result in their posting the same rate. Even including separate rates for different types of mortgages might be redundant since rates tend to move together—when the 15-year conventional rate is raised, so is the 30-year conventional rate. One explanatory variable representing a single rate or index of rates (the construction of which is discussed in a later chapter) might suffice.

Correlation among the explanatory variables can enter in many unsuspected ways. Two variables might appear to be measuring different things but yet be highly correlated. For instance, mortgage rates might be correlated among themselves and be correlated with measures of housing prices, which we also would want to include in an explanation of the demand for housing.

Fitting the model of new house demand $y = \beta_1 + \beta_2 x_2 + \beta_3 x_3 + \beta_4 x_4 + \varepsilon$, where x_2 might be the index of 15-year mortgage rates, x_3 might be the index of 30-year mortgages rates, and x_4 might be an index of house prices, could yield small t statistic values for b_2, b_3, and b_4, implying that $\beta_2 = 0$, $\beta_3 = 0$, or $\beta_4 = 0$ cannot be rejected. Yet, the F statistic might yield a high value, suggesting that at least one of these β's is not zero. These apparently contradictory results are an indication of correlation between the x's. In fact, the calculation of a simple correlation coefficient for the three combinations of x_2, x_3, and x_4 would confirm the magnitude of these relationships. High correlation among the explanatory variables is called multicollinearity.

Multicollinearity can force one or all of the standard errors of slope coefficients of the collinear explanatory variables to be overly large and thus cause the individual t statistics to be small. It is possible that all of the explanatory variables may be contributing to the explanation of the dependent variable but yet their contributions overlap so individual effects cannot be discerned.

We might be able to get an idea of the individual effects by sequentially removing each of the explanatory variables and then re-estimating the regression without the variable. If the omitted variable is correlated with the included variables, however, then the coefficient estimates of the included variables will be biased. That is, if collinear variables are included in a regression, then their standard errors will be overly large; but if they both belong in the model and one of them is excluded, then the coefficient of the included variable will be biased by its correlation with the error term.

Another possible way to address the issue of collinearity between two variables might be to create a new variable that is a ratio or difference of the two. The estimated regression would then include this new variable but not the two that were used to create the new variable. Getting more data will also help offset the effects of multicollinearity, if for no reason other than increasing the degrees of freedom. Short of specifying a different model and different estimation procedure or getting more sample data, there is no cure for multicollinearity. The important point is not to emphasize individual coefficient t tests before checking the overall fit with the F test.

Query 14.5:

"A Scorecard on the B-Schools" introduced in Query 14.3 provided three ratings on the top MBA programs in the United States. Michigan, the only public school among the five, had the lowest tuition but highest student/faculty ratio. In a least squares regression of tuition (y) on the student/faculty ratio and a (0-1) public school dummy variable, what is the significance of the overall fit of the regression versus the individual explanatory power of each regressor?

	Annual Tuition	Student/ Faculty Ratio
Chicago	\$13,500	9/1
Harvard	13,300	9/1
Michigan	11,312	18/1
Northwestern	13,200	9/1
Pennsylvania	13,354	8/1

ANSWER:

Using the MINITAB computer program the following data sets were created

$$y = \begin{bmatrix} 13500 \\ 13300 \\ 11312 \\ 13200 \\ 13354 \end{bmatrix} \qquad X = \begin{bmatrix} 1 & 9 & 1 \\ 1 & 9 & 1 \\ 1 & 18 & 0 \\ 1 & 9 & 1 \\ 1 & 8 & 1 \end{bmatrix}$$

Note that the column of ones shown in the X matrix need not be entered into MINITAB because this program, as with many others, will automatically insert this column of ones for the constant, x_1. As in Query/Answer 14.3, here tuition is y. It is stored in MINITAB's column C1, although it is not shown in the MINITAB printout. The student/faculty ratio is x_2 in MINITAB's column C2, and the private/public school variable is x_3 in MINITAB's C3.

THE REGRESSION EQUATION IS

$$Y = 11684 - 20.667X_2 + 1835.3X_3$$

	COLUMN	COEFFICIENT	ST. DEV. OF COEF.	T-RATIO COEF./S.D.
X_1	—	11684.0	3179.	3.68
X_2	C2	–20.6667	176.4	–.117
X_3	C3	1835.33	1640.	1.12

THE ST. DEV. OF Y ABOUT REGRESSION
LINE IS
S = 152.753
WITH (5 – 3) = 2 DEGREES OF FREEDOM
R-SQUARE = 98.6 PERCENT
R-SQUARE = 97.2 PERCENT, ADJUSTED
FOR D.F.

ANALYSIS OF VARIANCE

DUE TO	DF	SS	MS=SS/DF
REGRESSION	2	3285682.13	1642841.07
RESIDUAL	2	46666.67	23333.33
TOTAL	4	3332348.80	

To test the overall fit of the equation, and using the variable numbering system employed in this MINITAB printout, we have

$$H_O\text{: } \beta_2 = \beta_3 = 0$$

$$H_A\text{: at least one of } \beta\text{'s } (\beta_2 \text{ or } \beta_3) \text{ is nonzero}$$

At the 5 percent Type I error level, from Appendix Table A.5, the critical F is 19.00

$$\text{Prob}\left(F^{\text{df1}=2}_{\text{df2}=2} > 19.00\right) = 0.05$$

Using the analysis of variance values in the MINITAB printout, the calculated F ratio is

$$F = \frac{\text{RSS}/(k-1)}{\text{ESS}/(n-k)} = \frac{3285682.13/(3-1)}{46666.67/(5-3)} = \frac{642841.07}{23333.33} = 70.4$$

Since the calculated F exceeds the critical F, H_0 ($\beta_2 = \beta_3 = 0$) is rejected and H_A (at least one β, either β_2 or β_3, is nonzero) is not rejected. Because of the high degree of multicollinearity, however, neither one of the regressors appear significant. At the 5 percent level, the critical t is 4.303, for a two-tailed test with two degrees of freedom, while the calculated t for the student/faculty coefficient is –0.117 and for the school coefficient it is 1.12. Yet we know from the overall F test that if it were not for the multicollinearity, at least one of these coefficients would be significant.

As can be seen in the following MINITAB output, regressing tuition on the student/faculty ratio by itself suggests that the student/faculty ratio is a significant explanatory variable and that its coefficient estimate is highly sensitive to the inclusion of the dummy variable for school type. One more student per faculty member is now predicted to lower tuition by $216.90, with a standard error of only 19.12; before one more student per faculty member lowered tuition by only $20.67, with an extremely large standard error of $176.40. This is direct evidence of multicollinearity. Notice that the first column of x's differs from the third by only one element, and the second and third columns also appear related.

THE REGRESSION EQUATION IS

$Y = 15233 - 216.9X_2$

	COLUMN	COEFFICIENT	ST. DEV. OF COEF.	T-RATIO COEF./S.D.
X_1	—	15232.7	214.8	70.928
X_2	C_2	–216.931	19.12	–11.347

THE ST. DEV. OF Y ABOUT REGRESSION LINE IS
S = 159.0313
WITH (5-2) = 3 DEGREES OF FREEDOM
R-SQUARE = 97.72 PERCENT
R-SQUARE = 96.96 PERCENT, ADJUSTED FOR D.F.
ANALYSIS OF VARIANCE

DUE TO	DF	SS	MS=SS/DF
REGRESSION	1	3256575.9	3256575.9
RESIDUAL	3	75872.9	25290.9
TOTAL	4	3332348.8	

To determine if the multiple regression with both the student/faculty ratio and the school dummy is preferred to the simple regression with only the student/

faculty ratio as an explanatory variable, consider the population model being estimated. If our theory says that the true model is

$$y = \beta_1 + \beta_2 x_2 + \beta_3 x_3 + \varepsilon$$

but we do not include x_3, then the effect of x_3 is in the disturbance term. That is, if the implied population model is

$$y = \beta_1 + \beta_2 x_2 + \varepsilon^*$$

where $\varepsilon^* = \beta_3 x_3 + \varepsilon$, then as long as x_3 and x_2 are related, $E(x_2\varepsilon^*) \neq 0$, and coefficient estimates from the simple regression are biased.

There have been many cures proposed for multicollinearity. With the exception of getting a new and larger sample, they all require knowledge of the true population model—something few researchers have. Increasing the sample size always reduces the standard errors; a new and larger sample also may yield a sample with less correlation among the regressors.

Whether a more precise but biased estimator is better than a less precise but unbiased estimator is a matter of debate. If the specification of the true population model is known with certainty, econometricians tend to favor unbiasedness over precision. Unfortunately, here the true specification of the population model is not known. Because multicollinearity clearly is present and because there are so few observations, we should put little faith on the individual coefficient estimates.

14.14 Stepwise Regression (Optional)

Stepwise Regression A routine by which potential explanatory variables are selected on the basis of their contribution to reducing the error sum of squares.

Ideally we would have a theory that suggests the equational form and variables to include in the explanation of our dependent variable. This is seldom the case. Typically, we have a long list of potential explanatory variables. In the absence of theory, we may turn to a sorting or screening procedure known as **stepwise regression.** There are many forms of stepwise regression. The most popular routines appearing in computer programs choose explanatory variables for inclusion or exclusion on the basis of their contributions to reducing the error sum of squares, as reflected in an adjusted R^2, an F statistic, and the t statistic. In these programs, the user need only identify the independent variable, y, and the potential explanatory variables, the x's to be used in the regression. The search for the "best" explanatory variables is then automatic.

In assessing each x's relative potential to reduce the error sum of squares, the stepwise program may proceed from the smallest model to the largest or vice versa. In step one, for example, the computer program may fit all possible one explanatory variable models. The model with the highest F is declared the "best fitting" one variable model. Step two has the program search through the potential explanatory variables for the two "best" explanatory variables. In step three, the three best explanatory variables are sought and so on until the list of potential explanatory variables is exhausted. With many programs, at each step the user can get descriptive statistics (t, F, and R^2) for the "best fitting" regression and the runners-up.

Not all programs use the same criteria for "best fitting." Some programs add variables based on their individual t coefficients, others work off the F, while the better programs key off comparisons of several statistics. The best fitting regression may give different slope and standard error estimates depending on the step or number of other explanatory variables included. This is the consequence of collinearity between variables. How the computer program brings variables into and out of the regressions will affect descriptive statistics. There is no unique procedure or criteria for inclusion and exclusion. Thus, some programs give users the ability to interact with the programs in making these decisions.

Users need to know what criteria the program is using for automatic inclusion (called forward selection) and exclusion (called backward elimination) of potential explanatory variables. The sequence of selection and elimination can influence the statistics. For example, if the data are highly collinear, then including or excluding variables on the basis of the t statistic of the individual variable under consideration will give different results than looking at the overall F statistic for the regression. On the other hand, when comparing models with the same number of predictors, choosing the model with the highest R^2 is equivalent to choosing the model with the smallest error sum of squares. When comparing models of different size, choosing the model with the highest adjusted R^2 is equivalent to choosing the model with the smallest mean square error.

Program manuals will tell the user what criteria are being used in choosing the "best regression." Although numerous studies involving resampling from a known population (as described in Chapter 7) have been done, the only consensus is that the use of the t statistic in one-way forward or backward regression does not tend to identify the underlying population model as well as criteria based on the adjusted R^2 or F statistic (Lovell, 1984).

An example may be enlightening. Consider a search for the variables that best describe faculty salaries. The ASCII data file "um91yr-d.prn" contains 17 variables of faculty characteristics in a college at a large midwestern university, where the first field contains the identification number for each of 182 complete records and the second field contains current salaries (cursal). As shown at the bottom of Table 14.8, file "um91yr-d.prn" is a modified version of the file "UofM91yr.prn." The modification simply removed faculty members for whom data was missing and deleted the years from regular appointment variable (*ryr*) because of missing data. Using the data in Um91yr-d.prn, we seek a model to explain faculty salaries.

In the MINITAB commands "breg" at the top of Table 14.8, rank (c4) and department (c17) are absent from the list to be searched for potential explanatory variables. Neither variable has numerical values that are meaningful along a number line. That is, movement from assistant professor to associate professor is a one-unit change from 0 to 1; so is the movement from associate professor to full professor a one-unit change, from 1 to 2. But promotion from assistant to associate and promotion from associate to full are not equivalent; they may be associated with quite different salary rewards. The numbers placed on departments are totally arbitrary, as in the placement of numbers on baseball uniforms discussed in Chapter 1. The effect of different ranks could be searched if we represented the three ranks with two dummy variables.

Table 14.8 Stepwise Regression in MINITAB

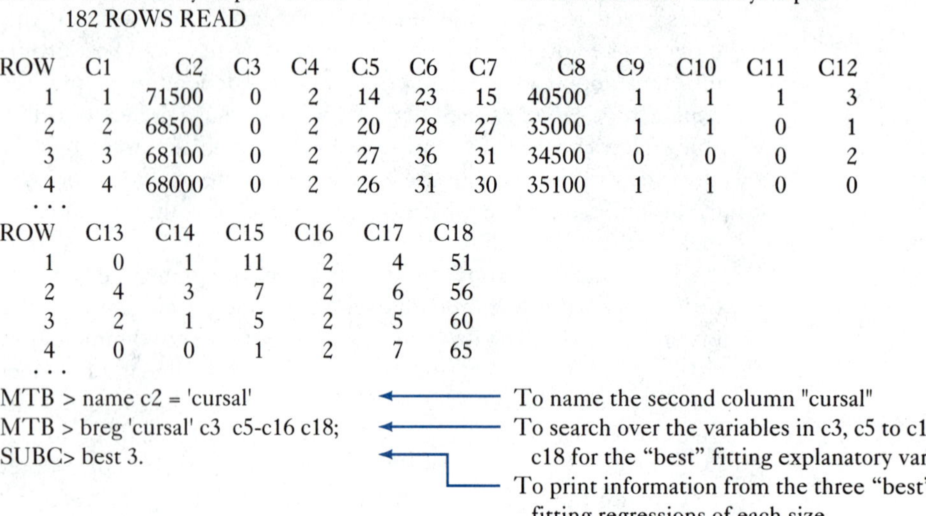

MTB > read 'um91yr-d.prn' c1-c18 ← To read data file "um91yr-d.prn"
182 ROWS READ

ROW	C1	C2	C3	C4	C5	C6	C7	C8	C9	C10	C11	C12
1	1	71500	0	2	14	23	15	40500	1	1	1	3
2	2	68500	0	2	20	28	27	35000	1	1	0	1
3	3	68100	0	2	27	36	31	34500	0	0	0	2
4	4	68000	0	2	26	31	30	35100	1	1	0	0
. . .												

ROW	C13	C14	C15	C16	C17	C18
1	0	1	11	2	4	51
2	4	3	7	2	6	56
3	2	1	5	2	5	60
4	0	0	1	2	7	65
. . .						

MTB > name c2 = 'cursal' ← To name the second column "cursal"
MTB > breg 'cursal' c3 c5-c16 c18; ← To search over the variables in c3, c5 to c16, and c18 for the "best" fitting explanatory variables.
SUBC> best 3. ← To print information from the three "best" fitting regressions of each size.

* WARNING * This may take a long time.

Best Subsets Regression of cursal

Vars	R-sq	Adj. R-sq	C-p	s	C3	C5	C6	C7	C8	C9	C10	C11	C12	C13	C14	C15	C16	C18
1	34.1	33.8	190.4	4477.2						X								
1	31.0	30.6	208.1	4583.6		X												
1	28.2	27.8	223.6	4674.6				X										
2	47.4	46.9	117.9	4010.1		X				X								
2	46.1	45.5	125.1	4059.3				X		X								
2	44.4	43.8	134.6	4122.9						X								X
3	53.0	52.3	88.6	3801.2		X				X						X		
3	52.2	51.4	93.2	3834.2		X				X	X							
3	52.2	51.3	93.5	3836.9				X	X	X								
4	57.1	56.1	67.9	3643.9				X	X	X						X		
4	57.1	56.1	68.0	3643.9		X				X	X					X		
4	56.4	55.5	71.5	3671.0				X	X	X	X							
5	61.0	59.9	47.9	3482.4				X	X	X	X				X			
5	60.8	59.7	49.3	3493.2				X	X	X	X					X		
5	60.2	59.1	52.4	3518.0		X		X		X	X					X		
6	63.5	62.3	36.0	3379.0				X	X	X	X		X		X			
6	62.9	61.6	39.6	3408.7				X	X	X	X				X		X	
6	62.6	61.3	41.0	3420.3				X	X	X	X					X	X	
7	65.4	64.0	27.5	3300.2				X	X	X	X		X		X		X	
7	64.5	63.1	32.3	3341.0		X		X	X	X	X		X		X			
7	64.4	63.0	33.0	3347.2		X		X	X	X	X					X	X	
8	66.5	65.0	23.2	3255.3		X		X	X	X	X		X		X		X	
8	66.2	64.6	25.1	3271.8				X	X	X	X		X		X	X	X	
8	66.1	64.6	25.4	3274.8		X		X	X	X	X		X			X	X	

(continued)

TABLE 14.8 Stepwise Regression in MINITAB *(continued)*

Vars	R-sq	Adj. R-sq	C-p	s	C3	C5	C6	C7	C8	C9	C10	C11	C12	C13	C14	C15	C16	C18
9	67.7	66.0	18.5	3206.0		X		X	X	X	X		X		X	X	X	
9	67.6	65.9	19.0	3210.4		X	X	X	X	X	X		X		X		X	
9	67.3	65.6	20.7	3225.9		X	X	X	X	X	X		X			X	X	
10	68.8	67.0	14.3	3159.3		X	X	X	X	X	X		X		X	X	X	
10	68.3	66.4	17.3	3187.1		X		X	X	X	X	X	X		X	X	X	
10	68.2	66.4	17.7	3189.9		X	X	X	X	X	X	X	X		X		X	
11	69.5	67.5	12.7	3136.5		X	X	X	X	X	X	X	X		X	X	X	
11	69.3	67.3	13.9	3147.5		X	X	X	X	X	X		X		X	X	X	X
11	68.9	66.9	15.8	3164.3	X	X	X	X	X	X	X		X		X	X	X	
12	70.0	67.9	11.7	3117.8		X	X	X	X	X	X	X	X		X	X	X	X
12	69.5	67.4	14.4	3142.2	X	X	X	X	X	X	X	X	X		X	X	X	
12	69.5	67.3	14.6	3144.9		X	X	X	X	X	X	X	X	X	X	X	X	
13	70.1	67.8	13.1	3121.5	X	X	X	X	X	X	X	X	X		X	X	X	X
13	70.0	67.7	13.6	3126.3		X	X	X	X	X	X	X	X	X	X	X	X	X
13	69.6	67.2	16.2	3150.4	X	X	X	X	X	X	X	X	X	X	X	X	X	
14	70.1	67.6	15.0	3129.7	X	X	X	X	X	X	X	X	X	X	X	X	X	X

MTB >
MTB > stop

Key to file "um91yr-d.prn," as a modification of file "uofm91yr.prn," with variable ryr deleted and cases 157, 164, 168, and 174 deleted because of missing values.

MINITAB COLUMN	VARIABLE NAME DESCRIPTION
c1:	code = Identification number (1 to 186, with 157, 164, 168, 174 deleted)
c2:	cursal = current salary
c3:	sex = 1, if the person is female and 0 if male
c4:	rank = 0, 1, or 2 for assistant, associate, or full professor
c5:	proyr = number of years from last promotion to 1991
c6:	phdoyr = number of years from receipt of PhD to 1991
c7:	appyr = number of years from appointment to 1991
Note:	ryr = years from regular appointment deleted because of missing data
c8:	stsal = starting salary
c9:	admin = 1, if person has been an administrator and 0 otherwise
c10:	serv = 1, if the person heavily involved in service activity; 0 otherwise
c11:	off = 1, if the person received an outside offer in past three years
c12:	pir = number of times person was principal investigator on a research grant
c13:	pit = number of times person was principal investigator on a teaching grant
c14:	book = number of books written by person
c15:	art = number of articles written by person
c16:	adv = number of PhD advisees finished
c17:	dept = 1, 2, 3, . . . 7 if the person was a member of the ith department
c18:	birth = age of the person in 1991

Representation of the seven departments would require the creation of six dummy variables.

The "breg" stepwise routine begins with a search for the one variable that best explains cursal. With the "best 3" subcommand we requested information from the "best" three models of each size. For the one variable model, the three best single explanatory variables are c9: admin, c5: proyr, and c7: appyr. The three best two explanatory variable models are

1. cursal as a function of c5: proyr and c9: admin, $R^2 = 47.4$
2. cursal as a function of c7: appyr and c9: admin, $R^2 = 46.1$
3. cursal as a function of c9: admin and c18: birth, $R^2 = 44.4$

and so on to the model with all 14 explanatory variables with $R^2 = 70.1$. The highest adjusted $R^2 = 67.9$ is for the model with 12 explanatory variables. In that model, the $C\text{-}p$ statistic is 11.7, which is calculated as

$$C\text{-}p = \frac{\text{ErrorSS}_p}{\text{MSE}_m} - \left(n - 2p\right)$$

where ErrorSS_p is the error sum of squares for the best model with p parameters (including the intercept, if it is in the equation), and MSE_m is the mean square error for the model with all m explanatory variables.

In general, we look for models with the highest adjusted R^2 and small $C\text{-}p$, which indicates a good fit and relative precision in estimation. Ideally, these models will not be complicated and involve many explanatory variables. In Table 14.8, the regression with the highest adjusted $R^2 = 67.9$ and lowest $C\text{-}p = 11.7$ has 12 explanatory variables plus an intercept. Some researchers might consider 12 regressors to be too many to be practical. From a practical standpoint, models should be "parsimonious"; that is, they should have only a small number of explanatory variables. For example, Tom Peters (Tribune Media Service, January 26, 1994) warns against complexity in model building and cites Digital Equipment Corp. founder Ken Olsen as saying "The smartest businesspeople keep a very simple model of their business in their heads . . . so simple you can meditate on it, modify it, adjust it, . . . while laying awake at night, driving in a car, leaning back with your eyes closed on an airplane, or keeping yourself occupied during a dull meeting."

Of some interest in looking over the results of Table 14.8 is that the gender variable (c3: sex) does not appear consistently in the best fitting models. Being an administrator, however, does enter the explanation of salary consistently. A question for further analysis might be whether or not these two variables are collinear, with only males tending to be administrators. This analysis is left to the student's discretion, as our intent here is only to describe the use of stepwise regression and not a full analysis of these data.

In closing our discussion, it is worth remembering that stepwise regression can be helpful at the early stages of an investigation or when there is no theory to guide the researcher. There are dangers in its use, however. First, because there is no underlying theory guiding selection or elimination of variables, chance relationships may be identified as the "best" when in fact they are flukes. Second, the automatic procedures are algorithms that might maximize or minimize one statistic while ignoring another. A program does not take account of specialized knowledge that a researcher might have about the data. In using stepwise procedures, it is essential that the "best fitting" regression make sense from a practical point of view, and only the researcher can make that decision.

Historical Perspective

Milton Friedman (1912 –) won the 1976 Nobel Memorial Prize in Economic Science in part for his regression analysis of the relationship between personal consumption and income. In the appendix to an *American Economic Review* paper he wrote with Anna Schwartz (1991), Friedman tells an intriguing story about his early experience with regression analysis during World War II.

In 1945–1946 Friedman was a member of the staff of the Statistical Research Group at Columbia University, a brain trust providing consulting service to the U.S. Armed Forces. One of his assignments was to assist in the development of an improved alloy for high-temperature use in airplane turbo-supercharger blades. Experimental turbine blades were put in a furnace, heated, and timed until they fractured. At one point, Friedman combined the test data from all the separate experiments and built a regression model that expressed time to fracture as a function of stress, temperature, and variables describing the composition of the alloy. Because this was physics and not economics, Friedman took great care to make sure that his regression was consistent with the metallurgical theory of the day.

Although trivial now, a major problem for Friedman was the computation of the coefficients of the regression and the associated test statistics. Harvard University had the only large-scale computer in the country that could perform the calculations. It was built from a large number of IBM card-sorting machines housed in an enormous air-conditioned gymnasium. Friedman received time on the machine to perform calculations; they took 40 hours per regression. Had the calculations been done on a desktop calculator by skilled operators, Friedman estimated that it would have taken three months to obtain the estimates. Today the same calculations could be done on a desktop microcomputer in seconds.

Friedman reported that he "was delighted with the calculated regression. It had a high multiple correlation, low standard error of estimate, and high t values for all of the coefficients, and it satisfied every other test statistic . . ." Based on his regressions, he constructed two new alloys, which he named F-1 and F-2. According to his regression, each should have taken several hundred hours to rupture at high temperatures; yet, his two alloys showed fractures in something like one to four hours. According to Friedman "F-1 and F-2 were never heard of again." In conclusion, Friedman stated:

> Ever since, I have been extremely skeptical of relying on projections from a multiple regression, however well it performs on the body of data from which it is derived; and the more complex the regression, the more skeptical I am. In the course of decades, that skepticism has been justified time and again. In my view, regression analysis is a good tool for deriving hypotheses. But any hypothesis must be tested with data or non-quantitative evidence other than that used in deriving the regression or available when the regression was derived. Low standard errors of estimate, high t values, and the like are often tributes to the ingenuity and tenacity of the statistician rather than reliable evidence of the ability of the regression to predict data not used in constructing it.

Exercises

14.29 Answer true or false to the following. Give the reason for your answer.

a. If multicollinearity is a problem, then the standard errors of the coefficients will be overly large.

b. If a theory suggests a multiple regression model, but regressors are independent (and uncorrelated in the sample), then an unbiased estimate of a β_j still may be obtained from a simple regression of y on x_j.

c. In simple regression, if the calculated t is larger than its critical value, then the coefficient and the model can be said to be significant; there is no need to calculate the overall F. In multiple regression, the overall fit of the regression can only be tested by calculating the F.

d. Adding another regressor to an equation lowers the explanatory power of the previous regressors because the variability in y is limited by the total sum of squares.

14.30 Under what condition would a regression of y on x_1 and a regression of y on x_1 and x_2 give the same value of b_1?

14.31 Using the data on B-schools in Exercise 14.3, specify and estimate an equation that you think best describes starting salary. Use only two explanatory variables. Why is your equation here a better (or worse) specification than that given in Exercise 14.3?

14.32 Using the data on houses in Exercise 14.8, construct your own housing price regression. Why do you like your specification better than that in Exercise 14.8?

14.15 Concluding Comments

Multiple regression provides a powerful tool of analysis, which was recognized by Yule at the turn of the century. As cautioned by economist Pigou in an assessment of Yule and as reported by Milton Friedman, multiple regression also can produce parameter estimates that are statistically significant but meaningless in application. Computer programs make regression estimation easy, but only the researcher can assess the practical usefulness of the regression estimated. It is the application of the regression to real-world situations and problems that determines its value.

Historically, the extension from simple regression models (in which y is regressed on x) to multiple regression analysis (in which y is regressed on many x's) required researchers to have familiarity with advanced topics in algebra. Today, however, computer regression programs automatically make these complex calculations with little or no knowledge of the intricacies of matrix algebra required on the part of the user. To use these programs, it is only essential that you understand the assumptions and notation of multiple regression.

By now you should recognize that the assumptions for multiple regression are the same as those for simple regression with one addition: The multiple explanatory variables cannot be correlated with each other. That is, we assume no multicollinearity. You should be able to articulate why each of the assumptions is required for least squares estimation and testing. Finally, you should be competent at reading and interpreting computer printouts for the purposes of predicting the value of the dependent variable and testing hypotheses about individual coefficients and the overall fit of a regression.

CHAPTER EXERCISES

14.33 Answer true or false to the following. Give the reason for your answer.

a. Adding another explanatory variable to a regression will raise the error sum of squares.

b. Estimates that are precise (small standard errors) are also unbiased (expected value equals parameter value).

c. Unlike simple regression with one regressor, in multiple regression the regressors are always independent of the disturbance ε because no explanatory variables need be omitted.

d. Unlike simple regression, in multiple regression the residuals need not add to zero. For example, if the residuals add to zero for a regression of y on x_1, then they could not add to zero when another regressor is added.

e. As in simple regression, as the calculated t for a slope coefficient in multiple regression increases in absolute value, the p-value decreases.

f. The interpretation of TSS, RegSS, and ErrorSS is the same for simple and multiple regression. The only difference is that in multiple regression, TSS is based on more than one regressor.

g. The 95 percent confidence interval for the expected value of y, conditional on the x's, is just another name for the 95 percent prediction interval of y, conditional on the x's.

14.34 What is the difference between R^2 and its adjusted value?

14.35 Below is the traffic death rate per 100 million vehicle miles traveled in the United States, for 1957 through 1984.

Year	Death Rate	Year	Death Rate	Year	Death Rate
57	5.71	67	5.25	77	3.26
58	5.32	68	5.17	78	3.25
59	5.17	69	5.02	79	3.35
60	5.06	70	4.72	80	3.30
61	4.92	71	4.44	81	3.20
62	5.08	72	4.32	82	2.80
63	5.18	73	4.11	83	2.60
64	5.39	74	3.52	84	2.60
65	5.30	75	3.35		
66	5.48	76	3.23		

In 1974, the 55 mph speed limit was imposed. Using the last two digits of the year as an explanatory variable (to control for any trend in the death rate) and a 0-1 dummy variable to reflect the 55 mph speed limit, test whether or not the enactment of this speed limit had an effect on the death rate. Do you feel that any of the least squares assumptions were violated in this regression?

14.36 If a human life is valued at $2 million, and if this value is used as the dependent variable in Exercise 14.35, how would your estimate of the effect of the 55 mph speed limit change?

14.37 The following computer display is based on the useable records of the 182 faculty members of the college at the large midwestern university referenced in this chapter. The variables are defined as:

CURSAL = 1987-80 salary of person, the dependent variable
SEX = 1, if person is female and 0 if male
PROYR = last two digits of year of last promotion
ADMIN = 1, if person has been an administrator and 0 otherwise
SERV = 1, if person heavily involved in service activity; 0 otherwise
ART = number of articles written by person
BOOK = number of books written by person
DEPTi = 1, if person was a member of the ith department within the college (i = 2, 3, 4, 5, 6, 7) and 0 otherwise

Dependent Variable . . . . CURSAL   Number of Observations = 182
Mean of Dependent Var. = 54278.55495   Std. Dev. of Dep. Var. = 5500.86761
Std. Error of Regression = 3462.5369   R-square = 0.63006

Variable	Coefficient	Std. Error	T-ratio(Sig.Lvl)	Mean of X	Std.Dev.of X
ONE	53974.9	3979.	13.565 (.00000)		
SEX	–1030.45	680.8	–1.514 (.12776)	.20879	.40757
PROYR	–463.486	55.56	–8.342 (.00000)	70.775	5.2326
ADMIN	5025.02	785.1	6.400 (.00000)	.17033	.37696
SERV	2706.64	606.6	4.462 (.00003)	.29670	.45806
ART	166.006	73.71	2.252 (.02426)	3.7363	4.3642
BOOK	745.127	383.5	1.943 (.05082)	.35714	.77881
DEPT2	631.320	847.1	.745 (.46359)	.14286	.35089
DEPT3	–897.412	940.3	–.954 (.34363)	.10989	.31362
DEPT4	3175.07	1014.	3.132 (.00221)	.87912E-01	.28395
DEPT5	901.833	858.3	1.051 (.29528)	.13187	.33928
DEPT6	1347.50	814.5	1.654 (.09578)	.15385	.36180
DEPT7	2513.94	1088.	2.310 (.02097)	.71429E-01	.25825

a. Interpret the coefficient estimate on the SEX variable, and construct a 95 percent confidence interval for the population parameter it estimates.
b. Which of the coefficients are significant at the 10 percent Type I error level?
c. Calculate the F ratio from the information provided.
d. What is the value of the error sum of squares and the total sum of squares?
e. Is the F significant at the 10 percent Type I error level?
f. There are seven departments in this college. Why isn't a dummy variable included for the department that is designated DEPT1?
g. Calculate the adjusted R-square.

14.38 The ASCII file "EX14-38.PRN" contains all 186 records on the faculty of a college at the large midwestern university referenced in this chapter. Every line contains a numerical value or missing observation (designated .). Observations on each of the 19 variables are separated by a space. In order, the variables are

CODE = Identification number (1 through 186)
CURSAL = 1988-89 salary of person
SEX = 1, if person is female and 0 if male
RANK = 0, 1, or 2 for assistant, associate, or full professor
PROYR = last two digits of year of last promotion
PHDYR = last two digits of year of receipt of Ph.D. degree
APPYR = last two digits of year of appointment

RYR = last two digits of year of regular appointment
STSAL = starting salary
ADMIN = 1, if person has been an administrator and 0 otherwise
SERV = 1, if person heavily involved in service activity; 0 otherwise
OFF = 1, if person received an outside offer in past three years
PIR = number of times person was principal investigator on a research grant
PIT = number of times person was principal investigator on a teaching grant
BOOK = number of books written by person
ART = number of articles written by person
ADV = number of Ph.D. advisees finished
DEPT = 1, 2, 3, 4, 5, 6, and 7 if person was a member of the ith department
BIRTH = last two digits of year of birth

a. Test for sex discrimination in department four for those 15 faculty members who are not administrators, using years since receiving the Ph.D., number of books published, and gender as explanatory variables of salary. Estimate a model in which you assume that raises are based on percentage increases and one in which you assume they are based on absolute increases. Which is the most appropriate for your test of discrimination in the salary determination process?
b. For all 182 useable records, estimate a semilog and a linear salary model that includes SEX, PROYR, ADMIN, SERV, ART, BOOK, and a zero-one dummy for the fourth department.
   1. Which model is better for testing hypotheses, and why?
   2. Is the fourth department significantly different from the others in explaining salary?
   3. Is the gender variable significant? What does this imply?

14.39 High school student learning of economics is of concern to economists and civic leaders. William Walstad and John Soper (*American Economic Review,* May 1988) report on a national study of student learning as measured by the *Test of Economic Literacy.* In a replication of their work, using a subset of their sample, the following results were obtained for the defined variables:

TELPOST, score on the test after the course is finished,
TELPRE, score on the test prior to the start of the course,
IQ, equals the ith student's score on the Quick Word Test,
MALE, equals 1 if the ith student is male and 0 otherwise,
SENIOR, equals 1 if the ith student is a senior and 0 otherwise,
BLACK, equals 1 if the ith student is black and 0 otherwise,
ECON, equal to 1, if course was economics and 0 otherwise,
CONECON, equals 1 if the ith student is in a consumer econ course,
SSECON, equals 1 if the ith student is in a social studies course with an economics component, and $i = 1, 2, \ldots n_1$.
TCOUR, equal to the number of economics credits taken by the teacher,
DEEP, equal to 1, if the school is participating in the Developmental Economic Education Program, and 0 otherwise,
SIZE, equal to the natural logarithm of school enrollment,
MIDINC and HIGHINC, equal 1 if the school is in a mid or high income SMSA district, respectively, and 0 otherwise,
SUBURB and URBAN, equal 1 if the school is in a suburb or urban community, respectively, and 0 otherwise,

NOREAST, SOUTH and WEST, equal 1 if the school is in the northeastern, southern, or western quadrant of the United States, respectively, and 0 otherwise.

Variable	Coefficient	Std. Error	T-ratio(Sig.Lvl)	Mean of X	Std.Dev.of X
ONE	−17.5855	1.951	−9.015 (.0000)		
TELPRE	.540466	.1868E-01	28.926 (.0000)	20.503	7.4713
IQ.	.170683	.9196E-02	18.561 (.0000)	59.870	15.198
MALE	.514706	.2299	2.239 (.0389)	.50886	.50003
SENIOR	.627319	.2679	2.341 (.0184)	.58228	.49329
BLACK	−.379560	.4001	−.949 (.3454)	.10549	.30724
ECON	5.15502	.3744	13.767 (.0000)	.55738	.49680
CONECON	1.41003	.4650	3.032 (.0026)	.15401	.36103
SSECON	2.53389	.5016	5.052 (.0000)	.12616	.33210
TCOUR	.514923	.6689E-01	7.698 (.0000)	4.4030	2.1498
DEEP	1.48799	.2914	5.107 (.0000)	.44599	.49718
SIZE	3.55161	.6239	5.693 (.0000)	3.0495	.22378
MIDINC	1.68947	.4180	4.042 (.0001)	.72489	.44666
HIGHINC	−.449207	.5039	−.891 (.3765)	.16793	.37389
SUBURB	.756856	.3441	2.199 (.0264)	.47679	.49957
URBAN	−.346618	.3685	−.941 (.3497)	.23207	.42224
NOREAST	.137861E-01	.3708	.037 (.9213)	.14388	.35104
SOUTH	−1.22757	.2888	−4.251 (.0001)	.37511	.48425
WEST	−.941348	.4989	−1.887 (.0561)	.11730	.32184

Number of Observations	2370
Mean of Dependent Variable	22.33418
Std. Dev. of Dep. Variable	9.04203
R-Squared	.63285
F-Statistic (18, 2351)	225.13248

a. What was the standard error of estimation for this regression?
b. What type of student, community, and region are represented by the constant term (that is, when all dummy variables equal 0)?
c. What is the effect of TELPRE on TELPOST? Is it significant?
d. What is the effect of the DEEP variable? Is it significant?
e. Is the overall fit of this regression significant?

14.40 You are to analyze beverage sales in the Tri-city area. There are 22 records on three possible explanatory variables of SALES—the dollar value of advertising (abbreviated AD), the number of retailers in the area served by your outlets (NUM), and the number of competing wholesale distributors in the area (COMP), as given below. In the Tri-city area, there are 10 retailers, four competing wholesalers, and $16,500 is to be spent on advertising, which is the only variable at your control.

a. Estimate the linear relationship between SALES and the other three variables; write it as an equation.
b. Predict sales in the Tri-city area and construct a 95 percent prediction interval.
c. At what level of significance is there a joint relationship between sales and the three explanatory variables?
d. At the 1 percent Type I error level, is advertising expense a significant positive explanatory variable?

SALES (in thousands of dollars)	AD (in hundreds of dollars)	NUM (No. of Retailers in area)	COMP (No. of Competitors in area)
514	165	12	2
398	109	9	7
422	146	8	4
613	192	10	5
378	128	9	4
541	150	8	3
704	185	11	2
487	134	10	5
532	158	9	6
295	124	6	5
336	160	5	4
668	175	9	5
403	142	8	6
467	170	10	7
606	181	10	4
380	159	8	8
618	163	8	6
584	155	8	2
521	131	7	3
289	156	7	6
467	160	8	4
599	180	9	2

14.41 A study of bank teller efficiency yielded the following:

Observation Period	Average $ Amount Transacted During Observation Period	Average Minutes To Complete Transaction	Teller Type 1 If only human 0 If some automation
1	2,314	2.0	1
2	1,178	1.0	0
3	897	.9	0
4	987	1.1	1
5	2,037	2.5	1
6	1,987	2.3	0
7	1,005	2.1	1
8	1,764	2.2	0
9	987	1.2	1
10	1,741	2.0	1
11	1,200	.9	0
12	876	.9	1
13	1,435	1.7	1
14	1,143	1.1	0
15	1,998	2.0	1
16	2,345	1.9	0
17	1,067	1.1	0
18	958	.9	1
19	2,978	1.5	0
20	1,591	1.0	0

a. Estimate a least squares regression in which average time to complete a transaction is a function of the average dollar amount of the transactions and the type of teller. Write out the equation.
b. Does this regression provide a good fit?
c. Are the automatic tellers faster than the humans?

14.42 Exercise 10.23 reported on a *WSJ* article that gave the death rates to be 12 in 368 bypass operations at Harrisburg Hospital, and six in 447 bypass operations at Lancaster General Hospital. A state health council predicted only 5.76 deaths for Harrisburg and 10.94 for Lancaster based on age, health, and the number of operations performed. Specify the multiple regression equation that the council might have used to make its predictions of death rates. Do you see any problems in fulfilling the assumptions for least squares estimation of this model?

14.43 The next table features a regression of profits that car dealers have made from each of their final offers given to "testers" who were sent into the dealerships as would-be buyers. It is from a study by Ian Ayers, "Fair Driving: Gender and Race Discrimination in Retail Car Negotiations," *Harvard Law Review* (February 1991).
a. What group is represented in the intercept term; that is, what group is not represented by a dummy variable?
b. Interpret each coefficient in terms of its influence on the profit.
c. Which variables are statistically significant at the 5 percent Type I error level?

Final Offer Profit Regression	Coefficients	t-ratio
Intercept	603.90	(1.17)
Tester-Salesperson Type:		
TWF (= 1 if tester white female, = 0 otherwise)	196.66	(1.37)
TBM (= 1 if tester black male, = 0 otherwise)	280.18	(1.84)
TBF (= 1 if tester black female, = 0 otherwise)	1029.67	(7.16)
SWF (= 1 if seller white female, = 0 otherwise)	83.21	(.41)
SBM (= 1 if seller black male, = 0 otherwise)	99.49	(.53)
Dealership Characteristics:		
SUB (= 1 if dealer in suburb,= 0 otherwise)	–84.59	(–.44)
WHITE (proportion of neighborhood white)	–271.53	(–.62)
RENT (average rent in neighborhood)	.83	(.42)
HOUSE (average house value in neighborhood)	–.002	(–.98)
INCOME (average income in neighborhood)	.013	(.97)
AGE (average age in neighborhood)	–10.6	(–.75)
Reliability Tests:		
PAIR (= 1 if the tester was the first member of the two testers to bargain at a particular dealership, = 0 otherwise)	–66.80	(–.67)
TEST (= number of tests that a tester completed)	.36	(.06)

R-squared	.388
Adjusted R-squared	.312
Standard Error	516.67
Degrees of Freedom	105

14.44 For Exercise 14.43 involving car dealership profits,

a. how many records (observations) were used in estimation?

b. what does the value of the coefficient of determination indicate?

14.45 Exercise 13.13 reported on an article in *The Wall Street Journal* (February 24, 1992) stating that "for Biotech (companies), pure genius isn't enough." The article suggested that both stock market capitalization and the number of drugs in clinical trial influence the number of drugs the company can bring to market. Listed with the data on capitalization were the number of drugs in clinical trials and the number of proprietary drugs for humans on the market. Fit the suggested multiple regression to these data.

a. What is the equation of your best fitting regression?

b. Are capitalization and the number of drugs in trial significant explanatory variables of the number of drugs on the market?

c. Are capitalization and number of drugs in trial individually significant explanatory variables of the number of drugs on the market?

d. How does your test in part c for the significance of the drug in trial differ from that in Exercise 13.13?

	Stock Market Capitalization ($ millions)	Number of Drugs in: Clinical Trials	Number of Drugs in: Market for Human Use
Amgen	$8,309.7	8	2
Genentech	3,080.2	10	3
Chiron	1,509.6	14	3
Centocor	1,296.8	5	0
Synergen	1,205.4	3	0
Genzyme	935.9	1	1
Biogen	879.8	3	2
Gensia	860.1	1	0
Immunex	791.8	4	2
Alliance	644.0	2	0
US Bioscience	570.8	4	1
Zoma	489.5	6	0
Medimmune	448.8	1	1
Affymax	409.6	0	0
Immune Response	395.2	1	0
Genetics Institute	379.6	3	1
Molecular Biosystems	373.8	0	0
Cytogen	352.3	1	0
Immunomedics	337.1	2	0
Liposome Co.	334.9	3	0

14.46 As discussed in this chapter, "Simpson's paradox," or more appropriately "Yule-Pigou effect," is a phenomenon in which matched pairs of y and x observations (say on salary and year of birth) from two groups (men and women) could show a negative slope for a regression within a group but a positive slope for a regression fit to the combined data. Would the inclusion of a dummy variable as an explanatory variable to reflect membership in a group cure the problem?

14.47 In Exercise 12.42, you were asked to estimate a least-squares regression of CEO compensation (y) on the number of headquartered companies in each country (x). Now add another 0-1 explanatory variable that indicates whether the country is English speaking. What does the coefficient on this 0-1 regressor indicate?

14.48 Using the *Business Week* data in Exercise 13.43 on inflation (y), unemployment (x_1), and political stability (x_2), estimate the parameters in the following model.

$$y = \beta_0 + \beta_1 x_1 + \beta_2 x_2 + \varepsilon$$

a. Write the sample regression, and interpret the coefficients.
b. Test $\beta_1 = 0$ against the alternative $\beta_1 < 0$, at the 0.05 Type I error level.
c. How do your results in part b above differ from those in Exercise 13.43? Why?

14.49 Carnegie Mellon University Professor M. Granger Morgan was quoted: "A simple linear relation, for instance, accurately describes the average cancer risk incurred by smokers: 10 cigarettes a day generally increase the chance of contracting lung cancer by a factor of 25; 20 cigarettes a day increase it by a factor of 50." He added, "For other risks, however, a simple dose-response function is not appropriate, and a more complex model must be used." (*Scientific American*, July 1993) Why might more complex models be needed?

14.50 A diagram in Exercise 14.12 shows the relationship between death and age for smokers and nonsmokers. Why is the death rate measured on a logarithm scale?

14.51 Frank's Poultry maintains a fleet of trucks. After reviewing the repair and maintenance bills for the trucks, the manager observed that the diesel trucks appeared to be more expensive to repair and maintain (R&M) than the gasoline-powered trucks; they also cost more to purchase. Frank's six diesel trucks averaged $1,400 and its eight gasoline-powered trucks averaged $1,374 in R&M expense, as shown below, along with the age and number of miles driven by each truck:

Truck	R&M Expenses (in dollars)	Mileage (in thousands of miles)	Age (in years, where 0 indicates not one year old)	Power Type (D = diesel, (G = gasoline)
1	1229	32.2	0	D
2	1550	34.8	1	D
3	1359	26.7	1	D
4	1350	24.8	2	D
5	1459	28.2	2	D
6	1480	27.5	2	D
7	1309	30.6	0	G
8	1310	29.4	0	G
9	1350	31.4	0	G
10	1360	28.3	1	G
11	1400	29.0	1	G
12	1310	26.2	1	G
13	1500	27.1	2	G
14	1460	25.8	2	G

The manager also found that there was no apparent relationship between a plot of R&M expenses and mileage driven, as seen in a scatter diagram. There appeared to be a negative correlation between R&M expenses and mileage, with the notable exception for the high expense observation for Truck 2. In previous years, there had always been a clear positive correlation between R&M expense and miles driven. The management at Frank's was disappointed that the diesel trucks had not produced savings in repair and maintenance expenses and it was puzzled by the breakdown in the relationship between R&M expense and mileage driven. Do you agree with the manager's analysis? As a consultant to Frank's management, write a memo explaining the problem, your analysis, and conclusions. Append printouts of your data set and the regression runs you performed. Refer to these appendices by number in the body of your memo. (This exercise is adapted from an unknown authored case that was given to W. Becker by a student in an accounting class.)

14.52 William Evans and W. Kip Viscusi in "Income Effects and the Value of Health" (*Journal of Human Resources*, Summer 1993 p. 505), hypothesized that the amount of money an individual will desire as compensation to accept a risk of injury (monetary loss equivalence) is positively related to the wealth of the individual. They state: "To allow for potential nonlinearity in this (monetary loss equivalence) relationship, we will test this hypothesis by assuming that L_i for injury type i is quadratic in wealth (W), where

$$L_i = \alpha_{i0} + \alpha_{i1}W + \alpha_{i2}W^2\text{''}$$

a. Draw three diagrams to show the effect of α_{i2} being positive, negative, or zero, assuming α_{i1} is positive.
b. Draw three diagrams to show the effect of α_{i2} being positive, negative, or zero, assuming α_{i1} is negative.

14.53 The data set in the computer file "EX14-53.PRN" has information on the day (numbered 41 to 46) a patient entered the emergency room of the hospital (first column), the time of entering (1 to 1440 minutes past midnight, second column), the time of leaving the emergency room, which could extend into the next day (third column), the person's age (fourth column), and other variables.

a. Create a variable to show the minutes that each person was in the emergency room. (Make sure to correctly calculate the time for those who stay past midnight.)
b. What is the mean time in the emergency room?
c. What is the standard deviation of emergency room time?
d. Estimate the following model: time in emergency room as a function of entering time, the age of the person, and an error term.

14.54 Are time of entry and age jointly significant explanatory variables of time in the emergency room in Exercise 14.53?

14.55 In Exercises 13.47 and 12.49, the data in the computer ASCII file "EX14-55.PRN" from *Car and Driver* (April 1993), was employed. In this data set, the first column identifies the car tested and the issue of *Car and Driver* in which the test appeared. The second column is the estimated price of the car tested. The third and fourth columns are 0 to 60 mph and 1/4 mile time in seconds. The fifth column is top speed. The sixth column is 70 to 0 braking distance in feet. EPA estimated fuel economy is in the seventh column. Finally, in column eight, road-holding ability (measured as g force) is provided as obtained from a 300-foot skidpad. Estimate the linear model:

price = function(TO60, quarter, top, braking, fuel, hold, error)

14.56 For the regression model estimated in Exercise 14.55, which car has a predicted price farthest above its actual price? Why might a comparison of predicted and actual price be of interest to a shopper?

14.57 For the regression model in Exercise 14.55, which regressor is the most significant contributor in explaining car price?

14.58 In the midwestern bank sex discrimination case, the bank's expert witness presented a salary regression that had a coefficient on the 0-1 male-female dummy variable that he claimed was insignificant. The plaintiff's expert stated that if the bank's expert argues that for this regression (where $b = -0.016$, with standard error $s_b = 0.01$) the null hypothesis that the population coefficient is equal to zero cannot be rejected, then a population coefficient as low as –0.032 cannot be rejected in a similar test, since –0.032 is the same number of standard errors below –0.016 as zero is above –0.016.

a. Draw the sampling distributions of b showing the two null hypotheses that cannot be rejected.
b. What is the relevant p-value for both null hypotheses?
c. What conclusion can you draw from this?
(Hint: Review the appendices to Chapter 9.)

14.59 Lloyd's Register of Shipping keeps track of minor, but costly, damage to the forward section of cargo vessels (from contact with large waves, debris at sea, docks, and the like) for purposes of setting insurance premiums and standards for hull construction. The data below identifies four types of ships, the last two digits of year of christening, months the ship was at sea, and the number of damage incidents.

Ship Type	Year of Christening	Months at Sea	Number of Damage Incidents
1	64	44992	39
1	65	17179	30
1	71	28603	58
1	69	20380	54
1	76	7044	14
1	74	13087	45
1	81	1000	2
1	84	7118	18
1	81	4510	10
2	64	128	1
2	70	1095	3
2	69	1094	5
2	78	1612	9
2	74	3353	18
2	79	50	0
2	80	2244	11
2	74	1352	10
2	65	578	6
3	64	251	0
3	65	162	1
3	69	297	0
3	70	201	1
3	74	543	2
3	78	1478	12
3	84	359	1
3	89	203	2
3	86	1345	3
4	64	1180	1

(continued)

Ship Type	Year of Christening	Months at Sea	Number of Damage Incidents
4	64	578	1
4	68	883	2
4	70	686	1
4	74	831	5
4	74	1958	3
4	78	1234	2
4	79	274	1
4	65	63	0

In a regression analysis of number of damage incidents, how many dummy variables are required to represent the four types of ships, assuming a constant term is estimated?

14.60 Estimate a regression to explain number of damage incidents with the dummy variables for the types of ships, year of christening, and months at sea, for the data in Exercise 14.59. For a type 2 ship, christened in 1973, at sea for 2000 months, what is the estimated expected number of damage incidents? Construct a 95 percent confidence interval for the expected number of damage incidents. What does this interval tell you?

14.61 An article in the *WSJ* (March 8, 1994) reported on "America's 20 Hottest White-Collar Addresses." Below are the data provided in the article. Estimate a regression of median household income on median age and the percentage increase in the population.

a. For a 0.1 probability of a Type I error, does an overall F test show either explanatory variable to be significant?
b. For a 0.1 probability of a Type I error, do the individual coefficient t statistics show either of the explanatory variables to have a significant positive influence on income?
c. What is the difference between the tests in part a and part b?

	1990 Population	% Increase 1980–90	Median Household Income	Median Age
Douglas, CO	60,391	140.1	$ 51,864	32.3
Fayette, GA	62,415	114.9	50,187	34.1
Fort Bend, TX	225,421	72.3	42,808	30.5
Howard, MD	187,328	58.0	54,407	32.2
Loudoun, VA	86,129	50.0	52,210	31.6
Shelby, AL	99,358	49.9	36,851	31.5
Prince Wm, VA	215,686	49.1	49,370	29.0
Chesterfield, VA	209,274	48.0	43,603	31.9
Dakota, MN	275,227	41.7	42,218	30.2
Williamson, TN	81,021	39.4	43,612	34.1
Hamilton, IN	108,936	32.8	45,747	32.9
Rockingham, NH	245,845	29.2	41,880	32.6
Washington, MN	145,896	28.5	44,120	31.5
Delaware, OH	66,929	24.3	37,895	33.4
Hunterdon, NJ	107,776	23.4	54,661	35.2
Chester, PA	376,396	18.9	45,642	33.8
Somerset, NJ	240,279	18.3	55,566	34.8
Jefferson, CO	438,430	17.9	39,084	33.3
Saratoga, NY	181,276	17.9	36,635	32.8
Olmsted, MN	106,470	15.7	35,788	31.6

14.62 In Exercise 14.61, why doesn't the 1990 population variable belong in the income regression as an explanatory variable along with age and the percentage increase in the population?

14.63 Exercises 12.52 and 13.50 provided Professor Ashenfelter's data for the analysis of the 1993 contested election in Pennsylvania's Second State Senatorial District in Philadelphia. Fit a quadratic regression to these data and comment on why it might be more appropriate than the linear regression fit in Exercises 12.52 and 13.50. Does your quadratic regression change the conclusions reached in Exercise 13.50?

Academic References

Finkelstein, M., and Bruce Levin. *Statistics for Lawyers.* New York: Springer-Verlag, 1990, pp. 363–364 and pp. 409–415.

Friedman, M., and Anna Schwartz. "Alternative Approaches to Analyzing Economic Data: Appendix." *American Economic Review*, March 1991, pp. 48–49.

Johnston, J. *Econometric Methods.* 3rd ed. New York: McGraw-Hill, Inc. 1984.

Lovell, M. "Data Mining." *Review of Economics and Statistics*, February 1983, pp. 1–12.

Simpson, E.H. "The Interpretation of Interaction in Contingency Tables." *Journal of the Royal Statistical Society*, 1951. B 13, pp. 238–241.

Stigler, S. *The History of Statistics: The Measurement of Uncertainty Before 1900.* Cambridge: The Belknap Press of Harvard University Press, 1986.

Endnotes

1. The normal equations resulting from a minimization of the Σe_i^2 with respect to b_1, b_2, and b_3 in a regression of y on x_2 and x_3 are:

$$nb_1 + b_2\,\Sigma x_2 + b_3\,\Sigma x_3 = \Sigma y$$

$$b_1\Sigma x_2 + b_2\Sigma x_2^2 + b_3\Sigma x_2 x_3 = \Sigma x_2 y$$

$$b_1\Sigma x_3 + b_2\Sigma x_2 x_3 + b_3\,\Sigma x_3^2 = \Sigma x_3 y$$

This system of three equations and three unknowns (b_1, b_2, and b_3) can be solved by normal simultaneous equation procedures. Similarly, if there were k parameters, there would be k normal equations from which the values of $b_1, b_2, \ldots b_k$ could be determined. This algebra of multiple regression, however, is better handled in matrix form, which is beyond the scope of this text, and is thus not pursued here. The interested reader will find the matrix algebra development of multiple regression in any one of the standard econometric textbooks, such as J. Johnston (1984).

2. The standard error of the regression plane associated with a regression of y on x_1 and x_2 if calculated by hand is $s_{\hat{y}*} =$

$$s_e\sqrt{\frac{1}{n}+\sum_{j=1}^{2}\frac{\left(x_{j*}-\bar{x}_j\right)}{\sum_{i=1}^{n}\left(x_{ij}-\bar{x}_j\right)^2\left(1-r_{12}^2\right)}+\frac{2\left(x_{1*}-\bar{x}_1\right)\left(x_{2*}-\bar{x}_2\right)\sum_{i=1}^{n}\left(x_{i1}-\bar{x}_1\right)\left(x_{i2}-\bar{x}_2\right)}{\sum_{i=1}^{n}\left(x_{i1}-\bar{x}_1\right)^2\sum_{i=1}^{n}\left(x_{i2}-\bar{x}_2\right)^2\left(1-r_{12}^2\right)}}$$

where r_{12} is the coefficient of correlation between x_1 and x_2, and

$$s_e = \frac{\sum_{i=1}^{n} e_i}{n-3} = \frac{\text{ErrorSS}}{n-3}$$

3. The standard error in predicting an individual y value with a regression of y on x_1 and x_2 is

$$\sqrt{s_{\hat{y}*}^2 + s_e^2}$$

using the definitions provided in Endnote 2.

Formulas

- The population model is

$$y_i = \beta_1 + \beta_2 x_{i2} + \beta_3 x_{i3} + \ldots + \beta_k x_{ik} + \varepsilon_i$$

where y_i is ith value of the dependent variable
x_{ij} is the ith value of the jth independent variable
β_j is the jth parameter to be estimated
ε_i is the population disturbance term

- The sample multiple regression equation is

$$\hat{y}_i = b_1 x_{i1} + b_2 x_{i2} + b_3 x_{i3} + b_4 x_{i4} + \ldots b_k x_{ik}$$

where x_{ij} is the ith value of the jth independent variable
y_i is the ith value of the dependent variable
b_j is the jth coefficient

- The sample error term is

$$e_i = y_i - \hat{y}_i$$

where $i = 1, 2, \ldots n$, so that

$$\begin{aligned} e_1 &= y_1 - \hat{y}_1 \\ e_2 &= y_2 - \hat{y}_2 \\ &\vdots \\ e_{n-1} &= y_{n-1} - \hat{y}_{n-1} \\ e_n &= y_n - \hat{y}_n \end{aligned}$$

- The t statistic for any coefficient estimator b_j is

$$t_{\text{df}=n-k} = \frac{b_j - \beta_j}{s_{b_j}}$$

where s_{b_j} is the standard error of the least squares estimator b_j.

- The error or residual sum of squares is

$$\text{ErrorSS} = \sum_{i=1}^{n} e_i^2$$

- The total sum of squares is

$$\text{TSS} = \sum_{i=1}^{n} (y_i - \bar{y})^2$$

- The regression sum of squares is

$$\text{RegSS} = \text{TSS} - \text{ESS}$$

- The multiple coefficient of determination is

$$R^2 = \frac{\text{RegSS}}{\text{TSS}} = 1 - \frac{\text{ErrorSS}}{\text{TSS}}$$

- The adjusted R^2 is

$$\bar{R}^2 = 1 - \frac{s_e^2}{s_y^2} = 1 - \frac{(n-1)\text{ErrorSS}}{(n-k)\text{TSS}}$$

where

$$s_e^2 = \frac{\text{ErrorSS}}{n-k} \text{ and } s_y^2 = \frac{\text{TSS}}{n-1}$$

- The calculated F ratio is

$$F = \frac{\text{RegSS}/(k-1)}{\text{ErrorSS}/(n-k)}$$

A 100 (1 – α) percent confidence interval for $E(y_* | x_{*1}, x_{*2} \ldots x_{*k})$, at the specified x_* value, is

$$\hat{y}_* \pm t_{(1-\alpha)/2,\, \text{df}=n-k} s_{\hat{y}*}$$

where $+t_{(1-\alpha)/2, \text{df}=n-k}$ is the value of t for which α percent of the area is in the upper tail of the t distribution and $-t_{(1-\alpha)/2, \text{df}=n-k}$ is the t value for which α percent of the area is in the lower tail.

- The 100 (1 – α) percent confidence interval for β_j is

$$b_j \pm t_{\alpha/2} s_{b_j}, \text{ for df} = n - k$$

where s_{b_j} is the standard error of b_j.

CHAPTER 15

TIME SERIES ANALYSIS AND FORECASTING

We have two classes of forecasters: Those who don't know—and —those who don't know they don't know.

John Kenneth Galbraith
b. 1908

15.1 Introduction

Time Series
A sequence of values collected on the same variable at regular and successive time intervals.

Data can be collected at a point in time or over time. When data are collected on the same variable at regular and successive time intervals, the resulting array or sequence of values is called a **time series.** In a time series, both the magnitude of an observation and its order in the series are important. For example, knowledge of the level of the Consumer Price Index (CPI) in a given year is necessary to calculate a "real wage" that reflects purchasing power in that year. The previous and following years' CPI's are needed to calculate changes in the CPI over time. In turn, the rate of change in the CPI may be used to calculate a "real rate of interest," which is an observed interest rate adjusted for an expected rate of inflation.[1]

Data on both the level and the rate of change in a time series are required to forecast future values. This chapter presents models that can be used in the analysis of past relationships and extrapolations from the past to the future. We will begin our formulation of these time series models with a decomposition of a time series into its component parts: a secular trend, cyclical and seasonal variations, and residual or random fluctuations. We will make use of the regression techniques developed in the past three chapters for this decomposition. We will also use regression to forecast future values based on the estimated past relationships uncovered. By the end of this chapter, you will be able to make predictions about future values in a time series by extrapolating from its trend and seasonal components and by building more elaborate econometric models that are based on economic theory and past relationships among variables.

15.2 Time Series Components

Trend
The long-run tendency of a time series to rise or fall as time passes.

The secular **trend** is the long-term or long-run tendency of a series to rise or fall as time passes. Some economic series, such as the prices of computers and other electronic equipment over the past 10 to 20 years, have decreased steadily. Other series, such as the prices of medical products, have increased dramatically. The value of sales of all manufacturing corporations in the United States has risen in almost a straight-line fashion since the 1970s. Table 15.1 gives the value of these total net sales by quarter, starting with the fourth quarter of 1973, when this series was first published. Figure 15.1 provides a line drawing of sales, where the upward movement can be seen.

Seasonal Variations
Deviations from the trend that occur within a year and tend to repeat regularly from year to year.

Seasonal variations represent deviations from a trend that occur within a year and tend to repeat regularly from year to year, such as the increase in sales associated with the holidays in the fourth quarter. Similarly, large expenditures on housing and automobiles in the spring and vacation expenditures in early summer tend to increase sales in the second quarter. These fourth and second quarter increases can be seen in the sawtooth pattern in Figure 15.1. Unlike other forms of variation, seasonal variations are usually quite easy to spot and rationalize.

Cyclical variations are the longer swings in a time series that may require many years to exhibit a repetitive pattern. The business cycle represents cyclical variations

TABLE 15.1 Net Sales of All Manufacturing Corporations in the United States (billions of dollars)

#	year	qt	sales	#	year	qt	sales	#	year	qt	sales	#	year	qt	sales
1	1973	4	236.6	18	1978	1	340.3	35	1982	2	521.9	52	1986	3	546.3
2	1974	1	242.0	19	1978	2	377.5	36	1982	3	508.0	53	1986	4	564.5
3	1974	2	269.4	20	1978	3	376.9	37	1982	4	506.6	54	1987	1	556.8
4	1974	3	272.1	21	1978	4	401.8	38	1983	1	490.8	55	1987	2	596.1
5	1974	4	277.0	22	1979	1	406.6	39	1983	2	527.1	56	1987	3	597.7
6	1975	1	247.1	23	1979	2	436.4	40	1983	3	534.7	57	1987	4	627.7
7	1975	2	265.8	24	1979	3	437.5	41	1983	4	561.6	58	1988	1	614.2
8	1975	3	271.0	25	1979	4	461.2	42	1984	1	566.1	59	1988	2	655.5
9	1975	4	281.3	26	1980	1	465.7	43	1984	2	597.9	60	1988	3	646.3
10	1976	1	284.2	27	1980	2	466.3	44	1984	3	577.1	61	1988	4	680.2
11	1976	2	307.6	28	1980	3	464.2	45	1984	4	594.0	62	1989	1	666.0
12	1976	3	301.6	29	1980	4	516.6	46	1985	1	565.3	63	1989	2	707.5
13	1976	4	309.8	30	1981	1	520.8	47	1985	2	594.1	64	1989	3	681.3
14	1977	1	311.5	31	1981	2	549.6	48	1985	3	578.0	65	1989	4	690.3
15	1977	2	338.6	32	1981	3	539.9	49	1985	4	593.9	66	1990	1	667.4
16	1977	3	331.7	33	1981	4	534.4	50	1986	1	544.0	67	1990	2	704.1
17	1977	4	346.2	34	1982	1	502.9	51	1986	2	566.2	68	1990	3	702.0

Source: Various issues of the *Economic Report of the President.*

FIGURE 15.1 Net Sales of All Manufacturing Corporations in the United States (billions of dollars)

Sales

707.500

589.775

472.050

354.325

236.600

.000 17.000 34.000 51.000 68.000

Trend

TIME PERIOD (t = 1, for 4th quarter 1973, to t = 68, for 3rd quarter 1990)

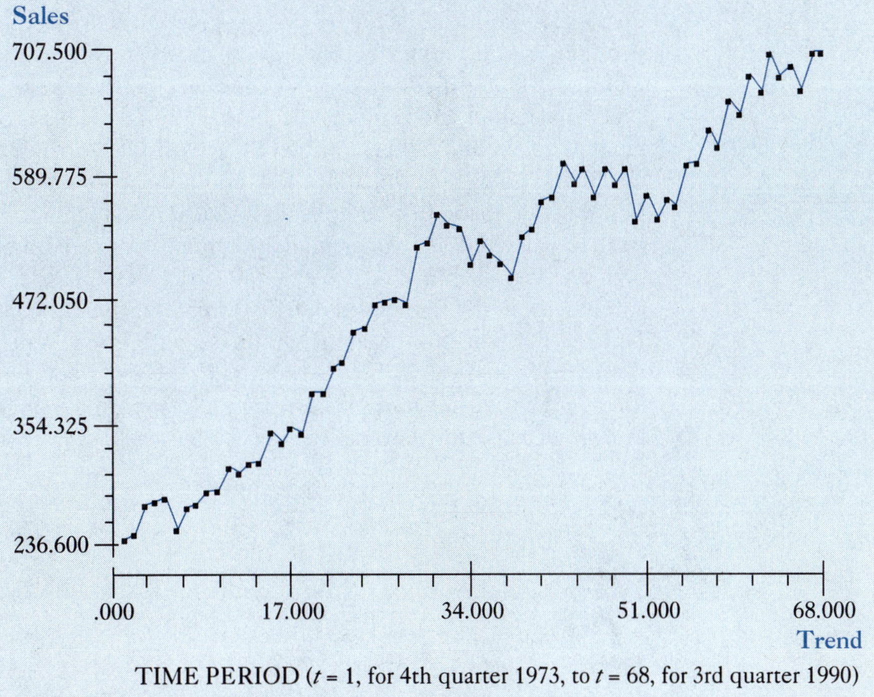

Cyclical Variations The longer swings in a time series that suggest a repetitive pattern over extended intervals of time.

in which business activity is below the trend for several months, but then rises above the trend line only to peak and then fall back below the trend after several months of being above it. In Figure 15.1, 1984 and 1985 show relative peaks in a business cycle, with lower sales in both 1983 and 1986.

Economic data may contain longer-term cycles that cannot be identified without more than the 17 years of data shown in Figure 15.1. For instance, the Kondratieff cycle (named after a Russian economist) is supposed to be evident in national income data every 48 to 60 years. By this theory of cycles, real growth in U.S. gross domestic product peaked in 1973 and has been falling ever since. Confirmation of this theory is difficult to establish because of the centuries of data required and the irregularities in cycles.

Theories for cycles in economic data are not lacking, and many of them have nothing to do with economics. The Elliot Wave theory of stock market prices, for example, is based on the alleged existence of patterns that follow Fibonacci number series (a series of numbers formed by the sum of the two preceding numbers: 1, 2, 3, 5, 8, 13, 21, . . .). According to newsletter publisher Robert Prechter, Elliot Wave stock market price analysis suggests "the end of a 'grand supercycle rise dating from 1784' that could send the stock market plunging 90% to 98% within 12 years" *Business Week* (June 29, 1992, p. 128). Other noneconomic theories of cycles in economic data can be seen in the 11-year sunspot cycles (for which turn-of-the-century economist Jevons was widely ridiculed) and planetary cycles (in which Henry Moore assigned a central role to Venus). Further consideration of these different theories of cycles is beyond the scope of this book.

Random (or residual) Fluctuations The irregular or unpredictable changes in a time series.

Random fluctuations are the irregular or unpredictable changes in a time series. Deviations from the trend caused by unexpected changes in the weather, by fads, or by unforeseen political unrest are examples of random fluctuations. Inherent in business decisions and economic activity are chance factors that are reflected in the outcome measures. Unfortunately, the outcome of chance factors is hard to distinguish from cyclical variation and requires elaborate residual analysis. Sources of both cyclical and random variability will be considered as parts of our residual analysis here.

Before going into the above four components of a time series in more detail, mention must be made of a complication that is often ignored. Observations on a time series are collected at regular points in time. Any variability between these collection points is lost. Table 15.1, for example, provides quarterly observations on sales. But surely sales in the first quarter of 1990 were not the same in each of its three months of January, February, and March. Within a month, there is no reason to believe that sales in the first week would be the same as in the last week. Variability within the quarter for quarterly observations and within the month for monthly observations are unknown, because they are not recorded.

Although the government provides quarterly and monthly data on many variables, variability that occurs over even shorter periods of time may be of greater importance to the decision maker. For instance, monthly variations in interest rates are not as relevant as daily fluctuations to someone shopping for a car or home loan. The time frame used for observing a time series must be related to the use to which it will be put.

QUERY 15.1:

Below is a diagram from *Newsweek* (January 6, 1992) showing net earnings of the Chrysler Corporation. Identify which components of a time series are apparent in this diagram, and state why the other components are not observable.

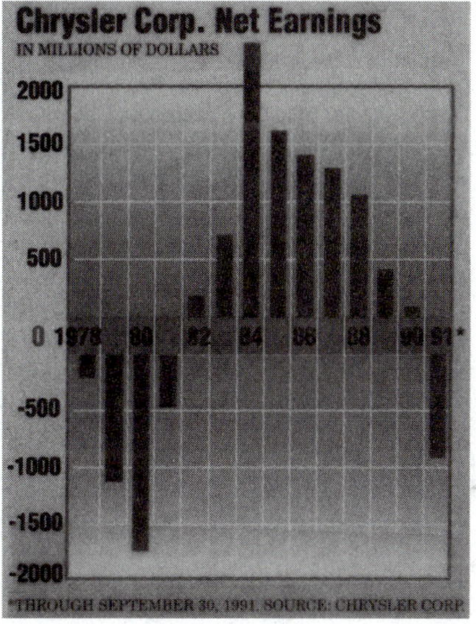

ANSWER:

There is little or no trend in this time series of earnings. The cyclical nature of Chrysler's earnings is apparent, with a trough showing a low point in 1980 and the upturn peaking in 1984. Without actually fitting a trend line and removing the cyclical component through a residual analysis, the randomness in Chrysler's earnings cannot be identified. There is no way to identify the seasonal nature of Chrysler's earnings from this chart because this is yearly data; all fluctuations within the year are thus lost.

15.3 Trend Analysis

Trend refers to the long-term tendency of a series to rise or fall over time. The description of this trend is of value to study historical patterns. For instance, changes in the rate of increase in prices associated with the imposition of price controls by President Richard Nixon would be of interest in assessing the effectiveness of these controls. Establishment of past trends also will enable the prediction of future trends. In the case of price controls, for example, the change in price trends associated with the Nixon controls might be used to predict future changes in trend should such controls ever be imposed again. Finally, once a trend has been estimated, its effects can be removed from the series, which enables the identification of seasonal, cyclical, and random variations. These three reasons for estimating a trend will be considered in later sections. First, however, the trend must be estimated.

Trend can be seen as a (straight or curved) line passing through the data with a positive or negative slope. This visually apparent trend can be estimated with least squares. In the case of a straight or linear trend line, estimation involves the fitting of the simple regression

$$\hat{y}_t = a + bt \tag{15.1}$$

where $\hat{y}_t$ is the predicted value of y_t at time t and a and b are least squares estimators of the intercept (a) and slope (b).

In Table 15.1, for example, the numbers 1, 2, . . . 68 are used to represent time, the explanatory variable.[2] In this case, $t = 1$ represents the fourth quarter of 1973, where $y_1 = 236.6$; $t = 2$ is the first quarter of 1974, where $y_2 = 242.0$; and so forth until the last time period $t = T = 68$, where $y_{68} = 702.0$.

Forecasting with a Trend Line

The intercept (a) and slope (b) can be calculated by hand, as in Chapter 12, or preferably with a computer program. The resulting least squares regression line is

$$\hat{y}_t = 247.86 + 6.89t$$

This equation indicates that sales are predicted to rise by \$6.89 billion each quarter, starting from a predicted \$247.86 billion in the third quarter of 1973, where t is zero. It is drawn on the scatterplot in Figure 15.2. This trend line equation can be used to predict or forecast future or past sales outside the time domain shown. In the third quarter of 1993, when t would equal 80, for instance, sales are estimated to be \$799.06 billion.

$$\hat{y}_t = 247.86 + 6.89(80) = 799.06$$

FIGURE 15.2 Scatter Diagram of Sales with Linear Trend Line

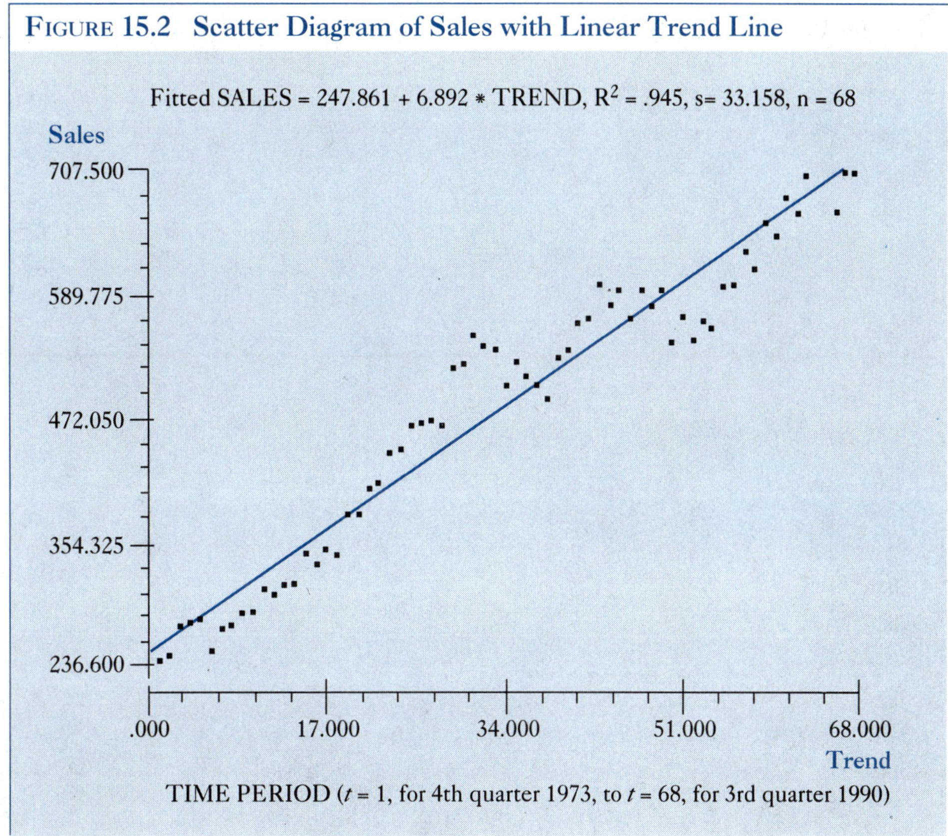

In the second quarter of 1973, prior to the publication of this series, sales can be predicted by setting $t = -1$. Predicted sales for the second quarter of 1973 are \$240.97 billion.

$$\hat{y}_t = 247.86 + 6.89(-1) = 240.97$$

Detrended Time Series

Detrending
Removing the trend from a time series, leaving only the seasonal, cyclical and random components.

In addition to predicting values, a trend line can be used to produce a series that has the effect of time removed. The residuals from the trend line $(y_t - \hat{y}_t)$ constitute such a **detrended** time series. Examination of the detrended residual series makes identification of other components possible. For instance, although the trend line appears to fit the sales data well, with an R^2 of .945, the residuals in Figure 15.3 show the influence of other factors. A serpentine pattern is apparent. A seasonal effect appears to cause an alternating pattern of rising and falling residuals. These time patterns in the residuals

FIGURE 15.3 Plot of Residuals after Removing Trend from Sales

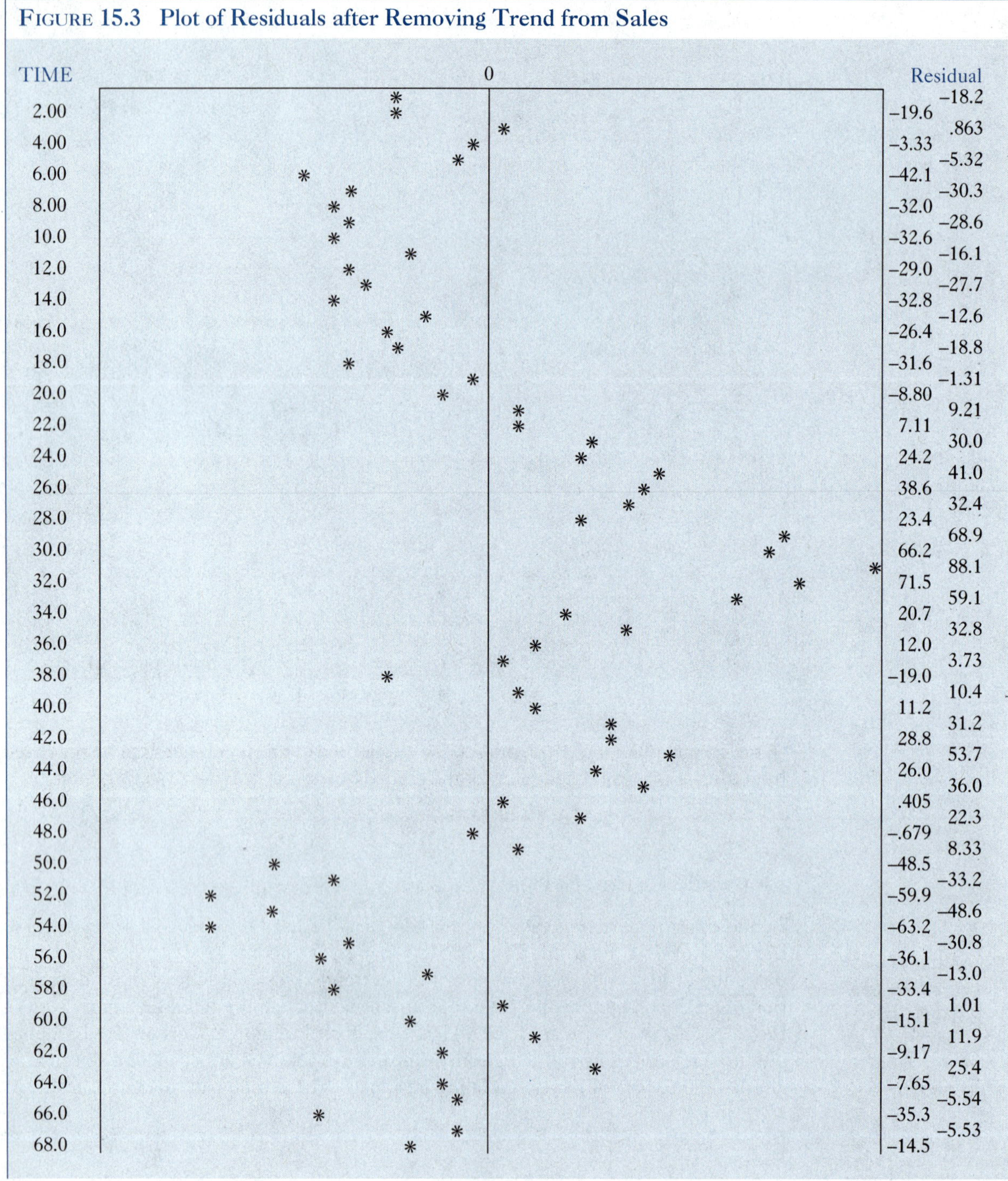

indicate possible violations of the assumption that the errors are independent of one another, as required for a statistical test of the slope of the trend line. Removal of these undesired patterns will be discussed in the following sections.

QUERY 15.2:
The diagram below is from a *Wall Street Journal* (December 12, 1991) article on U.S. champagne and sparkling-wine consumption. What is the trend in this data?

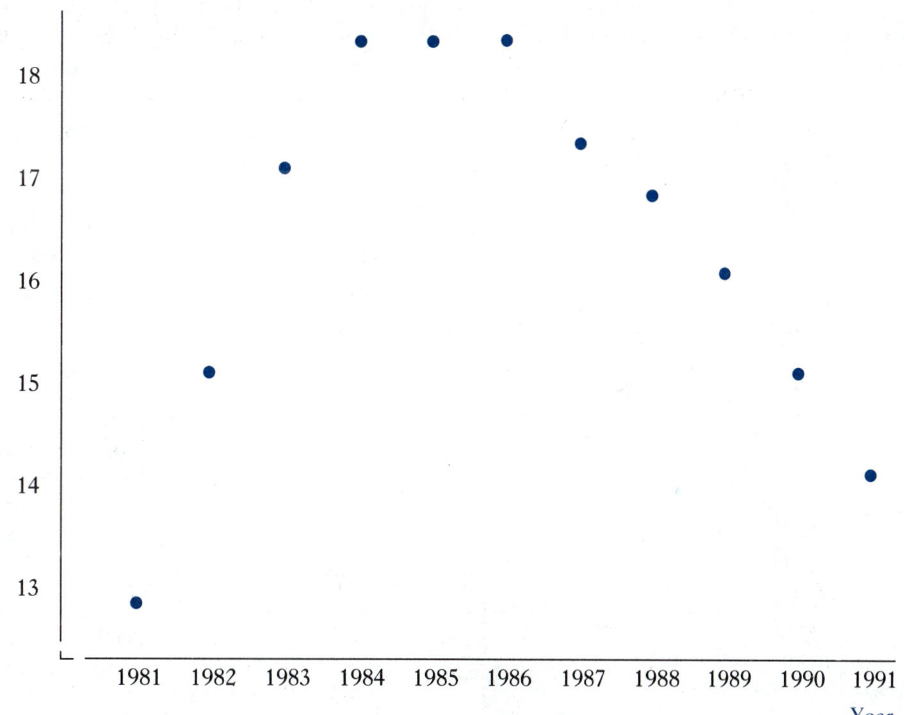

ANSWER:
There is no simple linear trend; the correlation between U.S. champagne and sparkling-wine consumption and time is zero between 1981 and 1991. Prior to 1985, however, consumption was positively related to time, but since 1985 this relationship has been negative.

15.4 SEASONAL VARIATION

Seasonal variations are exactly what the name implies: fluctuations in a time series that tend to repeat themselves from year to year on a regular basis. Unemployment rates, for example, tend to rise in the northern states in January as cold weather causes cutbacks in construction activity. Sales of houses tend to increase every spring, and decrease during the summer. Sales volume of most consumer products tends to repeat on a seasonal basis, peaking between Thanksgiving and Christmas. To make comparisons over time, these seasonal factors need to be identified and removed from the time series. That is, the series will need to be "seasonally adjusted" to take account of factors that are seasonally related or **deseasonalized** to totally remove the effects from the series. At a minimum, identification and use of information on seasonality will improve predictive power.

Deseasonalized Series
A time series with the seasonal effects removed, leaving only the trend, cyclical and random components.

Dummy variables can be used to capture the effect of seasonal components in a time series. Dummy variables are created by the researcher to reflect qualitative attributes or characteristics of the data. For instance, data for the fourth quarter may be different from data collected in the previous three quarters. Each of the four quarters can be represented by its own dummy variable, with the dummy variable equal to one if the observation is from the designated quarter, and zero if not. For the quarterly sales data, such a structure of dummy variables is shown in Table 15.2, where QT1, QT2, QT3, and QT4 represent the first through fourth quarters. Similar structures could be constructed for data collected on a monthly or weekly basis, where the number of dummy variables would be 12 or 52.

A least squares regression of sales on the trend and the dummy variables in Table 15.2 can be estimated, but care is needed to avoid perfect linear relationships among the regressors. As discussed in the previous chapter, the x data includes a column of

TABLE 15.2
Dummy Variable Representation of Quarterly Observations on Sales

TIME	SALES	QT1	QT2	QT3	QT4
1	236.60	.00000	.00000	.00000	1.0000
2	242.00	1.0000	.00000	.00000	.00000
3	269.40	.00000	1.0000	.00000	.00000
4	272.10	.00000	.00000	1.0000	.00000
5	277.00	.00000	.00000	.00000	1.0000
6	247.10	1.0000	.00000	.00000	.00000
7	265.80	.00000	1.0000	.00000	.00000
8	271.00	.00000	.00000	1.0000	.00000
9	281.30	.00000	.00000	.00000	1.0000
•	•	•	•	•	•
•	•	•	•	•	•
•	•	•	•	•	•
62	666.00	1.0000	.00000	.00000	.00000
63	707.50	.00000	1.0000	.00000	.00000
64	681.30	.00000	.00000	1.0000	.00000
65	690.30	.00000	.00000	.00000	1.0000
66	667.40	1.0000	.00000	.00000	.00000
67	704.10	.00000	1.0000	.00000	.00000
68	702.00	.00000	.0000	1.0000	.00000

ones for the constant term b_1. This column of ones would equal the sum of the four dummy variables if all four dummy variables were included as explanatory variables. Such a perfect linear relationship among the regressors makes least squares estimation impossible. To avoid this perfect collinearity, one of the dummy variables can be excluded from the regression equation. The effect of this omitted quarter is then captured in the constant term, and the effects of the other three included quarters are measured as additions to the constant term.

Which quarter is eliminated is irrelevant, but one must be omitted if the constant term is included in the regression. Excluding the first quarter dummy yields the following regression line for the prediction of sales ($\hat{y}_t$) based on the remaining three quarters and a time trend term t.

$$\hat{y}_t = b_1 + b_2(\text{QT2}) + b_3(\text{QT3}) + b_4(\text{QT4}) + b_5 t$$

The relevant regression results for the data in Table 15.1 are shown in Table 15.3. Notice that the residuals in Figure 15.4 no longer have a distinct quarterly pattern to them. Thus, the dummy variables have controlled for the seasonal variation. The least squares regression equation shown in Table 15.3 is

$$\hat{y}_t = 235.683 + 21.923(\text{QT2}) + 8.246(\text{QT3}) + 18.189(\text{QT4}) + 6.895t$$

The low test statistic value (0.734) and high p-value (.46586) for the coefficient of the dummy variable for the third quarter might be seen as suggesting that there is no difference between the first and third quarters in the population. The third quarter, however, is adding \$8.246 billion to predicted sales within the sample. From a practical standpoint, with a \$8.246 billion difference, it is difficult to conclude that the third quarter does not differ from the first quarter.

In addition to considering the coefficients' statistical significance and practical importance, the residuals should be checked to see if there is a noticeable pattern. Figure 15.4 provides the plot of those residuals. A comparison of the residuals in Figure 15.3 and Figure 15.4 shows that the seasonal sawtooth pattern has been removed but the serpentine pattern is still present. This serpentine pattern suggests that the sales in one quarter are related to that in the next; the residuals over time are not independent, as required for the legitimate use of the least squares test statistics. Although predictions can be made with this relationship, the tests of significance are unreliable.

TABLE 15.3 Regression with Three Quarter Dummies to Control for Seasonality

Dependent Variable	SALES	Number of records	68
Mean of Dep. Variable	485.6368	Std. Dev. of Dep. Var.	140.198740
Sum of Squared Residuals	.67555E+05	Sum of Squares Regression	.12494E+07
Sum of Squares Total	.13169E+07	Std. Error of Regr.	32.7460
R-squared	.94870	F(4, 63)	291.2840

Variable	Coefficient	Std. Error	T-ratio	Prob | t | > x
Constant	235.683	10.51	22.416	.00000
QT2	21.9230	11.23	1.952	.05544
QT3	8.24608	11.24	.734	.46586
QT4	18.1887	11.23	1.619	.11041
TIME	6.89461	.2026	34.023	.00000

FIGURE 15.4 Plot of Residuals after Removing Trend and Seasonality from Sales

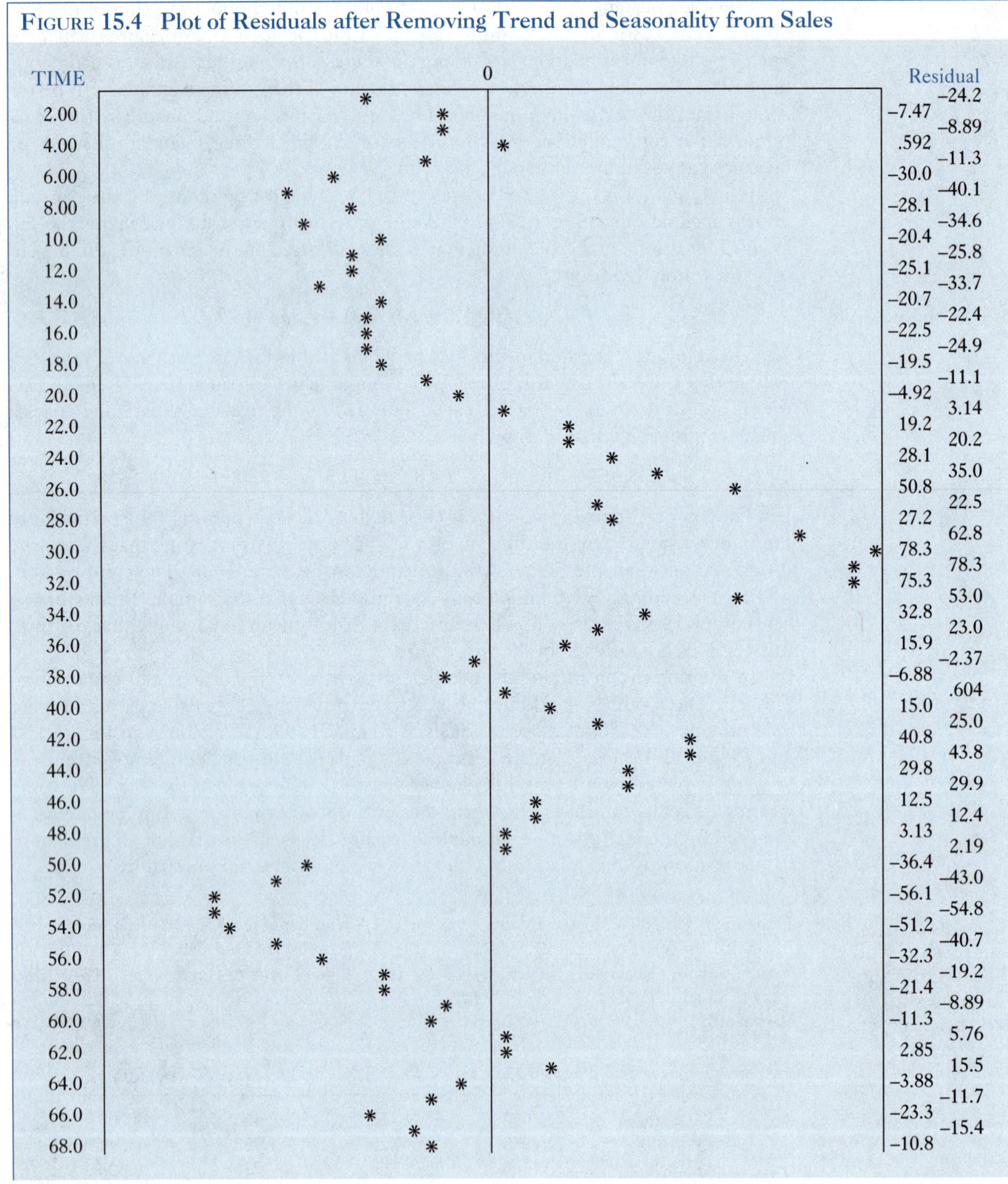

The regression in Table 15.3 represents four different trend lines. The line for each quarter is derived by substituting the dummy-variable values. For the first quarter, QT2 = 0, and QT3 = 0 and QT4 = 0.

$$\hat{y}_t = 235.683 + 21.923(0) + 8.246(0) + 18.189(0) + 6.895t$$
$$\hat{y}_t = 235.683 + 6.895t$$

For the second quarter, 21.923 is simply added onto the first quarter. That is, QT2 = 1 and QT3 = 0 and QT4 = 0.

$$\hat{y}_t = 235.683 + 21.923(1) + 8.246(0) + 18.189(0) + 6.895t$$
$$\hat{y}_t = 257.606 + 6.895t$$

In the third quarter, 8.246 is added onto the first quarter, since QT2 = 0 and QT3 = 1 and QT4 = 0.

$$\hat{y}_t = 235.683 + 21.923(0) + 8.246(1) + 18.189(0) + 6.895t$$
$$\hat{y}_t = 243.929 + 6.895t$$

And, finally, for the fourth quarter, 18.189 is added onto the first quarter (QT2 = 0 and QT3 = 0 and QT4 = 1).

$$\hat{y}_t = 235.683 + 21.923(0) + 8.246(0) + 18.189(1) + 6.895t$$
$$\hat{y}_t = 253.872 + 6.895t$$

These four trend lines are shown in Figure 15.5. To predict sales in the second quarter of 1990, for example, the time value for the quarter is $t = 67$, and the dummies are QT2 = 1, QT3 = 0, and QT4 = 0. Thus, the predicted sales are $719.6 billion.

$$\hat{y}_t = 235.683 + 21.923(1) + 8.246(0) + 18.189(0) + 6.895(67)$$
$$\hat{y}_t = 719.6$$

The observed sales were $704.1 billion, so the prediction error or residual is a negative $15.4 billion. The residuals for the other 67 quarters in the sample are shown down the right-hand column of Figure 15.4.

The seasonally adjusted trend line can also be used to forecast future values of sales. In the third quarter of 1993, when t would equal 80 and QT3 = 1, for instance, sales are predicted to be $795.5 billion.

$$\hat{y}_t = 235.683 + 21.923(0) + 8.246(1) + 18.189(0) + 6.895(80)$$
$$\hat{y}_t = 795.5$$

As stated earlier, some might argue that a dummy variable for the third quarter should not be included because its coefficient estimate is insignificant. This argument, however, overlooks the fact that this quarter adds $8.246 billion to the sales forecast. It also overlooks the fact that the assumptions for hypothesis testing may be violated; for example, the serpentine pattern in the residuals is not consistent with the assumption that errors are independently and identically distributed as normal random variables. Omitting the third quarter dummy from the regression and a re-estimation with the third quarter omitted may not be warranted.

FIGURE 15.5 Seasonally Adjusted Trend Lines for Sales

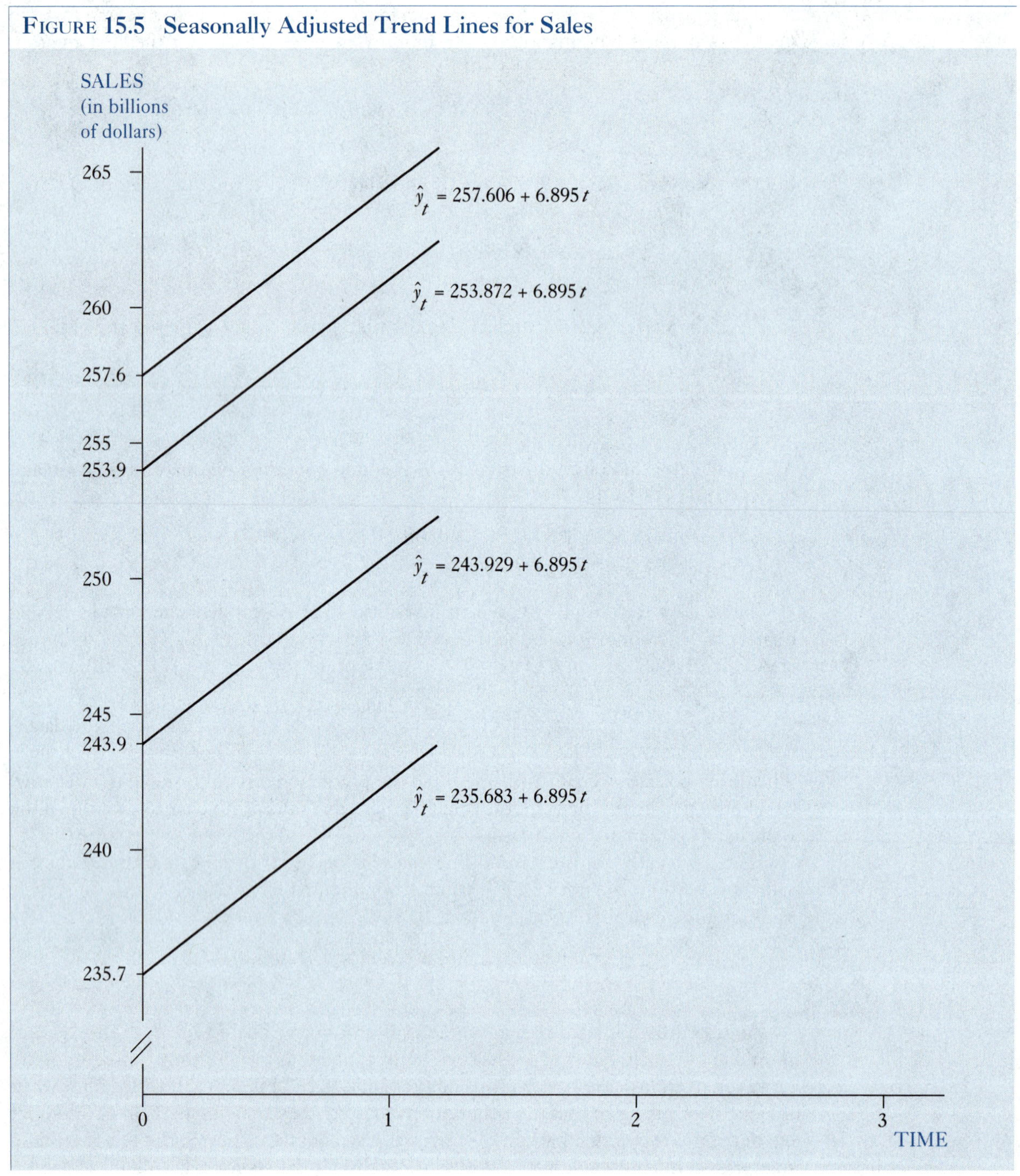

Historical Perspective

Much of what is done today in the name of economic forecasting and econometrics can be traced to three individuals: Francis Ysidro Edgeworth (1845–1926), Wesley C. Mitchell (1874–1948), and Trygve Haavelmo (who won the Nobel prize in economics in 1989). Edgeworth demonstrated the importance of economic indicators and time-series analysis. Mitchell institutionalized and advanced business and economic data analysis as we know it today. Haavelmo provided a framework for applications of the inferential procedures (of Ronald Fisher, Jerzy Neyman, and Egon Pearson) in econometric model building.

Although the idea of least squares was popularized by Galton, Edgeworth advanced the use of regression analysis and correlation with time series data in economic analysis. Edgeworth was a student of classical literature who also studied law at Oxford University; he learned economics, mathematics, and statistics on his own.

Edgeworth laid the foundation for what we now call causal models and structural equation models of behavior. His ideas were implemented in the United States by Mitchell, in the founding of the National Bureau of Economic Research (NBER) in 1920. The NBER continues to be a private organization and the recognized voice decreeing the cyclical turning points in economic activity, even though it has no explicit connection to the United States government. Through its studies, the NBER has identified variables that turn up (or down) prior to major changes in the business cycle. Leading indicators (variables that tend to lead the economy into and out of a recession or growth period) are now tracked by the United States Department of Commerce and summarized in an index that includes the average workweek, plant and equipment orders, unemployment claims, orders for consumer goods, slower deliveries, building permits, durable order backlog, material prices, money supply, and consumer expectations.

Articles on the behavior of the Commerce Department's index of leading indicators appear regularly in the *Wall Street Journal, Business Week,* and other business and economics publications. For example, the *WSJ* (April 27, 1992) presents scrap metal prices as a leading indicator of future economic strength. Federal Reserve Chairman, Allan Greenspan, is cited as watching scrap metal prices because they are the beginning of the production chain—rising scrap metal prices point toward a recovering economy. An article in *Business Week,* "Predicting the Market's Mood Swings" (June 29, 1992, p. 126), discusses predictions from eight different indicators.

The behavior of indicators that tend to lead, coincide, and lag turning points in the economy are followed closely by government and business economists. Unfortunately, in periods of rapid growth, the leading indicators occasionally turn down well before and with no apparent relationship to any following downturn in overall economic activity. Similarly, in recessions, they have failed to consistently signal upturns in a timely manner. Random behavior is also apparent in the coincidental and lagging indicators. Although no forecaster ignores these indicators, the advancements in computer technology and the building and estimation of multivariable econometric models, as promoted by Haavelmo and advanced earlier by Jan Tinbergen and Ragnar Frisch (who shared the first Nobel prize in economics in 1969), made econometric forecasting a more mathematically rigorous activity.

Haavelmo proposed the practice, still adhered to by many econometricians today, that economic research must start with a model from which the data are believed to be generated. Before looking at the data, however, economic theory and probability theory are used to justify the model specification and the estimation, testing, and forecasting procedures to be employed. Curiously, as stated by James Heckman in his *Journal of Economic Literature* (June 1992) review of Mary Morgan's book *The History of Econometric Ideas,* adherence to this Haavelmo paradigm may be the reason why mainstream statisticians and applied empirical scholars no longer look to econometricians for guidance on problems of exploratory data analysis. As discussed earlier in the historical perspective on hypothesis testing, problems associated with induction in Fisher-Neyman-Pearson hypothesis testing procedures have not been solved, and the Haavelmo paradigm of model specification, estimation, hypothesis testing, and forecasting is not universally accepted. In the world of forecasting, there is no one method and no one answer.

Exercises

15.1 Below are monthly changes in the Dow Jones Industrial Average

July	1.6%	August	1.5	January	1.4	December	1.3
April	1.1	June	0.7	March	0.5	November	0.4
February	0.1	May	0.0	October	–0.2	September	–0.8

Show the dummy variable structure (the variable and the arrangement of 0s and 1s) required to capture the four seasonal effects in these data.

15.2 An article in *The Wall Street Journal* (March 13, 1992) argued that the "dinner check growth slows" as fast-food chains are redefining their products. Below is the average dinner check of quick-service restaurants that was provided with the article. Does this graph suggest a slowing of growth?

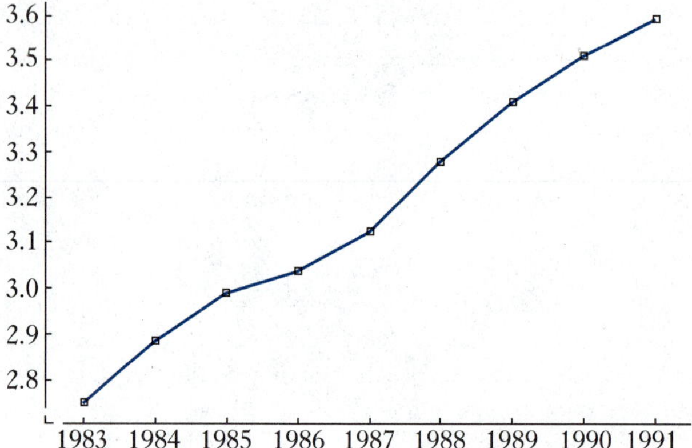

15.3 To the right is a diagram from *The Wall Street Journal* (October 6, 1992) that shows the "estimated number of convenience store robberies in the U.S., in thousands." Could a linear trend line be fit to the underlying data and then used to extrapolate "stickups" in 1992, 1993, and beyond? Explain?

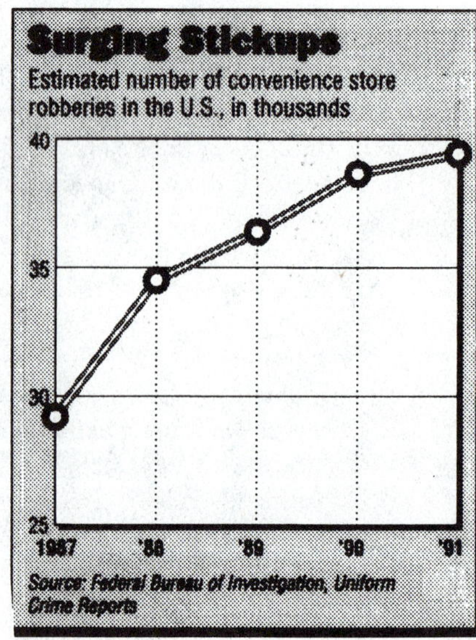

15.4 Students of the Federal Reserve System believe that there are three different trends in money growth: 1959–1973, 1974–1982, and 1983–1990. Investigate this in a model having both a linear trend component and two dummy variables to reflect the three periods. Use the data on the following page:

Year/Money stock (in billions):

1990	89	88	87	86	85	84	83	82	81	80	79	78	77	76
825	795	787	750	725	620	552	521	474	436	409	383	358	331	306

1975	74	73	72	71	70	69	68	67	66	65	64	63	62	61
288	274	263	249	228	214	204	197	183	172	168	160	153	148	145

1960	59
141	140

15.5 An article in the *Herald-Times* (March 3, 1992) discussed the downward trend in the number of residents of Monroe County who terminated a pregnancy. The data were as follows

Year	1985	1986	1987	1988	1989
Terminated	627	576	534	476	442

a. Fit a linear trend line to these data.

b. Why are extrapolations from the trend line in part a questionable?

15.6 A *Wall Street Journal* (January 31, 1992) article reported on the changing market shares for 386 microprocessor sales worldwide, as a percentage of total units sold. The accompanying diagram showed chip maker INTEL with 100 percent of the market in the first quarter of 1991, when chip maker AND had none. By the fourth quarter, however, AND's new clone of the INTEL 386 chip had about 27 percent of the market. In fitting linear trend lines to predict when AND's market share might equal that of INTEL, why isn't it necessary to fit two separate regressions?

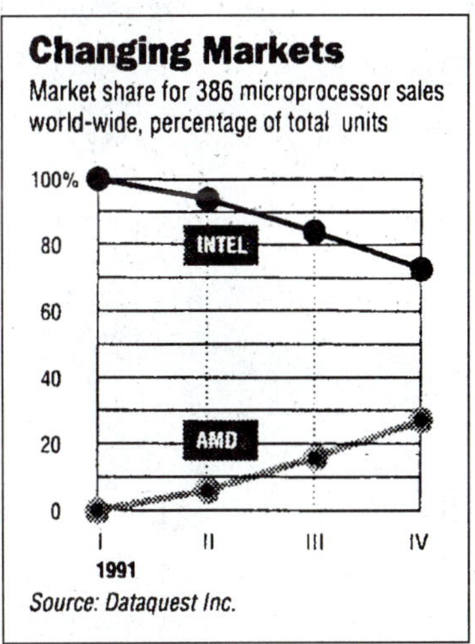

15.7 A *Wall Street Journal* (February 12, 1991) article stated, "Though the Texas economy began to recover in early 1987, housing prices in Houston continued to decline." The following data was provided as supporting evidence that there was a negative relationship between housing prices and time, and that the recovery that began in 1987 did not influence housing prices. Fit a regression of median price on the number of sales, a trend variable, and an intervention dummy for the 1987–1990 period. Did you find that the data and this regression support the alleged relationships? Comment.

Year	No. of Sales	Median Price
1982	13,091	$77,430
1983	13,275	80,260
1984	15,065	79,779
1985	16,495	78,324
1986	21,669	71,557
1987	29,938	65,784
1988	34,599	62,530
1989	31,167	68,176
1990	34,304	70,656

15.8 A *Wall Street Journal* (June 10, 1991) article quoted David Webb, head of real estate investment banking at Merrill Lynch & Co., as saying, "We believe there is a 12- to 24-month buying window in the domestic real estate markets" before prices start trending up again. Below are total annual returns on real estate investment based on when the investment was made, in percentages. Use these data to predict returns in the first quarter of 1992 and the first quarter of 1993, after which Webb believes the trend will turn up.

First Quarter:	1991	1990	1989	1988	1987	1986	1985	1984	1983	1982	1981	1980	1979
Annual Return:	–0.2%	0.0%	2.7%	4.1%	4.4%	4.8%	5.6%	6.5%	7.5%	7.6%	8.4%	9.0%	10.1%

15.9 Below is a time series plot of semester enrollments in the introductory business and economics statistics course at Indiana University. Fit a linear trend line and make seasonal adjustments to forecast enrollments in the fall of 1992 and spring of 1993.

Spring 1986	962	Fall 1986	758
Spring 1987	905	Fall 1987	761
Spring 1988	901	Fall 1988	640
Spring 1989	863	Fall 1989	839
Spring 1990	747	Fall 1990	712
Spring 1991	792	Fall 1991	489
Spring 1992	689		

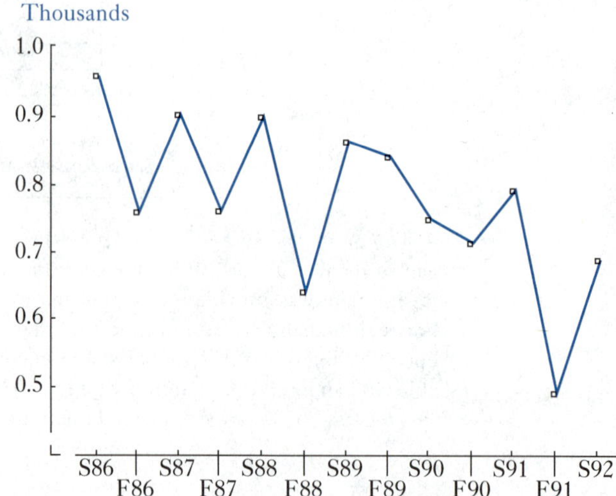

15.10 Technical analysts, also known as chartists, take their cues from the stock market itself by delving into trends and cycles. One line watched closely is the advance-decline line, which is a tally of the number of advancing stocks minus the number of declining ones per week on the New York Stock Exchange. Below is the advance-decline line from an article on the subject in *The Wall Street Journal* (May 1, 1992).

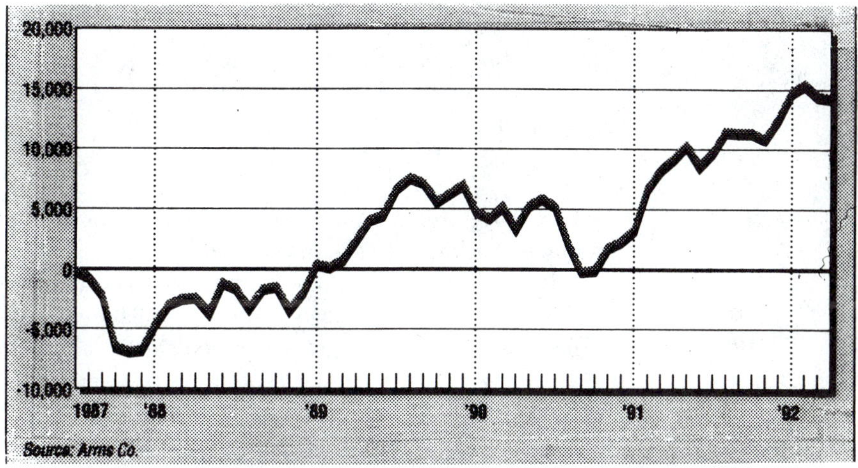

a. What does the trend in the advance-decline line suggest about the future of a bull market (prices rising) versus a bear market (prices falling)?
b. Is there anything in this advance-decline line that suggests something other than trend may be worth watching?

15.5 Residual Analysis

When observations around a trend line tend to remain above or below the line for an extended period of time and then slowly move to the other side of the trend line and continue there for a time, this movement may indicate cyclical variations. Cycles may show up in the residuals as the serpentine pattern in Figure 15.3. The tendency of the residuals to group alternatively into negative and positive clusters are evidence of correlation between time periods. The correlation of a variable with itself over time is a form of serial correlation. Conceptually, serial correlation of the residuals between two subsequent periods is defined by the **autocorrelation coefficient**

Autocorrelation Coefficient
A measure of the dependence of a time series on its own past values.

$$r = \frac{\sum_{t=1}^{T} e_t e_{t-1}}{\sqrt{\sum_{t=1}^{T} e_{t-1}^2 \sum_{t=1}^{T} e_t^2}} \tag{15.2}$$

where e_{t-1} and e_t are the residuals from a regression of y on a set of x's, estimated first at time $t-1$ and then again in the subsequent time period t, for a total of T periods. This autocorrelation coefficient poses calculation problems because a value prior to the starting value in the data set is always needed but missing. The net sales series in Table 15.1, for example, started in the fourth quarter of 1973, but the $t-1$ subscript requires its availability in the third quarter of 1973, which we do not have.

Random Walk
A form of serial correlation in which the value of a variable in this period is determined by its value in the previous period plus a random error.

There are many forms of serial correlation. The most studied is the **random walk.** It is likened to the path of a drunk trying to walk a straight line. The drunk may start out wandering to the left; catching his or her balance, the drunk then overcorrects and wanders back to the right but is unable to correct at the line and wanders past it, with an overcorrection on the other side.

A random walk is described mathematically as a first-order autoregressive process. For the population disturbance, it is described by a regression of ε_t on ε_{t-1},

$$\varepsilon_t = \rho\varepsilon_{t-1} + u_t \tag{15.3}$$

where u_t is a truly random perturbation and ρ is the population parameter relating the value of ε at time $t-1$ to its value at t.

Returning to our analogy, if the drunk is only slightly inebriated, then he or she may negotiate the line and finally end up oscillating around it in ever smaller movements. This is akin to an autoregressive process in which ρ is between –1 and +1, $|\rho| < 1$, a stationary process. In a stationary process, $|\varepsilon_t|$ approaches zero as t increases. If the drunk is extremely inebriated, then he or she may be unable to negotiate the line and make ever bigger oscillations around it. The drunk's path is now an autoregressive process in which $|\rho| > 1$, a nonstationary process. In an explosive process, $|\varepsilon_t|$ goes to infinity as t increases. We will first address the detection of stationary autoregressive error term processes.

15.6 Durbin-Watson Test

Conceptually, as with the autocorrelation coefficient, the parameter ρ could be estimated and tested by a least squares regression of e_t on e_{t-1}. A regression of e_t on e_{t-1}, however, is not that simple, because e_t is itself a byproduct of a regression of y on the x's. With T observations on y there can be only $T-1$ residuals for a regression of e_t on e_{t-1}. Rather than speculate about the strength of a first-order autoregressive process in the population disturbance term, the Durbin-Watson statistic can be calculated, from which ρ can be inferred and tested. The Durbin-Watson d statistic is the most well-known statistic in time series analysis; it is calculated by

$$d = \frac{\sum_{t=2}^{T}\left(e_t - e_{t-1}\right)^2}{\sum_{t=1}^{T} e_t^2} \tag{15.4}$$

Notice, unlike the autocorrelation coefficient, in the numerator of the d statistic the number of observations is only $T-1$, while in the denominator it is all T observations. The numerator is the sum of squared deviations in successive residuals $(e_t - e_{t-1})$, which necessitates the loss of one observation at the beginning of the series. The denominator is simply the sum of squared residuals (ErrorSS).

Table 15.4 shows actual sales y_t, predicted sales $\hat{y}_t$, and the calculated residual series e_t for the sales regression in Table 15.3. The lagged residual series e_{t-1} is also shown. From these values, the Durbin-Watson d statistic is calculated as

$$d = \frac{\sum_{t=2}^{T}\left(e_t - e_{t-1}\right)^2}{\sum_{t=1}^{T} e_t^2} = \frac{10759.406}{67554.906} = 0.1593$$

Most computer statistics packages routinely report the Durbin-Watson d along with other summary statistics whenever a regression is run. It is reported regardless of the type of data used—computers cannot tell time series from cross-sectional data. With few exceptions, however, the Durbin-Watson statistic only pertains to time series data.[3]

The exact probability distribution of d is difficult to derive. As discussed below, it depends in a complicated way on several assumptions about both the x's and the e's.

Table 15.4 Residuals and Lagged Residuals from the Regression of Sales on the Three Quarter Dummies and a Time Trend Variable

	year. quarter	sales	predicted sales	residual	lagged residual	residual²	(residual – lagresidual)²
		y_t	$\hat{y}_t$	$e_t = y_t - \hat{y}_t$	e_{t-1}	e_t^2	$(e_t - e_{t-1})^2$
1	1973.4	236.6	260.77	–24.167	n.a.	584.044	n.a.
2	1974.1	242.0	249.47	–7.473	–24.167	55.838	278.706
3	1974.2	269.4	278.29	–8.890	–7.473	79.036	2.010
4	1974.3	272.1	271.51	0.592	–8.890	0.351	89.915
5	1974.4	277.0	288.35	–11.345	0.592	128.709	142.496
6	1975.1	247.1	277.05	–29.951	–11.345	897.062	346.183
7	1975.2	265.8	305.87	–40.069	–29.951	1605.525	102.374
8	1975.3	271.0	299.09	–28.086	–40.069	788.823	143.592
9	1975.4	281.3	315.92	–34.624	–28.086	1198.821	42.745
10	1976.1	284.2	304.63	–20.429	–34.624	417.344	201.498
•	•	•	•	•	•	•	•
•	•	•	•	•	•	•	•
•	•	•	•	•	•	•	•
62	1989.1	666.0	663.15	2.851	5.757	8.128	8.444
63	1989.2	707.5	691.97	15.533	2.851	241.274	160.833
64	1989.3	681.3	685.18	–3.884	15.533	15.088	377.032
65	1989.4	690.3	702.02	–11.722	–3.884	137.405	61.430
66	1990.1	667.4	690.73	–23.327	–11.722	544.149	134.676
67	1990.2	704.1	719.55	–15.445	–23.327	238.548	62.126
68	1990.3	702.0	712.76	–10.763	–15.445	115.842	21.921
69	n.a.	n.a.	n.a.	n.a.	–10.763	n.a.	n.a.
						67554.906	10759.406
							$d = 0.1593$

The numerical limits of d, however, are relatively easy to derive because the relationship between d and ρ is[4]

$$d \approx 2(1 - \hat{\rho}) \tag{15.5}$$

If there is no autocorrelation in the residuals, then $d = 2$ and ρ is zero. On the other hand, if there is a positive one-for-one relationship between subsequent time period residuals, in which $\rho = +1$, then $d = 0$. A negative one-for-one relationship between time periods, in which $\rho = -1$, implies $d = 4$. In the case of the sales regression in Table 15.3, where the d statistic was calculated to be 0.1593, ρ is estimated to be 0.9203.

Unfortunately, there is no unique critical value for d that leads to the acceptance or rejection of the null hypothesis of no first-order autoregression in the disturbances (i.e., $\rho = 0$). There is only an upper bound d_U and a lower bound d_L for which more extreme d values define acceptance and rejection regions. These upper and lower bounds are determined by the number of observations T and the number of explanatory variables. These critical values for the d statistic are provided in Appendix Table A.7, for 6 to 200 observations on 1 to 20 explanatory variables, at the 0.05 and 0.01 Type I error levels.

The number of observations are given down the left-hand column of the Durbin-Watson Appendix Table A.7. There is no row for 68 observations; the closest rows are marked for 65 and 70. Across the top of the Durbin-Watson table, each column header identifies the number of explanatory variables. In the case of our sales regression, there were four explanatory variables (QT2, QT3, QT4, and TIME), which are designated by $k' = 4$.[5] For a 0.01 Type I error level, the intersection of the row marked 70 and column marked $k' = 4$ gives $d_L = 1.343$ and $d_U = 1.578$. Similarly, the intersection of the row marked 65 and the column marked $k' = 4$ gives $d_L = 1.315$ and $d_U = 1.568$. Thus, the calculated d = 0.1593, with 68 observations, in Table 15.4, is well below the critical lower bound d_L. The null hypothesis that there is no first-order autoregressive process in the disturbances must be rejected in favor of the alternative hypothesis that the disturbances are positive first-order autoregressive. Had the calculated d statistic been above the upper bound d_U, then the null hypothesis of no first-order autoregressive process would have been accepted. Calculated values of d between the lower and upper bound give indeterminate results. Table 15.5 summarizes these rules.

Table 15.5 Durbin-Watson *d* Statistic Decision Rules

NULL HYPOTHESIS: disturbances have	IF	THEN
no positive autocorrelation; i.e., H_O: $\rho = 0$ vs H_A: $\rho > 0$	$0 < d < d_L$	reject null
	$d_L < d < d_U$	no decision
	$d_U < d < 2$	accept null
no negative autocorrelation; i.e., H_O: $\rho = 0$ vs H_A: $\rho < 0$	$2 < d < 4 - d_U$	accept null
	$4 - d_U < d < 4 - d_L$	no decision
	$4 - d_L < d < 4$	reject null

where $d = \sum_{t=2}^{T}(e_{t-1} - e_t)^2 / \sum_{t=1}^{T} e_t^2$

and d_L and d_U are found in Appendix Table A.7.

Although widely used, the Durbin-Watson statistic is based on a set of assumptions that are seldom all true. These assumptions are

1. The regression equation must include an intercept term.
2. The x's must be nonstochastic. That is, only y and ε are random variables; the values of the x's are assumed to be the same under resampling.
3. There are no missing observations. That is, the y and x series are continuous.
4. There are no lagged dependent variables (such as y_{t-1}) included on the right-hand side of the regression as explanatory variables. For example, the regression equation cannot be as follows.

   $$y_t = \beta_0 + \beta_1 y_{t-1} + \beta_2 x_{t2} \ldots + \beta_k x_{tk} + \varepsilon_t$$

   Because y_{t-1} is included as a regressor, the above model is known as a dynamic model. It will be discussed in more detail later in this chapter.
5. The residuals are either unrelated or they are generated by the first-order autoregressive process, which is what we will be testing.

In addition, it is not uncommon to have the calculated d fall in the region for which the d statistic provides inconclusive or indecisive information. When the conditions for the Durbin-Watson test are not fulfilled, an alternative test that might be attempted is the nonparametric runs test of independence.[6]

15.7 Runs Test

The Durbin-Watson test is quite powerful when compared to other tests for detecting first-order autoregressive processes. Nevertheless, consideration must be given to an alternative nonparametric test for situations when the assumptions for the Durbin-Watson test are highly suspect.

As already stated, cyclical variations may show up in the residuals as a serpentine pattern in which the residuals tend to group alternatively into negative and positive clusters. This pattern is evident in Figure 15.4 and the residual column e_t in Table 15.4. If the residuals were purely random, no pattern would be apparent; the negative and positive values would not be clustered together in a continuous string or run of like signs. This is the intuition behind the nonparametric runs test.

The test begins with a count of the number of uninterrupted strings or continuous runs of pluses and minuses. Next, the number of pluses and minuses in a run is obtained. These counts are provided in Table 15.6 for the residuals in Table 15.4. The question is: Are the nine runs too few or too many as compared with the number of runs expected in a purely random sequence of 68 observations? Too many runs indicate negative serial correlation. (For instance, in the first-order autoregressive process

$e_t = \rho e_{t-1}$, with $\rho < 0$, e_t will be the opposite sign of e_{t-1}.) Too few runs indicate positive serial correlation. The expected number of runs is given by

$$E(\text{runs}) = \frac{2\,\#(+)\,\#(-)}{\#(+) + \#(-)} + 1 \tag{15.6}$$

where #(–) is the number of negative residuals

#(+) is the number of positive residuals

In our sales regression example, with 68 observations, the expected number of runs is

$$E(\text{runs}) = \frac{2\#(+)\#(-)}{\#(+) + \#(-)} + 1 = \frac{2(31)(37)}{(31)+(37)} + 1 = 34.7$$

but the actual number of runs is only nine. Apparently, there are too few runs in the residuals and positive serial correlation is suggested.

To test for serial correlation, the distribution of runs is needed along with the standard error. Under the null hypothesis that residuals are independent, the number of runs is approximately normally distributed, given a sufficiently large number of plus and minus residuals [#(+) > 10 and #(–) > 10]. The standard error is

$$\hat{\sigma}(\text{runs}) = \sqrt{\frac{2\#(+)\#(-)\left[2\#(+)\#(-) - \#(+) - \#(-)\right]}{\left[\#(+) + \#(-)\right]^2\left[\#(+) + \#(-) - 1\right]}}$$

which in the sales example yields

$$\hat{\sigma}(\text{runs}) = \sqrt{\frac{2(31)(37)\left[2(31)(37) - (31) - (37)\right]}{\left[(31) + (37)\right]^2\left[(31) + (37) - 1\right]}} = \sqrt{16.48} = 4.06$$

Table 15.6 Runs in the Residuals Found in Table 15.4

Number of Runs		Number of Pluses	Number of Minuses
1	– – –		3
2	+	1	
3	– – – – – – – – – – – – – – – –		16
4	+ + + + + + + + + + + + + + + +	16	
5	– –		2
6	+ + + + + + + + + + +	11	
7	– – – – – – – – – – –		11
8	+ + +	3	
9	– – – – –		5
		31	37

The calculated z is

$$z = \frac{\text{runs} - E(\text{runs})}{\sigma(\text{runs})} \approx \frac{9 - 34.7}{4.06} = -6.34$$

In a two-tailed test, the critical z is ± 1.645 for a 0.05 Type I error. Thus, the null hypothesis of independent population errors is rejected, and the alternative hypothesis of serially correlated residuals is accepted. The decision rule for any runs test of serially correlated residuals is

Decision Rule: Reject the null hypothesis of independent errors, and accept the alternative of serially correlated errors, if the absolute value of the calculated z exceeds the absolute value of the critical z, for the Type I error probability. If the absolute value of the calculated z does not exceed the absolute value of the critical z, then the null hypothesis of independence cannot be rejected.

Alternatively, reject the null hypothesis if the p-value is smaller than the Type I error probability, and do not reject the null hypothesis if the p-value is larger than the Type I error probability.

EXERCISES

15.11 When would it be inappropriate to use the Durbin-Watson test in residual analysis, but appropriate to use the runs test?

15.12 Below are quarterly profits (Profit, in thousands of dollars) for a national catalog sales firm between the years 1971 and 1991. Also shown are the firm's advertising (Adv, in thousands of dollars) budgets per quarter.

	Quarter							
	I		II		III		IV	
Year	Profit	Adv	Profit	Adv	Profit	Adv	Profit	Adv
1971	3101	100	3282	118	3802	239	5801	489
1972	3120	115	3348	144	4007	248	5980	513
1973	3241	125	3393	155	4673	278	5524	552
1974	3582	148	4110	185	5090	311	6195	572
1975	4070	177	4653	197	5725	332	8401	632
1976	4791	199	4984	210	6230	359	8880	651
1977	4913	222	5641	222	6706	401	9372	711
1978	5522	239	6029	245	7720	422	11704	788
1979	5850	257	6384	260	8105	455	11950	844
1980	6012	279	6510	284	7690	453	12136	893
1981	7841	282	8383	289	9403	481	13100	944
1982	8253	303	9087	300	10390	488	14122	972

(continued)

Quarter (*continued*)

Year	I Profit	I Adv	II Profit	II Adv	III Profit	III Adv	IV Profit	IV Adv
1983	8754	308	9502	313	10844	479	14960	999
1984	9550	316	10650	320	11630	493	15011	1001
1985	9285	333	10731	341	12487	525	16020	1112
1986	9011	478	10510	485	11691	652	15833	1194
1987	8843	483	10382	488	12404	683	16101	1245
1988	9253	504	11088	502	13391	689	17121	1271
1989	9755	507	11503	514	13845	678	17969	1298
1990	10551	515	12651	522	14631	695	18010	1009
1991	9386	517	9765	523	13899	703	18000	1000

a. Fit a regression of profits on advertising expenses, a linear time trend, and seasonal dummies.

b. Are the trend and individual seasonal dummies significant?

c. Is residual autocorrelation a problem?

15.13 An article in *The Wall Street Journal* (February 3, 1992), titled "Depressed Job Market Prompts Paradox: Upsurge of Interest in M.B.A. Programs," gave only the two diagrams below on the total number taking the Graduate Management Aptitude Test and applications at 10 business schools.

The MBA Boom

Total Test Takers
Registration for the Graduate Management Aptitude Test, the standardized exam for most business school applicants, in thousands

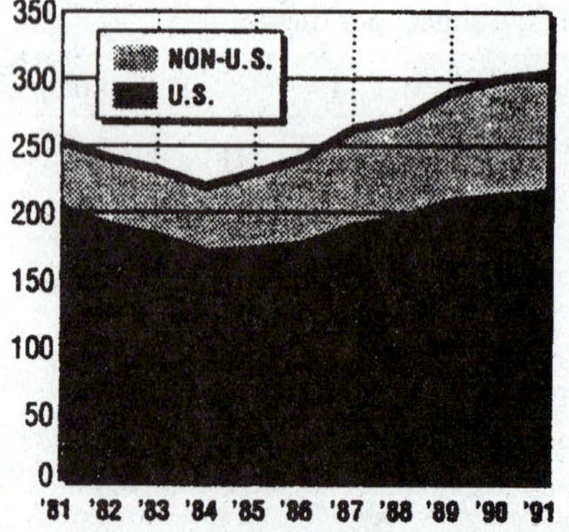

Applications at 10 Schools
Applicants seeking fall entrance to MBA programs*, in thousands

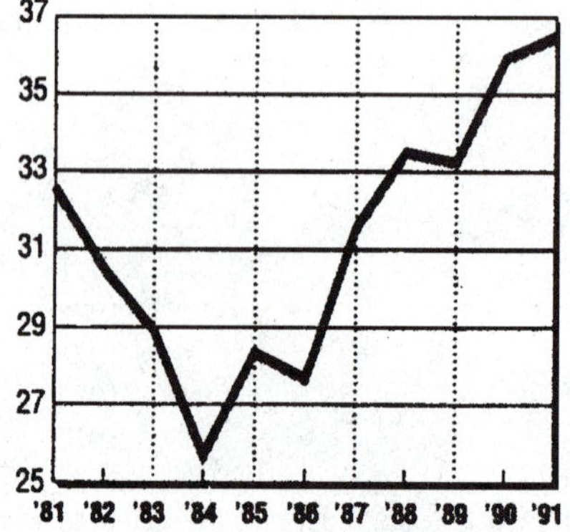

*Schools include: Dartmouth (Tuck), Harvard, Stanford, U. of Pennsylvania (Wharton), MIT (Sloan), Columbia, Northwestern (Kellogg), U. of Virginia (Darden), U. of Chicago, and UCLA.

Sources: Bill Brazenly, Graduate Management Admission Council, Los Angeles, CA. Sam T. Lundquist, Amos Tuck School Admissions Director, Dartmouth College, Hanover, NH.

From Bill Brazenly and Sam Lundquist, the author obtained the data that was used by the *WSJ* to construct the diagrams. Unemployment rates for experienced wage and salary workers, which were not cited in the *WSJ* article, were obtained from the *Economic Report of the President.*

Year	Applications for fall of identified school year, in thousands	Test takers in the previous school year, in thousands	Unemployment rate for experienced workers in prior calendar year, in percent
1981	32.578	252.531	6.9
1982	30.498	240.200	7.3
1983	28.916	231.274	9.3
1984	25.701	219.226	9.2
1985	28.365	229.379	7.1
1986	27.686	240.434	6.8
1987	31.569	262.567	6.6
1988	33.522	269.072	5.8
1989	33.357	291.378	5.2
1990	35.944	299.781	5.0
1991	36.531	305.072	5.3

a. Can the number of test takers be explained by the unemployment rate?
b. As reflected in the unemployment rate, what contribution does the job market make to explaining applications to the 10 schools cited in the *WSJ* article, after controlling for test takers?
c. Is autocorrelation a problem in your answer to part b? Explain.

15.14 Exercise 15.4 required the investigation of whether there is a difference in money growth. Using the runs test, is autocorrelation a problem in the regression that you fit?

15.15 Use the data on sales (x) and price (y) in Exercise 15.7 to generate a plot of residuals. Do they appear to be related over time? Test for autocorrelation via the runs test.

15.8 Dynamic Model

As introduced in the discussion of Assumption 4 for the Durbin-Watson statistic, in many time series the value of the dependent variable in the current period may be thought of as depending on the value of this variable in the prior period. For example, starting salaries for college graduates this year may be highly correlated with the starting salaries last year. Sales for this quarter may be accurately predicted from the value of sales in the previous quarter. Interest rates today might be related to interest rates yesterday. And even my rate of productivity this hour may be a function of my productivity in the past hour.

Building a Dynamic Model

Dynamic Model An autoregressive process of the dependent variable on itself plus other explanatory variables.

If y_t represents the value of the dependent variable at time t, then its lagged value is y_{t-1}. A **dynamic model** of y_t that is based on its immediate past value y_{t-1}, as well as other explanatory x variables, is given by

$$y_t = \beta_0 + \beta_1 y_{t-1} + \beta_2 x_{t2} \ldots + \beta_k x_{tk} + \varepsilon_t \qquad (15.8)$$

This dynamic model shows y to be a first-order autoregressive process of itself. A second-order autoregressive process would have y explained by y_{t-1} and y_{t-2}, as well as the values of the x's at time t. If the y autoregressive process was known perfectly, the resulting dynamic model would give y predictions that mirror its seasonal and cyclical variations.

Lagged Dependent Variable
A created variable that has as its current value a previous observation.

Lagging the dependent variable and including it along with the other explanatory variables provides a means to capture the growth or decay in a dependent variable. As long as the absolute value of the coefficient of the **lagged dependent variable** is less than one, y will approach an equilibrium (or target) value, for given values of the other explanatory x variables. Such a converging y process is shown in Figure 15.6, panels a and b. If the absolute value of the coefficient of the lagged dependent variables is

FIGURE 15.6 Dynamic Time Series Processes

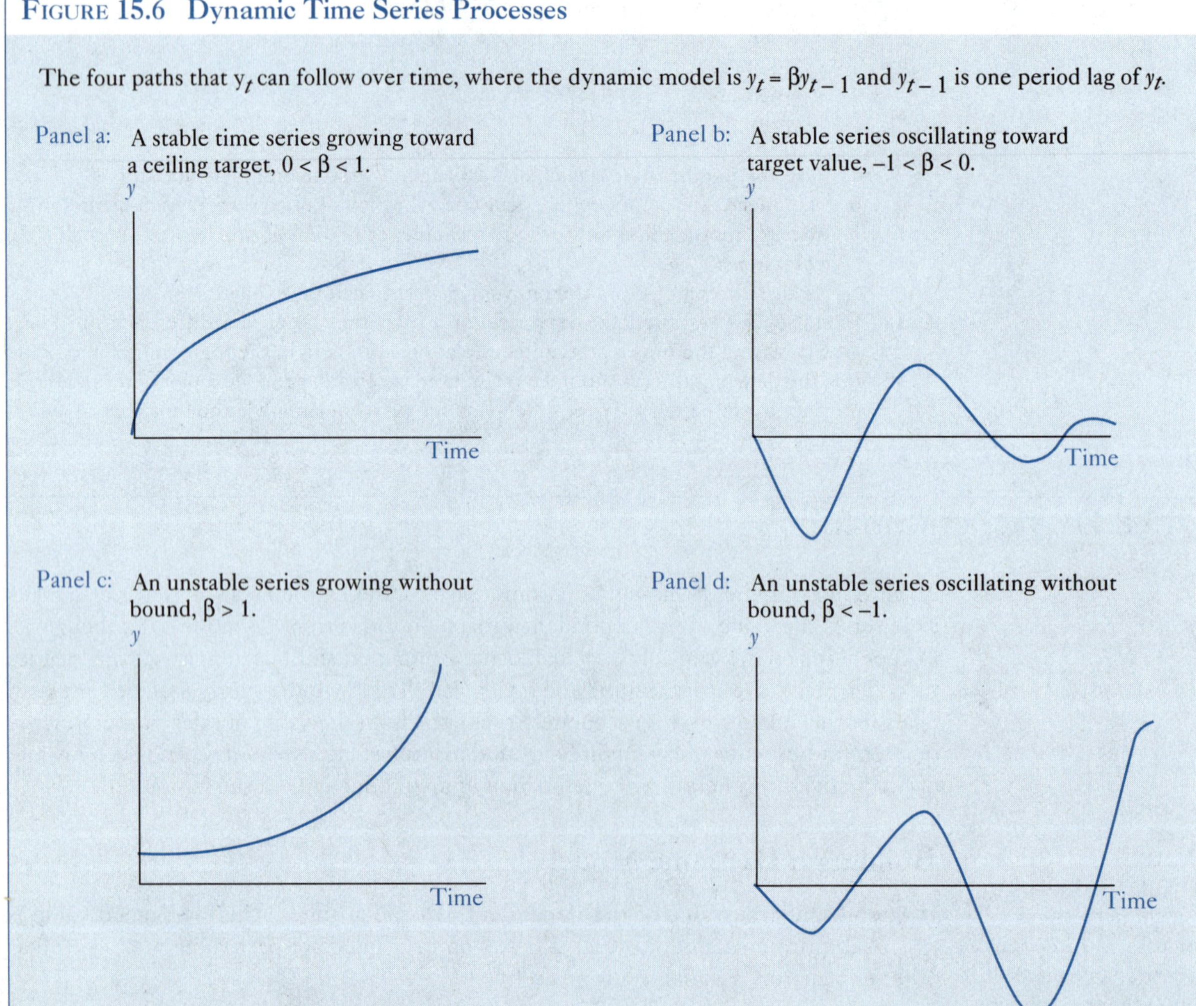

Table 15.7
Lag of Sales

	year. quarter	sales y_t sales$_t$	lagged sales y_{t-1} sales$_{t-1}$
1	1973.4	236.6	n.a.
2	1974.1	242.0	236.6
3	1974.2	269.4	242.0
4	1974.3	272.1	269.4
5	1974.4	277.0	272.1
6	1975.1	247.1	277.0
7	1975.2	265.8	247.1
8	1975.3	271.0	265.8
9	1975.4	281.3	271.0
•	•	•	•
•	•	•	•
•	•	•	•
61	1988.4	680.2	646.3
62	1989.1	666.0	680.2
63	1989.2	707.5	666.0
64	1989.3	681.3	707.5
65	1989.4	690.3	681.3
66	1990.1	667.4	690.3
67	1990.2	704.1	667.4
68	1990.3	702.0	704.1
69	n.a.	n.a.	702.0

greater than unity, however, growth will be unbounded (or explosive), as shown in Figure 15.6, panels c and d.

To illustrate a dynamic model, again consider the quarterly sales data in Table 15.1. There are 68 observations on sales. The first value, 236.6 (billion dollars), occurs in the fourth quarter of 1973 and the last value, 702.0 (billion dollars), occurs in the third quarter of 1990. Lagging this time series one quarter creates the new variable called "sales$_{t-1}$," shown in column four of Table 15.7. The first value of sales$_{t-1}$ is 236.6, but this value is now associated with the first quarter of 1974, while the last useable value is 704.1, which occurs in the third quarter of 1990.

Inclusion of a lagged variable necessitates an adjustment to the sample size. There is no value on the original sales series that matches the last observation (702.0, for the fourth quarter of 1990) on the lagged sales series. Similarly, there is no value on the lagged sales series that matches the first observation (236.6, for the third quarter of 1973) on the original sales series. Thus, there can only be $T-1$ matched pair observations on the y_t and y_{t-1} series. In this sales example, there are only 67 matched pairs on the dependent variable "sales" and its lagged series "sales$_{t-1}$." (Lagging a variable two periods results in only $T-2$ matched pairs because two observations are lost.)

Forecasting with a Dynamic Model

By regressing sales on sales$_{t-1}$, we can estimate the relationship between sales in consecutive periods and use information on sales in one period to forecast sales in the next. To control for seasonality, the regression again must include three quarterly dummy variables. A time trend variable also may be included to capture any linear growth in sales that is not captured in its autoregressive process. Thus, the sample regression specification is

$$\widehat{\text{sales}}_t = b_1 + b_2(\text{QT2}) + b_3(\text{QT3}) + b_4(\text{QT4}) + b_5(\text{TIME}) + b_6(\text{sales}_{t-1})$$

where again QTi is the ith quarter dummy that is equal to one for the identified quarter and zero otherwise, and TIME is the trend variable that takes on the value 2, 3, 4 . . . for each of the 67 quarters. (The first observation on each variable is sales_t = 242.0, sales_{t-1} = 236.6, QT2 = 0, QT3 = 0, QT4 = 0, TIME = 2 and the last is sales_t = 702.0, sales_{t-1} = 704.1, QT2 = 0, QT3 = 1, QT4 = 0, TIME = 68.) The least squares regression estimates for these data are given in Table 15.8.

Using the estimates in Table 15.8, future sales can be forecast. For instance, in the fourth quarter of 1990, sales are predicted to be $719.05355 billion.

$$\widehat{\text{sales}}_{69} = 10.8104 + 38.6459(0) + 4.91471(0) + 28.2447(1)$$
$$+ .526935(69) + .916866(702.0) = 719.05355$$

In making predictions about future values, remember that these forecasts are outside the sample time period. The further into the future one predicts, the less confidence one can have in the forecast. Predictions with a sample regression equation have meaning only within the range of sample values and the conditions under which the sample values were drawn. Regressions estimated under one set of economic conditions may have little meaning if those conditions change. Furthermore, before putting a lot of faith in any prediction, consideration should be given to the regression model specification, the magnitude and significance of coefficient estimates, and the fit of the regression line itself.

The coefficient of determination in Table 15.8 suggests that this dynamic specification fits well, R^2 = 0.99185, but high R^2's are not unusual for these models. Similarly, a high F statistic, (1483.8800) is to be expected in dynamic time series specification since sales in one quarter are highly related to sales in the next, with the lag coefficient

Table 15.8 Ordinary Least Squares Estimation of Dynamic Model of Sales

Dependent Variable	salest	Number of Observations	67
Mean of Dep. Variable	489.3537	Std. Dev. of Dep. Var.	137.839721
Durbin Watson statistic	1.5339	Estimated Autocorrelation	.23303
Std. Error of Regr.	12.9474	Sum of Squared Residuals	10225.8
Total variation	.12540E+07	Regression Variation	.12438E+07
R-squared	.99185	Adjusted R-squared	.99118
F(5, 61)	1483.8800	Prob. Value for F	.00000

Variable	Coefficient	Std. Error	T-ratio	Prob | t | > x	Mean of X	Std. Dev. of X
Constant	10.8104	12.99	.832	.40841		
QT2	38.6459	4.533	8.525	.00000	.25373	.43843
QT3	4.91471	4.448	1.105	.27351	.25373	.43843
QT4	28.2447	4.534	6.229	.00000	.23881	.42957
TIME	.526935	.3541	1.488	.14187	35.00000	19.48504
salest-1	.916866	.4986E-01	18.388	.00000	482.40746	138.68535

0.916866 being highly significant. A one billion dollar increase in sales in one quarter is predicted to raise sales in the next quarter by 0.916866 billion dollars, ignoring seasonal factors, which is close to a one-for-one relationship.

There also are strong seasonal factors, with the second and fourth quarters being significantly different from the first quarter. Although the third quarter dummy does not appear to be significantly different from the first quarter, its effect within the sample is not small. (There is a $4.91471 billion difference in predicted sales between the third and first quarter, holding other things constant.) Finally, the trend variable, although relatively small (0.526935), does show some significance with a test statistic of 1.488, implying significance at the 0.07094 Type I error level in a one-tailed test. Remember from our discussion of hypothesis testing in earlier chapters that a researcher is always wise to consider both practical significance and statistical significance before drawing a conclusion regarding importance.

15.9 Testing for Autocorrelated Errors in Dynamic Models

The above discussion of statistical significance assumes that the residuals behave randomly. The Durbin-Watson statistic (1.5339), however, is in the indeterminate region ($d_L = 1.438$, $d_U = 1.767$), for a 0.05 Type I error level. Furthermore, use of the Durbin-Watson statistic in a dynamic model is suspect because Assumption 4 is violated.

J. Durbin (1970) does provide a test for serially correlated residuals in dynamic specifications: the Durbin h statistic.

$$h = (1 - .5d)\sqrt{\frac{T}{1 - T(\text{variance of the lagged variable coefficient})}} \tag{15.9}$$

For large sample sizes ($T > 30$), Durbin has shown that under the null hypothesis of $\rho = 0$, h is distributed as an approximately standard normal random variable. That is, h is treated as if it were z in Appendix Table A.3. In particular, the test is

$$H_0\!: \rho = 0 \text{ versus } H_A\!: \rho > 0, \text{ where } \varepsilon_t = \rho\varepsilon_t + u_t$$

where the null hypothesis is rejected (not rejected) if the calculated h exceeds (does not exceed) the critical z value for the Type I error level selected.

Decision Rule: Reject the null hypothesis of independent residuals, and accept the alternative of positive autocorrelated residuals, if the calculated h exceeds the critical value of the z, for the Type I error selected. If the calculated value of h does not exceed the critical value of z, then the null hypothesis of independent residuals cannot be rejected.

Alternatively, reject the null hypothesis if the p-value is smaller than the Type I error probability, and do not reject the null hypothesis if the p-value is larger than the Type I error probability.

For the dynamic sales model estimated in Table 15.8, the calculated h statistic is

$$h = \left[1 - .5\left(1.5339\right)\right]\sqrt{\frac{67}{1 - 67\left(.04986\right)^2}} = 2.09$$

For a one-tailed test and a probability of a Type I error of 0.05, the critical z is 1.645. Thus, the null hypothesis is rejected in favor of the alternative. The presence of a first-order autoregressive process in the residuals of this dynamic specification is supported. The residuals are not independently distributed, as required for the t and F tests discussed above. More work is needed.

15.10 Models Involving First Differences

When the estimate of ρ in a first-order autoregressive disturbance process is close to unity, a procedure that might remove the effect of the autocorrelation in the residuals is to first difference the data. That is, if

$$y_t = \beta_1 + \beta_2 x_{t2} \ldots + \beta_k x_{tk} + \varepsilon_t$$

where $\varepsilon_t = \varepsilon_{t-1} + u_t$ and u_t is a purely random perturbation, then differencing the right- and left-hand variables gives

$$\begin{aligned}(y_t - y_{t-1}) &= \beta_2(x_{t2} - x_{t-12}) + \ldots + \beta_k(x_{tk} - x_{t-1k}) + \varepsilon_t - \varepsilon_{t-1} \\ (y_t - y_{t-1}) &= \beta_2(x_{t2} - x_{t-12}) + \ldots + \beta_k(x_{tk} - x_{t-1k}) + u_t\end{aligned} \tag{15.10}$$

Letting delta, Δ, represent the first difference in the following variable, this difference equation can be written more compactly as

$$\Delta y_t = \beta_2 \Delta x_{t2} \ldots + \beta_k \Delta x_{tk} + u_t \tag{15.11}$$

Although the β's in (15.11) are the same as in (15.10), first differencing the data removes the first-order autoregressive process from the residuals. A least squares regression of the differenced data will not suffer from autocorrelated residuals if the initial regression, specified on the levels, had autocorrelated residuals with a ρ close to unity.

First differencing the data cannot change the identity of the time period; thus, seasonal dummy variables included as explanatory variables should not be altered by first differencing—that is, seasonal dummies do not get first differenced. If β_2 through β_4 represented differences in quarterly effects in the initial regression, they must continue to represent these effects in the first differenced equation. With these three quarters identified, the first differenced regression is written

$$\Delta y_t = \beta_2(\text{QT2}) + \beta_3(\text{QT3}) + \beta_4(\text{QT4}) + \beta_5 \Delta x_{t5} \ldots + \beta_k \Delta x_{tk} + u_t$$

Notice that as a result of first differencing, the y intercept is eliminated, because $x_{t1} = 1$ for all t. If one of the x variables is a trend variable, however, then first differencing will reintroduce a constant term. For example, if x_{t5} was the trend variable in the initial regression, then the first differenced regression is

$$\Delta y_t = \beta_2(QT2) + \beta_3(QT3) + \beta_4(QT4) + \beta_5 + \beta_6\Delta x_{t6} \ldots + \beta_k\Delta x_{tk} + u_t$$

If the initial regression did not have a trend variable but did have the three dummy variables and a constant term to capture quarterly effects, then a fourth dummy would have to be introduced in the first differenced regression because the constant term is eliminated by first differencing.

We now have enough information to remove from the sales data the trend, seasonal variability, and the cyclical effects. To do this, the regression to be fit is

$$\Delta(\text{predicted sales})_t = b_2(QT2) + b_3(QT3) + b_4(QT4) + b_5$$

If the assumptions of the population model are correct, the residuals from this equation will be a series on sales that has no trend and has no seasonal and no cyclical variability.

$$\text{Residual}_t = \Delta(\text{sales})_t - \Delta(\text{predicted sales})_t$$

Table 15.9 provides the time series on first difference of sales. Again there are only 67 useable observations because one observation is lost in lagging the sales series. The

Table 15.9 First Difference in Sales

	year.quarter	sales y_t	lagged sales y_{t-1}	difference in sales Δy_t
1	1973.4	236.6	n.a.	n.a.
2	1974.1	242.0	236.6	5.4
3	1974.2	269.4	242.0	27.4
4	1974.3	272.1	269.4	2.7
5	1974.4	277.0	272.1	4.9
6	1975.1	247.1	277.0	−29.9
7	1975.2	265.8	247.1	18.7
8	1975.3	271.0	265.8	5.2
9	1975.4	281.3	271.0	10.3
10	1976.1	284.2	281.3	2.9
11	1976.2	307.6	284.2	23.4
12	1976.3	301.6	307.6	−6.0
•	•	•	•	•
•	•	•	•	•
•	•	•	•	•
57	1987.4	627.7	597.7	30.0
58	1988.1	614.2	627.7	−13.5
59	1988.2	655.5	614.2	41.3
60	1988.3	646.3	655.5	−9.2
61	1988.4	680.2	646.3	33.9
62	1989.1	666.0	680.2	−14.2
63	1989.2	707.5	666.0	41.5
64	1989.3	681.3	707.5	−26.2
65	1989.4	690.3	681.3	9.0
66	1990.1	667.4	690.3	−22.9
67	1990.2	704.1	667.4	36.7
68	1990.3	702.0	704.1	−2.1
69	n.a.	n.a.	702.0	n.a.

results from running a least squares regression on the three quarterly dummies and a constant term are shown in Table 15.10. The constant term –11.2941 indicates that in the first quarter, the change in sales (from the fourth quarter of the last year to the first quarter of this year) is predicted to fall by $11.2941 billion. In the second quarter, the change in sales (from the first quarter to the second quarter) is predicted to rise by $28.8177 billion (= 40.1118 – 11.2941). In the third quarter, the change in sales (from the second quarter to the third) is predicted to fall by $6.7823 billion (= 4.51177 – 11.2941), and in the fourth quarter, the predicted change in sales is $17.675 billion (= 28.9691 – 11.2941). In terms of billions of dollars all quarters are predicted to have large effects on sales. To check for their statistical significance, however, the residuals must first be considered for their compliance with the least squares assumptions.

The Durbin-Watson statistic 1.5870 is in the indeterminate zone, although the plot of residuals in Figure 15.7 shows little pattern. The runs test in Table 15.11 suggests that the residuals may in fact be random. The expected number of runs is 34.5, with a standard deviation of 4.06. With 27 runs of + and – signs, the calculated z is –1.85, which is well inside the critical value of ±1.96 for a 5 percent Type I error. Thus, the null hypothesis that the residuals are randomly distributed cannot be rejected. The residuals in Figure 15.7 represent a series of sales that is detrended, deseasonalized, and free of cycles. They are appropriate for testing hypotheses about seasonal effects.

Again, the third quarter does not appear to be significantly different from the first, even though its contribution in billions of dollars is large. The calculated t value of 1.007 is small and the p-value of 0.31775 is large relative to the 5 percent Type I error levels at which most tests are performed. Nonetheless, at a probability of a Type I error of 0.32, the third quarter would be accepted as being significant; that is, it does have some statistical significance. More important, the predicted drop in sales from the second to the third quarter of $6.7823 billion is not trivial and should not be ignored. Although some researchers advocate deleting variables that are not statistically significant at the 5 percent level, this author does not recommend deleting the third quarter dummy from this regression. Each of the four quarter seasonal effects in the sales data appears to be practically important.

TABLE 15.10 Regression of First Differenced Sales

Dependent Variable	DIFSALES	Number of Observations	67
Mean of Dep. Variable	6.9463	Std. Dev. of Dep. Var.	21.171060
Durbin Watson statistic	1.5870	Estimated Autocorrelation	.20651
Std. Error of Regr.	13.0616	Sum of Squared Residuals	10748.2
Total Variation	29582.	Regression Variation	18834.
R-squared	.63667	Adjusted R-squared	.61936
F (3, 63)	36.7981	Prob. Value for F	.00000

Variable	Coefficient	Std. Error	T-ratio	Prob|t|>x	Mean of X	Std.Dev. of X
Constant	–11.2941	3.168	–3.565	.00070		
QT2	40.1118	4.480	8.953	.00000	.25373	.43843
QT3	4.51177	4.480	1.007	.31775	.25373	.43843
QT4	28.9691	4.550	6.367	.00000	.23881	.42957

Figure 15.7 Plot of Residuals from the Regresssion in Table 15.10

Axis: −38.6 0 38.6

Time Period	Residual
2.00	16.7
3.00	−1.42
4.00	9.48
5.00	−12.8
6.00	−18.6
7.00	−10.1
8.00	12.0
9.00	−7.38
10.0	14.2
11.0	−5.42
12.0	.782
13.0	−9.48
14.0	13.0
15.0	−1.72
16.0	−.118
17.0	−3.17
18.0	5.39
19.0	8.38
20.0	6.18
21.0	7.22
22.0	16.1
23.0	.982
24.0	7.88
25.0	6.03
26.0	15.8
27.0	−28.2
28.0	4.68
29.0	34.7
30.0	15.5
31.0	−.177E−01
32.0	−2.92
33.0	−23.2
34.0	−20.2
35.0	−9.82
36.0	−7.12
37.0	−19.1
38.0	−4.51
39.0	7.48
40.0	14.4
41.0	9.22
42.0	15.8
43.0	2.98
44.0	−14.0
45.0	−.775
46.0	−17.4
47.0	−.177E−01
48.0	−9.32
49.0	−1.77
50.0	−38.6
51.0	−6.62
52.0	−13.1
53.0	.525
54.0	3.59
55.0	10.5
56.0	8.38
57.0	12.3
58.0	−2.21
59.0	12.5
60.0	−2.42
61.0	16.2
62.0	−2.91
63.0	12.7
64.0	−19.4
65.0	−8.67
66.0	−11.6
67.0	7.88
68.0	4.68

TABLE 15.11 Runs Test for Residuals in Figure 15.7

Runs	plus	minus
1	1	
2		1
3	1	
4		3
5	1	
6		1
7	1	
8		1
9	1	
10		1
11	1	
12		3
13	9	
14		1
15	3	
16		8
17	5	
18		9
19	5	
20		1
21	1	
22		1
23	1	
24		1
25	1	
26		3
27	2	
	33	34

H_O: Residuals are random
H_A: Residuals are not random

Expected number of runs is

$$E(\text{runs}) = \frac{2\#(+)\,\#(-)}{\#(+) + \#(-)} + 1$$

$$= \frac{2(33)(34)}{(33)+(34)} + 1$$

$$= 34.49$$

The standard deviation is

$$\sigma(\text{runs}) = \sqrt{\frac{2\#(+)\#(-)[2\#(+)\#(-) - \#(+) - \#(-)]}{[\#(+) + \#(-)]^2[\#(+) + \#(-) - 1]}}$$

$$\sigma(\text{runs}) = \sqrt{\frac{2(33)(34)[2(33)(34) - (33) - (34)]}{[(33) + (34)]^2[(33) + (34) - 1]}} = 4.06$$

The calculated z is

$$z = \frac{\text{runs} - E(\text{runs})}{\sigma(\text{runs})} = \frac{27 - 34.49}{4.06} = -1.85$$

15.11 Causal Model Building for Forecasting

Every manager must make forecasts: explicitly, at least once a year when proposing next year's budget for his or her department; and implicitly, every time he or she enters into a contract that will be fulfilled at a future date. These forecasts may be based on an implicit model in which a condition that exists today is simply assumed to hold tomorrow or as complex as the explicit multiple equation models used to simulate the operation of the entire economy over the next several years. Regardless of the sophistication employed, forecasts should be treated with caution, especially the more distant the date to which they apply. Remember the skeptic's question: "If forecasting is so exact, why are there so many forecasters?"

As already discussed, an effective and straightforward method of forecasting is to extrapolate values of a variable of interest from its past trend and seasonal components. For instance in the case of sales, we fit a regression to the time series components and then predict future values from this fitted regression. Once we move beyond this simple decomposition of the time series to include other variables (which theory suggests influence the variable of interest) we enter the world of causal model building. For example, to forecast the yield on U.S. Treasury bonds, we might start with Irving Fisher's theory of interest rate determination.

According to Fisher, the observed yield is determined by expectations about future rates, which might be influenced by past yields and expectations of future price inflation. Thus, in addition to a yearly trend, price inflation expectations should enter a dynamic model of yield:

$$(\text{YIELD})_t = \beta_1 + \beta_2(\text{YIELD})_{t-1} + \beta_3(t) + \beta_4(EP)_t + u_t$$

where $(\text{YIELD})_t$ is the yearly yield on 10-year maturity U.S. Treasury bonds in year t;

$(\text{YIELD})_{t-1}$ is the yield in year $t - 1$;

t is the year (1953, 1954, . . .). When used as a regressor, it is the trend term;

$(EP)_t$ is the expected rate of price inflation in year t; and

u_t is a population disturbance term.

But expected price inflation in year t is a theoretical concept that is unobservable. Even if the price expectation was observable in year t, this value would be unknown in the previous year $(t - 1)$. Herein lies a problem in building causal models for forecasting: Variables that are related concurrently cannot be used to predict each other because neither variable's value at time t is known in the previous period $t - 1$. A model must be built to predict the value of one of these variables that can then be used to predict the other. In the case of expected price inflation, for instance, the model might be that these expectations at time t are a function of actual price inflation in year $t - 1$ plus an error term; that is,

$$(EP)_t = \lambda p_{t-1} + v_t$$

where p_{t-1} is the price inflation, measured by change in the consumer price index, in year $t - 1$; and

v_t is a population disturbance term.

Although price expectations cannot be observed, they can be removed from the system by substituting for the deterministic and error term components in the dynamic model of yield; that is,

$$(\text{YIELD})_t = \beta_1 + \beta_2(\text{YIELD})_{t-1} + \beta_3(t) + \beta_4(\lambda p_{t-1}) + u_t + \beta_4 v_t$$

which upon simplification gives

$$(\text{YIELD})_t = \beta_1 + \beta_2(\text{YIELD})_{t-1} + \beta_3(t) + \beta_4^* p_{t-1} + \varepsilon_t$$

where β_4^* is $\beta_4\lambda$ and $\varepsilon_t = u_t + \beta_4 v_t$

Now future yields are determined by past values, trend, past rates of inflation, and an error that is independent of the regressors.

Table 15.12 provides the data necessary to estimate the dynamic model of yields; Table 15.13 provides the least squares parameter estimates and residual plot. This equation appears to fit well, with a correlation coefficient of 0.927, and thus, at first glance, may seem well suited for forecasting. The residual plot, however, shows signs of heteroscedasticity, with the variability in the residuals growing with time. This lack of stability in the errors may be caused by nonlinearity in the growth of yields that can

Table 15.12
Yields and Rates

	Year	Yield on 10 Year U.S. Bond	Rate of (CPI) Price Inflation
1	1953	2.85 %	0.80 %
2	1954	2.40	0.70
3	1955	2.82	−0.40
4	1956	3.18	1.50
5	1957	3.65	3.30
6	1958	3.32	2.80
7	1959	4.33	0.70
8	1960	4.12	1.70
9	1961	3.88	1.00
10	1962	3.95	1.00
11	1963	4.00	1.30
12	1964	4.19	1.30
13	1965	4.28	1.60
14	1966	4.92	2.90
15	1967	5.07	3.10
16	1968	5.65	4.20
17	1969	6.67	5.50
18	1970	7.35	5.70
19	1971	6.16	4.40
20	1972	6.21	3.20
21	1973	6.84	6.20
22	1974	7.56	11.00
23	1975	7.99	9.10
24	1976	7.61	5.80
25	1977	7.42	6.50
26	1978	8.41	7.60
27	1979	9.44	11.30
28	1980	11.46	13.50
29	1981	13.91	10.30
30	1982	13.00	6.20
31	1983	11.10	3.20
32	1984	12.44	4.30
33	1985	10.62	3.60
34	1986	7.68	1.90
35	1987	8.39	3.60
36	1988	8.85	4.10
37	1989	8.49	4.80
38	1990	8.55	5.40

Source: Various issues of the *Economic Report of the President.*

TABLE 15.13 Regression of Yield

Dependent Variable: YIELD,	Mean = 6.9165,	Std. Dev. = 3.033691,	N = 37
Durbin Watson statistic =	1.9334	Estimated Autocorrelation =	.03329
Std. Error of Regr. =	.8534	Sum of Squared Residuals =	24.0363
Sum of Squares Total =	331.32	Sum of Squares Regression =	307.28
R squared = .92745,	F (3, 33) = 140.6246,	Prob. Value for F =	.00000

Variable	Coefficient	Std. Error	T-ratio	Prob|t|>x	Mean of X	Std.Dev. of X
Constant	−92.8207	53.13	−1.747	.08992		
LAGYIELD	.617550	.1090	5.665	.00000	6.76243	3.09259
LAGINFLA	.219599	.5870E−01	3.741	.00070	4.30541	3.35087
YEAR	.479795E−01	.2723E−01	1.762	.08732	1972.00000	10.82436

Residual plot (scale −2.51 to 0 to 2.51)

YEAR	Residual
1954	−.467
1955	.205
1956	.499
1957	.282
1958	−.782
1959	.494
1960	.732E−01
1961	−.305
1962	.192E−01
1963	−.220E−01
1964	.233E−01
1965	−.520E−01
1966	.419
1967	−.160
1968	.235
1969	.608
1970	.324
1971	−1.38
1972	−.355
1973	.459
1974	.836E−01
1975	−1.03
1976	−1.31
1977	−.588
1978	.318
1979	.447
1980	.970
1981	1.64
1982	−.127
1983	−.612
1984	2.51
1985	−.425
1986	−2.14
1987	.715
1988	.316
1989	−.486
1990	−.406

TABLE 15.14 Regression of Logarithm of Yield

Dependent Variable: LnYIELD,	Mean = 1.8357,	Std. Dev. = .459097,	N = 37
Durbin Watson statistic	1.8895	Estimated Autocorrelation	.05523
Std. Error of Regr.	.1090	Sum of Squared Residuals	.392161
Sum of Squares Total	7.5877	Sum of Squares Regression	7.1955
R squared = .94832	F (3, 33) = 201.8331	Prob. Value for F =	.00000

Variable	Coefficient	Std. Error	T-ratio	Prob|t|>x	Mean of X	Std.Dev. of X
Constant	−20.2981	9.263	−2.191	.03560		
LAGLnY	.576059	.1299	4.434	.00010	1.80605	.47377
LAGINFL	.266825E−01	.8380E−02	3.184	.00316	4.30541	3.35087
YEAR	.106382E−01	.4798E−02	2.217	.03361	1972.00000	10.82436

−0.248 0 0.248

YEAR	Residual
1954	−.238
1955	.141E−01
1956	.601E−01
1957	.674E−01
1958	−.165
1959	.157
1960	.110E−03
1961	−.686E−01
1962	−.810E−02
1963	−.165E−01
1964	.406E−02
1965	−.121E−01
1966	.946E−01
1967	.842E−03
1968	.759E−01
1969	.139
1970	.956E−01
1971	−.153
1972	−.190E−01
1973	.943E−01
1974	.481E−01
1975	−.930E−01
1976	−.134
1977	−.533E−01
1978	.572E−01
1979	.606E−01
1980	.785E−01
1981	.913E−01
1982	−.133E−01
1983	−.335E−01
1984	.241
1985	−.230E−01
1986	−.248
1987	.619E−01
1988	.838E−02
1989	−.879E−01
1990	−.862E−01

be removed through a natural logarithmic transformation of the yields.[7] The transformed model is

$$\ln(\text{YIELD})_t = \beta_1 + \beta_2 \ln(\text{YIELD})_{t-1} + \beta_3(t) + \beta_4^* p_{t-1} + \varepsilon_t^*$$

The least squares estimates for this equation and the residual plot are in Table 15.14.

Unlike the residuals in Table 15.13, the variability in the residuals of the regression in Table 15.14 appears to be stable over time; the errors are homoscedastic. In addition, there is no apparent serpentine pattern; the errors are not serially correlated. In fact, the calculated Durbin h statistic of 0.548 is well below the critical z of 1.645 for a one-tailed, 5 percent Type I error test.

$$h = (1 - .5d)\sqrt{\frac{T}{1 - T(\text{variance of the lagged variable coefficient})}}$$

$$h = (.05525)\sqrt{\frac{37}{1 - 37(.1299)^2}} = 0.548$$

Each of the coefficients in the regression in Table 15.14 is significantly different from zero, at standard Type I error levels, with an overall fit that is extremely good. For forecasting the Treasury bond yield in 1991, this equation has the desired statistical properties. Table 15.12 provides the needed values on inflation and yield for 1990. The predicted logarithm of the yield in 1991 is thus

$$\begin{aligned}
\widehat{\ln(\text{YIELD})_t} &= -20.2981 + 0.5761[\ln(\text{YIELD})_{t-1}] + 0.0267p_{t-1} + 0.0106t \\
\widehat{\ln(\text{YIELD})_{1991}} &= -20.2981 + 0.5761[\ln(8.55)] + 0.0267(5.4) + 0.0106(1991) \\
&= -20.2981 + 0.576(2.1459) + 0.0267(5.4) + 0.0106(1991) \\
&= 2.2628
\end{aligned}$$

Taking the antilog of this predicted value gives the predicted yield on U.S. Treasury bonds in 1991.

$$\widehat{(\text{YIELD})_{1991}} = \exp(2.2628) = 9.61\%$$

QUERY 15.3:

An article in the Bloomington *Herald-Times* (January 6, 1992) had the headline "Hospital Rates Reflect Rising Demands" and stated, "Bloomington Hospital's (daily room) rates have more than doubled in the past decade, reflecting a national problem in health care: Patients can't pay for the type of service they demand." The following data from the hospital were available to the reporter. Do the data support the reporter's claim that daily room rates are rising because of increasing demand for hospital service?

Year	Average rate increase	Daily rate for 2-bed hospital room	Average daily census
1983	14.6	162.5	257.7
1984	9.6	178.0	227.7
1985	4.0	178.0	190.9
1986	9.9	204.5	179.8
1987	8.9	215.0	181.4
1988	8.0	248.1	186.8
1989	8.5	308.0	185.3
1990	9.0	336.0	190.7
1991	8.9	371.5	179.0
1992*	7.4	399.0	182.8

*Approximated by the hospital by some unspecified method

ANSWER:

The trend in rates is positive, but this says nothing about what is driving the rates up. It could be demand or supply. A regression of room rate on a yearly trend variable and daily census might suggest whether demand or supply concepts are relevant here.

Ordinary Least Squares			
Dependent Variable	PRICE	Number of Observations	10
Mean of Dep. Variable	260.0600	Std. Dev. of Dep. Var.	86.944634
Durbin Watson statistic	1.8963	Estimated Autocorrelation	.05184
Std. Error of Regr.	11.7977	Sum of Squared Residuals	974.304
Total Variation	68034.	Regression Variation	67060.
R-squared	.98568	Adjusted R-squared	.98159
F(2, 7)	240.9003	Prob. Value for F	.00000

Variable	Coefficient	Std. Error	T-ratio	Prob |t|>x	Mean of X	Std.Dev. of X
Constant	−66387.8	3676.	−18.057	.00000		
YEAR	33.4440	1.835	18.228	.00000	1987.50000	3.02765
CENSUS	.907006	.2151	4.217	.00395	196.21000	25.82723

This regression fits well, with a correlation coefficient close to one and high F statistic. Serial correlation does not appear to be a problem since the Durbin-Watson statistic is close to two. The census coefficient is positive, suggesting that as the census of patients in beds goes up the price charged for those beds rises, after controlling for the effect of yearly trend. We would have expected a negative census coefficient if this were a demand curve.

15.12 Tradeoffs in Modeling

Econometric Model
An equation or system of equations that is based on economic theory and observed relationships among variables.

Unfortunately, unlike making projections from only a trend, the inclusion of other explanatory variables requires knowledge of the value of these variables that is not always available contemporaneously. In the absence of this knowledge, multiple equation models must be built. These **econometric models** are based on economic theory and estimated from past observations on the key variables. Projections with these models can thus be wrong because of faulty economic theory or because of structural changes that make past relationships among variables obsolete. For example, World War II in the 1940s, the Korean War in the 1950s, the Vietnam conflict in the 1960s, and the oil embargo in the 1970s constitute major structural changes that may render data from these periods worthless in making predictions in the 1990s.

Ignoring data from the more distant past cuts the sample size; estimators lose efficiency, but the point estimates might be viewed as more valid than those obtained from the entire sample because the relationships represented by observations closer to the current period may be more like the immediate future simply by inertia. For example, in the case of bond yields, using only the most recent 11 observations in Table 15.12 gives the regression results in Table 15.15. Although the summary statistics for this regression are not as impressive as those for the 37 observations in Table 15.14, the predicted bond yield of 8.14 percent is closer to the actual yield (7.2 percent) that was realized after these projections were made.

In retrospect, even a simple trend line projection of Table 15.16 beats the more complicated econometric model of Table 15.14. The year 1991 proved to be a recession period in which the Federal Reserve eased credit greatly, enabling the negatively sloped trend line in Table 15.16 to hit the yearly average yield almost exactly, 7.19 percent. The accuracy of this simple trend line extrapolation is surprising since none of the 42 economists surveyed by *The Wall Street Journal* in the summer of 1991 "even came close to being right on how far down the Federal Reserve would drive short-term interest rates in 1991 . . . in mid-1990, 35 of the 40 economists surveyed predicted that the economy would continue expanding for at least another 12 months. Shortly thereafter, the recession began."(*WSJ*, January 2, 1992)

Although a simple trend line projection that is based on only a few past years is not eloquent, such trend line projections of the immediate next period are seldom far off. Unfortunately, trend line projections cannot predict turning points; they provide no information on what economic factors might be causing changes, and they often lead to absurd projections when made far into the future. For long-run forecasting, more complex multiequation econometric models cannot be avoided, although even they are not infallible—there is no way to predict the future with certainty.

Table 15.15 Regression of Logarithm of Yield, Based on 1980s

Dependent Variable: LnYIELD,	Mean = 2.3234,	Std. Dev. = .205053,	N = 11
Durbin Watson statistic	2.2085	Estimated Autocorrelation	.10424
Std. Error of Regr.	.1209	Sum of Squared Residuals	.102352
Sum of Squares Total	.42047	Sum of Squares Regression	.31812
R squared = .75658	F (3, 7) = 7.2521	Prob. Value for F =	.01491

Variable	Coefficient	Std. Error	T-ratio	Prob|t|>x	Mean of X	Std.Dev. of X
Constant	55.0143	42.86	1.284	.24011		
LAGLnY	.297943	.2510	1.187	.27388	2.33245	.19855
LAGINFL	.145044E–01	.1587E–01	.914	.39118	6.07273	3.83304
YEAR	–.269390E–01	.2140E–01	–1.259	.24843	1985.00000	3.31662

```
        -0.231                         0                           0.231
        ............................................................
YEAR    :                              :                           :  Residual
1980    :                    *         :                           :  -.690E-01
1981    :                              :        *                  :   .620E-01
1982    :                              : *                         :   .994E-02
1983    :                        *     :                           :  -.415E-01
1984    :                              :                   *       :   .190
1985    :                              : *                         :   .885E-02
1986    :*                             :                           :  -.231
1987    :                              : *                         :   .554E-02
1988    :                              :   *                       :   .349E-01
1989    :                             *:                           :  -.289E-02
1990    :                              :   *                       :   .333E-01
        ............................................................
```

$$\widehat{\ln YIELD_t} = 55.0143 + 0.2979[\ln(YIELD)_{t-1}] + 0.01450p_{t-1} - 0.0269t$$

$$= 55.0143 + 0.2979[\ln(8.55)] + 0.01450(5.4) - 0.0269(1991)$$

$$= 55.0143 + 0.639365 + 0.078323 - 53.5579 = 2.174088$$

$$\widehat{(YIELD)}_{1991} = \exp(2.174088) = 8.794\%$$

Table 15.16 Trend Line for Yield, Based on 1980s

Dependent Variable: YIELD,	Mean = 10.4082,	Std. Dev. = 2.140434,	N = 11
Durbin Watson statistic	1.7313	Estimated Autocorrelation	.13437
Std. Error of Regr.	1.2614	Sum of Squared Residuals	14.3191
Sum of Squares Total	45.815	Sum of Squares Regression	31.495
R squared = .68746	F (1, 9) = 19.7959	Prob. Value for F =	.00160

Variable	Coefficient	Std. Error	T-ratio	Prob|t|>x	Mean of X	Std.Dev. of X
Constant	1072.56	238.7	4.493	.00150		
YEAR	–.535091	.1203	–4.449	.00160	1985.00000	3.31662

```
            -2.19                          0                           2.19
YEAR  :                                    :                             :  Residual
1980  :      *                             :                             :  -1.62
1981  :                                    :                 *           :   1.36
1982  :                                    :            *                :   .987
1983  :                      *             :                             :  -.378
1984  :                                    :                   *         :   1.50
1985  :                                    :  *                          :   .212
1986  :*                                   :                             :  -2.19
1987  :                 *                  :                             :  -.948
1988  :                                    : *                           :  .471E-01
1989  :                                    :  *                          :   .222
1990  :                                    :          *                  :   .817
```

$$\widehat{\text{YIELD}}_t = 1072.56 - .535091t$$

$$\widehat{\text{YIELD}}_{1991} = 1072.56 - .535091\ (1991) = 7.19\%$$

Query 15.4:

A *Business Week* article, "When California Sneezes . . ."(December 30, 1991), provided the following data on homebuilding, in thousands of units for the years 1987 to 1991. A linear trend fit to the data suggested that by 1993, total homebuilding in California would be only slightly above half its 1991 level. Comment on this forecast.

CALIFORNIA HOMEBUILDING	1987	1988	1989	1990	1991
(thousands of units)	256.3	243.7	232.8	170.4	114.8

ANSWER:

Although not reported in the article, the linear trend and related statistics are

$$\widehat{\text{HOME}}_t = 71071.7 - 35.6300t \qquad R^2 = .892 \qquad F = 24.88$$

$$(\text{s.e.} = 14210)\ (\text{s.e.} = 7.143) \qquad \text{ErrorSS} = 1530.65 \qquad N = 5$$

PLOT OF RESIDUALS

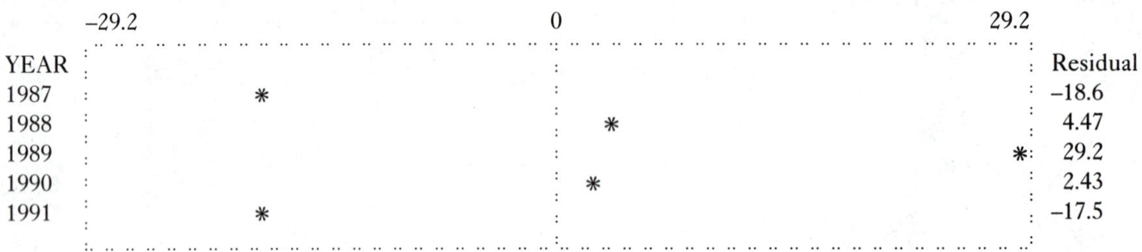

The overall fit of this regression is good, with a significant F and high R^2. Each of the coefficients is significant with calculated t ratios of approximately 5.0, impressive because the 0.05 two-tailed critical t value is 3.182, with three degrees of freedom. The predicted homebuilding for 1993 is 61.11 thousand units [= 71071.7 – 35.6300(1993)], as suggested. This forecast, however, is doubtful, because maintenance of the linear trend cannot be supported. First, the residuals show a nonlinear and random relationship for the 1987–1991 period. (Notice the ">" pattern in the plot of the residuals.) Second, building starts cannot be expected to approach zero linearly. If, in fact, building starts continue to fall, they will likely approach zero asymptotically (continuously approaching zero).

EXERCISES

15.16 Below are the number of professional degrees awarded between 1975 and 1989, (from the National Center for Education Statistics). In a dynamic model [$y_t = f(y_{t-1}, \varepsilon_t)$], with no trend, and in a model with only trend as an explanatory variable [$y_t = f(\text{trend}, \varepsilon_t)$, project the number of professional degrees for 1990 through the turn of the century for both men and women and the total.

Year	Total	Men	Women
1975	55,916	48,956	6,960
1976	62,649	52,892	9,757
1977	64,359	52,374	11,985
1978	66,581	52,270	14,311
1979	68,848	52,652	16,196
1980	70,131	52,716	17,415
1981	71,956	52,792	19,164

(continued)

Year	Total	Men	Women
1982	72,032	52,223	19,809
1983	73,136	51,310	21,826
1984	74,407	51,334	23,073
1985	75,063	50,455	24,608
1986	73,910	49,261	24,649
1987	72,750	47,460	25,290
1988	72,000	46,000	25,000
1989	72,200	46,400	25,800

15.17 Below are the projections of the number of professional degrees to be awarded as given in *Projections of Education Statistics* (National Center for Education Statistics). In a two-dimensional graph, plot your projections (for men, women, and the total) from Exercise 15.16 and those of the NCES. Why do your projections differ from those of the NCES? Which of your model estimates are closer to the NCES estimates?

Year	Total	Men	Women
1990	72,400	46,000	26,400
1991	72,300	45,700	26,600
1992	72,100	45,500	26,600
1993	72,700	45,600	27,100
1994	72,200	44,400	27,800
1995	70,600	43,000	27,600
1996	69,200	42,200	27,000
1997	68,300	41,500	26,800
1998	67,800	41,000	26,800
1999	67,600	40,800	26,800
2000	67,100	40,400	26,800

15.18 Exercise 15.8 has data on total annual returns on real estate investment based on when the investment was made, in percentages. Specify and estimate a dynamic model with these data to forecast returns in the first quarter of 1992 and the first quarter of 1993. What is the difference between your forecast here and that in 15.8?

15.19 Below are annual data on unemployment rates in Indiana and for 10 southern counties within Indiana. Fit an autoregressive model and a time trend model to the aggregate Indiana data for predicting future unemployment rates. Which model fits better? Why?

Year	Brown	Greene	Jack.	Lawr.	Martin	Monroe	Morgan	Orange	Owen	Wash.	Ind.
1972	4.9%	6.9%	4.8%	10.2%	8.7%	4.3%	n/a	13.9%	6.1%	5.6%	4.5%
1973	4.2%	6.7%	4.4%	7.9%	13.5%	4.0%	n/a	10.7%	6.2%	5.1%	4.3%
1974	4.3%	7.8%	4.2%	10.4%	10.7%	6.4%	4.4%	10.3%	8.9%	5.1%	5.2%
1975	4.9%	11.3%	12.9%	12.3%	10.0%	9.8%	6.0%	12.1%	10.4%	9.9%	8.6%
1976	4.5%	8.8%	7.8%	8.4%	10.1%	7.2%	6.5%	9.3%	8.6%	6.1%	6.1%
1977	4.9%	7.9%	6.7%	8.3%	9.3%	6.8%	5.8%	10.1%	7.9%	6.2%	5.7%
1978	4.7%	8.5%	5.8%	7.3%	8.5%	6.9%	5.3%	9.6%	6.1%	6.3%	5.7%
1979	4.4%	8.5%	5.7%	7.2%	8.5%	7.1%	5.1%	8.9%	6.1%	9.1%	6.4%
1980	6.3%	11.1%	11.5%	10.5%	11.3%	6.8%	7.2%	13.2%	8.3%	14.4%	9.6%

(continued)

Year	Brown	Greene	Jack.	Lawr.	Martin	Monroe	Morgan	Orange	Owen	Wash.	Ind.
1981	6.8%	12.1%	11.9%	13.0%	11.9%	8.3%	8.2%	13.7%	9.7%	12.5%	10.1%
1982	8.7%	12.8%	15.5%	16.1%	13.0%	9.4%	9.7%	15.2%	12.6%	16.4%	11.9%
1983	8.2%	12.7%	14.0%	13.4%	12.8%	7.6%	9.3%	14.5%	13.1%	13.3%	11.1%
1984	6.9%	10.9%	9.7%	11.0%	9.5%	5.2%	6.7%	11.4%	9.0%	9.2%	8.6%
1985	6.5%	11.2%	10.8%	11.1%	9.3%	5.5%	8.1%	10.4%	9.1%	8.2%	7.9%
1986	5.9%	10.1%	7.9%	8.8%	8.2%	3.9%	7.0%	9.3%	7.2%	6.3%	6.7%
1987	6.4%	9.5%	8.9%	8.7%	7.7%	4.3%	6.6%	8.3%	6.6%	6.7%	6.4%
1988	6.3%	8.4%	6.2%	7.2%	6.1%	3.4%	5.6%	8.1%	5.3%	5.3%	5.3%
1989	4.9%	6.6%	5.5%	6.5%	5.3%	3.2%	5.1%	7.2%	4.5%	5.0%	4.7%
1990	5.3%	7.0%	6.1%	8.8%	6.1%	3.3%	5.3%	9.7%	4.6%	6.4%	5.3%
1991	5.0%	7.3%	6.9%	9.1%	7.0%	3.7%	6.1%	10.2%	5.5%	7.7%	6.7%

15.20 Which Indiana county in Exercise 15.19 has an unemployment rate that most nearly follows the aggregate unemployment rate for the state of Indiana? On what basis are you making comparisons?

15.21 Use the data in Exercise 15.12 on quarterly profits to build a dynamic model of profits without employing a trend variable, but making use of the advertising data. Predict profits for each quarter in 1992, assuming an advertising budget of $600,000 per quarter.

15.22 For the dynamic model of annual returns in Exercise 15.18, why isn't the Durbin-Watson test appropriate?

15.13 Concluding Comments

Today, there are several widely used econometric models that are employed to forecast national aggregates, ranging from individual sector demand, as for motor vehicles, to the entire economy, as measured by gross domestic product. Models include the Wharton Model, Data Resources Inc. (DRI) Model, the Fair Model, the Chase Model, and many others routinely mentioned in the business press. Most of these models were developed by econometricians on the faculties at major United States universities. Many business schools now have bureaus of business and economic research that maintain econometric models to forecast state specific values. These forecasts are then sold to participating companies. Most large corporations also maintain their own models for forecasting product demand and costs. As eloquent as these models may be, they are all fallible: there is no sure way to predict the future.

Although any time series can be decomposed into its component parts (a secular trend, cyclical and seasonal variations, and residual or random fluctuations) this does not imply that previously observed patterns will continue into the future. Both the magnitude of observations and the order of observations in a time series provide information that can be exploited to forecast the future. But no real time series data has ever been found that perfectly reproduces itself. At best, we can attempt to identify the deterministic components of a time series, leaving only the random component to affect our predictions.

CHAPTER EXERCISES

15.23 Comment on the correctness of each of the following.

a. Time, by itself, cannot be an explanatory variable of anything.

b. Correlations between time and another variable may be a useful descriptive statistic, but they tell us nothing about what will happen in the future.

c. The idea of classical statistics when applied to time series analysis is silly. You cannot resample the year 1990; yet, classical statistics rests on the assumption that repeated sampling is possible.

15.24 Why would a researcher use a quadratic rather than a linear trend? What are the implications of each for future projections? Why would a defense lawyer question the use of a linear trend for projecting lost wages in a wrongful injury case?

15.25 According to the *New York Times* (March 4, 1990), economist Orley Ashenfelter's use of the Bordeaux equation, a time series regression to predict wine quality, caused a stir in the wine industry.

$$Q = -12.145 + 0.00117\text{WR} + 0.6164\text{TMP} - 0.00386\text{HR}$$

where Q = Logarithmic index of quality, with 1961 equal to 100
WR = Winter rain (October through March) in millimeters
TMP = Average temperature during growing season (April through September) in degrees centigrade
HR = Harvest rain (August through September) in millimeters.

a. To test the goodness of fit of Ashenfelter's model, what would you want to know?

b. To test the usefulness of predictions of wine quality from Ashenfelter's model, what would you want to know?

15.26 An Associated Press article (*HT* January 14, 1992) cited the *Evansville Courier* for uncovering data from the Indiana State Office of Medical Policy and Planning on increasing billing by chiropractors. The data supposedly showed chiropractors billing Indiana's Medicaid program more than nine times the sum billed five years earlier to treat welfare patients. The following data were provided as evidence of exponential growth in patient visits and amount billed.

Year	Medicaid patients visits	Amount billed in millions of dollars
1983	236	0.02
1984	674	0.10
1985	1,452	0.30
1986	2,466	0.80
1987	3,747	1.60
1988	4,745	2.40
1989	6,000	3.50
1990	7,626	5.50
1991	10,363	9.20

Which fits the data on billings better: a linear or an exponential trend? What are the differences in interpretation?

15.27 Use the chiropractor data in Exercise 15.26 to show the time series relationship between patient visits, a trend, and amount billed, making the amount billed the dependent variable. Does a linear or a semilog specification seem to do a better job of describing this relationship?

15.28 For the 1953–1991 time series data in the ASCII file named "EX15-28.PRN" estimate an ordinary least squares regression of the variable yield on the variable inflation. Is autocorrelation a problem? First difference these data and rerun the regression. Is there any autocorrelation in the regression of first differenced data?

15.29 For the 1957–1984 traffic death rate data in Exercise 14.35, repeat the test of the effect of the 1974 imposition of the 55 mph speed limit after removing any trend and serial correlation from the residuals.

15.30 A *Wall Street Journal* article (June 13, 1991) stated:

> Advertisers lauded NBC, a General Electric unit, for using a trend line, which they said at least acknowledges that audience levels have been steadily declining. ABC's plan in contrast, compares audience level with an average of the previous three years—and the average figure marks the steadily falling numbers. "At least NBC is more realistic than ABC; they are publicly acknowledging that there is, in fact, a trend of lower people-using-television," says John Mandel, senior vice president and director of national broadcasting for Grey Advertising.

Comment on the difference between ABC's and NBC's methods of time series analysis of audience levels.

15.31 A *Wall Street Journal* article (May 24, 1988) quoted Richard Belous, a Conference Board economist, as stating that leading econometric models, geared to the last recession (in the early 1980s) would underpredict unemployment in the next downturn. According to Belous, "The models ignore new flexible pay plans and more use of temporary and part-time workers." Go to the library and find an article that either confirms or dismisses this claim, based on the 1991 recession.

15.32 An article in *The Wall Street Journal* (February 1991) reported that "in eight of the 10 previous February auctions, prices of 30 year bonds have fallen, and yields have risen, ahead of the auction, . . . and in six of the 10 previous February refundings, Treasury bond prices rebounded (with yield falling) within two weeks following the auction." Below is the data on which this statement was based.

Yields on 30-Year Treasury Bonds

Year	Prior to Auction (Feb 1)	Average Yield at Auction	After Auction (Feb 20)
1981	12.46	12.68	12.53
1982	14.33	14.57	13.62
1983	10.96	11.01	10.60
1984	11.75	11.88	12.02
1985	11.32	11.27	11.50
1986	9.28	9.28	8.88
1987	7.48	7.49	7.47
1988	8.41	8.51	8.43
1989	8.84	8.91	9.05
1990	8.44	8.50	8.66

a. Fit a regression of the average yield at auction as a function of the yield prior to auction and a trend variable. Is there a positive relationship between the two yields?

b. Fit a regression to explain yields after the auction as a function of the average yield at auction and a trend variable. Is there a negative relationship between the two yields?

15.33 An article in *The Wall Street Journal* (March 16, 1992) showed how "interest paid on the savings component of universal life insurance policies has plunged."

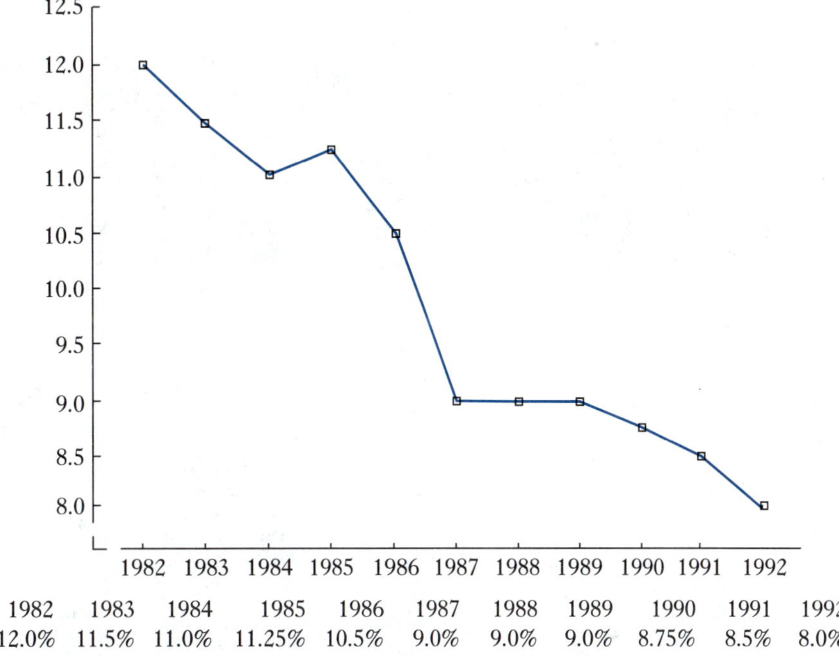

1982	1983	1984	1985	1986	1987	1988	1989	1990	1991	1992
12.0%	11.5%	11.0%	11.25%	10.5%	9.0%	9.0%	9.0%	8.75%	8.5%	8.0%

a. Fit an autoregressive one process to these data on yields and predict the yield in 1993.
b. Fit a linear trend to these data on yields and predict the yield in 1993.
c. What is the difference in the implications between parts a and b?

15.34 An article in *The Wall Street Journal* (May 18, 1992) stated, "Americans are most likely to move in the summer . . . . Fair weather is certainly a factor . . . children are on summer vacation . . . ." John Goodman, a senior economist for the Federal Reserve System is quoted as saying, "For buyers or renters, housing selection is greatest during this period." Create a seasonal dummy variable that could be included as a regressor in an equation designed to predict and explain housing availability using quarterly data.

15.35 An article by Padma Desai on "Reforming the Soviet Grain Economy" in the *American Economic Review* (May 1992) reported that "yields and outputs are estimated from the equation $y_t = AB^t u_t$, where y_t denotes either yield or output, t denotes time, A and B are the parameters, and u_t is the error term . . . the 1981–1989 period is characterized by the highest annual growth rate of yield, at 4.5 percent." Write the sample regression corresponding to this population model in its semilog specification, substituting the 4.5 percent for the letter it represents in the regression.

15.36 Edward S. Hayman, of the International Strategy & Investment Group, Inc., provided the author of this book with the data used in a *Newsweek* (June 8, 1992) graph showing that the annual percentage change in federal outlays is not increasing with a trend. The

Newsweek article stated, "George Bush may talk about budget cutting but in the end he's following the same pattern as his predecessor Ronald Reagan. When an election year approaches, the government shamelessly turns on the federal money spigots to boost the economy and attract voters."

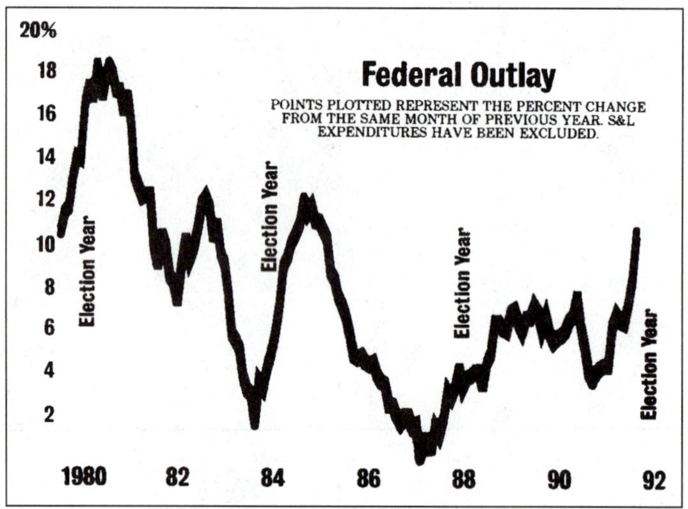

Year. Month	Fed Outlay	Year. Month	Fed Outlay	Year. Month	Fed Outlay	Year. Month	Fed Outlay
92.04	11.2	90.04	6.7	88.04	3.8	86.04	5.4
92.03	9.0	90.03	7.5	88.03	2.5	86.03	6.4
92.02	7.9	90.02	7.8	88.02	1.8	86.02	7.2
92.01	7.0	90.01	7.2	88.01	2.3	86.01	7.8
91.12	7.1	89.12	7.2	87.12	1.0	85.12	8.2
91.11	7.4	89.11	6.6	87.11	1.5	85.11	9.0
91.10	6.8	89.10	7.0	87.10	1.5	85.10	9.3
91.09	5.0	89.09	7.6	87.09	0.7	85.09	10.9
91.08	5.0	89.08	7.5	87.08	2.2	85.08	11.3
91.07	4.9	89.07	6.6	87.07	2.1	85.07	11.7
91.06	4.7	89.06	6.7	87.06	2.6	85.06	11.7
91.05	4.3	89.05	7.0	87.05	2.6	85.05	12.4
91.04	4.7	89.04	7.0	87.04	2.3	85.04	12.1
91.03	5.9	89.03	5.5	87.03	2.6	85.03	12.7
91.02	6.8	89.02	5.1	87.02	3.1	85.02	11.6
91.01	8.3	89.01	4.0	87.01	3.1	85.01	11.1
90.12	7.3	88.12	4.7	86.12	4.1	84.12	10.7
90.11	7.1	88.11	4.7	86.11	4.2	84.11	10.0
90.10	6.6	88.10	4.5	86.10	4.5	84.10	9.6
91.09	6.4	88.09	4.4	86.09	5.1	84.09	7.7
90.08	6.3	88.08	4.0	86.08	5.0	84.08	6.4
90.07	6.1	88.07	4.9	86.07	5.1	84.07	5.5
90.06	6.5	88.06	4.0	86.06	5.4	84.06	4.8
90.05	7.3	88.05	3.5	86.05	5.3	84.05	3.9

(continued)

Year. Month	Fed Outlay	Year. Month	Fed Outlay	Year. Month	Fed Outlay	Year. Month	Fed Outlay
84.04	4.2	82.11	10.4	81.06	17.6	80.01	11.2
84.03	2.3	82.10	9.9	81.05	16.8	79.12	10.2
84.02	3.6	82.09	10.9	81.04	17.8	79.11	11.2
84.01	3.8	82.08	9.7	81.03	17.8	79.10	10.2
83.12	4.4	82.07	7.9	81.02	18.6	79.09	10.0
83.11	5.7	82.06	8.8	81.01	19.0	79.08	9.9
83.10	6.1	82.05	9.1	80.12	18.1	79.07	9.9
83.09	6.4	82.04	10.5	80.11	17.5	79.06	10.3
83.08	8.4	82.03	11.1	80.10	19.1	79.05	10.3
83.07	9.7	82.02	9.6	80.09	17.4	79.04	9.8
83.06	10.2	82.01	10.5	80.08	18.1	79.03	10.1
83.05	11.6	81.12	12.9	80.07	17.2	79.02	10.7
83.04	10.9	81.11	12.9	80.06	14.6	79.01	11.0
83.03	12.3	81.10	12.8	80.05	14.7	78.12	12.1
83.02	12.8	81.09	13.3	80.04	13.7	78.11	11.1
83.01	12.5	81.08	13.8	80.03	12.3	78.10	10.9
82.12	11.3	81.07	16.3	80.02	12.1	78.09	12.0
						78.08	11.6

a. Use a dummy variable structure to capture any cyclical pattern around the trend in federal outlays.

b. Is the trend significantly different from zero (test at the 5 percent Type I error level)?

c. Is there a significant cyclical pattern consistent with the statement about outlays increasing prior to elections?

15.37 "For over a decade," according to Brian Jones, an economist at Salomon Brothers, Inc., "year to year changes in temporary jobs have been highly correlated with 12-month changes in payroll jobs a month or two later—with each gain or decline of 50,000 temporary positions foreshadowing equivalent shifts of about a million payroll jobs."(*Business Week*, June 29, 1992, p. 25) Write the equation of the regression line Jones is presenting. Make sure to specify the slope and the units of measurement for each variable.

15.38 In a wrongful injury lawsuit, there was a question regarding the effect of an accident on the plantiff's progress through school. Below is a listing of the plantiff's course completion rate before and after the accident:

School Term and Year		Credit Courses	GPA
2nd Summer	80	3	2.00
1st Semester	80–81	15	2.20
2nd Semester	80–81	12	1.93
1st Summer	81	6	1.00
2nd Summer	81	3	2.00
1st Semester	81–82	12	2.40
2nd Semester	81–82	0	
1st Summer	82	0	
2nd Summer	82	0	
1st Semester	82–83	0	

(continued)

School Term and Year		Credit Courses	GPA
2nd Semester	82 – 83	0	
1st Summer	83	0	
2nd Summer	83	0	
1st Semester	83 – 84	0	
2nd Semester	83 – 84	Withdrew	0.00
1st Summer	84	0	
2nd Summer	84	0	
1st Semester	84 – 85	0	
2nd Semester	84 – 85	0	
1st Summer	85	0	
2nd Summer	85	0	
1st Semester	85 – 86	0	
2nd Semester	85 – 86	9	3.33
1st Summer	86	4	2.43
2nd Summer	86	4	2.50
1st Semester	86 – 87	12	2.33
2nd Semester	86 – 87	3	3.30
1st Summer	87	0	
2nd Summer	87	0	
1st Semester	87 – 88	6	2.50
2nd Semester	87 – 88	6	2.85
1st Summer	88	0	
2nd Summer	88	0	
1st Semester	88 – 89	Withdrew	0.00
ACCIDENT - ACCIDENT			
2nd Semester	88 – 89	0	
1st Summer	89	0	
2nd Summer	89	0	
1st Semester	89 – 90	9	2.77
2nd Semester	89 – 90	10	3.60
1st Summer	90	3	4.00
2nd Summer	90	3	2.70
1st Semester	90 – 91	3	1.30
2nd Semester	90 – 91	6	2.85
1st Summer	91	0	
2nd Summer	91	0	
1st Semester	91 – 92	3	2.00

a. Fit a linear trend line to the credits completed, and include a dummy variable to capture the difference in the trend before versus after the accident.

b. Was there a change in the trend in credits completed before versus after the accident? Elaborate.

c. Why might a consultant want to consider other models in assessing the effect of the accident on course completion?

15.39 Below are data on Corvette sales, from *Corvette Quarterly* (Winter 1992), interest rates, and changes in hourly wages for the 1953–1991 period. To predict the production trend when the next Corvette body style is introduced later in the 1990s, estimate the time series relationship between Corvette production and the explanatory variables: trend, body style, interest rates, and changes in wages. For the body style, create and include four dummy variables. Write out your prediction equation in algebraic form. Did the different body styles influence production? Explain.

Year	Type of production: Convertible	Coupe	Total	Body style	Rate on 3yr U.S. T bills	% change in wages
1953	300	0	300	1	2.47	5.921
1954	3640	0	3640	1	1.63	2.484
1955	700	0	700	1	2.47	3.636
1956	3467	0	3467	2	3.19	5.263
1957	6339	0	6339	2	3.98	5.000
1958	9168	0	9168	2	2.84	3.175
1959	9670	0	9670	2	4.46	3.590
1960	10261	0	10261	2	3.98	3.465
1961	10939	0	10939	2	3.54	2.392
1962	14531	0	14531	2	3.47	3.738
1963	10919	10594	21513	3	3.67	2.703
1964	13925	8304	22229	3	4.03	3.509
1965	15376	8186	23562	3	4.22	4.237
1966	17762	9958	27720	3	5.23	4.065
1967	14436	8504	22940	3	5.03	4.688
1968	18630	9936	28566	3	5.68	6.343
1969	16633	22129	38762	4	7.02	6.667
1970	6648	10668	17316	4	7.29	6.250
1971	7121	14680	21801	4	5.65	6.811
1972	6508	20496	27004	4	5.72	7.246
1973	4943	25521	30464	4	6.95	6.486
1974	5474	32028	37502	4	7.82	7.614
1975	4629	33836	38465	4	7.49	6.840
1976	0	46558	46558	4	6.77	7.285
1977	0	49213	49213	4	6.69	8.025
1978	0	46776	46776	4	8.29	8.381
1979	0	53807	53807	4	9.71	8.260
1980	0	40614	40614	4	11.55	8.117
1981	0	40606	40606	4	14.44	8.859
1982	0	25407	25407	4	12.92	5.931
1983	0	43	43	4	10.45	4.427
1984	0	51547	51547	5	11.89	3.741
1985	0	39729	39729	5	9.64	3.005
1986	7315	27794	35109	5	7.06	2.217
1987	10625	20007	30632	5	7.68	2.511
1988	7407	15382	22789	5	8.26	3.341
1989	9749	16663	26412	5	8.55	4.095
1990	7630	16016	23646	5	8.26	3.727
1991	5672	14967	20639	5	6.82	3.194

style code: 1: introductory model
2: side panel insert model
3: original stringray model
4: shark shape model
5: wide stance model

15.40 From your answer in Exercise 15.39, has the trend in Corvette production been significantly upward? Discuss.

15.41 In Exercise 9.37, a finance analyst calculated a z test statistic suggesting that the Super Bowl Theory of Stock Prices is correct; that is, if an NFC team wins the Super Bowl, then a bull market (prices up) follows. If an AFC team wins, then a bear market (prices down) will prevail. The following diagram from *USA Today* (January 21, 1992) shows a time series that suggests why this test is suspect. Explain what it is about the stock market and Super Bowl wins that make any correlation artificial.

15.42 Exercise 12.50 cited a *Wall Street Journal* article in which Song Won Sohn stated that there was a negative relationship between bond yield and the Dow. In Exercise 13.48, you were to test this relationship at the 0.01 Type I error level. What time series problems would cause you to now call this test into question? Correct for these problems and retest for the negative relationship.

15.43 For the following data on average National Hockey League ticket prices from the *WSJ* (October 4, 1993), why wouldn't a linear trend be appropriate for predicting price in 1994? What type of trend line would be appropriate? Fit it, and predict price in 1994.

Year	Ticket Price
1983	\$12.50
1984	\$13.00
1985	\$14.00
1986	\$14.50
1987	\$15.50
1988	\$17.00
1989	\$18.50
1990	\$20.50
1991	\$22.50
1992	\$26.00

15.44 Writing in *Business Week* (November 27, 1989), Alan Blinder stated:

> Another myth that the 1980s exploded was the alleged death of the Phillips curve. An academic debate in the 1970s over the trade-off between inflation and unemployment left economists divided. By the late 1970s, the lack of consensus had emboldened policymakers of both the left and the right to deny that high unemployment would lower inflation.
>
> Then came the recession and disinflation of the 1980s, like a cold slap on the face. Lo and behold, high unemployment did indeed arrest an inflation that pessimists had thought unstoppable. And the quantitative dimensions of the trade-off were just what Phillips curves of the 1970s had predicted. Data from the 1980s verified an old rule of thumb that keeping unemployment one percentage point higher for a year will lower the inflation rate by one-half of a percentage point.

a. Specify a regression model, including dummy variables, a linear time trend, and variables besides unemployment variables that you believe might influence inflation, for a test of Blinder's claim about the old rule of thumb.
b. Go to the library and obtain the data to estimate your model in part a and to test Blinder's claims about the inflation and unemployment tradeoff for the 1960–1990 period.
c. Reconcile your work in parts a and b with the claims of Fischer and Dornbusch, as discussed in Chapter 13, Section 2.

15.45 Exercises 12.52, 13.50 and 14.63 provided Professor Ashenfelter's data and information for an analysis of the 1993 contested election in Pennsylvania's Second State Senatorial District in Philadelphia. Not yet checked in this analysis, however, is the possibility that there is a time trend in the explanation of absentee ballots. Add a time trend (as reflected in the last two digits of the year) to both the linear regression in Exercise 13.50, and to the quadratic regression in Exercise 14.63. Comment on the significance of trend in the explanation of absentee ballots. That is, does the inclusion of the time trend change the conclusions reached in Exercises 13.50 and 14.63?

Academic References

Box, G., and D. Pierce. "Distribution of Residual Autocorrelation in Autoregressive Moving Average Time Series Models." *Journal of the American Statistical Association* 65, 1970, pp. 1509–1526.

Durbin, J. "Testing for Serial Correlation in Least-Squares Regression when Some of the Regressors Are Lagged Dependent Variables." *Econometrica* 38, 1970, pp. 410–421.

King, M. "The Durbin-Watson Test for Serial Correlation Bounds for Regressions with Trend and/or Seasonal Dummy Variables." *Econometrica* 49, 1981, pp. 1571–1581.

Ljung, G., and G. Box. "On a Measure of Lack of Fit in Time Series Models." *Biometrika* 66, 1979, pp. 265–270.

Endnotes

1. As introduced in Chapter 14, and discussed in more detail in Chapter 16, a real wage (W_r) is defined by the nominal or observed money wage (W) divided by an index of prices (P) for the period; that is,

$$W_r = W/P$$

A real rate of interest (R_r) is typically defined by a nominal rate of interest (R) less a measure of inflationary expectations. One measure of those inflationary expectations might be the previous period change in the price index (Previous $\Delta P/P$). The real rate is then

$$R_r = R - (\text{Previous } \Delta P/P)$$

2. For linear trend, numerical values used to represent the time period are arbitrary as long as they increase by a fixed amount. For instance, instead of indexing the quarterly sales 1, 2, 3, 4, 5, 6, . . . 68, the coding could have been 73.75, 74.00, 74.25, 74.50, 74.75, 75.00 . . . 91.50. The estimated intercept and slope simply reflect these changes in units of measurement.
3. The Durbin-Watson statistic might have meaning for cross-sectional data if there is a justifiable order to the data; for example, if data are collected on household characteristics by a census taker moving from the poorest to wealthiest neighborhood. Typically, however, the order in which cross-sectional data is collected and organized is irrelevant.
4. The proof of $d \approx 2(1 - \hat{\rho})$ requires an expansion of d.

$$d = \frac{\sum_{t=2}^{T}\left(e_t - e_{t-1}\right)^2}{\sum_{t=1}^{T} e_t^2} = \frac{\sum_{t=2}^{T} e_t^2 + \sum_{t=2}^{T} e_{t-1}^2 - 2\sum_{t=2}^{T} e_t e_{t-1}}{\sum_{t=1}^{T} e_t^2}$$

Because the sum of e_t^2 in the numerator and denominator differs by only one term (the starting value at $t = 1$ versus $t = 2$), for a series with many observations (large T) these sums will be approximately equal, so

$$d \approx 2 - \frac{2\sum_{t=2}^{T} e_t e_{t-1}}{\sum_{t=1}^{T} e_t^2}$$

Finally, $\Sigma\, e_t^2$ and $\Sigma\, e_{t-1}^2$ will be approximately the same, giving

$$d \approx 2\left[1 - \frac{\sum_{t=2}^{T} e_t e_{t-1}}{\sum_{t=1}^{T} e_{t-1}^2}\right] = 2\left(1 - \hat{\rho}\right)$$

where $\hat{\rho}$ is the least squares estimator of ρ, in a regression of e_t on e_{t-1}.

5. The critical values in the Durbin-Watson table require that the regression contain a constant term. How additions to this constant term, in the form of seasonal dummies, are to be counted is known to influence the inconclusive region between the critical values. Here we are treating these dummies as explanatory variables. King (1981) provides an alternative interpretation and gives an additional set of critical values for regressions with quarterly seasonal dummy variables.

6. There are many alternative parametric tests of serial correlation. One of these tests is the Box and Pierce Q test

$$Q = T\sum_{j=1}^{L} r_j^2$$

where L is the maximum lag length considered and r_j is a generalization of the autocorrelation coefficient in Formula (15.2):

$$r_j = \frac{\sum_{t=j+1}^{T} e_t e_{t-j}}{\sqrt{\sum_{t=1}^{T} e_{t-j}^2 \sum_{t=1}^{T} e_t^2}}$$

where e_{t-j} is the residual from a regression of y on a set of x's, estimated at time $t-j$.

In practice, the denominator of r_j can be approximated by substituting e_t^2 for the product $e_{t-j}e_t$. The Q statistic is distributed as an approximate chi-square with L degrees of freedom. Thus, if the calculated Q is greater than the critical chi-square value, the null hypothesis of no autocorrelation is rejected. A refinement of the Q statistic proposed by Ljung and Box (1979) is

$$Q' = T(T-2)\sum_{j=1}^{L} r_j^2 \Big/ (T-j)$$

which adjusts for biases that may exist in Q.

These chi-square tests, as well as several others, have been found to be reasonably powerful in detecting autocorrelation. The arbitrary nature of the lag length L, however, has been criticized. With the exception of computer programs devoted explicitly to time series analysis, they are not routinely provided in computer output.

7. As discussed in previous chapters, many time series have an exponential trend rather than a straight line trend. Exponential growth is especially prevalent for financial variables that grow by compounding. The equation for continuous exponential growth in y is

$$y_t = \alpha e^{\beta t}$$

The natural logarithmic transformation gives

$$\ln(y_t) = \ln\alpha + \beta t$$

and the growth coefficient β can be estimated without bias by least squares. Unfortunately, $E(y_t)$ cannot be estimatated without bias by the antilog of the least squares line; that is,

$$E(y_t) \neq E[\text{antilog of } \widehat{\ln(y_t)}]$$

FORMULAS

- A straight trend line is

$$\hat{y}_t = a + bt$$

where $\hat{y}_t$ is the predicted value of y_t at time t and coefficients a and b are obtained from the least squares regression.

- Correlation of the residuals between two subsequent periods is defined by the autocorrelation coefficient

$$r = \frac{\sum_{t=1}^{T} e_t e_{t-1}}{\sqrt{\sum_{t=1}^{T} e_{t-1}^2 \sum_{t=1}^{T} e_t^2}}$$

where e_{t-1} and e_t are the residuals from a regression of y on a set of x's, estimated first at time $t - 1$ and then again in the subsequent time period t.

- The first-order autoregressive error term process is

$$\varepsilon_t = \rho \varepsilon_{t-1} + u_t$$

where u_t is a random perturbation and ρ is the parameter relating the error of ε_t to its previous value ε_{t-1}.

- The Durbin-Watson d statistic is

$$d = \frac{\sum_{t=2}^{T} \left(e_t - e_{t-1}\right)^2}{\sum_{t=1}^{T} e_t^2}$$

where e_t is the least squares regression residual at time t.

- The Durbin-Watson d statistic is related to the estimated autocorrelation coefficient ρ by

$$d \approx 2(1 - \hat{\rho})$$

- The expected number of runs in a series is given by

$$E\left(\text{runs}\right) = \frac{2\#(+)\#(-)}{\#(+) + \#(-)} + 1$$

where #(–) is the number of negative residuals
#(+) is the number of positive residuals

and its standard error is

$$\sigma\left(\text{runs}\right) = \sqrt{\frac{2\#(+)\#(-)\left[2\#(+)\#(-) - \#(+) - \#(-)\right]}{\left[\#(+) + \#(-)\right]^2 \left[\#(+) + \#(-) - 1\right]}}$$

- A dynamic model of y_t that is based on its immediate past value y_{t-1}, as well as other explanatory x variables, is given by

$$y_t = \beta_0 + \beta_1 y_{t-1} + \beta_2 x_{t2} \ldots + \beta_k x_{tk} + \varepsilon_t$$

- The Durbin h statistic

$$h = \left(1 - .5d\right)\sqrt{\frac{T}{1 - T\left(\text{variance of the lagged variable coefficient}\right)}}$$

- The model of first differenced data is

$$(y_t - y_{t-1}) = \beta_2(x_{t2} - x_{t-12}) + \ldots + \beta_k(x_{tk} - x_{t-1k}) + u_t$$

which by letting delta, Δ, represent the first differences can be written more compactly as

$$\Delta y_t = \beta_2 \Delta x_{t2} \ldots + \beta_k \Delta x_{tk} + u_t$$

Chapter 16

Index Numbers

Those who talk about the future are scoundrels. It is the present that matters. To evoke one's posterity is to make a speech to maggots.

Louis Ferdinand Celine
1894–1961

16.1 Introduction

Headlines reporting price increases within the United States are not unusual. Every month, the U.S. Labor department releases numbers that are published by all the major news services. For example, from *The Wall Street Journal* (March 17, 1994), we learned:

> The price consumers paid for goods and services increased 0.3% in February . . . a relatively low rate but enough to make some economists search for hints of inflation . . . . The increase was fueled by an increase in housing costs, a major component in the index. . . . New car and financing prices are up because of strong demand. . . .

And from *Business Week* (March 28, 1994) we also learned:

> . . . the price indexes tell us where inflation was, not where it's going, but the latest reports show that inflation was still falling in February. The producer price index rose 0.5% last month, reflecting higher energy prices, especially a 23.5% surge in heating-oil costs related to unusually heavy demand . . .

The questions addressed in this chapter are concerned with the meaning of the measures involved in these quotes: How are general price increases calculated from price changes in individual products? How can price indexes be used to assess price changes? How can a price index be created from a time series of prices? How do price indexes relate to the "real" purchasing power of money? By the end of this chapter, you will be able to answer questions such as these.

16.2 Measuring Price Change

The easiest way to measure general price changes is to "average" the observed changes in the prices of individual goods and services. For instance, if pork prices rose by 4 percent in a month, fruit prices declined by 6.8 percent, gas prices rose by 2.8 percent, and car prices fell by 0.9 percent, then an average price change for these four items could be calculated as

$$\text{average price change} = \frac{4.0 + \left(-6.8\right) + 2.8 + \left(-0.9\right)}{4} = -0.225$$

A major problem with this average is that the price change of each item depends on the units in which the item was measured and these units are not comparable across items. One car is not the same as one pound of pork. In addition, price changes of each item are given equal weight in the calculation of the average. A 1 percent change in the price of a car is far larger than a 1 percent change in the price of a pound of pork. On the other hand, a car is only purchased once every several years, while foodstuffs are consumed daily. In total, foodstuffs might make up a bigger proportion of a person's budget than would car costs. The trouble with a simple average of price changes is that it does not reflect the weights given to items in the representative person's budget or basket of things purchased.

16.3 Index Numbers

The best known index in business and economics is the Consumer Price Index (CPI). It is reported often in the newspapers as the measure for assessing price inflation. The CPI has a number of uses, including making adjustments to child support, Social Security payments, and wage contracts. Designed to measure the cost to consumers of a "typical" bundle or basket of goods and services, the CPI is a time series of numbers. Table 16.1 shows the CPI for selected years between 1950 and 1992. In essence, these numbers show that a basket of goods and services that cost about $100 in 1983 costs about $140 in 1992, approximately a 40 percent increase in prices.

The U.S. government tabulates many different indexes that are used widely throughout the business world. Examples include the Producer Price Index (PPI), the GNP (Gross National Product) Deflator, the GDP (Gross Domestic Product) Deflator, the Index of Industrial Production (IIP), the Export and Import Indexes, and the Employment Cost Index. The CPI itself is not a single and continuous series.

In the late 1970s and early 1980s, the consumer price indexes went through some major revisions; most notable was the introduction of the Consumer Price Index for all urban consumers (sometimes identified as the CPI-U). Before 1978, the Consumer Price Index was calculated only for purchases by urban wage earners (occasionally labeled the CPI-W). For comparison purposes, both of these indexes were used through the 1980s, but today a person who refers to the CPI is likely talking about the CPI for all urban consumers, as in Table 16.1.

Table 16.1
U.S. Consumer Price Index (Base is 1982–1984 equals 100)

Year	Index	Year	Index	Year	Index
1950	24.1	1975	53.8	1985	107.6
1960	29.6	1976	56.9	1986	109.6
1967	33.4	1977	60.6	1987	113.6
1968	34.8	1978*	65.2	1988	118.3
1969	36.7	1979	72.6	1989	124.0
1970	38.8	1980	82.4	1990	130.7
1971	40.5	1981	90.6	1991	136.2
1972	41.8	1982	96.5	1992	140.3
1973	44.4	1983**	99.6		
1974	49.3	1984	103.9		

* Data beginning 1978 are for all urban consumers; earlier data are for urban wage earners and clerical workers.

** Data beginning 1983 incorporate a rental equivalence measure for homeowners' cost and therefore are not strictly comparable with earlier figures.

Source: *Economic Report of the President.*

To demonstrate the importance of the CPI, Janet L. Norwood, senior fellow at the Urban Institute and U.S. Commissioner of Labor Statistics from 1979 to 1991, said the CPI affects the income of more than half the population of the country. It has enormous effects on the federal budget. In an article in the *Journal of Economic Education* (Summer 1994), she estimated that a one percentage point change in the CPI affects the federal government fiscal position by nearly $5 billion. Because of its importance in our daily lives, business decisions, and policy debates, an understanding of the CPI is essential.

The CPI represents the price of a basket of goods and services purchased by a typical household consumer. For example, in the case of the old CPI-W, the basket was the bundle of goods and services purchased by a typical urban wage earner. The new CPI-U is the basket purchased by an urban consumer regardless of whether he or she was a wage earner. This new basket, for example, includes a rental equivalence measure for homeowners' costs that was not measured in the old basket purchased by urban wage earners. The current market basket purchased by urban consumers can be broken down into eight broad classes: food and beverages, housing, apparel and upkeep, transportation, medical care, entertainment, energy, and other goods and services.

Each month the U.S. Bureau of Labor Statistics dispatches some 450 inspectors to 20,000 locations throughout the land to gather 80,000 prices on standardized items that Americans buy or rent. The cost of the basket is calculated by multiplying the price of each item by the amount the consumers purchased in the relevant time period. The total cost of the basket is the sum of all such costs for all the different goods and services purchased by consumers.

To see how a price index is constructed, consider the 1992 basket of food stuffs (consisting of meat, eggs, milk, bread, and butter) in Table 16.2. Suppose that in 1992, 100 pounds of meat were purchased at a price of $1.10 per pound. The following year (1993), this meat was priced at $1.15. The price paid for eggs was 90 cents per dozen in 1992, when 30 dozen were purchased, but eggs were 94 cents per dozen in 1993. In 1992, 75 gallons of milk were purchased at $1.40 per gallon, while in 1993 the price was $1.45. Bread was $1.30 per loaf in 1992, when 10 loaves were purchased; in 1993, bread was priced at $1.34. Finally, two pounds of butter, at $1.60 per pound, were purchased in 1992. The price of butter was $1.69 per pound in 1993.

The question: What is the total cost of the five-item basket as purchased in 1992, but as priced in both 1992 and 1993? As shown in Table 16.2, the total cost of the 1992

Table 16.2 Calculating a 1993 Price Index

	1992 Quantity	1992 Price	1993 Price	1992 Total Cost	1993 Total Cost
Meat	100 lbs	$1.10/lb	$1.15/lb	$110.00	$115.00
Eggs	30 doz	0.90/doz	0.94/doz	27.00	28.20
Milk	75 gal	1.40/gal	1.45/gal	105.00	108.75
Bread	10 lvs	1.30/lf	1.34/lf	13.00	13.40
Butter	2 lbs	1.60/lb	1.69/lb	3.20	3.38
				$258.20	$268.73

basket at 1992 prices is \$258.20. The total cost of this basket in 1993 is \$268.73. Thus, the 1993 price index for these five food items is

$$\text{1993 price index (base 1992)} = \frac{268.73}{258.20} \times 100 = 104.08$$

Laspeyres Price Index
The ratio of the total cost of a basket of goods and services purchased in a base year to the cost of the same basket purchased in a year of interest.

This is a **Laspeyres index.** In a Laspeyres index, the quantity purchased in the base year is held constant; 1992 is the base year and its price index value is 100. The index number 104.08 indicates that for this basket of food items, 1993 prices were 4.08 percent higher than 1992 prices.

The Consumer Price Index produced by the United States Bureau of Labor Statistics is an example of a Laspeyres index. All Laspeyres price indexes are defined as follows:

Laspeyres Price Index

$$\frac{\sum_{j=1}^{J} Q_{bj} P_{Ij}}{\sum_{j=1}^{J} Q_{bj} P_{bj}}(100)$$

where Q_{bj} is the base year quantity of *j*th item purchased
P_{bj} is the base year price of *j*th item
P_{Ij} is the *j*th item price in the year of interest

A problem with the Laspeyres index is that as tastes of consumers change, the quantities in the base year basket may become obsolete. For example, in the case of the CPI (1982–1984 base), cigarettes make up almost 2 percent of this particular Laspeyres index. Cigarettes constitute a larger share of the CPI than dairy products or men's clothing. An article in *Business Week* (May 10, 1993) reported that many analysts thus argue that a 10 percent drop in cigarette prices will slice 0.4 percentage points off CPI measured inflation, because cigarettes are overrepresented in the CPI today. As long as the quantity of cigarettes entering the CPI remains fixed, changes in cigarette prices will receive more weight than they should in subsequent index numbers.

One way that the U.S. Bureau of Labor Statistics attempts to avoid problems of outdated items in the Consumer Price Index basket is by adjusting items in the basket periodically. For example, in 1983, a rental equivalence measure for homeowners' cost was added. These changes in the basket, however, mean that the Consumer Price Index is not strictly comparable over long periods.

The base year of the Consumer Price Index is changed to reflect major changes in the basket. But these base year changes are made only once a decade or so. According to Harvard University economist Zvi Griliches, "Whole generations of new goods come and go before the data collectors can measure their impact on prices, productivity, and growth."(*Fortune*, March 8, 1993) The common pocket calculator, for example, was introduced in the late 1960s but didn't show up in the CPI market basket until the late 1970s, after its price had already fallen by 90 percent.

Paasche Price Index The ratio of the total cost of a basket of goods and services purchased in a year of interest to the cost of the same basket in a base year.

Another way to avoid the problem of an outdated basket is to use a **Paasche index.** In a Paasche index, the basket consists of current quantities. These current quantities are then priced at both current and base year prices. As can be seen in the following formula, calculations with a Paasche price index requires an updating of the basket in each new year, so both the numerator and the denominator must be recalculated every year.[1]

Paasche Price Index

$$\frac{\sum_{j=1}^{J} Q_{Ij} P_{Ij}}{\sum_{j=1}^{J} Q_{Ij} P_{bj}}(100)$$

where Q_{Ij} is the jth item consumed in year of interest
P_{bj} is the base year price of jth item
P_{Ij} is the jth item price in the year of interest

The Gross National Product Deflator is one example of a Paasche price index. As an example of calculations in a Paasche price index, suppose the 1993 basket of food stuffs is that in Table 16.3, where 115 pounds of meat, 37 dozen eggs, 71 gallons of milk, 11 loaves of bread and three pounds of butter were purchased. Again, the prices of meat were $1.10 per pound in 1992 and $1.15 per pound in 1993, and similarly for the other prices in 1992 and 1993, as in Table 16.2.

The total cost of the 1993 five-item basket as purchased in 1993 is $289.79. The total cost of this 1993 five-item basket, when priced at 1992 prices, is $278.30. Thus, the Paasche price index is calculated as

$$\text{1993 price index (base 1992)} = \frac{289.79}{278.30} \times 100 = 104.13$$

Table 16.3 Calculating a Paasche Price Index

	1993 Quantity	1992 Price	1993 Price	1992 Total Cost	1993 Total Cost
Meat	115 lbs	$1.10/lb	$1.15/lb	$126.50	$132.25
Eggs	37 doz	0.90/doz	0.94/doz	33.30	34.78
Milk	71 gal	1.40/gal	1.45/gal	99.40	102.95
Bread	11 lvs	1.30/lf	1.34/lf	14.30	14.74
Butter	3 lbs	1.60/lb	1.69/lb	4.80	5.07
				$278.30	$289.79

The base year is still 1992, with a price index value of 100, but now the Paasche price index in 1993 is 104.13. This Paasche index value indicates that in 1993 prices were 4.13 percent higher than 1992 prices.

16.4 Other Index Weighting Schemes (Optional)

As stated in Table 16.1, the base for the U.S. Consumer Price Index is not one year; it is a composite for the 1982–1984 period. The Consumer Price Index is a Laspeyres price index because it weights current prices by the quantities in a past market basket. The actual determination of those weights can be based on one period or some average of periods. For example, from Tables 16.2 and 16.3, an "average two-year 1992–1993 basket" can be formed as shown in the fourth column of Table 16.4. These average quantities will be the weights for the average-weight price index. The prices for the base period also can be formed as an average of the prices in the period, as shown in the seventh column in Table 16.4. The total cost of this average two-year basket, at average prices, is shown in column eight, $273.755. Using this total cost of the average two-year basket as the base, the price index can be calculated for both 1992 and 1993.

The value of the average weight price index in 1992 is

$$1992 \text{ price index (base } 1992-93) = \frac{268.25}{273.755} \times 100 = 97.989$$

Thus, prices in 1992 are 2.011 percent lower than the 1992–1993 period average (for which the base is 100). Similarly, the price index in 1993 is

$$1993 \text{ price index (base } 1992-93) = \frac{279.26}{273.755} \times 100 = 102.011$$

Prices in 1993 are 2.011 percent higher than the 1992–1993 average.

Table 16.4 Price Index for Average Basket and Average Base Prices

(1)	(2) 1992 Quantity	(3) 1993 Quantity	(4) 1992–93 Av. Qut.	(5) 1992 Price	(6) 1993 Price	(7) 1992–93 Av. Price	(8) 1992–93 Av. Total Cost	(9) 1992 Total Cost	(10) 1993 Total Cost
Meat	100 lbs	115 lbs	107.5	$1.10	$1.15	1.125	$120.9375	$118.25	$123.625
Eggs	30 doz	37 doz	33.5	0.90	0.94	0.920	30.8200	30.15	31.490
Milk	75 gal	71 gal	73.0	1.40	1.45	1.425	104.0250	102.20	105.850
Bread	10 lvs	11 lvs	10.5	1.30	1.34	1.320	13.8600	13.65	14.070
Butter	2 lbs	3 lbs	2.5	1.60	1.69	1.645	4.1125	4.00	4.225
							273.7550	268.25	279.260
							average-weight index =	97.9890	102.0109

The equation for the average-weight Laspeyres price index is summarized in the following formula.

Average-Weight Laspeyres Price Index

$$\frac{\sum_{j=1}^{J} \bar{Q}_{bj} P_{Ij}}{\sum_{j=1}^{J} \bar{Q}_{bj} \bar{P}_{bj}}(100)$$

where $\bar{Q}_{bj}$ is the average quantity of *j*th item purchased
$\bar{P}_{bj}$ is the average price of *j*th item in base period
P_{Ij} is the *j*th item price in the year of interest

Although the above average-weight price index is rather general, it is not a universally accepted measure of change. There is no universally accepted index for measuring change in an aggregate. Since 1922, when Yale University economist Irving Fisher wrote his authoritative book *The Making of Index Numbers: A Study of Their Varieties, Tests, and Reliability* (Boston: Houghton Mifflin Company), there has been little advancement in our knowledge of index numbers. The sponsors of Fisher's book "hoped that all those who are capable of understanding the subject will see their way clear to agreeing upon the Ideal Formula."(prefatory note by William Foster) Such agreement has never been realized because there is no unique way to measure discrete changes across diverse items.[2]

Exercises

16.1 What are the advantages and disadvantages of the Laspeyres versus the Paasche price indexes?

16.2 An article in *Business Week* (May 10, 1993) reported that the CPI is not an accurate measure of inflation because in

> HOUSING The CPI leaves out the impact of . . . mortgage rates on home ownership costs . . .
> DISCOUNTING . . . the CPI gives too much weight to full-price retailers . . .
> HOME COMPUTERS The CPI is based on spending patterns from 1982–84, when home computers were still scarce.

In recent years, what influence does each of these have on the CPI?

16.3 A price index for clothing for 1994 is to be constructed using 1992 as the base year. Determine both a Laspeyres and a Paasche index using the following data on purchases made in one year by a "representative" person. What is the difference in inflation as given by the two indexes?

	1992		1994	
Clothing item	Price	Quantity	Price	Quantity
Socks	5.00	4	6.63	5
Shoes	79.25	3	96.00	2
Pants	78.75	4	82.55	6
Shirts	33.45	7	36.85	8
Underwear	6.50	9	7.60	10

16.4 The prices and quantities of five products sold by a specialty shop are shown below for the years 1992 and 1993. Compute both a Laspeyres price index and a Paasche price index for these data. What is the difference in inflation as measured by each of these indexes?

	1992		1994	
Product:	Price	Quantity	Price	Quantity
1	4.56	300	5.55	333
2	12.90	237	14.95	245
3	25.45	79	36.00	84
4	28.32	70	32.98	74
5	31.97	55	41.00	64

16.5 Irving Fisher in *The Making of Index Numbers* (Formula 353, p. 482) called the following price index "ideal"

$$\sqrt{\frac{\Sigma p_1 q_0}{\Sigma p_0 q_0} \times \frac{\Sigma p_1 q_1}{\Sigma p_0 q_1}}$$

a. What are the components of this "ideal" index number?
b. Go read why Fisher thought this was an ideal index. What did he say? State why you agree or disagree with him.

16.5 Chaining of Index Numbers

In any discussion of index numbers, consideration must be given to chaining, or the creation of index numbers from a series of numbers with a specific unit of measurement (e.g., dollars, pounds, feet, etc.). Table 16.5 shows the average hourly dollar wages paid in the United States. To create an index for this time series of wages, the dollar wage value in any year of interest is divided by the dollar wage value in the year we wish to be the base; this ratio is then multiplied by 100 to form the price index for any year of interest.

The third column in Table 16.4 shows the wage index with a 1985 base of 100. All calculations in the third column were made with LOTUS 1-2-3, although any other spreadsheet program could have been used. In notational form,

$$I(85 \mid \text{base } 85) = (8.57/8.57)(100) = 100.00$$

TABLE 16.5 Index for Hourly Wages

Year	Wages	Wage Index (Fixed base, where 1985 = 100)	Wage Index (Shifting base, where previous year = 100)
1979	\$6.16	100 × 6.16/8.57 = 71.87	
1980	6.66	100 × 6.66/8.57 = 77.71	100 × 6.66/6.16 = 108.1168
1981	7.25	100 × 7.25/8.57 = 84.59	100 × 7.25/6.66 = 108.8588
1982	7.68	100 × 7.68/8.57 = 89.61	100 × 7.68/7.25 = 105.9310
1983	8.02	100 × 8.02/8.57 = 93.58	100 × 8.02/7.68 = 104.4270
1984	8.32	100 × 8.32/8.57 = 97.08	100 × 8.32/8.02 = 103.7406
1985	8.57	100 × 8.57/8.57 = 100.00	100 × 8.57/8.32 = 103.0048
1986	8.76	100 × 8.76/8.57 = 102.21	100 × 8.76/8.57 = 102.2170
1987	8.98	100 × 8.98/8.57 = 104.78	100 × 8.98/8.76 = 102.5114
1988	9.28	100 × 9.28/8.57 = 108.28	100 × 9.28/8.98 = 103.3407
1989	9.66	100 × 9.66/8.57 = 112.71	100 × 9.66/9.28 = 104.0948
1990	10.02	100 × 10.02/8.57 = 116.91	100 × 10.02/9.66 = 103.7267
1991	10.34	100 × 10.34/8.57 = 120.65	100 × 10.34/10.02 = 103.1936

Source for Wage Data: *Economic Report of the President.*

The wage index in 1991 is

$$I(91 \mid \text{base } 85) = (10.34/8.57)(100) = 120.65$$

which indicates that wages were 20.65 percent higher in 1991 than they were in 1985. In this wage index, all future dollar wages are measured relative to the 1985 base.

To quantify how wages vary from year to year, the dollar wage in the previous year is used as the base. Instead of a fixed year base, a sequentially shifting base may be employed to show year-to-year changes. For example, the shifting base index for 1985 is

$$I(85 \mid \text{base } 84) = (8.57/8.32)(100) = 103.005$$

which indicates that wages were 3.005 percent higher in 1985 than in 1984. The shifting, previous year based wage index in 1991 is

$$I(91 \mid \text{base } 90) = (10.34/10.02)(100) = 103.194$$

Chained Index Numbers
Index values based on a sequentially shifting base.

showing that wages are 3.194 percent higher than they were in 1990. Indexes based on a sequentially shifting base are sometimes called **chained index numbers.**

Both the chain index numbers and the fixed base index numbers need not be provided because, given one, the other can always be obtained. Let $p_{79}, \ldots p_{85}, p_{86}, \ldots p_{91}$ be the series of prices. With 1985 as the base, then the fixed base index number for 1991 is

$$I(91 \mid \text{base } 85) = (p_{91}/p_{85})(100)$$

This can be rewritten as

$$I(91 \mid \text{base } 85) = (p_{91}/p_{90})(p_{90}/p_{89})(p_{89}/p_{88})(p_{88}/p_{87})(p_{87}/p_{86})(p_{86}/p_{85})(100)$$

But since $(p_{91}/p_{90}) = I(91 \mid \text{base } 90)/100$, the above chain can be written as

$$I(91 \mid \text{base } 85) = [I(91 \mid \text{base } 90)/100][I(90 \mid \text{base } 89)/100] \ldots [I(86 \mid \text{base } 85)/100](100)$$

Notice that the chain index procedure can be used to update an index from year to year, provided that the previous year's fixed base index value is known and the current year price change is known. In particular, as soon as the dollar value for 1992 is available, the 1992 price index with 1985 as its base would be

$$I(92 \mid \text{base } 85) = (p_{92}/p_{91})[I(91 \mid \text{base } 85)]$$

That is, to calculate the index for 1992, multiply the index for 1991 by the ratio of 1992 values to 1991 values. In essence, this is the way that the Consumer Price Index, as well as other indexes, are updated by the United States government agencies responsible for their maintenance.

16.6 Splicing of Index Numbers

Changing the base year of a price index alters the level of the index, but unless there is a major change in the basket the rate of change in the series will not change. For example, when the U.S. Bureau of Labor Statistics rebased the CPI from the 1967 base of 100 to the 1982–1984 base of 100, the percentage change in prices indicated by each series was roughly the same even though major changes were made in the basket of goods and services represented. Table 16.6 shows both the old and new CPI indexes

Table 16.6 The 1967 and 1982–1984 Consumer Price Indexes

	CPI* 1967 base = 100	Percentage change (%)	CPI** 1982–1984 base = 100	Percentage change (%)
1980	246.8		82.4	
1981	272.4	10.3727	90.9	10.3155
1982	289.1	6.1306	96.5	6.1606
1983	298.4	3.2168	99.6	3.2124
1984	311.1	4.2560	103.9	4.3172
1985	322.2	3.5679	107.6	3.5611
1986	328.4	1.9242	109.6	1.8587
1987	340.4	3.6540	113.6	3.6496
1988	Not available		118.3	4.1373
1989	Not available		124.0	4.8183
1990	Not available		130.7	5.4032
1991	Not available		136.2	4.2081

* Source: *Economic Report of the President* (February 1988, p. 313).
** Source: *Economic Report of the President* (February 1992, p. 361).

from 1980 to 1987, the last year for which the 1967-based CPI was reported in the *Economic Report of the President.*

For instance, to find the percentage changes in 1987 in both of the Consumer Price Indexes in Table 16.6 requires calculating the ratios

$$\text{Change in 1987, 1967 base: } \frac{340.4 - 328.4}{328.4} = 0.036541, \text{ and}$$

$$\text{Change in 1987, 1982} - \text{1984 base: } \frac{113.6 - 109.6}{109.6} = 0.036496$$

Both of these ratios are then multiplied by 100 to obtain the percentage changes shown in Table 16.6. These percentage changes are the measures of price inflation that newspapers typically report. Regardless of the index used, the price inflation reported in 1987 would have been 3.65 percent.

Splicing
A method of continuing one index based on percentage changes in another index.

The closeness of the annual rates of inflation reflected in these two Consumer Price Indexes suggests a method for extending the old index. This **splicing** method involves increasing the old price index by the percentage increases indicated by the new index. To estimate the 1988 Consumer Price Index, for the 1967 base, the calculation is

$$\hat{I}(1988 \mid \text{base } 1967) = 340.4 + (0.041373)(340.4) = 354.4838$$

where 4.1373 is the percentage increase for 1988 found in the 1982–1984 price index. Similarly, the 1989 Consumer Price Index, for the 1967 base, is estimated by

$$\hat{I}(1989 \mid \text{base } 1967) = 354.4838 + (0.048183)(354.4838) = 371.5633$$

Because approximate percentage changes are maintained when indexes are rebased, the researcher can always forward or back splice different based indexes. As stated in the footnotes to the 1982–1984 based price index (Table 16.1), these splices will not always result in exactly comparable indexes. If the baskets differ greatly, then the indexes are providing different measures of purchasing power.

The use to which price indexes are put in controlling for changes in the purchasing power of money over time is best demonstrated through examples. Below, the calculations of the "real rate of interest" and the "real purchasing price" are considered in two queries involving the yield on Moody's Aaa index bonds and the price of gasoline in the United States.

QUERY 16.1:

In 1930, Irving Fisher published *The Theory of Interest* (New York: Macmillan Company) in which he defined a "real rate of interest" as the rate that would exist if inflationary expectations were zero. He approximated the real rate of interest by the observed, or nominal rate of interest, minus the rate of price inflation calculated from a price index. The real rate of interest is used in the comparison of investments over time. Calculate a real rate of interest from recent Moody's Aaa yields. How do changes in these nominal yields compare to changes in the corresponding real rates?

ANSWER:
From the *Economic Report of the President* (February 1993), the yield per annum on corporate bonds (Moody's Aaa) was selected as an arbitrary but representative interest rate. It is reported in the second column of the spreadsheet shown below. The yearly percentage changes in the Consumer Price Index are reported in the third column of this spreadsheet display. The real rate of interest (r) is now approximated by the observed or nominal rate of interest (i) minus the rate of price inflation ($\Delta p/p$); that is, $r = i - \Delta p/p$. The real rate of interest is given in the fourth column. Although the nominal rate of interest (i) fell throughout the 1980s, the real rate of interest rose and then slowly fell, as can be seen easily in the accompanying graph.

The Real Rate of Interest

Year	Corporate Bonds* (Moody's Aaa) Yield Per Annum i	CPI Percentage Change (%) $\Delta p/p$	Real Rate of Interest $r = i - \Delta p/p$
1981	14.17%	10.32%	3.85%
1982	13.79	6.16	7.63
1983	12.04	3.21	8.83
1984	12.71	4.32	8.39
1985	11.37	3.56	7.81
1986	9.02	1.86	7.16
1987	9.38	3.65	5.73
1988	9.71	4.14	5.57
1989	9.26	4.82	4.44
1990	9.32	5.40	3.92
1991	8.77	4.21	4.56
1992	8.14	3.01	5.13

* Source: *Economic Report of the President.*

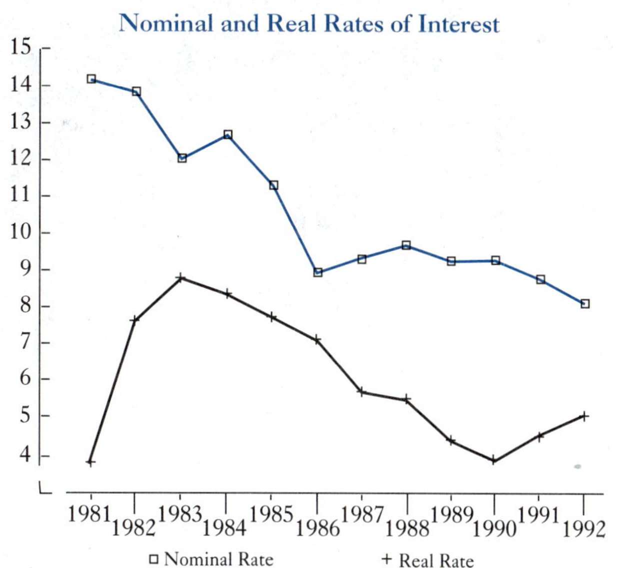

Query 16.2:

In the late 1970s, V8 engines were expected to become a thing of the past as the price of gasoline was projected to skyrocket. The price of gasoline did rise from 41 cents per gallon in 1972, to 61 cents per gallon in 1976, to \$1.38 per gallon in 1981, and then fell to \$1.13 per gallon in 1992, which was still higher than the price of gasoline in the 1970s. Production of big V8 engines, however, did not come to a halt. Why?

Answer:

Although the nominal or observed price of gasoline rose, a comparison with other prices shows that the "real" price of gasoline did not rise continuously from 1972 to 1992. A price index can be used to show how the price of gasoline changed relative to prices in general. Table 16.1 provides the Consumer Price Index for the relevant years. To compare the price of gasoline in terms of "real" prices, or "1982–1984 purchasing power," the 1972 to 1992 prices must be divided by the corresponding Consumer Price Index values (41.8 to 140.3), and then the ratios must be multiplied by 100. This arithmetic is done easily with a spreadsheet program. The "real" prices of gasoline in 1972 and 1992, in terms of 1982–1984 purchasing power (written as 1983 dollars) are

$$\text{1972 gas price in 1983 dollars} = \frac{\text{1972 gas price}}{\text{CPI}(1972)} = \frac{\$0.41}{41.8}(100) = 0.98$$

and

$$\text{1992 gas price in 1983 dollars} = \frac{\text{1992 gas price}}{\text{CPI}(1992)} = \frac{\$1.13}{140.3}(100) = 0.81$$

which indicates that the real price of gasoline decreased between 1972 and 1992. Although the nominal or observed price of gasoline is 176 percent higher in 1992 than it was in 1972 (from 41 cents to \$1.13), the Consumer Price Index rose by some 236 percent (41.8 to 140.3) over the same period. Thus, the real price of gasoline fell in terms of 1983 dollars. Similar calculations, provided in the spreadsheet display, show that the real price or 1983 constant dollar price of gasoline peaked in 1980. As seen in the graph, although the nominal price of gasoline rose through the 1970s and 1980s, the real price of gasoline rose only in the late 1970s and early 1980s; it fell in the late 1980s. For the entire 1972–1992 period, the use of gasoline for consumption in big V8 engines has become relatively less expensive.

The Real Price of Gasoline

Year	CPI 1982–1984 base = 100	Observed Current Price	Real Price in 1982–1984 dollars
1972	41.8	0.41	0.98
...	...	...	...
1976	56.9	0.61	1.07
1977	60.6	0.65	1.07
1978	65.2	0.66	1.01
1979	72.6	0.90	1.24
1980	82.4	1.26	1.53
1981	90.9	1.38	1.52
1982	96.5	1.29	1.34
1983	99.6	1.25	1.26
1984	103.9	1.22	1.17
1985	107.6	1.20	1.12
1986	109.6	0.93	0.85
1987	113.6	0.95	0.84
1988	118.3	0.95	0.80
1989	124.0	1.02	0.82
1990	130.7	1.16	0.89
1991	136.2	1.15	0.84
1992	140.3	1.13	0.81

Real and Nominal Gas Prices

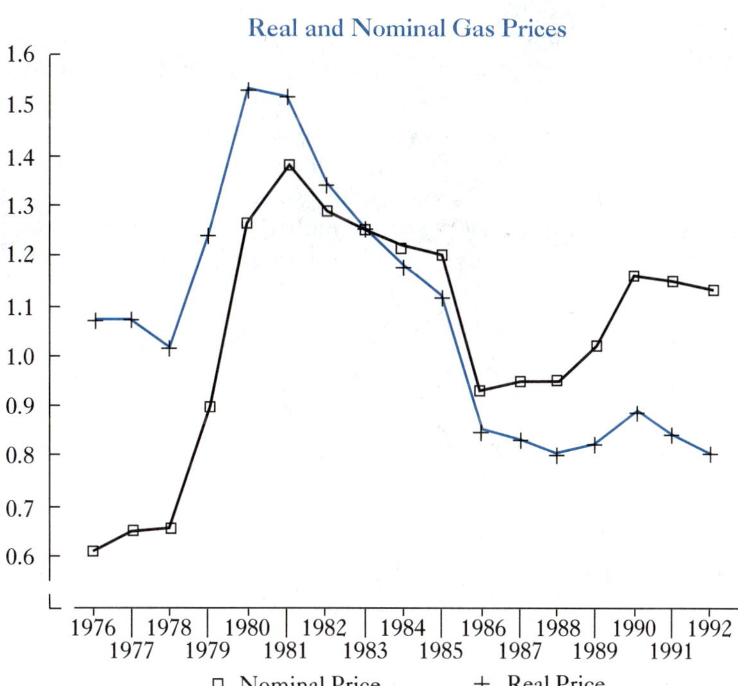

16.7 Seasonality and Smoothing of Index Numbers

In Chapter 15, we discussed the components of a time series. Indexes that are used to measure change over time may have trend, seasonal, and random components. The seasonal and random components in price indexes can produce highly variable values from month to month and quarter to quarter. The seasonal component, which is a regular peculiarity of the time period for which the measure pertains, can be removed in several different ways. The U.S. Bureau of Labor Statistics regularly studies these different methods of "deseasonalizing" index values. Together with the Commerce Department, BLS reports all key indexes both as "seasonally adjusted" and unadjusted. Here we will address only two different methods for deseasonalizing a time series of index numbers. Although we will restrict our attention to index numbers, the techniques demonstrated will work for "smoothing" any time series.

Comparable Period Differencing

The first way to remove the effect of seasonality is to measure changes in the index between comparable periods. When the business press reports monthly information on inflation, for example, it is typically given as a rate of change from the previous month and as a rate of change from 12 months previous. Monthly changes in the Consumer Price Index are magnified by seasonal components, but a 12-month difference removes the seasonal effect particular to that month.

The third column, titled "CPI," in Table 16.7 shows the Consumer Price Index by month for 54 months. Figure 16.1 provides a line graph of this monthly Consumer Price Index time series. Notice, for example, that there tends to be little increase from November to December, but from December to January there tends to be a large increase; this is a demonstration of seasonality. A comparison of these CPI percentage change from one month to the next can be meaningless. Thus, when a monthly change is reported, it is typically accompanied by its change from 12 months earlier—that is, from November to November, from December to December, and so on.

Moving Averages

Moving Average A method of smoothing data in which an observation at time t is replaced by the average of the series for a fixed number of past periods.

To smooth out the peaks and troughs in any time series, a **moving average** can be constructed. For the monthly Consumer Price Index, a T period moving average at time t may be defined by[3]

$$\overline{\text{CPI}_t} = \sum_{i=0}^{T-1} \text{CPI}_{t-i}/T, \text{ where } t - i > 0 \text{ for all } i$$

For example, a six-month moving average is shown in the fourth column of Table 16.7. The first possible value of this moving average appears in the sixth period ($T = 6$)

Table 16.7 Consumer Price Index by Month, by Six-Month Moving Average, and by 12-Month Moving Average

	Month/year	CPI	Six-month moving average	12-month moving average
1	1/1989	121.1		
2	2	121.6		
3	3	122.3		
4	4	123.1		
5	5	123.8		
6	6	124.1	122.7	
7	7	124.4	123.2	
8	8	124.6	123.7	
9	9	125.0	124.2	
10	10	125.6	124.6	
11	11	125.9	124.9	
12	12	126.1	125.3	124.0 = CPI 1989
13	1/1990	127.4	125.8	124.5
14	2	128.0	126.3	125.0
15	3	128.7	127.0	125.6
16	4	128.9	127.5	126.0
17	5	129.2	128.1	126.5
18	6	129.9	128.7	127.0
19	7	130.4	129.2	127.5
20	8	131.6	129.8	128.1
21	9	132.7	130.5	128.7
22	10	133.5	131.2	129.4
23	11	133.8	132.0	130.0
24	12	133.8	132.6	130.7 = CPI 1990
25	1/1991	134.6	133.3	131.3
26	2	134.8	133.9	131.8
27	3	135.0	134.3	132.4
28	4	135.2	134.5	132.9
29	5	135.6	134.8	133.4
30	6	136.0	135.2	133.9
31	7	136.2	135.5	134.4
32	8	136.6	135.8	134.8
33	9	137.2	136.1	135.2
34	10	137.4	136.5	135.5
35	11	137.8	136.9	135.9
36	12	137.9	137.2	136.2 = CPI 1991
37	1/1992	138.1	137.5	136.5
38	2	138.6	137.8	136.8
39	3	139.3	138.2	137.2
40	4	139.5	138.5	137.5
41	5	139.7	138.9	137.9
42	6	140.2	139.2	138.2
43	7	140.5	139.6	138.6

(continued)

TABLE 16.7 Consumer Price Index by Month, by Six-Month Moving Average, and by 12-Month Moving Average *(continued)*

	Month/year	CPI	Six-month moving average	12-month moving average
44	8	140.9	140.0	138.9
45	9	141.3	140.4	139.3
46	10	141.8	140.7	139.6
47	11	142.0	141.1	140.0
48	12	141.9	141.4	140.3 = CPI 1992
49	1/1993	142.6	141.8	140.7
50	2	143.1	142.1	141.1
51	3	143.6	142.5	141.4
52	4	143.9	142.9	141.8
53	5	144.2	143.2	142.2
54	6	144.4	143.6	142.5

FIGURE 16.1 Monthly Consumer Price Index

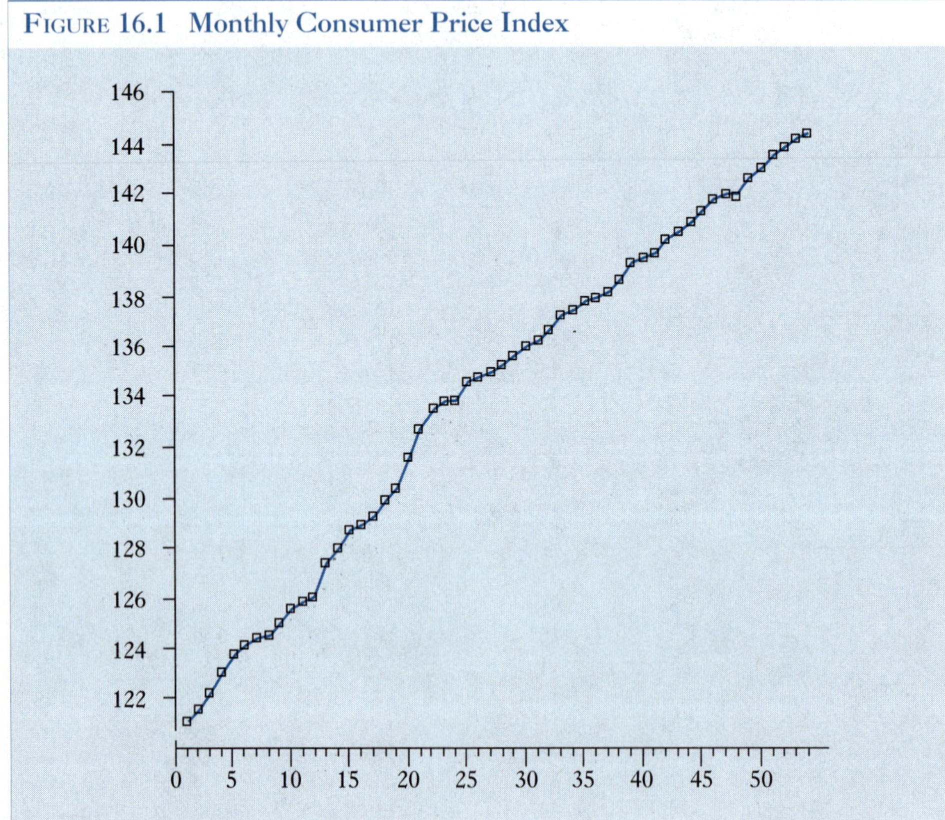

because $t - i$ must be greater than zero and the six-month period is the first period for which this is true. The first value of the six-month moving average is

$$\overline{\text{CPI}}_6 = \sum_{i=0}^{5} \text{CPI}_{6-i}/6 = (121.1 + 121.6 + 122.3 + 123.1 + 123.8 + 124.1)/6$$

$$= 122.7$$

The second value is

$$\overline{\text{CPI}}_7 = \sum_{i=0}^{5} \text{CPI}_{7-i}/6 = \left(121.6 + 122.3 + 123.1 + 123.8 + 124.1 + 124.4\right)/6$$

$$= 123.2$$

and so on, to the 49th value, where

$$\overline{\text{CPI}}_{54} = \sum_{i=0}^{5} \text{CPI}_{54-i}/6 = \left(142.6 + 143.1 + 143.6 + 143.9 + 144.2 + 144.4\right)/6$$

$$= 143.6$$

The effect of this six-month moving average has been to smooth the Consumer Index Price measurements, as apparent in Figure 16.2. The longer the moving average, the smoother will be the resulting series. For instance, a 12-month moving average is provided in the fifth column of Table 16.7 and graphed in Figure 16.3, along with the original monthly Consumer Price Index and the six-month moving average. The December values of this 12-month moving average are approximately the same as the annual Consumer Price Index values that are found in the *Economic Report of the*

FIGURE 16.2 Six-Month Moving Average of the Consumer Price Index

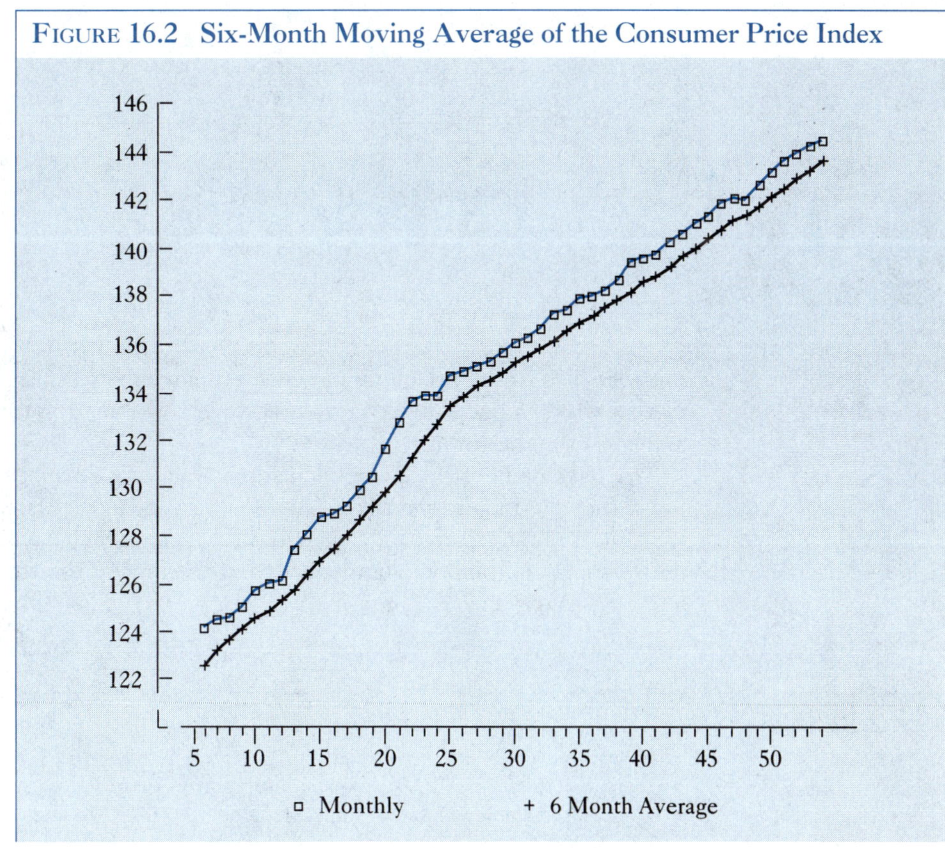

FIGURE 16.3 Twelve- and Six-Month Moving Average of the Consumer Price Index

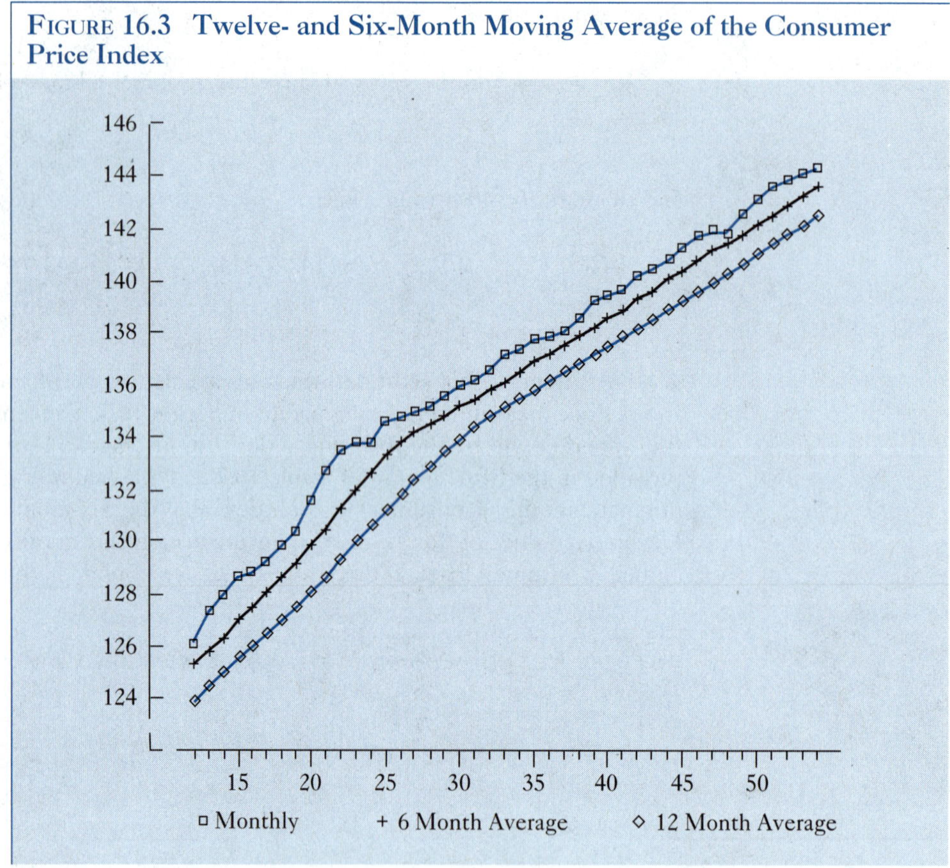

President. Twelve-month moving averages remove the seasonal and random effects that reoccur within a fixed period from year to year, making the trend and cyclical components in a time series more apparent.

We have demonstrated the calculation of a moving average for only the Consumer Price Index. Moving averages, however, can be constructed for any time series. Regardless of the time series considered, a moving average will always tend to remove the seasonal and random components and thus make the trend and cyclical components easier to see.

EXERCISES

16.6 The American Medical Association reported that the average income of physicians increased about $11,100 to some $155,800 between 1988 and 1989. By how much did this increase outpace increases in the Consumer Price Index in 1989?

16.7 In the early 1980s, when the 1982–1984 period was being established as the latest base for the Consumer Price Index, a rumor circulated that President Ronald Reagan wanted the CPI rebased to make it look as if inflation had been lowered. "After all," the rumor had him musing, "the CPI would be lower as a result of changing the base from 1967 to 1982–84." Comment on the substance of this rumor.

16.8 If a person's salary was tied to the CPI in the early 1980s, what would have been the effect of rebasing to the years 1982–1984?

16.9 Below is a diagram of real gasoline prices in the United States, with a base year of 1991 when gasoline was priced at \$1.15 per gallon. As can be seen, the real price of gasoline measured in terms of 1991 dollars peaked in 1980 at \$2.04 per gallon. Query 16.2 showed the real price of gasoline to be \$1.53 in 1980. Why is there a 51-cent difference between these two real prices of gasoline?

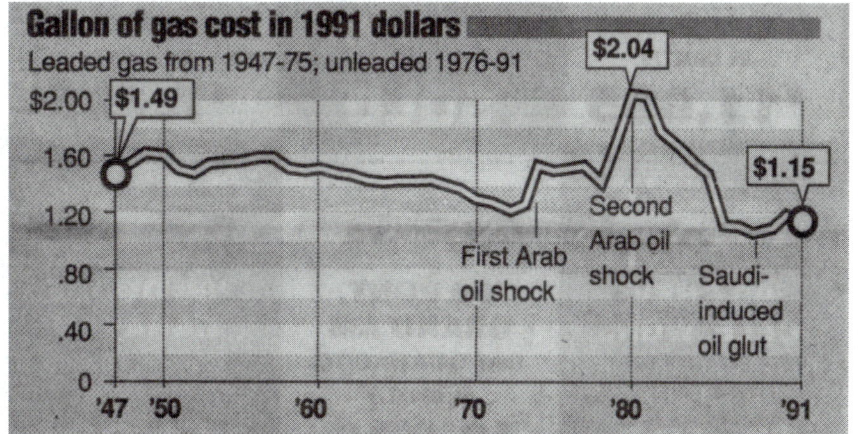

16.10 An article in *The Wall Street Journal* (February 12, 1991) stated that "At Brigham and Woman's (Hospital), salaries rose an average of about 8% last year; the consumer price index climbed 6.1%." What happened to real salaries at this hospital? (Give an answer involving a numerical value.)

16.11 A "real wage" is calculated as 100 times the ratio of the nominal or money wage to a price index. A "real interest rate" is calculated as the nominal or observed interest rate minus the rate of change in the price index. Why aren't these two "real values" calculated in the same way?

16.12 The Employment Cost Indexes for private industry, *Economic Report of the President* (February 1992, p. 347), are indexes for employer costs for wages, salaries, and benefits, which are

INDEX	1986	1987	1988	1989	1990	1991
Wages + salaries	91.1	94.1	98.0	102.0	106.1	110.0
Percent change		3.2930	4.1445	4.0816	4.0196	3.6757
Benefits	87.5	90.5	96.7	102.6	109.4	116.2
Percent change		3.4285	6.8508	6.1013	6.6276	6.2157
Compensation	90.1	93.1	97.6	102.3	107.0	111.7
Percent change		3.3296	4.8335	4.8155	4.5943	4.3925

a. During this period, what was the average percentage increase in wages and salaries, benefits, and total compensation?

b. Did increases in total compensation outpace price inflation during this period?

16.13 The late University of Minnesota Professor Walter Heller was also Chairman of the Council of Economic Advisors to U.S. President John F. Kennedy in the early 1960s. Heller was notorious for saying that if inflation continues at the current monthly rate, then the annual rate will be 12 times as much. That is, a monthly rate of 0.4 percent, if it persists, implies an annual rate of 4.8 percent. What is wrong with this inference?

16.14 In class, Walter Heller would infer from a monthly rate to an annual rate of inflation by the method stated in the Exercise 16.13. When corrected by the author of this textbook, Heller would shrug off the criticism and say his method gave an adequate approximation. In the early 1960s, was Heller's approximation adequate? In the late 1970s, was it adequate? Explain.

16.15 Will a nine- or a six-month moving average do a better job of "smoothing" a time series? Explain why.

16.16 For the annual money stock data in Exercise 15.4, calculate a three-year moving average. Plot your three-year moving average and the original data. Does either plot appear to be smoother? Why or why not?

16.8 Concluding Comments

Index numbers can be used as economic indicators for forecasting, as in projecting earnings, or as deflators, as in the calculation of real wages and real interest rates. They provide barometers of business changes over time and are thus needed in the consideration of historical magnitudes.

As demonstrated in this chapter, knowledge of the way price and quantity indexes are constructed is essential for their correct use. You now know some of the differences among the major indexes that are typically cited in the popular and business press as measures of economic activity. You also know how to construct, chain, and splice these indexes.

Chapter Exercises

16.17 The list price of a Chevrolet sedan changed as follows during the 1967–1992 period:

1967:	$ 3,395
1974:	$ 3,960
1977:	$ 4,901
1982:	$ 8,137
1987:	$11,770
1992:	$17,300

What happened to the "real" price of Chevrolet sedans over this period? To answer this question, what do you need in addition to the data provided on Chevrolet prices?

16.18 Are the "real" Chevrolet sedan prices calculated in Exercise 16.17 an adequate measure of the cost of personal transportation? Explain. What other factors should be taken into account in comparing car prices over time?

16.19 The "constant dollar" price of the top of the line IBM personal computer plummeted in the 1980s and early 1990s. Why isn't this an adequate measure of the true price of these computers?

16.20 Below is a *Business Week* (February 18, 1991) graph showing an index of mean physician income and average national wage between 1985 and 1989.

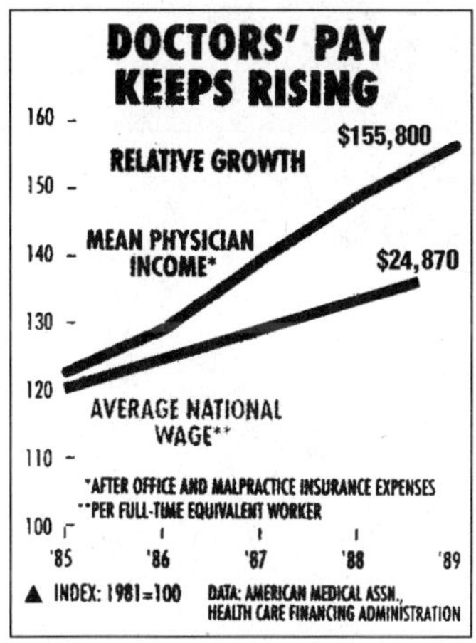

What were the approximate percentage and dollar increases in

a. physician income from 1981 to 1989?

b. national wage from 1981 to 1989?

16.21 From the *Business Week* graph in Exercise 16.20 determine the approximate percentage and dollar increase in

a. physician income from 1985 to 1989?

b. national wage from 1985 to 1989?

16.22 A reporter observed that the Consumer Price Index was 146.2 in January 1994 and 142.6 a year earlier. He observed that the CPI was 145.8 in December 1993 and 141.9 a year earlier. Thus, he concluded, "the rate of inflation continues to rise with the most recent monthly change showing a 0.4 percent increase" (i.e., 146.2–145.8). Comment on the correctness of his analysis and conclusion.

16.23 In finance, it is common to hear brokers refer to a "basis point" as 1/100th of a percentage point, as in "Freddie Mac's 30-day required yield on 30-year fixed rate mortgages averaged 9.89%, two basis points above the U.S. average rate of 9.87%." Are these basis points specific to a given year, as with an index number? Explain.

16.24 An Associated Press release (*HT*, April 18, 1991) stated that,

> People living along the East Coast enjoyed the fastest income growth. . . . At the top of the list were residents of New Jersey, whose per capita incomes jumped an average 8.0 percent annually, to $24,968, between 1980 and 1990.

Why is an arithmetic average growth rate of 8 percent not the rate of growth that was compounding annually between 1980 and 1990?

16.25 Below is the salary index for sales occupations. The index shows the earnings path that a salesperson s earnings might follow as he or she ages.

Age	Index	Age	Index
37	1.000000	57	1.110545
38	1.021484	58	1.098895
39	1.041249	59	1.085658
40	1.059301	60	1.070838
41	1.075648	61	1.054444
42	1.090295	62	1.036482
43	1.103249	63	1.016958
44	1.114518	64	0.995880
45	1.124106	65	0.973252
46	1.132022	66	0.949083
47	1.138271	67	0.923378
48	1.142860	68	0.896144
49	1.145796	69	0.867388
50	1.147085	70	0.837116
51	1.147000	71	0.805335
52	1.144748	72	0.772052
53	1.141136	73	0.737272
54	1.135903	74	0.701002
55	1.129056	75	0.663250
56	1.120601	76	0.624021

By this index, what percentage increase or decrease can a salesperson expect from

a. age 40 to 41?
b. age 50 to 51?
c. age 60 to 61?
d. age 70 to 71?

16.26 If a salesperson is making $42,692 at age 37, using the age-earnings index in Exercise 16.25, what is this person s expected earnings at ages 40, 50, 60, and 70?

16.27 For the data on federal outlays in Exercise 15.36, construct a five-period moving average. What is the effect of this moving average?

16.28 After seeing an article in *Business Week* (May 24, 1993) that showed slightly more than a 200 percent increase in college tuition between 1980 and 1992, a math major responded that this was more than an average 16.7 percent annual increase over this 12-year period. An economics/finance major said no way; it was more on the order of 10 percent per year. Who is correct and why?

16.29 On the top of the following page is a diagram showing an index of U.S. academic R&D and library expenditures (in constant 1982 dollars), where 1976 is the index base for each series (as circulated by Dr. Peter Shepherd, publishing director, Pergamon Press: August 1992). How were these indexes created?

16.30 Why is the vertical distance between the two index values in the diagram in Exercise 16.29 of R&D and library expenditures a misleading measure of differences in growth rates in the two series?

16.31 The bottom diagram on the following page was prepared by someone at the University of Minnesota Medical School. From this diagram, readers were to compare tuition increase vs. Consumer Price Index. Comment on the appropriateness of this diagram for the stated purpose.

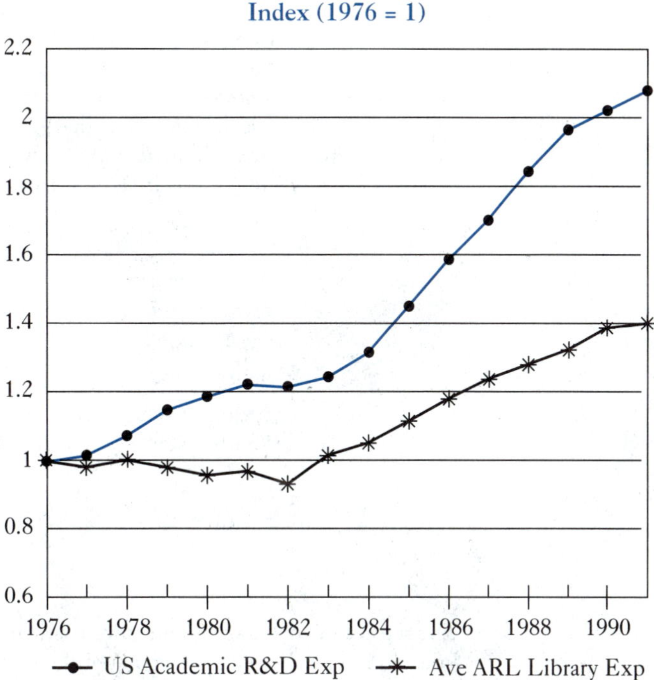

Sources: R & D - *Science & Engineering Indicators-1991*, National Science Board (Appendix Table 5.1) Library Expenditure-Association of Research Libraries (ARL) Statistics.

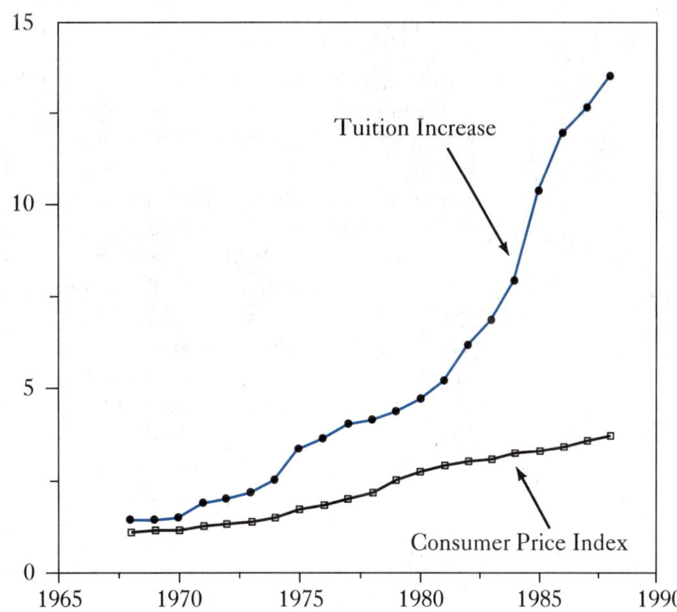

16.32 In a wrongful injury case, the economist hired by the plaintiff's lawyer to assess damages cited a 1979 study by W. H. Gauger and K. E. Walker, "The Dollar Value of Household Work" (The New York State College of Human Ecology, Revised 1980) for assigning a dollar value to household services the plaintiff lost as the result of his injuries. The economist stated, "the wage rates cited in 1979 were inflated to current levels using the Consumer Price Index." In reviewing this work for the defense, the author of this book obtained the following information on wages:

	1979	1991
Average hourly earnings in U.S.	$6.16	$10.34
Kitchen helper's hourly wage	$3.02	unavail.*
Cleaning person's hourly wage	$4.75	unavail.*
Minimum wage	$2.90	$ 4.25

* The U.S. Bureau of Labor Statistics "Job Bank" is unavailable but the *Occupational Outlook Handbook* lists the wage of "Private Household Workers" as $4.25/hr, for those covered by the federal minimum wage.

The author of this book then showed how inflating the 1979 wages by the Consumer Price Index would overstate actual observed wages in 1991. By how much would projecting 1979 wages based on changes in the Consumer Price Index overestimate actual wages?

16.33 To the right is a *Business Week* (August 23, 1993) diagram that reports to show an "annualized three-month moving average" of consumer inflation. State the different ways in which this could have been calculated.

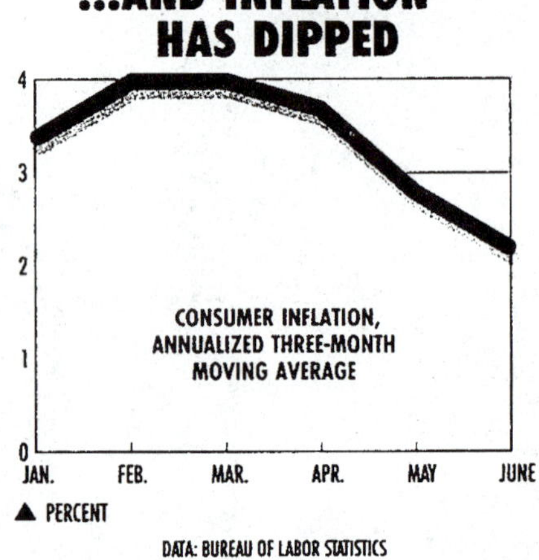

16.34 Use the information on the Consumer Price Index in this chapter, or from your library, to calculate an "annualized three-month moving average" of inflation. Graph your results as in the *Business Week* diagram in Exercise 16.33. Does your curve look like that in *Business Week?* (Why or why not?)

16.35 Economist L. Douglas Lee, of NatWest Washington Analysis, was cited in *Business Week* (July 19, 1993) as finding that gold prices tend to reflect overall inflationary trend, but they are far more erratic than changes in the Consumer Price Index. "Moreover, using three-month moving averages, he found no statistically significant relationship between gold prices and subsequent inflation, with time lags up to 12 months." Go to the library and retrieve data on monthly gold prices and the CPI for the past two decades (since the dollar was decoupled from gold convertibility). Create three-month moving averages of both series, and test for the lack of statistical significance reported by Lee. Do your results confirm or refute Lee's findings? Explain.

16.36 An article in *Business Week* (October 4, 1993) stated that Japan, with a 3.9 percent interest rate and 1 percent inflation rate, has a real rate of 2.9 percent on its 10-year bonds, which is "still startlingly high." Find a library source to check if this is truly a recent high for this real rate in Japan.

ENDNOTES

1. Students of economics may wish to demonstrate with utility analysis that a Laspeyres index tends to overstate price increases (as measured by a loss in utility) and understate price declines (as measured by a gain in utility). In contrast to a Laspeyres index, a Paasche index tends to understate price increases (as measured by utility loss) and overstate price declines (as measured by utility gain).
2. In addition to price indexes, quantity indexes can be defined. To create a quantity index, simply substitute price terms for quantity terms, and vice versa. In a quantity index, price becomes the weight. A Laspeyres quantity index is

$$\frac{\sum_{j=1}^{J} Q_{Ij} P_{bj}}{\sum_{j=1}^{J} Q_{bj} P_{bj}}(100)$$

and the Paasche quantity index is

$$\frac{\sum_{j=1}^{J} Q_{Ij} P_{Ij}}{\sum_{j=1}^{J} Q_{bj} P_{Ij}}(100)$$

3. There are no hard rules as to how long (T) the moving average should be, or from where the moving average should be calculated. As an alternative to calculating the moving average on only past values, we could center the moving average on past and future values. For a five-month period moving average, for example, the formula could be written:

$$\overline{\text{CPI}_t} = \sum_{i=-2}^{2} \text{CPI}_{t+i} / 5$$

Because two future values are required for this type of centering, a moving average calculated in this way can never be current.

FORMULAS

- Laspeyres price index

Ratio of the total cost in the year of interest of the quantities of each item consumed in the base year to the total cost of these quantities in the base year. In formula form:

$$\frac{\sum_{j=1}^{J} Q_{bj} P_{Ij}}{\sum_{j=1}^{J} Q_{bj} P_{bj}}(100)$$

where Q_{bj} is the base year quantity of jth item purchased
P_{bj} is the base year price of jth item
P_{Ij} is the jth item price in the year of interest

- Average-weight Laspeyres price index

$$\frac{\sum_{j=1}^{J} \bar{Q}_{bj} P_{Ij}}{\sum_{j=1}^{J} \bar{Q}_{bj} \bar{P}_{bj}}(100)$$

where $\bar{Q}_{bj}$ is the average quantity of jth item purchased
$\bar{P}_{bj}$ is the average price of jth item in base period
P_{Ij} is the jth item price in the year of interest

- Paasche price index

Ratio of the total cost in the year of interest of the quantities of each item consumed in the year of interest to what would have been the total cost of these items in the base year (i.e., at base year prices). In formula form

$$\frac{\sum_{j=1}^{J} Q_{Ij} P_{Ij}}{\sum_{j=1}^{J} Q_{Ij} P_{bj}}(100)$$

where Q_{Ij} is the jth item consumed in year of interest
P_{bj} is the base year price of jth item
P_{Ij} is the jth item price in the year of interest

- Moving average of CPI

For the monthly Consumer Price Index, a T period moving average at time t is defined by

$$\overline{\text{CPI}}_t = \sum_{i=0}^{T-1} \text{CPI}_{t-i} / T$$

where $t - i > 0$ for all i

Chapter 17

Decision Analysis

He is no wise man who will quit a certainty for an uncertainty.

Samuel Johnson
1709–1784

17.1 Introduction

What is the best course of action when

- selling a car: buying newspaper space to advertise the car for sale or waiting for someone to see the car for sale in your driveway?
- considering a move: building a new building, renovating an existing one, or continuing with the existing building as is?
- making any decision involving uncertainty: paying for more sample data or making a decision on available information?
- evaluating safety: facing new evidence that Genentech Inc.'s heart drug, TPA, "saves an additional one percentage point of patients, or one in 100, over streptokinase," but, "at $2,200 a dose, TPA costs 10 times more than streptokinase," (*Wall Street Journal*, April 3, 1993), health economists and cardiologists must decide if the benefits are worth the cost—that is, how much is a life worth?

Courses of Action The alternatives available to the decision maker.

These and other questions related to decision making under uncertainty are the focus of this chapter. By the end of this chapter, you will be able to use a decision tree to find the best **course of action** when confronted with sequential and alternative courses of action. You will be able to assess the importance of utility functions in the decision-making process. You will be able to update probabilities as newer information becomes available, and you will be able to determine when it is worthwhile to gather that additional information. Finally, you will learn of possible shortcomings in the decision tree/utility theory process of decision making and of ongoing research into statistical decision theory.

17.2 Decision Trees

Decisions involve uncertainty; when deciding among several courses of action, the outcomes of the actions are unknown. At best, the decision maker may know or be able to speculate about the likelihood of the outcome associated with each possible course of action. For example, the author of this book recently bought a new car without trading in his old one. A neighbor offered to buy the old car for $6,300, if the author could not sell it for the red book suggested retail price of $6,600. The local newspaper claimed that 80 percent of car ads result in sales; the ad cost $44. What was the appropriate action to be taken?

Decision Tree A diagram representing a series of choices as forks, with courses of action as the branches emanating from a fork.

A "decision tree" is helpful in visualizing the actions, outcomes, and the associated likelihoods in problems of the above type. A **decision tree** represents the series of choices as forks or branching points. Each fork represents a junction where the decision maker is in control of events or one in which the outcome is determined by chance. Figure 17.1 shows the decision tree facing the author in the sale of his car.

At the starting point, the left-hand side of the tree, there is a decision to be made. Either an ad can be placed in the newspaper or the car can be sold immediately to the neighbor. This is a decision point that is at the control of the decision maker; such

FIGURE 17.1 Decision Tree for Selling a Car

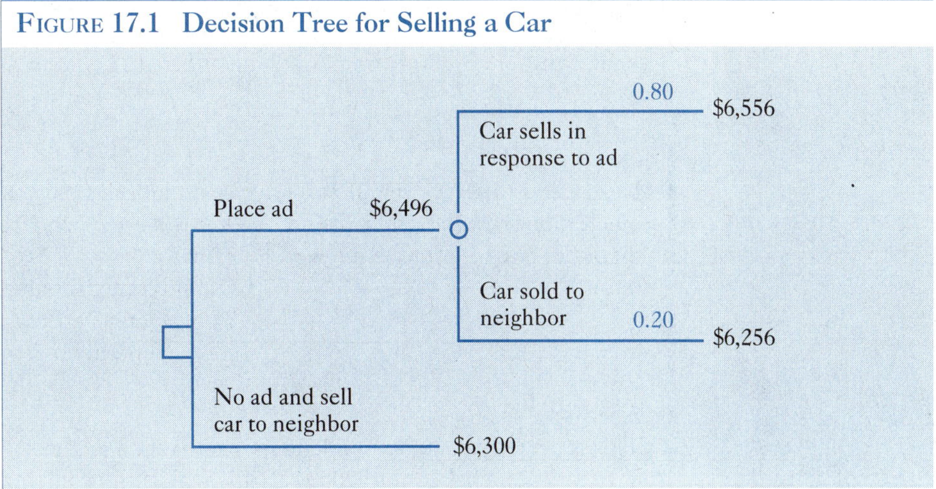

decision points are depicted as squares with extending branches showing the available outcomes. If no ad is placed and the car is sold to the neighbor, the consequence is certain—the author has $6,300 for his car, as shown at the right-hand end of the branch. If the ad is placed, the branch leads to a chance fork that is designated by a circle. The top branch following this chance fork represents the sale of the car for $6,600 less the cost of the ad ($6,556 = $6,600 – $44). The lower branch emanating from this chance fork depicts the sale of the car to the neighbor but with the added expense of the ad ($6,256 = $6,300 – $44). The probability of either of these two chance outcomes is shown at the end of the branch, along with the associated dollar amounts. In addition to being called outcomes, these dollar amounts are called the **payoffs.**

Payoff
The outcome of a combination of an action and the realization of a state of nature.

In Chapters 3 and 4, the idea of an expected value was introduced as a mean calculated from probability weighted numerical outcomes. The expected value at the chance fork is $6,496 [= (0.8)(6556) + (0.2)(6256)]. This is the expected value for advertising the car in the newspaper (at a price of $6,600, at a cost of $44 for running the ad, assuming that there is an 80 percent chance that it sells, and that if it does not, then the neighbor will buy it at $6,300). Moving back to the starting point, we can now see that the choice is between a certain $6,300 or the expected value of $6,496. This process of moving from right to left along the decision tree is known as **backward induction.**

Backward Induction
Working from right to left in a decision tree; reasoning from the payoffs to the course of action to be taken.

The decision tree analysis in Figure 17.1 led the author of this book to place the ad in the newspaper, because that action maximized expected monetary returns. As it turned out, the car did not sell through the newspaper ad and the car was eventually sold to the neighbor for $6,300. This outcome, however, is not inconsistent with the analysis because there was a nonzero (actually 20 percent) chance of this happening. Remember from Chapter 3 that expected values are means calculated with probabilities and not observed relative frequencies. In making decisions under the uncertainty reflected in probabilities, there are no assurances that an actual outcome will be realized as the result of any one decision. When working with expected values, we only know what could happen in the long run, with (infinitely) repeatable decisions.

THE BEST OF MARILYN ON QUERIES ABOUT PROBABILITY

The following are excerpts from Marilyn vos Savant's column, "Ask Marilyn," in *Parade Magazine.*

QUERY 17.1:

I am asked to select one of two envelopes and told only that one contains twice as much money as the other. I find $100 in the envelope I select. Should I switch to the other one to improve my worldly gains?

— Barney Bissinger
Hershey, PA
September 20, 1992

ANSWER:

This is a dandy paradox. While it appears as though you should switch, because you have an even chance for $200 vs. $50—which any gambler would grab—it actually makes no difference at all. Those even chances would apply only if you could choose one of two *more* envelopes, one with $200 and the other with $50.

As it is, there's just one more envelope sitting there, with either twice the amount you've already seen or half of it. And you knew that would be the case before you even started. So, when you opened the first envelope, you didn't gain any information to improve your chances. This can be illustrated by noting that the logic that causes you to switch (because you appear to have an even chance for $200 vs. $50) will lead you to switch every time (no matter what you find in the first envelope), making the second envelope just as randomly chosen as the first one!

— Marilyn vos Savant
September 20, 1992

QUERY 17.2:

Let's say we're playing roulette, and I offer you a bet. You can pick any triplet of black and red—say red/red/black or red/black/red. Then I'll pick a different one. At the starting point, we'll watch each spin of the roulette wheel until one of our triplets appears as a run. If yours comes first, you win. If mine comes first, I win. Even chances, right? But I'll give you 3 to 2 odds! When you win, I pay you $3; but when I win, you only pay me $2. We'll play as many times as you like, and you can always have the first choice. Will you take the bet?

— George Groth
Hendersonville, NC

ANSWER:

What a great "sucker" bet that would be! No—your chances of winning would range from 2/3 to 7/8, depending on what triplet I chose. You'd always be able to choose a triplet with a better chance of winning. I'll use the 7/8 chances as an example, because it's the most obvious:

There are eight different triplet combinations. Let's say I choose black/black/black. If it appears at the start, I win—and that'll happen 1/8th of the time. But before it appears anytime afterward, it would have to be preceded by a red. So if you choose red/black/black every time I choose black/black/black, you'll win 7/8ths of the time!

—Marilyn vos Savant
January 31, 1993

QUERY 17.3:

An article in the *Herald Times* (September 24, 1992) reported, "Consultant says library should be moved . . . it is time to look elsewhere for a new location which would house a new 80,000 to 90,000 square-foot library, according to the recommendation contained in the consultant's final report." The article provided the following cost estimates for adding on to and renovating the existing library versus building a new facility.

	Addition/renovation	New facility
Construction costs	$ 8,837,460	$10,490,550
Nonconstruction costs	$ 2,317,289	$ 2,659,450
Land acquisition costs	$ 2,240,000	$ 1,250,000
Revenue from land sale	– 0 –	$ 2,000,000
Total Cost	$13,394,749	$12,400,000

A local construction firm questioned these estimates, asserting that they were based on relatively low rates of price inflation. Alternative cost estimates, assuming a higher rate of inflation, suggested that the total cost of adding on and renovating was $13,497,830, and the total cost of a new facility was $13,987,345. Suppose a probability is two-thirds that inflation will be high and one-third that it will be low. What should the library board do—add on, renovate, or build a new facility?

ANSWER:

The payoff table is

	Assumed rate	Higher rate
New facility	–$12,400,000	–$13,987,345
Add on	–$13,394,749	–$13,497,830
Probability	1/3	2/3

where the negative signs indicate that each dollar amount constitutes a dollar outlay.

Assuming that the library board is only interested in expected values, then it should build a new library, even with a two-thirds chance that the higher rates of inflation will be realized. The expected values are

New facility:

$$\$\ 13.458226 \text{ million cost} = (-12.4)(0.333333) + (-13.987345)(0.666667)$$

Add on and renovate:

$$\$\ 13.463466 \text{ million cost} = (-13.394749)(0.333333) + (-13.497830)(0.666667)$$

Because the difference in expected values is not great, the board's consideration of other political interests should be anticipated in a final decision.

17.3 Risk in Decision Making

State of Nature
The environment or circumstances realized as the result of a chance event.

In choosing the action that maximizes the expected value, the only thing that is relevant is the average payoff. The variability in payoffs is not considered. But many people are uneasy about risk; they try to avoid it. For example, Table 17.1 presents a payoff table for the flip of a coin. The left-hand side of the table shows two actions: the first is a_1, which has payoffs of \$ 1.1 million if the coin comes up heads, or –\$1 million if tails. The second is a_2, which has payoffs of \$1.10 and –\$1. The payoffs for either action are determined by what is called the **state of nature,** which in this case is the outcome of a coin toss that results in "heads" or "tails" with equal probabilities.

With the notable exception of those who find risk exhilarating and who also have the funds to cover million-dollar losses, most people would select action a_2. Action a_1 has the higher expected value, but if this game is only played once, the possibility of losing \$1 million is overbearing. Action a_1 would be feasible and desired by most people only if this game could be repeated over and over and the decision maker had access to funding or credit to sustain occasional million-dollar losses. Of course, no one would offer such an ongoing game, since on average the payoff would be \$50,000; it is the existence of risk on the single play that makes games of chance appealing to both sides. The satisfaction or dissatisfaction that individuals derive from actions involving uncertainty is critical in decision making.

17.4 Expected Utility Theory

Utility
The amount of satisfaction or pleasure derived from a payoff.

Mathematician John von Neumann and economist Oskar Morgenstern, in their book *The Theory of Games and Economic Behavior* (1944), introduced the world to expected utility theory. Their approach to decision making emphasizes the satisfaction that a decision maker derives from a dollar gain versus a dollar loss when an action has an uncertain outcome. This satisfaction is called **utility,** which is the level of satisfaction

Table 17.1
Payoff Table for Coin Toss

	States of Nature		
	Heads	Tails	Expected Value
Action: a_1	\$1,100,000	–\$1,000,000	\$50,000
Action: a_2	\$1.10	–\$1.00	\$0.05
Probability	0.50	0.50	

Utility Function
An equation that determines the numerical value of utility derived from a payoff.

associated with an action. A **utility function** shows how much satisfaction a decision maker assigns to a monetary gain or loss. Figure 17.2 presents one such utility function.

In Figure 17.2, utility (measured on the vertical axis) rises with each additional dollar received (measured on the horizontal axis). In this figure, utility rises at a decreasing rate; that is, each additional dollar adds to overall satisfaction, but at a declining rate. For example, if $U(x)$ represents the utility of x dollars and the utility of \$50 is arbitrarily assigned a utility value of 100 [i.e., $U(\$50) = 100$], and if the utility of zero dollars is arbitrarily assigned a utility of zero [i.e., $U(0) = 0$], then the utility of \$25 would have to be assigned a utility greater than 50 but less than 100 to have utility increasing at a diminishing rate. In Figure 17.2, $U(\$25)$ is set at 70.

Expected Utility
The average utility derived from payoffs that occur by chance.

Von Neumann and Morgenstern extended the theory of utility to include risk. Instead of treating utility as a certainty, they introduced the idea of **expected utility.** Expected utility is the mean level of utility calculated from the utility associated with each possible outcome times its probability.

Risk Neutral Person
Someone who evaluates a gamble or a chance prospect in the same way using either expected utility or expected monetary payoff.

At least in theory, the shape of a utility function for outcomes involving risk could be derived by asking the decision maker a series a questions about the value of a prospect or game. For example: "What is the value to you of a coin toss where you win \$50 if heads comes up and zero if tails?" The expected monetary value of this prospect is \$25 [= 0.50(\$50) + 0.50(\$0)]. For a person who is indifferent to risk (risk neutral), the utility of a prospect with an expected value of \$25 is identical to the utility of \$25 with certainty. In Figure 17.3, panel a, this utility is shown at C, the point half-way along the straight line connecting points A and B. In panel a, the **risk neutral** person evaluates a gamble or the chance prospects in the same way using either expected utility or the expected monetary payoffs. That is, expected utility and the expected monetary value can be treated as one and the same for a risk neutral person.

FIGURE 17.2 A Utility Function

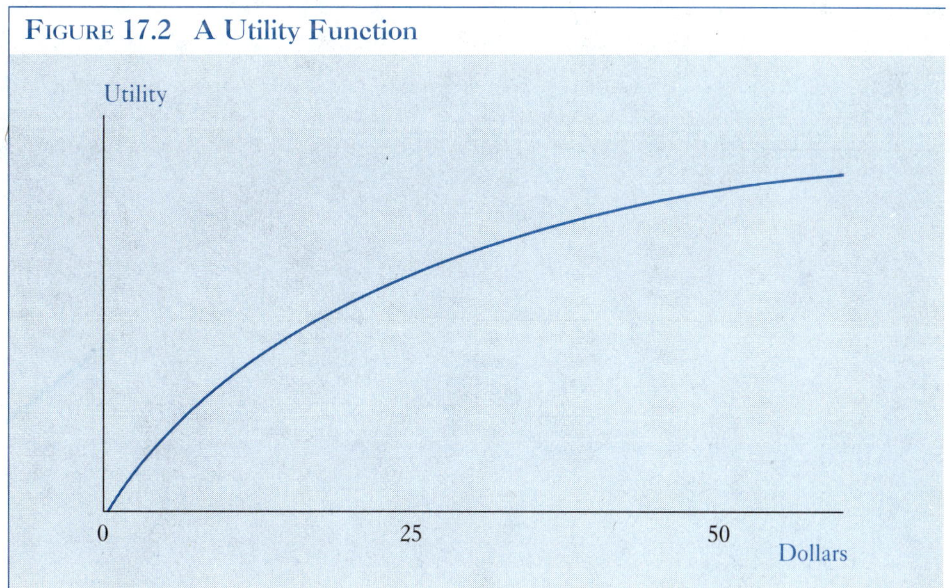

FIGURE 17.3 Three Utility Functions

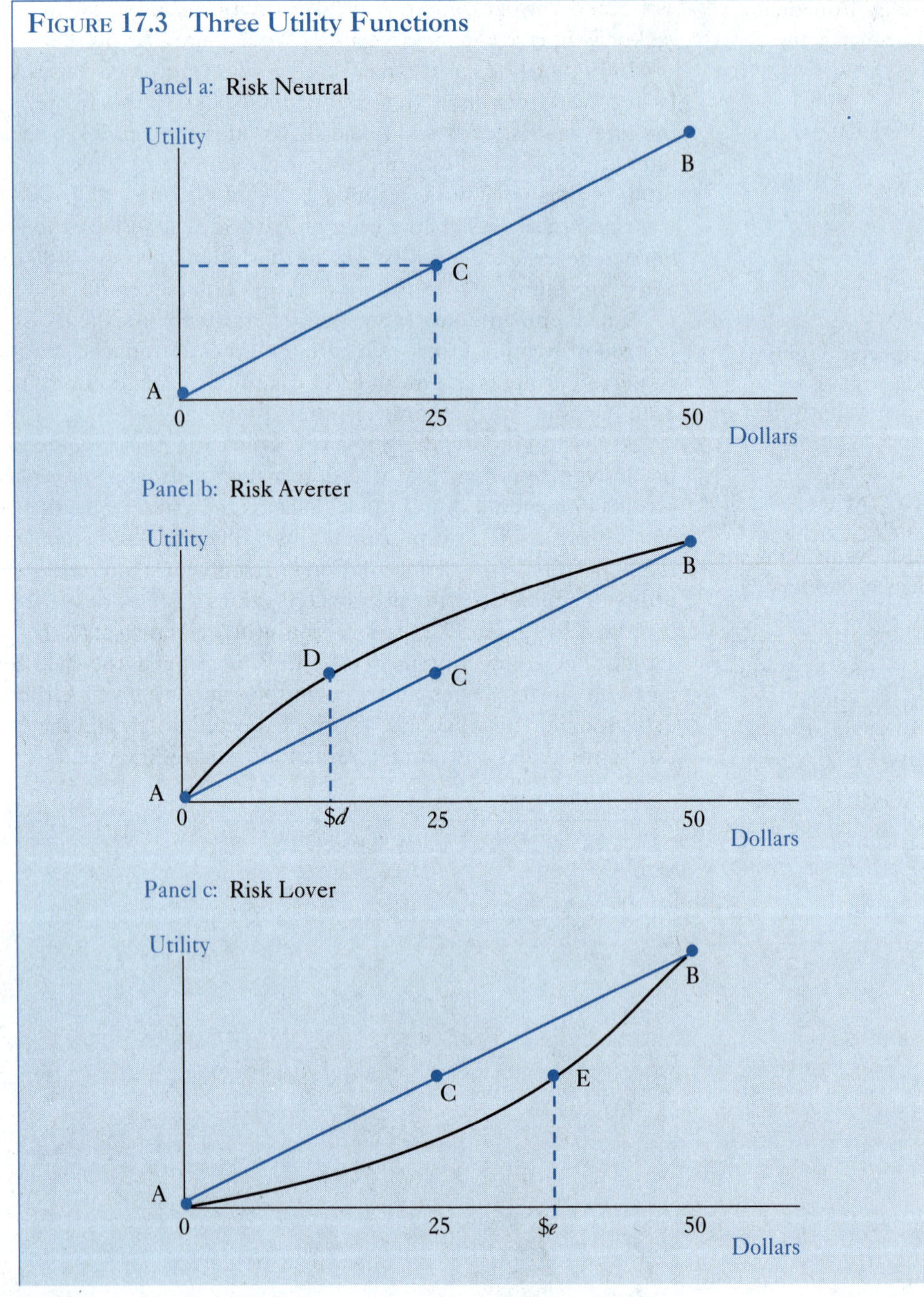

Risk Averter
Someone who prefers the more certain outcome to one with less certainty when both have equal expected monetary values.

Risk Lover (Risk Seeker)
Someone who prefers a more uncertain outcome to one with more certainty when both have the same expected monetary values.

If the decision maker prefers a dollar with certainty to the chance of a dollar involving the risk that the dollar will never be received, the utility function would be that in Figure 17.2 and Figure 17.3, panel b. In Figure 17.3, panel b, point D shows that a dollar amount ($*d*) less than $25 to be paid with certainty will yield satisfaction or utility equal to the prospect with an expected value of $25. A **risk averter** is a decision maker who prefers the certain outcome to a prospect involving uncertainty when the dollar value of the certain outcome equals the expected monetary value of the uncertain prospect. Generalizing, at the risk of over simplifying, a risk averter prefers the more certain outcome to one with less certainty, even though they may both have equal expected monetary values. A risk-averse person will demand a premium to accept risk.

If the decision maker prefers a gamble with an expected payoff of a dollar to a dollar with certainty, the utility function would look like that in Figure 17.3, panel c. In Figure 17.3, panel c, point *E* shows that a dollar amount ($*e*) greater than $25 with certainty will yield satisfaction or utility equal to the gamble that has an expected value of $25. A **risk lover** or **risk seeker** is a decision maker who prefers the more uncertain outcome to one with more certainty, if they have the same expected monetary value; a risk seeker will pay a penalty to take risk.

17.5 An Application of Expected Utility to Production

If an expected utility could be determined with objective probabilities, and if preferences could be well defined, the utility values reflected in this analysis could be treated as the payoffs on the right-hand side of a decision tree. Then, instead of calculating expected monetary values, the decision maker calculates expected utilities. That is, the process starts with the maximization of expected utility instead of expected dollar values. Consider the following quote from *Business Week* (September 7, 1992):

> The Hottest Thing Since the Flashbulb . . . the disposable camera—basically a roll of film with a cheap plastic case and lens . . . . Sales in the U.S. zoomed 50% last year . . . . Projected sales for 1992 are 22 million units in the U.S. or about $200 million at retail . . . . Just look at camera-happy Japan, where disposables now capture more than 10% of the film market, vs. 3% in the U.S.

In anticipation of increased demand for these cameras, the question, if you are a Kodak executive, is whether to build a new plant, modify existing facilities, or continue using existing facilities as they are. Assume the company has classified possible sales of the disposable camera as either "low," "medium," or "high" demand, with prior probabilities of 0.10, 0.50, and 0.40, and prepared the payoffs in Table 17.2. (Note: The assumed payoffs in Table 17.2 are to be interpreted as profits; that is, sales revenue less expenses.)

Sometimes, after displaying payoffs, one action is obviously inferior and should never be selected, no matter what the state of nature that may occur. If the monetary payoff of an action is inferior in every state of nature, then there is no need to admit it for further consideration. Continuing to use the existing plant unchanged can be seen

TABLE 17.2 Payoffs to Disposable Camera Production

	State of Nature Demand for the Disposable Cameras (in millions of dollars)		
	Low	Medium	High
a_1: Build new plant	–100	200	450
a_2: Renovate plant	100	200	300
a_3: Use plant as is	–100	100	200
Probability	0.10	0.50	0.40

to be such an inadmissible action in Table 17.2. Regardless of the state of nature (i.e., whether demand for disposable cameras is low, medium, or high), using the existing plant is always dominated by the other actions. That is, one of the other two actions will never result in a worse outcome than that achieved by producing in the unmodified existing plant.

An action a_i dominates action a_j if

1. for each state of nature, the payoff for action a_i is greater than or equal to the payoff for a_j, and
2. for at least one state of nature, the payoff for action a_i is greater than the payoff for a_j.

Dominated actions always result in inferior outcomes relative to the other actions and are thus inadmissible as solutions and may be eliminated from consideration.

So we eliminate action a_3: use the plant as is. The remaining two actions give rise to the expected monetary value decision tree in Figure 17.4. Ignoring the decision maker's behavior toward risk and using backward induction, Figure 17.4 suggests that the best action is to build a new plant. A new plant has the highest expected monetary value: \$270 million versus \$230 million for plant modification. But what if the decision maker is risk averse? Is this still the best action?

To assess alternative actions by the utility payoff criteria, a utility function must be specified. There is no unique utility function. For a risk averse person, one such utility function is

$$U_{ij} = \sqrt{(\text{dollar payoffs } j \text{ of } a_i)} \tag{17.1}$$

where U_{ij} is the utility of the jth dollar payoff of action a_i.

Utility function (17.1) has the property that, as the dollar payoff rises, the associated utility also rises but at a decreasing rate. For example, if the payoff is \$4, utility is 2; if the payoff is \$9, utility is 3; if the payoff is \$16, utility is 4; and so on, as shown in Figure 17.5. Using this utility function, the utility of each of the payoffs associated with the two admissible actions in Table 17.2 are shown in Table 17.3.

FIGURE 17.4 Decision Tree for Camera Plant: Expected Monetary Value Criteria

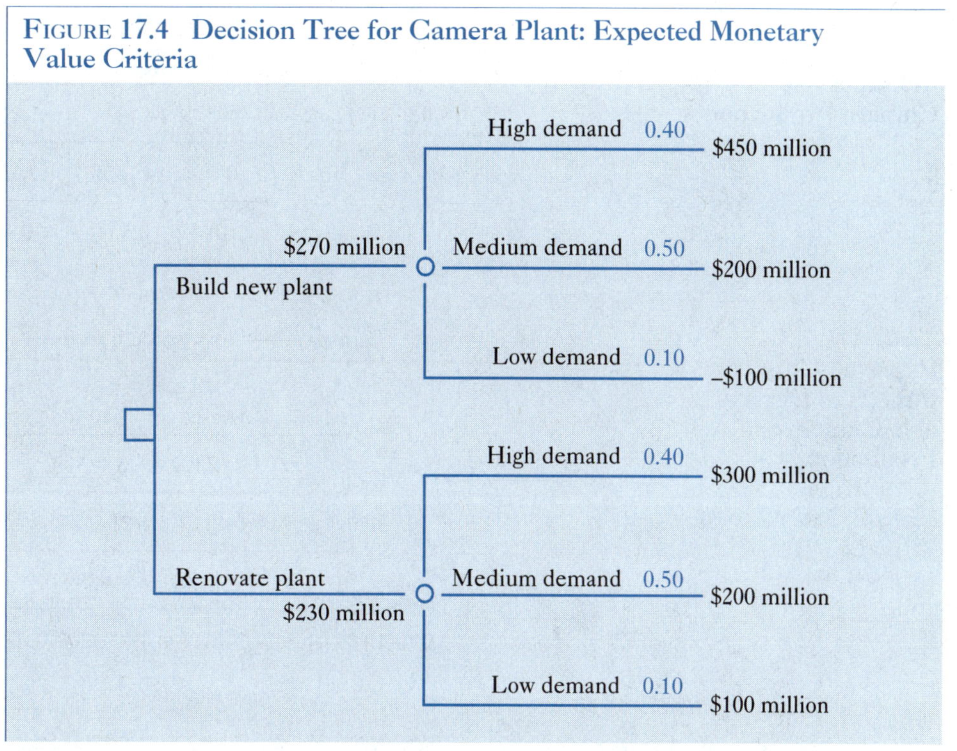

FIGURE 17.5 A Risk Averse Utility Function

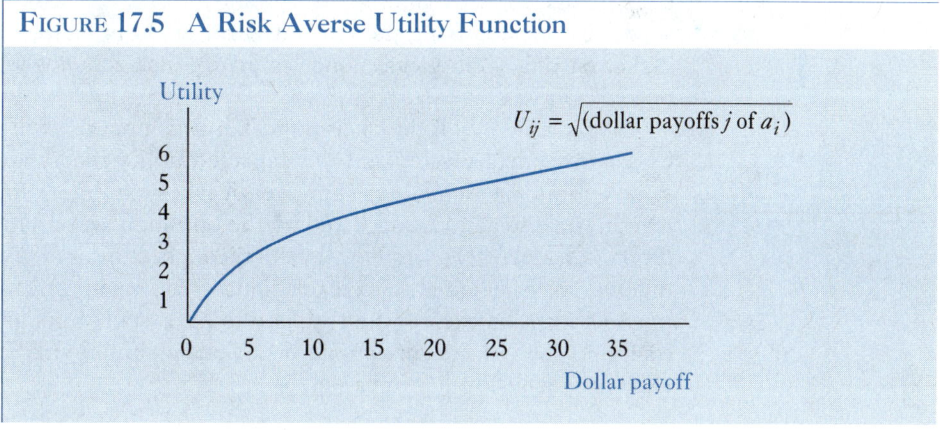

The expected utility is now calculated by multiplying the probability of each state of nature times the utility of the corresponding payoff. These expected utility values are shown in Table 17.4, along with the expected utility of each of the two admissible actions. The expected utility of renovating the existing plant (14.999) has a higher value than building a new plant (14.556). So renovating is the choice for a decision maker with this utility function.

TABLE 17.3
Utility of Payoffs to Disposable Camera Production

	Demand for Disposable Cameras		
	Low	Medium	High
a_1: Build new plant	−10.0	14.1421	21.2132
a_2: Renovate plant	10.0	14.1421	17.3205
Probability	0.10	0.50	0.40

TABLE 17.4
Expected Utility of Payoffs to Disposable Camera Production

	Demand for Disposable Cameras			
	Low	Medium	High	Expected utility of the action
a_1: Build new plant	−1.0	7.07107	8.48528	14.556
	0.10(−10.0) + 0.5(14.1421) + 0.4(21.2132) = 14.556			
a_2: Renovate plant	1.0	7.07107	6.92820	14.999
	0.10(10.0) + 0.5(14.1421) + 0.4(17.3205) = 14.999			

The utility function in Figure 17.5 represents someone who is so risk averse that, even though the expected monetary value of the new plant ($270 million) greatly exceeds that of modifying the existing plant ($230 million), the utility associated with higher payoffs is not great enough to justify such action when viewed in terms of expected utility.

Notice that only if the decision maker is risk neutral will decisions reached by the expected monetary value and the expected utility theory approach always imply the same optimum action. Often, however, it takes extremely strong risk aversion (or risk love) to reverse actions suggested by the optimum action under risk neutrality. In addition, a risk neutral position will yield the highest payoffs in the long run. While an individual may not be able to take a long view, organizations can. For these reasons, much of decision theory when applied to firms starts from an assumption of risk neutrality. As long as the appropriate behavioral assumption is risk neutrality, there is no need to specify a utility function.

QUERY 17.4:
The article in the *Herald-Times* (September 24, 1992) on library costs used in Query 17.3 suggested that a new facility should be built based on the expected monetary criteria. Is there a utility function for which the optimum course of action would be to add to and renovate the existing facility?

ANSWER:

A utility function that would suggest renovating and adding on to the existing building would be

$$U = (\text{Payoff})^3 \text{ for payoff less than zero,}$$

which gives the utility mapping in the following diagram.

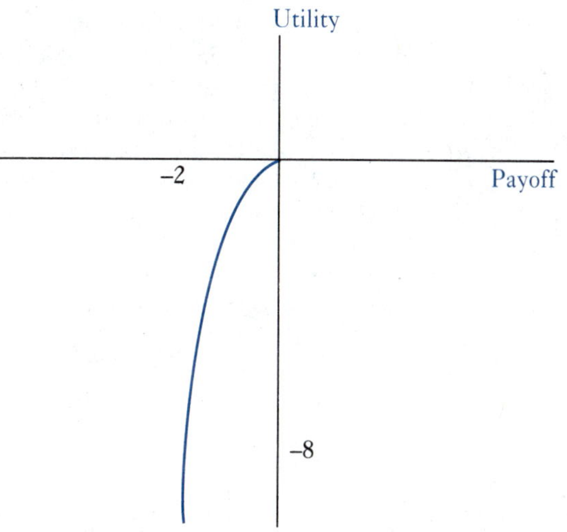

This is an extremely risk averse utility function. It gives the following expected utilities for each of the two courses of action.

New facility:

$$-2459.917845 = (-1906.624000)(0.333333) + (-2736.565584)(.666667)$$

Add on and renovate:

$$-2440.550609 = (-2403.276500)(0.333333) + (-2459.188743)(0.666667)$$

By this utility function, adding on to and renovating the existing facility is the optimum decision for the library board.

EXERCISES

17.1 Under what conditions might a decision maker be wise to behave as an expected monetary value maximizer rather than a risk averse expected utility maximizer?

17.2 Why can't a "best" course of action be defined without the specification of the criteria upon which optimization is to take place?

17.3 Identify each of the following as a risk lover, a risk averter, or a risk neutral person. A person who prefers a:

a. Guaranteed payment of $1,500, to an equally likely gamble between $3,000 and 0.

b. Gamble between $2,000 and $400, with equal odds, to a guaranteed payment of $1,200.

17.4 If a decision maker is indifferent between not playing a game (so there is no loss) or playing a game that costs $1 for an equally likely gamble between $502 and –$500, then this decision maker may be described as __________.

17.5 In the question to Marilyn in Query 17.1 about selecting one of two envelopes, what has Marilyn ignored or implicitly assumed in her answer?

17.6 An investor can place $100,000 into Treasury bills and get an annual net yield of 6 percent with certainty. Alternatively, the investor could place the entire $100,000 in a real estate venture that has the potential to yield a net return of 18 percent in a year (with probability of 0.50) or no net return.

a. If the investor is risk neutral, which investment is optimal?

b. If the investor is risk averse, with a utility function of $U = \text{Log}_e(\text{payoff})$, which investment is optimal?

17.7 A decision maker's utility function is

$$U(\text{payoff}) = 0.01(\text{payoff}) - 0.02(\text{payoff})^2 + 0.03(\text{payoff})^3$$

where payoffs are measured in thousands of dollars between $1,000 and $10,000 (i.e., payoff = 1, 2, . . . 10).

a. Graph U(payoff).

b. Is this utility function that of a risk lover or a risk averter?

17.8 For the decision maker described in Exercise 17.7, which of the following prospects or gambles would be selected?

a. A lottery in which a win of 0 or $10,000 is equally likely, or

b. $5,000 for certain.

17.6 Decision Analysis and New Sample Information

Prior Probability The marginal probability of a state of nature that exists before any new sample information is obtained.

The probabilities of alternative states of nature influence both the expected value of alternative actions and the expected utility of payoffs; thus, they affect the best decision. The accuracy of information on probabilities is critical to the decision process. Decision makers may have preliminary or **prior probability** estimates of the likelihood of each state of nature. These prior probability estimates, however, may be nothing more than hunches or they may be based on past frequency data. In any case, as time passes, the decision maker may be able to get additional information about the likely states of nature. New sample data can be used to revise or update the prior information so that the decisions are based on more accurate probability estimates.

Figure 17.4 provided the decision tree for the camera plant decision based on the expected monetary value criteria (or the risk neutral utility function). This tree shows a $270 million profit for building a new plant versus a $230 million profit for modifying the existing plant. The probabilities of low, medium, and high demand for disposable cameras are again 0.10, 0.50, and 0.40.

Posterior Probability The conditional probability of the state of nature given the indicator suggested by new sample information.

Indicator Sample information about the state of nature.

Suppose a market research firm is now hired to study the potential demand for disposable cameras. The market study will provide new sample data that can be used to update or revise probability estimates for the states of nature (low, 0.10; medium, 0.50; and high demand, 0.40). The updated probabilities are called **posterior probabilities.** They should be more accurate than the prior values because they are based on more and newer data. The new data are called the sample information. We will not use only this new sample information in the posterior probability calculations, because we do not know how accurate it is relative to the prior information. Typically, more information is better than less. Thus we combine the old with the new in the formation of posterior probabilities.

In the disposable camera case, suppose that the marketing firm will provide one of three types of sample information, which we will call **indicators:**

I_1 = a very favorable market indicator, which suggests an "extremely good" acceptance for the disposable cameras.

I_2 = a satisfactory indicator, which suggests a "medium" acceptance of the disposable cameras.

I_3 = a bad market indicator, which suggests "poor" acceptance of the disposable cameras.

The indicator can provide information that can be used to improve the accuracy of the probabilities of the three states of nature, where again the three states of nature are

s_1: high demand for disposable cameras,

s_2: medium demand for disposable cameras, and

s_3: low demand for disposable cameras.

The results of this probability updating process are posterior probabilities of the form $P(s_j \mid I_k)$, which is the conditional probability that the state of nature s_j occurs, given that the new sample information suggests indicator I_k. This updating process is summarized in Figure 17.6.

Unfortunately, the posterior probabilities typically cannot be obtained directly. They must be derived from knowledge of the probability relationships between the indicators and the states of nature. To find $P(s_2 \mid I_1)$, for example, we have to first answer

FIGURE 17.6 Probability Updating Based on Sample Information

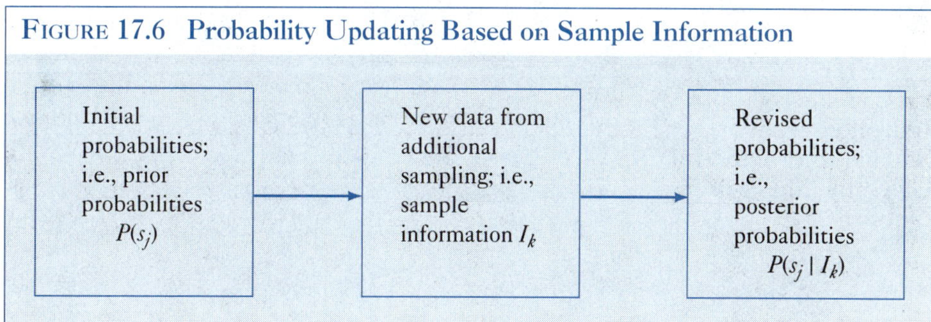

questions of the type: "If the state of nature turned out to be high demand, what is the probability that the market research will indicate an extremely favorable acceptance of the cameras?" That is, to find $P(s_2 | I_1)$, we must first find $P(I_1 | s_1)$, which is the conditional probability of indicator I_1, given that the state of nature is s_1. But from our discussion of conditional probability in Chapters 3 and 4, it can be seen that $P(I_1 | s_1)$ is only one piece in the determination of $P(s_2 | I_1)$. To determine $P(s_2 | I_1)$, we also need to find $P(I_1 | s_2)$, which is the probability of a highly favorable indicator given the state of nature turns out to be medium demand, and $P(I_1 | s_3)$, which is the probability of an highly favorable acceptance indicator given that the realized state of nature is a low demand.

Suppose that there are past records from which the probabilities of the indicators, conditional on the final state of nature, may be obtained. These past records give the $P(I_k | s_j)$ probabilities for I_1, I_2 and I_3 as shown in Table 17.5. In Table 17.5, for example, when the true state of nature was s_1 (high demand), the indicator gave a favorable signal 80 percent of the time. When the true state of nature was a medium demand (s_2), the indicator was correct 90 percent of the time; and when a low demand was realized as the state of nature, the indicator gave the correct signal 75 percent of the time.

So what is the optimum decision strategy that is to be followed by the decision maker? Build a new plant or renovate? As before, a decision tree can be employed to find the optimal strategy. Figure 17.7 shows the decision tree for the disposable camera case after the market research study is conducted. Again, a decision is made by backward induction—doing an analysis from right to left. Now, however, market indicators (I_1 or I_2 or I_3) are incorporated into the decision process regarding an action (a_1, a_2, or a_3) to be taken. By backward induction we move from the payoffs to the optimum course of action, taking the indicators into account.

The decision tree in Figure 17.7 introduces an indicator node at point number 1, from which three indicator branches (I_1, I_2, and I_3) emanate. The indicator node is a circle just like the state-of-nature nodes at 5, 6, 7, 8, 9, and 10; branches coming from the indicator and state-of-nature nodes are determined by chance and are not under the control of the decision maker. Only at the decision nodes 2, 3, and 4 is the decision maker in control and free to select actions a_1, a_2, or a_3.

Bayes' Rule
The formula used to update prior probabilities to obtain posterior probabilities. It yields a conditional probability for a state of nature given sample information.

We want the probability of a state of nature. But now that probability also depends on the probability of a branch emanating from the indicator node. We need to find the posterior probability of a state of nature s_j, given an indicator I_k, which is $P(s_j | I_k)$. This calculation will require the use of a mathematical formula known as **Bayes' rule.**

TABLE 17.5 Probabilities of Indicators from Sample Information Given the States of Nature

States of Nature		Market Research Report		
		Favorable I_1	Satisfactory I_2	Bad I_3
High demand	s_1	$P(I_1 \mid s_1) = .80$	$P(I_2 \mid s_1) = .15$	$P(I_3 \mid s_1) = .05$
Medium demand	s_2	$P(I_1 \mid s_2) = .05$	$P(I_2 \mid s_2) = .90$	$P(I_3 \mid s_2) = .05$
Low demand	s_3	$P(I_1 \mid s_3) = .05$	$P(I_2 \mid s_3) = .20$	$P(I_3 \mid s_3) = .75$

FIGURE 17.7 Disposable Camera Decision Tree Employing New Sample Information

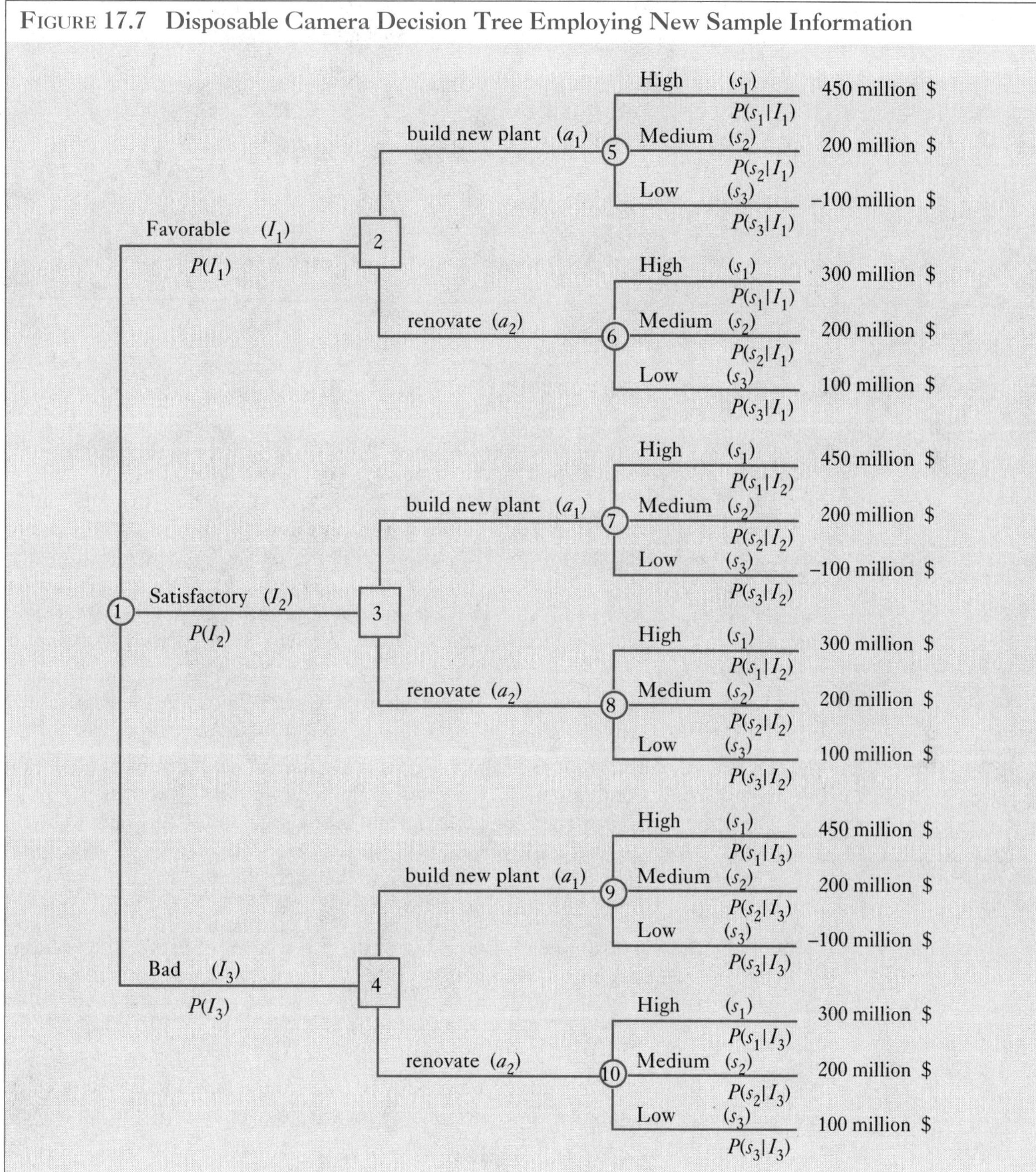

Bayes' Rule for Calculating Posterior Probabilities $P(s_1 | I_k)$

$$P(s_1|I_k) = \frac{P(I_k \cap s_1)}{P(I_k)} = \frac{P(I_k|s_1)P(s_1)}{\sum_{j=1}^{J} P(I_k|s_j)P(s_j)} \tag{17.2}$$

where s_j is the jth state of nature, for J such states
I_k is the kth indicator from sample information
$P(s_j)$ is the prior probability of the jth state of nature

17.7 Computing Probabilities with Bayes' Rule

To make posterior probability calculations with Bayes' rule, the prior probabilities for the states of nature are needed. In the disposable camera case, these prior probabilities are given in Table 17.2, $P(s_1) = 0.4$, $P(s_2) = 0.5$ and $P(s_3) = 0.1$. These marginal probabilities times the conditional probabilities of the indicators, given each state of nature, are needed to get the joint or intersection probabilities of the respective states of nature and the indicators. That is, the prior probabilities $P(s_j)$ and the conditional probabilities $P(I_k | s_j)$, as given in Table 17.5, are multiplied together to get the intersection probabilites $P(I_k \text{ and } s_j)$, which in turn are used to calculate the branch probabilities $P(I_k)$ and $P(s_j | I_k)$.

Starting with the probability, $P(I_1)$, notice that the indicator I_1 can occur with three different states of nature:

1. The sample information is favorable (I_1) and the state of nature turns out to be the high demand world (s_1), which is written as the intersection (I_1 and s_1).
2. The sample information is favorable (I_1) and the state of nature turns out to be the medium demand world (s_2), written as (I_1 and s_2).
3. The sample information is favorable (I_1) and the state of nature turns out to be the low demand world (s_3), written (I_1 and s_3).

Adding the three probabilities associated with the three states of nature and indicator I_1 yields the marginal branch probability for the indicator for high acceptability of the disposable cameras.

$$P(I_1) = P(I_1 \text{ and } s_1) + P(I_1 \text{ and } s_2) + P(I_1 \text{ and } s_3) \tag{17.3}$$

The multiplication property of probability provides the following method of calculating the intersection probabilities $P(I_1 \text{ and } s_1)$ and $P(I_1 \text{ and } s_2)$ and $P(I_1 \text{ and } s_3)$; namely,

$$P(I_1 \text{ and } s_1) = P(I_1 | s_1)P(s_1) \tag{17.4}$$

$$P(I_1 \text{ and } s_2) = P(I_1 | s_2)P(s_2) \tag{17.5}$$

$$P(I_1 \text{ and } s_3) = P(I_1 | s_3)P(s_3) \tag{17.6}$$

Substituting the above expressions for the probabilities of the intersections in Equation (17.3) gives

$$P(I_1) = P(I_1 \mid s_1)P(s_1) + P(I_1 \mid s_2)P(s_2) + P(I_1 \mid s_3)P(s_3) \tag{17.7}$$

This expression can be generalized to any indicator branch probability, $P(I_k)$, and J states of nature, $s_1, s_2, \ldots, s_J$:

$$P(I_k) = P(I_k \mid s_1)P(s_1) + P(I_k \mid s_2)P(s_2) + \ldots + P(I_k \mid s_J)P(s_J) \tag{17.8}$$

or

$$P(I_k) = \sum_{j=1}^{J} P(I_k \mid s_j)P(s_j) \tag{17.9}$$

which is the denominator of Bayes' rule.

In the disposable camera case, using the three prior probabilities in Table 17.2 and the nine conditional probabilities in Table 17.5, Equation (17.9) is used to compute the three indicator branch probabilities as follows:

$$\begin{aligned} P(I_1) &= P(I_1 \mid s_1)P(s_1) + P(I_1 \mid s_2)P(s_2) + P(I_1 \mid s_3)P(s_3) \\ &= (0.80)(0.4) + (0.05)(0.5) + (0.05)(0.1) = 0.35 \end{aligned}$$

and

$$\begin{aligned} P(I_2) &= P(I_2 \mid s_1)P(s_1) + P(I_2 \mid s_2)P(s_2) + P(I_2 \mid s_3)P(s_3) \\ &= (0.15)(0.4) + (0.90)(0.5) + (0.20)(0.1) = 0.53 \end{aligned}$$

and

$$\begin{aligned} P(I_3) &= P(I_3 \mid s_1)P(s_1) + P(I_3 \mid s_2)P(s_2) + P(I_3 \mid s_3)P(s_3) \\ &= (0.05)(0.4) + (0.05)(0.5) + (0.75)(0.1) = 0.12 \end{aligned}$$

Thus, the probability of the market indicator I_1, a favorable acceptance of the disposable cameras, is 0.35; the probability of I_2, a satisfactory acceptance, is 0.53; and the probability of I_3, a poor acceptance, is 0.12.

As shown in Formula (17.2), Bayes' rule requires the use of the three marginal indicator probabilities in computing the posterior state-of-nature branch probabilities $P(s_j \mid I_k)$. For example, the state-of-nature branch probability $P(s_1 \mid I_1)$, which is the probability the demand for disposable cameras is high (s_1), given that the indicator is favorable (I_1), is

$$P(s_1 \mid I_1) = \frac{P(I_1 \text{ and } s_1)}{P(I_1)} = \frac{P(I_1 \mid s_1)P(s_1)}{P(I_1)} = \frac{(0.80)(0.4)}{0.35} = 0.9143$$

The prior probability of a high demand for the disposable cameras was $P(s_1) = 0.40$. The state-of-nature branch probability calculation for $P(s_1 \mid I_1)$ says that if the sample information from the market research results in a highly acceptable rating, then the posterior probability of a high demand for the cameras is 0.9143.

FIGURE 17.8 Disposable Camera Decision Tree Employing New Sample Information to Revise Probabilities

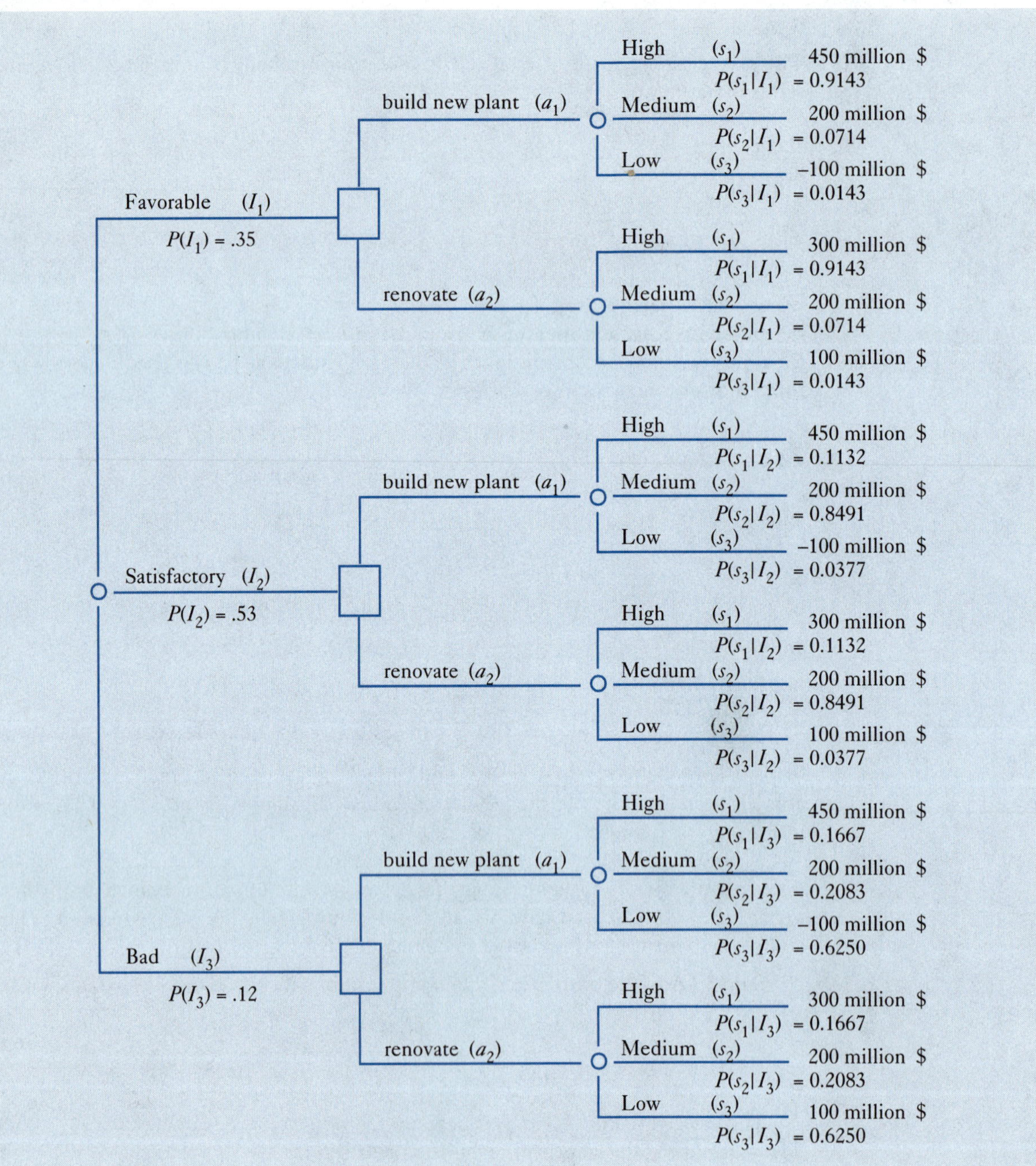

Using Bayes' rule, the remaining posterior state-of-nature branch probabilities can be determined. For example, the revised probability of medium demand, s_2, given the indicator is favorable, I_1, becomes

$$P(s_2|I_1) = \frac{P(I_1|s_2)P(s_2)}{P(I_1)} = \frac{(0.05)(0.5)}{0.35} = 0.0714$$

Figure 17.8 shows the disposable camera decision tree after all indicators and all posterior state-of-nature branch probabilities have been calculated. There is specialized computer software (such as DEMOS, which is presented in Section 17.10) that will do all of the calculations performed here. Section 17.8 shows how these probabilities are easily calculated with a microcomputer spreadsheet such as LOTUS 1-2-3, EXCEL, or QUATTRO.

17.8 Computing Probabilities with a Computer Spreadsheet

Although the above long-hand procedure can be used to compute branch probabilities, the calculations are made much easier using a computer spreadsheet and a tabular procedure. Within the spreadsheet, use the following matrix arrangement:

Column 1 States of nature s_j
Column 2 Prior probabilities $P(s_j)$
Column 3 Conditional probabilities $P(I_k|s_j)$
Column 4 Joint probabilities $P(I_k \text{ and } s_j)$
Column 5 Posterior probabilities $P(s_j|I_k)$

For the indicator I_1, the steps used to calculate $P(I_k)$ and the $P(s_j|I_k)$ are:

Step 1 In column 1, list the states of nature appropriate to the problem being analyzed—the s_j's.
Step 2 In column 2, enter the prior probability corresponding to each state of nature listed in column 1—the $P(s_j)$'s.
Step 3 In column 3, enter each value of $P(I_k|s_j)$ for the corresponding state of nature specified in column 1.
Step 4 To compute each probability of an intersection in column 4, multiply each entry in column 2 by the corresponding entry in column 3.
Step 5 Add the entries in column 4; the sum is the value of $P(I_1)$, which is shown below column 4.
Step 6 To compute each posterior probability $P(s_j|I_k)$ in column 5, divide each entry in column 4 by $P(I_k)$.

The result of these six steps will give the first block of four columns in Table 17.6; this block is conditional on indicator I_1. In the case of the disposable cameras, there are

TABLE 17.6 Computer Spreadsheet Calculation of Posterior Probabilities

State	P(s)	P(I1|s)	P(I1 and s)	P(s|I1)
s1	0.40	0.80	0.3200	0.9143
s2	0.50	0.05	0.0250	0.0714
s3	0.10	0.05	0.0050	0.0143
			0.3500 = P(I1)	

State	P(s)	P(I2|s)	P(I2 and s)	P(s|I2)
s1	0.40	0.15	0.0600	0.1132
s2	0.50	0.90	0.4500	0.8491
s3	0.10	0.20	0.0200	0.0377
			0.5300 = P(I2)	

State	P(s)	P(I3|s)	P(I3 and s)	P(s|I3)
s1	0.40	0.05	0.0200	0.1667
s2	0.50	0.05	0.0250	0.2083
s3	0.10	0.75	0.0750	0.6250
			0.1200 = P(I3)	

three possible indicators. Thus, two more blocks must be computed by the above six-step procedure. The difference between the blocks is the result of the different conditional probabilities $P(I_k | s_j)$ that are obtained from Table 17.6.

17.9 OPTIMAL DECISIONS BASED ON POSTERIOR PROBABILITIES

Once again, using backward induction, an optimum action can be derived. This first requires the calculation of expected values at each state-of-nature node, which is done using the sum of products formed by the payoffs to each of the states of nature and their respective posterior probabilities. For example:

The expected value of building a new plant, given sample information suggesting the favorable indicator I_1, is

$$E(a_1 | I_1) = (450)(0.9143) + (200)(0.0714) + (-100)(0.0143)$$
$$= \$424.285 \text{ million}$$

The expected value of renovating the existing plant, given sample information suggesting the satisfactory indicator I_2, is

$$E(a_2 | I_2) = (300)(0.1132) + (200)(0.8491) + (100)(0.0377)$$
$$= \$207.55 \text{ million}$$

and similarly for the other four expected values in Figure 17.9.

FIGURE 17.9 Disposable Camera Decision Tree Employing New Sample Information in Expected Value Calculations

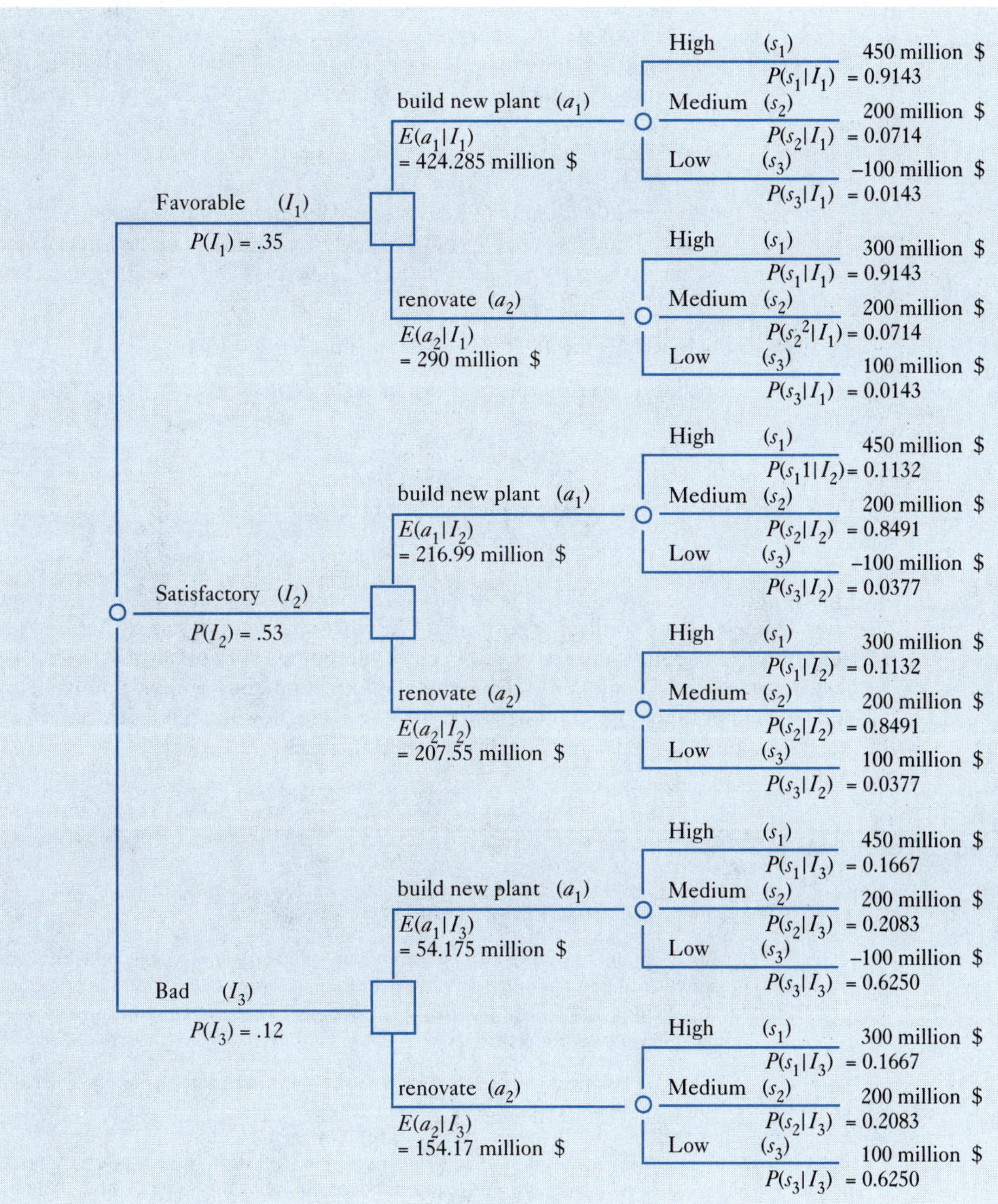

The decision maker can choose the action that maximizes expected return based on the indicator that results from the sample information. For instance, if the indicator turns out to be I_1 (favorable reaction to disposable cameras can be anticipated), then the best decision is to build a new plant (a_1) with expected profits of $424.285 million; if the existing plant is renovated, expected profits are only $290 million. If the indicator is I_2, then the optimum action is also to build the new plant. However, if the indicator is I_3 (a poor reaction to disposable cameras can be anticipated), then the optimum decision is to renovate the existing plant (a_2) with expected profits of $154.17 million; if a new plant is built, expected profits are only $54.175 million.

Notice that the final decision depends on the realization of the indicator I_k, which is a random variable. Random variables have expected values, so a natural question to ask is: What is the expected value of the indicator? The probabilities of the indicators I_1, I_2, and I_3 are found from Equation (17.9):

$$P(I_1) = 0.3500,\ P(I_2) = 0.5300 \text{ and } P(I_3) = 0.1200$$

The expected value of the optimum decision associated with each indicator is also known:

$$E(a_1 | I_1) = 424.285,\ E(a_1 | I_2) = 216.99 \text{ and } E(a_2 | I_3) = 154.17$$

Thus, the expected value of the indicators can be calculated as

$$E(I) = (0.35)(424.285) + (0.53)(216.99) + (0.12)(154.17) = 282.0048$$

That is, the mean value of the indicators is approximately $282.005 million. If we had never learned of the availability of sample information and the sample information were never obtained, then Figure 17.4 shows the optimum decision to yield an expected value of $270 million. The sample information thus added approximately $12.005 million to the optimum decision. As long as the new sample information does not cost more than $12.005 million, it is worth pursuing.

EXERCISES

17.9 An article in *The Wall Street Journal* (March 3, 1994) stated:

> Chrysler is trying to figure out how and where to make more cars and trucks, as the company is running out of capacity to meet demand for its most profitable products. Options include canceling a planned plant closing, expanding output in Mexico and adding shifts at existing plants.

If you were a Chrysler executive, what would you want to know to make this decision?

17.10 Under what conditions could the expected monetary value calculated from posterior probabilities be greater than that calculated from prior probabilities?

17.11 In planning to run for public office, Mrs. Becker has to decide on how many letters to mail out in the solicitation of contributions to support her campaign. Letters and follow-up telephone calls will be done by a clerical and word processing firm that requires orders in multiples of 50. The contributions to be received from potential contributors

Historical Perspective

Thomas Bayes (1701–1761) is credited with fathering work on the relationship between posterior and prior probabilities and advancing the idea that a conditional probability statement, such as $P(s \mid I)$, can be expressed in terms of its reverse conditional probability statement, $P(I \mid s)$. Bayes, an ordained Nonconformist minister, lived outside of London, England. His mathematical works were not published during his lifetime, but a posthumous essay, "An Essay Toward Solving a Problem in the Doctrine of Chance" (1764), made him the namesake of one of the most famous method of inference in all of science—"Bayesian inference"—and "Bayes' rule."

Mathematician Jansci (John) von Neumann and economist Oskar Morgenstern made Bayes' rule operational for economic analysis and business decision making. Von Neumann was a multifaceted scholar: he was multilingual, he invented game theory, he designed the internal binary logic used in computers, and he worked on the atomic bomb as part of the Manhattan Project during World War II. He was born in Hungary in 1903 and studied throughout Europe while enrolled at the University of Budapest. In 1930, he joined Princeton University's faculty, where he was a member of the Institute for Advanced Study, along with the likes of Albert Einstein. He served on the United States Atomic Energy Commission until his death in 1957.

Morgenstern (1902–1977) joined the Princeton University faculty in 1938. He was familiar with von Neumann's work with the theory of games and sought out von Neumann as a collaborator on a paper that grew into their famous book, *The Theory of Games and Economic Behavior* (Princeton University Press, 1944).

In his 1928 paper, von Neumann assumed that at the time of a decision each player in a game either knew everything about the previous moves of an opponent (as in chess) or knew nothing (as in matching coins). Using decision trees and partitioning the nodes of a tree into subsets, in 1944 von Neumann and Morgenstern relaxed this assumption. They assumed that players have only partial information about past moves at the time of a decision. This was a major extension that has not been greatly improved upon to this day.

Von Neumann also assumed only monetary payoffs. In their 1944 book, von Neumann and Morgenstern provided the axioms of utility theory. Daniel Bernoulli had advanced the notion that the utility of x dollars should be the natural logarithm of x, because an additional payment of Δlx to a rich person was worth less than the same additional payment to a poor person. But Bernoulli's rigid utility function was never accepted by economists. Von Neumann and Morgenstern provided a general framework for utility analysis that did not depend on any one utility function.

are uncertain, but from previous campaigns she knows that an average contribution of $50 per letter or more is a "large" return, an average return per letter of $25 to $50 is a "middle" contribution level, and less than $25 per letter is a "small" contribution return. The payoff table (net dollars) follows.

	Solicitation letter order size				
Contribution level	50	100	150	200	250
large	225	350	475	550	475
middle	225	350	300	225	175
small	225	175	125	100	0

Which solicitation letter order sizes are viable alternative actions? (Explain.)

17.12 In Exercise 17.11, suppose Mrs. Becker assigns probabilities to each contribution outcome as follows:

$$P(\text{large}) = 0.25,\ P(\text{middle}) = 0.55 \text{ and } P(\text{small}) = 0.20$$

a. What is the expected payoff of each act (letter order)?
b. If Mrs. Becker is risk neutral, how many letters should she order?

17.13 Following the removal of restrictions on advertising and cross-state practices by lawyers, many new partnerships were formed to capitalize on the new freedoms. Suppose a "big time New York lawyer" and a "hot-shot San Francisco lawyer" terminated their previous partnerships and started their own law firm with the intention of advertising as specialists in personnel practices and tax law. They are each being sued by their previous partners for client theft, however. The lawyers have not yet gone national with their advertising and are rethinking the advisability of doing so, given the threat of litigation. The payoff table of the two-lawyer partnership (in millions of dollars) is:

	Specialize and Advertise for			
Possible lawsuit outcomes:	Both tax & personnel	Tax only	Personnel only	Don't advertise
The two lawyers win both suits	1.600	1.535	1.320	1.189
They lose both suits	0.535	0.876	0.978	0.965
They win the suit involving the prior New York partner	1.487	1.555	1.345	1.123
They win the suit involving the prior San Francisco partner	0.835	0.860	0.912	0.900

a. Which of the four different courses of action should the lawyers consider as viable? (Explain.)
b. What else would be good to know here to make a decision? (Explain.)

17.14 After some legal research into the decision reached by the courts in previous cases involving alleged client theft, the two lawyers in Exercise 17.13 form the following subjective probabilities of victory:

Possible lawsuit outcomes:	Probability of outcome
The two lawyers win both suits	0.15
They lose both suits	0.05
They win the suit over the theft of personnel practice clients	0.35
They win the suit over the theft of tax clients	0.45

Assuming the lawyers are risk neutral, what course of action should they take? (Explain why.)

17.15 A computer firm is considering purchasing 0.25, 0.50, or 1.00 minute of advertising time during next year's NCAA basketball finals. The time will cost $500,000 for each half-minute. The firm does not know the number of prospective customers who watch the Final Four, but it arbitrarily classifies them as "a few," "most," and "almost all," with prior probabilities of 0.35, 0.50, and 0.15. The revenue associated with three states of nature and the three commercial times are

	Few (F)	Most (M)	Almost all (A)
0.25 min.	–150,000	550,000	2,100,000
0.50 min.	350,000	525,000	900,000
1.00 min.	400,000	750,000	1,500,000
Probability	0.35	0.50	0.15

a. Draw the decision tree for this problem, assuming risk neutrality.
b. What is the optimal decision, assuming risk neutrality?

17.16 The computer company considering the purchase of TV time in Exercise 17.15 has the option of purchasing a marketing study for $15,000. The marketing firm will provide sample information regarding the number of potential refrigerator buyers who watched last year's NCAA Final Four; the study will indicate either "small," "medium," or "large." From past experience, the marketing firm provides the following conditional probabilities:

P(small | few) = 0.450 P(medium | few) = 0.350
P(small | most) = 0.350 P(medium | most) = 0.550
P(small | almost all) = 0.150 P(medium | almost all) = 0.450

a. Draw the decision tree, assuming the sample information is pursued.
b. Is it worth paying for the sample study? (Explain.)

17.10 A Computer Based Decision-Making System

Our decision-making framework has developed to the point where we can answer questions about the worthiness of the options, the uncertainty associated with the outcomes, the worth of information from experts, and the value of seeking additional information. Doing the associated calculations for alternative assumptions is tedious, even for an experienced spreadsheet jockey. Highly specialized computer software is available to automatically assess the consequence of alternative sets of assumptions about the risks, costs, and benefits of particular actions. One such system is DEMOS, developed at Carnegie-Mellon University and now distributed by Lumina Decision Systems in Palo Alto, California.

DEMOS is a Macintosh-based program for use in risk analysis, decision analysis, cost-benefit analysis, and strategic planning. Users may build their own models from scratch or work with pre-existing templates. When the tutorial housing cost model is brought up, for example, the computer screen is filled with the general model structure shown in Figure 17.10. This diagram indicates the type of quantities involved in the

FIGURE 17.10 DEMOS Influence Diagram for Housing Costs

When a model is started up, DEMOS displays the top level of the model in an influence diagram window as shown below. A model contains a number of objects of different classes. Object classes include decision variables, chance variables and submodels. The shape of each node in the diagram indicates its object class.

The browser icon is grayed-out, indicating that you are in browse mode. This means you can examine the diagram but not change it.

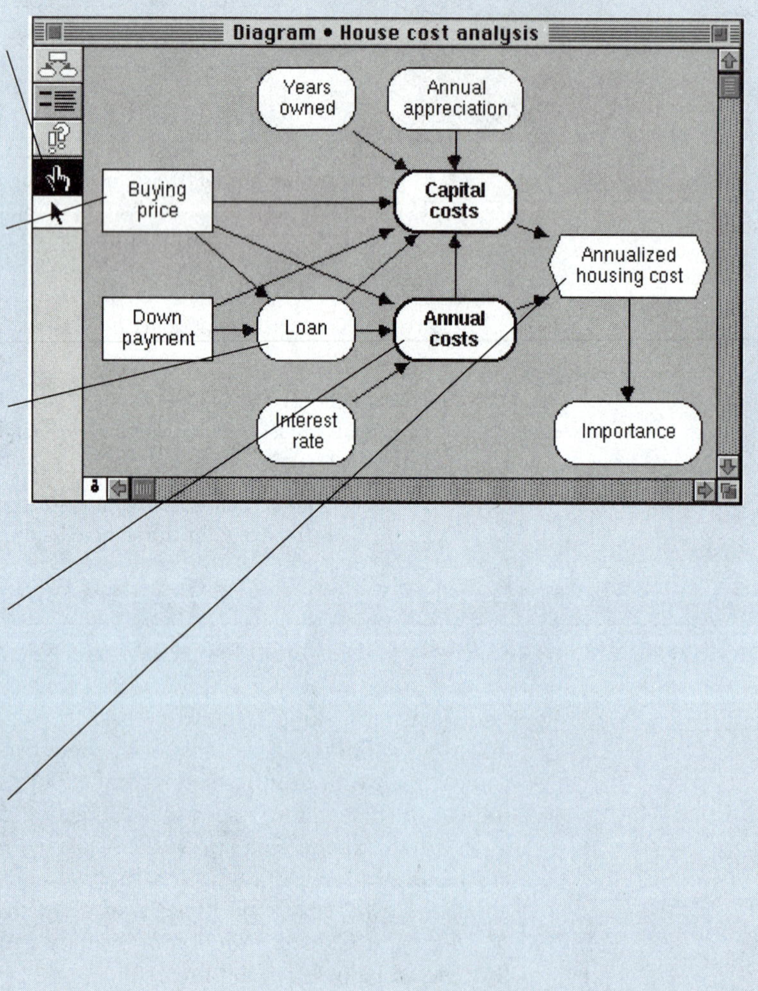

Decision variables are represented by rectangular nodes. Decisions are variables that are directly under the control of the decision maker.

Chance variables are represented by nodes with rounded corners. In general, chance variables are uncertain. They cannot be controlled directly by the decision maker.

Submodels are represented by thick-lined nodes with rounded corners. A submodel contains its own influence diagram, showing more detail.

The objective variable is represented by a hexagonal node. The objective is the variable that evaluates the overall value or desirability of outcomes. In this one, the goal is to minimize the *Annualized housing cost*. Most models contain a single objective node.

FIGURE 17.11 DEMOS Outcome Diagram of Annualized Housing Costs

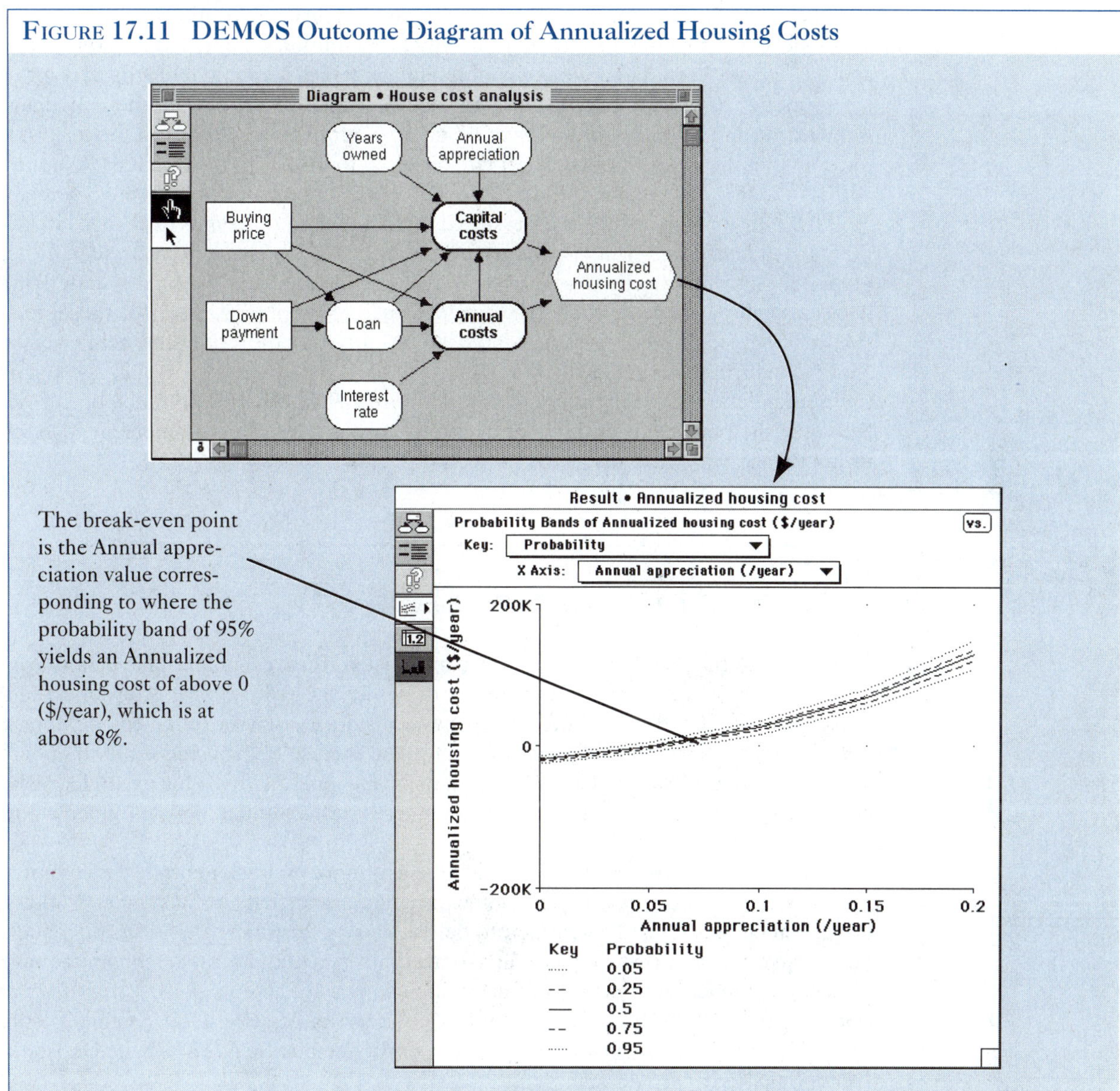

model by node shape—decision variables are rectangular nodes, nodes involving chance events have rounded corners, the objective to be achieved is a hexagonal node. In this example, the objective is to minimize housing cost, given uncertainty about years of ownership, annual appreciation, capital cost, annual maintenance costs, interest rates, and the size of the loan (which actually is predetermined once price and downpayment are specified). The choice or decision quantities are the buying price and the down payment amount.

Each chance node has its own underlying probability distributions or it involves a submodel consisting of one or more probability distributions. Thus, the structure of the analysis can be viewed as a hierarchy of models in which a submodel feeds data into a larger model. In these calculations, all chance variables are set at their medians. Results, the annualized housing costs, are given in probabilistic terms. In Figure 17.11, for example, the annual appreciation rate required to break even on a $400,000 house, with a $120,000 down payment, is about 8 percent for a probability band of 95 percent.

DEMOS is nicely suited to assessing the sensitivity of outcomes to changes in key inputs. Tradeoffs and switchpoints where the best options change can be seen either graphically or numerically. Users need only a basic understanding of the underlying decision-making framework and the model structure employed. Users of these decision-making programs, however, should be sensitive to the criticisms being voiced about actual practices in decision analysis, as raised by Rex Brown (1992), as well as their defenders, as articulated by Ronald Howard (1992). At issue in much of this debate over the practical usefulness of decision analysis are the basic tenets of expected utility theory, which are discussed in Section 17.12.

17.11 An Application of Expected Utility to the Value of Life (Optional)

In addition to the use of expected utility in decision trees, economists have employed expected utility theory (*EU*) in attempts to model the risky choices consumers and workers make and their willingness to pay to reduce risk. As reflected in Morgan (1993), engineers are now following economists in the use of expected utility theory to assess the value of safety. Computer programs such as DEMOS, for example, make child's play of risk analysis and the valuing of lives saved through policies aimed at environmental safety.

Unlike decision tree analysis in which the assumption of risk neutrality is common, to reflect the riskiness of a product, job or environmental change, consumers, workers, or general citizens must be assumed to be risk averse. Assuming that decision makers are risk averters and that they attempt to maximize expected utility, economists have a conceptual model for putting a value on life itself, as reflected in a *Wall Street Journal* (December 12, 1988) headline, "New Legal Theorists Attach a Dollar Value To the Joys of Living: Hedonic Damage Argument By Economist Stan Smith Stirs Debate in Death Suits."

Becker and Stout (1992), provided the following discussion of the expected utility gains that consumers may receive from the risk reduction associated with the purchase of fire alarms, as studied by Dardis (1980). From their willingness to pay for this risk reduction, Dardis derives the value consumers are placing on their lives. For illustrative purposes, Dardis's risk-averse utility function is reproduced in Figure 17.12, where utility is on the vertical axis and consumption (which is assumed to be equal to income) is on the horizontal axis. The utility function and notation are:

FIGURE 17.12 Dardis Utility Function

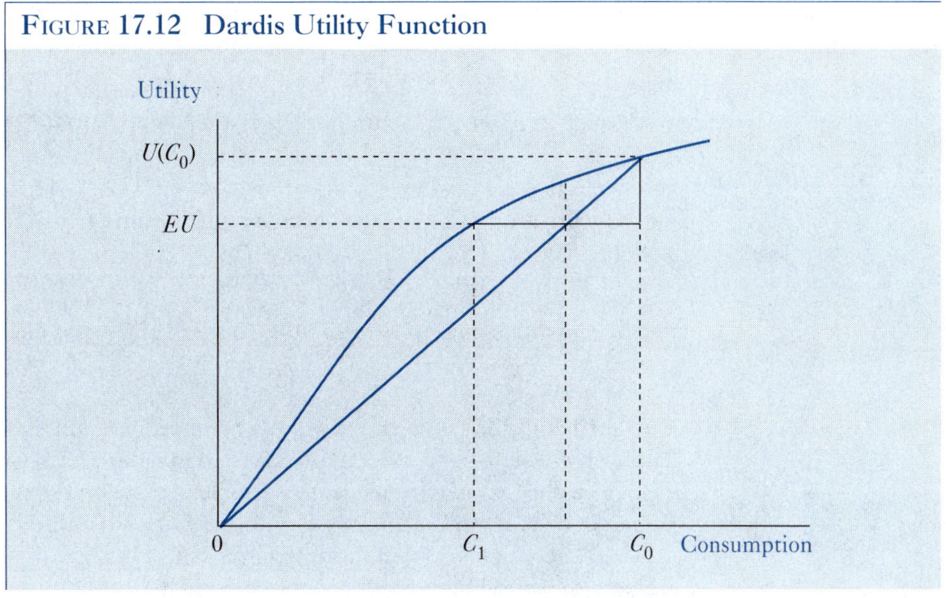

$U(C)$	is the utility of consumption (or income), where the marginal utility of consumption (change in utility associated with a unit change in consumption, $\Delta U/\Delta C$, or first derivative of the utility function with respect to consumption, dU/dC) is positive but decreases as consumption increases.	(17.10)
C_0	is the initial, risk-free lifetime consumption.	
C_1	is the certain value equivalent consumption; that is, the individual is indifferent to the choice of C_1 with certainty and a gamble that could result in death (the hazard).	
$U(D)$	is the "utility of death."	
p	is the subjective probability of death.	
$1 - p$	is the subjective probability of survival.	

In a risk-free state, the individual is assumed to earn and consume C_0 and enjoys lifetime utility $U(C_0)$ with certainty. The introduction of the hazard (death by fire) presents the individual with the following prospects: death with probability p or survival and C_0 with probability $1 - p$. Although criticized by Becker and Stout (1992), Dardis selected zero for both the utility of death and the utility of zero consumption and asserted that "the U(Death) = 0 is a convenient simplification." Thus, the expected utility of living with the hazard of zero consumption (death) is

$$EU = pU(D) + (1 - p)U(C_0) = (1 - p)U(C_0) \tag{17.11}$$

From the maximum risk-free consumption level and uncertain zero consumption level (death), a third point C_1 can be derived at which the individual is indifferent between

C_1 with certainty and zero consumption (death) with uncertainty. By the axioms of rational choice, the individual must consider EU equivalent to $U(C_1)$. The hazard or chance of death reduces utility by

$$dU = U(C_0) - U(C_1) = U(C_0) - EU = p[U(C_0) - U(D)] = pU(C_0) \qquad (17.12)$$

and the individual is willing to pay at most $dC = C_0 - C_1$ (which includes a risk premium) for reducing the risk. Thus, from the marginal utility

$$dU/dC = p[U(C_0) - U(D)]/(C_0 - C_1) = pU(C_0)/(C_0 - C_1) \qquad (17.13)$$

the change in income exchangeable for a small change in risk is obtained:

$$(C_0 - C_1)/p = [U(C_0) - U(D)]/(dU/dC) = U(C_0)/U' \qquad (17.14)$$

The left-hand side is the *ex ante* value of risk reduction. Division by the probability of death on the left-hand side implicitly extrapolates the payment $C_0 - C_1$ for a unit change in the probability of death. This leads Dardis (1980) and others to interpret Equation (17.14) and its analogs as the value of human life. That is, Equation (17.14) provides the rationale for the estimation of the value of a life by the amount of money an individual is willing to pay for risk reduction.

An average value of life is calculated from Equation (17.14) by assuming that all individuals make similar choices. Dardis uses the example of 1,000 people all requiring a compensation of $100 because of a decrease in survival probability of 0.001. The estimated value of a life is thus $100,000 for individuals in this community. This does not mean that any one individual is willing to sacrifice his or her life for $100,000; this is only a type of average or statistical artifact that can be interpreted via the rational choices and definitions assumed in Equation (17.14).

Query 17.5:

A *Wall Street Journal* (April 3, 1993) article, "Genentech Drug Raises Question On a Life's Value," reported that Genentech's TPA saves "one in 100, over streptokinase," but "at $2,200 a dose, TPA costs 10 times more than streptokinase." What is the implied value of life? Prior to this announcement, TPA and streptokinase were believed to be about equally effective. How should the stock market have responded to this new finding?

Answer:

Using Equation (17.14), $p = 0.01$, $C_0 = \$2,200$, and $C_1 = \$220$, and the implied value of a life at which cost equals benefit is $198,000 (or approximately $200,000), as stated in the *WSJ* article. Given that most other estimates of a value of life are much higher than this, it is not surprising that Genentech's stock price soared, going up $4.75 a share to close at $37.50 in composite New York Stock Exchange trading on April 30, 1993.

17.12 Empirical Observations on Expected Utility Theory

Critical to decision tree analysis and the use of utility theory generally is the manner in which individuals behave toward risk. Early research in the 1950s by Milton Friedman and Leonard Savage suggested that individuals are not consistently risk neutral, risk takers, or risk averters. A typical person may have no qualms about gambling small amounts of money, like playing Bingo, but may be very adverse to gambling large amounts of money, by going without house or auto insurance. Such a person will have a utility function that shows him or her to be a risk lover toward small losses and gains but a risk averter for large gains or losses. This utility function is depicted in Figure 17.13.

As discussed in Becker and Stout (1992), and reprinted in part here, prior to 1970 the notion that decision makers behave as expected utility maximizers was relatively unquestioned within and outside of economics. To behave as expected utility maximizers, however, individuals must have accurate perceptions of risks. Researchers have shown that subjective probabilities for low-probability outcomes tend to be biased above their objective probabilities and are frequently inconsistent with the axioms of probability.

Lichtenstein et al. (1978), in their study of people's estimates of the frequency of various causes of death, found that subjects overestimated small frequencies and un-

Figure 17.13 A Typical Utility Function

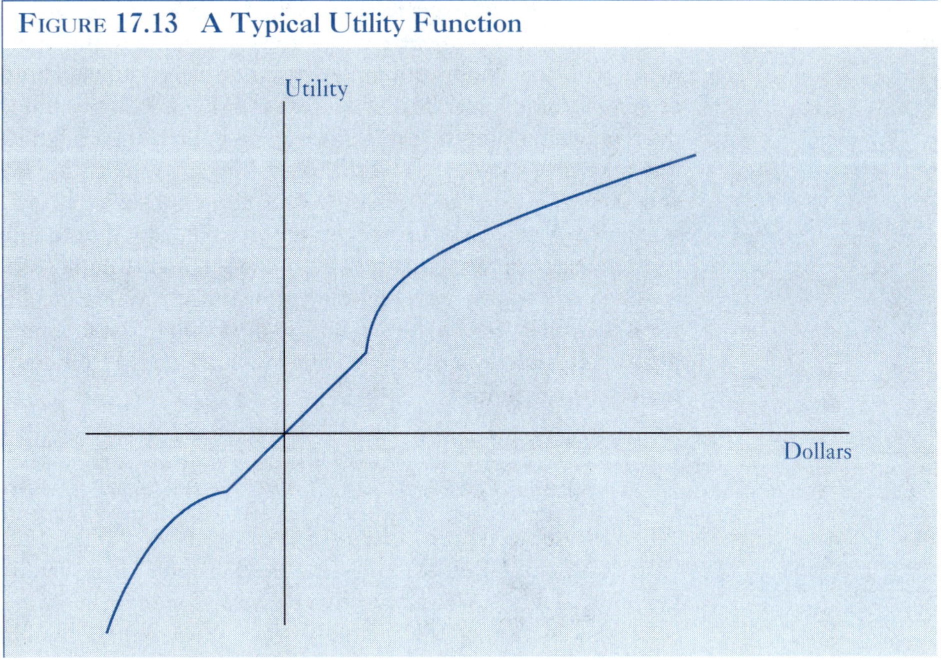

derestimated large frequencies. For example, subjects often assigned two events with the same mortality rate quite different risk estimates. They exaggerated the frequency of events that killed people they knew and events that received disproportionate media coverage. On the other hand, the suboptimal purchase of subsidized flood and earthquake insurance reported by Kunreuther et al. (1978) reflects a denial of the existence of the hazard. Even experts have biased subjective probabilities, as Christensen-Szalanski et al. (1983) reported for physicians asked to judge the mortality risks associated with a list of diseases.

In addition to problems inherent in people's use of subjective as opposed to objective probabilities, violations of the axioms of expected utility theory are well documented in the literature. Expected utility theory requires that a decision maker's preferences are **transitive;** that is, if outcome *A* is preferred to *B*, and if *B* is preferred to *C*, then *A* must also be preferred to *C*. Schoemaker (1982), however, discusses the experimental findings that show preferences to be intransitive.

Transitivity
The condition in expected utility theory that requires outcome A to be preferred to outcome C, if outcome A is preferred to outcome B and if B is preferred to C.

Expected utility theory also requires the independence axiom, which implies that the probabilities of the outcomes and the utilities of outcomes are determined separately and are not related. The **independence axiom** is required to give an expected value interpretation to the sum of products formed by the multiplication of utilities and probabilities. The empirical results of Kahneman and Tversky (1979) show subjects violating the assumption of independence. Early work showed that many subjects cannot even do the arithmetic to reduce a two-stage lottery to an equivalent one-stage lottery.

Independence Axiom
The condition that probabilities of action are determined independently.

Applications of expected utility theory require the assumption that market prices convey information about choices and well-defined and stable preferences. In accordance with the axioms of market-based models, rational choice should yield the same preference order regardless of context and the procedures used to elicit responses. If the decision maker prefers *A* to *B*, for example, then the cash equivalent or minimum selling price of *A* should exceed that of *B*. But, as reviewed by Tversky and Thaler (1990), and Slovic, Griffin, and Tversky (1990), there is a large body of evidence that shows that price orderings of risky prospects are systematically different from choice orderings.

The classic preference reversal phenomenon is demonstrated in a pair of gambles in which one option has a high probability of winning a small amount of money (the P-Bet), the other has a low probability of winning a large amount (the $-Bet), and each has roughly the same expected value. Goldstein and Einhorn (1987), for example, gave the following gambles.

P-Bet: Win $4, with $p = .97$ $-Bet: Win $16, with $p = .31$
Lose $1, with $p = .03$ Lose $1.50, with $p = .69$

When the subjects were asked for their preferred gamble, 50 to 60 percent chose the P-Bet over the $-Bet. When asked at what price they would offer the gamble, given they owned it, the $-Bet typically received a higher minimum selling price. Those subjects who chose the P-Bet and also gave the $-Bet the higher selling price exhibited **preference reversal.** Early work by Lichtenstein and Slovic (1973) demonstrated preference reversal even among professional gamblers at the Four Queens Casino in Las Vegas.

Preference Reversal
An inconsistency in someone's preference ordering.

Preference reversal has many potential sources. Agents who violate either the independence or other axioms of expected utility theory may exhibit preference reversal. As shown by Tversky, Slovic, and Kahneman (1990), however, violation of the axiom of expected utility theory is neither necessary nor sufficient for observing preference reversals; the way a question is stated may bring about preference reversal. The discomforting conclusion for standard choice theory is that preferences are not well defined. On the contrary, preferences are conditioned in the process of choice by many difference effects. Despite many attempts, no choice theorist has yet built a model of choice under uncertainty that reconciles anomalous descriptive realities with desired theoretical properties. Until this happens, expected utility theory, with all its faults, will continue to be advanced for decision making under uncertainty.

Chapter Exercises

17.17 Refer to the notions of subjective and objective probabilities in Chapters 3 and 4. Under what conditions would prior probability be considered objective, rather than subjective?

17.18 In response to the question ". . . would you please explain how gambling differs from buying insurance?" (Troy H. Main, Bartlesville, Oklahoma), Marilyn vos Savant, in her *Parade Magazine* column, wrote: "Life insurance isn't gambling; it's a responsibility. When you lose at the racetrack, your family loses too. But when you 'lose' at life insurance, they 'win'—even if only in a small way." (August 23, 1992) What has Marilyn ignored or implicitly assumed in her answer?

17.19 Because of "bank by phone" and "cash machines," according to *Business Week* (May 10, 1993), banks are assessing consumer demand for ATMs (automated teller machines) versus the cost of branch offices. Whether to close or not close your branch offices and expand ATMs is the question. Your marketing people have classified possible industry acceptance of the ATMs as either "low," "medium," or "high," with prior probabilities of 0.30, 0.50, and 0.20. They prepared the following net payoff table for going to ATMs and closing your banch offices.

	Demand (in millions of dollars)		
	Continue as is	Close branches and expand ATMs	Probability
High	300	500	.3
Medium	250	225	.5
Low	100	1	.2

a. By the expected value criteria, what is the optimum action?

b. If the utility function is $U = \text{Log}_e(\text{payoff})$, what is the optimum action?

17.20 In Exercise 17.19, involving the use of ATMs in banking, you now have the option of organizing a panel of experts who will form an opinion of industry acceptance of the new terminals. The panel's report will be either "favorable" or "unfavorable." Based on a variety of previous panel reports, you estimate the following conditional probabilities:

$$P(\text{fav.} \mid \text{low}) = 0.09 \quad P(\text{fav.} \mid \text{med}) = 0.50 \quad P(\text{fav.} \mid \text{hi}) = 0.75$$

$$P(\text{unfav.} \mid \text{low}) = 0.91 \quad P(\text{unfav.} \mid \text{med}) = 0.50 \quad P(\text{unfav.} \mid \text{hi}) = 0.25$$

If you are risk neutral, draw the decision tree for this problem. What is the optimum action?

17.21 Ceramic material may soon be used in automobile engines, although it is not yet widely employed. As a product developer, you are attempting to assess the demand for ceramics in engines. The decision facing your firm is whether to start a separate company devoted to ceramics or to continue with ceramic related projects within the existing structure. You have classified industry acceptance of the new material as "low," "medium," or "high," to which you attach prior probabilities of 0.25, 0.55, and 0.20. The payoffs associated with these probabilities are given in the following table.

	Demand (in millions of dollars)		
	Low	Medium	High
Continue as is	100	200	250
New startup	–100	200	450
Probability	0.25	0.55	0.20

By the expected value criteria, what is the optimum action?

17.22 For the decision in Exercise 17.21, you now have organized a consulting committee of experts who will advise you on industry acceptance of the new material. The committee's advice will be either "favorable" or "unfavorable." Based on previous advice, you estimate the following conditional probabilities:

$$P(\text{fav.} \mid \text{low}) = 0.10 \quad P(\text{fav.} \mid \text{med}) = 0.50 \quad P(\text{fav.} \mid \text{hi}) = 0.80$$

$$P(\text{unfav.} \mid \text{low}) = 0.90 \quad P(\text{unfav.} \mid \text{med}) = 0.50 \quad P(\text{unfav.} \mid \text{hi}) = 0.20$$

If you are risk neutral, draw the decision tree for this problem. What is the optimum action?

17.23 If a drug test is "99 percent accurate," meaning that 99 percent of drug users are accurately identified and 99 percent of nondrug users are accurately identified, and if we know that approximately 5 percent of the adult population of interest are drug users, what is the probability that someone who tests positive for drugs will in fact be a drug user? What is the relevance of this probability?

17.24 A small software house known as Alter-Becker Worklife has come out with a new program that can be used with the U.S. Bureau of Labor Statistics work life table to calculate expected earnings in wrongful death and personal injury litigation. This house can order disks with copy protected versions of its program in units of 1,000, 5,000, and 10,000 disks, at a cost of $1,000 for the initial order (fixed cost) and $5 per copy (variable cost). The house can produce the disks internally at a fixed start up cost of $3,000 and variable cost of $3.50 per copy. The house estimates that the probability of selling 1,000 copies is 0.3. The probability of selling 5,000 copies is 0.5 and the probability of selling 10,000 copies is 0.2. The finished disk will sell for $80.

a. If this software house wishes to maximize expected net return, what action should it take?
b. Draw the decision tree, indicating the optimal action suggested in part a.

17.25 In Exercise 17.24, if you learned that the owners of the software house were real "worrywarts," how would this affect the optimal decision, that is, would maximization of expected return be the appropriate criteria? Why or why not?

17.26 Under what alternative conditions might the same person be both a risk taker and a risk averter?

17.27 A decision maker's utility function is

$$U(\text{payoff}) = 0.01(\text{payoff}) + 0.55(\text{payoff})^2 - 0.03(\text{payoff})^3$$

where payoffs are measured in thousands of dollars between $1,000 and $10,000 (i.e., payoff = 1, 2, . . . 10).

a. Graph U(payoff).

b. Is this utility function that of a risk lover or a risk averter?

17.28 For the decision maker described in Exercise 17.27, which of the following prospects or gambles would be selected?

a. A lottery in which a win of 0 or $10,000 is equally likely, or

b. $5,000 for certain.

17.29 A lawyer for an insurance company can admit liability and pay a claim for $1 million or contest it in court. If it goes to court, there is a 50/50 chance of either verdict. The legal fees will be $500,000, regardless of the verdict. Below is the decision tree for these prospects. What course of action should the insurance company lawyer follow? What is the significance of the utility function in this problem?

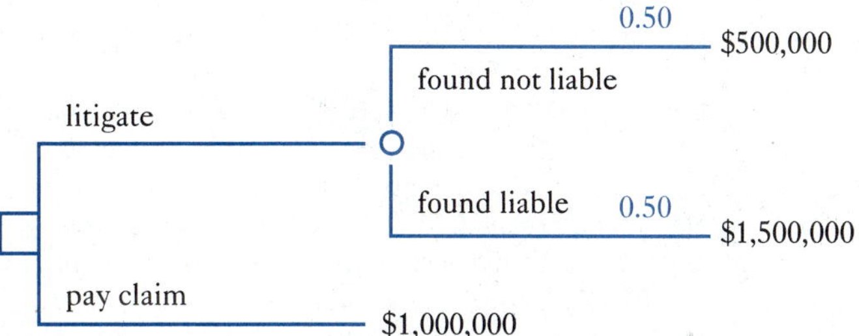

17.30 Sawdust accumulates at Jerald's Wood Shop. The sawdust could be sold at retail (a_1) or it could be sold to a wholesaler (a_2). If the sawdust does not sell at retail, it may be either discounted and sold by Jerald's staff or it will have to be sold wholesale. If sawdust builds up in the yard, Jerald can either try to unload the dust in small lots at substantial discounts (e_1) by advertising a sale (and burning whatever does not sell) or dispose of it all at once to a wholesaler (e_2). The decision tree for this problem is shown below (payoffs are expected future returns in thousands of dollars to Jerald).

a. What are the expected payoffs from alternative actions?

b. What is the optimal strategy, assuming risk neutrality?

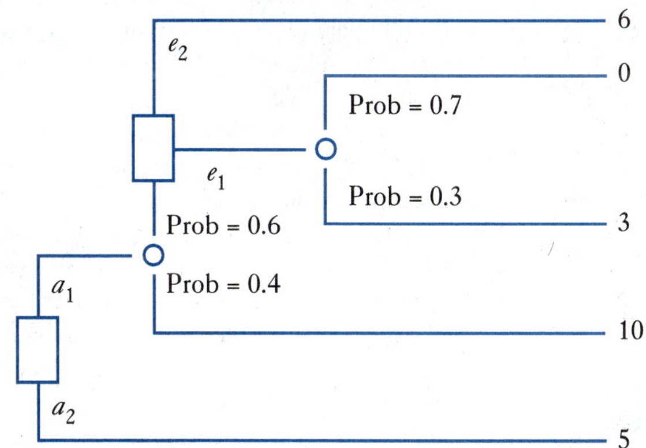

17.31 *The Wall Street Journal* (September 29, 1992) reported on fail-safe or highly reliable computing. "Every time Status Computer Inc. President William E. Foster shows off a new computer, he opens up its cabinet and yanks out a wire or a circuit board. It keeps on humming . . . 99.999% reliability." The article stated that many big computers these days can deliver 99.9 percent reliability. But when computer outages can cost $70,000 per minute in lost sales of airline reservations, "five-nines engineering" may be necessary. If there is a 0.001 percent chance of 8.8 hours of downtime a year on a big airline reservation system with a 99.9 percent reliable system while the probability for a Status computer is 99.999 percent, how much more per year should the airlines be willing to pay for Status' precision to five decimal places in nines?

17.32 In shopping for a used Corvette, Professor Becker was concerned that approximately 15 percent of owners responding to the survey for the annual auto issue of *Consumer Reports* reported electrical systems problems. Becker was particularly interested in one Corvette that seemed okay, but he took it to a "Corvette service specialist." In the past, the mechanic claims to have been 95 percent accurate in spotting faulty electronics on Vettes that later proved to be faulty. Of all the Corvettes he has examined that later proved to be okay, he correctly judged them as such 98 percent of the time.

a. Prior to having the specialist look at the Corvette, what is the probability that it is faulty?
b. What is the probability that the Corvette Becker is considering has faulty electronics, if:
    1. The specialist judges it to be okay?
    2. The specialist judges it to be bad?

17.33 If the Corvette service specialist in Exercise 17.32 charges $75 and the average cost of repairs to the Corvette electrical system over the period of ownership of interest is $950, should Becker hire the specialist, assuming he is risk neutral?

17.34 A computer manufacturer uses disk drives from two suppliers. The first supplier accounts for 80 percent of the drives and the second 20 percent. It is known that 6 percent of the drives supplied by the first supplier are defective and 1 percent of the drives supplied by the second supplier are defective. If a disk drive is found to be defective, what is the probability it came from the first supplier?

17.35 "A drug called DNase is a treatment for the nation's 20,000 CF (cystic fibrosis) victims . . . . Pharmaceutical executives, in an effort to defend price increases, often lament that it takes some $359 million and 12 years on average to develop a drug . . . . Arthur D. Levison, senior-vice president for research and development (Genentech Inc.), immediately put the chance that DNase would become a drug at 70%, an unheard of level at such an early stage even for Genentech's can-do gene jocks . . . . These statistics should help the company defend the drug's price, which analysts expect to be $10,000 to $15,000 per year." (*Business Week*, August 23, 1993) Assuming a steady state in which all the above values stay fixed with no time preference, for what life expectancy of CF victims could Genentech expect to break even at the $10,000 to $15,000 price?

17.36 If DNase, as described in Exercise 17.35, increases life expectancy of CF victims, then which of the steady state assumptions in Exercise 17.35 would change? Will this imply more profits for Genentech?

17.37 An editorial in *The Wall Street Journal* (June 24, 1992) on the U.S. Food and Drug Administration's ruling on silicone breast implants included Dr. Marcia Angell's remarks "that in breast implants, as always, the issue is a balance between risks and benefits, and in this case, 'the benefit has to do with personal judgments about the quality of life, which are subjective and unique to each woman.' A bureaucracy working with numbers can't measure this." Comment on the merits of a "risk and benefits" comparison in general and on its applicability in this case.

17.38 Ronald Howard, as cited in this chapter, reported that a distinguished decision theorist presented the following example involving a symmetric die with six equally likely sides, numbered 1 through 6, and the two prospects, labeled "Deal A" and "Deal B."

Die	1	2	3	4	5	6
Deal A	$600	700	800	900	1000	500
Deal B	$500	600	700	800	900	1000

In response to the question "do you prefer Deal A or Deal B?" Howard claimed the presenter said he preferred Deal A, because he would win more money in five of the six ways the die could fall. If the amount of money involved was hundreds of thousands of dollars rather than hundreds of dollars, then the presenter said he would prefer Deal B, because he would feel such regret if he chose Deal A and the number 6 appeared on the die. Why do you think Howard found this answer absurd.

17.39 In Exercise 17.38, assume Deals A and B each cost $750, and that a potential customer will play but is indifferent about selecting A or B. Now, in a new situation, assume that Deals A and B involve hundreds of thousands of dollars (so a 1 on the face of the die in Deal A paid $600,000) and the cost is $750,000 to play. If the potential customer refuses to play this second game, what would you conclude about this person's utility function?

Academic References

Becker, William, and Richard Stout. "The Utility of Death and Wrongful Death Compensation." *Journal of Forensic Economics,* (Fall 1992): pp. 197–208. Reprinted as "The Utility of Death and Wrongful Death Compensation." In J. Ward, ed. *Hedonics Primer for Economists and Attorneys.* Tucson, AZ: Lawyers and Judges Publishing Co., 1992, pp. 155–168.

Brown, Rex V. "The State of the Art of Decision Analysis: A Personal Perspective," and "Response." *Interfaces,* November-December, 1992, pp. 5–14 and 24–26.

Christensen-Szalanski, Jay J. J., Don E. Beck, Carlyn M. Christensen-Szalanski, and Thomas D. Koepsell. "Effects of Expertise and Experience on Risk Judgments." *Journal of Applied Psychology* 68, 1983, pp. 278–284.

Dardis, Rachel. "The Value of a Life: New Evidence from the Marketplace." *American Economic Review* 70, December 1980, pp. 1077–1082.

Goldstein, W. M., and H. J. Einhorn. "Expression Theory and the Preference Reversal Phenomena." *Psychological Review* 94, April 1987, pp. 236–254.

Howard, Ronald A. "Heathens, Heretics, and Cults: The Religious Spectrum of Decision Aiding." *Interfaces,* November–December 1992, pp. 15–27.

Kahneman, Daniel, and Amos Tversky. "Prospect Theory: An Analysis of Decision Under Risk." *Econometrica,* March 1979, pp. 263–91.

Kunreuther, H., R. Ginsberg, and L. Miller. *Disaster Protection Insurance: Public Policy Lessons.* New York: Wiley, 1978.

Lichtenstein, S. and P. Slovic. "Response-Induced Reversals of Preference Between Bids and Choices in Gambling Decisions: An Extended Replication in Las Vegas. *Journal of Experimental Psychology* 101, November 1973, pp. 16–20.

Lichtenstein, S., P. Slovic, P. Fischhoff, B. Layman, and B. Combs. "Judged Frequency of Lethal Events." *Journal of Experimental Psychology: Human Learning and Memory* 6, 1978, pp. 551–578.

Morgan, M. Granger. "Risk Analysis and Management." *Scientific American,* July 1993, pp. 32–41.

Schoemaker, Paul J. H. "The Expected Utility Model: Its Variants, Purposes, Evidence and Limitations." *Journal of Economic Literature* 20, June 1982, pp. 529–563.

Slovic, P., D. Griffin, and A. Tversky. "Compatibility Effects in Judgment and Choice." In R. M. Hogarth, ed. *Insights in Decision Making: Theory and Applications*. Chicago: University of Chicago Press, 1990.

Tversky, A., P. Slovic, and D. Kahneman. "The Causes of Preference Reversal." *American Economic Review* 80, March 1990, pp. 204–217.

Tversky, A. and R. Thaler. "Preference Reversals." *Journal of Economic Perspectives* 4, Spring 1990, pp. 201–211.

FORMULAS

- Bayes' Rule for Calculating Posterior Probabilities $P(s_1 \mid I_k)$

$$P(s_1 | I_k) = \frac{P(I_k \cap s_1)}{P(I_k)} = \frac{P(I_k | s_1)P(s_1)}{\sum_{j=1}^{J} P(I_k | s_j)P(s_j)}$$

where s_j is the jth state of nature, for J such states
I_k is the kth indicator from sample information
$P(s_j)$ is the prior probability of the jth state of nature

CHAPTER 18

STATISTICAL QUALITY CONTROL AND QUALITY MANAGEMENT[1]

The object of taking data is to provide a basis for action.

W. Edwards Deming
1900–1993

18.1 Introduction

The statistical techniques demonstrated in this book have had a major influence on business practices. Currently, that influence is highly visible in the area of quality control, where actions are taken as a direct result of data observations. To demonstrate some of the ways in which statistical techniques have been used in quality control, we consider the following situations.

- Critical in the production of adhesive bandages is the uniformity of the adhesive. A defective bandage has spotty adhesive. To assess quality, a manager of bandage production collects data over a period of 10 months on the proportion of adhesive strips that are defective. If the proportion of defects is judged to be beyond that assignable to chance, the manager must take action; that is, find the cause and correct it.
- "Material handlers" retrieve units exiting a conveyor belt; they transport the units via a forklift truck to their appropriate location in the finished products warehouse. The handlers claimed that the level of production (number of units coming off the belt in a given time period) had increased substantially to the point that another handler is now required to assist in the job. Data on production levels were available for the last nine days. Do these data support the handlers' contention that production levels are unstable?
- Hospital emergency rooms face a problem in treating trauma, acute illness, and less urgent ailments of patients who may not be covered by insurance. According to a March 2, 1994, U.S. Department of Health and Human Services news release, the government and private/commercial insurance cover only about 78 percent of those seeking emergency room care. A profit-oriented hospital may want to have the accounting department monitor emergency room practices to assess any change that might be occurring in its treatment of the uninsured. The question is how best to do this.
- An engineering graduate student designed a computer-based menu interface. A user's task involved searching the six sequential levels of the menu structure for a desired topic description. The graduate student was particularly interested in determining whether the incidence of retrieval errors (selecting the wrong icon) warranted an investigation into their causes.
- Many U.S. tool manufacturers have learned from the Japanese that suppliers of inputs must be viewed as part of the production process; purchased items must be scrutinized before they are used in the production of final goods. As part of the production process, questions need to be raised about the number of items to be inspected from an incoming shipment and when an entire shipment should be rejected or accepted based on the sample inspected.

The quality of a product or service is defined by specific dimensions of interest. In a market economy, the consumers determine which products and which attributes of those products define quality. In the sale of bandages, marketing experts working with sample focus groups and sample market data (as discussed in Chapter 7) found that bandages that stuck to human flesh, without pulling when purposefully removed, sold best. Engineers determined that uniformity in the adhesive was a physical property that influenced this desired attribute. In the case of a service, quality might be defined in terms of speed of delivery or absence of waiting time.

After completing this chapter, you should be able to define product quality in terms of measurable dimensions of interest. You should understand how to use both statistical process control and acceptance sampling. Finally, you should be able to identify the roles of the producer and the consumer along with the associated risks of each within the quality control framework.

18.2 A Perspective on Production and Quality Control

At the turn of the 20th century, Frederick W. Taylor (1856–1915) revolutionized production techniques by quantifying tasks in terms of their positions in the production sequence, difficulty, and time to complete. One of Taylor's (1947, p. 164) analyses started with subdividing a manual production task, shoveling dirt, into its various components: the time required to fill the shovel, the time required to walk with the load of dirt, and the time required to throw the shovelful of dirt. The least amount of time required to complete each subcomponent was then determined through the use of stopwatches and task observation. Taylor then redesigned the execution of the subcomponents in a manner that minimized overall time. One of Taylor's most enduring contributions to the field was the recognition that decisions to improve any process must be based on reliable and valid measurements—hard data.

Today, Taylor is regarded as the inventor of time and motion studies in manufacturing. Many executives consider Taylor the father of scientific management, while many union leaders consider him the scourge of the workplace. McGill University Professor Henry Mintzberg (1994) argues that Taylor and those that followed him in strategic planning have only succeeded in measuring what's measurable. Operating efficiency tends to be emphasized over product leadership that may entail nonmeasurable product attributes. Yet putting data to use and an emphasis on quality seem to have helped the Japanese become world leaders in several manufacturing sectors.

Quantitative quality control was introduced in Japan after World War II by two Americans, W. Edwards Deming and Joseph M. Juran. They revolutionized the quality and competitiveness of products and services produced by Japan and helped change the world's perception of Japanese products. Further, as a result of Japan's ability to manufacture superior products, the United States and other industrialized nations were forced to adopt a heightened sensitivity to the quality of products that were entering and leaving their factories.

These quality pioneers have left their mark on global manufacturing. In recognition of Deming's contributions to Japanese manufacturing, the Japanese Union of Scientists and Engineers created the prestigious Deming Prize, given to a company for creating outstanding products and services. One of Juran's many contributions is appropriately termed the Juran Trilogy, for the three phenomena of quality:

1. Quality improvement is a continuous process.
2. A structured managerial process must be developed for carrying out continued improvement, step by step, project by project.
3. Quality planning leads to new goods and services by identifying customer needs and meeting them. Quality improvement reduces deficiencies in exiting goods, services, or process. Quality control ensures that results are sustained.

Juran and the other gurus of quality-oriented management argue that an essential ingredient for a successful quality improvement program is that top executives be involved. In addition, Juran argues that firms will not recognize that high-quality products actually cost less to produce in the longer term until top executives recognize a crisis or a whopping opportunity.

It was not until the 1980s that American manufacturers looked to the "total quality philosophies" as a means to improve performance. At the time, for example, it was easy to point at the number of defects per vehicle produced by U.S. versus Japanese firms, as shown in Figure 18.1, as a telltale sign of why American manufacturing was in decline.

FIGURE 18.1 Automobile Quality

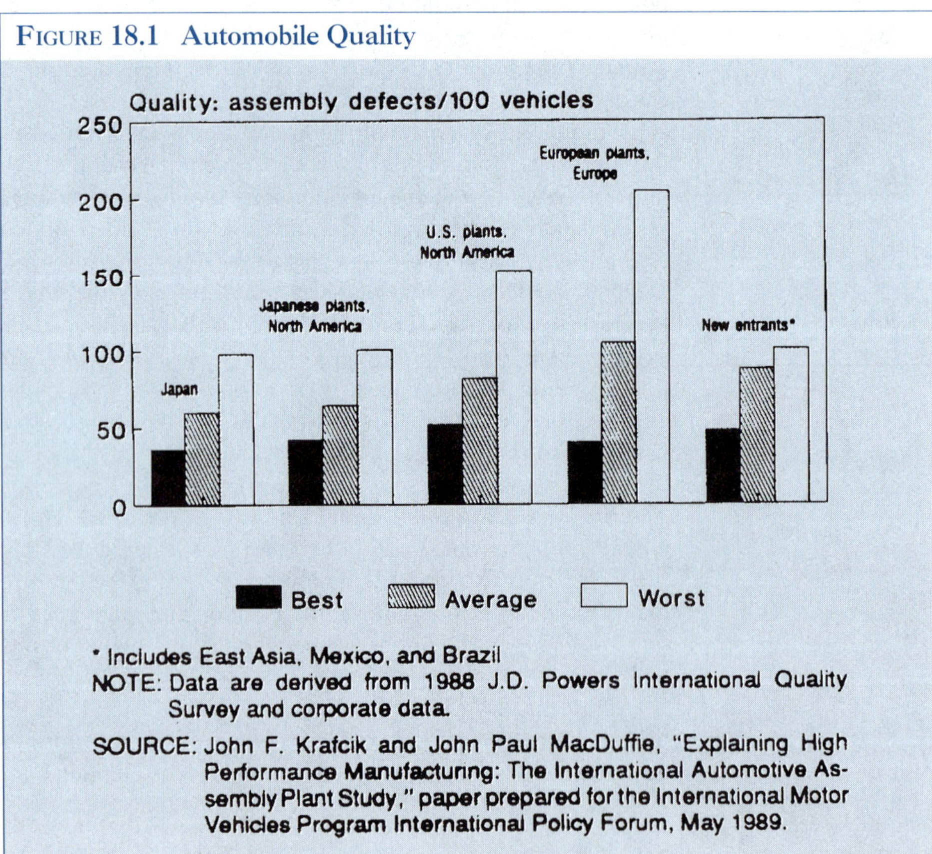

* Includes East Asia, Mexico, and Brazil

NOTE: Data are derived from 1988 J.D. Powers International Quality Survey and corporate data.

SOURCE: John F. Krafcik and John Paul MacDuffie, "Explaining High Performance Manufacturing: The International Automotive Assembly Plant Study," paper prepared for the International Motor Vehicles Program International Policy Forum, May 1989.

A perceived lack of competitiveness because of shoddy products led many business leaders and politicians to look for ways to stimulate interest in quality improvement. A total quality philosophy was viewed as a way to reduce the need for repairs, recalls, and retrofits. High quality was being viewed as a way to reduce cost and not add to it.

The Malcolm Baldrige National Quality Award was initiated as a stimulus to American industry to adopt quality improvement programs. The Baldrige Award was established by Congress in 1987 and has become the most visible award for quality in the United States. The seven criteria for the award are leadership, information and analysis, strategic quality planning, human resource utilization, quality assurance of products and services, quality results, and customer satisfaction; each of the criteria is weighted according to importance with customer satisfaction first.

Competition for the award has been intense, with only 19 of some 500 applying companies receiving the award between its inception and 1993. The burdens of the application process, however, may be leading to fewer applications, as evidenced by falling numbers in 1992 and 1993. Supporters continue to applaud the award, claiming that it serves as a vehicle through which American firms become more globally visible and competitive.

Critics of the award have asserted that "winners haven't necessarily solved their business problems and gone to capitalist heaven" (*Fortune,* July 1, 1991) and that the award judges the process, not the finished product. Evidence compiled by statistician William Schmidt of Michigan State University on behalf of the American Quality Foundation suggests that firms just getting started in total quality management (TQM) should not waste resources implementing formal statistical gauges of progress because gains will not be apparent. The business press has presented the argument that only firms with a quality-oriented infrastructure will benefit from trying to imitate the best of the best (benchmarking) and measuring incremental gains (*Business Week,* November 30, 1992).

Despite the critics, American manufacturers appear to be more willing today to use statistical methods in the assessment of product quality than in any previous time period. Pioneers like Taylor, Deming, and Juran deserve credit for the evolution. At the same time, the warnings of the critics cannot be ignored. That which is done in the name of total quality management must be sound on statistical as well as economic grounds. If it is just a mindless collecting of data, then it will be an addition to cost and not an efficient means of planning, control, and improving.

18.3 Statistical Quality Control and Graphical Tools

Final product quality is defined and judged by consumers in the marketplace. Thus, the level of quality desired by consumers must motivate the discussion and practice of statistical quality control at every level of an organization.

In statistical quality control, the term "control" refers to assuring product conformance to pre-established specifications that are set in accordance with ultimate consumer desires. The term "statistical" indicates that the assurance is based on numerical data and involves an analysis of variability in a specific manufacturing process.

Common Causes Causes of variability that are embodied in a system.

Assignable Causes Causes of variability that arise from influences outside of a process.

Typically, the variability in a process can be associated with a random variable and a specific probability distribution. This distribution is used to define the **common causes** of chance variability and variability **assignable** or specific to causes other than chance. Distinguishing between common and assignable causes is integral to the practice of quality control.

Common causes are embodied in the distribution of the random variable(s) represented in the process. Common causes are part of the process. They are found in the materials, methods, personnel, equipment, and environment that constitute the production process. Thus, eliminating common causes requires direct manipulation of features that are inherent to the process. Assignable causes arise from influences outside of the process; they are not part of the process design. Assignable causes can be either one-time events (a single wrench that falls into the gears) or persistent (every Friday afternoon around three o'clock a wrench falls into the gears). Assignable or special causes can typically be remedied without disrupting the entire process. Because they are localized, they can be dealt with on the spot and do not require a total reworking of the process.

Statistical Quality Control Identifying the causes of process variation and intervening to control that variation.

Process A series of activities or operations designed to yield a desired outcome.

Statistical quality control is the practice of identifying assignable and common causes present in a process and subsequently trying to control them to achieve the highest levels of performance. There are two major areas of study in statistical quality control: process control and acceptance sampling. Process control involves maintaining and improving product quality levels by monitoring a manufacturing process and intervening in the process when it is necessary. A **process** is a series of operations or activities that are employed in a specific manner to achieve a desired outcome that at least in part is determined by chance. A process thus can be viewed as a single or series of random variables. Getting a college education is an example of a process. It involves a series of activities that are conducted in a specific order that involves the uncertainty of course registration, test taking, and ultimately graduation. Process control might entail studies to make sure that things other than chance are not causing students to drop out after each semester.

Acceptance Sampling Evaluating a large product lot for conformance to standard by examining a sample.

Acceptance sampling involves an evaluation of a large product lot for conformance to a standard by examining a sample taken from the lot. For example, the acceptability of a magazine of bullets (the lot) might be determined by firing only a small sample of randomly selected bullets. Such tests have already been considered in detail in Chapter 9 and its appendices. Process control, on the other hand, was only considered briefly in Chapter 2. We will extend our earlier discussion of process control to reflect what was learned in Chapters 3 through 6 about probability and in Chapters 7 and 9 about sampling and estimation. We begin with some basic graphical tools that can be used to help identify the most influential factors critical to a process.

Flowcharts

A flowchart is a schematic used to graphically display a process of interest. It represents the flow of information and activity in a process through the use of symbols. Various symbols are linked by arrows that indicate the direction of information flow. Figure 18.2 is a flowchart for the registration process at a state college on the East Coast. Each of

FIGURE 18.2 Flowchart of the Registration Process

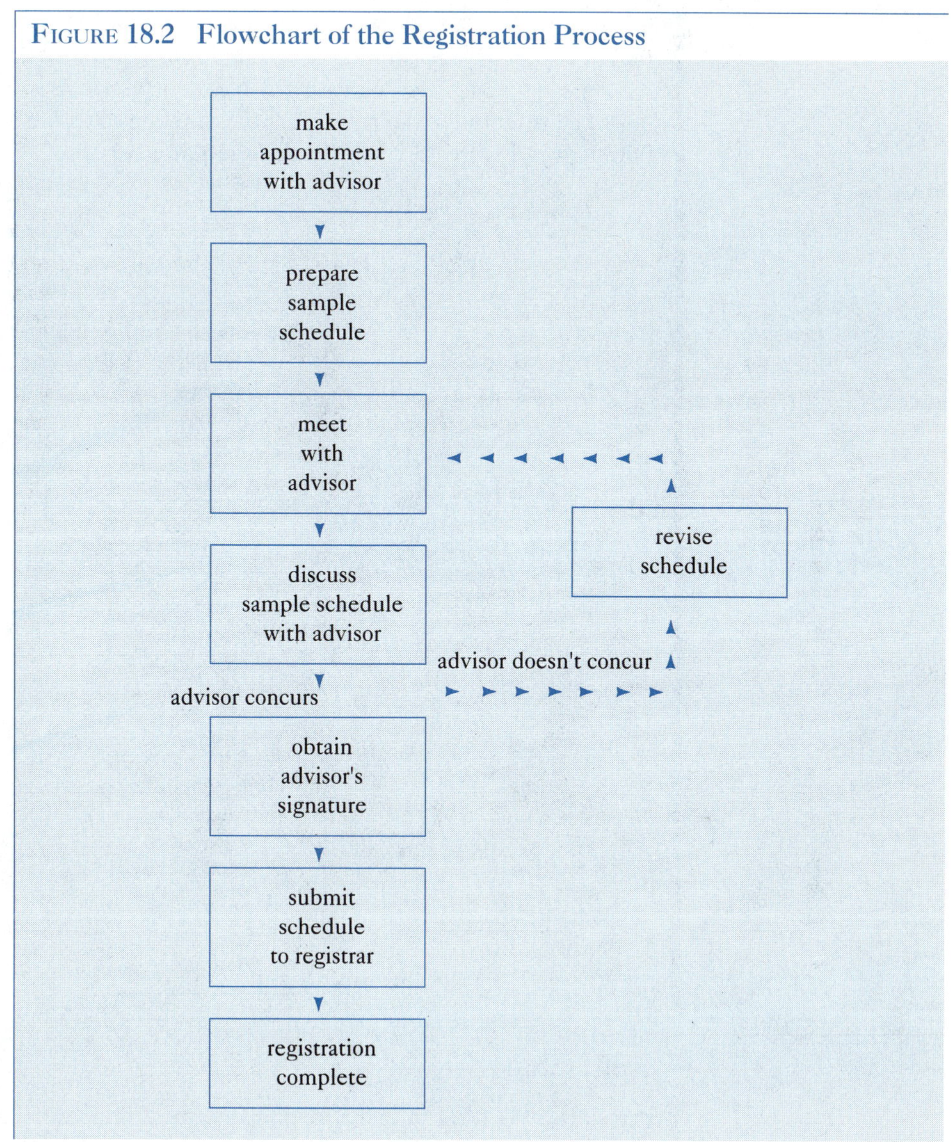

the rectangles on the flowchart represents a student's activity during the registration process. The arrows represent the order of activities the student encounters. More complex processes involving numerous participants and diverse objectives can be represented using symbols of multiple shapes and sizes.

The value of representing a process with a flowchart is that it forces the analyst to identify critical process operations. In addition, the flowchart illustrates the decision rules that are an integral part of achieving the desired outcome.

QUERY 18.1:

Before a student may submit a schedule to the registrar, tuition fees must be paid to the bursar for the next semester. A payment verification receipt from the bursar must be presented to the registrar at the time the schedule is submitted. Modify the flowchart in Figure 18.2 to reflect this requirement.

ANSWER:

Modified Flowchart of the Registration Process

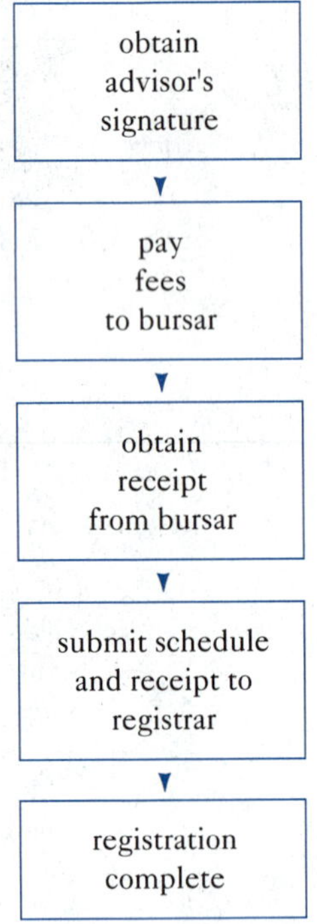

Pareto Chart

A Pareto chart is used to make evident the most important causes of a process deficiency. It consists of a succession of bars, arranged from largest to smallest, where the heights of the bars reflect the frequency of process deficiencies. Pareto charts help to visually separate the vital few from the trivial many.

Pareto charts are used to examine process deficiencies and are developed by first collecting process deficiency data. This data could be a measure or count of unusable product generated at the completion of a manufacturing process. Typically, this is referred to as process waste. It is classified according to type; each type of waste represents a percentage of the total waste generated. The percentage contribution to the total is then plotted for each waste type. If the percentages for each waste type are summed, they would total 1.0, because combining all waste types represents total waste generated.

The Pareto chart in Figure 18.3 represents how five different categories contribute to the total waste in a sanitary napkin manufacturing process. The process waste in this example is classified as one of five types: corregate, napkin, dust, clippings, or pulp waste. Corregate waste, napkin waste, and dust waste are the three major contributors to total process waste, while clippings and pulp waste are very small contributors.

The causes of corregate, napkin, and dust waste should be examined quite carefully. Eliminating or reducing those causes would have the most impact on the total amount of waste being produced by the manufacturing process. By the same token, completely eliminating clipping and pulp waste would have only a small effect on total waste. The value of the Pareto chart is its ability to visually differentiate between the heavy hitters and the less significant contributors.

FIGURE 18.3 Pareto Chart of Process Waste

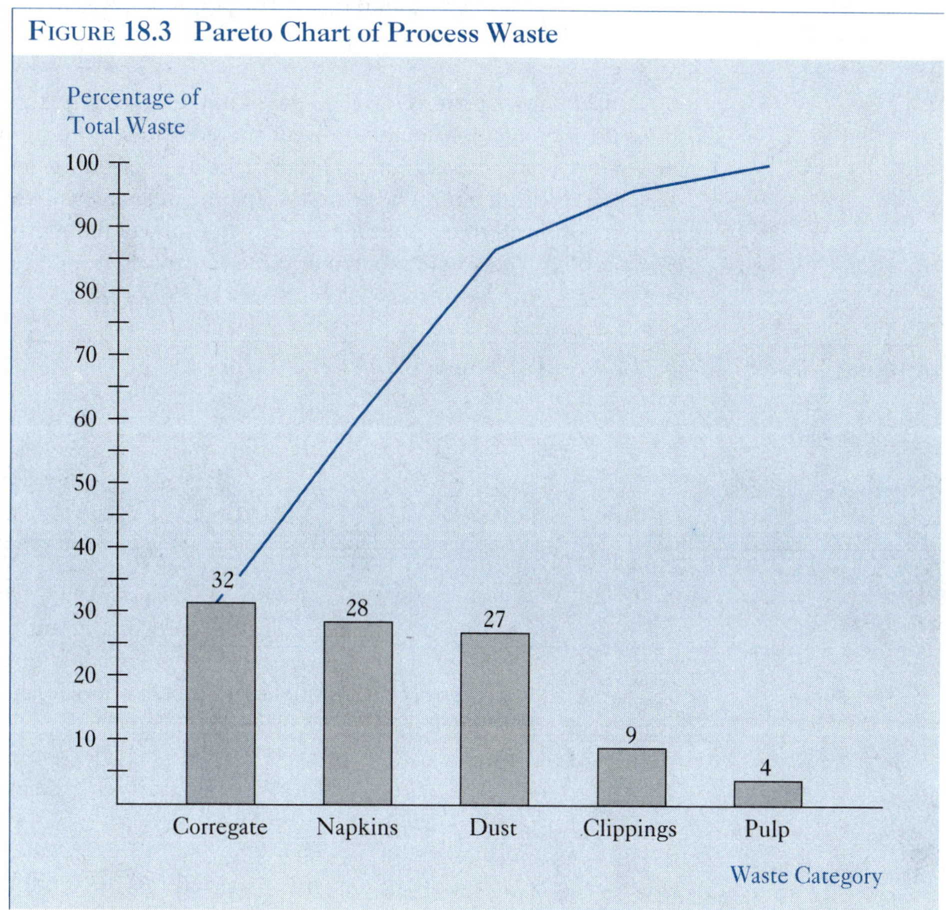

Fishbone Diagram

The fishbone diagram is referred to as the "cause and effect" diagram because it helps identify critical factors that affect a desired outcome. It can be used for studying processes, for examining situational problems, and for planning operations. The fishbone diagram must be used carefully, however. Banks indicates that simply constructing a fishbone diagram does not ensure that the process has been represented in a comprehensive manner or that all process deficiencies are automatically accounted for (1993, pp. 356–409).

Fishbone diagrams are developed by first identifying all the elements of a process. Each element is then examined more closely for identification of the components comprising that element. Figure 18.4 illustrates some of the critical elements that help generate shippable product. The angled lines show five main elements that influence shippable product: equipment, materials, methods, environment, and personnel. Along each angled line are the integral components corresponding to each of these five elements. For example, it is important that equipment be well-maintained. Many other factors could be incorporated into the diagram, such as noise, rodent, temperature, and ventilation control.

A fishbone diagram is a valuable tool to the process quality analyst because it provides a coherent representation of all the contributors to a process. Typically, these contributors can be classified under the labels of personnel, methods, machines, materials, and environment. These are the five main sources of variability in a production process that need to be controlled. This control will require quantification of variability that might be assigned to one of these process contributors.

FIGURE 18.4 Fishbone Diagram of Shippable Product Generation

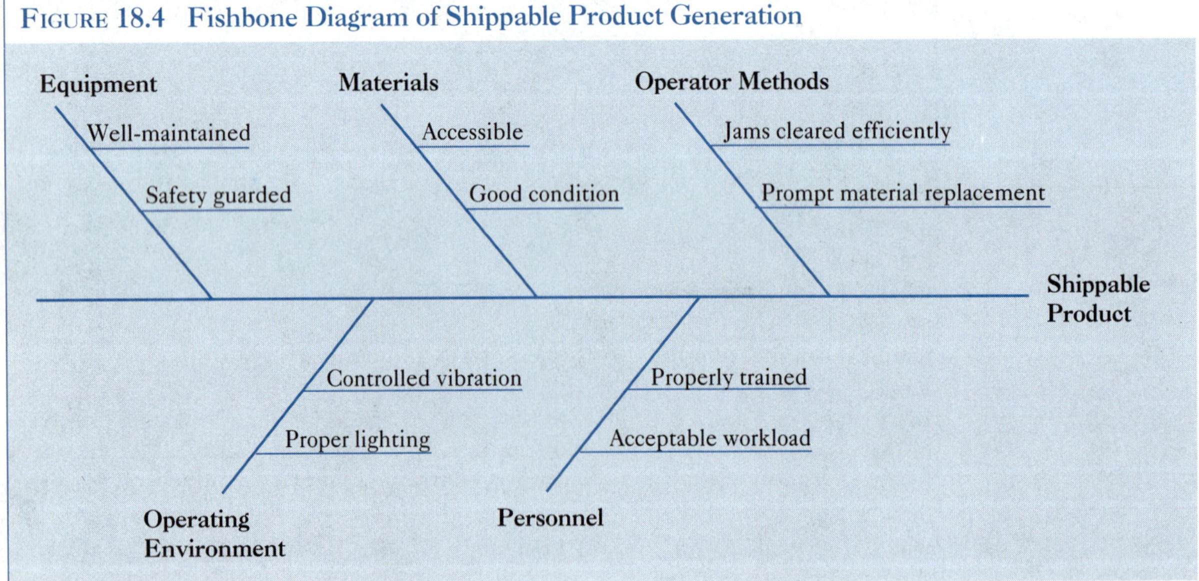

Histogram

The histogram was introduced and discussed in Chapters 1 and 2 as a means of summarizing quantitative information. A histogram displays the frequency or relative frequency of a measured quantity or variable outcome. The numerical outcomes of interest are typically plotted on the horizontal axis and the frequencies of each of these values are plotted on the vertical axis. When process data are incorporated into a histogram, the spread and the shape of a histogram of relative frequencies reveals the underlying distribution of the process. If the shape and spread of the relative frequency histogram is stable, with bar heights that do not jump abruptly, or a maximum (or minimum) that does not wander upward (or downward) as more observations are added, then the histogram is said to display the overall capability of a stable process. The construction of such a histogram from time-series data was demonstrated in Figure 13.11.

Exercises

18.1 An economist stated that "the Malcolm Baldrige National Quality Award may judge the process of production but it ignores the product and its value in the marketplace." What, if any, relevance does this quote have for quality control?

18.2 Are process quality and product quality mutually exclusive entities? Explain.

18.3 What is the difference between common causes and assignable causes?

18.4 Clegg said, "The only real way to improve quality is to make it right the first time" (1993, p. 16). What role does statistical process control play in achieving this?

18.5 Expand upon the fishbone diagram in Figure 18.4. Include any additional elements that contribute to the generation of shippable product in a manufacturing facility.

18.4 Statistical Process Control

Statistical Process Control
Involves maintaining and improving product quality levels by monitoring a process, operation or service and intervening when it is necessary.

Any process is a set of operations and activities that, when employed in certain ways, can be expected to lead to a desired outcome. A manufacturing process consists of components like materials, equipment, methods, and operators that contribute toward the generation of a finished product. **Statistical process control** provides the methodology to examine the stability of an ongoing process. Before employing statistical process control, the product features that define its quality must be identified. These features can be referred to as measurable dimensions or attributes of interest. Examples include weight, height, length, width, depth, clarity, brightness, and color. In the case of the adhesive bandages, the attribute of interest is uniformity in the adhesive, as measured by spottiness in the adhesive.

When a process generates finished product, samples may be taken and the specific dimension of interest assessed and recorded. The analyst can use the data to determine whether the process is producing too many defects to believe the process is stable. If the data indicate that the process is no longer stable, it is said to be "out of control," and intervention is necessary to alleviate the causes that are not due to chance, but rather to external sources.

In Control
A situation in which a measurable process dimension follows the same probability distribution over time.

A process is **in control** if measurements on the identified dimension follow the same probability distribution over time. If the dimension of interest involves a continuous measure, one way to assess whether the process is in control might be to monitor a relative frequency distribution of the results. As already stated, a process in control would exhibit a pattern of variation in the histogram that is stable over time. A stable histogram would suggest that only chance factors common to each occurrence are causing variability in observed measurements. As discussed in Chapter 7, common chance factors cannot be avoided in sampling. By definition, the random variable associated with a process will tend to give different values on each occurrence.

Out of Control
A situation in which both chance variation and assignable variation exist.

A process is said to be **out of control** if, in addition to chance variation, there is added variation caused by external sources. Some examples of external shocks that could contribute to added variation in the process include systematic tool breaks, a machine that is out of adjustment, an untrained operator, or imperfections in a new batch of raw material. Sources of variability assignable to external factors can be controlled and thus the variability in a process can be reduced with the removal of these external factors. Variability in a process could actually be increased, however, if it were constantly tampered with and overadjusted in response to chance occurrences.

18.5 Statistical Process Control: A Control Chart

Although a histogram could be used to monitor some types of processes over time, instead of constructing a new histogram each time another observation or set of observations is obtained, it is more efficient to plot the new information in a control chart. A control chart is used to spot outliers or peculiar measures as they occur. It also is used to monitor movements in the measures of central tendency and dispersion as they take place. With quick identification of the specific causes of these anomalies and trends, action can be taken immediately to remove the cause and thus reduce unwanted variability in the process.

As illustrated in Chapter 2 in the case of an $\bar{x}$ chart, development of the control chart involves taking samples of size n from the process that is being examined. Although a sample of one item ($n = 1$) can be taken, usually n is greater than one. As discussed in Chapter 6 and its appendix, these samples could be drawn by any one of several methods, depending on the nature of the process. As samples are taken, the product dimension of interest is measured, and the sample means are computed and recorded. The chart on which these sample means are recorded is the control chart.

To demonstrate the construction of the control chart, consider the material handlers' example described at the beginning of this chapter. The handlers retrieve units of product from a conveyor system and transport it to another location in the warehouse. They feel that production has been increasing and has reached an unacceptable level for their current work force. They want another handler assigned immediately if the production level increase is not the result of chance causes.

The first step in addressing this issue is to obtain and review data on the measures of interest, which come from data obtained while the material handlers are performing their jobs. Because handlers work one of four shifts per day, a shift is our time unit of observation (n = 4). The process data for nine days of production, and the sample means and ranges, are tabulated in Table 18.1.

The material handlers requested that another handler be assigned to help handle the increased production if the increase is due to assignable causes. An $\bar{x}$ chart will show how the amount of material handled has been changing. The $\bar{x}$ chart starts with a plot of the $\bar{x}$ values over time, as shown in Figure 18.5. Although the time series in Figure

TABLE 18.1 Production Process Data (nine days)

	Total Units Handled					
Day	Shift 1	Shift 2	Shift 3	Shift 4	$\bar{x}$	Range
1	104	127	124	100	113.8	27
2	127	131	121	121	125.0	10
3	126	130	126	103	121.3	27
4	110	131	90	122	113.3	41
5	112	128	99	97	109.0	31
6	90	134	118	126	117.0	44
7	80	140	100	98	104.5	60
8	97	138	105	145	121.3	48
9	105	145	130	128	127.0	40

FIGURE 18.5 Material Handling Production Levels Chart

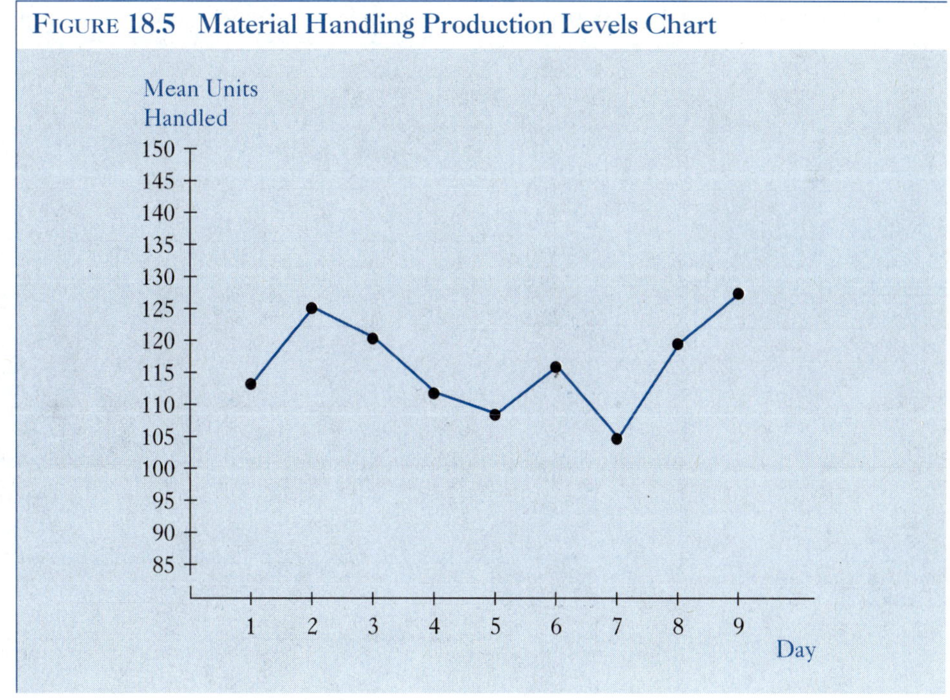

18.5 appears to end on a high reading, additional days may show the series turning down again as it has in the past. Figure 18.5 clearly shows variability in the sample means. The sources of that variability are difficult to identify, however. A standard is needed against which changes in $\bar{x}$ can be assessed. That standard could take many forms and be based on different time-series statistical tests, as discussed in Chapter 15. In actual day-to-day practices, where decisions often must be made on the spot by those with minimal education in statistical time-series analysis, the simplest of standards are employed.

As a standard for comparison, individual $\bar{x}$ values could be compared to the desired process population mean μ, if it were known. Ideally, the desired process population mean would be what customers want and what the firm can provide at a profit. Typically, an exact value for μ is unknown. By historical precedent in quality control work, μ is approximated by the grand mean $\bar{\bar{x}}$, which was defined in Chapter 11 as the mean of sample means. For k samples, where each is of size n, the grand mean is

$$\bar{\bar{x}} = \sum_{i=1}^{k} \bar{x}_i / k$$

For the nine samples in Table 18.1, the grand mean is 116.9.

$$\bar{\bar{x}} = (113.8 + 125.0 + 121.3 + 113.3 + 109.0 + 117.0 + 104.5 + 121.3 + 127.0) / 9$$
$$= 116.9$$

Centerline
A representation of the mean of an in control process.

The grand mean $\bar{\bar{x}}$ is a standard against which individual sample means can be compared. To facilitate this comparison, a chart showing $\bar{\bar{x}}$ as the centerline of the $\bar{x}$'s is constructed. The **centerline** of the control chart, denoted CL, represents the process mean when the process is in control. The centerline for the handlers' workload case, CL = 116.9, is shown in Figure 18.6. The nine sample means appear to deviate randomly

FIGURE 18.6 $\bar{x}$ Control Chart with Limits

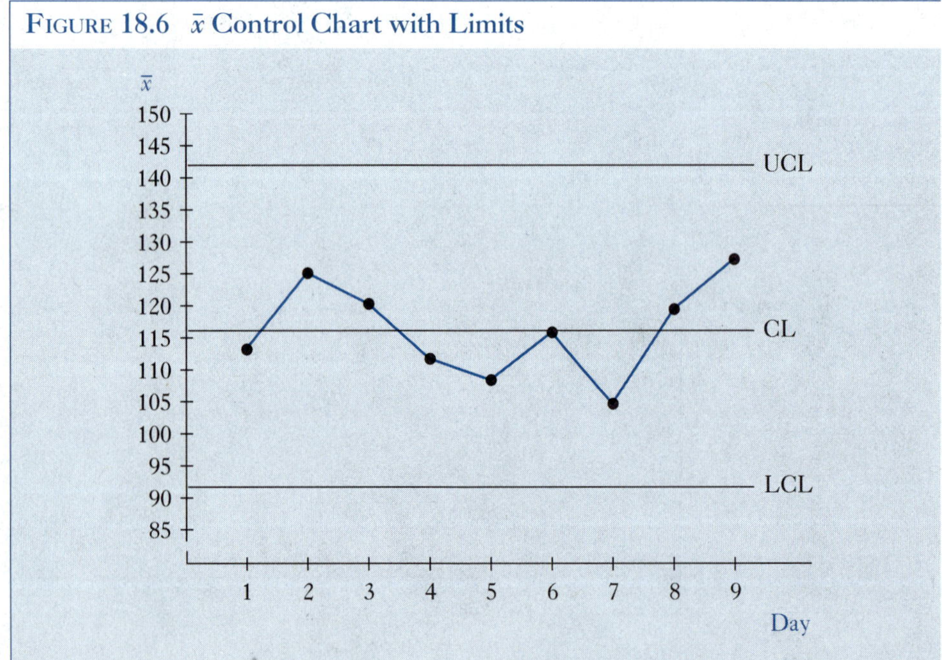

around this centerline with no particular pattern. If we had more data and sufficient time for analysis, the formal techniques of time-series analysis, as presented in Chapter 15, could be employed. Often in production work, as is the case here, such is not the case. A decision about process stability must be made with the data that is available.

A quick and easy way to assess whether a process is producing unusual observations is to establish control limits that should not be exceeded by purely chance events. For most distributions, almost all observations should be within three standard deviations of the mean. In the case of the sample mean, the relevant standard deviation is that of the sampling distribution of the sample mean, which was defined in Chapters 7 and 8 to be the standard error of the mean, $\sigma/\sqrt{n}$. The **control limits** beyond which observations may not be expected by chance are

Control Limits
The range of acceptable chance variation about the mean of a stable process.

$$\mu \pm 3\sigma/\sqrt{n}$$

Thus, there is an upper control limit, UCL, that equals $\mu + 3\sigma/\sqrt{n}$ and a lower control limit, LCL, that equals $\mu - 3\sigma/\sqrt{n}$. The quantity $\pm 3\sigma/\sqrt{n}$ can be thought of as the margin of error inherent in estimating the mean of the process.

The sample data are plotted over time, and their positions are assessed relative to the centerline and the control limits. The concepts of centerline and control limits contrast sharply with the concept of product **specifications.** Specifications are values established by manufacturing managers or product designers, not by the process itself. They represent tolerances to which product dimensions are designed to conform. Unfortunately, these exact specifications are seldom known with certainty, and we are forced to use statistics generated by the process itself. We have already shown that if the desired process population mean μ is unknown, as is typically the case, then the grand mean $\bar{\bar{x}}$ is used as an approximation. As an extension of this approximation, the upper and lower control limits are approximated by

Specification
Tolerances established for a measurable product dimension of interest.

$$\text{UCL} = \bar{\bar{x}} + 3\sigma/\sqrt{n} \tag{18.1}$$

and

$$\text{LCL} = \bar{\bar{x}} - 3\sigma/\sqrt{n} \tag{18.2}$$

As with the mean, usually σ is unknown and must be estimated. This could be done, as in Chapter 7, by calculating a sample standard deviation or by obtaining a pooled standard deviation from all the samples, as in Chapter 9. For ease in calculation, and by historical precedence in quality control work, we use an average measure of the range as an approximation of σ. As shown in Table 18.1, the range of a sample is the difference between the maximum and minimum value. For k samples the mean range is

$$\bar{R} = \sum_{i=1}^{k} R_i/k \tag{18.3}$$

which, for the nine samples of data in Table 18.1, the mean range is 36.4.

$$\bar{R} = (27 + 10 + 27 + 41 + 31 + 44 + 60 + 48 + 40) / 9$$

$$\bar{R} = 36.4$$

Ryan (1989) suggests that the mean range divided by the constant d_2 yields an unbiased estimator of σ. Thus, the upper and lower control limits are approximated by

$$\text{UCL} = \bar{\bar{x}} + 3[\bar{R}/d_2]/\sqrt{n} \tag{18.4}$$

$$\text{LCL} = \bar{\bar{x}} - 3[\bar{R}/d_2]/\sqrt{n} \tag{18.5}$$

Appendix Table C.2 gives d_2 as a value between 1.128, for a sample size $n = 2$, and 3.931, for sample size of $n = 25$. For a sample of $n = 4$, $d_2 = 2.059$. Calculation of the control limits are made easier with the constant A_2, where in Appendix Table C.2

$$A_2 = 3/d_2\sqrt{n} \tag{18.6}$$

Using the constant A_2, the control limits are

$$\text{UCL} = \bar{\bar{x}} + A_2\bar{R} \tag{18.7}$$

$$\text{LCL} = \bar{\bar{x}} - A_2\bar{R} \tag{18.8}$$

For a sample size of $n = 4$, A_2 can be calculated to be 0.729 or obtained directly from Appendix Table C.2. The control limits for the handlers' workload problem are thus

$$\text{UCL} = \bar{\bar{x}} + A_2\bar{R} = 116.9 + (0.729)(36.4) = 143.44$$

$$\text{LCL} = \bar{\bar{x}} - A_2\bar{R} = 116.9 - (0.729)(36.4) = 90.36$$

Plotting the $\bar{x}$ data, the centerline, and these control limits gives the control chart in Figure 18.6. All nine of the sample means lie well within the control limits. The idea that the process is out of control and in need of action (in the form of bringing in more handlers) cannot be accepted. Only chance factors may be causing the observed variations in the process.

In summary, the $\bar{x}$ control chart is used to assess variability as either that associated with chance factors or that assignable to specific causes other than chance. The goal of statistical process control is to bring the process into control, if it is out of control; once it is in control, the goal is to monitor the process to make sure it stays in control.

QUERY 18.2:

Suppose that on the 10th and 11th days you get average day workloads for the four shifts that give rise to the control chart additions given below. What would you conclude from this new information?

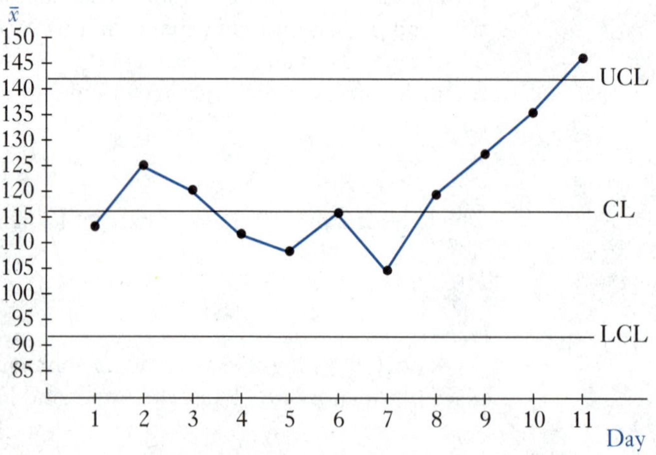

ANSWER:

Assuming the first nine observations define the parameters of the desired stable process, this chart shows that the process is becoming unstable, with the observation on the 11th day beyond the upper control limit. From this chart, it appears that the material handlers were correct. Production has reached an unacceptable level on the 11th day. A search for assignable causes is warranted. If assignable causes are present, an additional handler may be required to assist in the operation until the assignable cause is corrected.

18.6 STATISTICAL PROCESS CONTROL: EXAMINING PROCESS VARIABILITY

The $\bar{x}$ control chart is used to monitor process levels. It is usually coupled with another chart that displays process variability directly. An example of a process variability chart is the range chart. The following is a discussion of the construction and use of the *R* chart.

The range of each sample represents the outcome of a random variable whose value is determined as the difference between the highest and lowest value in a sample. It is more easily calculated and interpreted than the sample standard deviation and has thus come to dominate in quality control work. The mean of the range of the samples, $\bar{R}$, is the centerline of the *R* chart. The control limits of the *R* chart can be calculated through the use of the following expression:

$$\bar{R} \pm (3\bar{R}d_3) / d_2 \quad (18.9)$$

Again $\pm (3\bar{R}d_3) / d_2$ is a measure of the margin of error inherent in estimating the range. The control chart constant d_2 is the same constant used for the control limits of the mean. It, along with the new constant d_3, is found in Appendix Table C.2. To facilitate hand calculations, the control chart constants D_3 and D_4 are introduced

$$D_3 = 1 - 3d_3 / d_2 \quad (18.10)$$

$$D_4 = 1 + 3d_3 / d_2 \quad (18.11)$$

Thus, the *R* chart lower and upper limits are:

$$\text{LCL} = \bar{R}D_3 \quad (18.12)$$

$$\text{UCL} = \bar{R}D_4 \quad (18.13)$$

The first step in developing the *R* chart is the same as that for the $\bar{x}$ control chart: collecting the process data, which is already done in Table 18.1.

The second step is calculating the centerline of the *R* chart. The centerline for the *R* chart is the $\bar{R}$ value that was calculated in the $\bar{x}$ control chart example.

$$\text{CL} = \bar{R} = 36.4$$

The third step involves calculating the upper and lower control limits so an assessment of process control can be made for the data in terms of process variability. The control limits for the *R* chart are computed in the following way:

$$\text{LCL} = \bar{R}D_3 = (36.4)(0) = 0$$

$$\text{UCL} = \bar{R}D_4 = (36.4)(2.282) = 83.1$$

The fourth step is to plot the sample ranges against time. Additionally, the centerline and control limits must be represented on the plot. Figure 18.7 is the *R* chart with the centerline, control limits, and time-series data shown. The data that are plotted on this chart are the range values for each day.

The final step is interpreting the chart. Observation reveals that there are no data points outside the control limits; therefore, the process is said to be stable and in control. This evidence, coupled with the results of the $\bar{x}$ chart, provides an argument against the material handlers' claim. With only nine days of observations, however, the process should continue to be monitored closely for the development of upward trend in the range. Ideally, at least 20 days of observation would be involved before drawing a conclusion. Since a decision must be made immediately about adding another handler at this point, the decision is not to do it but to continue monitoring the process.

FIGURE 18.7 The *R* Control Chart

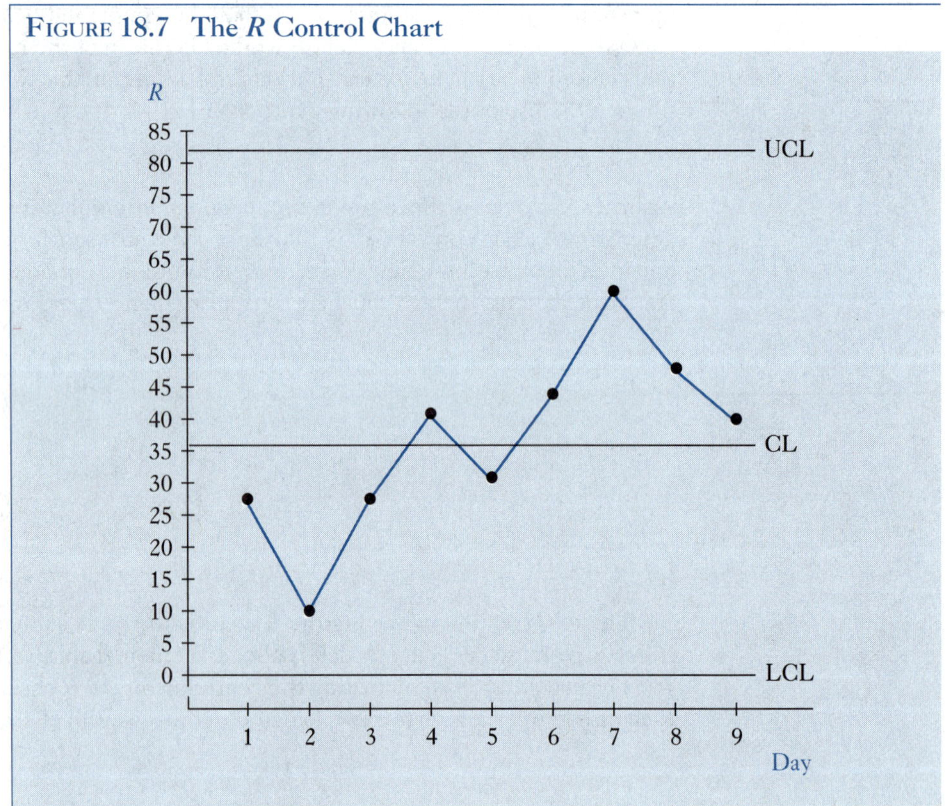

EXERCISES

18.6 What type of control chart is the one shown below? How was it constructed?

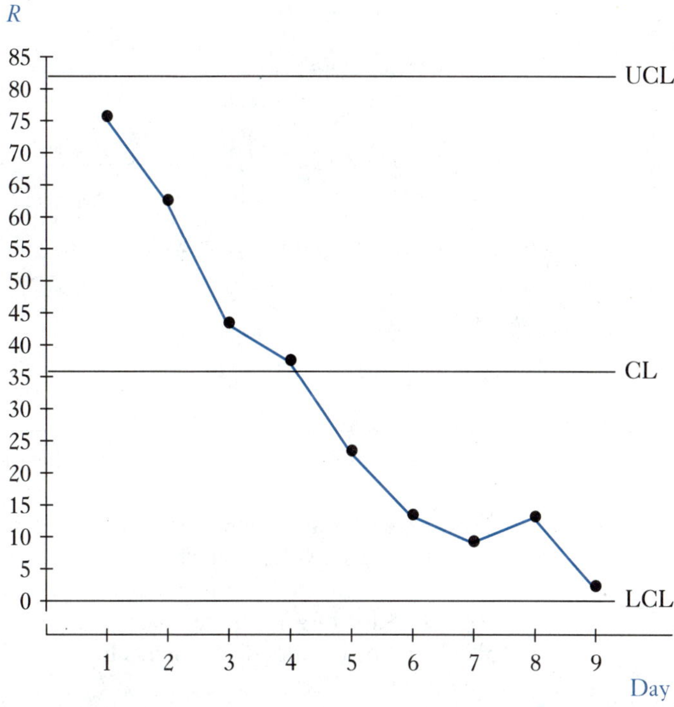

18.7 Does the control chart in Exercise 18.6 indicate that there is reason to suspect assignable causes of variability are present in the process? Is process intervention recommended for the process portrayed in Exercise 18.6? Why or why not?

18.8 The weekly data that follow are the average accounts receivable at a small law firm. The weekly average was constructed as a sample average with $n = 5$; that is, on each of the five working days in a week, the daily accounts receivable for the day were treated as the outcome of a random variable. Assuming that the process generating these data was initially in control:

a. Construct an $\bar{x}$ control chart showing the centerline and the control limits.
b. Construct an R-chart showing the centerline and the control limits.
c. What do the charts in parts a and b show?

Week	days/wk	$\bar{x}$	Range
1	5	$70,100	$81,060
2	5	$50,040	$60,700
3	5	$55,608	$56,109
4	5	$90,040	$98,060
5	5	$48,006	$62,908

(continued)

Week	days/wk	$\bar{x}$	Range
6	5	\$53,400	\$71,240
7	5	\$48,034	\$69,007
8	5	\$89,548	\$99,256
9	5	\$61,137	\$76,431

18.9 In Exercise 18.8, what would you conclude about the process, and what would you do:

a. If in the 10th week, the following accounts receivable information was recorded?

Week	days/wk	$\bar{x}$	Range
10	5	\$101,978	\$99,136

b. If in the 11th week, the following accounts receivable information was recorded?

Week	days/wk	$\bar{x}$	Range
11	5	\$134,782	\$157,894

18.7 Proportions and Their Application to Process Control

Measures of central tendency, such as means, medians, and modes will have little meaning for data sets involving attribute or qualitative data. For example, if we are assessing the hiring practices of a firm and observe that two women and five men are hired in a week, the idea of a mean sex has no use. The attribute "woman" and the attribute "man" cannot be added together. However, describing this data set of hires as two-sevenths female has meaning.

As discussed in Chapter 5, the proportion of a population or sample that has a given characteristic provides a summary measure for data involving a discrete attribute. The population proportion sharing an attribute has been denoted by the Greek letter $\pi = x/N$, where x is the number with the attribute in a population of size N. A sample proportion p is defined by x/n, where x is the number having the attribute in a sample of size n. If π is fixed from trial to trial (trials are independent), then for n identical trials x is distributed as a binomial random variable. The expected value of p is π, and the variance of p is $(1 - \pi)\pi/n$.

Defective
An item that will not perform its intended task.

Nonconforming
An item that does not meet its specifications.

In statistical quality control, practitioners are often interested in the proportion of defects generated by a process. An item is designated as **defective** if it does not perform the intended task. An item is designated as **nonconforming** if it does not meet its specifications. The distinction between the use of the words defective and nonconforming to represent an undesired outcome has been blurred in many practical applications; many practitioners treat these words as synonyms. Control charts used to display information on defects and nonconformities are called attribute control charts. There are several different types of attribute control charts. The p chart is one of the most well known.

Consider the hospital emergency room case described at the beginning of this chapter. From the accounting department's perspective, emergency room treatment of too many of those without insurance might be considered nonconforming practice.

From a social welfare point of view a certain amount of *pro bono,* indigent care is necessary, but too much will bankrupt the hospital. Table 18.2 shows 21 samples of 36 records each that were drawn on sequential weeks. In the first sample of 36, 10 patients were found to have no insurance. The second sample of 36 had four with no insurance, and so on to the 21st week.

We will take a closer look at these emergency room data to determine if there is a reason to suspect that assignable causes are contributing to nonconformity. Construction of a p chart will be helpful in this regard.

The second and third steps involve computing the components of the p chart. In the absence of a predetermined specification of the centerline of the control chart, the mean of the calculated sample proportions is obtained. If the proportion of those without insurance is initially believed to be in control, then this centerline will be a point of comparison for observed proportions. For k samples, where the jth sample is of size n_j with p_j proportion of nonconforming elements, calculation of the centerline is

$$\bar{p} = \frac{\sum_{j=1}^{k} n_j p_j}{\sum_{j=1}^{k} n_j} \text{ and for equal sample sizes } n,\ \bar{p} = \frac{\sum_{j=1}^{k} np_j}{kn} \tag{18.14}$$

TABLE 18.2 Hospital Emergency Room Data on Uninsured Patients

Week	Sample Size	Number Not Conforming	Proportion Not Conforming
1	36	10	0.2778
2	36	4	0.1111
3	36	9	0.2500
4	36	7	0.1944
5	36	5	0.1389
6	36	8	0.2222
7	36	11	0.3056
8	36	9	0.2500
9	36	5	0.1389
10	36	8	0.2222
11	36	7	0.1944
12	36	5	0.1389
13	36	9	0.2500
14	36	10	0.2778
15	36	6	0.1667
16	36	8	0.2222
17	36	11	0.3056
18	36	5	0.1389
19	36	9	0.2500
20	36	7	0.1944
21	36	10	0.2778
	756	163	0.2156

which is just the total number of nonconforming items in the k samples divided by the total number in the k samples. In the case of the 21 weeks (k) of 36 patient samples (n) from the emergency room, the centerline is

$$\bar{p} = \frac{\text{Total number with no insurance}}{\text{Total number of patients checked}} = \frac{163}{756} = 0.2156$$

$$\text{CL} = \bar{p} = 0.2156$$

From the U.S. Department of Health and Human Services, we know that nationally about 22 percent of those using the emergency room do not have insurance. Although we do not have a predetermined specification for the centerline of a process in control, our mean of sample proportions of 0.2156 is close to that reported by all emergency rooms, so, at least relative to the national norm, our emergency room mean proportion of 0.2156 suggests a stable process. The centerline value of 0.2156 seems an appropriate point of comparison.

For sufficiently large individual sample sizes, each p will be approximately normally distributed. The approximate margin of error that should include almost all of the sample proportions if the standard is being adhered to is thus $\pm 3\sqrt{\pi(1-\pi)/n}$, when centered on the true population proportion. Because the true population proportion is unknown, the margin of error is added into and subtracted from the mean of the sample proportions to form the upper and lower control limits.

For the upper control limit, we have

$$\begin{aligned} \text{UCL} &= \bar{p} + 3\sqrt{\left[\bar{p}\left(1-\bar{p}\right)\right]/n} \\ &= 0.2156 + 3\sqrt{0.2156\left(1-0.2156\right)/36} = 0.4212 \end{aligned} \tag{18.15}$$

For the lower control limit, we have

$$\begin{aligned} \text{LCL} &= \bar{p} - 3\sqrt{\left[\bar{p}\left(1-\bar{p}\right)\right]/n} \\ &= 0.2156 - 3\sqrt{0.2156\left(1-0.2156\right)/36} = 0.0100 \end{aligned} \tag{18.16}$$

For overly small sample sizes, it is not unusual to get a lower control limit that is less than zero. This is evidence that suggests that p should not be assumed to be approximately normal, and that control limits should be obtained from the binomial distribution. In practice, however, negative lower control limits are often ignored and arbitrarily set to zero. A better solution would be to increase n until it exceeds $9(1-\bar{p})/\bar{p}$.

The fourth step involves plotting the proportion of defective or nonconforming data, the centerline, and the control limits, as in Figure 18.8. The final step is interpretation of the control chart to consider whether process intervention is necessary. The p chart does not indicate that intervention is required nor should a search for assignable causes be necessary. The accounting department need not approach the emergency room personnel to inquire about chance in the proportion of uninsured patients being treated in the emergency room. This conclusion is reached because there are no sample proportions lying outside of the control limits.

FIGURE 18.8 The *p* Chart for Uninsured

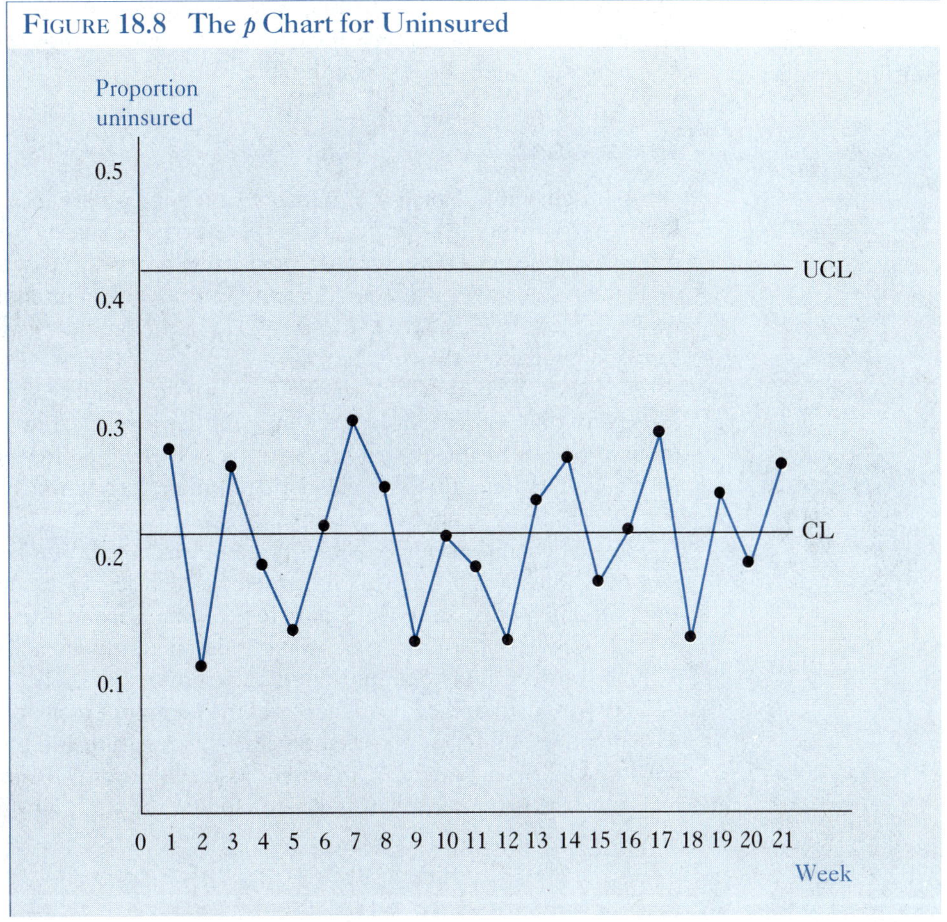

The hospital emergency room data on uninsured patients employed equal sample sizes of $n = 36$. But in practice, it is not always possible to obtain equal sample sizes. This is especially true when working with existing data that were not necessarily collected to be used in a control chart. If there are variable sample sizes, and the data are plotted on the same p chart, the data will share the same centerline. The control limits, however, require more thought. There is no unique way to set up the control limits for unequal sample sizes. One way is to use control limits based upon $\bar{n}$, which is the average sample size. Another approach is to use multiple limits for several sample size values spanning the expected range of the sample sizes.

Notice that although the conclusion is that the emergency room procedures of screening and treating patients appear to be in control for these 21 weeks, if in the 22nd week a sample of 36 yields 16 patients with no insurance, then there would be reason to further investigate the treatment process because the upper control limit of 0.4212 would be exceeded (i.e., $p = 16/36 = 0.4444$). If assignable causes are discovered, the action to be taken might require a new method of screening potential patients before they are allowed to enter the emergency room area. While the emergency room area might have comfortable chairs and a well-heated/air-conditioned environment, the pre-selection area might be less comfortable.

The *c* Chart

As discussed in Chapter 5, in many situations we are interested in the number of highly unlikely occurrences in a large unit of time or space. Where this unlikely occurrence is a nonconforming item in a production process, the *c* chart is used and the unit of time or space over which the nonconformities are counted is called "area of opportunity" (Burr, 1976, p. 133), and the Poisson probability distribution is the assumed function generating the count.

Consider the hierarchical computer menu described at the beginning of this chapter. Every time a user employs the menu to retrieve specific types of information, the menu provides an area of opportunity on which errors or nonconforming responses can be made. The *c* chart will provide a plot of the count of nonconforming responses and the control limits for stability of this process.

Several conditions must be met if the *c* chart is to be applied. First, the area of opportunity must be the same across samples, which seems reasonable in the hierarchical computer menu case since all menus are the same. Second, the nonconformities counted in each area of opportunity must occur independently of each other, which is assured if there is no recall or communication between uses or users.

To demonstrate the *c* chart, consider the computer menus during the design phase. The menu selection and retrieval process is not mature and the retrieval standards are not known. To determine if there are assignable causes responsible for poor performance on the menus, a *c* chart is constructed for the data collection per area of opportunity in Table 18.3.

Table 18.3 Data on Nonconforming Responses in Computer Menu Use

Subject	Nonconforming Responses
1	1
2	2
3	3
4	2
5	2
6	3
7	2
8	1
9	9
10	10
11	2
12	3
	40

After obtaining the data in step 1, steps 2 and 3 involve computing the components of the control chart. The centerline of the c chart is computed as follows:

$$CL = \bar{c} = \sum_{i=1}^{k} c_i \,/\, k \qquad (18.17)$$

which in the case of count data in Table 18.3 is

$$CL = 40/12 = 3.33$$

Assuming the process is initially stable, then CL = 3.33 is the standard against which sample counts are compared.

In Chapter 5, the standard deviation of a Poisson random variable was defined as the square root of the mean number of occurrences. Thus, if the mean number of occurrences is $\bar{c}$, and if the number of occurrences is approximately normal, then the upper control limit would be

$$UCL = \bar{c} + 3\sqrt{\bar{c}} \qquad (18.18)$$

which for the computer menu data in Table 18.3 is

$$UCL = 3.33 + 3\sqrt{3.33} = 8.80$$

We can see in Figure 18.9 that this upper control limit was exceeded for samples nine and 10 when the counts were nine and 10. Thus, there is evidence for assignable causes that need to be investigated for the nonconforming responses made while utilizing the hierarchical computer menu. Types of assignable causes that could be responsible include factors such as individual user abilities or interruptions encountered during the retrieval process. The process control analyst should intervene to eliminate these sources of variability.

In addition to looking for assignable causes that might account for the count exceeding the upper control limit, the analyst should recognize that the assumption of approximate normality may be inappropriate since the lower control limit is negative, which is impossible for a count. In particular, the lower control limit is

$$LCL = \bar{c} - 3\sqrt{\bar{c}} \qquad (18.19)$$

which for the computer menu data yields

$$LCL = 3.33 - 3\sqrt{3.33} = -2.14$$

While it is not unusual to see practitioners ignore this meaningless lower limit, it provides direct evidence that the normality assumption is inappropriate. The likelihood of occurrence in this case is best obtained with the Poisson distribution itself. For a random variable that follows the Poisson distribution, and has an expected value of 3.33, the probabilities of nine and 10 occurrences can be determined with a computer program such as ET or STATABLE to be respectively 0.005 and 0.002. Neither of these occurrences is as unlikely as suggested by the normal approximation, although both are still sufficiently unlikely that a search for assignable causes is in order.

FIGURE 18.9 *c* Chart for Nonconforming Responses

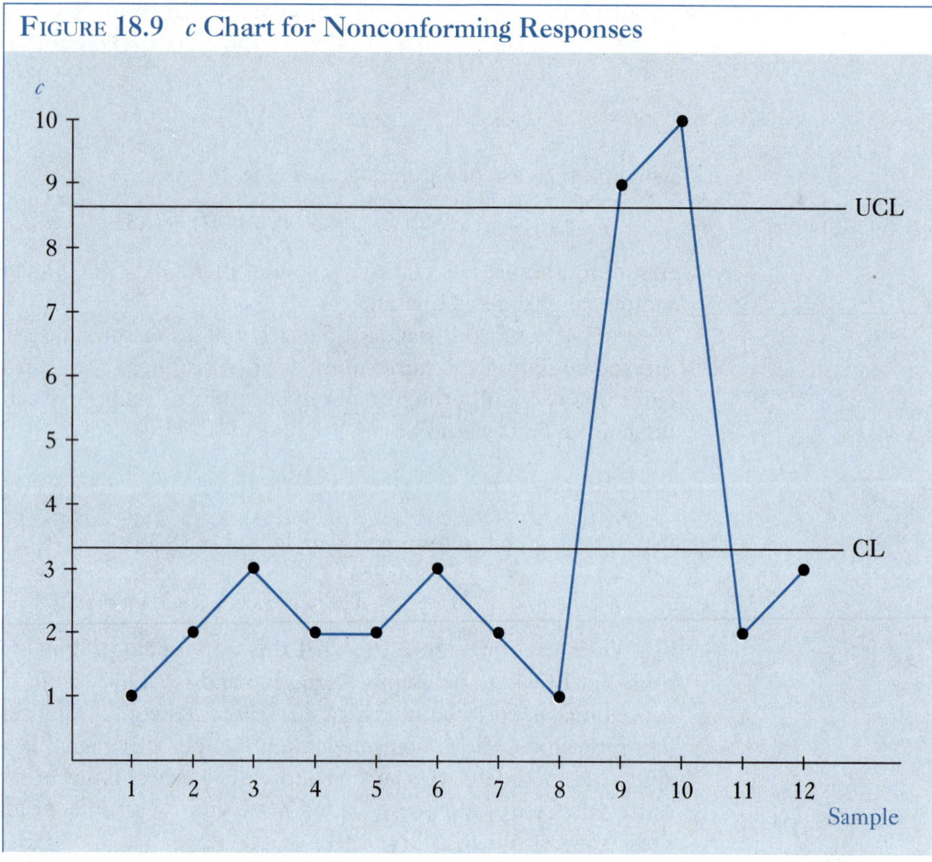

EXERCISES

18.10 What is the difference between the terms "defective" and "nonconforming." Provide examples of defective items and nonconforming items.

18.11 A mature manufacturing process generates laser printer paper. It is the job of a quality control inspector to examine the paper and count the number of instances of spotty pigmentation per page. The quality control inspector determined that use of the *c* chart is appropriate for this paper quality assessment. Which assumptions must be met to justify use of the *c* chart?

18.12 The following data are visual inspection results for defective soup cans. Twenty-five lots were inspected. The sample size per lot was 312. Construct a control chart that appropriately assesses soup can quality.

Lot	n	Defectives	Lot	n	Defectives
1	312	16	4	312	17
2	312	17	5	312	15
3	312	15	6	312	15

(continued)

Lot	n	Defectives	Lot	n	Defectives
7	312	14	16	312	16
8	312	15	17	312	33
9	312	22	18	312	20
10	312	16	19	312	18
11	312	17	20	312	15
12	312	17	21	312	19
13	312	16	22	312	15
14	312	16	23	312	21
15	312	16	24	312	25
			25	312	20

18.13 Does the control chart of Exercise 18.12 indicate that the process is in control? How did you conclude this?

18.14 Soup can labels were also visually inspected for label nonconformities. A label nonconformity was defined to be any place on the label where there was not smooth label adhesion. For example, a nonconformity could be a crease on the label, a bump on the label, or a place on the label where the can was visible through the label. From the following data, construct a control chart to assess the quality of the soup can labeling process.

Sample	Nonconformities	Sample	Nonconformities
1	12	13	11
2	11	14	12
3	11	15	11
4	10	16	11
5	12	17	10
6	13	18	11
7	12	19	10
8	11	20	11
9	11	21	10
10	12	22	12
11	11	23	10
12	11	24	11
		25	11

18.15 Does the control chart that was constructed in Exercise 18.14 indicate that the labeling process is in control? Why or why not?

18.8 Acceptance Sampling by Attributes

Acceptance sampling was described in Section 18.3 as the second of two elements within the field of statistical quality control. Acceptance sampling is the use of a sample drawn from a large shipment to determine the acceptability of the entire lot.

Acceptance sampling is different from statistical process control. Acceptance sampling occurs after a complete lot of the product has been manufactured and delivered to the consumer's doorstep. The acceptance sampling inspection does not occur "in-process"; that is, it is not part of the in-house production process. It typically occurs after some outsider's processing is completed and before the consumer's in-house use of the material begins. Acceptance sampling is conducted by a consumer. The consumer could be a separate business unit from the manufacturer, or the consumer might reside within the same organization as the manufacturer.

Quite often a consumer receives a quantity of finished product and must address the question of whether the entire lot is suitable for use. If the number of items in the shipment is small, and the items will not be destroyed in the testing, the consumer could inspect 100 percent of a shipment lot and only reject those items that are not acceptable. If the number of items in a shipment lot is large or if the items will be destroyed in testing, then only a sample of items may be inspected. That is, the consumer will have to determine whether an entire shipment lot should be accepted or rejected based upon the inspection of a subset of the lot.

Consider the tool manufacturer presented at the beginning of this chapter. Suppose that this tool manufacturer begins using fasteners from an outside supplier. To ensure the quality of the tools being produced at this facility, the quality control analysts can inspect a sample that was randomly drawn from a lot of fasteners received from the supplier. (Random selection was addressed in Chapter 6 and its appendix, so it will not be reviewed here.) Based on the results of the inspection, the tool manufacturer can determine whether to accept the entire lot of fasteners or reject the lot and return it to the supplier.

Producer's Risk
The probability that a good quality lot will be mistakenly rejected.

Consumer's Risk
The probability that a poor quality lot will be mistakenly accepted.

There are risks facing the producer and risks facing the consumer in using sample information to determine the acceptability of an entire lot. The **producer's risk,** denoted by α, is the probability that a good quality lot will be mistakenly rejected. The **consumer's risk,** denoted by β, is the probability that a poor quality lot will be mistakenly accepted. As discussed in Chapter 9 and its appendices, this is just a hypothesis testing situation in which the producer's risk is the probability of a Type I error and the consumer's risk is the probability of a Type II error, where the hypotheses are

H_0: Shipment lot fulfills specifications

H_A: Shipment lot does not fulfill specifications

Acceptance Number
The maximum number of defectives a buyer will tolerate and still accept a product lot.

Rejection Number
The minimum number of defectives required to reject a lot.

As employed in Chapter 9 and its appendices, single sampling plans utilize only one sampling stage, which means that just one sample of n pieces is drawn at random from the entire lot of product and is inspected. Based on the results of this single sample inspection, a decision is made whether to accept the lot or reject it. This accept or reject decision is based upon a predetermined **acceptance number,** a, and a predetermined **rejection number,** r. The acceptance number is the maximum number of defectives allowed in an acceptable lot. The rejection number is the minimum number of defectives that must be present for a decision to reject the lot. The rejection number is one plus the acceptance number. Suppose that the tool manufacturer had established an acceptance number of $a = 2$. Then it follows that the rejection number is $r = 3$.

The critical values for acceptance and rejection of a shipment provide the criteria for decision making about the acceptability of a shipment. They are determined by considering the tradeoff between Type I and Type II errors and constructing operating characteristics curves as presented in the appendices of Chapter 9. Tables showing producer (α) and consumer (β) risk under alternative sampling plans are available. One set of tables, known as the American Military Standard Tables, MIL-STD-105D, provides sampling plans for varying lot sizes and desired sample sizes to achieve preset acceptable quality (AQ) levels. Although still used in acceptance sampling, the Military Standard Tables do not incorporate the explicit and implicit costs of sampling. Alternative sampling plans, such as the Dodge-Romig plans, are based on the average outgoing quality limits (AOQL) or the lot tolerance percent defects (LTPD). Unlike the Military Standard 105D, these plans are based on a minimization of the cost of inspection for a given quality level. Entire books are devoted to the use of these tables, and the construction of multiple sampling plans is beyond the scope of this book.

In closing our discussion of acceptance sampling, recognize that the purpose of acceptance sampling may not be to immediately improve the quality of products received by the consumer, but rather to ensure future incoming product quality. This can be achieved by sending a signal to the supplier that an entire product lot is being rejected based on poor quality in samples taken from the entire quantity of product received. Although suppliers and consumers may work together to reduce the number of defects and nonconforming items in a shipment, decision criteria are needed for evaluating items leaving the producer's docks and arriving at the consumer's docks. This requires that

1. a concise, unequivocal definition of what comprises a defect or a nonconformity is agreed to by both the consumer and the producer.
2. samples must be taken in accordance with the random sampling procedures discussed in Chapter 7.
3. the inspection process must be conducted in an objective and accurate manner.
4. numerical criteria for acceptance or rejection of shipment lot is specified prior to inspection.

All four of these elements are necessary conditions for effective decision making when evaluating incoming product.

18.9 Concluding Comments

This chapter provided a perspective on the total quality movement in the United States through examination of those events that helped establish quality as a priority in manufacturing. The flowchart, Pareto chart, fishbone diagram, and histogram were presented as graphical means for representing processes and operations. The two primary components of statistical quality control were also presented: statistical process control and acceptance sampling.

Statistical process control is conducted concurrently with the manufacturing or design process. It is considered to be a preventive function for the producer. Acceptance sampling is conducted by the consumer after product lots have already been manufactured. Acceptance sampling is a protection mechanism for the consumer.

As stated throughout this chapter, statistical quality control requires the collecting of data, presenting of data, specifying of models, testing of hypotheses, and the consideration of uncertainty in decision making. It thus makes use of all of the concepts presented in this book.

Chapter Exercises

18.16 Your younger brother has asked you to teach him how to properly wash a car. As part of his instruction, construct a flowchart that represents the process required to produce a clean car.

18.17 Data compiled by the Cooperative Institutional Research Program of the American Council on Education and the Higher Education Research Center show the percentage of college freshman men and women planning to major in biology, engineering, physical sciences, social sciences, computer science, and math, respectively. Is the chart that displays this information (shown below) a legitimate Pareto chart? Explain.

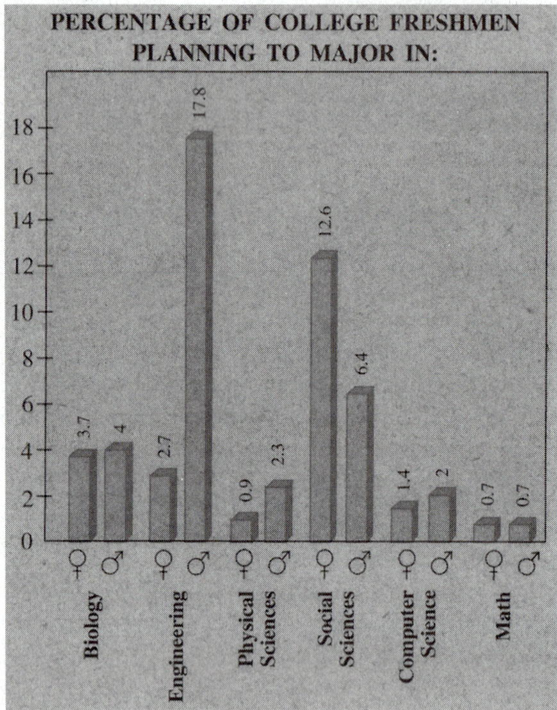

18.18 Construct a Pareto chart from the following data. The data represent errors committed by users of a hierarchical, computerized menu system. The errors are organized by type and frequency of occurrence

Error Type	Number Committed
Errors of omission	7
Selection errors	3
Sequence errors	9
Time errors	2
Qualitative errors	1
TOTAL	22

18.19 What action should be taken based upon the information displayed in the Pareto chart that was constructed in Exercise 18.18?

18.20 A tissueless maxi-pad was being produced on a trial basis at a leading feminine hygiene manufacturing facility. To assess the prevalence of downtime, downtime duration was recorded for a specific manufacturing machine. Downtime was defined as beginning at any point in time when there was an unscheduled machine stop. Downtime concluded when the machine resumed normal production. A total of 153.68 minutes of downtime was recorded. Construct a Pareto chart for downtime.

Downtime Classification	Duration
A	15.65
B	67.46
C	47.08
D	00.00
E	23.49
F	00.00
TOTAL	153.68

18.21 After examining the Pareto chart constructed in Exercise 18.20, determine which downtime classifications should be investigated. Why should they be investigated?

18.22 Sample means and ranges were acquired during examination of a new manufacturing operation. The measurable dimension of interest was the weight of a particular component of the product measured in grams. Each sample was of size $n = 20$. What would the sample range statistic represent? Why is it helpful to know the sample range when constructing an $\bar{x}$ chart?

18.23 The following data are the sample means and ranges for a new diaper manufacturing operation. Each sample is size $n = 20$. The weight of the inner lining was of interest; it was measured in grams.

Sample	Sample Mean, $\bar{x}$	Sample Range, R
1	32.27 grams	5.12 grams
2	23.31	2.37
3	19.21	4.31
4	40.90	8.93
5	37.94	6.52
6	25.34	3.95
7	14.39	7.29
8	26.40	4.09
9	51.53	9.59
10	28.57	6.41
11	19.48	7.10
12	36.45	5.34
13	42.46	7.17
14	56.31	8.83
15	68.28	8.43

If an $\bar{x}$ control chart was constructed from this data, what specific aspect of the process would the analyst be examining?

18.24 Calculate from the data of Exercise 18.23:

a. the centerline for the $\bar{x}$ control chart.

b. the lower and upper control limits for the $\bar{x}$ control chart.

18.25 Plot the $\bar{x}$ data from Exercise 18.23, the centerline, and the control limits calculated in Exercise 18.24 on an $\bar{x}$ control chart. What does the $\bar{x}$ control chart indicate about the control of the diaper manufacturing process?

18.26 The following data pertain to the amount of stock handled by storeroom clerks in a retail store. Fifteen samples were collected each day.

Day	Mean Product Handled (cases)	Range
1	79	5
2	86	6
3	85	8
4	85	9
5	84	7
6	88	7
7	86	4
8	89	8
9	87	8
10	82	7
11	81	6
12	84	6
13	78	6
14	80	5
15	82	7
16	85	4
17	89	4
18	86	7
19	83	7
20	83	4

Construct an R chart to assess the variability of this product handling operation. Would it be appropriate to also construct an $\bar{x}$ chart for this process? Why?

18.27 The following are data on exhaust valve assembly production. Each sample is size $n = 5$. Construct

a. An $\bar{x}$ chart. Comment on the process level control.

b. An R chart. Comment on the process variability.

c. Did you come to the same conclusion regarding overall process control in both parts a and b? Discuss.

Sample	$\bar{x}$	R
1	11	2
2	12	2
3	12	2
4	12	2
5	13	2
6	13	2
7	14	1

(continued)

Sample	$\bar{x}$	R
8	15	2
9	15	1
10	15	2
11	16	2
12	16	1
13	17	2
14	9	2
15	10	1
16	10	2
17	10	1
18	11	2
19	11	1
20	11	2

18.28 Consider the case of a process that stamps flat pieces of steel for use in a radiator. The measureable dimension of interest in this process is the length (inches) of the flat pieces of steel after stamping. The following data are average lengths and ranges for 10 samples of size $n = 10$.

Average Length (in)	Range
250.2	2.3
249.7	2.2
250.8	2.5
248.8	1.1
250.8	3.2
248.5	2.4
249.9	3.3
256.1	1.1
249.3	2.2
248.9	2.4

Construct an R chart from this data. What does this chart indicate about process quality?

18.29 Samples of the following sizes were collected at the start of a quality control study. Which value of n should be used for the construction of a p chart?

Day	Sample Size
1	21
2	25
3	22
4	20
5	18
6	29
7	34
8	22
9	18
10	20
11	24

(continued)

Day	Sample Size
12	19
13	22
14	21
15	25

18.30 From the information provided below, determine the values for the centerline, and upper and lower control limits of a p chart.

Day	Sample Size	Defectives	Proportion Defective
1	21	4	0.19
2	25	2	0.08
3	22	4	0.18
4	20	4	0.20
5	18	6	0.33
6	29	2	0.07
7	34	6	0.18
8	22	7	0.32
9	18	9	0.50
10	20	4	0.20
11	24	6	0.25
12	19	8	0.42
13	22	2	0.09
14	21	5	0.24
15	25	9	0.36

18.31 Construct the p chart by plotting the proportion of defective data, the centerline, and the control limits computed in Exercise 18.30. What does the p chart indicate about process control?

18.32 A mature manufacturing process generates laser printer paper. It is the job of a quality control inspector to examine the paper for nonconformities such as spotty pigmentation and uneven thickness. Prior continuous monitoring of the process has revealed that the standard number of nonconformities per piece of paper is 2.34. What is the appropriate control chart that should be examined to assess process control?

18.33 Calculate the centerline and control limits for the appropriate control chart designated in Exercise 18.32 based upon the standard number of nonconformities per piece of paper. What does this control chart indicate about process control?

18.34 Visual inspection of an electrical plug production process generated data on the number of nonconformities per plug. Inspection data on 21 days of production were recorded. The data follow:

Day	1	2	3	4	5	6	7	8	9	10
c	20	46	37	76	34	13	6	54	70	44

Day	11	12	13	14	15	16	17	18	19	20
c	63	55	66	59	43	48	20	81	80	26

Day	21
c	67

Construct a c chart from this data, and make a statement about the quality of the plug production process.

18.35 Acceptance sampling is generally considered to be a protection mechanism for the consumer. In what situations would acceptance sampling be employed by the producer as a protection mechanism?

18.36 How does acceptance sampling differ from statistical process control in terms of how it is employed, when it is employed, and who employs it?

18.37 One of the goals of just-in-time (JIT) management is to reduce set-up time, thereby permitting the customer to order very small lot sizes. Thus, if the customer ordered a lot size of 20 parts, JIT principles dictate that the customer should receive a lot of 20 good parts. The product received by the customer must be of high quality or else the customer will experience shortages. In practice, one solution to the problem of shortages is to "overorder" in anticipation of incoming product losses caused by defects. Theoretically, does successful JIT management preclude the need for acceptance sampling? Explain.

ENDNOTE

1. Julie A. Jacko wrote this chapter. It was edited by William E. Becker. Dr. Jacko is an Assistant Professor of Engineering Science at Trenton State College and received her Ph.D. in Industrial Engineering from Purdue University in 1993. While earning her bachelor's degree in Industrial Engineering from Purdue, Jacko worked in an industrial engineering capacity at Personal Products Company, a division of Johnson & Johnson. In addition, Jacko has been employed as a staff consultant with Andersen Consulting.

ACADEMIC REFERENCES

Banks, D. (1993). "Is Industrial Statistics Out of Control?" *Statistical Science, 8*, pp. 356–409.

Burr, I. W. (1976). *Statistical Quality Control Methods.* New York, NY: Marcel Dekker, Inc. p. 133.

Clegg, W. H. (1993). "The Role of Quality in JIT Management." *Industrial Engineering, 25*, November 1993, p. 16.

Deming, W. E. (1944). *Statistical Adjustment of Data.* New York, NY: John Wiley and Sons, Inc. pp. 1–2.

Holloway, M. (1993). "A Lab of Her Own. "*Scientific American*, November 1993, pp. 94–103.

Juran, J. M. (1992). *Juran on Quality By Design.* New York, NY: The Free Press. pp. 14–15.

Mintzberg, H. (1994). *The Rise and Fall of Stategic Planning.* New York: Free Press.

Ryan, T. P. (1989). *Statistical Methods for Quality Improvement.* New York: Wiley.

Shewhart, W. A. (1939). *Statistical Method From the Viewpoint of Quality Control.* Lancaster, PA: Lancaster Press. pp. 1–5.

Taylor, F. W. (1947). *Scientific Management.* New York, NY: Harper and Brothers Publishers. pp. 163–164.

U.S. Congress, Office of Technology Assessment (1990). "Making Things Better: Competing in Manufacturing," OTA-ITE-443 (Washington, D.C: U.S. Government Printing Office, February).

FORMULAS

- The $\bar{x}$ Chart, μ and σ known:

$$\text{CL} = \mu$$

$$\text{UCL} = \mu + \left(3\sigma/\sqrt{n}\right)$$

where μ and σ are known and n is the size of the samples taken.

$$\text{LCL} = \mu - \left(3\sigma/\sqrt{n}\right)$$

where μ and σ are known and n is the size of the samples taken.

- The $\bar{x}$ chart, μ and σ unknown:

$$\text{CL} = \bar{\bar{x}} = \sum_{i=1}^{k} x_i \,/\, k$$

where x_i is the value of the ith sample

$$\text{UCL} = \bar{\bar{x}} + A_2\bar{R}$$

$$\text{LCL} = \bar{\bar{x}} - A_2\bar{R}$$

where A_2 is a control chart constant

- p chart, process standards unknown:

$$\text{CL} = \bar{p}$$

$$\text{LCL} = \bar{p} - 3\sqrt{\left[\bar{p}\left(1-\bar{p}\right)\right]/n}$$

where n is the sample size

$$\text{UCL} = \bar{p} + 3\sqrt{\left[\bar{p}\left(1-\bar{p}\right)\right]/n}$$

where n is the sample size

- c chart, process standards unknown:

$$\text{CL} = \bar{c} = \sum_{i=1}^{k} c_i \,/\, k$$

$$\text{UCL} = \bar{c} + 3\sqrt{\bar{c}}$$

$$\text{LCL} = \bar{c} - 3\sqrt{\bar{c}}$$

APPENDICES

Appendix Table A.1 Binomial Distribution

The binomial mass function is defined by

$$P(x) = \frac{n!}{x!(n-x)!}\pi^x (1-\pi)^{n-x}$$

This is the probability of x successes in n independent trials, where the probability of success on each trial is π. The values of x in the left margin of each of the following sections are to be used in conjunction with the values of π at the top of that section. The values of x in the right margin of each section are to be used in conjunction with the values of π at the bottom of that section.

Example To evaluate $P(x)$ for $n = 4$, $x = 1$, and $\pi = 0.20$, locate the section of the table for $n = 4$, the column for $\pi = 0.20$, and the row for $x = 1$ and read $P(x) = 0.4096$

This table gives the binomial distribution for $n = 1$ through $n = 10$, $n = 20$, and $n = 50$.

$n = 1$

x	π	01	02	03	04	05	06	07	08	09	10		
0		9900	9800	9700	9600	9500	9400	9300	9200	9100	9000		1
1		0100	0200	0300	0400	0500	0600	0700	0800	0900	1000		0
		99	98	97	96	95	94	93	92	91	90	π	x
x	π	11	12	13	14	15	16	17	18	19	20		
0		8900	8800	8700	8600	8500	8400	8300	8200	8100	8000		1
1		1100	1200	1300	1400	1500	1600	1700	1800	1900	2000		0
		89	88	87	86	85	84	83	82	81	80	π	x
x	π	21	22	23	24	25	26	27	28	29	30		
0		7900	7800	7700	7600	7500	7400	7300	7200	7100	7000		1
1		2100	2200	2300	2400	2500	2600	2700	2800	2900	3000		0
		79	78	77	76	75	74	73	72	71	70	π	x
x	π	31	32	33	34	35	36	37	38	39	40		
0		6900	6800	6700	6600	6500	6400	6300	6200	6100	6000		1
1		3100	3200	3300	3400	3500	3600	3700	3800	3900	4000		0
		69	68	67	66	65	64	63	62	61	60	π	x
x	π	41	42	43	44	45	46	47	48	49	50		
0		5900	5800	5700	5600	5500	5400	5300	5200	5100	5000		1
1		4100	4200	4300	4400	4500	4600	4700	4800	4900	5000		0
		59	58	57	56	55	54	53	52	51	50	π	x

$n = 2$

x	π	01	02	03	04	05	06	07	08	09	10		
0		9801	9604	9409	9216	9025	8836	8649	8464	8281	8100		2
1		0198	0392	0582	0768	0950	1128	1302	1472	1638	1800		1
2		0001	0004	0009	0016	0025	0036	0049	0064	0081	0100		0
		99	98	97	96	95	94	93	92	91	90	π	x

x	π	11	12	13	14	15	16	17	18	19	20		
0		7921	7744	7569	7396	7225	7056	6889	6724	6561	6400		2
1		1958	2112	2262	2408	2550	2688	2822	2952	3078	3200		1
2		0121	0144	0169	0196	0225	0256	0289	0324	0361	0400		0
		89	88	87	86	85	84	83	82	81	80	π	x

x	π	21	22	23	24	25	26	27	28	29	30		
0		6241	6084	5929	5776	5625	5476	5329	5184	5041	4900		2
1		3318	3432	3542	3648	3750	3848	3942	4032	4118	4200		1
2		0441	0484	0529	0576	0625	0676	0729	0784	0841	0900		0
		79	78	77	76	75	74	73	72	71	70	π	x

x	π	31	32	33	34	35	36	37	38	39	40		
0		4761	4624	4489	4356	4225	4096	3969	3844	3721	3600		2
1		4278	4352	4422	4488	4550	4608	4662	4712	4758	4800		1
2		0961	1024	1089	1156	1225	1296	1369	1444	1521	1600		0
		69	68	67	66	65	64	63	62	61	60	π	x

x	π	41	42	43	44	45	46	47	48	49	50		
0		3481	3364	3249	3136	3025	2916	2809	2704	2601	2500		2
1		4838	4872	4902	4928	4950	4968	4982	4992	4998	5000		1
2		1681	1764	1849	1936	2025	2116	2209	2304	2401	2500		0
		59	58	57	56	55	54	53	52	51	50	π	x

$n = 3$

x	π	01	02	03	04	05	06	07	08	09	10		
0		9703	9412	9127	8847	8574	8306	8044	7787	7536	7290		3
1		0294	0576	0847	1106	1354	1590	1816	2031	2236	2430		2
2		0003	0012	0026	0046	0071	0102	0137	0177	0221	0270		1
3		0000	0000	0000	0001	0001	0002	0003	0005	0007	0010		0
		99	98	97	96	95	94	93	92	91	90	π	x

x	π	11	12	13	14	15	16	17	18	19	20		
0		7050	6815	6585	6361	6141	5927	5718	5514	5314	5120		3
1		2614	2788	2952	3106	3251	3387	3513	3631	3740	3840		2
2		0323	0380	0441	0506	0574	0645	0720	0797	0877	0960		1
3		0013	0017	0022	0027	0034	0041	0049	0058	0069	0080		0
		89	88	87	86	85	84	83	82	81	80	π	x

$n = 3$ *(cont.)*

x	π	21	22	23	24	25	26	27	28	29	30		
0		4930	4746	4565	4390	4219	4052	3890	3732	3579	3430		3
1		3932	4014	4091	4159	4219	4271	4316	4355	4386	4410		2
2		1045	1133	1222	1313	1406	1501	1597	1693	1791	1890		1
3		0093	0106	0122	0138	0156	0176	0197	0220	0244	0270		0
		79	78	77	76	75	74	73	72	71	70	π	x

x	π	31	32	33	34	35	36	37	38	39	40		
0		3285	3144	3008	2875	2746	2621	2500	2383	2270	2160		3
1		4428	4439	4444	4443	4436	4424	4406	4382	4354	4320		2
2		1989	2089	2189	2289	2389	2488	2587	2686	2783	2880		1
3		0298	0328	0359	0393	0429	0467	0507	0549	0593	0640		0
		69	68	67	66	65	64	63	62	61	60	π	x

x	π	41	42	43	44	45	46	47	48	49	50		
0		2054	1951	1852	1756	1664	1575	1489	1406	1327	1250		3
1		4282	4239	4191	4140	4084	4024	3961	3894	3823	3750		2
2		2975	3069	3162	3252	3341	3428	3512	3594	3674	3750		1
3		0689	0741	0795	0852	0911	0973	1038	1106	1176	1250		0
		59	58	57	56	55	54	53	52	51	50	π	x

$n = 4$

x	π	01	02	03	04	05	06	07	08	09	10		
0		9606	9224	8853	8493	8145	7807	7481	7164	6857	6561		4
1		0388	0753	1095	1416	1715	1993	2252	2492	2713	2916		3
2		0006	0023	0051	0088	0135	0191	0254	0325	0402	0486		2
3		0000	0000	0001	0002	0005	0008	0013	0019	0027	0036		1
4		0000	0000	0000	0000	0000	0000	0000	0000	0001	0001		0
		99	98	97	96	95	94	93	92	91	90	π	x

x	π	11	12	13	14	15	16	17	18	19	20		
0		6274	5997	5729	5470	5220	4979	4746	4521	4305	4096		4
1		3102	3271	3424	3562	3685	3793	3888	3970	4039	4096		3
2		0575	0669	0767	0870	0975	1084	1195	1307	1421	1536		2
3		0047	0061	0076	0094	0115	0138	0163	0191	0222	0256		1
4		0001	0002	0003	0004	0005	0007	0008	0010	0013	0016		0
		89	88	87	86	85	84	83	82	81	80	π	x

x	π	21	22	23	24	25	26	27	28	29	30		
0		3895	3702	3515	3336	3164	2999	2840	2687	2541	2401		4
1		4142	4176	4200	4214	4219	4214	4201	4180	4152	4116		3
2		1651	1767	1882	1996	2109	2221	2331	2439	2544	2646		2
3		0293	0332	0375	0420	0469	0520	0575	0632	0693	0756		1
4		0019	0023	0028	0033	0039	0046	0053	0061	0071	0081		0
		79	78	77	76	75	74	73	72	71	70	π	x

$n = 4$ *(cont.)*

x	π	31	32	33	34	35	36	37	38	39	40		
0		2267	2138	2015	1897	1785	1678	1575	1478	1385	1296		4
1		4074	4025	3970	3910	3845	3775	3701	3623	3541	3456		3
2		2745	2841	2933	3021	3105	3185	3260	3330	3396	3456		2
3		0822	0891	0963	1038	1115	1194	1276	1361	1447	1536		1
4		0092	0105	0119	0134	0150	0168	0187	0209	0231	0256		0
		69	68	67	66	65	64	63	62	61	60	π	x

x	π	41	42	43	44	45	46	47	48	49	50		
0		1212	1132	1056	0983	0915	0850	0789	0731	0677	0625		4
1		3368	3278	3185	3091	2995	2897	2799	2700	2600	2500		3
2		3511	3560	3604	3643	3675	3702	3723	3738	3747	3750		2
3		1627	1719	1813	1908	2005	2102	2201	2300	2400	2500		1
4		0283	0311	0342	0375	0410	0448	0488	0531	0576	0625		0
		59	58	57	56	55	54	53	52	51	50	π	x

$n = 5$

x	π	01	02	03	04	05	06	07	08	09	10		
0		9510	9039	8587	8154	7738	7339	6957	6591	6240	5905		5
1		0480	0922	1328	1699	2036	2342	2618	2866	3086	3280		4
2		0010	0038	0082	0142	0214	0299	0394	0498	0610	0729		3
3		0000	0001	0003	0006	0011	0019	0030	0043	0060	0081		2
4		0000	0000	0000	0000	0000	0001	0001	0002	0003	0004		1
		99	98	97	96	95	94	93	92	91	90	π	x

x	π	11	12	13	14	15	16	17	18	19	20		
0		5584	5277	4984	4704	4437	4182	3939	3707	3487	3277		5
1		3451	3598	3724	3829	3915	3983	4034	4069	4089	4096		4
2		0853	0981	1113	1247	1382	1517	1652	1786	1919	2048		3
3		0105	0134	0166	0203	0244	0289	0338	0392	0450	0512		2
4		0007	0009	0012	0017	0022	0028	0035	0043	0053	0064		1
5		0000	0000	0000	0001	0001	0001	0001	0002	0002	0003		0
		89	88	87	86	85	84	83	82	81	80	π	x

x	π	21	22	23	24	25	26	27	28	29	30		
0		3077	2887	2707	2536	2373	2219	2073	1935	1804	1681		5
1		4090	4072	4043	4003	3955	3898	3834	3762	3685	3601		4
2		2174	2297	2415	2529	2637	2739	2836	2926	3010	3087		3
3		0578	0648	0721	0798	0879	0962	1049	1138	1229	1323		2
4		0077	0091	0108	0126	0146	0169	0194	0221	0251	0283		1
5		0004	0005	0006	0008	0010	0012	0014	0017	0021	0024		0
		79	78	77	76	75	74	73	72	71	70	π	x

$n = 5$ *(cont.)*

x	π	31	32	33	34	35	36	37	38	39	40		
0		1564	1454	1350	1252	1160	1074	0992	0916	0845	0778		5
1		3513	3421	3325	3226	3124	3020	2914	2808	2700	2592		4
2		3157	3220	3275	3323	3364	3397	3423	3441	3452	3456		3
3		1418	1515	1613	1712	1811	1911	2010	2109	2207	2304		2
4		0319	0357	0397	0441	0488	0537	0590	0646	0706	0768		1
5		0029	0034	0039	0045	0053	0060	0069	0079	0090	0102		0
		69	68	67	66	65	64	63	62	61	60	π	x

x	π	41	42	43	44	45	46	47	48	49	50		
0		0715	0656	0602	0551	0503	0459	0418	0380	0345	0313		5
1		2484	2376	2270	2164	2059	1956	1854	1755	1657	1562		4
2		3452	3442	3424	3400	3369	3332	3289	3240	3185	3125		3
3		2399	2492	2583	2671	2757	2838	2916	2990	3060	3125		2
4		0834	0902	0974	1049	1128	1209	1293	1380	1470	1562		1
5		0116	0131	0147	0165	0185	0206	0229	0255	0282	0312		0
		59	58	57	56	55	54	53	52	51	50	π	x

$n = 6$

x	π	01	02	03	04	05	06	07	08	09	10		
0		9415	8858	8330	7828	7351	6899	6470	6064	5679	5314		6
1		0571	1085	1546	1957	2321	2642	2922	3164	3370	3543		5
2		0014	0055	0120	0204	0305	0422	0550	0688	0833	0984		4
3		0000	0002	0005	0011	0021	0036	0055	0080	0110	0146		3
4		0000	0000	0000	0000	0001	0002	0003	0005	0008	0012		2
5		0000	0000	0000	0000	0000	0000	0000	0000	0000	0001		1
		99	98	97	96	95	94	93	92	91	90	π	x

x	π	11	12	13	14	15	16	17	18	19	20		
0		4970	4644	4336	4046	3771	3513	3269	3040	2824	2621		6
1		3685	3800	3888	3952	3993	4015	4018	4004	3975	3932		5
2		1139	1295	1452	1608	1762	1912	2057	2197	2331	2458		4
3		0188	0236	0289	0349	0415	0486	0562	0643	0729	0819		3
4		0017	0024	0032	0043	0055	0069	0086	0106	0128	0154		2
5		0001	0001	0002	0003	0004	0005	0007	0009	0012	0015		1
6		0000	0000	0000	0000	0000	0000	0000	0000	0000	0001		0
		89	88	87	86	85	84	83	82	81	80	π	x

$n = 6$ *(cont.)*

x	π	21	22	23	24	25	26	27	28	29	30		
0		2431	2252	2084	1927	1780	1642	1513	1393	1281	1176		6
1		3877	3811	3735	3651	3560	3462	3358	3251	3139	3025		5
2		2577	2687	2789	2882	2966	3041	3105	3160	3206	3241		4
3		0913	1011	1111	1214	1318	1424	1531	1639	1746	1852		3
4		0182	0214	0249	0287	0330	0375	0425	0478	0535	0595		2
5		0019	0024	0030	0036	0044	0053	0063	0074	0087	0102		1
6		0001	0001	0001	0002	0002	0003	0004	0005	0006	0007		0
		79	78	77	76	75	74	73	72	71	70	π	x

x	π	31	32	33	34	35	36	37	38	39	40		
0		1079	0989	0905	0827	0754	0687	0625	0568	0515	0467		6
1		2909	2792	2673	2555	2437	2319	2203	2089	1976	1866		5
2		3267	3284	3292	3290	3280	3261	3235	3201	3159	3110		4
3		1957	2061	2162	2260	2355	2446	2533	2616	2693	2765		3
4		0660	0727	0799	0873	0951	1032	1116	1202	1291	1382		2
5		0119	0137	0157	0180	0205	0232	0262	0295	0330	0369		1
6		0009	0011	0013	0015	0018	0022	0026	0030	0035	0041		0
		69	68	67	66	65	64	63	62	61	60	π	x

x	π	41	42	43	44	45	46	47	48	49	50		
0		0422	0381	0343	0308	0277	0248	0222	0198	0176	0156		6
1		1759	1654	1552	1454	1359	1267	1179	1095	1014	0937		5
2		3055	2994	2928	2856	2780	2699	2615	2527	2436	2344		4
3		2831	2891	2945	2992	3032	3065	3091	3110	3121	3125		3
4		1475	1570	1666	1763	1861	1958	2056	2153	2249	2344		2
5		0410	0455	0503	0554	0609	0667	0729	0795	0864	0937		1
6		0048	0055	0063	0073	0083	0095	0108	0122	0138	0156		0
		59	58	57	56	55	54	53	52	51	50	π	x

$n = 7$

x	π	01	02	03	04	05	06	07	08	09	10		
0		9321	8681	8080	7514	6983	6485	6017	5578	5168	4783		7
1		0659	1240	1749	2192	2573	2897	3170	3396	3578	3720		6
2		0020	0076	0162	0274	0406	0555	0716	0886	1061	1240		5
3		0000	0003	0008	0019	0036	0059	0090	0128	0175	0230		4
4		0000	0000	0000	0001	0002	0004	0007	0011	0017	0026		3
5		0000	0000	0000	0000	0000	0000	0000	0001	0001	0002		2
		99	98	97	96	95	94	93	92	91	90	π	x

x	π	11	12	13	14	15	16	17	18	19	20		
0		4423	4087	3773	3479	3206	2951	2714	2493	2288	2097		7
1		3827	3901	3946	3965	3960	3935	3891	3830	3756	3670		6
2		1419	1596	1769	1936	2097	2248	2391	2523	2643	2753		5
3		0292	0363	0441	0525	0617	0714	0816	0923	1033	1147		4
4		0036	0049	0066	0086	0109	0136	0167	0203	0242	0287		3
5		0003	0004	0006	0008	0012	0016	0021	0027	0034	0043		2
6		0000	0000	0000	0000	0001	0001	0001	0002	0003	0004		1
		89	88	87	86	85	84	83	82	81	80	π	x

x	π	21	22	23	24	25	26	27	28	29	30		
0		1920	1757	1605	1465	1335	1215	1105	1003	0910	0824		7
1		3573	3468	3356	3237	3115	2989	2860	2731	2600	2471		6
2		2850	2935	3007	3067	3115	3150	3174	3186	3186	3177		5
3		1263	1379	1497	1614	1730	1845	1956	2065	2169	2269		4
4		0336	0389	0447	0510	0577	0648	0724	0803	0886	0972		3
5		0054	0066	0080	0097	0115	0137	0161	0187	0217	0250		2
6		0005	0006	0008	0010	0013	0016	0020	0024	0030	0036		1
7		0000	0000	0000	0000	0001	0001	0001	0001	0002	0002		0
		79	78	77	76	75	74	73	72	71	70	π	x

x	π	31	32	33	34	35	36	37	38	39	40		
0		0745	0672	0606	0546	0490	0440	0394	0352	0314	0280		7
1		2342	2215	2090	1967	1848	1732	1619	1511	1407	1306		6
2		3156	3127	3088	3040	2985	2922	2853	2778	2698	2613		5
3		2363	2452	2535	2610	2679	2740	2793	2838	2875	2903		4
4		1062	1154	1248	1345	1442	1541	1640	1739	1838	1935		3
5		0286	0326	0369	0416	0466	0520	0578	0640	0705	0774		2
6		0043	0051	0061	0071	0084	0098	0113	0131	0150	0172		1
7		0003	0003	0004	0005	0006	0008	0009	0011	0014	0016		0
		69	68	67	66	65	64	63	62	61	60	π	x

n = 7 (cont.)

x	π	41	42	43	44	45	46	47	48	49	50		
0		0249	0221	0195	0173	0152	0134	0117	0103	0090	0078		7
1		1211	1119	1032	0950	0872	0798	0729	0664	0604	0547		6
2		2524	2431	2336	2239	2140	2040	1940	1840	1740	1641		5
3		2923	2934	2937	2932	2918	2897	2867	2830	2786	2734		4
4		2031	2125	2216	2304	2388	2468	2543	2612	2676	2734		3
5		0847	0923	1003	1086	1172	1261	1353	1447	1543	1641		2
6		0196	0223	0252	0284	0320	0358	0400	0445	0494	0547		1
7		0019	0023	0027	0032	0037	0044	0051	0059	0068	0078		0
		59	58	57	56	55	54	53	52	51	50	π	x

n = 8

x	π	01	02	03	04	05	06	07	08	09	10		
0		9227	8508	7837	7214	6634	6096	5596	5132	4703	4305		8
1		0746	1389	1939	2405	2793	3113	3370	3570	3721	3826		7
2		0026	0099	0210	0351	0515	0695	0888	1087	1288	1488		6
3		0001	0004	0013	0029	0054	0089	0134	0189	0255	0331		5
4		0000	0000	0001	0002	0004	0007	0013	0021	0031	0046		4
5		0000	0000	0000	0000	0000	0000	0001	0001	0002	0004		3
		99	98	97	96	95	94	93	92	91	90	π	x

x	π	11	12	13	14	15	16	17	18	19	20		
0		3937	3596	3282	2992	2725	2479	2252	2044	1853	1678		8
1		3892	3923	3923	3897	3847	3777	3691	3590	3477	3355		7
2		1684	1872	2052	2220	2376	2518	2646	2758	2855	2936		6
3		0416	0511	0613	0723	0839	0959	1084	1211	1339	1468		5
4		0064	0087	0115	0147	0185	0228	0277	0332	0393	0459		4
5		0006	0009	0014	0019	0026	0035	0045	0058	0074	0092		3
6		0000	0001	0001	0002	0002	0003	0005	0006	0009	0011		2
7		0000	0000	0000	0000	0000	0000	0000	0000	0001	0001		I
		89	88	87	86	85	84	83	82	81	80	π	x

x	π	21	22	23	24	25	26	27	28	29	30		
0		1517	1370	1236	1113	1001	0899	0806	0722	0646	0576		8
1		3226	3092	2953	2812	2670	2527	2386	2247	2110	1977		7
2		3002	3052	3087	3108	3115	3108	3089	3058	3017	2965		6
3		1596	1722	1844	1963	2076	2184	2285	2379	2464	2541		5
4		0530	0607	0689	0775	0865	0959	1056	1156	1258	1361		4
5		0113	0137	0165	0196	0231	0270	0313	0360	0411	0467		3
6		0015	0019	0025	0031	0038	0047	0058	0070	0084	0100		2
7		0001	0002	0002	0003	0004	0005	0006	0008	0010	0012		1
8		0000	0000	0000	0000	0000	0000	0000	0000	0001	0001		0
		79	78	77	76	75	74	73	72	71	70	π	x

n = 8 *(cont.)*

x	π	31	32	33	34	35	36	37	38	39	40		
0		0514	0457	0406	0360	0319	0281	0248	0218	0192	0168		8
1		1847	1721	1600	1484	1373	1267	1166	1071	0981	0896		7
2		2904	2835	2758	2675	2587	2494	2397	2297	2194	2090		6
3		2609	2668	2717	2756	2786	2805	2815	2815	2806	2787		5
4		1465	1569	1673	1775	1875	1973	2067	2157	2242	2322		4
5		0527	0591	0659	0732	0808	0888	0971	1058	1147	1239		3
6		0118	0139	0162	0188	0217	0250	0285	0324	0367	0413		2
7		0015	0019	0023	0028	0033	0040	0048	0057	0067	0079		1
8		0001	0001	0001	0002	0002	0003	0004	0004	0005	0007		0
		69	68	67	66	65	64	63	62	61	60	π	x

x	π	41	42	43	44	45	46	47	48	49	50		
0		0147	0128	0111	0097	0084	0072	0062	0053	0046	0039		8
1		0816	0742	0672	0608	0548	0493	0442	0395	0352	0312		7
2		1985	1880	1776	1672	1569	1469	1371	1275	1183	1094		6
3		2759	2723	2679	2627	2568	2503	2431	2355	2273	2187		5
4		2397	2465	2526	2580	2627	2665	2695	2717	2730	2734		4
5		1332	1428	1525	1622	1719	1816	1912	2006	2098	2187		3
6		0463	0517	0575	0637	0703	0774	0848	0926	1008	1094		2
7		0092	0107	0124	0143	0164	0188	0215	0244	0277	0312		1
8		0008	0010	0012	0014	0017	0020	0024	0028	0033	0039		0
		59	58	57	56	55	54	53	52	51	50	π	x

n = 9

x	π	01	02	03	04	05	06	07	08	09	10		
0		9135	8337	7602	6925	6302	5730	5204	4722	4279	3874		9
1		0830	1531	2116	2597	2985	3292	3525	3695	3809	3874		8
2		0034	0125	0262	0433	0629	0840	1061	1285	1507	1722		7
3		0001	0006	0019	0042	0077	0125	0186	0261	0348	0446		6
4		0000	0000	0001	0003	0006	0012	0021	0034	0052	0074		5
5		0000	0000	0000	0000	0000	0001	0002	0003	0005	0008		4
6		0000	0000	0000	0000	0000	0000	0000	0000	0000	0001		3
		99	98	97	96	95	94	93	92	91	90	π	x

x	π	11	12	13	14	15	16	17	18	19	20		
0		3504	3165	2855	2573	2316	2082	1869	1676	1501	1342		9
1		3897	3884	3840	3770	3679	3569	3446	3312	2169	3020		8
2		1927	2119	2295	2455	2597	2720	2823	2908	2973	3020		7
3		0556	0674	0800	0933	1069	1209	1349	1489	1627	1762		6
4		0103	0138	0179	0228	0283	0345	0415	0490	0573	0661		5
5		0013	0019	0027	0037	0050	0066	0085	0108	0134	0165		4
6		0001	0002	0003	0004	0006	0008	0012	0016	0021	0028		3
7		0000	0000	0000	0000	0000	0001	0001	0001	0002	0003		2
		89	88	87	86	85	84	83	82	81	80	π	x

n = 9 *(cont.)*

x	π	21	22	23	24	25	26	27	28	29	30		
0		1199	1069	0952	0846	0751	0665	0589	0520	0458	0404		9
1		2867	2713	2558	2404	2253	2104	1960	1820	1685	1556		8
2		3049	3061	3056	3037	3003	2957	2899	2831	2754	2668		7
3		1891	2014	2130	2238	2336	2424	2502	2569	2624	2668		6
4		0754	0852	0954	1060	1168	1278	1388	1499	1608	1715		5
5		0200	0240	0285	0335	0389	0449	0513	0583	0657	0735		4
6		0036	0045	0057	0070	0087	0105	0127	0151	0179	0210		3
7		0004	0005	0007	0010	0012	0016	0020	0025	0031	0039		2
8		0000	0000	0001	0001	0001	0001	0002	0002	0003	0004		1
		79	78	77	76	75	74	73	72	71	70	π	*x*

x	π	31	32	33	34	35	36	37	38	39	40		
0		0355	0311	0272	0238	0207	0180	0156	0135	0117	0101		9
1		1433	1317	1206	1102	1004	0912	0826	0747	0673	0605		8
2		2576	2478	2376	2270	2162	2052	1941	1831	1721	1612		7
3		2701	2721	2731	2729	2716	2693	2660	2618	2567	2508		6
4		1820	1921	2017	2109	2194	2272	2344	2407	2462	2508		5
5		0818	0904	0994	1086	1181	1278	1376	1475	1574	1672		4
6		0245	0284	0326	0373	0424	0479	0539	0603	0671	0743		3
7		0047	0057	0069	0082	0098	0116	0136	0158	0184	0212		2
8		0005	0007	0008	0011	0013	0016	0020	0024	0029	0035		1
9		0000	0000	0000	0001	0001	0001	0001	0002	0002	0003		0
		69	68	67	66	65	64	63	62	61	60	π	*x*

x	π	41	42	43	44	45	46	47	48	49	50		
0		0087	0074	0064	0054	0046	0039	0033	0028	0023	0020		9
1		0542	0484	0431	0383	0339	0299	0263	0231	0202	0176		8
2		1506	1402	1301	1204	1110	1020	0934	0853	0776	0703		7
3		2442	2369	2291	2207	2119	2027	1933	1837	1739	1641		6
4		2545	2573	2592	2601	2600	2590	2571	2543	2506	2461		5
5		1769	1863	1955	2044	2128	2207	2280	2347	2408	2461		4
6		0819	0900	0983	1070	1160	1253	1348	1445	1542	1641		3
7		0244	0279	0318	0360	0407	0458	0512	0571	0635	0703		2
8		0042	0051	0060	0071	0083	0097	0014	0132	0153	0176		1
9		0003	0004	0005	0006	0008	0009	0011	0014	0016	0020		0
		59	58	57	56	55	54	53	52	51	50	π	*x*

$n = 10$

x	π	01	02	03	04	05	06	07	08	09	10		
0		9044	8171	7374	6648	5987	5386	4840	4344	3894	3487		10
1		0914	1667	2281	2770	3151	3438	3643	3777	3851	3874		9
2		0042	0153	0317	0519	0746	0988	1234	1478	1714	1937		8
3		0001	0008	0026	0058	0105	0168	0248	0343	0452	0574		7
4		0000	0000	0001	0004	0010	0019	0033	0052	0078	0112		6
5		0000	0000	0000	0000	0001	0001	0003	0005	0009	0015		5
6		0000	0000	0000	0000	0000	0000	0000	0000	0001	0001		4
		99	98	97	96	95	94	93	92	91	90	π	x

x	π	11	12	13	14	15	16	17	18	19	20		
0		3118	2785	2484	2213	1969	1749	1552	1374	1216	1074		10
1		3854	3798	3712	3603	3474	3331	3178	3017	2852	2684		9
2		2143	2330	2496	2639	2759	2856	2929	2980	3010	3020		8
3		0706	0847	0995	1146	1298	1450	1600	1745	1883	2013		7
4		0153	0202	0260	0326	0401	0483	0573	0670	0773	0881		6
5		0023	0033	0047	0064	0085	0111	0141	0177	0218	0264		5
6		0002	0004	0006	0009	0012	0018	0024	0032	0043	0055		4
7		0000	0000	0000	0001	0001	0002	0003	0004	0006	0008		3
8		0000	0000	0000	0000	0000	0000	0000	0000	0001	0001		2
		89	88	87	86	85	84	83	82	81	80	π	x

x	π	21	22	23	24	25	26	27	28	29	30		
0		0947	0834	0733	0643	0563	0492	0430	0374	0326	0282		10
1		2517	2351	2188	2030	1877	1730	1590	1456	1330	1211		9
2		3011	2984	2942	2885	2816	2735	2646	2548	2444	2335		8
3		2134	2244	2343	2429	2503	2563	2609	2642	2662	2668		7
4		0993	1108	1225	1343	1460	1576	1689	1798	1903	2001		6
5		0317	0375	0439	0509	0584	0664	0750	0839	0933	1029		5
6		0070	0088	0109	0134	0162	0195	0231	0272	0317	0368		4
7		0011	0014	0019	0024	0031	0039	0049	0060	0074	0090		3
8		0001	0002	0002	0003	0004	0005	0007	0009	0011	0014		2
9		0000	0000	0000	0000	0000	0000	0001	0001	0001	0001		1
		79	78	77	76	75	74	73	72	71	70	π	x

x	π	31	32	33	34	35	36	37	38	39	40		
0		0245	0211	0182	0157	0135	0115	0098	0084	0071	0060		10
1		1099	0995	0898	0808	0725	0649	0578	0514	0456	0403		9
2		2222	2107	1990	1873	1757	1642	1529	1419	1312	1209		8
3		2662	2644	2614	2573	2522	2462	2394	2319	2237	2150		7
4		2093	2177	2253	2320	2377	2424	2461	2487	2503	2508		6
5		1128	1229	1332	1434	1536	1636	1734	1829	1920	2007		5
6		0422	0482	0547	0616	0689	0767	0849	0934	1023	1115		4
7		0108	0130	0154	0181	0212	0247	0285	0327	0374	0425		3
8		0018	0023	0028	0035	0043	0052	0063	0075	0090	0106		2
9		0002	0002	0003	0004	0005	0006	0008	0010	0013	0016		1
10		0000	0000	0000	0000	0000	0000	0000	0001	0001	0001		0
		69	68	67	66	65	64	63	62	61	60	π	x

n = 10 *(cont.)*

x π	41	42	43	44	45	46	47	48	49	50	
0	0051	0043	0036	0030	0025	0021	0017	0014	0012	0010	10
1	0355	0312	0273	0238	0207	0180	0155	0133	0114	0098	9
2	1111	1017	0927	0843	0763	0688	0619	0554	0494	0439	8
3	2058	1963	1865	1765	1665	1654	1464	1364	1267	1172	7
4	2503	2488	2462	2427	2384	2331	2271	2204	2130	2051	6
5	2087	2162	2229	2289	2340	2383	2417	2441	2456	2461	5
6	1209	1304	1401	1499	1596	1692	1786	1878	1966	2051	4
7	0480	0540	0604	0673	0746	0824	0905	0991	1080	1172	3
8	0125	0147	0171	0198	0229	0263	0301	0343	0389	0439	2
9	0019	0024	0029	0035	0042	0050	0059	0070	0083	0098	1
10	0001	0002	0002	0003	0003	0004	0005	0006	0008	0010	0
	59	58	57	56	55	54	53	52	51	50	π *x*

n = 20

x π	01	02	03	04	05	06	07	08	09	10	
0	8179	6676	5438	4420	3585	2901	2342	1887	1516	1216	20
1	1652	2725	3364	3683	3774	3703	3526	3282	3000	2702	19
2	0159	0528	0988	1458	1887	2246	2521	2711	2828	2852	18
3	0010	0065	0183	0364	0596	0860	1139	1414	1672	1901	17
4	0000	0006	0024	0065	0133	0233	0364	0523	0703	0898	16
5	0000	0000	0002	0009	0022	0048	0088	0145	0222	0319	15
6	0000	0000	0000	0001	0003	0008	0017	0032	0055	0089	14
7	0000	0000	0000	0000	0000	0001	0002	0005	0011	0020	13
8	0000	0000	0000	0000	0000	0000	0000	0001	0002	0004	12
9	0000	0000	0000	0000	0000	0000	0000	0000	0000	0001	11
	99	98	97	96	95	94	93	92	91	90	π *x*

x π	11	12	13	14	15	16	17	18	19	20	
0	0972	0776	0617	0490	0388	0306	0241	0189	0148	0115	20
1	2403	2115	1844	1595	1368	1165	0986	0829	0693	0576	19
2	2822	2740	2618	2466	2293	2109	1919	1730	1545	1369	18
3	2093	2242	2347	2409	2428	2410	2358	2278	2175	2054	17
4	1099	1299	1491	1666	1821	1951	2053	2125	2168	2182	16
5	0435	0567	0713	0868	1028	1189	1345	1493	1627	1746	15
6	0134	0193	0266	0353	0454	0566	0689	0819	0954	1091	14
7	0033	0053	0080	0115	0160	0216	0282	0360	0448	0545	13
8	0007	0012	0019	0030	0046	0067	0094	0128	0171	0222	12
9	0001	0002	0004	0007	0011	0017	0026	0038	0053	0074	11
10	0000	0000	0001	0001	0002	0004	0006	0009	0014	0020	10
11	0000	0000	0000	0000	0000	0001	0001	0002	0003	0005	9
12	0000	0000	0000	0000	0000	0000	0000	0000	0001	0001	8
	89	88	87	86	85	84	83	82	81	80	π *x*

n = 20 *(cont.)*

x π	21	22	23	24	25	26	27	28	29	30	
0	0090	0069	0054	0041	0032	0024	0016	0014	0011	0008	20
1	0477	0392	0321	0261	0211	0170	0137	0109	0087	0068	19
2	1204	1050	0910	0783	0669	0569	0480	0403	0336	0278	18
3	1920	1777	1631	1484	1339	1199	1065	0940	0823	0716	17
4	2169	2131	2070	1991	1897	1790	1675	1553	1429	1304	16
5	1845	1923	1979	2012	2023	2013	1982	1933	1868	1789	15
6	1226	1356	1478	1589	1686	1768	1833	1879	1907	1916	14
7	0652	0765	0883	1003	1124	1242	1356	1462	1558	1643	13
8	0282	0351	0429	0515	0609	0709	0815	0924	1034	1144	12
9	0100	0132	0171	0217	0271	0332	0402	0479	0563	0654	11
10	0029	0041	0056	0075	0099	0128	0163	0205	0253	0308	10
11	0007	0010	0015	0022	0030	0041	0055	0072	0094	0120	9
12	0001	0002	0003	0005	0008	0011	0015	0021	0029	0039	8
13	0000	0000	0001	0001	0002	0002	0003	0005	0007	0010	7
14	0000	0000	0000	0000	0000	0000	0001	0001	0001	0002	6
	79	78	77	76	75	74	73	72	71	70 π	x

x π	31	32	33	34	35	36	37	38	39	40	
0	0006	0004	0003	0002	0002	0001	0001	0001	0001	0000	20
1	0054	0042	0033	0025	0020	0015	0011	0009	0007	0005	19
2	0229	0188	0153	0124	0100	0080	0064	0050	0040	0031	18
3	0619	0531	0453	0383	0323	0270	0224	0185	0152	0123	17
4	1181	1062	0947	0839	0738	0645	0559	0482	0412	0350	16
5	1698	1599	1493	1384	1272	1161	1051	0945	0843	0746	15
6	1907	1881	1839	1782	1712	1632	1543	1447	1347	1244	14
7	1714	1770	1811	1836	1844	1836	1812	1774	1722	1659	13
8	1251	1354	1450	1537	1614	1678	1730	1767	1790	1797	12
9	0750	0849	0952	1056	1158	1259	1354	1444	1526	1597	11
10	0370	0440	0516	0598	0686	0779	0875	0974	1073	1171	10
11	0151	1188	0231	0280	0336	0398	0467	0542	0624	0710	9
12	0051	0066	0085	0108	0136	0168	0206	0249	0299	0355	8
13	0014	0019	0026	0034	0045	0058	0074	0094	0118	0146	7
14	0003	0005	0006	0009	0012	0016	0022	0029	0038	0049	6
15	0001	0001	0001	0002	0003	0004	0005	0007	0010	0013	5
16	0000	0000	0000	0000	0000	0001	0001	0001	0002	0003	4
	69	68	67	66	65	64	63	62	61	60 π	x

n = 20 *(cont.)*

x	π 41	42	43	44	45	46	47	48	49	50	
1	0004	0003	0002	0001	0001	0001	0001	0000	0000	0000	19
2	0024	0018	0014	0011	0008	0006	0005	0003	0002	0002	18
3	0100	0080	0064	0051	0040	0031	0024	0019	0014	0011	17
4	0295	0247	0206	0170	0139	0113	0092	0074	0059	0046	16
5	0656	0573	0496	0427	0365	0309	0260	0217	0180	0148	15
6	1140	1037	0936	0839	0746	0658	0577	0501	0432	0370	14
7	1585	1502	1413	1318	1221	1122	1023	0925	0830	0739	13
8	1790	1768	1732	1683	1623	1553	1474	1388	1296	1201	12
9	1658	1707	1742	1763	1771	1763	1742	1708	1661	1602	11
10	1268	1359	1446	1524	1593	1652	1700	1734	1755	1762	10
11	0801	0895	0991	1089	1185	1280	1370	1455	1533	1602	9
12	0417	0486	0561	0642	0727	0818	0911	1007	1105	1201	8
13	0178	0217	0260	0310	0366	0429	0497	0572	0653	0739	7
14	0062	0078	0098	0122	0150	0183	0221	0264	0314	0370	6
15	0017	0023	0030	0038	0049	0062	0078	0098	0121	0148	5
16	0004	0005	0007	0009	0013	0017	0022	0028	0036	0046	4
17	0001	0001	0001	0002	0002	0003	0005	0006	0008	0011	3
18	0000	0000	0000	0000	0000	0000	0001	0001	0001	0002	2
	59	58	57	56	55	54	53	52	51	50 π	x

n = 50

x	π 01	02	03	04	05	06	07	08	09	10	
0	6050	3642	2181	1299	0769	0453	0266	0155	0090	0052	50
1	3056	3716	3372	2706	2025	1447	0999	0672	0443	0286	49
2	0756	1858	2555	2762	2611	2262	1843	1433	1073	0779	48
3	0122	0607	1264	1842	2199	2311	2219	1993	1698	1386	47
4	0015	0145	0459	0902	1360	1733	1963	2037	1973	1809	46
5	0001	0027	0131	0346	0658	1018	1359	1629	1795	1849	45
6	0000	0004	0030	0108	0260	0487	0767	1063	1332	1541	44
7	0000	0001	0006	0028	0086	0195	0363	0581	0828	1076	43
8	0000	0000	0001	0006	0024	0067	0147	0271	0440	0643	42
9	0000	0000	0000	0001	0006	0020	0052	0110	0203	0333	41
10	0000	0000	0000	0000	0001	0005	0016	0039	0082	0152	40
11	0000	0000	0000	0000	0000	0001	0004	0012	0030	0061	39
12	0000	0000	0000	0000	0000	0000	0001	0004	0010	0022	38
13	0000	0000	0000	0000	0000	0000	0000	0001	0003	0007	37
14	0000	0000	0000	0000	0000	0000	0000	0000	0001	0002	36
15	0000	0000	0000	0000	0000	0000	0000	0000	0000	0001	35
	99	98	97	96	95	94	93	92	91	90 π	x

n = 50 *(cont.)*

x π	11	12	13	14	15	16	17	18	19	20		
0	0029	0017	0009	0005	0003	0002	0001	0000	0000	0000		50
1	0182	0114	0071	0043	0026	0016	0009	0005	0003	0002		49
2	0552	0382	0259	0172	0113	0073	0046	0029	0018	0011		48
3	1091	0833	0619	0449	0319	0222	0151	0102	0067	0044		47
4	1584	1334	1086	0858	0661	0496	0364	0262	0185	0128		46
5	1801	1674	1493	1286	1072	0869	0687	0530	0400	0295		45
6	1670	1712	1674	1570	1419	1242	1055	0872	0703	0554		44
7	1297	1467	1572	1606	1575	1487	1358	1203	1037	0870		43
8	0862	1075	1262	1406	1493	1523	1495	1420	1307	1169		42
9	0497	0684	0880	1068	1230	1353	1429	1454	1431	1364		41
10	0252	0383	0539	0713	0890	1057	1200	1309	1376	1398		40
11	0113	0190	0293	0422	0571	0732	0894	1045	1174	1271		39
12	0045	0084	0142	0223	0328	0453	0595	0745	0895	1033		38
13	0016	0034	0062	0106	0169	0252	0356	0478	0613	0755		37
14	0005	0012	0025	0046	0079	0127	0193	0277	0380	0499		36
15	0002	0004	0009	0018	0033	0058	0095	0146	0214	0299		35
16	0000	0001	0003	0006	0013	0024	0042	0070	0110	0164		34
17	0000	0000	0001	0002	0005	0009	0017	0031	0052	0082		33
18	0000	0000	0000	0001	0001	0003	0007	0012	0022	0037		32
19	0000	0000	0000	0000	0000	0001	0002	0005	0009	0016		31
20	0000	0000	0000	0000	0000	0000	0001	0002	0003	0006		30
21	0000	0000	0000	0000	0000	0000	0000	0000	0001	0002		29
22	0000	0000	0000	0000	0000	0000	0000	0000	0000	0001		28
	89	88	87	86	85	84	83	82	81	80	π	*x*

x π	21	22	23	24	25	26	27	28	29	30	
1	0001	0001	0000	0000	0000	0000	0000	0000	0000	0000	49
2	0007	0004	0002	0001	0001	0000	0000	0000	0000	0000	48
3	0028	0018	0011	0007	0004	0002	0001	0001	0000	0000	47
4	0088	0059	0039	0025	0016	0010	0006	0004	0002	0001	46
5	0214	0152	0106	0073	0049	0033	0021	0014	0009	0006	45
6	0427	0322	0238	0173	0123	0087	0060	0040	0027	0018	44
7	0713	0571	0447	0344	0259	0191	0139	0099	0069	0048	43
8	1019	0865	0718	0583	0463	0361	0276	0207	0152	0110	42
9	1263	1139	1001	0859	0721	0592	0476	0375	0290	0220	41
10	1377	1317	1226	1113	0985	0852	0721	0598	0485	0386	40
11	1331	1351	1332	1278	1194	1089	0970	0845	0721	0602	39
12	1150	1238	1293	1311	1294	1244	1166	1068	0957	0838	38
13	0894	1021	1129	1210	1261	1277	1261	1215	1142	1050	37
14	0628	0761	0891	1010	1110	1186	1233	1248	1233	1189	36
15	0400	0515	0639	0766	0888	1000	1094	1165	1209	1223	35
16	0233	0318	0417	0529	0648	0769	0885	0991	1080	1147	34
17	0124	0179	0249	0334	0432	0540	0655	0771	0882	0983	33
18	0060	0093	0137	0193	0264	0348	0444	0550	0661	0772	32

$n = 50$ *(cont.)*

19		0027	0044	0069	0103	0148	0206	0277	0360	0454	0558	31
20		0011	0019	0032	0050	0077	0112	0159	0217	0288	0370	30
21		0004	0008	0014	0023	0036	0056	0084	0121	0168	0227	29
22		0001	0003	0005	0009	0016	0026	0041	0062	0090	0128	28
23		0000	0001	0002	0004	0006	0011	0018	0029	0045	0067	27
24		0000	0000	0001	0001	0002	0004	0008	0013	0021	0032	26
25		0000	0000	0000	0000	0001	0002	0003	0005	0009	0014	25
26		0000	0000	0000	0000	0000	0001	0001	0002	0003	0006	24
27		0000	0000	0000	0000	0000	0000	0000	0001	0001	0002	23
28		0000	0000	0000	0000	0000	0000	0000	0000	0000	0001	22
		79	78	77	76	75	74	73	72	71	70	π *x*

x	π	31	32	33	34	35	36	37	38	39	40	
4		0001	0000	0000	0000	0000	0000	0000	0000	0000	0000	46
5		0003	0002	0001	0001	0000	0000	0000	0000	0000	0000	45
6		0011	0007	0005	0003	0002	0001	0001	0000	0000	0000	44
7		0032	0022	0014	0009	0006	0004	0002	0001	0001	0000	43
8		0078	0055	0037	0025	0017	0011	0007	0004	0003	0002	42
9		0164	0120	0086	0061	0042	0029	0019	0013	0008	0005	41
10		0301	0231	0174	0128	0093	0066	0046	0032	0022	0014	40
11		0493	0395	0311	0240	0182	0136	0099	0071	0050	0035	39
12		0719	0604	0498	0402	0319	0248	0189	0142	0105	0076	38
13		0944	0831	0717	0606	0502	0408	0325	0255	0195	0147	37
14		1121	1034	0933	0825	0714	0607	0505	0412	0330	0260	36
15		1209	1168	1103	1020	0923	0819	0712	0606	0507	0415	35
16		1188	1202	1189	1149	1088	1008	0914	0813	0709	0606	34
17		1068	1132	1171	1184	1171	1133	1074	0997	0906	0808	33
18		0880	0976	1057	1118	1156	1169	1156	1120	1062	0987	32
19		0666	0774	0877	0970	1048	1107	1144	1156	1144	1109	31
20		0463	0564	0670	0775	0875	0956	1041	1098	1134	1146	30
21		0297	0379	0471	0570	0673	0776	0874	0962	1035	1091	29
22		0176	0235	0306	0387	0478	0575	0676	0777	0873	0959	28
23		0096	0135	0183	0243	0313	0394	0484	0580	0679	0778	27
24		0049	0071	0102	0141	0190	0249	0319	0400	0489	0584	26
25		0023	0035	0052	0075	0106	0146	0195	0255	0325	0405	25
26		0010	0016	0025	0037	0055	0079	0110	0150	0200	0259	24
27		0004	0007	0011	0017	0026	0039	0058	0082	0113	0154	23
28		0001	0003	0004	0007	0012	0018	0028	0041	0060	0084	22
29		0000	0001	0002	0003	0005	0008	0012	0019	0029	0043	21
30		0000	0000	0001	0001	0002	0003	0005	0008	0013	0020	20
31		0000	0000	0000	0000	0001	0001	0002	0003	0005	0009	19
32		0000	0000	0000	0000	0000	0000	0001	0001	0002	0003	18
33		0000	0000	0000	0000	0000	0000	0000	0000	0001	0001	17
		69	68	67	66	65	64	63	62	61	60	π *x*

n = 50 *(cont.)*

x π	41	42	43	44	45	46	47	48	49	50	
8	0001	0001	0000	0000	0000	0000	0000	0000	0000	0000	42
9	0003	0002	0001	0001	0000	0000	0000	0000	0000	0000	41
10	0009	0006	0004	0002	0001	0001	0001	0000	0000	0000	40
11	0024	0016	0010	0007	0004	0003	0002	0001	0001	0000	39
12	0054	0037	0026	0017	0011	0007	0005	0003	0002	0001	38
13	0109	0079	0057	0040	0027	0018	0012	0008	0005	0003	37
14	0200	0152	0113	0082	0059	0041	0029	0019	0013	0008	36
15	0334	0264	0204	0155	0116	0085	0061	0043	0030	0020	35
16	0508	0418	0337	0267	0207	0158	0118	0086	0062	0044	34
17	0706	0605	0508	0419	0339	0269	0209	0159	0119	0087	33
18	0899	0803	0703	0604	0508	0420	0340	0270	0210	0160	32
19	1053	0979	0893	0799	0700	0602	0507	0419	0340	0270	31
20	1134	1099	1044	0973	0588	0795	0697	0600	0506	0419	30
21	1126	1137	1126	1092	1030	0967	0884	0791	0695	0598	29
22	1031	1086	1119	1131	1119	1086	1033	0963	0880	0788	28
23	0872	0957	1028	1082	1115	1126	1115	1082	1029	0960	27
24	0682	0780	0872	0956	1026	1079	1112	1124	1112	1080	26
25	0493	0587	0684	0781	0873	0956	1026	1079	1112	1123	25
26	0329	0409	0497	0590	0687	0783	0875	0957	1027	1080	24
27	0203	0263	0333	0412	0500	0593	0690	0786	0877	0960	23
28	0116	0157	0206	0266	0336	0415	0502	0596	0692	0788	22
29	0061	0086	0118	0159	0208	0268	0338	0417	0504	0598	21
30	0030	0044	0062	0087	0119	0160	0210	0270	0339	0419	20
31	0013	0020	0030	0044	0063	0088	0120	0161	0210	0270	19
32	0006	0009	0014	0021	0031	0044	0063	0088	0120	0160	18
33	0002	0003	0006	0009	0014	0021	0031	0044	0063	0087	17
34	0001	0001	0002	0003	0006	0009	0014	0020	0030	0044	16
35	0000	0000	0001	0001	0002	0003	0005	0009	0013	0020	15
36	0000	0000	0000	0000	0001	0001	0002	0003	0005	0006	14
37	0000	0000	0000	0000	0000	0000	0001	0001	0002	0003	13
38	0000	0000	0000	0000	0000	0000	0000	0000	0001	0001	12
	59	58	57	56	55	54	53	52	51	50	π *x*

APPENDIX TABLE A.2 Poisson Distribution

The Poisson mass function is defined by

$$P(x) = \frac{e^{-\lambda}\lambda^x}{x!}$$

The probability of x successes, for various values of λ, are defined by the corresponding values in the body of the table.

Examples If $\lambda = 1.6$, then $P(1) = 0.3230$, $P(5) = 0.0176$

This table gives the Poisson distribution for $\lambda = 0.1$ through $\lambda = 10$.

	λ									
x	0.1	0.2	0.3	0.4	0.5	0.6	0.7	0.8	0.9	1.0
0	.9048	.8187	.7408	.6703	.6065	.5488	.4966	.4493	.4066	.3679
1	.0905	.1637	.2222	.2681	.3033	.3293	.3476	.3595	.3659	.3679
2	.0045	.0164	.0333	.0536	.0758	.0988	.1217	.1438	.1647	.1839
3	.0002	.0011	.0033	.0072	.0126	.0198	.0284	.0383	.0494	.0613
4	.0000	.0001	.0002	.0007	.0016	.0030	.0050	.0077	.0111	.0153
5	.0000	.0000	.0000	.0001	.0002	.0004	.0007	.0012	.0020	.0031
6	.0000	.0000	.0000	.0000	.0000	.0000	.0001	.0002	.0003	.0005
7	.0000	.0000	.0000	.0000	.0000	.0000	.0000	.0000	.0000	.0001

	λ									
x	1.1	1.2	1.3	1.4	1.5	1.6	1.7	1.8	1.9	2.0
0	.3329	.3012	.2725	.2466	.2231	.2019	.1827	.1653	.1496	.1353
1	.3662	.3614	.3543	.3452	.3347	.3230	.3106	.2975	.2842	.2707
2	.2014	.2169	.2303	.2417	.2510	.2584	.2640	.2678	.2700	.2707
3	.0738	.0867	.0998	.1128	.1255	.1378	.1496	.1607	.1710	.1804
4	.0203	.0260	.0324	.0395	.0471	.0551	.0636	.0723	.0812	.0902
5	.0045	.0062	.0084	.0111	.0141	.0176	.0216	.0260	.0309	.0361
6	.0008	.0012	.0018	.0026	.0035	.0047	.0061	.0078	.0098	.0120
7	.0001	.0002	.0003	.0005	.0008	.0011	.0015	.0020	.0027	.0034
8	.0000	.0000	.0001	.0001	.0001	.0002	.0003	.0005	.0006	.0009
9	.0000	.0000	.0000	.0000	.0000	.0000	.0001	.0001	.0001	.0002

	λ									
x	2.1	2.2	2.3	2.4	2.5	2.6	2.7	2.8	2.9	3.0
0	.1225	.1108	.1003	.0907	.0821	.0743	.0672	.0608	.0550	.0498
1	.2572	.2438	.2306	.2177	.2052	.1931	.1815	.1703	.1596	.1494
2	.2700	.2681	.2652	.2613	.2565	.2510	.2450	.2384	.2314	.2240
3	.1890	.1966	.2033	.2090	.2138	.2176	.2205	.2225	.2237	.2240
4	.0992	.1082	.1169	.1254	.1336	.1414	.1488	.1557	.1622	.1680
5	.0417	.0476	.0538	.0602	.0668	.0735	.0804	.0872	.0940	.1008
6	.0146	.0174	.0206	.0241	.0278	.0319	.0362	.0407	.0455	.0504
7	.0044	.0055	.0068	.0083	.0099	.0118	.0139	.0163	.0188	.0216
8	.0011	.0015	.0019	.0025	.0031	.0038	.0047	.0057	.0068	.0081
9	.0003	.0004	.0005	.0007	.0009	.0011	.0014	.0018	.0022	.0027
10	.0001	.0001	.0001	.0002	.0002	.0003	.0004	.0005	.0006	.0008
11	.0000	.0000	.0000	.0000	.0000	.0001	.0001	.0001	.0002	.0002
12	.0000	.0000	.0000	.0000	.0000	.0000	.0000	.0000	.0000	.0001

	λ									
x	3.1	3.2	3.3	3.4	3.5	3.6	3.7	3.8	3.9	4.0
0	.0450	.0408	.0369	.0334	.0302	.0273	.0247	.0224	.0202	.0183
1	.1397	.1304	.1217	.1135	.1057	.0984	.0915	.0850	.0789	.0733
2	.2165	.2087	.2008	.1929	.1850	.1771	.1692	.1615	.1539	.1465
3	.2237	.2226	.2209	.2186	.2158	.2125	.2087	.2046	.2001	.1954
4	.1734	.1781	.1823	.1858	.1888	.1912	.1931	.1944	.1951	.1954
5	.1075	.1140	.1203	.1264	.1322	.1377	.1429	.1477	.1522	.1563
6	.0555	.0608	.0662	.0716	.0771	.0826	.0881	.0936	.0989	.1042
7	.0246	.0278	.0312	.0348	.0385	.0425	.0466	.0508	.0551	.0595
8	.0095	.0111	.0129	.0148	.0169	.0191	.0215	.0241	.0269	.0298
9	.0033	.0040	.0047	.0056	.0066	.0076	.0089	.0102	.0116	.0132
10	.0010	.0013	.0016	.0019	.0023	.0028	.0033	.0039	.0045	.0053
11	.0003	.0004	.0005	.0006	.0007	.0009	.0011	.0013	.0016	.0019
12	.0001	.0001	.0001	.0002	.0002	.0003	.0003	.0004	.0005	.0006
13	.0000	.0000	.0000	.0000	.0001	.0001	.0001	.0001	.0002	.0002
14	.0000	.0000	.0000	.0000	.0000	.0000	.0000	.0000	.0000	.0001

	λ									
x	4.1	4.2	4.3	4.4	4.5	4.6	4.7	4.8	4.9	5.0
0	.0166	.0150	.0136	.0123	.0111	.0101	.0091	.0082	.0074	.0067
1	.0679	.0630	.0583	.0540	.0500	.0462	.0427	.0395	.0365	.0337
2	.1393	.1323	.1254	.1188	.1125	.1063	.1005	.0948	.0894	.0842
3	.1904	.1852	.1798	.1743	.1687	.1631	.1574	.1517	.1460	.1404
4	.1951	.1944	.1933	.1917	.1898	.1875	.1849	.1820	.1789	.1755
5	.1600	.1633	.1662	.1687	.1708	.1725	.1738	.1747	.1753	.1755
6	.1093	.1143	.1191	.1237	.1281	.1323	.1362	.1398	.1432	.1462
7	.0640	.0686	.0732	.0778	.0824	.0869	.0914	.0959	.1002	.1044
8	.0328	.0360	.0393	.0428	.0463	.0500	.0537	.0575	.0614	.0653
9	.0150	.0168	.0188	.0209	.0232	.0255	.0280	.0307	.0334	.0363
10	.0061	.0071	.0081	.0092	.0104	.0118	.0132	.0147	.0164	.0181
11	.0023	.0027	.0032	.0037	.0043	.0049	.0056	.0064	.0073	.0082
12	.0008	.0009	.0011	.0014	.0016	.0019	.0022	.0026	.0030	.0034
13	.0002	.0003	.0004	.0005	.0006	.0007	.0008	.0009	.0011	.0013
14	.0001	.0001	.0001	.0001	.0002	.0002	.0003	.0003	.0004	.0005
15	.0000	.0000	.0000	.0000	.0001	.0001	.0001	.0001	.0001	.0002

	λ									
x	5.1	5.2	5.3	5.4	5.5	5.6	5.7	5.8	5.9	6.0
0	.0061	.0055	.0050	.0045	.0041	.0037	.0033	.0030	.0027	.0025
1	.0311	.0287	.0265	.0244	.0225	.0207	.0191	.0176	.0162	.0149
2	.0793	.0746	.0701	.0659	.0618	.0580	.0544	.0509	.0477	.0446
3	.1348	.1293	.1239	.1185	.1133	.1082	.1033	.0985	.0938	.0892
4	.1719	.1681	.1641	.1600	.1558	.1515	.1472	.1428	.1383	.1339
5	.1753	.1748	.1740	.1728	.1714	.1697	.1678	.1620	.1632	.1606
6	.1490	.1515	.1537	.1555	.1571	.1584	.1594	.1656	.1605	.1606
7	.1086	.1125	.1163	.1200	.1234	.1267	.1298	.1301	.1353	.1377
8	.0692	.0731	.0771	.0810	.0849	.0887	.0925	.0926	.0998	.1033
9	.0392	.0423	.0454	.0486	.0519	.0552	.0586	.0662	.0654	.0688
10	.0200	.0220	.0241	.0262	.0285	.0309	.0334	.0359	.0386	.0413
11	.0093	.0104	.0116	.0129	.0143	.0157	.0173	.0190	.0207	.0225
12	.0039	.0045	.0051	.0058	.0065	.0073	.0082	.0092	.0102	.0113
13	.0015	.0018	.0021	.0024	.0028	.0032	.0036	.0041	.0046	.0052
14	.0006	.0007	.0008	.0009	.0011	.0013	.0015	.0017	.0019	.0022
15	.0002	.0002	.0003	.0003	.0004	.0005	.0006	.0007	.0008	.0009
16	.0001	.0001	.0001	.0001	.0001	.0002	.0002	.0002	.0003	.0003
17	.0000	.0000	.0000	.0000	.0000	.0001	.0001	.0001	.0001	.0001

x	6.1	6.2	6.3	6.4	6.5	6.6	6.7	6.8	6.9	7.0
0	.0022	.0020	.0018	.0017	.0015	.0014	.0012	.0011	.0010	.0009
1	.0137	.0126	.0116	.0106	.0098	.0090	.0082	.0076	.0070	.0064
2	.0417	.0390	.0364	.0340	.0318	.0296	.0276	.0258	.0240	.0223
3	.0848	.0806	.0765	.0726	.0688	.0652	.0617	.0584	.0552	.0521
4	.1294	.1249	.1205	.1162	.1118	.1076	.1034	.0992	.0952	.0912
5	.1579	.1549	.1519	.1487	.1454	.1420	.1385	.1349	.1314	.1277
6	.1605	.1601	.1595	.1586	.1575	.1562	.1546	.1529	.1511	.1490
7	.1399	.1418	.1435	.1450	.1462	.1472	.1480	.1486	.1489	.1490
8	.1066	.1099	.1130	.1160	.1188	.1215	.1240	.1263	.1284	.1304
9	.0723	.0757	.0791	.0825	.0858	.0891	.0923	.0954	.0985	.1014
10	.0441	.0469	.0498	.0528	.0558	.0588	.0618	.0649	.0679	.0710
11	.0245	.0265	.0285	.0307	.0330	.0353	.0377	.0401	.0426	.0452
12	.0124	.0137	.0150	.0164	.0179	.0194	.0210	.0227	.0245	.0264
13	.0058	.0065	.0073	.0081	.0089	.0098	.0108	.0119	.0130	.0142
14	.0025	.0029	.0033	.0037	.0041	.0046	.0052	.0058	.0064	.0071
15	.0010	.0012	.0014	.0016	.0018	.0020	.0023	.0026	.0029	0033
16	.0004	.0005	.0005	.0006	.0007	.0008	.0010	.0011	.0013	.0014
17	.0001	.0002	.0002	.0002	.0003	.0003	.0004	.0004	.0005	.0006
18	.0000	.0001	.0001	.0001	.0001	.0001	.0001	.0002	.0002	.0002
19	.0000	.0000	.0000	.0000	.0000	.0000	.0000	.0001	.0001	.0001

x	7.1	7.2	7.3	7.4	7.5	7.6	7.7	7.8	7.9	8.0
0	.0008	.0007	.0007	.0006	.0006	.0005	.0005	.0004	.0004	.0003
1	.0059	.0054	.0049	.0045	.0041	.0038	.0035	.0032	.0029	.0027
2	.0208	.0194	.0180	.0167	.0156	.0145	.0134	.0125	.0116	.0107
3	.0492	.0464	.0438	.0413	.0389	.0366	.0345	.0324	.0305	.0286
4	.0874	.0836	.0799	.0764	.0729	.0696	.0663	.0632	.0602	.0573
5	.1241	.1204	.1167	.1130	.1094	.1057	.1021	.0986	.0951	.0916
6	.1468	.1445	.1420	.1394	.1367	.1339	.1311	.1282	.1252	.1221
7	.1489	.1486	.1481	.1474	.1465	.1454	.1442	.1428	.1413	.1396
8	.1321	.1337	.1351	.1363	.1373	.1382	.1388	.1392	.1395	.1396
9	.1042	.1070	.1096	.1121	.1144	.1167	.1187	.1207	.1224	.1241
10	.0740	.0770	.0800	.0829	.0858	.0887	.0914	.0941	.0967	.0993
11	.0478	.0504	.0531	.0558	.0585	.0613	.0640	.0667	.0695	.0722
12	.0283	.0303	.0323	.0344	.0366	.0388	.0411	.0434	.0457	.0481
13	.0154	.0168	.0181	.0196	.0211	.0227	.0243	.0260	.0278	.0296
14	.0078	.0086	.0095	.0104	.0113	.0123	.0134	.0145	.0157	.0169
15	.0037	.0041	.0046	.0051	.0057	.0062	.0069	.0075	.0083	.0090
16	.0016	.0019	.0021	.0024	.0026	.0030	.0033	.0037	.0041	.0045
17	.0007	.0008	.0009	.0010	.0012	.0013	.0015	.0017	.0119	.0021
18	.0003	.0003	.0004	.0004	.0005	.0006	.0006	.0007	.0008	.0009
19	.0001	.0001	.0001	.0002	.0002	.0002	.0003	.0003	.0003	.0004
20	.0000	.0000	.0001	.0001	.0001	.0001	.0001	.0001	.0001	.0002
21	.0000	.0000	.0000	.0000	.0000	.0000	.0000	.0000	.0001	.0001

					λ					
x	8.1	8.2	8.3	8.4	8.5	8.6	8.7	8.8	8.9	9.0
0	.0003	.0003	.0002	.0002	.0002	.0002	.0002	.0002	.0001	.0001
1	.0025	.0023	.0021	.0019	.0017	.0016	.0014	.0013	.0012	.0011
2	.0100	.0092	.0086	.0079	.0074	.0068	.0063	.0058	.0054	.0050
3	.0269	.0252	.0237	.0222	.0208	.0195	.0183	.0171	.0160	.0150
4	.0544	.0517	.0491	.0466	.0443	.0420	.0398	.0377	.0357	.0337
5	.0882	.0849	.0816	.0784	.0752	.0722	.0692	.0663	.0635	.0607
6	.1191	.1160	.1128	.1097	.1066	.1034	.1003	.0972	.0941	.0911
7	.1378	.1358	.1338	.1317	.1294	.1271	.1247	.1222	.1197	.1171
8	.1395	.1392	.1388	.1382	.1375	.1366	.1356	.1344	.1332	.1318
9	.1256	.1269	.1280	.1290	.1299	.1306	.1311	.1315	.1317	.1318
10	.1017	.1040	.1063	.1084	.1104	.1123	.1140	.1157	.1172	.1186
11	.0749	.0776	.0802	.0828	.0853	.0878	.0902	.0925	.0948	.0970
12	.0505	.0530	.0555	.0579	.0604	.0629	.0654	.0679	.0703	.0728
13	.0315	.0334	.0354	.0374	.0395	.0416	.0438	.0459	.0481	.0504
14	.0182	.0196	.0210	.0225	.0240	.0256	.0272	.0289	.0306	.0324
15	.0098	.0107	.0116	.0126	.0136	.0147	.0158	.0169	.0182	.0194
16	.0050	.0055	.0060	.0066	.0072	.0079	.0086	.0093	.0101	.0109
17	.0024	.0026	.0029	.0033	.0036	.0040	.0044	.0048	.0053	.0058
18	.0011	.0012	.0014	.0015	.0017	.0019	.0021	.0024	.0026	.0029
19	.0005	.0005	.0006	.0007	.0008	.0009	.0010	.0011	.0012	.0014
20	.0002	.0002	.0002	.0003	.0003	.0004	.0004	.0005	.0005	.0006
21	.0001	.0001	.0001	.0001	.0001	.0002	.0002	.0002	.0002	.0003
22	.0000	.0000	.0000	.0000	.0001	.0001	.0001	.0001	.0001	.0001

	λ									
x	9.1	9.2	9.3	9.4	9.5	9.6	9.7	9.8	9.9	10
0	.0001	.0001	.0001	.0001	.0001	.0001	.0001	.0001	.0001	.0000
1	.0010	.0009	.0009	.0008	.0007	.0007	.0006	.0005	.0005	.0005
2	.0046	.0043	.0040	.0037	.0034	.0031	.0029	.0027	.0025	.0023
3	.0140	.0131	.0123	.0115	.0107	.0100	.0093	.0087	.0081	.0076
4	.0319	.0302	.0285	.0269	.0254	.0240	.0226	.0213	.0201	.0189
5	.0581	.0555	.0530	.0506	.0483	.0460	.0439	.0418	.0398	.0378
6	.0881	.0851	.0822	.0793	.0764	.0736	.0709	.0682	.0656	.0631
7	.1145	.1118	.1091	.1064	.1037	.1010	.0982	.0955	.0928	.0901
8	.1302	.1286	.1269	.1251	.1232	.1212	.1191	.1170	.1148	.1126
9	.1317	.1315	.1311	.1306	.1300	.1293	.1284	.1274	.1263	.1251
10	.1198	.1210	.1219	.1228	.1235	.1241	.1245	.1249	.1250	.1251
11	.0991	.1012	.1031	.1049	.1067	.1083	.1098	.1112	.1125	.1137
12	.0752	.0776	.0799	.0822	.0844	.0866	.0888	.0908	.0928	.0948
13	.0526	.0549	.0572	.0594	.0617	.0640	.0662	.0685	.0707	.0729
14	.0342	.0361	.0380	.0399	.0419	.0439	.0459	.0479	.0500	.0521
15	.0208	.0221	.0235	.0250	.0265	.0281	.0297	.0313	.0330	.0347
16	.0118	.0127	.0137	.0147	.0157	.0168	.0180	.0192	.0204	.0217
17	.0063	.0069	.0075	.0081	.0088	.0095	.0103	.0111	.0119	.0128
18	.0032	.0035	.0039	.0042	.0046	.0051	.0055	.0060	.0065	.0071
19	.0015	.0017	.0019	.0021	.0023	.0026	.0028	.0031	.0034	.0037
20	.0007	.0008	.0009	.0010	.0011	.0012	.0014	.0015	.0017	.0019
21	.0003	.0003	.0004	.0004	.0005	.0006	.0006	.0007	.0008	.0009
22	.0001	.0001	.0002	.0002	.0002	.0002	.0003	.0003	.0004	.0004
23	.0000	.0001	.0001	.0001	.0001	.0001	.0001	.0001	.0002	.0002
24	.0000	.0000	.0000	.0000	.0000	.0000	.0000	.0001	.0001	.0001

Appendix Table A.3 Standard Normal Distribution

The cumulative standardized normal distribution $F(z)$ is defined by

Example $P(z \leq 1.75) = F(1.75) = 0.9599$ $P(z \geq 1.75) = 0.0401$

$$F(z) = \int_{-\infty}^{z} \frac{1}{\sqrt{2\pi}} e^{-z^2/2} dz$$

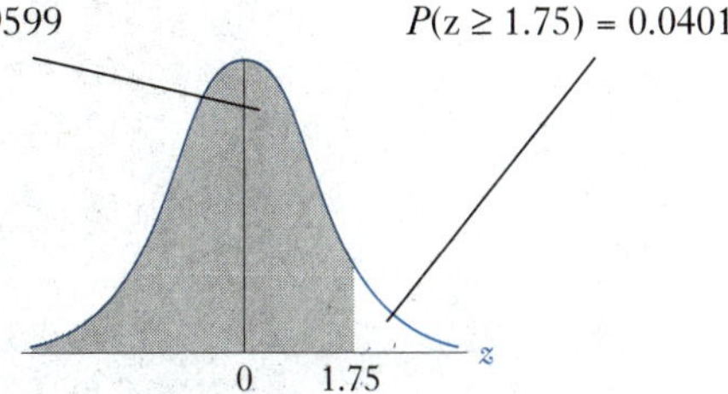

z	.00	.01	.02	.03	.04	.05	.06	.07	.08	.09
.0	.5000	.5040	.5080	.5120	.5160	.5199	.5239	.5279	.5319	.5359
.1	.5398	.5438	.5478	.5517	.5557	.5596	.5636	.5675	.5714	.5753
.2	.5793	.5832	.5871	.5910	.5948	.5987	.6026	.6064	.6103	.6141
.3	.6179	.6217	.6255	.6293	.6331	.6368	.6406	.6443	.6480	.6517
.4	.6554	.6591	.6628	.6664	.6700	.6736	.6772	.6808	.6844	.6879
.5	.6915	.6950	.6985	.7019	.7054	.7088	.7123	.7157	.7190	.7224
.6	.7257	.7291	.7324	.7357	.7389	.7422	.7454	.7486	.7517	.7549
.7	.7580	.7611	.7642	.7673	.7704	.7734	.7764	.7794	.7823	.7852
.8	.7881	.7910	.7939	.7967	.7995	.8023	.8051	.8078	.8106	.8133
.9	.8159	.8186	.8212	.8238	.8264	.8289	.8315	.8340	.8365	.8389
1.0	.8413	.8438	.8461	.8485	.8508	.8531	.8554	.8577	.8599	.8621
1.1	.8643	.8665	.8686	.8708	.8729	.8749	.8770	.8790	.8810	.8830
1.2	.8849	.8869	.8888	.8907	.8925	.8944	.8962	.8980	.8997	.9015
1.3	.9032	.9049	.9066	.9082	.9099	.9115	.9131	.9147	.9162	.9177
1.4	.9192	.9207	.9222	.9236	.9251	.9265	.9279	.9292	.9306	.9319
1.5	.9332	.9345	.9357	.9370	.9382	.9394	.9406	.9418	.9429	.9441
1.6	.9452	.9463	.9474	.9484	.9495	.9505	.9515	.9525	.9535	.9545
1.7	.9554	.9564	.9573	.9582	.9591	.9599	.9608	.9616	.9625	.9633
1.8	.9641	.9649	.9656	.9664	.9671	.9678	.9686	.9693	.9699	.9706
1.9	.9713	.9719	.9726	.9732	.9738	.9744	.9750	.9756	.9761	.9767
2.0	.9772	.9778	.9783	.9788	.9793	.9798	.9803	.9808	.9812	.9817
2.1	.9821	.9826	.9830	.9834	.9838	.9842	.9846	.9850	.9854	.9857
2.2	.9861	.9864	.9868	.9871	.9875	.9878	.9881	.9884	.9887	.9890
2.3	.9893	.9896	.9898	.9901	.9904	.9906	.9909	.9911	.9913	.9916
2.4	.9918	.9920	.9922	.9925	.9927	.9929	.9931	.9932	.9934	.9936
2.5	.9938	.9940	.9941	.9943	.9945	.9946	.9948	.9949	.9951	.9952
2.6	.9953	.9955	.9956	.9957	.9959	.9960	.9961	.9962	.9963	.9964
2.7	.9965	.9966	.9967	.9968	.9969	.9970	.9971	.9972	.9973	.9974
2.8	.9974	.9975	.9976	.9977	.9977	.9978	.9979	.9979	.9980	.9981
2.9	.9981	.9982	.9982	.9983	.9984	.9984	.9985	.9985	.9986	.9986
3.0	.9987	.9987	.9987	.9988	.9988	.9989	.9989	.9989	.9990	.9990
3.1	.9990	.9991	.9991	.9991	.9992	.9992	.9992	.9992	.9993	.9993
3.2	.9993	.9993	.9994	.9994	.9994	.9994	.9994	.9995	.9995	.9995
3.3	.9995	.9995	.9995	.9996	.9996	.9996	.9996	.9996	.9996	.9997
3.4	.9997	.9997	.9997	.9997	.9997	.9997	.9997	.9997	.9997	.9998

APPENDIX TABLE A.4 *t* Distribution

The cumulative *t* distribution is defined by

$$F(t) = \int_{-\infty}^{t} \frac{\left(\frac{v-1}{2}\right)!}{\left(\frac{v-2}{2}\right)!\sqrt{\pi n}\left(1+\frac{t^2}{v}\right)^{(v+1)/2}} dt$$

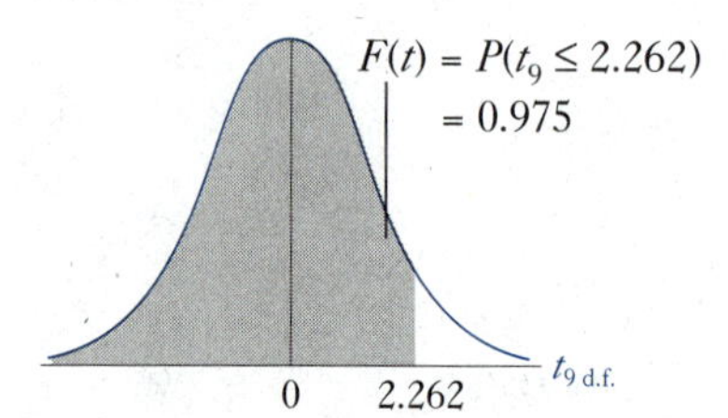

Example $n = 10$, df = 9

$F(t)$	.75	.90	.95	.975	.99	.995	.9995
1	1.000	3.078	6.314	12.706	31.821	63.657	636.619
2	.816	1.886	2.920	4.303	6.965	9.925	31.598
3	.765	1.638	2.353	3.182	4.541	5.841	12.941
4	.741	1.533	2.132	2.776	3.747	4.604	8.610
5	.727	1.476	2.015	2.571	3.365	4.032	6.859
6	.718	1.440	1.943	2.447	3.143	3.707	5.959
7	.711	1.415	1.895	2.365	2.998	3.499	5.405
8	.706	1.397	1.860	2.306	2.896	3.355	5.041
9	.703	1.383	1.833	2.262	2.821	3.250	4.781
10	.700	1.372	1.812	2.228	2.764	3.169	4.587
11	.697	1.363	1.796	2.201	2.718	3.106	4.437
12	.695	1.356	1.782	2.179	2.681	3.055	4.318
13	.694	1.350	1.771	2.160	2.650	3.012	4.221
14	.692	1.345	1.761	2.145	2.624	2.977	4.140
15	.691	1.341	1.753	2.131	2.602	2.947	4.073
16	.690	1.337	1.746	2.120	2.583	2.921	4.015
17	.689	1.333	1.740	2.110	2.567	2.898	3.965
18	.688	1.330	1.734	2.101	2.552	2.878	3.922
19	.688	1.328	1.729	2.093	2.539	2.861	3.883
20	.687	1.325	1.725	2.086	2.528	2.845	3.850
21	.686	1.323	1.721	2.080	2.518	2.831	3.819
22	.686	1.321	1.717	2.074	2.508	2.819	3.792
23	.685	1.319	1.714	2.069	2.500	2.807	3.767
24	.685	1.318	1.711	2.064	2.492	2.797	3.745
25	.684	1.316	1.708	2.060	2.485	2.787	3.725
26	.684	1.315	1.706	2.056	2.479	2.779	3.707
27	.684	1.314	1.703	2.052	2.473	2.771	3.690
28	.683	1.313	1.701	2.048	2.467	2.763	3.674
29	.683	1.311	1.699	2.045	2.462	2.756	3.659
30	.683	1.310	1.697	2.042	2.457	2.750	3.646
40	.681	1.303	1.684	2.021	2.423	2.704	3.551
60	.679	1.296	1.671	2.000	2.390	2.660	3.460
120	.677	1.289	1.658	1.980	2.358	2.617	3.373
$\infty(z_\alpha)$	.674	1.282	1.645	1.960	2.326	2.576	3.291

Appendix Table A.5 Upper Percentage Points of the *F* Distribution

Example

P ($F > 1.59$) = 0.25
P ($F > 2.42$) = 0.10
P ($F > 3.14$) = 0.05
P ($F > 5.26$) = 0.01
for df N_1 = 10
and N_2 = 9

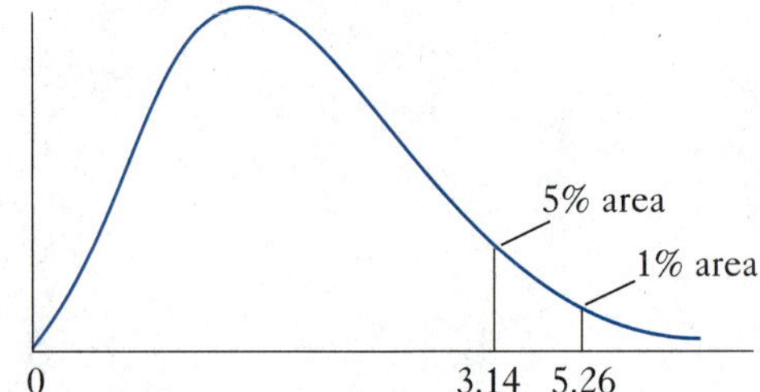

df for numerator N_1

df for denominator N_2	P	1	2	3	4	5	6	7	8	9	10	11	12
	.25	5.83	7.50	8.20	8.58	8.82	8.98	9.10	9.19	9.26	9.32	9.36	9.41
1	.10	39.9	49.5	53.6	55.8	57.2	58.2	58.9	59.4	59.9	60.2	60.5	60.7
	.05	161	200	216	225	230	234	237	239	241	242	243	244
	.25	2.57	3.00	3.15	3.23	3.28	3.31	3.34	3.35	3.37	3.38	3.39	3.39
2	.10	8.53	9.00	9.16	9.24	9.29	9.33	9.35	9.37	9.38	9.39	9.40	9.41
	.05	18.5	19.0	19.2	19.2	19.3	19.3	19.4	19.4	19.4	19.4	19.4	19.4
	.01	98.5	99.0	99.2	99.2	99.3	99.3	99.4	99.4	99.4	99.4	99.4	99.4
	.25	2.02	2.28	2.36	2.39	2.41	2.42	2.43	2.44	2.44	2.44	2.45	2.45
3	.10	5.54	5.46	5.39	5.34	5.31	5.28	5.27	5.25	5.24	5.23	5.22	5.22
	.05	10.1	9.55	9.28	9.12	9.01	8.94	8.89	8.85	8.81	8.79	8.76	8.74
	.01	34.1	30.8	29.5	28.7	28.2	27.9	27.7	27.5	27.3	27.2	27.1	27.1
	.25	1.81	2.00	2.05	2.06	2.07	2.08	2.08	2.08	2.08	2.08	2.08	2.08
4	.10	4.54	4.32	4.19	4.11	4.05	4.01	3.98	3.95	3.94	3.92	3.91	3.90
	.05	7.71	6.94	6.59	6.39	6.26	6.16	6.09	6.04	6.00	5.96	5.94	5.91
	.01	21.2	18.0	16.7	16.0	15.5	15.2	15.0	14.8	14.7	14.5	14.4	14.4
	.25	1.69	1.85	1.88	1.89	1.89	1.89	1.89	1.89	1.89	1.89	1.89	1.89
5	.10	4.06	3.78	3.62	3.52	3.45	3.40	3.37	3.34	3.32	3.30	3.28	3.27
	.05	6.61	5.79	5.41	5.19	5.05	4.95	4.88	4.82	4.77	4.74	4.71	4.68
	.01	16.3	13.3	12.1	11.4	11.0	10.7	10.5	10.3	10.2	10.1	9.96	9.89
	.25	1.62	1.76	1.78	1.79	1.79	1.78	1.78	1.78	1.77	1.77	1.77	1.77
6	.10	3.78	3.46	3.29	3.18	3.11	3.05	3.01	2.98	2.96	2.94	2.92	2.90
	.05	5.99	5.14	4.76	4.53	4.39	4.28	4.21	4.15	4.10	4.06	4.03	4.00
	.01	13.7	10.9	9.78	9.15	8.75	8.47	8.26	8.10	7.98	7.87	7.79	7.72
	.25	1.57	1.70	1.72	1.72	1.71	1.71	1.70	1.70	1.69	1.69	1.69	1.68
7	.10	3.59	3.26	3.07	2.96	2.88	2.83	2.78	2.75	2.72	2.70	2.68	2.67
	.05	5.59	4.74	4.35	4.12	3.97	3.87	3.79	3.73	3.68	3.64	3.60	3.57
	.01	12.2	9.55	8.45	7.85	7.46	7.19	6.99	6.84	6.72	6.62	6.54	6.47
	.25	1.54	1.66	1.67	1.66	1.66	1.65	1.64	1.64	1.63	1.63	1.63	1.62
8	.10	3.46	3.11	2.92	2.81	2.73	2.67	2.62	2.59	2.56	2.54	2.52	2.50
	.05	5.32	4.46	4.07	3.84	3.69	3.58	3.50	3.44	3.39	3.35	3.31	3.28
	.01	11.3	8.65	7.59	7.01	6.63	6.37	6.18	6.03	5.91	5.81	5.73	5.67
	.25	1.51	1.62	1.63	1.63	1.62	1.61	1.60	1.60	1.59	1.59	1.58	1.58
9	.10	3.36	3.01	2.81	2.69	2.61	2.55	2.51	2.47	2.44	2.42	2.40	2.38
	.05	5.12	4.26	3.86	3.63	3.48	3.37	3.29	3.23	3.18	3.14	3.10	3.07
	.01	10.6	8.02	6.99	6.42	6.06	5.80	5.61	5.47	5.35	5.26	5.18	5.11

df for numerator N_1													df for denominator
15	20	24	30	40	50	60	100	120	200	500	∞	P	N_2
9.49	9.58	9.63	9.67	9.71	9.74	9.76	9.78	9.80	9.82	9.84	9.85	.25	
61.2	61.7	62.0	62.3	62.5	62.7	62.8	63.0	63.1	63.2	63.3	63.3	.10	1
246	248	249	250	251	252	252	253	253	254	254	254	.05	
3.41	3.43	3.43	3.44	3.45	3.45	3.46	3.47	3.47	3.48	3.48	3.48	.25	
9.42	9.44	9.45	9.46	9.47	9.47	9.47	9.48	9.48	9.49	9.49	9.49	.10	2
19.4	19.4	19.5	19.5	19.5	19.5	19.5	19.5	19.5	19.5	19.5	19.5	.05	
99.4	99.4	99.5	99.5	99.5	99.5	99.5	99.5	99.5	99.5	99.5	99.5	.01	
2.46	2.46	2.46	2.47	2.47	2.47	2.47	2.47	2.47	2.47	2.47	2.47	.25	
5.20	5.18	5.18	5.17	5.16	5.15	5.15	5.14	5.14	5.14	5.14	5.13	.10	3
8.70	8.66	8.64	8.62	8.59	8.58	8.57	8.55	8.55	8.54	8.53	8.53	.05	
26.9	26.7	26.6	26.5	26.4	26.4	26.3	26.2	26.2	26.2	26.1	26.1	.01	
2.08	2.08	2.08	2.08	2.08	2.08	2.08	2.08	2.08	2.08	2.08	2.08	.25	
3.87	3.84	3.83	3.82	3.80	3.80	3.79	3.78	3.78	3.77	3.76	3.76	.10	4
5.86	5.80	5.77	5.75	5.72	5.70	5.69	5.66	5.66	5.65	5.64	5.63	.05	
14.2	14.0	13.9	13.8	13.7	13.7	13.7	13.6	13.6	13.5	13.5	13.5	.01	
1.89	1.88	1.88	1.88	1.88	1.88	1.87	1.87	1.87	1.87	1.87	1.87	.25	
3.24	3.21	3.19	3.17	3.16	3.15	3.14	3.13	3.12	3.12	3.11	3.10	.10	5
4.62	4.56	4.53	4.50	4.46	4.44	4.43	4.41	4.40	4.39	4.37	4.36	.05	
9.72	9.55	9.47	9.38	9.29	9.24	9.20	9.13	9.11	9.08	9.04	9.02	.01	
1.76	1.76	1.75	1.75	1.75	1.75	1.74	1.74	1.74	1.74	1.74	1.74	.25	
2.87	2.84	2.82	2.80	2.78	2.77	2.76	2.75	2.74	2.73	2.73	2.72	.10	6
3.94	3.87	3.84	3.81	3.77	3.75	3.74	3.71	3.70	3.69	3.68	3.67	.05	
7.56	7.40	7.31	7.23	7.14	7.09	7.06	6.99	6.97	6.93	6.90	6.88	.01	
1.68	1.67	1.67	1.66	1.66	1.66	1.65	1.65	1.65	1.65	1.65	1.65	.25	
2.63	2.59	2.58	2.56	2.54	2.52	2.51	2.50	2.49	2.48	2.48	2.47	.10	7
3.51	3.44	3.41	3.38	3.34	3.32	3.30	3.27	3.27	3.25	3.24	3.23	.05	
6.31	6.16	6.07	5.99	5.91	5.86	5.82	5.75	5.74	5.70	5.67	5.65	.01	
1.62	1.61	1.60	1.60	1.59	1.59	1.59	1.58	1.58	1.58	1.58	1.58	.25	
2.46	2.42	2.40	2.38	2.36	2.35	2.34	2.32	2.32	2.31	2.30	2.29	.10	8
3.22	3.15	3.12	3.08	3.04	2.02	3.01	2.97	2.97	2.95	2.94	2.93	.05	
5.52	5.36	5.28	5.20	5.12	5.07	5.03	4.96	4.95	4.91	4.88	4.86	.01	
1.57	1.56	1.56	1.55	1.55	1.54	1.54	1.53	1.53	1.53	1.53	1.53	.25	
2.34	2.30	2.28	2.25	2.23	2.22	2.21	2.19	2.18	2.17	2.17	2.16	.10	9
3.01	2.94	2.90	2.86	2.83	2.80	2.79	2.76	2.75	2.73	2.72	2.71	.05	
4.96	4.81	4.73	4.65	4.57	4.52	4.48	4.42	4.40	4.36	4.33	4.31	.01	

df for denominator N_2	P	df for numerator N_1: 1	2	3	4	5	6	7	8	9	10	11	12
10	.25	1.49	1.60	1.60	1.59	1.59	1.58	1.57	1.56	1.56	1.55	1.55	1.54
	.10	3.29	2.92	2.73	2.61	2.52	2.46	2.41	2.38	2.35	2.32	2.30	2.28
	.05	4.96	4.10	3.71	3.48	3.33	3.22	3.14	3.07	3.02	2.98	2.94	2.91
	.01	10.0	7.56	6.55	5.99	5.64	5.39	5.20	5.06	4.94	4.85	4.77	4.71
11	.25	1.47	1.58	1.58	1.57	1.56	1.55	1.54	1.53	1.53	1.52	1.52	1.51
	.10	3.23	2.86	2.66	2.54	2.45	2.39	2.34	2.30	2.27	2.25	2.23	2.21
	.05	4.84	3.98	3.59	3.36	3.20	3.09	3.01	2.95	2.90	2.85	2.82	2.79
	.01	9.65	7.21	6.22	5.67	5.32	5.07	4.89	4.74	4.63	4.54	4.46	4.40
12	.25	1.46	1.56	1.56	1.55	1.54	1.53	1.52	1.51	1.51	1.50	1.50	1.49
	.10	3.18	2.81	2.61	2.48	2.39	2.33	2.28	2.24	2.21	2.19	2.17	2.15
	.05	4.75	3.89	3.49	3.26	3.11	3.00	2.91	2.85	2.80	2.75	2.72	2.69
	.01	9.33	6.93	5.95	5.41	5.06	4.82	4.64	4.50	4.39	4.30	4.22	4.16
13	.25	1.45	1.55	1.55	1.53	1.52	1.51	1.50	1.49	1.49	1.48	1.47	1.47
	.10	3.14	2.76	2.56	2.43	2.35	2.28	2.23	2.20	2.16	2.14	2.12	2.10
	.05	4.67	3.81	3.41	3.18	3.03	2.92	2.83	2.77	2.71	2.67	2.63	2.60
	.01	9.07	6.70	5.74	5.21	4.86	4.62	4.44	4.30	4.19	4.10	4.02	3.96
14	.25	1.44	1.53	1.53	1.52	1.51	1.50	1.49	1.48	1.47	1.46	1.46	1.45
	.10	3.10	2.73	2.52	2.39	2.31	2.24	2.19	2.15	2.12	2.10	2.08	2.05
	.05	4.60	3.74	3.34	3.11	2.96	2.85	2.76	2.70	2.65	2.60	2.57	2.53
	.01	8.86	6.51	5.56	5.04	4.69	4.46	4.28	4.14	4.03	3.94	3.86	3.80
15	.25	1.43	1.52	1.52	1.51	1.49	1.48	1.47	1.46	1.46	1.45	1.44	1.44
	.10	3.07	2.70	2.49	2.36	2.27	2.21	2.16	2.12	2.09	2.06	2.04	2.02
	.05	4.54	3.68	3.29	3.06	2.90	2.79	2.71	2.64	2.59	2.54	2.51	2.48
	.01	8.68	6.36	5.42	4.89	4.56	4.32	4.14	4.00	3.89	3.80	3.73	3.67
16	.25	1.42	1.51	1.51	1.50	1.48	1.47	1.46	1.45	1.44	1.44	1.44	1.43
	.10	3.05	2.67	2.46	2.33	2.24	2.18	2.13	2.09	2.06	2.03	2.01	1.99
	.05	4.49	3.63	3.24	3.01	2.85	2.74	2.66	2.59	2.54	2.49	2.46	2.42
	.01	8.53	6.23	5.29	4.77	4.44	4.20	4.03	3.89	3.78	3.69	3.62	3.55
17	.25	1.42	1.51	1.50	1.49	1.47	1.46	1.45	1.44	1.43	1.43	1.42	1.41
	.10	3.03	2.64	2.44	2.31	2.22	2.15	2.10	2.06	2.03	2.00	1.98	1.96
	.05	4.45	3.59	3.20	2.96	2.81	2.70	2.61	2.55	2.49	2.45	2.41	2.38
	.01	8.40	6.11	5.18	4.67	4.34	4.10	3.93	3.79	3.68	3.59	3.52	3.46
18	.25	1.41	1.50	1.49	1.48	1.46	1.45	1.44	1.43	1.42	1.42	1.41	1.40
	.10	3.01	2.62	2.42	2.29	2.20	2.13	2.08	2.04	2.00	1.98	1.96	1.93
	.05	4.41	3.55	3.16	2.93	2.77	2.66	2.58	2.51	2.46	2.41	2.37	2.34
	.01	8.29	6.01	5.09	4.58	4.25	4.01	3.84	3.71	3.60	3.51	3.43	3.37
19	.25	1.41	1.49	1.49	1.47	1.46	1.44	1.43	1.42	1.41	1.41	1.40	1.40
	.10	2.99	2.61	2.40	2.27	2.18	2.11	2.06	2.02	1.98	1.96	1.94	1.91
	.05	4.38	3.52	3.13	2.90	2.74	2.63	2.54	2.48	2.42	2.38	2.34	2.31
	.01	8.18	5.93	5.01	4.50	4.17	3.94	3.77	3.63	3.52	3.43	3.36	3.30
20	.25	1.40	1.49	1.48	1.46	1.45	1.44	1.43	1.42	1.41	1.40	1.39	1.39
	.10	2.97	2.59	2.38	2.25	2.16	2.09	2.04	2.00	1.96	1.94	1.92	1.89
	.05	4.35	3.49	3.10	2.87	2.71	2.60	2.51	2.45	2.39	2.35	2.31	2.28
	.01	8.10	5.85	4.94	4.43	4.10	3.87	3.70	3.56	3.46	3.37	3.29	3.23

df for numerator N_1													df for denominator
15	20	24	30	40	50	60	100	120	200	500	∞	P	N_2
1.53	1.52	1.52	1.51	1.51	1.50	1.50	1.49	1.49	1.49	1.48	1.48	.25	
2.24	2.20	2.18	2.16	2.13	2.12	2.11	2.09	2.08	2.07	2.06	2.06	.10	10
2.85	2.77	2.74	2.70	2.66	2.64	2.62	2.59	2.58	2.56	2.55	2.54	.05	
4.56	4.41	4.33	4.25	4.17	4.12	4.08	4.01	4.00	3.96	3.93	3.91	.01	
1.50	1.49	1.49	1.48	1.47	1.47	1.47	1.46	1.46	1.46	1.45	1.45	.25	
2.17	2.12	2.10	2.08	2.05	2.04	2.03	2.00	2.00	1.99	1.98	1.97	.10	11
2.72	2.65	2.61	2.57	2.53	2.51	2.49	2.46	2.45	2.43	2.42	2.40	.05	
4.25	4.10	4.02	3.94	3.86	3.81	3.78	3.71	3.69	3.66	3.62	3.60	.01	
1.48	1.47	1.46	1.45	1.45	1.44	1.44	1.43	1.43	1.43	1.42	1.42	.25	
2.10	2.06	2.04	2.01	1.99	1.97	1.96	1.94	1.93	1.92	1.91	1.90	.10	12
2.62	2.54	2.51	2.47	2.43	2.40	2.38	2.35	2.34	2.32	2.31	2.30	.05	
4.01	3.86	3.78	3.70	3.62	3.57	3.54	3.47	3.45	3.41	3.38	3.36	.01	
1.46	1.45	1.44	1.43	1.42	1.42	1.42	1.41	1.41	1.40	1.40	1.40	.25	
2.05	2.01	1.98	1.96	1.93	1.92	1.90	1.88	1.88	1.86	1.85	1.85	.10	13
2.53	2.46	2.42	2.38	2.34	2.31	2.30	2.26	2.25	2.23	2.22	2.21	.05	
3.82	3.66	3.59	3.51	3.43	3.38	3.34	3.27	3.25	3.22	3.19	3.17	.01	
1.44	1.43	1.42	1.41	1.41	1.40	1.40	1.39	1.39	1.39	1.38	1.38	.25	
2.01	1.96	1.94	1.91	1.89	1.87	1.86	1.83	1.83	1.82	1.80	1.80	.10	14
2.46	2.39	2.35	2.31	2.27	2.24	2.22	2.19	2.18	2.16	2.14	2.13	.05	
3.66	3.51	3.43	3.35	3.27	3.22	3.18	3.11	3.09	3.06	3.03	3.00	.01	
1.43	1 41	1.41	1.40	1.39	1.39	1.38	1.38	1.37	1.37	1.36	1.36	.25	
1.97	1.92	1.90	1.87	1.85	1.83	1.82	1.79	1.79	1.77	1.76	1.76	.10	15
2.40	2.33	2.29	2.25	2.20	2.18	2.16	2.12	2.11	2.10	2.08	2.07	.05	
3.52	3.37	3.29	3.21	3.13	3.08	3.05	2.98	2.96	2.92	2.89	2.87	.01	
1.41	1.40	1.39	1.38	1.37	1.37	1.36	1.36	1.35	1.35	1.34	1.34	.25	
1.94	1.89	1.87	1.84	1.81	1.79	1.78	1.76	1.75	1.74	1.73	1.72	.10	16
2.35	2.28	2.24	2.19	2.15	2.12	2.11	2.07	2.06	2.04	2.02	2.01	.05	
3.41	3.26	3.18	3.10	3.02	2.97	2.93	2.86	2.84	2.81	2.78	2.75	.01	
1.40	1.39	1.38	1.37	1.36	1.35	1.35	1.34	1.34	1.34	1.33	1.33	.25	
1.91	1.86	1.84	1.81	1.78	1.76	1.75	1.73	1.72	1.71	1.69	1.69	.10	17
2.31	2.23	2.19	2.15	2.10	2.08	2.06	2.02	2.01	1.99	1.97	1.96	.05	
3.31	3.16	3.08	3.00	2.92	2.87	2.83	2.76	2.75	2.71	2.68	2.65	.01	
1.39	1.38	1.37	1.36	1.35	1.34	1.34	1.33	1.33	1.32	1.32	1.32	.25	
1.89	1.84	1.81	1.78	1.75	1.74	1.72	1.70	1.69	1.68	1.67	1.66	.10	18
2.27	2.19	2.15	2.11	2.06	2.04	2.02	1.98	1.97	1.95	1.93	1.92	.05	
3.23	3.08	3.00	2.92	2.84	2.78	2.75	2.68	2.66	2.62	2.59	2.57	.01	
1.38	1.37	1.36	1.35	1.34	1.33	1.33	1.32	1.32	1.31	1.31	1.30	.25	
1.86	1.81	1.79	1.76	1.73	1.71	1.70	1.67	1.67	1.65	1.64	1.63	.10	19
2.23	2.16	2.11	2.07	2.03	2.00	1.98	1.94	1.93	1.91	1.89	1.88	.05	
3.15	3 .00	2.92	2.84	2.76	2.71	2.67	2.60	2.58	2.55	2.51	2.49	.01	
1.37	1.36	1.35	1.34	1.33	1.33	1.32	1.31	1.31	1.30	1.30	1.29	.25	
1.84	1.79	1.77	1.74	1.71	1.69	1.68	1.65	1.64	1.63	1.62	1.61	.10	20
2.20	2.12	2.08	2.04	1.99	1.97	1.95	1.91	1.90	1.88	1.86	1.84	.05	
3.09	2.94	2.86	2.78	2.69	2.64	2.61	2.54	2.52	2.48	2.44	2.42	.01	

df for denom-inator N_2	P	1	2	3	4	5	6	7	8	9	10	11	12
		df for numerator N_1											
	.25	1.40	1.48	1.47	1.45	1.44	1.42	1.41	1.40	1.39	1.39	1.38	1.37
22	.10	2.95	2.56	2.35	2.22	2.13	2.06	2.01	1.97	1.93	1.90	1.88	1.86
	.05	4.30	3.44	3.05	2.82	2.66	2.55	2.46	2.40	2.34	2.30	2.26	2.23
	.01	7.95	5.72	4.82	4.31	3.99	3.76	3.59	3.45	3.35	3.26	3.18	3.12
	.25	1.39	1.47	1.46	1.44	1.43	1.41	1.40	1.39	1.38	1.38	1.37	1.36
24	.10	2.93	2.54	2.33	2.19	2.10	2.04	1.98	1.94	1.91	1.88	1.85	1.83
	.05	4.26	3.40	3.01	2.78	2.62	2.51	2.42	2.36	2.30	2.25	2.21	2.18
	.01	7.82	5.61	4.72	4.22	3.90	3.67	3.50	3.36	3.26	3.17	3.09	3.03
	.25	1.38	1.46	1.45	1.44	1.42	1.41	1.39	1.38	1.37	1.37	1.36	1.35
26	.10	2.91	2.52	2.31	2.17	2.08	2.01	1.96	1.92	1.88	1.86	1.84	1.81
	.05	4.23	3.37	2.98	2.74	2.59	2.47	2.39	2.32	2.27	2.22	2.18	2.15
	.01	7.72	5.53	4.64	4.14	3.82	3.59	3.42	3.29	3.18	3.09	3.02	2.96
	.25	1.38	1.46	1.45	1.43	1.41	1.40	1.39	1.38	1.37	1.36	1.35	1.34
28	.10	2.89	2.50	2.29	2.16	2.06	2.00	1.94	1.90	1.87	1.84	1.81	1.79
	.05	4.20	3.34	2.95	2.71	2.56	2.45	2.36	2.29	2.24	2.19	2.15	2.12
	.01	7.64	5.45	4.57	4.07	3.75	3.53	3.36	3.23	3.12	3.03	2.96	2.90
	.25	1.38	1.45	1.44	1.42	1.41	1.39	1.38	1.37	1.36	1.35	1.35	1.34
30	.10	2.88	2.49	2.28	2.14	2.05	1.98	1.93	1.88	1.85	1.82	1.79	1.77
	.05	4.17	3.32	2.92	2.69	2.53	2.42	2.33	2.27	2.21	2.16	2.13	2.09
	.01	7.56	5.39	4.51	4.02	3.70	3.47	3.30	3.17	3.07	2.98	2.91	2.84
	.25	1.36	1.44	1.42	1.40	1.39	1.37	1.36	1.35	1.34	1.33	1.32	1.31
40	.10	2.84	2.44	2.23	2.09	2.00	1.93	1.87	1.83	1.79	1.76	1.73	1.71
	.05	4.08	3.23	2.84	2.61	2.45	2.34	2.25	2.18	2.12	2.08	2.04	2.00
	.01	7.31	5.18	4.31	3.83	3.51	3.29	3.12	2.99	2.89	2.80	2.73	2.66
	.25	1.35	1.42	1.41	1.38	1.37	1.35	1.33	1.32	1.31	1.30	1.29	1.29
60	.10	2.79	2.39	2.18	2.04	1.95	1.87	1.82	1.77	1.74	1.71	1.68	1.66
	.05	4.00	3.15	2.76	2.53	2.37	2.25	2.17	2.10	2.04	1.99	1.95	1.92
	.01	7.08	4.98	4.13	3.65	3.34	3.12	2.95	2.82	2.72	2.63	2.56	2.50
	.25	1.34	1.40	1.39	1.37	1.35	1.33	1.31	1.30	1.29	1.28	1.27	1.26
120	.10	2.75	2.35	2.13	1.99	1.90	1.82	1.77	1.72	1.68	1.65	1.62	1.60
	.05	3.92	3.07	2.68	2.45	2.29	2.17	2.09	2.02	1.96	1.91	1.87	1.83
	.01	6.85	4.79	3.95	3.48	3.17	2.96	2.79	2.66	2.56	2.47	2.40	2.34
	.25	1.33	1.39	1.38	1.36	1.34	1.32	1.31	1.29	1.28	1.27	1.26	1.25
200	.10	2.73	2.33	2.11	1.97	1.88	1.80	1.75	1.70	1.66	1.63	1.60	1.57
	.05	3.89	3.04	2.65	2.42	2.26	2.14	2.06	1.98	1.93	1.88	1.84	1.80
	.01	6.76	4.71	3.88	3.41	3.11	2.89	2.73	2.60	2.50	2.41	2.34	2.27
	.25	1.32	1.39	1.37	1.35	1.33	1.31	1.29	1.28	1.27	1.25	1.24	1.24
∞	.10	2.71	2.30	2.08	1.94	1.85	1.77	1.72	1.67	1.63	1.60	1.57	1.55
	.05	3.84	3.00	2.60	2.37	2.21	2.10	2.01	1.94	1.88	1.83	1.79	1.75
	.01	6.63	4.61	3.78	3.32	3.02	2.80	2.64	2.51	2.41	2.32	2.25	2.18

df for numerator N_1													df for denom-inator
15	20	24	30	40	50	60	100	120	200	500	∞	P	N_2
1.36	1.34	1.33	1.32	1.31	1.31	1.30	1.30	1.30	1.29	1.29	1.28	.25	
1.81	1.76	1.73	1.70	1.67	1.65	1.64	1.61	1.60	1.59	1.58	1.57	.10	22
2.15	2.07	2.03	1 98	1.94	1.91	1.89	1.85	1.84	1.82	1.80	1.78	.05	
2.98	2.83	2.75	2.67	2.58	2.53	2.50	2.42	2.40	2.36	2.33	2.31	01	
1.35	1.33	1.32	1.31	1.30	1.29	1.29	1.28	1.28	1.27	1.27	1.26	.25	
1.78	1.73	1.70	1.67	1.64	1.62	1.61	1.58	1.57	1.56	1.54	1.53	.10	24
2.11	2.03	1.98	1.94	1.89	1.86	1.84	1.80	1.79	1.77	1.75	1.73	.05	
2.89	2.74	2.66	2.58	2.49	2.44	2.40	2.33	2.31	2.27	2.24	2.21	.01	
1.34	1.32	1.31	1.30	1.29	1.28	1.28	1.26	1.26	1.26	1.25	1.25	.25	
1.76	1.71	1.68	1.65	1.61	1.59	1.58	1.55	1.54	1.53	1.51	1.50	.10	26
2.07	1.99	1.95	1.90	1.85	1.82	1.80	1.76	1.75	1.73	1.71	1.69	.05	
2.81	2.66	2.58	2.50	2.42	2.36	2.33	2.25	2.23	2.19	2.16	2.13	.01	
1.33	1.31	1.30	1.29	1.28	1.27	1.27	1.26	1.25	1.25	1.24	1.24	.25	
1.74	1.69	1.66	1.63	1.59	1.57	1.56	1.53	1.52	1.50	1.49	1.48	.10	28
2.04	1.96	1.91	1.87	1.82	1.79	1.77	1.73	1.71	1.69	1.67	1.65	.05	
2.75	2.60	2.52	2.44	2.35	2.30	2.26	2.19	2.17	2.13	2.09	2.06	.01	
1.32	1.30	1.29	1.28	1.27	1.26	1.26	1.25	1.24	1.24	1.23	1.23	.25	
1.72	1.67	1.64	1.61	1.57	1.55	1.54	1.51	1.50	1.48	1.47	1.46	.10	30
2.01	1.93	1.89	1.84	1.79	1.76	1.74	1.70	1.68	1.66	1.64	1.62	.05	
2.70	2.55	2.47	2.39	2.30	2.25	2.21	2.13	2.11	2.07	2.03	2.01	.01	
1.30	1.28	1.26	1.25	1.24	1.23	1.22	1.21	1.21	1.20	1.19	1.19	.25	
1.66	1.61	1.57	1.54	1.51	1.48	1.47	1.43	1.42	1.41	1.39	1.38	.10	40
1.92	1.84	1.79	1.74	1.69	1.66	1.64	1.59	1.58	1.55	1.53	1.51	.05	
2.52	2.37	2.29	2.20	2.11	2.06	2.02	1.94	1.92	1.87	1.83	1.80	.01	
1.27	1.25	1.24	1.22	1.21	1.20	1.19	1.17	1.17	1.16	1.15	1.15	.25	
1.60	1.54	1.51	1.48	1.44	1.41	1.40	1.36	1.35	1.33	1.31	1.29	.10	60
1.84	1.75	1.70	1.65	1.59	1.56	1.53	1.48	1.47	1.44	1.41	1.39	.05	
2.35	2.20	2.12	2.03	1.94	1.88	1.84	1.75	1.73	1.68	1.63	1.60	.01	
1.24	1.22	1.21	1.19	1.18	1.17	1.16	1.14	1.13	1.12	1.11	1.10	.25	
1.55	1.48	1.45	1.41	1.37	1.34	1.32	1.27	1.26	1.24	1.21	1.19	.10	120
1.75	1.66	1.61	1.55	1.50	1.46	1.43	1.37	1.35	1.32	1.28	1.25	.05	
2.19	2.03	1.95	1.86	1.76	1.70	1.66	1.56	1.53	1.48	1.42	1.38	.01	
1.23	1.21	1.20	1.18	1.16	1.14	1.12	1.11	1.10	1.09	1.08	1.06	.25	
1.52	1.46	1.42	1.38	1.34	1.31	1.28	1.24	1.22	1.20	1.17	1.14	.10	200
1.72	1.62	1.57	1.52	1.46	1.41	1.39	1.32	1.29	1.26	1.22	1.19	.05	
2.13	1.97	1.89	1.79	1.69	1.63	1.58	1.48	1.44	1.39	1.33	1.28	.01	
1.22	1.19	1.18	1.16	1.14	1.13	1.12	1.09	1.08	1.07	1.04	1.00	.25	
1.49	1.42	1.38	1.34	1.30	1.26	1.24	1.18	1.17	1.13	1.08	1.00	.10	∞
1.67	1.57	1.52	1.46	1.39	1.35	1.32	1.24	1.22	1.17	1.11	1.00	.05	
2.04	1.88	1.79	1.70	1.59	1.52	1.47	1.36	1.32	1.25	1.15	1.00	.01	

Appendix Table A.6 Chi-Square Distribution

The cumulative chi-square distribution is defined by

$$F(\chi^2) = \int_0^{\chi^2} \frac{\chi^{(v-2)/2} e^{-\chi/2} d\chi}{2^{v/2}[(v-2)/2]!}$$

Example $P(\chi_9^2 \leq 16.9$ for df = 9

$P(\chi^2 \leq 30.1) = F(16.9) = 0.950$

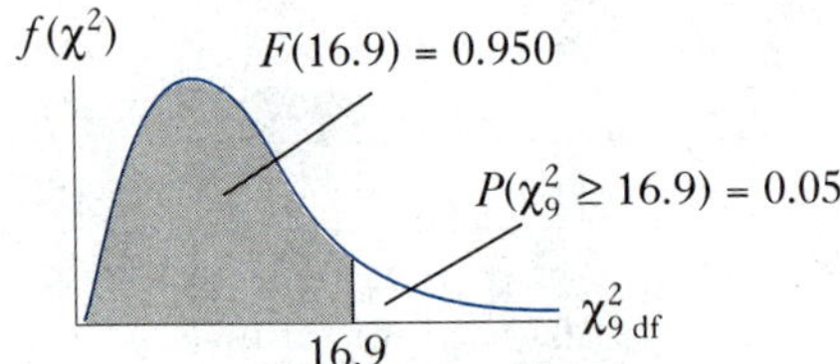

$F(\chi^2)$	.005	.010	.025	.050	.100	.900	.950	.975	.990	.995
1	$.0^4393$	$.0^3157$	$.0^3982$	$.0^2393$	0.158	2.71	3.84	5.02	6.63	7.88
2	.0100	.0201	.0506	.103	.211	4.61	5.99	7.38	9.21	10.6
3	.0717	.115	.216	.352	.584	6.25	7.81	9.35	11.3	12.8
4	.207	.297	.484	.711	1.06	7.78	9.49	11.1	13.3	14.9
5	.412	.554	.831	1.15	1.61	9.24	11.1	12.8	15.1	16.7
6	.676	.872	1.24	1.64	2.20	10.6	12.6	14.4	16.8	18.5
7	.989	1.24	1.69	2.17	2.83	12.0	14.1	16.0	18.5	20.3
8	1.34	1.65	2.18	2.73	3.49	13.4	15.5	17.5	20.1	22.0
9	1.73	2.09	2.70	3.33	4.17	14.7	16.9	19.0	21.7	23.6
10	2.16	2.56	3.25	3.94	4.87	16.0	18.3	20.5	23.2	25.2
11	2.60	3.05	3.82	4.57	5.58	17.3	19.7	21.9	24.7	26.8
12	3.07	3.57	4.40	5.23	6.30	18.5	21.0	23.3	26.2	28.3
13	3.57	4.11	5.01	5.89	7.04	19.8	22.4	24.7	27.7	29.8
14	4.07	4.66	5.63	6.57	7.79	21.1	23.7	26.1	29.1	31.3
15	4.60	5.23	6.26	7.26	8.55	22.3	25.0	27.5	30.6	32.8
16	5.14	5.81	6.91	7.96	9.31	23.5	26.3	28.8	32.0	34.3
17	5.70	6.41	7.56	8.67	10.1	24.8	27.6	30.2	33.4	35.7
18	6.26	7.01	8.23	9.39	10.9	26.0	28.9	31.5	34.8	37.2
19	6.84	7.63	8.91	10.1	11.7	27.2	30.1	32.9	36.2	38.6
20	7.43	8.26	9.59	10.9	12.4	28.4	31.4	34.2	37.6	40.0
21	8.03	8.90	10.3	11.6	13.2	29.6	32.7	35.5	38.9	41.4
22	8.64	9.54	11.0	12.3	14.0	30.8	33.9	36.8	40.3	42.8
23	9.26	10.2	11.7	13.1	14.8	32.0	35.2	38.1	41.6	44.2
24	9.89	10.9	12.4	13.8	15.7	33.2	36.4	39.4	43.0	45.6
25	10.5	11.5	13.1	14.6	16.5	34.4	37.7	40.6	44.3	46.9
26	11.2	12.2	13.8	15.4	17.3	35.6	38.9	41.9	45.6	48.3
27	11.8	12.9	14.6	16.2	18.1	36.7	40.1	43.2	47.0	49.6
28	12.5	13.6	15.3	16.9	18.9	37.9	41.3	44.5	48.3	51.0
29	13.1	14.3	16.0	17.7	19.8	39.1	42.6	45.7	49.6	52.3
30	13.8	15.0	16.8	18.5	20.6	40.3	43.8	47.0	50.9	53.7
z_α	−2.576	−2.326	−1.960	−1.645	−1.282	+1.282	+1.645	+1.960	+2.326	+2.576

Appendix Table A.7 Durbin-Watson d Statistic: Significance Points of d_L and d_U at 0.05 Level of Significance

	$k'=1$		$k'=2$		$k'=3$		$k'=4$		$k'=5$		$k'=6$		$k'=7$		$k'=8$		$k'=9$		$k'=10$	
n	d_L	d_U	d_L	d_U	d_L	d_U	d_L	d_U	d_L	d_U	d_L	d_U	d_L	d_U	d_L	d_U	d_L	d_U	d_L	d_U
6	0.610	1.400	—	—	—	—	—	—	—	—	—	—	—	—	—	—	—	—	—	—
7	0.700	1.356	0.467	1.896	—	—	—	—	—	—	—	—	—	—	—	—	—	—	—	—
8	0.763	1.332	0.559	1.777	0.368	2.287	—	—	—	—	—	—	—	—	—	—	—	—	—	—
9	0.824	1.320	0.629	1.699	0.455	2.128	0.296	2.588	—	—	—	—	—	—	—	—	—	—	—	—
10	0.879	1.320	0.697	1.641	0.525	2.016	0.376	2.414	0.243	2.822	—	—	—	—	—	—	—	—	—	—
11	0.927	1.324	0.658	1.604	0.595	1.928	0.444	2.283	0.316	2.645	0.203	3.005	—	—	—	—	—	—	—	—
12	0.971	1.331	0.812	1.579	0.658	1.864	0.512	2.177	0.379	2.506	0.268	2.832	0.171	3.149	—	—	—	—	—	—
13	1.010	1.340	0.861	1.562	0.715	1.816	0.574	2.094	0.445	2.390	0.328	2.692	0.230	2.985	0.147	3.266	—	—	—	—
14	1.045	1.350	0.905	1.551	0.767	1.779	0.632	2.030	0.505	2.296	0.389	2.572	0.286	2.848	0.200	3.111	0.127	3.360	—	—
15	1.077	1.361	0.946	1.543	0.814	1.750	0.685	1.977	0.562	2.220	0.447	2.472	0.343	2.727	0.251	2.979	0.175	3.216	0.111	3.438
16	1.106	1.371	0.982	1.539	0.857	1.728	0.734	1.935	0.615	2.157	0.502	2.388	0.398	2.624	0.304	2.860	0.222	3.090	0.155	3.304
17	1.133	1.381	1.015	1.536	0.897	1.710	0.779	1.900	0.664	2.104	0.554	2.318	0.451	2.537	0.356	2.757	0.272	2.975	0.198	3.184
18	1.158	1.391	1.046	1.535	0.933	1.696	0.820	1.872	0.710	2.060	0.603	2.257	0.502	2.461	0.407	2.667	0.321	2.873	0.244	3.073
19	1.180	1.401	1.074	1.536	0.967	1.685	0.859	1.848	0.752	2.023	0.649	2.206	0.549	2.396	0.456	2.589	0.369	2.783	0.290	2.974
20	1.201	1.411	1.100	1.537	0.998	1.676	0.894	1.828	0.792	1.991	0.692	2.162	0.595	2.339	0.502	2.521	0.416	2.704	0.336	2.885
21	1.221	1.420	1.125	1.538	1.026	1.669	0.927	1.812	0.829	1.964	0.732	2.124	0.637	2.290	0.547	2.460	0.461	2.633	0.380	2.806
22	1.239	1.429	1.147	1.541	1.053	1.664	0.958	1.797	0.863	1.940	0.769	2.090	0.677	2.246	0.588	2.407	0.504	2.571	0.424	2.734
23	1.257	1.437	1.168	1.543	1.078	1.660	0.986	1.785	0.895	1.920	0.804	2.061	0.715	2.208	0.628	2.360	0.545	2.514	0.465	2.670
24	1.273	1.446	1.188	1.546	1.101	1.656	1.013	1.775	0.925	1.902	0.837	2.035	0.751	2.174	0.666	2.318	0.584	2.464	0.506	2.613
25	1.288	1.454	1.206	1.550	1.123	1.654	1.038	1.767	0.953	1.886	0.868	2.012	0.784	2.144	0.702	2.280	0.621	2.419	0.544	2.560
26	1.302	1.461	1.224	1.553	1.143	1.652	1.062	1.759	0.979	1.873	0.897	1.992	0.816	2.117	0.735	2.246	0.657	2.379	0.581	2.513
27	1.316	1.469	1.240	1.556	1.162	1.651	1.084	1.753	1.004	1.861	0.925	1.974	0.845	2.093	0.767	2.216	0.691	2.342	0.616	2.470
28	1.328	1.476	1.255	1.560	1.181	1.650	1.104	1.747	1.028	1.850	0.951	1.958	0.874	2.071	0.798	2.188	0.723	2.309	0.650	2.431
29	1.341	1.483	1.270	1.563	1.198	1.650	1.124	1.743	1.050	1.841	0.975	1.944	0.900	2.052	0.826	2.164	0.753	2.278	0.682	2.396
30	1.352	1.489	1.284	1.567	1.214	1.650	1.143	1.739	1.071	1.833	0.998	1.931	0.926	2.034	0.854	2.141	0.782	2.251	0.712	2.363
31	1.363	1.496	1.297	1.570	1.229	1.650	1.160	1.735	1.090	1.825	1.020	1.920	0.950	2.018	0.879	2.120	0.810	2.226	0.741	2.333
32	1.373	1.502	1.309	1.574	1.244	1.650	1.177	1.732	1.109	1.819	1.041	1.909	0.972	2.004	0.904	2.102	0.836	2.203	0.769	2.306
33	1.383	1.508	1.321	1.577	1.258	1.651	1.193	1.730	1.127	1.813	1.061	1.900	0.994	1.991	0.927	2.085	0.861	2.181	0.795	2.281
34	1.393	1.514	1.333	1.580	1.271	1.652	1.208	1.728	1.144	1.808	1.080	1.891	1.015	1.979	0.950	2.069	0.885	2.162	0.821	2.257
35	1.402	1.519	1.343	1.584	1.283	1.653	1.222	1.726	1.160	1.803	1.097	1.884	1.034	1.967	0.971	2.054	0.908	2.144	0.845	2.236
36	1.411	1.525	1.354	1.587	1.295	1.654	1.236	1.724	1.175	1.799	1.114	1.877	1.053	1.957	0.991	2.041	0.930	2.127	0.868	2.216
37	1.419	1.530	1.364	1.590	1.307	1.655	1.249	1.723	1.190	1.795	1.131	1.870	1.071	1.948	1.011	2.029	0.951	2.112	0.891	2.198
38	1.427	1.535	1.373	1.594	1.318	1.656	1.261	1.722	1.204	1.792	1.146	1.864	1.088	1.939	1.029	2.017	0.970	2.098	0.912	2.180
39	1.435	1.540	1.382	1.597	1.328	1.658	1.273	1.722	1.218	1.789	1.161	1.859	1.104	1.932	1.047	2.007	0.990	2.085	0.932	2.164
40	1.442	1.544	1.391	1.600	1.338	1.659	1.285	1.721	1.230	1.786	1.175	1.854	1.120	1.924	1.064	1.997	1.008	2.072	0.952	2.149
45	1.475	1.566	1.430	1.615	1.383	1.666	1.336	1.720	1.287	1.776	1.238	1.835	1.189	1.895	1.139	1.958	1.089	2.022	1.038	2.088
50	1.503	1.585	1.462	1.628	1.421	1.674	1.378	1.721	1.335	1.771	1.291	1.822	1.246	1.875	1.201	1.930	1.156	1.986	1.110	2.044
55	1.528	1.601	1.490	1.641	1.452	1.681	1.414	1.724	1.374	1.768	1.334	1.814	1.294	1.861	1.253	1.909	1.212	1.959	1.170	2.010
60	1.549	1.616	1.514	1.652	1.480	1.689	1.444	1.727	1.408	1.767	1.372	1.808	1.335	1.850	1.298	1.894	1.260	1.939	1.222	1.984
65	1.567	1.629	1.536	1.662	1.503	1.696	1.471	1.731	1.438	1.767	1.404	1.805	1.370	1.843	1.336	1.882	1.301	1.923	1.266	1.964
70	1.583	1.641	1.554	1.672	1.525	1.703	1.494	1.735	1.464	1.768	1.433	1.802	1.401	1.837	1.369	1.873	1.337	1.910	1.305	1.948
75	1.598	1.652	1.571	1.680	1.543	1.709	1.515	1.739	1.487	1.770	1.458	1.801	1.428	1.834	1.399	1.867	1.369	1.901	1.339	1.935
80	1.611	1.662	1.586	1.688	1.560	1.715	1.534	1.743	1.507	1.772	1.480	1.801	1.453	1.831	1.425	1.861	1.397	1.893	1.369	1.925
85	1.624	1.671	1.600	1.696	1.575	1.721	1.550	1.747	1.525	1.774	1.500	1.801	1.474	1.829	1.448	1.857	1.422	1.886	1.396	1.916
90	1.635	1.679	1.612	1.703	1.589	1.726	1.566	1.751	1.542	1.776	1.518	1.801	1.494	1.827	1.469	1.854	1.445	1.881	1.420	1.909
95	1.645	1.687	1.623	1.709	1.602	1.732	1.579	1.755	1.557	1.778	1.535	1.802	1.512	1.827	1.489	1.852	1.465	1.877	1.442	1.903
100	1.654	1.694	1.634	1.715	1.613	1.736	1.592	1.758	1.571	1.780	1.550	1.803	1.528	1.826	1.506	1.850	1.484	1.874	1.462	1.898
150	1.720	1.746	1.706	1.760	1.693	1.774	1.679	1.788	1.665	1.802	1.651	1.817	1.637	1.832	1.622	1.847	1.608	1.862	1.594	1.877
200	1.758	1.778	1.748	1.789	1.738	1.799	1.728	1.810	1.718	1.820	1.707	1.831	1.697	1.841	1.686	1.852	1.675	1.863	1.665	1.874

Significance Points of d_L and d_U at 0.05 Level of Significance

	$k' = 11$		$k' = 12$		$k' = 13$		$k' = 14$		$k' = 15$		$k' = 16$		$k' = 17$		$k' = 18$		$k' = 19$		$k' = 20$	
n	d_L	d_U	d_L	d_U	d_L	d_U	d_L	d_U	d_L	d_U	d_L	d_U	d_L	d_U	d_L	d_U	d_L	d_U	d_L	d_U
16	0.098	3.503	—	—	—	—	—	—	—	—	—	—	—	—	—	—	—	—	—	—
17	0.138	3.378	0.087	3.557	—	—	—	—	—	—	—	—	—	—	—	—	—	—	—	—
18	0.177	3.265	0.123	3.441	0.078	3.603	—	—	—	—	—	—	—	—	—	—	—	—	—	—
19	0.220	3.159	0.160	3.335	0.111	3.496	0.070	3.642	—	—	—	—	—	—	—	—	—	—	—	—
20	0.263	3.063	0.200	3.234	0.145	3.395	0.100	3.542	0.063	3.676	—	—	—	—	—	—	—	—	—	—
21	0.307	2.976	0.240	3.141	0.182	3.300	0.132	3.448	0.091	3.583	0.058	3.705	—	—	—	—	—	—	—	—
22	0.349	2.897	0.281	3.057	0.220	3.211	0.166	3.358	0.120	3.495	0.083	3.619	0.052	3.731	—	—	—	—	—	—
23	0.391	2.826	0.322	2.979	0.259	3.128	0.202	3.272	0.153	3.409	0.110	3.535	0.076	3.650	0.048	3.753	—	—	—	—
24	0.431	2.761	0.362	2.908	0.297	3.053	0.239	3.193	0.186	3.327	0.141	3.454	0.101	3.572	0.070	3.678	0.044	3.773	—	—
25	0.470	2.702	0.400	2.844	0.335	2.983	0.275	3.119	0.221	3.251	0.172	3.376	0.130	3.494	0.094	3.604	0.065	3.702	0.041	3.790
26	0.508	2.649	0.438	2.784	0.373	2.919	0.312	3.051	0.256	3.179	0.205	3.303	0.160	3.420	0.120	3.531	0.087	3.632	0.060	3.724
27	0.544	2.600	0.475	2.730	0.409	2.859	0.348	2.987	0.291	3.112	0.238	3.233	0.191	3.349	0.149	3.460	0.112	3.563	0.081	3.658
28	0.578	2.555	0.510	2.680	0.445	2.805	0.383	2.928	0.325	3.050	0.271	3.168	0.222	3.283	0.178	3.392	0.138	3.495	0.104	3.592
29	0.612	2.515	0.544	2.634	0.479	2.755	0.418	2.874	0.359	2.992	0.305	3.107	0.254	3.219	0.208	3.327	0.166	3.431	0.129	3.528
30	0.643	2.477	0.577	2.592	0.512	2.708	0.451	2.823	0.392	2.937	0.337	3.050	0.286	3.160	0.238	3.266	0.195	3.368	0.156	3.465
31	0.674	2.443	0.608	2.553	0.545	2.665	0.484	2.776	0.425	2.887	0.370	2.996	0.317	3.103	0.269	3.208	0.224	3.309	0.183	3.406
32	0.703	2.411	0.638	2.517	0.576	2.625	0.515	2.733	0.457	2.840	0.401	2.946	0.349	3.050	0.299	3.153	0.253	3.252	0.211	3.348
33	0.731	2.382	0.668	2.484	0.606	2.588	0.546	2.692	0.488	2.796	0.432	2.899	0.379	3.000	0.329	3.100	0.283	3.198	0.239	3.293
34	0.758	2.355	0.695	2.454	0.634	2.554	0.575	2.654	0.518	2.754	0.462	2.854	0.409	2.954	0.359	3.051	0.312	3.147	0.267	3.240
35	0.783	2.330	0.722	2.425	0.662	2.521	0.604	2.619	0.547	2.716	0.492	2.813	0.439	2.910	0.388	3.005	0.340	3.099	0.295	3.190
36	0.808	2.306	0.748	2.398	0.689	2.492	0.631	2.586	0.575	2.680	0.520	2.774	0.467	2.868	0.417	2.961	0.369	3.053	0.323	3.142
37	0.831	2.285	0.772	2.374	0.714	2.464	0.657	2.555	0.602	2.646	0.548	2.738	0.495	2.829	0.445	2.920	0.397	3.009	0.351	3.097
38	0.854	2.265	0.796	2.351	0.739	2.438	0.683	2.526	0.628	2.614	0.575	2.703	0.522	2.792	0.472	2.880	0.424	2.968	0.378	3.054
39	0.875	2.246	0.819	2.329	0.763	2.413	0.707	2.499	0.653	2.585	0.600	2.671	0.549	2.757	0.499	2.843	0.451	2.929	0.404	3.013
40	0.896	2.228	0.840	2.309	0.785	2.391	0.731	2.473	0.678	2.557	0.626	2.641	0.575	2.724	0.525	2.808	0.477	2.892	0.430	2.974
45	0.988	2.156	0.938	2.225	0.887	2.296	0.838	2.367	0.788	2.439	0.740	2.512	0.692	2.586	0.644	2.659	0.598	2.733	0.553	2.807
50	1.064	2.103	1.019	2.163	0.973	2.225	0.927	2.287	0.882	2.350	0.836	2.414	0.792	2.479	0.747	2.544	0.703	2.610	0.660	2.675
55	1.129	2.062	1.087	2.116	1.045	2.170	1.003	2.225	0.961	2.281	0.919	2.338	0.877	2.396	0.836	2.454	0.795	2.512	0.754	2.571
60	1.184	2.031	1.145	2.079	1.106	2.127	1.068	2.177	1.029	2.227	0.990	2.278	0.951	2.330	0.913	2.382	0.874	2.434	0.836	2.487
65	1.231	2.006	1.195	2.049	1.160	2.093	1.124	2.138	1.088	2.183	1.052	2.229	1.016	2.276	0.980	2.323	0.944	2.371	0.908	2.419
70	1.272	1.986	1.239	2.026	1.206	2.066	1.172	2.106	1.139	2.148	1.105	2.189	1.072	2.232	1.038	2.275	1.005	2.318	0.971	2.362
75	1.308	1.970	1.277	2.006	1.247	2.043	1.215	2.080	1.184	2.118	1.153	2.156	1.121	2.195	1.090	2.235	1.058	2.275	1.027	2.315
80	1.340	1.957	1.311	1.991	1.283	2.024	1.253	2.059	1.224	2.093	1.195	2.129	1.165	2.165	1.136	2.201	1.106	2.238	1.076	2.275
85	1.369	1.946	1.342	1.977	1.315	2.009	1.287	2.040	1.260	2.073	1.232	2.105	1.205	2.139	1.177	2.172	1.149	2.206	1.121	2.241
90	1.395	1.937	1.369	1.966	1.344	1.995	1.318	2.025	1.292	2.055	1.266	2.085	1.240	2.116	1.213	2.148	1.187	2.179	1.160	2.211
95	1.418	1.929	1.394	1.956	1.370	1.984	1.345	2.012	1.321	2.040	1.296	2.068	1.271	2.097	1.247	2.126	1.222	2.156	1.197	2.186
100	1.439	1.923	1.416	1.948	1.393	1.974	1.371	2.000	1.347	2.026	1.324	2.053	1.301	2.080	1.277	2.108	1.253	2.135	1.229	2.164
150	1.579	1.892	1.564	1.908	1.550	1.924	1.535	1.940	1.519	1.956	1.504	1.972	1.489	1.989	1.474	2.006	1.458	2.023	1.443	2.040
200	1.654	1.885	1.643	1.896	1.632	1.908	1.621	1.919	1.610	1.931	1.599	1.943	1.588	1.955	1.576	1.967	1.565	1.979	1.554	1.991

Note: n = number of observations

k' = number of explanatory variables excluding the constant term.

Significance Points of d_L and d_U at 0.01 Level of Significance

	$k' = 1$		$k' = 2$		$k' = 3$		$k' = 4$		$k' = 5$		$k' = 6$		$k' = 7$		$k' = 8$		$k' = 9$		$k' = 10$	
n	d_L	d_U	d_L	d_U	d_L	d_U	d_L	d_U	d_L	d_U	d_L	d_U	d_L	d_U	d_L	d_U	d_L	d_U	d_L	d_U
6	0.390	1.142	—	—	—	—	—	—	—	—	—	—	—	—	—	—	—	—	—	—
7	0.435	1.036	0.294	1.676	—	—	—	—	—	—	—	—	—	—	—	—	—	—	—	—
8	0.497	1.003	0.345	1.489	0.229	2.102	—	—	—	—	—	—	—	—	—	—	—	—	—	—
9	0.554	0.998	0.408	1.389	0.279	1.875	0.183	2.433	—	—	—	—	—	—	—	—	—	—	—	—
10	0.604	1.001	0.466	1.333	0.340	1.733	0.230	2.193	0.150	2.690	—	—	—	—	—	—	—	—	—	—
11	0.653	1.010	0.519	1.297	0.396	1.640	0.286	2.030	0.193	2.453	0.124	2.892	—	—	—	—	—	—	—	—
12	0.697	1.023	0.569	1.274	0.449	1.575	0.339	1.913	0.244	2.280	0.164	2.665	0.105	3.053	—	—	—	—	—	—
13	0.738	1.038	0.616	1.261	0.499	1.526	0.391	1.826	0.294	2.150	0.211	2.490	0.140	2.838	0.090	3.182	—	—	—	—
14	0.776	1.054	0.660	1.254	0.547	1.490	0.441	1.757	0.343	2.049	0.257	2.354	0.183	2.667	0.122	2.981	0.078	3.287	—	—
15	0.811	1.070	0.700	1.252	0.591	1.464	0.488	1.704	0.391	1.967	0.303	2.244	0.226	2.530	0.161	2.817	0.107	3.101	0.068	3.374
16	0.844	1.086	0.737	1.252	0.633	1.446	0.532	1.663	0.437	1.900	0.349	2.153	0.269	2.416	0.200	2.681	0.142	2.944	0.094	3.201
17	0.874	1.102	0.772	1.255	0.672	1.432	0.574	1.630	0.480	1.847	0.393	2.078	0.313	2.319	0.241	2.566	0.179	2.811	0.127	3.053
18	0.902	1.118	0.805	1.259	0.708	1.422	0.613	1.604	0.522	1.803	0.435	2.015	0.355	2.238	0.282	2.467	0.216	2.697	0.160	2.925
19	0.928	1.132	0.835	1.265	0.742	1.415	0.650	1.584	0.561	1.767	0.476	1.963	0.396	2.169	0.322	2.381	0.255	2.597	0.196	2.813
20	0.952	1.147	0.863	1.271	0.773	1.411	0.685	1.567	0.598	1.737	0.515	1.918	0.436	2.110	0.362	2.308	0.294	2.510	0.232	2.714
21	0.975	1.161	0.890	1.277	0.803	1.408	0.718	1.554	0.633	1.712	0.552	1.881	0.474	2.059	0.400	2.244	0.331	2.434	0.268	2.625
22	0.997	1.174	0.914	1.284	0.831	1.407	0.748	1.543	0.667	1.691	0.587	1.849	0.510	2.015	0.437	2.188	0.368	2.367	0.304	2.548
23	1.018	1.187	0.938	1.291	0.858	1.407	0.777	1.534	0.698	1.673	0.620	1.821	0.545	1.977	0.473	2.140	0.404	2.308	0.340	2.479
24	1.037	1.199	0.960	1.298	0.882	1.407	0.805	1.528	0.728	1.658	0.652	1.797	0.578	1.944	0.507	2.097	0.439	2.255	0.375	2.417
25	1.055	1.211	0.981	1.305	0.906	1.409	0.831	1.523	0.756	1.645	0.682	1.776	0.610	1.915	0.540	2.059	0.473	2.209	0.409	2.362
26	1.072	1.222	1.001	1.312	0.928	1.411	0.855	1.518	0.783	1.635	0.711	1.759	0.640	1.889	0.572	2.026	0.505	2.168	0.441	2.313
27	1.089	1.233	1.019	1.319	0.949	1.413	0.878	1.515	0.808	1.626	0.738	1.743	0.669	1.867	0.602	1.997	0.536	2.131	0.473	2.269
28	1.104	1.244	1.037	1.325	0.969	1.415	0.900	1.513	0.832	1.618	0.764	1.729	0.696	1.847	0.630	1.970	0.566	2.098	0.504	2.229
29	1.119	1.254	1.054	1.332	0.988	1.418	0.921	1.512	0.855	1.611	0.788	1.718	0.723	1.830	0.658	1.947	0.595	2.068	0.533	2.193
30	1.133	1.263	1.070	1.339	1.006	1.421	0.941	1.511	0.877	1.606	0.812	1.707	0.748	1.814	0.684	1.925	0.622	2.041	0.562	2.160
31	1.147	1.273	1.085	1.345	1.023	1.425	0.960	1.510	0.897	1.601	0.834	1.698	0.772	1.800	0.710	1.906	0.649	2.017	0.589	2.131
32	1.160	1.282	1.100	1.352	1.040	1.428	0.979	1.510	0.917	1.597	0.856	1.690	0.794	1.788	0.734	1.889	0.674	1.995	0.615	2.104
33	1.172	1.291	1.114	1.358	1.055	1.432	0.996	1.510	0.936	1.594	0.876	1.683	0.816	1.776	0.757	1.874	0.698	1.975	0.641	2.080
34	1.184	1.299	1.128	1.364	1.070	1.435	1.012	1.511	0.954	1.591	0.896	1.677	0.837	1.766	0.779	1.860	0.722	1.957	0.665	2.057
35	1.195	1.307	1.140	1.370	1.085	1.439	1.028	1.512	0.971	1.589	0.914	1.671	0.857	1.757	0.800	1.847	0.744	1.940	0.689	2.037
36	1.206	1.315	1.153	1.376	1.098	1.442	1.043	1.513	0.988	1.588	0.932	1.666	0.877	1.749	0.821	1.836	0.766	1.925	0.711	2.018
37	1.217	1.323	1.165	1.382	1.112	1.446	1.058	1.514	1.004	1.586	0.950	1.662	0.895	1.742	0.841	1.825	0.787	1.911	0.733	2.001
38	1.227	1.330	1.176	1.388	1.124	1.449	1.072	1.515	1.019	1.585	0.966	1.658	0.913	1.735	0.860	1.816	0.807	1.899	0.754	1.985
39	1.237	1.337	1.187	1.393	1.137	1.453	1.085	1.517	1.034	1.584	0.982	1.655	0.930	1.729	0.878	1.807	0.826	1.887	0.774	1.970
40	1.246	1.344	1.198	1.398	1.148	1.457	1.098	1.518	1.048	1.584	0.997	1.652	0.946	1.724	0.895	1.799	0.844	1.876	0.749	1.956
45	1.288	1.376	1.245	1.423	1.201	1.474	1.156	1.528	1.111	1.584	1.065	1.643	1.019	1.704	0.974	1.768	0.927	1.834	0.881	1.902
50	1.324	1.403	1.285	1.446	1.245	1.491	1.205	1.538	1.164	1.587	1.123	1.639	1.081	1.692	1.039	1.748	0.997	1.805	0.955	1.864
55	1.356	1.427	1.320	1.466	1.284	1.506	1.247	1.548	1.209	1.592	1.172	1.638	1.134	1.685	1.095	1.734	1.057	1.785	1.018	1.837
60	1.383	1.449	1.350	1.484	1.317	1.520	1.283	1.558	1.249	1.598	1.214	1.639	1.179	1.682	1.144	1.726	1.108	1.771	1.072	1.817
65	1.407	1.468	1.377	1.500	1.346	1.534	1.315	1.568	1.283	1.604	1.251	1.642	1.218	1.680	1.186	1.720	1.153	1.761	1.120	1.802
70	1.429	1.485	1.400	1.515	1.372	1.546	1.343	1.578	1.313	1.611	1.283	1.645	1.253	1.680	1.223	1.716	1.192	1.754	1.162	1.792
75	1.448	1.501	1.422	1.529	1.395	1.557	1.368	1.587	1.340	1.617	1.313	1.649	1.284	1.682	1.256	1.714	1.227	1.748	1.199	1.783
80	1.466	1.515	1.441	1.541	1.416	1.568	1.390	1.595	1.364	1.624	1.338	1.653	1.312	1.683	1.285	1.714	1.259	1.745	1.232	1.777
85	1.482	1.528	1.458	1.553	1.435	1.578	1.411	1.603	1.386	1.630	1.362	1.657	1.337	1.685	1.312	1.714	1.287	1.743	1.262	1.773
90	1.496	1.540	1.474	1.563	1.452	1.587	1.429	1.611	1.406	1.636	1.383	1.661	1.360	1.687	1.336	1.714	1.312	1.741	1.288	1.769
95	1.510	1.552	1.489	1.573	1.468	1.596	1.446	1.618	1.425	1.642	1.403	1.666	1.381	1.690	1.358	1.715	1.336	1.741	1.313	1.767
100	1.522	1.562	1.503	1.583	1.482	1.604	1.462	1.625	1.441	1.647	1.421	1.670	1.400	1.693	1.378	1.717	1.357	1.741	1.335	1.765
150	1.611	1.637	1.598	1.651	1.584	1.665	1.571	1.679	1.557	1.693	1.543	1.708	1.530	1.722	1.515	1.737	1.501	1.752	1.486	1.767
200	1.664	1.684	1.653	1.693	1.643	1.704	1.633	1.715	1.623	1.725	1.613	1.735	1.603	1.746	1.592	1.757	1.582	1.768	1.571	1.779

	$k' = 11$		$k' = 12$		$k' = 13$		$k' = 14$		$k' = 15$		$k' = 16$		$k' = 17$		$k' = 18$		$k' = 19$		$k' = 20$	
n	d_L	d_U	d_L	d_U	d_L	d_U	d_L	d_U	d_L	d_U	d_L	d_U	d_L	d_U	d_L	d_U	d_L	d_U	d_L	d_U
16	0.060	3.446	—	—	—	—	—	—	—	—	—	—	—	—	—	—	—	—	—	—
17	0.084	3.286	0.053	3.506	—	—	—	—	—	—	—	—	—	—	—	—	—	—	—	—
18	0.113	3.146	0.075	3.358	0.047	3.357	—	—	—	—	—	—	—	—	—	—	—	—	—	—
19	0.145	3.023	0.102	3.227	0.067	3.420	0.043	3.601	—	—	—	—	—	—	—	—	—	—	—	—
20	0.178	2.914	0.131	3.109	0.092	3.297	0.061	3.474	0.038	3.639	—	—	—	—	—	—	—	—	—	—
21	0.212	2.817	0.162	3.004	0.119	3.185	0.084	3.358	0.055	3.521	0.035	3.671	—	—	—	—	—	—	—	—
22	0.246	2.729	0.194	2.909	0.148	3.084	0.109	3.252	0.077	3.412	0.050	3.562	0.032	3.700	—	—	—	—	—	—
23	0.281	2.651	0.227	2.822	0.178	2.991	0.136	3.155	0.100	3.311	0.070	3.459	0.046	3.597	0.029	3.725	—	—	—	—
24	0.315	2.580	0.260	2.744	0.209	2.906	0.165	3.065	0.125	3.218	0.092	3.363	0.065	3.501	0.043	3.629	0.027	3.747	—	—
25	0.348	2.517	0.292	2.674	0.240	2.829	0.194	2.982	0.152	3.131	0.116	3.274	0.085	3.410	0.060	3.538	0.039	3.657	0.025	3.766
26	0.381	2.460	0.324	2.610	0.272	2.758	0.224	2.906	0.180	3.050	0.141	3.191	0.107	3.325	0.079	3.452	0.055	3.572	0.036	3.682
27	0.413	2.409	0.356	2.552	0.303	2.694	0.253	2.836	0.208	2.976	0.167	3.113	0.131	3.245	0.100	3.371	0.073	3.490	0.051	3.602
28	0.444	2.363	0.387	2.499	0.333	2.635	0.283	2.772	0.237	2.907	0.194	3.040	0.156	3.169	0.122	3.294	0.093	3.412	0.068	3.524
29	0.474	2.321	0.417	2.451	0.363	2.582	0.313	2.713	0.266	2.843	0.222	2.972	0.182	3.098	0.146	3.220	0.114	3.338	0.087	3.450
30	0.503	2.283	0.447	2.407	0.393	2.533	0.342	2.659	0.294	2.785	0.249	2.909	0.208	3.032	0.171	3.152	0.137	3.267	0.107	3.379
31	0.531	2.248	0.475	2.367	0.422	2.487	0.371	2.609	0.322	2.730	0.277	2.851	0.234	2.970	0.196	3.087	0.160	3.201	0.128	3.311
32	0.558	2.216	0.503	2.330	0.450	2.446	0.399	2.563	0.350	2.680	0.304	2.797	0.261	2.912	0.221	3.026	0.184	3.137	0.151	3.246
33	0.585	2.187	0.530	2.296	0.477	2.408	0.426	2.520	0.377	2.633	0.331	2.746	0.287	2.858	0.246	2.969	0.209	3.078	0.174	3.184
34	0.610	2.160	0.556	2.266	0.503	2.373	0.452	2.481	0.404	2.590	0.357	2.699	0.313	2.808	0.272	2.915	0.233	3.022	0.197	3.126
35	0.634	2.136	0.581	2.237	0.529	2.340	0.478	2.444	0.430	2.550	0.383	2.655	0.339	2.761	0.297	2.865	0.257	2.969	0.221	3.071
36	0.658	2.113	0.605	2.210	0.554	2.310	0.504	2.410	0.455	2.512	0.409	2.614	0.364	2.717	0.322	2.818	0.282	2.919	0.244	3.019
37	0.680	2.092	0.628	2.186	0.578	2.282	0.528	2.379	0.480	2.477	0.434	2.576	0.389	2.675	0.347	2.774	0.306	2.872	0.268	2.969
38	0.702	2.073	0.651	2.164	0.601	2.256	0.552	2.350	0.504	2.445	0.458	2.540	0.414	2.637	0.371	2.733	0.330	2.828	0.291	2.923
39	0.723	2.055	0.673	2.143	0.623	2.232	0.575	2.323	0.528	2.414	0.482	2.507	0.438	2.600	0.395	2.694	0.354	2.787	0.315	2.879
40	0.744	2.039	0.694	2.123	0.645	2.210	0.597	2.297	0.551	2.386	0.505	2.476	0.461	2.566	0.418	2.657	0.377	2.748	0.338	2.838
45	0.835	1.972	0.790	2.044	0.744	2.118	0.700	2.193	0.655	2.269	0.612	2.346	0.570	2.424	0.528	2.503	0.488	2.582	0.448	2.661
50	0.913	1.925	0.871	1.987	0.829	2.051	0.787	2.116	0.746	2.182	0.705	2.250	0.665	2.318	0.625	2.387	0.586	2.456	0.548	2.526
55	0.979	1.891	0.940	1.945	0.902	2.002	0.863	2.059	0.825	2.117	0.786	2.176	0.748	2.237	0.711	2.298	0.674	2.359	0.637	2.421
60	1.037	1.865	1.001	1.914	0.965	1.964	0.929	2.015	0.893	2.067	0.857	2.120	0.822	2.173	0.786	2.227	0.751	2.283	0.716	2.338
65	1.087	1.845	1.053	1.889	1.020	1.934	0.986	1.980	0.953	2.027	0.919	2.075	0.886	2.123	0.852	2.172	0.819	2.221	0.786	2.272
70	1.131	1.831	1.099	1.870	1.068	1.911	1.037	1.953	1.005	1.995	0.974	2.038	0.943	2.082	0.911	2.127	0.880	2.172	0.849	2.217
75	1.170	1.819	1.141	1.856	1.111	1.893	1.082	1.931	1.052	1.970	1.023	2.009	0.993	2.049	0.964	2.090	0.934	2.131	0.905	2.172
80	1.205	1.810	1.177	1.844	1.150	1.878	1.122	1.913	1.094	1.949	1.066	1.984	1.039	2.022	1.011	2.059	0.983	2.097	0.955	2.135
85	1.236	1.803	1.210	1.834	1.184	1.866	1.158	1.898	1.132	1.931	1.106	1.965	1.080	1.999	1.053	2.033	1.027	2.068	1.000	2.104
90	1.264	1.798	1.240	1.827	1.215	1.856	1.191	1.886	1.166	1.917	1.141	1.948	1.116	1.979	1.091	2.012	1.066	2.044	1.041	2.077
95	1.290	1.793	1.267	1.821	1.244	1.848	1.221	1.876	1.197	1.905	1.174	1.934	1.150	1.963	1.126	1.993	1.102	2.023	1.079	2.054
100	1.314	1.790	1.292	1.816	1.270	1.841	1.248	1.868	1.225	1.895	1.203	1.922	1.181	1.949	1.158	1.977	1.136	2.006	1.113	2.034
150	1.473	1.783	1.458	1.799	1.444	1.814	1.429	1.830	1.414	1.847	1.400	1.863	1.385	1.880	1.370	1.897	1.355	1.913	1.340	1.931
200	1.561	1.791	1.550	1.801	1.539	1.813	1.528	1.824	1.518	1.836	1.507	1.847	1.495	1.860	1.484	1.871	1.474	1.883	1.462	1.896

Note: n = number of observations
k' = number of explanatory variables excluding the constant term.

Appendix Table B.1 Most Valuable Companies

Rank	Company	Market Value ($ Mil.)
1	Exxon	69086
2	Philip Morris	68770
3	General Electric	66969
4	Wal-Mart Stores	60954
5	Merck	57976
6	Coca-Cola	52411
7	American Telephone & Telegraph	49426
8	IBM	49322
9	Bristol-Myers Squibb	41431
10	Procter & Gamble	33558
11	Johnson & Johnson	32019
12	Du Pont	29535
13	GTE	26778
14	Abbott Laboratories	25410
15	Pepsico	25054
16	Pfizer	23941
17	American Home Products	23828
18	Mobil	23699
19	General Motors	22792
20	Bellsouth	21962
21	Amoco	21777
22	Chevron	21367
23	Eli Lilly	21105
24	Microsoft	21039
25	Waste Management	20916
26	Walt Disney	20229
27	Minnesota Mining & Mfg.	19805
28	Hewlett-Packard	18690
29	American International Group	18308
30	Southwestern Bell	17449
31	Ford Motor	17338
32	Federal National Mortgage Assn.	17062
33	Bell Atlantic	16731
34	Atlantic Richfield	16152
35	Boeing	15828
36	Ameritech	15798
37	Pacific Telesis Group	15790
38	Sears, Roebuck	15785
39	Anheuser-Busch	15619
40	Dow Chemical	15292
41	Texaco	14734
42	Nynex	14671
43	McDonald's	14308
44	Schlumberger	14308
45	Eastman Kodak	14084
46	US West	13835
47	Kellogg	13708
48	Home Depot	12957
49	Intel	12699
50	Pacific Gas & Electric	12423
51	Sara Lee	12203
52	Emerson Electric	11911
53	Schering-Plough	11452
54	RJR Nabisco Holdings	11217
55	General Mills	10964
56	J.P. Morgan	10922
57	Syntex	10627
58	Gillette	10483
59	Nationsbank	10319
60	American Express	10152
61	Kmart	10088
62	Berkshire Hathaway	9970
63	Motorola	9948
64	Baxter International	9877
65	Dun & Bradstreet	9859
66	Marion Merrell Dow	9843
67	Toys 'R' Us	9831
68	The Limited	9808
69	Southern	9627
70	Time Warner	9492
71	Bankamerica	9467
72	Union Pacific	9430
73	H.J. Heinz	9392
74	Campbell Soup	9309
75	International Paper	9021
76	Scecorp	8963
77	American Brands	8940
78	Archer Daniels Midland	8818
79	MCI Communications	8777
80	Warner-Lambert	8715
81	Banc One	8324
82	Monsanto	8311
83	Novell	8279
84	Norfolk Southern	8273
85	Kimberly-Clark	8244
86	General Re	8155
87	Amgen	8043
88	Texas Utilities	8007
89	Rhone-Poulenc Rorer	7929
90	Apple Computer	7583
91	Commonwealth Edison	7523
92	Capital Cities/ABC	7320
93	ITT	7307
94	Digital Equipment	7287
95	Loews	7237
96	J.C. Penney	7225

Rank	Company	Market Value ($ Mil.)	Rank	Company	Market Value ($ Mil.)
97	Federal Home Loan Mortgage Assn.	7202	147	Marsh & McLennan	5370
98	May Department Stores	7202	148	Houston Industries	5296
99	Food Lion	7132	149	PNC Financial	5277
100	Xerox	7091	150	Philadelphia Electric	5253
101	Upjohn	7009	151	Melville	5162
102	Allied-Signal	6769	152	Pitney Bowes	5160
103	Westinghouse Electric	6742	153	AMR	5157
104	Gannett	6738	154	Rubbermaid	5083
105	Weyerhaeuser	6649	155	Unocal	5073
106	AMP	6508	156	Honeywell	5039
107	Duke Power	6474	157	Entergy	5001
108	The Gap	6409	158	Caterpillar	4995
109	Cooper Industries	6381	159	CNA Financial	4913
110	Conagra	6376	160	Wachovia	4853
111	Corning	6345	161	Tenneco	4845
112	Colgate-Palmolive	6332	162	Air Products & Chemicals	4840
113	American TV & Communications	6314	163	Bankers Trust New York	4832
114	CPC International	6302	164	Suntrust Banks	4809
115	PPG Industries	6291	165	Aetna Life & Casualty	4802
116	United Technologies	6232	166	Central & South West	4801
117	Raytheon	6047	167	Quaker Oats	4792
118	Automatic Data Processing	6043	168	Security Pacific	4788
119	Public Service Enterprise Group	6036	169	Sprint	4769
120	Tele-Communications	5929	170	Burlington Resources	4765
121	Rockwell International	5914	171	Borden	4738
122	Aluminum Co. of America	5913	172	Dayton Hudson	4737
123	Phillips Petroleum	5883	173	Chrysler	4709
124	American Cyanamid	5875	174	Dillard Department Stores	4691
125	CSX	5874	175	Medtronic	4685
126	Ralston Purina	5864	176	Carnival Cruise Lines	4604
127	Merrill Lynch	5855	177	American General	4598
128	USX-Marathon Group	5842	178	Norwest	4588
129	Student Loan Marketing Assn.	5796	179	Detroit Edison	4556
130	UST	5749	180	Times Mirror	4530
131	McCaw Cellular Communications	5739	181	Turner Broadcasting System	4481
132	Consolidated Edison Co. of New York	5736	182	Walgreen	4446
133	Albertson's	5733	183	Newmont Gold	4431
134	Occidental Petroleum	5701	184	Primerica	4372
135	Chemical Banking	5679	185	Goodyear Tire & Rubber	4249
136	American Electric Power	5674	186	First Union	4222
137	Chubb	5673	187	Costco Wholesale	4204
138	Georgia-Pacific	5661	188	Morgan Stanley Group	4174
139	FPL Group	5656	189	Chemical Waste Management	4169
140	Reader's Digest Association	5650	190	Carolina Power & Light	4169
141	Pacificorp	5576	191	Sysco	4128
142	Citicorp	5535	192	Medco Containment Services	4105
143	Paramount Communications	5490	193	Viacom	4073
144	Dominion Resources	5480	194	Freeport-McMoran Copper & Gold	4052
145	U.S. Surgical	5383	195	Cigna	4016
146	Nike	5375	196	Masco	4002

Rank	Company	Market Value ($ Mil.)	Rank	Company	Market Value ($ Mil.)
197	International Flavors & Fragrances	3960	247	Knight-Ridder	3076
198	Deere	3916	248	Winn-Dixie Stores	3075
199	R.R. Donnelley & Sons	3913	249	Torchmark	3008
200	Humana	3881	250	Consolidated Natural Gas	2993
201	Lin Broadcasting	3881	251	Reebok International	2991
202	First Interstate Bancorp	3875	252	Sun Microsystems	2987
203	W.R. Grace	3799	253	Morton International	2971
204	Woolworth	3746	254	Halliburton	2969
205	Genuine Parts	3737	255	Whirlpool	2967
206	Hershey Foods	3709	256	St. Paul	2966
207	Great Lakes Chemical	3688	257	Sun	2958
208	Pennsylvania Power & Light	3668	258	Bausch & Lomb	2950
209	Centel	3617	259	Ethyl	2943
210	H&R Block	3556	260	Safeco	2941
211	UAL	3555	261	Federal Express	2931
212	Arco Chemical	3548	262	Temple-Inland	2904
213	Browning-Ferris Industries	3547	263	Unocal Exploration	2892
214	Union Camp	3530	264	Newmont Mining	2881
215	Illinois Tool Works	3489	265	Ohio Edison	2861
216	NBD Bancorp	3463	266	Roadway Services	2851
217	Enron	3429	267	McGraw-Hill	2838
218	Liz Claiborne	3428	268	Texas Instruments	2834
219	Delta Air Lines	3422	269	Wm. Wrigley Jr.	2828
220	Fluor	3404	270	Newell	2821
221	Fleet/Norstar Financial Group	3376	271	Lockheed	2819
222	Deluxe	3366	272	Tribune	2815
223	Rohm & Haas	3354	273	W.W. Grainger	2804
224	Reynolds Metals	3353	274	Washington Post	2803
225	Consolidated Rail	3351	275	General Public Utilities	2798
226	Avon Products	3345	276	Phelps Dodge	2798
227	Union Electric	3345	277	Hillenbrand Industries	2788
228	Wells Fargo	3341	278	Wheelabrator Technologies	2777
229	Burlington Northern	3332	279	Dresser Industries	2763
230	U.S. Healthcare	3309	280	Northeast Utilities	2758
231	Salomon	3305	281	Fifth Third Bancorp	2721
232	Amerada Hess	3253	282	AON	2707
233	Transamerica	3252	283	Storage Technology	2706
234	Geico	3241	284	Bank of New York	2701
235	Dow Jones	3223	285	Oracle Systems	2697
236	Textron	3196	286	Computer Associates International	2680
237	Nordstrom	3192	287	Baker Hughes	2669
238	Scott Paper	3189	288	TRW	2666
239	Ingersoll-Rand	3183	289	Capital Holding	2654
240	Alza	3162	290	Potomac Electric Power	2653
241	Duracell International	3153	291	Martin Marietta	2646
242	Union Carbide	3142	292	Tambrands	2629
243	Genentech	3130	293	Crown Cork & Seal	2620
244	Alltel	3085	294	United Healthcare	2620
245	Freeport-McMoran	3081	295	Baltimore Gas & Electric	2608
246	Chase Manhattan	3077	296	Long Island Lighting	2575

Rank	Company	Market Value ($ Mil.)	Rank	Company	Market Value ($ Mil.)
297	National Medical Enterprises	2573	347	Sigma-Aldrich	2201
298	Gerber Products	2570	348	Pet	2190
299	Becton, Dickinson	2565	349	Keycorp	2176
300	Imcera Group	2564	350	Chambers Development	2155
301	Clorox	2547	351	Ivax	2148
302	Centerior Energy	2540	352	McCormick	2141
303	General Dynamics	2520	353	Teco Energy	2134
304	Autozone	2519	354	Household International	2118
305	Northern States Power	2517	355	Cooper Tire & Rubber	2116
306	Compaq Computer	2511	356	Mead	2115
307	McDonnell Douglas	2505	357	Republic New York	2108
308	Fruit of the Loom	2504	358	Stanley Works	2104
309	National City	2482	359	Hasbro	2097
310	National Health Laboratories	2480	360	Travelers	2083
311	Golden West Financial	2476	361	Cincinnati Gas & Electric	2082
312	Sherwin-Williams	2476	362	Santa Fe Pacific	2081
313	Wisconsin Energy	2475	363	Hilton Hotels	2078
314	Price	2469	364	U.S. Bancorp	2074
315	Circus Circus Enterprises	2462	365	Brown-Forman	2063
316	Biomet	2458	366	Interpublic Group	2054
317	Champion International	2451	367	St. Jude Medical	2054
318	Unum	2450	368	Comcast	2043
319	American Stores	2441	369	Citizens Utilities	2027
320	Niagara Mohawk Power	2433	370	H.F. Ahmanson	2020
321	Eaton	2432	371	Dole Food	2019
322	Nalco Chemical	2421	372	First Bank System	2019
323	Corestates Financial	2418	373	Pioneer Hi-Bred International	2017
324	VF	2409	374	Mattel	2001
325	San Diego Gas & Electric	2408	375	Mellon Bank	1995
326	Tyson Foods	2405	376	Dial	1990
327	Hercules	2376	377	First Chicago	1970
328	AFLAC	2360	378	Maytag	1967
329	Pall	2359	379	Bear Stearns	1965
330	Coastal	2359	380	MBIA	1963
331	Great Western Financial	2352	381	Stone Container	1962
332	Dover	2352	382	Willamette Industries	1962
333	Allegheny Power System	2339	383	Nucor	1959
334	Louisiana-Pacific	2337	384	Super Valu Stores	1952
335	Florida Progress	2336	385	Contel Cellular	1949
336	Lincoln National	2334	386	Amdahl	1929
337	Tandy	2321	387	New England Electric System	1925
338	State Street Boston	2317	388	Cincinnati Financial	1909
339	Barnett Banks	2311	389	Litton Industries	1902
340	Westvaco	2295	390	Franklin Resources	1900
341	First Fidelity Bancorporation	2286	391	Paccar	1885
342	New York Times	2263	392	Southern New England Telecomms.	1879
343	Cisco Systems	2250	393	MBNA	1863
344	CBS	2247	394	E.W. Scripps	1854
345	Lubrizol	2218	395	Ashland Oil	1849
346	Blockbuster Entertainment	2215	396	Pennzoil	1846

Rank	Company	Market Value ($ Mil.)	Rank	Company	Market Value ($ Mil.)
397	Society	1841	447	Panhandle Eastern	1582
398	Jefferson-Pilot	1838	448	BMC Software	1576
399	Rite Aid	1832	449	BHC Communications	1573
400	Lyondell Petrochemical	1830	450	Sundstrand	1570
401	Comerica	1815	451	First Financial Management	1563
402	Stryker	1807	452	Oklahoma Gas & Electric	1551
403	Amax	1804	453	Whitman	1549
404	New York State Electric & Gas	1799	454	Homestake Mining	1542
405	Mapco	1789	455	Johnson Controls	1539
406	Northern Trust	1768	456	Charming Shoppes	1536
407	Unitrin	1768	457	Murphy Oil	1534
408	Coca-Cola Enterprises	1767	458	Nipsco Industries	1533
409	CMS Energy	1766	459	Black & Decker	1531
410	Borland International	1757	460	Premark International	1523
411	Flightsafety International	1753	461	Unisys	1516
412	Molex	1743	462	Circuit City Stores	1516
413	Kerr-McGee	1742	463	Owens-Corning Fiberglas	1515
414	Ryder System	1741	464	Williams	1509
415	Consolidated Papers	1727	465	Union Texas Petroleum Holdings	1509
416	Premier Industrial	1723	466	Healthtrust	1507
417	Illinois Power	1715	467	T^2 Medical	1504
418	Sonoco Products	1708	468	Policy Management Systems	1503
419	Tyco Laboratories	1706	469	A. Schulman	1502
420	Chiquita Brands International	1692	470	Pinnacle West Capital	1501
421	Safeway	1685	471	Russell	1496
422	DPL	1682	472	Shaw Industries	1495
423	James River Corp. of Virginia	1681	473	DQE	1492
424	Hubbell	1680	474	Kroger	1490
425	Engelhard	1675	475	Dana	1490
426	Loctite	1669	476	Office Depot	1482
427	FMC	1667	477	Dreyfus	1480
428	Bandag	1664	478	Public Service Co. of Colorado	1478
429	Avery Dennison	1664	479	American Greetings	1474
430	Alco Standard	1663	480	Continental Corp.	1467
431	Oryx Energy	1643	481	Teledyne	1462
432	Allergan	1640	482	Stride Rite	1461
433	Snap-On Tools	1636	483	Giant Food	1459
434	Beneficial	1631	484	Vulcan Materials	1448
435	Marriott	1630	485	Lowe's	1447
436	Firstar	1628	486	Manufacturers National	1439
437	General Cinema	1625	487	Advanced Micro Devices	1439
438	Boatmen's Bancshares	1624	488	Safety-Kleen	1437
439	Betz Laboratories	1623	489	Mercantile Stores	1437
440	Scana	1616	490	CUC International	1435
441	Southwest Airlines	1615	491	Geo. A. Hormel	1427
442	Kemper	1599	492	Jostens	1425
443	Cabletron Systems	1599	493	EG&G	1423
444	International Game Technology	1597	494	AES	1420
445	Parker Hannifin	1592	495	Polaroid	1419
446	Carter-Wallace	1587	496	C.R. Bard	1412

Rank	Company	Market Value ($ Mil.)
497	Bancorp Hawaii	1411
498	USX-U.S. Steel Group	1403
499	Bank of Boston	1401
500	Metro Mobile CTS	1392
501	Lotus Development	1388
502	Equifax	1386
503	Enron Oil & Gas	1385
504	Raychem	1374
505	Ameritrust	1368
506	Tandem Computers	1365
507	Service Corp. International	1361
508	Medical Care International	1354
509	Worthington Industries	1354
510	Pacific Enterprises	1351
511	Puget Sound Power & Light	1347
512	IMC Fertilizer Group	1340
513	Gulf States Utilities	1340
514	Bruno's	1331
515	Brunswick	1328
516	Triton Energy	1318
517	Sonat	1315
518	Charles Schwab	1314
519	Cintas	1310
520	Potlatch	1303
521	Forest Laboratories	1294
522	Telephone & Data Systems	1294
523	Surgical Care Affiliates	1293
524	Kansas City Power & Light	1288
525	International Specialty Products	1286
526	Mylan Laboratories	1282
527	Alexander & Baldwin	1271
528	McKesson	1271
529	Southwestern Public Service	1268
530	Centocor	1267
531	Montana Power	1266
532	Destec Energy	1261
533	Huntington Bancshares	1260
534	CBI Industries	1259
535	National Service Industries	1257
536	TJX	1248
537	P.H. Glatfelter	1247
538	Harris	1232
539	St. Joe Paper	1220
540	Valero Energy	1219
541	Smith's Food & Drug Centers	1217
542	B.F. Goodrich	1216
543	Federal Paper Board	1210
544	Ambac	1208
545	Northrop	1201
546	Ipalco Enterprises	1197
547	Great Atlantic & Pacific Tea	1194
548	Armstrong World Industries	1183
549	Computer Sciences	1182
550	Synergen	1170
551	Wendy's International	1169
552	Silicon Graphics	1159
553	Fina	1157
554	Vons	1157
555	Nicor	1156
556	Conner Peripherals	1155
557	Pep Boys-Manny, Moe & Jack	1150
558	Weis Markets	1145
559	Marshall & Ilsley	1138
560	Bemis	1134
561	Fleming	1132
562	Cincinnati Bell	1132
563	Perkin-Elmer	1131
564	Cracker Barrel Old Country Store	1130
565	Stratus Computer	1130
566	Meridian Bancorp	1126
567	A.G. Edwards	1123
568	Arkla	1121
569	U.S. Cellular	1117
570	National Semiconductor	1117
571	MGIC Investment	1116
572	Adobe Systems	1115
573	Bethlehem Steel	1113
574	E-Systems	1101
575	Allegheny Ludlum	1101
576	Unifi	1100
577	Century Telephone Enterprises	1096
578	Penn Central	1092
579	McDermott International	1089
580	Thermo Electron	1084
581	Asarco	1083
582	Wang Laboratories	1083
583	Old Kent Financial	1081
584	Delmarva Power & Light	1080
585	General Signal	1076
586	Critical Care America	1075
587	Perrigo	1074
588	Anadarko Petroleum	1073
589	Kelly Services	1069
590	Thomas & Betts	1067
591	Brinker International	1063
592	American National Insurance	1058
593	Petrie Stores	1052
594	Cray Research	1046
595	First of America Bank	1045
596	Loral	1043

Rank	Company	Market Value ($ Mil.)	Rank	Company	Market Value ($ Mil.)
597	Mercury Finance	1041	647	Crompton & Knowles	940
598	Ecolab	1040	648	Navistar International	938
599	Scimed Life Systems	1036	649	Pacific Telecom	938
600	Rochester Telephone	1035	650	Painewebber Group	932
601	Shoney's	1034	651	Bowater	931
602	Hannaford Brothers	1029	652	New Plan Realty Trust	926
603	Family Dollar Stores	1026	653	Wellman	923
604	Atlantic Energy	1024	654	Cadence Design Systems	923
605	Kansas Gas & Electric	1023	655	Avnet	922
606	Harley-Davidson	1021	656	Universal	920
607	Seagate Technology	1020	657	Nordson	918
608	Duty Free International	1018	658	KU Energy	917
609	Dean Foods	1015	659	Fleetwood Enterprises	916
610	Block Drugs	1012	660	Harsco	908
611	Provident Life & Accident	1012	661	J.M. Smucker	908
612	Illinois Central	1010	662	Novacare	905
613	Witco	1010	663	Minnesota Power	903
614	Foster Wheeler	1009	664	Cipsco	900
615	Midwest Resources	994	665	Peoples Energy	899
616	Millipore	992	666	Symantec	897
617	King World Productions	991	667	First Alabama Bancshares	896
618	20th Century Industries	989	668	York International	896
619	Progressive	984	669	Old Republic International	894
620	Ogden	983	670	Wilmington Trust	890
621	Olin	982	671	PSI Resources	888
622	West Point-Pepperell	978	672	Intergraph	882
623	Multimedia	977	673	Idaho Power	879
624	Shawmut National	977	674	United Investors Management	879
625	Omnicon Group	976	675	Tosco	878
626	Genzyme	975	676	Service Merchandise	876
627	Ball	974	677	Yellow Freight System	874
628	Utilicorp United	973	678	Caesars World	868
629	LG&E Energy	968	679	Calgon Carbon	866
630	Countrywide Credit Industries	965	680	Georgia Gulf	864
631	Conseco	962	681	Boise Cascade	863
632	Legent	960	682	Integra Financial	862
633	Hartford Steam Boiler Inspection	959	683	Enquirer/Star Group	861
634	Chiron	958	684	Dell Computer	861
635	Amsouth Bancorporation	958	685	Hawaiian Electric Industries	859
636	Fund American	958	686	Alexander & Alexander Services	859
637	Boston Edison	957	687	Cyprus Minerals	859
638	Ohio Casualty	956	688	Universal Foods	856
639	Columbia Gas System	954	689	Biogen	854
640	Cummins Engine	952	690	John H. Harland	850
641	Promus	947	691	First Hawaiian	850
642	Southtrust	944	692	First Empire State	848
643	Echlin	942	693	Lafarge	843
644	Crawford	942	694	Washington Federal Savings & Loan	839
645	Kansas Power & Light	942	695	Star Banc	838
646	Manor Care	940	696	Stewart & Stevenson Services	832

Rank	Company	Market Value ($ Mil.)	Rank	Company	Market Value ($ Mil.)
697	First Virginia Banks	832	747	Continental Medical Systems	751
698	Ogden Projects	832	748	Century Communications	749
699	Marquette Electronics	829	749	Quality Food Centers	746
700	Transatlantic Holdings	829	750	IBP	745
701	Longview Fibre	827	751	Vanguard Cellular Systems	742
702	Crane	826	752	Wausau Paper Mills	742
703	Intelligent Electronics	822	753	Portland General	742
704	WPL Holdings	822	754	Ametek	742
705	Commerce Clearing House	819	755	Tootsie Roll Industries	741
706	Bob Evans Farms	818	756	Questar	741
707	Brooklyn Union Gas	817	757	Public Service Co. of New Hampshire	740
708	Keystone International	814	758	Briggs & Stratton	738
709	Continental Bank	813	759	Affiliated Publications	738
710	UJB Financial	809	760	Atlanta Gas Light	735
711	USAir Group	808	761	Fingerhut	733
712	Signet Banking	807	762	First Tennessee National	732
713	Rollins Environmental Services	805	763	Home Shopping Network	732
714	Communications Satellite	795	764	Cabot	731
715	Edison Brothers Stores	791	765	Rochester Gas & Electric	730
716	Central Fidelity Banks	791	766	Gensia Pharmaceuticals	730
717	Centex	790	767	Mirage Resorts	728
718	Crestar Financial	790	768	Spiegel	728
719	Healthcare Compare	790	769	Exabyte	727
720	Healthsouth Rehabilitation	789	770	JWP	723
721	Mercantile Bankshares	786	771	Valspar	723
722	Kansas City Southern Industries	785	772	Banponce	722
723	Louisiana Land & Exploration	785	773	Acuson	722
724	National Fuel Gas	785	774	Longs Drug Stores	721
725	Washington Water Power	784	775	Sensormatic Electronics	717
726	Mentor Graphics	783	776	Adolph Coors	716
727	Maxus Energy	783	777	Immunex	715
728	Ferro	782	778	Argonaut Group	715
729	Hanover Insurance	779	779	Tecumseh Products	711
730	Alleghany	777	780	Amax Gold	711
731	Varian Associates	774	781	Valley National	710
732	Lance	774	782	Jones Apparel Group	710
733	Enserch	774	783	Associated Communications	709
734	Applied Materials	773	784	Micron Technology	709
735	Equitable Resources	772	785	American Water Works	708
736	Diebold	771	786	American Power Conversion	708
737	Rouse	771	787	Federal Signal	707
738	USF&G	769	788	Rollins	705
739	Timken	766	789	Mercantile Bancorporation	704
740	Kaiser Aluminum	766	790	Puget Sound Bancorp	702
741	Sotheby's Holdings	766	791	MNC Financial	699
742	Leggett & Platt	759	792	Commerce Bancshares	698
743	Topps	757	793	Autodesk	694
744	First Security	756	794	J.B. Hunt Transport Services	694
745	Tiffany	753	795	RPM	692
746	Jacobs Engineering Group	752	796	Southland	692

Rank	Company	Market Value ($ Mil.)	Rank	Company	Market Value ($ Mil.)
797	Mitchell Energy & Development	691	848	Fred Meyer	620
798	Quantum	691	849	BWIP Holding	619
799	American Medical Holdings	690	850	CML Group	618
800	Bergen Brunswig	688	851	Beckman Instruments	618
801	Royal Appliance Mfg.	687	852	Nevada Power	617
802	USLife	686	853	Linear Technology	616
803	Vicor	684	854	OEA	616
804	Morrison Knudsen	683	855	Meditrust	614
805	Valhi	683	856	Atlantic Southeast Airlines	613
806	Consolidated Stores	682	857	T. Rowe Price Associates	613
807	Alberto-Culver	681	858	Clayton Homes	611
808	Springs Industries	681	859	Sbarro	611
809	U.S. Shoe	679	860	Cypress Semiconductor	610
810	Chris-Craft Industries	678	861	BB&T Financial	608
811	Inland Steel Industries	677	862	Heilig-Meyers	608
812	Waban	675	863	Ryan's Family Steak Houses	608
813	Jenny Craig	674	864	Ballard Medical Products	607
814	IES Industries	671	865	Filene's Basement	606
815	PHM	668	866	Central Maine Power	604
816	Watts Industries	668	867	Value City Department Stores	603
817	Cablevision Systems	668	868	Overseas Shipholding Group	603
818	Trinova	667	869	Calmat	602
819	Washington Gas Light	666	870	Wisconsin Public Service	601
820	Central Bancshares of the South	665	871	Teleflex	601
821	A.H. Belo	665	872	Apache	597
822	Iowa-Illinois Gas & Electric	665	873	Alliance Pharmaceutical	597
823	Union Bank	663	874	Itel	597
824	Merchants National	659	875	McClatchy Newspapers	595
825	Stanhome	658	876	Standard Federal Bank	595
826	U.S. Bioscience	655	877	Kaufman & Broad Home	593
827	Comdisco	654	878	W.R. Berkley	591
828	Beverly Enterprises	654	879	Harnischfeger Industries	589
829	INB Financial	651	880	M.A. Hanna	588
830	Consolidated Freightways	649	881	Pittston	588
831	Lee Enterprises	644	882	Foundation Health	587
832	Trinity Industries	644	883	MCN	587
833	Michigan National	639	884	Picturetel	586
834	Rockefeller Center Properties	638	885	Leucadia National	582
835	Hon Industries	638	886	Synovus Financial	581
836	Paging Network	636	887	Neutrogena	580
837	Danaher	636	888	PHH	580
838	Thermo Instrument Systems	633	889	Catellus Development	580
839	Community Psychiatric Centers	633	890	Verifone	579
840	Pacificare Health Systems	633	891	Wallace Computer Services	578
841	Diamond Shamrock	631	892	First Brands	577
842	Xilinx	629	893	Central Newspapers	576
843	Battle Mountain Gold	628	894	Octel Communications	575
844	Grumman	628	895	International Dairy Queen	575
845	Genetics Institute	627	896	Holnam	573
846	BP Prudhoe Bay Royalty Trust	626	897	Teradyne	572
847	AST Research	621	898	Church & Dwight	570

Rank	Company	Market Value ($ Mil.)	Rank	Company	Market Value ($ Mil.)
899	Marvel Entertainment Group	570	950	Central Louisiana Electric	512
900	Pic 'N' Save	569	951	Control Data	510
901	Merry-Go-Round Enterprises	567	952	Altera	510
902	Sierra Pacific Resources	565	953	American Health Properties	509
903	National Presto Industries	564	954	Standard Register	509
904	Nerco	564	955	West One Bancorp	509
905	Media General	564	956	Handleman	507
906	Broad	563	957	Information Resources	507
907	Westmark International	563	958	Medimmune	505
908	Morrison	562	959	Shopko Stores	504
909	Structural Dynamics Research	562	960	Kimball International	502
910	Fisher-Price	561	961	Tektronix	502
911	H.B. Fuller	560	962	Health Care Property Investors	501
912	Dominion Bankshares	559	963	Neiman Marcus Group	500
913	Lawter International	559	964	Wesco Financial	498
914	Goulds Pumps	559	965	Lennar	498
915	Lands' End	557	966	Herman Miller	495
916	Mark IV Industries	554	967	Advanta	494
917	Cardinal Distribution	554	968	SPI Pharmaceuticals	494
918	Staples	551	969	Albany International	493
919	Amsco International	549	970	Arvin Industries	492
920	Diagnostek	547	971	System Software Associates	491
921	Dexter	543	972	Mid-American Waste Systems	491
922	Masco Industries	542	973	United Asset Management	491
923	Eastern Enterprises	541	974	Fisery	488
924	GATX	536	975	Sealed Air	487
925	Transco Energy	533	976	Lancaster Colony	486
926	Hechinger	531	977	Anthem Electronics	485
927	United Missouri Bancshares	531	978	Southern Indiana Gas & Electric	485
928	Armco	531	979	La-Z-Boy Chair	480
929	Flowers Industries	530	980	Buffets	478
930	Santa Fe Energy Resources	529	981	Greif Bros.	477
931	Savannah Foods & Industries	529	982	United Illuminating	477
932	Helmerich & Payne	527	983	MGM Grand	477
933	Noble Affiliates	525	984	Xoma	476
934	Henley Group	523	985	FHP International	476
935	Airborne Freight	523	986	Wellfleet Communications	476
936	Weingarten Realty	522	987	Stop & Shop	476
937	Precision Castparts	522	988	Dauphin Deposit	476
938	Liberty National Bancorp	519	989	Gibson Greetings	474
939	Artisoft	519	990	First American	474
940	Green Tree Acceptance	518	991	Wetterau	473
941	Magma Power	516	992	Symbol Technologies	472
942	Venture Stores	515	993	Public Service Co. of New Mexico	470
943	NCH	515	994	American President	470
944	Home Beneficial	515	995	Cilcorp	470
945	Delta Woodside Industries	514	996	Pentair	469
946	Tokos Medical	514	997	MDU Resources Group	468
947	Dollar General	513	998	Dibrell Brothers	466
948	Chesapeake	512	999	TCA Cable TV	466
949	International Multifoods	512	1000	Paychex	465

APPENDIX TABLE B.2 CEO Compensations

Company	Executive	1991 Salary/Bonus (In thousands)	%Change From 1990	Long-Term Compensation (In thousands)	Total Direct Compensation (In thousands)
BASIC MATERIALS					
Air Products	Dexter F. Baker	$1161.4	6	0.0	1161.4
Alcoa	Paul H. O'Neill	562.4	-32	644.5	1206.9
Amax	Allen Born	900.0	-23	606.7	1506.7
American Cyanamid	George J. Sella Jr.	1309.4	6	1164.2	2473.6
Armco	Robert L. Purdum	654.0	N/A	19.8	673.8
Asarco	Richard de J. Osborne	850.9	-23	59.4	910.3
Avery Dennison	Charles D. Miller	1065.0	12	501.2	1566.2
Bethlehem Steel	Walter F. Williams	742.6	-8	10.1	752.7
Boise Cascade	John B. Fery	625.4	-10	0.0	625.4
Cabot	Samuel W. Bodman	500.0	-23	0.0	500.0
Champion Int'l	Andrew C. Sigler	1144.0	-8	0.0	1144.0
Consolidated Papers	George W. Mead	447.9	6	0.0	447.9
Dow Chemical	Frank P. Popoff	1069.0	15	256.3	1325.3
Du Pont	Edgar S. Woolard Jr.	1356.0	-10	20.7	1376.7
Ethyl	Floyd D. Gottwald Jr.	967.4	3	30.0	997.4
Georgia-Pacific	T. Marshall Hahn Jr.	1500.0	0	703.7	2203.7
Goodrich (B.F.)	John D. Ong	723.2	-28	805.5	1528.7
Hanna (M.A.)	Martin D. Walker	790.3	-9	349.0	1139.3
Hercules	Thomas L. Gossage	994.0	N/A	0.0	994.0
Imcera	George D. Kennedy	1325.0	17	3848.5	5173.5
Inland Steel	Frank W. Luerssen	523.6	6	0.0	523.6
International Paper	John A. Georges	1356.6	-6	1551.0	2907.6
James River	Robert C. Williams	768.1	N/A	434.9	1203.0
Kimberly-Clark	Darwin E. Smith	1428.0	4	0.0	1428.0
Louisiana-Pacific	Harry A. Merlo	750.0	15	586.7	1336.7
Lubrizol	L.E. Coleman	880.0	-5	442.4	1322.4
Lyondell Petrochem	Bob G. Gower	915.1	4	957.7	1872.8
Mead	Burnell R. Roberts	560.0	2	111.9	671.9
Monsanto	Richard J. Mahoney	1530.0	25	993.0	2523.0
Morton Int'l	Charles S. Locke	1164.2	8	468.0	1632.2
Olin	John W. Johnstone Jr.	787.6	6	91.4	879.0
Phelps Dodge	Douglas C. Yearley	1068.2	5	1482.7	2550.9
Quantum Chemical	John Hoyt Stookey	600.6	-25	0.0	600.6
Reynolds Metals	William O. Bourke	1090.0	-24	0.0	1090.0
Rohm & Haas	J. Lawrence Wilson	976.7	N/A	0.0	976.7
Scott Paper	Philip E. Lippincott	600.4	-1	121.9	722.3
Sherwin-Williams	John G. Breen	1077.7	0	417.4	1495.1
Stone Container	Roger W. Stone	803.5	N/A	1206.4	2009.9
Texas Industries	Robert D. Rogers	418.5	-21	0.0	418.5
Union Camp	Raymond E. Cartledge	939.3	N/A	195.0	1134.3
Union Carbide	Robert D. Kennedy	807.5	-43	0.0	807.5
Westvaco	John A. Luke	948.3	-4	286.9	1235.2
Weyerhaeuser	Geo. H. Weyerhaeuser	515.1	N/A	527.7	1042.8
Willamette	William Swindells	580.1	5	0.0	580.1

ENERGY					
Amerada Hess	Leon Hess	300.0	N/A	0.0	300.0
Amoco	H. Laurance Fuller	1302.2	N/A	236.3	1538.5
Ashland Oil	John R. Hall	1038.0	-6	0.0	1038.0
Atlantic Richfield	Lodwrick M. Cook	1673.9	-26	2008.9	3682.8
Baker Hughes	James D. Woods	916.0	-27	393.1	1309.1
Chevron	Kenneth T. Derr	1385.6	2	363.8	1749.4
Coastal	James R. Paul	813.8	N/A	913.4	1727.2
Dresser Industries	John J. Murphy	1070.6	-3	578.7	1649.3
Exxon	Lawrence G. Rawl	1812.8	17	7453.0	9265.8
FINA	Ron W. Haddock	476.3	N/A	0.0	476.3
Halliburton	Thomas H. Cruikshank	1200.0	9	401.5	1601.5
Kerr-McGee	F.A. McPherson	542.0	-32	129.4	671.4
Mobil	Allen E. Murray	2040.0	9	2624.9	4664.9
Occidental	Ray R. Irani	2324.0	N/A	147.1	2471.1
Pennzoil	James L. Pate	650.0	-2	0.0	650.0
Sun Co.	Robert McClements Jr.	1001.1	3	0.0	1001.1
Texaco	James W. Kinnear	1667.0	-14	1658.8	3325.8
USX	Charles A. Corry	1313.2	-18	840.2	2153.4
Unocal	Richard J. Stegemeier	1054.7	N/A	0.0	1054.7
Valero	William E. Greehey	1097.4	12	352.5	1449.9
INDUSTRIAL					
AMP	Harold A. McInnes	617.7	7	0.0	617.7
Allied-Signal	Lawrence A. Bossidy	809.1	N/A	0.0	809.1
Armstrong	William W. Adams	467.6	-25	23.8	491.4
Ball	Delmont A. Davis	815.0	N/A	22.0	837.0
Browning-Ferris	Wm. D. Ruckelshaus	806.0	0	0.0	806.0
Burlington Northern	Gerald Grinstein	1180.4	-8	0.0	1180.4
CSX	John W. Snow	1235.2	28	1832.4	3067.6
Caterpillar	Donald V. Fites	522.5	19	0.0	522.5
Conrail	James A. Hagen	1127.8	41	0.0	1127.8
Cooper Industries	Robert Cizik	1124.1	-16	5175.4	6299.5
Cummins Engine	Henry B. Schacht	450.0	2	1691.8	2141.8
Deere	Hans W. Becherer	654.0	N/A	162.7	816.7
Donnelley (R.R.)	John R. Walter	644.0	-16	99.6	743.6
Dover	Gary L. Roubos	765.0	-1	195.2	960.2
Dun & Bradstreet	Charles W. Moritz	1629.5	3	789.9	2419.4
Eaton	James R. Stover	693.8	-34	1015.8	1709.6
Emerson Electric	Charles F. Knight	1803.6	2	0.0	1803.6
Engelhard	Orin R. Smith	2119.3	32	3710.4	5829.7
FMC	Robert H. Malott	1238.1	N/A	3771.0	5009.1
Federal Express	Frederick W. Smith	662.5	14	0.0	662.5
Fluor	Leslie G. McCraw	1110.9	N/A	2191.3	3302.2
Harnischfeger	William W. Goessel	974.7	1	66.5	1041.2
Hillenbrand	W. August Hillenbrand	966.7	24	198.5	1165.2
Ingersoll-Rand	Theodore H. Black	1120.0	-6	278.6	1398.6
Morrison Knudsen	William M. Agee	2026.7	N/A	441.5	2468.2
Navistar	James C. Cotting	450.0	0	90.8	540.8
Norfolk Southern	Arnold B. McKinnon	1735.0	N/A	1208.4	2943.4
Owens-Corning	Max O. Weber	840.0	N/A	1164.1	2004.1

PPG Industries	Vincent A. Sarni	1038.9	-34	701.1	1740.0
Parker-Hannifin	Paul G. Schloemer	774.7	-21	46.2	820.9
Pentair	D. Eugene Nugent	979.3	47	487.9	1467.2
Premark	Warren L. Batts	1364.9	45	1593.8	2958.7
Raychem	Robert J. Saldich	448.8	N/A	447.9	N/A
Ryder System	M. Anthony Burns	764.2	-2	0.0	764.2
Santa Fe Pacific	Robert D. Krebs	977.1	78	294.4	1271.5
Stanley Works	Richard H. Ayers	678.9	-11	0.0	678.9
Tenneco	James L. Ketelsen	1469.6	N/A	21.1	1490.7
Trinova	Darryl F. Allen	504.6	3	0.0	504.6
Tyco	John F. Fort	1170.0	0	3150.0	4320.0
USG	Eugene B. Connolly	475.0	18	0.0	475.0
Union Pacific	Drew Lewis	2025.0	6	2131.3	4156.3
Waste Management	Dean L. Buntrock	1100.0	-30	2647.0	3747.0
Westinghouse	Paul E. Lego	677.1	-60	0.0	677.1
CYCLICAL					
AMR	Robert L. Crandall	600.0	0	3446.4	4046.4
Arvin	James K. Baker	488.8	-16	0.0	488.8
Bergen Brunswig	Robert E. Martini	676.4	N/A	0.0	676.4
Berkshire Hathaway	Warren E. Buffett	100.0	0	0.0	100.0
Black & Decker	Nolan D. Archibald	1300.5	2	269.5	1570.0
Brunswick	Jack F. Reichert	736.0	-34	189.0	925.0
CBS	Laurence A. Tisch	1382.9	2	0.0	1382.9
Capital Cities/ABC	Daniel B. Burke	842.0	-7	2839.1	3681.1
Chrysler	Lee A. Iacocca	976.4	-36	1904.0	2880.4
Claiborne (Liz)	Jerome A. Chazen	1185.0	-3	0.0	1185.0
Dana	Southwood J. Morcott	574.0	N/A	0.0	574.0
Delta Air Lines	Ronald W. Allen	475.0	-35	562.5	1037.5
Disney (Walt)	Michael D. Eisner	5441.5	-52	0.0	5441.5
Dow Jones	Peter R. Kann	730.0	N/A	0.0	730.0
Eagle-Picher	Thomas E. Petry	575.0	-2	0.0	575.0
Echlin	Frederick J. Mancheski	642.1	12	97.5	739.6
Ford Motor	Harold A. Poling	1140.8	N/A	0.0	1140.8
GM	Robert C. Stempel	1000.0	N/A	311.3	1311.3
Gannett	John J. Curley	1300.0	0	779.4	2079.4
General Cinema	Richard A. Smith	1275.0	3	0.0	1275.0
Goodyear	Stanley C. Gault	735.0	N/A	0.0	735.0
Hartmarx	Harvey A. Weinberg	557.0	-1	0.0	557.0
Hasbro	Alan G. Hassenfeld	1411.1	23	2740.6	4151.7
Home Depot	Bernard Marcus	1586.9	0	0.0	1586.9
Johnson Controls	James H. Keyes	659.2	8	182.8	842.0
Kmart	Joseph E. Antonini	1545.8	56	0.0	1545.8
Knight-Ridder	James K. Batten	605.0	-5	629.1	1234.1
Kodak	Kay R. Whitmore	1146.2	-19	49.2	1195.4
Marriott	J.W. Marriott Jr.	1304.4	71	0.0	1304.4
Mattel	John W. Amerman	1364.4	12	886.3	2250.7
Maytag	Daniel J. Krumm	982.6	N/A	0.0	982.6
McDonald's	Michael R. Quinlan	1335.0	7	0.0	1335.0
McGraw-Hill	Joseph L. Dionne	834.3	-19	506.8	1341.1
Melville	Stanley P. Goldstein	1297.0	7	601.6	1898.6
Miller (Herman)	Richard H. Ruch	405.0	-9	0.0	405.0
New York Times	Arthur Ochs Sulzberger	943.9	90	141.6	1085.5
Nike	Philip H. Knight	653.6	20	0.0	653.6

Outboard Marine	James C. Chapman	412.0	8	203.3	615.3
Paramount	Martin S. Davis	2750.0	-25	2280.8	5030.8
Penney (J.C.)	William R. Howell	937.2	-6	387.3	1324.5
Polaroid	I. MacAllister Booth	528.4	N/A	150.5	678.9
Price	Robert E. Price	270.0	12	0.0	270.0
Reebok	Paul B. Fireman	2007.2	-86	0.0	2007.2
Rubbermaid	Walter W. Williams	825.6	N/A	1594.1	2419.7
Sears Roebuck	Edward A. Brennan	979.8	0	1508.4	2488.2
Service Merchandise	Raymond Zimmerman	754.6	1	0.0	754.6
Shaw Industries	Robert E. Shaw	1334.0	9	0.0	1334.0
Smith (A.O.)	Robert J. O'Toole	442.4	-21	0.0	442.4
Springs Industries	Walter Y. Elisha	816.3	74	59.6	875.9
Tandy	John V. Roach	722.9	-17	0.0	722.9
Times Mirror	Robert F. Erburu	909.4	3	577.5	1486.9
Timken	Joseph F. Toot Jr.	536.0	-18	0.0	536.0
Tribune	Charles T. Brumback	869.2	N/A	1147.8	2017.0
UAL	Stephen M. Wolf	575.0	0	2235.0	2810.0
USAir	Seth E. Schofield	466.8	N/A	425.1	891.9
Vons	Roger E. Stangeland	1009.4	14	0.0	1009.4
Walgreen	Charles R. Walgreen III	858.8	8	207.3	1066.1
Washington Post	Donald E. Graham	375.0	N/A	0.0	375.0
Wendy's Int'l	James W. Near	947.4	30	147.0	1094.4
Whirlpool	David R. Whitwam	1590.4	60	0.0	1590.4
Zenith	Jerry K. Pearlman	523.7	16	0.0	523.7
NONCYCLICAL					
Abbott Labs	Duane L. Burnham	1340.5	15	2867.0	4207.5
Albertson's	Gary G. Michael	701.2	N/A	1710.0	2411.2
Alco Standard	Ray B. Mundt	742.9	-29	880.8	1623.7
American Brands	William J. Alley	1826.3	12	89.2	1915.5
American Home Prods	John R. Stafford	1840.0	8	3790.0	5630.0
Anheuser-Busch	August A. Busch III	1878.8	8	4776.1	6654.9
Archer-Daniels-Midland	D.O. Andreas	2000.0	3	0.0	2000.0
Avon	James E. Preston	1192.9	0	2230.5	3423.4
Bausch & Lomb	Daniel E. Gill	1777.0	N/A	3720.3	5497.3
Baxter	Vernon R. Loucks Jr.	1785.5	2	2517.1	4302.6
Becton Dickinson	Raymond V. Gilmartin	757.4	4	180.2	937.6
Borden	R.J. Ventres	1513.5	N/A	0.0	1513.5
Bristol-Myers Squibb	Richard L. Gelb	2051.1	-2	10126.9	12178.0
CPC International	Charles R. Shoemate	920.8	N/A	374.9	1295.7
Campbell Soup	David W. Johnson	1444.3	N/A	0.0	1444.3
Clorox	C.R. Weaver	869.5	N/A	240.3	1109.8
Coca-Cola Co.	Roberto C. Goizueta	2962.0	0	943.6	3905.6
Coca-Cola Enterprises	Brian G. Dyson	786.6	N/A	4224.5	5011.1
Colgate-Palmolive	Reuben Mark	1747.6	N/A	0.0	1747.6
ConAgra	Charles M. Harper	2463.0	33	4748.8	7211.8
Dean Foods	Howard M. Dean	779.0	18	211.8	990.8
Dial	John W. Teets	2160.6	6	1415.5	3576.1
Fleming	Dean Werries	763.9	12	314.3	1078.2
General Mills	H. Brewster Atwater Jr.	1786.9	8	3100.1	4887.0
Gillette	Alfred M. Zeien	1111.7	N/A	2724.4	3836.1
Heinz (H.J.)	Anthony J.F. O'Reilly	1075.8	4	73723.9	74799.7
Hershey Foods	Richard A. Zimmerman	815.0	6	562.5	1377.5
Hormel	Richard L. Knowlton	1662.3	4	2337.5	3999.8

Humana	David A. Jones	1366.4	20	5045.0	6411.4
Johnson & Johnson	Ralph S. Larsen	1253.0	N/A	1160.9	2413.9
Kellogg	W.E. LaMothe	1852.2	18	7873.7	9725.9
Kroger	Joseph A. Pichler	600.4	-21	91.4	691.8
Lilly (Eli)	Richard D. Wood	2217.9	N/A	3856.5	6074.4
McCormick	Charles P. McCormick Jr.	1035.7	4	1591.5	N/A
Merck	P. Roy Vagelos	3344.5	N/A	0.0	3344.5
National Medical	Richard K. Eamer	1397.6	5	15515.6	16913.2
National Service	Sidney Kirschner	756.9	-7	0.0	756.9
PepsiCo	D. Wayne Calloway	1911.9	1	0.0	1911.9
Pfizer	William C. Steere Jr.	1113.0	N/A	842.6	1955.6
Philip Morris	Hamish Maxwell	1740.8	N/A	13936.6	15677.4
Procter & Gamble	Edwin L. Artzt	2180.9	N/A	371.8	2552.7
Quaker Oats	William D. Smithburg	1404.6	30	2211.8	3616.4
Ralston Purina	William P. Stiritz	1197.8	3	12614.9	13812.7
Sara Lee	John H. Bryan Jr.	1750.8	25	0.0	1750.8
Schering-Plough	Robert P. Luciano	1674.0	12	1634.5	3308.5
Sysco	John F. Woodhouse	1067.4	-5	0.0	1067.4
Tyson Foods	Leland Tollett	886.8	N/A	0.0	886.8
U.S. Surgical	Leon C. Hirsch	1351.3	N/A	21211.9	22563.2
Universal	Henry H. Harrell	533.1	11	0.0	533.1
Warner-Lambert	Melvin R. Goodes	1286.7	N/A	1665.2	2951.9
Whitman	James W. Cozad	1209.6	24	2012.0	3221.6
Winn-Dixie	A. Dano Davis	646.4	8	1115.1	1761.5
Wrigley	William Wrigley	729.3	1	403.5	1132.8
TECHNOLOGY					
AM International	Merle H. Banta	450.0	-38	0.0	450.0
AT&T	Robert E. Allen	2061.2	2	788.8	2850.0
Amdahl	John C. Lewis	637.3	N/A	0.0	637.3
Apple	John Sculley	1338.0	-39	14531.0	N/A
Boeing	Frank A. Shrontz	1188.0	9	0.0	1188.0
Compaq Computer	Joseph R. Canion	737.3	N/A	1195.6	1932.9
Control Data	Lawrence Perlman	756.0	-16	168.7	924.7
Corning	James R. Houghton	981.6	7	690.7	1672.3
Data General	Ronald L. Skates	817.5	76	1225.0	2042.5
Digital Equipment	Kenneth H. Olsen	995.0	1	0.0	995.0
E-Systems	E. Gene Keiffer	957.9	8	91.9	1049.8
GenCorp	A. William Reynolds	870.0	6	0.0	870.0
General Dynamics	William A. Anders	1500.0	N/A	2950.0	4450.0
General Electric	John F. Welch Jr.	3207.0	8	1893.8	5100.8
General Signal	Edmund M. Carpenter	962.2	27	187.8	1150.0
Grumman	Renso L. Caporali	595.0	37	64.5	659.5
Harris	J.T. Hartley	868.8	-19	79.8	948.6
Hewlett-Packard	John A. Young	1519.4	2	1491.9	3011.3
Honeywell	James J. Renier	1276.3	-13	1291.9	2568.2
IBM	John F. Akers	1575.0	-40	921.1	2496.1
Intel	Andrew S. Grove	1115.4	36	0.0	1115.4
Litton	Orion L. Hoch	1406.5	N/A	0.0	1406.5
Lockheed	Daniel M. Tellep	1225.0	21	89.8	1314.8
Martin Marietta	Norman R. Augustine	1261.5	12	2981.5	4243.0
McDonnell Douglas	John F. McDonnell	850.8	47	0.0	850.8
Minn. Mining & Manuf.	Allen F. Jacobson	1300.9	N/A	1216.1	2517.0
Motorola	George M.C. Fisher	1118.4	20	0.0	1118.4

National Semiconductor	Charles E. Sporck	730.7	23	0.0	730.7
Northrop	Kent Kresa	1261.3	12	945.7	2207.0
Perkin-Elmer	Gaynor N. Kelley	389.6	N/A	0.0	389.6
Pitney Bowes	George B. Harvey	1166.3	51	0.0	1166.3
Rockwell Int'l	Donald R. Beall	1517.0	2	695.2	2212.2
Rohr Industries	Robert H. Goldsmith	625.2	94	384.1	N/A
Seagate	Alan F. Shugart	1179.4	-16	890.6	2070.0
Sun Microsystems	Scott G. McNealy	1571.6	12	0.0	1571.6
Sundstrand	Harry C. Stonecipher	945.8	2	0.0	945.8
TRW	Joseph T. Gorman	1098.8	7	141.1	1239.9
Tandem Computers	James G. Treybig	452.8	-17	0.0	452.8
Tektronix	Jerome J. Meyer	791.6	N/A	0.0	791.6
Teledyne	William P. Rutledge	683.0	N/A	0.0	683.0
Texas Instruments	Jerry R. Junkins	604.1	1	5.3	609.4
Textron	B.F. Dolan	2859.7	N/A	175.1	3034.8
Unisys	James A. Unruh	990.0	70	16.9	1006.9
United Technologies	Robert F. Daniell	1166.3	-15	0.0	1166.3
Varian	J. Tracy O'Rourke	1772.5	N/A	229.0	2001.5
Wang	Richard W. Miller	1000.0	N/A	0.0	1000.0
Western Digital	Roger W. Johnson	538.5	-8	0.0	538.5
Xerox	Paul A. Allaire	1083.4	-9	506.8	1590.2
FINANCIAL					
Aetna	James T. Lynn	1086.5	2	0.0	1086.5
Aflac	Daniel P. Amos	1108.9	N/A	1264.9	2373.8
Ahmanson (H.F.)	Richard H. Deihl	1214.4	-1	229.4	1443.8
Alleghany	F.M. Kirby	979.4	1	855.8	1835.2
American Express	James D. Robinson III	1625.0	-10	0.0	1625.0
American General	Harold S. Hook	1766.3	32	2182.9	3949.2
American Int'l Group	Maurice R. Greenberg	1950.0	11	0.0	1950.0
Anchor Bancorp	James M. Large Jr.	556.2	6	0.0	556.2
Banc One	John B. McCoy	1782.6	43	387.1	2169.7
Bank of Boston	Ira Stepanian	650.0	0	0.0	650.0
Bank of New York	J. Carter Bacot	809.6	-7	68.5	878.1
BankAmerica	Richard M. Rosenberg	1600.0	0	2350.0	3950.0
Bankers Trust	Charles S. Sanford Jr.	1850.0	23	2197.1	4047.1
Bear Stearns	Alan C. Greenberg	5299.1	26	0.0	5299.1
Beneficial	Finn M.W. Caspersen	1991.8	3	0.0	1991.8
Chase Manhattan	Thomas G. Labrecque	1180.0	N/A	0.0	1180.0
Chemical Banking	Walter V. Shipley	1715.8	132	762.0	2477.8
Chubb	Dean R. O'Hare	1241.3	8	997.9	2239.2
Cigna	Wilson H. Taylor	1058.0	3	510.4	1568.4
Citicorp	John S. Reed	1150.0	1	0.0	1150.0
Dime Savings	Richard D. Parsons	525.0	20	0.0	525.0
First Bank System	John F. Grundhofer	1009.5	N/A	0.0	1009.5
First Chicago	Barry F. Sullivan	770.1	5	0.0	770.1
First Interstate	Edward M. Carson	665.2	N/A	145.1	810.3
First Union	Edward E. Crutchfield Jr.	650.0	8	291.1	941.1
Fleet/Norstar Financial	Terrence Murray	1402.0	85	493.6	N/A
General Re	Ronald E. Ferguson	1257.3	N/A	540.4	1797.7
Glenfed	Norman M. Coulson	596.7	-12	0.0	596.7
Golden West	Herbert M. Sandler	774.6	7	0.0	774.6
	Marion O. Sandler	774.4	7	0.0	774.4
Great Western	James F. Montgomery	1719.4	27	0.0	1719.4

Household Int'l	Donald C. Clark	1094.6	-23	0.0	1094.6
ITT	Rand V. Araskog	2800.0	-27	1080.0	3880.0
Kemper	Joseph E. Luecke	685.7	-6	245.7	931.4
Loews	Laurence A. Tisch	339.3	N/A	0.0	339.3
	Preston R. Tisch	1324.2	6	0.0	1324.2
Mellon	Frank V. Cahouet	1176.8	N/A	1753.0	2929.8
Merrill Lynch	William A. Schreyer	5850.0	256	5586.2	11436.2
Midlantic	Garry J. Scheuring	574.4	N/A	0.0	574.4
Morgan (J.P.)	Dennis Weatherstone	2043.2	30	2069.1	4112.3
Morgan Stanley	Richard B. Fisher	7620.0	N/A	0.0	7620.0
NationsBank	Hugh L. McColl Jr.	2000.0	186	1353.0	3353.0
Norwest	Lloyd P. Johnson	1628.3	18	1439.3	3067.6
PNC Financial	Thomas H. O'Brien	1294.4	75	0.0	1294.4
PaineWebber	Donald B. Marron	3600.0	279	0.0	3600.0
Primerica	Stanford I. Weill	3106.6	21	13672.3	16778.9
Republic New York	Walter H. Weiner	1019.6	7	0.0	1019.6
Safeco	Bruce Maines	649.0	13	1068.0	1717.0
Schwab (Charles)	Charles R. Schwab	3176.2	73	0.0	3176.2
St. Paul	Douglas W. Leatherdale	954.5	0	238.7	1193.2
SunTrust	James B. Williams	601.7	37	5354.1	5955.8
Transamerica	Frank C. Herringer	1100.1	N/A	162.5	1262.6
Travelers	Edward H. Budd	923.1	32	0.0	923.1
U.S. Bancorp	Roger L. Breezley	623.3	5	0.0	623.3
USF&G	Norman P. Blake Jr.	1346.9	N/A	0.0	1346.9
Unum	James F. Orr III	1067.2	4	1377.4	2444.6
Wells Fargo	Carl E. Reichardt	775.0	-47	0.0	775.0
UTILITIES					
American Elec. Power	Richard E. Disbrow	618.0	N/A	0.0	618.0
Ameritech	William L. Weiss	1270.0.	-2	849.2	2119.2
Bell Atlantic	Raymond W. Smith	1457.2	14	1293.6	2750.8
Bell South	John L. Clendenin	1355.5	-1	39.6	1395.1
Centerior Energy	Richard A. Miller	504.6	8	41.7	546.3
Commonwealth Edison	James J. O'Connor	671.0	N/A	0.0	671.0
Con Edison	Eugene R. McGrath	802.7	N/A	0.0	802.7
Detroit Edison	John E. Lobbia	522.0	18	0.0	522.0
Duke Power	William S. Lee	624.9	4	0.0	624.9
Entergy	Edwin Lupberger	641.6	-15	38.9	680.5
FPL	James L. Broadhead	1120.5	29	0.0	1120.5
GPU	William G. Kuhns	537.4	N/A	0.0	537.4
GTE	James L. Johnson	2167.8	24	709.3	2877.1
Lilco	William J. Catacosinos	455.0	15	0.0	455.0
MCI Communications	William G. McGowan	1594.4	N/A	2594.4	4188.8
Nynex	William C. Ferguson	1214.5	16	338.7	1553.2
Ohio Edison	Justin T. Rogers Jr.	526.3	28	0.0	526.3
PacifiCorp	A.M. Gleason	1055.3	10	1279.2	2334.5
Pacific Enterprises	James R. Ukropina	594.0	N/A	440.0	1034.0
Pacific Telesis	Sam Ginn	1113.7	16	726.7	1840.4
Philadelphia Electric	Joseph F. Paquette Jr.	696.0	38	0.0	696.0
SCEcorp	John E. Bryson	810.0	N/A	125.6	935.6
Southern Co.	Edward L. Addison	971.8	N/A	0.0	971.8
Southwestern Bell	Edward E.Whitacre Jr.	1379.8	0	1088.2	2468.0
U S West	Richard D. McCormick	660.0	N/A	740.7	1400.7
Union Electric	William E. Cornelius	660.0	19	0.0	660.0

Appendix Table C.1 Random Numbers

53535	04260	77609	93799	92171	45524	10968	30231	70864	29908
41292	15201	66342	59155	46163	69248	31029	62034	21855	27863
07320	22682	09595	44805	54593	53350	61354	14029	10195	18644
77676	67772	45072	08940	02592	45976	82099	90739	77072	42081
43227	20568	16309	23841	53173	39475	27282	82699	00022	96419
90712	41695	67474	27567	93269	10163	94190	36188	41491	71217
88103	21514	60787	33170	58215	89951	01634	98155	05154	08971
72252	35791	84125	31962	81093	93068	41197	57779	88515	48002
51702	49516	69510	19678	47298	11355	68459	96360	13436	66314
63055	86998	22187	59898	96371	61370	35937	34292	00678	33505
32373	57889	85880	66515	37489	37854	72926	23437	62233	38651
71996	16525	25618	56577	69130	25035	93551	54394	81572	90624
26912	70619	22576	22780	99118	18487	58801	36063	32886	60453
74589	82677	13353	67658	17080	43212	34585	17179	86980	81899
56041	53072	19912	47466	32585	41414	07564	80712	27286	07966
09286	68067	84883	10023	78195	84711	85988	31545	39904	14984
33610	84843	07145	38437	06148	06094	89601	96751	49124	55092
14113	06396	59084	02534	09360	81918	77118	91640	92978	24815
56302	89765	63857	42747	28592	41784	00822	60356	96389	11728
06362	94540	29532	09994	55277	43897	63268	40481	00312	46039
48568	34412	84939	54850	84317	92032	60130	49071	68962	28953
65975	60965	77679	95782	67541	50654	09482	56111	98710	35803
66686	32977	48472	30226	54226	72490	18395	37338	88279	79089
51610	13000	73849	46654	30324	78000	72852	28934	83197	59003
47600	86103	25788	08774	72020	04543	25849	88887	41159	30131
34860	67572	83116	99579	81303	41889	56577	64142	51596	25329
76649	50908	67006	29332	29689	68786	98987	34815	53512	20620
78321	54309	85956	04976	37863	06711	72679	03405	28770	08515
35775	21295	39621	02339	16537	42246	06571	81193	94930	05376
06783	21338	89886	78826	02303	37886	70453	11021	62887	36855
25887	53024	71881	51208	95739	98572	01903	68043	62661	71273
37784	42100	70838	78963	10927	05448	25759	74051	47577	30196
02120	59536	82996	22671	89267	65924	46725	69179	15182	59158
55292	03836	28883	71134	08547	93204	09656	11671	29735	59573
66186	43648	97926	80469	66412	73647	36779	84688	96862	51937
55010	11479	55036	82146	37120	62328	56276	28906	45311	61818
02322	18679	18478	30052	05666	84405	47513	09244	78978	91819
78056	67836	82582	25809	20198	37222	62629	75733	77420	58746
69812	88260	83519	10062	60865	35038	14665	18163	59351	25794
84904	66864	26982	37928	32988	87652	81415	24416	93778	20391
83143	47631	79772	08576	10311	17597	71049	63326	47168	05737
44423	71197	91081	40781	72403	76245	31881	55716	89255	71997
59882	58479	59609	80115	91569	23152	51781	85744	78640	80172
74890	90405	75945	31645	61008	24448	42249	84909	29013	12529
52174	64334	77631	19855	17723	02897	80427	20700	92210	92091
41361	24347	53420	33639	83765	97935	83630	33765	21502	15589
94585	84798	98480	08335	08728	60428	22282	76784	37316	08624
36020	71966	61443	12554	67446	08676	46177	22422	87471	27283
08112	59807	28404	60316	49676	52901	90604	48379	85233	52060
05853	69681	52034	77617	78644	57321	14162	01849	94684	14628

Appendix Table C.2 Control Chart Constants

Number of Observations in Subgroup, n	A_2	d_2	d_3	D_3	D_4
2	1.880	1.128	.853	.000	3.267
3	1.023	1.693	.888	.000	2.574
4	.729	2.059	.880	.000	2.282
5	.577	2.326	.864	.000	2.114
6	.483	2.534	.848	.000	2.004
7	.419	2.704	.833	.076	1.924
8	.373	2.847	.820	.136	1.864
9	.337	2.970	.808	.184	1.816
10	.308	3.078	.797	.223	1.777
11	.285	3.173	.787	.256	1.744
12	.266	3.258	.778	.283	1.717
13	.249	3.336	.770	.307	1.693
14	.235	3.407	.762	.328	1.672
15	.223	3.472	.755	.347	1.653
16	.212	3.532	.749	.363	1.637
17	.203	3.588	.743	.378	1.622
18	.194	3.640	.738	.391	1.608
19	.187	3.689	.733	.403	1.597
20	.180	3.735	.729	.415	1.585
21	.173	3.778	.724	.425	1.575
22	.167	3.819	.720	.434	1.566
23	.162	3.858	.716	.443	1.557
24	.157	3.895	.712	.451	1.548
25	.153	3.931	.709	.459	1.541
More than 25	$3/\sqrt{n}$				

Short Answers to Selected Even-Numbered Exercises

Chapter 1

1.2 Yes. The set of all students taking a course in a given semester can be viewed as the population in research on those who take the course that semester. It can be viewed as a sample in a study where the population is the set of all students attending the university.

1.4

less than 25,000	0.033
25,000–99,999	0.773
100,000–199,999	0.168
200,000–299,999	0.021
300,000–399,999	0.004
400,000+	0.002

1.6 Quarterbacks (1–19) tend to weigh less than running backs (20–49), who tend to weigh less than offensive linemen (50–79).

1.8 The former is a cross-sectional analysis; the latter is time series.

1.10 Depends on the data source.

1.12 The average age for deaths in a region in a given year tells us nothing about life expectancies at that age.

1.14 That more than 1 percent of those 30 percent who refused the interview may have had sex exclusively with other men in the past ten years.

1.16 All frequency distributions can be constructed in a form of bar chart, but not all bar charts represent frequency distributions.

1.22 To analyze each group of owners' relative share, the pie chart would be best. If we are interested in the number of owners in each group, the bar chart would be better. If we want to visualize the amount of the distribution less than the upper boundary housing value, the ogive would be best.

1.24 The following frequency distribution can be used to draw the chart.

	f_i/N
Banks	0.31
Insurance	0.39
Mutual Fund	0.18
Other	0.12
Column Sum	1.00

1.28 The ogive shows the amount of distribution that is less than the upper boundary value on the horizontal axis. This is not shown in the pie chart or in the histogram for the relative frequency distribution.

1.30 Plaintiffs want the pool to be the entire workforce because that will give the highest proportion of women or other minority. If a firm does not have a high proportion of women or other minority, a censored pool might be to its advantage for comparison purposes.

1.32 You might question whether this report collected data for all female-owned businesses.

1.34 This diagram only tells us the productivity figures for the service and manufacturing sectors. To know the effect on the overall economy, we need more information about their relative shares within the overall economy and other sectors' productivity figures and their shares.

1.36 Those 561 trucks stopped can hardly be regarded as a representative sample or even a random sample.

1.38 Answer depends on the data source.

1.40 a. Could have been as high as 95 percent.
b. Could have been as low as 33 percent.

1.44 Killington, 10%; Sugarbush, 15%; Sunday River, 24%; Sugarloaf, 25%; Okemo, 28%; Smuggler's Notch, 45%.

1.46 WSJ diagram is wrong because it does not correctly show the relative shares of Pinkerton and Conwood.

1.50 The ogive shows the amount of the distribution that is less than the upper boundary value on the horizontal axis.

Chapter 2

2.2 Consumer satisfaction is a qualitative variable. It is only meaningful to compare the order of the rankings and not the relative magnitudes.

2.4 $4.114 million

2.6 a. 1.85
b. 23,400

2.8 When a distribution is unimodal, continuous and symmetric.

2.10 a. 67.78
b. 70
c. Any value can be a mode.

2.12 a. The mean is 16.47; the median is 16.50.
b. The distribution is left skewed.

2.14 The medians for three years are approximately 15, 14.375, and 22.5 weeks, respectively.

Not all means can be calculated because the class 30+ is open.

2.16 The medians for three spreadsheets are 7.15, 5.825, and 6.34, respectively. The median is a better measure of central tendency because the values in this data set are qualitative.

2.18 He is probably referring to the median because he is comparing an average survival of 7.3 months with his own result that "half" of the patients have survived 23 months or longer.

2.20 Sample mean = 52.09(= 520.9/10 : sum of values divided by number of values). Sample standard deviation = 32.5438($=\sqrt{1059.10}$: square root of sample variance). Sample variance = 1059.10(= 9531.9189/9: sum of squared deviations divided by degrees of freedom $(n-1)$). Standard error of the mean = 10.29125 = 32.5438/3.16 (the sample standard deviation divided by the square root of the sample size).

The other advertisers bought $1,069 million of commercial time.

2.22 The mean of data set B is 100 times larger than the mean of data set A. The variance is 10,000 times larger for data set B than for data set A. These relationships would hold even if these were population data.

2.24 The approximate sample mean is $9,520.

The approximate sample standard deviation is $190.595.

2.26 The highest degree of skewness can be found in the distribution for FY 90.

2.28 CEO compensation tends to deviate from its mean in the same direction as the number of headquartered companies deviates from its mean. This implies that the CEO compensation and the number of headquartered companies are positively related to each other.

2.30 The smallest fraction of calls that are handled within 25 and 85 seconds is about 0.95.

2.32 "At most" 44 percent of total students could have been paid above $140,000.

2.34 The mean delivery time is 156 minutes. The exact median and the mode cannot be calculated.

2.36 5th percentile = 3.90; 15th = 9.25; 25th = 11.70; 35th = 13.50; 45th = 15.35; 55th = 17.70; 65th = 20.65; 75th = 24.90; 85th = 33; 95th = 59.50.

2.40 An approximate median is 17.6 percent.

To calculate the median, we have to assume the observations are evenly spread within the median class.

2.42 The median numbers of times of play per year are 12 for skiers, 25 for tennis players, and 25 for golfers, respectively. The means and modes cannot be determined.

2.44 It would be possible to observe that 54.2 percent are above average if the distribution of the cost is left skewed.

2.46 The mean and median are not equal in each year because the distribution of automobile output is not symmetric. Mean in 1992 = 144524; mean in 1991 = 142188.6. Median in 1992 = 62000; median in 1991 = 49000.

2.48 The percentage change in average is a better measure of change than the percentage change in median.

2.50 a. No
b. No
c. Yes
d. Yes

2.52 For Austin-Los Angeles, Continental Airline has the median price of $258. For the remaining two cities, the medians cannot be clearly identified.

2.54 Mean = 41,914.8; median = 37,750; mode = 38,000 or 46,000; range =182,460; variance = 1,332,754,040; standard deviation = 36,506.904

2.56 a. The means cannot be calculated because the class 11+ is open. Approximate medians are 3.46 days and 5.63 days.
b. The data is not a proper form for a frequency distribution because there is an open class. It is not possible to calculate the mean with this data set.
c. The approximate median is a better measure because it can be determined without any trouble caused by the open class.

2.58 The sample mean is $21.618. The sample median is $18.42. The sample variance is 24.886 and hence the standard deviation is $4.989.

2.60 Average yearly production has increased with the introduction of new body styles, except for the last several years. Although the model average may show a trend in Corvette popularity, if the number of years that each model is produced is too different, then model average may not be comparable.

2.62 Average capital/asset ratio = 16.21 percent; the ratio of (average capital)/(average asset) = 14.64 percent

The average capital/asset ratio is the simple arithmetic average of capital/asset ratios of 100 banks while the ratio of (average capital)/(average asset) is nothing but the sum of 100 weighted capital/asset ratios(using the asset/(sum of asset) ratio as each weight).

2.64

	Sales	Profit
U.S. firms:	6.77%	−7.73%
Netherlands/U.K. firms:	4.69%	2.94%
Japan firms:	6.43%	−12.22%
Germany firms:	12.75%	20.00%
France firms:	13.00%	6.00%
Italy firm:	16.00%	N.A.
Switzerland firms:	11.00%	26.00%
Australia firm:	16.00%	9.00%

2.66 Women are receiving less than the men. The median salaries are $33,150 versus $39,850. Even if the ten highest paid women each received $50,000 raises, nothing changes in the median salary for women.

The mean is a better measure when assessing the cost of removing salary differences because both means would be exactly the same if a total of $1,894,439.838(= (218)(43331.449 – 34641.358)) is raised for salaries for those 218 women.

2.68 Time of arrival seems to affect the range of wait time. Range of wait time tends to be small in early hours of the day with the exception of 0 – 2 hour time interval.

Chapter 3

3.2 An outcome is one of possible results in a statistical experiment. An event is a subset of the sample space that is formed by outcomes that share an attribute of interest.

3.4 Although the chance of failure is minuscule, we can observe it actually happens in less than 300 years, and we did. The Challenger accident could be understood as an event with a minuscule probability that was actually observed. The occurrence of an uncertain event is not tied to a fixed number of trials.

3.6 As long as there is a limited number of trials we cannot guarantee that the relative frequency of 0.2 will be actually realized.

3.8 They are different because one statement is referring to a fixed number of trials while the other is referring to an infinite number of trials.

3.12 720

3.14 1/3

3.16 3/10

3.18 42

3.20 a. False; if $P(A_1|A_2) = P(A_1)$, then they are independent.
b. True; $P(A_1|A_2) = 0$ implies $P(A_1|A_2) = P(A_1 \text{ and } A_2)/P(A_2) = 0$ and thus implies $P(A_1 \text{ and } A_2) = 0$.
c. False; two probability do not need to be the same for independent events.

3.22 0.0675

3.24 Letting P(A) be the probability that an institutional investor would favor the creation of a compensation committee, P(B) be the probability that an institutional investor would favor a compensation committee that has the power to hire and pay compensation consultants, P(A) = 0.66 and P(B) = 0.50 are given. P(B|A), however, is not given. Thus, we cannot calculate P(A and B), i.e., the probability that an institutional investor would favor the creation of a compensation committee that has the power to hire and pay compensation consultants.

3.26 0.0532

3.28 0.15

3.30 0.96

3.32 If properly interpreted, they convey the same thing. For example, if odds of a team winning are "3 to 4" then on the average, 3 out of 7 times this team will win; thus the probability, of this team winning is approximately 0.42857(=3/7).

3.34 The favorites to win (1st), place (2nd), and show (3rd) were Prairie Bayou, Sea Hero, and Personal Hope, respectively.

3.36 They are similar because both measure the likelihood of an outcome. However, a frequency distribution is constructed based on what has occurred while a probability is based on what would occur in an infinite number of trials.

3.38 4200

3.40 1/5

3.42 a. 0.2143
b. 0.0357
c. 0.4643
d. 0.1429

3.44 0.9995

3.46 a. 0.35 for the 9-year old and 0.70 for the 12-year old.
b. 0.245
c. 0.805
d. We assume they understand the TV commercial independently. This assumption will be reasonable if they are not allowed to communicate with each other.

3.48 See the table at the bottom of the page.

3.50 a. 0.1
b. 0.7
c. 0.1429
d. 0.3333

3.52 0.1875

3.54 0.99; We need to assume that two detectors are operating independently.

3.56 P(the series will last 4 games) = 0.0625
P(the series will last 5 games) = 0.125
P(the series will last 6 games) = 0.15625
P(the series will last 7 games) = 0.15625

Table for Exercise 3.48

	Percentage within the computer users	Percentage in total-professional workforce
Hermits	8%	(0.08)(0.75) = 6.00%
Collaborators	8%	(0.08)(0.75) = 6.00%
Globetrotters	10%	(0.10)(0.75) = 7.50%
Small Site Bosses	11%	(0.11)(0.75) = 8.25%
Corporate Wanderers	12%	(0.12)(0.75) = 9.00%
Corridor Cruisers	15%	(0.15)(0.75) = 11.25%
Solo Practitioners	16%	(0.16)(0.75) = 12.00%
Road Warriors	20%	(0.20)(0.75) = 15.00%
Total	100%	75.00%

Thus, P(a person is either a "Globetrotter" or "Road Warrior") = 0.225.

Chapter 4

4.2 Approximately 0.5

4.4 The expected value is calculated with probabilities. Thus, the expected value is not based on what has occurred but on what would occur over an infinite number of trials.

4.6 Because the expected payoff ($50) is less than the cost of the policy ($210), you would be better off not buying it.

4.8 The reason someone buys the service insurance policy might be that they calculate their expected payoff to be greater than the insurance cost (i.e., they subjectively estimate the probability of total destruction to be greater than 0.19192).

4.10 a. The profit to be made on sale of a 27-inch Schwinn.
b. $40.50
c. 62.25 square dollars
d. $4,050

4.12 a. 3.00
b. 1.24
c. 96 percent
d. The result is not consistent with the "Rule of Thumb" because the given distribution is not bell-shaped although it is symmetric.

4.14 The risk neutral would prefer the investment C because it has the largest expected net return. The risk neutral does not care about the risk, but is only interested in the long run net return.

4.16 The risk neutral would prefer the investment B because it has a larger expected net return.

4.18 a. Marginal probabilities are calculated as follows:

	A_1	A_2	A_3	marginal
B_1	0.12	0.03	0.10	0.25
B_2	0.21	0.02	0.18	0.41
B_3	0.11	0.09	0.14	0.34
marginal	0.44	0.14	0.42	1.00

b. 0.12
c. 0.64
d. 0.48
e. not independent

4.20 $5,038.6

4.22 1500($)

4.24 478.71

4.26 a. 2.3
b. 2.3
c. 0

4.28 a. 1.41
b. 1.41
c. 0

4.30 –6.67($)

4.32 What is important to the insurance company is the amount of claim per accident. If the amount of claim is large, the expected loss could also be large, although the probability of accident is not high.

4.34 The word "expects" implies the expected value of Pepsi's market share.

4.36 He meant that the expected payoff for the lottery buyer is less than the cost to buy it.

4.38 a. Because of the increase in cost of administrating lotteries.
b. There would have been more educated people and thus an increase in numbers of those educated people who are not willing to buy lottery tickets might have led to future reduction in lottery revenues.

Chapter 5

5.2 0.016. If 0.016 is viewed as "large" then we might conclude that less than the majority of customers would prefer the new product and be willing to pay up to 20 percent more for it.

5.4 0.987

5.6 0.410. We have to assume that five male friends should be a random sample from the large population of American men who use a hair dryer.

5.8 0.999967 ~ 1.0

We have to assume that each member independently decides to uphold a guilty verdict, which does not seem to be a reasonable assumption.

5.10 0.251. This probability could be viewed as "large." Thus, it would not be unusual to observe six big down days in October.

5.12 a. 1.6; 1.26.
b. 4.25; 2.06.
c. 5.625; 2.37.

5.14 a. 2.1
b. 0.027

5.16 0.172. This probability could be viewed as "large." Thus, we might again question NordicTrack's ad.

5.18 a. Yes
b. 0.0038

5.20 0.021978

5.22 If a person is once selected he or she cannot be returned to the pre-selection pool.

5.24 a. 0.2240
b. 0.9434
c. 0.4480

5.26 0.00000 (not zero, but very close to zero).

The Poisson distribution approach is appropriate because the number of possible cardiac arrests in a 15-month period can be large while the number of expected realization, 5, is small. The calculated probability is extremely "small" and thus it would be very unusual to observe 57 or more cardiac arrests when 5 is the expected number. We might conclude that the nurse may have been a contributing factor.

5.28 0.0498

5.30 a. 0.1251
b. 0.0378
c. 0.0000 (not zero, but close to zero)

5.32 0.185

5.34 a. 36
b. 0.097
c. The probability is so small that we cannot believe that only guessing was involved.

5.36 $\pi = 0.95$ and $n = 11$ are given where π is the probability of an expert arriving at the right verdict. By a "simple majority" the author considers X is 6 where X is the number of experts who are in favor of the right choice. Thus, using the binomial, $P(X = 6|n = 11, \pi = 0.95) = 0.0001$ which is approximately odds of 10,000 to 1.

5.38 Majority could be any percentage above 50 percentage points. Thus, 7, 8, 9, 10, 11, as well as 6, could be included in the probability calculation. Now, $P(X \geq 6|n = 11, \pi = 0.95) = 0.9999$ (almost one). The odds are changed to 1 to 1.

5.40 0.0000

5.42 a. 2.4
b. $E(X) = \Sigma x_1 p(x_1) + \Sigma x_2 p(x_2) + \Sigma x_3 p(x_3)$
$= 1(0.8) + 0(0.2)$ —— for the first trial
$+ 1(0.8) + 0(0.2)$ —— for the second trial
$+ 1(0.8) + 0(0.2)$ —— for the third trial
$= 2.4$

where x_1, x_2 and x_3 denote the number of error free items in the first, second, and third trial, respectively.

c. The average number of error free items per sample, over infinite number of random samples of each size three, is 2.4.
d. 0.48
e. $Var(X) = \Sigma(x_1 - \mu_1)^2 p(x_1) + \Sigma(x_2 - \mu_2)^2 p(x_2) + \Sigma(x_3 - \mu_3)^2 p(x_3)$
$= (1 - 0.8)^2(0.8) + (0 - 0.8)^2(0.2)$ — for the first trial
$+ (1 - 0.8)^2(0.8) + (0 - 0.8)^2(0.2)$ — for the second trial
$+ (1 - 0.8)^2(0.8) + (0 - 0.8)^2(0.2)$ — for the third trial
$= 0.16 + 0.16 + 0.16 = 0.48$

where x_1, x_2, x_3 denote the number of error free items and μ_1, μ_2, μ_3 ($\mu_1 = \mu_2 = \mu_3 = 0.8$) denote the expected (average) number of error free items in the first, second, and third trials, respectively.

5.44 The probability of getting one totally error free item is $P(X = 1|n = 3, \pi = 0.8) = 0.096$. The probability 0.096 can be viewed to be "large." Thus, it would not be unusual to observe one error free item when 2.4 is the expected number. We might conclude that the assembly line is running in a way that the probability of any given item being error free is a 0.8.

5.46 0.3125

5.48 a. 0.077 with 55 mph limit; 0.001 without 55 mph limit.
b. 0.868 with 55 mph limit; 0.185 without 55 mph limit.
c. 0.999 with 55 mph limit; 0.863 without 55 mph limit.
d. We need to assume the probability of having a death must be same for all the time intervals (100 million miles). This assumption might not be appropriate because road conditions are not the same for all interstate highways.

5.50 Not all accidents result in death. We need to know the value of lost earning from injuries. Off-setting this added cost of injuries at the higher speed is the benefit of saved time from traveling at the higher speed.

5.52 0.03125

Chapter 6

6.2 The standard normal distribution is the normal distribution with zero mean and unit standard deviation. A normal random variable can take any value as its mean, and any positive value as its standard deviation. Once a distribution is normal, it can be easily converted into the standard normal distribution using the converting formula in calculating probabilities.

6.4 a. −2.55
b. 0.91
c. 3.43
d. 0.98
e. −2.00
f. 0.43

6.6 a. 53.8
b. 39.4
c. 48.04
d. 25.72
e. there is no such value of b
f. 17.8

6.8 0.9210

6.10 It does not seem to be reasonable to assume that the distribution is normal because the distribution is highly right skewed with a very large value pulling up the mean.

6.12 $154.60

6.14 The probability that an agent is overcharged more than $585 is less than 0.01, which is extremely unlikely. We typically do not observe unlikely events, so given the agent is truthful, either the assumption of normality or the assumed $25 undercharge mean must be called into question.

6.16 62.47 percent

6.18 0.0001

6.20 The implied value of σ should lie between 3.8 and 4.4.

6.22 0.0287

6.24 0.2946

6.26 0.9216

Because the line speed is a continuous random variable that can be expected to vary symmetrically around its mean, with values closer to the mean occurring most often, it seems reasonable to assume that the average line speed (X) is normally distributed.

6.28 In order to satisfy the rule of normality, the distance between the mean and the upper end point and the distance between the mean and the lower end point should be the same. The distance between the mean and the upper end point is $248,760(= 330000 − 81240) while the distance between the mean and the lower end point is $61,240(= 81240 − 20000); the distribution cannot be symmetric, but is right skewed with a long right tail. It is reasonable to assume that the distribution is lognormal, because right skewed distributions can be transformed into lognormal distributions.

Chapter 7

7.2 Suppose the mean time devoted to prayer in the United States is to be estimated for Catholics, Jews, Muslims, and Protestants. Because these groups are not of equal size, simple random sampling might get none or too few of the smallest group. To avoid this, a number proportional to each group's representation in the population could be selected at random; thus, variability between the groups (strata) is controlled and precision is increased.

7.4 For a sample to be a random sample each and every element or set of elements in the population must have an equal chance to be selected into a sample. Those who valued their time at more than $1.00 were less likely to respond.

7.6 Answer depends on the random numbers used.

7.8 Decreases; 1/2.

7.10 The standard error of the mean is the standard deviation of the sampling distribution of the mean, which Altman does not acknowledge. It is a measure of the average variability in the sample means around the population mean and it is equal to $\sigma/\sqrt{n}$, where σ is the standard deviation of the values in the population and n is the sample size. The standard error of the mean is estimated by $s/\sqrt{n}$, where s is the standard deviation of values in a sample. Altman's definition does not distinguish clearly between σ and s; his statement about the standard deviation pertains to s and not σ.

7.12 Yes. The standard error of an estimator varies inversely with the sample size but increasing the sample size does not result in proportional reductions in the standard error.

7.14 Whenever the sample size is large enough (i.e., $n \geq 30$), we can assert that the sampling distribution of the sample mean is approximately normally distributed according to the Central Limit Theorem.

7.16 a. 0.1587
b. 0.0228
c. The sample size is 1 in a and 4 in b. If n = 1, the sampling distribution of the sample mean is just the population distribution itself. With a larger size sample, sampling distribution of the sample mean has smaller standard deviation than that in the population.

7.18 0.1056

We need to assume that the population of time in household work is normally distributed in order to have a normally distributed sampling distribution of the sample mean of time. We don't need additional assumptions about the distribution of the mean time and about the sample of 4 families.

7.20 0.9332

7.22 a. False. There is no reason to believe that any one value of the sample proportion would equal to the population proportion.
b. False. The sample size has nothing to do with whether the population proportion is equal to a sample proportion.
c. Yes. See above.
d. False. The variance of the distribution of the sample proportion is $\pi(1 - \pi)/n$ where π is the population proportion, while the variance of X is σ^2.
e. False. s^2/n is an estimator of the variance of the sample mean. It has nothing to do with the variance of the sampling distribution of the sample proportion.
f. Yes. The variance of the sample proportion for samples of size n equals $\pi(1 - \pi)/n$ where π is the population proportion.

7.24 a. 0.206
b. 0.1446

7.26 Because division managers *handpicked* the 35 employees from the pool of 1,700 employees, it cannot be considered to be a random sample.

7.28 a. mean of 320 lbs., standard deviation of 5 lbs.
b. 0.0013
c. 10 lbs.

7.30 Since the sample size n = 16 is not large, we need to assume that the population distribution is normal to answer the question. Under this assumption, the sample mean width will be normally distributed with a mean of 0.25 inches and a standard deviation of 0.0000025 inches. The probability that a sample mean is 0.25007 or larger will be 0.0000.

7.32 11.504

7.34 Even if the population mean (μ) is negative, we might observe a sample mean ($\bar{x}$) of 5.5 percent or greater. Thus, it could be possible to observe 5.5 percent or greater as a monthly change in sales of new homes when the sales in fact decreased.

7.36 $2.85 < \frac{\sigma}{\sqrt{4}} < 3.3$

7.38 Because the distribution of the time to be "up and running" is highly left skewed, it might be appropriate to use the binomial distribution instead of the normal distribution even though the sample size is relatively large. We can compare the following probabilities:

If the binomial random variable X denotes the number of "successes" with $\pi = 0.98$ when 30 days are randomly drawn, then the probability of getting 29 or fewer "successes" is

$$P(x \le 29 \mid n = 30, \pi = 0.98) =$$
$$\sum_{x=0}^{29} \binom{30}{x} 0.98^x (1-0.98)^{30-x} = 4.55$$

using the binomial distribution table.

Or as an alternative method, we could try a calculation with the normal distribution. The sampling distribution of the sample proportion (p) is approximately normal with a mean of 0.98 and a standard deviation of

$$0.0256 \left(= \sqrt{\frac{(0.98)(0.02)}{30}} \right)$$

Thus, the probability that the sample proportion (p) is 0.9667(= 29/30) or less will be

$$P(p \le 0.9667) = P\left(z \le \frac{0.9667 - 0.98}{0.0256}\right)$$
$$= P(z \le -0.52) = 0.3015$$

These two probabilities do not seem to agree with each other. The probability 0.455, calculated using the binomial distribution, can be regarded to be more accurate.

7.40 The probability of getting the mean difference 7.83 or greater is 0.0000.

7.42 a. 0.0392. Because the line speed is a continuous random variable that can be expected to vary symmetrically around its mean, with values closer to the mean occurring most often, it seems reasonable to assume that the average line speed (X) is normally distributed.
b. 0.000. Because n = 36 is large enough to invoke the Central Limit Theorem, the sample average line speed will be approximately normally distributed with a mean of 11,500 cans per minute and a standard deviation of 23.67.

7.44 Kilpatrick was questioning whether the items actually measured what McDowell claimed they measured. That is, the words in the items may have meant different things to different people and thus not assessed an identifiable trait.

Chapter 8

8.2 They are interchangeable before sampling, when a sample statistic has not yet been calculated. But once a confidence interval has been constructed based on a sample, the words are no longer interchangeable because the value of the random variable has been realized and thus there is no chance to which probability applies.

8.4 a. No, because they take their own samples independently, they are likely to have different samples and different sample means.
b. It depends on the situation. If σ is unknown and thus they must use their own sample standard deviations (s) instead of σ, then the margin of error will not be the same.

8.6 99.8 percent

8.8 2.58

8.10 The margin of error depends on the value of the sample proportion p. Thus, the margin of error for each question would not be the same as long as the sample proportion for each question differs.

8.12 The t distribution has more area beyond the standardized value of 1.00. The degrees of freedom determine the magnitude of the difference in these areas.

8.14 Because the t distribution has more variability, the confidence interval using the t distribution is wider than the interval using the z distribution, other things being equal.

8.16 If the distribution of $\bar{x}$ is normal and the population standard deviation (σ) is known, the z distribution has to be used. But even in such a situation, t distribution can also be used instead of z, if the sample size is large enough.

8.18 0.242. Under the assumption of normality of the population of the car prices, the distribution of $\bar{x}$ is normal with a mean of 21,750 and a standard deviation of 849.618 $(= \sigma/\sqrt{n} = 2081.131/\sqrt{6})$assuming $\sigma \sim s$.

8.20 a. 6.28 cents per dollar of sales.
b. 0.2794
c. $5.70 < \mu < 6.86$

8.22 $0.2286 < \pi < 0.4114$

8.24 $0.798 < \pi < 0.922$

8.26 $0.52 < \pi < 0.70$

8.28 a. n = 173 is required.
b. We need a sample size four times as large as what was originally required.
c. n = 40 is required.

8.30 139 or more.

8.32 a. Neither is appropriate.
b. z is appropriate.
c. Neither is appropriate.
d. t is appropriate.
e. Either may be used.
f. z is appropriate.

8.34 The pay distribution is usually highly right skewed with a few high paying jobs pulling up the mean.

8.36 This confidence interval does not say that chances are 95 percent that population average salary is within ± $410 of the survey result. We can say only that under repeated sampling, 95 percent of similarly constructed intervals would contain the true population average and the interval 29,700 ± 410 may (or may not) be one of those intervals.

8.38 984 or more.

8.40 a. 54.48
b. $46.26 < \mu < 62.70$
c. There is a truncation or censoring problem in the calculation of the mean and the confidence interval if there are bills that are never paid.

8.42 $339.961 < \mu < 2020.529$

8.44 2.7. We must consider a random sample of size 3 or more.

8.46 $0.61 < \pi < 0.69$

8.48 a. $0.62 < \pi < 1.04$

b. The interval constructed in part a is wrong because a proportion cannot exceed 1. We get this strange result because the sample size 12 is not large enough to assume that the sampling distribution of the sample proportion (p) is approximately normal.

8.50 The "unbiasedness" used in this quote implies that the opinion of a real estate appraiser is not biased for a specific interest. The "unbiasedness" in statistical estimation means that the mean of an estimator is equal to the corresponding population parameter.

8.52 The point estimate of the population proportion is the sample proportion, which is 0.07, and the point estimate of the standard deviation of the distribution of the sample proportion is 0.012.

8.54 a. 0.429

b. We need to assume that the population distribution is normal so that the sampling distribution of sample mean can be normal.

c. At a typical confidence level of 95 percent, a confidence interval for the population mean receipt is $465.88 < \mu < 576.52$. This interval includes the Healthy's claimed value $517.40. This suggests the claimed value could be consistent with the true average receipt of the company and thus the IRS does not need to pursue the company.

8.56 95 percent.

8.58 $0.46 < \pi < 0.62$. This interval includes not only the majority (more than 50 percent) of judges but also 50 percent or less. This implies that the majority of judges does not necessarily think sexual bias is widespread in California state courts.

8.60 Because the distribution of the time to be "up and running" is highly left skewed, it would not be appropriate to use the normal distribution in a confidence interval construction even though the sample size (30) is relatively large.

8.62 The margin of error is 5 percentage points in either direction of the poll result. Using p as the sample proportion (poll result), a 95 percent (as implied by the statement "chances are 19 to 20") classical confidence interval for π is $p \pm 5\%$. This confidence interval of $p \pm 5\%$, however, does not say that chances are 19 to 20 (95 percent) that population proportion π (proportion of black-owned business with annual revenue of at least $100,000) is within 5 percent of the poll result. We can say again that in classical statistics the population proportion is a parameter so there is no probability or chance associated with this interval estimate.

Chapter 9

9.2 A small p-value indicates an unlikely sample result when the null hypothesis is assumed to be correct. The observation of sample data that is highly unlikely casts doubt on the correctness of the null hypothesis.

9.4 To test the claim, we need to state the following set of hypotheses; H_O: $\mu \le 1{,}000$ vs. H_A: $\mu > 1{,}000$ where μ is the average or expected life (hours) of the lightbulb. We have to randomly draw a sample of n packages ($n \ge 30$, so that we have an approximately normal distribution of sample mean). Using the sample mean $\bar{x}$ from this sample, we can calculate the test statistic to test the correctness of the claim.

$z = \dfrac{\bar{x} - \mu_0}{\sigma/\sqrt{n}} = \dfrac{\bar{x} - 1000}{50/\sqrt{n}}$ where $\bar{x}$ and n are obtained from the sample.

If this calculated test statistic is greater than the critical value of z at a certain level of α, then we can reject the null hypothesis in favor of the alternative hypothesis.

9.6 a. H_O: $\mu = 87.4$ vs. H_A: $\mu \ne 87.4$, where μ is average number of applicants for a training program in sales at IBM.

b. H_O: $\mu = 0.05$ vs. H_A: $\mu \ne 0.05$, where μ is the inflation rate (%) in the area.

c. H_O: $\mu \ge 3$ vs. H_A: $\mu < 3$, where μ is the average sick days per employee per month at GM.

d. H_O: $\pi \le 0.1$ vs. H_A: $\pi > 0.1$, where π is the defect rate (%) for the machine.

e. H_O: $\mu \le .018$ vs. H_A: $\mu > .018$, where μ is the average stretch (%) of the new line at 15 percent of its working load.

f. H_O: $\mu = 378$ vs. H_A: $\mu \ne 378$, where μ is the average cost ($) of piece work.

g. H_O: $\mu \le 1000$ vs. H_A: $\mu > 1000$, where μ is the average breaking strength (lbs.) of the new composite per square inch.

h. H_O: $\pi \le 0.07$ vs. H_A: $\pi > 0.07$, where π is the proportion (%) of unemployment in the next recession.

i. H_O: $\pi \ge 0.06$ vs. H_A: $\pi < 0.06$, where π is the proportion (%) of overdue accounts receivable.

j. H_O: $\mu \ge 7$ vs. H_A: $\mu < 7$, where μ is the average whitening power of the new bleach evaluated on a 0 to 10 scale where 10 is the whitest.

9.8 a. $z_{critical} = \pm 2.33$
b. $t_{critical} = 1.753$
c. $z_{critical} = \pm 1.96$

9.10 A Type I error occurs if the null hypothesis is rejected in favor of the alternative when in fact the null hypothesis is true. Thus, 1 – P(Type I error) equals the probability of accepting the null hypothesis when it is correct. This does not equal the probability of making a Type II error.

9.12 4.56 percent

9.14 If the hypothesized value of the population parameter in the null hypothesis is found to be inside the $(1 - \alpha)$ 100 percent confidence interval constructed based on a sample result, this implies that the null hypothesis value could not be rejected at a level of the probability of Type I error as high as α.

9.16 a. H_O: $\mu \geq 10{,}000$ vs. H_A: $\mu < 10{,}000$, where μ is the average mileage of the company's cars.
b. The null hypothesis cannot be rejected (i.e., the company's claim cannot be accepted as true) unless the level of α is set at 24 percent or higher, which is hardly usual.

9.18 We can conclude that the new process is better than the old.

9.20 To test the hypothesis H_O: $\mu \leq 1.0$ vs. H_A: $\mu > 1.0$, where μ is the average number of defects per car, we need to know the sample size and the sample standard deviation (or the population standard deviation) for the sample that produced a sample mean of 5.1 in order to conduct a z or t test.

9.22 a. H_O: $\pi = 0.5$ vs. H_A: $\pi \neq 0.5$, where π is the winners' chance of repeating their success.
c. 0.049
d. The value of the estimate could have been between 0.405 (40.5 percent) and 0.595 (59.5 percent). With any estimated value in this range, the p-value cannot be smaller than 0.05 and thus the null hypothesis cannot be rejected.

9.24 We need to test the following hypothesis:

H_O: $\pi \leq 1/3$ vs. H_A: $\pi > 1/3$, where π is the proportion of consumers who prefer bottle A. The p-value is 0.2843.

Because this p-value is greater than the α level 0.1, H_O cannot be rejected. Or equivalently, the test statistic $z = 0.65$ is not beyond the critical value $z_{0.1} = 1.28$ and thus H_O cannot be rejected. We cannot conclude that more than one third of the customers prefer bottle A.

9.26 We need to test the following hypothesis:

H_O: $\mu = \mu_O$ vs. H_A: $\mu \neq \mu_O$ where μ is the composition of vapors of the batch of alloys and μ_O is the norm.

Type I error would be rejecting the idea that composition of vapors does not deviate from the norm when in fact it does not deviate from the norm.

Type II error would be saying that the composition of vapors does not deviate from the norm and thus accept the batch when in fact it deviates from the norm.

9.28 a. Assuming the population distribution is normal, we have to use the t distribution for testing because the population standard deviation is unknown and the sample size is not large.
b. The p-value is 0.245 where the degrees of freedom of the t is 15. Since the p-value is greater than the α level 0.05, we cannot reject the null hypothesis. We cannot conclude that the average is more than six defects per car.

9.30 We need to test the following hypothesis:

H_O: $\pi = 0.5$ vs. H_A: $\pi \neq 0.5$, where π is the proportion of companies that showed a stock price falling after move.

The p-value is 0.4654. Because this p-value is greater than a typical α level 0.05, H_O cannot be rejected. The test result does not contradict the *WSJ*'s hypothesis.

9.32 We need to test the following hypothesis:

H_O: $\mu = 42$ vs. H_A: $\mu \neq 42$, where μ is the average passenger miles flown.

The p-value is 0.816 (=2(0.408)). Because this p-value is greater than a typical α level 0.05, H_O cannot be rejected. We can conclude that the regulators' claim has been realized.

9.34 a. We need to test the following hypothesis:

H_O: $\mu \leq 0.19$ vs. H_A: $\mu > 0.19$, where μ is Folger's average market share.

b. Using the t distribution, the p-value will be 0.146 where the degrees of freedom of the t is 9. Because this p-value is greater than a typical α level 0.05, H_O cannot be rejected. We cannot conclude that the market share exceeds 19 percent this year.

9.36 We need to test the following set of hypotheses: H_O: $\pi \leq 0.5$ vs. H_A: $\pi > 0.5$, where π is the proportion of GE 60-watt lightbulbs lasting longer than 1,000 hours. The p-value will be 0.0475.

Because this p-value is less than the given α level 0.05, H_O can be rejected. Or equivalently, the test statistic $z = 1.68$ is beyond the critical value $z_{0.05} = 1.645$ and thus H_O can be rejected. We might conclude that 50 percent of all GE 60-watt bulbs will last longer than 1,000 hours.

9.38 The new sample proportion is 0.8889(= 24/27) and thus the test statistic z is calculated to be 4.04. With this z statistic, the test conclusion would be the same as in the previous exercises.

9.40 a. If some consumers refuse to answer, the sample of those consumers who participate cannot be interpreted as random.

b. We need to test the following set of hypotheses:

H_O: $\pi = 0.36$ vs. H_A: $\pi \neq 0.36$, where π is the proportion of American consumers who would not respond to questions.

To test this, a large ($n \geq 30$) sample should be obtained to have a sample proportion and thus calculate the test statistic z. If the test statistic is beyond the critical value of z, then we can reject H_O in favor of H_A.

9.42 a. The following set of hypotheses needs to be tested:

H_O: $\mu \geq 500{,}000$ vs. H_A: $\mu < 500{,}000$ where μ is the average amount ($) of commission. The p-value will be 0.0000 where the degrees of freedom of the t is 24. Because this p-value is much smaller than a typical α level 0.05, H_O can be rejected. We can conclude that the average commission is smaller than the claimed $500,000.

b. The competitor's sample cannot be viewed to be random because the sample only included those brokers who applied for the competitor's fictitious position. This lack of randomness may have caused a lower sample average and thus a false conclusion in the test in part a.

9.44 We need to test the following hypothesis:

H_O: $\mu = 12$ vs. H_A: $\mu \neq 12$, where μ is the average percentage of health care cost.

Using the t distribution, the test statistic will be $t = -0.178$.

Because this calculated t statistic -0.178 is not beyond the critical value $-t_{(\alpha = 0.025,\ df = 9)} = -2.262$ at the given α level 0.05, the null hypothesis cannot be rejected. We can conclude that this sample data support a population average of 12 percent.

9.46 We need to test the following set of hypotheses: H_O: $\pi = 0.5$ vs. H_A: $\pi \neq 0.5$, where π is the proportion of law firms having marketing specialists. The p-value will be 0.4238. Because this p-value is greater than a typical α level 0.05, H_O cannot be rejected. We can conclude that the observed 48 percent is not significantly different from the hypothesized 50 percent.

9.48 We need to test the following set of hypotheses: H_O: $\pi = 0.5$ vs. H_A: $\pi \neq 0.5$, where π is the proportion of decisions in favor of plaintiffs. The calculated z statistic will be 2.36. Because the test statistic $z = 2.36$ is beyond the critical value $z_{0.025} = 1.96$, H_O can be rejected at a typical α level 0.05. We can conclude that there is a significant difference between Han's sample and the national norm.

9.50 The observation that five (39 percent) of the 13 died implies nothing about the hypothesis that drinking prior to riding increases the risk of falling and dying. To test this hypothesis we need to know the proportion of those who drank and fell and the proportion of those who drank alcohol and did not fall.

9.52 Stone's observation that 5 percent of those in prison are innocent does not imply "95 percent accuracy." From this 5 percent number we can infer nothing about the probability of a guilty person being found innocent or an innocent person being found guilty. We only know that for those in jail 5 percent are innocent and 95 percent are guilty.

Chapter 10

10.2 H_O: $\mu_2 - \mu_1 = 0$ vs. H_A: $\mu_2 - \mu_1 \neq 0$ where μ_1 is the average gas mileage of domestic cars and μ_2 is the average gas mileage of imported cars. The calculated $z = 0.995$ is smaller than the critical $z_{0.025} = 1.96$. Thus, H_O cannot be rejected.

10.4 a. H_O: $\mu_2 - \mu_1 = 0$ vs. H_A: $\mu_2 - \mu_1 \neq 0$ where μ_1 is the average reading time of older people and μ_2 is the average reading time of younger people. The calculated $z = -2.04$ is beyond the critical $-z_{0.025} = -1.96$. Thus, H_O can be rejected at $\alpha = 0.05$.

10.6 H_O: $\mu_2 - \mu_1 \geq 0$ vs. H_A: $\mu_2 - \mu_1 < 0$ where μ_1 is the average income ($) of academic engineers and μ_2 is the average income ($) of industry engineers. The calculated $t = -3.43$ is beyond the critical value of $-t_{(0.05,\ df = 23)} = -1.714$. Thus, H_O can be rejected in favor of H_A at the level of $\alpha = 0.05$.

10.8 To test the validity of the assumption of equal variances, the set of hypotheses should be H_O: $\sigma_1^2 = \sigma_2^2$ vs. H_A: $\sigma_1^2 \neq \sigma_2^2$. The calculated $F = 1.65$ is smaller than the critical $F_{(0.025,\ df = 11,12)} = 3.322$. Thus, H_O cannot be rejected. The assumption of equal variances is reasonable.

10.10 To test the equality of variances, the set of hypotheses should be H_O:$\sigma_1^2 = \sigma_2^2$ vs. H_A: $\sigma_1^2 \neq \sigma_2^2$. The calculated $F = 1.60$ is smaller than the critical $F_{(0.005,\ df = 12,12)} = 4.907$. Thus, H_O cannot be rejected. The assumption of equal variances is reasonable in this case. To test the difference in two population means, H_O: $\mu_2 - \mu_1 = 0$ vs. H_A: $\mu_2 - \mu_1 \neq 0$ where μ_1 is the average size (in billions of dollars) of projected deficit and μ_2 is the average size (in billions of dollars) of actual deficit. Thus, under the assumption of normal populations and equal population variances, the test statistic is calculated to be 1.446.

This calculated t-value is smaller than the critical $t_{(0.005,\ df = 24)} = 2.797$. Thus, H_O cannot be rejected at the level of $\alpha = 0.01$.

10.12 We had to make the assumption of normality of the populations from which the samples were drawn. However, this assumption of normality is not satisfied for the income distribution. Income distribution is usually highly right skewed with some large values pulling up the mean; thus, a violation of normality. We might consider performing a non-parametric test such as Wilcoxon-Mann-Whitney rank sum test to test the difference in population means in this situation. But it is impossible to perform a Wilcoxon-Mann-Whitney test because it requires that every data value be known to calculate the test statistic. The original data values are not available in Exercise 10.6.

10.14 H_O: $\mu_2 - \mu_1 = 0$ vs. H_A: $\mu_2 - \mu_1 \neq 0$ where μ_1 is the average size (in billions of dollars) of projected deficit and μ_2 is the average size (in billions of dollars) of actual deficit. To test the hypothesis, we perform a Wilcoxon-Mann-Whitney rank sum test.

The calculated $t = 1.217$ is smaller than the critical value of $t_{(0.005,\ df = 24)} = 2.797$. Thus, H_O cannot be rejected at the level of $\alpha = 0.01$ as used in Exercise 10.10. We cannot conclude that there is a significant difference between two means.

10.16 The paired t test requires that a sample of matched pairs should be randomly drawn from the population. That implies every pair in the population should have equal chance of being drawn. If there are any pairs whose posttesting scores are missing, one has to discard those incomplete pairs from the sample in order to make the calculation of t statistic possible. Those incomplete pairs actually have no chance to be in a sample. This weakens the randomness of the sampling process and thus may result in a biased outcome.

10.18 Because the data for old and new pizza prices were collected as a paired observation at each outlet, a paired t test is possible. The paired t test is clearly better in this situation. Outlets vary from one to the other. Once we use a paired data set we can remove some of the variability caused by individual outlet characteristics.

To perform the paired t test, the set of hypotheses should be H_O: $\mu_2 - \mu_1 \leq 0.50$ vs. H_A: $\mu_2 - \mu_1 > 0.50$ where μ_1 is the average price($) of old pizza and μ_2 is the average price($) of new pizza.

The calculated $t = 0.505$ is smaller than the critical $t_{(0.05,\ df = 9)} = 1.833$. Thus, H_O cannot be rejected at the typical level of $\alpha = 0.05$.

10.20 a. The set of hypotheses should be H_O: $\mu_2 - \mu_1 = 0$ vs. H_A: $\mu_2 - \mu_1 \neq 0$ where μ_1 is the average gasoline price in Dec. 1990 and μ_2 is the average gasoline price in Dec. 1991.

The calculated $t = -2.17$ is beyond the critical $-t_{(0.025, df = 17)} = 2.110$. Thus, H_O can be rejected in favor of H_A at the typical level of $\alpha = 0.05$. We can conclude that there was a significant change in the average price of gasoline between two periods.

b. The F statistic to test the assumption of equal variances is calculated to be 1.058, which is smaller than the critical $F_{(0.025,\ df = 17,17)} = 2.674$. Thus, H_O: $\sigma_1^2 = \sigma_2^2$ cannot be rejected at $\alpha = 0.05$. The assumption of equal variances is reasonable in this case.

10.22 If the difference in two proportions is within the margin of error, we could not reject the null hypothesis of equal population proportions of car defects in two models. We could not decide that the Buick LeSabre has fewer defects than the Ford Crown Victoria.

10.24 We need to test the hypothesis H_O: $\pi_2 - \pi_1 = 0$ vs. H_A: $\pi_2 - \pi_1 \neq 0$. The pooled estimate of π is calculated to be $p = (20 + 70)/(40 + 183) = 0.40$.

The calculated value $z = -1.40$ is not beyond the critical $-z_{0.025} = -1.96$. Thus, we cannot reject H_O at the level of $\alpha = 0.05$. We can conclude that there is no significant difference in the population proportions in two surveys.

10.26 We need to test the hypothesis H_O: $\pi_2 - \pi_1 \geq 0$ vs. H_A: $\pi_2 - \pi_1 < 0$ where π_1 is the population proportion of law firms having marketing directors in 1990 and π_2 is the population proportion of law firms having marketing directors in 1991. Thus, the pooled estimate of π is calculated to be $p = (264 + 192)/(400 + 400) = 0.57$.

The calculated value $z = -5.14$ is far beyond the critical $-z_{0.05} = -1.645$ at a typical α level 0.05. Thus, we can reject H_O at the level of $\alpha = 0.05$. We can conclude that there was a significant drop in the proportion of law firms with marketing directors. In fact, the null hypothesis can be rejected in favor of the alternative hypothesis at any α level as low as 0.0000 (not zero but very small), because the p-value can be calculated to be 0.0000. Thus, the test result is not sensitive to the practical α level.

10.28 The set of hypotheses should be H_O: $\pi_2 - \pi_1 = 0$ vs. H_A: $\pi_2 - \pi_1 \neq 0$ where π_1 is the population proportion of blacks who never play the lottery and π_2 is the population proportion of whites who never play the lottery. The statement "those numbers are statistically equal" might imply that we cannot reject the null hypothesis of equal proportions. Thus, it would be better to say "those numbers are not statistically significantly different."

10.30 The set of hypotheses is H_O: $\mu_2 - \mu_1 = 0$ vs. H_A: $\mu_2 - \mu_1 \neq 0$ where μ_1 is the average objective probability and μ_2 is the average subjective probability. We need to use the distribution free Wilcoxon-Mann-Whitney rank sum test.

The calculated t = 0.043 is smaller than the critical value of $t_{(0.025,df=16)} = 2.120$. Thus, H_O cannot be rejected at the level of $\alpha = 0.05$. We cannot conclude that there is a significant difference between two means.

10.32 To test the equality of the population variances for typing speed the set of hypotheses should be H_O: $\sigma_1^2 = \sigma_2^2$ vs. H_A: $\sigma_1^2 \neq \sigma_2^2$.

The calculated F = 1.105 is smaller than the critical $F_{(0.005,\ df = 5,6)} = 11.464$. Thus, H_O cannot be rejected at $\alpha = 0.01$. The assumption of equal variances is reasonable in this case.

Similarly, to test the equality of the population variances for typing accuracy the set of hypotheses should be H_O: $\sigma_1^2 = \sigma_2^2$ vs. H_A: $\sigma_1^2 \neq \sigma_2^2$.

The calculated F = 2.362 is smaller than the critical $F_{(0.005,\ df = 6,5)} = 14.514$. Thus, H_O cannot be rejected at $\alpha = 0.01$. The assumption of equal variances is also reasonable for typing accuracy.

10.34 The matched pairs approach could be proposed if pool 1 and pool 2 involve the same secretaries. This approach can be free from the possible problem that by chance less able secretaries could have been in pool 1 and more able in pool 2 or vice versa.

10.36 We can test the following hypothesis: H_O: $\pi_2 - \pi_1 = 0$ vs. H_A: $\pi_2 - \pi_1 \neq 0$ where π_1 is the population proportion of blacks who terminated the training class and π_2 is the population proportion of whites who terminated the training class.

The calculated value z = –3.47 is beyond the critical $-z_{0.025} = -1.96$ at $\alpha = 0.05$. Thus, we can reject H_O at the level of $\alpha = 0.05$. We can conclude that there is a significant difference in termination rates between blacks and whites.

10.38 The process that determines who terminates before testing may be related to the process that determines who passes. If so, then who takes the test is not randomly determined.

10.40 The sample size $n(= n_1 = n_2)$ must be greater than 3578.

10.42 We should test the following hypothesis: H_O: $\pi_2 - \pi_1 = 0$ vs. H_A: $\pi_2 - \pi_1 \neq 0$ where π_1 is the population proportion of CEOs who agreed with the proposition and π_2 is the population proportion of board members who agreed with the proposition.

The calculated value z = 1.22 is not beyond the critical $z_{0.025} = 1.96$ at $\alpha = 0.05$. Thus, we cannot reject H_O at the level of $\alpha = 0.05$. We can conclude that these results show their agreement on the issue.

10.44 An individual SAT score would not necessarily fall within the range of 18 points when one takes the SAT repeatedly. We can say that one's mean SAT score would be within the range of 18 points of the true population mean SAT score with a probability of 0.68 or within the range of 36 points of the true population mean SAT score with a probability of 0.95 and so on, assuming the population distribution is normal.

10.46 We need to test the following hypothesis:

H_O: Population mean difference in rank in market value between 1992 and 1991 is zero.

H_A: Population mean difference in rank in market value between 1992 and 1991 is not zero.

The Wilcoxon signed-ranks test statistic t is calculated to be –0.26.

This calculated t-value is not beyond the critical value of $-t_{(0.025,df = 83)} = -1.99$. Thus, H_O cannot be rejected at the level of $\alpha = 0.05$.

10.48 The calculated t = –0.99 is not beyond the critical value of $-t_{(0.025,df = 192)} = -1.96$. Thus, H_O cannot be rejected at the level of $\alpha = 0.05$. We cannot conclude that there was a significant difference in two population means of profits between these two years.

10.50 The calculated value z = 1.37 is not beyond the critical $z_{0.025} = 1.96$ at $\alpha = 0.05$. Thus, we cannot reject the null hypothesis of no difference at the level of $\alpha = 0.05$. This test result, however, does not seem to be reliable enough to state that GM trucks performed as well as competitors because each sample size (11) is too small to apply the normal distribution in testing.

10.52 The set of hypotheses are H_O: $\mu_2 - \mu_1 \leq 0$ vs. H_A: $\mu_2 - \mu_1 > 0$ where μ_1 is the average retirement age last year and μ_2 is the average retirement age this year. Because the population standard deviations are assumed to be the same, we can calculate the z statistic as follows:

$$z = \frac{(\bar{x}_2 - \bar{x}_1) - D_0}{\sqrt{\frac{\sigma_1^2}{n_1} + \frac{\sigma_2^2}{n_2}}} = \frac{(62-61)-0}{\sqrt{\frac{\sigma^2}{1000} + \frac{\sigma^2}{1000}}}$$

For the null hypothesis of no difference to be rejected at $\alpha = 0.05$, this z value should be greater than $z_{0.05} = 1.96$. Thus, σ^2 must be less than or equal to 130. This implies that the standard deviation σ must be 11.4 or less.

10.54 The arithmetic mean is not an appropriate measure of central tendency because the values in this data set are qualitative. Their relative magnitudes have no meaning. Median might be a better measure for the purpose of comparison.

Chapter 11

11.2 Because the p-value 0.377 is greater than 0.05, the null hypothesis of no difference cannot be rejected. We can conclude that the factor is not significant.

11.4 Because the p-value 0.308 is greater than 0.05, the null hypothesis of no difference cannot be rejected. Between variability is not making a significant contribution to the total sum of squares.

11.6 The p-value 0.712 is greater than 0.05, so the null hypothesis of no difference cannot be rejected. To do this test we have to assume that each population of change scores is normally distributed with equal variances and that the observations are independent.

11.8

Source	SS	df	MS	F	p
FACTOR1	52	2	26	1.00	0.402
FACTOR2	88	3	29.33	1.13	0.383
ERROR	260	10	26		
TOTAL	400	15			

11.10

Source	SS	df	MS	F	Sig.
FACTOR1	61	2	30.5	3.05	0.092
FACTOR2	664	3	221.3	22.13	0.000
ERROR	100	10	10		
TOTAL	825	15			

11.12 The F statistic for C3 is 35.57 and the p-value is 0.000. Because the p-value 0.000 is smaller than 0.05, the null hypothesis of no difference can be rejected. We can conclude that the factor C3 makes a significant contribution to the total sum of squares. Similarly, the F statistic for C4 is 8.17 and the p-value is 0.019. Because the p-value is smaller than 0.05, the null hypothesis of no difference can be rejected. We can also conclude that the factor C3 makes a significant contribution to the total sum of squares.

11.14 We can construct the following two-way ANOVA table:

Source	SS	df	MS	F	p
FACTOR1	530.19	3	176.73	30.52	0.000
FACTOR2	44.19	3	14.73	2.54	0.122
ERROR	52.06	9	5.79		
TOTAL	626.44	15			

The p-value of factor 1 is 0.000, which is smaller than 0.01, and thus the null hypothesis of no difference can be rejected. We can conclude that the country in which the person lives makes a significant contribution to the total sum of squares. However, the p-value of factor 2 is 0.122, which is greater than 0.01. Thus, the null hypothesis of no difference cannot be rejected. We can conclude that the job classification does not make a significant contribution to the total sum of squares.

11.16 The expected frequency is 15 for a sample of 200.

11.18 We need to test the following hypothesis:

H_O: Days and type of worker are independent.

H_A: Days and type of worker are not independent.

The χ^2 test statistic is 2.96908. This is smaller than the critical value of 15.5073 at a level of $\alpha = 0.05$. Thus, we cannot reject H_O. We can conclude that days and type of worker are independent.

11.20 We need to test the following hypothesis:

H_O: Type of business and the period of observation are independent.

H_A: Type of business and the period of observation are not independent.

The χ^2 test statistic is calculated to be 2,095.698, which is much greater than the critical value of 21.0261 at $\alpha = 0.05$. Thus, we can reject H_O. We can conclude that type of business is related to the period of observation in which failures occurred.

11.22 a. 1.151
b. The between sample variance = 0.034 and the within sample variance = 0.107.

11.24 The following ANOVA table can be constructed:

Source	SS	df	MS	F	p
BETWEEN	24.0	4	6.000	0.517	0.725
WITHIN	116.0	10	11.600		
TOTAL	140.0	14			

Because the p-value 0.725 is greater than 0.05, the null hypothesis of no difference cannot be rejected. We can conclude that the five individuals are equal producers.

11.26 The following ANOVA table can be constructed:

Source	SS	df	MS	F	p
BETWEEN	211.1	4	52.8	0.523	0.720
WITHIN	2521.7	25	100.9		
TOTAL	2732.8	29			

Because the p-value is greater than 0.05, the null hypothesis of no difference cannot be rejected. We can conclude that there is no difference in word processing speed among the five days.

11.28 We can construct the following two-way ANOVA table:

Source	SS	d	MS	F	p
FACTOR1	2235	5	447	31.3	0.000
FACTOR2	211	4	53	3.7	0.021
ERROR	287	20	14		
TOTAL	2733	29			

The p-value of factor 1 is smaller than 0.05 and thus the null hypothesis of no difference can be rejected. Similarly, the p-value of factor 2 is smaller than 0.05. Thus, the null hypothesis of no difference can be rejected. We can conclude that both the day and the person are making a significant contribution to explaining total variability.

11.30 The classical F test requires that the population distribution be normally (or approximately normally) distributed with the same (or similar) variances. However, the difference in the sample standard deviations for those four methods suggests that the equal variance assumption may not be fulfilled.

11.32 The following ANOVA table can be constructed, assuming the population variances are equal:

Source	SS	df	MS	F	p
BETWEEN	4799.9	3	1600.0	14.09	0.000
WITHIN	3633.8	32	113.6		
TOTAL	8433.7	35			

Because the p-value is smaller than 0.05, the null hypothesis of no difference can be rejected. We can conclude that there is a significant difference between damage incidents among the four ship types.

11.34 Using the ASCII data file EX11-33.PRN or the LOTUS spreadsheet file EX11-33.WK1 provided on the computer disk, the following two-way ANOVA table can be constructed (excluding 16 observations missing ER DOCTOR data):

Source	SS	df	MS	F	p
FACTOR1	273629	6	45605	2.60	0.019
FACTOR2	12772	2	6386	0.36	0.698
ERROR	3002992	171	17561		
TOTAL	3289393	179			

The p-value of factor 1(Day) is 0.019, which is smaller than the α level 0.05, and thus the null hypothesis of no difference can be rejected. However, the p-value of factor 2 (Doctor) is 0.698 which is greater than the α level 0.05. Thus, the null hypothesis of no difference cannot be rejected. We can conclude that the day is significant in explaining time in the emergency room while the doctor is not.

11.36 Students differ in their abilities to learn. When there is a source of nuisance variation, one way to separate the effect of the variation due to the individual differences is to use a randomized-block design. This design needs a single sample of students (20 blocks each of size three) to be assigned to each treatment (class). Three students in a block can be believed to share about the same abilities. Thus, if we "randomly" assign the students in each block to one of three classes, we will be able to separately identify the variability resulting from the individual block differences. This requires a two-way ANOVA test.

11.38 Answer depends on the sample student selects.

11.40 Answer depends on the sample selected.

Chapter 12

12.2 If r = +1 or –1, then r^2 = 1. If r = 0, then r^2 = 0.

12.4 b. The mean is 56.
d. The average change score for those who scored below the average (56) on the pretest is 4.41.
e. The average change score for those who scored above the average (56) on the pretest is 4.04.
f. The average change scores for the two groups of students are approximately equal. We cannot conclude that the special effort resulted in the larger gain scores for those below the average. We could expect that the same thing would have happened from the middle to the end of the program if those below the average had received special attention and assistance in the second part of the course. But if all trainees received special attention and assistance in the second part of the course, the change in score for those below the average would be lower than for those above the average.

12.6 For positive (negative) deviations from the mean of the pretest score, there tends to be positive (negative) deviations from the mean of the midtest score and vice versa.

12.8 The correlation coefficient r is 0.28799. This value is too low to say those two items are highly correlated.

12.10 r = –0.99752

12.12 r = 0.73321 for those selected for executive positions
r = 0.75903 for those not selected for executive positions
r = 0.88809 for the total group

The two correlation coefficients for the subsamples are less than the correlation for the entire sample. If we divide the entire data set into two subsamples, those wide ranged observations play crucial roles in determining the correlation coefficients for both subsamples and thus result in smaller correlation coefficients.

12.14 The correlation coefficient r is calculated to be 0.569. There exists a positive, but not strong linear relationship between the number of votes obtained by candidates in the primary and the general election.

12.16 a. True
b. False. The slope of the sample regression line indicates how the predicted value of y changes as x changes.
c. True
d. False. A residual can be positive or zero or negative. The sum of all residuals is zero.
e. False. The least square method minimizes the sum of squared residuals.
f. True
g. True
h. True
i. False. The residuals always sum to zero whether the y intercept is zero or not.

12.18 a. b = –1.06098, a = 0 and the regression is $\hat{y} = -1.06098$ x
b. Both the mean of x and mean of y are zero and thus the intercept is zero.
c. The slope coefficient –1.06098 means that a one unit increase in x is associated with a 1.06098 unit decrease in y.

12.20 a. The slope coefficient is 0.5019 and the intercept coefficient is 20.8247. As a result, we have the regression equation $\hat{y}$ = 20.8247 + 0.5019 x
b. $\hat{y}$ = 20.8247 + 0.5019 (40) = 40.9007

12.22 The slope coefficient is 0.4698 and the intercept coefficient is 4.2885. As a result, we have the regression equation $\hat{y}$ = 4.2885 + 0.4698 x

12.24 The regression equation GENÊRAL = –375.634 + 1.45015 PRIMARY implies that each additional vote in the primary is associated with 1.45015 votes in the general election.

12.26 Considering a ~ 0 and b ~ 27, Fisher's regression can be written $\hat{y}$ = 0 + 27x where *y* is the TV station advertising revenue and x is the audience size. Using the relationship, a = $\bar{y}$ – b$\bar{x}$ the slope coefficient b can be calculated as

$$b = \frac{\bar{y}}{\bar{x}} - \frac{a}{\bar{x}} = \frac{\bar{y}}{\bar{x}} = \frac{\sum y/n}{\sum x/n} = \frac{\sum y}{\sum x} \quad \text{since a} \sim 0.$$

Σy is nothing but the sum of all television revenues and Σx is the total audience size. The total audience size equals the number of television households multiplied by the fraction (60 percent) of prime time that the average household watches television. Thus, the estimation of b should result in the same coefficient value, 27.

12.28 a. The slope coefficient is 1.0220 and the intercept coefficient is 2.9866.
b. The total sum of squares is 380. 905 and the residual sum of squares is 115.490.

12.30 a. $17,546.6
b. $103,670.4
c. 0.56047
d. 11584.0

12.34 b. The least-squares regression equation is

$\widehat{\text{SPEED}}$ = 33.713 + 0.727536 EXPERIENCE and r^2 is 0.93852

c. 35.90 mph

12.36 Approximately 83.9 percent.

12.38 a. The least-squares regression equation is

$\widehat{\text{SCORE}}$ = –20.6842 + 0.410106 DAY

b. An increase of 8.2 points.
c. 0.76467

12.40 a. Spearman's rho is 0.9833
b. Spearman's rho is 0.9333.
c. There exists a stronger positive relationship between the number of industrial failures in 1985 and 1987 than in 1987 and 1990.
d. Because the magnitude of the number of failures has less importance than the relative ranking, the comparison of simple correlation coefficient would not be meaningful.

12.42 0.77044

12.46 The predicted MPG is 37.65. The regression equation gives a better prediction than just using the average mileage of the two cars because the regression makes use of relationship between the mileage and weight of all the cars.

12.48 The implied least-squares regression equation is $\widehat{\text{CANCER-FACTOR}}$ = 2.5 CIGARETTE where $\widehat{\text{CANCER-FACTOR}}$ is the incidence of lung cancer and CIGARETTE is the number of cigarettes consumed by a smoker a day.

12.50 Answer will depend on data and years selected.

12.52 a. The least-squares regression equation is

$\widehat{\text{ABSDIF}}$ = –125.904 + 0.0127 MACHDIF

where ABSDIF is the difference in absentee votes and MACHDIF is the difference in machine votes.

The slope coefficient 0.0127 indicates that an additional machine vote is associated with 0.0127 additional absentee votes.

c. 0.699. There is a positive and somewhat strong relationship between the difference in absentee votes and the difference in machine votes.

d. At MACHDIF = –564, the predicted ABSDIF is –133(= –125.904 + 0.0127(–564)); the difference between the actual and the predicted ABSDIF in 1993 is 1158(= 1025 –(–133)).

12.54 a. $\sum(y_i-\bar{y})^2=\sum(y_i^2-2y_i\bar{y}+\bar{y}^2)$

$$=\sum y_i^2-2\bar{y}\sum y_i+n\bar{y}^2$$
$$=\sum y_i^2-2n\bar{y}^2+n\bar{y}^2=\sum y_i^2+n\bar{y}^2$$

b. $RSS=\sum(\hat{y}_i-\bar{y})^2=\sum(a+bx_i-\bar{y})^2$

$$=\sum(\bar{y}-b\bar{x}_i+bx_i-\bar{y})^2$$
$$=\sum[b(x_i-\bar{x}]^2=b^2\sum(x_i-\bar{x})^2$$

c. $ESS=TSS-RSS=\sum(y_i-\bar{y})^2-b^2\sum(x_i-\bar{x})^2$

d. $=\sum(y_i-\bar{y})^2-b\dfrac{\sum(x_i-\bar{x})(y_i-\bar{y})}{\sum(x_i-\bar{x})^2}\sum(x_i-\bar{x})^2$

$$=\sum(y_i-\bar{y})^2-b\sum(x_i-\bar{x})(y_i-\bar{y})$$
$$=TSS-b\sum(x_i-\bar{x})(y_i-\bar{y})$$

e. $r=\dfrac{\sum(x_i-\bar{x})(y_i-\bar{y})}{\sqrt{\sum(x_i-\bar{x})^2\sum(y_i-\bar{y})^2}}$

$$=\frac{\sum(x_i-\bar{x})(y_i-\bar{y})}{\sum(x_i-\bar{x})^2}\frac{\sqrt{\dfrac{\sum(x_i-\bar{x})^2}{n-1}}}{\sqrt{\dfrac{\sum(y_i-\bar{y})^2}{n-1}}}=b\frac{s_x}{s_y}$$

Chapter 13

13.2 As shown in the confidence interval formula, when there exists a lot of variability in the independent variable x, the value of the denominator $\Sigma(x_i-\bar{x})^2$ tends to be large and thus the margin of error tends to be small.

13.4 $\hat{y}$ = 17.88 at x = 5.

A 95 percent confidence interval for the expected value of y at x = 5 is 17.88 ± 1.68.

A 95 percent prediction interval for an individual value of y at x = 5 is 17.88 ± 4.01.

The prediction interval for an individual y value at given x is wider than the confidence interval for E(y|x) at the same value of x, because the prediction interval reflects additional variability of y around E(y|x).

13.6 A higher (positive or negative) correlation between X and Y implies a higher r^2 and thus a smaller sum of squared residuals (ESS), which in turn results in a smaller value of standard error of regression(s_e). With a smaller value of s_e, we have small value of margin of error and thus smaller confidence interval for the expected value of y at an x.

13.8 a. The standard error of regression (estimation) is 11.7813. This is the estimate of the standard deviation of population error term ε and is calculated to be

$$s_e = \sqrt{\frac{ESS}{n-2}} = \sqrt{\frac{1110.3953}{8}} = 11.7813$$

b. The t statistic is to test the hypothesis H_O: ß = 0 vs. H_A: ß ≠ 0 and is calculated as follows:

$$t = \frac{b - \beta_0}{s_b} = \frac{0.7747 - 0}{0.4685} = 1.654$$

where s_b = 0.4685

c. prob = 0.13678 is the p-value to test the hypothesis H_O: ß = 0 vs. H_A: ß ≠ 0 and is calculated as follows:

p-value = 2P(b ≥ 0.7747) = $2P(t_{(df=8)} \geq 1.654)$ = 2(0.068) = 0.136

d. To test hypothesis H_O: ß ≥ 1 vs. H_A: ß < 1, the t statistic is -0.4809.

The calculated t value is not beyond the critical $-t_{(0.05,\ df=8)}$ = -1.860; H_O cannot be rejected at α = 0.05 (Notice this requires a one-tailed test).

13.10 a. The least-squares regression equation is

$\widehat{\text{JOBLESS-RATE}}$ = 4.38196 + 0.300121 DROP-IN-OUTPUT

b. The p-value is 0.00805 for the slope coefficient. We can reject the null hypothesis of zero slope value at an α level as low as 0.00805; we can conclude that there is a significant relationship between the jobless rate and the drop in industrial output.

13.12 The least-squares regression equation is

$\widehat{\text{YEAR}}$1988 = –8488.10 + 1.01264 YEAR1986

To test the hypothesis H_O: ß ≤ 0 vs. H_A: ß > 0, the test statistic is 241.1.

The calculated t value is much greater than the critical $t_{(0.05,\ df=23)}$ = 1.714; H_O is rejected at α = 0.05. We cannot conclude that there is a significant drop in populations between the two years.

This conclusion agrees with the test result in Exercise 10.19 where we concluded no significant change in mean population between two years. Further, it also agrees with the conclusion in Exercise 12.35 where we found a positive relationship between two years' populations.

13.14 As n is increased, s_b is decreased and thus t is increased. This implies that increasing the sample size results in an increase in the power of the test, because rejection of the null hypothesis becomes easier with a large t value.

13.16 Although the coefficient of determination (r^2 = 0.9802) is very high and the slope coefficient has a relatively large t ratio (21.1), the residual plot shows a "<" pattern and suggests that the year be not linearly related to the return.

13.18 If we use the natural logarithmic transformation for both sides of the given relationship, we get

$\ln(y_t) = \ln(A) + \ln(B)\, t + \ln(u_t)$. This can be rewritten to be $y_t^* = A^* + B^*\, t + u_t^*$ where $y_t^* = \ln(y_t)$, $A^* = \ln(A)$, $B^* = \ln(B)$ and $u_t^* = \ln(u_t)$.

Thus, A* and B* correspond the intercept coefficient and the slope coefficients, respectively, in the transformed relationship.

13.20 Only the regression y1 on x seems to satisfy the required assumptions. The regression y2 on x violates the assumption of linearity; the second residual plot suggests a second order polynomial relationship between y2 and x. The regression y3 on x violates the assumption of no correlation between the disturbance term and the independent variable x; the third residual plots suggests a patternized relationship between the disturbance term and x. The regression y4 on x4 violates the assumption of homoscedastic disturbances; the variance of disturbance terms seems to be unequal over the range of x values.

13.22 a. True

b. True

c. True

d. False. This confidence interval of 10 < E(y|x = 20) < 30 does not say that the mean of y is between 10 and 30 with a probability of 0.95. We can say only that under repeated sampling, 95 percent of similarly constructed intervals would contain the true mean of y and the interval 20 ± 10 may (or may not) be one of those correct intervals.

e. False. A 95 percent confidence interval for E(y|x) could not be wider than a 95 percent prediction interval for an individual value of y.

f. False. The decision to reject depends on the α level.

g. False. There is no reason to believe that an estimated b equals the parameter ß. We, however, can say that the mean of b equals ß, i.e., E(b) = ß.
h. True
i. True
j. True
k. False. The purpose of regression analysis is not necessarily limited to the prediction of the dependent variable.
l. False. The standard deviation of b is calculated based on the fact that b is an unbiased estimator of ß. The derivation of the standard deviation of b does not depend on any specific value of the parameter ß.

13.24 To test the hypothesis H_O: $ß = 0$ vs. H_A: $ß \neq 0$, the test statistic is –8.347. This t value is beyond the critical $-t_{(0.025, df = 8)} = -2.306$; H_O is rejected at $\alpha = 0.05$. We can conclude that there is a significant relationship between two variables.

13.26 DISKSEEK(Disk seeking time) is defined as the independent variable and FLOAT (Floating point calculation time) is defined as the dependent variable.

To test the hypothesis H_O: $ß \leq 0$ vs. H_A: $ß > 0$, the test statistic t is 0.951, which is not significant at a typical α level. We cannot reject H_O and thus we cannot conclude that there is a statistically significant positive relationship between two variables. We have the same conclusion as in Exercise 12.8.

13.28 The standard error of regression is 0.2226 and the standard error of b is 0.02588. We could use these quantities in the calculation of test statistic t.

13.30 A 95 percent confidence interval for the expected value of y (number of votes in general election) is 1365 ± 597. This interval means that, for a candidate who receives 1,200 votes in the primary, if we repeat the construction of a confidence interval for the mean number of votes that the candidate will receive in the general election, 95 percent of similarly constructed intervals would contain the true mean number of votes in the general election.

13.32 a. The slope coefficient estimate 0.727536 is significant at $\alpha = 0.1$. We can conclude that there is a significant relationship between horse speed and jockey experience.
b. 35.9 ± 0.49
c. 35.9 ± 1.16

13.34 We need to test the following set of hypotheses:

H_O: The rankings in the two populations are not correlated.

H_A: The rankings in the two populations are correlated.

Spearman's rho test statistic is $t = 3.321$.

This calculated t is greater than the critical $t_{(0.025, df=16)} = 2.120$; H_O can be rejected. The two rankings are correlated.

13.36 a. The slope coefficient implies that a one percent increase in industrial production is associated with a 0.5317 percent increase in employment.
b. We can conclude that there is a significant relationship between a change in industrial production and change in employment in the population.

13.38 5.5943; 0.9657; 0.8274

13.40 The slope coefficient estimate 4902.26 is significant at $\alpha = 0.05$; we can conclude that there is a significant relationship between CEO compensation (y) and the number of headquartered companies (x).

13.42 The slope coefficient estimate 1.06059 is significant at a typical α level 0.05; we can conclude that there is a significant relationship between two variables.

13.44 The slope coefficient estimate 5.52941 is significant at an α level as low as 0.00000 (not zero, but very small), which means that there is a significant relationship between the misery index and political stability.

13.46 The t value for the slope coefficient is 85.762, which is significant at $\alpha = 0.01$. We can conclude that there is a statistically significant relationship between two variables.

13.48 Answer will depend on data selected.

13.50 a. The MACHDIF coefficient estimate is significant at an α level as low as 0.00042; we can conclude that there is a significant relationship between absentee votes and machine votes.
b. –133.07 ± 721.64
c. The observed absentee votes 1025 is not inside this 95 percent of prediction interval, which may suggest a fraud in the election.

Chapter 14

14.2 a. If a jockey's experience increases by one year, then the horse speed increases by b miles per hour, holding all other things (jockey's sex, weight carried on horse and horse's condition) fixed.
b. When a jockey is male (=1), then the horse speed is c miles per hour faster or slower (depending on the sign of c) than at a female jockey, holding all other things (jockey's experience, weight carried on horse and horse's condition) fixed.

14.4 The least-squares regression equation for a first degree polynomial is $\hat{y} = 27.1112 - 0.791624\ x$, $r^2 = 0.09567$ and the slope coefficient estimate is not significant at $\alpha = 0.1$

The least-squares regression equation for a second degree polynomial is $\hat{y} = 21.2096 + 2.91584\ x - 0.51833\ x^2$, $R^2 = 0.15540$ and neither slope coefficient estimate is significant at $\alpha = 0.1$

The least-squares regression equation for a third degree polynomial is $\hat{y} = 62.2093 - 38.1160\ x + 11.7792\ x^2 - 1.13818\ x^3$, $R^2 = 0.54575$ and all three slope coefficient estimates are significant at $\alpha = 0.1$; this can be viewed to be the best regression equation to fit the given data set.

14.6 a. 1. $\hat{y} = 8130.65 + 29.0514\ x$ where y is real earnings and x is age.

3. This regression equation does not appear to be a good representation of the relationship between two variables. r^2 is only 0.017781 and the residual plot shows a ">" pattern in the following plot of residuals. We suspect that there is a nonlinear relationship between x and y.

b. 1.

$\hat{y} = -12103.3 + 1138.43\ x - 18.1007\ x^2 + 0.0876612\ x^3$

This regression equation appears to provide better representation of the relationship between two variables. R^2 is 0.96023 and the residual plot does not show any special pattern.

14.8 a. The least-squares regression equation is

$\widehat{\text{PRICE}}$ = 2719.54 – 4.94027 ROOMS – 0.125848 TOTAL –1.33721 AGE where TOTAL = MAIN-FL + UPPER-FL + BASEMENT

b. $236,540

14.10 a. For batters, the implied regression equation is

$\widehat{\text{SALARY}}$ = a + 9000 HOMERUN + 6000 RUN where SALARY is the batter's next year salary, HOMERUN is the number of homeruns and RUN is the number of runs. Silk did not supply the value of intercept coefficient.

b. For pitchers, the implied regression equation is

$\widehat{\text{SALARY}}$ = a + 38000 VICTORY + 16000 SAVE + 3000 INNING + 12000 ERA where SALARY is the pitcher's next year salary, VICTORY is the number of victories, SAVE is the number of saves, INNING is the number of additional innings pitched and ERA is the decrement in e.r.a. measured in the number of one-tenths of point. Silk did not supply the value of intercept coefficient.

14.12 The death rate of smokers is higher than that of non-smokers of the same age, but since the dependent variable is measured in natural logarithm units (where the base is e = 2.71828) assessment of this difference requires the use of the antilog.

14.14 Those intervals get wider, because the standard error of $\hat{y}_*$ in the formula $\hat{y}_* \pm (\alpha/2, df = n-k) s_{y*}$ becomes larger as the x values are farther from their respective means.

14.16 At ROOMS = 8, TOTAL = 2500 and AGE = 0, a 95 percent prediction interval for the individual price is (1695.6, 3035.2) and a 95 percent confidence interval for the expected price is (2075.9, 2654.9). The prediction interval is wider than the confidence interval.

14.18 In Exercise 14.5, the estimated least-squares regression equation was $\hat{y} = 3.84521 - 0.277241\ x + 4.42645$ DUMMY where y is inflation rate and x is unemployment rate. The dummy variable coefficient has the t value of 5.078, which is significant at an α level as low as 0.00009. The coefficient for inflation rate, however, has the t value –0.826, which is not significant at a typical α level. Further, a plot of residuals shows a "<" pattern, which suggests that the relationship between the inflation rate and the unemployment rate be adjusted.

14.20 Only the income coefficient is significant at $\alpha = 0.01$ because the p-value is 0.00113.

14.22 We need to test the hypothesis H_O: $\beta_2 \le 0$ vs. H_A: $\beta_2 > 0$, where β_2 is the SALES coefficient. We can reject H_O at an α level as low as 0.1772.

14.24 We need to test the following set of hypotheses:

H_O: $\beta_2 = \beta_3 = 0$

H_A: not H_O (at least one β is nonzero)

The F value is 13.45. The p-value is 0.00086 and thus we can reject H_O at $\alpha = 0.05$; at least one of β's is nonzero.

The conclusion is that the multiple regression relationship is significant, i.e., we conclude that at least one β is nonzero. We need to conduct individual coefficient significance tests to see which individual coefficient is significant.

14.26 The statement is incorrect. Type I error level of 5 percent means that if the computed F exceeds the critical F, we reject the null hypothesis (all β's are zero). If we do not reject the null hypothesis, we may be correct in that decision with a probability of 95 percent.

14.28 R and thus R^2 are not related to the percentage of correct predictions except the extreme case where R = 1. R = 0.9 and thus $R^2 = 0.81$ means that 81 percent of the total variation in y values is explained by the fitted regression line.

14.30 If x_1 and x_2 are not correlated.

14.32 Answer will depend on the regression equation selected.

14.34 The R^2 is simple ratio of RSS to TSS, while the adjusted R^2 is its degrees of freedom adjusted value, i.e.,

$$\bar{R}^2 = 1 - \frac{n-1}{n-k}(1 - R^2)$$

14.36 The dummy coefficient estimate –0.987623 means that the enactment of 55 mph speed limit results in 0.987623 units decrease in death rate implying $1,975,246 savings per 100 million vehicle miles (= 0.987623(2,000,000)).

14.38 a. First Model:

$\widehat{\text{INCR\%SAL}}$ = –0.0062 + 0.0275 PHDYR – 0.0060 BOOK – 0.0076 SEX where INCR%SAL = (CURSAL – STSAL)/STSAL. The SEX coefficient estimate is negative, but is not significant at a typical α level.

Second Model:

$\widehat{\Delta \text{SAL}}$ = 511.460 + 1042.74 PHDYR – 198.069 BOOK – 582.326 SEX where ΔSAL = CURSAL – STSAL. The SEX coefficient estimate is negative, but is not significant at a typical α level.
In both models, we cannot find a significant evidence for a sex discrimination in the department for those 15 faculty members who are not administrators. Model choice does not seem to affect the conclusion.

b. Semilog Linear Model:

$\widehat{\text{LCURSAL}}$ = 10.792 – 0.028 SEX + 0.008 PROYR + 0.090 ADMIN + 0.049 SERV + 0.003 ART + 0.013 BOOK + 0.045 DUMMY where LCURSAL = Log(CURSAL) and DUMMY = 1 for a person from the department 4, 0 otherwise.

Simple Linear Model:

$\widehat{\text{CURSAL}}$ = 48397.3 – 1367.8 SEX + 466.0 PROYR + 5185.9 ADMIN + 2685.1 SERV + 165.5 ART + 741.4 BOOK + 2693.9 DUMMY

1. The semilog salary model is considered to be better than the salary level model because salary raises from year to year tend to be based on percentage increases, which yields a nonlinear relationship between salary levels and time dependent regressors.
2. In the semilog model, the DUMMY coefficient estimate has a t value of 2.522, which is significant at an α level as low as 0.01256; we can conclude that the fourth department is significantly different from the others in explaining salary.
3. In the semilog model, the SEX coefficient estimate is negative and has a t value of –2.281, which is significant at an α level 0.05; at $\alpha = 0.05$, we can conclude that the gender variable plays a significant role in explaining salary, which may imply a sex discrimination.

14.40 a. $\widehat{\text{SALES}}$ = –30.5857 + 2.4199 AD + 26.6538 NUM – 18.2615 COMP
b. At AD = 165, NUM = 10 and COMP = 4, a 95 percent prediction interval is (383.7, 740.7).
c. An α level as low as 0.00078
d. The AD variable is not significant at $\alpha = 0.01$

14.42 One example of a regression model that might be estimated is death = $\beta_1 + \beta_2$(age) + β_3(health) + β_4(operation) + ε, where age is the average age of those operated on in a year at each of the n hospitals in Pennsylvania; health is an average measure of a health index of those operated on in a year at each of the n hospitals in Pennsylvania; and operation is the number of operations per year performed at each of the n hospitals in Pennsylvania.

We assume that ε is distributed normally but death is a count (0, 1, 2, •••• 12, ••••); thus, it is not continuous as implied by the assumption that ε is normal.

14.44 a. n = 119
b. The multiple coefficient of determination (R^2) of 0.388 means that 38.8 percent of the total variability in the profit is explained by the fitted regression line.

14.46 Yes, the inclusion of a dummy variable would be able to cure the problem. If we use the dummy variable of DUMMY = 1 for women and 0 for men, we could separate the gender effect on salary.

14.48 a. The least-squares regression equation is

$\widehat{\text{INFLATION}}$ = –0.735 – 0.530 UNEMPLOYMENT + 4.203 POLITICS

The UNEMPLOYMENT coefficient indicates that a one percent increase in the unemployment rate is associated with a 0.530297 percent decrease in the inflation rate. The POLITICS coefficient indicates that a one unit increase in the political stability index is associated with a 4.20284 percent increase in the inflation rate.

b. The calculated t value −2.280 is beyond the critical $-t_{(0.05,df=17)} = -1.740$. We can conclude that β_1 is significantly less than zero.

c. We have a different conclusion about the significance of the variable UNEMPLOYMENT.

14.50 If the death rate is measured on a logarithm scale, a linearized form ln(Death Rate) = ln(α) + rt, can be applied to estimate the given relationship.

14.54 The F statistic is reported as 4.4789, which is significant at an α level as low as 0.01255. For example at α = 0.05, we can conclude that two variables are jointly significant.

14.56 Ferrari F40 shows the largest residual, which is about 207,000. The shopper might question the actual price of the car because the actual price is much higher than what is expected according to the regression equation.

14.58 b. For the null hypothesis of $\beta \geq 0$, the relevant p-value is 0.055. For the null hypothesis of $\beta \leq -0.032$, the relevant p-value is also 0.055.

c. Not rejecting the null hypothesis does not imply its acceptance. As long as the estimated coefficient is different from the value in the null hypothesis, but the null hypothesis is not rejected, there is another value in the opposite direction of that implied by the null hypothesis that likewise cannot be rejected in a similar Type I error level.

14.60 The least-squares regression equation is

$\widehat{\text{NUMBER-DAMAGE}}$ = −2.152 + 11.943 DUMMY1 + 4.689 DUMMY2 + 0.800 DUMMY3 + 0.043 YEAR + 0.001 ATSEA where DUMMY1 = 1 for type 1, 0 otherwise, DUMMY2 = 1 for type 2, 0 otherwise, and DUMMY3 = 1 for type 3, 0 otherwise.

At DUMMY1 = 0, DUMMY2 = 1, DUMMY3 = 0, YEAR = 73, ATSEA = 2000, the predicted NUMBER-DAMAGE is 7.8.

At DUMMY1 = 0, DUMMY2 = 1, DUMMY3 = 0, YEAR = 73, ATSEA = 2000, a 95 percent confidence interval for the expected number of damage incidents is (2.33, 13.25). That is, we are predicting from 2.33 to 13.25 damage incidents with 95 percent confidence interval.

14.62 If we include the 1990 population variable together with the percentage increase in the population as explanatory variables, then it implies that we include two years' population in the regression equation, which could cause a multicollinearity problem.

Chapter 15

15.2 The given diagram shows the average dinner check. Thus, the dinner check growth is defined to be the value of the first derivative of the slope; the diagram shows that the first derivative has decreased since 1987.

15.4 $\widehat{\text{MONEY-STOCK}}_t$ = −631.534 + 15.3573 t − 197.716 DUMMY1_t −204.227 DUMMY2_t

where t is the last two digits of the year, DUMMY1 = 1 for 1959 to 1973, 0 otherwise and DUMMY2 = 1 for 1974 to 1982, 0 otherwise.

All coefficient estimates are significant at the 0.01 α level; we can conclude that the students' belief is reasonable.

15.6 There is no need to fit two regressions because AND's market share is one minus INTEL's share in this diagram.

15.8 At t = 1992(1st quarter of 1992) and 1993(1st quarter of 1993), the predicted values of ANNUAL-RETURN are

$\widehat{\text{ANNUAL-RETURN}}_{1992}$ = 1619.60 − 0.813187(1992) = −0.27%

$\widehat{\text{ANNUAL-RETURN}}_{1993}$ = 1619.60 − 0.813187(1993) = −1.08%

15.10 a. The advance-decline line shows an upward trend in general. This trend might suggest a bull market in future.

b. We can notice some other fluctuations in addition to the trend. This may result from seasonal factors and certain events that affect the stock market.

15.12 a. $\widehat{\text{PROFIT}}t$ = 3286.9 + 5.5621 ADVt + 96.5468 t −2052.6 QT1_t −1352.3 QT2_t − 848.628 QT3_t where t = 1 for 1st quarter 1971, 2 for 2nd quarter 1971, ••••••, 84 for 4th quarter 1991; QT1 = 1 for 1st quarter, 0 otherwise; QT2 = 1 for 2nd quarter, 0 otherwise; and QT3 = 1 for 3rd quarter, 0 otherwise.

b. The coefficients for the variables t and QT1 are significant at α = 0.05, but not for the variables QT2 and QT3.

c. The Durbin-Watson statistic is reported as 1.0541, which is less than the critical lower bound $d_L = 1.525$ at α = 0.05; the null hypothesis of no autocorrelation is rejected.

15.14 The z statistic is

$$= \frac{\text{runs} - E(\text{runs})}{\hat{\sigma}(\text{runs})} = \frac{5 - 16.9375}{2.7712} = -4.31$$

This z value is beyond the critical $-z_{0.025} = -1.96$ at $\alpha = 0.05$. We can conclude that there is an autocorrelation problem.

15.16 See table below.

15.18 The least-squares regression equation is

$$\widehat{RETURN}_t = 1.25243 + 0.921703\ RETURN_{t-1}$$

Thus, at t = 1992(1st quarter of 1992) and 1993(1st quarter of 1993), the predicted values of RETURN are

$$\widehat{RETURN}_{1992} = 1.25243 + 0.921703\ RETURN_{1991}$$
$$= 1.25243 + 0.921703(-0.2) = 1.07\%$$

$$\widehat{RETURN}_{1993} = 1.25243 + 0.921703\ RETURN_{1992}$$
$$= 1.25243 + 0.921703(1.07) = 2.24\%$$

The predicted values are higher than those in Exercise 15.8.

15.20 Excluding Morgan county data because of missing observations, we can conclude that Martin and Greene counties have unemployment equations that most nearly follow the aggregate equation for the State of Indiana.

15.22 The Durbin-Watson test requires that the regressors be fixed, but in a dynamic model, y_{t-1} is correlated with ε_{t-1}.

15.24 If the time trend cannot be represented as the slope in a simple linear regression, we need to consider alternative specifications and more complex regressions. A quadratic trend is among possible approaches. A defense lawyer would question the use of a linear trend in a wrongful injury or death case because the lifetime income does not follow a simple linear relationship with the time, but follows a curvilinear (possibly a quadratic) relationship.

15.26 Linear Trend Model:

$$\widehat{BILL}_t = -2014.87 + 1.01533\ t$$

Exponential Trend Model:

$\widehat{LBILL}_t = -1409.54 + 0.709316\ t$ where LBILL = ln(BILL)

The model with exponential trend is better; it has a higher r^2 and its coefficient estimates have larger t ratios.

This exponential model assumes that the log of billings is linearly related to the trend and thus, that billings is exponentially related to the trend, while the simple linear trend model assumes a linear trend.

15.28 Simple Model:

$\widehat{YIELD}_t = 4.0605 + 0.63425\ \text{INFLATION-RATE}_t$. The Durbin-Watson statistic is reported as 0.2978, which is less than the critical lower bound $d_L = 1.427$ at $\alpha = 0.05$; the null hypothesis of no autocorrelation is rejected.

Table for 15.16

Predicted number of degrees

	Dynamic model with no trend			Model with trend		
Year	Total	Men	Women	Total	Men	Women
1989	72200	46400	25800	72200	46400	25800
1990	72506	46517	26242	77434	47558	29640
1991	72713	46626	26628	78398	47177	30963
1992	72852	46728	26966	79361	46796	32286
1993	72947	46823	27262	80324	46415	33609
1994	73011	46911	27521	81288	46034	34932
1995	73054	46994	27747	82251	45653	36255
1996	73083	47071	27945	83215	45272	37578
1997	73103	47144	28118	84178	44891	38901
1998	73116	47211	28270	85141	44510	40224
1999	73125	47274	28403	86105	44129	41547
2000	73132	47333	28519	87068	43748	42870

* The values for 1989 are actual.

First Difference Model:

$\widehat{DIFYIELD}_t = -0.45694 + 0.13793$ INFLATION-RATE$_t$ where DIFYIELD$_t$ = YIELD$_t$ – YIELD$_{t-1}$. In this first difference model, a serial correlation problem no longer exists.

15.30 ABC simply compared the current audience level to the average of previous three years, while NBC used a time series regression analysis to derive a trend line.

15.32 a. AVERAGE$\hat{-}$YIELD$_t$ = –4.419 + 1.0297 PRIOR-YIELD$_t$ + 0.0021 t. There is a positive relationship between the average and prior yields.
b. PRIOR$\hat{-}$YIELD$_t$ = –77.108 + 0.957 AVERAGE-YIELD$_t$ + 0.039 t. There is not a negative relationship between the yield after and the average yield.

15.34 We can create a seasonal dummy variable as DUMMY = 1 for the summer, 0 otherwise.

15.36 a. We can create relevant dummy variables as: DUMMY = 1 for the months in election years, i.e., the months in 1980, 1984, 1988 and 1992 and DUMMY = 0 otherwise
b. The least-squares regression equation is $\widehat{OUTLAY}_t = 66.2862 - 0.679863\, t - 0.124783$ DUMMY$_t$ The trend coefficient estimate is negative and is significant at $\alpha = 0.05$
c. The DUMMY coefficient estimate is negative as opposed to the cited statement, but is not significant at a typical α level.

15.38 a. The least-squares regression equation is $\widehat{CREDIT}_t = 4.44395 - 0.094276\, t + 2.45757$ DUMMY$_t$ where t = 1 for 2nd summer 1980, 2 for 1st semester 1980–81, ••••••, 46 for 1st semester 1991–92, and DUMMY = 1 for the semesters after the accident, 0 otherwise.
b. The DUMMY coefficient 2.45757 is positive although it is not significant at a typical α level. This implies that the credits completed increased after the accident.
c. The regression with a simple trend does not seem to support the plaintiff's position. A consultant might want to consider other models to help the plaintiff.

15.40 The trend coefficient estimate is negative; the Corvette production shows a downward trend over time. But the coefficient estimate is not significant at a typical α level; we cannot reject the null hypothesis of zero coefficient.

15.42 Answer will depend on data selected.

15.44 a. 1/INFLATION$_t$ = $\beta_0 + \beta_1$ UNEMPLOYMENT$_t$ + β_2 t + ???
b. Requires library visit.
c. Answer will depend on data selected.

Chapter 16

16.2 Leaving mortgage rates out of home ownership costs implies that the CPI would overestimate price increases when mortgage rates fell as they did in the early 1990s. Leaving out discount store prices implies that the CPI would overestimate price increases as consumers move from buying at retail stores to discounters. Leaving out home computers implies that this improved technology is not reflected in the CPI and price increases are overestimated.

16.4 According to a Laspeyres Index, the inflation rate is 24.0 percent. According to a Paasche Index, the inflation rate is 24.1 percent.

16.6 2.72 percent

16.8 Changing the base would have lowered his/her salary. Tying a person's salary to the CPI, however, usually implies that increases or decreases in salary are tied to the percentage change in the CPI. Rebasing the CPI does not affect percentage changes.

16.10 The real salaries rose about 1.9 percent(= 8 – 6.1).

16.12 a. Wages+salaries, 3.8429; Benefits, 5.8448; Compensation, 4.3931.
b. No

16.14 His approximation could be adequate for early 1960s, when inflation was low, but not for late 1970s, when it was in the double digits.

16.16 Both series appear to be equally smooth, because the original series does not have much fluctuation.

16.18 A 1967 Chevrolet was not as reliable, rust resistant, fuel efficient, comfortable, maneuverable, crash damage proof, maintenance free, or passenger and environmentally safe as a 1992 Chevrolet.

16.20 a. 57 percent; \$56,600.
b. 37 percent; \$6,700.

16.22 The rate of inflation was 2.7 percent in December 1993 and 2.5 percent in January 1994; the inflation rate fell from December 1993 to January 1994. We must calculate the percentage change from the same month of previous year in CPI to compare the inflation rates.

16.24 If the annual growth rate was 8.0 percent between 1980 and 1990, then the compounded growth rate during the same period would have been 115.9 percent, which implies an average annual growth rate of 11.59 percent.

16.26

Age	Earnings
37	\$42,692
40	\$45,224
50	\$48,971
60	\$45,716
70	\$35,738

16.28 $A_0 (1 + r)^{12} = 3A_0$ is implied where r is the annual increase rate, A_0 is the tuition in 1980 and $3A_0$ is the tuition in 1992. Thus, the annual increase rate r is 0.096. More than 9.6 percent is the correct calculation of the annual increase rate.

16.30 The difference in growth rates is visualized as the difference in the degree of steepness of both slopes, not the vertical distance between two series.

16.32 The average hourly earnings in U.S. is overestimated by \$1.22 and the minimum wage is overestimated by \$1.19.

16.34

Year.Month	Annualized percent change
93. 1	0.0314
2	0.0488
3	0.0369
4	0.0311
5	0.0225

16.36 Requires library visit.

Chapter 17

17.2 The "best course of action" will not be the same for a risk averse utility maximizer and a risk neutral utility maximizer.

17.4 Risk neutral

17.6 a. The investor chooses the real business venture.
b. The investor chooses the Treasury bills as optimal.

17.8 The risk lover would select a.

17.10 If one can obtain new additional sample information, then the new information can be used to obtain posterior probabilities and thus the resulting expected monetary value could be greater than the old one.

17.12 a.

	Solicitation letter order size			
Contribution level	50	100	150	200
Expected payoff	225	315	309	281

b. 100

17.14 Risk neutral lawyers should choose the "tax only" action.

17.16 b. The expected payoff is the same as the payoff in Exercise 17.15; the sample information provides no additional payoff and thus is not worth paying to buying.

17.18 If we view human life as a piece of capital equipment that has an income stream attached to it, then from the standpoint of the estate there is no difference between the decision to buy or not buy life insurance and any other investment, including betting on a horse race.

17.20 The best course of decision is choosing the action "close branches and expand ATMs" if the indicator is "favorable," and choosing the action "continue as it is" if the indicator is "unfavorable."

17.22 The best course of decision is choosing the action "new startup" if the indicator is "favorable," and choosing the action "continue as it is" if the indicator is "unfavorable."

17.24 a. The action "produce 10,000 copies internally" provides the highest expected net return; thus, that action is optimal.

17.26 For gambles involving a likely small loss and unlikely big gain, an individual may be a risk taker but for equally likely big losses and big gains this same person might be risk averse.

17.28 The decision maker would select the option a.

17.30 b. The optimal strategy is initially to choose selling at retail and, if the sawdust does not sell at retail, then choose to dispose of it all at once to a wholesaler.

17.32 a. 0.15
b. P(faulty|judge okay) = 0.0089. P(faulty|judge faulty) = 0.8934

17.34 0.96

17.36 As users live longer then the stock of users will tend to increase with time. Both the increased life expectancy and the increase in their number will increase Genentech profits.

17.38 If, for any reason, the presenter chooses the deal A in the first game (involving hundreds of dollars), his decision in the second game (involving hundreds of thousands of dollars) should be consistent with the decision in the first game because both games are virtually the same with the exception of dollar amounts in payoffs.

Chapter 18

18.2 No. Process quality is a necessary and sufficient condition for product quality.

18.4 Statistical process control allows in-process verification of process quality. It enhances the manufacturer's ability to meet specifications initially, instead of accepting rework as a necessary part of production.

18.6 The control chart shown is an R chart. It was constructed by computing the ranges of the samples taken per day. These range values were plotted for each day and lines were drawn to connect each data point. The centerline of the chart, denoted CL, was computed by determining the average of all nine days' range values, $\bar{R}$. The lower control limit, LCL, was computed by multiplying the average of all nine days' range values, $\bar{R}$, and the control chart constant D_3. The upper control limit, UCL, was computed by multiplying the average of all nine days' range values, $\bar{R}$, and the control chart constant D_4. The CL, LCL, and UCL were then plotted on the graph.

18.8 a. CL = 62,879.2; $\bar{R}$ = 74,974.6; UCL = 106,139.5; LCL = 19,618.9
b. CL = $\bar{R}$ = 74,974.6; UCL = $\bar{R}D_4$ = 158,571.3; LCL = $\bar{R}D_3$ = 0.0
c. Both charts indicate that the process is in control as all data points reside within their respective control limits.

18.10 A defective item does not perform its intended task. An example is a diaper whose velcro fastener won't hold the diaper together. A nonconforming item doesn't meet its specifications. An example is a bandage which is 1/8" wider than its specifications.

18.12 The p chart is required in this case.

CL = 0.0572. UCL = 0.097. LCL = 0.018.

18.14 The appropriate control chart for this problem is a c chart. CL = 11.12. UCL =21.12. LCL = 1.12.

18.18

Error type	% of total
omission	32
selection	14
sequence	41
time	9
qualitative	4

The Pareto chart shows the errors organized from largest contributor to smallest contributor.

18.20

Downtime classification	% of total
A	10
B	44
C	31
D	00
E	15
F	00

The Pareto chart shows the classifications organized by largest contributor to smallest contributor.

18.22 The sample range statistic represents the range of component weight values (maximum weight – minimum weight) for each sample of size n = 20.

The sample range statistic is useful for construction of the $\bar{x}$ chart because it's used as an approximator of σ.

18.24 a. CL = 34.85
b. UCL = 35.99
LCL = 33.71

18.26 CL = $\bar{R}$ = 6.25
UCL = $\bar{R}D_4$ = 10.33
LCL = $\bar{R}D_3$ = 2.18

18.28 CL = $\bar{R}$ = 2.27
UCL = $\bar{R}D_4$ = 4.03
LCL = $\bar{R}D_3$ = .5062

The R chart indicates that the process is in control.

18.30 CL = $\bar{p}$ = 0.2294
$\bar{n}$ = 22.7
UCL = .4941
LCL = –0.0353

18.32 The c chart.

18.34 CL = 48.0. UCL = 68.8. LCL = 27.2.

The c chart indicates that the process is out of control. The data repeatedly exceed both the upper and lower control limits.

18.36

	SPC	Acceptance Sampling
When?	in-process	post-processing
Who?	producer	consumer
How?	process sampling	large lot sampling

Index